D0583483

Advances and Innovations in Systems, Computing Sciences and Software Engineering

Advances and Innovations in Systems, Computing Sciences and Software Engineering

Edited by

Khaled Elleithy
University of Bridgeport
CT, USA

A C.I.P. Catalogue record for this book is available from the Library of Congress.

ISBN 978-1-4020-6263-6 (HB)
ISBN 978-1-4020-6264-3 (e-book)

Published by Springer,
P.O. Box 17, 3300 AA Dordrecht, The Netherlands.

www.springer.com

Printed on acid-free paper

To my wife and sons.

Table of Contents

Preface

This book includes Volume I of the proceedings of the 2006 International Conference on Systems, Computing Sciences and Software Engineering (SCSS). SCSS is part of the International Joint Conferences on Computer, Information, and Systems Sciences, and Engineering (CISSE 06). The proceedings are a set of rigorously reviewed world-class manuscripts presenting the state of international practice in Advances and Innovations in Systems, Computing Sciences and Software Engineering.

SCSS 06 was a high-caliber research conference that was conducted online. CISSE 06 received 690 paper submissions and the final program included 370 accepted papers from more than 70 countries, representing the six continents. Each paper received at least two reviews, and authors were required to address review comments prior to presentation and publication.

Conducting SCSS 06 online presented a number of unique advantages, as follows:

- All communications between the authors, reviewers, and conference organizing committee were done on line, which permitted a short six week period from the paper submission deadline to the beginning of the conference.
- PowerPoint presentations, final paper manuscripts were available to registrants for three weeks prior to the start of the conference.
- The conference platform allowed live presentations by several presenters from different locations, with the audio and PowerPoint transmitted to attendees throughout the internet, even on dial up connections. Attendees were able to ask both audio and written questions in a chat room format, and presenters could mark up their slides as they deem fit.
- The live audio presentations were also recorded and distributed to participants along with the power points presentations and paper manuscripts within the conference DVD.

The conference organizers are confident that you will find the papers included in this volume interesting and useful.

Khaled Elleithy, Ph.D.
Bridgeport, Connecticut
June 2007

Acknowledgements

The 2006 International Conference on Systems, Computing Sciences and Software Engineering (SCSS) and the resulting proceedings could not have been organized without the assistance of a large number of individuals. SCSS is part of the International Joint Conferences on Computer, Information, and Systems Sciences, and Engineering (CISSE). I had the opportunity to co-found CISSE in 2005, with Professor Tarek Sobh, and we set up mechanisms that put it into action. Andrew Rosca wrote the software that allowed conference management, and interaction between the authors and reviewers online. Mr. Tudor Rosca managed the online conference presentation system and was instrumental in ensuring that the event met the highest professional standards. I also want to acknowledge the roles played by Sarosh Patel and Ms. Susan Kristie, our technical and administrative support team.

The technical co-sponsorship provided by the Institute of Electrical and Electronics Engineers (IEEE) and the University of Bridgeport is gratefully appreciated. I would like to express my thanks to Prof. Toshio Fukuda, Chair of the International Advisory Committee and the members of the SCSS Technical Program Committee, including: Abdelaziz AlMulhem, Alex A. Aravind, Ana M. Madureira, Mostafa Aref, Mohamed Dekhil, Julius Dichter, Hamid Mcheick, Hani Hagras, Marian P. Kazmierkowski, Low K.S., Michael Lemmon, Rafa Al-Qutaish, Rodney G. Roberts, Sanjiv Rai, Samir Shah, Shivakumar Sastry, Natalia Romalis, Mohammed Younis, Tommaso Mazza, and Srini Ramaswamy.

The excellent contributions of the authors made this world-class document possible. Each paper received two to four reviews. The reviewers worked tirelessly under a tight schedule and their important work is gratefully appreciated. In particular, I want to acknowledge the contributions of the following individuals: Yongsuk Cho, Michael Lemmon, Rafa Al-Qutaish, Yaser M. A. Khalifa, Mohamed Dekhil, Babar Nazir, Khaled Hayatleh, Mounir Bousbia-Salah, Rozlina Mohamed, A. Sima Etner-Uyar, Hussein Abbass, Ahmad Kamel, Emmanuel Udoh, Rodney G. Roberts, Vahid Salmani, Dongchul Park, Sergiu Dumitriu, Helmut Vieritz, Waleed Al-Assadi, Marc Wilke, Mohammed Younis, John Zhang, Feng-Long Huang, Natalia Romalis, Hamid Mcheick, Minkoo Kim, Khaled Rasheed, Chris Panagiotakopoulos, Alex Aravind, Dinko Gichev, Dirk Mueller, Andrew Vincent, Ana Madureira, Abhilash Geo Mathews, Yu Cai, Spyros Kazarlis, Liu Xia, Pavel Osipov, Hamad Alhammady, Fadel Sukkar, Jorge Loureiro, Hemant Joshi, Hossam Fahmy, Yoshiteru Ishida, Min Jiang, Vien Ngo Anh, Youming Li, X. Sheldon Wang, Nam Gyu Kim, Vasso Stylianou, Tommaso Mazza, Radu Calinescu, Nagm Mohamed, Muhammad Ali, Raymond Wu, Mansour Tahernezhadi, Trevor Carlson, Sami Habib, Vikas Vaishnav, Vladimir Avdejenkov, Volodymyr Voytenko, Vygantas Petrauskas, Shivakumar Sastry, U. B. Desai, Julius Dichter, Hani Hagras, Giovanni Morana, Mohammad Karim, Thomas Nitsche, Rosida Coowar, Anna Derezinska, Amala Rajan, Aleksandras Vytautas Rutkauskas, A. Ismail, Mostafa Aref, Ahmed Abou-Alfotouh, Damu Radhakrishnan, Sameh ElSharkawy, George Dimitoglou, Marian P. Kazmierkowski, M. Basel Al-Mourad, Ausif Mahmood, Nawaf Kharma, Fernando Guarin, Kaitung Au, Joanna Kolodziej, Ugur Sezerman, Yujen Fan, Zheng Yi Wu, Samir Shah, Sudhir Veerannagari, Junyoung Kim and Sanjiv Rai.

Khaled Elleithy, Ph.D.
Bridgeport, Connecticut
June 2007

An Adaptive and Extensible Web-based Interface System for Interactive Video Contents Browsing

Adrien Joly **Dian Tjondronegoro**

Faculty of Information Technology,
Queensland University of Technology
GPO Box 2434, Brisbane 4001, Queensland

adrien.joly@gmail.com, dian@qut.edu.au

Abstract - With the growing popularity of mobile devices (including phones and portable media players) and coverage of Internet access, we tend to develop the need of consuming video content on the move. Some technologies already allow end-users to watch TV and listen to news *podcasts* or download music videos on their devices. However, such services are restricted to a provider's selection of pre-formatted and linear content streams. Hence, we propose a web-based interface system that supports interactive contents navigation, making it possible for end-users to "surf" on video content like they are used to on the Web. This system is extensible to any specific domain of video contents, any web-enabled platform, and to any browsing scheme. In this paper, we will explain the architecture and design of this system, propose an application for soccer videos and present the results of its user evaluation.

Keywords: architecture, multimedia, content, browse, hypermedia, navigation, video, semantic, XML, pipeline

I. INTRODUCTION

Nowadays, most of us carry latest-technology mobile devices, such as mobile phones, PDA, pocket game consoles and portable media players, allowing to play video contents wherever we go. Along with the increasing coverage and bandwidth of wireless Internet access, today's consumers expect a richer, easier and more rewarding experience from their video-enabled devices to find, select, retrieve and consume video content on the move. The popular solution of cellular network providers is to propose a restrictive range of videos to download and/or access to certain TV channels. But this approach is very restrictive as these providers "push" their own content instead of leaving the consumer browse any content from any source on the Internet.

While streaming any video from the Internet is becoming possible on mobile devices like on desktop computers, their technical and ergonomical constraints bring new issues to consider. Firstly, according to [1], usage of mobile devices is not as exclusive as using a desktop computer at home. Mobile users can be distracted at any time by their context, hence they want an adaptive and flexible way to access the information they are expecting at a time. Secondly, mobile devices are technically limited: battery life, memory capacity, computing power, screen size, input interfaces, etc… Hence, the browsing experience and the display of contents must be adapted to ensure their usefulness on the mobile device. Thirdly, wireless access to the internet is too expensive for users to afford wasting long and bandwidth-consuming connections as they could at home using unlimited broadband Internet access or TV.

In order to bring the multimedia web to our mobile devices while satisfying these constraints, we need to adapt the retrieval of multimedia content for these specific platforms and their usage. Our approach is to allow users to browse inside the video content without having to transfer it integrally on the device and to personalize the content. This is made possible by video indexing, as long as a browsing interface can be generated from the resulting indices, providing direct links to access its most relevant segments.

In this paper, we propose a web-based interface system relying on a SEO-indexed video library to bring rich and personalized video content efficiently (by focusing on the information that matters for the user) and adaptively (to the platform's constraints) at anytime (on demand) and anywhere (on the move or at home) to the average end-user (with ease of use). As we are aware that new devices and new types of video contents are constantly appearing on the market, this system is adaptive to new devices and extensible to new video domains. Moreover, its modular architecture makes it possible to integrate new components allowing browsing content in an intuitive, precise and enjoyable manner.

This paper is structured as follows: Section 2 describes our previous work and the proposed application of the system; Section 3 specifies the workflow of the expected application; Section 4 outlines the architecture of the proposed interface system; Section 5 explains the design of the dynamic interface generation; Section 6 describes the implementation of our application on the system; Section 7 discusses the success of our approach by evaluating the application; and finally, Section 8 and 9 describe the conclusions and future work.

II. APPLICATION AND PREVIOUS WORK

As an application of such a system, we have proposed navigation on soccer matches with adaptation to user preferences [2]. We have identified use cases which mobile soccer enthusiasts would benefit from. The idea is to browse

K. Elleithy (ed.), Advances and Innovations in Systems, Computing Sciences and Software Engineering, 1–6.

soccer matches on the move after having recorded them at home. Users could then use their mobile device (e.g. during their daily bus journey) to browse the highlights of a match in a constrained/noisy environment, browse the latest exciting events concerning one's favorite players, list matches with interesting specificities, etc...

For this application, we need to extract some valuable metadata (structure and semantics) from the soccer videos recorded on TV. Inspiring from MPEG-7, we have proposed a video indexing scheme and developed tools permitting to segment and semi-automatically analyze soccer videos in order to generate these indices from TV-recorded soccer matches [2]. The extraction process is out of the scope of this paper, but we will summarize the SEO indexing scheme.

SEO is a semi-schema and object-relational based indexing scheme allowing a flexible and efficient way of annotating and indexing video content in XML. This work has been focusing on soccer videos essentially but the same paradigm can be used for different domains as well. Moreover, it is based on MPEG-7 concepts and ideas, which have been widely-accepted as the video description standards by video professionals and institutions. A *SEO* index consists of the following components:

- *Segments*: A segment is an individual spatio-temporal range of a video document that contains indexable contents such as whistle, slow motion replay, and close-up on players' face and near goal area.
- *Events*: An event is a special type of video segment that contains a particular theme or topic such as a soccer goal and foul. It usually embeds annotations and possibly linking to objects.
- *Objects:* An object can be a person, a place, or any other material or immaterial entity that is involved with an event.

Thus, any video item (e.g. a soccer match) contains a definition of its belonging segments, events and objects.

In the SEO model, access to content is brought by the use of domain-specific queries (expressed in the XQuery language) that generate hierarchical summaries from the XML metadata. Some of those queries are depending on the user's preferences in order to personalize the results.

III. WORKFLOW OF THE SOCCER APPLICATION

In order to ensure interactivity and intuitivity for end-users, we have designed the system as a website, allowing users to browse the content by jumping from page to page using hyperlinks. Moreover, this paradigm is easily portable to devices that have limited input capabilities (e.g. few keys, stylus) and can be supported on mobile phones using the WAP technology.

The interface system is session-aware, that means that it keeps the context for each connected user and adapts the content according to his/her preferences. As shown on Figure 1, connecting on the system leads to a login page. Once authenticated, the user is brought to the homepage which proposes links to queries and video segments for registered domains and latest matches, as shown on Figure 2. For each match, a keyframe and some metadata (e.g. place and date of the match) are displayed. At all times, the user can go back to this page or to the User Preferences management pages shown on Figure 3. These pages allow the user to select his/her favorite players, event and segment types for each sport, using 2-list interfaces.

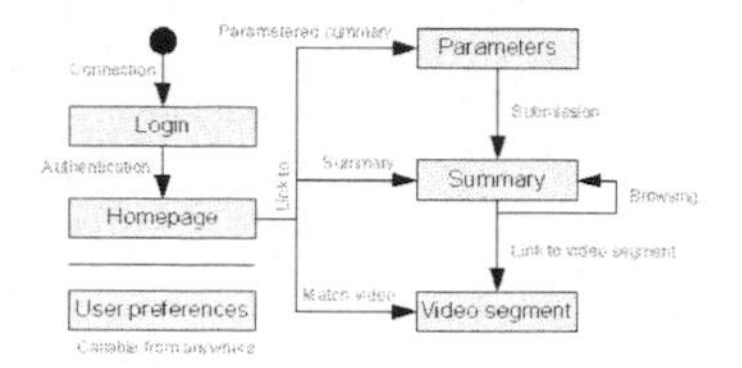

Figure 1: Workflow of the soccer application

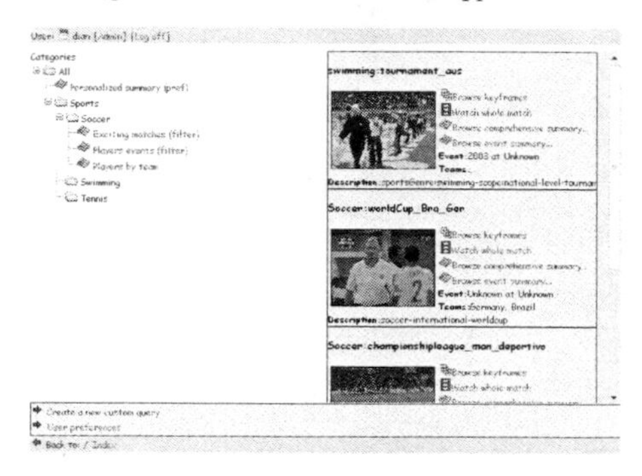

Figure 2: Homepage on PC

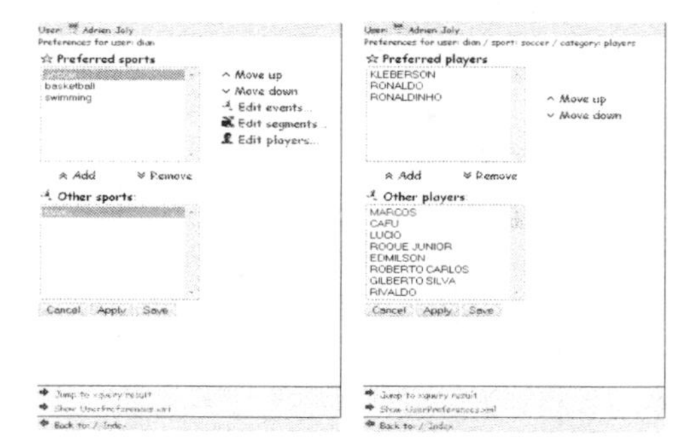

Figure 3: User preferences management pages

Two types of queries are proposed on the main page: (i) domain queries return summaries which scope a given domain only (e.g. soccer), whereas (ii) media queries return summaries which scope a given video (e.g. a match). Some queries are parameterized, making it possible for the user to customize the results. As seen on Figure 4, when the user selects such one, he/she will be invited to fill in a form of parameters (a) before displaying the resulting summary (b). This form is already populated in order to propose predefined (and existing) options to the user instead of letting him/her type the corresponding information. The summary is an interactive page containing a hierarchical browsing pane at the top and a details pane at the bottom showing data about the currently selected item. Depending on the type of item, links to video segments or to other summaries can be proposed.

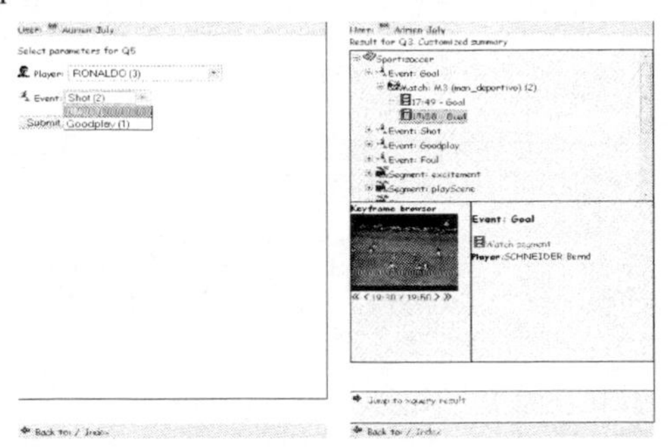

Figure 4: (a) Parameters form and (b) summary

More details about this application and its user evaluation are given in Sections VI and VII. We are now describing the architecture and design of the interface system on which relies the soccer application.

IV. System Architecture

In order to easily support many kinds of client devices by generating and delivering adapted user interfaces to each of them, we decided to adopt a web-based architecture that consists in "light" clients using a web browser to interact with an applicative server called the *Retrieval Server*. Figure 5 depicts the architecture of the proposed system.

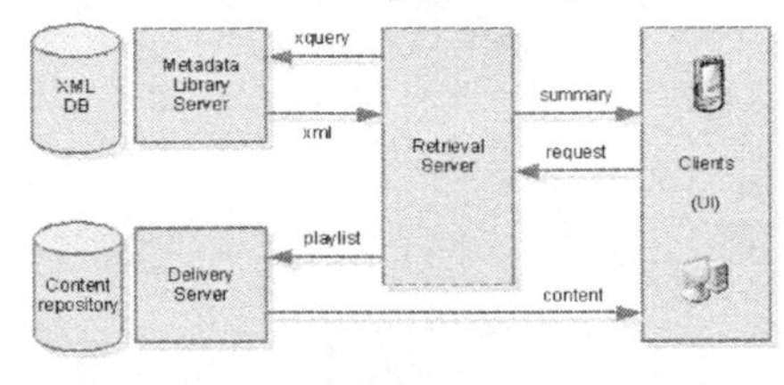

Figure 5: Architecture of the system

Because the metadata library is heavily solicited by this system, access to the metadata has to be handled by a XML database server that ensures reliable and efficient retrieval using XQueries [3]. For that purpose, we have chosen to use eXist [4], a popular and robust open source solution.

The actual delivery of video content must be handled by one or several streaming server(s), according to the type and format of content and the final application. We chose not to focus on streaming issues, hence the proposed system was designed to support any streaming server by extension.

A. Design of the retrieval server

Our system is expected to be extensible to new domains, browsing schemes, platforms and delivery servers. In order to satisfy these specifications, the *Retrieval Server* was designed with the layers listed on Figure 6: (i) the *data access* layer contains the queries that feed the system with data from the metadata library, (ii) the *representation* layer defines the templates that transform raw metadata into their human-readable representation, (iii) the *user interface* layer proposes components that implement platform-specific browsing schemes, (iv) the *services* layer handles the calls for delivery of summaries and content to the user that will be actually processed by (v) the *servers* layer.

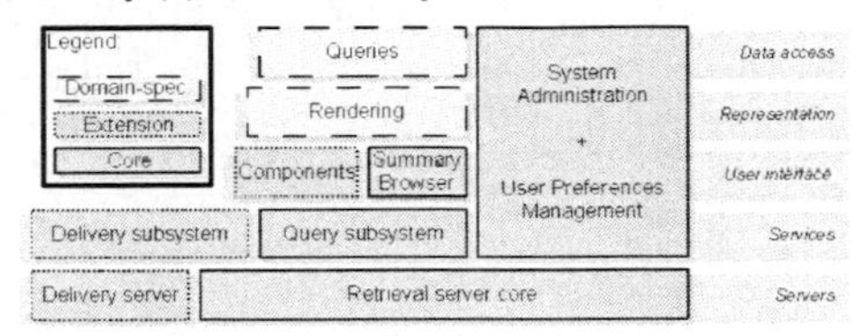

Figure 6: Extensibility of the retrieval server

As depicted on Figure 6, the layers (i) and (ii) can be extended with domain-specific queries and rendering templates. New browsing schemes and platforms can be supported by extending the layer (iii) with components. And content delivery servers can be plugged by adding a handler to the layer (iv) services.

B. Implementation of the retrieval server

The *Retrieval Server Core* located on the layer (v) is a J2EE application based on Java servlets which handle general web logic (e.g. sessions). On the layer (iv), the *Query Subsystem* consist in servlets implementing the GUI (Graphical User Interface) generation engine which drives the data flow from the metadatabase to the final user interface. This will be explained further in Section V. On the same layer, the *Delivery Subsystem* can be extended by servlets to give access to the content from specific delivery servers. In the current implementation of the system, we use the HTTP server as a *Delivery Server*, providing direct download access to content. Hence we implemented two servlets: the first returns URLs to keyframes which are identified by the video identifier and the timestamp, and the second returns URLs to video segments. Both layers (iii) and (ii) rely on XSL transformations. This will be explained further in Section V. At last, the layer (i) consists in XQuery files.

V. Dynamic GUI generation pipeline

In order to generate summaries adaptively to an extensible pool of queries, domains, browsing schemes and client platforms, the GUI must be dynamically generated. As our queries return XML data that must be processed to generate platform-adapted HTML summaries, we chose to follow the "XML/XSLT pipeline" approach presented in [5]. This approach consists in processing the XML input data with different XSLT transformations in order to obtain platform-adapted documents at the end of the pipeline. Furthermore, because the numerous extensible layers result in many different chain combinations, the pipeline has to be generated dynamically. In order to achieve this, the GUI subsystem generates the pipeline by building a graph in which transformations are connected according to their specified input and output formats.

Browsing Schemes are defined as GUI components that transform high-level results of queries into platform-adapted interactive summaries allowing the user to browse the contents. The system natively includes a generic GUI component called the *Summary Browser*, which consists of a hierarchical browser composed of a *tree pane* and a *content pane*, as seen previously on Figure 4. It is generic since this tree can match the structure of any XML output from the queries. In this browsing scheme, users can select nodes in the tree of results to display their corresponding metadata and content (e.g. details, keyframes and other related data). For example, if a query returns a set of soccer matches, clicking on a match will show the location and date of that match, some keyframes, hyperlinks to match-specific queries and to the whole match video.

Note that a "*Keyframe Browser*" is natively included in the system as another GUI component. It enables the user to browse the keyframes of a segment.

A. Case of the Summary Browser

The generation of a summary using the *Summary Browser* component is a particular scenario of the pipeline. As it is the only browsing scheme natively included with the system, we will now describe the process of generating a summary from the results of a query.

As depicted on Figure 7, this process consists in 3 transformations:

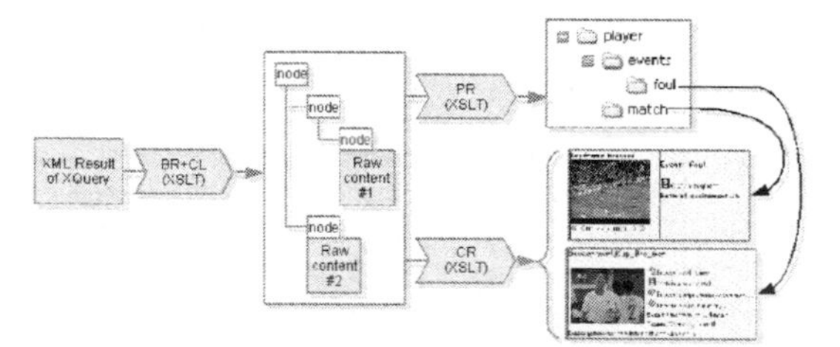

Figure 7: Data flow from the XML query result to the final HTML page using the Summary Browser

1. **Browsing document Rendering (BR) with Content Location (CL)**: This query-specific XSL transformation builds up the structure of the summary that will be shown on the final HTML page for browsing from the results of the query. The output of this transformation respects the input format of the *Summary Browser* component; it is a high-level XML format with no platform-specificities. Moreover, this transformation calls a generic "content location" transformation that will embed the content (e.g. reference to keyframes and textual details) associated with nodes of the tree structure, in order to leave their rendering for later. This transformation permits to separate the structure from the content while keeping linkage information using identifiers for late binding.

2. **Page Rendering (PR)**: This transformation renders the final platform-adapted DHTML page from the high-level tree structure resulting of the previous transformation. This page contains a *treeview* component and the scripts driving the browsing logic for the final summary. This transformation is actually the instantiation of the *Summary Browser* component.

3. **Content Rendering (CR)**: This transformation is a second pass on the output of the BR+CL transformation. It extracts the embedded content elements generated by the *Content Locator* and delegates their rendering to domain-specific templates defining the representation of those elements in the expected output format (HTML).

Then, the rendered content is integrated to the page (by merging them) to obtain the final platform-adapted summary that will be returned to the client. The DHTML code of the page drives the browsing interactivity, content being bound to their corresponding tree nodes thanks to the identification applied by the CL transformation.

The main strengths of this design are that: firstly, we abstract the platform specificities in the first transformation, as it only defines the way to structure the results of a given query in a high-level format. Thus, it is easy to add support for new client platforms or change the GUI layout by implementing different versions of the *Page* and *Content Rendering* transformations only. Secondly, the rendering of content is delegated to domain-specific templates. Hence, it's easy to add support for new domains; the main transformation engine remains unchanged. Thirdly, the separation of the tree and the linked content makes it possible to implement a "PULL" version of the *Summary Browser* that would download content on-demand from the server instead of downloading everything ("PUSH" of the contents from the server).

B. System configuration model

This section will describe the configuration model which the extensible GUI generation pipeline relies on. This model defines the entities that the system deals with in order to generate adapted interfaces dynamically. It is stored as a XML document in the database. Extending the system consists in merging extensions in this common system configuration XML tree.

As depicted on Figure 8, in the *system* configuration tree are defined *domains* and *platforms*. For both of them, instances are hierarchically linked using a reference to the *parent* identifier. In the domain hierarchy, each *domain* node inherits the *queries* that are defined for its ancestors. In the *platform* hierarchy, each *platform* node inherits the *transformations* that are defined for its ancestors.

A *domain* is defined by:

- *Queries* are proposed on the main page to provide summaries that scope on the domain,
- *Media-queries* are defined like *queries* but are proposed for each video items of the corresponding domain (e.g. the "event summary" and "comprehensive summary") and,
- Its *renderer*: a XSL *transformation* that renders the metadata and associated content from the SEO components (segments, events and objects) that are defined for this domain and returned by the queries.

A *Query is* defined by:

- Its *presentation* consisting in XSL *transformations* that will convert the XML result of the query to the input format of the rendering component that will be used to generate the final interactive summary. Note that different transformations can be proposed, depending on the *target* platform.
- An optional *form* consisting in a *query* and its *presentation* (defined as above), for parameterized queries only. The query will ensure the retrieval of data that will be used to populate the form generated by a XSL *transformation*.

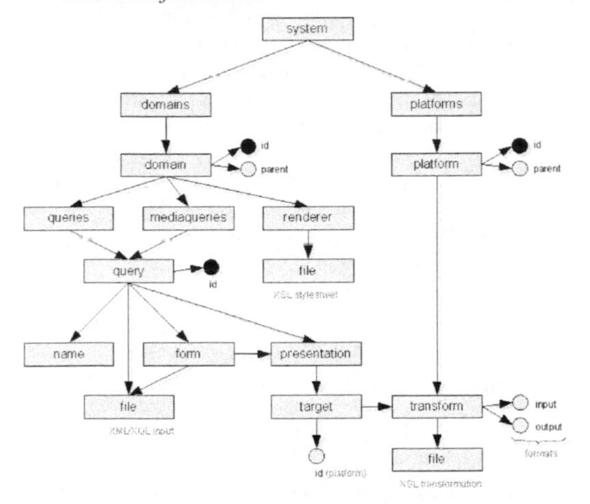

Figure 8: System configuration scheme

Platforms are defined as a set of platform-specific XSL *transformations* for given input and output *formats*. Each transformation is actually the implementation of a GUI component (like the *Summary Browser*) that will generate the final user interface from formatted query results. The platform hierarchy enables the GUI generation engine to elect higher-level (and thus, less precisely adapted) platforms for the case where a perfect match between a required *target* platform and an existing platform definition could not be found. As an example, we can consider the following platform hierarchy: *pfm.pc, pfm.ppc, pfm.ppc.vga, etc*, where *pfm* denotes platform; *pc* corresponds to Personal Computers (desktop), *ppc* stands for Pocket PC and *vga* is a special kind of *ppc* with a high-definition screen. If we are using a Pocket PC client with VGA screen, but that no version of the *Summary Browser* has been implemented specifically for VGA screens, the standard Pocket PC version will be used instead.

The *Format* identifiers are used to match inputs and outputs in order to build transformation graphs automatically. Like platforms, formats are also identified hierarchically, where the descendants specialize their ancestors, giving more flexibility for the GUI generation process. Note that, contrary to *platforms*, *formats* are not explicitly declared. They are implicitly defined by their hierarchical identifier. As an example, we can consider the following format hierarchy: *fmt.browser, fmt.html.basic, fmt.html.v4,* where *fmt* denotes format, *browser* denotes the "*Summary Browser*" input format, and *html* format could be *basic* or advanced (*v4*) depending on the target web browser.

A *transformation* file is specified given its *input* and *output* formats. During use, the system will build the transformation pipeline required to render the GUI as a chain of transformations in which the inputs are optimally corresponding to the outputs for the given platform.

VI. IMPLEMENTATION WITH SOCCER VIDEOS

The system has been implemented and deployed on Tomcat as a web application. As seen on Figure 9, the *Summary Browser* GUI component has been implemented in DHTML (HTML + JavaScript) for both modern browsers on PC and Pocket Internet Explorer on Pocket PC.

In order to evaluate the system, we have implemented the "soccer" domain (queries and templates) and added 3 SEO-indexed soccer matches in the video library. This application supports 6 summaries, including 2 which are parameterized, and 1 which is based on user preferences. We will now describe the summaries Q1, Q2, Q3, Q4, Q5 and Q6.

Q1. Comprehensive Summary: This summary lists the « play-break tracks » of a given match. Each track consists in a "play" and a "break" segment, the "play" phase being interrupted by an event (e.g. foul, goal…). A "play" segment describes the cause of an event whereas a "break" segment describes its outcome.

Q2. Events Summary: This summary chronologically lists the events happening in a given match.

Q3. Players by Team: This summary lists all video segments in which a players appears as part of one of his teams, for every player of any team.

Q4. Player Summary: After selecting a player, this query returns personal and strategic details (e.g. position) and event segments related to this player. Segments are grouped by event type and match.

Q5. Exciting Matches: This summary provides a list of matches filtered by their custom number of goals and fouls.

Q6. Personalized Summary: Basing on the user's preferences, this summary shows the latest video segments related to his/her favorite players, types of events and segments.

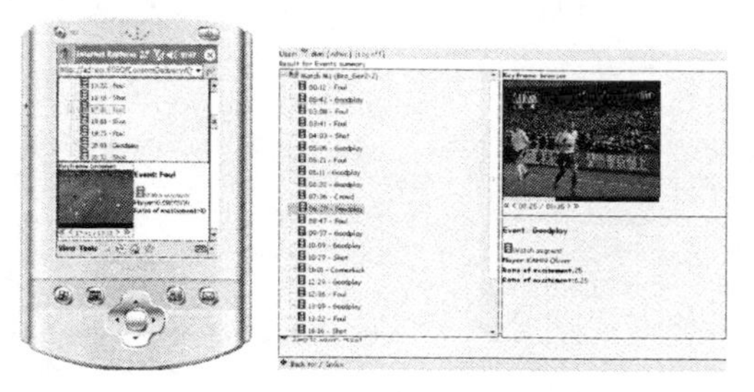

Figure 9: Web-based Video Retrieval System Accessible for Desktop and Mobile Devices

VII. USER EVALUATION

To prove the effectiveness of our proposed interface system, we have conducted a user evaluation on the soccer application relying on this system with a group of 53 university students. The aim of this survey is to gather users' feedback on the effectiveness, intuitiveness and enjoy-ability of the system for this application. Each criterion is measured uniformly by quantified "strongly agree, agree, disagree, and strongly disagree".

As shown on Figure 10, an average of 90% of the surveyed users agreed with the effectiveness, intuitiveness and enjoy-ability of the system for the proposed soccer application, and about 15% of them strongly agreed. The most appreciated summaries were Q2 (Event Summary) and Q6 (Personalized Summary).

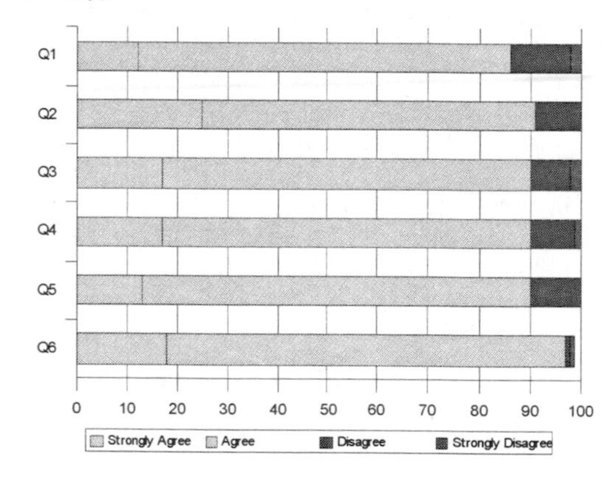

Figure 10: User acceptability of summaries

VIII. CONCLUSION

In this paper, we have proposed the architecture and design of an extensible video retrieval and browsing system. The architecture is intended for use with a XML-based video content indexing scheme, and we have shown its effectiveness with SEO-indexed soccer videos. The design is highly modular, allowing adaptation to various domains, browsing schemes and any web-enabled platform/device.

The major contributions proposed in this paper are: (i) the extensible multi-layered architecture of the system, (ii) the *Summary Browser* as a hierarchical browsing scheme allowing to browse the results of virtually any query in a rich and user-friendly manner, (iii) the graph-based GUI generation pipeline, and (iv) the system configuration scheme including domain description and platform adaptation.

Our approach expands the hyper-navigation paradigm that is used on today's websites (pages made of text and images) to video content in order to browse it in a non linear manner. Moreover, our evaluated implementation demonstrated that such an approach is realizable and effective.

This system could be the platform for new services brought to end-users for video hyper-navigation and summarization. It is suitable for many specific applications that could bring profit to content providers by selling more content and/or by proposing highly-targeted advertising to their clients.

IX. FUTURE DIRECTIONS

As the architecture and design of the system proposed in this paper has already shown its robustness in the scope of its application in a soccer video library, we propose some future directions to improve it further: (i) add new browsing schemes (e.g. using thumbnails or chronological representation), (ii) add new data rendering tools (e.g. charts), (iii) consider user's location and environment for delivery of targeted and adapted content, (iv) add user community and exchange features (e.g. forums, and tagging), and (v) support custom query creation.

REFERENCES

[1] H. Knoche and J. McCarthy, "Design requirements for mobile TV," *Proceedings of the 7th international conference on Human computer interaction with mobile devices & services*, pp. 69-76, 2005.

[2] D. Tjondronegoro, *Content-based Video Indexing for Sports Applications using Intergrated MultiModal Approach.* Melbourne: Deakin University, 2005.

[3] S. Boag, D. Chamberlin, M. F. Fernández, D. Florescu, J. Robie, and J. Siméon, "XQuery 1.0: An XML Query Language," *W3C Candidate Recommendation*, 2006

[4] W. Meier, "eXist: An Open Source Native XML Database," *Web, Web-Services, and Database Systems. NODe*, pp. 7-10, 2002

[5] M. Butler, *Current Technologies for Device Independence.*: Hewlett-Packard Laboratories, 2001.

Design and Implementation of Virtual Instruments for Monitoring and Controlling Physical Variables Using Different Communication Protocols†

A. Montoya [1], D. Aristizábal [2], R. Restrepo [3], N. Montoya [4], L. Giraldo[5]
Scientific and Industrial Instrumentation Group
Physics Department
National University of Colombia Sede Medellín
AA 3840, Medellin, Colombia

***Abstract* – In this Project were developed software components (Java Beans) which have the capability of communication through different communication protocols with hardware elements interconnected to sensors and control devices for monitoring and controlling different physical variables, conforming a hardware-software platform that obeys the virtual instruments design pattern. The implemented communication protocols are RS232, 1-Wire and TCP/IP with all of its annexed technologies like WiFi (Wireless Fidelity) and WiMax (Worldwide Interoperability for Microwave Access); also these elements communicate with a database and have the capability of sending text messages to cell phones through a GSM modem. For the development of software were used the object-oriented programming paradigm (OOP), Java programming language, LINUX OS and database server MySQL. As hardware, were used sensors and control devices in a 1-Wire network, a TINI (Tiny InterNet Interfaces) embedded system and a PIC (Peripheral Interface Controller) microcontroller.**

***Key words*: Virtual Instrumentation, Java, Monitoring, Control, Communication Protocols.**

I. INTRODUCTION

The virtual Instrumentation (VI) is a concept introduced by the National Instruments (NI) company. In 1983, Truchard and Kodosky of the NI, decided to face the problem of creating a software that would allow to use the personal computer (PC) as an instrument to make measurements, as result of this, they obtained the software denominated Laboratory Virtual Instrument Engineering Workbench (LabVIEW). Thus, the VI concept is conceived as "an instrument that is not real, it's executed in a computer and has its functions defined by software" [1].

A traditional instrument is characterized for performing one or several specific functions that cannot be modified. A virtual instrument is a hardware-software combination through a PC that fulfils the same functions of a traditional instrument [2]. Besides the VI are very flexible and their functions may be changed by modifying software. For the construction of a VI, it's required a PC, a data acquisition board and appropriated software.

This article describes a way to design and implement the necessary software to create virtual instruments with the property of being capable to communicate in a transparent way with the outside (to acquire data) through different communication protocols, such as: RS232, 1-Wire, TCP/IP, WiFi, WiMax, among others. This kind of software (Java Beans) allows the VI to be implemented in many kinds of applications like remote monitoring and control of physical variables and distributed automatism [4], [5],[6],[7].

II. MATERIALS AND METHODS

For the design and implementation of VI there were used the object-oriented programming paradigm (OOP), UML software modelling language (Unified Modelling Language) with ArgoUML tool , Java programming language of Sun Microsystems, IDE (Integrated Development Environment) NetBeans 5.0 of Sun Microsystems too and database server (DB) MySQL 5.0. All of them are free distributed.

The dynamic polymorphism, implemented through inheritance, is applied in the VI design in a way that allows the extension of the instruments in a transparent way and with an efficient code reutilization. The classes responsible of the communication are totally disconnected of the instruments code; this makes possible the adaptation of any new communication protocol with minimum effort. The VI are made fulfilling all the demanded requirements of Java language for them to be Java Bean components, accelerating control boards development through an IDE.

† This article is result of the investigation project "Remote monitoring and control of the physical variables of an ecosystem", which is financed by the DIME (project # 20201006024) of National University of Colombia Sede Medellin.
[1] e-mail: amontoy@unalmed.edu.co
[2] e-mail: daristiz@unalmed.edu.co
[3] e-mail: rrestrep@unalmed.edu.co
[4] e-mail: namontoy@unalmed.edu.co
[5] e-mail: lgiraldo@unalmed.edu.co

K. Elleithy (ed.), Advances and Innovations in Systems, Computing Sciences and Software Engineering, 7–11.

The developed VI are put under tests using communication through serial ports and inlaying them into 1-Wire sensor networks through the TINI embedded system of MAXIM / DALLAS Semiconductors [3], allowing remote monitoring and control through Internet. For this task, there were designed and implemented VI boards for remote monitoring and control of physical variables (temperature, relative humidity, luminosity) of a greenhouse dedicated to the production of flowers.

III. RESULTS

Virtual Instruments Software

The project establishes a main mother class which is the base of the virtual instruments. This class has all the fundamental methods that each virtual instrument should have like the method in charge of the instrument's thread and the data acquisition method, besides it defines the basic variables required to draw the object in an adequate and standard way for all the child instruments. The designed mother class is denominated *InstrumentoVirtual*, it extends *JComponent* and it's abstract.

The methods responsible of the alarm and to adjust the measurement scale are abstracts and should be defined by the child classes (this depend whether the instrument handles or not alarms or measurement scales, like the particular case of an on/off switch). The method in charge of repainting the objects on the screen is defined, but has to be rewritten by each child class due to the different characteristics of each instrument.

By having the Java Beans as the main idea of the project, all variables that directly affect the performance of the final element like variables in which are defined the colors of the final object, the measurement scales or the specific communication protocol that the particular Java Bean will use, should have its methods *set* and *get* to modify the variable and to give the actual value of it.

For the acquisition of data that will be shown on the screen, each protocol has its own capture method, nevertheless all these methods have the same name (*setLectura*), in other words, the method is recharged. The *setLectura()* method is responsible of creating the object that makes possible the communication. This method is recharged for the different protocols (RS-232, TCP/IP, 1-Wire, among others) desired to use. Each communication protocol is encapsulated in a class responsible of the configuration of all the parameters for an adequate communication. Therefore, in the *InstrumentoVirtual* class there are four different *setLectura()* methods where one of them is the main and it's the one that is going to be executed by defect by the child class that will be born from *InstrumentoVirtual*. This method has a switch/case block that decides the kind of connexion that has been defined for the child instrument. If for any reason it hasn't been defined the kind of connexion, the program assigns the connexion option to database. The connexion definition is made by *selectConexion*, which is an *int* variable in charge of telling the switch/case block the kind of connexion the user has defined for the specific object.

The group of methods in charge of communication are completely defined on the *InstrumentoVirtual* class because each object of a child class should have the capability of using any communication protocol, depending of the characteristics of the particular system that is being developed with software.

There's also a child class of *InstrumentoVirtual*, which is abstract too, this class is called *InstrumentoVirtualLinealCircular*, and has the only purpose of making possible the implementation of circular geometry objects (like for example a manometer). The implementation of a particular class for circular geometry objects is due to the complexity on the measurement scale transformation, the animation on data visualization and the transformation of the data acquired by the instrument to be correctly drawn on circular geometry.

The performance idea of the program for a specific Java Bean of the virtual instrumentation project, for example the thermometer is the next one: The element is drawn and the instruction to begin the thread that governs the Java Bean thermometer is executed. This thread remains constantly running, and to avoid PC recharging on the same process, there's a variable *tiempoMuestreo*, which indicates the thread to sleep for an entire *tiempoMuestreo* after executing the assigned routine, optimising in that way the program performance. Therefore, for velocity exigent programs, these threads may sleep less than 20 milliseconds, and being actualizing data, or in cases like the serial port, each time the data arrives an interruption of the thread's dream is generated, paying attention to the serial port and acquiring the data without any lost. After the thread has slept for a defined time and if there isn't any problem, the thread executes all its routine again, repainting the object on the screen, actualizing the lecture value to the last taken data and returns again to its dream.

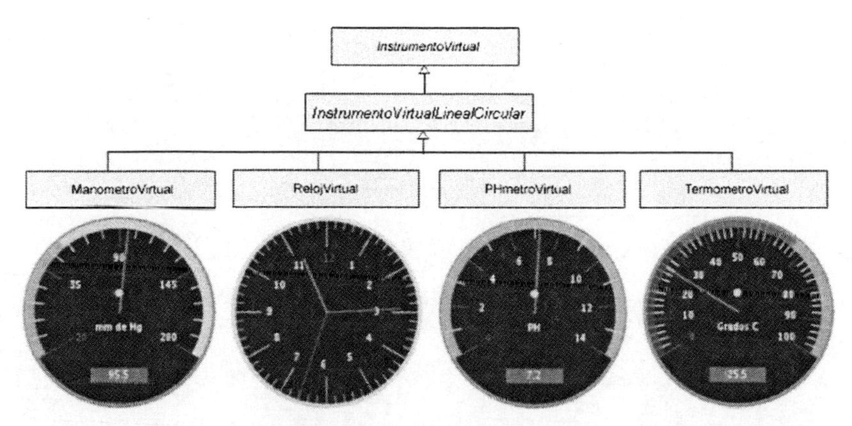

Fig. 1: Virtual Instruments and its inherency.

Fig.1 illustrates some of the designed VI: a thermometer (*TermometroVirtual*), a manometer (*ManometroVirtual*), a phmeter (*PHmetroVirtual*) and a clock (*RelojVirtual*). It can be clearly seen they inherit from *InstrumentoVirtual* class or from the *InstrumentoVirtualLinealCircular* child class, depending of the specific instrument geometry.

Fig. 2 shows the structure of *InstrumentoVirtual* class and some of its child classes, like *PHmetroVirtual* and *TermometroVirtual* in detail. There is shown the great complexity of the mother class in comparison of the simplicity of child classes, this clearly shows the huge potential of OOP and one of its great characteristics, the inheritance.

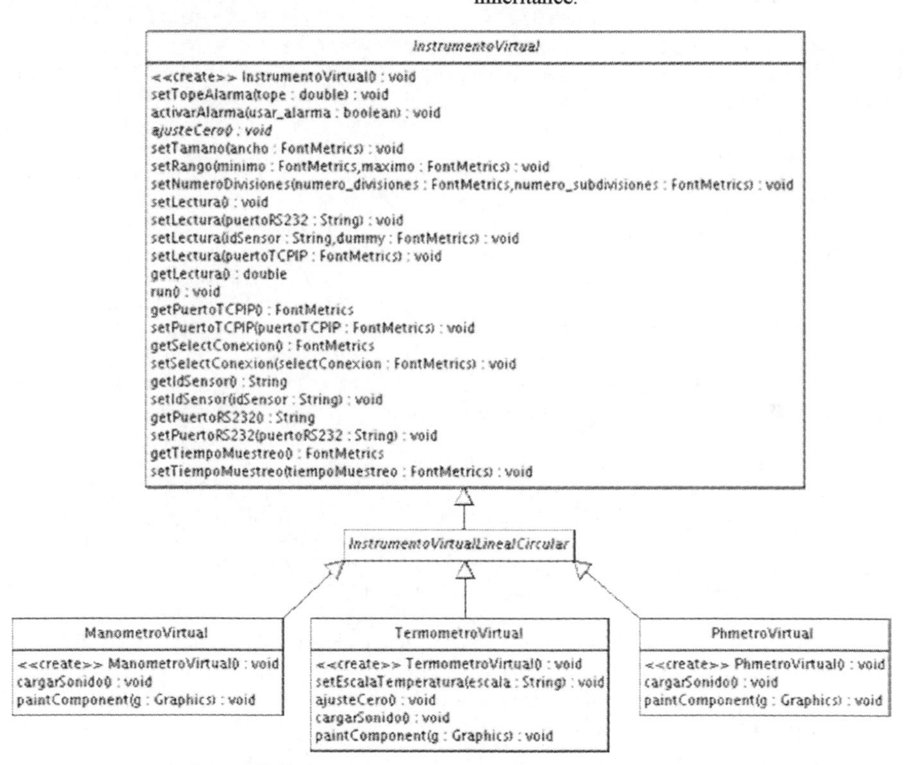

Fig. 2: Small UML diagram that show the huge differences between mother and child classes.

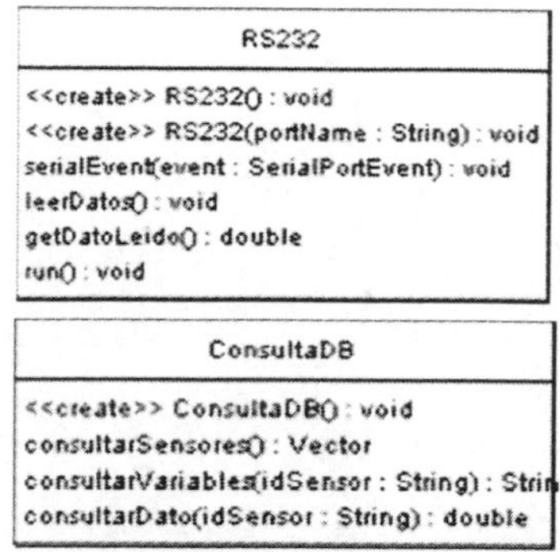

Fig. 3: Some classes in charge of communication protocols. RS232 is the class that manages the RS232 communication protocol. ConsultaDB is the class that manages the database connexion

Fig 3. shows two of the classes that manage the different communication protocols. As explained before, notice that these classes are totally disconnected of the virtual instruments code.

Implemented Hardware

Due to the fact the designed VI handle different communication protocols; there were implemented two different hardware assemblies to carry out all the pertinent tests.

A) Greenhouse Remote Monitoring and Controlling System.

The greenhouse remote monitoring and controlling system used in this project is a hardware-software platform whose principal idea is the monitoring and control of the physical variables of a greenhouse through Internet. For this a new type of embedded system called TINI was used, which has a lot of potentialities like being able to program it with JAVA language and to support a lot of communication protocols, like TCP/IP, RS232, CAN, 1-Wire, among others. The TINI embedded system is in charge of the sensor/control devices network (this network uses 1-Wire communication protocol which is a communication protocol specialized on sensor networks), the sending of data to a central server and the reception of different kinds of requests from the server. The central server saves data into a database and allows it visualization. Fig. 4 shows a general scheme of the entire system.

A very important fact is that communication between TINI embedded systems and the server may be made in several ways: through a wire connexion using TCP/IP communication protocol or through a wireless connexion using WiFi or WiMax connexion devices. Moreover, it's worth to notice the TINI's great capacity of data acquisition from different communication protocols and to hand it over in TCP/IP data packages; this gives great flexibility to the system.

The joint between the system and the VI may be made in two ways: The first one consist in that the data arriving from the sensor/control devices network are saved on database, therefore the VI directly communicate with the DB for acquiring their respective data and showing it on the screen

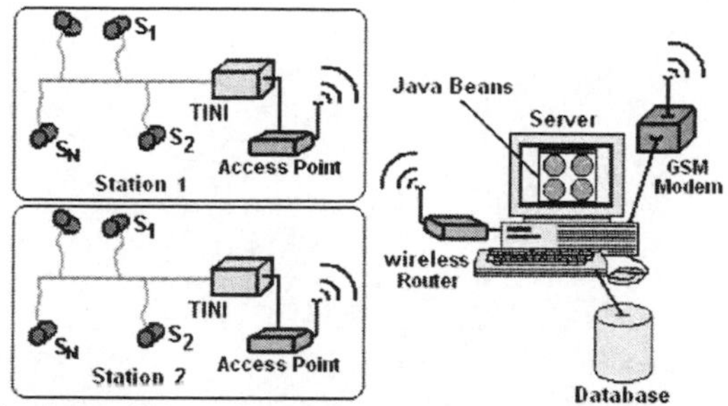

Fig. 4: General scheme of the greenhouse remote monitoring and controlling system.

The second way consist in the direct data acquisition made by the VI, in this way each VI handles a connexion with the TCP/IP port, only receives the data of concern (data sent specially to the VI to a specific virtual port) and shows the receive data on the screen. In this way, there's not necessary a DB.

B) Acquisition board – RS232.

On the development of the project, there was made a board set up by a PIC 16F877 microcontroller, a LM35 temperature sensor and a MAX-232 device (this one was used to transform PIC TTL voltages to RS232 communication protocol required voltages), which is in charge of acquiring the temperature sensor data and sending the acquired data, by way of RS232 protocol, to a server in which they are directly captured by a VI element. An important fact is that the board has the capability of acquiring data from eight sensors, however in this case, there's the limitation that only one VI occupies the serial port and denies the other seven VI to read the port and therefore to present data on screen. This problem may be resolved by an agent responsible of taking possession of the port and saving the arriving data on DB, in this way the VI would directly connect to DB without any lost of information.

The problem of occupying a physical PC port by a VI doesn't apply on TCP/IP communication protocol because this protocol handles "virtual" ports, whose (in theory) may be infinite (they are truly limited by the working PC). All these ports have the same physical input channel which is never occupied for particular software, unlike it happens on serial ports.

C) GSM MODEM – Alert SMS Messages.

The third implemented hardware device is a Samsung X486 cell phone which is used as a GSM MODEM for the delivery of alerts in form of SMS messages. For this purpose, there were used AT commands to handle the cell phone as a GSM MODEM; all this process is carried out by serial communication with the device. All the VI have the capacity of using this MODEM, however, a limitation of this system is that the MODEM may only be used by a VI at the time (due to the limitations of serial ports, explained before).

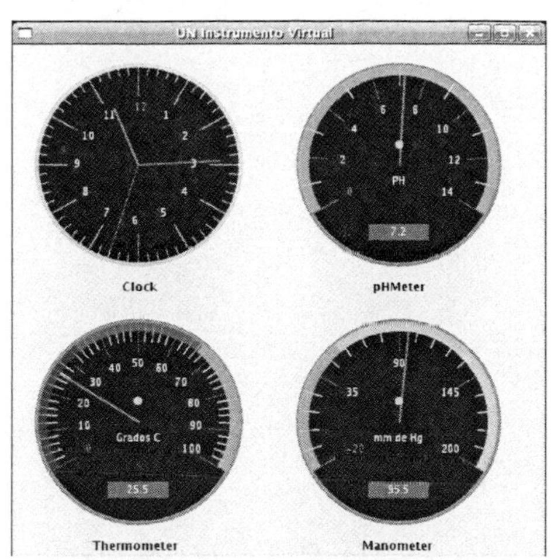

Fig. 5: An image of some virtual instruments acquiring data

The complete system was proved using four VI acquiring data from different communication protocols (Fig 5): two of them connected directly to the DB, another one acquired its data directly from the TCP/IP port (1414 port) and the last one connected directly to the serial port (ttyS0). Each one of these VI presented the acquired data on screen in a very satisfactory way. Furthermore, the GSM MODEM was connected to the USB port, being registered as a serial port identified as ttyUSB0. The data sent to the VI were modified in a way that alerts would be thrown; these ones were satisfactory sent as text messages by the module that handles the GSM MODEM.

IV. Discussion and Conclusions

As a result of this project, there were obtained virtual instruments capable of acquiring data in an autonomous way using different communication protocols like RS232, TCP/IP, among others. The carefully used software engineering for the VI design allows them to be easily adapted to different situations and, in this way, having instruments that may be implemented on monitoring and control software development for the small and medium national companies in a fast, economic and reliable way.

The manufacture applications require software to be reliable, of high performance and great adaptability. The virtual instruments designed and implemented as described in this article, bring all those advantages through the integration of characteristics such as alerts, security and network management. The great adaptability of this technology allows its incorporation in different kinds of environment like a house, a laboratory, a forest, a greenhouse or an industry.

Virtual instruments bring significant advantages on each stage of the engineering process, from the design and investigation to the manufacture test.

One of the biggest advantages of software is the great versatility it has due to its implementation on Java language, which allows it to have a successful performance on diverse operating systems. Furthermore, being Java and all its used tools (for design and development) of free distribution, it's very possible the implementation of this kind of systems at a very low cost.

V. Recommendations and Development Fronts

If there are introduced more sophisticated graphics visualization systems, statistical analysis and distributed network implementation into virtual instruments, they will bring capacity for robust company applications. By having well structured software, the VI may have intelligence and decision making capabilities.

Actually, the Scientific and Industrial Instrumentation Group of National University of Colombia Sede Medellín, is working on the design and implementation of VI on distributed wireless sensor networks, using the autonomous sensors MICA Mote of the University of California, Berkeley.

This technology will allow the realization of fundamental investigations for the Nation, like the monitoring of physical variables of forests and monitoring of civil structures.

VI. References

[1] Nacional Instruments, *Our History* [web] http://www.ni.com/company/history.htm [Last access, 23 June 2006].

[2] House, R., *Choosing the right Software for Data Acquisition*, IEEE Spectrum, 1995 p. 24-34.

[3] MAXIM - Dallas Semiconductor, *Tech Brief 1: MicroLAN Design Guide*, [web] http://pdfserv.maxim-ic.com/en/an/tb1.pdf, [Last access, 4 June 2006].

[4] Pulido, J., Serrano, C. , Chacón, R., *Monitoreo y Control de Procesos a través de la Web*. III Simposio de Control Automático. Habana, Cuba, 2001 p. 73 -78

[5] N. Montoya, L. Giraldo, D. Aristizábal, A. Montoya, *Remote Monitoring and Control System. of Physical Variables of a Greehouse through a 1-Wire Network,* International Conference on Telecommunications and Networking, University of Bridgeport, diciembre 10-20, 2005, Sobh, Tarek ,Elleithy, Khaled, *Circuits & Systems Advanced in Systems, Computing Sciences and Software Engineering* , **XIV**, 437 p., Springer, 2006, Hardcover ISBN: 1-4020-5262-6.

[6] A. F. Muñetón, J..M. Saldariaga A., A. Montoya, D. Aristizábal, *A Module for Monitoring and Remote Control of Physical Variables Using Mobile Devices,* International Conference on Telecommunications and Networking, University of Bridgeport, diciembre 10-20, 2005, Elleithy, K., Sobh, T., Mahmood, A., Iskander, M. Karim, M. (Eds.), *Circuits & Systems Advanced in Systems, Computing Sciences and Software Engineering*, **XV**, 489 p., Springer,2006, Hardcover ISBN: 1-4020-5260-X .

[7] Vidal, A., Pérez, F., Calle, A., Valdés, E., *Estrategias para una Instrumentación Virtual de Bajo Costo con Aplicaciones Médicas.*, Memorias V Congreso de la Sociedad Cubana de Bioingeniería, Habana, Cuba, 2003., Articulo T_0060.

Online Decision Support System for Dairy Farm

A. Savilionis, A. Zajančkauskas, V. Petrauskas, S. Juknevičius
Lithuanian University of Agriculture
Studentų g. 11 LT-53361
Akademija, Kauno r. Lithuania

***Abstract* – online decision support system for dairy farm was created for helping Lithuanian dairy farmers, scientists, dairy technology producers, students and other peoples interesting in dairy business. It enable they use newest information and technology for planning own business**

I. INTRODUCTION

Dairy business is one of underlying branch of Lithuanian agriculture. In creation or development dairy farm it is very important optimally select livestock breed, feeding ration, technology for livestock keeping, milker system and other equipment. All of them must secure good quality and biggest quantity of produced milk, low producing cost and biggest profitability.

With new information technologies we have more possibility using mathematical models for optimization economical and technological parameters of dairy farm.

In dairy farming sector of Europe countries are used some computer based models for creating feeding ration subject to individual cow productivity, cows breed and other factors. Commonly those models are created for feed, which are produced by models owner.

By this time in Lithuania wasn't created multipurpose mathematical models for dairy farm. Such models will be useful for farmers, scientifists, students and dairy adviser. It could be effective to make decision in selecting cows breed, feeding ration, livestock keeping and milking technology. It will be very helpful for all in dairy business.

Goal and subject of investigation was creating multipurpose mathematical model of dairy farm. Model is based on data bases which are accessed by World Wide Web. Data bases could be renewed and updated online by administrator grants for information suppliers. These suppliers are feed producing companies, dairy technology suppliers or scientists and agriculture advisers. Users of this information would be farmers, students and other men's interesting in dairy business.

Mine goal of creating online decision support system for dairy farm are:

1. Stimulate using information technology and mathematical models in dairy business for modeling and prognoses economical processes.
2. Make ability for scientific research and discoveries immediately apply in practice.
3. Improve information attainability.
4. Cheeping consulting cost.

Keywords: mathematical modeling, agriculture, dairy farm.

II. METHODOLOGY OF RESEARCH

Methodology of research are based on mathematical modeling of technological processes by using statistical methods and mathematical equations, estimated correlations in scientific research and POWESIM software

III. RESULTS OF RESEARCH

In creating of dairy farm, modernizing or expanding it mine question are time for getting dividends of madden investments

In this mathematical model are evaluated such kind of expenses:

KISKG – variable expenses in farm for producing 1 kg of milk;

KTIK – expenses for one cow kept in farm, Lt/cow;

IKR – part of expanses for producing 1 kg of milk;

PRDM – average milk yield in farm per year kg/year, calculated by equation;

$$PRDM = PRDN \cdot MK \cdot 365 . \quad (1)$$

PRDN – milk yield from one cow, kg.

After evaluation costs of feed, energy, labor and other expenses we estimate milk cost, i.e. total expenses for producing 1 kg of milk

$$ISKG = BPSKAIN + KISKG, \text{ Lt/kg.} \quad (2)$$

Process of investment in model from financial point of view, could be divided in two separate processes – capitalization and getting constant income from cumulated capital. Those processes are even in particular time interval. Both processes in time could have different distributions. In model distribution in time form (especially profit) are very

K. Elleithy (ed.), Advances and Innovations in Systems, Computing Sciences and Software Engineering, 13–15.

important, because money flows are planed on end of every year

In separate sectors of model are formed necessary investments for project implementation requirement. Investments usually are written in project fund requirement and sources table, in which are information how much and for which purpose necessary money for project implementation. Main part of project outlay is for long-term means (building and maintenance expenses). Necessary fund for project implementation in model consist of:

1) Long-term means = { FAV;MIKN}, sum of expenses for farm buildings (FAV) and milking equipment (MIKN).

2) Circulating capital = ISKGP, sum of expenses milk producing.

3) Unexpected expenses ((1)+(2))*0,01, it is 1% from sum of expenses for long-term means and circulating capital.

$$\text{Necessary fund} = (1) + (2) + (3) + \text{GLVK}, \quad (3)$$

GLVK-price of cow.

In this model object of analysis are payment flows which characterize both processes. Models elements of flow are described by profit and investment.

Net profit – it is total profit collected until end of year after all necessary payments are done. These necessary payments include all real expenses for produced production.

Profit flow element (R_t) in model could be estimated in such way:

Calculating most important economical parameter – profit. Profit can be raised by increasing production volume and decreasing outlay. For this reason in model are set coefficient of soled milk from all produced (PPK), i.e. proportion between produced and soled milk. Then soled amount of milk per year will be calculated (PRK):

$$PRK = 365 \cdot PRDN \cdot PPK \quad (4)$$

In this model part, after determination price of raw milk (SUPK), we can calculate incomes for soled milk (PA):

$$PA = PRK \cdot SUPK \quad (5)$$

Profit (P) is calculated by subtract outlay from incomes:

$$P = PA - S, \quad (6)$$

Outlay can be calculated:

$$S = ISKG - PRK \quad (7)$$

ISKG – outlay for producing one kg milk.

After subtracting taxes (IMK) from profit (P) we will get net profit (GRPL):

$$GRPL = P - IMK \quad (8)$$

Then, after estimation net profit for every year and depreciation expenses (AMR) we get total profit flow element for every year:

$$R_t = GRPL_t + AMR_t, \quad t = \overline{0,n} \quad (10)$$

R_t - Total profit flow in t year;

$GRPL_t$ - Net profit in t year;

AMR_t - Depreciation expenses in t year.

In this model for evaluation of efficiency of investments are used index of pay dividend time. This index is calculated by compare in time moment sum of incomes and investments (I_t). Our investigated process of investment is presented like continuous flow of payments in every year.

Time of investment pay dividends in this model are calculated like time interval in which investment become equal net profit after subtract all outlay for producing milk, i.e. sum of investment will be equal in same time (n) sum of incomes:

$$TI + \sum_{t=0}^{n} I_t = \sum_{t=0}^{n} R_t \quad (11)$$

TI – sum of all means used for project implementation.

Time interval (n) in which equality show us period of investments start to pay dividends.

Sum of investments ($\sum_{t=0}^{n} I_t$) can be calculated in this way:

a. we determine sum of loan;
b. determine interest rate, %;
c. determine project term in years, n;
d. By number of year's n and interest rate (%), we find paid sum with interest rate for creditors in every year (Lt.). This sum (RGSUM) are calculated by equation (12):

$$I_t = \frac{PASK \cdot \frac{i}{100} \cdot \left(1 + \frac{i}{100}\right)^t}{\left(1 + \frac{i}{100}\right)^t - 1}, \quad t = \overline{0,n} \quad (12)$$

i – Interest rate, %;
t – Number of year for project implementation;
PASK – sum of loan;
I_t - t – paid sum for creditors every year.

Period (n) after which investment start to pay dividend are calculated:

$$TI + \sum_{t=0}^{n} I_t - \sum_{t=0}^{n} R_t \leq 0 \qquad (13)$$

Period after which investment start to pay dividend are calculated by summing discounted interest rate and continuously played credit return in every year until sum become equal investment sum.

IV. CONCLUSIONS

1. Created multipurpose mathematical model of dairy farm can be used in Europe Union, because it takes account of EU law for environment and other requirements.
2. Created model can improve IT using in agriculture, stimulate using new research data in practice and increase information attainability for dairy business.

REFERENCE

1. Lietuvos mokslo ir technologijų baltoji knyga. – http://193.219.137.48/mokslas/mbk.htm
2. Attonaty J.-M., Chatelin M.-H., Garcia F. Interactive simulation modeling in farm decision-making // Computers and Electronics in Agriculture. -1999, vol.22, p. 157-170.
3. Brodersen C., Kuhlmann F. Nutzung der EDV in der Landwirtschaft // Agrarwirtschaft. – 1999, Bd. 48, H. ¾, S. 122-129.
4. Petrauskas V. POWERSIM modeliavimo programa:Metodinė priemonė. –Kaunas-Akademija, 1997- 41 p.
5. Halachmi I.; Maltz E.; Edan Y.; Metz J.H.M.; Devir S. The body weight of the dairy cow - II. Modeling individual voluntary food intake based on body weight and milk production. Livestock Production Science, Volume 48, Number 3, June 1997, p. 244-244(1).
6. John J. Glen. Mathematical Models in Farm Planning: A Survey. Operations Research, Vol. 35, No. 5 (Sep. - Oct., 1987), p. 641-666.
7. David J. Pannell, Bill Malcolm, Ross S. Kingwell. Are we risking too much? Perspectives on risk in farm modeling. Agricultural Economics, Vol. 23 Issue 1, June 2000, p. 69.

Decision Making Strategies in Global Exchange and Capital Markets

Aleksandras Vytautas Rutkauskas, Viktorija Stasytyte
Vilnius Gediminas Technical University
Sauletekio al. 11
LT-10223 Vilnius-40, Lithuania

Abstract. **The main objective of this paper is to present the investment decision management system in exchange and capital markets – the Double Trump model. The main problems being solved with this model are named as quantitative decision search problems. Computer-imitational methods are also analysed as the main solving means for the mathematical models viewed as stochastical programming tasks in order to reflect the problems characteristics. Attention is paid to the revealing of the analytical possibilities of the decision management system and to decision methods identification, analyzing such non-traditional problems of financial engineering as three-dimensional utility function maximization in the adequate for investment decisions reliability assessment portfolio possible set of values, searching for investment decisions profitability, reliability and riskiness commensuration concept and mathematical decisions methods. Solving of the problems named above ensures sustainable investment decisions development in capital and exchange markets.**

I. Introduction

Strategy, in cybernetics opinion, is any rational rule determining certain actions in any decision making situation. Formally, strategy is the function of the information obtained, which takes values in all possible set of alternatives at the given moment. This rule must include the whole decision making period and all possible situations.

Determined rules and situations are named as simple strategies. Their using result is usually described as strategy implementation or non-implementation. Strategies, which compose simple strategies' possibilities probability distributions, are named as mixed strategies, and about their realization we can say by these categories: by mean, with probability 1, by probability, etc.

Nowadays strategy's category more often goes together with the adjective "sustainable". There is no difference - is it a global atmosphere pollution reduction problem, or is it a small firm energy supply problem. And this is explained not only in terms of intellectual development, but also in terms of behaviour economy. Naturally, with the beginning of the broad exploitation of the category "sustainable strategy", its contents vary a lot. However, almost unambiguous trend is noticed – sustainable strategy more often is described quantitatively, i.e. by finding quantitative indicators allowing to identify strategy sustainability. There is no doubt that the core grating for sustainability grounding is the reliability of the analysed strategy's separate elements or their certain combinations.

Investment strategy is the set of investment decisions. Implementation of these decisions allows investor to get the best profitability and reliability composition. Here the possibility reliability is a very important factor and towards its evaluation the idea of adequate investment decisions reliability assessment portfolio was directed.

The sustainable investment decisions, or simply investment in currency and capital markets strategy, can be called such a strategy, which allows to secure not lower than market generated profitability, as well as invested capital value increase. Considering exchange and capital market riskiness degree and risk variety, the attempt to develop such a strategy can seem as intention to swim through the Atlantic ocean with simple boat. So, the real solution of such complex problem as sustainable investment strategy development is possible only with the adequate means for this problem solving.

Adequate portfolio, retaining profitability and risk commensuration possibilities reveals also profitability and reliability, as well as reliability and risk commensuration possibilities. Space of adequate portfolio values is the set of survival functions, constructed on every level of riskiness. This allows estimating how many units of profitability need to be denied in order to increase reliability by one unit. However, space of adequate portfolio values is the space of izoguarantees. This allows not only to commensurate profitability and risk, but also performs reliability and risk commensuration. Actual utility function provides a possibility for an investor to perform these actions individually, i.e. considering investor's own willingness to undertake certain risk level.

Recall that efficient investment decisions' in the exchange market strategy will further be perceived as the mix

K. Elleithy (ed.), Advances and Innovations in Systems, Computing Sciences and Software Engineering, 17–22.

of the actions and means, which would allow to select such invested or speculative capital management, which would guarantee advantage over all the existing investment means of respective duration and risk in the market.

Double Trump model, as investment possibilities analysis', goal formation and decision-making means, has an adequate structure:

- exchange rate forecasting subsystem;
- goal formation and achievement means subsystem;
- decision-making in exchange market model system characteristic identification and its quantitative decision-making methods subsystem;
- efficiency evaluation of the decisions being made and efficient decisions possibilities and conditions assessment.

II. THE STRUCTURE AND CHARACTERISTICS OF THE DOUBLE TRUMP MODEL

The main principles of an adequate forecasting system

Further we will illustrate the main utilization principles of a one-step currency rate and stock price forecasting system. The core of the forecasting system consists in the probability distribution selected parameter regression dependence of the forecasted index value at a (t+1) moment on the probability distribution certain parameters value of the index under analysis at a t-th and previous moments:

$$\overset{t+1}{y} = f\left(\overset{t}{x_1}, \overset{t}{x_2}, \dots, \overset{t}{x_n}; \Theta(0,t)\right) \tag{1}$$

where in general we can say that:

$\overset{t+1}{y}$ - probability distributions of the forecasted currency rate or stock price possible values at (t+1) moment;

$\overset{t}{x_i}$- i-th factor's possible values probability distribution's vector at a t-th and previous moments;

$\Theta(0,t)$ - the resultant of the influence of the other factors on the factor under analysis at (t+1)-th moment;

f - regression.

Practical results of the forecasting system application

One of the authors of this paper [3] developed the new decision management system for exchange and capital markets – the Double Trump model. A wide experiment was performed with the model, which gave valuable results.

The essence of the Double Trump model is that the two currencies were selected as the basic – EUR and USD, while in general there were analysed 7 currencies: EUR, USD, GBP, CHF, CAD, AUD, JPY [5], [8].

The rebalancing of the portfolio, i. e. the selection of an optimal portfolio, is carried out step by step. The scheme of every step of portfolio management strategy looks like this:

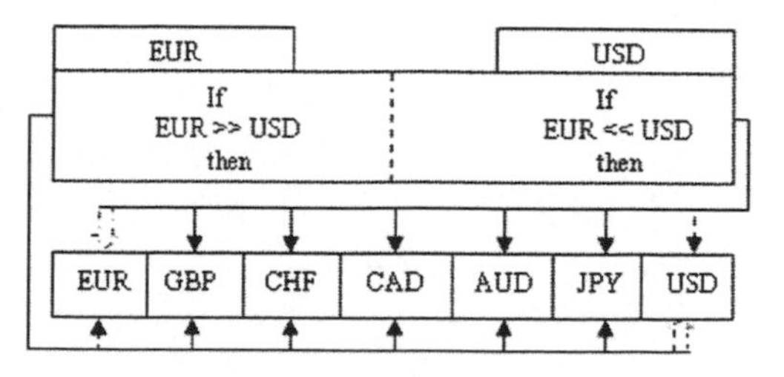

Fig. 1. Double Trump (EUR and USD) portfolio decision management in currency (EUR, GBP, CHF, CAD, AUD, JPY, USD) market model

- we choose EUR and USD as trump currencies;
- make prognoses of the EUR/USD, EUR/GBP, EUR/CHF, EUR/CAD, EUR/AUD, EUR/JPY and USD/GBP, USD/CHF, USD/CAD, USD/AUD, USD/JPY rates, or evaluate them on the basis of gathered FOREX historical data;
- If the EUR exchange rate increases (>>), then EUR is considered to be the trump currency, the diversification of a portfolio is performed on the basis of prognoses of EUR and exchange rates of other currencies. If EUR<<USD, the USD is chosen as the trump currency.
- After we have chosen the trump currency, we choose the currency portfolio which makes it possible to maximize the profitability of the subject at the end of each step, in the particular case – which makes it possible to maximize the purchasing power of the portfolio both in euros and dollars (Fig. 1).

Peculiarities of proposed forecasting system and its comparison with methods of technical analysis

As one can see from the presented model structure (Fig. 1), the preparation of our system of decision making in exchange markets formally begins with selecting the methods of currency rate forecast. Since in this system, as in technical analysis, a particular research object is historical currency rate indices, the suggested methods of forecast should be compared with the forecasting methods already in use in technical analysis, which are numerous and various. Here, next to traditional methods of forecast used in all areas of activities (various traditional models, regression models, moving averages models, etc.), the principle of pattern identification is also intensively used. The essence of the latter method is that particular patterns are being tried to identify, according to which the changes of future indices should repeat changes of historical data [5], [6].

Knowing that the set of technical analysis forecasting models is wide and diverse, it would be negligent to specify the summarized characteristics of this set. Therefore, even though many technical analysis forecasting methods are theoretically suitable for currency rate and stock price forecast and have a long-time practice of utilization not only in this area, we have to admit that they do not satisfy all the main attributes necessary for forecasting methods. Adequate forecasting system should poseess the following characteristics:

- **Adaptivity.** A currency rate forecasting method must be adaptive, i.e. it should help in considering in each point of variation of currency rate both the set and importance of the factors, as well as the functional dependence of currency rates and the factors, when the factors themselves and the forms of interdependence of these factors are being modified.
- **Flexibility.** The forecasting methods of currency rates and stock prices must be flexible, i.e. they must be applicable in every forecasting system.
- **Consistency.** Actions and results in the forecasting method must be clearly separated, i. e. they must be consistent. It is very important when determining and using the analytical interrelation between the result and the factor as well as among the factors themselves.
- **Correctness.** The diagrams of reliability zones could be a good illustration to explain the correctness of the models.
- **Accuracy.** The dislocation of the historical parameters in confidence zones indicates that the behavior of the currency rates and stock prices not only is compatible with the consistent patterns of the behavior of stochastic variables in their confidence intervals, but also these confidence intervals have much greater confidence levels.
- **Reliability.** Employment of one or another method of forecast should allow to measure quantitatively the reliability of the obtained results.
- **Constructiveness.** Forecasting methods must be constructive, i.e. they should allow selecting the most probable values of forecast variables or processes.

III. Development Of Adequate Portfolio For The Investment Decisions Reliability Assessment

Anatomy of adequate portfolio

In order to reveal the portfolio of investment decisions reliability role mechanism in details, we will briefly take a look over adequate investment decisions reliability assessment portfolio anatomy.

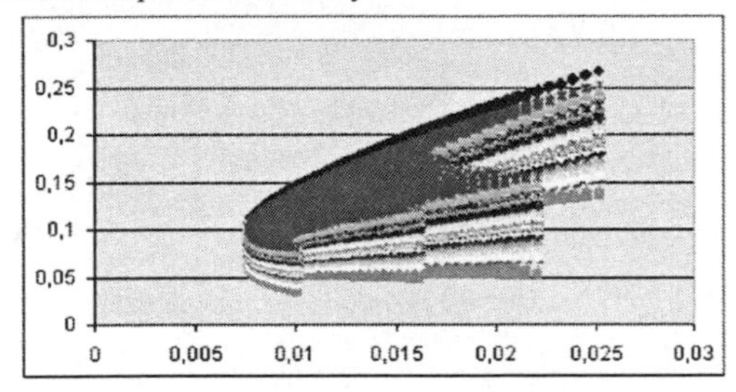

a. Bunch of "quintiles (percentiles) – standard deviation" portfolios

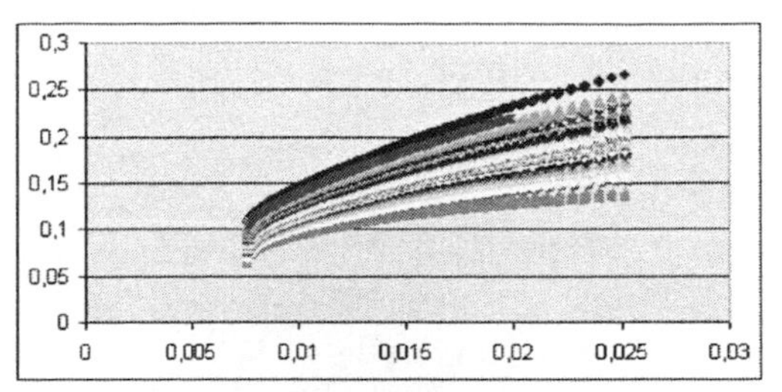

b. The confidence zone of adequate portfolio

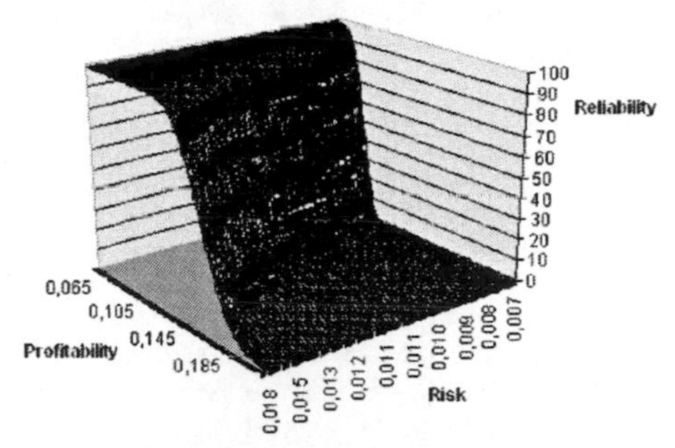

c. Three-dimensional view of the investment portfolio

Fig. 2. Elements of the adequate portfolio

Fig. 2 present adequate portfolio for investment decisions reliability assessment. "Mean – standard deviation" portfolio (modern, or Markowitz portfolio) is a portfolio formed for independent values, having normal probability distributions [1], [2]. Next, a bunch of the possible values of all possible "quintiles – standard deviation" portfolios (Fig. 2, section a) is formed. More precisely speaking, not all the quintiles were used for this bunch, but only percentiles. In turn, the efficiency zone - all portfolios' "quintile – standard deviation" set of values for each quintile efficiency lines is presented in Fig. 2, section b. Fig. 2, section c presents the three-dimensional view of the investment portfolio.

There is no doubt that investor is interested not only in quantitative indicators of investment profitability possibilities, but also in guarantee of each possibility, i.e. the probability investment efficiency (return). In case of modern stock portfolio, the guarantees of investment profit possibilities are usually not discussed, although in case when portfolio returns possibilities probability distribution is a normal one there is a direct possibility to evaluate these guarantees, if mean value and standard deviation are known [5].

Fig. 3 (left side) shows set of values of the adequate portfolio and utility functions' (right side) interaction possibilities searching for the most useful portfolio values for the subject whose interests reflects the utility function.

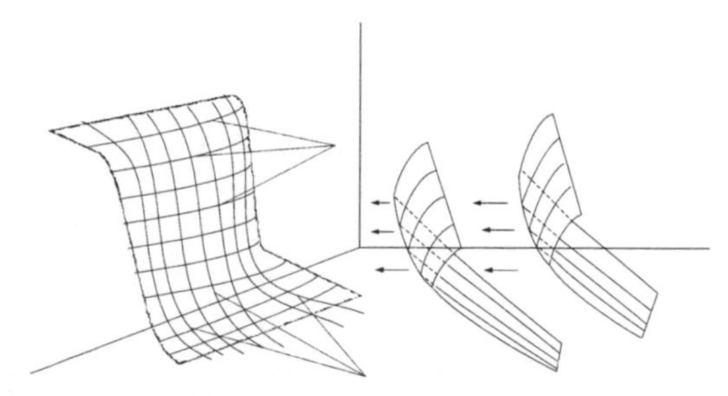

Fig. 3. Schematic view of portfolio possibilities and utility functions

On Fig. 3 applicate and ordinate axes reflect reliability and riskiness, and abscissa shows required profitability levels, which would guarantee selected utility level.

Imitative computer system for the solution of the forecasting and stochastical programming problems

The first step of the system can be viewed the data mining in order to maintain system capacity in general. The data collected should be able to give the possibility to qualitatively implement the main decision management objectives in currency and capital market.

Data mining is the continuous process constantly accessing with new data and – which is no less important – with new requirements for permanent data mining. Considering the variety of the data collectable and constantly changing data requirements, data mining process is automatizated and data sources are all accessible internet databases of the exchange and capital markets being discussed.

Performed complex forecasting and stochastical programming problem solving methods play a big role on the integral data mining, data analysis, forecasting and investment strategies search scheme. They can be called as formalized intelligence roots of the whole system.

Technically this general forecasting and stochastical programming problem can be solved in such a way:

To find such portfolio structure matrix

$$W_{\overline{t+1,T}} = \left(W_j^s\right) \tag{2}$$

which would maximize the gross utility of investor

$$U = U_{t+1} + U_{t+2} + \ldots + U_T \tag{3}$$

when forecasted currency rates values probability distributions:

$$\begin{array}{c} D_{(r_1)}^{t+1}, D_{(r_2)}^{t+1}, \ldots, D_{(r_n)}^{t+1} \\ ----------- \\ D_{(r_1)}^{T}, D_{(r_2)}^{T}, \ldots, D_{(r_n)}^{T} \end{array} \tag{4}$$

and currency rates statistical inter-dependence matrix

$$C = (C_{r_1, r_2, \ldots, r_n}^{\overline{t+1,T}}) \tag{5}$$

are assumed as true.

Schematically, according strategy performing steps, this problem can be presented as integrated data mining, data analysis, forecasting and investment strategies search system.

In the separate integrated system steps mentioned above and in formulated problems the following notes are used:

In data mining step –

r_j^t – selected currency rates for other currencies in t-year;

In data analysis and summation step –

$D\left(r_j, \alpha_{1,j}^t, \ldots, \alpha_{k,j}^t\right), j = 1, \ldots, n$ – distributions, identifying reporting data of the first t-years as random variables.

In forecasting step –

$D(r_j; a_{1,j}^s, \ldots, a_{k,j}^s)$, j = 1,2,...n; s = t+1,...,T – probability distributions of the forecasted currency rates possibilities;

$c = (\text{cov}(r_i^s, r_j^s))i, j = 1, \ldots, n; s = t+1, \ldots, T$ - forecasted correlation relations among the same year currency rates, as well as among different years forecasted currency rates;

In investment strategy search step –

$(W_j^s), \sum_{j=1}^{n} w_j = 1$

s = t + 1,..., T - yearly investment (diversification) portfolio structures maximizing accumulated utility U = U1 + U2 +....+Un (Us – the utility of the s-year) of the [t+1,...,T] period.

Discussing forecasting and stochastical programming problem solving results it is worth noticing that:

- forecasting system is based only on historical data, which is treated as random variables' realization; the general principals of the system are presented in the beginning of the paper;
- distribution forms identified on the basis of the historical data and determined parameters, as well as correlation dependencies are only information for the change determination of these distributions forms,

parameters and correlation dependencies in the further steps;
- performed currency rates' and their correlation dependencies' forecasts are the unique information, along with historical data, for selecting portfolio structure coefficients W_j^s;
- optimal selection of the structure coefficients is based on the three-parametrical (P – profitability, E – reliability, R – riskiness) criteria – U (P;E;R), among which there were utility functions.

The solution of this stochastical programming set of tasks is performed by the following steps:
- First, all continuous probability distributions are approximated to the desired accuracy by their discreet analogues and all mathematical operations are performed as with discreet stochastical variables.
- In order to perform almost all mathematical operations, including continuous variables and their dependencies approximation with discreet variables, Monte Carlo and other imitational methods are used.

IV. Analytical Possibilities And Results Of The Experimental System Implementation

The main goal of the paper is to reveal the system possibilities to prepare sustainable strategies of investment decisions. There is no doubt that the power and constructiveness of every decision management system primarily depends on the functional-analytical possibilities of this system and its orientation towards practical objectives. Thus, searching for the full reasoning of the objective reaching possibilities, further we will discuss three problems:
- sustainable investment strategy in exchange and capital market concept and its concretization;
- system analytical possibilities,
- experimental system implementation results and broad monitoring organization requirement and possibilities.

Analytical possibilities of the system

In the description of the adequate portfolio the particular portfolio, as well as its main characteristics is presented, the investment is made in four nondependent and having normal probability distributions assets:

N(x; a1 = 0,06, σ1 = 0,01), N (x; a2 = 0,09, σ2 = 0,016), N (x; a3 = 0,11, σ3 = 0,022), N (x; a4 = 0,2, σ4 = 0,025).

It is, of course, idealized case, having probably one advantage that allows evaluating practically all portfolio characteristics which we are interested in, using classical analytical methods. However, in reality assets as a rule have complex interdependence, and often their probability distributions are expressed not in classical probability distributions, but in empirically expressed information. While the most important thing in portfolio management, as well as in finance management, is the possibility of managing different dependences, including assets interdependence, as well as the forms of distributions' using.

The method of computer imitation technologies, created by one of the authors [7], allows determining characteristics of adequate portfolio at any level of accuracy and evaluating such parameters as investment assets interdependence or their profitability possibilities probability distributions form impact on the final results.

Sustainability of decision making strategies

Adequate portfolio, its components and space utility function discussed in the paper are just separate, though very important elements of the decision management system in financial markets. In order to use decision management system in financial markets for sustainable investment strategies' search it is required to use all its subsystems step by step [5]:
- strategy objectives and tasks;
- historical data analysis and forecasts;
- decision making methods selection;
- monitoring.

Using all possibilities of the developed system, the sustainable strategy search experiment is performed (see the A.V.Rutkauskas experiment in the Internet). It emphasizes possibilities to choose such strategies in a series of markets. Here (Fig. 4) we will present examples of London and Vilnius markets.

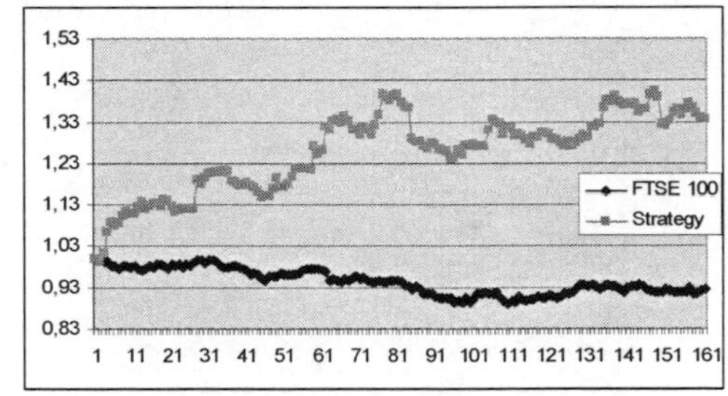

a. FTSE 100 (London) and the strategy. Daily step

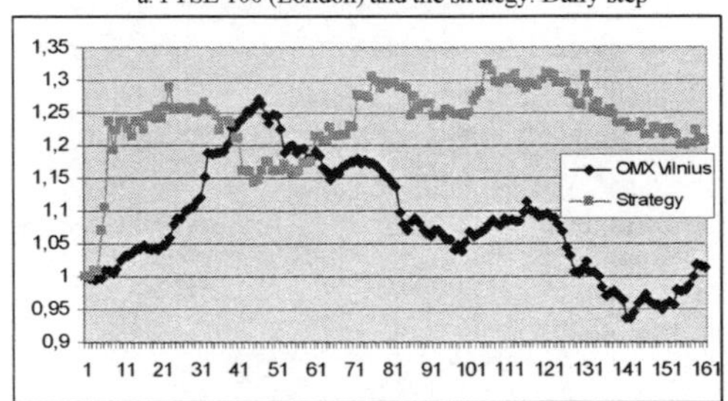

b. OMX Vilnius and the strategy. Daily step

Fig. 4. Comparison of indices and strategy using results in capital markets

Thus we can make a conclusion that in London and Vilnius the use of sustainable investment decisions strategy gives higher than the average results, because on Fig. 4 the strategy result curve exceed the average market indicators' – FTSE 100 and OMX Vilnius index curves.

The system of sustainable investment decisions strategy development can be also implemented in exchange market [4]. Such application has given results that are further presented.

The main goal of the sustainable investment strategy development in currency market is the formation of the effective portfolio of currencies giving high return [3]. The main statements and organizing principles of the strategy are:

- forecasting the probability distribution of rate change for the t+1 step using the historical currency rates data for the $[t_0, t]$ period;
- choosing a new currency portfolio for the t+1 step on the basis of the current portfolio and the forecasts;
- evaluating the effect of the decision made as soon as the historical data for the t+1 period appears;
- combining the t+1 period data with the historical database and performing forecasts and making a portfolio for the t+2 period;
- continuing the process.

Fig. 5 presents the results of the sustainable investment decisions strategy implementation in currency market. Section a shows the increase of invested capital and section b – accumulated capital sum, if we compose the portfolio suggested by the strategy.

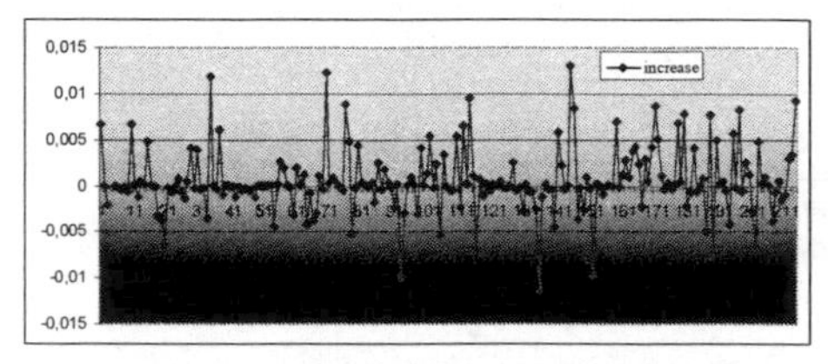

a) graphical view of the increase

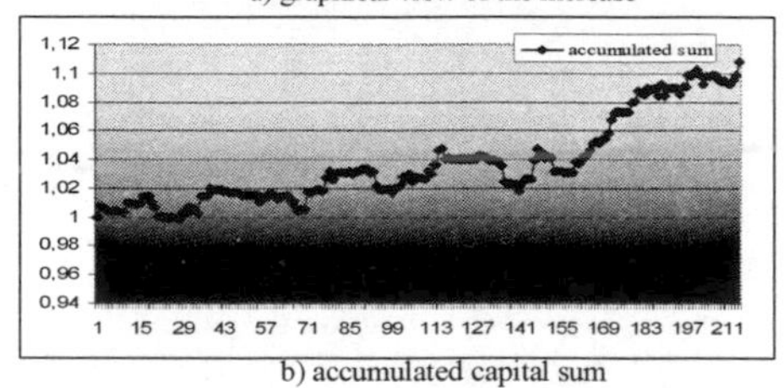

b) accumulated capital sum

Fig. 5. Results of one of the strategies application

V. Conclusions

While analysing decision management system in capital and exchange markets practical implementation possibilities with the help of adequate investment portfolio formation and imitative technologies application, the following conclusions have been made:

- Sustainable investment strategy in the exchange or capital market can be called a strategy allowing exceeding market generated effect in rather long period.
- One of the most suitable means for sustainable investment strategy development is the adequate for investment decisions reliability assessment portfolio. It differs from the modern portfolio, because modern portfolio operates only with two categories – profitability and risk. On the other hand, adequate portfolio commensurates profitability, riskiness and reliability.
- The system of decision management in currency and capital markets was used for the development and implementation of sustainable investment strategy. Strategy implementation results show that the strategy is capable of achieving higher than the average utility.
- The Double Trump model allows to form, leaning only on historical information, practically nonrisky investment strategies, allowing to gain higher investment effect than the effect, insured by the market investment instruments of the corresponding term.

References

1. Frank J. Fabozzi and Harry M. Markowitz, *The Theory and Practice of Investment Management*, John Willey & Sons, 2002, p. 894.
2. Frank K. Reilly and Keith C. Brown, *Investment Analysis and Portfolio Management*. Seventh edition. Thompson: South-Western, 2003, p. 1162.
3. A. V. Rutkauskas, "The double-trump decision management model in global Exchange", *Economics*, Vol. 72, 2005, pp. 84-104.
4. A. V. Rutkauskas and V. Stasytyte, "The double trump portfolio as the core of sustainable decision making strategy in currency markets," *The 10th World Multi-Conference on Systemics, Cybernetics and Informatics. July 16-19, 2006, Orlando, Florida, USA. International Institute of Informatics and Systemics*, pp. 57-62.
5. A. V. Rutkauskas, "Formation of adequate investment portfolio for stochastic possibilities", *Property Management*, Vol. 4, No. 2. Vilnius: Technika, 2000, pp. 100–116.
6. A. V. Rutkauskas, "Isoguarantees as instrument of portfolio decision making", *Proceedings of international conference "Modelling and simulation of business systems", May 13–14, 2003, Vilnius, Lithuania. Kaunas University of Technology Press*, 2003, pp. 239–243.
7. A. V. Rutkauskas, „Computerized imitative technologies for risk and return trade-off," *Materials of international conference "Strategie zarządzania ryzykiem w przedsiębiorstwie – elementy wiedzy teoretycznej i praktycznej"*. Bydgosczcz, 2000, p. 115-141.
8. The Double-Trump Model Experiment. Available http://www.vgtu.lt/usr/rutkauskas/en/eksperimentas.html.

A Simple and Efficient Solution for Room Synchronization Problem in Distributed Computing

Alex A. Aravind, *Member, IEEE*
University of Northern British Columbia,
Prince George, BC, CANADA V2N 4Z9.
csalex@unbc.ca

***Abstract* – Room synchronization problem was first introduced by Joung in 1998 and widely studied subsequently. The problem arises in various practical applications that require concurrent data sharing. The problem aims at achieving exclusive access to shared data while facilitating suitable concurrency.**

This paper presents a new algorithm to solve room synchronization problem in shared memory systems. The algorithm is simple and easy to prove. The main appeal of our algorithm is that, in some sense, it closely emulates the traditional queuing system commonly used in practice. Also, our algorithm is distributed and offers unbounded concurrency, while assuring bounded delay, which was conjectured as unattainable in [8].

I. INTRODUCTION

A. Background

Mutual exclusion and concurrency are two fundamental issues in distributed computing. Mutual exclusion problem is well known for more than four decades since it was posed as early as 1962 [4]. The problem is to basically ensure consistency when accessing a shared resource. Room synchronization or group mutual exclusion problem is a generalization of mutual exclusion problem introduced by Joung in 1998 [8] and it aims at achieving exclusive access to shared resource while facilitating suitable concurrency.

Consider a system consisting of n philosophers (processes) who spend their time thinking alone and talking in a forum. The philosophers may be interested in attending different forums, but only one forum can be held at a time. However, in any forum, any number of philosophers interested in that forum can participate simultaneously. The problem is to design an algorithm to ensure that (i) at any time only one forum is active, (ii) any philosophers attempting to attend a forum will eventually succeed, and (iii) if several philosophers are interested in the same forum then they may be able to attend the forum simultaneously.

A solution to room synchronization problem essentially has two logical components: one to achieve mutual exclusion among the processes interested in different forums (*mutual exclusion* (ME) *component*) and the other to facilitate concurrent participation of philosophers interested in the same forum (*concurrent entering* (CCE) *component*). With different combinations of ME and CCE components, we may get different room synchronization algorithms. The complexity and the properties like fairness, efficiency, simplicity, etc. of a room synchronization algorithm depend on its ME and CCE components.

B. Motivation

As resource integration and resource sharing are becoming increasingly common, room synchronization is an important issue in many such situations. For illustration, we recall three example applications of room synchronization from the literature [2], [6], [8].

Consider a CD "juke box" shared by multiple client processes. Any number of processes can simultaneously access the currently loaded CD, but processes wishing to access a CD other than the currently loaded one must wait. Here the forums are CDs in the juke box. The second example considers a server that can cache locally one chunk of large distributed database (e.g. the web). The clients of this server interested in the currently cached chunk can read it simultaneously; clients interested in a different chunk must wait. In this example the forums are the database chunks. The number of forums may be known or unknown. If all the processes are interested in distinct forums always then the problem becomes the traditional mutual exclusion problem.

The last example is about implementing a parallel stack using an atomic *move(k)* instruction, which moves the stack pointer *sp* (up if $k > 0$, down if $k < 0$) by k positions and returns the old address of *sp*. The stack operations *push* and *pop*, respectively, can be implemented using the instructions *move(1)* and *move(-1)*. Interleaving of two or more pushes or two or more pops is safe, because the order does not matter when the stack pointer *sp* is moving in the same direction. Consider an interleaving of *push* and *pop* on the stack *ST* as follows:

```
i = move(1) ;      // for push
j = move(-1);      // for pop
x = ST[j-1];       // for pop
ST[i] = y;         // for push
```

Here *pop* position will return an inconsistent value and therefore interleaving of *push* and *pop* is not safe. Such parallel access to shared resources is becoming important component in modern programming, due to the availability of multiple processors within a systems and widespread use of thread programming.

An existing mutual exclusion algorithm may be tuned appropriately and used as ME component to solve room synchronization problem. Knuth's algorithm [10], Lamport's algorithm [13], and Peterson's algorithm [14] are three widely studied in concurrent programming and operating systems contexts.

Joung refined Knuth's algorithm and used it as ME component to design his room synchronization algorithm [8].

K. Elleithy (ed.), Advances and Innovations in Systems, Computing Sciences and Software Engineering, 23–28.

Takamura and Igarashi's room synchronization algorithm used Lamport's bakery algorithm as its ME component. In [1], many variations to Peterson's algorithm are proposed and we use one such variation with improved performance (given in Figure 2) as the ME component to design our room synchronization algorithm in this paper.

C. *Literature Review*

We review the room synchronization algorithms presented for shared memory systems. In 1998, Joung introduced room synchronization problem (which he referred as Congenial Talking Philosophers problem and also as group mutual exclusion problem) and gave three solutions: first one centralized, second is semi-distributed, and the final one is distributed. The distributed version uses the CCE component similar to the CCE component of one of our algorithms [8].

Hadzilacos proposed an algorithm assuring FIFO fairness with respect a segment of code in the entry section called doorway [6] and its space efficiency improved version was presented subsequently in [7] and [15]. In [16], Takumara and Igarashi modified Lamport's bakery algorithm to solve room synchronization problem. Although the algorithm inherits the attractive properties of bakery algorithm, it requires unbounded shared space for token values.

A scalable solution to room synchronization problem was presented in [2] and [3] using higher level atomic instruction fetch-and-add. Assumption of a higher level atomic instruction limits its applicability to the systems where such support is available.

Kean and Moire designed a local-spin algorithm [11], [12] that is suitable for the systems where remote memory access is costly and the algorithm does not satisfy one of the important properties called concurrent entering. It assures bounded remote memory accesses. The advantage of Keane and Moire's algorithm is its generality that the exclusive lock can be implemented by any of the known mutual exclusion algorithms. They also discuss wait-free implementation that is costly. In [5], Danek and Hadzilacos proposed four local-spin algorithms, three for DSM multiprocessors and one for CC multiprocessors with varying degrees of fairness and concurrency.

A generalized version of room synchronization problem called mutual *l*-exclusion problem and its solution based on Peterson's algorithm are presented in [17].

D. *Contributions*

In this paper we propose a new algorithm to solve room synchronization problem in shared memory systems. The algorithm is simple and easy to prove. The main appeal of our algorithm is that, in some sense, it closely emulates the traditional queuing system commonly used in practice. Also, our algorithm offers unbounded concurrency, while assuring bounded delay, which was conjectured as unattainable in [8]. The proofs of our algorithm are also simpler and intuitive than the proofs of the algorithms given in [6], [8], [9], and [15].

E. *Organization*

The rest of the paper is organized as follows. Section II presents the system model and problem statement. Peterson's algorithm and its improved version given in [1] are reviewed in Section III. The new room synchronization algorithm and its correctness are presented in Section IV and Section V sketches some improvements. Section VI concludes the paper.

II. System Model And Problem Statement

We assume a **system** of *n* independent cyclic processes, numbered as *1,2, ..., n*, competing to attend forums of their interest. Forum ids are assumed to be positive integers. The code segment of a competing process can be divided into two parts: the part in which the process attends the forum (*Critical Section* (CS)) and the remaining part (*Non-critical Section* (NCS)). Following are the assumptions made for any room synchronization algorithm:

A1: No static priority is assumed on any process.

A2: The execution speed of any process is finite but unpredictable.

A3: Forum execution time of any process is finite but unpredictable.

A4: Each process begins its execution and subsequently halts, after possibly many cycles, only in its non-critical section.

The **room synchronization problem** is to design an algorithm that assures the following properties:

- *(P1) Safety:* If some process is attending a forum then no other process can be in a different forum at the same time [8].
- *(P2) Liveness:* When one or more processes have expressed their intentions to attend forums, one of them eventually enters its forum [8].
- *(P3) Concurrent Entering*: If some processes are interested in a forum and when no process is interested in a different forum, then the processes should be allowed to eventually enter the current forum concurrently [11], [12].

The property *P3* is to facilitate safe concurrency. The concurrent entering property first characterized in [8] states as follows: "If some processes are interested in a forum and no process is interested in a different forum, then the processes can attend the forum concurrently." As indicated by Hadzilacos in [6] that the phrase "can attend the forum concurrently" although suggestive, is not precise. So, we use a more precise definition given in [11] and [12] for concurrent entering property *P3* (which was renamed as *concurrent occupancy* in [6]).

In [6], another definition for concurrent entering was proposed. This definition enforces an additional requirement that the number of steps required to reach the CS to be bounded. To avoid confusion, we call this property as *bounded concurrent entering.*

Maximal resource utilization can be achieved if high degree of concurrency is allowed when a forum is active. In [8], it is conjectured that unbounded concurrency cannot be achieved concomitantly with bounded delay, if there is no centralized mechanism to schedule the forums. Our algorithm refutes this claim by providing unbounded concurrency. Hence, we introduce an additional desirable property referred as *unbounded concurrency*.

In addition to the properties *P1, P2*, and *P3*, the following are desirable properties of a room synchronization algorithm:

- *(P4) Freedom from Starvation:* Any process that expresses its intention to enter the forum will be able to do so in finite time [7].
- *(P5) Bounded Exit:* If a process exits the forum, then it enters the non-critical section within a bounded number of its own steps [6].
- *(P6) Bounded Concurrent Entering*: If a process *i* is interested in a forum and when no process is interested in a different forum, then the process *i* enters the current forum within a bounded number of its own steps [6].
- *(P7) Unbounded Concurrency:* The number of entry to a forum is not bounded.

Structurally, a **room synchronization algorithm** has two parts: *Entry Section* (the code to be inserted before entering the forum) and *Exit Section* (the code to be inserted after exiting the forum). These components have to be inserted appropriately in the code of all competing processes, as shown in Figure 1, to ensure the above mentioned properties.

```
Repeat
        NCS;
        Entry Section;
        CS;
        Exit Section;
forever
```

Figure 1

III. MUTUAL EXCLUSION COMPONENT

As indicated earlier, we inherit the ME component from an algorithm presented in [1], which is an improvement over Peterson's algorithm. The basic idea behind Peterson's ME algorithm is that each process passes through n-1 stages before entering the CS. These stages are designed to block one process per stage so that after n-1 stages only one process will be eligible to enter the CS. The algorithm uses two integer arrays *stage* and *pos* (for position) of sizes n-1 and n and it is given for process *i* in Figure 2.

Many variations to Peterson's algorithm are presented in [1]. The variation we choose to use as our ME component has both fairness and performance improvement over Peterson's algorithm. The performance improvement was to make the algorithm "stage-adaptive". That is, the number of stages to be crossed by a competing process is based on the competition that it observes in the beginning of its competition, rather than crossing n-1 stages always. For example, if a process i observes Ci as the number of processes competing for CS, then it enters the stage n-Ci+1 straightaway to cross only the last Ci-1 stages to reach the CS. This can be viewed in another way as that, when a process observes the competition as Ci, it enters the stage (n-Ci+1) leaving the first n-Ci stages for the n-Ci non-competing processes.

The fairness improvement was to avoid the unbounded overtakes possible in Peterson's algorithm. For that, the condition ($\forall k \neq i$, $pos[k] < j$) (which allows the processes in the highest stage j to move up) has been replaced by the condition (n-$Ci \geq j$) (which allows the process in the lowest stage j to move up when a process leaves the CS). Also, this improved fairness makes the variation, in some sense, to closely emulate traditional queuing system. The stage-adaptive mutual exclusion algorithm is reproduced in Figure 3. We introduce the following terminology which will be used subsequently.

- A process i is said to be *at stage* j, $1 \leq j \leq n$-1, if $pos[i] = j$.
- A process is said to be *blocked at stage* j, $1 \leq j \leq n$-1, if it is waiting (in line 5) for a condition to become true at stage j. A process is said to be *blocked*, if it is blocked at some stage j.
- A process is said to have *crossed stage* j, if it is at stage j or higher, and has finished executing line 5 in the jth iteration.
- A process i is said to have *bumped from stage* j, if it has crossed the stage j by observing the condition ($stage[j] \neq i$) as true.
- A stage j is said to be *a hole* if there is no process blocked at stage j and there exists two processes blocked at stages i and k such that $i < j < k$.

```
Process i

Entry Section:

1. for j := 1 to n-1 do
2. begin
3.    pos[i] := j;
4.    stage[j] := i;
5.    wait until ((stage[j] ≠ i) ∨ (∀ k≠i, pos[k] < j))
6. end;

Exit Section:

1. pos[i] := 0;
```

Figure 2: Peterson's Algorithm [14]

The algorithm given in Figure 3 assures that at any time the number of processes that have crossed stage j is at most n-j, for $1 \leq j \leq n$-1, $n > 1$.

Theorem 1: *The mutual exclusion algorithm given in Figure 3 assures safety, liveness, and freedom from starvation properties [1].*

```
Process i

Entry Section:

1.  for j := n-Ci+1 to n-1 do
2.  begin
3.     pos[i] := j;
4.     stage[j] :=i;
5.     wait until ((stage[j] ≠ i) ∨ (n-Ci ≥ j))
6.  end;

Exit Section:

1.  pos[i] := 0;
```

Figure 3: Stage-Adaptive Algorithm [1]

IV. ROOM SYNCHRONIZATION ALGORITHM

A. *Introduction*

As we mentioned in the background, mutual exclusion (ME) and concurrent entering (CCE) are two logical components of any room synchronization algorithm. The performance of room synchronization algorithm depends on the performance of its ME and CCE components. Although the underlying ideas of ME and CCE components appear simple and intuitive, gluing together to create a room synchronization algorithm normally involves many subtleties. We already discussed the ME component. Next we deal with CCE component before combining both to create our room synchronization algorithm.

B. *CCE Component*

In this section we explore the ways of designing concurrent entering component. We use an integer array F of size n to hold the ids of the forums for the processes. The aim of concurrent entering component is to allow the processes interested in the same forum to attend concurrently. Obviously, freedom from starvation cannot be guaranteed if the algorithm allows the processes to join an ongoing forum any time. (For example, two processes i and j may be in a forum f, i may get out, and start competing again for f, join f, then j may get out, and start competing again for f, join f, then i may get out and join again, and so on.) That is, a necessary condition for freedom from starvation is that every forum should be "active" only for a finite period of time if another forum is in demand. By tuning CCE components suitably, we can control various levels of concurrency in the system.

By Assumption A3, each process will be in a forum only for a finite number of times. So to ensure freedom from starvation we need only that processes can enter a forum only for a finite period of time if another forum is in demand. Based on this requirement, we describe the functionality of a concurrent entering component as follows:

- Opening the forum,
- Facilitating entry to the forum for a finite period of time if another forum is in demand, and
- Closing entry to the forum if another forum needs to be opened.

These three tasks may be accomplished in various ways. We start with designing the CCE component which requires minimum change in the ME component. Recall that the ME component closely emulates the queuing system traditionally used in practice. In addition to the choices of crossing a stage in the ME component (ie., $(stage[j] \neq i)$ and $(n\text{-}Ci \geq j)$), the CCE component introduces the following third choice to facilitate concurrency among the processes interested in same forum.

- *A process at stage j can move towards the CS if no process at stages above j requests a different forum.*

That is, between the stages 1 and n-1, inclusive, if j is the highest stage with a process requesting different forum, then all the processes at stages above j can attend the current forum concurrently. Formally, a process i interested in forum $F[i]$ at stage j can move further if the condition $(\forall\ k \neq i,\ (pos[k] < j) \vee (F[k]=F[i]))$ is true.

If some processes are interested in a forum and no process is interested in a different forum, then the CCE component allows the processes to attend the current forum concurrently. In other words, the resulting room synchronization algorithm clearly satisfies the condition *P3*.

C. *Algorithm*

Combining the ME and CCE components described above, we get the first room synchronization algorithm given in Figure 4, and we refer it as *RSA*. The basic idea behind the algorithm is that a competing process first enters a stage based on the current competition that it observes. Then it follows the logic used in ME component, given in Figure 3, when it observes a process in a higher stage interested in different forum. Otherwise, it moves up in the stages towards CS as long as it does not see any process at a higher stage interested in different forum. So all the processes interested in a same forum clustered in the higher stages can enter and attend the forum concurrently. The main appeal of this algorithm is that it achieves significant concurrency with a simple change in the original mutual exclusion algorithm. Now the condition $(n\text{-}Ci \geq j)$ at line 6 might appear as redundant, but it is required to assure freedom from starvation. Next we prove the correctness of *RSA*.

Lemma 1: *At any stage j,* $1 \leq j \leq n\text{-}1$, *if the number of forums requested by processes at stage j or higher is greater than* 1, *then the room synchronization algorithm RSA assures that the number of processes that can cross stage j and enter stage j+*1 *is at most n-j.*

Proof: We prove by induction on j.

<u>*Base Step:*</u> For j =1, we need to prove only for the case when all processes are competing for CS. Among n competing processes, the process which enters the stage 1 last will

observe all three conditions at line 6 as false and therefore blocked at stage 1. That is, the number of processes that can cross stage 1 and enter stage 2 is at most n-1.

Induction Step: Assume that the assertion is true for some m, $1 < m < n$. We need to consider the case where the number of forums requested by processes at stage m+1 or higher is greater than 1. This implies that the number of forums requested by processes at stage m or higher is also greater than 1. By induction hypothesis, the number of processes that can cross stage m and enter the stage m+1 is at most n-m. Among these n-m processes, the process which enters the stage m+1 last will observe all three conditions at line 6 as false and therefore blocked at stage m+1. That is, the number of processes that can cross stage m+1 and enter stage m+2 is at most n-(m+1). The assertion follows by induction. □

```
Process i

Entry Section:

1. Set F[i];
2. for j:= n-Ci+1 to n-1 do
3. begin
4.    pos[i] := j;
5.    stage[j] :=i;
6.    wait until ((stage[j]≠i) ∨ (n-Ci≥j) ∨
                 (∀k ≠i, (pos[k] <j)∨(F[k]=F[i]))
7. end

Exit Section:

1. pos[i] := 0;
```

Figure 4: ***RSA***

Theorem 2: *The room synchronization algorithm RSA assures safety.*

Proof: By contradiction. Assume that two or more processes interested in different forums are in CS simultaneously. These processes interested in more than 1 forum must have crossed the stage n-1 and entered CS. This contradicts Lemma 1. □

Theorem 3: *The room synchronization algorithm RSA assures liveness.*

Proof: The condition ($stage[j] \neq i$), for every process i, will enable them to eventually spread across the stages to occupy at most one process per stage. Then, the condition ($\forall k \neq i$, ($pos[k] < j$)$\vee$($F[k]=F[i]$)) will enable the process i in the highest stage to reach the CS in a finite time. □

Theorem 4: *The room synchronization algorithm RSA assures freedom from starvation.*

Proof: It is enough to show that no process will be blocked at a stage j, $1 \leq j \leq n$-1, forever. We prove this by induction on j.

Base step: We prove the assertion true for position j=1. That is we need to show that no process will be blocked at stage 1 forever. Assume that a process, say i, is blocked at stage 1, starting from time t. That means the conditions at line 6 in *RSA* are false, whenever i checks them. By assumption A3, the processes which access the CS currently will leave the CS in a finite time. After t, let t_1 be the time at which the first process leaves the CS. After t_1, let t_2 be the earliest time at which i checks the conditions at line 5 and observes them as false. Particularly it observes all n processes as competing. Assume that j is the latest process to compete for the CS in the interval [t_1, t_2]. The process j should have observed all n processes as competing and therefore should have entered the stage 1, which will make the condition ($stage[1] \neq i$) true for the process i at time t_2. This is a contradiction. Hence the assertion is true for the base step.

Induction Step: Assume as induction hypothesis for some stage m, $1 < m < n$, that no process will be blocked forever at m. We need to prove that no process will be blocked forever at stage m+1. Assume that a process, say i, is blocked at stage m+1, starting from time t. That means the conditions at line 5 are false, whenever i checks them at line 5 after t. That is, at least n-m-1 processes are competing whenever i checks the condition at line 5 after t. The assertion follows immediately if at least one competing process is at a stage between 1 and m, after t. That is, in this case, by induction hypothesis, no process will be blocked at the stages between 1 and m forever. Therefore, the process at a stage between 1 and m will eventually reach the stage m+1 and that will make ($stage[m+1] \neq i$) true for the process i at some time after t.

We need to prove for the case where none of the competing process is at a stage between 1 and m after t. By assumption A3, the processes which access the CS currently will leave the CS in a finite time. After t, let t_1 be the time at which the first process leaves the CS. After t_1, let t_2 be the earliest time at which i checks the conditions at line 5 and observes them as false. Particularly it observes n-m-1 processes as competing. Assume that j is the latest process to compete for the CS in the interval [t_1, t_2]. The process j should have observed n-m-1 processes as competing and therefore should have entered the stage m+1, which will make the condition ($stage[m+1] \neq i$) true for the process i at time t_2. This is a contradiction. Hence the assertion is true for the induction step. Hence, by induction, the assertion is true for all j. □

Theorem 5: *The room synchronization algorithm RSA satisfies bounded concurrent entering and bounded exit properties.*

Proof: The condition ($\forall k \neq i$, ($pos[k] < j$)$\vee$($F[k]=F[i]$)) for a process i at line 5 will be true as long as it observes the number forums requested as 1. This condition has to be executed to cross each stage and that clearly requires only a bounded number of steps. □

Theorem 6: *The room synchronization algorithm RSA satisfies unbounded concurrency property.*

We illustrate the possibility of unbounded concurrency by a simple example. Four processes 1, 2, 3, and 4 are competing, in that the process 1 is slow and the processes 2, 3, and 4 are fast. Assume that the process 1 is interested in forum f and the processes 2, 3, and 4 are interested in g. Consider the following scenario:

1. Processes 2, 3, and 4 are attending the forum g and the process 1 is blocked at stage n-3.
2. Processes 3 and 4 leave the forum and this enables the process 1 to move up.
3. Process 1 is still at stage n-3, however, due to its low speed.
4. Process 3, again interested in g, comes back, observes the competing processes as 3, and therefore enters the stage n-2.
5. Since the process 2 is attending the forum g, and no other processes interested in a different forum at a stage above n-2, the process 3 also joins the ongoing forum.
6. Now, the process 2 leaves the forum and comes back quickly again with the interest in attending the forum g and the process 1 is still at stage 1. This is identical to the step 4.

This scenario, although rare, can repeat any number of times allowing unbounded concurrency. □

V. FURTHER IMPROVEMENTS

Achieving high level of concurrency and maintaining fairness are often two conflicting goals. In *RSA*, a process interested in a forum f at a stage j is not allowed to enter the CS, even when f is active, if there is another process interested in different forum at a stage higher than j. Allowing such processes to enter the forum as long as the forum is active would increase the concurrency and the resulting fairness is justifiable. Because the processes interested in different forums cannot anyway attend the current forum. The key would be controlling the duration of the forums in order to avoid unfair wait. This is achieved by usually by identifying some process(es) as **captain(s)** and other processes as **followers** of the forums. The captains control the durations of the forums.

We consider two different ways of performing concurrent entry to the forum. The first approach (referred as **automatic join based**) allows one captain per session and in the second approach (referred as **invited join based**) many captains are possible.

In the automatic join based approach, the captain opens the forum and other processes interested in the same forum becomes as followers to enter the forum directly. This happens as long as the forum is open. Immediately after leaving the forum, the captain closes the forum. In the invited join based approach, since many processes can become captain, every captain of a single forum explicitly invites other processes in its forum to join.

Due to space constraint, these improved algorithms, their correctness proofs, and detailed analysis are given in an expanded version.

VI. CONCLUSION

In this paper, we have presented an efficient room synchronization algorithm. The algorithm uses an improved version of Peterson's algorithm as the mutual exclusion component. Also, the concurrent entering component is simple and transparent in our algorithm, and that makes the algorithm intuitive. The algorithm is easy to understand and has many nice properties such as unbounded concurrency, which was believed to be unattainable in [8]. The proofs are also simpler and intuitive than the proofs of algorithms given in [6], [8], [9], and [15].

REFERENCES

[1] K. Alagarsamy and K. Vidyasankar, "Fair and Efficient Mutual Exclusion Algorithms," *Lecture Notes in Computer Science*, vol. 1693, pp. 166-179, 1999.

[2] G.E. Blelloch, P. Cheng, and P.B. Gibbons, "Room synchronizations", *Proceedings of the ACM Symposium on Parallel Algorithms and Architectures*, pp. 122-133, 2001.

[3] G.E. Blelloch, P. Cheng, and P.B. Gibbons, "Scalable room synchronizations", *Theory of Computing Systems*, vol. 36, pp. 397-430, 2003.

[4] E.W. Dijkstra, "Solution of a problem in concurrent programming control," *Communications of the ACM*, vol. 8, no. 9, pp. 569, 1965.

[5] R. Danek and V. Hadzilacos, "Local-spin Group Mutual Exclusion Algorithms", *Lecture Notes in Computer Science*, vol. 3274, pp. 71-85, 2004.

[6] Hadzilacos, "A note on group mutual exclusion," *Proceedings of the ACM symposium on PODC*, pp. 100-106, 2001.

[7] P. Jeyanti, S. Petrovic, and K. Tan, "Fair group mutual exclusion," *Proceedings of the ACM Symposium on PODC*, pp. 275-284, 2003.

[8] Y. Joung, "Asynchronous group mutual exclusion,", *Proceedings of the ACM symposium on PODC*, pp. 51-60, 1998

[9] Y. Joung, "Asynchronous group mutual exclusion," *Distributed Computing*, vol. 13, pp. 189-206, 2000.

[10] D. E. Knuth, "Additional comments on a problem in concurrent programming problem," *Communications of the ACM*, vol. 9, no. 5, pp. 321-322, 1966.

[11] P. Kean and M. Moir, "A simple local-spin group mutual exclusion algorithm," *Proceedings of the ACM Symposium on PODC*, pp. 23-32, 1999.

[12] P. Kean and M. Moir, "A simple local-spin group mutual exclusion algorithm," *IEEE Transactions on Parallel and Distributed Systems, vol. 12, no. 7,* pp. 23-32, 2001.

[13] L. Lamport, "A new solution of Dijkstra's concurrent programming problem," *Communications of the ACM*, vol. 17, no. 8, pp. 453-455, 1974.

[14] G.L. Peterson, "Myths about the mutual exclusion problem," *Information Processing Letters*, vol. 12, no. 3, pp. 115-116, 1981.

[15] S. Petrovic, "Space-efficient FCFS group mutual exclusion", *Information Processing Letters*, vol. 95, pp. 343-350, 2005.

[16] M. Takamura and Y. Igarashi, Highly concurrent group mutual exclusion algorithms based on ticket orders, *IEICE Transactions on Information and Systems*, vol. E87-D(2), pp. 322-329, 2004.

[17] K. Vidyasankar, "A Simple group mutual l-exclusion algorithm", *Information Processing Letters*, vol. 85, pp. 79-85, 2003.

Improving Computer Access For Blind Users

Amina Bouraoui, Mejdi Soufi
UTIC
5 Av. Taha Hussein, BP 56, Bab Menara
hannibal.a@topnet.tn; hannibal.a@wanadoo.tn

Abstract-**This paper discusses the development of applications dedicated to the blind users, with the help of reusable components. The methodology relies on component based development. For this purpose, braille-speech widgets adapted from classical widgets, have been studied, specified and implemented. The developed components can be used by developers to implement software for blind users. The contribution of this work in the field of assistive technology is valuable, because there are no existing tools that facilitate the creation of interfaces for the blind users, and it may considerably improve computer access for this category of users.**

I. Introduction

The existing integrated developing environments (IDE) can be extended to deal with specific implementations [1]. In order to create easily non visual applications, existing IDEs have to integrate the following modules : libraries to hold communication with specific I/O devices, functions to deal with all the incoming events, and the production of outcoming events, functions that implement the different I/O modalities, and a set of non visual controls to facilitate the development of adapted representation and interaction.
For instance, a GUI developer, has a set of controls (usually called widgets) s/he manipulates visually by "drag and drop", giving them the position, shape, and content they will have at runtime.
It would be interesting if a non visual application developer has a matching set of controls that can be represented on adequate I/O devices.

Sun Microsystems furnishes a toolkit for accessibility, but it works only on Unix and Linux systems. Most of the personal computers today use Windows systems. Microsoft has also an accessibility program but it gives a set of developed tools to be used with existing applications.

Some attempts have been made to create non visual controls. These non visual widgets used only sound modality[2, 3], and/or cannot be used to facilitate the development of interfaces for the blind users. Some attempts failed because of the immature technology. A lot of projects [4] existed in this field but no tools were created to help developing non visual interfaces and interaction objects.

The objective of this work is to help creating stand alone applications and to facilitate the development of specific applications for blind people. The advantage is that no screen reader is needed, and that the proposed controls can be integrated in any existing IDE. Accessibility features are built inside of them and they can be used for both purposes : classic applications and non visual applications.
The following paragraphs present the specification, the design, the implementation and the evaluation of the non visual components. The final section of this paper present some applications that have been built using our components. All these sections are illustrated with examples in order to facilitate the comprehension for the newbie reader in the field of assistive technology.

II. Non visual controls specification

The developer of a GUI uses languages, toolkits, resource editors that facilitate the task. For example, when an error message is needed, the developer has the choice between a control ready message box, a window with a label inside it, or a text displayed in a text box. All s/he has to do is to pick up the needed control and to integrate the message inside it (by programming, or by visual association).
When the developer wants to give the user a set of commands or functionalities, s/he can propose them inside a menu. The resource editor gives the possibility to create a menu, to give it a general appearance, to add items in it, and to associate these items to functions inside the core of the application.
We propose to implement some comparable controls that can be used on non visual output devices such as a Braille edit box, a Braille and speech edit box, a Braille list box, a speech edit box…etc.
Let's consider an example : some choices are to be given to the visually impaired user, and the output device is a Braille display coupled with a speech synthesizer. The developer picks up the control called NonVisualMenu in the component gallery provided by the authoring tool, and then decides for its general proprieties (modality to be used, name, items, events…).

The developer enters the different items of the menu and attaches them to the possible events given for the control : interactive key click followed by enter, or interactive key double-click, or standard keyboard event. Then s/he defines the possible functionality (callback) for each menu item.

K. Elleithy (ed.), Advances and Innovations in Systems, Computing Sciences and Software Engineering, 29–34.

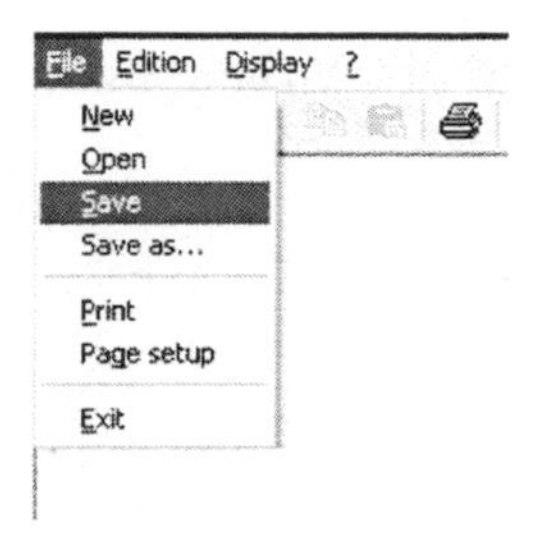

Figure 1. : An example of menu

The fig. 1 is an example of menu, and fig. 2 is its Braille adaptation.

The fig. 2 shows an example of menu adaptation using a Braille terminal. The lower part of the figure shows four keys that allow the user to navigate in the menu. Four of the Braille terminal function keys can be used and are programmed to navigate in the menu : up, down, next group, previous group. If the Braille terminal does not have function keys, it is possible to use the standard keyboard.

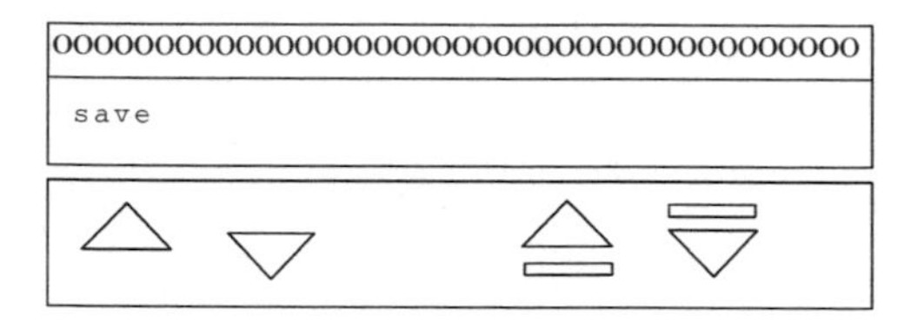

Figure 2. : Braille adaptation of the menu

In the same way we studied the example of a menu, other controls have been adapted and implemented that offer an equivalent for a developer of non visual applications, as they exist for GUI developers.

The controls proposed are the following : edit box, list box, menu, button.

The creation of a non visual application using these controls creates at the same time a matching graphical but very simple application. This is necessary for visual feedback when the developer is still developing, testing or maintaining the application. It is also useful to create dual interfaces, used by both visually impaired and sighted users [5].

III. ARCHITECTURE OF THE NON VISUAL CONTROLS

The controls proposed respect a four-level architecture. The control level, the non visual control level, the functionality level, and the physical device level. All these levels respect defined models.

The control level is composed of the graphical interface given to the developer. For the menu example this level contains the graphical tools that allow the developer to create the menu, its items and its basic features : title, aspect,...etc. This component is the same whether the developer chooses to implement a Braille control, a sound control, a speech control or a multimodal control.

The non visual control level is composed by the behaviour of the control on the chosen devices. For example, how will the menu behave on a Braille display, what will be displayed first, how the items are displayed since the principal menu, when are the items read by a speech synthesizer, etc.

The functionality level is composed of the functions that implement the different modalities of a control. The Braille modality, or speech modality are examples for this level.

The device level defines models of I/O peripherals that can be used, and how communication is established between them and the application, and how events are received, interpreted, and produced [6].

Eurobraille is an example of Braille terminal device, and Microsoft Speech API is an example of speech synthesis (in this case the device is software).

The fig. 3 shows the interaction between the four control levels in the case of a menu represented on a Braille terminal. The event is received by the lower physical level -the Braille terminal-, interpreted and then sent to the non visual control level. The non visual control asks the menu control for the associated callback when receiving this event, the menu control responds to the query. The non visual control asks then for a Braille support from the functionality level and produces the output event.

For example, an event is received by the Braille terminal and interpreted as a click on an interactive key. The event is sent accompanied by parameters such as the key position to the non visual control. The non visual control associates the click to the menu "File" which is displayed at the given position, and asks the general menu control about the callback that must be associated to this event.

The menu control returns the needed callback which is the replacing of the actual display by the items related to "File" menu. The non visual control receives the action accompanied by the text to be displayed, asks the Braille support to translate and format the text, and then asks the device to display it.

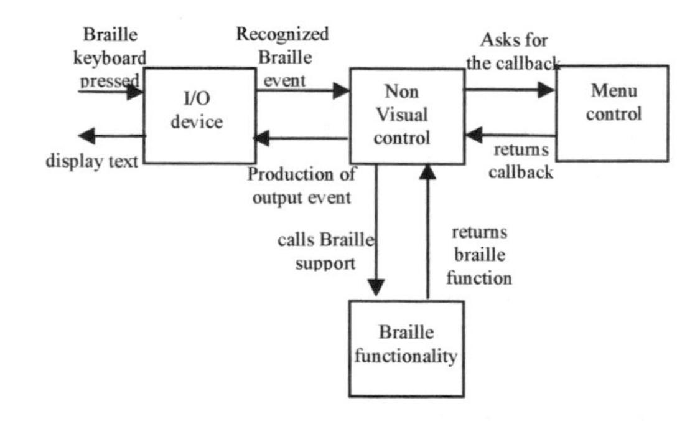

Figure 3. : Managing an input event for a control.

IV. DESIGN

We propose the use of a model driven approach to develop the proposed components. MDA[7] is an approach for application specification and interoperability which is based on the separation of concerns between domain knowledge and platform specificities. It relies on the use of MOF (Meta Object Facility) meta-models and UML (Unified Modelling Language) models for any step of the application life cycle.

This approach depends on the definition of :

- a specification model called Computational Independent Model or CIM,
- a conception model called Platform Independent Model or PIM,
- an implementation model called Platform Specific Model or PSM,
- a set of model transformations (also called mappings).

The control components are designed using the UML language. MOF (Meta Object Facility) formalism has been used to define the component meta model shown in the fig. 4.

All the developed components must correspond to this meta model :

- the component inherits from a classic control which inherits from a class such as CEdit, CButton…; a control is defined with attributes and methods;
- the component can define its own attributes and methods;
- the component is accessible via a modality or composition of modalities; Braille and speech modalities are given as example;
- the component is used by the developer via a graphic interface;
- the component generates and responds to events; the latter can be internal (application events) or external (device events).

After the definition of the general meta-model, we have to create a Platform Independent Model (or PIM) which describes the different components without considering any implementation techniques, or languages or platforms. Then, one or more Platform Specific Models (PSM) can be defined. Each PSM describes how the PIM is adapted for a different platform. When a new technology, device, platform emerges, it is only necessary to create the related PSM.

The passages from PIM to PSM and from PSM to application code are based on transformation models which need to be specified.

For instance, the device level model consists in a Braille terminal package, a sound package, and a speech synthesizer package. The control model consists in the description of menu, edit box, button, list box, window, and dialog box controls.

Once the PIM created, the next step is the description of the platform related models, and the transformation models and rules.

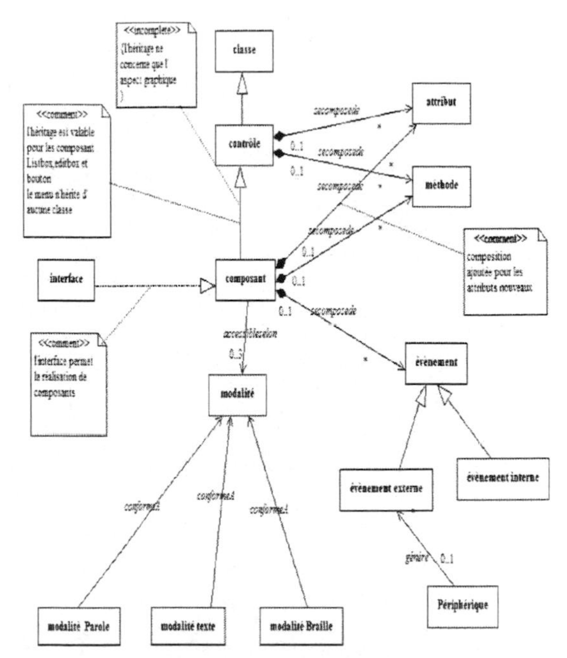

Figure 4 : The component meta-model.

The advantage of this approach is the permanence of the conception models. If we need to add a modality, to change the device for a component, or if a new technology appears, the models are modified, and the code is generated. Another advantage is that the conception models become productive.

At present, we have adopted this methodology in our piece of work, and we used Objecteering/UML environment to support the model driven approach.

We have specified the MOF meta-model of fig.4, and the UML platform independent models corresponding to the four levels of the non visual controls.

The source code is generated partly in C++. However, we have to make some enhancements to the mapping models in order to generate the complete components source code.

V. IMPLEMENTATION

The non visual controls can be used since C++ or Basic or Java languages in a Windows environment. from .net or php or other platforms. For these reasons we proposed the Microsoft COM technology to implement them. This technology (integrating OLE, Active X, COM and DCOM) creates object components that offer one or more interfaces that can be called from different languages. A COM client application can use one or more COM objects as if they were part of it. The COM components have a unique identifier, and are used via their exposed interface. MDA is not yet mature for the generation of components source code. There is a lot of work to do to create transformation tools adapted to our needs.

A. The component architecture

The partial source code generated by Objecteering/UML starting from the UML models has been intensely modified using MSVC++ IDE. Other IDEs or languages can be used if they are more adapted to the task.

A COM component has public methods and events which are its exposed interface: the way that allows the developer or other applications to use it.

The controls can respond at the same time to user defined events (Braille, speech recognition…) and standard events (keyboard, mouse…). This facilitates the creation of a non visual application that can also be used by a sighted user.

The four levels of our proposed architecture are implemented as independent COM components, so they can be reused for other applications.

B. The non visual control implementation

The device components which manage the input and the output events of assistive devices are implemented as ActiveX dynamic link libraries (DLL), and so are the functionality components which define the Braille modality and the speech modality.

The non visual controls are implemented as OCXs which are components written in C++ language.

At present, the developed components are an edit box, a list box, a button, and a menu. They are respectively called EditBoxNV, ListBoxNV, ButtonNV, and MenuNV.

Below is a series of figures that the developer will encounter when using our components.

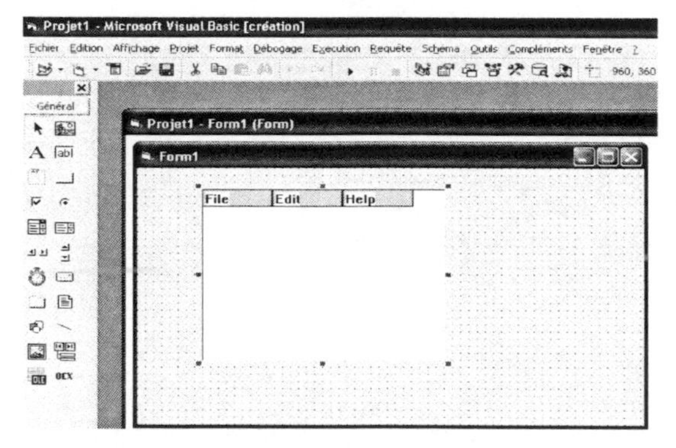

Figure 5 : Inserting MenuNV into an application.

Fig. 5 represents the insertion of a non visual menu for the developing of a new application. MenuNV OCX appears among the classic available controls.

After that the developer has to give a content to the menu, as shown in fig. 6, the menu is called "File" and it contains various items such as "New, Open, Edit…". The OCX gives the developer the opportunity to generate code related to the menu. The generation of source code works actually with four languages : Java, Visual Basic, MFC/C++ and C#. Other languages may be considered.

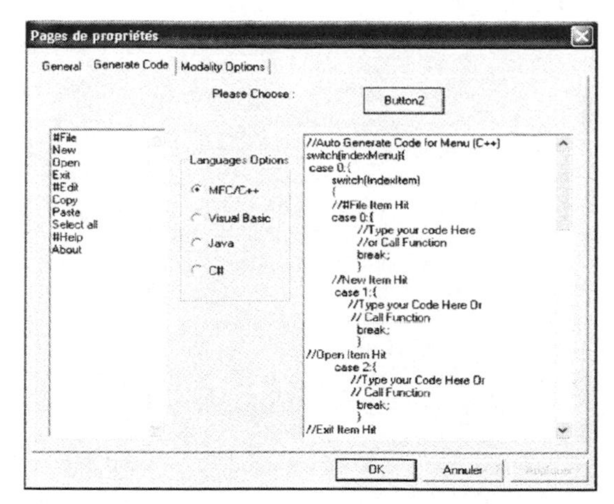

Figure 6 : Generating VB code for MenuNV.

The next step is to choose the modalities for the menu control (Fig.7) , for instance Braille and speech modalities can be used. The developer enters some keys to stop or to restart speech synthesis. Various options are available such as the possibility to repeat the reading of menu items for an indefinite period when a specific key is pressed by the user; or the possibility to stop the speech synthesis when the chosen key in the corresponding list box is pressed, etc.

For the Braille modality two choices can be made : a vertical and horizontal display, or a specific item display. The horizontal display mode displays the application entire menu on the Braille display when the user moves horizontally using direction keys. And it displays menu specific items if the user moves up and down. The second option allows the display of the current item only.

Figure 7 : Modality selection for MenuNV.

Fig. 8 shows an example of a non visual menu into a container application at runtime, and the focus is on the menu item "New". For instance this menu is represented horizontally. The window on the right of the figure simulates what is displayed on the Braille terminal. This is useful for the developer in order to have a visual feedback. The developer does not necessarily understand the Braille language.

VI. EVALUATION

The evaluation process has been made at two levels : the use of the components for a developing purpose by a developer, and the use of the components in a container application by a blind user.

A. The developing mode

The components have been tested using Visual Basic IDE, MSVC++ IDE, PHP and .NET platform. Developers in our research unit, that have not been involved in the components project, have used them for implementation purposes in Visual Basic IDE. They spent a short time in understanding their functionalities and integrating them into an application. Their feedback is that they found no differences in use between the non visual components and other OCX and VBX components they used before.

B. The user mode

We have used the non visual components to develop a word processor having basic features such as text entering, saving, and printing (fig. 8). The upp²er part of the application main window contains the non visual menu component MenuNV. The toolbar in the middle of the window has been built using several times the component ButtonNV. The lower part of the screen contains the two other components : ListBoxNV on the left, and EditBoxNV on the right. The toolbar represents the most frequently used functionalities of the application. The list box displays the most recently opened files, and the edit box displays the file content. The speech synthesis states the name of the component having the focus, while its content is written on the Braille display (that is simulated on the second window of fig. 8).

The application was given to a blind user without further explanations. He found it usable, and very friendly as it was faster than what he was accustomed to. (He usually use a screen reader under Microsoft environment). The user has also proposed some enhancements we are planning to integrate to our developments, such as multilingual speech synthesis.

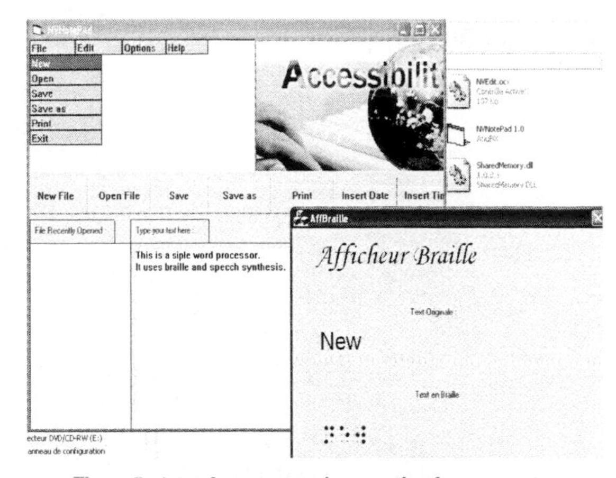

Figure 8 : A word processor using non visual components.

VII. UTILITY OF THE NON VISUAL COMPONENTS

In this section we present another simple applications that have been developed with our components, and discuss the possibility to use the components with Java platform.

A. A Web application

This example shows the use of the non visual components into a simple dynamic website developed with PHP and MySQL (Fig. 9). This website gives the user the opportunity to search and to buy a book. The components used are the button and the edit box.

The user can move through the different components using the tabulation key. Other keys allow the user to have extra information.

The fig. 10 shows how to setup ButtonNV control.

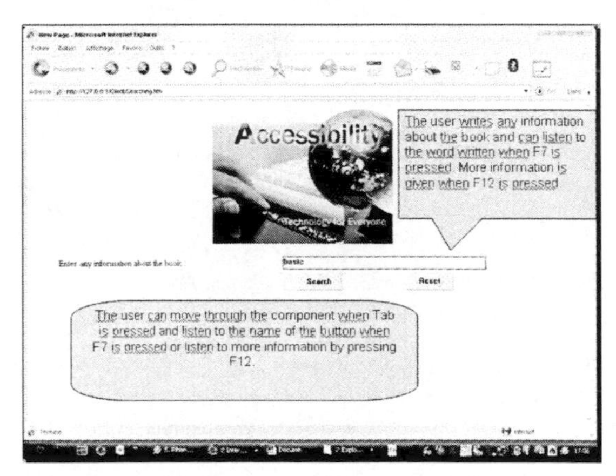

Fig 9 : The start up page of the website

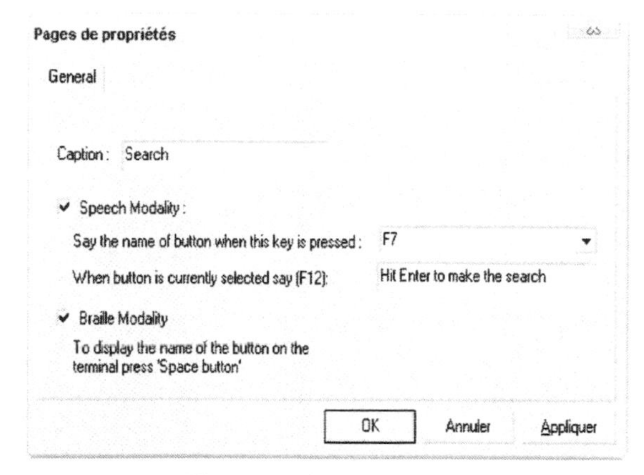

Figure 10 : Setup for ButtonNV

When the user launches the search, the results page is displayed (fig. 11).

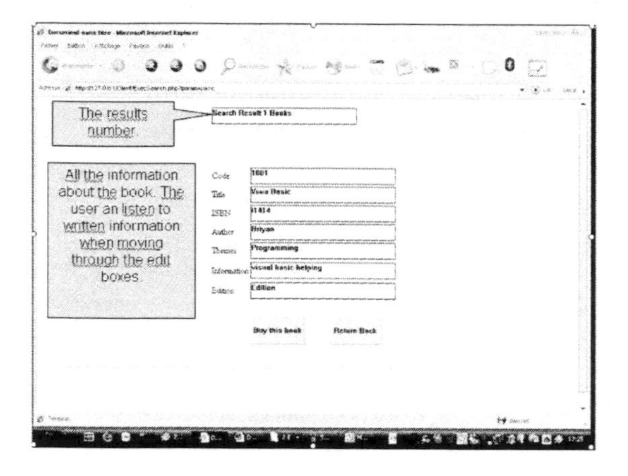

Figure 11: The website results page.

The results page uses several times the non visual edit box and button. The user listens to page contents when s/he moves trough the different controls. The Braille display is updated with each control change.

B. Non visual components and Java Platform

Using the components from Java platform is possible but more difficult because COM components can not be used without being converted to JavaBeans. There are tools that can help the developer to do this conversion, such as the Migration Assistant, or Bridge2Java which help to convert OCX files to Java files.

VIII. Conclusion and Further Work

The proposed controls are used to extend a programming environment [1,8], and evaluation proved that they can be easily used by developers. The four level architecture is composed by reusable components that can be useful for other purposes.
However, this work can be improved if we apply completely the OMG's Model Driven Architecture [7].
Our future work consists mainly in applying the OMG's MDA approach explained above to create the non visual controls.
We also want to integrate more I/O devices for the controls such as the speech recognition; define and implement more controls; define new controls that are not adapted from existing controls. We are also planning to apply multilingual functionalities to speech synthesis and to adapt Arabic language to the Braille modality.
Another perspective is now under study, to create components with zoom, contrast and colour enhancements for the visually impaired users.
At the end of this work we will be in presence of a reusable control repository.
The main advantage of our piece of work is the developing of specific applications for the blind users in a rapid speed and at a low cost. For instance, these controls are valuable in the education field for blind children in Tunisia. By installing applications containing non visual components on each computer in a classroom, we can imagine the gain compared to the use of a screen reader at a prohibitive price.

References

[1] Bouraoui A., Extending an existing IDE to create non visual interfaces. *Proc. Cisse 2005 International Conference on Systems, Computing Sciences and Software Engineering*, Springer, New York, 2005.

[2] Brewster S, The Design of Sonically-Enhanced Widgets, *Interacting with computers, 11* (2), 1998, 211-235.

[3] Edwards A.D.N, Mitsopoulos E.N, A Principled Methodology for the Specification and Design of Non-Visual Widgets, *ACM Transactions on Applied Perception*, *2*(4), 2005, 442-449.

[4] Emiliani P.L, Stephanidis C, From Adaptations to User Interfaces for All, *Proc. The 6th ERCIM workshop on "User interfaces for all"*, Italy, 2000.

[5] Savidis, A., Stephanidis, C., The HOMER UIMS for Dual User Interface Development: Fusing Visual and Non-Visual Interactions. *Interacting with Computers, 11* (2) 1998, 173-209.

[6] Wang A.J.A, Diaz Herrera, J.L, Device drivers as reusable components, *Proc .Software Engineering and Applications*, Marina Del Rey, USA, 2003.

[7] OMG Model Driven Architecture resource page. Needham, MA: Object Management Group. Internet : http://www.omg.org/mda/index.htm. 2003.

[8] Stephanidis, C, Emiliani , P..L , Universal Access to Information Society Technologies: Opportunities for People with Disabilities, *Proc. Computer Helping People with Special Needs : 8th International Conference*, Linz, Austria, 2002.

Developing a Multi-Agent System for Dynamic Scheduling trough AOSE Perspective

ANA MADUREIRA[α] JOAQUIM SANTOS[α] NUNO GOMES[β] ILDA FERREIRA[χ]

GECAD – Knowledge Engineering and Decision Support Group
Institute of Engineering – Polytechnic of Porto
Porto, Portugal

[α]{anamadur, filipe}@dei.isep.ipp.pt
[β]gomesn@gmail.com
[χ]ilda.ferreira@sbn.pt

Abstract – Agent-based computing can be considered as a new general purpose paradigm for software development, which tends to radically influence the way a software system is conceived and developed, and which calls for new agent specific software engineering approaches. This paper presents an architecture for distributed manufacturing scheduling and follows Agent Oriented Software Engineering (AOSE) guidelines trough specification defined by Ingenias methodology. This architecture is based on a Multi-Agent System (MAS) composed by a set of autonomous agents that cooperates in order to accomplish a good global solution.

I. Introduction

A major challenge in the area of global market economy is the development of new techniques for solving real world scheduling problems. Indeed, any industrial organization can only be economically feasible by maximizing customer services, maintaining efficient, low cost operations and minimizing total investment.

Traditional scheduling methods, encounter great difficulties when they are applied to some real-world situations. The interest in optimization algorithms for dynamic optimization problems is growing and a number of authors have proposed an even greater number of new approaches, and as a result, the field lacks a general understanding as to suitable benchmark problems, fair comparisons and measurement of algorithm quality [1][2][3][4].

Current practices and newly observed trends lead to the development of new ways of thinking, managing and organizing in enterprises, where autonomy, decentralization and distribution are some of the challenges. In manufacturing, a new class of software architectures, and organizational models appeared to give form to the Distributed Manufacturing System concept [5].

In the recent years, the characteristics and expectations of software systems have changed dramatically having as result that a variety of new software engineering challenges have arisen [6][7][8].

In this work we have two main purposes. Firstly, the resolution of more realistic scheduling problems in the domain of manufacturing environments, known as Extended Job-Shop Scheduling Problems [9][10], combining Multi-Agent Systems (MAS) and Meta-Heuristics technologies. Secondly, to demonstrate that is essential for MAS development the integration of Software Engineering concepts like the AOSE paradigm.

The proposed Team-based architecture is rather different from the ones found in the literature; as we try to implement a system where each agent (Machine Agent) is responsible to achieve a near optimal solution to schedule operations related with one specific machine through Tabu Search or Genetic Algorithms. After local solutions are found, each Machine Agent is required to cooperate with other Machine Agents in order to achieve a global optimal schedule.

The remaining sections are organized as follows: Section II summarizes some related work and the research on the use of multi-agent technology for dynamic scheduling resolution. In Section III are introduced some terms and definitions in Multi Agent Systems. This section presents some Agent-Oriented Methodologies and describes some considerations regarding Software Architectures and Multi-Agent Systems. In section IV the scheduling problem under consideration is defined. Section V presents the Team-Work based Model for Dynamic Manufacturing Scheduling and a proposal by Ingenias methodology.

K. Elleithy (ed.), Advances and Innovations in Systems, Computing Sciences and Software Engineering, 35–40.

Finally, the paper presents some conclusions and puts forward some ideas for future work.

II. Related Work

Dynamic scheduling is one that is receiving increasing attention amongst both researchers and practitioners. In spite of all previous contributions the scheduling problem is still known to be NP-complete [2]. This fact incites researchers to explore new directions. Multi-Agent technology has been considered as an important approach for developing industrial distributed systems.

In [11] Shen and Norrie presented a state-of-the-art survey referencing a number of publications that attempted to solve distributed dynamic scheduling problems. According to these authors, there are two distinct approaches in the mentioned work. The first is based on an incremental search process that may involve backtracking. The second approach is based on systems in which an agent represents a single resource and is therefore responsible for scheduling that resource. Agents then negotiate with other agents in order to accomplish a feasible solution.

For further works developed on MAS for dynamic scheduling, see for example, [9][12].

The characteristics and expectations of software systems have changed dramatically in the last few years, with the result that a range of new software engineering challenges have arisen [6][7]. First, most software systems are concurrent and distributed, and are expected to interact with components and exploit services that are dynamically found in the network. Second, software systems are becoming "always-on" entities that cannot be stopped, restored, and maintained in the traditional way. Finally, current software systems tend to be open, because they exist in a dynamic operating environment where new components can join and existing components can leave the system on a continuous basis, and where the operating conditions themselves are likely to change in unpredictable ways.

From the literature we can conclude that Agent-based computing is a promising research approach for developing applications in complex domains. However, despite the great research effort [8][13][14], there still exists a number of challenges before making agent-based computing a widely accepted paradigm in software engineering practice. In order to realize an engineering change in agent oriented software, it is necessary to turn agent oriented software abstractions into practical tools for facing the complexity of modern and current application areas.

III. Multi-Agent Systems

Agents and multi-agent systems (MAS) have recently emerged as a powerful technology to deal with the complexity of current Information and Communication Technologies environments. In this section we will describe some issues and considerations regarding the developing of the MAS following a software engineering perspective.

A. *Terms and Definitions*

The development of multi-agent systems requires powerful and effective modelling, architectures, methodologies, notation techniques, languages and frameworks. Agent-based computing can be considered as a new general purpose paradigm for software development, which tends to radically influence the way a software system is conceived and developed, and which calls for new, agent specific software engineering approaches [8].

The main term of Multi-Agent based computing is an Agent. However the definition of the term Agent has not common consent. In the last few years most authors agreed that this definition depends on the domain where agents are used. In Ferber [15] is proposed a definition: "*An agent is a virtual or physical autonomous entity which performs a given task using information gleaned from its environment to act in a suitable manner so as to complete the task successfully. The agent should be able to adapt itself based on changes occurring in its environment, so that a change in circumstances will still yield the intended result.*"

An agent can be generally viewed as a software entity with characteristics [16] such as:

- Autonomy - where an agent has its own internal thread of execution, typically oriented to the achievement of a specific task, and it decides for itself what actions it should perform at what time.
- Situatedness - agents perform their actions while situated in a particular environment.
- Proactivity - in order to accomplish its design objectives in a dynamic and unpredictable environment the agent may need to act to ensure that its set goals are achieved and that new goals are opportunistically pursued whenever appropriate.
- Sociability - agents interact (cooperate, coordinate or negotiate) with one another, either to achieve a common objective or because this is necessary for them to achieve their own objectives.

A Multi-Agent System (MAS) can be defined as "*a system composed by a population of autonomous agents, which cooperate with each other to reach common objectives, while simultaneously each agent pursues individual objectives*" [15]. According to Russell and Norving [17] multi-agent systems "[...] *solve complex problems in a distributed fashion without the need for each agent to know about the whole problem being solved*".

We can see MAS like a society of agents that cooperates to work in the best way possible. With this we gain the ability of solve complex problems like dynamic and distributed scheduling. Considering the complexity inherent to the manufacturing systems, the dynamic scheduling is considered an excellent candidate for the application of agent-based technology. In many implementations of multi-agent systems for manufacturing scheduling, the agents model the resources of the system and the tasks scheduling is done in a distributed way by means of cooperation and coordination amongst agents [8]. There are also approaches that use a single agent for scheduling (centralized scheduling algorithm) that defines the schedules that the resource agents will execute [8][16]. When responding to disturbances, the distributed nature of multi-agent systems can also be a benefit to the rescheduling algorithm by involving only the agents directly affected, without disturbance to the rest of the community that can continue with their work.

The main advantages of a Multi-Agent system are the abilities of coordination and cooperation in order to accomplish a common objective.

B. Agent-Oriented Methodologies

Several methodologies for the analysis and design of MAS have been proposed in the literature, however only few of them focus on organizational abstractions.

MASE Methodology [18] provides guidelines for developing MAS based on a multi-step process. In analysis, the requirements are used to define use-cases and application goals and sub-goals, and eventually to identify the roles to be played by the agents and their interactions. In design, agent classes and agent interaction protocols are derived from the outcome of the analysis phase, leading to a complete architecture of the system.

MESSAGE methodology [19] exploits organizational abstractions that can be mapped into the abstractions identified by Gaia. In particular, MESSAGE defines an organization in terms of a structure, determining the roles to be played by the agents and their topological relations (i.e., the interactions occurring among them). In addition, in MESSAGE, an organization is also characterized by a control entity and by a workflow structure.

GAIA methodology described in Zambonelli [20] is an extension of the version described in Wooldridge et al. [21]. The first version of GAIA, provided a clear separation between the analysis and design phases. However, as already noted in this paper, it suffered from limitations caused by the incompleteness of its set of abstractions. The objective of the analysis phase in the first version of GAIA was to define a fully elaborated role model, derived from the system specification, together with an accurate description of the protocols in which the roles will be involved. This implicitly assumed that the overall organizational structure was known a priori (which is not always the case). In addition, by focusing exclusively on the role model, the analysis phase in the first version of GAIA failed to identify both the concept of global organizational rules (thus making it unsuitable for modelling open systems and for controlling the behaviour of self-interested agents) and the modelling of the environment (which is indeed important, as extensively discussed in this paper). The new version of GAIA overcomes these limitations.

The TROPOS methodology first proposed in [22], adopts the organizational metaphor and an emphasis on the explicitly study and identification of the organizational structure. TROPOS recognizes that the organizational structure is a primary dimension for the development of agent systems and that an appropriate choice of it is needed to meet both functional and non-functional requirements.

PASSI (Process for Agent Societies Specification and Implementation) [23] is a methodology for MAS development, that integrates the definition of MAS philosophy, modelling and the orientation to objects using UML. This is composed by five models that address different visions and twelve steps during the development process (http://mozart.csai.unipa.it/passi/).

The described methodologies have different proposals to model agents and MAS, but they share some characteristics. All of them model agents like autonomous entities and address the interaction between agents in an agent society. A comparative study can be found in [24].

MESSAGE, MaSE and PASSI are more adaptable to industrial scenarios, because they consist on evolutions of UML that is a common standard in this kind of environments.

C. Software Architectures and Multi-Agent Systems

Research in the area of agent-oriented software engineering has expanded significantly in the past few years. Several groups have started addressing the problem of modelling agent systems with appropriate abstractions and defining methodologies for MAS development [25].

Traditional object-based computing promotes a perspective of software components as functional or service-oriented entities that directly influences the way that software systems are architected.

Usually, the global design relies on a rather static architecture that derives from the decomposition and modularisation of the functionalities and data required by the system to achieve its global goals and on the definition of their inter-dependencies [8].

IV. PROBLEM DEFINITION

Most real-world multi-operation scheduling problems can be described as dynamic and extended versions of the classic or basic Job-Shop scheduling combinatorial optimization problem. The general Job-Shop Scheduling Problem (JSSP) can be generally described as a decision-making process on the allocation of a limited set of resources over time to perform a set of tasks or jobs. In this work we consider several extensions and additional constraints to the classic JSSP, namely: the existence of different job release dates; the existence of different job due dates; the possibility of job priorities; machines that can process more than one operation in the same job (recirculation); the existence of alternative machines; precedence constraints among operations of different jobs (as quite often, mainly in discrete manufacturing, products are made of several components that can be seen as different jobs whose manufacture must be coordinated); the existence of operations of the same job, on different parts and components, processed simultaneously on different machines, followed by components assembly operations (which characterizes the Extended Job-Shop Scheduling Problem (EJSSP)[9][10]).

Moreover, in practice, scheduling environment tend to be dynamic, i.e. new jobs arrive at unpredictable intervals, machines breakdown, jobs are cancelled and due dates and processing times change frequently.

V. Multi-Agent System for Distributed Manufacturing Scheduling with Genetic Algorithms and Tabu Search

This section describes the architecture proposed for dynamic and distributed scheduling and proposes a methodology trough Ingenias for its specification.

A. *MASDScheGATS Architecture*

Distributed environment approaches are important in order to improve scheduling systems flexibility and capacity to react to unpredictable events. It is accepted that new generations of manufacturing facilities, with increasing specialization and integration, add more problematic challenges to scheduling systems. For that reason, issues like robustness, regeneration capacities and efficiency are currently critical elements in the design of manufacturing scheduling system and encouraged the development of new architectures and solutions, leveraging the MAS research results. The work presented in this paper describes a system where a community of distributed, autonomous and often conflicting behaviours, cooperating and asynchronously communicating machines tries to solve scheduling problems. Global system behaviour can emerge with requested abilities of reactivity and flexibility to accommodate all the external perturbations.

The main purpose of MASDScheGATS (Multi Agent System for Distributed Manufacturing Scheduling with Genetic Algorithms and Tabu Search) is to create a Multi-Agent system where each agent represents a resource (Machine Agents) in a Manufacturing System. Each Machine Agent is able to find an optimal or near optimal local solution trough Genetic Algorithms or Tabu Search meta-heuristics, to change/adapt the parameters of the basic algorithm according to the current situation or even to switch from one algorithm to another.

In our case the dynamic scheduling problem is decomposed into a series of Single Machine Scheduling Problems (SMSP)[9][10]. The Machine Agents obtain local solutions and cooperate in order to overcome inter-agent constraints and achieve a global schedule.

Agents agree to work together in order to solve a problem that is shared by all agents in the team. Such approach allows for the resolution of large-scale problems that a single agent would not be able to solve. Moreover, Team-based architecture has the ability to meet global constraints given the capability that agents possess to act in concert. As we shall see later, this characteristic is critical for the problem treated in this work and defined in section IV.

The proposed architecture is based on three different types of agents. In order to allow a seamless communication with the user, a User Interface Agent is implemented. This agent, apart from being responsible for the user interface, will generate the necessary Task Agents dynamically according to the number of tasks that comprise the scheduling problem and assign each task to the respective Task Agent.

The Task Agent will process the necessary information regarding the task. That is to say that this agent will be responsible for the generation of the earliest and latest processing times, the verification of feasible schedules and identification of constraint conflicts on each task and the decision on which Machine Agent is responsible for solving a specific conflict. Finally, the Machine Agent is responsible for the scheduling of the operations that require processing in the machine supervised by the agent. This agent will implement meta-heuristic and local search procedures in order to find best possible operation schedules and will communicate those solutions to the Task Agent for later feasibility check.

B. *Proposal methodology trough Ingenias*

The development cycle that is proposed by INGENIAS (http://grasia.fdi.ucm.es/ingenias/) methodology sees MAS like a computational representation of a set of models. Each of these models has a partial view of the system: definition of the autonomous agents that compose the system, interaction between agents, system organization, domain, tasks and objectives.

In order to specify these models, the definition of meta-models is needed. One meta-model is a representation of all types of entities that can exist in a model, their relations and application restrictions.

The meta-models used in this methodology are an evolution of MESSAGE methodology work [19]. This methodology uses five different kinds of meta-models that describe the correspondent diagrams:

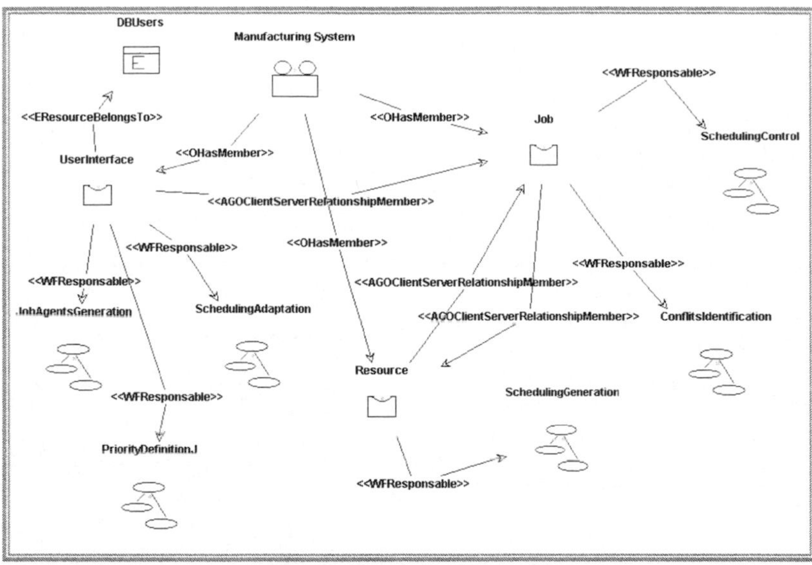

Figure 1 - Organization Meta-model

1. Organization meta-model: defines groups of agents, system functionality and restrictions to agent's behaviour. Is equivalent to system architecture in MAS. The value of these models is the definition of workflows.

2. Interaction meta-model: details how agents coordinate and communicate among them. The definition of systems interaction allows the identification of dependencies among components.

3. Agent meta-model: describes agents, excluding interactions with other agents, and the mental states that they have in their life cycle. This meta-model is centred in agent functionality and in is control drawing. It gives information about the responsibilities or tasks that an agent is able to perform.

4. Tasks and Objectives meta-model: is used to attach an agent mental state to the task that it executes. Aims at the collection of MAS motivations, to define the identified actions in the organization, interaction or agents models, and how these actions influence the agents.

5. Environment meta-model: Defines everything that is present in the environment and the way that each agent understands it. Its main function is to identify all environment elements and define a relation with the other entities.

C. *Illustrative example*

It seems now relevant to exemplify the methodology through the illustration of one meta-model applied to the MASDScheGATS problem. Given the system architecture description provided above, we think that, from the five meta-models, the Organization meta-model is the most suited to be described (Fig. 1).

As already mentioned, the Organization meta-model defines the groups (or types) of agents and their functionalities, adding up the objectives and mental states for each agent. Therefore, our Manufacturing system consists of three types of agents: User Interface, Task and Machine.

The *User Interface agent* is responsible for the definition and attribution of priorities to the Machine Agents. It is also responsible for the creation of the Task Agents and for the introduction of changes in the scheduling plan when required by the user, as it belongs to this agent the task of communication with the external environment. Its objective is to perform a seamless interface between the user and the system and its initial mental state is the reception of user requests and the final mental state is achieved when the results of the requests are received.

The *Task agent* is responsible for the scheduling control, particularly for the generation of information regarding the task (earliest processing times, due dates, etc) and for the feasibility check of the scheduling plans. Also, this agent identifies the conflicts between operations of the same task and which machine is responsible to solve them. The objective of the Task agent is to achieve a feasible scheduling plan for the task. Its initial mental state is the reception of all operations concerning the task and its final mental state is the communication of the viability of the scheduling plan.

Finally, the *Machine agent* is responsible for scheduling the operations of the different tasks that must be processed in the machine portrayed by the agent. The objective of this agent is the minimization of the processing time in the scheduling plan generated. The initial mental state of this agent is the reception of data about the operations to be scheduled and its final mental state is the scheduling plan proposal.

VI. Conclusions and Future Work

The Team-Work based architecture for distributed scheduling that we have proposed in this paper seems to be a good way to solve real world scheduling problems, because a good global solution may emerge from a set of autonomous agents that cooperate through a communication mechanism to accomplish a common goal. Coordination seems to be the edge in MAS, because it is not possible for all autonomous agents to work together in a effective way even if one intervenient in the system is not an active part of the system.

We consider that the AOSE paradigm can perform an important role when a MAS in being developed, because with this definition it becomes easier to model the system and to find problems observing global system structure. When a structure problem is discovered in the middle of systems procedure implementation, often such situation implies an important loss of time.

Work still to be done in the MASDScheGATS system includes the testing of the system and negotiation mechanisms under dynamic environments subject to several random perturbations. We realize, however, that this is not an easy task because it is difficult to find test problems and computational results for the considered dynamic environment where the jobs to be processed have release dates, due dates and different job assembly levels (parallel/concurrent operations).

The proposed AOSE approach needs to be refined in order to support dynamic environments with unexpected disruptions that can not be strictly considered in the modelling because they can happen without any specific warning. Despite of this, in our opinion this kind of work can be very significant in order to turn MAS development a structured process that doesn't go from modelling to implementation without any intermediate test and validation.

ACKNOWLEDGMENTS

The authors would like to acknowledge the FCT, FEDER, POCTI, POCI 2010 for their support to R&D Projects and the GECAD Unit.

REFERENCES

[1] H. Aytug, M.A. Lawley, K. McKay, S. Mohan and U. Reha, "Executing production schedules in the face of uncertainties: a review and some future directions," European Journal of Operat. Research, vol. 16 (1), pp. 86- 110, 2005.

[2] J. Blazewicz, K. H. Ecker, E. Pesch, G. Smith, J. Weglarz, *Scheduling Computer and Manufacturing Processes,* Springer. 2nd edition. New York, 2001.

[3] P. Cowling, and M. Johansson, "Real time information for effective dynamic scheduling," European Journal of Operational Research, 139 (2), pp. 230-244, 2002.

[4] R. N. Jennings, Coordination Techniques for Distributed Artificial Intelligence, in Foundations of Distributed Artificial Intelligence (eds. G. M. P. O'Hare and N. R. Jennings), Wiley, pp. 187-210, 1996.

[5] P. Bresciani, A. Perini, P. Giorgini, F. Giunchiglia and J. Mylopoulos, "A knowledge level software engineering methodology for agent oriented programming," Proceedings of the 5th Int. Conf. on Autonomous Agents, pp. 648–655, ACM Press, Montreal (CA), 2001.

[6] P. Bresciani, A. Perini, P. Giorgini, F. Giunchiglia and J. Mylopoulos, "A knowledge level software engineering methodology for agent oriented programming," Proceedings of the 5th International Conference on Autonomous Agents, pp. 648–655, ACM Press, Montreal (CA), 2001.

[7] M. Wooldridge, and R. Jennings, "Agent Theories, Architectures, and Languages: A Survey,". Workshop on Agent Theories, Architectures and Languages, 11th European Conf. on Artificial Intelligence, Amsterdam, The Netherlands, 1994.

[8] F. Zambonelli and H. Parunak, "Toward a change of paradigm in computer science and software engineering: A synthesis," Knowledge Engineering Review, 18, 2004.

[9] A. Madureira, "Meta-Heuristics Application to Scheduling in Dynamic Environments of Discrete Manufacturing," PhD Dissertation, University of Minho, Braga, Portugal (in portuguese), 2003.

[10] A. Madureira, C. Ramos, and S. Silva, "Toward Dynamic Scheduling Through Evolutionary Computing," WSEAS Transactions on Systems, Issue 4. vol. 3, pp. 1596-1604, 2004.

[11] W. Shen, and D. Norrie, "Agent-based systems for intelligent manufacturing: a state of the art survey," Int. J. Knowledge Inf. Systems, vol. 1, no. 2, pp. 129– 156, 1999.

[12] W. Shen, and D. Norrie, "Agent-based systems for intelligent manufacturing: a state of the art survey," Int. J. Knowledge Information Systems, vol. 1, no. 2, pp. 129– 156, 1999.

[13] J. Lind, "A Process Model for the Design of Multi-Agent Systems," Research Report TM- 99-03, German Research Center for AI (DFKI), 1999.

[14] F. Zambonelli, N. Jennings, and M. Wooldridge, "Developing Multiagent Systems: The Gaia Methodology," ACM Transactions on Software Engineering and Methodology vol. 12(3), pp. 317–370, 2003.

[15] J. Ferber, *Les Sístemes multi-agents: versune intelligence collective*, Interedition, 1995.

[16] M. Wooldridge, *An Introduction to Multiagent Systems*, John Wiley and Sons, 2002.

[17] S. Russel, and P. Norvig, *Artificial Intelligence: A Modern Approach*, Prentice Hall/Pearson Education Intern.: Englewood Cliffs (NJ), 2nd ed., 2003.

[18] M. Wood, A. DeLoach, and C. Sparkman, "Multiagent system engineering," International Journal of Soft. Engineering and Knowledge Engineering, 11(3), pp.231–258, 2001.

[19] G. Caire, et al., "Agent Oriented Analysis using MESSAGE/UML," Proceedings of the 2nd International Workshop on Agent-oriented Software Engineering (AOSE 2001), Montreal, 2001.

[20] F. Zambonelli, N. Jennings, and M. Wooldridge, "Developing Multiagent Systems: The Gaia Methodology," ACM Transactions on Software Engineering and Methodology vol. 12(3), pp. 317–370, 2003.

[21] M. Wooldridge, R. Jennings, and D. Kinny, "The Gaia Methodology for Agent-Oriented - Analisys and Design," Journal of Autonomous Agents and MAS", 15, 2000.

[22] S. DeLoach, "Multiagent Systems Engineering: A Methodology And Language for Designing Agent Systems," Agent-Oriented Information Systems, AOIS, 1999.

[23] M. Cossentino and C. Potts, "PASSI: a Process for Specifying and Implementing MAS Using UML", 2002.

[24] A. Lindoso, "Uma Metodologia baseada em Ontologias para a Engenharia de Aplicações Multiagente," Msc. Dissertation, Univ. Fed. Maranhão, Brasil, (in Portuguese), 2006.

[25] F. Zambonelli, and A. Omicini, "Challenges and Research Directions in Agent-Oriented Software Engineering," Autonomous Agents and Multi-Agent Systems, 9, Kluwer Academic Publishers, pp. 253–283, 2004.

Criminal Sentencing, Intuition and Decision Support

Andrew Vincent,[1,3] Tania Sourdin,[2] John Zeleznikow[1]
[1]School of Information Systems, Victoria University
[2]School of Law, La Trobe University
[3]School of Historical and European Studies, La Trobe University
Melbourne, Australia

I. Introduction

In criminal sentencing it is virtually impossible to predict the outcome of a particular case, even in jurisdictions that employ grid guideline sentencing regimes the prediction of sentence will only be in terms of a range. In some jurisdictions like the Australian state of Victoria where judges are allowed wide latitude or discretion in the manner they sentence offenders, the prediction of sentencing tariffs is extremely difficult. There are fundamental elements that come together in criminal sentencing namely, the effect of wide discretion and the effect of intuitive decision making on the final tariff. In constructing systems that support decision makers in the criminal justice system it is of fundamental importance to not only represent sentencing knowledge in a manner that empowers the decision maker given the discretionary nature of criminal sentencing but that also allows decision makers to hone their intuitive skills. As part of on going research being conducted at both Victoria University and La Trobe University we attempting to build a decision support system for legal aid so that they may have more success for their clients.

II. Sentencing in Victoria

In the Australian state of Victoria and in other Australian states sentencing is governed by acts of parliament. In Victoria sentencing is conduct under the *Sentencing Act 1991*. The act does not define crimes (crimes are defined in various other acts, but primarily in the *Crimes Act 1958*) but rather the procedural rules and guidelines that are incumbent on judges when sentencing convicted criminals. Sentences in Victoria are not defined in terms of minimum periods of incarceration but rather maximums.[1]

Judges in Victoria are delegated the responsibility by the state for the selection of the appropriate purpose for sentencing these are:[2]

(a) to punish the offender to an extent and in a manner which is just in all of the circumstances; or
(b) to deter the offender or other persons from committing offences of the same or a similar character; or
(c) to establish conditions within which it is considered by the court that the rehabilitation of the offender may be facilitated; or
(d) to manifest the denunciation by the court of the type of conduct in which the offender engaged; or
(e) to protect the community from the offender; or
(f) a combination of two or more of those purposes.

Judges are also then required to take in to account the following matters:[3]

(a) the maximum penalty prescribed for the offence; and
(b) current sentencing practices; and
(c) the nature and gravity of the offence; and
(d) the offender's culpability and degree of responsibility for the offence; and
(daa) the impact of the offence on any victim of the offence; and
(da) the personal circumstances of any victim of the offence; and
(db) any injury, loss or damage resulting directly from the offence; and
(e) whether the offender pleaded guilty to the offence and, if so, the stage in the proceedings at which the offender did so or indicated an intention to do so; and
(f) the offender's previous character; and
(g) the presence of any aggravating or mitigating factor concerning the offender or of any other relevant circumstances.

It can be seen from the two lists why wide discretion has attracted fierce detractors, especially regarding the purposes of sentencing [1] [2]. Indeed as indicated [3], there has for a long time existed a

> "kind of cafeteria system, in which judges and magistrates have been encouraged to choose a rationale from several ... with relatively little constraint in the choice (331)."

There can be no doubt that sentencing in Victoria favours individualsation over other forms of sentencing which have solely at their core the promotion of consistency.[4]

---

[1] Sentences are classified into nine levels of severity each relating to a *maximum* penalty. A person guilty of armed robbery (*Crimes Act 1958*, s. 75A), for example, is liable to a level two imprisonment, which corresponds to a twenty-five year maximum. There are *no* minimums specified. There is a statutory requirement for the judiciary to set non-parole periods (the period of time the offender must spend in prison before being eligible for release on parole), which should not be considered minimum sentences.

[2] *Sentencing Act 1991*, s. 5(1).

[3] *Sentencing Act 1991*, s. 5(2).

[4] Methods of achieving consistency usually involve mandatory minimum sentences, guideline grid sentencing or guideline sentences. It is however

K. Elleithy (ed.), Advances and Innovations in Systems, Computing Sciences and Software Engineering, 41–46.

III. Intuition

In Victoria the idea of instinctive or intuitive synthesis has been the accepted methodology of judicial decision making. It was put in to currency in a Court of Criminal Appeal judgment in 1975 (*R v. Williscroft* [1975] VR) where it was stated

> "...ultimately every sentence imposed represents the sentencing judge's instinctive synthesis of all the various aspects involved in the punitive process ... it is profitless ... to attempt to allot to the various considerations their proper part in the assessment of the particular punishments presently under consideration ... (300)."

This view of the "right" approach to judicial decision making has recently been reinforced in a High Court of Australia decision *Markarian v The Queen* ([2005] HCA 25). Broadly speaking, sentencing method has fallen in to one of two camps. One the one hand there is the "logical", "rational" approach that has stages and levels in argument which would be made transparent with full and proper reasons [4], this is known as the two-tiered approach and was ultimately rejected as the appropriate method for arriving at criminal sentences. At the other end of the spectrum is the instinctive or intuitive synthesis camp. The instinctive synthesis method involves, as indicated above, the weighing of all the circumstances of the offence and offender to arrive at an appropriate sentence. All the factors are evaluated in reaching the decision but no one is given priority or weighted more than another.

The "two-tiered" approach is where on the first tier the judge considers the objective circumstances of the offence (factors associated to the gravity of the criminal activity) in order to gauge the seriousness of the offence. On the second tier the judge considers the subjective factors which usually relate to the offender (both aggravating and mitigating) and then the sentence is decided. Judges will often suggest a tariff they regard as proportionate to the crime and then adjust the tariff by specific amounts by reference to particular factors. It is obvious that this method should encourage the judiciary to be more explicit in their reasoning by declaring the weight given to individual factors. This procedure is more overtly mathematical than the so-called "black box" approach and at first blush more likely to be able to be predicted and especially modelled for the construction of decision support systems. This though cannot be further from the truth. This approach also has its critics, whose argument usually revolves around the "mathematical" nature of the process. One of the main criticism that seems to be overlooked by commentators is that there is still an intuitive decision to be made, this involves the selection of an appropriate starting point for the various calculations of aggravation and mitigation. It has been suggested [5] that

> "we see no harm in retaining a two-tiered sentencing methodology wherein the sentencing judge or magistrate determines the upper, and sometimes lower, limits of an appropriate sentence based on the notion of offence seriousness or objectivity (the outer range of proportional punishment) and then proceeds to fine-tune the sentence by reference to other considerations."

Given that sentences in Australia are bounded only by maximums it is still an instinctive or intuitive decision about what the starting point should be. It is further suggested that in some cases the quantum for individual factors could be disclosed. This then leads to more intuitive decisions about when to disclose specific discounts and then also explanations as to the reasons. The guilty plea is the most common mitigating factor leading to sentence reduction that is indicated in by judges in their reasons and the judicial annunciation of the specific discount in a particular case could lead to further legal wrangling about the appropriateness of that figure.

There can be little doubt that the instinctive synthesis rationale that judges advocate as correct most accurately reflects their decision making practices. Intuition plays a very large part in the sentencing process. Mackenzie [6] in her important study of judges in Queensland indicates that many judges describe the process as an intuitive one. This is an important finding as it adds to the list of studies that have found that judges use their intuition to assist in their decision making [7]. Brest [8] suggests that intuition and analysis are involved in a complicated dance and it is all but the most simple decisions that are analytical in nature.

In a very useful discussion of the importance role of intuition and decision support, Sauter [9] has suggested the paucity of information required for very complicated decisions means that detailed analyses by decision makers are unfeasible. In most jurisdictions, judges are under increasing pressure with respect to their case loads and given the often parlous state of information available on the effectiveness of the various sentencing options available to them, there is a great reliability on intuition in determining the appropriate sentence.

Intuition is variously described as "a sense of feeling of pattern or relationship" [9] or intuitive responses "are reached with little apparent effort, and typically without conscious awareness" [10]. It usually involves a decision that is complicated, all the information required might not be present and the outcome is not certain. In sentencing, intuition is educated by experience and by providing information relating to the interactions of the factors that can be taken in to account judges can hone their intuition. The provision of recent and relevant information to decision makers is also of critical importance.

Building systems to support the various parties involved in the sentencing process is fraught with difficulties. Tata [11] has detailed the effort in the construction of the Scottish Sentencing Information System and discusses some of the reasons why judicial decision support systems are not well received by the judiciary they are made to support. One of the

---

obvious even to the most causal of observers that mandatory sentences have the capacity to produce extremely unjust outcomes, most recently US v. Hungerford, No. 05-30500 (9th Cir. Oct. 13, 2006) (http://www.ca9.uscourts.gov/ca9/newopinions.nsf/CA706AAB28F6B229882572050076D19A/$file/0530500.pdf?openelement (last viewed 14/10/2006), where a 52 year old mentally ill women was sentenced to 159 years in jail.

primary reasons for judicial ambivalence is the fact that most systems do not accurately reflect either the manner in which judges reach their decision or are so over-burdened with complication that they are virtually useless. So far in this short paper we have not discussed the link between how a sentencing decision is reached and how the reasons for the sentence are articulated. In Australia written decisions are provided by the judiciary the more serious crimes. The opaqueness of the process is further exacerbated by the articulation of reasons, this though will not be discussed in any further detail.

In the remainder of the paper the work to date on a system designed to assist criminal defence lawyers will be presented. The approach used is also a method of attempting to delve in to the depths of intuitive decision making.

IV. THE TOULMIN ARGUMENT MODEL

The approach to modelling the discretionary and intuitive domain of sentencing is based on the model of argument proposed by Toulmin [12]. It is a method of structuring an argument that is not mathematical in nature. The Toulmin model is jurisprudential. It is concerned with showing that logic can be seen as a kind of jurisprudence rather than science. The jurisprudential nature of the Toulmin argument structure means that it is process focused and more useful in structuring an argument after it has been articulated. It is able to capture arguments regardless of content. The procedural nature and simplicity of the Toulmin model means that argument chains can be constructed by linking together single argument units. The claim of one argument can be used as the data item for the next. According to Toulmin, an argument is made up of a combination of five components: a claim, some data (grounds), a warrant, some backing, and a qualifier. Claims are ideas that the arguer would like the audience to believe. The data lends support to the claim and makes it more likely that the audience will believe it. The warrant, on the other hand, is the logic of the argument: the rules of inference that lead the claimant to conclude the claim, given one ground or a set of grounds. Backings usually give reasons why the audience should believe the warrant. Modal qualifiers modify the claim by indicating a degree of reliance on, or scope of generalisation of, the claim, given the grounds, warrants, and backing available. Rebuttals are the possible exceptions to the conditions under which a claim holds.

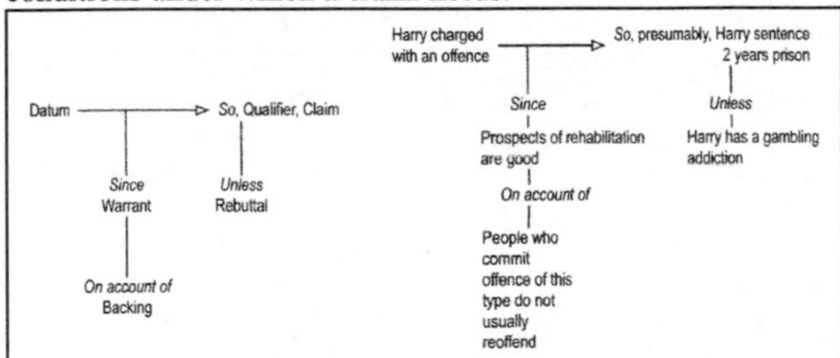

Figure 1. Toulmin argument structure

The Toulmin argument structure offers those interested in knowledge engineering a method of structuring domain knowledge. It also enables the reasoning behind certain claims to be made explicit. In any system that will be of use to decision makers, reasons for decisions are important, especially for transparency. Yearwood and Stranieri [13] present a list of other attempts to use the Toulmin argument model as a method of structuring reasoning and modelling discourse. Knowledge representation has fallen into two main areas, dialectical and non-dialectical. The Toulmin argument model is used in structuring the reasoning in judicial sentencing in a non-dialectical manner, as it is delivered by the judiciary. In a sentencing verdict the judges does not enter in to a dialogue in an effort to explain the reasons. A sentence verdict stands like a tombstone unless challenged by an appeal. Stranieri, Yearwood and Meikle [14] have suggested that discretion is intimately associated with the way knowledge is represented. They suggest that the way in which discretion is operationalised in particular knowledge representations is important for the design of computer-based systems that support decision makers. While this may be particularly obvious to policy makers who construct complex legislation, it has not been recognised or articulated very often in information systems research. A rule-based system offers no discretionary action to a user; similarly, a mandatory sentencing scheme offers little discretion to a sentencing judge. It would be inappropriate to attempt to capture the complicated discretionary area of sentencing by a rule-based system, since there would be too many rules and so the system would be virtually useless to all but the most patient users. The approach is very beneficial in attempting to model the sentencing domain due to the ability of the system to provide information to a sentencing judge or defence lawyer. Information can be stored around clusters of arguments.

V. A TOULMIN BASED MODEL OF SENTENCING

The Generic/Actual Argument Model (GAAM) represents a variation of the original Toulmin argument structure described briefly above. The variations made to the Toulmin structure in the construction of the GAAM have been described in great detail in [15] [16]. The GAAM has been used to model other legal domains most notably family law in the Split-Up system which advised disputing partners regarding property distribution at divorce [17].

Arguments are represented at two levels of abstraction: the generic and actual. The main changes to the original structure include: (1) a variable-value representation of claim and data items, (2) a certainty variable associated with each variable-value rather than a modality associated with the entire argument, (3) reasons for relevance of the data items in place of the warrant, (4) a list of inference procedures that can be used to infer a claim value from data values in place of the warrant, (5) reasons for appropriateness of each inference procedure, (6) context variables, (7) absence of the rebuttal feature and (8) the inclusion of a claim value reason. The following diagram represents the GAAM template.

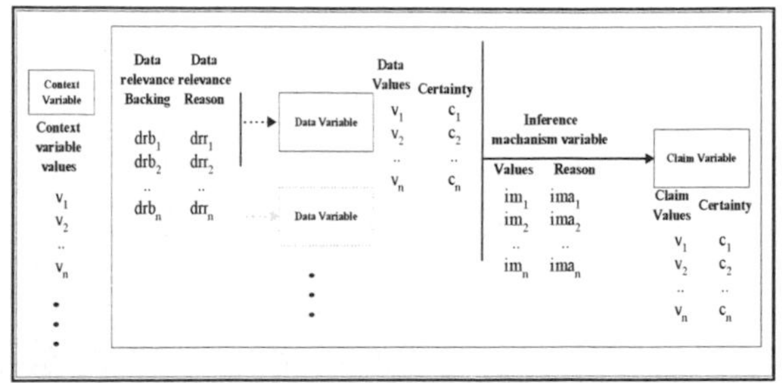

Figure 2. GAAM template.

The generic argument is a representation of the particular domain being modelled, where the following components are set: (1) claim, data, and context variables are specified but no values are assigned, (2) relevance reason statements and backing statements are specified, (3) inference procedures are listed, (4) reason for inference procedure is specified, (5) claim and data variables are not assigned certainty values. The generic argument is general enough to model the discretionary behaviour of a judge in deciding a sentence. It is contended by Stranieri, Zeleznkiow and Yearwood [15] that this method of representing knowledge corresponds to a non-dialectical perspective. It does not model the direct exchange of views between discursive participants, but instead describes assertions made from premises and a way that multiple claims can be organised. One of the most important variations to the original Toulmin structure is the removal of the rebuttal element, which makes the GAAM model firmly non-dialectical in character.

These changes to the original Toulmin model facilitate a machine-based implementation of knowledge representation. Context variables allow the actual argument to be contextualised, for example a sentence for client (VLA client) John Doe. The Toulmin warrant has been replaced by two components: an inference procedure and a reason for relevance. Yearwood and Stranieri [16] show that the warrant can indicate a reason for the relevance of a data item and on the other hand the warrant can be interpreted as a rule which, when applied to the data items, leads to a claim inference. The inference method can be an algorithm or some other method used to infer a claim.

The modelling framework integrates two techniques: decision trees and argument trees which derive from the GAAM. A decision tree is an explicit representation of all scenarios that can result from a given decision. The root of the tree represents the initial situation, while each path from the root corresponds to one possible scenario [18]. Discretion is operationalised as the selection of alternate ways to combine existing factors and to include or ignore new factors, and is therefore appropriate for modelling reasoning in 'bounded discretion' fields such as sentencing. *Figure 3* illustrates a decision tree. Nodes represent decision points and the possible outcomes of each decision are captured in arcs emerging from the node and ending in leaf nodes.

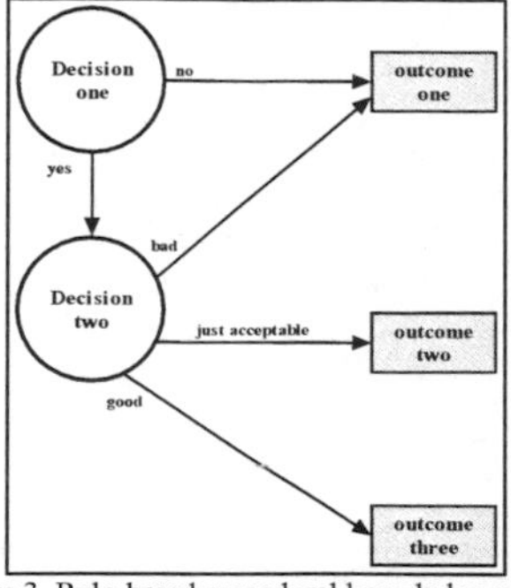

Figure 3. Rule-based procedural knowledge model.

Decision one in *Figure 3* has two possible outcomes: *'no'* leading to a conclusion (outcome one) and *'yes'* leading to a second decision. Decision two in *Figure 3* has three possible outcomes: *'good'*, *'bad'* and *'just acceptable'* with no explicit rules for deciding between them. The shadow indicates that further information about this decision is available in a second diagrammatic model. Such decisions with discretionary elements are modelled using an argumentation technique.

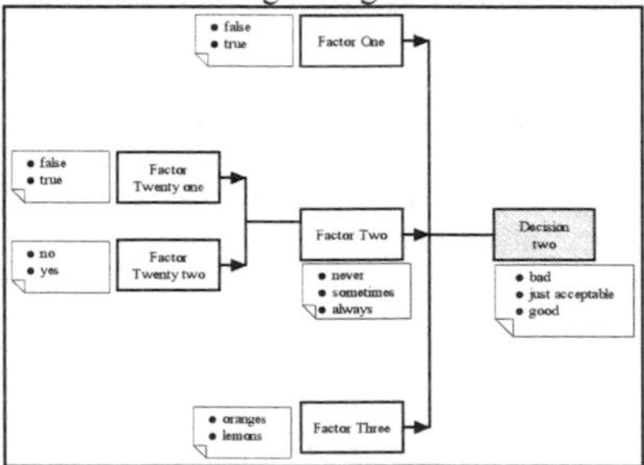

Figure 4. Argument tree

Argument trees derive from a model of structured reasoning called the GAAM advanced in [15]. As indicated earlier the trees are hierarchies of relevant factors where the root node or culminating factor is a decision tree node. When discretion is present, argument trees are used to further refine the knowledge depicted as directed graph nodes as for example, Decision two in *Figure 3* is further elaborated in *Figure 4*.

Figure 4 illustrates an argument tree with nodes representing factors that are relevant for inferring nodes higher in the tree. For example, *Figure 4* shows that '*Factor One*', '*Factor Two*' and '*Factor Three*' are all relevant for inferring '*Decision Two*'. However, how these three factors combine is left unspecified. Further, the value of '*Factor Two*' is inferred in some discretionary way from the values of '*Factor Twenty One*' and '*Factor Twenty Two*'. The values of '*Factor One*', '*Factor Two*' and '*Factor Three*' are used to infer whether the '*Decision Two*' outcome is '*bad*', '*good*' or '*just acceptable*'. This in turn is fed back to the decision tree depicted in *Figure 3*.

The argument tree provides a diagrammatic representation of the structure of reasoning. The tree can be elicited from experts in a bounded discretion field of law. Once the structure is explicated a variety of methods can be used to model the way in which factors are combined to infer values at the next level. Stranieri and his colleagues trained neural networks, a machine learning technique from artificial intelligence, from past cases in family law [19] [17].

The modelling phase was undertaken by knowledge engineers in conjunction with domain experts to establish the practical nature of the sentencing environment in Victoria. After reading the relevant parliamentary acts governing the Victorian sentencing system, both knowledge engineers and domain experts developed the decision and argument trees. The modelling procedures and steps are more fully discussed in [20]. One of the decision trees for the project are present below (*Figures 5*), along with two of the argument trees (*Figures 6* and *7*).

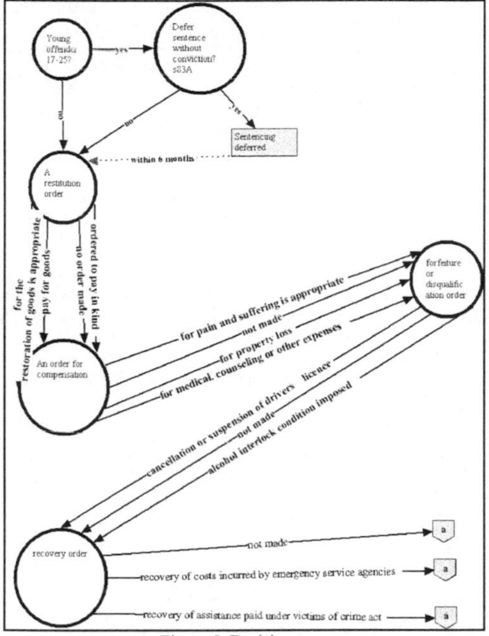

Figure 5. Decision tree.

Figures 5 represents the procedural decision trees used in the project. The 'impose sentence' area is a discretionary element and the factors which influence the decision are shown below (*Figures 6* and *7*).

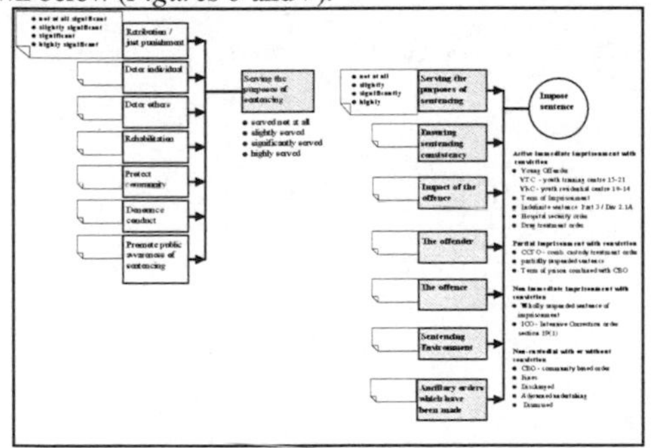

Figure 6. Top level argument tree with expansion to the first node.

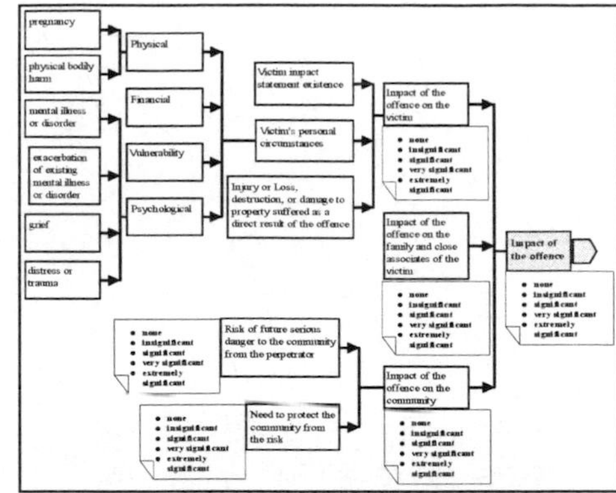

Figure 7. Argument tree for the 'impact of offence' node.

In each of *Figures 6* and *7* there are data values for only a few of the nodes. The argument trees are the most contentious elements of the modelling process and require many iterations of refinement involving both knowledge engineers and legal experts. A working prototype system without weights has been constructed but is being reviewed to take account of legislative change. It was constructed using an expert system shell justReason.[5] This is open source software designed for encoding knowledge as decision and argument trees for the rapid generation of web programs. The open source version of the shell program has a built-in weighted sum mechanism for implementing argument tree inferences. GetAid is a system that utilises the same software has been constructed for VLA by JustSys and has been implemented for use by VLA assessors. Below in *Figure 8* is a one of the screens from the prototype system. It can be seen in the figure below that the argument tree represented in *Figure 6* (above) is represented as the '*not sure*' button. Selection of this button drills the user down into the list of prompts that derive from *Figure 6*.

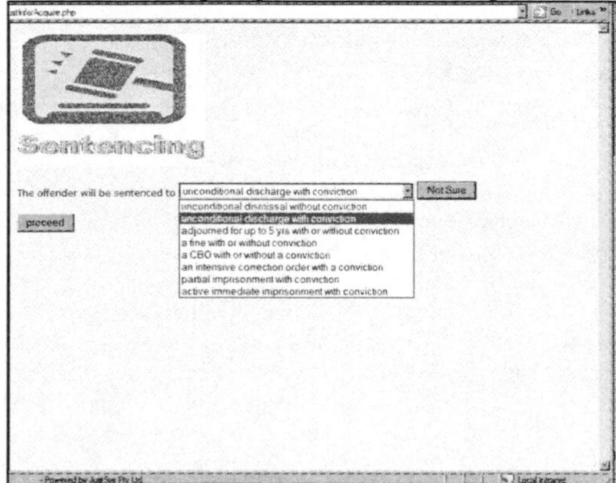

Figure 8. Screen shot of justReason constructed sentencing prototype

The figure below shows the screen that the knowledge engineer uses in conjunction with the domain experts to set weights for the weighted sum formula that the program uses to model the way decision makers combine factors in sentencing.

[5] http://www.justsys.com.au (last viewed, 14/10/2006).

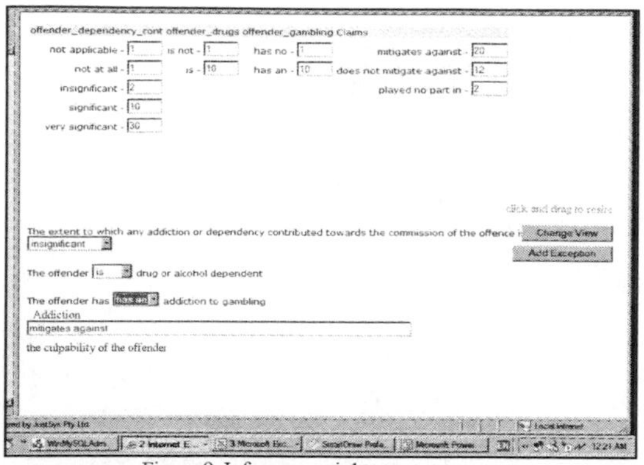

Figure 9. Inference weights capture page.

At present weights and thresholds cannot be derived from databases of past sentencing decisions because the data is difficult to obtain. This is especially the case in the lower courts in Victoria where, even though a sentence is handed-down by a magistrate, the reasons for the sentence are often not recorded. The only ways to capture these reasoning behind the sentences in these situations is to be present at sentence pronouncement or request the transcripts which then must be transcribed from recording and then purchased. The data for this system is to be gleaned from written sentencing decisions delivered in the County Court, they are not appellate cases, rather they are everyday cases. Besides ensuring legislative consistency for the information represented, establishing the weights and thresholds will be the next major hurdle. Following from the successful use of neural networks in the family law domain [19] [17] it is anticipated that similar results will be obtain in criminal sentencing.

VI. CONCLUSION

The Toulmin model, although probably intended as a method of exploring arguments in a more theoretical setting, is finding itself used more and more in representing knowledge in different types of decision support whether computerized or not. The GAAM presented above shows that it is possible to represent complex knowledge in a non-dialectical manner and that decisions which are discretionary in nature requiring intuitive action can be modelled and hopefully predicted. The great benefit of this type of system comes about as it begins to make the intuitive part of sentencing more transparent and open to scrutiny. Even though the system is not designed for the judiciary to use for decision support for their own sentencing requirements, it could be used in helping to train judges and magistrates. This system, once in use by VLA, will hopefully provide a method for lawyers, both experienced and inexperienced, to make better arguments for sentences for client before the bench.

REFERENCES

[1] M. E. Frankel, *Criminal Sentences: Law without Order*, New York: Hill and Wang, 1972.

[2] M. Bagaric and R. Edney, "The Sentencing Advisory Commission and the hope of smarter sentencing," *Current Issues in Criminal Justice* vol. 16, pp. 125-139, 2004.

[3] A. Ashworth, *Sentencing and Criminal Justice*, 2nd ed, London: Butterworths, 1995.

[4] R. Edney, "Still plucking figures out of the air?: Markarian and the affirmation of the instinctive synthesis," *High Court Quarterly Review*, vol. 1, pp. 50-57, 2005.

[5] S. Traynor and I. Potas, "Sentencing Methodology: Two-tiered or Instinctive Synthesis," *Sentencing Trends and Issues*, vol. 25, 2002, New South Wales Judicial Commission (http://www.judcom.nsw.gov.au/st/st25/index.html, last viewed 15/10/2006).

[6] G. Mackenzie, *How Judges Sentence*, Annandale: The Federation Press, 2005.

[7] A. Ashworth, E. Genders, G. Mansfield, J. Peay and E. Player, *Sentencing in the Crown Court: Report of an exploratory study*, Centre for Criminological Research, Oxford, 1984.

[8] P. Brest, "The critique of pure reason: The role of intuition in judgment and decision making," *International Conference on Artificial Intellgience and Law*, 2005, keynote paper.

[9] V. L. Sauter, "Intuitive decision-making," *Communications of the ACM*, vol. 42(6), pp. 109-115, June 1999.

[10] R. M. Hogarth, *Educating Intuition*, Chicago: University of Chicago Press, 2001.

[11] C. Tata, "Resolute ambivalence: Why judiciaries do not institutionalise their decision support systems," *International Review of Law, Computers and Technology*, vol 14, pp. 297-316, 2000.

[12] S.E. Toulmin, *The uses of argument*, Cambridge: Cambridge University Press, 1958.

[13] J. Yearwood and A. Stranieri, "The Generic Actual Argument Model of practical reasoning," *Decision Support Systems*, vol. 41, pp. 358-379, 2006.

[14] A. Stranieri, J. Yearwood, and T. Meikle, "The dependency of discretion and consistency on knowledge representation," *International Review of Law Computers and Technology*, vol. 14, pp. 325–340, 2000.

[15] A. Stranieri, J. Zeleznikow & J. Yearwood, Argumentation structures that integrate dialectical and non-dialectical reasoning, *Knowledge Engineering Review*, vol. 16, pp. 331-348, 2001.

[16] J. Yearwood and A. Stranieri, "The Generic Actual Argument Model of practical reasoning," *Decision Support Systems*, vol. 41, pp. 358-379, 2006.

[17] A. Stranieri, J. Zeleznikow, M. Gawler, and B. Lewis, Bryn, "A hybrid-neural approach to the automation of legal reasoning in the discretionary domain of family law in Australia," *Artificial Intelligence and Law*, vol. 7, pp. 53-183, 1999.

[18] J. Zeleznikow and G. Hunter, *Building intelligent legal information systems: knowledge representation and reasoning in law* Dordrecht: Kluwer Academic Publishers, 1994.

[19[J. Zeleznikow, A. Stranieri, and M. Gawler, "Split-Up: A legal expert system which determines property division upon divorce," *Artificial Intelligence and Law*, vol. 3, pp. 267-275, 1996.

[20] J. Hall, D. Calabro, T. Sourdin, A. Stranieri, and J. Zeleznikow, "Supporting discretionary decision making with information technology: a case study in the criminal sentencing jurisdiction," *University of Ottawa Law and Technology Journal*, vol. 2(1), pp. 1-36, 2005.

An Approach for Invariant Clustering and Recognition in Dynamic Environment

Andrey Gavrilov, Sungyoung Lee
Computer Engineering Department, Kyung Hee University,
1 Seocheon-dong, Giheung-gu, Yongin-si, Gyeonggi-do, 446-701,
Republic of Korea

Abstract - **An approach for invariant clustering and recognition of objects (situation) in dynamic environment is proposed. This approach is based on the combination of clustering by using unsupervised neural network (in particular ART-2) and preprocessing of sensor information by using forward multi-layer perceptron (MLP) with error back propagation (EBP) which supervised by clustering neural network. Using MLP with EBP allows to recognize a pattern with relatively small transformations (shift, rotation, scaling) as a known previous cluster and to reduce producing large number of clusters in dynamic environment, e.g. during movement of robot or recognition of novelty in security system.**

I. Introduction

Most important tasks for actions of mobile robot in unknown environment using behavior based approach [1] are the clustering and recognition of objects and situations. Application of Adaptive Resonance Theory (ART) [2] (in particular ART-2) is rather appropriate for solving these tasks, because this model combines properties of plasticity and stability, and also it does not require a priori knowledge of the fixed quantity of necessary classes. Many different modifications of this model and its combinations with other neural networks are known [3], [4, [5] and others. However, most of them (except FANNC [5]) are supervised learning and demand a teacher. Thus, the most important feature of model ART is refused. As to model FANNC, this one aims to classify static images and is not oriented on recognition of novelty in flow of images many of which may be similar and differ only by some gradual transformations. So we focus on model ART-2 to provide fast unsupervised learning.

However, model ART-2 assumes usage of only one layer of neurons (not including entry, associated with sensors). It results in that the neural network works only with the metrics of primary features, e.g. for visual images these features are pixels. At that similarity between images (for classification or creation of a new cluster - output neuron), are calculated usually using Euclidean distance. As result, even any small transformation of input vector can cause recognition of it as belonging to new cluster. It leads to the result that the model ART-2 is nearly never used for real applications. For example, considering the clustering and pattern recognition of a mobile robot, [6], [7] it is required to recognize the object in different foreshortenings which may allocated in different parts of a field of vision, i.e. recognition should be invariant concerning transformation of the input pattern, such as shifts, rotations and others.

The ability of invariant recognition of objects (situations) in dynamic environment is one of most important capabilities of natural brain [8]. This ability is closely connected with another ability which is to detect novelty. These are "two sides of one same medal".

In this paper we suggest and investigate one hybrid model of neural network based on ART-2 and multi layer perceptron (MLP) with error back propagation (EBP) training algorithm [9]. In this model we keep the unsupervised learning and remove one major disadvantage of ART – the sensitivity to transformations of input patterns. Multi layer perceptron provides preprocessing of patterns for invariant recognition because its hidden layers form secondary features during learning. It could be said that in MLP each hidden layer provides conversion of any feature space to another one.

There are many other approaches to achieve invariant recognition by neural networks, for example proposed in [10], [11], [12], [13], [14], [15], [16]. But each of them is either too complex or specialized for determined any kind of images and transformations.

We suggest potentially universal approach for invariant recognition which can be implemented in real time systems, because it not requires long time processing as in usual applications of EBP.

For the first time the paradigm of MLP-ART2 model was proposed by us in [17], [18]. In that paper the idea was suggested and some previous experiments and problems were shown. In current paper this paradigm and algorithm of model get continue developments. Some new ideas, experiments and conclusions are discussed.

II. Main Concepts and Algorithm of Hybrid Neural Network MLP-ART2

In our suggested model the first several layers of neurons are organized as MLP. Its outputs are the inputs of model ART-2. MLP provides conversion of primary feature space to secondary feature space with lower dimension. Neural network ART-2 classifies images and uses secondary features to do it. Training of MLP by EBP (with limited small number of iterations) provides any movement of an output vector of MLP to centre of recognized cluster of ART-2 in feature space. In this case the weight vector (center) of recognized

K. Elleithy (ed.), Advances and Innovations in Systems, Computing Sciences and Software Engineering, 47–52.

cluster is desired output vector of MLP. It could be said that the recognized class is a context in which system try to recognize other images as previous, and in some limits the system "is ready to recognize" its by this manner. By other words neural network "try to keep recognized pattern inside corresponding cluster which is recognizing now".

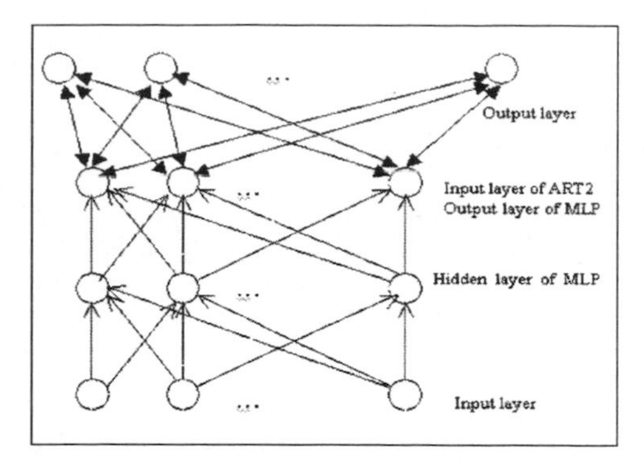

Fig. 1. Structure of hybrid neural network.

Action of the suggested model is described by the following unsupervised learning algorithm:

1. In MLP let the weights of connections equal to *1/n*, where *n* is quantity of neurons in the previous layer (number of features for first hidden layer). The quantity of output neurons N_{out} of ART-2 is considered equal zero.

2. The next example from training set is presented to inputs of MLP. Outputs of MLP are calculating.

3. If N_{out}=0, then the output neuron is formed with the weights of connections equal to values of inputs of model ART-2 (the outputs of MLP).

4. If N_{out}> 0, in ART-2 the algorithm of calculation of distances between its input vector and centers of existing clusters (the weight vectors of output neurons) is executing using Euclidian distance:

$$d_j = \sqrt{\sum_i (y_i - w_{ij})^2}$$

Where: y_i – i^{th} feature of input vector of ART-2, w_{ij} – i^{th} feature of weight vector of jth output neuron (the center of cluster). After that the algorithm selects the output neuron-winner with minimal distance. If the distance for the neuron-winner is more than defined a vigilance threshold or radius of cluster *R*, the new cluster is created as in step 3.

5. If the distance for the neuron-winner is less than *R*, then in model ART-2 weights of connections for the neuron-winner are recalculated by:

$$w_{im} = w_{im} + (y_i - w_{im})/(1 + N_m)$$

Where: N_m – a number of recognized input vectors of mth cluster before. In addition, for MLP a recalculation of weights by algorithm EBP is executing. In this case a new weight vector of output neuron-winner in model ART-2 is employed as desirable output vector for EBP, and the quantity of iterations may be small enough (e.g., there may be only one iteration).

6. The algorithm repeats from step 2 while there are learning examples in training set.

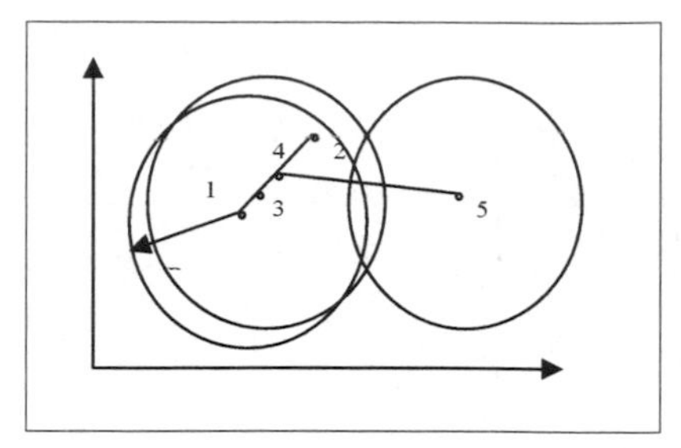

Fig. 2. Explanation of action of hybrid model.

Action of the suggested model is explained in a figure 2. Here we see the space of secondary features in which by points are represented output vector of perceptron (input vector of model ART-2) and centers of clusters. In this figure the following points are represented:

1) - a new (first) image for which new cluster with radius *R* is created,

2) - the next image recognized as concerning to this cluster,

3) - the new centre of a cluster calculated in step 5 of algorithm,

4) - a new output vector of perceptron, attracted to centre of a cluster as a result of executing of algorithm "error back propagation",

5) - the new image recognized as inhered to other cluster.

Note that in this algorithm EBP aims at absolutely another goal different from that in usual MLP-based systems. In those systems EBP reduces error-function to very small value. But in our algorithm EBP is needed only for some decreasing distance between actual and desirable output vectors of MLP. So in our case the long time learning of MLP is not required.

Note that EBP and forming of secondary features are executed only when image "is captured" by known cluster. So selection of value for vigilance threshold is very important. Intuitively obvious that one must be depending on transformation speed of input images and may be changed during action of system.

In our experiments we used three variants for calculation of vigilance threshold:

1) it is fixed value selected empirically,

2) it is calculated for every image by formulas $S/(N_a^2)$, where S – average input signal, N_a – number of output neurons of MLP (input neurons of ART-2),

3) it is calculated as kD_{min}, where D_{min} – minimal distance between input vector of ART2 and weight vector in previous image, $k>1$ – coefficient.

For experiments we developed program for simulation in which we process series of visual images. Every image consists of 100x100 pixels. The model MLP-ART2 consists of one hidden layer of MLP. Program is implemented in Delphi.

III. Experiments

A.. *Experiments with Images of Moving Objects*

We use series of multi-colored images for experiments which obtained from video. This series simulates the sight of mobile robot during its movement. There are images of shifting couple of chairs. Figure 3 shows the images 1, 8 and 15 (last) of this series.

Fig. 3. First image of series from video

Figures 4 and 5 show results of clustering and recognition at repeating of this series four times. In these experiments we used following parameters of neural network:

- the number of hidden neurons of MLP – 20,
- the number of output neurons of MLP (or input neurons of ART-2 – 10,
- exponential sigmoid with parameter a=1,
- the number of iterations of EBP - 1,
- fixed value of vigilance threshold $R=0,05$.

Figure 4 is the evidence of stability of our model at constant value of vigilance threshold. Cluster 2 is corresponding to situation when the robot sees only one chair; and cluster 3 is corresponding to situation when robot sees basically already empty wall. Figure 5 shows distance between input vector of ART-2 and center of recognized cluster 1 for first 7 images after 4 times of repeated series.

In case of variable vigilance threshold the stability is less. Figure 6 shows results with calculation of one by method 3 with $k=4$ and number of iterations of EBP equals to 5.

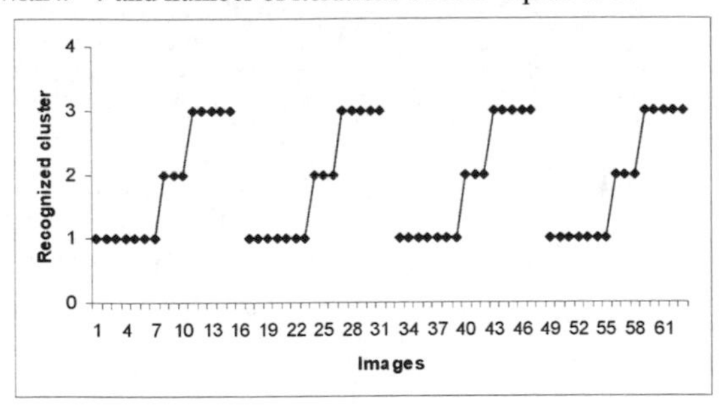

Fig. 4. Recognized clusters at repeating of series of images

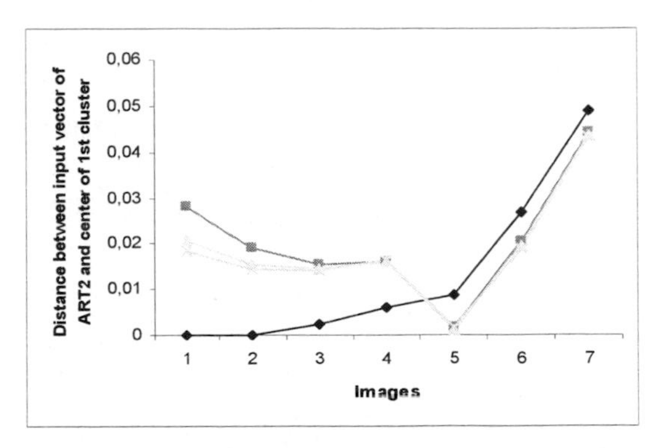

Fig. 5. Distance between recognized cluster 1 and input vector of ART-2 (output vector of MLP).

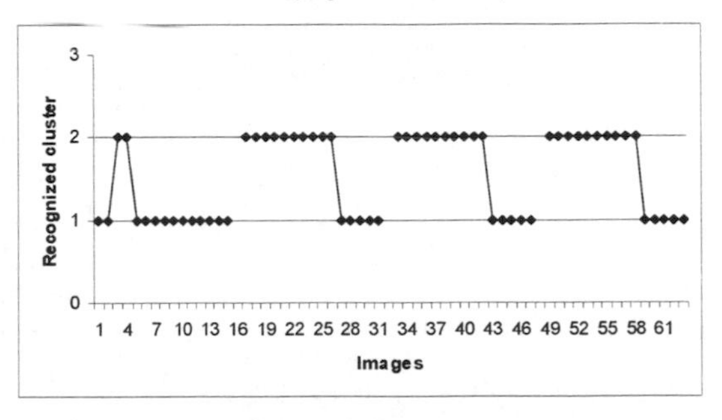

Fig. 6. Recognized clusters at usage of calculation of vigilance threshold by method 3 with k=4

Note that in repetitions 2, 3 and 4 the results of recognition are more intuitively well-taken than in first time.

Figure 7 shows results of clustering and recognition in case when vigilance threshold is calculated by method 3 with $k=2$ and number of iterations of EBP equals to 1.

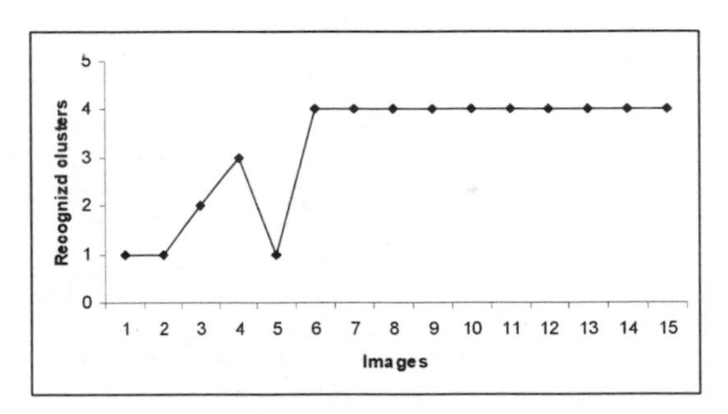

Fig. 7. Recognized clusters at usage of calculation of vigilance threshold by method 3 with k=2.

Figure 8 shows distance between input vector of ART-2 and center of cluster 4 during its recognition for same parameters and number of iterations of EBP equals to 1, 3, 5 and 10. In this figure lower lines are corresponding to large number of iterations of EBP. Thus we can see that the distance is decreasing in accordance with increasing of number of iterations.

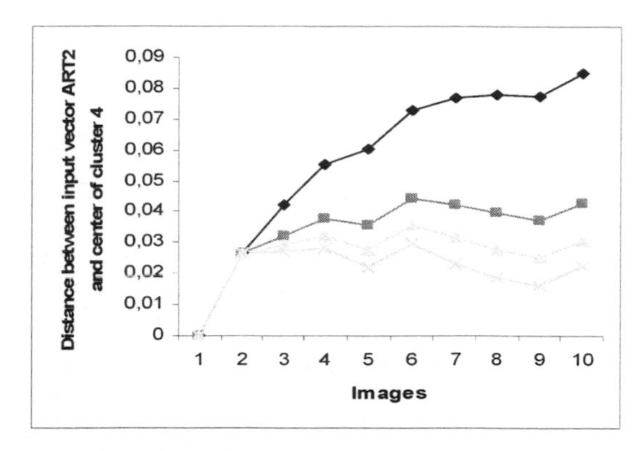

Fig. 8. Distance between input vector of ART-2 and center of recognized cluster 4 with different number of iterations of EBP

B. Experiments with Series of Faces

For these experiments we use collection of faces from Olivetti and Yale. Series of images used in experiments are shown in figure 9. This series consists of set of gray-scaled pictures of certain person with different expressions (from 1 to 8), with different illumination (from 9 to 11), with transformation by turning (from 12 to 18), another 3 faces (from 19 to 21), faces of last person with different expressions and position (from 22 to 29), repetition of face 19 (30).

In these experiments following parameters of model are used:

- the vigilance threshold is calculated for every image by method 2, described above,
- the number of iterations of EBP - 1,
- the number of inputs of ART-2 = 10,
- the number of hidden neurons of MLP – 20,
- exponential sigmoid with parameter a=1,

Figure 10 shows results of clustering and recognition of two repeating of series. We can see from this picture that the model demonstrates stable results. Figure 11 partially explains the result shown in Figure 10. The images 9, 10, 11 with other illuminations in contrast to previous images are difficult for recognition and cause creation of new clusters. Images 19-21 are recognized as new persons and images 22-29 - as 2 different faces in accordance with its mimic and position features. Image 30 is recognized like previously looked face 19.

Figure 12 shows distance between input vector of ART-2 and center of recognized cluster 1 with different number of iterations of EBP: 1, 3, 5 and 10. The line beginning from lowest point is corresponding to value 1, and the line from highest point is corresponding to value 3 and further increasing number of iterations causes decreasing of first point and increasing of last point in this graphic.

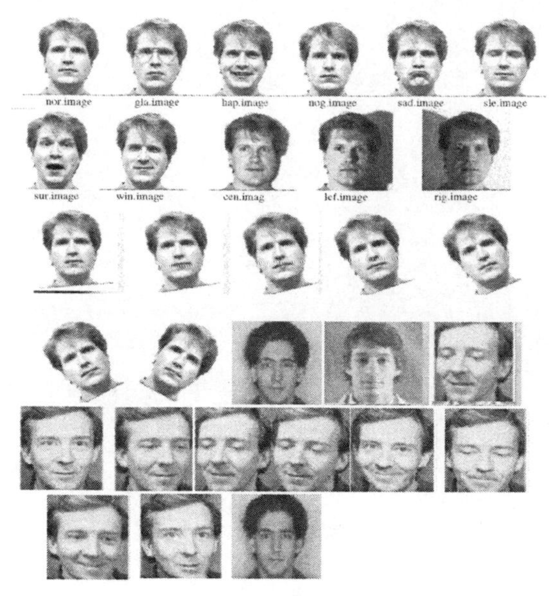

Fig. 9. Series of faces

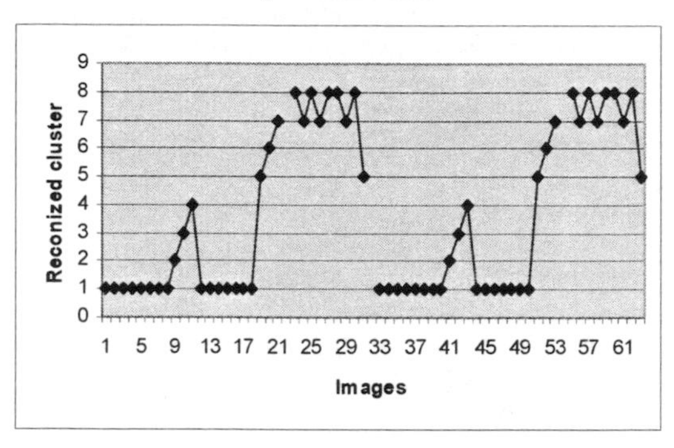

Fig. 10. Recognized clusters for 2 repetitions of series of faces

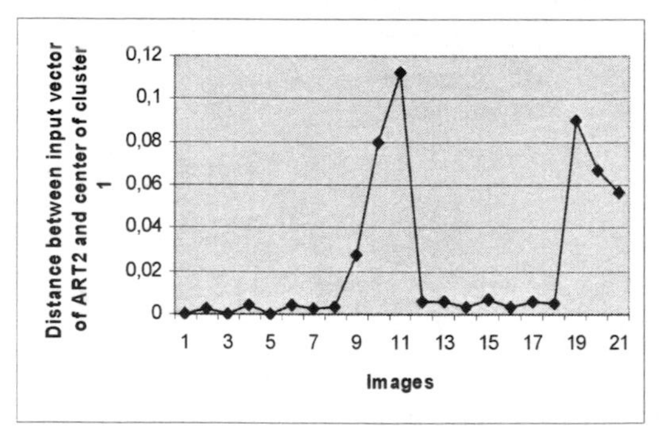

Fig. 11 Distance between input vector of ART-2 and center of cluster 1

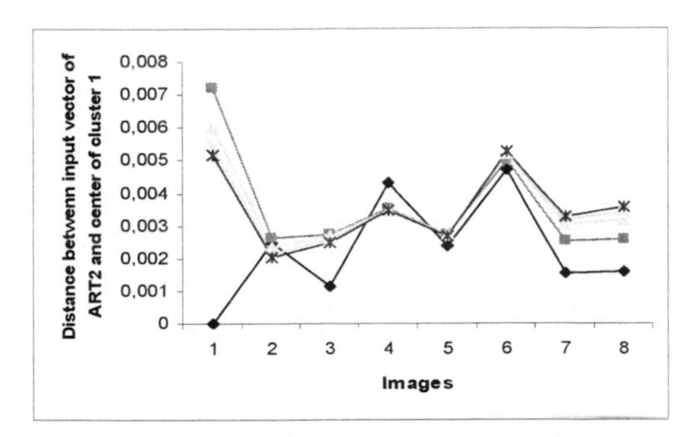

Fig. 12. Distance between input vector of ART-2 and center of recognized cluster 1 for first 8 images of series for different number of iterations for EBP

Figure 13 shows results of clustering and recognition in case with calculation of vigilance threshold by methods 3 with $k=4$, number of iterations equals to 5 and number of repetitions of series equals to 4.

From these results we can see that in this case every new repetition of series improves accuracy of recognition. At beginning all images of first person with other illumination are recognized as one cluster 2. Almost all images of other persons are recognized as one cluster 3 (may be interpreted as "other person"). In next repetition different variants of illuminations are recognized as different clusters. In third repetition other persons are recognized as different persons. But in this case one of these persons is recognized as person 1 with other illumination (cluster 6). In last repetition this single mistake is removed. It could be said that repetition causes more detailed clustering and recognition that is very similar to well known features of human learning.

IV. Discussions

The demonstrated above and other obtained results of experiments show that proposed and investigated model MLP-ART2 of hybrid neural network demonstrates adaptation to relatively small transformations of visual image and capability to recognize new images in flow of ones. This adaptation coexists with keeping ability to recognize previously occurred images. For all that this model demonstrates some features of perception of humans, such as "we see what we expect to see" [8] and sometimes effect of some difficulty to recognize following new images immediately after adaptation to recognition of any previous one (new different images are recognized as one cluster). Also experiments show stability and capability to improve accuracy (in human-like respect) of recognition and clustering during repetition of same series of images.

The suggested hybrid model of the neural network can be used in the mobile robot when it is necessary to watch sequence of images visible by the robot during its movement, and to extract its novelty, i.e. essential changes in the scene visible by the robot. Also this model may be used in security systems for recognition of new objects or persons in sight of camera. Another possible application of this approach is recognition of novelty and dangerous situations in different monitoring systems, e.g. in health care systems.

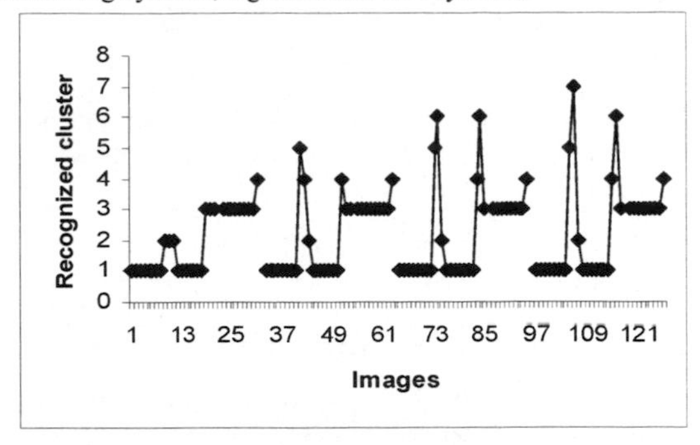

Fig. 13. Recognized clusters at calculation of vigilance threshold by method 3 with $k=4$

Modification of this algorithm may be algorithm in which the quantity of created clusters is limited. In this case, if the quantity of clusters (output neurons) has achieved a limitation, it is needed to decide what to do with images which are not recognized, i.e. which cannot be related to any existing cluster. In this case following method may be offered. At first, increasing of parameter R (radius of clusters) and then trying to apply algorithm of recognition again until the new image will not be related to one of clusters. After that, it is necessary to reduce quantity of clusters (output neurons) to unify clusters with the centers which are occurred in one cluster with new value R, and to change weights of connections between outputs of MLP and outputs neurons-clusters.

Also one modification of this algorithm may be algorithm of training with the teacher in which before creating new cluster the system requests teacher what to do in order to create a new cluster or to increase vigilance threshold R. Moreover in this case may be not only to ask from user but also to employ any additional procedure for recognition of novelty of the image.

Similar approach for building hybrid neural network from supervised and unsupervised neural networks may be applied to self-organized maps [19].

V. Conclusions

In this paper we suggested and investigated hybrid model of neural network MLP-ART2 for unsupervised clustering and invariant recognition of images in dynamical unknown environment. The suggested hybrid model of the neural network can be used in the mobile robot and security systems when it is necessary to watch sequence of the visible images and to recognize similar or novel objects (situations).

This approach may be applied to self-organized maps of T.Kohonen.

The following further researches of the suggested hybrid model of the neural network are planned:

- continue research of influence of MLP and ART-2 parameters on action of the neural network,

- testing and investigation of the suggested model on simulated mobile robot and the real robot,
- development of hierarchical system based on proposed hybrid neural network which may be basis for building of model of mind,
- development of hybrid architecture of intelligent system based on proposed model and knowledge based approach for perception and processing of multi-modal information.

ACKNOWLEDGEMENT

This work was supported by MIC Korea (IITA Visiting Professorship Program). Dr. S.Y.Lee is the corresponding author.

REFERENCES

[1] R.A. Brooks, "Intelligence without representation," *Artificial Intelligence J.*, vol. 47, 1991, pp. 139-159.

[2] G.A. Carpenter and S. Grossberg, "*Pattern recognition by self-organizing neural networks*," MIT Press, MA, Cambridge, 1991.

[3] R.A. Baxter, "Supervised adaptive resonance networks," In *Proc. of the Conf. on Analysis of Neural Network Applications*, Fairfax, VA, 1991, ACM, pp. 123-137.

[4] G.A. Carpenter, S. Grossberg. N. Markuzon, J.H. Reynolds and D.B. Rosen, "Fuzzy ArtMap: neural network architecture for incremental supervised learning of analog multidimensional maps," *IEEE Trans. on Neural Networks,* 1992, vol. 3, pp. 698-713.

[5] Zhihua Zhou, Shifu Chen and Zhaoqian Chen, "FANNC: a fast adaptive neural network classifier," *Knowledge and Information Systems, vol.* 2(1), 2000, pp. 115-129.

[6] A.V. Gavrilov, V.V. Gubarev, K.-H. Jo and H.-H. Lee, "Hybrid neural-based control system for mobile robot," In *Proc. of 8th Korea-Russia Int. Symp. on Science and Technology KORUS-2004, vol. 1,* Tomsk, Russia, June, 2004, TPU, pp. 31-35.

[7] An-Min Zou, Zeng-Kuang Hou, Si-YaoFu and Min Tan, "Neural networks for mobile robots navigation: a survey," In *Proc. of Third Int. Symp. on Neural Networks ISNN-2006*, part 2, Chengdu, China, May, 2006, Springer, Advances in Neural Networks ISNN-2006, LNCS 3972, pp. 1218-1226.

[8] J. Hawkins and S. Blakeslee, "*On intelligence*," Times Books, 2004.

[9] D.E. Rumelhart, "Parallel distributed processing," In *Mcclelland J.L. (Eds.), Explorations in the Microstructure of Cognition,* vol.. I, II, MIT Press, MA, Cambridge, 1986.

[10] K. Fukushima, "Neocognitron: a self organizing neural network model for a mechanism of pattern recognition unaffected by shift in position," *Biol. Cybern.,* 1980, vol. 36, pp. 193-201.

[11] V. Petridis and V.G. Caburlasos, "Fuzzy lattice neural network (FLNN): a hybrid model for learning," *IEEE Trans. on neural networks*, vol. 9(5), 1998, pp. 877-890.

[12] Satoshi Suzuki and Hiroshi Ando, "A modular network scheme for unsupervised 3D object recognition," *Neurocomputing,* vol. 31, 2000, pp. 15-28.

[13] L. A. Torres-Méndez, J. C. Ruiz-Suárez, Luis E. Sucar, and G. Gómez, "Translation, rotation, and scale-invariant object recognition," *IEEE Tran. on Systems, Man and Cybernetics - Part C: Applications and Reviews*, vol. 30(1), 2000, pp. 125-130.

[14] Jian Huang Lai, Pong C. Yuen and Guo Can Feng, "Face recognition using holistic Fourier invariant features," *Pattern Recognition,* vol. 34, 2001, pp. 95-109.

[15] Jung-Hua Wang, Jen-Da Rau, Wen-Jeng Liu, „Two-stage clustering via neural networks," *IEEE Trans. on Neural Networks*, vol. 14(3), 2003, pp. 606-615.

[16] Hahn-Ming Lee, Chih-Ming Chen, Yung-Feng Lu, "A self-organizing HCMAC neural-network classifier," *IEEE Trans. on Neural Networks*, vol. 14(1), 2003, pp. 15-27.

[17] A.V. Gavrilov, "Hybrid neural network based on models of multi-layer perceptron and adaptive resonance theory," In *Proc. of 9th Korean-Russian Int. Symp. on Science and Technology KORUS-2005,* Novosibirsk, Russia, June, 2005, NSTU, pp. 119-122.

[18] A.V. Gavrilov, Y.-K. Lee and S.-Y. Lee, "Hybrid neural network model based on multi-layer perceptron and adaptive resonance theory," In *Proc. of Third Int. Symp. on Neural Networks ISNN-2006*, part 1, Chengdu, China, May, 2006, Springer, Advances in Neural Networks ISNN-2006, LNCS 3971, pp. 707-713.

[19] T. Kohonen, "Self-organized formation of topologically correct feature maps," Biological Cybernetics, vol. 43(1), 1982, pp. 59-69.

Modelling non Measurable Processes by Neural Networks: Forecasting Underground Flow Case Study of the *Cèze* Basin (*Gard - France*)

A. Johannet*, P.A. Ayral**, B. Vayssade*
Ecole des Mines d'Alès
*Centre des matériaux de Grande Diffusion
**Laboratoire de Génie de l'Environnement Industriel
6 avenue de Clavières / 30319 Alès Cedex / FRANCE

Abstract: After a presentation of the nonlinear properties of neural networks, their applications to hydrology are described. A neural predictor is satisfactorily used to estimate a flood peak. The main contribution of the paper concerns an original method for visualising a hidden underground flow Satisfactory experimental results were obtained that fitted well with the knowledge of local hydrogeology, opening up an interesting avenue for modelling using neural networks.

I INTRODUCTION

During the last twenty years there has been considerable research devoted, on the one hand, to the field of nonlinear and adaptive modelling, and on the other hand to the study of neural networks in order to perform such tasks. Nevertheless, the idea of using neural networks' ability to model nonlinear and non-stationary behaviours in hydrological systems emerged only about ten years ago. Currently, several theoretical results and many different learning schemes have proven that neural networks are becoming a very effective tool in hydrological applications.

The *Gard* Region, in the South-East of France, has a specific geographical position which makes it particularly vulnerable to flash floods: each autumn, storms formed over the sea and pushed by southerly winds provoke extreme rainfall events. Flash floods are very important natural hazards for the Mediterranean Region. Unfortunately, scientific knowledge about them is insufficient.

In that context, this paper has two objectives: the first is to present how and why the neural methods are appropriate for solving such environmental problems. The second objective of this paper is to present how the neural *black box* can be changed into a *grey box* in order increase our knowledge.

The paper contains four parts: part one introduces neural networks including nonlinear properties. The second part is devoted to presenting the principal neural architecture and learning rules. The third part presents the problematic and the fourth the results.

II NEURAL NETWORKS FOR IDENTIFICATION

System identification is the modelling of systems. It is useful for the knowledge it gives about the system and in that it provides ways to control it, and to predict or forecast its behaviour.

Neural networks are devices capable of learning. In the case of signal processing, or system identification, the set of examples consists of sampled input and output signals. The second fundamental property of neural networks is that they can implement non linear functions. This property is a necessary one for systems such as catchment areas which may have different responses even when the input is the same (for example, the behaviour during summer or winter is very different).

The Model of Neuron and Multilayer Network

An artificial neuron is a mathematical operator which generally computes two actions: first the linear weighted sum of its inputs, and second the non-linear evaluation of its output. Various models of neurons have been proposed depending on the evaluation function. The formula is:

$$o_l = f\left(\sum_{example\_m} c_{lm}.i_m\right)$$

where o_l is the output of the neuron l, i_m is one of its inputs, c_{lm} is the synaptic coefficient linking this input to the neuron under consideration, and f(.) is the evaluation function. For example it is possible to choose f(.) = tanh(.). Linear neurons may exist, they have an Identity function.

A neural network is a set of interconnected neurons. These connections (defined by the set of coefficients c_{lm}) are computed during the learning phase.

It has been demonstrated that any non linear, smooth function can be identified by such a network [1]. The accuracy of the identification depends on the number of hidden neurons. This result is of course very important, but it only constitutes a proof of the existence of the solution; therefore the difficulty is to find the solution using the appropriate learning method operating on an architecture which includes a sufficient number of neurons. In this study we firstly consider the well known two-layer perceptron, and secondly an *ad hoc* network, coding in its architecture the function we want to implement.

Learning

The neural network learning phase is the computation of the synaptic weights in order to minimise a *"goal function"*. Different learning rules can be derived taking into account different *goal functions* and different minimising methods. Let

K. Elleithy (ed.), Advances and Innovations in Systems, Computing Sciences and Software Engineering, 53–58.

us consider only identification and forecasting applications; principally two types of *goal functions* have been proposed: supervised or unsupervised one. Amongst the unsupervised methods, the Reward-Penalty learning algorithm [2] is very interesting because it enables interpretation as the gradient descent of a *goal function* (redo or undo past action) and does not need a comprehensive modelling of the environment. We have implemented this method for solving a robotic task. The aim was to make a hexapod robot learn gait and obstacle avoidance. Results obtained both in simulations and with the real plant [3] were very satisfactory; the robot learnt its task without explicit modelling of the actuator-environment relations. This application highlights the fact that neural networks find their field of excellence when they are applied to model the real world or natural environment.

On the other hand, the cost function G is more understandable in the case of supervised learning, since this function is generally the sum of the squared errors between the measured outputs and the computed values, for each input-output couple of interest. It is possible to consider this "cost" function J as follows (only one output neuron):

$$J(C,k)=\frac{1}{2}\sum_{\{k\}}\left(o^k(C)-d^k\right)^2$$

where {k} is the set of input-output couples taken for k past values, and C is the set of synaptic coefficients.

Starting from this *cost function*, several learning rules have been proposed depending on the chosen minimising method. The most popular method has been the backpropagation learning rule introduced by D. Rumelhart [4] which uses the steepest gradient descent. However, other more efficient rules have been proposed, for example a descent inspired by second order minimisation methods [5] [6]. Amongst these second order methods the "Levenberg-Marquardt" learning rule [7] is at present the most powerful and leads in a few iterations to a very satisfactory solution.

Backpropagation learning rule

The backpropagation learning rule provides a method for modifying the network's synaptic weights according to the gradient of the quadratic error. It was the first learning rule which enabled learning on nonlinear networks, and which could also efficiently operate on multilayer networks.

Let us consider the network shown in Figure 3. An input-output couple is presented to the network which has to associate the input vector i^k $\{i^k_1, i^k_2, i^k_3, \dots\}$ to the desired output d^k (scalar value in case of one output neuron). It can be noticed that the intermediate, or hidden, neurons have no desired value. After computation of the network's output o^k, the modification to apply to the coefficients, at time t, using a gradient method with a constant step μ is:

$$c^k_m(t+1)=c^k_m(t)-\mu\frac{\partial J(C;t)}{\partial c^k{}_m} \quad (4)$$

Therefore, using the backpropagation learning rule, the synaptic coefficients of a multilayered neural network can be computed. Its principal drawbacks are the sensitivity of the result to the initialisation of the synaptic weights, and the slowness of the convergence rate toward a minimum of the *cost function*.

Levenberg-Marquardt Learning Rule

Because of its efficiency, the Levenberg-Marquardt rule should be used whenever possible. Nevertheless, the Levenberg-Marquardt learning rule suffers from two drawbacks: first it has to invert a matrix which is an approximation of the Hessian: the second order derivative of the *cost function* relative to the synaptic coefficients, *i.e* a matrix whose dimension is equal to nc.nc if nc is the number of synaptic coefficients. Sometime this matrix is too huge to be inverted; sometimes this Hessian matrix may be non-invertible [8]. We will see later that in case of hydogeological modelling, the data are very noisy and lead to difficult problems for which Levenberg–Marquardt algorithm may be inefficient. In such cases, the backpropagation algorithm provides adequate results.

In some words (see [7][8] for full presentation), Levenberg-Marquardt algorithm starts, as bakpropagation, from a problem of cost function minimization. The principle of the rule is to apply to the coefficients an increment taking into account the first and second order of the Taylor decomposition of the *cost function* (notes that Levenberg-Marquardt addresses the *cost function*, taking into account the whole set of learning couple at the same time t). Noting that the second term of the Taylor decomposition needs the computation of the Hessian Matrix, Levenberg–Marquardt method considers an approximation of the Hessian:

$H=\Delta^T\Delta$, where Δ is the vector composed of the first order derivative of the *cost function* (computed by the backpropagation), the formula is:

$$[H]_{ij,lm}\cong\sum_{\{k\}}\frac{\partial J}{\partial c_{ij}}\frac{\partial J}{\partial c_{lm}}$$

The Levenberg-Marquardt rule assumes that at each presentation t of the whole set of learning couples $\{i^k,o^k\}$, an increment to the coefficients is computed in the direction of the gradient: Δ, with amplitude $\mu(C,t)$ such that:

$$\mu(C,t)=\left(\Delta^T\Delta+\lambda(t).\mathrm{Id}\right)^{-1}$$

where Id is the Identity matrix.

The interpretation is the following: at the beginning of the learning process, a high value of factor $\lambda(t)$ is chosen in order to lead the matrix $\mu(C)$ to be diagonal dominant. The rule is therefore close to a first order gradient descent rule.

The factor $\lambda(t)$ is then decreased in order to be neglected in relation to the approximation of the Hessian part : $\Delta^T\Delta$. At the end of learning, the computation essentially uses the second order information and in a few iterations comes close to the *cost function* minimum.

This presentation of the Levenberg-Marquardt rule shows that backpropragation is necessarily computed in order to estimate the derivatives Δ.

A SYSTEM IDENTIFICATION

Starting from the previous considerations, the identification of a dynamic system can be addressed by neural networks in computing learning with input-output couples. It is well known that the behaviour of a dynamic system depends not only on external inputs but also on some internal variables that represent the "state" of the system. Under the condition of observability of the system, these state variables are assumed to be past outputs of the real process. However expertise may indicate that another choice may be to select the most relevant state variables (see S. Narendra in [9] for further considerations).

Learning of a discrete-time feedback network

Considering a network at a given instant, learning is performed using the previous external inputs: {i(t)} plus the state variables: the previous output or complementary state variables. Learning on recurrent networks can be performed in at least two ways: the first one consists in taking into account all the previous values using a recurrent method, see for example K. Narendra [9] and P. J. Werbos [10]; the second way takes into account only a few time events, and formulates the backpropagation on a small window of time as proposed by L. Personnaz [11]. The second way was chosen in this study because of its simplicity.

Schemes of identification

Two strategies are possible in order to implement the learning: in the first one the objective is to capture the dynamics of the process. Then the errors coming from the network are taken into account during the learning. The looped input is initialised with the past estimated value of the network. This scheme of identification is called "non directed".

The second way of learning uses measured values coming from the system. This mode is termed "directed".

It is immediately clear that in the case of a neural model with feedback operating on non measurable state variables, the previous discussion is not relevant; the only solution is the non directed model. The identification of the underground flow of water was approached in this way [12].

III NEURAL NETWORKS FOR HYDROLOGY

Because of its complexity there are many models dealing with the rainfall-runoff relation. Usually the models can be classified as: deterministic or statistical; local, global or distributed; static or dynamic; empiric, physics-based or conceptual. Flash flood forecasting is usually addressed by physics-based, conceptual and statistical models. For example, TopModel [13] is a physics-based model used for flood forecasting, while conceptual models have been developed for the same objective: ALHTAÏR [14], MARINE [15] or SCS [16]. For real time use, the models generally have to be distributed [14] [15] [17] or semi-distributed.

Clearly, Neural Networks are statistical models. They have been used for about ten years in an increasing number of applications for elaborate rainfall-runoff models using RBF networks [18], or multilayer networks [19]. Other approaches are also used, such as fuzzy logic [20] or sequential automata [21].

Because of the lack of knowledge about fast floods [22] we hope that neural network models may significantly improve not only flash flood forecasting, but also scientific knowledge about them. This point is at the heart of this work.

A CONTEXT PRESENTATION

The target of our study is the river *Cèze* (fig. 1), a tributary of the *Rhône*. The *Cèze* is 112 km long and its catchment area is about 950 km^2 [23].

The upper part of the river flows on antestephanian schists and gneiss which are impervious; then, on carboniferous deposits (schists, sandstone and coal); in this part of the river, galleries of former coal mines bring water from the neighbouring catchment of the *Avène* river to the *Cèze via the Auzonnet river*. This underground flow is quite low, and had been estimated at 0.5 m^3/s [24]. Then, the *Cèze* flows on oligocen sediments composed of conglomerates which are impervious, before crossing cretaceous deposits, mainly limestones. In this area, the valley flows in a canyon and the plateau has a typically karstic relief. In short, the adjective karstic comes from Slovenia and is generally used for limestone in which the water has eroded galleries. In this part of the river, the relation between the karstic network and the river is not very well understood, so the limit of the area which contributes to outflow in the *Cèze* is not precisely known. The *Cèze* joints the *Rhône* on its right bank at an altitude of 26 meter above sea level, flowing on impervious and semi-pervious deposits.

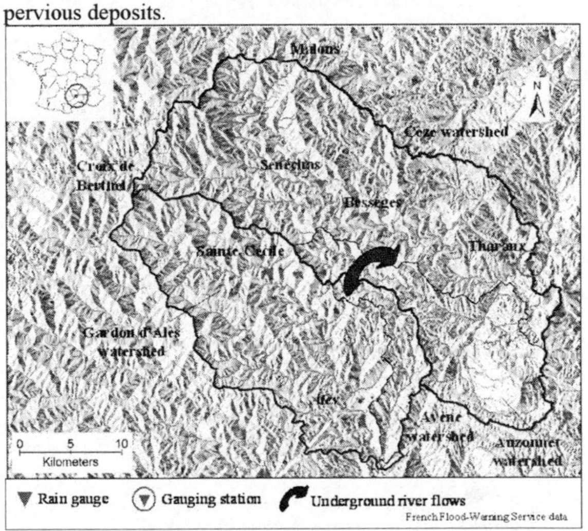

Fig. 1: Location of study area

The main two features of the *Cèze* are its very irregular outflow, which reflects the irregularity of the meteorology, and the contrast between the part of the river on impervious rocks and the part located in a karstic region.

The particular feature of the *Cèze* is that its flows are not fully explained by the rainfall on its catchment. Several hypotheses have been proposed in order to find another definition of its catchment. One of these hypotheses is that some water could arrive in the river via its affluent the *Auzonnet*, coming via underground circuits from a neighbouring, but different, catchment: the *Gardon d'Alès* river. Two explanations can be found: the karstic network, or galleries of former coal mines. Coal Mines galleries may be neglected because of crumbling in the galleries which limits the flow. We therefore propose in this study to explore the first hypothesis of karstic communication from the *Gardon d'Alès* catchment to the *Cèze* catchment (Fig. 1).

B HYDROLOGICAL DATA

For this study, five floods of the *Cèze* were selected. The following figure shows these events.

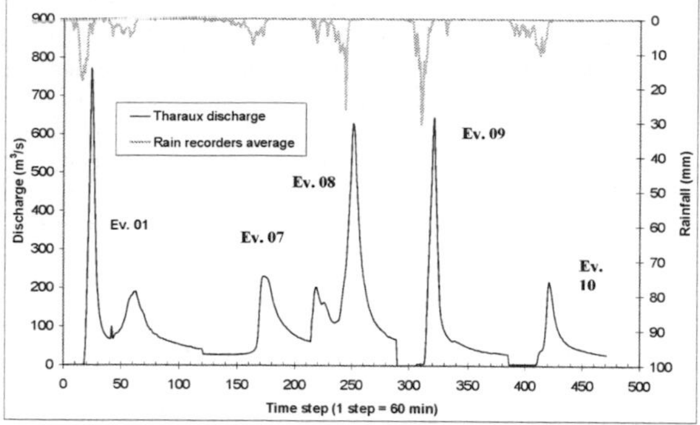

Fig. 2: Floods Hydrographs of the *Cèze* (*Tharaux outlet*)

These data were collected by the flood warning service of the *Gard* Region. The discharge was measured at *Tharaux* gauging station (fig. 1) and the rainfall measured by five rain recorders which located in the *Cèze* and *Gardon d'Alès* catchment areas. For both the discharge and the rainfall, the sampling period was 60 minutes.

TABLE I
Hydrological Database Used

Event	Date	Peak discharge (m^3/s)	Rainfall (mm)
1	22th / 26th October 1993	770	205
2	25th / 28th November 1995	230	105
3	11th / 14th November 1996	630	145
4	06th / 09th October 1997	640	175
5	28th September / 22th October 2000	215	145

Table I contains more details on these events. It can be noticed that because of the difficulty of accurately measuring the rainfall and flows, only five events are available, due to a lack of data for other events.

C FLOOD SIMULATIONS

As usual in the neural network field, the first approach is the multilayered perceptron with one hidden layer. We applied rainfall as inputs and runoff as output. We chose the rainfall measured by 3 rain gauges in the *Cèze* catchment and two in the *Gardon* catchment in order to observe whether the latter input has a major influence on the forecast (Figure 3). An input bias is necessary in order to represent the base flow. Its value is not 1, as is usually applied, but a lower value due to the great number of very low values of the flow during the flood recording. This adjustment is necessary in order not to saturate the sigmoids during learning. The mean of the inputs was shown to be a good value. As shown in Figure 3 we apply the rainfall to the network in a temporal window. This temporal window is essential in order to capture the temporal behaviour of the catchment. Thirty time steps were chosen for rain recorders near the *Tharaux* outlet, and forty five for remore recorders, situated far upstream, in order to take into account a longer propagation time.

At the output of the network we measured the quality of the response using a criterion used in hydrology and called the Nash criterion [25]. The Nash criterion is analogous to the coefficient of determination and is calculated as:

$$\text{Nash} = 1 - \frac{\sum_k \left(o^k - d^k\right)^2}{\sigma^2}$$

where σ is the standard deviation of the test signal.

The Nash criterion takes into accounts the quadratic error and normalises this error by the variance of the signal. The closer the criterion to the value 1, the better the model. If forecasting is limited to predicting the mean value, the criterion is equal to zero; negative values are very bad. Table II contains the values obtained from learning with validation on an example which was not taken into account during learning.

TABLE II
SYNTHETIC RESULT OF THE BEST NASH CRITERIA

Network architecture	Nash criteria computed after test on:					
	Event 1		Event 7		Event 8	
	BP	LM	BP	LM	BP	LM
A	0.68	0.68	0.86	0.92	0.68	0.67
A with Loop	0.75	0.89	0,64	0,88	0,42	0,56
B	0.70	0.74	0.86	0,94	0.62	0.61
C	0.75	0.8	0.89	0.92	0.63	0.66
D	0,75	0.77	0.87	0.92	0.63	0,62

BP is Backpropagation, LM is Levenberg-Marquardt.

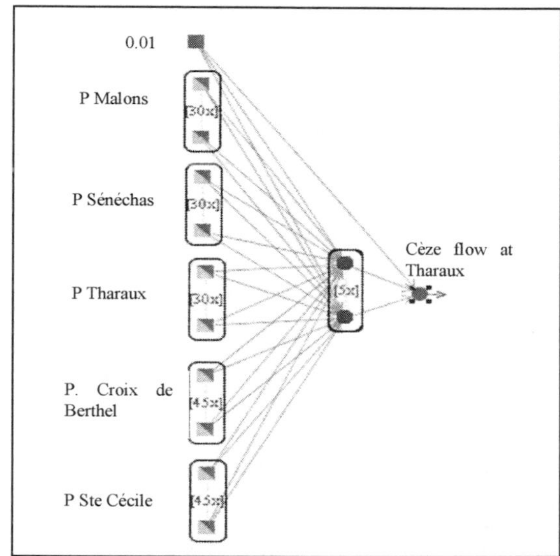

Fig. 3: *Cèze* flow modelling using multilayer perceptron as black box model (Network A.). Each input has a temporal window of 30 or 45 delays

Because of the bad results obtained by this simple static network, we applied the same function, with the same external input, but with a recurrent network. As suggested above, we chose non directed scheme in order better to capture the dynamics of the system. In this case, the Levenberg-Marquardt rule works well and significantly improves the Nash criterion (Table II). It can be noticed that the important estimation of the peak value is not improved by the Levenberg Marquardt. The advantage of the recurrent non directed network is that the input window may be smaller than in the static case. The window of temporal values applied to the networks is ten hours for all the rain recorders. And it is well known that in this type of configuration, the lower the number of coefficients, the better the learning.

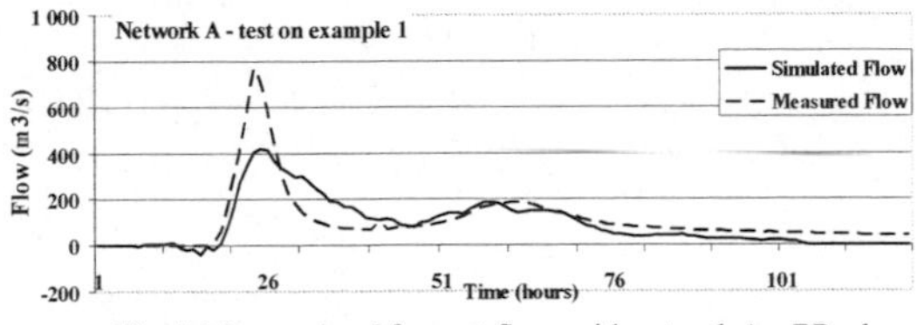

Fig. 4: Measured and forecast flows with network A – BP rule.

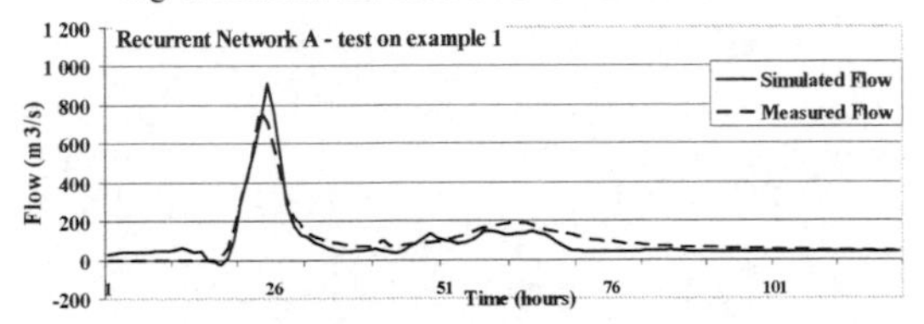

Fig. 5: Measured and forecasted flows with looped network A - LM rule.

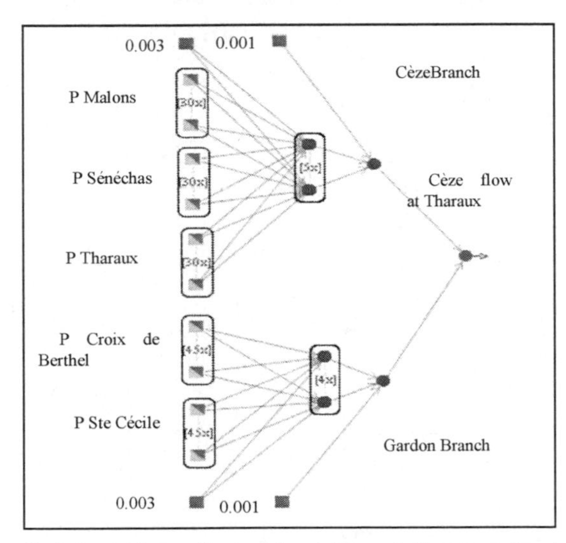

Fig. 6: the contribution of each catchments is computed in a separate branch of the network, and the addition is performed by the output neuron (Network B).

These first simulations allow us to predict the flows accurately but not to estimate the contribution of the *Gardon* catchment. The problem of quantifying the impact of a particularly input on the quality of the prediction is well identified for non-linear models but has no satisfactorily solution. Thus, in order to estimate the contribution of the *Gardon*'s catchment we propose an original architecture which separates the two catchments into two branches: one for the *Gardon* and one for *Cèze* catchment, as shown in Figure 6.

In the proposed architecture, the flow coming from the *Cèze* catchment is computed by a classical network devoted for identification: one hidden layer and one linear output network. Another network devoted to the *Gardon* catchment is computed in parallel. At the end, both networks, called the *Cèze* branch and the *Gardon* branch, are integrated in a single network using a supplementary linear neuron which is the output neuron of the whole network. The flow at *Tharaux* is then computed.

The interest of this network is that the flow coming from each catchment can be obtained by observing the value of the neurons of the second hidden layer: we only need to multiply the output of the neurons by the coefficient linking this neuron to the output neuron. Figure 7 shows the flows obtained in this way with architecture B. In order to illustrate the richness of this approach, we have plotted the estimation of flows on the learning examples, because it is interesting to observe how each sub-basin contributes to the whole flow, for each event.

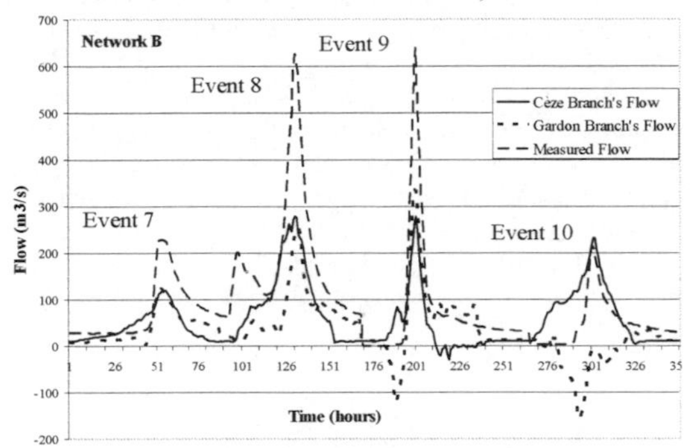

Figure 7: Learning curves obtained for architecture B.

One can note that the *Cèze* branch and the *Gardon* branch contribute equally to the final flow.

The result obtained is that, with architecture B, the contributions of the sub-basins are similar except for event 10. From the hydrological point of view, this result is not possible: the flows coming from the *Gardon* catchment cannot be equal to those from the *Cèze*. Moreover they cannot be negative, as shown for event 10. Thus the architecture is not realistic: the *Croix de Berthel* rain collector should not be inputted to the *Gardon* branch.

Starting from these considerations, we inputted the *Croix de Berthel* rain collector to the *Cèze* branch and left only the *Ste Cécile* rain collector inputting the *Gardon d'Alès* branch, thus building network C.

After learning, we obtained the following results: the forecasting has a Nash criterion of the same order as for network B, but the water now essentially comes from the *Cèze* catchment and only marginal and limited flows come from the

Gardon (Figure 8). We also obtained an estimation of 6.2m^3/s for the maximum flow and 2.4m^3/s for the mean value.

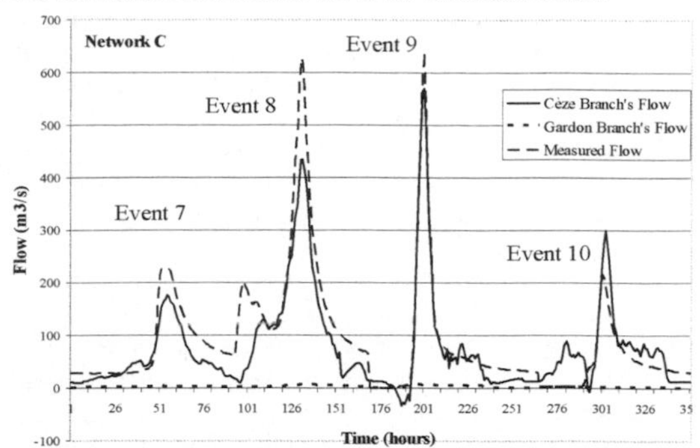

Figure 8: Learning curves obtained for architecture C. On can note that the *Cèze* branch contribute predominately to the final flow. The *Gardon* contribution is very small.

The interpretation is that the flow of the *Cèze* river is well explained by including the *Croix de Berthel* rain collector in the *Cèze* catchment without taking into account a huge karstic flow from the *Gardon*. This interpretation is confirmed by the bibliography: karstic inputs are diffuse along the stream with only one perennial spring: the *Peyrouse* spring. The very interesting property of the neural network is that it estimates this hidden flow.

In fact, the *Croix de Berthel* is near to the frontier of the catchments. Thus, complementarily, we tried an extra network where the *Croix de Berthel* rain collector was connected to both catchments (network D) in order to take into account the possibility of water going to *Tharaux* via the *Céze* or *Gardon* branch. It appears in this case that the results are the same as above (Figure 8): no major quantity of water comes from the Gardon network.

IV CONCLUSION

We have shown in this paper that neural networks can usefully be applied to very complex problems in hydrogeology. We first showed that because of their ability to identify non linear dynamical models, recurrent non-directed neural networks are good candidates for simulating fast floods. Moreover, using a specific architecture we showed that static models can be interpreted in terms of hydrogeology and provide an estimation of hidden variables. This last property is really innovative and opens up a wide field of fruitful research in earth science.

ACKNOWLEDGEMENTS

We are very grateful to Miss S. Eurisouké for the simulations performed during her Master professional training course.

We also wish to thank M. D. Bertin (from SAPI) for his highly affective collaboration in the conception and realization of the Neural Network simulation tool: RnfPro.

We also thank the *Gard* Warning Service for the data used.

REFERENCES

[1]. G. Cybenko"Approximation by Superposition of a Sigmoidal Function". Math. Ctrl Signal Syst, 2 (1989) 293-342.

[2]. A. G. Barto et P. Anandan,."Pattern Recognition Stochastic Learning Automata". IEEE Trans. System Man and Cybernetics. SMC-15 : 360-375 (1985).

[3].Johannet A. and Sarda I. "Goal-Directed Behaviours by Reinforcement Learning". Neurocomputing, Special Issue on Neurap'98. May 1999.

[4].D. Rumelhart, G. Hinton, R. Williams. "Learning Internal Representation by Error Propagation. In "PDP, MIT Press, 1988.

[5] E.A. Bender. "Mathematical Method for Artificial Intelligence".IEEE Computer Society Press. 1996.

[6] A.J. Shepherd. «Second-Order Methods for Neural Networks». Springer 1997.

[7] D.W. Marquardt. Journal of the Society for Industrial and Applied Mathematics, vol 11, pp 431-441.

[8] W.H. Press, S.A.Teukolsky, W.T. Vetterling, B.P. Flannery. «Numerical recipies in C». Cambridge University Press 1992.

[9] Narendra K. S., Parthasarathy K. "Gradient Methods for the Optimization of Dynamical Systems Containning Neural Networks". IEEE trans. neur.net., vol 2, n°2, pp. 252-262, 1991.

[10]Werbos P.J. "Backpropagation Throught Time : What it Does and How to Do It". PROC. IEEE, 78, N°10, PP1550-1560, 1990.

[11].Nerrand O., Roussel P., Personnaz L., Dreyfus G. & Marcos S. (1992) "Neural Networks and Non-Linear Adaptive Filtering : Unifying Concepts and new Algotithms". Neural computation **5**, 165-199.

[12]. A. Johannet, A. Mangin, D. D'Hulst. "Subterranean Water Infiltration Modelling by Neural Networks: Use of Water Source Flow". In Proc. of ICANN 1994; pp1033-1036. M. Marinaro and P.G. Morasso eds, Springer.

[13] K. J. Beven and M. J. Kirby, "A physically based, variable contributing area model of basin hydrology." *Hyd. Sciences*, vol. 24, 1979, pp. 43-69.

[14] P-A. Ayral, S. Sauvagnargues-Lesage, F. Bressand and S. Gay, "Forecasting flash floods with an operational model: application in the South-East of France (Gard)", Flood risk management in Europe: Innovation in Policy and practice". ANTHR, vol. 25. Begum S., Stive Marcel JF, James W (Eds), Springer 2006, 600 p.

[15] V. Estupina-Borrell, J. Chorda, D. Dartus, « Flash floods anticipation." *C. R. Geosciences*, vol. 337, 2005, pp.1109-1119.

[16] E. Gaume, M. Livet, M. Desbordes and J-P. Villeneuve, "Hydrological analysis of the river Aude, France, flash flood on 12 and 13 November 1999". *J. Hydrol.*, vol. 286, 2004, pp. 135-154.

[17] C. Bouvier, P-A. Ayral, P. Brunet, A. Crespy, A. Marchandise and C. Martin, "Recent advances in modelling extreme floods in Mediterranean French catchments",(FRIEND-Amhy) – Extreme events – UNESCO Project, Cosenza (Italy), 3-4 May 2006, 11p.

[18] "H. Moradkhani, K. Hsu, H. V. Gupta, S. Sorooshian. "Improved Streamflow Forecasting Using Self-Organizing Radial Basis Function Artificial Neural Networks. Journal of Hydrology 295 (2004) 246-262.

[19] I. N. Daliakopoulos, P. Coulibaly, I. K. Tsanis. "Groundwater Level Forecasting Using Artificial Neural Networks". Journal of Hydrology 309 (2005) 229-240.

[20] A. P. Jaquin, A. Y. Shamseldin. "Development of rainfall-runoff models using Takagi-Sugeno fuzzy inference systems". Journal of Hydrology 329 (2006) 145-173.

[21] A.-L. Courbis; E. Touraud and B. Vayssade. "Water balance diagnosis based on a simulation tool". ENVIROSOFT'98, 10-12th November 1998, Las Vegas USA,pp199-208.

[22] E. Gaume, P-A. Ayral, C. Bouvier, J.D. Creutin, G. Delrieu and M. Livet, "Hydrological analysis of the Gard River (France) extraordinary flood: 8 and 9 September 2002", 5th Plinius Conference, "Mediterranean storm", 1 – 3 October 2003, 7 p.

[23]E. Dumas: « Statistique géologique, minéralogique, Métallurgique et paléontologique du département du Gard », Paris, ARTHUS BERTRAND, 1875

[24] A. Blachère : « Rapport pour Charbonnages de France », CESAME, 2003.

[25]J.E. Nash , J. V. Sutcliffe. «River Flow Forecasting through Conceptual Models. Part I – A Discussion of Principles » Journal of Hydrologie 10 (1970) 282-290.

Significance of Pupil Diameter Measurements for the Assessment of Affective State in Computer Users

Armando Barreto[1], Jing Zhai[1], Naphtali Rishe[2], and Ying Gao[1]

Electrical and Computer Engineering Department[1] and School of Computer and Information Sciences[2]
Florida International University
Miami, FL 33174 USA

Abstract- **The need to provide computers with the ability to distinguish the affective state of their users is a major requirement for the practical implementation of Affective Computing concepts. The determination of the affective state of a computer user from the measurement of some of his/her physiological signals is a promising avenue towards that goal. In addition to the monitoring of signals typically analyzed for affective assessment, such as the Galvanic Skin Response (GSR) and the Blood Volume Pulse (BVP), other physiological variables, such as the Pupil Diameter (PD) may be able to provide a way to assess the affective state of a computer user, in real-time. This paper studies the significance of pupil diameter measurements towards differentiating two affective states (stressed vs. relaxed) in computer users performing tasks designed to elicit those states in a predictable sequence. Specifically, the paper compares the discriminating power exhibited by the pupil diameter measurement to those of other single-index detectors derived from simultaneously acquired signals, in terms of their Receiver Operating Characteristic (ROC) curves.**

I. Introduction

New developments in human-computer interaction technology seek to close the communication gap between the human and the machine. A key component needed to meet these requirements is the ability of computer systems to address user affect. Picard and others have described the importance of the emotional and affective factors in human-computer interaction [1]. The knowledge of a user's affect can provide useful feedback regarding the degree to which a user's goals are being met, enabling dynamic and intelligent adaptation. Since physiological variables in humans are inherently controlled by their autonomic nervous system, these expressions of emotion are less susceptible to environmental interference or voluntary masking than others, such as, for example, facial expression or speech activity. Previous attempts to recognize emotions from physiological changes have analyzed a variety of autonomic activities such as the Electroencephalogram (EEG), the Electrocardiogram (ECG), the Electromyogram (EMG), Blood Pressure (BP), Blood Volume Pulse (BVP), Galvanic Skin Response (GSR), Skin Temperature (ST), Heart Rate Variability (HRV), etc. Many of these physiological variables have been chosen because they can be monitored in non-invasive and non-intrusive ways. However, one physiological variable that has not been studied extensively for the purpose of affect recognition is the pupil dilation. In an isolated fashion, it has been verified that variations of the Pupil Diameter (PD) reflect the emotional changes driven by auditory emotional stimulation [2].

From human physiology studies, it is known that the Sympathetic Division of the Autonomic Nervous System (ANS) significantly influences these physiological variables. The sympathetic division prepares the body for heightened levels of somatic activity. When fully activated, this division readies the body for a crisis that may require sudden, intense physical activity, which is known as the "fight or flight" response. Generally, an increase in sympathetic activity stimulates tissue metabolism and increases alertness. The heart rate, skin resistance, blood pressure and pupil diameter are all affected by branches of the sympathetic division of the ANS. In this study, we monitored four physiological variables (GSR, BVP, ST and PD) simultaneously and compared the significance of signals derived from these measurements towards the detection of sympathetic activation associated with a multifaceted emotional state — 'Stress'.

When a subject experiences stress and nervous tension, the palms of his/her hands become moist. Increased activity in the sympathetic nervous system will cause increased hydration in the sweat duct and on the surface of the skin. The resulting drop in skin resistance (increase in conductance) is recorded as a change in electrodermal activity (EDA), also called Galvanic Skin Response (GSR). So, in everyday language, electrodermal responses can indicate 'emotional sweating'. The GSR is measured by passing a small current through a pair of electrodes placed on the surface of the skin and measuring the conductivity level. In spite of its simplicity, GSR measurement is currently considered one of the most sensitive physiological indicators of psychological phenomena. GSR is also one of the signals used in the polygraph or 'lie detector' test. A GSR2 module, by Thought Technology LTD (West Chazy, New York) was used in our research to measure GSR. The resistance found in between its two electrodes determines the oscillation frequency of a square-wave oscillator inside the device. We have used a "frequency-to-voltage-converter" integrated circuit (LM2917N) to obtain output voltages that are proportional to instantaneous skin conductance. This modified device was calibrated by connecting several resistors of known resistance to it and measuring the output voltage of the frequency-to-voltage converter in each case.

K. Elleithy (ed.), Advances and Innovations in Systems, Computing Sciences and Software Engineering, 59–64.

The measurements of Blood Volume Pulse (BVP) in this project were obtained using the technique called photoplethysmography (PPG), to measure the blood volume in skin capillary beds, in the finger. PPG is a non-invasive monitoring technique that relies on the light absorption characteristics of blood. Traditionally, the Blood Volume Pulse has been used to determine the heart rate only. However, if measured precisely enough, it can be used to extract estimates of the heart rate and its variability. In our experiment, the sampling rate used to record the BVP signal was 360 samples/second.

Changes of acral skin blood flow are also a commonly used indicator for sympathetic reflex response to various stimuli. When sympathetic stimuli are applied to a person, the blood volume in the finger vessels is expected to decrease due to the vasoconstriction in the hairless areas of the hand but not in the hairy skin of the hand. If this assumption is true, the finger temperature should transiently decrease according to this effect. A thermistor can be attached to the subject's finger to sense the temperature changes. In our experiment, the subject's skin temperature was measured with an LM34 integrated circuit that provided a linear output between –50 and 300 degrees Fahrenheit. The output of the sensor was buffered and fed into a differential amplifier (with a gain of 31 V/V) to amplify the temperature changes in the range of 75-100 °F. This sensor was attached to the distal phalanx of the left thumb finger with the help of Velcro. The signal was recorded at the sampling rate of 360 samples/second. The experiments were performed in an air-conditioned room, to minimize the potential impact of environmental temperature changes on this experimental variable.

The diameter of the pupil is determined by the relative contraction of two opposing sets of muscles within the iris, the sphincter and dilator pupillae, and is determined primarily by the amount of light and accommodation reflexes [3]. The pupil of the human eye can constrict and dilate such that its diameter can range from 1.5 to more than 9mm. The pupil dilations and constrictions in humans are governed by the ANS. Several researchers have established that pupil diameter increases due to many factors. Anticipation of solving difficult problems, or even thinking of performing muscular exertion will cause slight increases in pupil size. Hess [4] indicated that other kinds of anticipation may also produce considerable pupil dilation. Previous studies also have suggested that pupil size variation is also related to cognitive information processing. This, in turn, relates to emotional states (such as frustration or stress) since the cognitive factors play an important role in emotions [5]. Partala and Surakka have found, using auditory emotional stimulation, that the pupil size variation can be seen as an indication of affective processing [2]. All these previous results found in the literature prompted us to attempt to use the pupil size variation to detect affective changes during human-computer interactions. There are several techniques available to quantify pupil size variations [5]. Currently, automatic instruments, such as infrared eye-tracking systems, can be used to record eye-related information, including pupil diameter and point of gaze. In our study, the subject's left eye was monitored with an Applied Science Laboratories series 5000 eye tracking system running on a PC computer to extract the values of pupil diameter. The sampling rate of the system was 60 samples/second. To minimize the potential impact of illumination changes on the subject's pupil diameter, the lighting of the experimental environment and the average brightness of the stimulus computer were kept constant during the complete experimental sequences and across all the subjects.

II. METHODOLOGY

A. Stress Elicitation

Our aim in this research is the detection of mental stress, as physical stressors occur far less frequently in the context of human-computer interaction. Therefore, in order to elicit mental stress at controlled intervals a computerized "Paced Stroop Test" was used. The Stroop Color-Word Interference Test [6], in its classical version, requires that the font color of a word designating a different color be named. In our research, the classical Stroop Test was adapted into an interactive version that requires the subject to click on the correct answer rather than stating it verbally. Since adding task pacing to the Stroop Test might intensify the physiological responses [7], each trial was designed to only wait 3 seconds for a user response. If the subject could not make a decision within 3 seconds, the screen automatically changed to the next trial. This modified version was implemented with Macromedia Flash® and also programmed to output bursts of sinusoidal tones through the sound system of the laptop used for stimulation, at selected timing landmarks through the protocol to time-stamp the recorded signals at those critical instants.

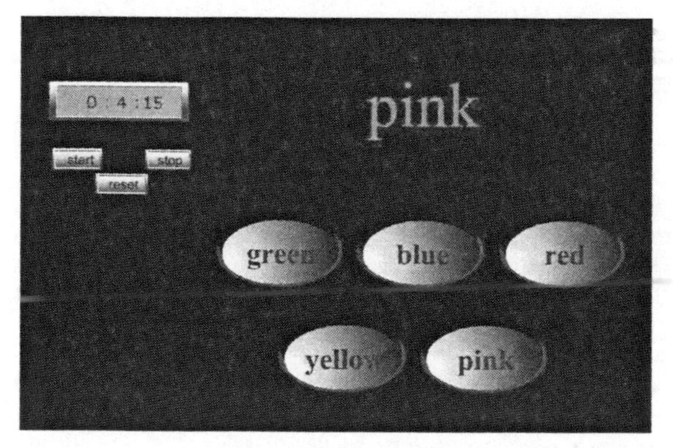

Figure 1. Sample Stroop Test interface

Figure 2 is the audio output schedule for the experiment, from the beginning of the session to its end. The complete experiment comprises three consecutive sections. In each section, we have four segments including: 1) 'IS' - the Introductory Segment to let the subject get used to the task environment, in order to establish an appropriate initial level for his/her psychological state, according to the law of initial values (LIV) [8]; 2) 'C' – is a Congruent segment, comprising 45 Stroop congruent word presentations (font color matches the meaning of the word), which are not expected to elicit significant stress in the subject; 3) 'IC' – is an Incongruent segment of the Stroop Test in which the font color and the meaning of the 30 words presented differ, which is expected to induce stress in the subject; 4) 'RS' – is a Resting Segment to let the subject relax for some time. The binary numbers shown in Figure 2 represent the de-multiplexed output of the audio signaling used in the system to time-stamp the four physiological signals, BVP, GSR, PD and ST. Our previous report on the instrumental setup [9] provides more details on this audio scheme.

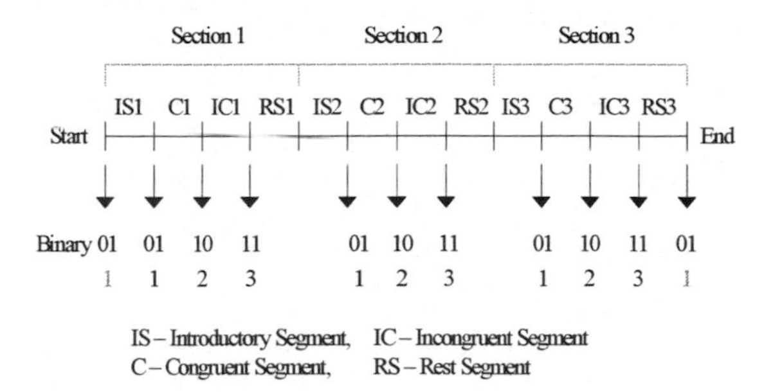

Figure 2. Audio output schedule.

B. Physiological Recording Setup

The complete instrumental setup developed for this research is illustrated in Figure 3. The stimulus program (interactive paced Stroop Test) described above runs in a laptop PC. While performing the Stroop Test, the subject has the GSR, BVP and ST sensors attached to his/her left hand. These three signals are digitized, using a multi-channel data acquisition system, NI DAQPad-6020E for USB, a product of National Instrumentation Corp, and the samples are read into Matlab® directly at rate 360 samples/sec. Additionally, the eye gaze tracking system (ASL-504) records PD data to a file on its own interface PC, at a rate of 60 samples/sec. The software for this system allows the extraction of selected variables (in this case the pupil diameter and the marker channel) to a smaller file, which in turn can be read into Matlab® also, where it can be aligned with the BVP, GSR and ST signals, thanks to their common timing marks for the start and stop events. At this point the pupil diameter data can be upsampled (interpolated) by six, to achieve a common sampling rate of 360 samples/sec for all four measured signals.

Figure 4 shows an example of the four signals recorded from a subject through the complete length of the experimental session, after all of the signals have been synchronized (at a sampling rate of 360 samples per second). The gaps in the pupil diameter signals, due to blinking, have been compensated by automatic interpolation.

Signals from 32 experimental subjects were collected and divided into 192 data entries, since each participant generated data under three relaxed (congruent Stroop) segments and three stressed (incongruent Stroop) segments.

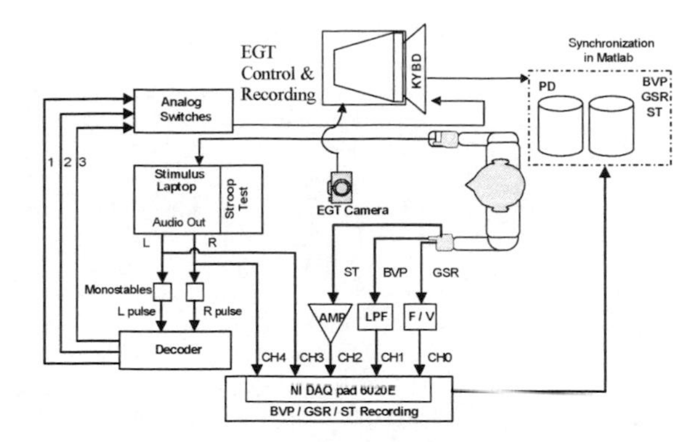

Figure. 3. .Instrumental Setup.

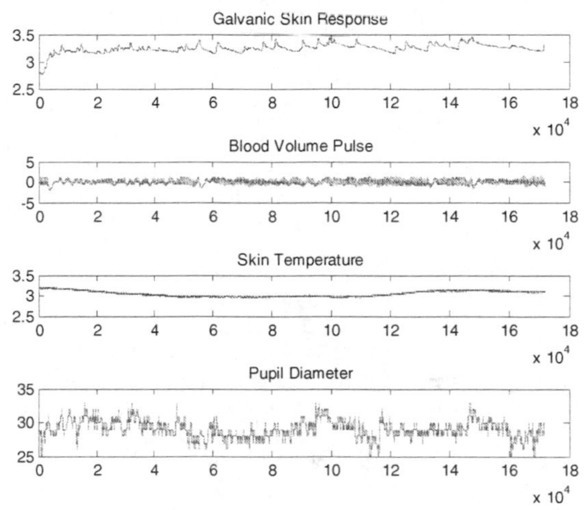

Figure. 4. Four physiological signals after synchronization. From top to bottom: GSR, BVP, ST, and PD.

C. Definition and Normalization of Individual Detection Signals

In this study our goal is to compare the potential of single-index indicators derived from the four physiological signals measured in terms of their individual discriminating power to differentiate between the congruent Strop segments (associated with a "relaxed" affective state in the user) and the incongruent Stroop segments (which are assumed to have caused a "stressed" state in the subject). The following paragraphs describe how the sample values of each of the signals were consolidated in a single feature value for each congruent or incongruent segment in the test.

The average value of the GSR samples collected during the whole extent of a congruent or incongruent Stroop segment was used as a representative response for this variable for each segment: *GSRmean*. Increased sweat production during "stressed" segments would predict a noticeable change of this average value during those segments.

From the BVP signal the interbeat interval (IBI), defined as the time in milliseconds between two normal, consecutive peaks in the BVP signal was defined for each two consecutive beats. The inverse of the IBI, expressed in beats per minute (BPM) is the heart rate, which is known to be altered by autonomic activation. The single index value defined from the BVP signal for each (congruent or incongruent) segment was the average of the IBI values in the segment: *BVPIBImean* .

For the skin temperature signal, it was expected that the temperature in the finger surface would display transient decreases when the stressor stimuli occur. To extract this information, the amplified ST signal was first filtered to remove recording noise. The slope of the filtered skin temperature in each segment was then used as a feature element of this signal. We found that the patterns of temperature slope provided more indicative information than the patterns of mean value of this signal. One possible explanation for this finding is that the skin temperature seems to obey much longer time constants in its variation, and, as such, its instantaneous value does not necessarily reflect well the "current" affective status of the subject, at any given time. In a protocol that included alternation between two types of stimuli (congruent and incongruent Stroop), the ST level during one given interval may still reflect the response to the previous interval. However, the derivative of the changing signal showed an interesting pattern. When the stressor stimuli occur, the slope of the temperature signal was generally negative. The slope of the ST signal was estimated using the digital low pass differentiation algorithm 1f3, as defined in [10], to yield the detection signal *STslope*.

The raw pupil diameter (PD) signal was recorded separately, as previously described. The artifact gaps due to blinking were automatically detected and filled by interpolation. The single-index signal extracted from the pupil diameter samples in each segment was simply the average value of PD, which we have labeled: *PD*. According to previous knowledge from the literature, we expected the mean PD should increase during the stress segments.

Prior to attempting to use these single-index signals to identify "stressed" (incongruent) and "relaxed" (congruent) experimental segments, they underwent a process of normalization. Let Xs represent the feature value for any of the raw features defined from the signal sample values during congruent and incongruent segments of the experiment. Let Xr represent the corresponding feature value extracted from the signals samples that were recorded during the relaxation period, prior to the first congruent Stroop segment. To eliminate the initial level due to the individual differences, Equation (1) was first applied to get the corrected feature signals (Y_s) for each of the subjects.

$$Y_s = \frac{X_s}{X_r} \tag{1}$$

For each subject, there were three congruent segments and three incongruent segments. Therefore, six values of any of the features were obtained from the signals recorded during these segments. Equation (2) normalizes each feature value dividing it by the sum of all six segment values.

$$Y_s' = \frac{Y_{si}}{\sum_{i=1}^{6} Y_{si}} \tag{2}$$

These two stages of normalization aimed at minimizing the impact of individual subject responses on the affective state identification process. After this pre-processing, all features (GSRmean, BVPIBImean, STslope and PD) were normalized to the range of [0, 1] using max-min normalization, as shown in Equation (3), to be considered as detection signals and compared against a threshold that spanned a uniform range of possible values: [0,1], for a fair comparison.

$$Y_{norm} = \frac{Y'_s - Y'_{s\min}}{Y'_{s\max} - Y'_{s\min}} \tag{3}$$

III. RESULTS AND DISSCUSION

A. Comparison of single detection signals

The four physiological signals monitored in our experiments are expected to exhibit different characteristics during the intervals when the subject was not under stress (i.e., during congruent Stroop segments) and during the intervals in which the subject was being stressed by incongruent Stroop word presentations. It is possible to summarize the information contained in each physiological signal by extracting one or several numerical features from each. In previous studies we have developed affective state classifiers that combine the information from several features extracted from each of the four physiological signals monitored, by means of machine learning systems [11][12][13].

In this study, however, our goal was to compare the discriminant power of information derived from the Pupil Diameter mean in a given interval (PD), with respect to other single detection signals GSRmean, BVPIBImean and STslope). Therefore, these three signals, as well as the PD measurements, were normalized as indicated by equations (1), (2) and (3).

B. Receiver Operating Characteristic (ROC) Curves

Next we present the comparison of the Receiver Operating Characteristic (ROC) curves for the four chosen normalized detection signals (PD, GSRmean, BVPIBImean and STslope), as a way to compare the levels of affective state discrimination power associated with them.

Receiver Operating Characteristic (ROC) curves show graphically the trade-off that a classifier must make between its "false positive rate" (which reflects the false alarm level, i.e., fraction of negative cases incorrectly classified as positive) and its "true positive rate" (i.e., the fraction of all positive cases correctly classified), by means of adjusting a threshold. The ROC is a plot of false positive rate vs. true positive rate for a classifier as different settings for the threshold are considered. At the lowest sensitivity level (i.e., setting the threshold at the highest possible value of the detection signal) the classifier produces no false alarms but detects no positive cases. This is represented by the origin of the coordinate axes in the ROC plot. As the sensitivity is increased, (i.e., as the threshold is lowered) the classifier detects more positive examples but may also start generating false alarms (false positives). Eventually the sensitivity may become so high (threshold set at the lowest possible value of the detection signal) that the classifier always claims each case is positive. So the classifier gets all positive cases right (true positive rate = 1), but it gets all the negative cases wrong, because it raises a false alarm on each negative case (false positive rate = 1). This corresponds to the top right-hand corner of the ROC. While all ROC curves "start" at the coordinate origin, (0,0) and "end" at the upper-right corner (1,1), the trajectory between these points followed by a given ROC, and consequently, the "Area under the ROC" are indicators of the discriminant power of the classification signal being thresholded. A "random classifier" (i.e., a process that produces uniformly distributed random numbers, without any relation to the input which is supposedly being classified) would display a ROC that follows approximately a 45° diagonal ascent from (0,0) and (1,1). The "area under the ROC" (AUROC) would, therefore, be close to 0.5 (half the area of the 1.0-by-1.0 square). On the other hand, a classification system that produces a highly discriminating detection signal will have one or several threshold levels that map close to the upper-left corner of the ROC, at (0,1) indicating a high sensitivity (large true positive rate) and also a high specificity (low false positive rate). If that is the case, the AUROC will come close to encompassing the full 1.0-by-1.0 square. That is, the AUROC will approach the ideal value of 1.

C. ROC comparison for Pupil Diameter and other signals

In the light of the background provided by the previous sub-section, our interest is to compare the discriminant power of the Pupil Diameter with those of the other 3 normalized physiological measures chosen to represent each (congruent or incongruent) interval (GSRmean, BVPIBImean and STslope).

The ROC curve for each of these detection functions has been estimated using the values for the 6(segments) x 32(subjects) = 192 segments analyzed in our experiments. Only half of these segments correspond to "stressed" states induced by incongruent Stroop stimulation (ideal classifier output = 1), while the other half are known to be associated with "relaxed" (congruent Stroop) intervals (ideal classifier output = -1). Each point of the ROC curves is determined by comparing the value of the detection signal to a specific threshold level and determining which portion of the "positive classifier outputs" match the ideal (1) and which portion of the "negative classifier outputs" are in disagreement to the ideal output (-1). These "portions", expressed as fractional numbers, yield the coordinates of the ROC point (false positive rate, true positive rate) for the threshold value tested. The process was carried out using the ROC Matlab ® scripts provided by Dr. Gavin C. Cawley (University of East Anglia, Norwich, UK) in his web site http://theoval.sys.uea.ac.uk/matlab/default.html. These scripts not only sweep the complete range of normalized threshold values, [0,1], and draw the ROC, but additionally estimate a "convex hull" that fits the actual ROC points calculated. The convex hull is shown with dashed lines in the following plots.

Figures 5 through 8 show the ROC computed for the GSRmean, BVPIBImean, STslope and PD detection signals, respectively. The area under the ROC curve computed in each case is stated in the caption for each figure. It should be noted that both the ROC curves for GSRmean and for BVPIBImean show a convexity that makes them depart from the random classification diagonal to some extent. However, their areas under the ROC are only moderately better than 0.5: AUROC_GSRmean = 0.6519 and AUROC_BVPIBImean = 0.6455 .

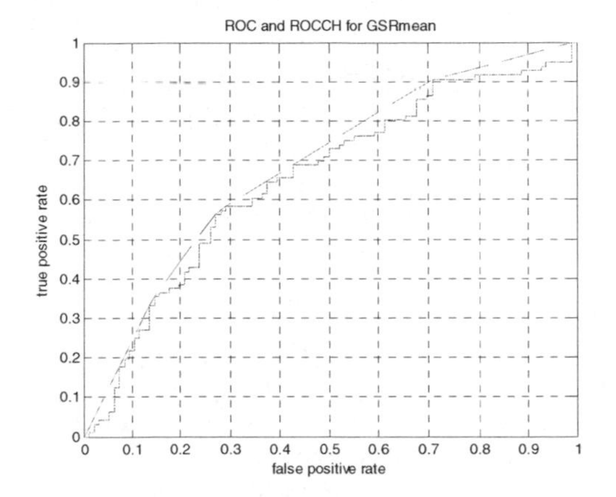

Figure 5. GSRmean ROC curve (AUROC = 0.6519)

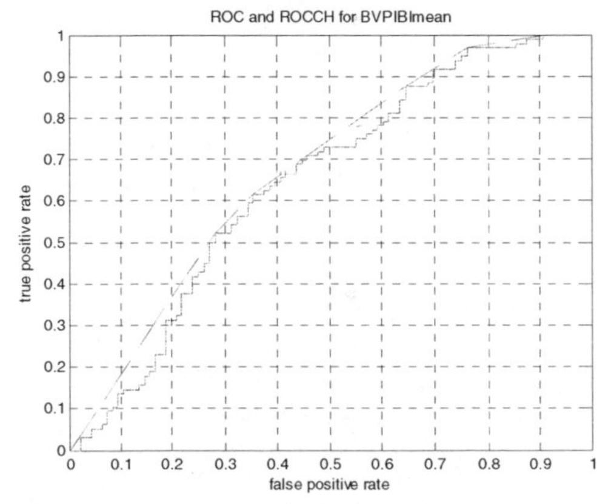

Figure 6. BVPIBImean ROC curve (AUROC = 0.6455)

The ROC curve for STslope shown in Figures 7 is actually very close to the random classification diagonal and, in general terms, follows a straight line at an angle just slightly larger than 45°. As such, the area under this curve is not much higher than ½ : AUROC_STslope = 0.5849 .

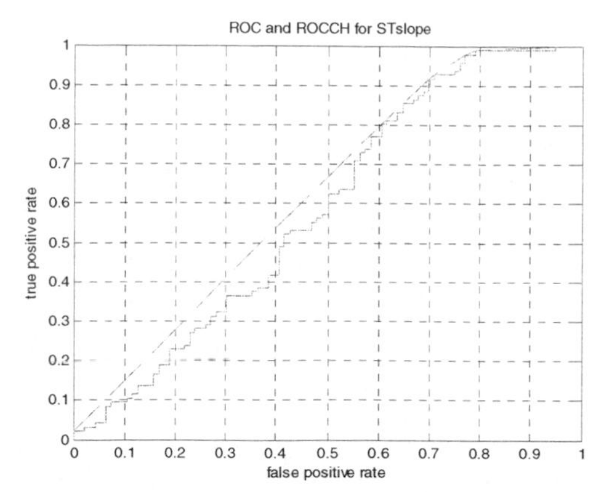

Figure 7. STslope ROC curve (AUROC = 0.5849)

In contrast, the ROC curve for PD, shown in Figure 8, exhibits a sharp slope from the coordinate origin, almost immediately reaching into high levels of true positive rate. Then the curve exhibits a number of intermediate points (threshold levels) for which the true positive rate is better than 0.8 while simultaneously having a false positive rate of less than 0.2. Accordingly, the area under this curve is large: AUROC_PD = 0.9647.

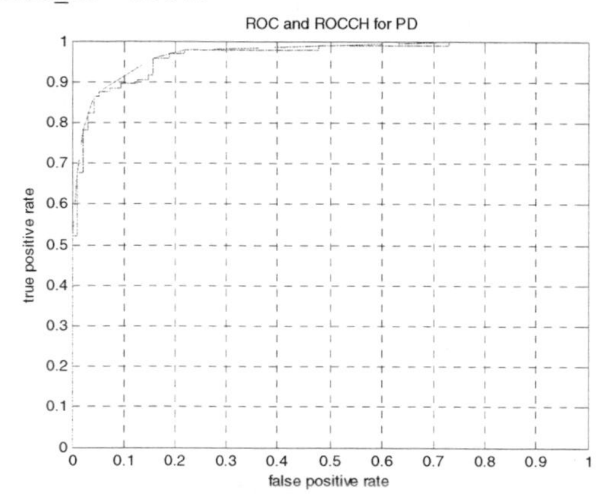

Figure 8. PD ROC curve (AUROC = 0.9647)

If a figure of merit indicating the *Discriminating Potential* of a given detection signal, X, calculated as:

$$DP = (AUROC(X) - 0.5) \times 200\% \quad (4)$$

is considered (such that a random classifier will yield 0% and a detection signal for which AUROC approaches 1.0 will yield approximately 100%), we would find that: DP(GSRmean) = 30.38%; DP(BVPIBImean)=29.10%; DP(STslope)=16.98% and, significantly, DP(PD)=92.94% . This indicates that, at least in terms of its ROC curve, the PD detection signal has significantly more potential to help identify one state from the other, while STslope shows particularly limited discriminating potential.

IV. CONCLUSION

We have investigated the potential of four detection signals derived from physiological measurements, GSRmean, BVPIBImean, STslope and PD, to act as individual classification signals for the differentiation between stress and relaxation in computer users. Our results indicate that two of the signals (GSRmean and BVPIBImean) derived from two of the physiological variables most commonly monitored for affective sensing exhibited only moderate discriminating potential. The STslope signal exhibited limited potential for this detection problem. In contrast, the mean value of the pupil diameter, PD, displayed a strong potential for single-signal discrimination between relaxed and stressed user states. Of course, this analysis has only addressed the *potential* of these signals for discrimination, and the actual performance of a detector based on any of these signals will be strongly influenced by the definition of the actual threshold used.

ACKNOWLEDGMENT

This work was partially sponsored by NSF grants IIS-0308155, CNS-0520811, HRD-0317692 and CNS-0426125.

REFERENCES

[1] Picard, R.W. 1997. Affective computing. MIT Press, Cambridge, Mass.

[2] Partala, T., Surakka, V. 2003. Pupil size variation as an indication of affective processing. Int. J. of Human-Computer Studies 59:185-198

[3] Beatty, J., Lucero-wagoner, B. 2000. The Pupillary System. In: Handbook of Psychophysiology. J.T. Cacioppo, L.G. Tassinary, and G.G. Berntson, editors. pp. 142-162. Cambridge University Press

[4] Hess, E.H. 1975. The tell-tale eye : how your eyes reveal hidden thoughts and emotions. Van Nostrand Reinhold, New York

[5] Grings, W.W., Dawson, M.E. 1978. Emotions and Bodily Responses A psychophysiological Approach. Academic Press, Inc.

[6] Stroop, J.R. 1935. Interference in serial verbal reactions. Journal of Experimental Psychology 18:643-661

[7] Renaud, P. and Blondin, J.-P., "The stress of Stroop performance: physiological and emotional responses to color-word interference, task pacing, and pacing speed," International Journal of Psychophysiology, vol. 27, pp. 87-97, 1997.

[8] Stern, R.M., Ray, W.J., Quigley, K.S. 2001. Psychophysiological Recording. Oxford University Press.

[9] Barreto, A. and Zhai, J., "Physiological Instrumentation for Real-time Monitoring of Affective State of Computer Users," WSEAS Transactions on Circuits and Systems, vol. 3, pp. 496-501, 2003.

[10] Usui, S. and Amidror, I., "Digital Low-Pass Differentiation for Biological Signal Processing," IEEE Trans. BME, 29: 686-693, 1982.

[11] Zhai J., and Barreto A., "Stress Detection in Computer Users Through Noninvasive Monitoring of Physiological Signals ", Biomedical Science Instrumentation, vol. 42, pp. 495-500, 2006..

[12] Zhai, J., and Barreto, A., "Stress Detection in Computer Users Based on Digital Signal Processing of Noninvasive Physiological Variables", Proceedings of the 28th IEEE EMBS Annual International Conference New York City, USA, Aug 30-Sept 3, 2006, pp. 1355 – 1358.

[13] Zhai, J., and Barreto, A., "Stress Recognition Using Non-invasive Technology", Proceedings 19th Int. Florida Artificial Intelligence Research Society Conference (FLAIRS 2006), pp. 395 – 400, 2006.

A Novel Probing Technique for Mode Estimation in Video Coding Architectures

Ashoka Jayawardena
School of Mathematics, Statistics and Computer Science,
University of New England.
Australia.

Abstract- **Video compression standards operate by removing redundancy in the temporal, special, and even frequency domains. Temporal redundancy is usually removed by motion compensated prediction resulting in Inter-, Intra-, Bidirectional- frames. However, video coding standards do not specify the encoding process but the bit stream. Thus one of the key tasks of any implementations of such standards is to estimate the modes of frames as well as macro blocks. In this article we propose a novel technique for this purpose.**

I. INTRODUCTION

Video compression has become a key component in modern digital broadcast and entertainment media. The MPEG series of standards and H.26X series of standards have been developed for this purpose [1,2,3,4,5]. The latest of this is the MPEG-4 part 10 and H.264 overlapping standard [6,7]. Yet these standards only specify the encoding bit stream leaving the innovation of decision logic for implementers. Thus for the same standard and for the same video clip the performance may change from implementation to implementation. In this article we will propose an algorithm for this decision logic.

The article is organized as follows. In section II, I will briefly discuss motion vector search for macro blocks. In section III, I will explain the philosophy of mode probing and its relevance to motion vector search. In section IV, the mode estimation algorithm will be given. Finally the conclusion is given in section V.

II. MOTION VECTOR SEARCH

The techniques used to analyze the 2D motion across various applications can be classified in to three techniques: block matching, pel recursion, and optical flow [8]. The block matching techniques are the most relevant for existing video coding applications. However the other techniques can be used for various techniques of video compression. For example affine mesh based motion models [9,10] can use pel recursive and optical flow techniques. Pel recursive and optical flow field are usually invertible while block matching motion field are not invertible.

All these motion estimation methods are iterative. It is attempted to improve the motion estimation in each iteration. Some of these methods usually require a good initial guess and may have varying rates of convergence.

III. MODE PROBING

Our philosophy for mode decisions of macro blocks is that if modes can be decided early it must be decided as early as possible and if it needs to be delayed it must be delayed as late as possible. Mode decisions can be made before we reach optimal points of motion vector search. If mode decisions cannot be made early, it is an indication that we should be prepared to allocate more computational time for those decisions. Thus if the mode decisions cannot be made early, we must delay the computation of motion vector until such time that the actual modes are known thus saving precious computational time. On the other hand, if mode decisions can be made with confidence early we should simply set those modes.

Thus mode probing is about determining prediction modes without calculating the actual motion vectors. Thus we expect it to take less computational time. However video coding is computationally demanding already and hence we need to minimize computational resources allocated to mode probing. An efficient approach is to view that mode probing consists of the first few iterations of a motion vector search algorithm until it is clear that the iteration would lead to an acceptable motion vector or it is clear that the iterations may not lead to an acceptable motion vector.

Whenever the motion vector search is delayed we simply save the status of the iteration so that we can resume it at a later time to reach the optimal motion vector. This approach requires that the motion vector the search space is sufficiently stable and smooth.

IV. MODE ESTIMATION LOGIC

In H.264, for example, each macro block in a B slice may be predicted from one or two reference frames, before or after the current frame in temporal order. Thus the decoding order in the decoder does not have to be the temporal order of frames. These B macro blocks must be handled carefully.

However for simplicity, we assume that the principal modes of the macro blocks as forward predicted, backward predicted or intra-coded. We further identify that the principle modes of macro blocks can be arranged in the order of preference: forward predicted, backward predicted and Intra-coded. Forward prediction is preferred since it does not enforce a frame delay in the decoder.

K. Elleithy (ed.), Advances and Innovations in Systems, Computing Sciences and Software Engineering, 65–66.

The backward prediction can introduce a frame delay in the decoder. The Intra-coded blocks require higher bit rates.

The mode estimation of macro blocks for each frame is done in two stages. In the first stage all the forward prediction modes are decided whenever possible. In the second stage we attempt to decide backward prediction modes after the first stage of the next frame. If both forward prediction and backward prediction fails the macro block will be intra-coded.

We also need to avoid motion vector loops, i.e. prediction of a pixel can be traced back to itself. This can occur when trying to set backward prediction. The simplest way to avoid such motion loops is to only allow forward prediction in frames in reference (or decoding) order. Whenever there is backward prediction from a frame to its previous in temporal order, we will insert the frame just before it's previous in reference order. The algorithm is as follows:

```
BPCurrentFrame ← false
For all the frames F_i in the sequence
        BPLastFrame ← BPCurrentFrame.
        FPTemporalCount = 0.
        For M_j = M_first to M_last in F_i
                If Probe(j,F_i,F_i-1)
                        FPTemporalCount++;
                        TemporalMode(j,F_i) = FP;
                Else if Probe(j,F_i,F_i+1)
                        BPCurrentFrame ←true;
                        TemporalMode(j,F_i) = BP;
                If (BPLastFrame)
                        F_r ← previous frame of F_i-1 in reference buffer
                        FPReferenceCount++;
                        If Probe(j,F_i',F_r)
                                FPReferenceCount++;
                                ReferenceMode(j,F_i) = FP
                        Else if TemporalMode(j,F_i) != BP
                                If Probe(j,F_i,F_i+1)
                                        BPCurrentFrame ←true;
                                        TemporalMode(j,F_i) = BP;
      FPReferenceCost = FPTemporalCount –
                                        FPReferenceCount
      BPCostLastFrame = number of macro blocks of F_i-1
                        such that TemporalMode(j,F_i-1) = BP
      If (BPCostLastFrame – FPReferenceCost) > 0
              Place F_i behind F_i-1 in the reference buffer
              Complete the backward prediction for all the macro blocks of
                        F_i-1 with mode BP.
      Else
              Place F_i in front of F_i-1 in reference buffer
              Complete the forward prediction for all the macro blocks of
                        F_i-1 with mode BP.
    Complete the forward prediction for all the macro blocks of
                        F_i with mode FP.
```

The descriptions of the functions and variables used in the algorithm are as follows:

TemporalMode(j,F_i): It is variable which stores the mode of jth macro block of ith frame. This mode could be either FP (forward prediction) or BP (backward prediction) and calculated relative to previous (or next) frame in temporal order (display order).

ReferenceMode(j,F_i): It is variable which stores the mode of jth macro block of ith frame.The modes are calculated relative to previous frame in reference order. Only forward prediction is used to avoid motion loops.

Probe(j,F_i,F_{i-1}): Returns true if j^{th} macro block of F_i can be successfully predicted from F_{i-1}.

V. CONCLUSION

We have proposed an algorithm which can be used to estimate modes for macro blocks for existing video compression standards. However, these standards specify larger number of modes than we have given. The algorithm can be easily generalized to implement those modes after identifying the order of preference among those modes.

Our future research of this project will concentrate on a comprehensive mode decision framework within a resource constrained environment. Also we will work on hybrid motion vector search and mode probing algorithms with parameters optimized for compression performance.

VI. REFERENCES

[1] "Coding of Audio-Visual Objects-Part 2: Visual," ISO/IEC JTC1, ISO/IEC 14496-2 (MPEG-4 visual version 1), 1999.

[2] "Video Coding for Low Bit rate Communication, Version 1," ITU-T, ITU-T Recommendation H.263, 1995.

[3] "Information Technology – coding of moving pictures and associated audio for digital storage media at up to about 1.5Mbit/s, ISO/IEC 11172, 1993 (MPEG-1).

[4] "Information Technology: generic coding of moving pictures and associated audio information, ISO/IEC 13818, 1995 (MPEG-2).

[5] "Video CODEC for audiovisual services at px64 kbit/s", ITU-T Recommendation H.261, 1993.

[6] T. Wiegand, G. J. Sullivan, G. Bjontegaard, and A. Luthra, "Overview of the H.264/AVC video coding standard," IEEE Trans. Circuits Systems and Video Technology, vol. 13, pp. 560-576, July 2003.

[7] I. E. G. Richardson, "H.264 and MPEG-4 Video Compression: Video Coding for Next-generation Multimedia", ISBN 0470848375, John Wiley & Sons, 2003.

[8] Y. Q. Shi and H. Sun, "Image and Video Compression for Multimedia Engineering: Fundamentals, Algorithms, and Standards", ISBN 0849334918, CRC, December 1999.

[9] A. Secker and D, Taubman, "Motion-compensated highly scalable video compression using an adaptive 3D wavelet transform based lifting," Proc. IEEE International Conference on Image Processing, pp. 1029-1032, Oct 2001.

[10] Y. Nakaya and H. Harashima, "Motion compensation based on spatial transformations," IEEE Trans. Circuits Systems and Video Technology, vol. 4, pp. 339-367, June 1994.

The Effects of Vector Transform on Speech Compression

B.D.Barkana & M.A.Cay
Eskişehir Osmangazi University, Electrical-Electronics Engineering Department,
26480 Eskişehir, TURKEY
bdbarkana@aol.com
acay@ogu.edu.tr

Abstract

Until now, many techniques have been developed for speech compression. In this study, firstly, linear vector quantization is used to compress speech signals. Then, Linde-Buzo Gray (LBG) algorithm, which is used for image compression, is adapted to speech signals for compressing process. Before this process, vector transform (VT) that is defined at second chapter is applied on speech signals. After the VT, speech vectors are coded using vectoralized LBG algorithm. Inverse VT is applied to decoded data. Obtained compression results are evaluated using graphics and SNR values.

1. Introduction

Signals such as speech and image data are stored and processed in computers as files or collections of bits. Signal compression is concerned with reducing the number of bits required to describe a signal to a given accuracy. The compression of speech signals has many practical applications such as digital cellular technology. Compression allows more users to share the system than otherwise possible. Also, for a given memory size, compression allows longer messages to be stored than otherwise in digital voice storage such as answering machines. Speech compression is called speech coding. During the coding process, after high-rate data transforms low-rate data, loss or accuracy will be increased, relatively. The aim of the coding is to find the best possible accuracy for a given compression rate. Such as transform coding, predictive coding, or pulse code modulation techniques give good accuracy, relatively [1-2]. All this techniques are scalar.

For all compression techniques, modeling has three steps as shown Figure 1 [7]. First step is signal processing. At this step, original speech signals are transformed into a different domain. Second step is quantization. Third step is lossless coding. Speech compression and data loss are occurred in the second step, which is quantization. This step is the most important part of the coding. At this step, transforms during signal processing may be based on scalar or vectoral [3-8].

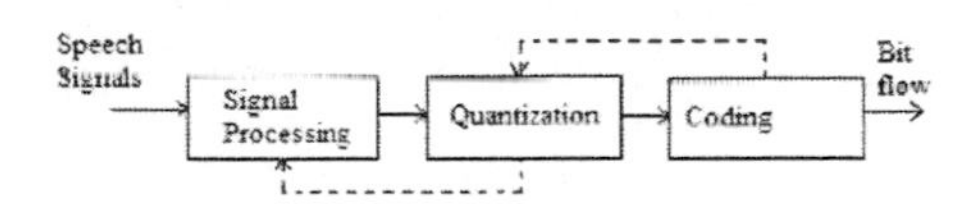

Figure 1. A model for compression.

In this study, we use vector transform (VT) which is presented second part, to prepare speech signals for quantization [4-5]. After speech signals are divided blocks, VT is applied to them. Compression is completed after these vectors are quantized.

2. Vector Transform and Speech Coding

2.1. Vector Transform

The vector transformation does not reduce the inter vector correlation as much as DCT does, but it preserves the intra vector correlation much better, which makes VQ more efficient. Vector quantization is shown using VT in Figure 2. After speech signals are divided into blocks, each block is considered like a vector. Each vector is represented with a code vector in the codebook.

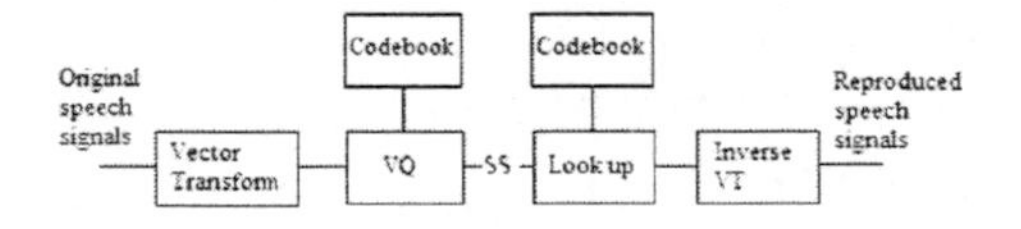

Figure 2. Vector Quantization using VT.

Codebook is given to the receiver. After look up, each set of code vectors is applied to the inverse vector transform. At the end of this process, speech signal are reproduced [5].

K. Elleithy (ed.), Advances and Innovations in Systems, Computing Sciences and Software Engineering, 67–70.

Firstly, speech signal data is divided frames and each frame is treated as a vector. These vectors can be defined;

$$X = \left[x_0, x_1, \ldots, x_{N-1}\right] \tag{1}$$

$$X = \begin{bmatrix} x_{0,0} & x_{0,1} & . & . & x_{0,N-1} \\ x_{1,0} & x_{1,1} & . & . & x_{1,N-1} \\ . & . & . & . & . \\ . & . & . & . & . \\ x_{\frac{N}{2}-1,0} & x_{\frac{N}{2}-1,1} & . & . & x_{\frac{N}{2}-1,N-1} \end{bmatrix} \tag{2}$$

$$x_0 = \begin{bmatrix} x_{0,0} \\ x_{1,0} \\ . \\ . \\ x_{\frac{N}{2}-1,0} \end{bmatrix} \quad x_1 = \begin{bmatrix} x_{0,1} \\ x_{1,1} \\ . \\ . \\ x_{\frac{N}{2}-1,1} \end{bmatrix} \quad x_{N-1} = \begin{bmatrix} x_{0,N-1} \\ x_{1,N-1} \\ . \\ . \\ x_{\frac{N}{2}-1,N-1} \end{bmatrix} \tag{3}$$

x_n signal is defined generally as equation 4.;

$$x_n = \left[x_{0,n} \quad x_{1,n} \quad . \quad . \quad x_{\frac{N}{2}-1,n}\right]^T \quad n = 0,1,\ldots,N-1 \tag{4}$$

Vector transformation is applied to these vectors to prepare for quantization. There is no data loss during the transformation. Since VT is applied to signals at the beginning, inverse VT has to be applied to signals at the end of the compression. Vector transformation and inverse VT are defined with equation 5 and 6, relatively.

$$x_k = \tfrac{1}{\sqrt{N}} \sum_{n=0}^{N-1} x_n^T W^{nk} \tag{5}$$

$$x_n = \tfrac{1}{\sqrt{N}} \sum_{k=0}^{N-1} x_k^T W^{-nk} \tag{6}$$

W is a skew-circulant matrix that its dimension is $\frac{N}{2} \times \frac{N}{2}$. W is defined as;

$$W = \begin{bmatrix} 0 & 1 & 0 & . & 0 \\ 0 & 0 & 1 & . & 0 \\ . & . & . & . & . \\ 0 & 0 & 0 & . & 1 \\ -1 & 0 & 0 & . & 0 \end{bmatrix}_{\frac{N}{2} \times \frac{N}{2}} \tag{7}$$

2.2. Bit Allocation

After VT is applied to the vectors, each vector has different energy density. Because of the different energy densities, different sized codebooks are used for vectors during the quantization. While vectors, which have higher energy density, are coded with bigger codebooks, vectors, which have low energy densities, are coded with small codebooks. To obtain codebook size and design is important for vector transform coding.

We used the "vector bit-allocation algorithm [5]" to find the bits for the vectors after the VT. For following equations, D_k; mean square distortion for k^{th} transform vector x_k.

$$D_k = Ae^{-\frac{4}{N}b_k} \quad \text{k=0,1,2,...,N-1} \tag{8}$$

A is a constant. b_k is a bit size that is allocated for k^{th} vector. Using the Eq.8, total distortion on transform vectors,

$$D = \sum_{k=0}^{N-1} \frac{\sigma_k^2}{N} D_k = \frac{A}{N} \sum_{k=0}^{N-1} \sigma_k^2 e^{-\frac{4}{N}b_k} \tag{9}$$

σ_k^2 is variance for x_k. Allocated bit number for each kth vector is given with Eq.10b. B is a total bit number for a speech block. R is an average bit number for each block.

$$R = \frac{B}{N} \quad \text{(bit/dimension)} \tag{10a}$$

$$b_k = \frac{N}{2} R + \frac{N}{4} \left(\log_2 \sigma_k^2 - \frac{1}{N} \sum_{k=0}^{N-1} \log_2 \sigma_k^2 \right) \tag{10b}$$

Calculated bits are not an integer. For this reason, they are rounded to the nearest integer. The first term in the bit-allocation equation, $\frac{N}{2} \times R$, is average bit number for each vector. The term in the parenthesis can be positive or negative depends on the variance value. There are three important things in the bit-allocation equation. These are;

1. b_k is generally not an integer.
2. If σ_k^2 is very small, b_k can be a negative.
3. If σ_k^2 is very big, b_k can be a positive integer [5].

In addition, in this study, we design a codebook using vectorilized LBG algorithm. Therefore, each speech vector are coded using different size codebook. Before the LBG algorithm [6], we used vector transformation for speech signals. The reason for using the VT before the LBG algorithm is so that vectors having a higher energy density will be represented with a larger codebook. Speech vectors are less correlated after the VT.

The signal to noise ratio of the reproduced speech signals is given with Eq.11.

$$SNR(dB) = 10\log_{10}\frac{\sum_{n=0}^{V-1} x[n]^2}{\sum_{n=0}^{V-1}\left(x[n]-\hat{x}[n]\right)^2} \qquad (11)$$

$x[n]$ is an original speech signal, $\hat{x}[n]$ is a reproduced speech signals.

3. Compression Results

In this section, results of the coding techniques in this study are presented and compared. Original speech signals that are studied are given with Figure 3.

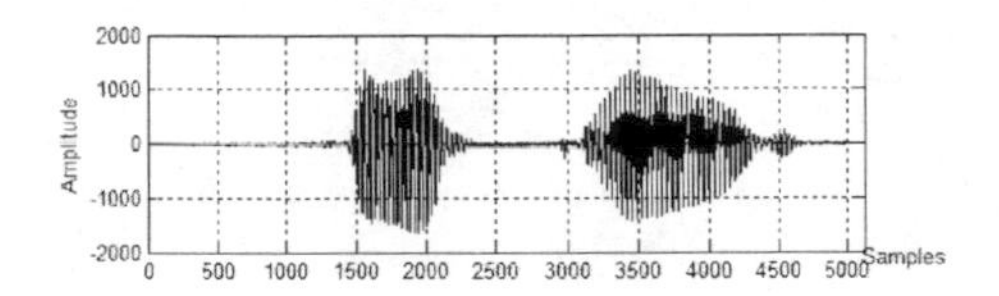

Figure 3. Original speech signals.

Original signals have 5120 data that is 40 frames, which have been studied in this paper. These signals were sampled at 9600 Hz.

3.1. Vector Quantization

Speech signals, which have 40 frames, are coded with VQ method. For 256 vectors included 20 elements, 512 vectors included 10 elements, and 1024 vectors included five elements, are coded with 1:16 compression rate. Codebook sizes are 16, 32, and 64 relatively. These coded signals are given with Figure 4. SNR values are 5.3907 for 256 vectors, 8.7052 for 512 vectors, and 14.1594 for 1024 vectors.

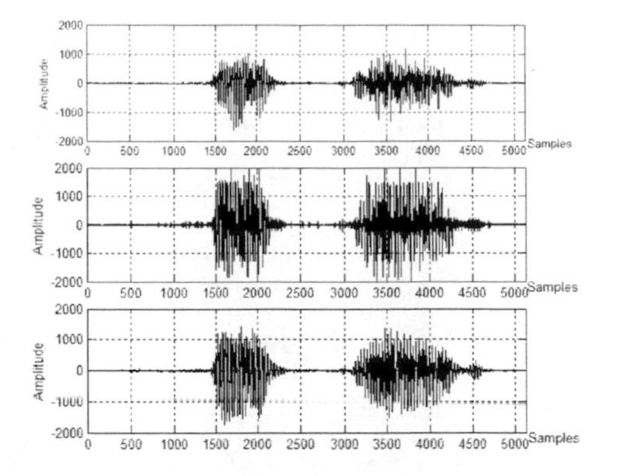

Figure 4. (a) Compression rate is 1:16 for 20x256, SNR=5.3907 (b) rate is 1:16 for 10x512, SNR=8.7052 (c) rate is 1:16 for 5x1024, SNR=14.1594.

3.2. LBG Algorithm

Speech signals are coded with 1:122, 1:64, 1:32, and 1:16 compression rate using LBG algorithm, which does not include VT, are given with Figure 5. N value is chosen as 4. SNR values are 14.7612, 15.9155, 18.0576, and 22.0292 relatively.

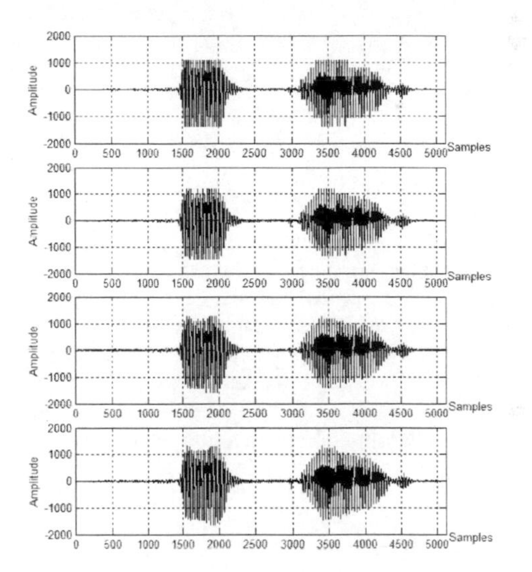

Figure 5. Compression results for LBG algorithm. N=4. (a) Compression rate 1:122, SNR=14.7612 (b) 1:64, SNR=15.9155 (c) 1:32, SNR=18.0576 (d) 1:16, SNR=22.0292.

3.3. Vectorilized LBG Algorithm

Speech signals are coded with 1:122, 1:64, 1:32, and 1:16 compression rate using LBG algorithm, which includes VT, are given with Figure 6. N value is chosen as 4. SNR values are 9.9798, 11.6234, 15.3492, and 19.1503 relatively.

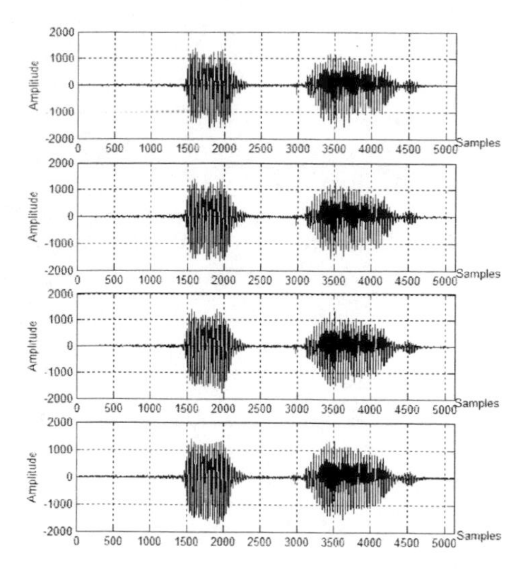

Figure 6. Compression results for Vectorilized LBG algorithm (a) Compression rate 1:122, SNR=9.9798 (b) 1:64, SNR=11.6234 (c) 1:32, SNR=15.3492 (d) 1:16, SNR=19.1503.

4. Conclusion

For linear quantization, when speech signals are represented with more vectors, reproduced signals are much closer to the original speech signals. In addition, SNR values are increased. Compression using LBG algorithm has much better results than VQ. The reason for this, each speech vector is coded using different size codebook. Surprisingly, after VT is an applied speech signal, LBG algorithm's compression results are not better than the without VT. However, SNR values and compression rate are still better than VQ.

Vectorilized LBG algorithm has advantages for speech signal compression despite of VQ.

5. References:

[1] J.A. Fuemmeler, R.C. Hardie, and W.R. Gardner, "Techniques for The Regeneration of Wideband Speech from Narrowband Speech", EURASIP Journal on Applied Signal Processing, Vol. 2001 (2001), Issue 4, pp. 266-274.

[2] S. Van de Par, A. Kohlrausch, R. Heusdens, J. Jensen, and S.H. Jensen, "A Perceptual Model for Sinusoidal Audio Coding Based on Spectral Integration", EURASIP Journal on Applied Signal Processing , Vol.2005 (2005), Issue 9, pp.1292-1304

[3] J. Makhoul, S. Roucos, and H. Gish, "Vector quantization in speech coding", IEEE, Vol.73, pp.1551-1558, November 1985.

[4] W. Li, "On Vector Transformation", IEEE Transactions on Signal Processing, Vol.41, No.11, pp.3114-3126, November 1993.

[5] W. Li, "Vector Transform and image coding" IEEE Transactions on Circuit and Systems for Video Tech., Vol.1, No.4, pp. 297-307, December 1991.

[6] Y. Linde, A. Buzo, and R.M. Gray, "An algorithm for vector quantizer design", IEEE Transactions on Communications, Vol.28, No.1, January 1990.

[7] W. Li and Y. Zhang, "Vector-based signal processing and quantization for image and video compression", Processing of the IEEE, Vol.83, No.2, February 1995.

[8] R.M. Gray, "Vector Quantization", IEEE ASSP Magazine, Vol.1, pp.4-29, April 1984.

Software Development Using an Agile Approach for Satellite Camera Ground Support Equipment

D. dos Santos, Jr.
Opto Electronics S/A
1071, Joaquim A. R. de Souza
São Carlos, SP, 13563-330, Brazil
daniel@opto.com.br

I. N. da Silva
University of São Paulo
400, Av. Trabalhador Sancarlense
São Carlos, SP, 13566-590, Brazil
insilva@sel.eesc.usp.br

R. Modugno, H. Pazelli, A. Castellar
Research and Development (R&D)
Opto Electronics S/A
1071, Joaquim A. R. de Souza
São Carlos, SP, 13563-330, Brazil

***Abstract* - This work presents the development of the software that controls a set of equipments, called Ground Support Equipment (GSE), which verifies requirement fulfilling and helps integration procedures of the CBERS-3 and 4 satellites' Multispectral Camera (MUXCAM). The software development followed an iterative spiral model, with agile methods characteristics that were originally used at Opto Electronics in industrial and medical equipment's projects. This approach allowed a small team, constituted by only four engineers, to fast create the first software version, even sharing time with GSE's hardware development, and to keep the project on schedule, in spite of some requirement changes.**

I. Introduction

Opto Electronics S/A won in 2004 the auction for developing and producing the Multispectral Camera (MUXCAM, Fig. 1) of the China and Brazil Earth Resource Satellites (CBERS) 3 and 4. The MUXCAM Ground Support Equipment (MUX-GSE) is a complex set of test equipments that also integrates the project. It is composed by several measurement equipments and an optical bench that allows evaluation of requirements and helps during assembly stage. The MUX-GSE is also the responsible for approving MUXCAM for launching, performing the last automated tests, after camera and satellite integration

The MUX-GSE was specified to work in the most automated possible way, using an industrial computer as general controller. This computer should be equipped with the necessary software to command all the systems' components and to perform all the specified test procedures. The development of this and other auxiliary software needed to be performed in the available short term, without prejudicing the quality and reliability of the system. As it was the first Opto's aerospatial project, the development strategy applied to medicals' and industrials' areas products had to be adapted, however, it was avoided to perform big changes, since the practices were known for all the team and had already proved its performance in the past.

Figure 1. MUXCAM – Structural and Thermal Model

In this section we will briefly present the CBERS program and its importance. Section II contains some of the characteristics of the MUX-GSE. In Section III, we will present information about the agile methodologies of software development. Section IV shows the strategy applied to develop MUX-GSE's software, comparing with the agile development's techniques. Finally, we will show the obtained results and the conclusions about the performed work, emphasizing the qualities and flaws of the applied methods.

A. The CBERS Program

The China and Brazil Earth Resource Satellite (CBERS) is a cooperative pacific program signed in 1988, when was accorded the development of two advanced remote sensing satellites. Two satellites were produced in this first program stage. CBERS-1 was launched on October 14th 1999 and operated successfully until August 13th 2003, almost two years beyond its projected lifetime, which was only two years. The CBERS-2, for its time, was launched on October 21st 2003 and is working until now, in spite of some limitations. It was also projected for a two years lifetime.

These two satellites are identical, equipped with three cameras that together allow acquiring images in spectral bands[1] ranging from blue to thermal infrared and four different spatial resolutions[2], from 260m up to 20m. They allowed a large number of earth resource researches in Brazil. For example, only in the XII Brazilian Remote Sensing Symposium [1] more than seventy of the presented works applied CBERS images.

The Brazilian Government considers the CBERS' images as a powerful instrument to scientific community developing

[1] Spectral band is the spectral electromagnetic range, comprising the set of frequencies (or wavelengths) for that each individual camera sensor operates.
[2] Spatial resolution is the size that each image pixel represents in real world.

K. Elleithy (ed.), Advances and Innovations in Systems, Computing Sciences and Software Engineering, 71–76.

works about water quality, cartography, agricultural, deforestation, etc. Thereby, it decides to free distribute the images to national users and to charge only operational costs to international users, increasing significantly the number of viable applications. This strategy makes Brazil the bigger satellite's images distributor in the world. Other countries have also demonstrated interests about the CBERS' images. One proof is that the United States Geological Survey (USGS) has already accomplished reception tests to acquire the images of CBERS-2 and it is possible that they become users of the CBERS satellites in the next program stages, receiving its data directly.

Due to the great success obtained by CBERS-1 and 2 and the large number of users depending on its images, Brazil and China decided to continue the program, developing three new satellites. The CBERS-2B is a copy of CBERS-2, except by one of the cameras, the Infra-Red Multispectral Scanner, which will be replaced by a panchromatic camera with better spatial resolution (2.5m). The CBERS-3 and 4, however, will incorporate a lot of technological innovations. Besides those innovations, there is also other important change to Brazilian industry: The work share is now 50% to 50%. Brazil will produce more systems than did to the first three satellites. Improving the capabilities of spatial technology's Brazilian industries was one of the main objectives of the CBERS program since its beginning.

The change in the work share brings to Brazil the opportunity of developing the MUXCAM. It is the first time that the Brazilian industry will produce a camera with its characteristics. Its success is necessary not only to the CBERS images' users, but to the whole CBERS program and, specially, to the Brazilian spatial industry, to assure the continuity of government investments.

II. MUX-Ground Support Equipment

The MUXCAM has rigid requirements to assure its reliability and image quality. It demands a lot of tests during the assembly stages and verification of several characteristics after the camera integration. These test procedures requires a large set of specialized equipment that can accomplish optical, mechanical, electrical and logical evaluation. The MUX-GSE also contains cables, tools, mechanical cases and supports, software and accessories. Some of the MUX-GSE requirements, from [2] and [3], are:

- Generate optical patterns for calibration and test procedures.
- Perform quantitative measurements of specified optical, mechanical and electronic characteristics, as modular transfer function, focal plane, radiometric parameters and image distortion.
- Receive, decode, display in real-time and store the MUXCAM images, verifying the data format and rate.
- Obtain information about focus, collimation and centering of the MUXCAM optical system and about the CCD positioning, acting as auxiliary equipment during the integration procedures.
- Perform calibration procedures and stores its data.
- Evaluate mechanical parameters as dimensions, mass and mass center.
- Evaluate electrical parameters as power consumption, operational limits and circuit characteristics.
- Command and certify the MUXCAM working in all of its working modes.
- Check the properly working of all telemetries and telecommands.

Besides that, the MUX-GSE shall execute all the verification tests proposed to the MUXCAM. These tests are determined in the requirements definition and design stages and shall be executed during its manufacture, assembly and after the system integration, to assure that the desired characteristics are reached. The test procedures shall be as automated as possible, mitigating errors caused by the operator, however, all the MUX-GSE measurement equipments shall be available to manual control, so it is possible to accomplish measurements that were not originally planned. These necessities led the development team to choose virtual instruments, instead of real bench equipments.

This way, the MUX-GSE is composed by an industrial computer and several PCI boards that allows to emulate an oscilloscope, a multimeter, a digital waveform generator, a logic analyzer and make available a bunch of digital and analogical I/O ports, used to develop the necessary interface circuits to control the MUXCAM and acquire its data.

This approach is more flexible, making the system more adaptable to possible changes in its requirements. It also permits the automation of the system and the development of software interfaces to command the PCI boards emulating the bench instruments functions. However, the use of virtual instruments increases strongly the needed software development effort and the high level of system's automatism demand more robust and reliable software, once that any measurement error can cause from project delays to damages to the MUXCAM. Worst than that, it can approve to launch a system that may have not reached the specified requirements.

III. Agile Development

The rise of the agile methodologies in the last years has shown that the production of less documentation, the transference of part of the management responsibility to developers, the creation of more simplified software, do not necessarily mean reducing the quality of the product. Even though it looks like the produced software will be less reliable and specialists are yet discussing the evolved issues [4], it was reported that, applying suitable methodology, the system could be produced faster, with complete reliability and more satisfaction for customers, developers and employers.

Agile methods as eXtreme Programming (XP), Scrum and Crystal have common characteristics, most of them summarized in the Agile Manifesto[3]. These methods valorize

[3] 17 experts in software development that worked with lightweight methodologies created the Agile Manifesto in 2001, after discussing their strategies for creating software. It defines the principles of agile development. It can be found at <http://agilemanifesto.org/>

the individuals' characteristics, the software quality, direct communication and collaboration, and the system's adaptability to changes in the requirements. Each method has its own way of reaching these qualities and each one is more appropriated to certain projects, however all of them asserts that the method should be adapted to specific companies' and projects' characteristics. The Crystal, for instance, is a family of methodologies, which states that the most suitable Crystal method shall be selected in agreement with the project's complexity level.

The agile methods are based on an iterative development model. The iterations are composed by communication stage, design, codification, tests and deployment. The time spent in each iteration changes from method to method and sometimes is not fixed, however it is always a short time, so the customer can interact with the system and give opinion as feedback. Shorter iterations assure that the system is more adaptable to changes. The definition of what shall be implemented in the next iteration generally belongs, in commercial applications, to the client. It is also possible to keep this decision in the manager hands, following a risk-driven sequence in specific applications, as originally proposed by Boehm[4].

The developers' importance is emphasized in agile development. They are responsible by all the technical decisions as the better algorithm to produce a solution, the code's parts each one will write and, sometimes, even the time needed to implement a solution. Some researchers think that this last activity may due to the manager, which can decide when to pressure the team to obtain the maximum performance [5], however most of the agile methods states that developers are the only that known exactly the time effort needed to create a piece of software and employers should rely on them to makes this definition, because it is intrinsic to them that they shall do a good job, developing the best possible software in the shortest possible time.

Communication is one of the bases of agile development. Researchers have reported that direct speaking is the most effective communication way, while diagrams and text are the less effective one [6]. The main mode of sharing information in agile methods is by meetings. They are used to get customers' ideas and feedback, to take team decisions, to delivery results, to estimate the project progress and term, etc. Meetings shall be frequent, short, and direct. They can not take most time than the needed to allow ideas discussion. Once again [5] defends that the manager should participate of most meetings, leaving the developers free to work while decisions that do not relate directly to them are taken.

Besides, developer team communication shall be encouraged, by taking out physical barriers between the teammates. Thereby they shall work in the same room, using close and large tables, which allow people sitting together to talk. XP also propose pair-programming, so the furniture in this case shall also admit two programmers sitting together, sharing the same keyboard. Close working motivates the teammates to help themselves to solve difficult problems, to use the same programming style, to share their code and to do their best all the time. It also promotes knowledge sharing and continuous learning, what tends to make the team more homogeneous and reduce the difficulties of a teammate replacement.

As agile methods are suitable for changes in the project requirements, they do not demand extensive documentation. On the contrary, they propose that only the general requirements are registered and the needed document for the current iteration shall be created. These files do not need to be complex since they will be used just during this iteration. After it finishes, the files lose its value, since the program can change anytime and the documents will became out of date. The general requirements shall be revised every iteration to absorb customer feedback and market changes. There are some agile patterns for software design that can be applied when considered valid.

System quality is assured by the selected strategies for codification and testing, together with the commitment of the programming team. While XP defend test-driven development, other methods let the team free to decide its own strategy. An alternative is to evaluate the systems' risks and select what has to test and how these tests will be executed. In any case, the program structure must be clear, keeping only the structures really needed, because next iterations probably will demand changes and the system must continue working. As the development model requires the software running at the end of each iteration, it is also necessary to accomplish a test set at the end of each cycle, to assure that the last changes do not prejudiced the older functions. If these tests are correctly defined and performed, the system keeps high level of reliability all over its life.

IV. Development at Opto

Opto Electronics is a Brazilian company that develops high technology medical and industrial equipments since 1985. Most of its products demand development of embedded software and some other projects require software graphical interfaces that run on standard PCs. The first products did not demand complex software and one or two programmers, free to apply their development strategies, were responsible by accomplishing them. Software quality was assured by the product risk analysis and the definition of a set of acceptance tests and its execution.

As projects increase in size, price, relevance and risk, more complex strategies had to be defined. Software engineers always had defined Opto's development strategies, so they were listened and a set of practices were selected, creating not a complete method, but general rules that should guide any development project. Involuntarily, the selected rules have a lot in common with agile methods. It is probably due to characteristics of most Opto's projects: Generally software teams are composed by less than six developers that share time creating also the hardware together with other engineers, and the products are medical or industrial equipments, which requirements vary all the time due to customer feedback as of market changes.

[4] Barry Boehm proposed the spiral software development model in 1985, which is the most widely known iterative model.

For that reason, software development at Opto (Fig. 2) follows the iterative Boehm's spiral model, with some adaptations. Company's direction defines the releases and the R&D director discusses possibilities and necessities with programmers. The development team defines the next iteration execution jointly with the R&D director, using information from Marketing and Sells Departments, which are directly in contact with customers.

Communication between teammates has ever been instigated. Not only software developers share the same room, but also as Opto's projects generally incorporate mechanical, electrical, optics and software systems interacting, all the researchers work together in a large room. People from the same area have their desks closer, but everybody in the project is available all the time. It stimulates people to share their problems and ideas, discuss integration issues and find the best solutions in a multidisciplinary point of view.

Electrical engineers are responsible by software development, because they have a more complete view about the system and Opto's products generally have intense hardware/software interaction. Of course, they have to prepare themselves, expanding knowledge about software engineering, algorithms, data structures, programming languages, etc. Sometimes they share time developing also the hardware, managing projects or developing more than a project at the same time. It is common in Opto's R&D structure, because all the researchers have the same hierarchical level, below the R&D manager, being selected to projects according with their competences, skills and company's necessities.

The R&D manager coordinates directly all projects. Him and all researchers have a weekly meeting, unless in exceptional cases, where project progress is evaluated, next steps are defined and major problems are discussed at system level. The software developers can show their necessities, correct requirements and expected time. If necessary, a specific software meeting can be requested. Also, the R&D manager is always available at his room and frequently is present at the development room.

The software manager is a member of the team, generally the most experienced programmer, or the one who has more

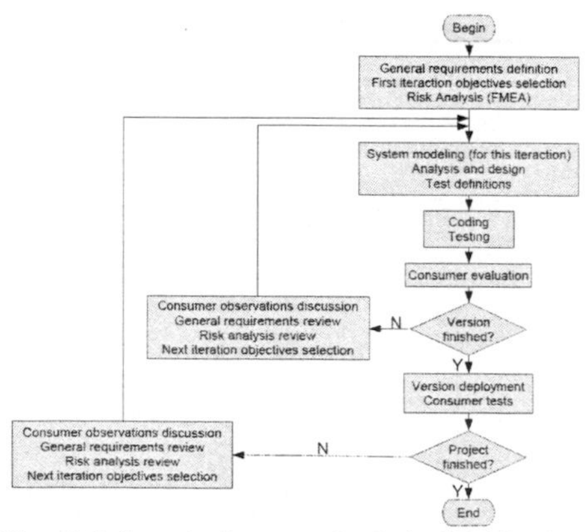

Figure 2. Software development at Opto's simplified flowchart

knowledge about the project. His responsibility is to keep the project in track, availing time, necessities and quality. Developers can organize meetings every time they consider it necessary. It occurs more frequently in planning stages and rarely during codification, however, as everybody is closer, they never stop to comment about the generated code. Anytime someone is in trouble to solve a problem, other teammates are called to help.

Customers are not ever available, as recommended by most agile methods. The Marketing and Sells Departments, however, provide R&D with information about products under development. Market researches, customer visitation and prototypes evaluations are frequently performed to provide the necessary data. Since a release is done, Marketing personal evaluates its characteristics and, if possible, some key consumers are selected to test and opine about the equipment. If the software has a user interface, it is designed in early stages and consumer ideas are incorporated as soon as possible, because changing the way the user controls the system generally implicates in hardware and software deep modifications that are quite expensive, even to agile projects.

Except in very simple programs, a software developer never works alone. Complex functions are pair-programmed and the other routines are revised and tested by someone other than its programmer. This strategy allows tracking most of errors, since experienced developers can easily detect software fragilities. However, it is not enough. Software risks are availed and a Failure Mode And Effects Analysis (FMEA) [6] is generated. From the FMEA is defined a set of tests that have to be accomplished before every release to guarantee the software reliability.

The program code is created using a unified style, defined by the developers in the project beginning. It assures that everyone can understand and modify other's code if necessary. An integration path is defined, so routines created isolated can be linked to the entire system and its working reviewed. Code comments are used when can help algorithm understanding and avoided when unnecessary, keeping the code clean. Other documents are created only to help the execution of the current iteration. Manuals, helps, reports and other documentation are generated only after a version release and are not updated until the next one. It avoids unnecessary efforts during intermediary iterations. Each developer has its own project book, which he can take its individual notes registering meetings decisions, test results, ideas, problems detected, etc. Opto's quality management system defines some standards to project book notes, so in spite of being personal books, its information can be shared and understood by others teammates.

V. MUX-GSE's Software Development

The development of MUX-GSE's electronics and software started in September 2005. Four electrical engineers, authors of this work, were chosen to carry out the project. The first meetings were used to define methodology, development platform, hardware characteristics and its implications in software. The optical bench equipments were already being studied, so it was necessary to discuss project's details with the physicists.

After that, we decided to use virtual instrumentation as presented in Section II. We then evaluate PCI boards for data acquisition, instruments simulation and data I/O and decided to use National Instruments' products. This, due to the necessity of fast development and the low programming experience of two of the teammates (which is a risk factor to software quality [7]), took us to decide by applying LabView as development platform. It allows fast learning, prototyping and virtual instrumentation development. It has a lot of libraries and sample programs that can increase the programming speed.

The same time we discussed the more suitable development method. Classical waterfall model could be used, since it is a project with well-defined requirements, and [6] shows that this kind of project can reach success with this strategy. However, we considered that the defined requirements ([2] and [3]) were just about the system, not specific for software, besides we had no experience developing this kind of product and had never applied LabView for high level software development. More than that, it was the first time that this kind of equipment was developed in Brazil, so nobody had a clear idea of what had to be done. At the same time, the MUXCAM was also an experimental product, without a Brazilian predecessor, and could need different tests from MUX-GSE than the originally defined. Considering all these information we agreed that an iterative method was more suitable to the project and then, to avoid more risk factors than the already existent, we decided to adopt the same methodology used successfully to develop medical and industrial equipment, with the fewer possible adaptations.

After that key decision, we determined that the first step should be to deeply learn the LabView programming. For that, we selected four independent modules like LAN communication and digital waveform generator software interface to develop using a code and fix strategy, each of the developers working independently. After the first week we had a meeting to discuss possibilities and problems. We then defined programming style based on [9], integration strategies and path, project parts and initial work share, and restarted the software applying our agile method. All the code created during the first week needed refactoring, because of its low quality, since it was created without the necessary planning.

The next iterations were used to develop independent equipment interfaces as digital waveform generator, logic analyzer, power supply (controlled through GPIB interface) and digital and analog I/O. These equipments were created with essential controls, but without excessive options, since it was not much relevant, and could be done in late stages if necessary. These first routines, called VIs in LabView development, were tested and approved. At the same time the MUXCAM's development team and we evaluated the oscilloscope and digital multimeter software front panels supplied by National Instruments. They were considered enough for our necessities and consequently we didn't need to create control software for these virtual instruments.

Once the programmers were finishing their first work modules, we were deciding the next steps. This time we considered ourselves ready to create more complex VIs, and then started the development of important prototypes, as routines to acquire MUXCAM images and store the data in files. The data rate demands the maximum of hardware capabilities and software could not cause delays. Various different algorithms were tested and most of them failed because of its time wasting, we then decided to pair-programming this routine. The performed tests were important because made us understand a lot about LabView memory management and execution speed ([10] also helped a lot). After several tests, we had to create a Dynamic Link Library (DLL) in C++ and access it through LabView to obtain the best performance.

At the same time, the main interface was created and started the iterative process for refining its working and appearance. This project has the advantage that the main customers are the MUXCAM developers, so they were available all the time, sharing our room. It allowed fast feedback and the interface was quickly created. The interface for the MUXCAM images has also caused some problems because of the needing for real time storing camera data, however, after using the DLL to process the acquiring data, the interface worked properly.

Next stage was the creation of automatic test VIs. It was a large number of tests, more than thirty, and some of then were composed by sequences of procedures. We decided to create a generic interface that could match all of them. The screen was designed and availed by MUXCAM developers. After finished, developers selected some tests and started the VIs creation. Optical evaluations were ignored for this time, because their hardware would not be available in the next months. Most of the implemented tests could not be immediately verified since electronic circuit was not yet completely integrated. The generic interface needed some changes, but due to system's modularity they could be easily implemented.

Software development had then an interval, because developers concentrate efforts in debugging hardware. There were thirteen interface boards, so their tests took some time. When it was accomplished, hardware commandment by software was tested. Again, debugging the system required considerable time. The next step was to execute the automated test routines using the interface hardware and track its bugs. The fact of electrical engineers had developed the software helped a lot during this stage, the most difficult until now, because they had full knowledge about the system and could better detect error causes.

Simultaneously, routines not originally previewed had to be created to allow MUXCAM's circuit tests. It happened because the MUX-GSE requirements do not include routines to test internal boards, but creating the signals necessary to run each one of the boards individually could make possible to isolate its problems. These new routines were rapidly codified and tested, because of their urgency. Their development followed the better sequence for the MUXCAM's electrical engineers. Sharing the same space favored communication between the teams, accelerating the solution of this necessity.

After accomplishing the integration hardware/software to allow the execution of automated tests, excepting that to optical evaluation, the first release could be prepared. The whole software was tested using the FMEA data. Version 1.0 allows the verification of all required electrical parameters, telemetries, telecommands, operation modes, MUXCAM imaging acquiring, evaluation, visualization and storing and manual control by software interface of all required measurement equipments. Extra routines allows MUXCAM internal boards independent running, for easy debugging. As planned, this first release is not the complete software, but is useful. MUXCAM are applying it while next release is prepared.

Next development step is to create control interfaces for the optical testing equipments, even before having them available, and then will be possible to create the automated test routines that were not yet coded. After generating this code, integrating with hardware, which is planned to be ready in November 2006, and testing the whole system, will be possible to release the second version.

Printing, log information recording and calibration routines are other requirements that were not implemented in version 1.0. There is no prevision yet for when they will be done, because they are not very important in this stage of MUXCAM project, when the first laboratory models are being integrated and tested. Probably these requirements will only be fulfilled in the third or fourth release.

VI. Results

MUX-GSE software version 1.0 is useful, reliable and was created in short term, by only four engineers that had to share their time with hardware development. MUXCAM engineers are applying the MUX-GSE software in their tests successfully. If this version could not be done in time, their works would be truly harder, and the MUXCAM project could suffer some delays.

Small errors in version 1.0 were detected and quickly corrected, in spite of they can not prejudice test results. To avoid its repetition, new risks were added to software FMEA, and verification procedures were defined.

The users are giving opinions about software that are being collected to improve the next release. Important modifications, needed to allow immediate test execution are implemented, tested and added to version 1.0.

Next release is scheduled to November 2006. It is important to not have delays since the optical characteristics of MUXCAM's test models need to be availed. It had to be delayed since some MUX-GSE's optical equipments require clean room, with particle and temperature control. This laboratory is under construction and will be completely operational in October's last week.

VII. Conclusions

The agile approach helped to select the more important routines to be created, supporting the MUXCAM engineers in their works. The applied methodology allowed modifying development sequence to absorb unexpected changes, as delays in the clean room construction and equipment availability. It did not stop the software development because we could concentrate efforts in other routines. Besides, as we use modular development, we could create graphical interface and automated test procedures before having available hardware, just low-level interface routines were left to next stages, when the hardware will be complete.

Pair-programming is a good choice when developing complex routines, but until now, in spite of works showing its value [11], we could not implement it all the time. However, having a developer to test other work also brings good results. We found most of algorithm and implementation errors in this stage.

Communication is the key for our development method. All the R&D team sharing the same room strongly contributed to the success at the project. The constant presence of the R&D director and of the main consumer allowed the developers to better understand their desires and produce software that fits exactly their needs.

Test-driven development has not yet been applied at Opto. We want to try it, but this project did not allow us, because of its extremely short and fixed term. We believe that it can improve even more the quality of produced software.

The freedom allowed the software developers to take most of technical decisions, which deeply influenced the project quality. It brings to project different ideas, proposed by people with distinct points of view. The constant learning stimulated by Opto's direction also incorporates to the project the most recent results, which is fundamental since it is a high technologic product.

References

[1] Brazilian Remote Sensing Symposium, Goiânia, Brazil, São José dos Campos: INPE, 2005. CD-ROM.

[2] National Institute for Space Research (INPE), *CBERS 3&4 MUX Ground Support Equipment Preliminary Specification*, September 2004.

[3] Opto Electronics, *MUX GSE Specification Review*, rev. 2, June 2006.

[4] T. DeMarco, B. Boehm, "The agile methods fray", Computer, v. 35, n. 6, pp. 90-92, June 2002

[5] J. A. Blotner, "It's more than just toys and food: Leading agile development in an enterprise-class start-up" in Proceedings of the Agile Development Conference (ADC'03), 2003, pp. 81-92

[6] A. A. R. Cockburn, "Characterizing People as Non-Linear, First-Order Components in Software Development" in Proceedings of the 4th International Multi-Conference on Systems, Cybernetics and Informatics, June 2000.

[7] N. Snooke, "Model-based failure modes and effects analysis of software" in Proceedings of the 15th International Workshop on Principles of Diagnosis, June 2004.

[8] J. M. Beaver, G. A. Schiavone, "The effects of development team skill on software product quality", ACM Sigsoft Software Engineering Notes, v. 31, n3, pp. 1-5, May 2006.

[9] National Instruments, *Labview Development Guidelines*, Apr., 2003

[10] National Instruments, *Labview Performance and Memory Management*, March 2004.

[11] J. Aiken, "Technical and human perspectives on pair-programming", ACM Sigsoft Software Engineering Notes, v. 29, n.5, pp. 1-14, September 2004.

Priming the Pump: Load Balancing Iterative Algorithms

Assistant Professor David J. Powers
Department of Mathematics and Computer Science
Northern Michigan University
Marquette, MI 49855
phone: 906/227-2501
fax: 906/227-2010
dpowers@nmu.edu

12-October-2006

Abstract

Load balancing iterative algorithms is an interesting problem in resource allocation that is useful for reducing total elapsed processing time through parallel processing. Load balancing means that each processor in a parallel processing environment will handle about the same computational load. It is not sufficient to allocate the same number of processes to each processor since different processes or tasks can require different loads [1]. For iterative algorithms, load balancing is the process of distributing the iterations of a loop to individual processes [2]. This paper will analyze different methods used for load balancing. Each method will be measured by how well it reduces the total elapsed time and by algorithm complexity and overhead. Measured data for different load balancing methods will be included in this paper.

I. INTRODUCTION

The goal of load balancing a parallel processing algorithm is to minimize the total elapsed processing time. To accomplish this goal a mapping or allocation of tasks to processors is done in such a way as to minimize processor idle time and to minimize inter-process communication. There are different ways or methods that can be used in assigning tasks to processors. This paper presents details on the effectiveness of different methods for load balancing iterative algorithms.

A. Problem Assumptions

- Each iteration may generate a different computational load.
- Each iteration is an independent task that can be computed without dependency on other iterations.
- Each iteration or task does not generate additional iterations or tasks to process.
- Each processor in the parallel processing environment may have a different speed or computational capacity.
- The computational load of each task is high compared with the volume of data transferred between tasks.

B. Problem Considerations

A master-slave or boss-worker model will be used to allocate tasks to worker processors. The model uses a centralized approach since the master or boss will provide tasks to the workers. If there are $p+1$ processors in the cluster, then one processor will be used for the master or boss task and p processors are available for the worker or slave tasks. Static or dynamic allocation can be used.

- **Static scheduling**

The master or boss task evenly divides the total number of iterations or tasks between the available worker processors. This is a simple algorithm with negligible inter-process communication. This method will provide the minimum total elapsed processing time if all processors in the cluster run at the same speed and each iteration generates the same computational load. If each iteration takes elapsed time, e, to process, then the ideal total elapsed time, t_p, will be equal to

$$t_p = (e * n) / p$$

where n = total number of iterations
and p = number of processors (workers) available

Other methods of static scheduling include round robin, randomized, recursive, and genetic algorithms [3]. These static methods may provide better load balancing, but in all cases the total iterations or tasks are divided equally among the available worker processors. If the cluster is not homogeneous and the elapsed time, e, for each iteration is not the same, then processor idle times can occur waiting for longer tasks or slower processors to finish. Processor idle times will increase total elapsed processing time. In the case of different processor speeds and/or different task loads, static scheduling will not be able to divide the workload evenly among available processors. A more sophisticated mapping of tasks to processors is needed and can be obtained by dynamic scheduling.

K. Elleithy (ed.), Advances and Innovations in Systems, Computing Sciences and Software Engineering, 77–81.

- **Dynamic scheduling**

Instead of mapping all tasks to processors before processing begins, tasks are dynamically allocated to processors during execution of the parallel algorithm. The master or boss task *primes the pump* by giving each processor (worker) an initial chunk of tasks or range of iterations. When a worker completes the current chunk of tasks, the worker requests an additional chunk of tasks from the master or boss. This process continues until all tasks or iterations are completed. The chunk size can be varied from 1 task or iteration up to n / p

where n = total number of tasks or iterations
and p = total number of slave or worker processors

When the chunk size is equal to 1, then the maximum inter-process overhead occurs, since the master and slave tasks must communicate on each iteration.

When the chunk size reaches n / p, then the allocation is static and not dynamic since all the tasks are mapped to all the processors in the beginning.

If we consider the case where are processors run at the same speed, and each iteration may have a different elapsed time, e_i, then the ideal total elapsed time, t_p, will be equal to

$$t_p = (\Sigma e_i) / p$$

where p = number of processors (workers) available
and $1 \leq i \leq n$, n = total number of iterations

This ideal elapsed time assumes that the computational load is evenly divided among available processors (workers).

As with static scheduling, the different worker processors may not complete at the same time. However, when chunks are small, the idle time of the worker processors at the end of the process will also be small as compared with static scheduling (maximum chunk size). Reducing the chunk size will minimize the chance of processor idle time but will increase the amount of inter-process communication overhead [3].

Parallelization of loops by chunk scheduling is an example of the boss-worker model with centralized mapping when the tasks are statically available [4].

With dynamic scheduling, it is usually best to hand out the longer tasks first. This assumes that this information is easily known a priori. For some iterative algorithms this is the case. For example, in a prime number determination algorithm computational loads generally increase as the numbers or iteration increase. If a longer task were handed out late in the whole process, then workers that have completed shorter tasks would then be idle while waiting for longer tasks to be completed [3].

C. Performance Metrics

If we consider that total execution time can be broken down into computation and communication overhead, then

t_s = sequential execution time with no communication overhead

c = communication overhead
= actual elapsed time – ideal elapsed time

t_p = parallel execution time
$= (t_s / p) + c$

Speedup Factor = t_s / t_p

Computation/communication Ratio =
Parallel computation time / Communication Overhead =

$$(t_s / p) / c = t_s / (p * c)$$

The goal is to find the chunk size that provides the minimum parallel execution time, t_p. High values for speedup factor and computation/communication ratio are also desirable.

Note that overhead, c, may also include idle time, time for termination detection, time contending for shared resources, as well as, communication time. Communication overhead is caused by work requests and work transfers [4].

D. Test Cases

The following three (3) test cases will be implemented and analyzed:

1. Static scheduling with a chunk size of n / p
   where n = total number of tasks or iterations
   and p = total number of slave or worker processors

2. Dynamic scheduling with a chunk size of s
   where $1 < s < n / p$

3. Dynamic scheduling with a chunk size of 1.

II. IMPLEMENTATION

A. Static Scheduling

The Static Scheduling logic is quite simple and is shown below:

Static scheduling program logic:

Master Process (P_0):

```
Spawn p worker tasks.
Send message to each worker task
  containing first number and chunk size.
  (All work is assigned at this time)
Receive message from each worker that
  contains results.
Save results for each worker.
Display results for all workers.
Display elapsed time.
Exit program.
```

Slave Process P_i ($1 <= i <= p$)

```
Receive message from Master containing
  first number and chunk size.
Process all data in the range.
Send message to Master which contains
  results.
Exit program.
```

B. Dynamic Scheduling

The Dynamic Scheduling logic more complex and is shown below:

Dynamic scheduling program logic:

Master Process (P_0):

```
Spawn p worker tasks.
// priming the pump
Send initial message to each worker task
  containing first number and chunk size.
While more chunks left to do
{
     Receive message from worker that
       results.
     Save results.
     Send another chunk to the worker.
}
Receive final message from each worker
   task that contains results.
Send termination message to each worker.
Display results.
Display elapsed time.
Exit program.
```

Process P_i ($1 <= i <= p$)

```
Receive message from Master containing
  first number and chunk size.
While more work to do
{
      Process all data in range.
      Send message to Master which
             contains results.
      Receive next chunk from Master.
}
Exit program.
```

Compared to static scheduling, the dynamic scheduling algorithm is not only more complex with additional loops required but also requires much more inter-task communication between master and slave tasks.
Each algorithm requires two (2) messages for each chunk of data or range of iterations processed. Figure 1 below shows the master/slave communication required.

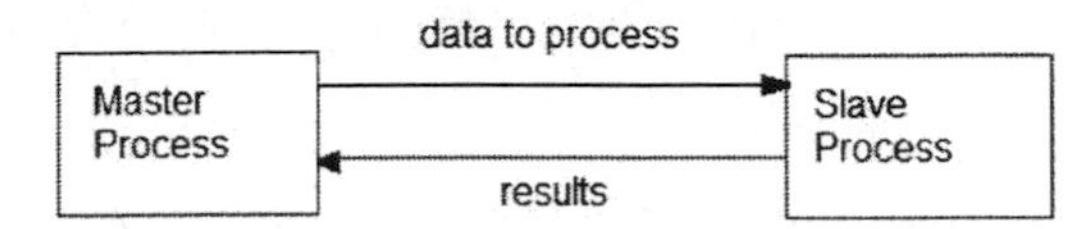

Fig. 1. Master/Slave communication for each chunk of data.

For static scheduling with *p* slave processors, only 2 * *p* messages are required since only *p* chunks of data are processed. For dynamic scheduling, the number of messages depends on the chunk size. For example, Table 1 below lists the number of messages required when p = 2 and n = 2,000 where n = number of iterations.

chunk size	number of chunks	number of messages
1	2,000	4,000
2	1,000	2,000
4	500	1,000
10	200	400
20	100	200
40	50	100
100	20	40
200	10	20
1000	2	4

Table 1. Number of messages required as a function of chunk size.

C. Iterative Algorithm

The iterative algorithm selected for testing is the prime number routine. This algorithm tests each number in a range to determine if it is prime. Larger numbers usually take more processing time, but non-primes are found much quicker than primes. So there is a certain randomness to the amount of computation time required per iteration. Also, each iteration can be processed independently of any other iteration so that no task dependencies exist. Even though the prime number routine is a very specific algorithm it represents a general class of iterative algorithms where each iteration generates a different computational load and where there are no task dependencies. For these reasons, the prime number routine is an appropriate algorithm for testing.

III. TEST RESULTS

This paper presents different methods used to allocate tasks or iterations to available worker processors. The best method is the one that minimizes the total elapsed time. The test results are displayed in Figure 2 below. An analysis of the results is summarized as follows:

The sequential execution time, t_s, is determined for 2000 iterations of the prime number algorithm:

$$t_s = 176 \text{ seconds}$$

So, for p = 2 processors the ideal parallel processing time with no overhead would be 88 seconds, and the speedup factor would be 2.0, since

$$t_p = t_s / p = 176 / 2 = 88 \text{ seconds}$$

$$\text{speedup factor} = t_s / t_p = 176 / 88 = 2.0$$

A. Method 1
Static scheduling with a chunk size of 1000 produces an elapsed time of 96 seconds which means there are 8 seconds of overhead (96 – 88). In this case, the overhead is due to idle time of one of the processes waiting for the other process to finish the chunk of iterations.
A minimal amount of communication overhead occurs when the chunk size equals 1000 (refer to Table 1).

B. Method 2
Dynamic scheduling with a chunk size greater than 1. Chunk sizes 2-10 (refer to Figure 2) provide for an ideal parallel processing time of 88 seconds and a speedup factor of 2.0. Even with a high amount of communication, there is negligible overhead incurred since the computation per iteration is large compared to the communication.

C. Method 3
Dynamic scheduling with a chunk size equal to 1 incurs a slight communication overhead of 1 second due to the large number of messages that must be transmitted between the master and slave tasks.

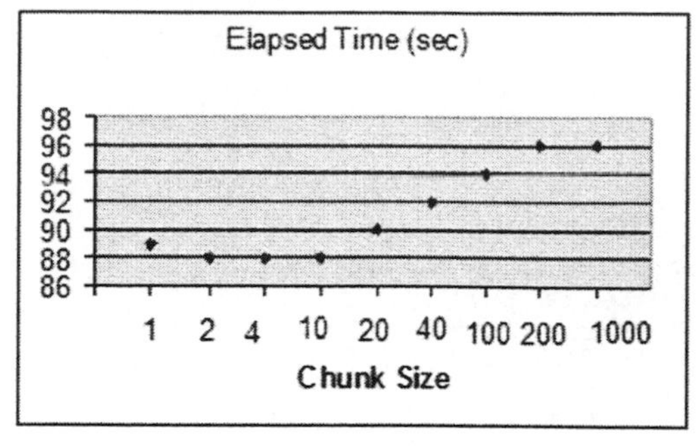

Fig. 2. Elapsed Time in seconds for different chunk sizes using a 2-processor cluster and prime number algorithm

IV. CONCLUSIONS

- Method 1, static scheduling works well if worker processors are homogeneous (same speed) and each iteration of the iterative algorithm requires about the same amount of computation. The load can then be balanced by sending the same number of iterations to each processor. More generally, static scheduling will work well if the chunk size is apportioned by processor speed and the amount of computation per iteration averages out over the large chunks used for static scheduling.

- Method 2, dynamic scheduling with chunk sizes greater than 1. This method reduces the amount of inter-task communication as compared to method 3 with a chunk size equal to 1. In the test case above, Method 2 provides the best parallel elapsed time even though a relatively high number of messages are required.

- Method 3, dynamic scheduling with a chunk size equal to 1. This method requires the most inter-task communication and overhead. For this reason, Method 3 has slightly more overhead than Method 2 which requires less than half the messages that Method 3 requires.

- The algorithms for mapping tasks to processors when task dependencies exist would be much more complex than the master/slave algorithms presented in section II. of this paper.

- Static and dynamic scheduling methods are unique for classes of or individual iterative algorithms. The methods presented in this paper and the results obtained only apply to the specific iterative algorithm tested and any other similar iterative algorithms.

- The centralized dynamic scheduling model used here works well if there are few work processors and the tasks are computationally intensive. The potential exists for a bottleneck when one master task is servicing many worker processors. In this case it may be more appropriate to use a decentralized or distributed model for dynamic load balancing.

- The two main components of overhead in parallel processing are idle time and inter-task communication. For large chunk sizes inter-task communication is reduced and the chance of idle time increases. For small chunk sizes inter-task communication increases and the chance of idle time decreases. The impact of communication overhead depends on the computational load per iteration compared to the communication time. For a particular iterative algorithm a chunk size is desired that will provide the minimum parallel elapsed time.

ACKNOWLEDGMENT

I would like to thank Professor Barry Peterson for his advice and suggestions.

REFERENCES

[1] Gregory R. Andrews, *Concurrent Programming, Principles and Practice*, Addison-Wesley, 1991.

[2] Scott Oaks, Henry Wong, *JAVA Threads*, O'Reilly, 1999.

[3] B. Wilkinson, M. Allen, *Parallel Programming*, Pearson-Prentice Hall, 2005.

[4] Grama, Gupta, Karypis, Kumar, *Introduction to Parallel Computing*, Addison-Wesley, 2003.

[5] C. Hughes, T. Hughes, *Parallel and Distributed Programming Using C++*, Addison-Wesley, 2004.

An Ontology for Modelling Flexible Business Processes

Denis Berthier
GET/INT (Groupe des Écoles des Télécommunications)
9 rue Charles Fourier, 91011 Evry Cedex, France

Abstract: With the aim of developing information systems better fitted to the main challenges raised by globalisation, we propose an ontology for the modelling of interoperable and flexible business processes. We distinguish three types of Activities a Process can be made of: whereas Procedures are defined by a sequence of Tasks and Services by a Service Description, Interactions are specified by a Goal and ruled by a Social Convention. Correlatively an Actor can have three different Statuses: Performer, Provider or Agent.
Keywords: information system, business process, interoperability, agent, modelling, ontology

I. Introduction

In the general context of globalisation and in a resulting highly competitive and rapidly changing world, information systems (IS) are facing two major tightly interconnected challenges: interoperability (IS of different institutions, or separately developped parts of the IS of the same institution, must communicate or even cooperate) and flexibility (IS must be able to be easily adapted to the frequent changes a reactive institution must make in its business processes).

In the last decades, the process view has been playing an increasing role both in organizational theories and in the IS area. Process modelling is recognized as a key element when representing the behaviour of an IS [1]; an IS process is considered to be an information oriented view of a business process [2]; and successful IS design starts with business process modelling [3]. However, most process models were initially motivated by operationnal processes and, based on the subsequent generic definition of a process [4], they give a central place to the concept of an activity; moreover the way activities are considered in this classical context limits the reuse of process models and it hides the dimension of communication between the actors. But, this dimension tends to become more and more important as one needs to model more and more complex processes, which are also more and more interactive (project management, decision processes, innovation processes,...).

In previous work ([5] and [6]), we explained why, when one wants to facilitate the modelling of business processes better fitted to the two challenges mentionned at the start, it is natural to take inspiration in the multi agent systems (MAS) paradigm and we reviewed some of the existing work in this direction ([7, 3, 8, 9, 10, 11, 12, 13, 14]). We also developed the first sketch of an ontology, represented as an UML metamodel, coherent with classical approaches (in the sense that it is an extension of them) such as described in [15].

The present paper aims at being as self sufficient as possible. Part II gives a general overview of the ontology. Part III distinguishs three types of Activities a Process can be made of: whereas Procedures are defined by a sequence of Tasks and Services by a Service Description, Interactions can be specified by a Goal and ruled by a Social Convention; correlatively, we distinguish three Statuses an Actor can have: Performer, Provider and Agent. Part IV concentrates on what are probably the most significant concepts of this paper for flexible IS: the concepts of an Interaction and of a Social Convention, that allow for much flexibility.

Conventions used troughout the paper:

1) in order to differentiate them from the same words used with their natural meanings, formal concepts introduced in our ontology are consistently written with capital letters;

2) where it is defined (which may not be the first time it appears in the text), a concept of our ontology is in bold fonts; of course, this does not mean that definitions are independent of each other;

3) unless otherwise stated (arrows on the arcs), UML graphs are read from left to right and from top to bottom.

II. Graphical representation of our ontology

In our ontology, the notion of a **Process** occupies the highest level. In the most classical tradition[1], we consider it as a coordinated set of interoperating **Activities**, which are assigned to Actors; a Process is motivated by a **Purpose**, the full meaning of which resides at the organization level, where it corresponds to strategic orientations of the institution. An Actor is an active element (human being, organizational entity or software component) involved in some of the Activities of a Process (we therefore assume from the start a distinction between active elements and passive ones – that can be input

[1] "A process is a partially ordered set of activities of a business executed so as to reach a business goal" ([4]).

K. Elleithy (ed.), Advances and Innovations in Systems, Computing Sciences and Software Engineering, 83–88.

or outpout to processes, resources, ...). An Actor can be internal or external to the institution and a Process can be executed by one or several Actors.

We distinguish three different types of Activities, with three corresponding different Statuses for the Actors involved in them. The Status is related to the kind of autonomy expected from the Actor while he performs this Activity; it should not be confused with the much more specific notion of the Role(s) an Actor can have in some Activity(ies) (see later). A given Actor can have multiple Statuses if he participates in Activities of various types: this should allow for more flexibility in the organization of the institution. Finally, a special fourth Status for an Actor is introduced: Pilot; a Process has a unique Pilot, in charge of its management.

Before entering into more details, let us display our ontology in the UML "metamodel" of Fig. 1 (next page).

III. ACTIVITIES AND ACTORS

A. *Type of an Activity: Procedure, Service or Interaction*

A.1 Procedure

A **Procedure** is a kind of Activity defined by the Tasks it is composed of. Generally, it is an *ad hoc* Activity, designed for a specific Process, and it has therefore very low reusability. The Tasks sequence (including possible predefined variants) of a Procedure may be specified by an UML sequence diagram or by a task graph or by a Petri net. An Actor to whom a Procedure is assigned has (in the context of this Activity) the Status of a mere Performer: he has no autonomy for modifying this Procedure.

Procedures are adapted for the modelling of operational Processes; they are not adapted for more complex Processes.

A.2 Service

An Activity of type **Service** is described by a Service Description. An Actor who provides a Service has (in the context of this Activity) the Status of a Provider. In the context of the Process being modelled, an Activity of type Service is not supposed to be described with more detail than: 1) what is specified by its Service Description, 2) the designation of the Provider responsible for providing it (he can be internal or external to the institution) and 3) the Contract to which this Provider is submitted. The **Service Description** enunciates fixed constraints, that are definitory of the Service. The **Contract** defines constraints specific to a given instance of the Service and the Provider, that can vary from one instance to another (delays for deliverables, prices,...). A typical example of a Service, in case it is a software component, might be a Web service (in which case, the Service Description might ultimately, at the technical level, be formalized in the WSDL language). Notice that, even though he has no option for modifying the Service Description, a Provider has much more autonomy than a Performer: he is totally responsible for the means and the methods he uses to provide the Service and satisfy his Contract. A Service is thus the analogue of a black box.

A.3 Interaction

An **Interaction** is an Activity defined by its *unique* Goal (we insist on uniqueness – see further on for explanations); this unique Goal can be satisfied only with the participation of several Actors, whose Statuses are then that of Agents. Although an Interaction is not specified procedurally by rigid Tasks that would define it in detail, it is nevertheless constrained by an organizational framework, i.e. the actions of the various participating Agents are regulated, both externally and internally:

– externally: since the Interaction occurs, as any other type of Activity, in the framework of a given Process, it is steered via **Steering Indicators** (for instance: a planning of deliverables, indices of quality for the output of the Interaction,...), that will be checked by the Pilot of the Process;

– internally: the relationships and interactions between the Agents participating in the Interaction are defined, in a way that allows for much flexibility, via the concept of a Social Convention, which will be discussed in section IV.C.

The concept of a **Goal** allows to describe some finality, more limited than the finality attached to the concept of a Purpose (at the level of the Process). Given the Purpose of the Process, the various Activities that compose it and the various Goals associated to Interactions in it correspond to a first level of choice in the way the Process definition can be elaborated so as to reach this Purpose.

The Goal of an Interaction defines fixed general constraints common to any instance of the given Interaction. Nevertheless, contrary to a Service Description that strictly defines the expected result, a Goal can be much less precise. For instance: the Goal of Interaction "prepare an answer to a call for proposal" is not of the same nature as the Service Description "manage invoices"; it leaves open lots of possibilities and it grants the participating Agents much more autonomy (both individual and collective) than a Provider can have.

A Goal characterizes both an Interaction (an Interaction aims at reaching a Goal) and the participating Agents (a group of Agents is collectively able to reach a Goal). It should be noted that, since an Interaction is entrusted to a group of Agents, the (unique) Goal that such an Activity aims at is the Goal common to this group as a whole. The way this common Goal is decomposed into subgoals, the way those are assigned to the participating Agents and the way all this is coordinated by the

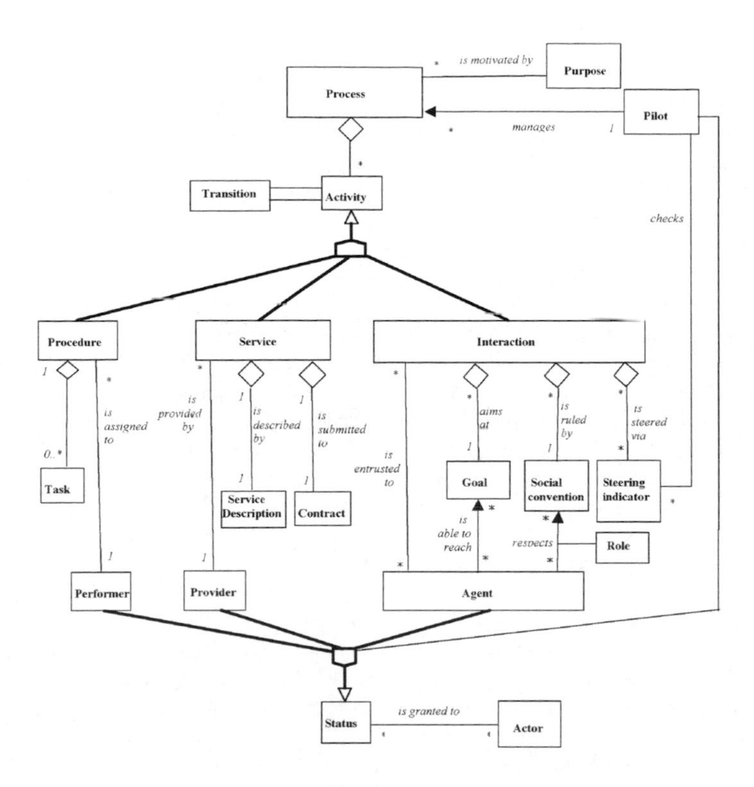

Figure 1: UML representation of our ontology

group itself depends on the Social Convention ruling the Interaction (see section IV.C). Notice that (this is a classical result) a group of Agents having a common goal means much more than each Agent having this goal as its individual goal[2] [17]. In particular, having a common goal supposes that various mutual beliefs must be held and that various types of communication, coordination and cooperation must take place for the group to reach the goal.

B. Status of an Actor: Performer, Provider, Agent or Pilot

As in the classical conception, an **Actor** is an active element that may play a role in the definition and the unfolding of a Process. Once the Activities of a Process have been defined, they are assigned to Actors. An Actor can be internal or external to the institution, and a Process can thus be executed by several cooperating partners.

[2] This must be compared with the no less classical result in logic that a proposition P being a common belief of a group means much more than each element of the group believing P individually [16].

To the three types of Activities correspond three different possible **Statuses** for an Actor: Performer, Provider or Agent. Whatever his Status, an Actor can be either a human person, a group of such persons, an organizational entity or a software component; our description at the organization level makes no difference as to this point.

Hereafter, we give some typical examples of how an Actor (in the Process model at the organization level) with a given Status can be implemented at the technical level (if it has been decided that it should be a software component), but this does not imply that there should be a systematic correspondance. For instance, for lots of reasons, there may be cases when an Agent will actually be implemented by techniques of lower level than allowed by MAS or AI.

A **Performer** is defined as an Actor with no autonomy: he is expected to accomplish the Tasks assigned to him in total conformance with their procedural description. He necessarily belongs to the institution (because an institution is not supposed to have control on *how* things are done ouside).

A **Provider** is defined as an Actor with total autonomy relative to *how* he manages to provide the Servie defined by the

Service Description, as far as he also satisfies the further constraints specified in its Contract. His autonomy is an autonomy of *means*, not of *goals*. A Provider is typically external to the institution; but it may also be an internal department, providing some peripheral services (mailing department, accounting department,...).

An **Agent** is defined as an Actor with the ability to execute Activites in an "autonomous" way, i.e. he is not told how to operate, but only which Goal he must contribute to reach within an Interaction; even his contribution in the Interaction may be formulated in very general terms. Thus, although the autonomy of an Agent is not a full autonomy of its *goals*, it is not only a mere autonomy of *means*, as is the case for a Provider. An Agent can be internal or external to the institution (since the institution is entitled to have some control on the Interaction via the Steering Indicators and via the final satisfaction of the Goal). Given the autonomy implicit in the notion of an Agent, in case one (or several) of the Agents participating in an Interaction is a software component, its implementation at the technical level may require MAS and/or AI techniques.

A special Status is introduced as a fourth posibility, that of a **Pilot**. Whereas the first three Statuses where attached to Activities, the Status of a Pilot is attached to the Process as a whole. This Actor is unique for each Process; he is in charge of managing the Process (it is highly unlikely that such a Status can be granted to a software component in the near future).

IV. SOCIAL CONVENTION

A **Social Convention** is a set of clauses that regulate the communication, coordination and cooperation among the Agents participating in an Interaction in order to reach its Goal. It is a set of rules and constraints that tie these Agents together in the context of this Interaction – excluding rules specific to each Agent's internal behaviour (such as his reasonning processes, his personnal motives,...). A Social Convention thus defines constraints internal to the Interaction (whereas external constraints are taken care of via the Steering Indicators). Much more flexible than a Procedure, a Social Convention has therefore much more reusability. Moreover, it grants the participating Agents much more autonomy.

The concept of a Social Convention is a very general one, which can have many specializations. What follows has no claim to exhaustivity.

A. Standard Conversation

A seemingly degenerate case of a Social Convention is a **Standard Conversation**.

In a Standard Conversation, the communication events between the participating Agents are strongly guided, although the actions they have to accomplish between these events may be very complex and require much autonomy on their part.

This concept is interesting for discussing a potential diffculty of our approach and for precising the limits of our "inspiration from the MAS paradigm". With such inspiration, it might be tempting to rely on the notion of a Standard Conversation as it is standardized by FIPA (Foundation for Intelligent Physical Agents, the international association in charge of standardizing MAS). For FIPA, a standard conversation can be defined starting from the more elementary concept of a message:

– either as an *ad hoc* graph of the possible sequences of messages between agents, using for instance AUML diagrams (Agent UML, an extension of UML – see [17]);

– or by chosing among a list of standardized communication protocols such as: Contract-net, English-auction, Brokering, cfp (i.e. call for proposal), ... (where each protocol is predefined with AUML diagrams).

But we consider it would be a major failure for us if we had to define a Standard Conversation in this way. The concept of a message, as it appears in so central a position in the MAS paradigm is much too "atomic" to mean anything at the organization level. We want the organization level concepts to appear directly in our ontology with no reference to concepts of lower levels. Concretely, one can specify the notion of a call for proposal (for instance) without having to specify in detail all the possible paths such a conversation can follow (if needed, such detail can be added in subsequent modelling phases).

Therefore, we posit in our ontology a Standard Conversation as a subtype of an Interaction. We think that some of the FIPA standard conversations can be abstracted from their technical definition in terms of sequences of messages and asserted in our ontology as specializations of the concept of a Standard Conversation.

Notice that, although a Standard Conversation might look like a Procedure, there is a major difference: a Procedure is decomposed into Tasks, that are in turn decomposed into Tasks,... In a Standard Conversation, what is defined with some rigidity is only the possible sequences of communication events between Agents. Neither the internals of each Agent's activities between such events nor the exact content of their outputs is specified.

For instance, in a cfp (call-for-proposal) Standard Conversation, the initiator of the conversation issues a call for proposal. An Agent may answer with a proposal. But, in between, he may have very complex activities, supposing a great deal of autonomy, to elaborate his answer (see IV).

B. Social Conventions based on roles

Some Social Conventions can be based on the attribution of

specific roles to each participating Agent (one could posit a Role-Based-Social-Convention as a general subtype of Social-Convention).

Consider again the following typical case of an Interaction: "prepare an answer to a call for proposal". Assume the participating Agents are the Commercial Department, the Technical Department and the Legal Department. In a company, when preparing such an answer, each of these (institutionnal) Agents naturally assumes one or more specific roles in its contribution, corresponding to its competencies. Moreover, there is naturally a special role for internal coordination of the Interaction, which will typically be assumed by the Commercial Department, in addition to its purely commercial role. This last role must be distinguished from the Status of the Pilot of the Process; the coordinator of the Interaction will generally be responsible for interacting with the Pilot of the Process. Defining the associated Social Convention of this example consists in part in formalizing all these natural roles.

Although each participating Agent has a fixed role in it, the Interaction cannot be defined by a fixed sequence of actions from them. This is a main difference with a Standard Conversation. The actual actions will be determined dynamically from the Goal. For instance, the Technical Department may be confronted to a major difficulty that requires lots of work to assess the feasability (by its company or by subcontractors) of some parts of the requirements.

C. Social Conventions in general

More generally, a Social Convention can be based on a large panel of social interaction models: negociation, game theory, planification, free collaboration between agents, remuneration of the agents according to certain economical models,... In practice, in an IS, Social Conventions will be more restrictive and more specific than such general models.

In our example "prepare an answer to a call for proposal", in addition to the attribution of a specific role to each agent as in section III.B, a Social Convention might specify that whenever a participating agent finds that he won't be able to fulfill his part of the job in time, he must warn the others immediately (and not let them discover it at the last moment); to whoever this seems too obvious to be explicited: have you have ever worked in collaboration? A Social Convention might also specify the format in which the results of each agent are delivered to the others, so as to minimize subsequent assembly work. More generally, it may refer to general rules of the institution.

These simple examples show that there is a link between the notion of a Social Convention and what has been elaborated for now three decades under the name of Quality Management. Many "quality procedures" are not actual procedures, but rules implicitly describing how various participants must cooperate to reach a common goal.

A Social Convention also defines, in a more or less direct way, the degree of autonomy of the participating Agents; actually, it is only through this concept that the notion of autonomy, that we have until now used in a very vague manner, can acquire a precise meaning: the autonomy degrees of the various Agents participating in an Interaction are defined by the Social Convention that rules their social behaviour.

There is currently a lot of research activity on social organization of agents in MAS. Simply listing them would require a full paper. Just as a brief illustration of the notion, let us cite Jennings GRATE* model ([18]). Jennings distinguishes several levels on which the various clauses of a Social Convention can bear:

- a set of rules for information communication between the Agents and for assessing the advancement of a common action plan: 1°) what type of information each agent must exchange with which other agents, in which conditions (for instance in a reactive or proactive way); 2°) what kind of reporting he has to do, to whom, when; 3°) what kind of tasks he can delegate, to whom, in which conditions. Such a convention may be enough in a hierarchically organized system, when tasks planning and delegation is done from the top (when there is no cooperation to build an action plan).
- a set of rules defining how the commitments of the participating agents (towards common goals, plans, distribution of tasks, planning,...) can be taken, re-assessed or dropped.

V. Conclusion

We have proposed a general purpose ontology for modelling flexible and interoperating business processes. On the theoretical side, it defines the top level elements of a modelling language, in words that have a meaning for the organization and its managers. Given the vagueness or the non standardization of the vocabulary in the domain, one should not underestimate the usefulness of having precise definitions for all the terms we use. But is is also important on the practical side, since this ontology becomes a guide for the analyst when he tries to model (or to re-enginner) a business process: one of the first very concrete choices he must make, for every Activity, is: will it be described in a classical procedural way, like a strictly organized set of Tasks, or as a Service or as an Interaction? Thus, our ontology introduces several new practical facilities for the modelling of processes:

- Inter-organizational processes: one can represent Processes resorting to Services that are external to the corporation. Such an Activity entrusted to an external Provider will not be described by detailed Tasks, but only by its Service Description and by the Contract with the Provider. This case typically includes such examples as Web Services and e-commerce.

- Modelling "systemic" ([19, 20]) processes: our ontology allows modelling Processes some parts of which are not structured with precise Tasks, but in which Agents can communicate and cooperate freely within the limits of their only constraints: their Goal, their Social Convention and their Steering Indicators. This is another element for building flexible Processes.

In this last case, it seems that we have specified only the "static" part of our ontology. For Interactions, one of the main problems is specifying their dynamics; at the level of detail of this paper, it may seem that our model is of little help and we still have to resort to the classical representations: UML activity or sequence diagrams, A-UML ([17]), Petri nets, BPML (Business Process Modelling Language: a language based on finite state automata, that now tends to be supplanted by BPEL4WS),... Nevertheless, the concepts of a Goal and of a Social Convention may open the door to alternatives to such semi-graphical representations – which are often very cumbersome for complex processes. For instance, for specifying complex interactions, one could use a rule or a constraint language. On the technical side, the point behind such possibilities is that, in the agent paradigm, as the communicationnal aspects of the agents are based on the pragmatico-linguistic speech acts theory, they are at a level of abstraction strongly coherent with the possibility of providing these agents with "cognitive" aspects, in the sense of Artificial Intelligence (AI). That such an approach is feasible has already been illustrated by Jennings [18] in a technical domain. We have not yet evocated the "cognitive" aspects of the agents, because they have become very classical. Given this coherency at the technical level, both aspects can be consistently transposed at the organization level of interest in this paper, combining our ontology with Newell's notion of a knowledge level and with any of the subsequent modelling techniques of AI (such as KADS: [23], [24]).

REFERENCES

[1] B. Curtis, M.I. Kellner et J. Over: Process modelling, *Communications of the ACM*, vol 35, n° 9, pp. 75-90, 1992.

[2] S. Alter: *Information Systems: A Management Perspective*, 3e éd., Addison-Wesley, 1999.

[3] K.A. Butler, C. Esposito et R. Hebron: Connecting the design of software to the design of work, *Communications of the ACM*, vol. 42, n° 1, pp. 39-46, 1999.

[4] CEN/CENELEC: ENV12204, Advanced Manufacturing Technology – Systems Architecture – Constructs for enterprise modelling, 1995.

[5] D. Berthier, C. Morley et M. Maurice-Demourioux: Using the Agent Paradigm to Improve Business Process Modelling, *Colloque ICSSEA 2005*, Paris, 29 Novembre - 1er Décembre 2005.

[6] D. Berthier, C. Morley et M. Maurice-Demourioux: Enrichissement de la modélisation des processus métiers par le paradigme des systèmes multi-agents, *Systèmes d'Information et Management*, Vol. 3, n° 10, 2005, pp. 25-45.

[7] E. Aurimaki, E. Lehtinen E. et K. Lyytinen, A speech-act-based office modeling approach, *ACM transaction on Office Information Systems*, vol. 6, n° 2, 1988.

[8] J. Castro, M. Kolp et J. Mylopoulos: Towards requirements-driven information systems engineering: the Tropos project, *Information Systems*, vol. 27, pp. 365-389, 2002.

[9] Y. Hoffner, H. Lwdwig, C. Gulcu et P. Grefen: An architecture for cross-organisational business processes, *Advanced Issues of E-Commerce and Web-Based Information Systems*. WECWIS, 2000.

[10] R. Kishore, H. Zhang et R. Ramesh: Enterprise integration using the agent paradigm: foundations of multi-agent-based integrative business information systems, *Decison Support Systems*, sept. 2004.

[11] G. Mentzas, C. Halaris et S. Kavadias: Modelling business processes with workflow systems: an evaluation of alternative approaches, *International Journal of Information Management*, 21, 123-135, 2001.

[12] I. Paik, S. Takami et F. Watanabe: Intelligent agent to support design in supply chain based on semantic web services, *Proceedings of the 4th International Conference on Hybrid Intelligent Systems*, 2004.

[13] G. Wagner: The Agent-Object-Relationship Metamodel: towards a unified view of state and behavior, *Information Systems*, vol. 28, pp. 475-504, 2003.

[14] H. Weigand et W.J. van den Heuvel: Cross-organizational workflow integration using contracts, *Decision Support Systems*, vol. 33, n° 3, pp. 247-265, juillet 2002.

[15] C. Morley, J. Hugues, B. Leblanc et O. Hugues: *Processus métiers et systèmes d'information*, Dunod, 2005.

[16] J.P. Dupuy: *Introduction aux sciences sociales, Logique des phénomènes collectifs*, Ellipses, Paris, 1992.

[17] B. Bauer et J. Odell: UML 2.0 and agents: how to build agent-based systems with the new UML standard, *Engineering Applications of Artificial Intelligence*, 2004.

[18] N. Jennings: *Cooperation in Industrial Multi-Agent Systems*, World Scientific, Singapore, 1994.

[19] P. Vidal et S. Nurcan: Coordination des actions organisationnelles et modélisation des processus, *in* [21]

[20] N. Melão et M. Pidd: A conceptual framework for understanding business process and business process modeling, *Information Systems Journal*, vol. 10, pp. 105-129, 2000.

[21] M. Wooldridge: Semantic Issues in the Verification of Agent Communication Languages, *Autonomous Agents and Multi-Agent Systems*, vol. 3, n° 1, pp. 9-31, 2000.

[22] Newell Allen, The Knowledge Level, *Artificial Intelligence*, Vol 59, pp 87-127, 1982.

[23] Schreiber Guus, Wielinga Bob & Breuker Joost, *KADS: A Principled Approach to Knowledge-Based System Development*, Academic Press, 1993.

[24] Breuker Joost & Van de Velde Walter, *CommonKADS Library for Expertise Modelling*, IOS Press, Amsterdam, 1994.

Routing Free Messages Between Processing Elements in a Hypercube with Faulty Links

Dinko Gichev

Centre of Informatics and Technical Sciences
Burgas Free University
San Stefano str.62, Burgas 8000, Bulgaria

Abstract - **An algorithm for routing free messages between processing elements in a multiprocessor system is proposed. As a basic architecture an n-dimensional hypercube is applied. Only one of the processors in the hypercube is connected with an external user. The external machine is called host processor. Bidirectional one-port links, some of them faulty at same time are applied. The algorithm can be applied on an arbitrary connected multiprocessor system.**

I. INTRODUCTION

Here we shall describe an algorithm of type "wave" for routing free communication messages between processing elements (PE's) in a multiprocessor system. Some of the links in the system can be temporary or permanent faulty. Like a basic configuration, an *n*-dimensional hypercube is used. Multiprocessor systems of hypercube type (HC-architectures) are well known and a number of useful algorithms is known, solving different optimization aspects /[1-3]/. Many authors consider the problem for describing a new type of architecture [4-6] or for applying a communication strategy [7-9] which tolerates the presence of faulty components (links and/or nodes).

In [10] routing and broadcasting algorithms are presented for an *incomplete hypercube,* i.e., for a hypercube (HC) that is missing certain of its nodes. These algorithms can be used to solve a problem on a system whose number of PE's is not a power of two, but they imposed some restrictions on the interconnection topology of the multiprocessor system. A similar idea is realized in [11].

In a more general case of an *injured hypercube* [7,8], i.e., an HC, a part of whose components are faulty, we shall note *Algorithm A1* from [7] - an algorithm for routing messages between PE's which uses only local information which any PE contains about its own faulty links. The basic restriction on the HC is that the number of faulty components in all the HC must be smaller than the dimension of the HC.

In [12] this restriction has been reduced by constructing the algorithm which can always be performed when the HC is a connected graph and without using more than the local information which any PE contains about its own faulty links. A similar configuration is considered in [13] too.

II. TERMINOLOGY

At first we shall consider a system whose number of PE's is a power of two. It is not necessary for the PE's to form a unique HC, i.e. PE's may take place at different research centers. PE's are numbered in a common manner.

Let $N = 2^n$ be the number of PE's numbered from 0 to $N-1$. Let PE_i be the PE having number *i*, $i=0,1,\dots,N-1$ and let $i=(i_{n-1},i_{n-2},\dots,i_0)$ be the binary representation of *i*. With $\oplus$, as is usual, we shall note the bitwise EXCLUSIVE-OR operation, namely

$$i \oplus j = k = (k_{n-1}k_{n-2}\dots k_0),$$

where

$$k_m = i_m \oplus j_m = \begin{cases} 1, \text{ for } i_m \neq j_m; \\ 0, \text{ for } i_m = j_m. \end{cases}$$

Definition 1. $(i_{n-1}i_{n-2}\dots i_0)$ is called *the address of* PE_i.

Definition 2. $i \oplus j = k = (k_{n-1}k_{n-2}\dots k_0)$ is called *the relative address of* PE_i *with respect to* PE_j (or vice versa). It is clear that the relative address of PE_i, with respect to PE_0 is equal to its address.

Definition 3. *The Hamming distance between i and j* (between PE_i and PE_j) is defined as the number of bits in which *i* and *j* differ, i.e.

$$\text{Hamming}(i,j) = \sum_{m=0}^{n-1} i_m \oplus j_m$$

Definition 4. $\{ PE_i, i=0,\dots,N-1 \}$ *form an n-dimensional hypercube* (n-dimensional HC) when PE_i and PE_j are connected if and only if $\text{Hamming}(i,j)=1$.

Definition 5. If $\text{Hamming}(i,j)=1$ and *i* and *j* differ only in k^{th} bit, then the link between PE_i and PE_j is called *the link on the* k^{th} *direction.*

If all the links between PE's in an HC are nonfaulty, then every PE is connected with exactly *n* of the other PE's

K. Elleithy (ed.), Advances and Innovations in Systems, Computing Sciences and Software Engineering, 89–92.

having one link on any of the directions from the 0th to the $(n-1)^{st}$.

The total number of links in an n-dimensional HC is $n2^{n-1}$, and there are 2^{n-1} links on each of the n directions.

III. DESCRIPTION OF THE CONFIGURATION

A multiprocessor system with N PE's is applied for routing free messages. The result obtained must be sent to an external user - master (host) processor which is connected with only one of the PE's (e.g., PE_0).

The class of auxiliary problems which produces messages can be described as follows. The PE's of the multiprocessor system perform some (parallel) program having for an object to register the obtaining of some result R (e.g., the coming into existence of a condition which makes the subsequent work senseless or unnecessary). It is possible to obtain the result R by any of the PE's (i.e., everywhere in the system), and it has to be sent to the external user (resp. to PE_0) as fast as is possible.

The algorithm in [12] solves a similar problem in the case when a number of links (0 or more) in the system is faulty. The maximal number of faulty links

$$MF = n2^{n-1} - 2^n + 1$$

which are tolerated by the algorithm in [12] is the maximal number of links that can be removed from the n-dimensional HC and keep the connectivity.

In view of the fact that the presence of (2^n-1) nonfaulty links does not make us sure that the HC is always a connected graph, the connectivity condition is only necessary but not sufficient. But as there exists a connected HC with only (2^n-1) links, this condition can be used as an upper bound for MF.

The corresponding value when Algorithm Al in [7] is applied is

$$MF = n - 1.$$

It is clear that even if n is not too big, the difference between both upper bounds is sensible.

Let us point out that both the algorithms tolerate up to $(n-1)$ faulty links anywhere in the HC. This is due to the fact that removing up to $(n-1)$ links cannot destroy the connectivity of an n-dimensional HC.

Now let us consider the following configuration. Let we have two (or more) HC's in two (or more) research centres having one (or more) link between their host processors (respectively between their PE_0's). Let the dimensions of HC's are $n_1, n_2, \ldots, n_k$. In a more general sense this configuration can be considered as a HC with faulty links having dimension n where $n = max\{n_1, n_2, \ldots, n_k\} + k - 1$. All of the PE's in all of the HC's are used to solve the upper mentioned auxiliary problem.

Every n-dimensional HC can be easy divided into two subcubes with dimensions (n-1) by "removing" connections between nodes on fixed direction. The following table

Dimension	Nodes	Links into the HC	Links into the first subcube	Links into the second subcube	Links between the subcubes
n	2^n	$n\,2^{n-1}$	$(n-1)\,2^{n-2}$	$(n-1)\,2^{n-2}$	2^{n-1}

shows us that even if n is not too big, the significant number of links are *into subcubes*. In that sense the proposed configuration when HC's in different centres are considered as subcubes of a *common HC* with faulty links is reasonable.

IV. DESCRIPTION OF THE ALGORITHM

A. *Operational environment*

We shall use bidirectional one-port communications between the PE's in the HC /[1]/, i.e., a processor can only send *and* receive data on *one* of its *I/O* ports at any given time. Of course, the corresponding link must be a nonfaulty one. If there is more than one claim for communication at the same time, the PE first serves the link whose direction has a lesser number (see Section II).

In the local memory of every PE for the needs of the algorithm, two n-dimensional binary working arrays - F and PF, and two binary working fields - WF and FIRST, are reserved. Their purposes are given below.

In the array F, information about links of the PE is kept, i.e., the jth bit of F is equal to zero if the link of the PE on the jth direction is nonfaulty. It is equal to one in the opposite case.

In the array PF, information about "pseudofaulty" links of PE is kept, i.e., about the directions which have already been used (see Subsection IV.B.2).

WF is equal to one if the algorithm has already been started on this PE. It is equal to zero in the opposite case.

FIRST is equal to one if the result R has been obtained by this PE and it is equal to zero in the opposite case. There may exist more than one PE with field FIRST being equal to one.

When a new auxiliary problem is received from the external user, arrays PF and fields WF and FIRST in all the PE's filled up with zeroes.

As the idea of "faulty link" can be considered in a more broad sense, namely, we can assume that the link may be faulty just for a time. If so and if PE needs a reasonable time for testing its own links, then the information in the array F may be renewed periodically (e.g., after receiving a new problem from the external user or in a period of time which is necessary for restoration of the faulty link).

B. *Basic idea of the algorithm*

1. *Structure of messages*

When some PE (e.g., PE_r) in some of the subcubes obtains the result R, it forms a message which must be sent to its own PE_0 (resp. to the other PE's in the other research centres) having the following structure

$$(d, d_1, d_2, \ldots, d_d, mess),$$

where

- d is the dimension of the smallest subcube that contains both PE_r and PE_0, so called the *spanning subcube* [7] of PE_r and PE_0. In particular, d = Hamming (r,0).
- $d_1, d_2, \ldots, d_d$ are all the directions in the spanning subcube of PE_r and PE_0, i.e., d_i, $i = 1,\ldots, d$ are positions of 1's in binary representation of r; $0 \leq d_i \leq n - 1$ for $i = 1,\ldots, d$.
- *mess* contains the result R to be sent to the external user.

After forming the message, PE_r changes the value of its field FIRST, making it equal to one and starting the algorithm.

2. *Moving messages*

After receiving a free message from one of its neighbors (i.e., this neighbor has already started the algorithm), te corresponding PE immediately stops working on the current auxiliary problem and starts the program which realized the algorithm for subsequent sending of the message.

The field WF is changed to one and thus, all the possible subsequent messages from the other neighbors are ignored.

The link on which the message is received is marked in the array PF as "pseudofaulty" to avoid sending back the message.

In view of the fact that the operational environment is "one-port" one, the algorithm uses the "wave" strategy, i.e., it sends the message in a consecutive order on all its links which are not marked either in the array F (i.e., they are not faulty) or in the array PF (i.e., they have not been used yet).

At first, PE tries to send the message on the directions which are marked in the message, i.e., the directions in the spanning subcube of *this* PE and its PE_0.

If it is possible to do that (i.e., if the corresponding link is not marked either in F or in PF), then in the message to be sent the following changes are made - the dimension of the spanning subcube decreases by d to $(d - 1)$, and the direction on which the message is sent is excluded from the message.

If not (i.e., the message at this step of the algorithm is sent on a direction out of the spanning subcube), then the dimension increases by d to $(d + 1)$, and the direction on which the message is sent is included in the message.

After sending the message on some direction, the corresponding link is marked as "pseudo-faulty" in the array PF to avoid repeated sending of the message.

Formal description of the algorithm performed by PE after obtaining the result R (FIRST = 1) or after receiving the message (FIRST = 0) from the neighbour given in [12]. It not depends on the new configuration that is proposed here.

V. PERFORMING OF THE ALGORITHM

In view of the connectivity of the *every* subcube, it is clear that after a finite number of steps, the free message will arrive from PE_r to PE_0. Then this PE_0 send the message to the other PE_0's to start the next auxiliary problem.

As the operational environment is "one-port" one, we cannot be sure that the number of steps is equal to the length of the shortest path between PE_r and PE_0. This is due to the fact that having only local information about its own faulty links, it is impossible for PE to preliminarily determine which direction (or directions) is part of the shortest path (or paths) to the PE_0.

This inconvenience can be overcome by using different methods described in [12].

VI. BASIC SHORTCOMING OF THE ALGORITHM

The basic shortcoming of the algorithm is its impossibility to determine if the free message has been sent over the shortest path and if there are any other free messages of the same type. It is possible for more than one equivalent messages to "travel" towards the PE_0 (of course, on different paths) at the same time. So, when one message is received by PE_0, there is not a reasonable possibility to stop other messages. They will "travel" in the system towards the PE_0 until being received by PE, which has already received another such message, or by "final" PE, i.e., PE with only one link on which the message is received. Of course, this "travelling" is a finite process. In a worst case, when the "external user" starts a new auxiliary problem on the multiprocessor system, all the earlier programs will be aborted.

VII. CONCLUSIONS

When the number of faulty links increases, the HC-architecture loses its advantages. Therefore, we have used the HC-architecture only like an example which is convenient and easy for formal describing of the algorithm. In a general case, the algorithm can be applied on an arbitrary *connected* configuration. But then the program realizations will be different for different PE's and will depend on the number of *I/O* ports. Format of the message will be different too, e.g., only (mess). All parts of the program which tolerate directions in the spanning subcube must be omitted. Analysis in Section V and Section VI is valid in the general case of an arbitrary connected configuration.

REFERENCES

[1] S.L. Johnsson and C.T. Ho, Optimum broadcasting and personalized communication in hypercubea, *IEEE Trans. Comput.* **C-38** (9), 1249-1268 (Sept. 1989).

[2] Y. Saad and M.H. Shultz, Data communication in hypercubes, *Journal of Parallel and Distributed Computing* 6 (1989).

[3] Y. Saad and M.H- Shultz, Topological properties of hypercubes, *IEEE Trans. Comput.* **C-37** (7), 867-872 (July 1988).

[4] L.N. Bhuyan and D.P. Agrawal, Generalized hyper-cube and hyperbus structures for a computer network, *IEEE Trans. Comput.* **C-33** (4), 323-333 (Apr. 1984).

[5] D.K. Pradhan, Fault-tolerant multiprocessor link and bus network architectures, *IEEE Trans. Comput.* **C-34** (1). 33-45 (Jan. 1985).

[6] D.K. Pradhan and S.M. Reddy, A fault-tolerant communication architecture for distributed systems, *IEEE Trans. Comput.* **C-31** (9), 863-870 (Sept. 1982).

[7] M.S. Chen and K.G. Shin, Adaptive fault-toterant routing in hypercube multicomputers, *IEEE Trans. Compui.* **C-39** (12), 1406-1416 (Dec. 1990).

[8] M.S. Chen and K-G. Shin, Subcube allocation and task migration in hypercube multiprocessors, *IEEE Trans. Comput.* **C-39** (9.), 1146-1155 (Sept. 1990).

[9] J.R. Armstrong and F-G- Gray, Fault diagnosis in a Boolean n-cube array of microprocessors, *IEEE Trans. Comput.* **C-30** (8), 587-590 (Aug. 1981).

[10] H. Katseff, Incomplete hypercube, *IEEE Trans. Comput.* **C-37** (5), 604-608 (May 1988).

[11] G.H. Chen, D.R. Duh and C.C. Hsu, An algorithm paradigm for incomplete hypercubes, *Intern. J. Comput. Math. Appl.* **22** (6) (1991).

[12] D.Gichev, An algorithm for routing messages between processing elements in a multiprocessor system which tolerates a maximal number of faulty links, *Math. Copmput. Modelling* 16 (12), 143-148, 1992.

[13] D.Gichev, Vignesh K., *Downloading Data into a Hypercube plus Star Configuration*, Proc. of ObCom-2004 Asia Pacific Conference on Parallel & Distributed Computing Technologies, VIT, Vellore, India,2004

OPTGAME: An Algorithm Approximating Solutions for Multi-Player Difference Games

Doris A. Behrens and Reinhard Neck
Department of Economics
Klagenfurt University
Universitaetsstrasse 65–67, A-9020 Klagenfurt, Austria / Europe

Abstract- **We present a new numerical tool to determine solutions of non-zero-sum multi-player difference games. In particular, we describe the computer algorithm OPTGAME (version 2.0) which solves affine-quadratic games and approximates solutions for nonlinear games iteratively by using a local linearization procedure. The calculation of these solutions (open-loop and feedback Nash and Stackelberg equilibrium solutions) is sketched, as is the determination of the cooperative Pareto-optimal solution.**

I. Introduction

Engineers and mathematicians have performed an impressive job in developing tools and methods for the control of physical, chemical, biological and other systems. Although the number of applications of dynamic optimization techniques (with one or more decision makers) in economics has grown during the last years, a success story comparable to the one that happened in natural sciences has not yet materialized. One of the reasons may be found in the lack of appropriate tools – such as easy to use software for determining solutions for dynamic non-zero-sum games – designed and formulated for dynamic systems especially appearing in economic applications.

In this paper, we describe the OPTGAME algorithm, a tool that we have developed for solving discrete-time LQ games and for determining the solutions of multi-player nonlinear-quadratic difference games arising especially in economics (cf. [1]). These are dynamic games where the objective functions are polynomials of degree two in multi-variable state and control vectors, and are to be optimized over a pre-specified period of time by each player (decision maker, controller), individually or jointly, subject to a nonlinear autonomous discrete-time system. An example for using an earlier version of the OPTGAME computer algorithm (version OPTGAME 1.0; see [2]) for analyzing strategic interdependence between two decision makers in an actual economic policy problem can be found in [3].

As in the literature on dynamic games (see [4]), we consider both noncooperative and cooperative solution concepts for dynamic games. Among noncooperative solutions, we distinguish between Nash equilibrium solutions, where no player can improve his performance by one-sided deviations from the equilibrium strategy, and Stackelberg equilibrium solutions, where the players exhibit asymmetric hierarchical roles. For both the Nash and the Stackelberg solutions of a dynamic game, we consider open-loop information patterns, where the players' strategies depend only on the initial state of the dynamic system, and feedback information patterns, where the strategies depend on the current state of the system but not on the initial conditions.

The core of the OPTGAME algorithm is the "combination" of a local numerical linearization procedure along a reference path with elements taken from the theory of affine-quadratic games, extended to meet the requirements of economic problems. For a more detailed description of the version OPTGAME 2.0, see [5].

II. The Dynamic Game Problem

In the dynamic game-theoretic problems considered by the OPTGAME algorithm, each of the decision-makers $i=1,\ldots,N$ aims at minimizing an individual intertemporal quadratic loss function of states and controls:

$$\min_{u_1^i,\ldots,u_T^i} J^i = \sum_{t=1}^{T} L_t^i\left(x_t, u_t^1, \ldots, u_t^N\right), \quad i=1,\ldots,N, \tag{1}$$

with

$$L_t^i\left(x_t, u_t^1, \ldots, u_t^N\right) := \tfrac{1}{2}[X_t - \widetilde{X}_t^i]' \Omega_t^i [X_t - \widetilde{X}_t^i], \; i=1,\ldots,N \tag{2}$$

The parameter T denotes the terminal period of the players' finite time horizon, i.e. the duration of the interaction (the game).

X_t ($t=1,\ldots,T$) in (2) denotes the stacked state vector

$$X_t := [\, x_t \quad u_t^1 \quad u_t^2 \quad \cdots \quad u_t^N \,]' \tag{3}$$

and consists of an n_x-vector of state variables

$$x_t = [x_t^1 \quad x_t^2 \quad \ldots \quad x_t^{n_x}]', \tag{4}$$

an n_1-vector of control variables determined by player 1, an n_2-vector of control variables determined by player 2, etc., i.e.

$$\begin{aligned} u_t^1 &= [u_t^{11} \quad u_t^{12} \quad \ldots \quad u_t^{1n_1}]', \\ u_t^2 &= [u_t^{21} \quad u_t^{22} \quad \ldots \quad u_t^{2n_2}]', \\ &\vdots \\ u_t^N &= [u_t^{N1} \quad u_t^{N2} \quad \ldots \quad u_t^{Nn_N}]'. \end{aligned} \tag{5}$$

Thus, the dimension of the "stacked" state vector is given by $r := n_x + n_1 + n_2 + \ldots + n_N$.

Financial support from the Austrian Science Foundation (FWF) (project no. P12745-OEK and project no. P14060-OEK) and from the Ludwig Boltzmann Institute for Economic Analyses, Vienna, is gratefully acknowledged.

K. Elleithy (ed.), Advances and Innovations in Systems, Computing Sciences and Software Engineering, 93–98.

For any player i=1,...,N, the terms $\widetilde{X}_t^i$ store the desired (target) values for all variables of the game for t=1,...,T. Equation (2) contains $r \times r$-dimensional symmetric penalty matrices Ω_t^i, weighting the deviations of states and controls from their desired levels, $X_t - \widetilde{X}_t^i$, in any time period t. The matrices

$$\Omega_t^i := \begin{bmatrix} Q_t^i & 0 & \cdots & 0 \\ 0 & R_t^{i1} & 0 & \vdots \\ \vdots & 0 & \ddots & 0 \\ 0 & \cdots & 0 & R_t^{iN} \end{bmatrix}, \quad i=1,...,N, \tag{6}$$

are of block-diagonal form, where the blocks Q_t^i and R_t^{ij} (i,j=1,...,N) are symmetric. $Q_t^i \geq 0$ are positive semi-definite, R_t^{ij} are positive semi-definite for $i \neq j$ but positive definite for $i = j$.

For determining a cooperative solution of the dynamic game, we have to define a joint objective function of all the players J as a convex combination of the individual cost functions

$$J = \sum_{t=1}^{T}\sum_{i=1}^{N} {}^{i}\, L_t^i\left(x_t, u_t^1, \ldots, u_t^N\right), \qquad \sum_{i=1}^{N} {}^{i} = 1. \tag{7}$$

The dynamic system, which constrains the choices of the decision-makers, is represented in state-space form by a first-order system of nonlinear autonomous difference equations,

$$x_t = f\left(x_{t-1}, x_t, u_t^1, ..., u_t^N, z_t\right), \qquad x_0 = \bar{x}_0, \tag{8}$$

where the initial state value, $\bar{x}_0$, is given. The $n_x \times 1$-dimensional vector z_t contains exogenous variables, i.e., variables not subject to control by any player. f is a vector-valued function, where f^{α} (α=1,...,n_x) denotes the αth component of f. For the algorithm, we require that the first and second derivatives of the system function f with respect to $x_{t-1}, x_t, u_t^1, \ldots, u_t^N$ exist and are continuous. The particular type of dynamic representation used in (8) appears frequently in economic models. It is equivalent to the discrete-time forms used in other scientific disciplines like engineering.

III. Input and Output of the OPTGAME Algorithm

A simple way to describe the OPTGAME algorithm is to treat it as a black box, where input parameters are inserted and outputs (tables, graphics) are derived.

Input of the algorithm for $t = 1,...,T$ and $i, j = 1,...,N$:

- N number of players
- T length of planning horizon
- f system function
- $x(0) = \bar{x}_0$ initial value of the multi-variable state vector
- z_t path of exogenous variables not subject to control
- $\bar{x}_{t-1}$ historical values for state variables
- $\bar{u}_{it}$ initial tentative (possibly historical) control paths
- Ω_t^i weighting matrices for the quadratic terms in the objective functions
- μ_i weights in the joint objective function (7) (for Pareto solution)
- $\widetilde{u}_t^{ij}$ values of each player i's desired levels for all control variables for $t = 1,...,T$
- $\widetilde{x}_t^i$ values of each player i's desired levels for all state variables for $t = 1,...,T$

Output of the algorithm for $t = 1,...,T,\ i, j = 1,...,N$ (for each of the five solution concepts considered):

- u_{it}^* optimal (i.e. equilibrium or solution) paths of the control variables
- x_t^* optimal (i.e. equilibrium or solution) paths of the state variables
- J_i^*, J^* quadratic loss functions evaluated along the optimal (i.e. equilibrium or solution) paths

IV. Description of the Algorithm

The OPTGAME algorithm can be summarized as follows; more details can be found in [5].

<u>Step 1. Initialize the OPTGAME program</u>: Initialize state variables with $x(0) = \bar{x}_0$, define non-controlled exogenous variables z_t for $t = 1,...,T$. Specify penalty matrices and individual targets. Define initial tentative paths for the lagged state and the control variables $\bar{x}_{t-1}, \bar{u}_{1t}, ..., \bar{u}_{Nt}$ ($t = 1,...,T$), e.g. by using historical values.

<u>Step 2. Compute a tentative state path</u>: Use the Gauss-Seidel or the Newton-Raphson algorithm, the given exogenous non-controlled variables z_t, the lagged tentative state variables $\bar{x}_{t-1}$, the tentative policy paths $\bar{u}_{it}$ ($i = 1,...,N$, $t = 1,...,T$), and the system equation f to calculate the tentative state path $\bar{x}_t$ according to

$$\bar{x}_t - f\left(\bar{x}_{t-1}, \bar{x}_t, \bar{u}_{1t}, ..., \bar{u}_{Nt}, z_t\right) = 0. \tag{9}$$

<u>Step 3. Apply the nonlinearity loop</u>: Starting with the tentative state and control paths, repeat steps A to C as defined below until the number of iterations exceeds a pre-specified number or "convergence" is reached, i.e. until the state and the control variables do not change by more than a small amount between two successive iterations.

A. Compute the time-dependent parameters of the linearized system equations.

Linearize the nonlinear autonomous system (8) at the reference values $\bar{x}_{t-1}, \bar{x}_t, \bar{u}_{1t}, \ldots, \bar{u}_{Nt}, z_t$ ($t = 1,...,T$) in a similar way as in [6] and replace it by a linear nonautonomous system as given by

$$x_t = A_t x_{t-1} + s_t + \sum_{j=1}^{N} B_{jt} u_{jt}, \quad (10)$$

with

$$A_t := (I - F_{x_t})^{-1} F_{x_{t-1}}, \quad (11)$$

$$B_{it} := (I - F_{x_t})^{-1} F_{u_{it}}, \quad (12)$$

$$s_t := \bar{x}_t - A_t \bar{x}_{t-1} - \sum_{j=1}^{N} B_{jt} \bar{u}_{jt}, \quad (13)$$

for $i = 1,...,N$, where I denotes the $n_x \times n_x$–dimensional identity matrix. The $n_x \times n_x$–dimensional matrix A_t, the $n_x \times n_1$–dimensional matrix B_{1t}, the $n_x \times n_2$–dimensional matrix B_{2t}, etc., and the n_x–dimensional vector s_t are time-dependent functions of the reference paths along which they are evaluated. The $n_x \times n_x$–dimensional matrices $F_{x_{t-1}}$ and F_{x_t} are defined by

$$(F_{x_{t-1}})_{ij} := \frac{\partial f^i}{\partial x_{t-1,j}}, \quad i,j = 1,\dots,n_x, \quad (14)$$

$$(F_{x_t})_{ij} := \frac{\partial f^i}{\partial x_{t,j}}, \quad i,j = 1,\dots,n_x, \quad (15)$$

and the $n_x \times n_k$–dimensional matrix (for $k = 1,...,N$) $F_{u_{kt}}$ is defined by

$$(F_{u_{kt}})_{ij} := \frac{\partial f^i}{\partial u_{kt,j}}, \quad \begin{matrix} i = 1,\dots,n_s, \\ j = 1,\dots,n_k, \end{matrix} \quad k = 1,\dots,N. \quad (16)$$

Here $x_{t-1,j}$ denotes the j-th element of the lagged state vector x_{t-1}, etc.

B. Perform the optimization of the quadratic objective functions under Nash, Pareto and/or Stackelberg strategies with open-loop and/or feedback information structure.

The sequence of procedures necessary to approximate the open-loop and feedback Nash and Stackelberg equilibrium solutions and the Pareto-optimal solution proceeds as follows. In each case, we start by

- *deleting the old and creating new matrices* for storing the $\sigma\times\sigma$– and $\sigma\times 1$–dimensional Riccati matrices, H_{it} and h_{it} for $i = 1,...,N$ and H_t and h_t for all time periods $t \in \{1,...,T\}$ where $\sigma = Nn_x$ for the open-loop Stackelberg equilibrium solution, and $\sigma = n_x$ for all other types of solutions discussed here. We proceed by
- *deleting the old and creating new matrices* for storing the $n_i \times n_x$– and $n_i \times 1$–dimensional feedback matrices, G_{it} and g_{it} for $i = 1,...,N$ for all time periods $t \in \{1,...,T\}$, for both types of feedback solutions and the cooperative Pareto-optimal solution. Then we proceed by
- *allocating the terminal conditions* for the Riccati matrices, H_{iT} and h_{iT} for $i = 1,...,N$ and H_T and h_T, in a similar manner as in linear-quadratic optimum control theory. Then we are ready for
- *obtaining the Riccati matrices (and – in the case of the feedback and the cooperative solutions – the feedback matrices) by backward iteration.* Starting from the terminal conditions and stepping backwards in time towards the initial node, we derive (and store) the Riccati matrices, H_{it} and h_{it} for $i = 1,...,N$, and the feedback matrices, G_{it} and g_{it} for $i = 1,...,N$, for the feedback and the Pareto-optimal solutions. We use these stored matrices and the initial conditions for the state vector, $x(0) = \bar{x}_0$, for
- *calculating the optimal control and state vectors of the underlying affine-quadratic games* by forward iteration, i.e. for $t = 1,...,T$ for each of the solution concepts. This optimal state vector can be calculated as

$$x_t^* = K_t x_{t-1}^* + k_t \quad (17)$$

for $x_0^* = \bar{x}(0)$, where K_t and k_t are defined below.

Only for the feedback equilibrium solutions and the Pareto-optimal solution, we are able to express the control variables u_{it}^* $(i = 1,...,N)$ for each player directly using the feedback matrices, G_{it} and g_{it} for $i = 1,...,N$ according to

$$u_t^{i*} = G_t^i x_{t-1}^* + g_t^i, \quad i = 1,\dots,N. \quad (18)$$

Additionally, for the feedback and the cooperative solutions, the matrices K_t and k_t can be computed as

$$K_t := A_t + \sum_{j=1}^{N} B_t^j G_t^j, \quad (19)$$

$$k_t := s_t + \sum_{j=1}^{N} B_t^j g_t^j. \quad (20)$$

Storing the values of *the control vectors* allows us to

- *calculate the state vector* according to the nonlinear system equation (8).

For each of the five different strategies we now report the details of the computation of the solutions of the underlying affine-quadratic difference game.

1. The Open-Loop Nash Equilibrium Solution

At the beginning of the open-loop Nash game, each of the simultaneously acting players makes a binding commitment to stick to a chosen policy. As long as these commitments hold, the solution is an equilibrium in the sense that none of the players can improve individual welfare by one-sided deviations from the equilibrium path.

The Riccati matrices for all players $i = 1,...,N$ and for all time periods $t \in \{1,...,T\}$ are derived by backward iteration according to the following system of recursive matrix equations:

$$H_{t-1}^i = Q_{t-1}^i + A_t' H_t^i [\Lambda_t]^{-1} A_t, \quad H_T^i = Q_T^i, \quad (21)$$

$$h_{t-1}^i = -Q_{t-1}^i \tilde{x}_{t-1}^i + A_t' [H_t^i [\Lambda_t]^{-1} \eta_t + h_t^i], \; h_{iT} = -Q_T^i \tilde{x}_T^i, \quad (22)$$

where

$$\Lambda_t := I + \sum_{j=1}^{N} B_t^j [R_t^{jj}]^{-1} B_t^{j\prime} H_t^j, \quad (23)$$

$$\eta_t := s_t + \sum_{j=1}^{N} B_t^j [\tilde{u}_t^{jj} - [R_t^{jj}]^{-1} B_t^{j\prime} h_t^j]. \quad (24)$$

With these Riccati matrices, H_{it} and h_{it} computed for all time periods $t \in \{1,...,T\}$, the optimal controls for the open-loop Nash game for all players are computed by forward iteration according to the following rule:

$$u_t^{i*} = \tilde{u}_t^{ii} - [R_t^{ii}]^{-1} B_t^{i'} [H_t^i x_t^* + h_t^i], \quad (25)$$

for $i = 1,...,N$, together with the optimal state vectors as

$$x_t^* = [\Lambda_t]^{-1}[A_t x_{t-1}^* + \eta_t] \ . \quad (26)$$

2. The Open-Loop Stackelberg Equilibrium Solution

The open-loop Stackelberg equilibrium solution assumes that the so-called leader (player $i = 1$) makes binding commitments about future policy actions, where the rational reaction functions of the so-called followers (players $i = 2,...,N$) are taken into consideration.

The Riccati matrices for all players $i = 1,...,N$ and for all time periods $t \in \{1,...,T\}$ are derived by backward iteration according to the following system of matrix equations:

$$H_{t-1} = \underline{Q}_{t-1} + \underline{A}_t' [\Lambda_t]^{-1} H_t \underline{A}_t, \quad H_T = \underline{Q}_T, \quad (27)$$

$$h_{t-1} = -\underline{q}_{t-1} + \underline{A}_t' [\Lambda_t]^{-1} [H_t d_t + h_t], \quad h_T = -\underline{q}_T, \quad (28)$$

where

$$\underline{A}_t := \begin{bmatrix} A_t & 0 & 0 \\ 0 & \ddots & 0 \\ 0 & 0 & A_t \end{bmatrix}, \quad (29)$$

$$\underline{Q}_t := \begin{bmatrix} Q_t^1 & Q_t^2 & \cdots & Q_t^N \\ Q_t^2 & 0 & \cdots & 0 \\ \vdots & \vdots & \ddots & \vdots \\ Q_t^N & 0 & \cdots & 0 \end{bmatrix}, \quad (30)$$

$$\underline{q}_t := [Q_t^1 \tilde{x}_t^1 \ \cdots \ Q_t^N \tilde{x}_t^N]', \quad (31)$$

$$d_t := [s_t + \sum_{j=1}^{N} B_t^j \tilde{u}_t^{jj} \ \ 0 \ \cdots \ 0]', \quad (32)$$

$$\Lambda_t := I + H_t C_t, \quad (33)$$

$$C_t := \begin{bmatrix} B_t^1 [R_t^{11}]^{-1} B_t^{1'} & B_t^2 [R_t^{22}]^{-1} B_t^{2'} & \cdots & B_t^N [R_t^{NN}]^{-1} B_t^{N'} \\ B_t^2 [R_t^{22}]^{-1} B_t^{2'} & 0 & \cdots & 0 \\ \vdots & \vdots & \ddots & \vdots \\ B_t^N [R_t^{NN}]^{-1} B_t^{N'} & 0 & \cdots & 0 \end{bmatrix}. \quad (34)$$

With the Riccati matrices, H_t and h_t computed for all time periods $t \in \{1,...,T\}$, the supplementary state vector, ξ_t, which consists of the state vector x_t and the costate vectors of player 1 corresponding to the costate equations of players $i = 2,...,N$, is given by

$$\xi_t^* = [\Lambda_t]^{-1}[\underline{A}_t \xi_{t-1}^* - C_t h_t + d_t]. \quad (35)$$

Then the control vector is determined by

$$u_t^* := \begin{bmatrix} u_t^{1*} \\ \vdots \\ u_t^{N*} \end{bmatrix} = \begin{bmatrix} \tilde{u}_t^{11} \\ \vdots \\ \tilde{u}_t^{NN} \end{bmatrix} - \underline{R}_t [H_t \xi_t^* + h_t], \quad (36)$$

where

$$\underline{R}_t := \begin{bmatrix} [R_t^{11}]^{-1} B_t^{1'} & 0 & 0 \\ 0 & \ddots & 0 \\ 0 & 0 & [R_t^{NN}]^{-1} B_t^{N'} \end{bmatrix}. \quad (37)$$

3. The Feedback Nash Equilibrium Solution

The feedback Nash equilibrium solution is generated by performing the following procedure: We derive the feedback matrices, G_{it} and g_{it} for $i = 1,...,N$, alternating with the Riccati matrices, H_{it} and h_{it} for $i = 1,...,N$, starting with the terminal conditions for the Riccati matrices, according to the solutions of the following set of linear matrix equations for $i = 1,...,N$:

$$H_{t-1}^i = Q_{t-1}^i + K_t' H_t^i K_t + \sum_{j=1}^{N} G_t^{j'} R_t^{ij} G_t^j, \quad H_T^i = Q_T^i, \quad (38)$$

$$h_{t-1}^i = Q_{t-1}^i \tilde{x}_{t-1}^i - K_t' [H_{it} k_t - h_{it}] + \sum_{j=1}^{N} G_{jt}' R_{it}^j [\tilde{u}_{ijt} - g_{jt}], \quad h_T^i = Q_T^i \tilde{x}_T^i, \quad (39)$$

$$D_t^i G_t^i + B_t^{i'} H_t^i \sum_{\substack{j=1 \\ j \neq i}}^{N} B_t^j G_t^j + B_t^{i'} H_t^i A_t = 0, \quad (40)$$

$$D_t^i g_t^i + B_t^{i'} H_t^i \sum_{\substack{j=1 \\ j \neq i}}^{N} B_t^j g_t^j + v_t^i = 0, \quad (41)$$

where

$$D_t^i := R_t^{ii} + B_t^{i'} H_t^i B_t^i, \quad (42)$$

$$v_{it} := B_t^{i'} [H_t^i s_t - h_t^i] - R_t^{ii} \tilde{u}_t^{ii}. \quad (43)$$

K_t is defined by equation (19) and k_t by equation (20). Using $x_0^* = \bar{x}(0)$, the state equations (17) and the feedback rules (18), the optimal state paths and all players' control paths can be determined by forward iteration.

4. The Feedback Stackelberg Equilibrium Solution

The feedback Stackelberg equilibrium solution assumes the following interaction: The leader (player 1) announces his decision rule, $u_{1t} = \phi_1(x_{t-1})$, whereas the followers (players $i = 2,...,N$) base their actions on the current state and on the decision of the leader according to the reaction function $u_{it} = \phi_i(x_{t-1}, u_{1t})$ for $i = 2,...,N$. The leader in turn considers the reaction coefficients, $\partial u_{it}/\partial u_{1t}$ $(i = 2,...,N)$, as rational reactions of the followers $i = 2,...,N$ in the optimization process. At the time of optimizing his performance, the leader considers the reaction coefficients,

$$\Psi_t^i := \frac{u_t^i}{u_t^1}, \quad i = 2,...,N, \quad (44)$$

as rational reactions of the followers $i = 2,...,N$. These reaction coefficients, Ψ_t^i $(i = 2,..., N)$, are determined as the solutions of the following set of $N-1$ linear matrix equations:

$$B_t^{i'} H_t^i B_t^1 + D_t^i \Psi_t^i + B_t^{i'} H_t^i \sum_{j=2}^{N} B_t^j \Psi_t^j = 0, \quad i = 2,...,N, \quad (45)$$

where the H_t^i denote the Riccati matrices of the feedback Stackelberg game. The associated functional relationship in recursive form is given by (38) and the matrix D_t^i is given by (42).

Given that the matrix

$$\overline{\Lambda}_t := R_t^{11} + \overline{B}_t' H_t^1 \overline{B}_t + \sum_{j=2}^{N} \Psi_t^{j'} R_t^{1j} \Psi_t^j \tag{46}$$

is invertible, for

$$\overline{B}_t := B_t^1 + \sum_{j=2}^{N} \Psi_t^j B_t^j \tag{47}$$

we can derive the Riccati matrices, H_t^i and h_t^i for $i = 1,...,N$, by backward iteration according to the Riccati equations given by (38) and (39), augmented, however, with different coefficient matrices. The feedback matrices are determined by

$$G_t^1 := -[\overline{\Lambda}_t]^{-1}[\overline{B}_t' H_t^1 A_t + \sum_{j=2}^{N} \overline{D}_t^j W_t^j], \tag{48}$$

$$g_t^1 := -[\overline{\Lambda}_t]^{-1}[v_t^1 + \overline{v}_t + \sum_{j=2}^{N} \overline{D}_t^j w_t^j], \tag{49}$$

$$G_t^i := W_t^i + \Psi_t^i G_t^1, \quad i = 2,\ldots,N, \tag{50}$$

$$g_t^i := w_t^i + \Psi_t^i g_t^1, \quad i = 2,\ldots,N, \tag{51}$$

where

$$\overline{D}_t^i := \Psi_T^{i'} R_t^{1i} + \overline{B}_t' H_t^1 B_t^i, \quad i = 2,\ldots,N, \tag{52}$$

$$\overline{v}_t := \sum_{j=2}^{N} \Psi_t^{j'} [B_t^{j'} H_t^1 s_t - B_t^{j'} h_t^1 - R_t^{1j} \widetilde{u}_t^{1j}]. \tag{53}$$

The matrices W_t^i and w_t^i (for $i = 2,...,N$) necessary for determining the feedback matrices, G_t^i and g_t^i for $i = 1,...,N$, as given by (48)–(51), are determined as solutions of the following set of N–1 linear matrix equations:

$$D_t^i[W_t^i + \Psi_t^i G_t^1] + B_t^{i'} H_t^i [A_t + B_t^1 G_t^1 + \sum_{j=2}^{N} B_t^j [W_t^j + \Psi_t^j G_t^1]] = \mathbf{0}, \tag{54}$$

$$D_t^i[w_t^i + \Psi_t^i g_t^1] + v_t^i + B_t^{i'} H_t^i [B_t^1 g_t^1 + \sum_{j=2}^{N} B_t^j [w_t^j + \Psi_t^j g_t^1]] = \mathbf{0}. \tag{55}$$

Using the Riccati matrices, H_t^i and h_t^i for $i = 1,..., N$, formally given by (38) and (39), and the feedback matrices, G_t^i and g_t^i for $i = 1,..., N$, (48)–(51), for all time periods, we can compute the matrices K_t and k_t as formally given by (19) and (20). Then, the approximate feedback Stackelberg equilibrium values of the states and the controls for the game for all players ($i = 1,...,N$) are determined by forward iteration according to the functional relationships determined by (17) and (18).

5. The Pareto-Optimal Solution

The solution of the cooperative Pareto-optimal game with N players is given by solving a classical optimum control problem: We derive the feedback matrices, G_{it} and g_{it} for $i = 1,...,N$, and the Riccati matrices, H_t and h_t, according to the solutions of the following set of linear matrix equations:

$$B_t^{j'} H_t A_t + [R_t^j + B_t^{j'} H_t B_t^j] G_t^j + B_t^{j'} H_t \sum_{\substack{k=1 \\ k \neq j}}^{N} B_t^k G_t^k = \mathbf{0}, \tag{56}$$

$$[R_t^j + B_t^{j'} H_t B_t^j] g_t^j + B_t^{j'} H_t \sum_{\substack{k=1 \\ k \neq j}}^{N} B_t^k g_t^k + B_t^{j'} H_t s_t - B_t^{j'} h_t - r_t^j = \mathbf{0}, \tag{57}$$

$$H_{t-1} = Q_{t-1} + K_t' H_t K_t + \sum_{j=1}^{N} G_t^{j'} R_t^j G_t^j, \quad H_T = Q_T, \tag{58}$$

$$h_{t-1} = q_{t-1} - K_t' [H_t k_t - h_t] + \sum_{j=1}^{N} G_t^{j'} [r_t^j - R_t^j g_t^j], \quad h_T = q_T, \tag{59}$$

where

$$Q_t := \sum_{i=1}^{N} {}^iQ_t^i, \tag{60}$$

$$q_t := \sum_{i=1}^{N} {}^iQ_t^i \widetilde{x}_t^i, \tag{61}$$

$$R_t^j := \sum_{i=1}^{N} {}^iR_t^{ij}, \quad j = 1,\ldots,N, \tag{62}$$

$$r_t^j := \sum_{i=1}^{N} {}^iR_t^{ij} \widetilde{u}_t^{ij}, \quad j = 1,\ldots,N. \tag{63}$$

Using the Riccati matrices, H_t and h_t for $i = 1,..., N$, and the feedback matrices, G_{it} and g_{it} for $i = 1,..., N$, for all time periods, the state equations (17), starting with $x_0^* = \overline{x}(0)$ and augmented with K_t and k_t as defined by (19) and (20), and the feedback rules (18), the optimal state and control paths of all players ($i = 1,...,N$) can be determined by forward iteration.

Having stored all (five times N rows of) control variables, u_{it}^* for $i = 1,...,N$ and for all time periods $t = 1,...,T$, resulting from the derivation of the equilibrium solutions of the affine-quadratic open-loop and feedback Nash and Stackelberg game and from solving the cooperative Pareto game, we compute the associated (five rows of) state vectors, x_t^* for all time periods $t=1,...,T$, according to the nonlinear system dynamics (8).

C. Compute new tentative values for state and control variables.

The "new" tentative state and control paths for the next iteration are determined from the *optimized* state and control variables from the current iteration (see B), i.e.

$$\overline{x}_t := x_t^*, \quad t = 1,\ldots,T, \tag{64}$$

$$\overline{u}_{it} := u_{it}^*, \quad i = 1,\ldots,N, \quad t = 1,\ldots,T, \tag{65}$$

which will (hopefully) converge towards the "optimized solution" returned by the OPTGAME algorithm. Using the Gauss-Seidel or the Newton-Raphson algorithm, the given exogenous non-controlled variables z_t, the lagged tentative state variables $\overline{x}_{t-1}$, the tentative policy paths $\overline{u}_{it}$ ($i = 1,...,N$, $t = 1,...,T$), and the system equation f (provided in a separate input file), the tentative state path $\bar{x}_t$ is calculated according to (9).

Step 4. Compute the welfare loss determined by equation (1) and (7), evaluated along the five types of optimal control and state paths.

Step 5. Output the results in graphical form and list the optimal states, controls and associated losses.

V. Concluding Remarks

In this paper, we have described an algorithm which has been developed to deal with dynamic game problems for economic policy questions, though it is not limited to economics. The OPTGAME algorithm approximates the solutions of multi-player difference games assuming a quadratic structure for the objective functions and using a numerical linearization procedure. This numerical tool was already used for problems of international policies, where several decision makers (governments and central banks) from different countries were involved. Applications to dynamic oligopoly and other forms of imperfect competition in microeconomics, to national or international policy coordination problems can easily be implemented. Applications from other fields of research include international technology transfer, pollution control, and international drug control, to mention only a few. Some promising and challenging extensions of the algorithm, which are topics for further research, include stochastic systems, rational expectations, and the possibility of coalition formation.

References

[1] M.L. Petit, *Control Theory and Dynamic Games in Economic Policy Analysis*. Cambridge: Cambridge University Press, 1990.

[2] D.A.Behrens, M. Hager, and R. Neck, "OPTGAME 1.0: A numerical algorithm to determine solutions for two-person difference games," in *Modeling and Control of Economic Systems,* R. Neck, Ed. Oxford: Elsevier, 2003, pp. 47–58.

[3] M. Hager, R. Neck, and D.A. Behrens, "Solving dynamic macroeconomic policy games using the algorithm OPTGAME 1.0," *Optimal Control Applications and Methods,* vol. 22, pp. 301–332, 2001.

[4] T. Başar and G.J. Olsder, *Dynamic Noncooperative Game Theory*, 2nd ed. Philadelphia, PA: SIAM, 1999.

[5] D.A. Behrens and R. Neck, "Approximating solutions for multi-decision-maker tracking problems," Working Paper. Klagenfurt: Klagenfurt University, 2006.

[6] J. Matulka and R. Neck, "OPTCON: An algorithm for the optimal control of nonlinear stochastic models," *Annals of Operations Research,* vol. 37, pp. 375–401, 1992.

Rapid Development of Web Applications with Web Components

Dzenan Ridjanovic
Université Laval
Pavillon Palasis-Prince
Québec (Québec) G1K 7P4, Canada

Abstract-**This paper provides a brief overview of Domain Model RAD, a web framework, which is used for developing dynamic web applications with a minimum amount of programming. Domain Model RAD uses Domain Model Lite to represent a domain model of a web application. Domain Model Lite is a framework that facilitates the definition and the use of domain models in Java. Domain Model RAD uses Wicket for web application pages and page sections. Wicket is a web framework that provides basic web components, to construct, in an object oriented way, more advanced web components. Domain Model RAD interprets the application model and creates default web pages from its web components that are based on the domain model.**

I. INTRODUCTION

There are many Open Source Java web frameworks [1]. The most popular is Struts [2] from the Apache Jakarta Project. Struts relies more on external configuration files and less on Java code to speed up web application development. It is an action based framework. As a consequence, the control part of Struts is rather elaborate for developers and is not suitable for rapid development of web applications.

There is a new web component based framework called Wicket [3]. A web component, such as a web page or a page section, is in the center of preoccupation of Wicket developers. The control part of Wicket is largely hidden from the developers.

A web application, as any other software has two major parts: a domain model and views. Although, a Wicket component requires a model for the component data, the actual model is outside of the Wicket realm.

I have developed a domain model framework, called Domain Model Lite or dmLite [4, 5, 6], to provide an easy support for small domain models, which are usually used in rapid web development. Its name reflects the framework objective to provide an easy to learn and easy to use framework. In addition, I have developed a web component framework, called Domain Model RAD or dmRad, to produce rapidly a web application based on the given domain model. Domain Model RAD creates more advanced web components from basic components provided by Wicket. Both frameworks are developed in Java.

II. DOMAIN MODEL LITE

A domain model is a model of specific domain classes that describe the core data and their behavior. The heart of any software is a domain model. When a model is well designed and when it can be easily represented and managed in an object oriented language, a developer can then focus more rapidly on views of the software, since they are what users care about the most.

There is a class of small projects where there is no need to elaborate on different design representations, such as sequence and collaboration diagrams in UML [7]. In a small application, a domain model is the core part of the application software. Domain Model Lite has been designed to help developers of small projects in representing and using application domain models in a restricted way. The restrictions minimize the number of decisions that a domain model designer must make. This makes Domain Model Lite easy to learn.

A domain model must be configured in an XML configuration file. This configuration reflects Java classes of the model. However, it provides more information about the domain model default behavior used heavily in Domain Model RAD. The model XML configuration is loaded up-front and converted into meta entities. Those meta entities are consulted by Domain Model Lite and Domain Model RAD quite often.

In most cases, domain model data must be saved in an external memory. If a database system is used for that purpose, the software requires at least several complex installation steps. Since Domain Model Lite uses XML files to save domain model data, there is no need for any special installation. In addition, Domain Model Lite allows the use of a db4o object database [8]. The upgrade of an application from XML data files to a database does not require a single line of programming code to be changed. It is enough to change a domain model configuration and an application configuration, both in XML, and, of course, create an empty database. The migration of XML data to database objects, and vice versa, is also provided.

A domain model is a representation of user concepts, concept properties and relationships between concepts. The easiest way to present a domain model is through a graphical representation [9]. The Fig. 1. displays a domain model of

K. Elleithy (ed.), Advances and Innovations in Systems, Computing Sciences and Software Engineering, 99–103.

slide show presentations, called Public Point Presentation.

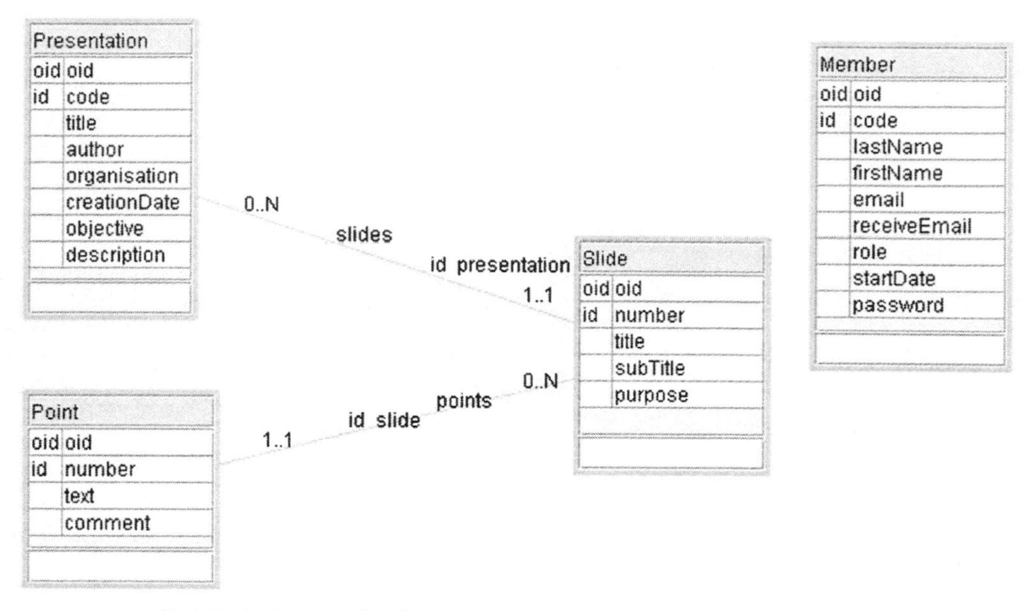

Fig. 1. Graphical representation of a domain model of the Public Point Presentation web application.

A domain model is represented in Domain Model Lite as the `DomainModel` Java class that implements the `IDomainModel` interface.

```
public interface IDomainModel
{
  public void setInitialized(
    boolean initialized);
  public boolean isInitialized();
  public void setModelConfig(
    ModelConfig modelConfig);
  public ModelConfig getModelConfig();
  public IEntities getEntry(
    String entryCode);
  public List<IEntities> getEntries();
  public boolean isSession();
  public Session getSession();
  public ModelMeta getModelMeta();
  public void notifyObservers(Object arg);
}
```

A domain model concept is described by its properties and neighbors. The oid property is mandatory. It is used as an artificial identifier and is managed by Domain Model Lite. In addition, a concept may have at most one user oriented identifier (id) that consists of the concept properties and/or neighbors.

The model from Fig. 1. has four concepts. The Member concept is used for the application login. The Presentation concept has several properties and one neighbor, the Slide concept. A relationship between two concepts is represented by two neighbor directions, displayed together as a line. A neighbor direction is a concept neighbor property, with a name and a range of cardinalities. For example, a presentation may have from 0 to N slides. A slide belongs to exactly one presentation.

A concept is represented in Domain Model Lite as two Java classes, one for `Entity` and the other for `Entities` (or `OrderedEntities` if an order of entities is important). The `Entity` class implements the `IEntity` interface, and the `Entities` class implements the `IEntities` interface.

A domain model has a few concepts that are entry points. The access to concept entities is provided only through the entry points. In the PPP domain model from Fig. 1., the entry points are the Presentation and Member concepts. A certain point of a certain slide can only be accessed through the slide's presentation, then through the slide itself.

```
public interface IEntity extends Serializable
{
  public IDomainModel getDomainModel();
  public ConceptConfig getConceptConfig();
  public void setOid(Oid oid);
  public Oid getOid();
  public void setCode(String code);
  public String getCode();
  public Id getId();
  public void setProperty(
    String propertyCode, Object property);
  public Object getProperty(
    String propertyCode);
```

```
  public void setNeighborEntity(
    String neighborCode,
    IEntity neighborEntity);
  public IEntity getNeighborEntity(
    String neighborCode);
  public void setNeighborEntities(
    String neighborCode,
    IEntities neighborEntities);
  public IEntities getNeighborEntities(
    String neighborCode);
  public boolean update(IEntity entity)
    throws ActionException;
  public IEntity copy();
  public boolean equalContent(
    IEntity entity);
  public boolean equalOids(IEntity entity);
  public boolean equalIds(IEntity entity);
}

public interface IEntities extends
  Serializable
{
  public IDomainModel getDomainModel();
  public ConceptConfig getConceptConfig();
  public Collection<IEntity> getCollection();
  public IEntities getEntities(
    SelectionCriteria selectionCriteria)
    throws SelectionException;
  public IEntities getSourceEntities();
  public void setPropagateToSource(
    boolean propagate);
  public boolean isPropagateToSource();
  public IEntities union(IEntities entities)
    throws SelectionException;
  public IEntities intersection(
    IEntities entities)
    throws SelectionException;
  public boolean isSubsetOf(
    IEntities entities)
    throws SelectionException;
  public boolean add(IEntity entity)
    throws ActionException;
  public boolean remove(IEntity entity)
    throws ActionException;
  public boolean update(
    IEntity entity, IEntity updateEntity)
    throws ActionException;
  public boolean contain(IEntity entity);
  public boolean containCode(String code);
  public IEntity retrieveByOid(Oid oid);
  public IEntity retrieveByCode(String code);
  public IEntity retrieveByProperty(
    String propertyCode,
    Object paramObject);
  public IEntity retrieveByNeighbor(
    String neighborCode,
    Object paramObject);
  public Iterator<IEntity> iterator();
  public int size();
  public boolean isEmpty();
  public Errors getErrors();
}
```

III. Domain Model RAD

A. Web Components

Domain Model Lite has a companion rapid web application development framework, called Domain Model RAD, which can be used to make a default Wicket application out of a domain model. Domain Model RAD uses the domain model configuration to find the model entry points and to provide a web page for each entry point, either for the display or for the update of data. An entry point is a collection of entities and it is presented in a web page as a table, a list, or a slide show of entities. This choice and other view presentation properties are defined in the XML configuration of the domain model. The traversal of the domain model is done by navigating from an entry entity to neighbor entities following the parent-child neighbor directions.

A default application may help developers validate and consequently refine the domain model. In addition, Domain Model RAD has a collection of web components that may be easily reused in specific web applications to display or update entities. For example, the component called `EntityDisplayTablePanel` in Fig. 2. is used to display entities as a table with a link to update the selected entity and a link to display child entities of the selected entity.

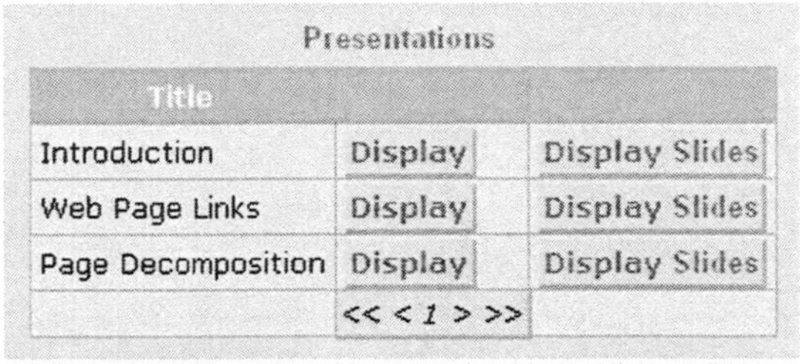

Presentations

Title		
Introduction	Display	Display Slides
Web Page Links	Display	Display Slides
Page Decomposition	Display	Display Slides
	<< < 1 > >>	

Fig. 2. Page table section with presentation titles and links to slides.

The following is a list of the most important web components that belong to Domain Model RAD.

```
LoginPanel
NewMemberConfirmPanel
CountryLanguageDropDownChoicePanel
EntryUpdateTablePanel
EntityAddFormPanel
EntityConfirmRemovePanel
EntityEditFormPanel
EntryDisplayTablePanel
EntityDisplayTablePanel
EntityDisplayListPanel
EntityDisplaySlidePanel
EntityDisplayPanel
EntityDisplayMinPanel
EntityPropertyKeywordSelectPanel
EntityLookupTablePanel
EntitySlideNavigatePanel
EntityPropertyDisplayListPanel
ParentChildPropertyDisplayListPanel
```

The `LoginPanel` component is used by a member to sign in. The `NewMemberConfirmePanel` component allows a new member to register. The registration is valid only if the new member confirms the registration based on the information in a confirmation email sent by the web application. The `CountryLanguageChoicePanel` component enables the support for international application versions. The EntryUpdateTablePanel component shows the entry concepts in a table, for the purpose of updates. Web components with the Add, Edit and ConfirmRemove actions in their names are used to update entities. The EntryDisplayTablePanel component shows the entry concepts in a table, for the purpose of displays. Web components with Display in their names serve to present entities in different ways. The EntityPropertyKeywordSelectPanel component provides a keyword search on a single property. The EntityLookupTablePanel component enables a selection of single entity in a table of parent entities. The EntitySlideNavigatePanel component provides the first, previous, next and last actions on entities. The EntityPropertyDisplayListPanel component displays single property values in a list. The ParentChildPropertyDisplayListPanel component displays a list of parent-children sublists, with a parent property and a child property in a sublist.

B. Public Point Presentation

Public Point Presentation (PPP) is a web application developed in three spirals: ppp00 [10], ppp01 and ppp02 [11]. Each spiral is a separate web application. The ppp00 spiral has been created with almost no programming [12], while the ppp02 spiral has only a few specific classes.

The ppp00 spiral has three packages: org.ppp.model.xml, org.ppp.view and org.ppp.view.home. The first package provides the domain model context, the second package determines the web application, while the third package describes the application home page. The real programming is done only in the last package that has two classes: HomePage and HomePageMenuPanel.

The HomePage class has two web components, one specific – HomePageMenuPanel, and another generic – EntityDisplayTablePanel.

```
public class HomePage extends WebPage
{
  public HomePage() {
    add(new
      HomePageMenuPanel("homePageMenuPanel",
      this));
    App app = (App) getApplication();
    Presentations presentations =
      (Presentations)
      app.getEntry("Presentations");
    Presentations orderedPresentations =
      (Presentations)
      presentations.getOrderByCode();
    ModelContext pppDisplayModelContext =
      new ModelContext();
    pppDisplayModelContext.setDomainModel(
      app.getDomainModel());
    pppDisplayModelContext.setEntities(
      orderedPresentations);
    ViewContext pppDisplayViewContext =
      new ViewContext();
    pppDisplayViewContext.setApp(app);
    pppDisplayViewContext.setWicketId(
      "pppDisplayPanel");
    pppDisplayViewContext.setContextPage(
      this);
    pppDisplayViewContext.setPage(this);
    add(new EntityDisplayTablePanel(
      pppDisplayModelContext,
      pppDisplayViewContext));
  }
}
```

The generic web component requires two parameters, one for the model context and another for the view context. The model context has access to the domain model and to the entities used to support the component view. The view context has access to the application, to the view context page and to the view page. The Wicket id is used to link the web component with the HTML div tag in the HomePage.html file that resides in the same package as its corresponding HomePage.java file.

```
<div wicket:id="pppDisplayPanel">
  To be replaced dynamically by a
  table of presentations.
</div>
```

The ppp02 spiral has one additional package, org.ppp.view.component.slide.display, to create a specialized version of the generic EntityDisplaySlidePanel web component. This new version displays a slide in a way more appropriate for a slide show presentation.

In this spiral, the home page uses three new generic web components:

```
LoginPanel,
EntityPropertyDisplayListPanel,
ParentChildPropertyDisplayListPanel.
```

The LoginPanel component, in Fig. 3., allows a member to sign in, in order to create or update a presentation.

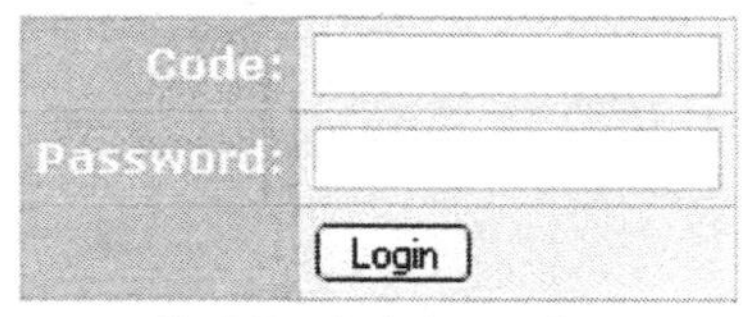

Fig. 3. Page login form section

EntityPropertyDisplayListPanel, in Fig. 4., shows email addresses of those members that have accepted to receive email messages (here only one member).

Member E-mail

dzenan.ridjanovic@fsa.ulaval.ca

Fig. 4. Page email list section

`ParentChildPropertyDisplayListPanel`, in Fig. 5., lists only one presentation title with its slide titles. In general, this component displays a list of list sections. Each list section shows one presentation title together with its slide titles.

Presentations with Slide Titles

Introduction

Objective

Frameworks

Dynamic Web Applications

Configurations

Domain Model

Model Classes

View Classes

Next Spiral

Fig. 5. Page list section with presentation title and slide titles

After a successful login, a member may use the *PPP Administration* menu to create a new presentation or update an existing one. This part of application is completely handled by Domain Model RAD. It starts with the generic `EntityUpdateTablePage` web component. The component shows a table of presentation titles with links to add, edit and remove a presentation. The link to update slides of the selected presentation is also provided.

IV. Conclusion

I have developed a web component framework, called Domain Model RAD, which makes it possible to produce quickly a web application based on the given domain model expressed in Domain Model Lite. Domain Model RAD uses Wicket for advanced web components that are based on Wicket's base components. Advanced web components are tightly integrated with Domain Model Lite and its meta entities that represent the configuration of a domain model. The configuration has both model and view properties. Even if view properties are not indicated, Domain Model RAD uses default values, so that a domain model may become a default web application without specific view configurations and without view programming. However, for customizing a web application, Domain Model RAD provides generic web components that can be easily specialized. In this way, a designer can focus on how to divide a page into sections, where each section is supported by a generic or a specific web component.

There are several web applications [4, 5] developed by using both generic and specific web components. The source code for the frameworks [4] and the web applications [4, 5] is in the public domain.

References

[1] http://java-source.net/open-source/web-frameworks

[2] http://struts.apache.org/

[3] http://wicketframework.org/

[4] http://drdb.fsa.ulaval.ca/dm/

[5] http://drdb.fsa.ulaval.ca/urls/

[6] D. Ridjanovic, "Domain Model Driven Development of Web Applications", *the 6th OOPSLA Workshop on Domain-Specific Modeling, Object-Oriented Programming, Systems, Languages and Applications - OOPSLA 2006*, Portland, Oregon, USA, October 22 - 26, 2006.

[7] http://www.uml.org/

[8] http://www.db4o.com/

[9] http://drdb.fsa.ulaval.ca/mmLite/

[10] http://drdb.fsa.ulaval.ca/ppp00/

[11] http://drdb.fsa.ulaval.ca/ppp02/

[12] http://drdb.fsa.ulaval.ca/ppp00/code.html

Mesh-adaptive methods for viscous flow problem with rotation

E. Gorshkova
Department of Mathematical Information Technology,
P.O. Box 35 (Agora). FIN-40014,
University of Jyvaskyla, Finland

P. Neittaanmaki
Department of Mathematical Information Technology,
P.O. Box 35 (Agora). FIN-40014,
University of Jyvaskyla, Finland

S.Repin
V.A. Steklov Institute of Mathematics at St. Petersburg,
191023, Fontanka 27, St.Petersburg, Russia

In this paper, new functional type a posteriori error estimates for the viscous flow problem with rotating term are presented. The estimates give guaranteed upper bounds of the energy norm of the error and provide reliable error indication. We describe the implementation of the adaptive finite element methods (AFEM) in the framework of the functional type estimates proposed. Computational properties of the estimates are investigated on series of numerical examples.

INTRODUCTION

To obtain good numerical solution of a problem in Computational Fluid Dynamics on regular meshes is almost impossible. Complicated behavior of fluids (such as turbulence) requires specially adapted meshes in the respective regions. Typically such meshes are constructed with help of a posteriori error estimation or a posteriori error indicator.

Two main aims of a posteriori error estimation are:
a) to provide a guaranteed upper bound of the error (a quality used for this purpose is usually called "error estimator")
b) to present an adequate indicator that shows the distribution of local errors (such a quantity is usually called "error indicator")

Ability to provide a guaranteed upper bound of the error is what differs error estimator from error indicator.

For finite element approximations, a posteriori error indicators started receiving attention in the late 70th (see [5],[6]). First investigations in the area were focused on linear elliptic problems. Later, a lot of work have been done for some other linear and nonlinear problems. We refer to the monographs [1], [7], [14], [16] for surveys in the area.

Modern outlook on the mesh adaptive numerical methods can be expressed by the logical sequence (see, e.g. [8])

SOLVE $\Rightarrow$ INDICATE $\Rightarrow$ MARK $\Rightarrow$ REFINE

A posteriori error estimators (indicators) are required to perform the second step in the previous sequence.

In this paper, we present and numerically investigate a posteriori error estimator for a basic stationary model of viscous fluid with the rotating term.

$$\begin{aligned} -\nu\Delta u+\varpi\times u &= f-\nabla p \quad \text{in } \Omega \\ \operatorname{div} u &= 0 \quad \text{in } \Omega \\ u &= u_0 \quad \text{on } \partial\Omega \end{aligned} \tag{1}$$

Here $\nu>0$ is the viscosity parameter, $f\in L_2(\Omega,R^n)$ is a given vector-valued function, p is the pressure function and $u_0\in H^1(\Omega,R^n)$ defines the Dirichlet boundary conditions on $\partial\Omega$. It is assumed that $\operatorname{div} u_0=0$ in Ω

Such equations are physically motivated by the geophysical flow problem. It models the movement of the atmosphere or the oceans at mid-latitude. Those equations have been studied by a number of authors (see [2], [3], [4], [10], [11]).

To the best of our knowledge, a posteriori error estimators for a system with Coriolis term have not been investigated seriously yet. Certainly, we consider one of the simplest stationary models that does not take into account nonlinear effects. However, we believe that the present work is natural and necessary step in a posteriori error control for more complicated system in the theory of fluids.

A POSTERIORI ERROR ESTIMATES

Consider v to be some approximate solution of the problem obtained by any numerical method. We consider v to be some finite element approximation, but generally speaking, it can be obtained by e.g. finite difference method or some other method. The difference between v and the exact solution can be estimated as follows:

$$\begin{aligned} \nu\|\nabla(u-v)\| \le \|\nu\nabla v-\tau\| + C_\Omega\|\operatorname{div}\tau-\varpi\times v+f-\nabla q\| + \\ +\frac{1}{C_{LBB}}(2\nu+|\varpi|C_\Omega)\|\operatorname{div} v\| \end{aligned} \tag{2}$$

Here and later on we call the right-hand side of (2) ***the functional type error majorant***. This estimator is valid for arbitrary tensor-function

$$\tau\in\{L_2(\Omega,M^{n\times n})\,|\,\operatorname{div}\tau\in L_2(\Omega,R^n)\}$$

(here, by $M^{n\times n}$ we denote the space of real symmetric $n\times n$ matrices) and scalar-function $q\in L_2(\Omega)\cap H^1(\Omega)$. The constant C_Ω comes from Friedrichs-Poincare inequality, C_{LBB} is the constant that appears in Ladyzhenskaya-Babuska-Brezzi inequality (inf-sup inequality).

If the rotation parameter ϖ is equal to zero, then the error majorant (2) coincides with the functional majorant for the

K. Elleithy (ed.), Advances and Innovations in Systems, Computing Sciences and Software Engineering, 105–107.

Stokes problem, derived in [15] and numerically investigated in [12] and [13].

For the divergence-free approximations, that exactly satisfy the condition $\operatorname{div} v = 0$, he error estimate has a simpler form:

$$\nu\|\nabla(u-v)\| \le \|\nu\nabla v - \tau\| + C_\Omega\|\operatorname{div}\tau - \varpi\times v + f - \nabla q\| \quad (3)$$

In (3) and (4) the functions τ and q are "arbitrary" and, therefore, in our disposal. There exist several possible ways of choosing them.

From the computational viewpoint, the cheapest way is to define τ as a certain averaging of the tensor function $\nu\nabla v_h$, where v_h is a velocity constructed with help of FEM on a mesh of the character size "h". The error majorant will provide a guaranteed upper bound, which, of course, can be rather overestimated. To improve it, we can use gradient averaging from the finer mesh (e.g., next adaptation step).

However, sharp upper bound can be achieved if the majorant is minimized with respect to τ and q. In this case, we minimize (3) over some finite dimensional space. By the squaring of the (3) and implementing the Young's inequality, we rewrite it in the following form:

$$\nu^2\|\nabla(u-v)\|^2 \le (1+\alpha)\|\nu\nabla v - \tau\|^2 + \\ +(1+\frac{1}{\alpha})C_\Omega^2\|\operatorname{div}\tau - \varpi\times v + f - \nabla q\|^2 \quad (4)$$

Now, minimization of the right-hand side of (4) is a problem of minimization of quadratic functional, that is reduced to a system of linear simultaneous equations.

Note, that it is not necessary to find the exact minimizer of this system. Starting from the initial approach of gradient averaging $\nu\nabla v_h$ we can use some iterative numerical procedure. On the each iteration step, the value of the emajorant give an upper bound, and, therefore solves the problem (a).

A function τ^* obtained at the end of minimization process yield an error indicator $\|\nu\nabla v - \tau^*\|^2$ that is used to solve problem (b). It is worth nothing, that if τ^* is close to $\nu\nabla u$ then the second term of the error majorant (4) is small, while the first one perfectly represents the error.

MESH ADAPTATION STRATEGIES

Certainly, the best possible adaptive algorithm can be constructed on the basis of the true error distribution obtained by comparing the true and approximate solutions. Denote by E_i the normalized local contribution of the error on the element, by M_i the normalized local contribution of error indicator. To compare them not only qualitatively but also quantitatively we introduce a special coefficient

$$p_{shape} = 1 - \frac{\Sigma|M_i - E_i|}{N}, \quad (5)$$

Which is equal to one only in the ideal case when the normalize error indicator coincides with the normalized true error, what means that may they differ by factor only.

If we use two color marking, then with help of a marking strategy we assert "one" or "zero" to each element, i.e. we construct an element-wise Boolean function.

$$R(M_i) = \{0,1\}$$

"Zero" value of such a function means that the element will not be refined and value "one" means that the element should be subject to further subdivision.

In our numerical test we use two refinement strategies:

The first strategy follows the "maximum principle": within its framework we mark the element to be refined if

$$R^{\max}(M_i) = \begin{cases} 1, & if\ M_i \ge \theta^{\max} M^{\max} \\ 0, & otherwise \end{cases}$$

where $M^{\max}$ is the maximum local contribution over all elements in the triangulation. Typically, $\theta^{\max} = 1/2$ what means that element is refined, if the error is bigger then one half of the maximum error (see, e.g., [16]).

The second strategy is the so-called "bulk criterion". Here, the elements are ranked by the values of the local errors. For the refinement, we take those, that contain maximum errors and jointly give some certain part θ^{bulk} of the total error (see [9]). In other words

$$R^{bulk}(M_i) = \begin{cases} 1, & if\ \Sigma M_i \ge \theta^{bulk}\Sigma M_i \\ 0, & otherwise \end{cases} \quad (6)$$

In our tests we follow the principle $\theta^{bulk} = 0.6$

Let us define p_{eff} which shows the per cent of the elements, marked in the same way as in the etalon marking, i.e.

$$p_{eff} = 1 - \frac{\Sigma|R_i - R_i^{etalon}|}{N}, \quad (7)$$

where N is a number of elements in triangulation.

If p_{eff} is close to 1, then the error indicator produces almost optimal mesh adaptation.

NUMERICAL EXAMPLES

Consider the flow in a container depicted on the figure1. The container is rotated around the vertical axis i_z with the angular velocity ϖ. Taking into account axial symmetry of the problem, we use cylindrical coordinate system. The income and outcome boundary conditions define as follows:

$$u_r = 0$$

$$u_\phi = \varpi r$$

$$u_z = \begin{cases} (R_{top}^2 - r^2)/R_{top}^3 & on\ \ input \\ (R_{bottom}^2 - r^2)/R_{bottom}^3 & on\ \ output \end{cases}$$

In this case, exact solution is not known, as a "true error" we understand the difference between an approximate solution and the solution, obtained on the very fine mesh.

We conducted a series of numerical tests. We select an initial mesh and solve the problem by the finite element method using quadratic approximations for the velocity and linear ones for the pressure. Further, we project the approximate solution obtained to the space of divergence-free functions. Then we implement a posteriori error control. Namely, we find the tensor-valued functions τ and scalar function q (we use quadratic approximations for them) by minimization of the right-hand side of (4). We calculate guaranteed error bound and evaluate an error indicator. Then with help of some adaptation criterion we mark the zones with excessively high errors and refine the corresponding elements.

In the table 1 we collect the results of several iteration steps. The parameters are $R_{top} = R_{bottom} = 0.6$, $\varpi = 1$.

$I_{eff} = \sqrt{\frac{M}{E}}$ denotes ration between error majorant and true error, this quantity is greater then one by definition and shows the quality of the error estimate. Quantities $p_{eff}^{\max}$ and p_{eff}^{bulk} refers to two different refinement strategies described above. The actual refinement is done following the "bulk" strategy.

CONCLUSION

We observe, that in all the cases on the each iteration step, the error majorant provide realistic error indication and efficient upper bound of the true error.

It is important to emphasize, that mesh adaptation, based on the functional type error majorant is very close to those, which would be obtained on the basis of the exact knowledge of the error distribution.

REFERENCES

[1] M. Ainsworth, J. T. Oden, "A posteriori error estimation in finite element analysis", Wiley, New York, 2000.

[2] A. Babin, A. Mahalov, and B. Nicolaenko, "Global splitting, integrability and regularity of 3D Euler and Navier-Stokes equations for uniformly rotating fluids", European Journal of Mechanics, 15, (1996), pp. 291-300.

[3] A. Babin, A. Mahalov, and B. Nicolaenko, "Resonances and regularity for Boussinesq equations", Russian Journal of Mathematical Physics, 4, (1996), pp. 417-428.

[4] A. Babin, A. Mahalov, and B. Nicolaenko, "Global regularity of 3D rotating Navier-Stokes equations for resonant domains", Indiana University Mathematics Journal, 48, (1999), pp. 1133-1176.

[5] I. Babuska, W.C. Rheinboldt, "A-posteriori error estimates for the finite element mthod" Int. J. Numer. Methods Eng., Vol. 12. 1978 pp.1597-1615

[6] I. Babuska, W.C. Rheinboldt, "Error estimates for adaptive finite element computations". SIAM J.Numer. Anal., 1978 V. 15. pp. 736-754

[7] I. Babuska, T. Strouboulis, "The finite element method and its Reliability", Clarendon Press, New York, 2001.

[8] C. Carstensen, R. Hoppe, "Error reduction and cinvergence for an adaptive mixed finite element method". Math. of Comput, 75(255), 2006, pp. 1033-1042

[9] W. Doerfler, "A convergent adaptive algorithm for Poisson's equation". SIAM J. Numer. Anal., vol. 33 (3), 1996, pp. 1106-1124

[10] P. Embid and A. Majda, Averaging over fast gravity waves for geophysical ows with arbitrary potential vorticity, Communications in Partial Diferential Equations, 21 (1996), pp. 619-658.

[11] I. Gallagher, L. Saint-Raymond, "Weak convergence results for inhomogeneous rotating fluid equation". http://www.math.jussieu.fr/~gallagher/articles/debut.pdf

[12] E. Gorshkova, S. Repin, "On the functional type a posteriori error estimates of the Stokes problem", Proceedings of the 4th European Congress in Applied Sciences and Engineering ECCOMAS 2004 (eds. P.Neittaanmaki, T. Rossi, S. Korotov, E. Onate, J. Periaux, and D. Knorzer), CD-ROM.

[13] E. Gorshkova, P. Neittaanmaki, S. Repin "Comparative study of the a posteriori error estimators for the Stokes Problem", in the proceedings of the The Sixth European Conference on Numerical Mathematics and Advanced Applications (ENUMATH 2005), Santiago de Compostela, Spain, pp. 254--259,Springer Berlin, Heidelberg, New York, 2006.

[14] P. Neittaanmaki, S. Repin, "Reliable methods for computer simulation. Error control and a posteriori estimates". ELSEVIER, New York, London, 2004.

[15] S.I Repin, "A posteriori estimates for the Stokes problem". J. Math. Sciences, vol. 109 (5), 2002, pp.1950--1964

[16] R. Verfurth, "A review of a posteriori error estimation and adaptive mesh–refinement techniques", Wiley; Teubner, New York, 1996.

Table 1 Numerical results.

Iteration	N	E	M	I_{eff}	$p_{eff}^{\max}$	p_{eff}^{bulk}
1	312	0.0087	0.0129	1.22	0.96	0.94
2	472	0.0066	0.0086	1.14	0.96	0.90
3	643	0.0059	0.0081	1.17	0.93	0.90
4	692	0.0051	0.0061	1.09	0.94	0.89
5	786	0.0033	0.0050	1.23	0.95	0.93

Figure 1. Computational domain.

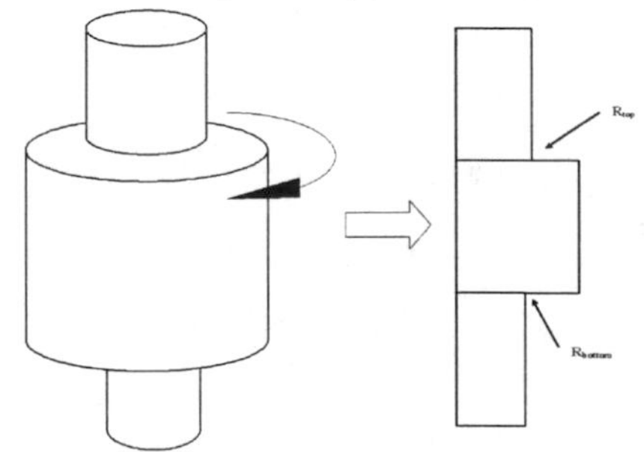

Metamodel-based Comparison of Data Models

Erki Eessaar
Department of Informatics, Tallinn University of Technology, Raja 15, 12618 Tallinn, ESTONIA
eessaar@staff.ttu.ee

Abstract-**A data model specifies the building blocks of databases, the rules how to assemble these blocks and operations that can be performed based on the built-up structures. We have to create increasingly complex applications. Properties of the underlying data model of a Database System (DBMS) determine how easy is to create an application that uses a database. There are many different data models. We have to choose a DBMS, the underlying data model of which best fulfils the needs of an application. Existing comparisons of data models are mostly based on the experiences of using one or another DBMS. This paper explains how to perform non-empirical comparison of data models by using the metamodels, which describe abstract syntax of these data models. We also present some results of the comparison of the underlying data model of SQL:2003 and the data model that is proposed in The Third Manifesto.**

I. INTRODUCTION

The concept "data model" has two different meanings:

- Meaning 1: "An abstract, self-contained, logical definition of the data structures, data operators, and so forth, that together make up the abstract machine with which users interact." [1]
- Meaning 2: "A model of persistent data of some particular enterprise." [1]

In this work, we use the concept "data model" in the sense of meaning 1.

Examples of data models are hierarchical, network, relational, object-relational, TransRelational and object-oriented. Some of them are general names because there are different proposals about the exact nature of these models. Examples of proposals about object-relational data model are [2], [3], [4], [5]. In addition, SQL:2003 [6] follows object-relational paradigm [7] and its underlying data model is also one interpretation of object-relational data model.

Applications that use databases become increasingly complex and they demand more and more from the DBMSs. A very important selection criterion of a DBMS is its underlying data model. How should we compare data models? The work [8] is an example of thorough and methodical comparison of two data models. It presents similarities and differences of relational and network data model in the form of discussion and examples. The authors even had to work out definitions of concepts of the network data model based on CODASYL DBTG language proposals in order to do it properly. Additional examples are the comparison of the prescriptions, proscriptions and suggestions of The Third Manifesto with SQL and with ODMG proposal of object model and associated database language [4].

However, many comparisons or judgments of the data models that are presented in the literature are based on the experiences and the intuitive understandings of the researchers and developers. One reason of the prevalence of informal descriptions and empirical observations is that there is often no precise specification of a particular data model. Instead, there is a set of research papers and textbooks that reflect their authors understanding of the model and a set of DBMSs that implement the model with their own limitations and extensions. Inadequacies and shortcomings of the DBMSs as well as lack of understanding what are the parts of a data model can cause unfair criticism of a data model. For example, research [9] shows that some of the criticism towards relational data model is caused by the exactly these reasons. A more precise method for evaluating data models is needed.

Metamodeling is well-known activity in software engineering. Metamodel "makes statements about what can be expressed in the valid models of a certain modeling language." [10] Data model is also kind of abstract language.

The goal of this paper is to present a non-empirical metamodel-based method for comparing data models. We also present some results of comparison of the underlying data model of SQL:2003 standard ("OR_{SQL}") [6] and the data model ("OR_{TTM}") that is described in The Third Manifesto [5].

According to The Third Manifesto, all the good features that are expected from object-relational data model can actually be implemented within the framework of the relational model. In particular, the support to complex data types is already present in the relational model in the form of domains [1].

The rest of the paper is organized as follows. Section 2 gives an overview of important concepts and presents a possible improvement of CIM Database Model. Section 3 discusses different comparison methods of data models and proposes a new method. Section 4 presents some results of the metamodel-based comparison of two data models (OR_{SQL} and OR_{TTM}). Finally, we draw some conclusions.

Participation in the conference was supported by the Doctoral School in ICT of Measure 1.1 of the Estonian NDP.

II. IMPORTANT CONCEPTS OF DATA MODELS

CIM (Common Information Model) Database Model [11] is a conceptual model that describes common database management concepts. However, it models data model only as an experimental property *DataModelType* of class *CommonDatabase*.

K. Elleithy (ed.), Advances and Innovations in Systems, Computing Sciences and Software Engineering, 109–114.

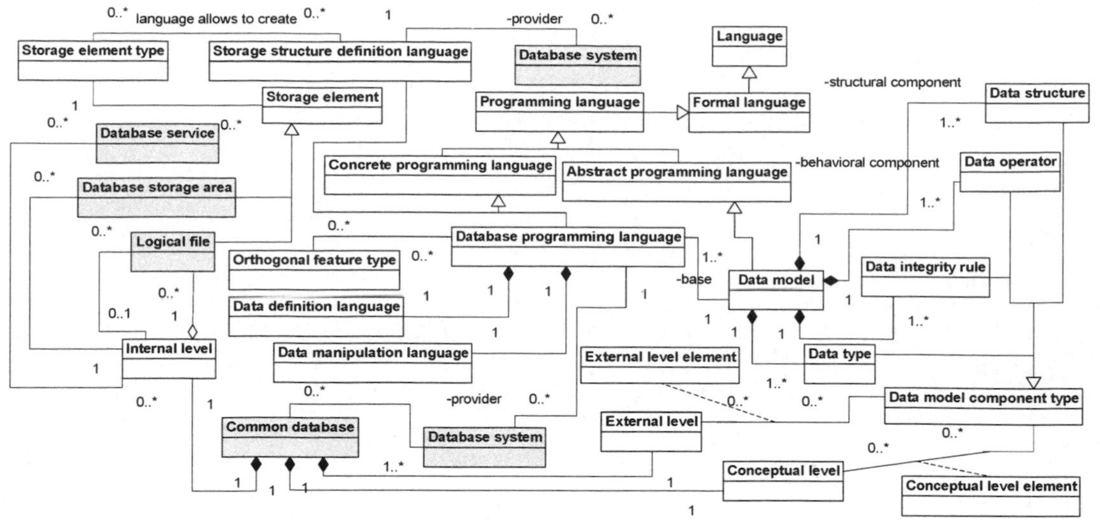

Figure 1. Domain model of data models (desired state of affairs).

We think that it is necessary to model this concept more precisely and present the domain model (see Fig. 1). The classes with grey background are already present in CIM Database Model. The new classes are with white background.

A programming language is a formal language designed specifically for machine processing [12]. A data model is a kind of an abstract programming language [1] that specifies the data structures and operators, which are its structural and behavioural components, respectively (see Fig. 1). In addition, a data model specifies "a collection of general integrity rules, which implicitly or explicitly define the set of consistent database states or changes of state or both" [13].

A specification of a formal language, like modeling or programming language, must contain specifications of abstract syntax, semantics and concrete- and serialization syntaxes [12]. A Database System (DBMS) is a software system used for managing databases. A user can interact with it by using a database programming language that is designed according to some data model. The data model is the basis for the abstract syntax of this language. A database programming language has two sublanguages – a Data Definition Language (DDL) and a Data Manipulation Language. Statements of a DDL are used in order to create data types, structures, operators and integrity rules. Statements of a DML are used in order to perform operations with data.

A database can be divided into conceptual, external and internal levels according to ANSI/SPARC architecture [1]. *Ideally*, a data model specifies elements that belong to the logical levels - conceptual and external level (and not elements at the internal level). In addition, a DBMS should provide a Storage Structure Definition Language (SSDL) for managing storage structures at the internal level [4]. In practice, there is often no separate SSDL. Instead, it is possible to specify elements of the internal level (indexes, tablespaces, clusters, segments, files etc.) and other properties of data storage by using DDL statements.

Database programming languages provide also features that are independent of a data model. The existence of these orthogonal features does not depend on the underlying data model of a database language and they could be present in many languages that have different underlying models. Examples of these orthogonal features are the support to the nested transactions [4] or security mechanisms (for example, a possibility to specify roles, users and their privileges). A data model can have more than one corresponding database programming languages. Different languages could provide support to different orthogonal features. For example, The Third Manifesto that is a proposal for future database systems uses the language name "D" in order to refer to any language that follows its principles. The manifest book also presents Tutorial D language that is "a computationally complete programming language with fully integrated database functionality" [4]. Nevertheless, the authors acknowledge that their proposed language is a "toy" language that must support learning. Industrial-strength languages would need additional features.

If we want to compare data models and reason about them, then we must have their specifications at our disposal. The relational model is an example of the data model that was formally specified before the appearance of systems that implemented it [13]. Sometimes a data model is formally specified only after its implementations (DBMSs) have been created. This is for example true in case of hierarchic and network data models [13].

Nowadays object-relational data models are of major interest. SQL:1999 and SQL:2003 standards specify the object-relational database programming language. However, these specifications do not contain a clear and compact description of an object-relational data model. The Third Manifesto on the other hand specifies the data model *and* the database programming language (Tutorial D) that is created based on this model.

An abstract syntax of a language describes its elements and rules about their interconnections [12]. It is possible to use context-free grammars or metamodels in order to describe the abstract syntax [12]. For example, context-free grammars are used in order to present the syntax of SQL and Tutorial D. The syntax is expressed by using a form of Backus-Naur Form (BNF) notation. Reference [14] writes: "Understanding semantics of SQL (not even of SQL-92), covering all combinations of nested (and correlated) subqueries, null values, triggers, ADT functions, etc. is a nightmare." We need better ways how to present the data model to the interested readers.

People can benefit from a visual presentation of a concrete syntax of a programming language [15]. Is it possible to specify an abstract syntax of a language by using visual means? Specification of the Unified Modeling Language (UML) [16] is an example of using a metamodeling approach in order to define an abstract syntax of a language. If we use UML in order to create a metamodel, then "a metamodel characterizes language elements as classes, and relationships between them using attributes and associations." [12] Other examples of using metamodeling approach are the metamodel-based comparison of workflow management systems [17], ontologies [18] and description of Object Constraint Language (OCL) [19]. Reference [20] presents the metamodel of the object-oriented DBMS. Reference [7] presents the ontology of SQL:2003 Object-Relational Features by using UML class diagrams and well-formedness rules written in OCL.

What are the advantages of creation of metamodel of a data model? Metamodel of a data model visualizes underlying concepts of a data model and their interconnections. It is possible to get overview about the data model with the help of much more compact document compared to purely textual specification. For example, a foundation part of SQL:2003 standard [6] is 1332 pages long.

A metamodel can be used for the teaching purposes, in order to give visual overview of the concepts of a data model and its overall complexity.

It is possible to compare data models based on their metamodels. Creation of a metamodel requires thorough study of a data model and therefore can contribute to finding inconsistencies and other problems in the existing specifications and in the data model themselves.

A metamodel is a basis for creating a database catalog and metadata management systems that manage metadata about the various data sources.

The metamodels of data models help to work out the language for interchanging the management information between management systems and applications. CIM [11] is a step towards this direction. CIM v. 2.13 specifies some SQL Schema elements, but this specification is not complete. The metamodels of different data models are potentially important source that helps to extend CIM.

III. COMPARISON METHODS OF THE DATA MODELS

Reference [21] introduces and classifies the methods for evaluating existing information modeling methods. The ideas behind these comparison methods can be used in order to compare different data models. A comparison method is either empirical or non-empirical. Examples of empirical methods are surveys, laboratory and field experiments, case studies and action research. Next, we describe possible non-empirical methods.

Feature comparison. Data models can be compared with each other based on the features that they provide to the database designers. One could also create a checklist of the desired features and compare data models with this list. Reference [13] names components of a data model and notes that comparisons of data models often ignore operators and integrity rules and therefore "run the risk of being meaningless".

The following methods require metamodels of data models.

Comparison based on metamodels. Data models can be compared by finding common metamodel elements as well as elements that have no counterpart in another metamodel or have more than one counterpart.

Comparisons based on the metrics values that are calculated based on the metamodels. For example, reference [22] proposes the set of metrics for comparing systems development methods and techniques.

Ontological evaluation. Ontological evaluation of a language means comparison of the concrete metaclasses of a language metamodel (language constructs) with the concepts of an ontology in order to find ontological discrepancies: construct overload, construct redundancy, construct excess and construct deficit [23]. For example, reference [23] contains a comparison of UML and Bunge–Wand–Weber (BWW) model of information systems.

Are there any ontologies about the databases? CIM Database Model [11] is a conceptual model that describes common database management concepts. These concepts correspond mainly to the internal (storage) level of ANSI/X3/Sparc Framework. Examples of the classes in this model are *LogicalFile*, *SystemResource*, *DatabaseServiceStatistics*. A data model specifies constructs that are used in order to build up a conceptual and external level of a database. In addition, the CIM Database model specifies SQL Schema. It is part of the specification of one data model but not databases in general.

References [1], [4], and [5] use a set of core concepts (type, value, variable) as a basis of the description of OR_{TTM}. They do not present the research results as ontology but we believe that these core concepts should be part of an ontology that describes the most basic concepts of conceptual and external level of a database, independent of any specific data model.

A. Proposed Approach for Comparing Data Models

We propose to do the comparison in terms of the components of data models – data structures, integrity rules and data operators. In addition, this comparison should cover the data types. There are different viewpoints whether the specification of data types is one of the components of a data model. According to one school of thought, one of the main differences between the relational model and the object-relational model is that the former supports only simple predefined data types but the latter supports also complex types and allows users to create new types. One the other hand, reference [4] writes: "The question as to what data types are supported is orthogonal to the question of support for the relational model." Even the founder of relational model E. F. Codd acknowledges the possibility of the non-simple domains (types), the permitted values of which are relations [24].

The comparison of two data models should consist of the following parts:

1. Metamodels of the data models in the form of UML class diagrams

If we create the metamodel of a data model based on the database language description, then we first have to decide which parts of the language are relevant in terms of data model and which are orthogonal to it (and therefore have no corresponding constructs in the metamodel of the data model).

We propose to use packages in order to control complexity and create groupings of logically interrelated classes. These packages are – *Data types*, *Data structures*, *Data integrity* and *Data operators*. In some cases, it is necessary to add the stereotype <<singleton>> to a class as [19] does. This stereotype indicates that there is exactly one instance of this class. An example is metaclass *Boolean type* that belongs to package *Data types*. Some attributes of the classes could have type *Enum*. A value of this kind of attribute is one of an enumerated set of values.

2. Mapping between the metaclasses of the metamodels of the data models

For each metaclass in a metamodel, we have to try to find one or more corresponding metaclasses from another metamodel. We have a pair of metaclasses in the mapping if the constructs behind these metaclasses have exactly the same semantics or they are semantically quite similar. Whether or not the constructs are semantically so similar that the mapping can be created depends on the opinion of the persons who perform the comparison. This comparison is a kind of framework that allows us to reason about semantic similarity of different constructs.

3. Discrepancies between the data models

We have to consider at least the constructs that are represented as metaclasses in the metamodels.

Let us assume that we compare two data models A and B. If we decide that data model A has much clearer and much more precise specification than the other data model B, then we can think about A as a kind of ontology. Then we can perform an ontological evaluation of data model B in order to find its construct redundancy, construct overload, construct excess and construct deficit problems.

Generally, we do not prefer one data model and want to compare them without prejudice. Based on the mapping between metaclasses of two data models A and B we can find:

- Cases when a metaclass of A/B has more than one corresponding metaclass of B/A.
- Cases when a metaclass of A/B does not correspond to any metaclass of B/A.

We could use the same names as [23] in order to refer to different cases of discrepancies. If a metaclass of metamodel of A has *more than one* corresponding metaclass of metamodel of B, then its reason could be:

- Data model B (and therefore its metamodel as well) is too complex. Data model A pays more attention to the orthogonality principle of language design. Its one requirement is that a language should provide a comparatively small set of primitive constructs [4]. The metaclasses of B that correspond to the metaclass of A have a common supertype or it is at least possible to create that supertype. We say that there is a construct redundancy in B.
- The construct of A is the counterpart of two or more constructs of B, the semantic of which is very different (in the metamodel of B their corresponding metaclasses do not have a common superclass and it is not possible to create that). We say that there is a construct overload in A. For example, metaclass "table" in OR_{SQL} has corresponding metaclasses "relation variable" and "relation value" in OR_{TTM}.

If a metaclass of the metamodel of A has *no corresponding* metaclass of the metamodel of B, then B has construct deficit A has *construct excess*. Its reasons could be:

- Data model B is less powerful than data model A because it does not provide an important and necessary construct.
- Metamodel of data model B does not have a clearly corresponding metaclass, but it could be created in the metamodel (for example, by creating a common superclass of some existing metaclasses).
- Creators of data model B think that a construct is orthogonal to the data model and therefore it is missing from the specification of B.
- The construct is not in B because creators of B think that a similar effect can be achieved by using other constructs that are already present in B.

In the latter two cases, the authors of B might explicitly argue against a construct. For example, OR_{SQL} allows us to create reference types and typed tables. However, The Third Manifesto argues explicitly against these constructs in the section "OO Prescriptions" because they increase the complexity of a data model without providing clear advantages.

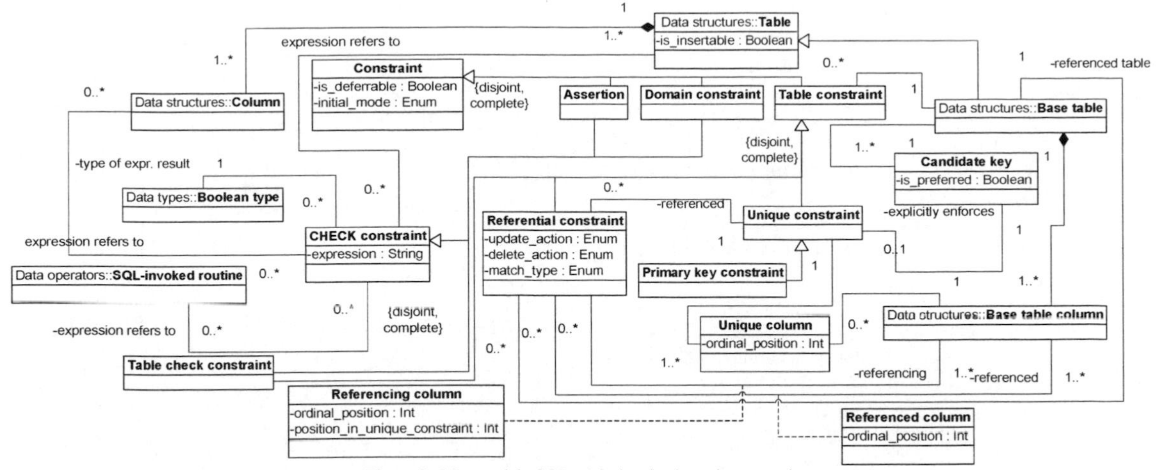

Figure 2. Metamodel of OR_{SQL} declarative integrity constraints.

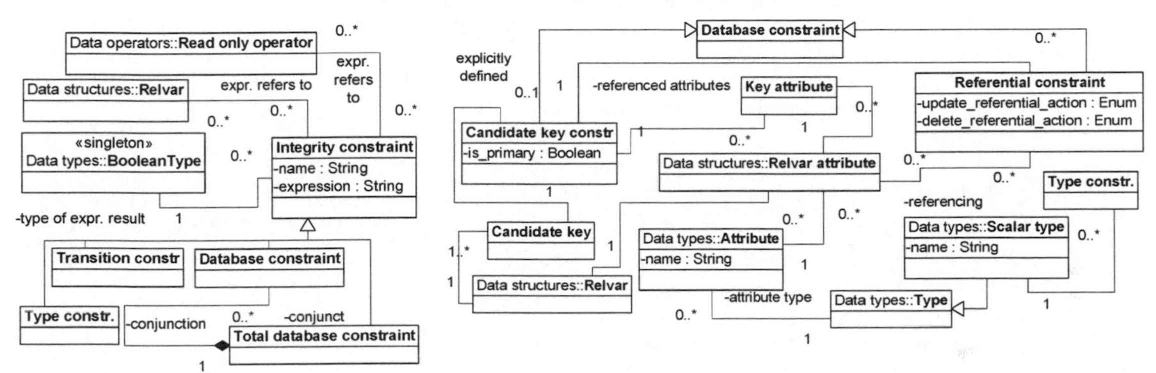

Figure 3. Metamodel of OR_{TTM} declarative integrity constraints.

4. Mapping between the metaclasses does not mean that the constructs behind them have exactly the same semantics. Therefore, we need an additional section that contains the textual description of the differences.
5. Metrics values

For each data model, we propose to calculate at least the amount of metaclasses and the amount of their attributes.

It is sometimes difficult to decide whether to model something by using a class or using an attribute in a UML class diagram. "If in doubt, define something as a separate conceptual class rather than as an attribute." [25] Therefore, we also present the sums of these two values. The resulting values characterize the relative complexity of the data models. We propose to calculate metrics values in case of each package of a metamodel as well as in general for the entire data model.

IV. RESULTS OF COMPARISON OF DATA MODELS

In this section, we present some of the results of comparison of the OR_{SQL} and OR_{TTM} data models.

Due to the space restrictions, we present only the parts of the OR_{SQL} and OR_{TTM} metamodels that specify *declarative* integrity constraints (see Fig. 2 and Fig. 3).

Table I presents mapping of OR_{SQL} and OR_{TTM} metaclasses that belong to package *Data integrity*. We have omitted the part of the OR_{SQL} metamodel that specifies triggers.

TABLE I
MAPPING OF OR_{SQL} AND OR_{TTM} METACLASSES

OR_{SQL} metaclass	OR_{TTM} metaclass
Assertion	Database constraint
Candidate key	Candidate key
CHECK constraint	Database constraint
Constraint	Integrity constraint
Primary key constraint	Candidate key constraint
Referential constraint	Referential constraint
Table check constraint	Database constraint
Table constraint	Database constraint
Unique column	Key attribute
Unique constraint	Candidate key constraint

Construct redundancy in OR_{SQL}:

a) *Assertion*, *CHECK constraint*, *Table check constraint* and *Table constraint* in OR_{SQL} vs. *Database constraint* in OR_{TTM}.

b) *Primary key constraint* and *Unique constraint* in OR_{SQL} vs. *Candidate key constraint* in OR_{TTM}.

Construct deficit in OR_{TTM}: *Domain constraint*, *Referenced column*, *Referencing column*.

Construct deficit in OR_{SQL}: *Transition constraint*, *Type constraint*. The OR_{TTM} metaclass *Total database constraint* does not have one clearly corresponding metaclass in the current OR_{SQL} metamodel. However, it is possible to create it as an abstraction without violating the principles of OR_{SQL}.

For example, construct redundancy (a) points that OR_{SQL} is overly complicated. The possible result of distinguishing table constraints and assertions is that we currently cannot use assertions in any DBMS [26]. OR_{TTM} does not support domain constraints (as in OR_{SQL}) because it considers the concepts "type" and "domain" as synonyms. It does not support the notion of domain as a reusable specification of column properties (as in OR_{SQL}). On the other hand, OR_{TTM} advocates the use of declarative type constraints.

Our current complete metamodel of OR_{SQL} contains 109 metaclasses and 94 attributes (total 203). The current complete metamodel of OR_{TTM} contains 94 metaclasses and 18 attributes (total 112). It shows that OR_{SQL} is more complex compared to OR_{TTM}.

V. CONCLUSIONS

Metamodel-based comparison of modeling methods or ontologies is well known. We applied metamodeling in a new context and proposed the creation of metamodels of data models and a metamodel-based comparison method of data models. Part of this method is the creation of mapping between the metaclasses, identification of discrepancies and calculation of metrics values. We also presented some results of comparison of two data models in order to prove the concept.

A big challenge of the metamodel-based comparison method is the creation of the mapping of metamodel elements. We need clear definitions of the data model constructs in order to reduce subjectivity of this process. Therefore, the use of such comparison method could trigger the creation and improvement of these definitions.

REFERENCES

[1] C.J. Date, *An Introduction to Database Systems*, 8th ed., Boston: Pearson/Addison Wesley, 2003.

[2] M. Stonebraker, L.A. Rowe, B. Lindsay, J. Gray, M. Carey, M. Brodie, P. Bernstein, and D. Beech "Third-generation database system manifesto," Computer Standards and Interfaces, vol. 13, no. 1-3, pp. 41-54, Oct. 1991.

[3] P. Seshadri, "Enhanced abstract data types in object-relational databases," The VLDB Journal, vol. 7, no. 3, 1998, pp. 130-140.

[4] C.J. Date, and H. Darwen, *Foundation for Future Database Systems: The Third Manifesto*, 2nd ed. Massachusetts: Addison-Wesley, 2000.

[5] C.J. Date and H. Darwen, *Databases, Types and the Relational Model*, 3rd edn, Addison Wesley, 2006. Chapter 4 – The Third Manifesto. Retrieved August 13, 2006, from http://www.dcs.warwick.ac.uk/~hugh/TTM/CHAP04.pdf

[6] J. Melton, ISO/IEC 9075-2:2003 (E) Information technology — Database languages — SQL — Part 2: Foundation (SQL/Foundation). August, 2003. Retrieved December 26, 2004, from http://www.wiscorp.com/SQLStandards.html

[7] C. Calero, F. Ruiz, A.L Baroni, F.B. Abreu F, and M. Piattini, "An Ontological Approach to Describe the SQL:2003 Object-Relational Features," Journal of Computer Standards & Interfaces. vol. 28, issue 6. pp. 695-713, 2005.

[8] E.F. Codd, C.J. Date, "Interactive support for non-programmers: The relational and network approaches," In: Proceedings of the 1975 ACM SIGFIDET (now SIGMOD) workshop on Data description, access and control, 1975, pp. 11-41.

[9] E. Eessaar, "Using Relational Databases in the Engineering Repository Systems," In: Proceedings of the Eighth International Conference on Enterprise Information Systems, Paphos, Cyprus, May 23 -27, 2006, Databases and Information Systems Integration, pp. 30 – 37.

[10] E. Seidewitz, "What models mean," IEEE Software, vol. 20, issue 5, pp. 26-31, Sept.-Oct. 2003.

[11] DMTF Common Information Model (CIM) Standards. CIM Schema Ver. 2.13. Database specification. Retrieved October 16, 2006 from http://www.dmtf.org/standards/cim/cim_schema_v213/

[12] J. Greenfield, K. Short, S. Cook, and S. Kent, Software Factories: Assembling Applications with Patterns, Models, Frameworks, and Tools. USA: John Wiley & Sons, 2004.

[13] E.F. Codd, "Data models in database management," SIGART Bull., 74, pp. 112-114, Jan. 1981.

[14] S. Chaudhuri, and G. Weikum, "Rethinking Database System Architecture: Towards a Self-tuning RISC-style Database System," In: Proceedings of Int. Conf. on Very Large Data Bases, 2000, pp. 1-10.

[15] L.M Braz, "Visual syntax diagrams for programming language statements," In: Proceedings of the 8th Annual international Conference on Systems Documentation. New York: ACM Press, 1990, pp. 23-27.

[16] OMG Unified Modeling Language Specification formal/03-03-01. March 2003. Version 1.5.

[17] M. Mühlen, "Evaluation of Workflow Management Systems Using Meta Models," In: Proceedings of the 32nd Hawaii International Conference on System Sciences (HICSS'99), 1999, vol. Track5, pp. 1-11.

[18] I. Davies, P. Green, S. Milton, and M. Rosemann, "Using Meta Models for the Comparison of Ontologies," In: Proceedings Evaluation of Modeling Methods in Systems Analysis and Design Workshop - EMMSAD'03, Klagenfurt/Velden, Austria, 2003.

[19] M. Richters, and M. Gogolla, "A Metamodel for OCL," In: Lecture Notes In Computer Science, issue 1723, Springer, 1999, pp. 156-171.

[20] P. Habela, M. Roantree, and K. Subieta, "Flattening the Metamodel for Object Databases," In: Proceedings of the 6th East European Conference on Advances in Databases and Information Systems, Bratislava, Slovakia, Sept. 8-11, 2002. LNCS 2435, Springer, 2002, pp. 263-276.

[21] K. Siau, and M. Rossi, "Evaluation of Information Modeling Methods -- A Review," In: Proceedings of the 31st Annual Hawaii International Conference on System Sciences, vol. 5, 1998, p. 314.

[22] M. Rossi, and S. Brinkkemper, "Complexity Metrics for Systems Development Methods and Techniques," Information Systems, vol. 21, no. 2, 1996, pp. 209-227.

[23] A.L. Opdahl and B. Henderson-Sellers, "Ontological Evaluation of the UML Using the Bunge–Wand–Weber Model," Software and Systems Modeling, vol. 1, issue 1, pp. 43 – 67, Sept. 2002.

[24] E.F. Codd, "A relational model of large shared data banks," Comm. ACM, vol. 13, no. 6, 1970, pp. 377-387.

[25] C. Larman, *Applying UML and Patterns: An Introduction to Object-Oriented Analysis and Design and the Unified Process*, 2nd edn, Upper Saddle River, USA: Prentice Hall, 2002.

[26] C. Türker, and M. Gertz, "Semantic integrity support in SQL:1999 and commercial (object-) relational database management systems," The VLDB Journal, vol. 10, no. 4, pp. 241–269, 2001.

BEMGA: A HLA Based Simulation Modeling and Development Tool

Ersin Ünsal, Fatih Erdoğan Sevilgen

Gebze Institute of Technology, Department of Computer Engineering,
41400, Kocaeli, Türkiye
eunsal@havelsan.com.tr,
sevilgen@gyte.edu.tr

Abstract- **High Level Architecture (HLA) is a general purpose architecture, developed to support reuse and interoperability across a large number of different types of distributed simulation projects.**

HLA-compliant simulation development is a complex and difficult engineering process. This paper presents a case tool, named BEMGA, which aims to decrease the complexity of the process. Using BEMGA, one can easily model a distributed simulation, generate the simulation software and produce the documentation files from the model.

I. Introduction

The High Level Architecture (HLA) is a general purpose distributed simulation architecture. The HLA was developed under the leadership of the Defense Modeling and Simulation Office (DMSO) to support reuse and interoperability across a large number of various simulation projects developed or maintained by the United States Department of Defense [1]. The HLA was approved as an open standard through the Institute of Electrical and Electronic Engineers (IEEE) - IEEE Standard 1516 - in 2000 [2]. Currently, the HLA is used extensively in modeling and simulation projects [12, 13].

Developing HLA-compliant simulations requires a complex and time-consuming process [4, 5, 6]. Even for a typical "Hello World" application, the required software contains about 1500 lines of code and only the 2% of the code is about simulation logic. The 98% of the code is developed to initialize necessary services and to make the simulation run on a distributed environment. Moreover, developers need a long education period about the HLA specifications, to be able to develop HLA-compliant simulations. There are various tools to overcome the complexity issues about the HLA [10, 11]. But those tools are in their infant ages and a lot has to be done in the field.

BEMGA is a new tool developed to make HLA-compliant simulation development process easier and faster. BEMGA provides a modeling interface to build visual models for simulations and supports code generation for these visual models. Moreover, BEMGA can generate XML based tabular documents describing the model. Platform independence and multi-language (Turkish and English, currently) support are important distinguishing features of BEMGA.

The organization of the paper is as follows. First, there is a short section introducing the HLA specification. Next, BEMGA is introduced in details and a sample simulation development process using BEMGA is provided. Finally, the results will be discussed.

II. High Level Architecture

The High Level Architecture is a general purpose architecture, developed to support reuse and interoperability among distributed simulations. In the HLA, simulations are formed by means of federations. Each federation is a set of cooperating federates. A federate is the simulation component that models a real world domain in a federation. Once a federate is developed, it can be used as part of various distributed simulation systems and it can operate together with other federates.

The HLA specification consists of three elements: HLA Rules, Interface Specification and Object Model Template. HLA Rules define the relationships among federation components. These rules should be obeyed by each federate and federation to be regarded as HLA-compliant. There are totally ten rules, first five rules are for federations and the other five are for federates.

Interface Specification defines the interface between federates and the Run-time Infrastructure (RTI) is the

K. Elleithy (ed.), Advances and Innovations in Systems, Computing Sciences and Software Engineering, 115–119.

implementation of this interface. The RTI provides necessary services for a HLA-compliant simulation to run. RTI services are divided into six management areas:

- Federation Management: Includes tasks such as creating federations, destroying federations, joining federates to federations and resigning federates from federations.
- Declaration Management: Includes tasks to publish and subscribe objects.
- Object Management: Includes object management tasks such as object registration, object update, discovering objects and reflecting objects.
- Ownership Management: Includes tasks for managing the ownership of objects, attributes and interactions.
- Data Distribution Management: Includes tasks for distributing the data efficiently within the distributed simulation.
- Time Management: Includes tasks for managing and enforcing the simulation time policy.

Object Model Template (OMT) defines a common format for definitions of federates and federations. To achieve reusability and interoperability, a standard should define how the objects and interactions should be referred.

Main components of OMT are object classes and interaction classes. Object classes refer to the simulated entities persisting for some interval of simulation time. Object classes have attributes which hold data describing the characteristics of the object class. Interaction classes are simulated entities that do not persist. Interactions are used to represent an occurrence or an event in the simulation. Interaction classes have parameters holding the interaction data. Both object classes and interaction classes form hierarchical structures. Each object class or interaction class has exactly one immediate ancestor or super class. All the object classes extend from ObjectRoot and all the interaction classes extend from InteractionRoot.

OMT defines the Federation Object Model (FOM), the Simulation Object Model (SOM) and the Management Object Model (MOM). For each federation, a FOM introducing all the shared information should be developed. SOM focuses on internal federate details and introduces objects and interactions which can be used externally. Finally, MOM is a universal definition describing the objects and interactions which are designed to manage the federation.

III. BEMGA

BEMGA, which is a modeling and development tool for the HLA-compliant simulations, is being developed for academic interests. The name "BEMGA" is the Turkish abbreviation for "Benzetim Modelleme ve Geli tirme Aracı" which means "Simulation Modeling and Development Tool". The motivation for BEMGA and a preliminary version of BEMGA is presented in a previous work [7]. BEMGA is being developed in the Java programming language. It uses the open source JGraph visual graph library for modeling interface. Basic functionalities of BEMGA are listed below:

- User-friendly Modeling Interface: BEMGA has an elegant and functional modeling interface for building FOM and SOM models. Modeling interface supports drawing and editing object classes and interaction classes. Interaction hierarchies between classes can be modeled by using inheritance arrows. Simulation time policy can be specified within the model. Most of the standard functionalities appearing in other modeling tools such as cut, copy, paste, redo and undo exists in BEMGA. The BEMGA toolbar provides quick access to modeling functionalities.
- Source Code Generation: BEMGA's code generating capabilities allow the developers to build the simulation software easily. Generated software code contains all the necessary initialization and handling functions for communicating with RTI and data-type declarations for the object classes and interaction classes. The generated code is well-documented. Since, BEMGA depends on templates for source code generation; the structure of generated code can be changed easily or code generation for another programming language can be supported without any difficulty.
- Generating Federation File: BEMGA generates federation file which is used by the RTI. Federation file introduces the structure of the objects and interactions in the simulation to the RTI.

- Document Generation: BEMGA generates XML based tabular documents (OMT tables) describing the model.

- Data Interchange Format (DIF) Support: BEMGA supports standard DIF format so, models developed with BEMGA can be used in other case tools supporting the HLA Data Interchange Format.

BEMGA makes use of object oriented techniques and design patterns. MVC, factory and command design patterns are extensively used in BEMGA [8]. Inheritance, polymorphism and delegation are applied where appropriate. Tree algorithms and recursion are employed for processing the model. BEMGA is designed to have a layered architecture. The layered architecture of BEMGA is illustrated in Figure 1.

User interface and modeling layer provides a functional and elegant graphical user interface to the developer. Modeling interface is based on JGraph library [9]. JGraph is an open source graph library supporting modeling diagrams, workflows, flowcharts and organizational charts.

User interface layer does not contain business logic code. User interactions are managed by controller package. Controller package employs command design pattern. Action classes handle user requests. Generation tasks are delegated to the generation package by the controller package. Code generation is based on source code templates. Currently BEMGA can generate only C++ codes. Generation package can also generate tabular documents according to OMT.

File system package have classes to handle file system access. Generation package and controller package delegate I/O operations to file system package. Exceptions and logging is managed in the exception package.

BEMGA's architecture is extensible and maintainable since object oriented principles and design patterns are used. Furthermore, BEMGA is developed by using the Eclipse IDE and various Eclipse plug-ins such as PMD and JMeter, to improve the quality of the software.

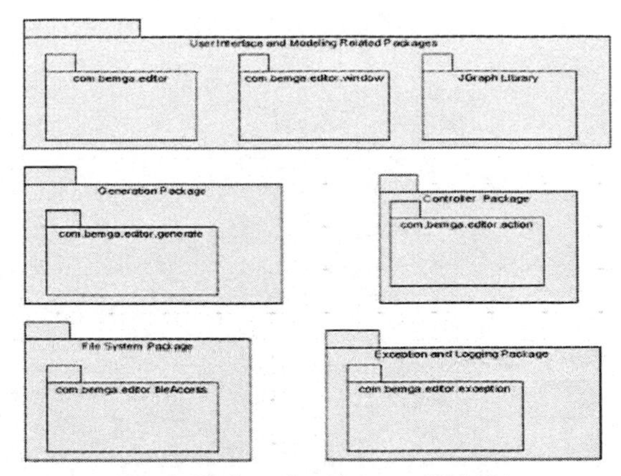

Fig. 1. The layered architecture of BEMGA

A. Developing a HLA-Compliant Weapon Management Simulation

To introduce BEMGA, a HLA-compliant Weapon Management Simulation (WMS) is developed with BEMGA. Fighters, reconnaissance planes (recce) and a control center are the main players of the WMS. A recce caps over a predefined orbit and sends an "EnemyAppeared" interaction whenever an enemy fighter is detected. When the control center receives an "EnemyAppeared" interaction message, the control center sends an "EmergencyCase" interaction. The fighters will approach to the enemy fighters when "EmergencyCase" interaction is received.

B. Modeling the Simulation with BEMGA

The model developed for the WMS simulation is shown in Figure 2. In the model, the Fighter and the Recce object classes inherit from the super class, Aircraft. The Aircraft has four attributes: Position, Speed, Direction and Waypoints. The other object class in the model is the ControlCenter object class. There are two interaction classes, EnemyAppeared and EmergencyCase. Both interaction classes have one attribute named Position which is used to specify the position of the event.

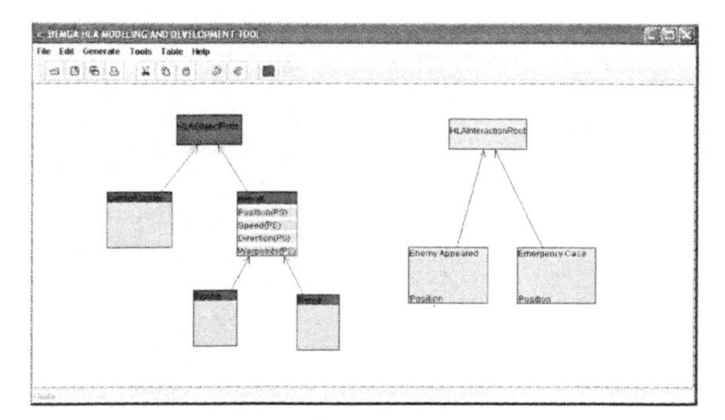

Fig. 2. Federation Object Model

C. Generating Federation File, Simulation Software and Documentation

To develop the simulation, Simulation Object Models for the AirControl federate, AirDefense federate and Viewer federate were built using BEMGA. Building a SOM is very straightforward with BEMGA. BEMGA asks for the related FOM when a new SOM is started. Afterwards, the developer can use "InsertFromFOM" pop-up menu item to select the object classes and interaction classes from the specified FOM. BEMGA will automatically insert the selected classes into the SOM. Then the developer can generate federation files and necessary source code files for the simulation.

BEMGA generates C++ source code files for the related SOM. These files can be compiled and an executable federate can be produced. However, generated source code does not include any simulation logic, so the execution of the federate just builds the federation and iterate on an empty while loop. Simulation logic implementation should be inserted properly into the generated source code. In Table 1, an outline of generated source code files and changes to the source code files for the AirDefense federate is given. The last column of Table 1 shows the percentage of generated source code over total source code for the working federate: 96 percent of the total simulation source code is automatically generated by BEMGA. Similar results are obtained for other federates. It should be noted that the simulation logic for this federate is very basic and minor changes have been done to the generated source code. For a complex federate, the size of the simulation logic code and the percentages given in Table 1 will change dramatically. Here the aim is to show that BEMGA generates all the necessary source code to build up a basic federate and allows the simulation developer to focus on the simulation logic code.

TABLE I Outline of the source code files for AirDefense federate

Federate	Source File	Generated LOC	Inserted LOC	%
AirDefense	Aircraft.h	130	0	100%
	Aircraft.cpp	766	0	100%
	Fighter.h	132	0	100%
	Fighter.cpp	766	6	99%
	Recce.h	132	0	100%
	Recce.cpp	766	0	99%
	EnemyAppeared.h	55	0	100%
	EnemyAppeared.cpp	45	0	100%
	AirDefense.h	51	3	94%
	AirDefense.cpp	329	103	76%
	AirDefenseFedAmb.h	361	0	100%
	AirDefenseFedAmb.cpp	552	0	100%
	AirDefenseMain.cpp	17	0	100%
	Globals.h	9	16	36%
	Total	4113	4241	97%

BEMGA can also generate documentation for the FOM. Documentation includes XML based tabular OMT tables: Object Model Identification Table, Object Class Structure Table, Interaction Class Structure Table, Attribute Table and Parameter Table. Table 2, 3 and 4 show some of the OMT tables generated for WMS FOM model. Object Model Identification Table contains basic information about the model and model developer. Object Class Structure Table and Interaction Class Structure Table contains the objects and interactions in the model. These tables also shows the interaction hierarchy between the classes. Finally, Attribute Table and Parameter Table contains detailed information about the attributes of object classes and parameters of the interaction classes respectively.

TABLE II Object Model Identification Table for WMS

Object Model Identification Table	
Category	Information
Name	Weapon Management Simulation
Type	FOM
Version	1.0
Modification Date	28-Aug-06
Purpose	Sample
Application Domain	Air Defense Operations
Sponsor	-
POC	Ersin ÜNSAL
POC Organization	GYTE
POC Email	eunsal@havelsan.com.tr
References	CIS2E 2006

TABLE III Object Class Structure Table for WMS

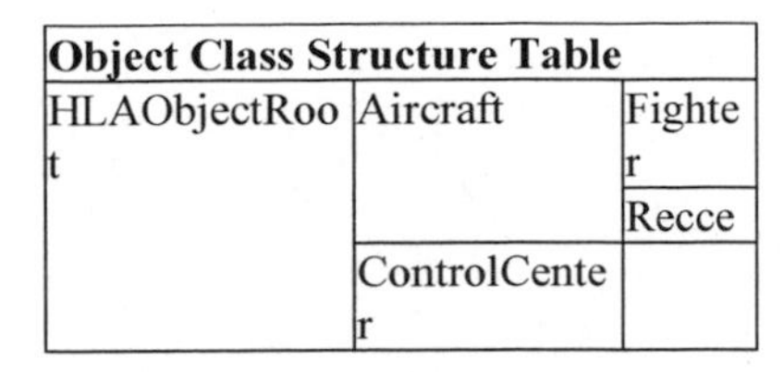

Object Class Structure Table		
HLAObjectRoot	Aircraft	Fighter
		Recce
	ControlCenter	

Table IV Interaction Class Structure Table for WMS

Interaction Class Structure Table	
HLAInteractionRoot	EnemyAppeared
	EmergencyCase

D. Running the WMS Simulation

A screenshot of the Viewer federate can be seen in Figure 3. The screenshot is taken when the "EnemyAppeared" interaction is sent by the AirControl federate. To run the simulation the rtiexec process should be started, firstly. Then, AirControl, AirDefense and Viewer federates are started from the command line. The simulation starts to run after the federates build up and join to the the federation. The sample simulation runs on the RTI1.3NG-V3.2 [3]. Viewer federate has a GUI developed as a windows dll. When an event occurs, the Viewer federate updates its GUI.

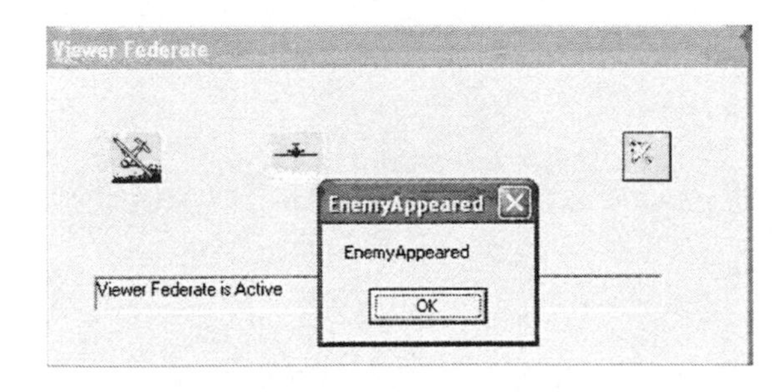

Fig. 3. Screenshot of the Viewer Federate

IV. Conclusions and Future Work

BEMGA can speed up the simulation development process by reducing the complexity of this process. BEMGA allows visual modeling and generates code for these visual models. Generated code supports federation management, declaration management, object management and time management services.

The visual modeling interface of BEMGA improves understandability and maintainability of the object models. Developers can update models and regenerate the simulation code and documentation. Federation file generation and documentation are important aspects of HLA based simulation development process and BEMGA provides necessary support in these steps.

BEMGA's current features address most of the complexities of HLA-compliant simulation development process. However, BEMGA can be improved by adding following new capabilities:

- Support for optimistic time management policy,
- Support for Data Distribution Management and Ownership Management services of Interface Specification,
- Code generation support for other programming languages such as Java,
- Providing necessary interfaces to manage the code generation process,
- Support for behavioral modeling and generation of simulation logic code for behavioral models.

References

1. Defense Modeling and Simulation Office, http://www.dmso.mil/
2. Institute of Electrical and Electronics Engineers, Document Number: 1516-2000
3. RTI 1.3-Next Generation Programmer's Guide Version 3.2, DMSO, 2000
4. Kuhl, F., Weatherly, R., Dahmann, J.: "Creating High Level Computer Simulation Systems", Prentice Hall PTR, 1999
5. Parr, S., Radeski, A., Whitney, R.: "The Application Of Tools Support in HLA", SimTect, 2002
6. Radeski, A., Parr, S.: "Towards a Simulation Component Model for HLA", SISO SW, Fall 2002
7. Ünsal, E., Sevilgen, F.E.: "Yüksek Seviyeli Mimari İçin Modelleme ve Uygulama Geliştirme Aracı", SAVTEK, 2006
8. Gamma, E., Helm, R., Johnson, R., Vlissides, J.: Design Patterns Elements of Reusable Object-Oriented Software Components, Addison Wesley Longman, 1998
9. JGraph Graph Library, www.jgraph.com
10. Simplicity, http://www.calytrix.com/default.php
11. Visual OMT 1516, http://www.pitch.se/visualomt/default.asp
12. RAMOS Project, http://www.modsim.metu.edu.tr/ramos.htm
13. DS-GRID Project, http://www.cs.bham.ac.uk/research/projects/dsgrid/

Comparison of Different POS Tagging Techniques (n-gram, HMM and Brill's tagger) for Bangla

Fahim Muhammad Hasan
Email: stealth_310@yahoo.com
Naushad UzZaman
Email: naushad@bracuniversity.nct
Mumit Khan
Email: mumit@bracuniversity.net
Center for Research on Bangla Language Processing
BRAC University
Bangladesh

Abstract-**There are different approaches to the problem of assigning each word of a text with a parts-of-speech tag, which is known as Part-Of-Speech (POS) tagging. In this paper we compare the performance of a few POS tagging techniques for Bangla language, e.g. statistical approach (n-gram, HMM) and transformation based approach (Brill's tagger). A supervised POS tagging approach requires a large amount of annotated training corpus to tag properly. At this initial stage of POS-tagging for Bangla, we have very limited resource of annotated corpus. We tried to see which technique maximizes the performance with this limited resource. We also checked the performance for English and tried to conclude how these techniques might perform if we can manage a substantial amount of annotated corpus.**

Keywords: POS tagging, POS tagger, Bangla, Bengali, Bangla, n-gram, HMM, Brill's transformation based tagger.

I. Introduction

Bangla is among the top ten most widely spoken languages [1] with more than 200 million native speakers, but it still lacks significant research efforts in the area of natural language processing.

Part-of-Speech (POS) tagging is a technique for assigning each word of a text with an appropriate parts of speech tag. The significance of part-of-speech (also known as POS, word classes, morphological classes, or lexical tags) for language processing is the large amount of information they give about a word and its neighbor. POS tagging can be used in TTS (Text to Speech), information retrieval, shallow parsing, information extraction, linguistic research for corpora [2] and also as an intermediate step for higher level NLP tasks such as parsing, semantics, translation, and many more [3]. POS tagging, thus, is a necessary application for advanced NLP applications in Bangla or any other languages.

We start this paper by giving an overview of a few POS tagging models; we then discuss what have been done for Bangla. Then we show the methodologies we used for POS tagging; then we describe our POS tagset, training and test corpus; next we show how these methodologies perform for both English and Bangla; finally we conclude how Bangla (language with limited language resources, tagged corpus) might perform in comparison to English (language with available tagged corpus).

II. Literature Review

Different approaches have been used for Part-of-Speech (POS) tagging, where the notable ones are rule-based, stochastic, or transformation-based learning approaches. Rule-based taggers [4, 5, 6] try to assign a tag to each word using a set of hand-written rules. These rules could specify, for instance, that a word following a determiner and an adjective must be a noun. Of course, this means that the set of rules must be properly written and checked by human experts. The stochastic (probabilistic) approach [7, 8, 9, 10] uses a training corpus to pick the most probable tag for a word. All probabilistic methods cited above are based on first order or second order Markov models. There are a few other techniques which use probabilistic approach for POS Tagging, such as the Tree Tagger [11]. Finally, the transformation-based approach combines the rule-based approach and statistical approach. It picks the most likely tag based on a training corpus and then applies a certain set of rules to see whether the tag should be changed to anything else. It saves any new rules that it has learnt in the process, for future use. One example of an effective tagger in this category is the Brill tagger [12, 13, 14, 15].

K. Elleithy (ed.), Advances and Innovations in Systems, Computing Sciences and Software Engineering, 121–126.

All of the approaches discussed above fall under the rubric of supervised POS Tagging, where a pre-tagged corpus is a prerequisite. On the other hand, there is the unsupervised POS tagging [16, 17, 18] technique, and it does not require any pre-tagged corpora.

The following tree figure demonstrates the classification of different POS tagging schemes.

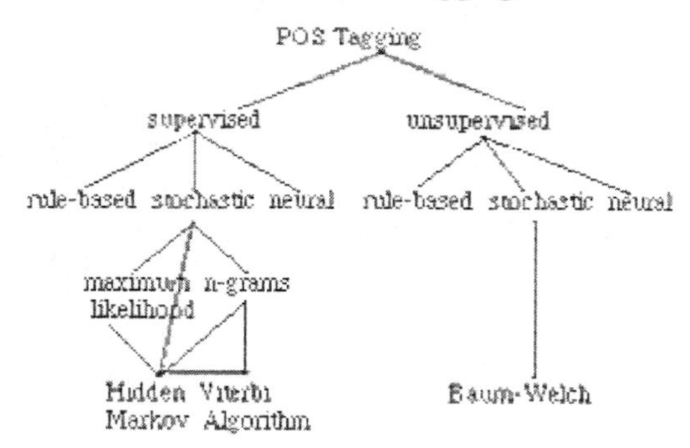

Figure: Classification of POS tagging models [19]

For English and many other western languages many such POS tagging techniques have been implemented and in almost all the cases, they show a satisfying performance of 96+%. For Bangla work on POS tagging has been reported by [20, Chowdhury et al. (2004) and Seddiqui et al. (2003).

Chowdhury et al. (2004) implemented a rule based POS tagger, which requires writing laboriously handcrafted rules by human experts and many years of continuous efforts from many linguists. Since they report no performance analysis of their work, the feasibility of their proposed rule based method for Bangla is suspect. No review or comparison of established work on Bangla POS tagging was available in that paper; they only proposed a rule-based technique. Their work can be described as more of a morphological analyzer than a POS tagger. A morphological analyzer indeed provides some POS tag information, but a POS-tagger needs to operate on a large set of fine-grained tags. For example, the [23] for English consists of 87 distinct tags, and Penn Treebank's [24] tagset consists of 48 tags. Chowdhury et al.'s tagset, by contrast, consists of only 9 tags and they showed only rules for nouns and adjectives for their POS Tagger. Such a POS-tagger's output will have very limited applicability in many advanced NLP applications.

For English, researchers had tried this rule-based technique in the 60s and 70s [4, 5, 6]. Taking into consideration of the problem of this method, researchers have switched to statistical or machine learning methods, or more recently, to the unsupervised methods for POS tagging.

In this paper we compare the performance of different tagging techniques such as Brill's tagger, n-gram tagger and HMM tagger for Bangla; such comparison was not attempted in [20, 21, 22].

III. METHODOLOGY

NLTK [25], the Natural Language Toolkit, is a suite of program modules, data sets and tutorials supporting research and teaching in computational linguistics and natural language processing. NLTK has many modules implemented for different NLP applications. We have experimented unigram, bigram, HMM and Brill tagging modules from NLTK [25] for our purpose.

Unigram Tagger

The unigram (n-gram, n = 1) tagger is a simple statistical tagging algorithm. For each token, it assigns the tag that is most likely for that token's text. For example, it will assign the tag `jj` to any occurrence of the word `frequent`, since `frequent` is used as an adjective (e.g. a `frequent` word) more often than it is used as a verb (e.g. I `frequent` this cafe).

Before a unigram tagger can be used to tag data, it must be trained on a training corpus. It uses the corpus to determine which tags are most common for each word.

The unigram tagger will assign the default tag `None` to any token that was not encountered in the training data.

HMM

The intuition behind HMM (Hidden Markov Model) and all stochastic taggers is a simple generalization of the "pick the most likely tag for this word" approach. The unigram tagger only considers the probability of a word for a given tag *t*; the surrounding context of that word is not considered.

On the other hand, for a given sentence or word sequence, HMM taggers choose the tag sequence that maximizes the following formula:

```
P (word | tag) * P (tag | previous n tags)
```

Brill's transformation based tagger [15]

A potential issue with nth-order tagger is their size. If tagging is to be employed in a variety of language technologies deployed on mobile computing devices, it is important to find ways to reduce the size of models without overly compromising performance. An nth-order tagger with backoff may store trigram and bigram tables, large sparse arrays, which may have hundreds of millions of entries. A consequence of the size of the models is that it is simply impractical for nth-order models to be conditioned on the identities of words in the context. In this section we will examine Brill tagging, a statistical tagging method which performs very well, using models that are only a tiny fraction of the size of nth-order taggers.

Brill tagging is a kind of transformation-based learning. The general idea is very simple: guess the tag of each word, then go back and fix the mistakes. In this way, a Brill tagger successively transforms a bad tagging of a text into a good one. As with nth-order tagging this is a supervised learning method, since we need annotated training data. However, unlike nth-order tagging, it does not count observations but compiles a list of transformational correction rules.

The process of Brill tagging is usually explained by analogy with painting. Suppose we were painting a tree, with all its details of boughs, branches, twigs and leaves, against a uniform sky-blue background. Instead of painting the tree first then trying to paint blue in the gaps, it is simpler to paint the whole canvas blue, then "correct" the tree section by overpainting the blue background.

In the same fashion we might paint the trunk a uniform brown before going back to overpaint further details with a fine brush. Brill tagging uses the same idea: get the bulk of the painting right with broad brush strokes, then fix up the details. As time goes on, successively finer brushes are used, and the scale of the changes becomes arbitrarily small. The decision of when to stop is somewhat arbitrary.

In our experiment we have used the taggers (Unigram, HMM, Brill's transformation based tagger) described above. Detailed descriptions of these taggers are available at [2, 26].

IV. POS TAGSET

For English we have used the Brown Tagset [23]. And for Bangla we have used a 41 tag-sized tagset [28]. Our tagset has two levels of tags. First level is the high-level tag for Bangla, which consists of only 12 tags (Noun, Adjective, Cardinal, Ordinal, Fractional, Pronoun, Indeclinable, Verb, Post Positions, Quantifiers, Adverb, Punctuation). And the second level is more fine-grained with 41 tags. Most of our experiments are based on the level 2 tagset (41 tags). However, we experimented few cases with level 1 tagset (12 tags).

V. TRAINING CORPUS AND TEST SET

For our experiment for English, we have used tagged Brown corpus from NLTK [25]. For Bangla, we have a very small corpus of around 5000 words from a Bangladeshi daily newspaper Prothom-alo [27]. In both cases, our test set is disjoint from the training corpus.

VI. TAGGING EXAMPLE

Bangla (Training corpus size: 4484 tokens)

Untagged Text:

1. দ্বিতীয় বিশ্বযুদ্ধে মিত্র বাহিনীর নেতা ব্রিটিশ প্রধানমন্ত্রী উইন্সটন চার্চিলকে গত সপ্তাহের শুরুতে টপকে বেয়ার এ তালিকায় স্থান লাভ করেন ।
2. তবে তিনি যদি আবার নির্বাচন করেন এবং জয়ী হন তাহলে হয়তো এ রেকর্ডও ভাঙতে পারবেন ।

Tagged output:

Level 2 Tagset (41 Tags)

Brill:

1. দ্বিতীয়/NC বিশ্বযুদ্ধে/NC মিত্র/NC বাহিনীর/NC নেতা/NC ব্রিটিশ/ADJ প্রধানমন্ত্রী/NC উইন্সটন/NP চার্চিলকে/NP গত/ADJ সপ্তাহের/NC শুরুতে/ADVT টপকে/NP বেয়ার/NP এ/DP তালিকায়/NC স্থান/NC লাভ/NC করেন/VF ।/PUNSF
2. তবে/INDO তিনি/PP যদি/INDO আবার/ADVM নির্বাচন/NC করেন/VF এবং/CONJC জয়ী/NC হন/VE তাহলে/INDO হয়তো/OTHER এ/DP রেকর্ডও/NC ভাঙতে/NC পারবেন/VF ।/PUNSF

Unigram:

1. দ্বিতীয়/NP বিশ্বযুদ্ধে/NP মিত্র/NP বাহিনীর/NC নেতা/NC ব্রিটিশ/ADJ প্রধানমন্ত্রী/NC উইন্সটন/NP চার্চিলকে/NP গত/ADJ সপ্তাহের/NC শুরুতে/ADVT টপকে/NP বেয়ার/NP এ/DP তালিকায়/NC স্থান/NC লাভ/NP করেন/VF ।/PUNSF
2. তবে/INDO তিনি/PP যদি/INDO আবার/ADVM নির্বাচন/NC করেন/VF এবং/CONJC জয়ী/NP হন/VE তাহলে/INDO হয়তো/OTHER এ/DP রেকর্ডও/NP ভাঙতে/NP পারবেন/NP ।/PUNSF

HMM:

1. দ্বিতীয়/DP বিশ্বযুদ্ধে/NC মিত্র/NC বাহিনীর/NC নেতা/NC ব্রিটিশ/ADJ প্রধানমন্ত্রী/NC উইন্সটন/NP চার্চিলকে/NP গত/ADJ সপ্তাহের/NC শুরুতে/ADVT টপকে/ADVT বেয়ার/NP এ/NP তালিকায়/NC স্থান/NC লাভ/NC করেন/VF ।/PUNSF
2. তবে/INDO তিনি/PP যদি/INDO আবার/ADVM নির্বাচন/NC করেন/VF এবং/CONJC জয়ী/NC হন/VF তাহলে/PUNSF হয়তো/OTHER এ/DP রেকর্ডও/NC ভাঙতে/VNF পারবেন/VF ।/PUNSF

Level 1 Tagset (Reduced Tagset: 12 Tags)

Brill:

1. দ্বিতীয়/NN বিশ্বযুদ্ধে/NN মিত্র/NN বাহিনীর/NN নেতা/NN ব্রিটিশ/ADJ প্রধানমন্ত্রী/NN উইন্সটন/NN চার্চিলকে/NN গত/ADJ সপ্তাহের/NN শুরুতে/ADV টপকে/NN বেয়ার/NN এ/PN তালিকায়/NN স্থান/NN লাভ/NN করেন/VB ।/PUNC
2. তবে/IND তিনি/PN যদি/IND আবার/ADV নির্বাচন/NN করেন/VB এবং/IND জয়ী/NN হন/VB তাহলে/IND হয়তো/OTHER এ/PN রেকর্ডও/NN ভাঙতে/VB পারবেন/VB ।/PUNC

Unigram:

1. দ্বিতীয়/NN বিশ্বযুদ্ধে/NN মিত্র/NN বাহিনীর/NN নেতা/NN ব্রিটিশ/ADJ প্রধানমন্ত্রী/NN উইন্সটন/NN চার্চিলকে/NN গত/ADJ সপ্তাহের/NN শুরুতে/ADV টপকে/NN বেয়ার/NN এ/PN তালিকায়/NN স্থান/NN লাভ/NN করেন/VB ।/PUNC
2. তবে/IND তিনি/PN যদি/IND আবার/ADV নির্বাচন/NN করেন/VB এবং/IND জয়ী/NN হন/VB তাহলে/IND হয়তো/OTHER এ/PN রেকর্ডও/NN ভাঙতে/VB পারবেন/NN ।/PUNC

HMM:

1. দ্বিতীয়/PN বিশ্বযুদ্ধে/NN মিত্র/NN বাহিনীর/NN নেতা/NN ব্রিটিশ/ADJ প্রধানমন্ত্রী/NN উইন্সটন/NN চার্চিলকে/NN গত/ADJ সপ্তাহের/NN শুরুতে/ADV টপকে/ADV বেয়ার/NN এ/NN তালিকায়/NN স্থান/NN লাভ/NN করেন/VB ।/PUNC

2. তবে/IND তিনি/PN যদি/IND আবার/ADV নির্বাচন/NN করেন/VB এবং/IND জয়ী/NN হন/VB তাহলে/IND হয়তো/OTHER এ/PN রেকর্ডও/NN ভাঙতে/VB পারবেন/VB ।/PUNC

VII. PERFORMANCE

We have experimented POS taggers (Unigram, HMM, Brill) for both Bangla and English. For Bangla we experimented in both tag levels (level 1 – 12 tags, level 2 – 41 tags). Experiment results are given below in form of table and graph.

Tokens	HMM Accuracy	Unigram Accuracy	Brill Accuracy
0	0	0	0
60	15.4	51.2	50.4
104	18	51.1	44.6
503	34.2	60.7	56.3
1011	42.3	64.2	62.6
2023	45.8	69.1	67.8
3016	49.4	70.1	70.9
4484	45.6	71.2	71.3

Table 1: Performance of POS Taggers for Bangla [Test data: 85 sentences, 1000 tokens from the (Prothom-Alo) corpus; Tagset: Level 1 Tagset (12 Tags)]

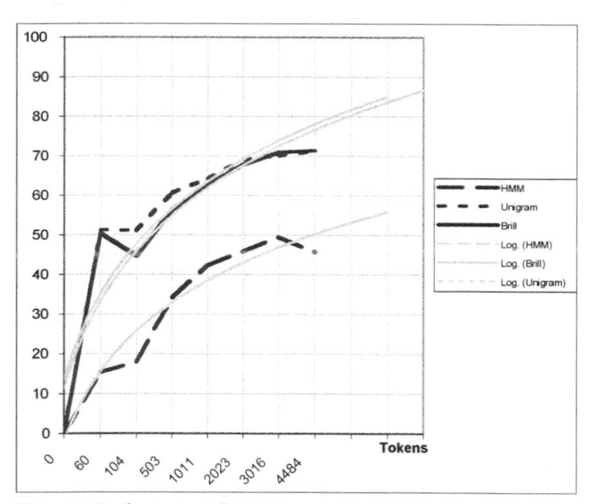

Figure 1: Performance of POS Taggers for Bangla [Test data: 85 sentences, 1000 tokens from the (Prothom-Alo) corpus; Tagset: Level 1 Tagset (12 Tags)]

Tokens	HMM Accuracy	Unigram Accuracy	Brill Accuracy
0	0	0	0
60	19.7	17.2	38.7
104	18.1	17.4	26.2
503	28.8	26.1	46.1
1011	32.8	30	51.1
2023	40.1	36.7	49.4
3016	44.5	39.1	51.9
4484	46.9	42.2	54.9

Table 2: Performance of POS Taggers for Bangla [Test data: 85 sentences, 1000 tokens from the (Prothom-Alo) corpus; Tagset: Level 2 Tagset (41 Tags)]

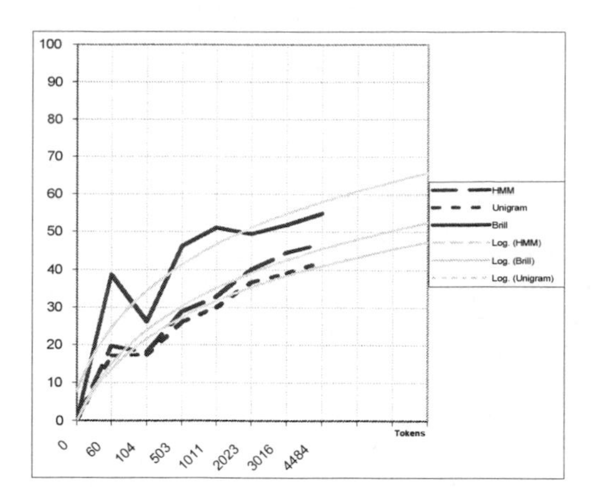

Figure 2: Performance of POS Taggers for Bangla [Test data: 85 sentences, 1000 tokens from the (Prothom-Alo) corpus; Tagset: Level 2 Tagset (41 Tags)]

Tokens	HMM Accuracy	Unigram Accuracy	Brill Accuracy
0	0	0	0
65	36.9	28.7	33.6
134	44.2	34	42.9
523	53.4	41.6	53.7
1006	62	47.7	58.3
2007	66.8	52.4	62.9
3003	68.2	55.1	66.1
4042	70	57.2	67.5
5032	71.5	59.2	70.2
6008	71.9	60.8	71.4
7032	74.5	61.5	71.8
8010	74.8	62.1	72.4
9029	76.8	63.5	74.5
10006	77.5	65.2	75.2
20011	80.9	69.5	79.8
30017	83.1	71.7	78.8
40044	84.7	73.3	79.8
50001	84.6	74.4	80.4
60022	85.3	75.2	80.8
70026	86.3	75.8	81
80036	87.1	77.1	81.6
90000	87.8	78.1	82.4
100057	87.5	78.9	83.4
200043	91.7	83	86.8
300359	89.5	84.2	87.3
400017	89.7	84.8	88.5
500049	90.3	85.6	
600070	90	85.9	
700119	90.3	86.1	
800031	90.2	86.2	

900073	90.3	86.6	
1000107	90.3	86.5	

Table 3: Performance of POS Taggers for English [Test data: 22 sentences, 1008 tokens from the Brown corpus; Tagset: Brown Tagset]

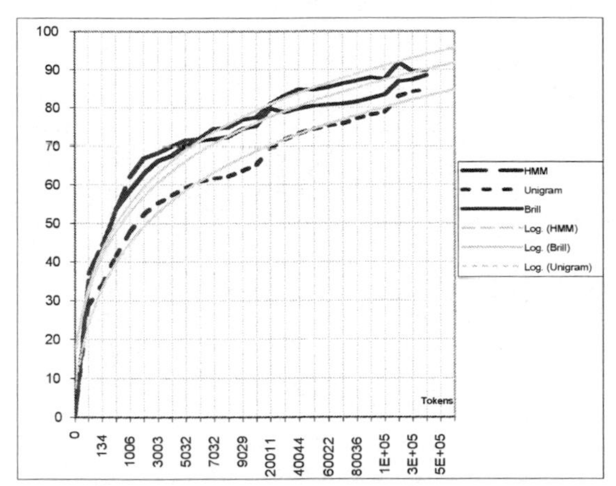

Figure 3: Performance of POS Taggers for English [Test data: 22 sentences, 1008 tokens from the Brown corpus; Tagset: Brown Tagset]

VIII. Analysis Of Test Result

English POS taggers report high accuracy of 96+%, where the same taggers did not perform the same (only 90%) in our case. This is because others tested on a large training set for their taggers, whereas we tested our English taggers on a maximum of 1 million sized corpus (for HMM and unigram) and for Brill, we tested under training of 400 thousand tokens.

Since our Bangla taggers were being tested on a very small-sized corpus (with a maximum of 4048 tokens), the resulting performance by them was not satisfactory. This was expected, however, as the same taggers performed similarly for a similar-sized English corpus (see Table 3). For English we have seen that performance increases with the increase of corpus size. For Bangla we have seen it follows the same trend as English. So, it can be safely hypothesized that if we can extend the corpus size of Bangla then we will be able to get the similar performance for Bangla as English.

Within this limited corpus (4048 tokens), our experiment suggests that for Bangla (both with 12-tag tagset and 41-tag tagset), Brill's tagger performed better than HMM-based tagger and Unigram tagger (see Tables 1, 2). Researchers who are studying a sister language of Bangla and want to implement a POS tagger can try Brill's tagger, at least for a small-sized corpus.

IX. Future Work

Unsupervised POS tagging is a very good choice for languages with limited POS tagged corpora. We want to check how Bangla performs using unsupervised POS tagging techniques.

In parallel to the study of unsupervised techniques, we want to try a few other state of the art POS tagging techniques for Bangla. In another study we have seen that in case of n-gram based POS tagging, backward n-gram (considers next words) performs better than usual forward n-gram (considers previous words).

Our final target is to propose a hybrid solution for POS tagging in Bangla that performs with 95%+ as in English or other western languages and use this POS tagger in other advanced NLP applications.

X. Conclusion

We showed that using n-gram (unigram), HMM and Brill's transformation based techniques, the POS tagging performance for Bangla is approaching that of English. With the training set of around 5000 words and a 41-tag tagset, we get a performance of 55%. With a much larger training set, it should be possible to increase the level of accuracy of Bangla POS taggers comparable to the one achieved by English POS taggers.

XI. Acknowledgment

This work has been supported in part by the PAN Localization Project (www.panl10n.net) grant from the International Development Research Center, Ottawa, Canada, administrated through Center for Research in Urdu Language Processing, National University of Computer and Emerging Sciences, Pakistan.

References

[1] The Summer Institute for Linguistics (SIL) Ethnologue Survey (1999).
[2] Daniel Jurafsky and James H. Martin, *Chapter 8: Word classes and Part-Of-Speech Tagging*, Speech and Language Processing, Prentice Hall, 2000.
[3] Yair Halevi, *Part of Speech Tagging*, Seminar in Natural Language Processing and Computational Linguistics (Prof. Nachum Dershowitz), School of Computer Science, Tel Aviv University, Israel, April 2006.
[4] B. Greene and G. Rubin, *Automatic Grammatical Tagging of English*, Technical Report, Department of Linguistics, Brown University, Providence, Rhode Island, 1971.
[5] S. Klein and R. Simmons, *A computational approach to grammatical coding of English words*, JACM 10, 1963.
[6] Z. Harris, *String Analysis of Language Structure*, Mouton and Co., The Hague, 1962.
[7] L. Bahl and R. L. Mercer, *Part-Of-Speech assignment by a statistical decision algorithm*, IEEE International Symposium on Information Theory, pages: 88 - 89, 1976.
[8] K. W. Church, *A stochastic parts program and noun phrase parser for unrestricted test*, In

proceeding of the Second Conference on Applied Natural Language Processing, pages: 136 - 143, 1988.

[9] D. Cutting, J. Kupiec, J. Pederson and P. Sibun, *A practical Part-Of-Speech Tagger*, In proceedings of the Third Conference on Applied Natural Language Processing, pages: 133 - 140, ACL, Trento, Italy, 1992.

[10] S. J. DeRose, *Grammatical Category Disambiguation by Statistical Optimization*, Computational Linguistics, 14 (1), 1988

[11] Helmut Schmid, *Probabilistic Part-Of-Speech Tagging using Decision Trees*, In Proceedings of The International Conference on new methods in language processing, page 44 - 49, Manchester, UK, 1994.

[12] Eric Brill, *A simple rule based part of speech tagger*, In Proceedings of the Third Conference on Applied Natural Language Processing, ACL, Trento, Italy, 1992.

[13] Eric Brill, *Automatic grammar induction and parsing free text: A transformation based approach*, In proceedings of 31st Meeting of the Association of Computational Linguistics, Columbus, Oh, 1993.

[14] Eric Brill, *Transformation based error driven parsing*, In Proceedings of the Third International Workshop on Parsing Technologies, Tilburg, The Netherlands, 1993.

[15] Eric Brill, *Some advances in rule based part of speech tagging*, In Proceedings of The Twelfth National Conference on Artificial Intelligence (AAAI-94), Seattle, Washington, 1994.

[16] Robbert Prins and Gertjan van Noord, *Unsupervised Pos-Tagging Improves Parsing Accuracy And Parsing Efficiency*, In Proceedings of the International Workshop on Parsing Technologies, 2001.

[17] Mihai Pop, *Unsupervised Part-of-speech Tagging*, Department of Computer Science, Johns Hopkins University, 1996.

[18] Eric Brill, *Unsupervised Learning of Disambiguation Rules for Part of Speech Tagging*, In Proceeding of The Natural Language Processing Using Very Large Corpora, Boston, MA, 1997.

[19] Linda Van Guilder, *Automated Part of Speech Tagging: A Brief Overview*, Handout for LING361, Fall 1995, Georgetown University.

[20] Sandipan Dandapat, Sudeshna Sarkar and Anupan Basu, *A Hybrid Model for Part-Of-Speech Tagging and its Application to Bengali*, In Proceedings of the International Journal of Information Technology, Volume 1, Number 4.

[21] Md. Shahnur Azad Chowdhury, Nahid Mohammad Minhaz Uddin, Mohammad Imran, Mohammad Mahadi Hassan, and Md. Emdadul Haque, *Parts of Speech Tagging of Bangla Sentence*, In Proceeding of the 7th International Conference on Computer and Information Technology (ICCIT), Bangladesh, 2004.

[22] Md. Hanif Seddiqui, A. K. Muhammad Shohel Rana, Abdullah Al Mahmud and Taufique Sayeed, *Parts of Speech Tagging Using Morphological Analysis in Bangla*, In Proceeding of the 6th International Conference on Computer and Information Technology (ICCIT), Bangladesh, 2003.

[23] Brown Tagset, available online at: http://www.scs.leeds.ac.uk/amalgam/tagsets/brown.html

[24] Mitchell P. Marcus, Beatrice Santorini and Mary Ann Marcinkiewicz, *Building a Large Annotated Corpus of English: The Penn Treebank*, Computational Linguistics Journal, Volume 19,Number 2, Pages: 313-330, 1994. Available online at: http://www.ldc.upenn.edu/Catalog/docs/treebank2/cl93.html

[25] NLTK, The Natural Language Toolkit, available online at: http://nltk.sourceforge.net/index.html

[26] NLTK's tagger documentation, available online at: http://nltk.sourceforge.net/tutorial/tagging.pdf

[27] Bangla Newspaper, Prothom-Alo. Online version available online at: http://www.prothom-alo.net

[28] Bangla POS Tagset used in our Bangla POS tagger, available online at http://www.naushadzaman.com/bangla_tagset.pdf

Real-Time Simulation and Data Fusion of Navigation Sensors for Autonomous Aerial Vehicles

Francesco Esposito, Domenico Accardo, and Antonio Moccia
Department of Space Science and Engineering "Luigi G. Napolitano"
University of Naples "Federico II"

U. Ciniglio, F. Corraro, and L. Garbarino
Italian Aerospace Research Center (CIRA)

***Abstract*—This paper presents an integrated navigation tool developed in the framework of an advanced study on navigation of Unmanned Aerial Vehicles. The study aimed at testing innovative navigation sensor configurations to support fully autonomous flight even during landings and other critical mission phases. The tool is composed of sensor simulation and data fusion software. The most important navigation sensors that are installed onboard an unmanned aircraft have been modeled: i.e. inertial, GPS, air data, high accuracy altimeter, and magnetometer. Their model included every non negligible error source that has been documented in the literature. Moreover, a specific sensor data fusion algorithm has been developed that integrates inertial sensor measurements with GPS and radar altimeter measurements. The paper reports on numerical testing of sensor simulator and data fusion algorithm. The algorithm was coded for real time implementation to perform hardware–in-the-loop validation and in flight tests onboard a small Unmanned Aerial Vehicle.**

I. INTRODUCTION

The effort for realizing fully autonomous and operative Unmanned Aerial Vehicles (UAVs) induced the need of developing innovative techniques for integrating measurements derived from different aircraft navigation systems. Since no human aid is available onboard UAVs, navigation hardware must attain larger capabilities than the ones of manned platforms. In particular, the most important features that shall be considered are autonomy, safety, compatibility with previously developed flight standards, and whole mission coverage [1]. This latter feature means that adequate and reliable navigation must be accomplished even during critical phases such as takeoff and landing. During these phases, the required positioning accuracy should be better than 1 meter for the vertical channel [2] and from a minimum of 2 meters for runway large less than 13 meters to a maximum of 7 meters for runway large more than 50 meters for the horizontal channel.

The research activity described in this paper refers to the development of a navigation system for UAVs, based only on on-board systems and without any external support, such as Differential GPS or Instrument Landing System.

Currently, no single sensor is capable of reliably realizing the required performance without relying on some ground measurement. Hence, UAV navigation requirements can be fulfilled only by integration of measurements from multiple sensors. In particular, configurations that integrate inertial sensor measurements with GPS, altimeters, air data sensors, and magnetometers are very frequent [3-5], and resulting performance and reliability depends both on sensor accuracy and on adopted integration techniques. The most common integrated navigation techniques involve the adoption of Kalman filtering [6-8]. Unfortunately, available data fusion methods do not allow reliable analytical methods for accuracy determination to be applied. Only numerical and statistical methods have been developed to measure integrated system performances such as Covariance Propagation and Montecarlo Analysis [9-10]. Numerical simulation is required to perform statistical determination of navigation system performances in terms of accuracy, reliability, and mission coverage. Moreover, real time simulations with hardware in the loop can demonstrate that fusion algorithm implementation and overall system latency are suitable for UAV control, before flight tests are carried out.

The need of simulations in UAV development has been pointed out by many authors [11-13]. Several simulation tools have been developed for performance verification: such as, NAVSIM™ developed by CAST™ [14], GPS/INS Toolboxes™ produced by GPSOFT™ [15-16] and GPS Simulator™ produced by Navsys™ [17]. However, no tool is currently available for testing GPS/INS integrated navigation in a real time simulated environment, which needs both algorithms under test and simulation models to be implemented and executed on suitable real-time machines. Simulink™ by Mathworks™ [18] is a widely used software package for modeling and simulating dynamical systems. Its Real Time Workshop™ toolbox allows automatic real-time code generation starting from Simulink™ models. As a consequence, sophisticated integration algorithms, involving multiple sensor aiding and Kalman filtering, may be first implemented as Simulink™ models and then automatically coded for real time implementation. In the same manner the sensor models used for numerical validation of the algorithms could be quickly used for real-time simulation of the environment. In any case, the starting simulation models shall be adequately designed in order to become suitable for a real-time implementation. This method also allows one to run on the flight computer, the same software version tested in simulations. In this paper, the above

K. Elleithy (ed.), Advances and Innovations in Systems, Computing Sciences and Software Engineering, 127–136.

guidelines for development and validation of the proposed integrated navigation system have been adopted and are briefly below described.

All sensor models and the proposed navigation algorithms have been developed and tested by means of numerical simulations. Subsequently, they have been automatically coded and downloaded, respectively, on a hardware UAV simulator (Iron Bird Facility) and a duplication of the flight computer. The resulting test rig, realized by CIRA (Italian Center for Aerospace Research), mainly reproduces the actual avionics architecture with simulated aircraft dynamics and sensor outputs. This facility is mounted on a hexapod mobile platform for reproducing actual sensor input, at least concerning aircraft attitude. The same software tested with the above facility is installed on-board a small UAV prototype (the CIRA FSSD flying laboratory), for in-flight validation [19].

In the first part of the paper the models for sensor simulation are described. Next, the data fusion algorithm is described in detail and, finally, ground based real-time simulation and partial flight test validation results are briefly presented and discussed by comparing the navigation outputs of the proposed filtering procedure with very accurate position and velocity measurements performed by a Differential GPS.

II. The Flight Small Scale Demonstrator (FSSD)

The FSSD is a low-cost lightweight flying laboratory designed and integrated by CIRA in order to be used for in flight testing and validation of Guidance, Navigation and Control system prototypes. Similar experiences performed in past years around the world both at universities and at Aerospace Research Centers, have shown that using scaled flying platforms does not affect the experiments. The only limitation concerns the available weight and space to install on board required avionics equipment. As shown later, these conditions allowed in any case to perform an in flight demonstration and validation of the proposed advanced navigation sensor system. In the following there is a brief description of the system main components.

The flying platform (fig. 1): is a one third scaled model of the piper PA Superclub remotely controlled via dedicated radio link. It is an off-the-shelf vehicle belonging to the Giant Class with a wing span of about 4 m, an empty weight of about 20 kg, a maximum speed of 30 m/s and a ceiling altitude of about 200 m.

Ground Control Station (GCS): It is a portable ground control station (fig. 2) designed to collect telemetry data and present it to the flight test engineers through a dedicated Human Machine Interface. It can be used also for remote reconfiguration of the on board avionics system and for controlling the aircraft via a virtual cockpit.

On board avionics systems: It includes all devices needed to perform the in flight experimental validation of advanced guidance, navigation and control functionalities. The devices have been selected among the Commercial-Off-The-Shelf (COTS) ones.

Fig. 1. FSSD flying platform.

Fig. 2. Ground Control Station.

They are:

a. A Flight Control Computer (FCC) based on a PowerPC processor provided by the supplier together with a tool based on the most advanced control system rapid prototyping methodologies. The tool allows for automatic real-time coding directly from Simulink™ diagrams;

b. A navigation sensor suite including a DGPS-RTK L1/L2 system capable to provide position measurements with an accuracy of few centimeters, a solid state Attitude Heading Reference System (AHRS) and two dedicated sensors for altitude measurements respectively using radar and laser technology. The altimeter sensors can be alternatively mounted because of weight limitations;

c. Digital electromechanical servos to command both aerodynamic surfaces and throttle, driven by the FCC via PWM signals;

d. A digital data-link system able to exchange data between on board FCC and the Ground Control Station with a maximum bit-rate of 9600 bit/sec;

e. An on board camera integrated with a dedicated radio-link able to send on board view to the ground control station with a maximum range of 2 km.

III. Sensor Models

The development of a sensor model must take into account all known errors or degradations of the signal output. Usually, they can be divided into two main categories: the external and the internal noise sources. The models presented in this section refer to the standard configuration of the navigation sensors installed onboard CIRA FSSD, and simulate the basic operations of the following sensors: Inertial Measurement Unit (IMU), Global Positioning System (GPS), Magnetometer, Air Data System (ADS), Laser/Radar Altimeter. The software architectures have been accurately designed to guarantee the flexibility of the system to varying user requirements and/or flight phases.

A. Inertial Measurement Unit Model

The first model refers to the IMU and is aimed at computing a realistic estimate of its outputs. Its Simulink™ layout is shown in fig. 3. The box named IMU simulates a six-axis measurement system that calculates linear acceleration and angular velocity

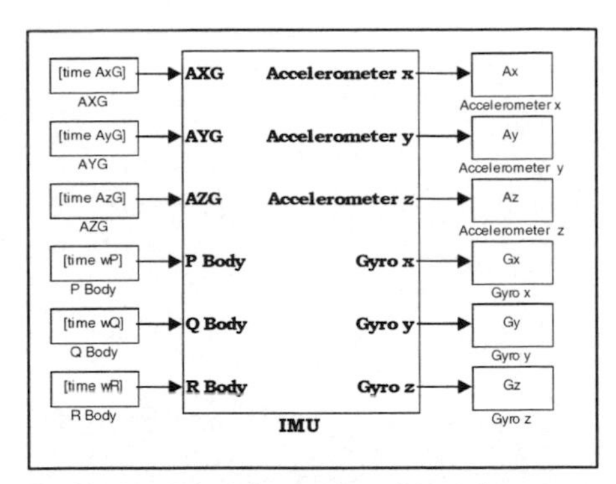

Fig. 3. IMU simulation model overall block diagram.

components along three orthogonal axes. This configuration allows complete determination of vehicle dynamics. On the left side of figure are the inputs to the model: the acceleration components in Body Reference Frame (BRF), expressed in m/s^2, and the angular rate components in BRF, expressed in rad/s. The model generates the error corrupted components of acceleration and angular rate in the Sensor Reference Frame (SRF). The crucial task is to corrupt the inputs by accounting for all the characteristics sources of error concurring in the determination of inertial sensors measures [6-9]: bias, scale factor, random walk, misalignment, thermal drift, and lever arm.

The IMU should be mounted as close as possible to the centre of gravity (CG) of the system. This will minimize any "lever arm effect". If it is not mounted close enough to the CG, then rotations about the CG will cause the accelerometers to measure a non negligible contribution to acceleration due to relative motion.

Moreover, errors in accelerometers and rate sensors alignment with SRF will contribute directly to errors in measured acceleration and angular rate relative to any selected system axis.

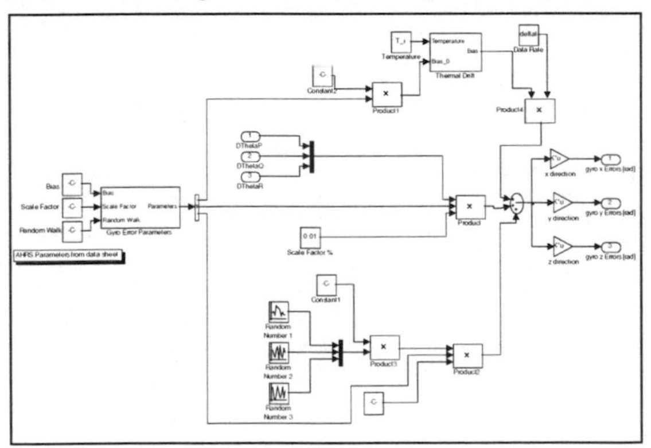

Fig. 4. Bias, scale factor, random walk, and thermal drift.

On the base of these considerations, misalignment has been treated by performing co-ordinate transformation of vectors, and lever effect has been accounted by computing the centrifugal terms in the equation of relative motion, expressed as:

$$\omega \times (\omega \times \delta r) \tag{1}$$

where ω is the angular rate vector and δr is the IMU CG position vector with respect to UAV CG.

Bias, scale factor, and random walk, have been considered as depicted in fig. 4 for rate sensors, where also the block that computes thermal drift effects appears. In particular, thermal drift effect simulation is based on a stochastic model describing the bias variation with the internal temperature of the inertial unit, that has been deduced from a series of laboratory tests conducted on the embarked sensor: the IMU VG400CC produced by Crossbow™. Experimental results have shown that thermal drift effect can be satisfactory modeled by means of second order polynomials, fitting bias changes in temperature [20-21]:

$$\text{bias}(T) = a \cdot T^2 + b \cdot T + c \tag{2}$$

where the coefficients a, b, c can be determined experimentally.

An example of simulated sensor measurements, referring to accelerations and angular rates generated by the flight profile of fig. 5, is reported in fig. 6 where the rate sensor outputs computed by the model are plotted. For the sake of simplicity only bias, scale factor, and random walk have been considered as error sources in the simulation.

B. GPS Receiver Model

The second sensor model is a very sophisticated and detailed simulator of a 12 channels GPS receiver. Its Simulink™ layout is shown in fig. 7, where model inputs, outputs, and a box that performs receiver computations are highlighted. The actual user position, in ENU co-ordinate frame, is given as input to the receiver box. Before starting simulations, the model requires that broadcast almanac orbital parameters had been loaded into global memory. YUMA formatted almanac files may be obtained at the U.S. Coast Guard Navigation Center web site.

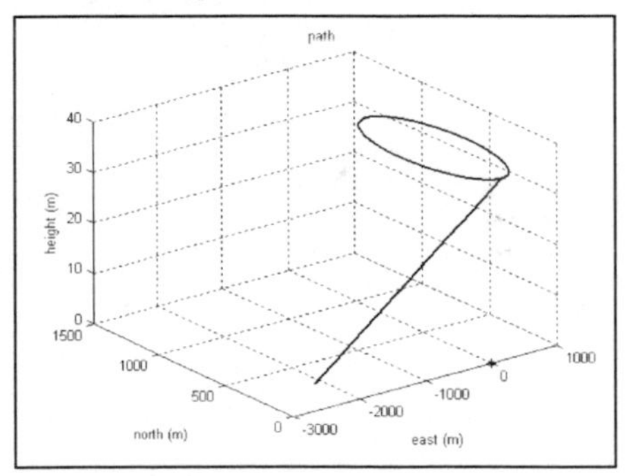

Fig. 5. Simulated flight profile, consisting in an ascent phase and a closed loop turn.

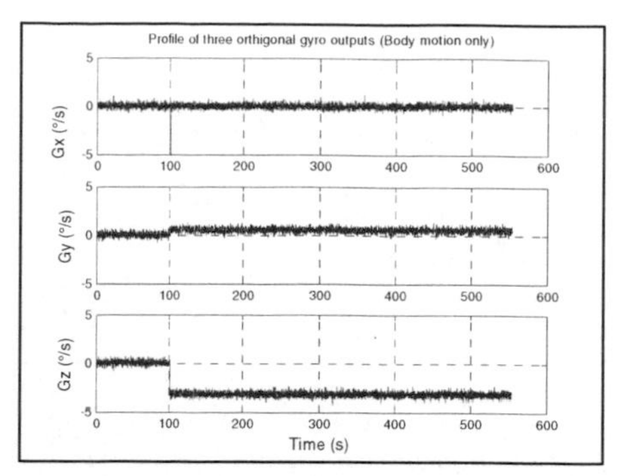

Fig. 6. Rate sensor estimated outputs.

The first step of the process is the generation of the positions of the satellites that are visible at user location. It can be done by using information extracted from the broadcast almanac. A block calculates the combination of four satellites that minimize one of Dilution Of Precision (DOP) parameters (to be set by the user) [8]. The second step is the computation of the pseudoranges to all visible satellites. The most referenced error sources are taken into account to produce measured pseudoranges: thermal noise of the receiver; tropospheric and ionospheric delays [8], Selective Availability (SA), and multipath. The last step is the estimation of user co ordinates and user clock offset by means of a least squares solution. User co ordinates are provided for both ECEF Cartesian and geodetic Latitude, Longitude, and Altitude (LLA) co ordinates. GPS time can be opportunely set, in order to obtain a realistic scenario of visible satellites at a given location, and at a specified epoch of GPS week. Giving in input to the model the flight profile of fig. 5 and considering as sources of error the thermal noise of the receiver and the tropospheric and ionospheric delays, block outputs the estimated user ENU co-ordinates of fig. 8.

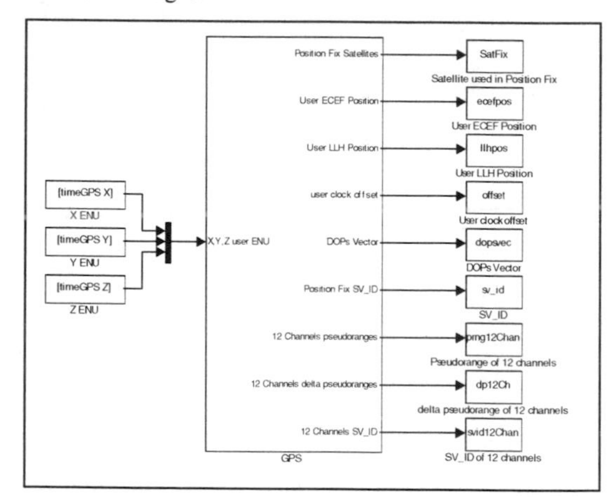

Fig. 7. GPS receiver simulation model overall block diagram.

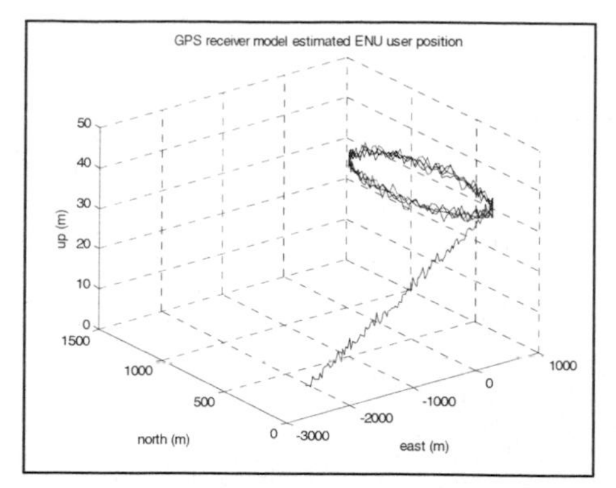

Fig. 8. Receiver model computed position.

TABLE 1. Error budget for SPS [22].

Parameter	Value (m)	
Signal-in-space ranging error (rms)	3.1	
Ionospheric errors (rms)	7.3	
Selective availability (rms)	23.0	
Tropospheric errors (rms)	0.2	
Receiver errors due to noise	0.7	
Multipath (rms)	1.2	
Total User Equivalent Range Error (rms)	SA on	SA off
	24.4	8

The models adopted for GPS data errors are reported in [8]. An error budget for Standard Positioning Service (SPS) with SA on and off is shown in Table 1.

Although quite accurate, this model cannot operate in real time due to requested inputs. As a consequence, a simplified, but adequate to real-time applications, model of GPS receiver has been developed taking into account the requirements of a sensor fusion strategy based on Kalman filtering [23]. Such model is quite different from the one just described and will be presented in the sensor fusion section of the paper.

C. Magnetometer Model

The model simulates a three-axial magnetometer, whose output is an estimate of the Earth magnetic field in SRF. Fig. 9 shows its layout in Simulink™.

It is worth pointing out the two boxes representing the external environment and the internal dynamics of the sensor. The external environment is the Earth magnetic field, whose most referenced model is the International Geomagnetic Reference Field (IGRF). The model implements the equations reported in [24-27] and includes Gauss coefficients for the 1995 IGRF. The magnetic field vector at a point Q above the Earth's surface can be expressed as the sum:

$$B = \sum_{n=1}^{\infty} \sum_{m=0}^{n} B_{n,m} \tag{3}$$

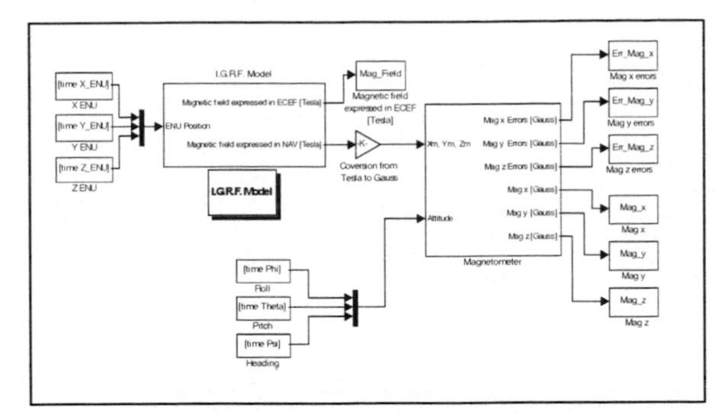

Fig. 9. Magnetometer simulation model overall block diagram.

where B can be derived from a magnetic scalar potential which is expanded into terms of spherical harmonics [26].

The IGRF model needs as inputs the position at which the field must be evaluated, in ENU co-ordinates, and the Schmidt-normalized Gauss coefficients [27]. Hence, it gives as outputs the ECEF components of magnetic field exerted at a specific point, computed assuming the field as a gradient of a potential function that, in turn, can be described as an infinite series of spherical harmonics. Computation of the field vector is efficiently performed by means of recursion formulas, and the computations do not suffer from singularities when evaluated at points that lie on the polar axis.

More generally, this Matlab™ Simulink™ package can be used in simulations of aircraft and spacecraft motion when a planetary magnetic field model is required. For example, it could be used for evaluating the torque produced by interaction of the system's magnetic moment with the external field, or by magnetic actuators for attitude control. Then, the model calculates a magnetic field vector.

The magnetic field vector obtained in ECEF frame is then converted in NED and SRF reference frames by means of co-ordinate transformations. Subsequently, the principal sources of error are included in the output measurements, namely: bias, scale factor, random noise, misalignment. Fig. 10 shows the Simulink™ block that makes these computations and provides the magnetic field at user position, as it would be measured by the magnetometer.

D. Air Data System Model

An Air Data System (ADS) estimates parameters useful for aircraft flight control starting from the measurements of quantities connected to the air flow around the vehicle. The layout of the simulation model of an ADS, developed in Simulink™, is presented in fig. 11. It involves dynamic and thermodynamic quantities that take part in ADS measures and, of course, it has been developed accounting for the incompressible flight regime of the CIRA FSSD UAV prototype.

The box in the left side of figure computes some of the quantities usually measured by an ADS, while the box on the right side generates the noisy output of the sensors. Four parameters must be given as inputs to the first box: aircraft speed in North-East-Up (NEU) components [m/s]; air density [kg/m^3]; wind speed in NEU axes [m/s]; attitude angles [rad]. The box outputs three of the four quantities which are directly measured by ADS components: differential pressure [N/m^2], sideslip angle [rad], and angle of attack [rad]. The developed model does not estimate air data parameters such as the static pressure ps using aircraft altitude and standard atmosphere model because this function is already provided by MathWorks™ Aerospace Blockset [28]. The angle of attack and sideslip are derived from the components of aircraft speed in the body-fixed co-ordinate frame as shown in [29].

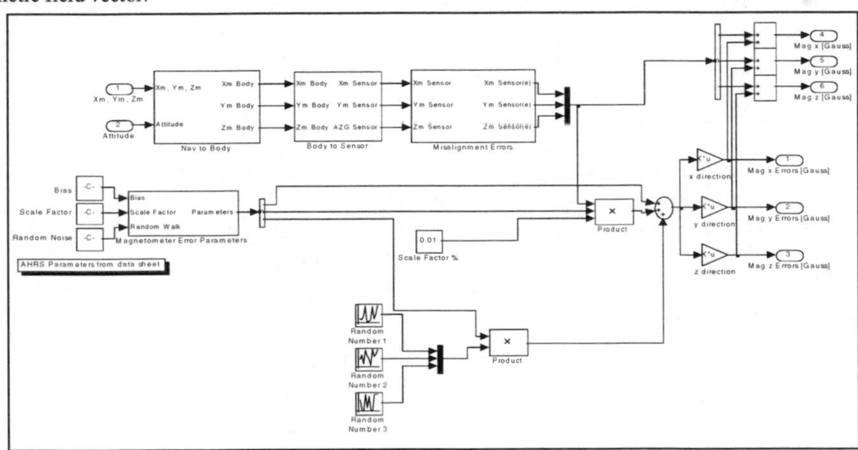

Fig. 10. Block diagram of magnetometer dynamics simulation and relevant output.

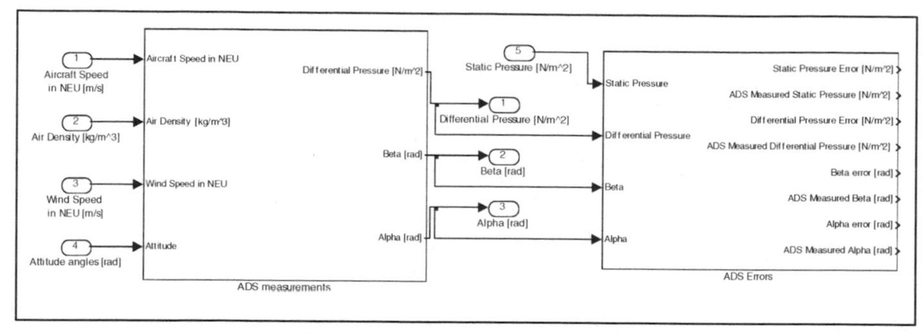

Fig. 11. ADS simulation model block diagram.

The second box in fig. 11 estimates the error contributions in ADS measures. Again, the error sources taken into account are: bias, scale factor accuracy, and random noise. For example, in sideslip angle measure estimate, the three contributions of error are computed as displayed in fig. 12.

D. Laser/Radar Altimeter Model

The simulation block modeling a Laser or a Radar Altimeter sensors is depicted in fig. 13. In particular, two main boxes are shown: a range computer and a block introducing instrument dynamics. The first box outputs the true slant range information, which represents the quantity an altimeter measures. It receives in input: the altitude above the Earth reference ellipsoid in meters and the attitude, in terms of body to navigation frames Euler angles, expressed in radians. Of course, the lever arm effect has been taken into account, because the altimeter must be necessarily installed on the bottom of fuselage, hence supposedly relatively far from aircraft CG. So the altitude input to the range computer represents the true quantity at the location of the firing point of the instrument.

The equation defining the true range in terms of altitude and attitude angles is reported in flow chart form in fig. 14. The range comes from the rearrangement of the following expression relating the altitude q to the measured range r and instrument tilt angles (pitch θ and roll φ) [2]:

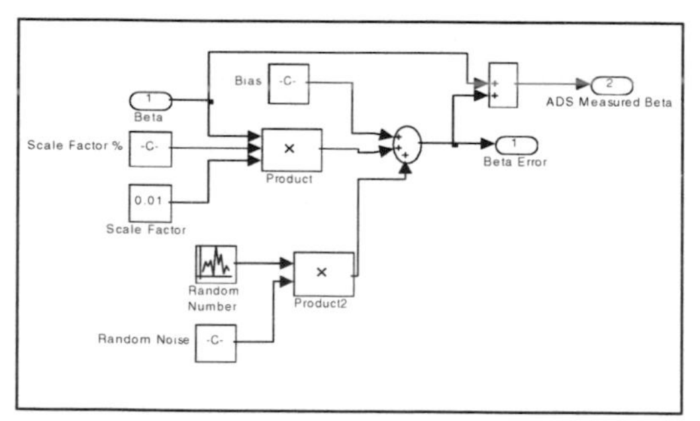

Fig. 12. Error contributions in sideslip angle estimation.

$$q = r\sin\beta_{Alt}\sin\theta + r\cos\beta_{Alt}\sin\alpha_{Alt}\cos\theta\sin\varphi + r\cos\alpha_{Alt}\cos\beta_{Alt}\cos\theta\cos\varphi \quad (4)$$

where α_{Alt}, β_{Alt} are the altimeter mounting angles.

The computed true range is then given as input to the box in the right side of fig. 13, which simulates the typical measurement error sources. This box has two outputs: the simulated altimeter range and range error, both expressed in meters. The main referenced error sources that have been considered for degrading the true range are bias, scale factor, and random noise. The bias is expressed in meters and the scale factor in percentage. The scale factor error can be estimated by multiplying the scale factor accuracy by the true range. With reference to the random noise, relevant error is obtained by randomly extracting from a Gaussian distribution with zero mean and error standard deviation furnished by instrument data sheet.

IV. Sensor Fusion

Following the sensor fusion strategy described in [7], a Simulink™ model has been implemented and successfully tested to integrate the measures of various navigation sensors and to provide a more accurate attitude, position, and velocity information, capable of satisfying guidance and control requirements for fully autonomous UAV navigation. It has been designed to operate both in off line simulations and in real time applications with Matlab™ Real Time Workshop®.

Fig. 15 reports the main layout of the Simulink™ block diagram that had run on the CIRA FSSD onboard computer during flight test campaign performed in May 2004. The model inputs are listed on the left side of the computational boxes in figure. Basically, they are the outputs of the IMU, the GPS receiver, and the altimeter sensor. Whereas the outputs of the procedure are reported on the right side of figure and refer to attitude, position, and velocity estimates, computed by the sensor fusion algorithm. It is worth dealing in more details the two main steps in the sensor fusion algorithm, consisting in: GPS raw data elaboration and INS/GPS/Altimeter Kalman navigation.

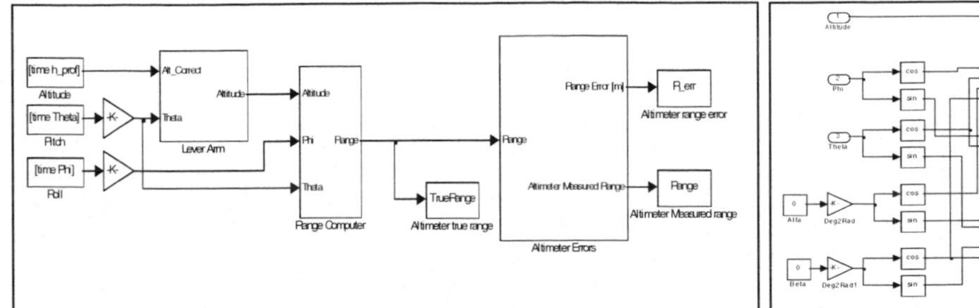

Fig. 13. Altimeter simulation model block diagram.

Fig. 14. Range determination block diagram.

A. GPS Receiver Model

Since the GPS receiver Simulink™ module must now be supported by Real-Time Workshop®, it is totally different from the one described in the previous section. It produces GPS fixes from GPS raw measurements according to the strategy of determining the position with the best Horizontal Dilution of Precision (HDOP) [22] in the ENU reference frame, a function that is not directly implemented in the GPS receiver. Inertial navigation aiding will benefit of these measures according to the strategy reported in [7]. The layout of the developed block is displayed in fig. 15. GPS block input ports will accept raw measurements from the GPS receiver. Specifically, block input ports are:

7. CP – Carrier phase to pseudorange difference. It is a vector reporting for each satellite in view the difference between ranges from satellite to receiver, computed with carrier phase and standard measurements;

8. SI – Satellite IDs. It is a vector reporting the Satellite Vehicle Identification Numbers (ID) for each satellite in view;

9. ID – Ionospheric delays. It is a vector that contains the GPS signals Ionospheric Delays (ms) estimated by the receiver for each satellite in view;

10. RC – Pseudorange. It is a vector with all measured pseudoranges (ms);

11. RT – Receiver Time. It is a scalar with the reference Receiver Time of Week (s) for GPS measurements;

12. TO – Time Offset. It is a scalar with the estimated Time Offset (ms) between Receiver Time and the GPS System Time;

13. GE – Satellite ephemeris. It is a 29 x 31 matrix that reports data broadcast from each satellite about its ephemeris, as described in [8], necessary to perform precise orbit determination;

14. NSat – Number of Satellites. It is a scalar reporting the total number of satellites in view;

15. Quaternary – It is a matrix containing all the sets of four satellite IDs that can be formed by all the satellites in view. The row of the matrix reports the IDs in ascending order of HDOP, so that the first set is the one with the minimum HDOP;

16. Dim_quaternary – It is the number of full rows of the matrix Quaternary. It is equal to the binomial coefficient $\binom{N_{Sat}}{4}$.

These inputs are returned as a result of proper queries that must be made to the on-board GPS receiver, except for input port 15 and 16 that are computed offline on the basis of the GPS almanac. However, future versions of the code could be provided with real time computation of Quaternary matrix. The matrix GE is formed by refreshing the values of a column that represents a satellite ID each time a new answer to a "GE" query is received from GPS receiver. The output ports are:

Out – It outputs the position fix with best HDOP, when raw GPS measurements are available for at least four satellites;

Nsat<4 – It outputs 1 when less than four satellites are in view;

Nsat_corr<4 – It outputs 1 when ephemeris data are available for less than four satellites in view;

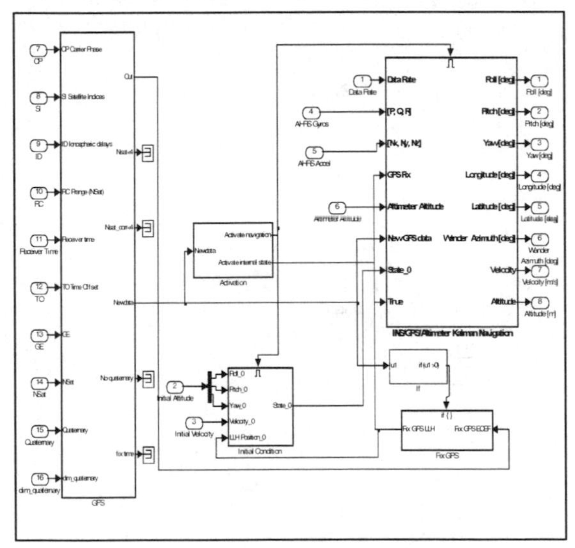

Fig. 15. Sensor fusion model basic operations.

Newdata – It outputs 1 when a new position fix is available;

No quaternary – It outputs 1 when no set of four satellites in the Quaternary matrix is available;

Fix time – It outputs GPS fix time;

The second, third and fifth output ports report all conditions that prevent from realizing a correct position fix.

From the computational point of view, GPS module is formed by four sequential sub-blocks, each performing a specific function in successive levels. The first level sub-block checks if at least four satellites in view are available and the current value of the receiver time (RT) is greater than the immediate past value, but of less than two seconds. If above conditions are verified, it enables the second level sub-block, which checks if ephemeris data are available for at least four satellites in view. This step has been introduced since the GPS receiver does not guarantee that ephemeris of a satellite are available before pseudoranges are output. The third level sub-block first of all selects the set of four satellites in view with best HDOP by browsing the Quaternary matrix, then performs the position fix. If one of above conditions is not met, as an example if no set of four satellites in view is recognized, the sub-blocks exit without enabling successive operations.

The fourth level sub-block first of all determines satellite ECEF positions at starting time of GPS message, for the four satellites reported in the input port Quaternary, and corrects pseudorange values by subtracting receiver time offset and relativistic error [2]. Then, user ECEF position fix and receiver time offset are estimated by using the satellite positions and previously determined corrected ranges. The algorithm adopted for satellite precise positioning is the one proposed by [8]. It is worth noting that the algorithm requires solution of Kepler equation. Since no analytical solution is available, it is solved by means of an iterative algorithm [29], which requires a limited number of iterations, because it is particularly effective for small orbital eccentricity. The position fix block is based on the standard GPS equation and adopts a standard iterative least-square algorithm [7] based on the determination of the inverse of the linear GPS observation matrix. Also in this case the total number of iteration is quite limited, usually up to 16.

B. Integrated Kalman Navigation Simulation Model

The INS/GPS/Altimeter Kalman Navigation block on the right side of fig. 15 performs data fusion of IMU (accelerometers and gyros), GPS, and altimetric (Laser or Radar) measurements by means of Kalman filtering. It is activated by a specific block, and receives the initial conditions calculated in another block, both represented in fig. 15. As regard the eight outputs, they are the solution of the navigation equations (attitude, position, and velocity) with Kalman filter corrections.

The model works according to the main steps of the adopted sensor fusion algorithm [23, 30]. Equations describing navigation state have been implemented according to the formulation presented in [9]. Wander azimuth mechanization has been chosen for the navigation frame [6]. The differential equations have been numerically integrated using the mathematical techniques described in [9]. The linearization approach is based on the perturbation representation for position, velocity, and attitude errors according to the "phi" formulation [9]. The position, attitude, and velocity error equations have been implemented in a sub-block that receives in input the actual state vector, solution of the navigation equations, and the state error vector, computed by Kalman filtering, and outputs the corrected state vector. Then, a specific block performs the quaternion to Direction Cosine Matrix (DCM) transformation [9] and provides the corrected attitude, longitude, latitude, and wander angles.

The Kalman filter architecture adopted is the feedback implementation of the EKF (Extended Kalman Filter) [7] that is illustrated in the scheme of fig. 16. The INS provides the reference trajectory, whereas the EKF estimates the INS state errors, which are fed back to correct INS internal state. Every time this operation is accomplished, the corresponding EKF error state is reset to zero. At the next measurement epoch, the INS produces again the full navigation state estimate. GPS concurs to the correction of EKF state estimate when a new position fix is available.

Altitude measurements provided by the altimetric system are also given in input to the EKF, independently from the availability of a new GPS receiver fix. The logic of the developed algorithm is that the altimeter sensor can be engaged continuously in the data fusion procedure or its measurements can be used to perform aiding only during specific mission phases, i.e. limited instrument tilt angles and/or altitudes.

Indeed, it is expected that altimeter aiding is particularly useful and its measurement reliable and effective during runway approach and landing.

Finally, the output of the block that performs Kalman filtering are the estimated state error vector and covariance matrix of estimation uncertainty [7].

V. Flight Test: Sensor Fusion Results

During the flight tests of CIRA FSSD UAV prototype an IMU, a GPS receiver, and a laser altimeter were installed on-board. Fig. 17 shows the UAV path calculated by the differential GPS Real Time Kinematik RTK™ fixed mode measurements, that has a rms accuracy of 4 cm when available [31]. This is a differential mode that requires a ground station to perform measurements. Indeed, the maximum accuracy is available as a result of an

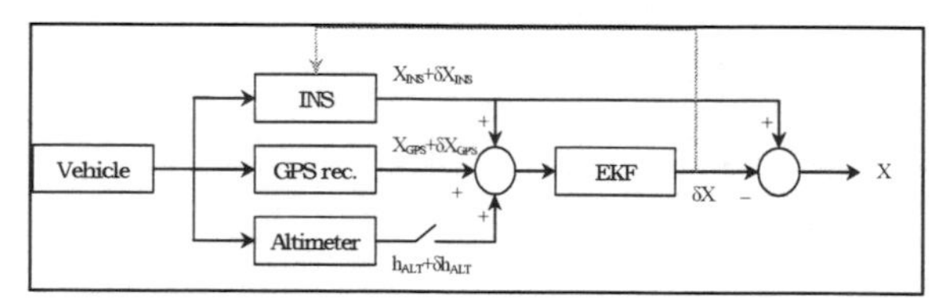

Fig. 16. Feedback implementation of Kalman filter.

iterative procedure that depends on many parameters such us satellite geometry, aircraft attitude, aircraft and ground station relative position. As a consequence, fully accuracy may be not available for a not negligible part of the trajectory. Since altimeter are slant range measurement system, the RTK™ measurements are very useful to test altimetric measurement performances.

The results of the flight data processing by means of the previously described sensor fusion model, integrating all the onboard navigation units, are reported in fig. 18, that depicts the estimated longitude-latitude path. It is worth noting the effect of repeated inclusions of UAV positions computed by using the stand-alone GPS receiver in the fusion procedure, graphically represented by plot discontinuities. In fig. 19 are depicted three runway flyovers of sensor fusion solution for path.

These results allow definition of an error budget for the sensor fusion algorithm with respect to RTK™ solution, assumed as the most accurate available reference during the time intervals where the maximum precision of the RTK™ algorithm solution is guaranteed. It was found that the horizontal position rms error is 6.5 m. The same analysis has been performed for altitude (fig. 20), where an rms error of 0.8 m was found during runway flyovers with engaged altimeter, when a comparison between measured and estimated altitude makes sense, thanks to the coincidence of altimeter slant range and GPS RTK™ altitude, apart from known mounting and attitude angles. Figures 21 and 20 plot the position error distributions during a runway flyover. The negative error peak shown in fig. 22 is generated by laser slant range measurement discontinuities, supposedly due to variable reflection characteristics of the test runway surface, along with altimeter not exactly nadiral pointing angle, hence causing a variation in signal input to the instrument. This effect resulted more significant during approaching phase, when the UAV flew over differently vegetated areas with more perturbed attitude.

In summary, proposed HW/SW integrated navigation measurement system allowed for continuous and autonomous dynamics determination, attaining an adequate accuracy both in cruise and during runway approach.

CONCLUSIONS

The development of an integrated navigation software has been presented. The tool included both sensor simulators and data fusion algorithms. The aim of the study was to develop an autonomous strategy capable to provide an adequate estimate of the aircraft dynamical state even during the most critical mission phases such as takeoff and landing. This process was comprehensively described in the paper. The relevant software has been tested with numerical simulations, ground hardware-in-the-loop experimentations, and in flight verification. A low-cost configuration composed by a solid-state inertial unit, a standalone GPS receiver, and a laser altimeter resulted enough accurate to output vertical position estimates with a rms uncertainty of less than 1 meter, whereas the horizontal positioning accuracy resulted in the order of 6 meters. The study showed that the available sensor set attains adequate accuracy for the vertical position whereas sensors that are more accurate than the standalone GPS are needed to attain an adequate horizontal positioning accuracy to support autonomous landing on narrow runways.

ACKNOWLEDGMENTS

This research has been carried out in the framework of the technology studies of the Italian Government research program PRORA UAV, carried out by the Italian Aerospace Research Center (CIRA) with the objective to develop a High Altitude Long Endurance (HALE) UAV for civil applications.

Mr. Esposito's participation to the research activity was sponsored by a grant for a Ph.D. degree in Aerospace Engineering co-funded by the European Union.

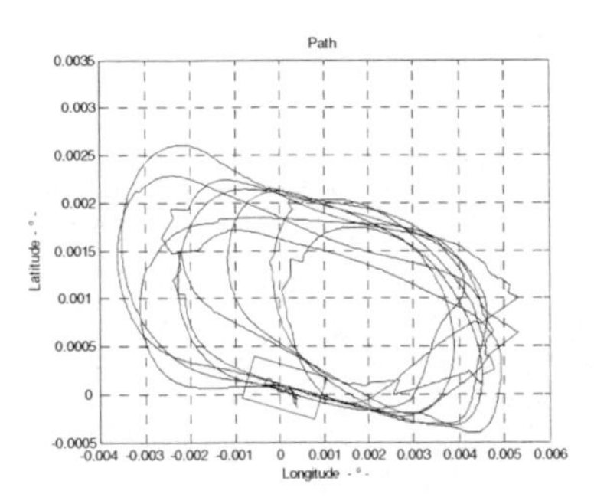

Fig. 17. GPS RTK™ path (Geodetic angles are referred to the center of the runway).

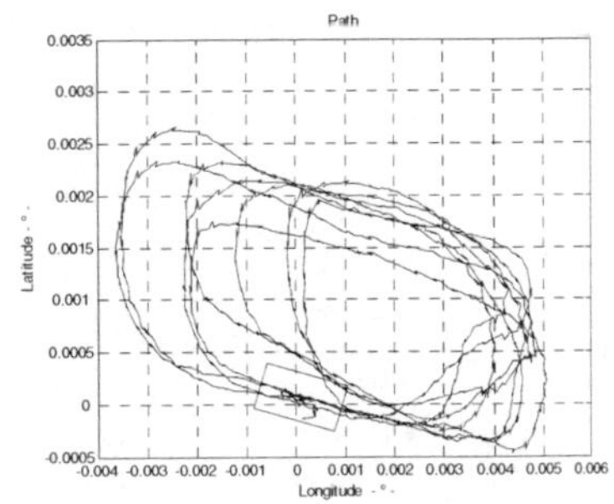

Fig. 18. Sensor Fusion path (Geodetic angles are referred to the center of the runway).

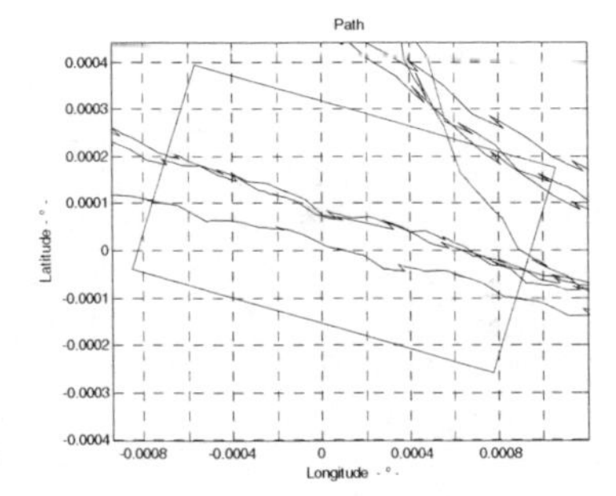

Fig. 19. Runway flyovers in Sensor Fusion solution for path (Geodetic angles are referred to the center of the runway).

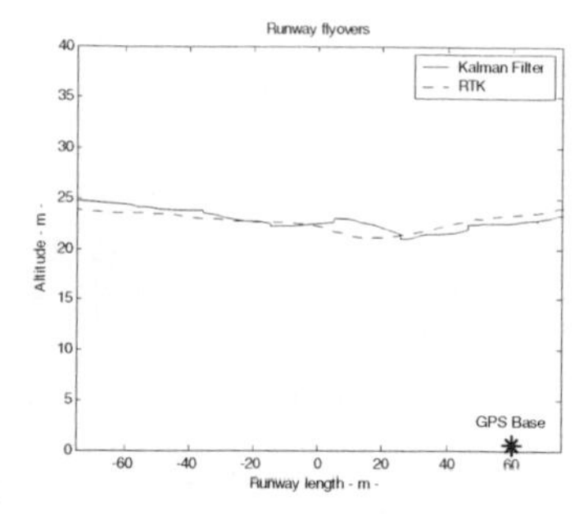

Fig. 20. Altimetric measurements comparison during runway flyovers.

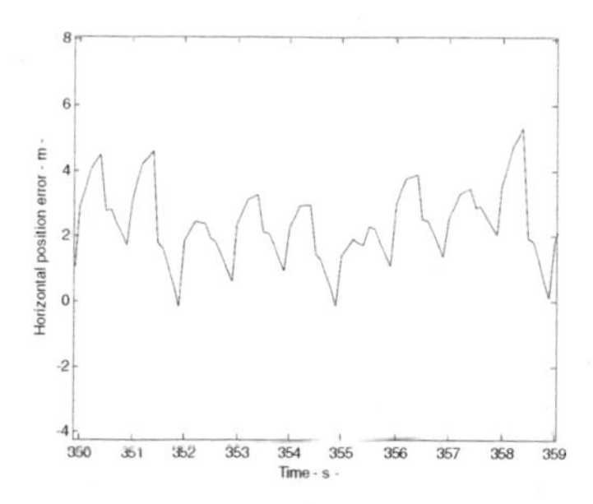

Fig. 21. Horizontal position error during a runway flyover.

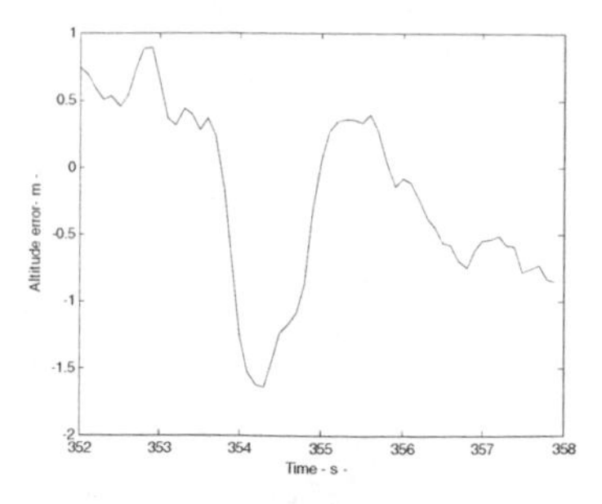

Fig. 22. Altitude error during a runway flyover with engaged altimeter.

REFERENCES

[1] Office of the Secretary of Defense, "UAV Roadmap 2002-2007," Department of Defense, Washington DC, 2002, pp. 153-164.

[2] Kayton, M. and Fried, W.R., "Avionics Navigation Systems," 2nd ed., Wiley-Interscience, New York NY, 1997, pp. 600-607.

[3] Kaminer, I., Yakimenko, O., Dobrokhodov, V., and Jones, K., "Rapid Flight Test Prototyping System and the Fleet of UAV's and MAVs at the Naval Postgraduate School," AIAA-2004-6491, AIAA, 3rd "Unmanned Unlimited" Conference, Chicago IL, 2004.

[4] Kingston, D., and Beard, D., "Real-Time Attitude and Position Estimation for Small UAVs Using Low-Cost Sensors," AIAA-2004-6488, AIAA, 3rd "Unmanned Unlimited" Conference, Chicago IL, 2004.

[5] Johnson, E., Schrage, D., Prasad, J., and Vachtsevanos, G., "UAV Flight Test Programs at Georgia Tech," AIAA-2004-6492, 3rd "Unmanned Unlimited" Conference, Chicago IL, 2004.

[6] Savage, P., "Strapdown Analytics," Strapdown Associates Inc., Minneapolis MN, 2002, Chap. 15.

[7] Farrell, J.A., and Barth, M., "The Global Positioning System & Inertial Navigation," McGraw Hill Professional, New York NY, 1998, pp 135-139, pp 241-257.

[8] Grewal, M.S., Weill, L.R., and Andrews, A.P., "Global Positioning System, Inertial Navigation and Integration," John Wiley & Sons, New York NY, 2002, pp 14-37, pp 103-130.

[9] Rogers, R.M., "Applied Mathematics in Integrated Navigation Instruments," AIAA Education Series, AIAA, Washington DC, 2000, pp 18-94, pp 163-177.

[10] Chatfield, A.B., "Fundamentals of High Accuracy Inertial Navigation," AIAA Press, Washington DC, 1997, pp. 267-271.

[11] Johnson, E.N., Proctor, A.A., Ha, J., and Tannenbaum, A.R., "Development and Test of Highly Autonomous Unmanned Aerial Vehicles," AIAA, Journal of Aerospace Computing, Information and Communication, Vol. 1, Issue 12, 2004,pp. 485-501.

[12] Walter, B.E., Knutzon, J.S., Sannier, A.V., and Oliver, J.H., "Virtual UAV Ground Control Station," AIAA-2004-6230, AIAA, 3rd "Unmanned Unlimited" Conference, Chicago IL, 2004.

[13] Evans, J., Inalhan, G., Jang, J.S., Teo, R., and Tomlin, C.J., "Dragonfly: a Versatile UAV Platform for the Advancement of Aircraft Navigation and Control," IEEE, Proceedings of Digital Avionics Systems, Vol. 1, Daytona Beach, FL, 2001, pp. 1C3/1 - 1C3/12.

[14] CAST LLC., "Navigation Simulator (NAVSIM) Product Brochure," Billerica, MA, URL: http://www.castnav.com/cast_pdf/cast_navsim.pdf, [cited 27 July 2005].

[15] GPSoft LLC., "Inertial Navigation System TOOLBOX Users's Guide," Athens, OH, 1998.

[16] GPSoft LLC., "Satellite Navigation TOOLBOX Users's Guide," Athens, OH, 1998.

[17] Gold, K., and Brown, A., "Architecture and Performance Testing of a Software GPS Receiver for Space-based Applications," IEEE, Proceedings of Aerospace Conference, Big Sky MT, 2004.

[18] The Mathworks Inc., "Using Matlab (version 6) ," Natick, MA, 2002.

[19] Amendola, A., Pecora, M., Mingione, G., and Mercurio, U., "CRX3 – The CIRA High Altitude Long Endurance UAV," AIDAA, Proc. of 17th Congress of Italian Association of Aeronautics and Astronautics, Rome, Italy, 2003.

[20] Dauderstadt, U.A., Sarro, P.M., and Middelhoek, S., "Temperature Dependence and Drift of a Thermal Accelerometer," IEEE, Proc. of International Conference on Solid-State Sensors and Actuators, Chicago IL, 1997.

[21] Esposito, F., Accardo, D., and Moccia, A., "An Integrated GPS/INS System for Mini-UAVs Autonomous Navigation," AIDAA, Proc. of 17th Congress of Italian Association of Aeronautics and Astronautics (AIDAA), Vol. 3, Rome, Italy, 2003, pp-1615-1623.

[22] McDonald, K.D., "The Modernization of GPS: Plans, New Capabilities and the Future Relationship to Galileo," Journal of Global Positioning Systems [online journal], Vol. 1, Issue 1, Paper 1, URL: http://www.gmat.unsw.edu.au/wang/jgps/ [cited 27 July 2005].

[23] Accardo, D., Esposito, F., and Moccia, A., "Low-cost Avionics for Autonomous Navigation Software/Hardware Testing," IEEE, Proc. of 2004 International Aerospace Conference, Big Sky MT, USA, 2004.

[24] Haymes, R.C., "Introduction to Space Science," John Wiley & Sons, New York NY, 1971.

[25] Lundberg, J.B., and Schutz, B.E., "Recursion Formulas of Legendre Functions for Use with Nonsingular Geopotential Models," AIAA, Journal of Guidance, Control, and Dynamics, Vol. 11, 1988, pp 32-38.

[26] Mueller, A.C., "A Fast Recursive Algorithm for Calculating the Forces Due to the Geopotential," NASA JSC Internal Note No. 75-FM-42, 1975.

[27] Roithmayr, C., "Contributions of Spherical Harmonics to Magnetic and Gravitational Fields," EG2-96-02, NASA Johnson Space Center, Houston TX, 1996.

[28] The Mathworks Inc., "Simulink™ Aerospace Blockset Manual", Natick MA, 2002.

[29] Collinson, R.P.G., "Introduction to Avionics Systems," 2nd ed., Springer, London, UK, 2002, pp. 355-392.

[30] Accardo D., Esposito F., Moccia A., and Russo M., "Performance Evaluation of Different Sensor Configurations for Autonomous Navigation of Unmanned Aerial Vehicles," 10th Saint Petersburg International Conference on Integrated Navigation Systems, Saint Petersburg, Russia, 2003

[31] Allison., T., "An Introduction to Carrier Phase and RTK Initialization," Proc. of Trimble User Conference, San Jose CA, 1998.

Swarm-based Distributed Job Scheduling in Next-Generation Grids

Francesco Palmieri[1] and Diego Castagna[2]

[1]Università degli Studi di Napoli Federico II, Centro Servizi Didattico Scientifico, Via Cinthia 45, 80126 Napoli, Italy
[2]Stazione Zoologica Anton Dohrn, Villa Comunale, Napoli, Italy

Abstract - **The computational Grid paradigm is now commonly used to define and model the architecture of a distributed software and hardware environment for executing scientific and engineering applications over wide area networks. Resource management and load balanced job scheduling are a key concern when implementing new Grid middleware components to improve resource utilization. Our work focuses on an evolutionary approach based on swarm intelligence and precisely on the ant-colony based meta-heuristic, to map the solution capability of social insects to the above resource scheduling and balancing problem, achieving an acceptable near-optimal solution at a substantially reduced complexity. The Grid resource management framework, will be implemented as a multi-agent system where all the agents communicate each other through the network and cooperate according to ant-like local interactions so that load balancing and Grid makespan/flowtime optimization can be achieved as an emergent collective behaviour of the system. We showed, by presenting some simulation results, that the approach has the potential to become really appropriate for resource balanced scheduling in Grid environments.**

Keywords – **GRID, Management Agents, Swarm Intelligence, Ant-Colony Optimization**

I. INTRODUCTION

A growing number of high-performance scientific and industrial applications, ranging from real-time particle physics or radio astronomical experiments to complex weather forecast, and financial modeling, are increasingly taking advantage from large geographically distributed computing, network infrastructure and data management resources, commonly referred to as "Grids". A Grid offers a uniform and often transparent interface to its resources such that an unaware user can submit jobs to the Grid just as if he/she was handling a large virtual supercomputer, so that large computing endeavors, consisting of one or more related jobs, are then transparently distributed over the network on the available computing resources, and scheduled to fulfill requirements with the highest possible efficiency. The management and scheduling of dynamic grid resources in a scalable way requires new and smarter technologies to implement a next generation intelligent grid environment and is now a key concern when implementing new grid middleware components to improve resource utilization. In traditional approaches, very common in early production grids, there was a centralized manager, often called resource broker, that was the only entity with a complete view of the resources available on the whole Grid. The broker selected computing resources based on actual job requirements and a number of criteria identifying the available resources and their location with the aim to minimize the total time to delivery for the individual application, and performed job distribution on them. This is clearly not applicable in modern grid computing where both the network and the computing infrastructure itself lack of a fixed structure. Accordingly, in this paper we present a new distributed approach, based on swarm intelligence, to efficiently map the jobs submitted on a computational grid by running them on the resources available on the network so that they can be completed as soon as possible. Swarm intelligence [1] is the collective behavior from a group of social insects, namely ants, that communicate interactively either directly or indirectly in a distributed problem-solving manner. More specifically, the simple observation of the phenomenon of a group of ants in a natural environment, which can dynamically and adaptively find and collect foraging objects into their nest without any master or central authority driving them, gives a significant clue to solve our problem. The ants work together to achieve an optimal solution and move towards the optimal solution by sharing their own knowledge with their neighbors. According to the above paradigm, an ant-like self-organizing mechanism can be used to perform efficient resource management on the Grid nodes through a collection of very simple local interactions. This can be achieved by heuristically determining a scheduling solution that distributes the jobs on the Grid resources minimizing the overall Grid *Makespan*, that is the maximum completion time of all the job instances in the schedule, and *Flowtime*, referring to the response time to the user petitions for task executions. The above process can result in an indirect coarse-grained load balancing effect since each task tends to be dispatched to a grid resource that has less workload and can meet the application execution deadline. It can be modeled as a multi-agent system where all the agents communicate each other through the network and cooperate through stigmergy according to ant-like local interactions. Each task is carried by an ant, represented by an agent. Ants cooperatively search the less-loaded nodes with sufficient available resources, and transfer the tasks to be executed to these nodes. The proposed self-organizing mechanism does not need a centralized control, which otherwise might act as a potential bottleneck, and is an attractive solution for very large, dynamic and computationally intensive Grid infrastructures because it is inherently parallel and easy distributable, with each node running one or more agents performing search in the solution space or directly managing the resources available on it. Adding more agents (or "ants") generally increases the solution quality at the cost of a very limited additional workload. Extensive simulation results obtained upon different experimental Grid topologies clearly indicate that our ant-colony based approach is highly adaptive, robust and effective in handling the above scheduling/load

K. Elleithy (ed.), Advances and Innovations in Systems, Computing Sciences and Software Engineering, 137–143.

balancing problem.

II. THE AGENT-BASED FRAMEWORK

Distributed Agent technology can be thought of as the next step in the evolution of swarm system modelling techniques. In a distributed agent framework, we conceptualize a dynamic community of agents, where multiple agents contribute services to the community by cooperating like individuals in a social organization. The appeal of such architectures depends on the ability of populations of agents to organize themselves and adapt dynamically to changing circumstances without top-down control from a central control logic.

A. Implementation details

The proposed resource management framework can be implemented as a multi-agent system where each agent is a representative of a local grid resource (resource management agent) or of an independent execution request from an user (search agent), associated with explicit computational requirements. In our model we consider the following basic assumptions: the jobs being submitted to the grid are independent and are not preemptive, that is, they cannot change the resource they has been assigned to once their execution is started, unless the resource is dropped from the grid. Examples of this scenario in real life grid applications arise typically when independent users send their tasks to the grid, or in case of applications that can be split into independent tasks. Such applications are frequently encountered in scientific and academic environments. They also appear in intensive computing applications and data intensive computing, data mining and massive processing of data, etc. Initially, search and resource management agents are generated from nodes on the grid according to their specific roles, that is, the system consists of a certain number of agents which either handle computational resources in various nodes or wander on the network, searching for available resources. When a group of tasks is generated on the grid, then a group of search agents, whose number can be less or equal to that of the tasks, is generated. Each search agent is associated to a task, or to a set of related tasks, not yet assigned to a computing resource for execution, so that we can regard each pending task waiting for execution as an agent, or better, in our swarm-based model as an "ant". The agents wander on networks and search proper nodes to join and queuing by interacting with the corresponding resource management agents. Agents can be generated from every node on networks. Each node on the networks supplies the same service and each search agent must be served by one node through one of its resource management agents. At a higher level of the global Grid, agents cooperate with each other according to ant-like local interaction to manage the overall workload according to the above agent-based self-organization performing complementary scheduling of the available resources on the network-distributed Grid.

B. The agent structure

In our environment, a specific agent will be implemented at the Grid middleware level for managing computing resources on each local grid node and scheduling incoming tasks to achieve local load balancing, and it provides a high-level representation of the corresponding grid resources. It also characterises these resources as high performance computing service providers in a wider grid environment. Agents can be structured according to a layered model to better define their architectural characteristics and ease the overall implementation tasks, as defined in the following schema [2].

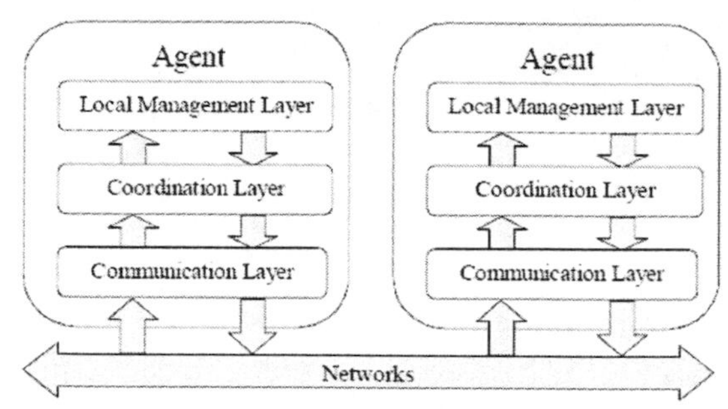

Fig. 2. The agent multi-layer architecture

- *Communication Layer.* Agents in the Grid system must be able to communicate with each other or with users using common data models and communication protocols. The communication layer provides an agent with an interface to heterogeneous networks and operating systems.
- *Coordination Layer.* The request an agent receives from the communication layer should be explained and submitted to the coordination layer, which decides how the agent should act on the request, during the search space exploration, according to its own knowledge. For example, if an agent receives a service request, it must decide whether it has related service information. The ant-driven resource scheduling process does not aim to find the best service for each request, but endeavours to find an available service provided by a neighbouring agent. While this may decrease the optimal load balancing effect, the trade-off is reasonable as grid users prefers to find a satisfactory resource as fast and as local as possible.
- *Local Management Layer.* This layer performs functions of an agent for local and grid resource management, that is participating to scheduling decisions to handle load distribution between the local resources (i.e. local processors) and the overall Grid resources. A local grid resource is considered to be a cluster of tightly coupled workstations, i.e. in a blade enclosure, operating according to a common scheduling policy, or a multiprocessor system. An agent takes its local available computational resources as one of its capabilities. It is also responsible for submitting local service information to the coordination layer for agent decision making. Within each agent, its own service provided by the local grid resource is evaluated first. Of course, if the

requirement can be met locally, the job execution can be handled successfully without interaction with other external agents/nodes.

III. THE ANT-COLONY OPTIMIZATION PARADIGM

The ant colony optimization (ACO) techniques are a subset of swarm intelligence meta-heuristics inspired by the foraging behaviour of real biological networks [3] in finding the paths to food sources and route around obstacles and consider the ability of simple individuals to solve complex problems by cooperation. In biological "networks" or colonies of ants consisting of thousands and in some cases tens of thousands of dynamic elements, each ant alone has relatively little intelligence, while the collective emergent behaviour of the "network" exhibits a great deal of global intelligence capable of dynamic near-global optimization of certain tasks. Engineering models and algorithms based on these biological systems have the potential to leverage the tremendous gains made in this century in understanding their individual and collective colony-based behaviour. Initial work in swarm intelligence has revealed a great deal of synergy between the routing requirements of communication networks and certain tasks that exist in biological swarms. For instance, a key characteristic of swarm intelligence is the ability of search agents (ants) to find optimal (or near optimal) routing (in food gathering operations for example), where intelligent behaviour arises through indirect communications between the agents. The latter point is the most interesting: the ants do not need any direct communication for the solution process, instead they communicate by stigmergy. The notion of stigmergy means the indirect communication of individuals through modifying their environment. In detail, any ant that leaves its colony in search of food leaves a trail of chemical called pheromones on the path that it takes. When an ant returns from the food source to its nest, it reinforces the pheromones on the path that it has used. Pheromones acts to attract other ants to follow a particular path. When a large number of ants forage for food, the shortest or however the best available path to the food source will eventually contain the highest concentration of pheromones, thereby attracting all the ants to use that path. Thus, the concentration of pheromone on a certain path is an indication of its usage. With time the concentration of pheromone decreases due to diffusion effects. This property is important because it is integrating dynamic into the path searching process. The following figure shows a scenario with two routes from the nest to the food place. At the intersection, the first ants randomly select the next branch. Since the upper route is shorter than the lower one, the ants which take this path will reach the food place first. On their way back to the nest, the ants again have to select a path. After a short time the pheromone concentration on the shorter path will be higher than on the longer path, because the ants using the shorter path will increase the pheromone concentration faster. The shortest path, that in this case is straightforwardly the best one, will thus be identified and eventually all ants will only use it.

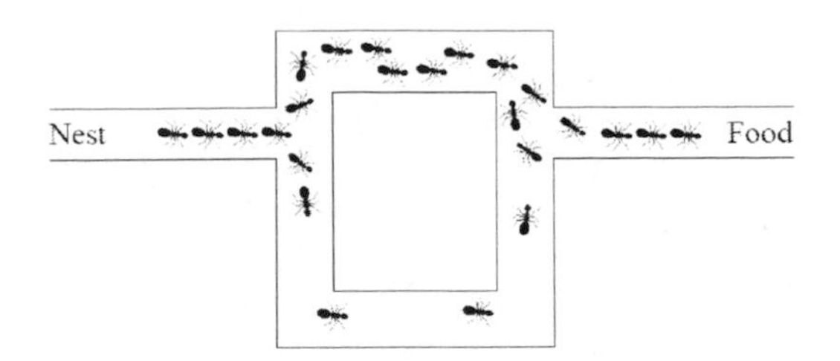

Fig. 2. The ant-colony path searching behaviour

Algorithms modelled according to this behaviour, have historically been used to solve shortest path problems and problems that can be reduced to a kind of shortest path problems but have also been successful in attacking various popular combinatorial optimization problems such as the travelling salesman problem (TSP), the quadratic assignment problem (QAP) and the job-shop scheduling problem (JSP) [4]. Furthermore, almost all the results obtained to date in developing swarm-based path searching or combinatorial optimization algorithms exhibits the following potential benefits:

- Dynamic "online" optimization using local information
- No or very limited exchange of global information for solution determination
- Inherent scalable nature, resulting in graceful builds and degradations
- Characteristics leading to adaptivity and robustness (fault-tolerance) under most contingencies

The solution to a combinatorial optimization problem is a set $S = \{c_1, c_2, \ldots c_n\}$, where c_i, with $1 \leq i \leq n$, are known as the solution components. The Ant Colony-based optimization techniques are inherently iterative in their behaviour. During each iteration, an ant constructs a solution starting from the empty solution $S=\varnothing$. Solution components are incrementally added, one at a time, to the partial solution until a complete, feasible solution is formed. At the end of an iteration, the candidate solutions constructed by the ants will be evaluated using a problem specific objective function f. For example in TSP, the objective function is the length of a complete tour. After that, each ant will update the artificial pheromone associated to each component that is found within its constructed solution. Components that appear frequently in good candidate solutions will have higher pheromone values. The search process in subsequent iterations will then be biased towards these favourable components. The search process can stop if for example the number of maximum iterations has been met. In any case, the five main characterizing elements of the above techniques are identified below:

- A heuristic function η, which will guide the ants' search with problem specific information.
- A pheromone trail definition, which states what information is to be stored in the pheromone trail. This allows the ants to share information about good solutions.

- The pheromone update rule, this defines the way in which good solutions are reinforced in the pheromone trail.
- A fitness function which determines the quality of a particular ant's solution.
- A construction procedure that the ants follow as they build their solutions (this also tends to be problem specific).

Although the ACO method is a fairly novel approach, it can be viewed as something of an amalgam of other techniques. The ants essentially perform an adaptive greedy search of the solution space. Greedy search methods have been used for many problems, and are often good at finding reasonable results fast. The greedy search takes into account information provided in the pheromone trail, and this information is reinforced after each iteration. Because a population of ants is used, ACO also has similarities with evolutionary computation approaches, such as genetic algorithms (GAs). GAs similarly use a population of solutions which share useful information about good solutions, and probabilistically use this information to build new solutions. They do this in a rather different way, however, since a GAs "stores" its information in the population at each iteration whereas the ants store it in the pheromone trail. So, while the specific details of an ACO algorithm are new, and the techniques used are combined in a novel way, the ACO approach implicitly uses several proven problem solving strategies.

IV. THE PROPOSED ALGORITHM

Our objective is to determine an efficient solution to the problem of scheduling a set J of m independent jobs $J=\{j_1, \dots j_m\}$ each of which have an associated predicted running time $t(j_i)$ onto a set V of n Grid nodes, $V=\{v_1, \dots v_n\}$, for each $1 \le i \le n$, with a specific computing capacity C_i, such that the load will be fairly balanced on the available computing resources and consequently all the jobs are completed as quickly as possible, by optimizing the overall Grid performance. Really, this is a multi-objective optimization, the two most important objectives being the minimization of the makespan and the flowtime of the overall grid system. This problem, widely simplified, is closely related to a commonly known NP-hard combinatorial optimisation problem, the bin packing problem, where the items to be packed are viewed as the jobs and the bins as the Grid nodes with their capacity represented by the available computing power. Consequently, also our multiple variable resource scheduling problem, that is inherently more complex, will be NP-hard and thus a reasonable heuristic solution, achieving near-optimal results is strongly desirable. Accordingly we propose an highly adaptive approach that employ a collection of ant agents that collaborate to explore the search space. A stochastic decision making strategy is proposed in order to combine global and local heuristics to effectively conduct this exploration. As the algorithm proceeds in finding better quality solutions, dynamically computed local heuristics are utilized to better guide the searching process. In our schema an ant's "life" begins at the originating node of each task execution demand. It proceeds until it finds the corresponding destination node resource management agent offering adequate resources available for task execution. At the completion of its search, each ant deposits pheromone to mark the detected solution and perform its stigmergy-based interaction with the other socially-related individuals. Thus, the most important factor that influences each ant's decision, and hence the overall heuristic search behaviour, will be the pheromone.

A. The pheromone trail

There is not an immediately clear definition for the pheromone trail in this problem, and so the information that will be stored in the pheromone trail must be carefully selected. As the problem is essentially to allocate jobs to the grid nodes, intuitively it seems that the trail could store the degree of success associated to assigning a particular job to a particular node. The pheromone value $\tau(i,j)$ can be therefore selected to represent the overall success in finding a feasible solution implied by scheduling a particular job i onto a particular node j. The pheromone matrix will thus have a single entry for each job-node pair in the problem. Furthermore, a policy for updating the pheromone trail has been established according to the following equation:

$$\tau(i,j) = \rho \cdot \tau(i,j)$$

where ρ is a parameter which defines the pheromone evaporation rate.

B. The ant-driving heuristic

The heuristic information that the ants use when building their solution is also very important, it guides their search with problem specific information. However, because the ACO approach relies on multiple ants building solutions over several "generations" the heuristic information must be quick to establish, and so only fairly simple heuristic values can be used. The heuristic value used by the ants for each job j has been defined in our model as:

$$\eta(j) = \frac{1}{\min_i(\log_{10}(t(j_i)))}$$

That is inversely proportional to the minimum completion time for the job j on all the available nodes, or better stated its completion time on the best available node on the grid. To allow this value to be effectively controlled with the β parameter, determining the extent to which heuristic information is used by the ants, it is necessary to "scale" the heuristic value up. Therefore in the implementation of this function all the $\eta(j)$ values are computed for each job and then the job list is sorted into descending order of these values.

C. *Finding a solution*

The set of jobs and nodes will serve as components from which each ant will use to incrementally construct a solution during each iteration of the algorithm. For each ant, an iteration consists of a finite number of steps. At step r of iteration t, an ant k will select a specific job j to be executed on a node vi to be included in its partially constructed solution $Sk(t,r)=\{s1,\dots sj\}$ according to a stochastic decision making strategy properly driven by the above heuristic and pheromone trail. More precisely, let b_j the node on which the job j can be executed in the minimum completion time, and α the extent to which pheromone information is used as the ants build their solution, the probability $p(j)$ of selecting job j to schedule next is given by equation:

$$p(j) = \frac{\tau(j,b_j)^{\alpha} \cdot \eta(j)^{\beta}}{\sum_{i=1}^{n}(\tau(i,b_i)^{\alpha} \cdot \eta(i)^{\beta})}$$

A job is then selected based on this value, and the chosen job j_c is preferably allocated, on the b_{jc} node. The pheromone trail update procedure is then used on the iteration best ant. This process is repeated until all jobs have been scheduled, a complete solution has been built and there is no further improvement in the fitness value of the solution. The fitness value is determined according to a properly-crafted fitness function, whose goal is essentially to help the algorithm in discerning between high and low quality solutions that in our case means obtaining that all the jobs are completed as quickly as possible. In other words it implicitly means best balancing the load on the available nodes, that is, minimising the makespan and flowtime of the solution itself. Makespan and flowtime are both strong fitness indicators of the grid system; their relation is not trivial at all, in fact they are contradictory in the sense that trying to minimize one of them could not suit to the other, especially for plannings close to optimal ones. Note that makespan is an indicator of the general productivity of the grid system. Small values of makespan mean that the scheduler is providing good and efficient planning of tasks to resources. On the other hand, minimizing the value of flowtime means reducing the average response time of the grid system. For better results, the value of mean flowtime, flowtime/M (where M is the number of machines in the grid), can be used instead of flowtime. Essentially, we want to maximize the productivity (throughput) of the grid environment through an intelligent load balancing and at the same time we want to obtain plannings that offer a quality of service acceptable to the users. Consequently the fitness function for our assignment and balancing problem could simply be the inverse of the sum of makespan mks_s and mean flowtime ft_s of the solution s, weighted by a properly crafted parameter λ, a priori fixed to 0.75 to give more priority to makespan, as it is the most important parameter. The fitness function equation is reported below.

$$f_s = \frac{1}{\lambda \cdot mks_s + (1-\lambda) \cdot ft_s}$$

The proposed algorithm is conceived to be flexible in the sense that the number of ants can be adjusted by the launching probability of ants to achieve a good performance.

V. PERFORMANCE EVALUATION

To show that the approach has the potential to become an acceptable distributed framework for self-management of computational resources in the next generation grids, extensive simulation has been conducted upon four different Grid scenarios built in a random way upon some sample grid dimensional characteristics (small: 32 hosts and 512 tasks; average: 64 hosts and 1024 tasks; large: 128 hosts and 2048 tasks; and very large: 256 hosts and 4096 tasks). The grid networking topologies, modelled as non-oriented graphs, have been created using Waxman's method [5], [6]. In this method, the probability of the existence of link between two nodes u and v is given by:

$$P(u,v) = \varphi e^{-\frac{d}{\chi L}}$$

where $0 < \varphi,\chi \leq 1$ are model parameters, d is the Euclidean distance between u and v and L is the maximum Euclidean distance between any two vertices of the graph. The available computational resources have been assigned randomly on all the grid nodes In all the experiments, we used a dynamic model in which the job execution requests arrive at the grid according to a Poisson process with an arrival rate ε (jobs/second). The predicted job execution time is exponentially distributed with mean μ (1800 seconds in our tests).

A. *Building the system model*

The ACO meta-heuristic algorithm has been implemented in Java using the RePast multi-agent simulation framework. We used the Java version of Repast in order to take advantage of its great extensibility, ease of modifiability, portability, strict math and type definitions (to guarantee duplicatable results), and object serialization (to checkpoint out simulations). Repast is a free open source toolkit that was originally developed at the University of Chicago [7] and is now managed by the non-profit volunteer Repast Organization for Architecture and Development (ROAD). Repast seeks to support the development of extremely flexible models of living social agents, but is not limited to modelling living social entities alone. In short, our RePast simulation is primarily a collection of agents of both the search and resource management type and a model that sets up and controls the execution of these agents' behaviours according to a schedule. This schedule not only controls the execution of agent behaviours, but also actions within the model itself, determined by the ACO meta-heuristic paradigm.

B. Simulation results

All the results presented are taken from 100 iteration runs on 2GHz HP Proliant DL380 machines running Linux, and each run was performed 10 times to collect the average makespan and flowtime values, that are the most interesting performance parameter for our evaluation. The proposed framework takes a comparatively long time to build solutions, approximately more then 10 seconds per iteration, so that the whole simulation took some hours. The ACO algorithm has been allowed to run for so long because this gives it reasonable time to build up a useful pheromone trail. The ants need some more running time to find solutions which significantly improve on the other solutions. The efficiency of the proposed solution can be easily observed from the graph in Fig. 3 below where the Makespan and Flowtime values measured as the results of the ACO meta-heuristic in our four typical Grid scenarios have been compared with the same values obtained by applying the classic Tabu Search (TS) approach as showed in [8].

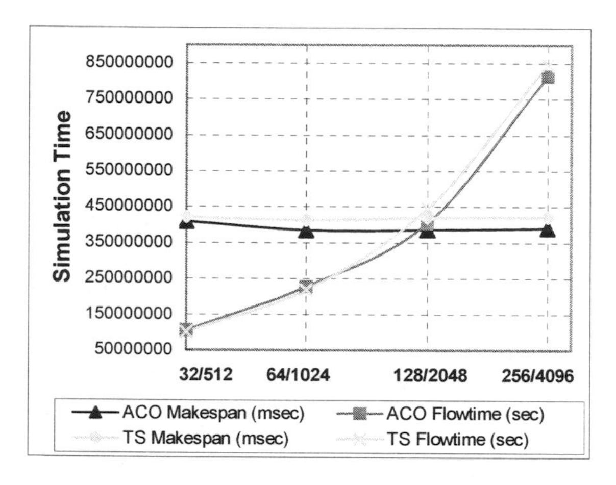

Fig. 3. Makespan and flowtime measured in typical grid scenarios

When representing both makespan and flowtime values simultaneously in the same graph we have to take into account that even though makespan and flowtime are measured in the same unit (seconds), the values they can take are in incomparable ranges, due to the fact that flowtime has a higher magnitude order over makespan, and its difference increases as more jobs and machines are considered. Consequently, in the above graph the makespan values have been scaled properly in msecs to be representable together with the flowtime ones. As can be seen, our ACO approach performs slightly better than the Tabu Search heuristic, especially in presence of larger problems, with more grid nodes and presented jobs, since in these cases the number of ants, or in other words the agents associated to tasks and nodes greatly increases. The setting of the ACO parameters in our model will also affect the performance of the whole framework. Due to the time taken for a decent sized run of the ACO algorithm, and to the inbuilt stochasticity of the approach, finding the optimal values for these parameters has been a complex and very time-consuming task. For each topology used in our evaluation, the best observed values for the parameters for pheromone control α and β and ρ have been empirically determined through experiences on simulation results. At first, the pheromone evaporation parameter ρ, defining how quickly the ants "forget" past solutions has been always set to be in the range $0 \leq \rho \leq 1$. A higher value makes for a more aggressive search, in our tests a value of around 0.8 gave good results. The value of α determining the extent to which pheromone information is used as the ants build their solution, showed to be very critical for the success of the ACO search, and having tested values between 1-30, it seems that the ant-based algorithm works best with a relatively high value of 15 for all the topologies. Again, also for the parameter β, determining the extent to which heuristic information is used by the ants, all the values in the range 1-30 were tested, and a value near 10 worked well for all the tests and topologies. Once again, we observed that an higher β value may provide good solutions quickly, but a lower value may provide better results after a longer period of time. The best-performing values for the ACO model parameters used in our simulation, are reported in the table 1 below.

Table 1. Best ACO parameters used in the simulation

Parameter name	Parameter Value
α	15
β	10
ρ	0.8

The above values as experimentally determined in our simple tests work well enough, as the observed results show, but there is undoubtedly room for further improvement.

VI. CONCLUSIONS

The next generation grid computing environment must be intelligent and autonomous to meet requirements of smart self-management. Accordingly, we presented in this work a new adaptive strategy to efficiently distribute the jobs submitted on a grid on the available computational resources. The proposed approach is based on swarm intelligence and precisely on the ant colony optimization meta-heuristic implemented in a multi-agent system scenario. These fascinating family of algorithms try to apply the ability of swarms to mathematical problems and were applied successfully to several optimization problems, so that they are widely recognised as of the major self-organizing search mechanisms used in nowadays adaptive applications. One of the most interesting features of ant colony optimization-based approaches is that it may allow enhanced efficiency when the representation of the problem under investigation is spatially distributed and changing over time. The agent-based ACO approach used in our work can be conceived as an initial attempt towards a distributed framework for building the next generation intelligent grid environments.

REFERENCES

[1] J. Kennedy, Y. Shi and R.C. Eberhart, "Swarm Intelligence" , Morgan Kaufmann Publishers, San Francisco, 2001.

[2] J. Cao , D. P. Spooner , S. A. Jarvis , G. R. Nudd , "Grid load balancing using intelligent agents", Future generation computer systems , vol. 21, n. 1, pp. 135-149, Elsevier, 2005

[3] M. Dorigo and G. Di Caro, "The Ant Colony Optimization Metaheuristic" In D. Corne, M. Dorigo and F. Glover (eds), New Ideas in Optimization, McGraw-Hill, 1999.

[4] M. Dorigo and L.M. Gambardella, "Ant colony System: Optimization by a colony of cooperating agents", IEEE Transactions on Systems, Man, and Cybernetics – Part B, vol. 26, no. 1, pp 29-41, 1996.

[5] E. W. Zegura, K. L. Calvert, and S. Bhattacharjee, "How to model an internetwork," in IEEE Infocom, vol. 2. San Francisco, CA: IEEE, pp. 594–602, 1996.

[6] B. M. Waxman, "Routing of multipoint connections," IEEE Journal on Selected Areas in Communications, vol. 6, no. 9, pp. 1617–1622, 1988.

[7] E. Tatara, M.J. North, T.R. Howe, N.T. Collier, and J.R. Vos, "An Introduction to Repast Modeling by Using a Simple Predator-Prey Example", Proceedings of the Agent 2006 Conference on Social Agents: Results and Prospects, Argonne National Laboratory, Argonne, IL USA

[8] T. D. Braun, H. J. Siegel, N. Beck, L. L. Boloni, M. Maheswaran, A. I. Reuther, J. P. Robertson, M. D. Theys, B. Yao, D. Hensgen and R. F. Freund, "A comparison of eleven static heuristics for mapping a class of independent tasks onto heterogeneous distributed computing systems", Journal of Parallel and Distributed Computing, 61(6):810–837, 2001

Facial Recognition with Singular Value Decomposition

Guoliang Zeng
Phone: (480) 727-1905; Fax: (480) 727-1723; E-mail: gzeng@asu.edu
Motorola DSP Laboratory, Division of Computing Studies
Arizona State University Polytechnic Campus
7001 East Williams Field Rd., ISTB3, Mesa, AZ 85212

ABSTRACT:
This paper implements a real-time system to recognize faces. The approach is essentially to apply the concepts of vector space and subspace to face recognition. The set of known faces with $m \times n$ pixels forms a subspace, called "face space", of the "image space" containing all images with $m \times n$ pixels. This face space best defines the variation of the known faces. The basis of the face space is defined by the singular-vectors of the set of known faces. These singular-vectors do not necessarily correspond to the distinct features like ears, eyes and noses. The projection of a new image onto this face space is then compared to the available projections of known faces to identify the person. Since the dimension of face subspace is much less than the whole image space, it is much easier to compare projections than origin images pixel by pixel. Based on the above idea, a Singular Value Decomposition (SVD) approach is implemented in this paper. The framework provides our system the ability to learn to recognize new faces in a real-time and automatic manner.

KEYWORDS:
Image processing, Face recognition, Singular value decomposition.

I. INTRODUCTION:

Over the last ten years or so, face recognition has become an active area of research in computer vision, neuroscience, and psychology. It is the general opinion that advances in computer vision research will provide useful insights to neuroscientists and psychologists into how human brain works, and vice versa.

Some of the applications of the facial recognition technology include:

- Security
- Computer-human interaction
- System (key) access based on face/voice recognition.
- Tracking people, either spatially with a large network of cameras or temporally by monitoring the same camera over time.
- Locating people in large image data.

Several approaches to face recognition have been proposed for the 2-dimensional facial recognition. Each has its own advantages and limitations. Much of the work has focused on detecting individual features such as eyes, nose, mouth, and head outline, and defining a face model by the position, size, and relationships among these features [1][2]. The face recognition strategies have modeled and classified faces based on normalized distances and ratios among feature points. Such approaches have proven difficult to extend to multiple views, and have often been quite fragile. Like Principal Component Analysis (PCA), our SVD approach treats a set of known faces as vectors in a subspace, called "face space", spanned by a small group of "base-faces" [3]. Recognition is performed by projecting a new image onto the face space, and then classifying the face by comparing its coordinates (position) in face space with the coordinates (positions) of known faces. But the SVD approach has better numerical properties than PCA.

II. PRINCIPALES:

K. Elleithy (ed.), Advances and Innovations in Systems, Computing Sciences and Software Engineering, 145–148.

Assume each face image has $m \times n = M$ pixels, and is represented as an $M \times 1$ column vector $\mathbf{f}_i$. Then, a 'training set' S with N face images of known individuals forms an $M \times N$ matrix:

$$S = [\mathbf{f}_1, \mathbf{f}_2, \ldots\ldots, \mathbf{f}_N] \tag{1}$$

The mean image $\bar{\mathbf{f}}$ of set S, is given by

$$\bar{\mathbf{f}} = \frac{1}{N}\sum_{i=1}^{N}\mathbf{f}_i \tag{2}$$

Subtracting $\bar{\mathbf{f}}$ from the original faces gives

$$\mathbf{a}_i = \mathbf{f}_i - \bar{\mathbf{f}},, i = 1,2,\ldots N \tag{3}$$

This gives another $M \times N$ matrix A:

$$A = [\mathbf{a}_1, \mathbf{a}_2, \ldots, \mathbf{a}_N] \tag{4}$$

Assume the rank of A is r, and $r \le N << M$. It can be proved that A has the following Single Value Decomposition (SVD):

$$A = U\Sigma V^T \tag{5}$$

Here, Σ is an $M \times N$ diagonal matrix

$$\Sigma = \begin{bmatrix} \sigma_1 & 0 & \cdots & 0 & 0 & \cdots & 0 \\ 0 & \sigma_2 & \cdots & 0 & 0 & \cdots & 0 \\ \vdots & \vdots & \ddots & \vdots & \vdots & \ddots & \vdots \\ 0 & 0 & \cdots & \sigma_r & 0 & \cdots & 0 \\ 0 & 0 & \cdots & 0 & \sigma_{r+1} & \cdots & 0 \\ \vdots & \vdots & \ddots & \vdots & \vdots & \ddots & \vdots \\ 0 & 0 & \cdots & 0 & 0 & \cdots & \sigma_N \\ 0 & 0 & \cdots & 0 & 0 & \cdots & 0 \end{bmatrix} \tag{6}$$

For i = 1, 2, …, N, σ_i are called Singular Values (SV) of matrix A. It can be proved that

$$\sigma_1 \ge \sigma_2 \ge \cdots \ge \sigma_r > 0, \text{ and}$$
$$\sigma_{r+1} = \sigma_{r+2} = \cdots = \sigma_N = 0. \tag{7}$$

Matrix V is an $N \times N$ orthogonal matrix:

$$V = [\mathbf{v}_1, \mathbf{v}_2, \ldots \mathbf{v}_r, \mathbf{v}_{r+1}, \ldots, \mathbf{v}_N] \tag{8}$$

That is the column vectors $\mathbf{v}_i$, for i = 1, 2, …, N, form an orthonormal set:

$$\mathbf{v}_i^T \mathbf{v}_j = \delta_{ij} = \begin{cases} 1,, i = j \\ 0,, i \ne j \end{cases} \tag{9}$$

Matrix U is an $M \times M$ orthogonal matrix:

$$U = [\mathbf{u}_1, \mathbf{u}_2, \ldots \mathbf{u}_r, \mathbf{u}_{r+1}, \ldots, \mathbf{u}_M] \tag{10}$$

That is the column vectors $\mathbf{u}_i$, for i = 1, 2, …, M, also form an orthonormal set:

$$\mathbf{u}_i^T \mathbf{u}_j = \delta_{ij} = \begin{cases} 1,, i = j \\ 0,, i \ne j \end{cases} \tag{11}$$

The $\mathbf{v}_i$'s and $\mathbf{u}_i$'s are called right and left singular-vectors of A.
From (5),

$$AV = U\Sigma \tag{5a}$$

We get

$$A\mathbf{v}_i = \begin{cases} \sigma_i \mathbf{u}_i,,, i = 1,2,\ldots,r \\ 0,,,,,,, i = r+1,\ldots,N \end{cases} \tag{12}$$

It can be proved that $\{\mathbf{u}_1, \mathbf{u}_2, \ldots, \mathbf{u}_r\}$ form an orthonormal basis for $R(A)$, the range (column) subspace of matrix A. Since matrix A is formed from a training set S with N face images, $R(A)$ is called a 'face subspace' in the 'image space' of $m \times n$ pixels. Each $\mathbf{u}_i$, $i = 1,2,\ldots,r$, can be called a 'base-face'.

Let $\mathbf{x}$ $(= [x_1, x_2, \ldots, x_r]^T)$ be the coordinates (position) of any $m \times n$ face image $\mathbf{f}$ in the face subspace. Then it is the scalar projection of $\mathbf{f} - \bar{\mathbf{f}}$ onto the base-faces:

$$\mathbf{x} = [\mathbf{u}_1, \mathbf{u}_2, \ldots, \mathbf{u}_r]^T (\mathbf{f} - \bar{\mathbf{f}}) \tag{13}$$

This coordinate vector $\mathbf{x}$ is used to find which of the training faces best describes the face **f**. That is to find some training face $\mathbf{f}_i$, $i = 1,2,\ldots,N$, that minimizes the distance:

$$\varepsilon_i = \|\mathbf{x} - \mathbf{x}_i\|_2 = [(\mathbf{x} - \mathbf{x}_i)^T (\mathbf{x} - \mathbf{x}_i)]^{1/2} \quad (14)$$

where $\mathbf{x}_i$ is the coordinate vector of $\mathbf{f}_i$, which is the scalar projection of $\mathbf{f}_i - \bar{\mathbf{f}}$ onto the base-faces:

$$\mathbf{x}_i = [\mathbf{u}_1, \mathbf{u}_2, \ldots, \mathbf{u}_r]^T (\mathbf{f}_i - \bar{\mathbf{f}}) \quad (15)$$

A face **f** is classified as face $\mathbf{f}_i$ when the minimum ε_i is less than some predefined threshold ε_0. Otherwise the face **f** is classified as "unknown face".

If **f** is not a face, its distance to the face subspace will be greater than 0. Since the vector projection of $\mathbf{f} - \bar{\mathbf{f}}$ onto the face space is given by

$$\mathbf{f}_p = [\mathbf{u}_1, \mathbf{u}_2, \ldots \mathbf{u}_r]\mathbf{x} \quad (16)$$

where **x** is given in (13).

The distance of **f** to the face space is the distance between $\mathbf{f} - \bar{\mathbf{f}}$ and the projection $\mathbf{f}_p$ onto the face space:

$$\varepsilon_f = \|(\mathbf{f} - \bar{\mathbf{f}}) - \mathbf{f}_p\|_2 = [(\mathbf{f} - \bar{\mathbf{f}} - \mathbf{f}_p)^T (\mathbf{f} - \bar{\mathbf{f}} - \mathbf{f}_p)]^{1/2} \quad (17)$$

If ε_f is greater than some predefined threshold ε_1, then **f** is not a face image.

Notes:

1). In practice, a smaller number of base-faces than r is sufficient for identification, because accurate reconstruction is not a requirement. This smaller number of significant base-faces is chosen as those with the largest associated singular values.

2). In the training set, each individual could have more than one face images with different angles, expressions, and so on. In this case, we can use the average of them for the identification.

III. STEPS:

1. Obtain a training set S with N face images of known individuals. An example of S is shown in Fig. 1.

2. Compute the mean face $\bar{\mathbf{f}}$ of set S by (2). The mean face of the set in Fig. 1 is shown in Fig. 2. Form matrix A in (4) with the computed $\bar{\mathbf{f}}$.

3. Calculate the SVD of A as shown in (5). The result base-faces are shown in Fig. 3.

4. For each known individual, compute the coordinate vector $\mathbf{x}_i$ from (15).

5. Choose a threshold ε_1 that defines the maximum allowable distance from face space. Determine a threshold ε_0 that defines the maximum allowable distance from any known face in the training set S.

6. For a new input image **f** to be identified, calculate its coordinate vector **x** from (13), the vector projection $\mathbf{f}_p$, the distance ε_f to the face space from (17). If $\varepsilon_f > \varepsilon_1$ the input image is not a face.

7. If $\varepsilon_f < \varepsilon_1$, compute the distance ε_i to each known individual from (14). If all $\varepsilon_i > \varepsilon_0$, the input image may be classified as unknown face, and optionally used to begin a new individual face. If $\varepsilon_f < \varepsilon_1$, and some $\varepsilon_i < \varepsilon_0$, classify the input image as the known individual associated with the minimum ε_i ($\mathbf{x}_i$) in (14), and this image may optionally added to the original training set. Steps 1-5 may be repeated. This can update the system with more instances of known faces.

IV. CONCLUSIONS:

We tested our system with the following data:

- Image Size: $M = 92 \times 112 = 10{,}304$.
- Number of known individuals: N = 20 -40.
- Number of face images per person: C = 1-5.
- Different Conditions: All frontal and slight tilt of the head, different facial expressions.

Essentially, a face image is of M (say 10,000) dimension. But the rank r of matrix A is less than or equals N (say 40). For most applications, a smaller number of base-faces than r is sufficient for identification. In this way, the amount of computation is greatly reduced. The Golub-Reinsch Algorithm is used for computing the singular value decomposition of a matrix. This can be done offline as part of the training. On machines with a 1.2 GHz clock rate, the recognition will take a time much less than 33 msec, a frame rate for real-time system.

The SVD approach is robust, simple, and easy and fast to implement which works well in a constrained environment. It provides a practical solution to the recognition problem. Instead of searching a large database of faces, it is better to give a small set of likely matches. By using base-faces, this small set of likely matches for given images can be easily obtained.

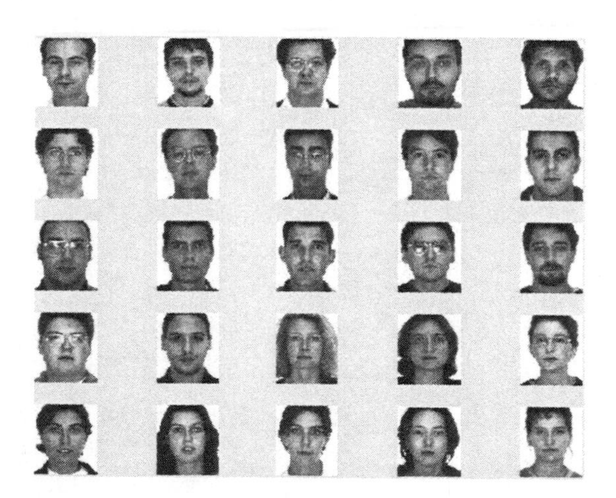

Fig. 1 Training set S

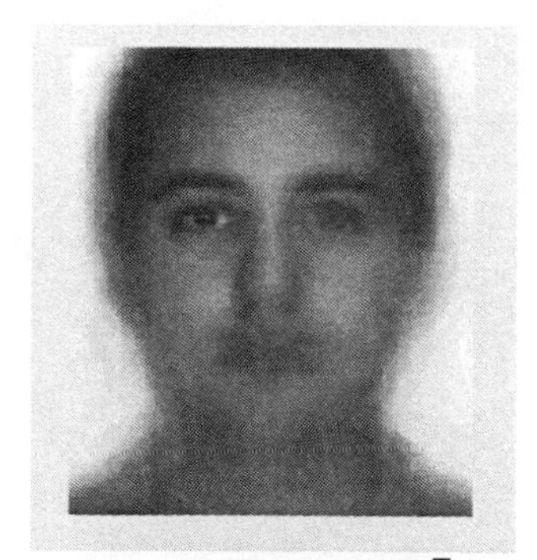

Fig. 2 Mean face $\bar{\mathbf{f}}$

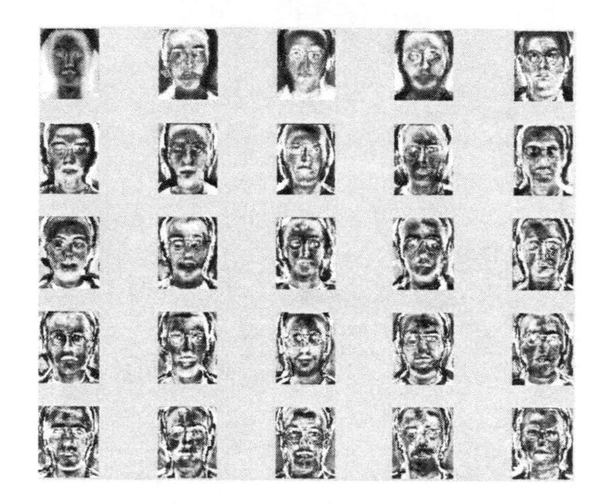

Fig. 3 Base-faces $\mathbf{u}_i$

REFERENCES:

[1] T. Kanade, "Picture Processing System by Computer Complex and Recognition of Human Faces," Department of Information Science, Kyoto University, Nov.1973.

[2] A. L. Yuille, D. S. Cohen, and P. W. Hallinan, "Feature Extraction from Faces Using Deformable Templates," *Proc. CVPR*, San Diego, CA June 1989.

[3] Matthew A. Turk, Pentland P. Alex(1991), "Face Recognition using Eigenface method", *IEEE Conference on Computer Vision and Pattern Recognition*, pp.586-591, 1991.

The Application of Mobile Agents to Grid Monitor Services

Guoqing Dong[1,2], Weiqin Tong[1]
[1] School of Computer Engineering and Science
Shanghai University, Shanghai 200072, China
[2] School of Computer and Communication Engineering
China Petroluem University, Dongying, China
E-mail: donggq@hdpu.edu.cn, wqtong@shu.edu.cn

Abstract—**Grids provide a uniform interface to a collection of heterogeneous, geographically distributed resources. In recent years, the research on the Grid Monitoring System gets increasingly essential and significant. In this paper we put forward a novel Mobile Agent-based Grid Monitoring Architecture (MA-GMA), which is based on the GMA from GGF and introduces the mobile agents and cache mechanism of MDS. Based on the Open Grid Service Architecture (OGSA) standard, we merge the intelligence and mobility characteristic of mobile agent into the current OGSA to constructing a dynamic and extensible monitoring system. In the end, we do some experiments under different environments. As the results shown, this MA-GMA is proved to be effective and improve the monitoring performance greatly.**

I. Introduction

Grids provide an infrastructure that allows for flexible, secure, coordinated resource sharing among dynamic collections of individuals, resources, and organizations. Each and every resource has its own specific static properties (e.g. processor speed, total amount of memory, total amount of disk space, networking capabilities) and dynamic status information (e.g. processor usage, available memory, available disk space, network usage). Furthermore, these resources are dynamic in nature: resources can join/part from the Grid, hardware failures can occur, etc. In order to make intelligent Grid resource management and scheduling decisions, accurate resource property and state information is required. Consequently, the users and developers of Grid must usually monitor the system running to discover unknown performance problems. As a result, people can find out the root failure in time, analyze the system performance bottleneck, resume and adjust the system in shortest time. Therefore, it is a hotspot in grid researches to develop Grid Monitoring System.

Grid environments have many common aspects with distributed and parallel systems, but are different than the latter for some features. However, the need for high performances causes the most common distributed programming paradigms not to be always suitable for Grids. On the other hand, the existing parallel programming tools and techniques are not always appropriate for coping with the heterogeneous Grid systems. These remarks show how a new generation of techniques, mechanisms and tools need to be considered, that can cope with the complexity of such system and provide the performances suitable for scientific computing applications [1].

In this scenario we think that mobile agent technology [2] can play a central role since their capability to cope with systems' heterogeneity and to deploy user customized procedures on remote sites seems to be very adequate to Grid environments. By interacting with a remote host after migrating on it, an agent is able to make complex operations on remote data without transferring them: the basic idea under this paradigm is that of moving the application logic near the data it needs. This may produce (in general) a significant saving in bandwidth on one hand and the possibility to analyze remote performance data with user-customized algorithms encapsulated in an agent.

This paper applies mobile agents to traditional GMA [3], [4], and puts forward a Mobile Agent-based Grid Monitor Architecture (MA-GMA).

The rest of this paper is organized as follows: Section 2 provides an introduction to related work. Section 3 contains a description of the MA-GMA, which is the architecture designed. Section 4 describes the principle of information collecting. Section 5 provides experiments and the performance analysis of MA-GMA. Finally, Section 6 presents the conclusions of this paper.

II. Related Work

A. Similar Work

The Grid Monitoring Architecture, as defined by the GGF [3], [4], is a reference architecture for feasible Grid monitoring systems and consists of three important components: producers, consumers and a directory service (see Fig. 1). The directory service stores the location and type of information provided by the different producers, while consumers typically query the directory to find out which producers can provide

K. Elleithy (ed.), Advances and Innovations in Systems, Computing Sciences and Software Engineering, 149–154.

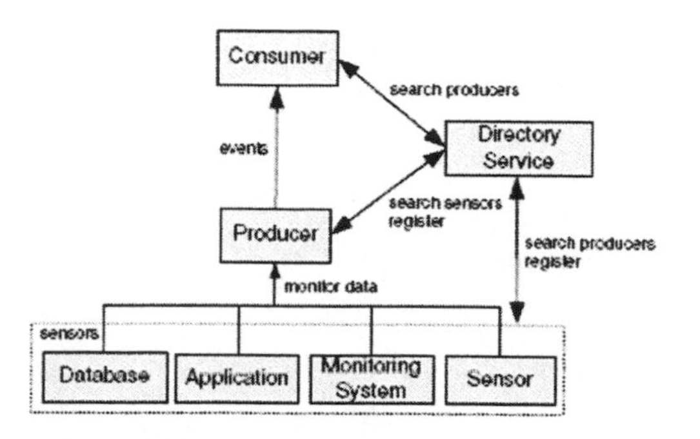

Fig. 1 Grid Monitoring Architecture overview

their needed event data (after which they contact the producers directly). Producers in turn can receive their event data from a variety of providers (software/hardware sensors, whole monitoring systems, databases, etc.). The GMA does not specify the underlying data models or protocols that have to be used.

Multiple monitoring architectures for distributed computing systems have already been successfully deployed. Below we present some notable Grid monitoring platforms with architecture similar to our framework, and point out the differences with our implementation.

GMA-compliant Grid monitoring systems include the European DataGrid's [5] Relational Grid Monitoring Architecture R-GMA [6] and GridRM [7], [8]. R-GMA offers a combined monitoring and information system using a Relational Database Management System as directory service and monitoring data repository. GridRM is an open source two-layer Grid monitoring framework, the upper layer being structured according to the GMA. This upper layer connects the per-site monitoring systems in a scalable way. Currently, GridRM's directory service can be a bottleneck and/or single point of failure, but work is under way to remedy this problem.

Network Weather Service [9], [10] is a framework for measuring the performance of distributed systems in processing intense environments. It can predict network and processing load in the near future based on monitored historical data.

Java Agents for Monitoring and Management (JAMM [11]) is a monitoring architecture fully implemented in Java. JAMM is mainly based on the GMA architecture and offers automated deployment of sensor agents on hosts from a central HTTP server. These sensors are actually wrappers for popular UNIX/Linux system utilities such as netstat, iostat and vmstat and are therefore badly deployable on other operating systems.

MDS2 is the Globus Toolkit (version 2) Monitoring and Discovery Service [12], and although MDS development was started before the GMA architectural reference appeared, it can still be seen as a GMA implementation. MDS2 only supports latest-state queries, making it mandatory for the consumers to actively retrieve status information from the GRIS. An extensive comparison of MDS2 against other monitoring frameworks has already been carried out in [13], [14]. It was shown that MDS2 outperforms the other frameworks mentioned in most use cases.

B. The Mobile Agent

With the agent technologies and applications developing, we think that a new approach using technologies based on mobile agents might be the key for dealing with some of the above mentioned problems adequately.

A mobile agent is a software module able to migrate among the hosts of a network, in order to carry on a specific task [2]. The agent is not linked to the system where it starts its execution. After being created in an execution environment, an agent can carry its state and code to another execution environment in another host of the network, where the execution can be restarted or continued. By "state" of the agent we mean a set of values of the agent's attributes, which allow it to determine what to do when the execution is restarted in another host.

The mobile agent programming paradigm overcomes some of the limits of traditional distributed processing techniques, which are typically based on the client/server paradigm [15]. In fact, in a mobile agent approach, the agent moves close to the data to be processed, thus eliminating the network traffic due to messages (excluding the initial migration), and allowing the execution of operations dynamically defined by the user. A typical case concerns client/server applications in which the client must retrieve some data from the server and operate complex filtering operations on such data; by moving an agent containing the procedures that deal with filtering, only the data that actually concern the client are sent through the network, with a considerable reduction of communication costs [16]. Besides, a permanent connection between client and server is not necessary in such scheme; the agent, once it is sent to the site of destination, can continue doing its operations and can communicate the results as soon as the client connects to the network again.

Mobile Agent has following characteristics besides common characteristics of simple agent [17]: mobility, durative, off line computing, etc. Therefore it has following advantages:

- The computing unit is close with source data; consequently the network traffic is reduced.
- Rule-based intelligentized migration policy can avoid blindness of resource access.
- Mobile agent can interact and collaborate with another one and fulfill tasks on different levels together.
- The whole asynchronous computing environment is offered.
- Demands of network bandwidth and communication devices are reduced greatly.
- Mobile Agent can make a reply and migrate itself according to the environment and events.

Considering above advantages of Mobile Agent, we put

forward a Mobile Agent-based Grid Monitor Architecture (MA-GMA).

III. Mobile Agent-based Grid Monitor Architecture (MA-GMA)

The MA-GMA is based on the traditional GMA. Our major

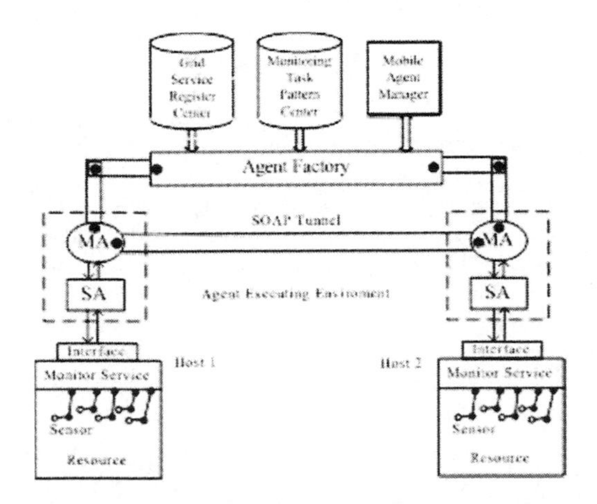

Fig. 2 Information-collecting module

contribution is to improve its performance through adding mobile agent-based collecting module and the cache mechanism to the producer of GMA.

The architecture of information collecting module in this MA-GMA is illustrated as Fig. 2, which is designed according to OGSA standard with characteristics of mobile agent. After handling this module, the dynamic and expandable grid monitoring system with higher QoS can be developed well.

A. Monitoring Service

This service holds a set of interfaces launched according to grid service rules. Other than interfaces of common grid services, these interfaces response the SA's (Stationary Agent) requests, collects information of grid resources via sensors, and then send the information to SA. The Monitoring Service main purpose is to use the data which is retrieved from the Sensors. And the Sensors are used to monitor a specific type of resources shown as follows:

Host: sensors monitoring hosts should be able to monitor CPU load, total and available memory and some other host related parameters;

Network: sensors monitoring network should be able to retrieve at least SNMP information from network devices and to retrieve other information about network links characteristics;

Storage: sensors monitoring storage resources should be able to measure the typical I/O parameters such as read/write throughput, access time, etc, for a mass data storage system;

Process/application: these sensors monitor the life cycle of a process, report all its interested information; they should also be able to monitor user-defined signals and/or events coming from the application.

B. Monitoring Task Pattern Center

Two libraries are available in this center: one is the system monitoring library, which provides basic monitoring task patterns; and the other is the user monitoring library, which is used to customize monitoring tasks by users according to special demands. The user monitoring library can improve flexibility and expandability of Grid Monitoring System.

C. Mobile Agent Manager

The Mobile Agent Manager is responsible for all kinds of management of the agents. It provides the necessary infrastructure for agents' transferring and receiving in the grid, and manages agents' life cycle to provide a complete environment for agents' execution. It is composed of a Mobile Agent Manager (MAM), a Mobile Agent Transportation (MAT), a Mobile Agent Naming Service (MANS), a Mobile Agent Communication (MAC), a Mobile Agent Security Manager (MASM), and Collaboration Service (CS).

MAT accomplishes the migration of agents, and do packaging, compressing and calculating on the agent that need to migrate.

MANS manages Mobile Agent's naming service, traces the location of the agent which moves frequently, handles the communication among mobile agents and controls the creator of the agent to remote agent.

MAC manages the communication among agents and notifies events to and from agents.

MASM provides bilateral security mechanism. It is responsible for distinguishing users, authenticating their agents, protecting servers' resources, ensuring the security and integrity of the agents and their data objects when moving in the network.

CS provides collaboration and resources-share services among mobile agents. Besides these functions, it also provides a mechanism for developers making it possible to add some related service on the destination server when agents move to it.

D. Mobile Agent Factory

In this module, the agents are generated and are send to the Grid. In order to combine mobile agent technology with OGSA standards smoothly, we employ SOAP as Agent Transfer Protocol and implement a simple and lightweight mechanism for exchanging structured information using XML [18], [19]. The factory extracts monitoring objects from grid services, searches corresponding patterns from Monitoring Task Pattern Center according to monitoring requirements, puts the patterns with additional information about migration path and destination into mobile agent process logic, and finally a new mobile agent was born.

E. SOAP Tunnel

In our architecture, all the interaction information, including mobile agents themselves, is encapsulated and transferred in SOAP message format in launcher, and is parsed in receiver by SOAP engine. It acts as a tunnel connecting all hosts.

F. Agent Executing Environment

In order for agents to perform any given task, an environment is required for the agents to execute on -- an Agent Executing Environment (AEE). An AEE is a software system that provides the necessary runtime components for mobile agent execution. It provides a common interface for interaction and communication between agents and other agents, or agents and hosts. The AEE also provides services for creation, migration and termination of agents.

IV. The Principle of Information Collecting

Two methods are mainly used in monitoring information collecting: passive method and active method. By passive method, the manager node collects information of sub nodes by turns at intervals. By active method, sub nodes notify the manager node that its information has changed, and the manager node is supposed to collect its information.

By active method, when a user sends requests to Grid Service Register Center, Grid Service Register Center provides detailed information about the grid service and the host. Agent Factory searches the Grid Service Register Center for the corresponding MMA (Monitor Mobile Agent) of the grid service. If the MMA cannot be found, Agent Factory will send the detailed information of the service to Monitoring Task Pattern Center, and embed the monitoring object provided by Monitoring Task Pattern Center in the logical unit of MMA. When the MMA is created by SA (Stationary Agent), the information of the grid service must be registered in the center. After MMA is deployed in the executable environment of the host, the agent will be instantialized instantly and added to the CodeServer. Via the CodeServer, Mobile Agent Manager can obtain the location information of mobile agent instance. When the grid service is consumed again, location information of the corresponding agent and instance will be found in the Mobile Agent Manager.

By passive method, the consumer sends request, and Agent Factory searches the Grid Service Register Center for the corresponding MMA of the grid service. If the MMA cannot be found, Agent Factory will send the detailed information of the service to Monitoring Task Pattern Center, and embed the monitoring object provided by Monitoring Task Pattern Center in the logical unit of MMA. Then Agent Factory creates the instance of the MMA. The MMA goes through all the grid nodes along the optimization path of Mobile Agent's migration and goes back to the manager node at last. The MMA gathers complete information of grid resources and then sends it to Consumer.

Mobile agent-based grid monitoring system has many merits compared with traditional one. For instance, migrating agents to sub nodes can decrease dependence on performance of the manager node. Services provided by sub nodes can be reduced because mobile agents have resources monitoring code; Network can be disconnected when a mobile agent runs after it is migrated.

To shorten the delay of the system, cache mechanism should be set up in every grid nodes. The MDS adopted the cache mechanism to improve the resources monitoring efficiency greatly. The MA-GMA inherits the cache mechanism and defines the conception of Period of Validity of grid resource information. The Period of Validity is an additive time property of resource information. If Tg is the creation time of information and Te is the end time that the information can be received by users, then (Te-Tg) is the Period of Validity of this resource information. The resource information in the cache will be invalid when its Period of Validity expires.

Resource information in the cache has different period of validity. We designed two cache policies to deal with different period of validity -- longer one or shorter one, which are described as following:

If the resource information has long period of validity, such as memory capacitance, it will be collected by the Sensor after the resource startups, then the cache will be updated to modify the information. When its period of validity expires, the Sensor collects the information and updates the cache over again.

If the resource information has short period of validity, such as CPU load, using above mentioned cache policy would worsen the efficiency of system, because its period of validity expires frequently and consequently the updating of cache is frequent. For this reason, checking its validity only when a user queries this information can reduce the updating of cache.

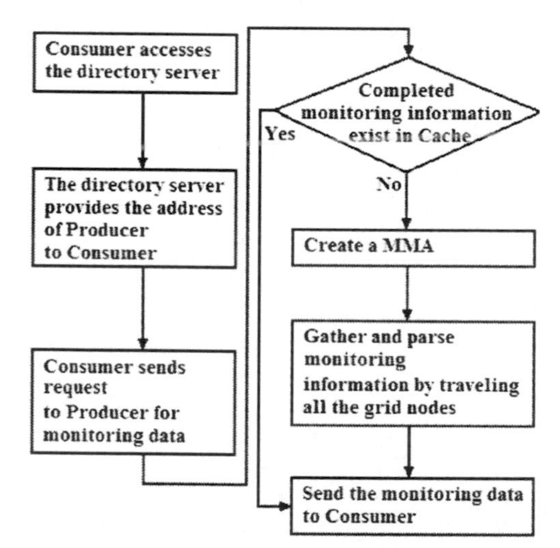

Fig. 3 Working process of the Grid Monitoring System based on MA-GMA

Because the information with a short period of validity changes quickly, the meaning of its instantaneous value is not important generally. The average value in recent period of time is more meaningful. The average value can be used instead to prolong the period of validity, and to enhance the cache hit rate.

The working process of the Grid Monitoring System based on MA-GMA is illustrated in Fig. 3.

V. Experiments and Performance Analysis

Here the performance test of MA-GMA refers to the test policy of MDS. The manage grid node is a Linux server, and the directory server is another Linux server with the same configuration. The nodes to be monitored include eight servers with different configuration. They are divided into four groups. The Globus Toolkit, grid services and grid applications are available in every server. All resources are configured by the policy of short period of validity and cache not to be modified automatically. The client is a personal computer installed with Windows 2000 OS and works via a remote login. A special

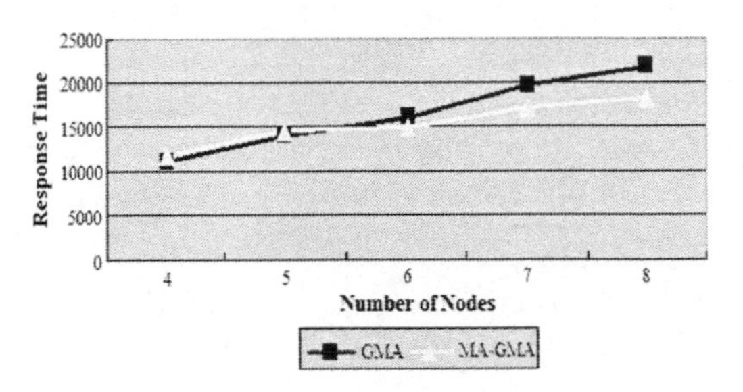

Fig. 4 Response time of GMA and MA-GMA without cache

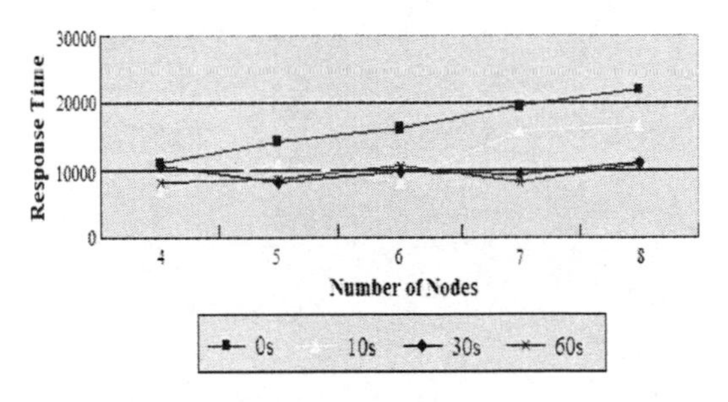

Fig. 5 Response time of MA-GMA with cache

program that was written for performance test runs on the client.

We experimented with different period of validity of information about grid resources. The value of period of validity is set to 0s(close the cache), 10s, 30, and 60s respectively in different test. The results of experiments are shown in Fig. 4 and Fig. 5.

From Fig. 4 and Fig. 5, the following facts can be drawn:

(1) If the number of nodes is small in a grid, the response time of grid monitoring system based on MA-GMA is higher than that based on GMA. But the response time of grid monitoring system based on MA-GMA is lower greatly than that based on GMA if the number of nodes is large.

(2) The response time tends to low with the increasing of the period of validity. It is due to the cache mechanism. The increasing of the period can lead a higher rate of hit the target in cache.

(3) When the period of validity is increased to 60s from 30s, there is no obvious improvement in the response time. It shows that after the period of validity is increased to a certain extent, its influence on the rate of hit the target in cache and whole performance becomes powerless.

VI. Conclusion

In this paper, the cache mechanism used in MDS and the technology of mobile agent are introduced into the grid monitoring system and we provide a novel idea in grid monitoring researches by merging the intelligence, mobility characteristic of mobile agent into current OGSA. A Mobile Agent-based Grid Monitor Architecture (MA-GMA) based on GMA is put forward, and its performance analyses are performed in the end. In our future work, we would like to investigate hierarchical mechanism of grid monitoring in more details and to strengthen the mechanism of security and authorization for grid monitoring.

References

[1] Ian Foster, and Carl Kesselman, *The Grid2: Blueprint for a New Computing Infrastructure*. Morgan Kaufmann, 2004.

[2] K. Rothermel and R.Popescu-Zeletin Eds., Mobile Agents. *Lecture Notes in Comp. Science*, LNCS1219, 1997.

[3] *Global Grid Forum*, http://www.gridforum.org/.

[4] B. Tierney, R. Aydt, D. Gunter, A Grid Monitoring Architecture(2002); http://www-didc.lbl.gov/GGF-PERF/GMA-WG/papers/GWD-GP-16-2.pdf.

[5] *The DataGrid Project*. http://eu-datagrid.web.cern.ch /eu-datagrid/.

[6] A. Cooke, A.Gray, L. Ma, W. Nutt, J. Magowan, P. Taylor et al, "R-GMA: An information integration system for grid monitoring," In *Proc. of the 11th International Conference on Cooperative Information Systems*, 2003. 2-26 GRID MONITORING.

[7] M.A. Baker and G. Smith. "GridRM: a resource monitoring architecture for the Grid," In *Springer-Verlag, editor, Proc. of the 3rd International Workshop on Grid Computing*, 2002.

[8] M.A. Baker and G. Smith, "GridRM: an extensible resource monitoring system," In *proceedings of the IEEE Cluster Computing Conference*, 2003.

[9] R. Wolski, N. Spring, and Jim Hayes, "The network weather service: a distributed resource performance forecasting service for metacomputing," *Journal of Future Generation Computing Systems*, 15(5-6), 1999.

[10] Rich Wolski. *Experiences with Predicting Resource Performance On-line in Computational Grid Settings*. ACM SIGMETRICS Performance Evaluation Review, 30:41–49, 2003.

[11] Brian Tierney, Brian Crowley, Dan Gunter, Mason Holding, Jason Lee, and Mary Thompson, "A monitoring sensor management system for grid environments," In *Proc. of High Performance Distributed Computing'00*, 2000.

[12] K. Czajkowski, S. Fitzgerald, I. Foster, and C. Kesselman, "Grid information services for distributed resource sharing," In *Proc. of the 10th IEEE International Symposium on High-Performance Distributed Computing*, 2001.

[13] X. Zhang, J.L. Freschl, and J. Schopf, "A performance study of monitoring and information services for distributed systems," In *Proc. of the 12th IEEE International Symposium on High-Performance Distributed Computing*, 2003.

[14] X. Zhang and J. Schopf, "Performance analysis of the Globus Toolkit Monitoring and Discovery Service, MDS2," In *Proceedings of the International Workshop on Middleware Performance (MP 2004)*, 2004.

[15] A. Fuggetta, G.P. Picco, and G. Vigna, "Understanding code mobility," *IEEE Transaction on Software Engineering*, 24(5), May 1998.

[16] A. Puliafito and O. Tomarchio, "Using Mobile Agents to implement flexible NetworkManagement strategies," *Com-puter Communication Journal*, 23(8):708–719, April 2000.

[17] Todd Papaioannou, *On the Structuring of Distributed Systems: The Argument for Mobility*, Doctoral Thesis, Loughborough University, 2000.

[18] W3C Recommendation, *SOAP: Simple Object Access Protocol* 1.2,24 June 2003.

[19] Artail and E. Kahale, "MAWS: a platform-independent framework for mobile agents using web services," *Journal of Parallel and Distributed Computing*, vol. 66, no. 3, pp. 428 - 443, March 2006.

Expanding the Training Data Space Using Bayesian Test

Hamad Alhammady
Etisalat University College, UAE
Email: hamad@euc.ac.ae

Abstract - Expanding the training dataset is a new technique proposed recently to improve the performance of classification methods. In this paper, we propose a powerful method to conduct the previous task. Our method is based on applying the Bayesian test based on emerging patterns to evaluate and improve the quality of the new data instances used to expand the training data space. Our experiments on a number of datasets show that our method outperforms the previous proposed methods and is able to add additional knowledge to the space of data.

I. INTRODUCTION

Many classification methods have been developed recently. However, the performance of these classifiers is proportional to the knowledge obtained from the training data. As a result, traditional classifiers can not perform very well when the training data space is very limited. In this paper, we propose a new approach to expand the training data space using the Bayesian test and emerging patterns (EPs). We combine the power of the Bayesian theorem and EPs to expand the training data space before applying standard classifiers.

The Bayesian theorem [11] is based on computing the probability scores of a test instance in each class in the dataset. The test instance is assigned to the class with the highest probability. For example, the probability of a test instance, t, in class C_j is as follows.

$$P(C_j \mid t) = P(C_j)\frac{\prod_{l=1}^{m} P(a_l \mid C_j)}{P(C_j)^{m-1}} \quad (1)$$

where $P(x \mid y)$ denotes the probability of x given y, m is the number of attributes, and a_l is the value of the lth attribute in t. The above score is calculated for all the classes, and t is assigned to the highest score class.

EPs are a new kind of patterns introduced recently [2]. They have been proved to have a great impact in many applications [3] [4] [5] [6] [9] [10]. EPs can capture significant changes between datasets. They are defined as itemsets whose supports increase significantly from one class to another. The discriminating power of EPs can be measured by their growth rates. The growth rate of an EP is the ratio of its support in a certain class over that in another class. Usually the discriminating power of an EP is proportional to its growth rate.

For example, the Mushroom dataset, from the UCI Machine Learning Repository [7], contains a large number of EPs between the poisonous and the edible mushroom classes. Table 1 shows two examples of these EPs. These two EPs consist of 3 items. e1 is an EP from the poisonous mushroom class to the edible mushroom class. It never exists in the poisonous mushroom class, and exists in 63.9% of the instances in the edible mushroom class; hence, its growth rate is ∞ (63.9 / 0). It has a very high predictive power to contrast edible mushrooms against poisonous mushrooms. On the other hand, e2 is an EP from the edible mushroom class to the poisonous mushroom class. It exists in 3.8% of the instances in the edible mushroom class, and in 81.4% of the instances in the poisonous mushroom class; hence, its growth rate is 21.4 (81.4 / 3.8). It has a high predictive power to contrast poisonous mushrooms against edible mushrooms.

TABLE I
EXAMPLES OF EMERGING PATTERNS

EP	Support in poisonous mushrooms	Support in edible mushrooms	Growth rate
e1	0%	63.9%	∞
e2	81.4%	3.8%	21.4

e1 = {(ODOR = none), (GILL_SIZE = broad), (RING_NUMBER = one)}
e2 = {(BRUISES = no), (GILL_SPACING = close), (VEIL_COLOR = white)}

II. RELATED WORK

The approach of expanding the training data space has been proposed recently [1]. The aim of this approach is to increase the amount of knowledge (hence, the classification accuracy) in the training data by creating additional data instances. The creation process is based on the characteristics of the original data instances. These characteristics can be expressed as EPs (definitions and terminologies are given in section 3).

In [4], the same idea has been used to expand the space of rare-class data. This approach has a powerful impact on the quality of rare-class classification.

In [1], the authors proposed four methods for creating additional instances. These methods are based on EPs and some genetic algorithms.

The first method is based on superimposing EPs to create additional data instances. In this method, the set of EPs is divided into a number of groups such that EPs in each group have attribute values for most of the elements in the attribute set. The new instances are generated by combining the EPs in each group. If a value for an attribute is missing from a group, it is substituted by the value that has the highest growth rate for the same attribute.

The second method is based on the crossover genetic algorithm. For two original instances, the attribute values are

K. Elleithy (ed.), *Advances and Innovations in Systems, Computing Sciences and Software Engineering*, 155–159.

switched before a random breaking point to generate two new instances.

The third method is based on the mutation genetic algorithm. A random binary number is chosen and overlapped over the original instance. All attribute values in the instance that match 1's in the random binary number are replaced by the values that have the highest growth rates.

The fourth method is based on EPs and the mutation genetic algorithm. That is, all values in the original instance are replaced with their matched values in the EP to generate a new instance.

Each of the above methods can improve the performance of different classifiers. The most powerful method is the union of them. That is, choosing the best instances resulted from the four methods and adding them to the training data space.

III. EMERGING PATTERNS AND CLASSIFICATION

Let $obj = \{a_1, a_2, a_3, \dots a_n\}$ is a data object following the schema $\{A_1, A_2, A_3, \dots A_n\}$. $A_1, A_2, A_3 \dots A_n$ are called attributes, and $a_1, a_2, a_3, \dots a_n$ are values related to these attributes. We call each pair (attribute, value) an item.

Let I denote the set of all items in an encoding dataset D. *Itemsets* are subsets of I. We say an instance Y contains an itemset X, if $X \subseteq Y$.

Definition 1. Given a dataset D, and an itemset X, the support of X in D, $s_D(X)$, is defined as

$$s_D(X) = \frac{count_D(X)}{|D|} \tag{2}$$

where $count_D(X)$ is the number of instances in D containing X.

Definition 2. Given two different classes of datasets D_1 and D_2. Let $s_i(X)$ denote the support of the itemset X in the dataset D_i. The growth rate of an itemset X from D_1 to D_2, $gr_{D_1 \to D_2}(X)$, is defined as

$$gr_{D_1 \to D_2}(X) = \begin{cases} 0, & \text{if } s_1(X) = 0 \text{ and } s_2(X) = 0 \\ \infty, & \text{if } s_1(X) = 0 \text{ and } s_2(X) \neq 0 \\ \frac{s_2(X)}{s_1(X)}, & \text{otherwise} \end{cases} \tag{3}$$

Definition 3. Given a growth rate threshold $\rho > 1$, an itemset X is said to be a ρ-emerging pattern (ρ-EP or simply EP) from D_1 to D_2 if $gr_{D_1 \to D_2}(X) \geq \rho$.

Let $C = \{c_1, \dots c_k\}$ is a set of *class labels*. A *training dataset* is a set of data objects such that, for each object *obj*, there exists a class label $c_{obj} \in C$ associated with it. A *classifier* is a function from attributes $\{A_1, A_2, A_3, \dots A_n\}$ to class labels $\{c_1, \dots c_k\}$, that assigns class labels to unseen examples.

IV. EXPANSION USING BAYESIAN TEST

The key idea to expand the training data space is to generate more training instances. Figure 1 explains this idea. Suppose that we have a dataset consisting of three classes. The complete data space of this dataset can be identified if and only if we have all the data points (instances). In real life, we know a certain portion of the data space. This portion represents the training data space (the black areas in figure 1). The aim of expanding the training data space is to generate additional training instances (the striped areas in figure 1) that support classification.

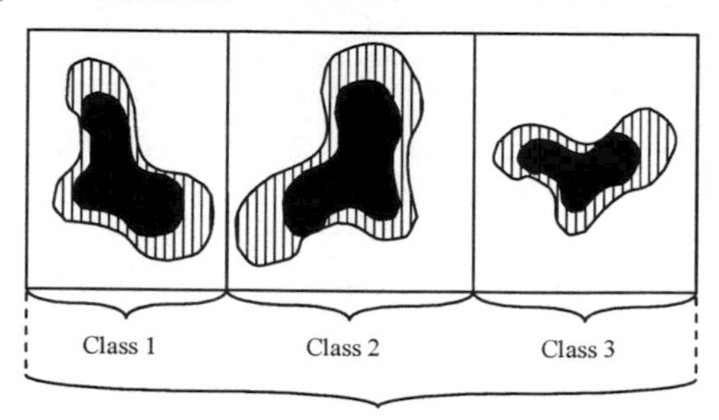

Figure 1. Expanding the training data space

In this paper, we propose a sophisticated method to expand the training data space. Our method consists of two phases. In the first phase, a new data instance is generated by planting strong attribute values from two original instances in an EP. In the second phase, the Bayesian theorem is used with the aid of EPs to evaluate the new instance generated in the first phase and correct its attribute values according to its connection with the intended class.

The generation process in the first phase is conducted as follows. The missing attribute values in the input EP is substituted by the highest growth rate values from the two input instances. Suppose that we have a dataset consisting of seven attributes $\{A_1, A_2, A_3, A_4, A_5, A_6, A_7\}$, the input instances are presented as following: $i_1 = \{a_1, a_2, a_3, a_4, a_5, a_6, a_7\}$ and $i_2 = \{b_1, b_2, b_3, b_4, b_5, b_6, b_7\}$, and the input EP is $e_1 = \{(A_2 = v_2), (A_6 = v_6)\}$. Notice that five attributes are not included in e_1. These attributes are A_1, A_3, A_4, A_5, and A_7. These attributes are planted in e_1 using attribute values in i_1 and i_2. That is, the attribute values that have the highest growth rates in the input instances are added to the input EP to fill the place of the excluded attributes. Figure 2 clarifies the first phase of our approach assuming that the growth rates of a_1, b_3, b_4, a_5, and a_7 are higher than the growth rates of b_1, a_3, a_4, b_5, and b_7, respectively.

	A_1	A_2	A_3	A_4	A_5	A_6	A_7
i_1	a_1	a_2	a_3	a_4	a_5	a_6	a_7
e_1		v_2				v_6	
i_2	b_1	b_2	b_3	b_4	b_5	b_6	b_7
i_3	a_1	v_2	b_3	b_4	a_5	v_6	a_7

Figure 2. Instance generation by attribute planting

In the second phase, we propose a new test to evaluate the connection between the resulted instance, i_3, and the intended class. Our proposed test is based on the Bayesian theorem and EPs. The idea of combining the advantages of the Bayesian theorem and EPs was first introduced as a classification method in [5].

In the Bayesian classification by emerging patterns (BCEP), EPs are used with the Bayesian theorem to predict class labels for the test instances. Given a test instance t, BCEP combines the evidence provided by the subsets of t that are present in F_i to approximate $P(t,C_i)$, where F_i denotes the final high quality EPs from class C_i for classification and $P(t,C_i)$ determines the probability that t belongs to class C_i given the evidence. The evidence which is selected from F is denoted as B, where $F = \sum_{i=1}^{c} F_i$, and c is the number of classes in the training dataset. The EPs of B are used sequentially to construct the product approximation of $P(t,C_i)$. The result is the class with the highest value of $P(t,C_i)$.

We convert this powerful classification method to an evaluating test for the new generated instances as follows. In the first stage, $P(t,C_i)$ is calculated for the new instance, i_3, in all the classes associated with the dataset. If the probability that i_3 belongs to the intended class is lower than the probability that i_3 belongs to another class, then i_3 is discarded as this is evidence that it is far away from being a member of the intended class. On the other hand, if the probability that i_3 belongs to the intended class is higher than that for the other classes, then i_3 undergoes a second stage of testing, $P(t,C_i)$ is calculated for all instances in the intended class. The average $P(t,C_i)$ is found for the intended class, avg($P(t,C_i)$). If $P(t,C_i)$ of i_3 is higher than avg($P(t,C_i)$), then i_3 is accepted as a new member of the intended class as this proves the strong connection between them . In contrast, if $P(t,C_i)$ of i_3 is lower than avg($P(t,C_i)$), then this indicates that i_3 has a reasonable connection with the intended class but this connection needs to be strengthened. As a result, i_3 undergoes a third stage of testing. In this stage, the attribute value which has the lowest growth rate in i_3 is replaced with the highest growth rate value in the intended class. $P(t,C_i)$ is recalculated for i_3 and the process is repeated till $P(t,C_i)$ of i_3 is larger than avg($P(t,C_i)$). When this condition is true, i_3 is accepted as a new member of the intended class. Figure 3 shows the flowchart of our proposed Bayesian test.

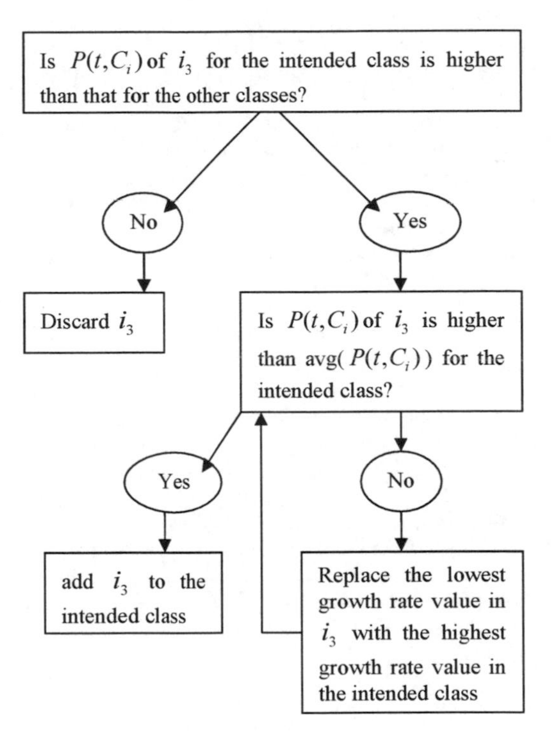

Figure 3. The flowchart of the Bayesian test

We present the following advantages of our approach:

- The new instances are basically EPs with strong planted attribute values. This represents an excellent source of knowledge.
- The Bayesian test employs the advantage of EPs to evaluate the new classes. It makes decisions on whether to discard the new instances, keep them, or improve them.

V. EXPERIMENTAL EVALUATION

We conduct experiments on 30 datasets from UCI repository of machine learning databases [7]. The testing method is stratified 10-cross-validation. We use C4.5 decision tree [8] as the base classifier (M). We compare our proposed method of generating new instances using the Bayesian test (GBT) with the five methods proposed in [1]. These methods are generation by superimposing EPs (M1), generation by crossover (M2), generation by mutation (M3), generation by mutation and EPs (M4), and the union of the previous four methods (M*).

The results are shown in table 2. We summarize the results as follows:

- The average accuracy of GBT is higher than that of the other methods.
- GBT outperforms methods M, M1, M2, M3, and M4 on all datasets. It also outperforms method M* on 29 datasets out of 30 datasets.
- GBT improves the performance of the base classifier extensively on some datasets (German, Hayes, and Vehicle).

The results show that our proposed method (GBT) is superior over the other methods. Moreover, it is much simpler than the nearest method (M*) which involves choosing the best instances resulted from the first four methods.

TABLE II
EXPERIMENTAL RESULTS

Dataset	M	M1	M2	M3	M4	M*	GBT
Adult	86.1	88.3	87.2	87.7	86.4	87.8	**89.8**
Aust	84.3	84.9	85.7	84.4	83.9	86.9	**87.5**
Breast	94.6	96.3	94.6	95.1	95.9	95.9	**97.8**
Cleve	73.8	72.6	73.2	74.9	74.6	74.8	**78.3**
CC	85.3	86.4	85.2	86.7	87.2	86.9	**87.8**
Crx	85.3	85.8	85.1	83.7	84.6	85.5	**86.9**
Diabetes	73.4	71.6	71.8	72.8	72.2	74.7	**79.1**
Flags	57.5	59.3	59.7	57.9	57.1	59.3	**62.1**
German	69.6	69.9	70.2	71.3	70.6	72.6	**77.9**
Glass	64.7	66.2	64.9	64.1	65.3	66.1	**67.8**
Hayes	70.2	70.7	72.5	75.3	75.8	75.5	**78.5**
Heart	80.6	81.1	80.4	81.3	80.9	81.1	**88.2**
Hepa	81.8	80.9	80.2	81.7	82.6	82.4	**87.1**
Horse	85.2	85.4	85.4	85.9	85.7	85.6	**87.9**
Hypo	99.3	97.2	98.1	96.8	98.4	**99.4**	99.3
Iono	89.4	88.6	89.5	89.1	89.8	91.1	**94.9**
Labor	76.9	76.2	77.3	77.8	76.9	77.5	**84.4**
Liver	58.1	59.3	59.8	58.5	61.3	60.6	**65.1**
Machine	87	87.5	89.5	89.2	88.4	89.2	**89.9**
Pima	74	76.2	74	73.4	74.7	76.1	**79.7**
Segment	93.5	93.6	92.7	94.8	94.1	94.4	**96.4**
Sick	98.7	97.5	97.7	98.9	98.1	98.6	**99**
Sonar	75.3	76.8	76.3	77.6	75.4	77.4	**80.3**
Staimage	85.2	84.7	85.5	86.1	84.9	85.8	**88.9**
TTT	84.2	86.5	88.1	84.6	87.1	87.8	**90.3**
Vehicle	71.2	72.8	69.5	70.2	70.7	72.9	**84.4**
Votes	77.8	78.9	77.4	78.5	78.1	78.8	**83.1**
Wine	84.2	85.6	83.1	84.8	85.2	85.9	**86.9**
Yeast	49.9	48.1	48.5	49.4	48.6	49.7	**51**
Zoo	93	91	89	91	90	92	**95**
Average	79.7	80	79.7	80.1	80.2	81.1	**84.2**

VI. CONCLUSIONS

Recently, the approach of expanding the training data space has been proposed to improve the accuracy of traditional classifiers. In this paper, we proposed a sophisticated and powerful technique to conduct the previous task. Our technique consists of two phases. In the first phase, new instances are generated by planting high quality values from two original instances in an EP. In the second phase, we combine the advantages of the Bayesian test and the power of EPs to evaluate and enhance the new instances, Our technique show a very high performance compared to the other methods.

REFERENCES

[1] H. Alhammady, and K. Ramamohanarao. Expanding the Training Data Space Using Emerging Patterns and Genetic Methods. In Proceeding of the 2005 SIAM International Conference on Data Mining, New Port Beach, CA, USA.

[2] G. Dong, and J. Li. Efficient Mining of Emerging Patterns: Discovering Trends and Differences. In Proceedings of the 1999 International Conference on Knowledge Discovery and Data Mining, San Diego, CA, USA.

[3] H. Alhammady, and K. Ramamohanarao. The Application of Emerging Patterns for Improving the Quality of Rare-class Classification. In Proceedings of the 2004 Pacific-Asia Conference on Knowledge Discovery and Data Mining, Sydney, Australia.

[4] H. Alhammady, and K. Ramamohanarao. Using Emerging Patterns and Decision Trees in Rare-class Classification. In Proceedings of the 2004 IEEE International Conference on Data Mining, Brighton, UK.

[5] H. Fan, and K. Ramamohanarao. A Bayesian Approach to Use Emerging Patterns for Classification. In Proceedings of the 14th Australasian Database Conference (ADC'03), Adelaide, Australia.

[6] Guozhu D., Xiuzhen Z., Limsoon W., and Jinyan L.. CAEP: Classification by Aggregating Emerging Patterns. In Proceedings of the 2nd International Conference on Discovery Science (DS'99), Tokyo, Japan.

[7] C. Blake, E. Keogh, and C. J. Merz. UCI repository of machine learning databases. Department of Information and Computer Science, University of California at Irvine, CA, 1999.
http://www.ics.uci.edu/~mlearn/MLRepository.html.

[8] I. H. Witten, E. Frank. Data Mining: Practical Machine Learning Tools and Techniques with Java Implementations. Morgan Kaufmann, San Mateo, CA., 1999.

[9] H. Alhammady & K. Ramamohanarao. Mining Emerging Patterns and Classification in Data Streams. In Proceedings of the IEEE/WIC/ACM International Conference on Web Intelligence (WI), Compiegne, France (Sep 2005), pp. 272-275.

[10] H. Alhammady & K. Ramamohanarao. Using Emerging Patterns to Construct Weighted Decision Trees. In IEEE Transactions on Knowledge and Data Engineering. Volume 18, Issue 7 (July 2006), pp. 865-876.

[11] Duda, R., & Hart, P. (1973). Pattern Classification and scene analysis. New York: John Wiley & Sons.

A Multi-Agent Framework for Building an Automatic Operational Profile

Hany EL Yamany and Miriam A. M. Capretz
Department of Electrical and Computer Engineering
Faculty of Engineering
University of Western Ontario
London, Ontario, N6A 5B9, Canada
{helyaman, mcapretz}@uwo.ca

Abstract **– Since the early 1970s, researchers have proposed several models to improve software reliability. Among these, the operational profile approach is one of the most common. Operational profiles are a quantification of usage patterns for a software application. The research described in this paper investigates a novel multi-agent framework for automatically creating an operational profile for generic distributed systems after their release into the market. The operational profile in this paper is extended to comprise seven different profiles. Also, the criticality of operations is defined using a new composed metrics in order to organise the testing process as well as to decrease the time and cost involved in this process. The proposed framework is considered as a step towards making distributed systems intelligent and self-managing.**

I. INTRODUCTION

Since the last century, software systems have become a vital component for human life. They have controlled most aspects such as industry and education. Nowadays, millions of users all over the world depend entirely on these applications to conduct their daily activities such as flight booking and banking transactions. Any failure or breakdown in these programs would result in substantial financial loss or even the loss of one or more human lives. Thus, software reliability is essential in order to improve the operation of software to save money and lives.

Software reliability refers to the probability of execution without failure for some specified interval of natural units or time [1]. The reliability measurement process is shown in Fig. 1 [2]. The most important step in this process is efficiently constructing an operational profile, which refers to the set of operations or processes for a software application, and the probabilities of occurrence of those operations or processes [3]. Identifying an operational profile is an efficient approach because it detects failures, and hence the faults causing them in the order of how often failure occurs [1]. However, as software systems become larger, being composed of thousands of operations and processes, the operational profile in [3] may not be an accurate reflection of the real use of the system shown in [2].

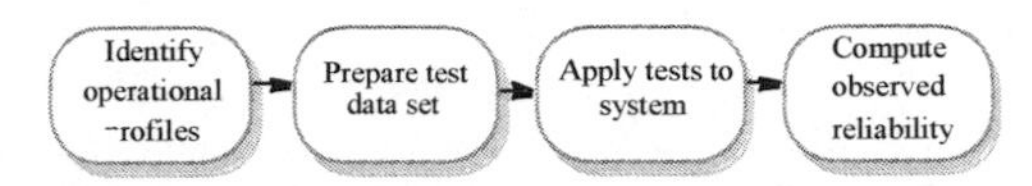

Fig. 1: The reliability measurement process [2]

Also in distributed systems, software testing presents two fundamentally difficult problems: choosing test cases and evaluating test results. Choosing test cases is challenging because there is an astronomical number of possible test inputs and sequences, yet only a few of those will succeed in revealing a failure. The other problem, evaluation requires generating an expected result for each test input and comparing this expected result to the actual output of a test run [4].

Moreover, current software systems are aiming to be more intelligent and self-managing. This requires more automated and reliable testing in order to keep costs within an acceptable and reasonable range. Another challenge involves computing the operations criticality in a distributed system. Operation criticality refers to the importance of an operation in terms of the safety or the value added by satisfactory execution; it also considers the risk to human life, the cost, or the damage resulting from failure [1]. The testing process should be focused on the operations that have high criticality value. This leads to a decrease in the time and cost of the testing process by guaranteeing that the essential operations are working well and then ensuring that the whole product is also efficient. Furthermore, the operations criticality can be used as a metric to organize the testing of the different paths in a distributed system instead of selecting them randomly as in [1].

The major goal of this paper is to introduce a novel multi-agent framework to automatically regenerate the operational profile for distributed systems after their release into the market. Furthermore, this paper proposes new composed metrics to determine the operations criticality.

The paper is structured as follows: Section 2 presents research related to building operational profiles for distributed systems. Section 3 describes the overall model for the proposed multi-agent framework, whereas Section 4 depicts the architecture of the proposed framework and Section 5 explains the generation of an operational profile using this proposed framework. Next, Section 6 determines the operations criticality, and Section 7 includes an implementation for determining the operations criticality. Finally, Section 8

K. Elleithy (ed.), Advances and Innovations in Systems, Computing Sciences and Software Engineering, 161–166.

provides conclusions and future work.

II. RELATED WORK

The operational profile is a quantitative characterization of how a system will be used and is applied to guide test selection [3]. Developing an operational profile involves progressively breaking systems use down into more details. The operational profile is applied to guide test selection. It consists of five steps as follows [3]: Find the customer profile, establish the user profile, define the system-mode profile, determine the functional profile, and finally determine the operational profile itself. Some steps may not be necessary in a particular application.

Testing driven by an operational profile is efficient, especially in communication software systems, because it identifies failures, and hence, the faults causing them. This approach rapidly increases the reliability and reduces the intensity of failures per unit of execution time, because the failures that occur most frequently are caused by the repetition of faulty operations. However, the performance of this testing technique could be further improved by adjusting many vital factors, such as selecting critical operations and reducing the number of operations. Controlling these factors would efficiently increase the software reliability.

Whittaker *and Voas.* [5] indicate that *a simple distribution of inputs from a human user does not come close to describing the situation.* The operating environment of the software can affect the operation of the software even if the user follows the tested traditional operational profile. The operating system enforces the limits on memory and makes decisions based on the requirements of the other applications in the operating environment [5, 6]. Thus, aspects such as the nature of the data structures, data size and data types are important issues to be considered when executing test cases that deal with the operations and thus have to be taken into account [7, 8].

Therefore, the operational profile must include information about the operating environment; information about other applications in the operating environment and external data that is used by the application. As a result, an extended operational profile can be built [9].

In addition to the normal operational profile, the extended operational profile includes two additional profiles; the structure profile and the data profile. Firstly, the structural profile contains both the structure of the system on which the application is running and the configuration or structure of the actual application itself. Data structures can often be characterized by attributes that have numerical value and may change over time. Secondly, the data profile consists of an application's input values from many users.

Furthermore, the extended operational profile depicts a higher level of refection than the normal operational profile for any applications in the software market. This extended profile will help the different organizations validate their systems, and consequently, improve their reliability. However, the selection of test cases is not addressed and also, there is no specified identification for the operation criticality. Moreover, the automatic regeneration of the operational profile is not considered in that work [9].

The greatest challenges that face organisations in validating their applications are those of choosing tests and evaluating test results; this is due to the great variation of test cases and the high cost of the assessment process. The automated model-based testing approach described in [4] could assist in solving those issues. The automated system is designed to support the rapid incremental development of complex distributed systems. It is able to revise the profile, regenerate it, and then run different test suites. In this latter technique, the most important issue involves generating fresh test suites every time the testing is running using discrete event simulation and AI-based meta-techniques. These fresh test cases are more likely to discover newer defects since re-running the same profile-based test suite is inefficient and useless. As high volume automated testing is generated, the output of the system should be monitored. The output checking requires the development and evaluation of expected results, so that, along with fresh test inputs, the system is extended to automatically produce new and expected results to evaluate the test run output. On the other hand, AI-based meta- programming architecture in this approach did not scale well imposed a high maintenance cost.

This paper proposes a novel technique for building a distributed operational profile (DOP) for generic distributed systems using multi-agent based framework. The DOP will consist of seven steps utilising the benefits of the normal operational profile in [3] and the extended operational profile in [9]. The first version of DOP will be built statically as suggested in [3, 9]. Consequently, the multi-agent system will automatically modify and regenerate the DOP according to the changes in the distributed system. This regeneration will be done, after releasing the software product in market; it will be accomplished by monitoring its usage and the changes that might occur at the vendor site due to the modification in requirements and the system development.

III. SYSTEM OVERVIEW

The major goal of this research is to automate the process of monitoring customers' software usage in order to establish a connection between the defects with the software usage in the customer environment. Specifically, there are two major sets of data necessary for collection. The first set is related to the static and dynamic information associated with the customer environment. This includes information on the machine, the operating system, the software configuration, the data held within the database, the movement of data, and the overall operations. This information will assist in revising and regenerating a new DOP that will represent a true reflection of the running distributed system. The second set of information is on the defects that are found by the customers themselves.

As mentioned, the DOP will consist of seven steps; each one will include a different profile. The first five steps are derived from [3] (1. find the customer profile, 2. establish the user profile, 3. define the system-mode profile, 4. determine the functional profile, and finally 5. determine the operational profile itself), the sixth one is complied of the profiles suggested in [9] (6. the structure profile and the data profile)

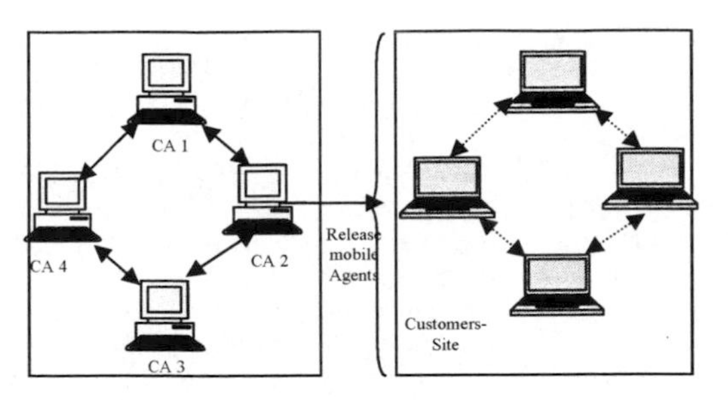

Fig. 2: A framework of a multi-agent system for building DOP.

and the last one describes the influence of the surrounding environment, including the hardware and software. Consequently, the descriptions of the modified profiles are:

1. *Determine the data profile:* A definition of the types or patterns of data and an analysis of its structure.
2. *Determine the machine profile:* A specification of each environment in which the system will run. This profile would help the vendor to track the system behaviour in different operating environments.

After the release of the product into the market, a multi-agent based system will start monitoring the product's behaviour. A software agent, such as this in the proposed system, is a piece of software that can be viewed as perceiving its environment through sensors and acting upon that environment through effectors [10]. Changes in a software system and its environment may require changes in the testing strategy. Software agents [10-12] are adaptive and can adjust their behaviours based on changes in their environment. They are also autonomous; they can continuously monitor customer's system usage and thus report errors as they are found. The agents in this research are also goal-oriented (pro-active), they can generate an operational profile and calculate the required statistics to track the total number of failures as well as the interval between failures. As a result, we believe

TABLE 1. Agents Functions.

Agent's Name	Goal	Perception	Action	Output
CA	Automatically generate the operational profile. Determine the operation's criticality. Determine the appropriate testing type.	1st step of operational profile. The vendor's data and the frequency of user usage for the distributed system's functionality	Rebuild the operational profile. Calculate the criticality for each operation. Select the best strategy for removal of defects.	A new version of the operational profile. A table of criticality value for each operation.
RMA	Monitor the user's usage.	User's log file.	Calculate the frequency of user's usage for each function in a distributed system. Reclaim for any found error.	The frequency value of user's usage for each function in the system.

that a software agent framework is the best solution for automatically revising and regenerating the operational profile for distributed systems, especially after their release into the market.

The software agents will be performing four major tasks: monitoring, detection, diagnosis and repair. Monitoring involves observance at both the point of sale and at the customer's location leading to the generation of reports, and conception and/or enhancement of the customer's operational profile. The second task deals with building the agent's capabilities to detect defects by making use of the operational profiles generated by data collected at different customers' sites. Subsequently, after detecting an error, agents work on the diagnosis and attempt to estimate the required actions and tests to fix the error. Eventually, with accumulated knowledge acquired from the environment, agents should be able to not only implement the determined test in the former step but also to automatically repair automatically the found errors.

IV. THE MULTI-AGENT FRAMEWORK ARCHITECTURE

The proposed multi-agent framework consists of two main types of agents: the Centralized Agent (CA) and the Release Mobile Agent (RMA), as shown in Fig. 2.

The following sections depict the agents' functionality as well as the mechanism of the communication between them.

A. The Agents' Functionality

To exhibit the agents' functionality in this framework, we have to demonstrate their goals, perceptions, actions and outputs according to the previous agent description. Table 1 describes these activities.

The CA (Centralized Agent) would be able to perform tasks such as determining the appropriate testing type such as regression testing. This testing could be based on statistical calculations such as the sum of the total number of failures and the interval between failures. Also, it could be based on the criticality degree of the different operations inside a distributed system.

B. The Communication Mechanism between the Agents

Because of the diversity of environments where the distributed systems are running, the agents use SOAP messages to communicate with each other. The data in these SOAP messages varies depending on the information required to rebuild the operational profile. For example, the CA sends SOAP messages to the RMAs requesting them to send the usage frequency for each function in a distributed system.

Furthermore, the CAs at the different servers of the vendor exchange SOAP messages including data such as the calculated criticality values of operations, about the behavior of the distributed system to effectively build the new version of the operational profile.

V. DEVELOPING THE OPERATIONAL PROFILE

The major aim of this work is to rebuild new versions of an operational profile for a distributed system after its release into the market. The new versions of the operational profile should realistically reflect the distributed system as it is running in a given environment. The first version of the operational profile is built statically as mentioned in [3, 9]. This version contains the basic knowledge for the agents in the proposed framework which will be used to generate the subsequent versions.

After the release of the product into the market, the proposed framework will start its work monitoring the vendor and the user sites. At the vendor site, the CA will register any changes that might occur in the distributed system's functionalities. Also, the vendor can update the agents' database by any modifications of the distributed system's components. At the user site, RMAs monitor customer usage by analyzing the user's log file. For example, the RMA can reveal a change in the environment where the distributed system is running and send this information in a SOAP message to the CA in order to modify the machine profile; this is the 7th step in defining the operational profile. Consequently, the CA combines the data from the vendor's server, the vendor's developers and the customer usage and utilizes it to regenerate a new operational profile.

VI. OPERATIONS CRITICALITY

There are two other weaknesses of current operational profiles, such as the one in [3]. The first of these is that the efficiency of operational testing decreases as the testing progresses, since more and more parts of the software code have already been examined [13]. Secondly, the random order of executed test cases further reduces the efficiency of testing, because it requires more navigation between different parts of the program [14]. Therefore, there needs to be a new approach that guarantees that the parts of a software product are covered in an efficient order become a necessary requirement in software testing. We believe that determining the criticality of operation will assist in this issue.

As previously mentioned, determining the criticality value for the operations of a distributed system helps to decrease the time and cost involved in testing by focusing on these operations to ensure effectiveness of the distributed system. This does not mean that the operations that have a low critical value will never be tested; however, the testing starts with the operations of high criticality value and finishes with the ones of low value. This technique assures that the maximum number of software's parts will be tested in an organized order. In our work, we use composed metrics to determine the function criticality, which will be calculated by the tester and the user, or in other words, by the agents that monitor the user usage and the changes that might occur in the system functionality. Since, according to [1], particular functions comprise each operation, it will be easier to measure the operations criticality when the functions criticality is known. The proposed metrics include function Complexity (C), Size (S), the Number of Input States (IS) and the Frequency (F) of the function usage. We have selected these metrics based on the criteria mentioned in [15] including the fact that they are quantified, continuous and defined on the basis of the function definition. The description of each metric is as follows:

1. *Complexity:* It evaluates the complexity of an algorithm in a function. A function with a high complexity might be considered a high critical function due to the fact that it may contain a greater number of faults [16].
2. *Size:* It can be measured in a variety of ways including the number of all physical lines of code, the number of statements, and the number of blank lines. In this work, we measure the size by the physical lines of code. A function of large size might be considered as a critical function.
3. *Input States:* It is the set of the input values of variables associated with a function and either used by it or affected by it.
4. *Frequency:* This is the number of times that a function is executed during a period of time.

To begin, the tester feeds the CA by using the values of the first three metrics: complexity, size and input states. Then, the CA will automatically recalculate these values after any changes occur in the system. On the other hand, the RMA automatically calculates the last metric, the frequency, for each function in the system by monitoring the user usage and analyzing the log file. Finally, the RMA sends these values to the CA which will produce the criticality value for each function by adding the values of the metrics and then computing the criticality value for each operation in the system.

The last issue in this framework involves defining a critical defect. When the RMA finds an error during its monitoring of the user usage, it sends a SOAP message to the CA reclaiming this error. The SOAP message contains the name of the function(s) and the module(s) where the error occurs or is influenced. The CA checks these functions and examines if they are critical functions or not. The defect that causes to halt one critical function or more is called a critical defect. In the case that a critical defect is found, the CA will suggest performing a testing to gather more information about this defect; otherwise CA will ignore this defect. However, the CA will register the non-critical defect in its database and create a report for the tester to deal with this defect in subsequent versions of the system.

VII. IMPLEMENTATION

Our interest in this work is to demonstrate that the suggested composed metrics are efficient enough to compute the operations criticality. We studied a simple financial-based distributed system that belongs to a small company working in the electronics field. Their financial system is composed of many operations; three of them are considered the most important operations. The first operation is creating invoices; each invoice operation is composed of many functions, which include customer data selection, product data retrieval, product price, product stock availability, invoice computation and invoice data storing. The second operation is a product manufacturing, whereby a product is built from many parts. This process includes some functions such as parts selection, parts stock availability, parts ordering and product registration. The third operation is creating purchase order for parts. The operation's functions are: supplier's data selection, parts' data selection, parts' price retrieving, producing and storing purchase order data.

We implemented an application to compute the complexity value for each function and then produced the criticality degree for each operation in that system, which was considered high, medium or low. Fig. 3 is a snapshot of this application. First, we calculated the values for the first three metrics (C, S and IS) for each previous function, so that the metrics' values are stored in the CA's database. Also, F is computed by analyzing the users' log files for this system and then it is stored in the same database. Each user log file includes an Operation Id, a Function Id and the frequency value for the function. F for a function is equal to the average of all users' usage for this function.

Consequently, the CA adds all the metrics' values separately for each function to compute its criticality. The operation criticality value is calculated by taking the average of the criticality values of its own functions. In this work, the range between the maximum and the minimum values of the operations criticality is computed and then divided into three distinct levels to produce the operations criticality degree for all operations in that system. The results show that the three operations mentioned above have a higher criticality degree as compared to other operations in the system. Fig. 4 shows the results.

VIII. CONCLUSIONS & FUTURE WORK

In this paper, we have proposed a novel multi-agent framework to automatically regenerate the operational profile for distributed systems after their release into the market. The automation in building the operation profile helps to decrease time and cost required for testing. The framework includes intelligent agents that monitor the changes and the user usage at the site of the vendor and client in order to efficiently build new versions of the operational profile that represent a more accurate reflection of the distributed systems' behaviour. Overall, the new operational profiles will increase the reliability and the performance of distributed systems. With these operational profiles, a framework is described and the communication system between them is also defined.

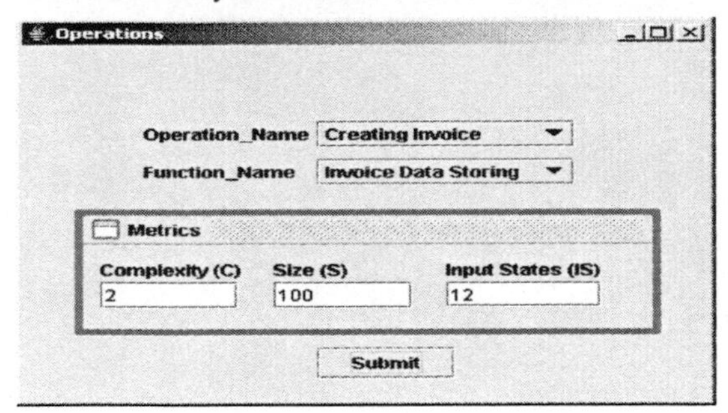

Fig. 3: A Snapshot of Operations' Metrics Calculation

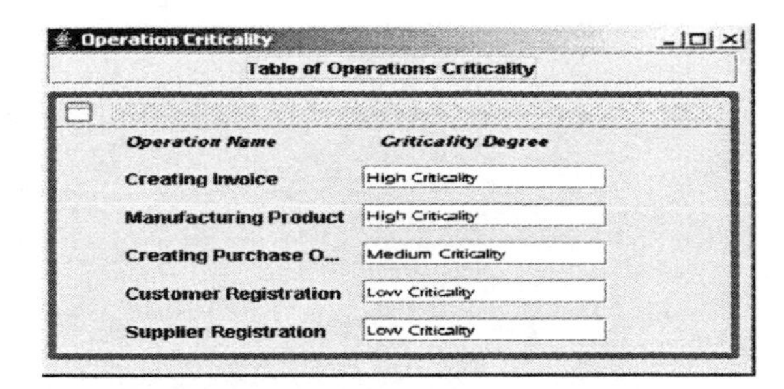

Fig. 4: A Snapshot of Operations' Criticality Results

Furthermore, this paper proposes new composed metrics to determine the operations criticality. The metrics are complexity, size, input states and frequency. Using these metrics to determine the criticality value for the distributed system's operations using these metrics involves focusing the testing process on the operations that have high criticality value. This is done in order to decrease the time and cost involved in testing and to guarantee that the essential operations and system as a whole are working effectively. The evaluation of operations criticality could be used to organize the testing process starting by testing the high critical operations and finishing by testing the lowest critical operations; this guarantees that the testing process will cover all paths of a distributed system.

The proposed framework in this paper is considered as a step toward making distributed systems intelligent and self-managing. Since this study has proposed the basic framework and related functionalities, our future work will develop a tool to further validate this model as well as conduct additional testing on other distributed systems to optimize the model's functionality.

REFERENCES

[1] J. Musa, *Software Reliability Engineering: More Reliable Software Faster and Cheaper*, McGraw-Hill. 2004.

[2] I. Sommerville, *Software Engineering*, Addison-Wesley, 7th Edition, Chapter 24, 2004.

[3] J. Musa, "Operational Profile in Software Reliability Engineering," *IEEE Software*, Vol. 10, No. 2, Mar. 1993, pp. 14-32.

[4] R. V. Binder, "Automated Testing with an Operational Profile", *The Software Tech News*, Vol. 8, No. 1. Dec. 2004, pp. 7-10.

[5] J. A. Whittaker and J. Voas, "Toward a more reliable theory of software reliability", *IEEE Computer*, Vol. 33, No. 12, Dec. 2000, pp. 36-42.

[6] J. Voas, "Will the real operational profile please stand up", *IEEE Software*, Vol. 17, No. 2, Mar./Apr. 2000, pp. 87-89.

[7] D. M. Woit, "Specifying operational profile for modules". *In Proceedings of the ACM International Symposium on Software Testing and Analysis*, ACM, 1993.

[8] D. M. Woit, "Operational profile specification, test case generation, and reliability estimation for modules", *Technical report*, Queen's University, Kingston, Ontario Canada, 1994

[9] M. Gittens, H. Lutfiyya, and M. Bauer, "An Extended Operational Profile Model", *In the proceedings of the Fifteenth International Symposium on Software Reliability Engineering*, Nov. 2004.

[10] N. R. Jennings, K. Sycara, M. Wooldridge, "A Roadmap of Agent Research and Development," *Journal of Autonomous Agents and Multi-Agent Systems*, Vol. 1, No. 1, 1998, pp. 5-38.

[11] J. Lind, "Patterns in agent-oriented software engineering," *in Proceedings of AOSE Workshop*, 2002, pp. 47-58.

[12] M. Wooldridge, "Agent-based software engineering," *IEE Proceedings Software Engineering*, Vol. 144, 1997, pp. 26-37.
[13] Mitchell, B.; Zeil, S. J.: A Reliability Model Combining Representative and Directed Testing, *Technical Report TR 95-18*, Old Dominon University, 1995.
[14] M. Grottke and K.D-Zieger, "Systematic vs. Operational Testing: The Necessity for Different Failure Models," in *Proc. of the 5th Conference on Quality Engineering in Software Technology*, 2001, pp. 59 - 68.
[15] 16 Critical Software Practices for Performance-Based Management: <http://www.spmn.com/16CSP.html>.
[16] V. R. Basili, and W. L. Melo, "A Validation of Object Oriented Design Metrics as Quality Indicators", *IEEE Transactions on Software Engineering*, Vol. 22, No. 10, Oct. 1996, pp. 751-761.

An Efficient Interestingness based Algorithm for Mining Association Rules in Medical Databases

Siri Krishan Wasan†, Vasudha Bhatnagar††, and Harleen Kaur†

†Deptt. of Mathematics,
Jamia Millia Islamia,
New Delhi-110 025,
India

††Deptt. of Computer Science,
University of Delhi,
New Delhi-110 007
India

Abstract— **Mining association rules is an important area in data mining. Massively increasing volume of data in real life databases has motivated researchers to design novel and efficient algorithm for association rules mining. In this paper, we propose an association rule mining algorithm that integrates interestingness criteria during the process of building the model. One of the main features of this approach is to capture the user background knowledge, which is monotonically augmented. We tested our algorithm and experiment with some public medical datasets and found the obtained results quite promising.**

Keywords— Association rules, algorithm, data mining, knowledge discovery in databases (KDD), interestingness

I. INTRODUCTION

Association rule discovery often generate very large number associations with support and confidence as measures. This imposes a large burden on the analysts to determine which of these associations are of interest and which are of not interest. Researchers therefore have been strongly motivated to propose techniques and ability to detect interesting associations between fields in a database. The association rules algorithms that reflect the changing data trends and the user beliefs are attractive in order to make the over all KDD process more effective and efficient [1-3].

We propose an algorithm based on the premise that unless the underlying data generation process has changed dramatically, it is expected that the rules discovered from one set are likely to be similar (in varying degrees) to those discovered from another set [4, 5, 6, 7].

Interestingness measures can be used as an effective way to filter the rule set discovered from the target data set thereby, reducing the volume of the output. We propose to extend the approach presented in [18]. The proposed approach is a self-upgrading filter that keeps known knowledge rule base updated as new interesting rules are discovered. The proposed filter quantifies interestingness of the discovered knowledge on the basis of deviation of the newly discovered items with respect to the known knowledge.

The proposed association rule algorithm operates on the current training set and induces the model. During the model building, the algorithm computes the interesting factor for each pass of Apriori like algorithm against domain expert knowledge.

Although novelty, actionability and unexpectedness of the discovered knowledge are the basis of the subjective measures, their theoretical treatment still remains a challenging task [7-10]. Rules are interesting if:

(i) They are unknown to the user or contradict the user's existing knowledge or expectations (this is referred as Unexpectedness).

(ii) Users can do something with them to their advantage (this is referred as Actionability).

(iii) They add knowledge to the user prior knowledge (this is referred as novelty).

Our algorithm helps to discover interesting patterns at current time with respect to the previously discovered patterns (candidates), rather than exhaustively discovering all patterns. This will minimize the search space for frequent items during the process. It further helps in reducing the size of the induced model and in maintaining the model. One of the features of the approach is to capture the user background knowledge, which is monotonically augmented and can be determined using a mathematical technique.

Our approach differs from existing approaches. We generate the set of candidate itemsets of length (k+1) against expert knowledge from the set of frequent itemsets of length k (for $k \geq 1$) and check their corresponding interesting factor against user threshold in the database.

II. RELATED WORKS

There has been several approaches for developing algorithms for association rules mining [11-13] and mining of the frequent itemsets [13- 17]. The main approach for mining association rule in general derived by Agrawal et al [16] and

K. Elleithy (ed.), Advances and Innovations in Systems, Computing Sciences and Software Engineering, 167–172.

is called Apriori algorithm with support and confidence for association rules. Algorithm Apriori is an effective algorithm for mining association rules. It uses prior knowledge of frequent itemsets. It uses iterative process to explore k+1 itesets from k-itemsets according to a user-defined minimum support. Apriori is based on an important property that all non-empty subsets of frequent ietmsets are frequent because if I is not frequent itemset i.e. P(I) < minSup, then for any set I ∪ {A}, A being an additional item, P (I ∪ {A}) < minSup. Thus I ∪ {A} is also not frequent. Apriori is a two-step process consisting of join and prune actions. Let L_k be the set of k-frequent itemsets. We generate L_k from the set of C_k of k-itemsets defined by

$$C_k = L_{k-1} \times L_{k-1} = \left\{ (a_i, a_j) \mid a_i \neq a_j;\ a_i, a_j \in L_{k-1} \right\}$$

A scan of the database to determine the count of each candidate in C_k, then

$$L_k = \left\{ a_i \in C_k \mid P(a_i) \geq minSup \right\}$$

It may be noted that if any (k-1) subset of a candidate k-itemset in C_k is not in L_{k-1} then this candidate can not in L_k and so can be removed from C_k.

Similar to Apriori, another well known iterative algorithm DHP [14], generates candidate k-itemsets from L_{k-1}. However, DHP employs a hash table, which is built in the previous pass, to test the eligibility of a k-temsets not for all k-itemsets from $L_{k-1} \times L_{k-1}$ into C_k. DHP adds a k-itemsets into C_k only if k-temsets represents in a hash table whose value is ≥ minSup. Such a filtering technique reduces the size of database, candidate set C_k and prune the number of transactions in the database.

Kaur H et al [18] described a unified approach for discovery of interesting associtaion rules by computing deviation of recently discovered rules from the known rule and then comparing this deviation against the user threshold value. Rule deviation is obtained by combing conjunct level deviation. We may compare the candidate itemsets in the database DB against the candidate itemsets in domain expert knowledge base (KB) with a meeting user threshold in each case of classical Apriori. This will further enable us to prune the frequent itemsets and the discovery of interesting rules.

III. MINING OF INTERESTING ASSOCIATION RULES FROM DATABASES

Most association rule mining algorithms employ support confidence thresholds to exclude uninteresting rules. But many rules satisfying minimum thresholds and minimum confidence still may not be interesting to experts. Ultimately, experts can judge if a rule is interesting or not.

First we introduce some basic concepts, using the formalism presented in [16]. Let I={i_1,i_2...,i_n} be a set of items. Let D, the task-relevant data, be a set of database transactions where each transaction T is a subset of items I i.e. T ⊂ I. Each transaction is associated with an identifier, called TID. Let 'A' is a set of items. A transaction T is said to contain A if and only if A ⊆ T. An association rule is an implication of the form A=>B, where A, B ⊆ I, A ∩ B = Ø and A, B ≠ Ø. The rule A ⇒ B holds in the transaction set D with support s, if s is the percentage of transactions in D that contain A ∪ B (i.e., both A and B). This is taken to be the conditional probability, p(A ∪ B). The rule A ⇒ B has confidence 'c' in the transaction set D if 'c' is the percentage of transactions in D containing A that also contain B. This is taken to be the conditional probability, p(B / A). That is, Support (A ⇒ B) = p(A ∪ B) and Confidence (A ⇒ B) = p(B / A). Rules that satisfy both a minimum support threshold (minSup) and a minimum confidence threshold (min-conf) are called strong rules.

The problem of mining association rules is to find all itemsets whose interestingness level is greater than the user specified minimum support and user specified threshold. The proposed algorithm utilizes the interestingness criterion to construct interesting model.

At each iteration (pass) of **I**nterestingness based **A**priori-like **R**ule **M**ining (IARM) algorithm, IARM performs the following tasks. The pruning is done after every pass, while building the association rule model:

1. Find the set N of frequent itemsets using Apriori-like.
2. Compute on interestingness factor ($\chi_{KB}(I)$), of set of candidate itemsets of length (k+1) from the set of frequent itemsets of length k (for k ≥ 1) with respect to user-specified knowledge base itemsets, which results in generation of interesting candidate itemsets based on user specified minimum support and user threshold at each pass. If the largest frequent itemset is a j-itemset, then an proposed algorithm may need to scan the database up to (j+1) times. Such a factor is particularly powerful in reducing the number of frequent itemsets.
3. Generate interesting association rules from frequent itemsets. For any pair frequent itemsets A and B satisfying A ⊂ B, if $\frac{Supp(A)}{Supp(B)} \geq \alpha$, where α is the confidence of the rule, then A=> B is a valid association rule.

IV. INTERESTINGNESS BASED APRIORI LIKE RULE MINING (IARM) ALGORITHM

This **I**nterestingness based **A**priori-like **R**ule **M**ining (IARM) algorithm, effectively and efficiently generates interesting association rules in the database by applying the Apriori property. However, IARM employs a interestingness factor, which is built in the previous pass, to test the occurrence of k-itemset. IARM adds a k-itemset into C_k only if k-itemset is interesting and whose value is greater than equal to minimum confidence threshold.

The problem of mining association rules is decomposed into two sub-problems. Firstly, generating all the frequent

X is an symbol for concatenation of L_{k-1} pass

items that satisfy the interestingness constraint, i.e. compute interestingness factor, $\chi_{KB}(I)$ *for each* $I \in DB$, I is the set k-items at each pass with respect to the domain expert knowledge base (KB). Computed value of the interestingness factor is compared with the threshold provided by the user. Computed interestingness value, $\chi_{KB}(I)$ reduces the number of frequent items resulting in time savings during the mining process. When all large itemsets are found, generating association rules is straight forward. Secondly, generating interesting association rules in the database with the required confidence and user threshold.

An association rule mining algorithm usually generates too many itemsets including a lot of uninteresting ones. Most interestingness criteria have been proposed to prune those interesting itemsets. Our model prunes the items in the itemsets generation process instead of post pruning. Algorithm computes the interestingness factor, $\chi_{KB}(I)$ between the database itemsets and expert knowledge base (KB) frequent itemsets as computed below:

Definition 3.1.1:

Let KB = Expert knowledge base of frequent itemsets as per his/her experience

DB = Current database of itemsets

For each $I \in DB$, I is the k-itemset where I and I_j are all possible items in DB and KB, we compute

$$\chi_{KB}(I) = \max_{I_j \in KB,\, I_j \text{ is } kth-itemset} \{\psi(I, I_j)\}$$

where

$$\psi(I,I_j) = \begin{cases} 0, & \text{if } I \cap I_j = \varnothing \quad \text{(Disjoint itemsets)} \\ \theta = \left\{ \dfrac{n\,(I \cap I_j)}{\text{No. of Items to be compared}} \right\}, & \text{where } I \cap I_j \neq \varnothing \text{ but } I \neq I_j \\ 1, & \text{if } I = I_j \quad \text{(Identical itemsets),} \end{cases}$$

Note:- If $\chi_{KB}(I) \geq \Phi$ *(user specified threshold) and* I *satisfies minSup (minimum support) then it will go to the next iteration* L_k,

3.2 Proposed Algorithm

The problem of mining association rules is decomposed into two sub-problems. Firstly, generating all large itemsets in the database and secondly, generating association rules according to the large itemsets generated in the first step The proposed algorithm is similar to Apriori algorithm except that after each frequent itemsets generation, the interestingness measure is computed with respect the existing model T_i and prune uninteresting candidate itemsets that are not relevant in the current training set. Fig. 1 explains the proposed algorithm.

During the process of generation of items from the candidates, the items is not to be extracted if the computed degree of interestingness is less than a user threshold or its accuracy level or less than the rule accuracy level.

Algorithm: Find the frequent itemsets using an iterative level-wise approach based on candidate generation meeting minimum support and interestingness constraint

- Input: Database DB of transactions (t); minimum Support (minSup), user threshold (Φ); interestingness factor ($\chi_{KB}(I)$).
- Output: L_k frequent itemsets in DB meeting minSupp and interestingness constraints.

KB: Expert knowledge base of frequent itemsets
DB: Current database of itemsets
t : transaction in database DB
C_k: Candidate itemset of size k
L_k : Frequent itemset of size k

L_1 = {frequent items};
for ($k = 2$; $L_k \neq \varnothing$; k++) **do begin**
  C_k= candidates generated from L_{k-1};
  for each transaction $I \in C_k$, candidate set of k-itemsets **do** generated from the join of L_{k-1} with itself
  Compute $\chi_{KB}(I)$ for each $I \in C_k$ /**explained in section3.1**/
    if $\chi_{KB}(I) \geq \Phi$ for I satisfies minimum support then it will go to next iteration L_k
  end
return $\cup_k L_k$;

- Use the interesting frequent itemsets to generate interesting association rules.

Fig. 1. Algorithm IARM

Example 1. Consider the transaction database (*DB*) and user-specified knowledge base (*KB*) given in Fig. 2. In each pass, L_k algorithm IARM constructs a candidate itemsets of large itemsets, count the number of occurrences of each candidate itemset, and compute interestingness factor ($\chi_{KB}(I)$) as given in definition 3.1.1 with respect to user-specified knowledge base (*KB*). It will be based on predetermined minimum support (minSup) and user threshold (Φ). In the first pass, proposed algorithm simply scans all the transaction to count the number of occurrences of each items. Assuming minSup=20% (i.e. 2) and user threshold (Φ), the set of large 1-itemsets (L_1) composed of candidate 1-itemsets with the minSup, can then be determined using Interestingness factor, $\chi_{KB}(I)$ with respect to knowledge base. The pruning is done after every stage of the algorithm.

	t_1	A
KB	t_2	B C
	t_3	A D

DB	T_1	A B C
	T_2	B D
	T_3	B C
	T_4	A B D
	T_5	A C
	T_6	B C
	T_7	A C
	T_8	A B C E
	T_9	A B C

Fig. 2. An illustrative known knowledge base (KB) and transaction database (DB)

To discover the set of large 2-itemsets, we consider L_1 X L_1 to generate a candidate set C_2. C_2 consists of 2-itemsets. Next, the nine transactions in *DB* are scanned and the support C_2 is counted. Fig. 3 shows the generation of candidate itemsets and frequent itemsets using algorithm IARM.

Proposed algorithm ends the process of discovering frequent itemsets and computing interesting factor ($\chi_{KB}(I)$), when there is no candidate itemsets to be constituted from L_{k+1}. Such a filtering technique prunes the frequent itemsets. This improves the computing efficiency while the number of candidate itemsets to be checked is reduced.

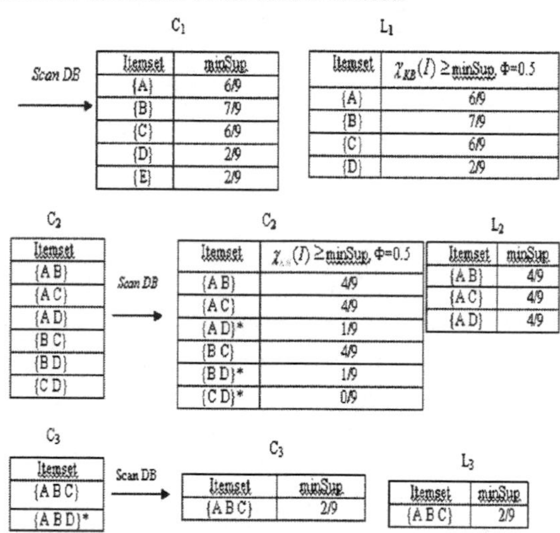

C1 (Scan DB)

Itemset	minSup
{A}	6/9
{B}	7/9
{C}	6/9
{D}	2/9
{E}	2/9

L1

Itemset	$\chi_{KB}(I) \geq$ minSup, Φ=0.5
{A}	6/9
{B}	7/9
{C}	6/9
{D}	2/9

C2

Itemset
{A B}
{A C}
{A D}
{B C}
{B D}
{C D}

C2 (Scan DB)

Itemset	$\chi_{KB}(I) \geq$ minSup, Φ=0.5
{A B}	4/9
{A C}	4/9
{A D}*	1/9
{B C}	4/9
{B D}*	1/9
{C D}*	0/9

L2

Itemset	minSup
{A B}	4/9
{A C}	4/9
{A D}	4/9

C3

Itemset
{A B C}
{A B D}*

C3 (Scan DB)

Itemset	minSup
{A B C}	2/9

L3

Itemset	minSup
{A B C}	2/9

Fig. 3. Generation of candidate itemsets and frequent itemsets using algorithm IARM

3.3 Discovery of Selecting Interesting Association Rules from Large Itemsets

There is no universally accepted definition for interesting rules. However, we prefer to have the following description for interesting rules. A rule is interesting for several reasons.

(i) if it is unknown to the user or contradicts the user's existing knowledge (unexpectedness) or

(ii) if it can do something to the advantage of the user (actionability) or

(iii) if it adds to the user prior knowledge (novelty).

The problem we consider in this paper is essentially how to find the most interesting rules. Once all interesting large itemsets and their support are determined, the rules can be generated in a straightforward manner as follows: if L_i is large interesting itemset, then for every subset X of L_i ($X \subset L_i$) such that $X \neq \emptyset$, the support. L_i / support. X is computed. If the ratio is at least greater then equal to the user specified minimum confidence, then the rule $R = R \cup (X \Rightarrow (L_i - X)$ and this is an interesting association rule.

IV. QUANTIFICATION OF THRESHOLD USING MATHEMATICAL TECHNIQUES

Many algorithms and techniques have been developed for discovering association rules [11-17]. The most important is these Apriori or Apriori like algorithms which are based on the assumption that user can specify the minimum support threshold. In real world problems, it is almost impossible for the user to give suitable thresholds if he has no understanding of the database. In Psychology, the study of expertise has gained impetus because of the advent of the expert systems and new technologies for preserving knowledge [22]. Knowledge Elicitation (KE) is not easy, Expert systems are intended to assist experts. Various methods of Knowledge Elicitation experts (KE) have been used by experimental psychologists and developers of expert systems. Analysis of tasks and interview methods (structured and unstructured) are some of the important KE techniques. Experts often reason in terms of their experiences. Test cases can be used in task analysis. Unstructured interviews are equally important in which open ended questions are put before the expert. Initially, this may provide an overview of the domain knowledge and subsequently interview can be structured for precise knowledge extraction [21]. Domain specific probe questions can be prepared to generate rules and procedures. By using monotonicity of some sort and Hansel chain we can reduce the number of questions to be put before the expert [22].

Zang et al [20] have proposed a Fuzzy approach for identifying Association Rules with database independent minimum support (FARDIMS). This provides a good man-machine interface that allows user to take commonly used interval [0,1] for specifying minimum-support. User threshold is appropriate to a database to be mined only if the distributions of deviations are known to the user. For a given database, one must try to find different set of interesting cases by giving different values to user threshold. We may allow user to specify relative threshold and develop algorithms to convert relative threshold to true threshold. To find interestingness in R_1 with respect to R_2 is compared with user threshold, we may apply Zhang et al [20] technique of Fuzzy logic.

V. EVALUATION

This evaluation tries to answer the question "Does the discovered interesting itemsets are accurate"? A widely used scenario within the *IF* community consist of using a small part of database D^{LM} (10%) from a dataset for learning model and the remaining part of D^{EM} (90%) for evaluation [3]. Then, we measure the error between the discovered interesting itemsets and expert knowledge itemsets in D^{EM}. The method consists of computing the Mean Absolute Error (MAE) between the real discovered interesting candidate itemsets and expert itemsets. In fact, MAE a widely used accuracy metric in the evaluation of IF systems, measures the absolute average deviation between the discovered interesting itemsets $S(I_i, I_j)$ and expert itemsets say $E(I_i, I_j)$ in the evaluation of data set. Where I_i and I_j are the discovered and expert itemsets. This absolute deviation for a particular user is computed As

$$|MAE| = \frac{\sum_{i=1}^{Mtest} |S(I_i, I_j) - E(I_i, I_j)|}{M_{test}}$$

We employed the following metric in our evaluation. In this framework, accuracy is defines as 'How good the model mimics knowledge contained in the association rule'. Accuracy is measured as original value by the Mean Absolute Error (MAE), in the following way. As expected, the accuracy and user threshold (Φ) decreases with minimum support (minSup) and the number of interesting rules increases.

$$acc = \frac{Orginal\ value}{|MAE|}$$

VI. IMPLEMENTATION AND EXPERIMENTATION

To demonstrate the applicability of our algorithm, data sets were identified from the UCI knowledge discovery and machine learning repository [23,24] with 0.1% and 1% to indicate minimum confidence and minimum support respectively. All the experiments were performed on a 1.5GHz Pentium IV PC with 640 MB of main memory, running under Windows 2000 professional. For the performance comparison experiment, we used the same datasets as in [18].

6.1 Experiment I

The objective of this experiment is to show the effectiveness of the approach in reducing the number of unnecessary itemsets. It is expected that the number of discovered rules from itemsets that are interesting keeps on decreasing over the time. Six different datasets were used for comparison and assume that the user threshold value (Φ) = 0.5. The datasets are Heart, Hepatitis, Sick, Lymph, Diabetic and Breast Cancer. The results are shown at time T_1, T_2 and T_3 respectively are given in Table I. Fig. 4 shows the results of frequent itemsets generated from IARM algorithm with the standard Apriori.

TABLE I. The discovered frequent itemsets and medical rules at time T_1, T_2, and T_3

Dataset	Time	APRIORI		IARM	
		Total No. of Freq. Itemsets	Discovered AR's	Total No. of Freq. Itemsets	Discovered AR's
Heart	T_1	347	987	273	798
	T_2	439	566	218	374
	T_3	88	207	45	130
Hepatitis	T_1	987	1207	478	856
	T_2	344	980	230	467
	T_3	286	626	120	320
Sick	T_1	238	4502	56	1208
	T_2	789	2709	290	469
	T_3	388	986	110	398
Lymph	T_1	1560	32000	856	13080
	T_2	780	28562	320	10900
	T_3	998	26781	439	12080
Diabetic	T_1	430	3678	175	1100
	T_2	367	2890	96	1080
	T_3	238	1400	57	569
Breast Cancer	T_1	192	802	49	405
	T_2	173	725	60	140
	T_3	120	540	39	78

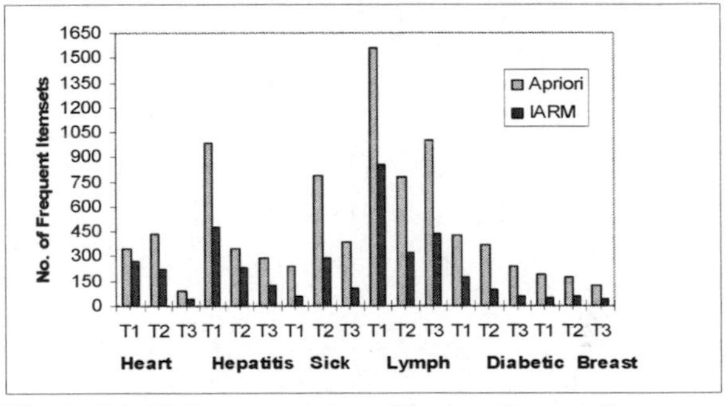

Fig. 4 . Graphical representation of frequent itemsets for different datasets

VII. CONCLUSION

With the use of experience of medical experts, we may develop data mining tools/models to discover more interesting, novel and actionable rules that may provide guidelines for better diagnosis and management of a disease. There is no universally accepted definition of interesting rules.

We prefer that interesting rules are important in the medicine and healthcare system. It is possible that there are contradictions between medical-expert diagnostic rules and rules discovered by data mining techniques. Most association rule mining algorithms employ support confidence thresholds to exclude uninteresting rules. But many rules satisfying minimum thresholds and minimum confidence still may not be interesting to medical experts. Ultimately, medical experts can judge if a rule is interesting or not.

We have integrated the framework into the mining algorithm. IARM generates interesting rules which is based on computation of the interestingness factor of discovered itemset with respect to known domain knowledge. The main feature of this algorithm is to capture user subjectivity. This helps in reduction in I/O overhead and CPU time. We have implemented and experimented with the medical datasets, and it has proved useful. Currently we are trying to apply the proposed methodology in a general case.

ACKNOWLEDGMENT

We are grateful to Dr Mrs. Manju Kanga, Associate Specialist, Wycompe Hospital, London, U.K. for her valuable comments and suggestions.

REFERENCES

[1] Han, J. and M. Kamber, 2001. Data Mining: Concepts and Techniques. San Francisco, Morgan Kauffmann Publishers.

[2] Dunham M. H.: Data Mining: Introductory and Advanced Topics. 1st Edition Pearson Education (Singapore) Pte. Ltd. 2003

[3] Hand, D., Mannila, H. and Smyth, P. (2001), Principles of Data Mining, Prentice-Hall of India Private Limited, India.

[4] Bronchi, F., Giannotti, F., Mazzanti, A., Pedreschi, D.: Exante: Anticipated Data Reduction in Constrained Pattern Mining. In Proceedings of the 7th PAKDD03. 2003.

[5] Bronchi, F., Giannotti, F., Mazzanti, A., Pedreschi, D.: ExAMiner: Optimized Level-wise Frequent pattern Mining with Monotone Constraints. In Proceedings of the 3rd International Conference on Data Mining (ICDM03). (2003).

[6] Bronchi, F., Giannotti, F., Mazzanti, A., Pedreschi, D.: Adaptive Constraint Pushing in Frequent Pattern Mining. In Proceedings of the 17th European Conference on PAKDD03. 2003

[7] C Freitas, A. A.: On Rule Interestingness Measures. Knowledge-Based Systems. 12:309-315. 1999

[8] Liu, B., Hsu, W., Chen, S., Ma, Y.: Analyzing the Subjective Interestingness of Association Rules. IEEE Intelligent Systems. (2000).

[9] Silberschatz, A., Tuzhilin, A.: On Subjective Measures of Interestingness in Knowledge Discovery. In Proceedings of the 1st International Conference on Knowledge Discovery and Data Mining, 1995

[10] B. Liu, W. Hsu, S. Chen, and Y. Ma.: Analyzing the Subjective Interestingness of Association Rules. IEEE Intelligent Systems, 2000.

[11] M.Zaki and C. Hsiao. Charm: An efficient algorithm for closed itemset mining. In Proceeding of the 2nd SIAM International Conference on Data Mining, Arlington, USA, 2002.

[12] S. Zhang, C. Zhang, Estimating itemsets of interest by sampling, in: Proceedings of the 10th IEEE International Conference on Fuzzy Systems, Melbourne, Australia, December, 2001.

[13] R. Agrawal, T. Imielinski, A. Swami, Mining association rules between sets of items in large databases, in: Proceedings of the ACM SIGMOD Conference on Management of Data, 1993, pp. 207–216.

[14] J. Park, M. Chen, P. Yu, Using a hash-based method with transaction trimming for mining association rules, IEEE Transactions on Knowledge and Data Engineering 9 (5) 813–824, 1997.

[15] Klemetinen, M., Mannila, H., Ronkainen, P., Toivonen, H., Verkamo, A. I.: Finding Interesting Rules from Large Sets of Discovered Association Rules. In Proceedings of the 3rd International Conference on Information and Knowledge Management. Gaithersburg, Maryland. (1994).

[16] C. Aggarawal, P. Yu, A new framework for itemset generation, in: Proceedings of the ACM PODS, pp. 18–24, 1998.

[17] J. Han, J. Pei, Y. Yin, Mining frequent patterns without candidate generation, In: Proceedings of the ACM SIGMOD International Conference on Management of Data, 2000, pp. 1–12.

[18] Kaur H et al,. A Unified Approach for Discovery of Interesting Association Rules in Medical Databases. 'Advances in Data Mining', LNAI -4065, Springer-Verlag, Berlin Heidelberg. 2006, pp. 53-63.

[19] Wasan S K., and Kaur H, A Hybrid Approach in Medical Decision Making, in : Special short paper proceedings, Industrial Conference on Data Mining, Germany. ICDM -2006, Springer-Verlag, Berlin Heidelberg. [accepted].

[20] Shichao Zhang et al. Fuzzy logic based method to acquire user threshold of minimum support for mining association rule. Information Sciences, (1-16). 2003.

[21] Kovalerchuk, B., Triantaphyllou, E., Despande, A. and Vtyaev, E. 1996. Interactive Learning of Monotone Boolean Function. Information Sciences, 94 (1-4):87-118.

[22] R. Hoffman, N. Shadbolt, Eliciting knowledge from experts: a methodological analysis, Organizational and Human Decision Process 62 (2), 129–158, 1995.

[23] S. D. Bay. The UCI KDD archive. [http://kdd.ics.uci.edu] University of California, Department of Information and Computer Science, Irvine,C.A; 2001.

[24] Blake and C. J. Merz. UCI repository of machine learning databases. [Machine-readable data repository]. .University of California, Department of Information and Computer Science, Irvine,C.A; 2001.

NeSReC: A News meta-Search Engines Result Clustering Tool

Hassan Sayyadi
Web Intelligence Laboratory
Sharif University of Technology
sayyadi@ce.sharif.edu

Sara Salehi
Web Intelligence Laboratory
Sharif University of Technology
sarasalehi@ace.tju.ir

Hassan AbolHassani
Web Intelligence Laboratory
Sharif University of Technology
abolhassani@sharif.edu

Abstract—**Recent years have witnessed an explosion in the availability of news articles on the World Wide Web. In addition, organizing the results of a news search facilitates the user(s) in overviewing the returned news. In this work, we have focused on the label-based clustering approaches for news meta-search engines, and which clusters news articles based on their topics. Furthermore, our engine for NEws meta-Search REsult Clustering (NeSReC) is implemented along. NeSReC takes queries from the users and collect the snippets of news which are retrieved by The Altavista News Search Engine for the queries. Afterwards, it performs the hierarchical clustering and labeling based on news snippets in a considerably tiny slot of time.**

Index Terms— **News, Clustering, Labeling, News Retrieval, News Mining**

I. INTRODUCTION

CURRENTLY, there exist a considerable number of online news sites and traditional news agencies provide electronic version of news on their web sites. To find news more effectively, special tools and search engines have been developed recently. For example News Feeder software, RSS standard and Google News site (which uses near 4500 news sources) can be mentioned. On the other hand, online news is specific type of public information available on the Web with unique features which result in different processing demands for gathering, searching and exploration on them in comparison to the ordinary web contents. Of those features, we can mention of the trustworthiness of news sources as well as the rapid update of them.

One of the problems of many search engines which is also true for many news search engines is the lack of the ability to categorize the search results before showing to a user. In fact results are ranked and displayed as a long list. When the user query is very specific, the results are not so much and then the user can rapidly find the relevant items. Nevertheless, unfortunately most of the time the query is general and ambiguous, resulting in a considerable number of items to display to the user. Average number of terms in a query is near 3 and most of the times a user only checks the first 3 results [2]. For an example when we use Google and search for "jaguar", pages talking about feline mammal are displayed in ranks 10, 11, 32 and 71. Therefore, it is more appropriate to cluster the results before showing to a user. Such clustering helps in resolving two important issues. Firstly, it helps user to have an overall view of the results and to refine the query needed more appropriately. Secondly, one of the pitfalls of the link-analysis based algorithms is when results are from different categories. In such a case the top results are normally, belong to one of the categories and considering the fact that users usually only check top ranked results they may miss many relevant results.

Van Rijsbergen [12] was the first one that investigated the effect of clustering hypothesis in Information Retrieval for query-based systems: documents similar to each other are relevant to similar queries with high probability. In fact the main point is that relevant documents are more similar to each other compared to non-relevant ones. Based on this hypothesis we can use clustering in two different ways:

1) Before retrieval which has a long history and we can mention Scatter as a famous example,
2) After retrieval which is what we are interested on in this paper.

It should be noted that the web search result clustering has important differences with traditional text clustering. One of the main differences is the existence of links between pages. The other important difference is the need to do fast processing and dealing with a multi-line abstract instead of whole document. According to [10] desire characteristics for clustering of search results are:

- **There is no need for all pages to be clustered:** Not all pages should be clustered, because some pages can be not related to any generated clusters.
- **Overlapping:** Clusters may have some overlaps, because one page can point to several topics. Consequently, it can fall in several clusters. Furthermore, clusters overlapping should be as few as possible. Because two clusters, which have great overlaps generally, point to the same topic, they should be merged.

Manuscript received October 13, 2006. This work was supported in part by the Web Intelligence Research Laboratory, Sharif University of Technology.

K. Elleithy (ed.), Advances and Innovations in Systems, Computing Sciences and Software Engineering, 173–178.

- **Incremental Clustering:** Finally, because of handling time complexity, incremental clustering methods are more satisfactory.

One of the reasons that commercial search engines are not doing this is the high runtime complexity of it. It should be noted that some search engines like Altavista[1] does aggregation in very simple form. In NorthernLight[2] results are divided in some Custom folders. Such division is not intelligent and is based on attributes like page type (personal page, product page, etc.), language, domain, site and so on. Better commercial examples are Kartoo[3], Grokker[4], Mooter and specifically Vivisimo which apply simple clustering algorithms. They also have very user-friendly interfaces. In addition, Clusty[5] uses Vivisimo[6] but its internal clustering algorithm is not known. The result of Vivisimo for query "Iran" is shown in figure 1.

In section 2 we have a review of the related works in which we are mainly focused on clustering and labeling tasks. Then in section 3 our proposed method is introduced and elaborated. Section 4 discusses on evaluation results on the implemented system (NeSReC). Conclusions and future works is given in the final section.

Fig. 1. Vivisimo result for query "Iran"

II. Related Works

For the first time clustering and its effects on the retrieval was reported in [13]. Its main purpose is to create a directory for documents, which facilitates users' access to them. It first divides the repository to a small number of clusters. Then a user selects some of them and they are combined and a sub-repository is constructed. Such operations are repeated until user is satisfied. To reduce the cost of algorithm for large repositories a sampling approach is taken. The study in [14] is a good but rather old study on it.

A. Clustering

Clustering is done in different levels of a text for different purposes. As noted above one usage of clustering is for providing better navigation. However, a clustering also can be used to do summarization in the level of paragraph [4] or even sentences [3] for event detection. Approaches for better review of results can be categorized in the following two [1]:

1) Document based approach
2) Label based approach

Document based approaches are those they does clustering based on similarity of texts like their terms vector similarity. After clustering is done, some words or phrases of the texts in a cluster are combined to build a label for the cluster like [7,8]. Clusters made this way have no overlap and the quality of their labels is highly dependent on the quality of the clusters themselves. Works in the document based clustering methods are different from the following two aspects:

1) The clustering algorithm they use
2) Measurement of distances between texts or clusters

Since human better understands hierarchical structures, the tendency for creation of clusters in a hierarchical structure is high. In addition, specification of parameters like number of clusters and the similarity threshold is a difficult task. Therefore, normally the generated labels do not satisfy users. For this reason, such techniques are not used anymore.

One of the document-based methods is what reported in [5], which first eliminates the stop-words and does stemming to unify words like "went" and "go". Then n words having higher TF values are selected from news. If S1 and S2 are words extracted from two news then similarity value of them is intersection of S1 and S2 divide by n. In addition, distance measure is easily computed from the similarity value:

$$Sim(n_1, n_2) = \frac{|s_1 \cap s_2|}{n}$$

$$Dissim(n_1, n_2) = 1 - Sim(n_1, n_2)$$

Using such a measure k-nearest neighbor algorithm is applied on the news. Furthermore, the single-link algorithm is reported in the paper and it is claimed that a combination of them produces best results.

STC test is also a linear incremental algorithm which instead of treating a document as a set of words considers it as a sequence of words and then combines clusters based on their

[1] Http://www.altavista.com/
[2] Http://www.northernlight.com/
[3] Http://www.kartoo.com/
[4] Http://www.grokker.com/
[5] Http://www.clusty.com/
[6] http://www.vivisimo.com/

intersection of postfixes of their documents. It allows overlaps of clusters that can handle noise efficiently and let documents themselves specify the number of clusters.

In [10] link structure of pages is used as a characteristic for clustering retrieved results of a search. In this method, firstly, the pages are loaded and their output links are extracted. In-links are also gathered using a standard search engine. Such two vectors of in-links and out-links are used to compute the similarity of two documents.

Authors of [9] as like [11] used classification instead of clustering. Statistical analysis is used to learn some specifications of some classes (for example from DMOZ[7] categories) and then the trained system is used for classification of new data. This method is highly dependent on the train set and therefore can't be extended for the whole web.

To create clusters in [15] in the first step a hierarchy for concepts is created and then assigns documents to those concepts. [16] uses a special tool (which is based on a large repository of documents) to find concepts related to a user query and uses fuzzy C-Mean for clustering.

In [6] Lycos search engine is used and the performance of two standard indexing methods (N-gram and vector model) in clustering are compared and it is shown that the first method is resistant to noise. In addition, it is mentioned that automatically created clusters even when the performance of algorithm is high has more categories than what a human may build. For example, it is possible for a human to put all news of a news agency in a cluster while an automatic system makes many smaller clusters for such news based on different topics they cover. Using N-grams results in fewer numbers of clusters and from this point of view acts like a human user. During the clustering fuzzyfication is used to reduce the unwanted results for points in boarders in both indexing and labeling phases.

One the other hand, in label based approach, words and informative phrases are first extracted using some statistical analysis like word occurrences and then create clusters base on selected labels. Vivisimo and Mooter, which have satisfactory results, use label-based approaches.

[11] at first discovers important words based on a training set and then to each of them assigns related words. It uses a within cluster similarity measure to evaluate the quality of output. For naming a cluster, also important words are used. Nevertheless, because of the difference between real words and N-grams the quality of naming is not so good.

Furthermore, Authors in [1] proposed a label-based method used named entities but with defining some new measures have reached to a more effective model. As they noted it is shown that TF-IDF cannot remove high frequent un-important words. Their new measures try to overcome this problem. Scoring of a phrase comes from two factors: local factor and global factor. In the traditional TF-IDF scoring TF is the local and IDF is the global factor. In the new proposal two new local factor named LRDF and OLF as defined below is used:

$$LF_i^{LRDF} = \log(1 + DF_{R,i})$$

$$LF_i^{OLF} = DF_{R,i} * \log(\frac{|R|}{DF_{R,i}})$$

Also a new global factor named OGF as below is defined:

$$DF_i^{OGF} = \frac{DF_{r,i} / |R|}{DF_{D,i} / |D|}$$

In their experiments, it is shown that OLF-OGF produces very good results. In the context of news the uses of named entities is very effective since unlike ordinary web pages a news is about an event in a specific location for a given person or group in a specific time. As results extraction of those entities like time and places is very beneficial.

B. Labeling

For the labeling two points are important [1]:

1) readability of labels which facilitate understanding by users, and
2) their conciseness in representing related documents.

In fact mislabeling results in reduction of both Precision and Recall measures. It is also mentioned that the stemming quality has a strong effect on labeling [14]. In Grouper [7] labeling are based on the same phrases which are used during clustering. In [17] for cluster naming super concepts and words in the title of news are used but it is only applicable for Japanese language.

III. PROPOSED METHOD

Our approach is for clustering as well as labeling of news which is architected as a meta-search engine for news. In the other words, it is proposed for engines that by accepting user queries forward them to a search engine and process the results to cluster them and make labels for those clusters based on the snippets and title of news. As mentioned in [1] labeling based clustering makes more efficient clusters than the traditional methods, so our proposed method follows this approach. The process contains following tasks:

1) receiving initial results (texts, links)
2) extraction of candidate titles
3) ranking of the titles
4) clustering
5) display

The initial results are made by using an ordinary search engine. Display of the results is also discussed when we explain the experimental results. Therefore, in this section we focus on the following two important tasks:

[7] Http://dmoz.org/

- Extraction of titles and ranking them
- Clustering

A. Extraction of Titles

In this phase at first, the candidate titles should be extracted from news' snippets. Then such titles are ranked based on some factors which is explained shortly and finally some of them which satisfy selection criteria are selected. Since the best titles are noun phrases in our method, we only extract noun phrases from the snippets. In this process, unimportant words are eliminated. However, as it is shown in the experimental results, stemming or rooting has not much effect on the quality of clustering.

To score titles we use three factors. As like many other methods, our first two factors are local and global factors. Their sole purpose is to eliminate very frequent or very rare words since they can't be good titles for clusters. Therefore, the local factor is a score for frequency of a term and the global factor scores a term's rareness. The third factor is the length of a phrase which is used to score longer titles since they better represent a cluster than shorter ones. The final score for each phrase (or word) is computed as the multiplication of these factors:

$$R(t_i) = LF(t_i) * GF(t_i) * SF(t_i)$$

Here $R(t_i)$ represent the score of title t_i and *LF*, *GF* and *SF* are local, global and length factors, respectively.

- Local factor is what presented in [1]. To balance its effect with other factors we also use a logarithmic form of term frequencies (DF) as below:

$$LF(t_i) = DF_i * \log \frac{|R|}{DF_i}$$

  Here |R| is the number of extracted news and DF_i is the document frequency of term i which expresses the number of documents that contain term i.

- Our global factor is:

$$GF(t_i) = \log \frac{|R|}{DF_i}$$

- For the length factor we propose three function of linear, exponential and logarithmic forms:

$$SF^{linear}(t_i) = \max(\alpha, sizeof(t_i))$$

$$SF^{power}(t_i) = \sqrt{sizeof(t_i)}$$

$$SF^{log}(t_i) = \log(sizeof(t_i)) + 1$$

  α parameter is a threshold value to control the maximum value obtained from the first equation and the *sizeof(ti)* is the number of words constructing term *i*.

After computation of scores of titles, they are sorted. Now we should select K titles among them that also satisfy following condition:

- If a title is sub-phrase of another title and their size ratio is higher than a ß threshold then the shorter title is eliminated from the list of candidates.

This condition is applied to prevent the selection of almost similar titles for different clusters.

B. Creation of Clusters

As said before in the clustering by labeling, clusters are made according to the selected label. Therefore, for each label t_i a cluster *Ci* is created and news having such a title is put on that cluster. It is clear that news can be put in several clusters. Since news may point out to different subjects this behavior is rational and is also in the interest of users. In our approach a news is belonged to a title when:

- it completely has that title
- it has a sub-phrase of the title with the length ratio of more than ß threshold

After completion of this process for a level of clustering, it can be applied to each cluster to obtain a hierarchical clustering structure. To control the level of hierarchy number of documents in a cluster is used and when such number is less than σ it is not anymore sub-clustered. In addition, we can define another h threshold to control the height of the tree and reduce complexity of having a very tall tree of clusters.

IV. EXPERIMENTAL RESULTS

In this section, our meta-search engine named NeSReC which is developed for the evaluation of our proposed method. It is a meta-search engine that clusters the results of a search for news. To have an initial result set Altavista news search is used. When a user searches some terms, Altavista is called to receive titles, addresses and snippets of the relevant news. Then candidate titles are selected from snippets. As mentioned, titles are noun phrases and to distinguish them we use JMontylingua[8] tool. Noun phrases as well as words in them are considered as candidate titles. A title with less than three letters or being a stop-word is eliminated. Our stop-word list contains 1000 words.

Stemming or rooting has not so much effects on the titles selection. For example, in one of our experiments without stemming, we reached to 760 titles while stemming reduced to 670 words. Considering its rather high execution cost it is not so much effective to be considered. In addition, it is noted that such a words gains low ranks by our ranking model and ultimately they are not selected so the existence of them has no harm on the quality of results. The main point is that without stemming, the execution time is less than 1 second in a typical run but when stemming is considered by

[8] http://web.media.mit.edu/~hugo/montylingua/

WORDNET tool, it raises to 20 seconds.

The reasons why stemming has not much effect on news clustering is clear. It is that the news agencies has many cross references to each other a news which is published by different sources has a main source which publishes it at first and others use very similar wording to the original source.

We also see that the logarithmic function for size factor leads to the best results for label selection. After extraction of label, the top 10, which also satisfy condition, are considered as the labels for the clusters in the given level. In our experiments, we set ß to 0.5.

After titles and clusters are constructed, those clusters having more than σ news are divided to lower level clusters with the same method. We set σ to the number of extracted news divided by 10 times clustering level:

$$\sigma = \frac{ResultSize}{20 * ClusterLevel}$$

Additionally, we limit the height of the tree to two levels. In other words, the threshold of the number of news for clusters in level 1 is 10 and this threshold for level 2 is 5. To reduce the complexity of the algorithm, we restrict our levels into two. However, the method can be customized for each user based on his preferences. The first level of result clustering and labeling in NesRec for user query "Iran" is shown in the figure 2.

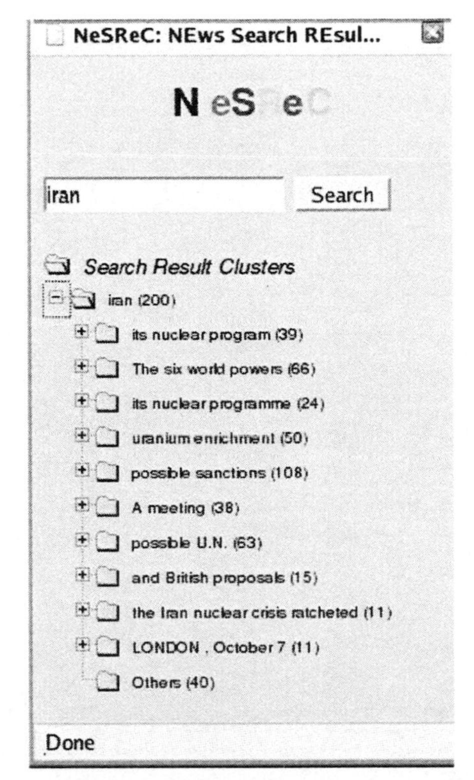

Fig. 2. NeSReC result clusters for query "Iran"

V. Conclusion

The main merit of our method is its simplicity while at the same time produces high quality results. The difference of our method with what reported in [1] is that our proposed method is used for meta-search engines and its clustering based on snippets, while their method is used for ordinary search engines. Furthermore, they limit labels to named entities for person names, places, organizations, and artifacts. It is true that such entities are important for news but limiting labels to them is not a good idea. In our method the only limitation is that the selected title should be noun phrase which seems a logical assumption. The other advantage of our method is that the created clusters are balanced which means they have almost same number of news.

In addition, the number of un-categorized news is low and is near 25% of total size of retrieved news in average which is shown in diagram in fig. 3. Furthermore, these non-clustered news have low ranks in search process. Such results confirm this points that the size of cluster for news is important for its rank. Moreover, Inasmuch as NeSReC engine generate 10 clusters in each level, the average size of first level clusters is near 15% of total size of retrieved news as shown in fig. 4. Although we escape stemming and cluster overlapping criteria, our less complicated model contributes to conspicuous results.

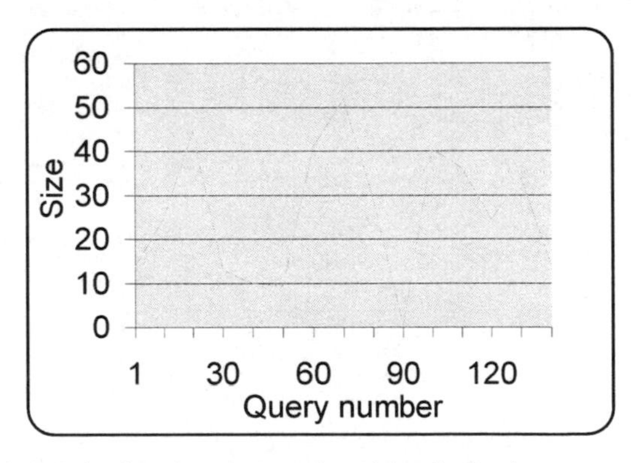

Fig. 3. Ratio of non-clutered news to the total size of retrieved news

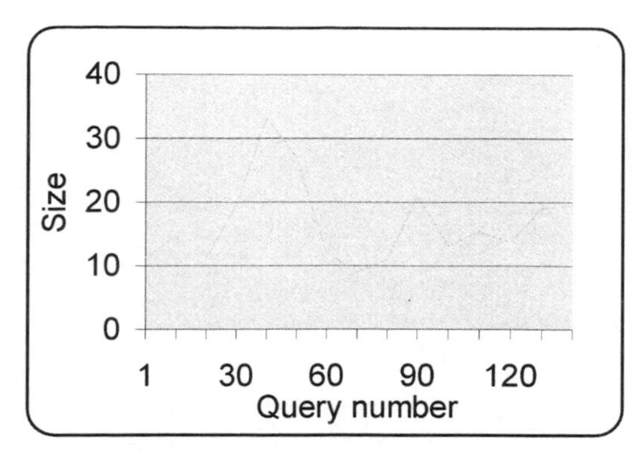

Fig. 4. Ratio of average size of first level clusters to the total size of retrieved news

In the future, we will work on some criteria for clusters overlaps to avoid similar clusters with different names. In addition, considering clusters overlapping avoid further simple problems which may arise from escaping stemming.

REFERENCES

[1] H. Toda, R. Kataoka, *"A search result clustering method using informatively named entities"*, In: WIDM '05: Proceedings of the 7th annual ACM international workshop on Web information and data management, New York, NY, USA, ACM Press (2005) 81–86

[2] C. Silverstein, M. Henzinger, H. Marais, and M. Moricz, *"Analysis of A Very Large Web Search Engine Query Log"*, SIGIR Forum, 33(1), 1999.

[3] M. Naughton, N. Kushmerick, J. Carthy, *"Clustering sentences for discovering events in news articles"*, In: ECIR. (2006) 535–538

[4] V. Hatzivassiloglou, J. Klavans, M. Holcombe, R. Barzilay, M. Kan, K. McKeown, *"Simfinder: A flexible clustering tool for summarization"*, (2001)

[5] N. A. Shah, E. M. ElBahesh, *"Topic-based clustering of news articles"*, In: ACM-SE 42: Proceedings of the 42nd annual Southeast regional conference, New York, NY, USA, ACM Press (2004) 412–413

[6] Z. Jiang, A. Joshi, R. Krishnapuram, and L. Yi., *"Retriever: Improving Web Search Engine Results Using Clustering"*, In Managing Business with Electronic Commerce 02.

[7] O. Zamir and O. Etzioni, *Grouper: A Dynamic Clustering Interface to Web Search Results*". In Proceedings of the Eighth International World Wide Web Conference, Toronto, Canada, May 1999.

[8] M. A. Hearst and J. O. Pedersen, *"Reexamining the Cluster Hypothesis: Scatter/Gather on Retrieval Results*", in Proceedings of the 19th International ACM SIGIR Conference on Research and Development in Information Retrieval (SIGIR96), 1996, pp 76-84.

[9] H. Chen and S. Dumais, *"Bringing Order to The Web: Automatically Categorizing Search Results*", in Proceedings of the CHI 2000 Conference on Human Factors in Computing Systems, pp. 142–152, 2000.

[10] Y. Wang and M. Kitsuregawa, *"Link Based Clustering of Web Search Results*", in Second International Conference on Advances in Web-Age Information Management (WAIM), 2000.

[11] H. Zeng, Q. He, Z. Chen, W. Ma, and J. Ma, *"Learning to Cluster Web Search Results*". In Proceedings of ACM SIGIR '04, 2004.[19] Douglas Cutting, Jan O. Pedersen, David Karger, and John W. Tukey. "Scatter /Gather: A Cluster-Based Approach to Browsing Large Document Collections". In Proceedings of SIGIR'92, pages 318-329, Copenhagen, Denmark, June 21-24 1992.

[12] V. Rijsbergen, C. J., *"Information Retrieval*". London: Butterworths; 1979.

[13] M. A. Hearst and J. O. Pedersen, "Reexamining the Cluster Hypothesis: Scatter/Gather on Retrieval Results" , in Proceedings of the 19th International ACM SIGIR Conference on Research and Development in Information Retrieval (SIGIR96), 1996, pp 76-84.

[14] A. V. Leouski and W. B. Croft, "An Evaluation of Techniques for Clustering Search Results". Technical Report IR-76, Department of Computer Science, University of Massachusetts, Amherst, 1996.

[15] K. Kummamuru, R. Lotlikar, S. Roy, K. Singal, and R. Krishnapuram, "A Hierarchical Monothetic Document Clustering Algorithm for Summarization and Browsing Search Results". In Proceedings of the 13th international conference on World Wide Web, pages 658-665, New York, NY, USA, 2004.

[16] O. Hoeber and X. D. Yang, "Visually Exploring Concept-Based Fuzzy Clusters in Web Search Results", In Proceedings of the Fourth International Atlantic Web Intelligence Conference, 2006.

[17] T. Noda, H. Ohshima, T. Tezuka, S. Oyama, and K. Tanaka, "Automatic Extraction of Topic Terms for Web Search Result Clustering", The 1st China-Kyoto Student Workshop on Digital Content and Web Computing (CKSW2006), Beijing, China, March 2006.

Automatic Dissemination of Text Information using the EBOTS system

Hemant Joshi* and Coskun Bayrak†
* Department of Applied Science
† Computer Science Department
University of Arkansas at Little Rock
{hmjoshi, cxbayrak}@ualr.edu

Abstract

World Wide Web contains 170 Terabytes of information [1] and storage estimates show that the new information is growing at a rate of over 30% a year. With the quanta of information growing exponentially, it is important to understand the information semantically to know what concepts are relevant and what are irrelevant. The Evolutionary Behavior Of Textual Semantics (EBOTS) system being developed at University of Arkansas at Little Rock [2] aims at the quantitative reasoning aspect of textual information. In the automatic decision-making mode, the EBOTS system can distinguish between relevant and irrelevant information, discarding irrelevant documents and accepting only relevant information to develop expertise in a particular field. This paper discusses the usefulness of Information Theory in the development of relevance criteria and the results obtained in the context of textual information.

Keywords: Information Theory, Entropy, Absolute Information Gain, Quantitative Reasoning of Text data, Textual Information Relevance

1. Introduction

Shannon's Information Theory [3] has made a tremendous impact on modern communication methodologies. It is human nature to try to measure anything quantitatively, in terms of units that the human brain can parse in objective numbers. Measuring information not only reveals quantity but also makes it possible to compare and draw conclusions based on such comparisons. Quantitative reasoning is a promising and novel approach that enables the *EBOTS system* not only to gain new knowledge, but also to reason about how much textual information has been gained over a certain period of time. This also makes it possible for the *EBOTS system* to learn and acquire expertise in a particular field or area as needed. Using entropy [4] as the measure of information content, the relevance of the new information acquired by the *EBOTS system* can be measured.

EBOTS is a contextually aware system that helps form and identify correlations among text data. Text documents consist of a number of sentences and each sentence in turn, consists of a set of words appearing together to convey a particular meaning. The core philosophy of the *EBOTS system* exploits this natural hierarchy to yield a context-aware system [5]. The methodology behind the EBOTS system is explained in detail in the following sections. In section 2, basic formalism behind context aware framework of the EBOTS system is presented. Section 3 will discuss mathematical foundation for the quantitative reasoning. Section 4 primarily addresses Information Theory aspects that are applicable to the EBOTS model, prototype experiments and results. Section 5 draws a logical conclusion to this discussion and also throws light upon possible and future enhancements that can be contributed to this approach.

2. Background

The *EBOTS system* uses the meanings of different words in identified contexts to represent acquired knowledge. Once the words are identified in a particular context, the system forms the hierarchy of words in a tree-like structure. Each tree represents the direct context of a single sentence. These trees can form paragraphs and documents when extended.

The *EBOTS system* employs knowledge acquisition techniques that are dynamic compared to those deployed by *N-gram* systems. *N-gram* systems [6] [7] are a commonly used technique in information retrieval [8] [9] to find word pattern combinations using co-occurrence, but are not sufficient to determine context. Also, the window for a particular context depends upon user selection. Thus a window of *100* words while appropriate in one case may not be useful for another dataset. The *EBOTS system* works with sentences and can be considered as a natural sentence-gram model. Each sentence changes in its number of words and so N-gram models proposed variable *N-grams*. Variable *N-gram* models are not only complex but also less effective in retrieving the intuitive meaning of the text. While N-gram models serve the basic purpose of context dissemination, they are either too divergent or convergent for a system to attain human-like intuitive results.

The theoretical foundation of the *EBOTS system* is composed of domains, reference domains, and correlation types (strong, weak and/or no correlation)

K. Elleithy (ed.), Advances and Innovations in Systems, Computing Sciences and Software Engineering, 179–184.

between the reference domains [10]. A brief definition of each of these follows.

Domain

The domain of information represents a concept [11]. A concept can be generic and abstract or it can be specific when expressed as a context. Domains can consist of more sub-domains and are related to other domains. If two domains represent two different concepts, then the relationship between two domains gives rise to a new concept altogether.

Reference Domain

A Reference domain represents the context of text data. They follow sentence boundaries. The words form the elements of the reference domain. A single domain consists of at least one reference domain. When two reference domains have common element(s), they are more likely to share the same context. Hence the correlations can be defined among the related reference domains. These correlations can be of three different types.

Strong Correlation

A single document consists of a number of reference domains. If an element 'a' of one reference domain A is common with an element of reference domain *B*, and if *A*, *B* belong to the same document, a Strong Correlation is said to exist among the two reference domains. This is logical because if the two sentences appear in the same document, and if both of the sentences share the common (partial) context, then both sentences are likely to share same meaning as well.

Weak Correlation

Consider the case in which two documents consist of two reference domains *A* and *B* respectively. If these two reference domains share one or more elements, there is a Weak Correlation between the two reference domains. When two documents share common elements among their reference domains, they are likely to have partially common context. Sharing of the common context is a necessary but not a sufficient condition to determine the correlation between the two documents. There is a level of fuzziness involved whether having common elements in two documents is a sufficient condition for the two documents to have same meaning. This fuzziness is well represented with a Weak Correlation.

No Correlation

If there is no common element between the two reference domains, it cannot be predicted whether they share the same context or not. Irrespective of whether the two reference domains exist in the same or different documents, no commitment can be made about their correlation and No Correlation is said to exist in this case.

Reference domains consist not only of synonyms [12] but they also include contextual information for every word in the sentence. This is shown in the reference domain structure formed for a simple sentence like "*Mary had a little lamb*". Only three words, namely *Mary, little* and *lamb* are considered after filtering common words and stop words to find word roots [13]. The resulting structure is shown in Figure 1. The *EBOTS system* uses a Machine Readable Dictionary [12] to obtain synonyms and glossary meanings of each word.

Mary had a little lamb		
mary	Virgin Mary Madonna The Virgin	the mother of Jesus ; Christians refer to her as the Virgin Mary; she is especially honored by Roman Catholics
little	fiddling	small and of little importance
	brief	of short duration or distance
	lilliputian	Small in size
	younger	Younger by age
	miniscule	Small in size and or shape
	little	small in a way that arouses feelings
lamb	lamb	Younger sheep
	Charles Lamb, Elia	English Essayist
	lamb	a person easily deceived or cheated
	dear	a sweet innocent mild-mannered person
	lamb	the flesh of a young domestic sheep eaten as food
	lamb	give birth to a lamb

Figure 1 Detailed Reference Domain structure for a simple sentence "Mary had a little lamb"

3. Formal Model of Correlation

Theoretically, correlations are helpful in determining direct or indirect association with a particular concept, but it is important to measure its quality mathematically. Based on the correlation theory, a formal model is presented.

Consider T to be the total number of correlations that are referenced to a particular reference domain, A. Out of a total T correlations, T_w is the number of weak correlations of another reference domains with A while T_s is the number of strong correlations that are established by other reference domains with A.

$$T = T_s + T_w \tag{1}$$

Correlation Delta

The correlation delta of any given reference domain, A, is a formal representation of its significance to the other reference domains in the system as well as its relevance to the query context. The correlation delta is computed for every reference domain in the *EBOTS system* with respect to the given query context, Q. It is also a useful representation to find out how many reference domains share a common context. Correlation delta Δ_i, for a reference domain i, is defined as follows:

$$\Delta_i = \left[\frac{T}{1+T} \right] * \left(\frac{1+T_s}{1+T} \right) \tag{2}$$

The above equation achieves normalized correlation delta values in the range between *0* and *1* (inclusive),

i.e., $0 \leq \Delta_i \leq 1$

The formula in equation (2) is significant in determining the strength or weakness of the reference domain with respect to the query context. The correlation delta value of zero refers to an isolated reference domain that does not share a concept with any other reference domain of the system. A correlation delta value of *1* indicates the presence of strong correlations for that particular reference domain. The Highest correlation delta value *1* is possible only when all correlations with the reference domain under consideration are strong.

As the number of weak correlations increase, the strength of its correlation to the query context decreases. In this case, the value of the correlation delta lies between *0* and *1* (excluding *0* and *1*) i.e., $0 < \Delta_i < 1$. It should also be noted that the correlation delta is directly proportional to the number of strong correlations and inversely proportional to the number of weak correlations. The weak correlation delta values allow fuzziness of correlation strength among reference domains for a particular query context.

4. Information Theory in EBOTS system

According to equation 2, the value of correlation delta always ranges between 0 and 1 (inclusive). Correlation delta can be considered as the likelihood of a particular reference domain being associated to a query Q. Hence correlation delta becomes a good representation of the statistical probability of the event in which that reference domain would be correlated (strongly / weakly) to the given query *Q*. When no query is given, the probability of the event of whether the aforementioned reference domain is associated or not, is equally likely for any hypothetical qucry. Thus, when knowledge *X* is initially acquired and no query is provided, i.e., in the equilibrium state, one cannot determine correlations and hence, the *EBOTS system* assumes equal likelihood of both possibilities.

If *p* is the probability of the event that a particular reference domain is associated to a given query context, then (*1-p*) is the probability that this reference domain is not associated or irrelevant to the query. According to information theory, entropy or self-information is the average uncertainty for a particular event. If p_i is the probability of a reference domain *i* and the acquired initial knowledgebase consists of *D* documents ($D > 0$) for a total of *N* reference domains ($N \geq D$), then the entropy, *H* of the *EBOTS system* can be defined as

$$H = -\sum_{i=1}^{N} p_i * log_2 p_i \qquad (3)$$

Figure 2 shows the steps in chronological manner that the *EBOTS system* learns in. This ever-iterative process is evolutionary and self-guided, accepting or rejecting new knowledge based on certain criteria that are discussed briefly in the following part of the document.

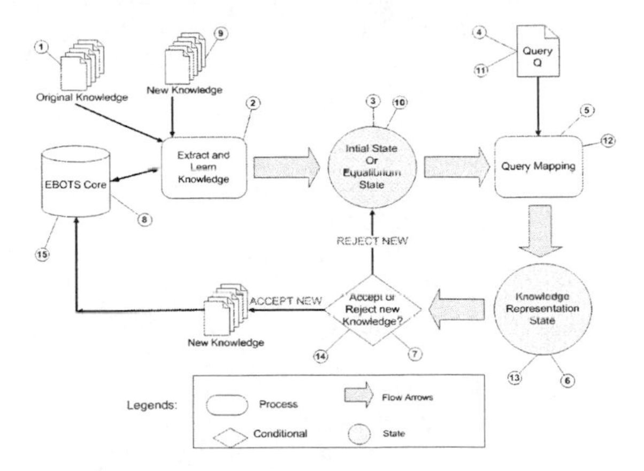

Figure 2 **Steps 1 through 15 (marked with circle around) in chronological manner for the EBOTS system knowledge acquisition and learning process**

From Figure 2, when initial knowledge is presented at step 1 (number 1 marked with circle around it), the knowledge parsing and acquisition process takes control and creates an equilibrium state as shown in step 3. Entropy *H* at the equilibrium state can also be calculated by taking into account the probability of association for each reference domain. Each p_i in the equilibrium state resembles equal likelihood of being associated or not associated to any given generic query. Hence using equation (3), the probability of each reference domain is

$$P_i = 0.5 \qquad (4)$$

Equal probability of each reference domain to be associated or not associated to the generic query is justified because in the equilibrium state, the *EBOTS system* does not express bias over reference domains. Using equations (3) and (4), in equilibrium state, for the initial knowledge *X*, entropy H_0 at step 3, can be calculated as

$$H_0 = [-0.5 * log_2(0.5)] + \dots$$
$$\dots + [-0.5 * log_2(0.5)]$$

$$H_0 = 0.5 + \dots + 0.5 = \sum_{1}^{N} 0.5$$

$$\therefore H_0 = \frac{N}{2} bits \tag{5}$$

When query Q is introduced to the system at step 4, it is carefully chosen to determine and orient the expertise in that particular field or area. At this stage, the information content of the *EBOTS system* in response to the query Q can be measured as the entropy H_1 of the represented knowledge as associated to query Q. Using correlation delta (Δ_i) as the probabilistic measure for entropy calculation, entropy H_1 is calculated for the query state as

$$H_1 = -\sum_{i=1}^{N} \Delta_i * \log_2(\Delta_i) \tag{6}$$

Where, Δ_i is the correlation delta value of each reference domain entailing correlation to the query Q.

Initial knowledge is always assumed to be a good starting point and accepted in steps 7 and 8. Chronologically, at step 9, new knowledge is to be acquired. The decision has to be made about the relevance of this new knowledge pertaining to the its relevance with respect to the existing or initial knowledge. Suppose the new knowledge X' combined with initial knowledge X constitutes knowledgebase Y. This allows the EBOTS system to determine the entropy H_2 in equilibrium state (step 10) as well as the query state (step 13) for the same query Q. If the knowledgebase Y now consists of D' documents ($D' >> D$) and total N' reference domains ($N' >> N$), then the entropy H_2 of the *EBOTS system* at equilibrium state will be

$$H_2 = -\sum_{i=1}^{N'} p_i * \log_2 p_i \tag{7}$$

Similar to equation (3), the entropy can be calculated using p_i value to be *0.5*

$$\therefore H_2 = \frac{N'}{2} bits \tag{8}$$

Now for the same query Q, at step 13, it needs to be determined how the new knowledge fairs in relevance to the query. The entropy value H_3 can be computed for query Q using correlation delta values Δ_i for each reference domain i of the system. So the entropy with new knowledge in the query state would be

$$H_3 = -\sum_{i=1}^{N'} \Delta_i * log_2(\Delta_i) \tag{9}$$

With all the four entropy values H_0, H_1, H_2 and H_3 thus computed, a naïve simplistic graphical representation of the knowledgebase can be summarized as shown in Figure 4.

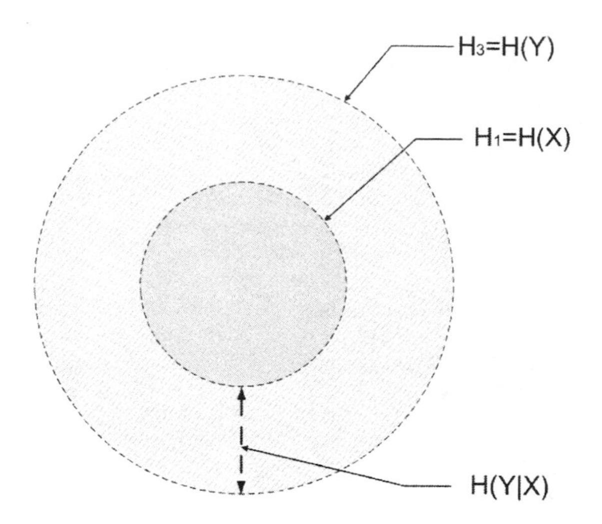

Figure 3 **Simplistic graphical representation of the EBOTS knowledgebase.**

Using Figure 3, we can compute the difference between the entropies of initial knowledge $H(X)$ and new knowledge $H(Y)$ in response to the given query Q. The difference between H_3 and H_1 is indicated as conditional entropy $H(Y|X)$ and can be represented as

$$H(Y \mid X) = H(Y) - H(X)$$
$$H(Y \mid X) = H_3 - H_1 \tag{10}$$

Here, $H(Y|X)$ indicates conditional entropy Y given prior establishment of X.

Please note that the difference between H_3 and H_1 indicates how relevant the newer knowledge is with respect to the initial knowledge. The least value of $H(Y|X)$ would be zero, which indicates that nothing relevant can be learnt from new knowledge for the given query Q. To calculate the absolute relevance or *Absolute Information Gain* (AIG) of the *EBOTS system* for the new knowledge in proportion to acquired knowledge, the ratio of conditional entropy $H(Y|X)$ to initial knowledge is computed.

$$AIG = \frac{H(Y \mid X)}{H(X)}$$
$$\therefore AIG = \frac{H_3 - H_1}{H_1} \tag{11}$$
$$\therefore AIG = \frac{H_3}{H_1} - 1 \tag{12}$$

The AIG values range between *0* to positive infinity (inclusive). The AIG value would be *1* if the entropy value represented by H_3 is zero, i.e., if the new knowledge is completely irrelevant to the given query. If the initial knowledge entropy H_0 is *0* (i.e., the initial knowledge is irrelevant to the query *Q*) and the new knowledge entropy is a positive value, then Absolute Information Gain would be infinite. In another case, if H_3 and H_1 represent exactly the same entropy values, there would be no gain i.e., AIG=*0*.

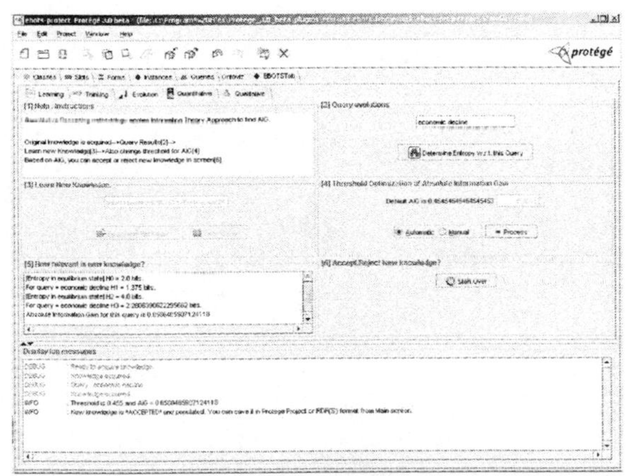

Figure 4 **Screenshot of the EBOTS system running as Quantitative Reasoning Module in Protégé Ontology environment to automatically accept/reject new knowledge for the query *economic decline***

The *EBOTS system* runs through the Protégé ontology editor [14] developed by the Stanford University's SMI laboratory. Though ontological framework is not necessary for the scope of this paper, the *EBOTS system* also uses qualitative learning techniques through ontologies. As shown in the Figure 4, the automatic dissemination process decides whether to accept or reject new knowledge based whether it is relevant to existing knowledge. In order to automatically decide this relevance, it is critical to define threshold λ dynamically for given Absolute Information Gain, thus making the EBOTS system operate closer to intuitive and human-like learning. Above the decided threshold value, new knowledge will be accepted and below λ, it will automatically be rejected. Based on the initial knowledge, the λ value can be dynamically calculated as

$$\lambda = H_0 - H_1 \tag{13}$$

Threshold λ is the quantitative expectation from the new knowledge to be relevant to existing knowledge. If the system has learnt and acquired knowledge expertise pertaining to a certain field, threshold λ automatically relaxes the expectation from new knowledge. However, in the initial learning cycles, when the *EBOTS system* is establishing its expertise, the threshold expectation from new knowledge for relevance is relatively high. Threshold λ is also a good measure of how well query *Q* defines the expert context that the *EBOTS system* is trying to learn. If the query *Q* is regarding music and the initial knowledge is completely irrelevant to music, then threshold λ would be equal to H_0 as entropy H_1 value would be zero.

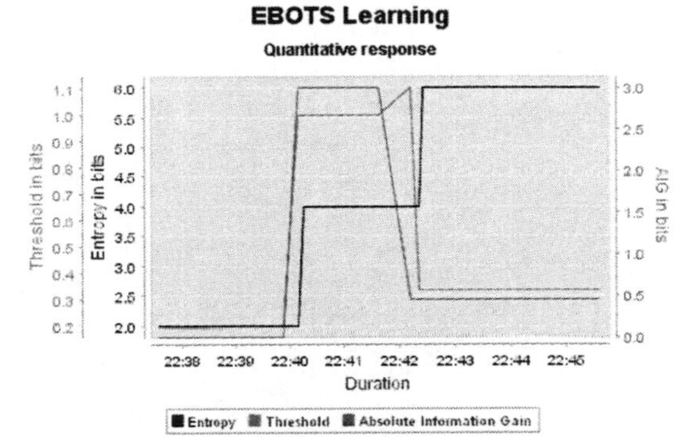

Figure 5 **Knowledge based quantitative Learning Response of the EBOTS learning system in quantitative reasoning mode during automatic dissemination of 14 documents about query concept economic decline**

In Figure 5, a typical learning trend of the *EBOTS system* is shown. The *X* axis indicates time domain and values on *Y* axis represent the AIG (blue), threshold (red) as well as entropy (black) of the *EBOTS system*. The only increase in the entropy of the system occurs when new knowledge is accepted and added to the existing knowledgebase. At this time, the expectation threshold value also drops. Out of a total of *14* documents, only *10* relevant documents to the concept economic decline were accepted but the remaining *4* documents were rejected, due to their being irrelevant to the given concept *economic decline*.

In a separate experiment (See Figure 6) we used Information Retrieval Classic3 MEDLINE dataset [15].

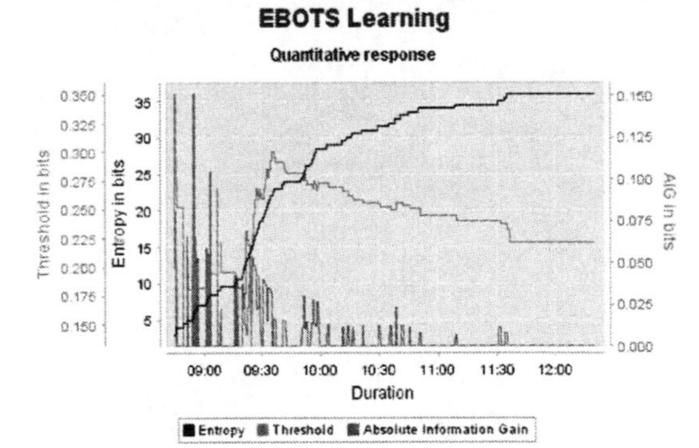

Figure 6 **Quantitative learning response of the EBOTS system using MEDLINE data for query "the**

crystalline lens in vertebrates, including humans" for over 3 hours.

Medline dataset has 1024 medical documents and the queries and retrieval results provided. We used the query "the crystalline lens in vertebrates, including humans" and determined the relevance of all documents when the EBOTS system was given few documents to start with (namely document ids *13, 14, 15, 72, 79* and *171*). The system achieved 63 documents as relevant to the given knowledge with respect to given query.

Typical Information Retrieval search engine would take user query to search through its database. Lot of research is being made to understand user preferences and personalized searching. Using the proposed approach the EBOTS system can actually learn user preferences and customize the search results (or knowledge accepted / rejected) accordingly.

5. Conclusion and Future Work

This paper demonstrates the ability and potential of the *EBOTS system* to reason quantitatively regarding acquired knowledge and the system automatically accepting or rejecting new knowledge based on a dynamic relevance threshold. The computed Absolute Information Gain of the system is a good measure of knowledgebase relevance to an expert query context. The *EBOTS system* can operate and disseminate information quantitatively in automatic as well as manual modes of operation. Manual intervention by a domain expert is required to define a query concept accurately.

Future enhancements to the quantitative reasoning module include the use of a database-backed knowledgebase to automatically disseminate knowledge and sense disambiguation [16] to determine the correct sense of textual semantics before decision-making. Another potential use of the *EBOTS system* may be to link a proposed reasoning engine to web search and web crawler programs that will acquire expertise in the distributed manner dynamically, given a particular concept. Currently, better threshold affecting factors are being investigated. We also plan to use TREC Novelty track [17] dataset which focuses on similar interests for user based feedback profiling. The important contribution of this paper is the framework for the quantitative modeling of the intuitive information seeking process.

6. Acknowledgements

Authors would like to acknowledge Mr. Subhashish Dutta Chowdhury for his valuable feedback.

7. References

[1] Lyman, P., Varian, H., Charles, P., Good, N., Jordan, L. L., Pal, J.: *How much Information? 2003*. SIMS Lab at University of Berkeley on the web at http://www.sims.berkeley.edu/research/projects/how-much- info-2003/ Jan 18, 2005

[2] *Evolutionary Behavior Of Textual Semantics*, online http://bayrak.ualr.edu/symsel/hemant/research_hemant.htm Mar 14, 2005

[3] Shannon C., *A Mathematical Theory of Communication*, Bell System Technical Journal, July-October 1948

[4] Gray R., *Entropy and Information Theory*, Springer Verlag, New York, 1990 online at http://www-ee.stanford.edu/~gray/it.pdf Mar 14, 2005

[5] Toda, M., 1956, "*Information-Receiving Behavior of Man*", The Psychological Review, Vol. 63, No. 3, pp. 204-212

[6] Lewis, D.: *Evaluating and Optimizing Autonomous Text Classification Systems*. International ACM-SIGIR Conference on Research and Development in Information Retrieval, July, 1995, pp. 246–254.

[7] Cavnar, W., Trenkle, J.: N-Gram-Based Text Categorization. Proceedings of SDAIR-94, 3rd Annual Symposium on Document Analysis and Information Retrieval (1994) 171-180

[8] Salton, G., Buckley, C.: Term-weighting Approaches in Automatic Text Retrieval. Information Processing & Management, (1988), 513–523.

[9] Salton, G., McGill, M.: Introduction to Modern Information Retrieval. McGraw-Hill, (1983).

[10] Joshi, H., Bayrak, C., 2005, "*Learning Contextual Behavior of Text Data*", International Conference on Machine Learning and Applications, ICMLA 2005

[11] Joshi, H., Bayrak, C., 2004, "Semantic Information Evolution" Artificial Neural Networks in Engineering Conference, Nov 07, 2004

[12] *WordNet, lexical resource system* developed by Princeton University, Cognitive Science Laboratory, online at http://wordnet.princeton.edu/

[13] Porter, M.F., *An algorithm for suffix stripping. Program*, 1980. 14(3): p. 130-137.

[14] *Protégé Ontology Editor*, Stanford University, online at http://protege.stanford.edu Mar 14, 2005

[15] MEDLINE Classic3 dataset online at ftp://ftp.cs.cornell.edu/pub/smart/med

[16] Lesk, M. E., 1986, "*Automatic sense disambiguation using machine readable dictionaries: how to tell a pine cone from an ice cream cone*", In proceedings of Special Interest Group for Documentation Conference , Toronto, Canada, pp. 24-26

[17] TREC Novelty Track online at http://trec.nist.gov/data/novelty.html

Mapping State Diagram To Petri Net : An Approach To Use Markov Theory For Analyzing Non-Functional Parameters

H. Motameni
Department of Computer Islamic Azad University, Sary, Iran
Motameni@iausari.ac.ir

M. Siasifar
Department of Computer University of Science and Technology of Mazandaran, Babol, Iran
Mahsa.Siasifar@gmail.com

H. Montazeri
Department of Computer University of Science and Technology of Mazandaran, Babol, Iran
H.Montazeri@ece.ut.ac.ir

A. Movaghar
Department of Computer Sharif University of Technology,Tehran,Iran
Movaghar@sharif.edu

M. Zandakbari
Department of Computer University of Science and Technology of Mazandaran, Babol, Iran
domanz111@yahoo.com

***Abstract*-The quality of an architectural design of a software system has a great influence on achieving non-functional requirements to the system.**

Unified Modeling Language (*UML*), which is the industry standard as a common object oriented modeling language needs a well-defined semantic base for its notation. Integrating formal methods Petri nets (*PNs*) with object oriented design concepts *UML* is useful to benefit from the strengths of both approaches. Formalization of the graphical notation enables automated processing and analysis tasks. In this paper we use a method to converting State Diagram to Generalized Stochastic Petri Net (*GSPN*) and then we derive the embedded Continues Time Markov Chain from the *GSPN* and finally we use Markov Chain theory to obtain performance parameters.

Keywords: *Unified Modeling Language (UML), State Diagram (SD), Generalized Stochastic Petri Net (GSPN), Continues Time Markov Chain (CTMC), Non-Functional Parameter.*

I. Introduction

Modeling of business processes is an important area in software engineering, and, given that it typically occurs very early in a project, it is one of those areas where model driven approaches definitely have a competitive edge over code driven approaches [1] modeling needs a language that is intuitive and easy to use. The *UML* specifies a modeling language that incorporates the object oriented community's consensus on core modeling concepts [2].

PNs are a formal and graphical appealing language, appropriate for modeling systems with concurrency [2]. So we translate *SD* to *GSPN* and then analyze it.

A *PN* is an abstract, formal model of information flow. The major use of *PNs* has been the modeling of systems of events in which it is possible for some events to occur concurrently but there are constraints on the concurrence, precedence, or frequency of these occurrences.

There are three general characteristics of *PNs* that make them interesting in capturing concurrent, object-oriented behavioral specifications. First, *PNs* allow the modeling of concurrency, synchronization, and resource sharing behavior of a system. Secondly, there are many theoretical results associated with *PNs* for the analysis of such issues as deadlock detection and performance analysis. Finally, the integration of *PNs* with object oriented software design architecture could provide a means for automating behavioral analysis [3].

We present at first a brief talk about *UML* and *SD*, and the method we used to transform *SD* to *GSPN*. And then we discuss about *GSPN* to introduce its fundamentals and history.

At the next step we derive a *CTMC* from the *GSPN*. And at the last step we do a performance evaluation on the derived *CTMC*.

II. RELATED WORK

Many studies and researches are being done in order to evaluate performance parameters from *UML* models which some include transformation of the *UML* model to a formal model like *PN* extensions. Saldhana and Shatz [4] transformed *UML* Diagrams to Object *PN* Models for performance valuation. A group of works are devoted to transforming the software model to *Color Petri Net* for evaluation (which seems to be more related to software properties than the other *UML* extensions) [5], [6], [7], [8], and [9]. Using *Stochastic Petri Net (SPN)* and its extensions has been discussed in several papers. Trowitzsch et al have transformed the software *UML* diagrams to *SPN* models for performance evaluation of real-time systems. In [10] the use of *UML* sequence diagram and statechart for the validation and the performance evaluation of systems is presented. For this purpose, the sequence diagrams are automatically transformed into *GSPN* models. Merseguer et al used the derived *SPN* from the *UML* model to evaluate performance of internet based software retrieval systems. Derivation of an executable *GSPN* model from a description of a system, expressed as a set of *UML* state machines (*SMs*) is reported in [11]. They have other works

K. Elleithy (ed.), Advances and Innovations in Systems, Computing Sciences and Software Engineering, 185–190.

on different *UML* diagrams [12], and [10]. Also we have some previous works in these areas in [13]. In [14] we transformed the *UML* model to *CPN* and then analyzed the *CPN* by the means of simulation. And in [15] we derived non-functional parameters through numerical analyses. In this paper we have focused on the evaluation of the final performance model in the way that leads us to gain meaningful parameters of the system like availability and performance efficiency.

III. UML AND STATE DIAGRAM

The *UML* [16] is a semi formal language developed by the Object Management Group (OMG) [17] to specify, visualize and document models of software systems and non-software systems too. *UML* defines three categories of diagrams: static diagrams, behavioural diagrams and diagrams to organize and manage application modules. Being the objective of our works the performance evaluation [18] of software systems at the first stages of the software development process, as proposed by the *software performance engineering* (*SPE*) [19], behavioural diagrams play a prominent role, since they are intended to describe system dynamics. These diagrams belong to five kinds: State diagram (SD), Use Case diagram (UC), Sequence diagram (SED), Collaboration diagram and Statechart diagram (SC) [12].

At last, it has to be mentioned that huge parts of software do not handle its intended usage but errors that may occur. This fact leads to the demand for explicit error handling support in *UML* [20].

A. State Diagram

A simple state indicates a condition or situation of an element. For example, the project management system may be in one of the following simple states:

Inactive: Indicates that he project management system is not available to its users, because it is not started or has been shut down

Suspended: Indicates that the project management system has encountered some server error, perhaps because it is running low on secondary storage and requires user intervention before becoming active again.

An initial state indicates the state of an element when it is created. In the UML, an initial state is shown using a small solid filled circle surrounding a small solid filled circle.

IV. GSPN

GSPN [21], [22], and [23] are ideally suited to model a large class of performance and reliability problems, but their numerical analysis requires the solution of a very large *CTMC*. The size of the transition rate for the *CTMC* is the main obstacle, since its memory requirements can easily exceed the capacity machines even when sparse storage techniques are employed.

*GSPN*s were originally proposed in [24], which is a well-known technique for describing complex stochastic systems in a compact way [22], and [24]. The state space, or *reachability graph*, of a *GSPN* is a semi-Markov process, from which states with zero time are eliminated to create a Markov chain. The Markov chain can be solved numerically, yielding probability values which can be combined to provide useful information about the net.

A *GSPN* is a directed bipartite graph with nodes called *places*, represented by circles, and *transitions*, represented by rectangles. Places may contain tokens, which are drawn as small black circles. In a *GSPN*, two types of transitions are defined, *immediate transitions*, and *timed transitions*. If there is a token in each place that is connected to a transition by an *input arc*, then the transition is said to be *enabled* and may *fire* after a certain delay, causing all these tokens to be destroyed, and creating one token in each place to which the transition is connected by an *output arc*. The markings of a *GSPN* are classified into two types. A marking is *vanishing* if any immediate transition is enabled in the marking. A marking is *tangible* if only timed transitions or no transitions are enabled in the marking. Conflicts among immediate transitions in a vanishing marking are resolved using a *random switch* [21].

As mention above, *GSPN* models comprise two types of transitions:

1. *Timed transitions*, which are associated with random, exponentially distributed firing delays, as in *SPN*, and
2. *Immediate transitions*, which fire in zero time, with priority over timed transitions.

Furthermore, different priority levels of immediate transitions can be used, and weights are associated with immediate transitions.

A *GSPN* is thus an eight-tuple

$$\text{GSPN} = (P, T, \Pi, I, O, H, M_0, W), \quad (1)$$

where $(P, T, \rho, I, O, H, M_0)$ is the underlying untimed *PN* with priorities, that comprises

- the set P of places,
- the set T of transitions,
- the input_ output and inhibitor functions I,O,H: $T \rightarrow N$,
- the initial marking M.

Additionally, the *GSPN* definition comprises the priority function Π: $\Pi : T \rightarrow N$ which associates lowest priority (0) with timed transitions and higher priorities (≥ 1) with immediate transitions:

$$\Pi(t) = \begin{cases} 0 & \textit{if t is timed} \\ \geq 1 & \textit{if t is immediate} \end{cases}$$

Finally, the last item of the *GSPN* definition is the function $W : T \rightarrow \Re$, that associates a real value with transitions. W(t) is:

- the parameter of the negative exponential *PDF* of the transition firing delay, if t is a timed transition,
- a weight used for the computation of firing probabilities of immediate transitions, if t is an immediate transition.

In the graphical representation of *GSPN*, immediate transitions are drawn as segments, and exponential transitions as white rectangular boxes.

When a marking is entered, it is first necessary to ascertain whether it enables timed transitions only, or at least one immediate transition. Markings of the former type are called *tangible*, whereas markings of the latter type are called *vanishing*.

In the case of a tangible marking, the timers of the enabled timed transitions either resume their decrement, or are re-initialized and then decremented, until one timed transition fires, exactly as in the case of *SPN*.

In the case of a vanishing marking, the selection of which transition to fire cannot be based on the temporal description, since all immediate transitions fire exactly in zero time. The choice is thus based on priorities and weights. The set of transitions with concession at the highest priority level is first found, and if it comprises more than one transition, the further selection, of probabilistic nature, is based on the transition weights according to the expression:

$$P\{t\} = \frac{w(t)}{\sum_{t' \in E(M)} w(t')}, \quad (2)$$

where E(M) is the set of enabled immediate transitions in marking M, i.e., of the transitions with concession at the highest priority level.

From a modeller point of view it may be difficult to specify transitions weights if they are then normalized over the whole net. However, it was proved in [26] that if no confusion is present in the net (that is to say if no interplay exists between conflict and concurrency), it is possible to determine at a structural level the sets of possibly conflicting transitions, called "extended conflict sets (*ECS*)," and the normalization of weights can be done only among transitions that belong to the same *ECS*.

Note that the semantics of a *GSPN* model always assumes that transitions are fired one by one, even in a vanishing marking comprising nonconflicting enabled immediate transitions. The equivalence of this behavior with the one resulting from the simultaneous firing of some immediate transitions in the model can be exploited to reduce the complexity of the solution algorithms [27].

The analysis of a *GSPN* model requires the solution of a system of linear equations comprising as many equations as the number of reachable *tangible* markings. The infinitesimal generator of the *CTMC* associated with a *GSPN* model is derived with a contraction of the reachability graph labeled with the rates or weights of the transitions causing the change of marking [28].

A different approach to the analysis of *GSPN* models, which also implies a different semantics, is presented in [29].

V. MARKOV CHAIN

A Markov process is a mathematical model that describes, in probabilistic terms, the dynamic behavior of certain type of system over time. The change of state occurs only at the end the time period and nothing happens during the time period chosen. Thus, a Markov process is a stochastic process which has the property that the probability of a transition from a given state $P_i(t)$ to a future state $P_i(t+1)$ is dependent *only* on the present state and not on. This is also called the Markovian property. If a Markov process meets the following conditions: the manner in which the current state was reached.

1. The system can be described by a set of finite states and that the system can be in one and only one state at a given time.
2. The transition probability P_{ij}, the probability of transition from state i to state j, is given from every possible combination of i and j fincluding i = j) and the transition probabilities are assumed to be stationary (unchanging) over the time period of interest and independent of how state i was reached. and
3. Either the initial state of the system or the probability distribution of the initial state is known.

Then, it is called a *finite-state first order Markov chain.*

As mentioned in the introduction, to completely specify the model, all we need to know are the initial state or probability distribution of the intial state) of the system p(0) = [p_1, p_2,.... p_n] and the transition probability matrix P.

$$P = \begin{bmatrix} P_{11} & P_{12} & P_{13} & \cdots & P_{1n} \\ P_{21} & P_{22} & P_{23} & \cdots & P_{2n} \\ \cdots & \cdots & \cdots & \cdots & \cdots \\ \cdots & \cdots & \cdots & \cdots & \cdots \\ P_{n1} & P_{n2} & P_{n3} & \cdots & P_{nn} \end{bmatrix}$$

Here, P_{ij} represent the constant probability (finite state first order Markov chain) of transition from state $X_i(t)$ to state $X_i(t+1)$ for any value of t. The Markovian property makes *P, time invariant*. Knowing *P*, we can also construct a transition diagram to represent the system.

Given the initial distribution **p**(0),

$$\mathbf{p}(1) = \mathbf{p}(0) \,.\, \mathbf{P}$$

$$\mathbf{p}(2) = \mathbf{p}(1).\mathbf{P} = \text{p}(0).\mathbf{P}.\mathbf{P} = \text{p}(0).\mathbf{P}^2$$

thus, for any k,

$$\mathbf{p}(\text{k}) = \mathbf{p}(0) \,.\, \mathbf{P}^{\text{k}}$$

We also note that the elements of *P* must satisfy the following conditions:

$$\sum_{j-1}^{j-n} P_{ij} = 1 \quad \text{For all i (row sum)}$$

and P_{ij} >= 0 for all i and j [30].

VI. TRANSFORMING SD TO CTMC

In this section we transform *SD* to *GSPN* first, and then we derive the *CTMC* from *GSPN* to analyze some parameters like availability and performance efficiency.

While transforming *SD* to *GSPN* we need some other information. Because there are some other parameters in *GSPN* like time information and *ECS*. We use the method which has been described in [12] which has recommended including two aspects in the annotation: time and probability. This method uses tagged values. Annotations will be attached to both transitions and states [12].

A formal semantics for the *SD* is achieved in terms of stochastic *PNs* that allow to check logical properties as well as to compute performance indices. Obviously, this formal semantics represents an interpretation of the "informally" defined concepts of the *UML SD*.

So we use this method but by using [15] that changes a little the use of the algorithm is the method to relate the ratio assigned to the *UML SD* to *GSPN* elements. These ratio are then included in the *GSPN* together with firing rates of transitions and the weights of immediate transitions.

It is very important to [22] check the *ECS* these results ensure that the net is suitable for a numerical evaluation yielding the steady-state probabilities of all its markings.

In this section we show how *GSPN* systems can be converted into Markov chains and how their analysis can be performed to compute interesting performance indices [31]. The construction of the Markov chain associated with a *SPN* system is described first, to set the ground for the subsequent derivation of the probabilistic model associated with a *GSPN*.

The *CTMC* associated with a given *SPN* system is obtained by applying the following simple rules [31]:

1. The *CTMC* state space S $=\{s_i\}$ corresponds to the reachability set RS (M_0) of the *PN* associated with the $SPN(M_i \leftrightarrow s_i)$.
2. The transition rate from state s_i (corresponding to marking M_i) to state $s_j(M_j)$ is obtained as the sum of the firing rates of the transitions that are enabled in M_i and whose firings generate marking M_j.

Based on the simple rules listed before, it is possible to devise algorithms or the automatic construction of the infinitesimal generator (also called the state transition rate matrix) of the isomorphic *CTMC*, starting from the *SPN* description. Denoting this matrix by Q, with wk the firing rate of T_k, and with:

$$E_j(M_i)=\{h: T_h \in E(M_i) \wedge M_i[T_h\rangle M_j\}$$

The set of transitions whose firings bring the net from marking M_i to marking M_j, the components of the infinitesimal generator are:

$$q_{ij}=\begin{cases}\sum T_k \in E_j(M_i)\ w_k & i \neq j \\ -q_i & i=j\end{cases} \tag{3}$$

$$q_i=\sum_{T_k \in E(M_i)} w_k \tag{4}$$

Let the row vector η represent the steady-state probability distribution on markings of the *SPN*. If the *SPN* is ergodic, it is possible to compute the steady-state probability distribution vector solving the usual linear system of matrix equations:

$$\begin{cases}\eta Q=0 \\ \eta 1^T=1\end{cases}, \tag{5}$$

where 0 is a row vector of the same size as η and with all its components equal to zero and 1^T is a column vector (again of the same size as η) with all its components equal to one, used to enforce the normalization condition typical of all probability distributions.

Calculation the probability and sojourn time the probability that a given transition $T_k \in E(M_i)$ fires (first) in marking M_i has the expression:

$$P\{T_k \mid M_i\}=\frac{w_k}{q_i} \tag{6}$$

Using the same argument, we can observe that the average sojourn time in marking M_i is given by the following expression:

$$SJ_i=\frac{1}{q_i} \tag{7}$$

When transitions belonging to the same *ECS* are the only ones enabled in a given marking, one of them (transition t_k) is selected to fire with probability:

$$P\{t_k \mid M_i\}=\frac{w_k}{w_k(M_i)}, \tag{8}$$

where $w_k(M_j)$ is the weight of ECS (t_k) in marking M_i and is defined as follows:

$$w_k(M_i)=\sum_{t_j \in [ECS(t_k) \wedge E(M_i)]} w_j \tag{9}$$

Gaining the transition probability matrix

The transition probability matrix U of the *Embedded Markov Chain (EMC)* can be obtained from the specification of the model using the following expression:

$$u_{ij}=\frac{\sum_{T_k \in E_j(M_i)} w_k}{q_i} \tag{10}$$

By ordering the markings so that the vanishing ones correspond to the first entries of the matrix and the tangible ones to the last, the transition probability matrix U can be decomposed in the following manner:

$$U=A+B=\begin{bmatrix} C & D \\ 0 & 0 \end{bmatrix}+\begin{bmatrix} 0 & 0 \\ E & F \end{bmatrix} \tag{11}$$

The elements of matrix A correspond to changes of markings induced by the firing of immediate transitions; in particular, those of sub matrix C are the probabilities of moving from vanishing to vanishing markings, while those of D correspond to transitions from vanishing to tangible markings. Similarly, the elements of matrix B correspond to changes of markings caused by the firing of timed transitions: E accounts for the probabilities of moving from tangible to vanishing markings, while F comprises the probabilities of remaining within tangible markings.

The solution of the system of linear matrix equations.

$$\begin{cases} \psi = \psi\, U \\ \psi\, 1^T = 1 \end{cases} \tag{12}$$

In order to restrict the solution to quantities directly related with the computation of the steady-state probabilities of tangible markings, we must reduce the model by computing the total transition probabilities among tangible markings only, thus identifying a *Reduced Embedded Markov Chain (REMC)*.

The following example illustrates the method of reducing the *EMC* by removing the vanishing Markings.

Example:

Consider the example of *Fig1.* From the initial marking $M_i = P_1$, the system can move to marking $M_j = P_4$ following three different paths. The first corresponds to the firing of transition t_2, that happens with probability $\beta/(\alpha+\beta+\gamma)$, and that leads to the desired (target) marking M_j in one step only. The second corresponds to selecting transition t_1, T_1 with probability $\alpha/(\alpha+\beta+\gamma) * \mu_1/(\mu_1+\mu_2)$, and the third with probability $\gamma/(\alpha+\beta+\gamma) * \mu_3/(\mu_3+\mu_4)$. The total probability of moving from marking M_i to marking M_j is thus in this case:

$$u_{i,j} = \frac{\alpha}{(\alpha+\beta+\gamma)}\frac{\mu_1}{(\mu_1+\mu_2)} + \frac{\beta}{(\alpha+\beta+\gamma)} + \frac{\gamma}{(\alpha+\beta+\gamma)}\frac{\mu_3}{(\mu_3+\mu_4)}$$

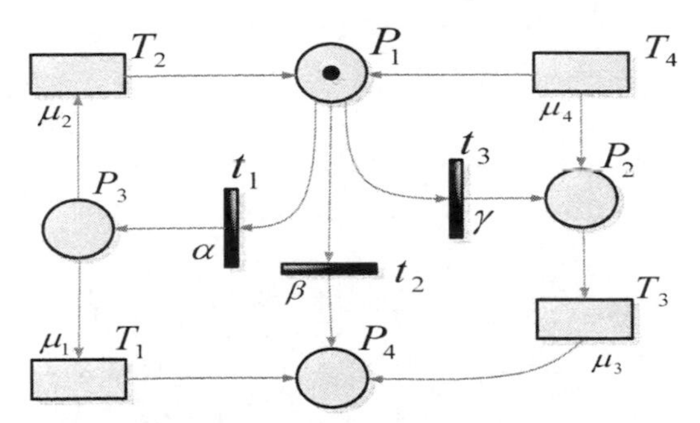

Fig. 1. A *GSPN* system with multiple paths between tangible markings

In general, recalling the structure of the U matrix, a direct move from marking M_i to marking M_j corresponds to a non-zero entry in block F $(f_{ij} \neq 0)$, while a path from M_i to M_j via two intermediate vanishing markings corresponds to the existence of

1. a non-zero entry in block E corresponding to a move from M_i to a generic intermediate marking M_r.
2. a non-zero entry in block C from this generic state M_r to another arbitrary vanishing marking M_s.
3. a corresponding non-zero entry in block D from M_s to M_j.

These informal considerations are precisely captured by the following formula:

$$u_{ij}' = f_{ij} + \sum_{r\,:\,M_r \in V\ RS} e_{ir} P\{r \to s\} d_{sj}, \tag{13}$$

where $P\{r \to s\} d_{sj}$ is the probability that the net moves from vanishing marking M_r to tangible marking M_j in an arbitrary number of steps, following a path through vanishing markings only.

In order to provide a general and efficient method for the computation of the state transition probability matrix U_0 of the *REMC*, we can observe that equation (13) can be rewritten in matrix notation in the following form:

$$U' = F + E\,G\,D, \tag{14}$$

where G can be computed from the formula:

$$G = \sum_{k=0}^{\infty} C^k = [I - C]^{-1} \tag{15}$$

We can thus write:

$$H = \begin{cases} (\sum_{k=0}^{n_0} C^k)\, D \\ [I - C]^{-1}\, D \end{cases} \tag{16}$$

That the first equation represents for times which there is no loops among vanishing states. While the second represents for times which there is loop among vanishing states.

Now we can get some parameters which are important in analyzing the behavior of a net: CY (M_i) represents the mean cycle time of the *GSPN* with respect to state M_i, i.e., the average amount of time that the *GSPN* takes to return to marking M_i. CY (M_i) has the following expression:

$$CY(M_i) = \sum_{k:M_k \in T} u_{ik}\, SJ_k \tag{17}$$

And another parameter which can be analyzed is stationary probability distribution associated with the set of tangible markings is thus readily obtained by means of their average sojourn times eq. 8 using the following formula:

$$\eta_j' = \frac{u_{ij}\, SJ_j}{CY(M_i)} \tag{18}$$

We can compute some other parameters like those computed above for example we can gain a metric for comparing the availability of different architectures but because availability is usually related to the places of the system. Thus, the availability factor is usually associated to the places than transitions. We can gain it by the equation:

$$Avalibility_{Net} = \frac{(\sum_{p \in P} a_p * T_p)}{\sum_{p \in P} SJ_p} \tag{19}$$

Which A_p is the availability associated to the transition t, t_p is the expected time in which there is a token in place P.

This is similar to our previous work using simulation. Another parameter that we can gain a metric is performance efficiency because performance efficiency is usually related to the transitions of the system it is associated to the transitions than places and we can gain it by the equation:

$$Performanc\ e\ Efficiency_{Net} = \frac{(\sum_{t \in T} P_t * f_t)}{\sum_{t \in T} f_t}, \quad (20)$$

where P_t is the performance efficiency that associated to the transition t and f_t is the firing rate of t.

VII. CONCLUSION AND FUTURE WORKS

In this paper, we presented a method to support formal validation of *UML* specifications. The main idea is to derive non-functional parameters from *UML SD*. There are some key activities to achieve this goal: Transforming *SD* to *GSPN*, and gaining the *CTMC* from *GSPN;* which is described in this paper extensively, and analyzing the *CTMC* and obtaining the non-functional parameters.

One area for future research is to obtain other non-functional parameters. Another direction for future research is to mapping other *UML* diagrams to *GSPN* models and obtaining non-functional parameters. In future work, we attempt to create a case tool. We will also evaluate non-functional parameters by its ability.

REFERENCES

[1] Storrle H. " Towards a Petri-net Semantics of Data Flow in UML 2.0 Activities", *IFI-PST, Universit¨at M¨unchen, Oettingenstr., M¨unchen, Germany.*

[2] J. A. Saldhana, S. M. Shatz, and Z. Hu. "Formalization of Object Behavior and Interactions From UML Models". *International Journal of Software Engineering and Knowledge Engineering, 11(6), 2001, pp. 643-673.*

[3] Robert G. Pettit IV and Hassan Gomaa, "Validation of Dynamic Behavior in UML Using Colored Petri Nets," *UML 2000 Dynamic Behavior Workshop*, York, England, October, 2000.

[4] Saldhana, J. and Shatz, S. M. "UML Diagrams to Object Petri Net Models: An Approach for Modeling and Analysis" *Proc. of the Int. Conf. on Software Eng. And Knowledge Eng. (SEKE), Chicago (2000) 103- 10.*

[5] Elkoutbi, M. and Rodulf K. Keller: "Modeling Interactive Systems with Hierarchical Colored Petri Nets" *1998 Advanced Simulation Technologies Conf., Boston, MA (1998) 432- 437.*

[6] Eshuis, R. "Semantics and Verification of UML State Diagrams for Workflow Modelling" *Ph.D. Thesis, University of Twente (2002).*

[7] Fukuzawa, K. et al "Evaluating Software Architecture by Colored Petri Net" Dept. of Computer Sience,Tokyo Institute of Technology Ookayama 2-12-1, Meguro-uk, Tokyo 152-8552, Japan 2002.

[8] Pettit, R. G. and Gomaa, H. "Validation of dynamic behavior in UML using colored Petri nets" *UML'00 (2000).*

[9] Shin, M., Levis, A. and Wagenhals, L."Transformation of UML-Based System Model into CPN Model for Validating System Behavior" *In: Proc. of Compositional Verification of UML Models, Workshop of the UML'03 Conference, California, USA, Oct. 21, 2003.*

[10] Bernardi, S. Donatelli, S. and Merseguer, J. "From UML Sequence Diagrams and Statecharts to Analysable Petri Net Models" *ACM Proc. Int'l Workshop Software and Performance, pp. 35-45, 2002.*

[11] Merseguer, J., Bernardi, S., Campos, J. and Donatelli, S."A Compositional Semantics for UML State Machines Aimed at Performance Evaluation" M. Silva, A. Giua and J. M Colom (eds.*), Proc. of the 6th Int. Workshop on Discrete Event Systems (WODES'02), Zaragoza, Spain (2002) 295-302.*

[12] Merseguer, J. , L´opezGrao, J. P., Campos J."From UML State Diagrams To Stochastic Petri Nets:Application To Software Performance Engineering" *ACM, WOSP 04 January 1416, 2004.*

[13] Motameni, H et al. "Evaluating UML State Diagrams Using Colored Petri Net" *SYNASC'05.*

[14] Motameni, H et al. "Verifying and Evaluating UML State Diagram by Converting to CPN" *SYNASC'05.*

[15] Motameni, H. Movaghar, A. and Zandakhbari, M. "Deriving Performance Parameters from The State Diagram Using GSPN and Markov Chain" *Proc. of the 4th Int. Conf. on Computer Science And Its Application (ICCSA), San Diego ,California (June 2006).*

[16] Object Management Group, http:/www.omg.org. *OMG Unified Modeling Language Specification*, March 2003. version 1.5.

[17] Object Management Group. http://www.omg.org.

[18] M.K. Molloy. *Fundamentals of Performance Modelling*. Macmillan, 1989.

[19] C. U. Smith. "*Performance Engineering of Software Systems".* The Sei Series in Software Engineering. Addison–Wesley, 1990.

[20] Kirsten Berkenkötter, "Using UML 2.0 in Real-Time Development.A Critical Review", *SVERTS, workshop hold in conjunction with UML 2003.*

[21] M. Ajmone Marsan, G. Balbo, and G. Conte, "A class of Generalized Stochastic Petri Nets for the performance evaluation of multiprocessor systems". *ACM Trans. Comp. Syst. 2, 2 (May 1984), 93-122.*

[22] Ajmone Marsan, M., Balbo, G., Conte, G., Donatelli, S., and Franceschinis, G. "Modelling with generalized stochastic Petri nets". *John Wiley & Sons, 1995.*

[23] G. Ciardo, A. Blakemore, P. F. J. Chimento, J. K. Muppala, and K. S Trivedi .Automated generation and analysis of Markov reward models using Stochastic Reward Nets. In Linear Algebra, Markov Chains, and Queueing Models, C. Meyer and R. J. Plemmons, Eds., vol. 48 of IMA Volumes in Mathematics and its Applications. Springer-Verlag, 1993, pp. 145-191.

[24] M. Ajmone Marsan, G. Balbo, and G. Conte, "Performance models of multiprocessor systems". MIT Press, 1986.

[25] G. Ciardo, J. Muppala, and K. Trivedi. SPNP: "Stochastic Petri net package". In Proc. Int. Workshop on Petri Nets and Performance Models, *pages 142-150, Los Alamitos, CA, Dec. 1989. IEEE Computer Society Press.*

[26] M. Ajmone Marsan, G. Balbo, A. Bobbio, G. Chiola, G. Conte, and A. Cumani. "The effect of execution policies on the semantics and analysis of stochastic Petri Nets". *IEEE Transactions on Software Engineering, SE-15:832-846, 1989.*

[27] G. Balbo, G. Chiola, G. Franceschinis, and G. Molinar Roet. "On the efficient construction of the tangible reachability graph of Generalized Stochastic Petri Nets". In Proceedings International Workshop on Petri Nets and Performance Models – PNPM87. *IEEE Computer Society, 1987.*

[28] M. Ajmone Marsan, A. Bobbio, and S. Donatelli "Petri Nets in performance analysis: an introduction", *lecture notes in computer science, 1998 - springer verlag kg*

[29] H. H. Ammar and R. W. Liu. "Analysis of the generalized stochastic Petri nets by state aggregation", In Proceedings *International Workshop on Timed Petri Nets, pages 88-95, Torino (Italy), 1985. IEEE Computer Society Press no. 674.*

[30] "school of mechanical, manufacturing & medical engineering men170: systems modelling and simulation", *online, available at http://www.cs.utexas.edu/~browne/cs380ns2003/Papers/Markov Chains.pdf, Sep 25,2006*

[31] Ajmone Marsan M et al "Modeling with Generalized Stochastic Petri Nets" *Universitiá degli studi di Torino.*

A Distributed Planning & Control Management Information System for Multi-site Organizations

Ioannis T. Christou
Athens Information Technology
19Km Markopoulou Ave. PO Box 68
Paiania 15669 GREECE
ichr@ait.edu.gr

Spyridon Potamianos
Sinastria Ltd.
52 D. Ralli str.
Maroussi 15124 GREECE

Abstract- **This paper describes the design, development and deployment challenges facing an implementation of an enterprise-wide distributed web-based *Planning, Budgeting and Reporting Control Management Information System* for a large public utility organization. The system serves the needs of all departments of the company's General Division of Production. The departments of the division are situated all over the Greek state with many geographically remote plants under the control of the division. To speed-up the exchange of Management Information between the various levels of the hierarchy regarding daily, monthly, or longer-term reports on the operational level, a portal was set-up that enabled all levels of management personnel to have controlled access to constantly updated information about operations, strategic goals and actions. A new planning and budgeting system for controlling operational, investment, and personnel expenses based on the Activity-Based Costing (ABC) & Budgeting model was then integrated into the portal to provide a web-based Planning Budgeting & Reporting Control MIS. The system is capable of handling many thousands of requests per hour for internal reports, graphs, set goals etc. and allows the seamless collaboration and coordination between all departments in the organizational hierarchy.**

I. Introduction

In recent years, there has been a growing tendency to integrate most Management Information Systems (MIS) in various departments of an organization into unified enterprise-wide portals that act as gateways for finding and accessing information from every department independent of its geographic location [7]. Such portals, at a minimum, act as integrators of various data sources and present appropriate views to different users accessing them with single sign-on ability. In more advanced situations the portal integrates applications or services allowing users to perform the business processes that they have been designed to automate or support.
This paper, presents the architectural and design decisions made in response to the challenges facing an implementation of a portal acting as a division-wide MIS, enabling users to manage approved or permanent reports as well as on-line charts of production related indices in real-time; and to plan the operational –referred to as exploitation-, investment, and personnel related expenses for the next fiscal year as well as prepare next year's budgets, from plant level to division level. The tool enables managers to handle the division budgets in accordance with the strategic directions of the company. In addition, the system enforces and automates the business processes of the company as an integrated workflow.

The portal was implemented in J2EE, using the model 2 [10] architectural pattern. Special emphasis is placed on the underlying content management mechanisms. The main advantages of the introduction of the new system from the users view point can be summarized as (I) immediate access to reports providing appropriate level of information for all cost centers no matter of geographic cost-center location, (II) secure access to the system through incorporation of Virtual Private Networks (VPNs) and single sign-on secure authentication mechanisms using Secure Socket Layer infrastructure (SSL), (III) viewing real-time charts of production related indices, and (IV) automation of many processes regarding data entry for the budgets and yearly expense planning, automatic consolidation of lower-level cost-centers' budgets into higher-level cost-centers' budgets, preparation of drill-down or roll-up reports showing costs broken down to different levels of activity granularity or account granularity or both. Prior to the introduction of the new MIS, budget planning was an error-prone and time consuming process, often misaligned with strategic goals of the company. After the introduction of the new system, it is no longer possible to use outdated forms, and the consolidation and report generation processes have been fully automated. Further, functions such as comparing how expensive an activity is between different cost centers are now possible at the click of a button enabling management to do a much better job controlling and planning expenses.

II. System Architectural Design

A. General Architecture

As mentioned before, the MIS has an appearance of a company-wide portal that is implemented as a combination of J2EE applications. From a functionality perspective, there are six major thematic categories –or areas- for the portal:

- Reports Management
- News and Announcements
- On-line Production-related Key Performance Indicator (KPI) Charts
- Planning
- Budgeting
- User Management

The overall MIS system architecture is shown in the following diagram.

K. Elleithy (ed.), Advances and Innovations in Systems, Computing Sciences and Software Engineering, 191–195.

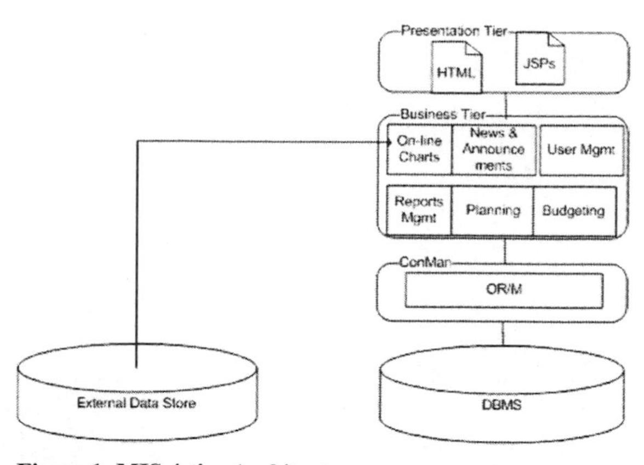

Figure 1: MIS 4-tier Architecture

The bottom DBMS tier is based on the standard relational model whereas the top tier, presentation, is built as a thin browser client, via standard JSP technology. The most challenging components are the two intermediate layers, the Content Management layer (ConMan), and the business logic layer, which are going to be the focus of this paper.

B. Content Management Layer

In order to handle data input and output from the database, we developed ConMan, a Content Management back-end that handles structured or semi-structured content easily and efficiently using as underlying mechanism standard relational databases. Data type definitions and content can be imported to the system -"ingested"- via XML files, or via a system-native Java API. Content retrieval (search) is performed via a Query-By-Example (QBE) API. The system allows for great flexibility in the data dependencies through a reference mechanism via which persistent objects may have references to other objects of other data types. The system also supports single inheritance of data types as well as array objects.

ConMan provides facilities to define *hierarchies* of structured data. "Hierarchies" means that content resides in the leafs of a conceptual tree called the *content-tree*, the paths of which correspond to an effective name for that content. Categorical similarities between different contents are represented by the grouping the tree representation corresponds to. "Structured data", means that content is defined as the aggregation of more elementary content, this regression stopping at trivial, non-decomposable data types, like integer, text or MPEG movie. The abstract form of these structures, which serve as rubber-stamps for actual content, is named "Asset". The elementary data types supported in ConMan that do not decompose any further, include the following:

- Boolean
- Numerical: Integer, Double
- Times: Time, Date, Timestamp
- Textual:Text, URL
- External File: RichText, Image, Audio, Video, VRML, XML, HTML
- REF TO another Asset.

All other assets in the system are composite, user-defined assets that are aggregations of named attributes of type any of the above data types and/or any other composite asset. The type of such assets is "COMPOSITE".

The system allows for array attributes of maximum or indefinite size of objects of type some asset. Finally, constraints of various types have been implemented and may be imposed on instances of any asset. These constraints include MIN/MAX constraints as well as ONE-OF type constraints allowing an attribute to take on a value from a restricted set of possibilities.

As a quick example of an Asset definition, the following asset defines a Document:

```
Asset Document {
  did: INT;
  url: TEXT[MAX=255];
  type: REF TO DocType;
  name: REF TO DocName;
  dept: REF TO DeptNode;
  year: INT;
  month: INT[MIN=1, MAX=12];
  author: TEXT[MAX=255];
  ...
};
```

Figure 2: Asset Definition in ConMan

Each line in the definition of the Document asset defines a named attribute for that composite asset. For example, the attribute "`did`" is of type integer, while the attribute "`type`" is a reference to a content instance of type `DocType`.

C. Reports Management

The general division of production of the company is composed of a number of departments, each of which usually has several cost centers underneath it, and so on, forming a hierarchical tree structure. Each cost center produces every month a number of reports showing various aspects of its operation. In particular, it produces a number of *technical reports* with production levels related data, a number of *reports showing financial data* regarding its operation, and a number of *techno-economic reports.*

Each report therefore belongs to one of the above-mentioned general categories, but is also characterized by a number of other attributes comprising its meta-data description. Of these attributes, only five, namely *department, month, year, type and report title* are searchable and allow the use to navigate through the digital library. Each user of the system "belongs" to a particular node in the cost centers hierarchy, and can view or modify –depending on their role privileges- only reports related to the cost centers in which they belong and all cost centers beneath them.

Reports are produced by authorized personnel using standard office automation tools, such as Microsoft Excel spreadsheets or PowerPoint presentations, which they upload to the system once they have been approved by the manager responsible for the cost center. However, reports are also automatically produced in PDF format by another tool that has been installed in the division's servers and is capable of extracting all necessary data from the division's databases (once they are approved) and automatically compute reports showing all appropriate production related technical, financial and techno-economic indices. This tool –which is external to the system we describe- is guided by a human user with appropriate privileges. Each PDF report produced is then automatically uploaded to the system's databases via yet another connector tool that performs batch updates/uploads of documents to the web-based MIS. This process minimizes human errors –since all indices are automatically computed and formatted in a report- while at the same time maximizing throughput; it currently uploads reports with an average of more than 1000 reports per minute.

It is also possible to browse through and retrieve news and announcements regarding the division. The news and announcements are managed by users having administrator privileges for the site. The announcements are usually related to the operation of the MIS itself, for example announcing when the system will be open to budget submissions from the various cost-centers. Part of the underlying data model in terms of the ConMan persistence API is shown in the following figure, which doesn't show the *Document* Asset.

```
Asset DocType {
  did: INT;
  name: TEXT;
  belongs: REF TO DocArea;
};
```

```
Asset DocKind {
  did: INT;
  name: TEXT;
};
```

```
Asset DeptNode {
  id: INT;
  name: TEXT;
  parent: REF TO DeptNode;
  ...
};
```

```
Asset DocArea {
  id: INT;
  name: TEXT;
  ...
};
```

Figure 3: ConMan Definitions of Reports Related Assets

The component's static design is shown in the next figure. There are a number of coordinator classes that ensure proper synchronization when different users concurrently view and modify the digital library. This coordination allows for the maximum possible degree of concurrency in the system and has enabled us to maximize throughput and response times, a major challenge in the project.

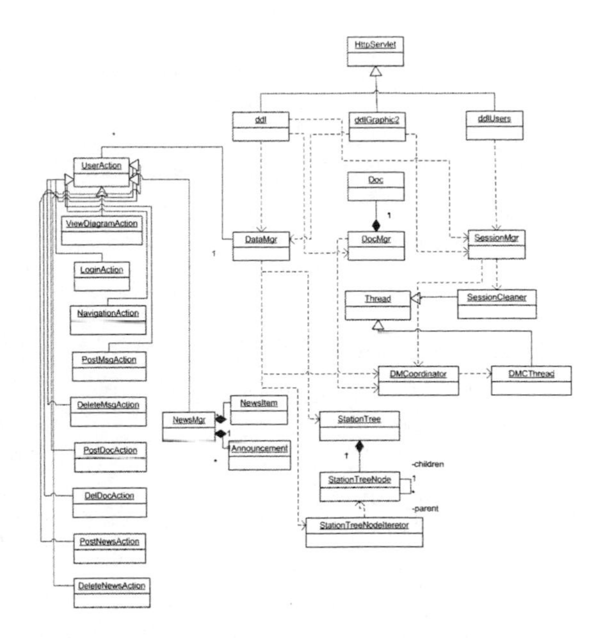

Figure 4: Reports Management Class Design

The user interacts with the system via a set of dynamically generated DHTML pages containing all the appropriate forms and buttons. The User Interface allowing the users to navigate through the digital library of reports contained in the system is shown in Figure 5.

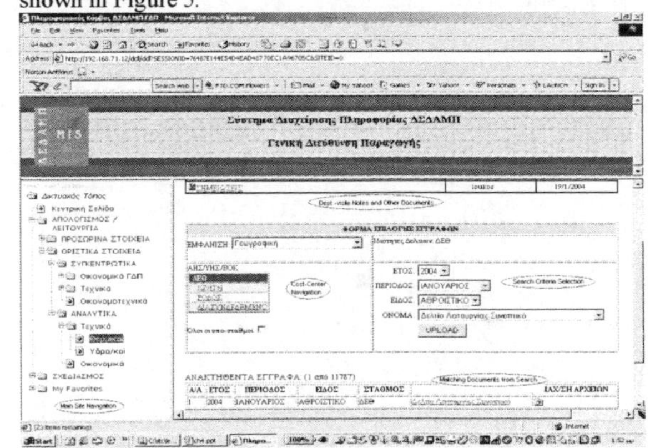

Figure 5: Digital Library User Interface

There exists yet another real-time data collection system that continuously collects and stores various performance index data regarding the operations of each cost center in the division. These indices can be shown in bar charts alone or in groups. In the case of groups, various indices are grouped together in meaningful groups and shown in comparative bar charts, where the horizontal axis shows the days from the beginning of the month to the current day, and the vertical axis shows performance level. In another dimensionality slice, the user can select a specific day of the month, and the horizontal axis then displays each of the cost centers directly beneath the currently selected cost-center, while the vertical axis still shows performance levels, allowing management to compare the performance of various cost centers for the same day.

The rendering of the charts was done with the help of the open-source library JOpenChart, a source-forge project [8] written entirely in Java. Certain instances of these charts are shown in Figure 6.

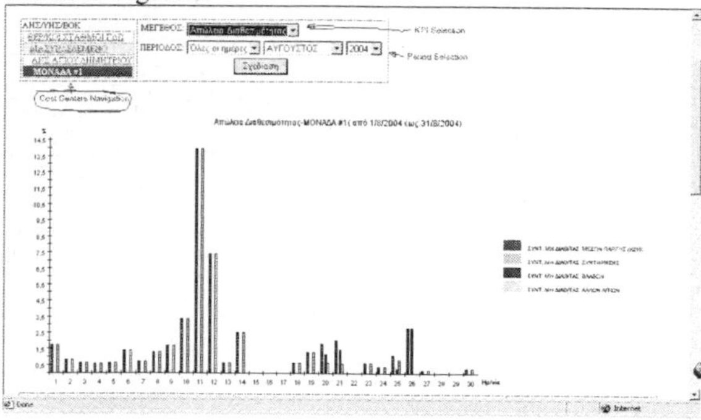

Figure 6: Live KPI Charts

D. Planning & Budgeting

Among the (five) basic responsibilities of management are the control of expenses, and planning. Both of these activities are aided by our planning and budgeting portal. The division adopted the *Activity-Based Costing and Budgeting model* [6] for monitoring, controlling and planning their expenses. In this model, costs are broken down by activity (in the vertical axis), and by account (horizontal axis). Both the activities and the accounts form hierarchical structures and they have slight variations among different cost-centers. Different cost centers have different allowed activities and accounts to which they may charge expenses. The expenses are divided among three large categories:

- Exploitation (Operational) Expenses
- Investment Expenses
- Personnel Expenses

The Exploitation Expenses category allows managers to track, monitor, and plan at a yearly or even monthly level the operational day-to-day expenses of the company that are related to production. The Investment Expenses category on the other hand allows managers to monitor, track and plan the long-term projects (investments) that the company has undertaken in order to improve production or lower costs. As such, this category implements a "project tracking and management" function that allows managers to keep track of all financial elements of all projects approved for any of the company's cost-centers. From the time of inception of a project, to project completion, the system keeps track of the project status plus all finance-related figures of the project and produces upon request detailed reports of these figures. Finally, the Personnel Expenses category keeps track of all expenses related to personnel –including of course payrolls, paid overtime, etc.

The Department of "Production Output Management & Planning" determines the cycles when the planning and budgeting processes open and close. At the end of each such cycle, the *planning and/or budgeting data are characterized with a new version number*. This versioning functionality allows all parties involved to have a better log of their planning decisions and results of negotiations.

A typical data-entry screen (using minimal JavaScript code for entering comments to justify a particular planning expense) is shown in Figure 7. Notice that the aesthetics of a spreadsheet are maintained in this screen due to the heavy familiarity of users with Excel usage. As mentioned before, planning used to be the process of filling out various fields in properly designed Excel spreadsheets. To minimize the learning curve of the users and their chances of making mistakes, plus to maintain a familiar look-and-feel in the way they work, we opted for spreadsheet-like screen forms for these tasks. The results were that all users –even though not very familiar with web technologies or computers in general- felt at home with the new system as soon as it was introduced, forming no resistance to the changes that the new system introduced.

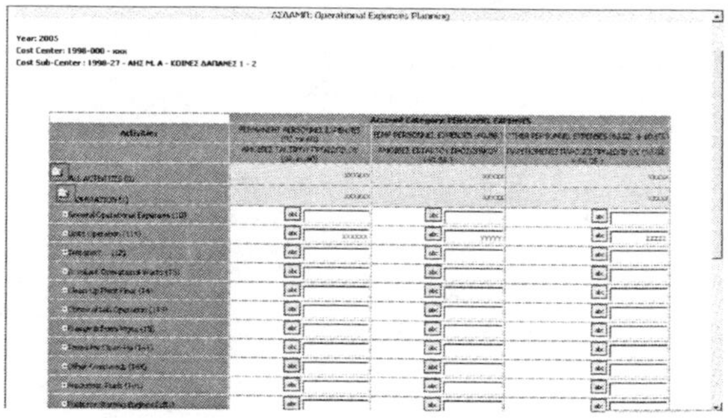

Figure 7: Operational Expenses Data Entry Form

Similarly, there are a number of OLAP style operations that can be performed in real-time by the system to produce various roll-up or drill-down reports on the various categories of expenses so that management can have a complete picture of how the budget is running at any moment.

E. User Management

The system is a totally integrated portal, and so it *offers a true single sign-on functionality*. Each user belongs to a node in the cost centers hierarchy. Also, each user may have a number of roles in different parts of the system, and each role is associated with different access levels, such as no-permissions, reader, writer, admin, etc. Unlike other approaches, role privileges are associated with a whole thematic area, so for example a user may be a reader for the techno-economics area of the reports systems, but cannot be a reader for the area's cost-center and a writer for the area's children cost-centers, which are also known as cost-center positions. We followed this specific approach to privileges because the client organization set it as a requirement which allowed their system administrators to more easily manage their many users. Typically there are more than 150 simultaneous active users of the system. The management of roles, access levels, user privileges etc. collectively called User Management, is done via a set of dynamically generated web-pages. Non-admin users typically are only allowed to change their login name and/or their password.

III. PROJECT MANAGEMENT LESSONS

Because the client had a number of stakeholders that did not have the time to articulate clearly all project requirements up-front, we decided to follow an iterative development approach, and prototype even the requirements [1]. As it turned out, an evolutionary approach worked best, because in all areas someone involved had a reasonable understanding of the requirements. After decomposing the project in the thematic areas mentioned in this paper, and settling upon a multi-tiered web architecture, each component was separately developed –on top of the ConMan OR/M infrastructure- with relative ease. ConMan itself was built on ideas from popular persistence schemes [2-3], [5], [9]. Essentially, only the user-interfaces went through more than 2 cycles of modification and enhancements (i.e. the presentation layer). Certain parts of the business logic –related to month-by-month budgeting estimations- went through two iterations to improve the efficiency with which aggregate and composite reports are produced.

Project size is not easy to measure in any project, as many different factors come into play. However, measured in terms of screen forms, the project has more than 25 different forms, and approximately 10 summary-level use cases that were rigorously documented in the inception and elaboration phases of the project. With the help of the ConMan architecture, a very small team with significant experience in J2EE development, delivered the project within time and budget constraints in less than 6 months.

IV. CONCLUSIONS

We described the design and implementation of a division-wide portal for a large organization with many geographically separated plants and offices that allows the production division's managers to perform two major functions: *controlling and planning*. Plans have a yearly horizon, even though this is a parameter that can easily change should the company decide to use the system to help in long-term strategy formulation and implementation [4]. The main thematic areas of the portal are reporting control, planning, and budgeting. Of course, the system also provides the ability to compare a specific year's plans with the outcome of that year and compute various indices such as how much the plan deviated from the actual implementation, or to what extent targets were met etc.

Recently, we complemented the system with yet another component, namely Inventory Expenses Planning & Budgeting. Using the same philosophy as before, each cost center maintains a record of their supplies –mostly spare parts- and plans in financial terms for their future needs for each recorded item in the inventory. Upon finalizing these figures, specialized spreadsheet-style reports can be produced on-the-fly regarding the use of any category of items etc.

Among the major advantages of the introduction of the system, we would name the ability of all managers, independently of their location, to have instant access to the information they need, to monitor production indices and compare results in dynamically generated real-time charts, and to plan and track their expenses in a consistent and coherent way throughout the division's many departments.

REFERENCES

[1] D. Alur, J. Crupi, and D. Malks, *Core J2EE Patterns: Best Practices and Design Strategies*. Prentice-Hall, 2nd ed., Upper Saddle River, NJ, 2003.
[2] C. Bauer, and G. King, *Hibernate in Action*. Manning, Greenwich, CT, 2004.
[3] R. Bourret, Java Packages for Transferring Data between XML Documents and Relational Databases, 2005 http://www.rpbourret.com/xmldbms/readme.htm.
[4] J. Johnson, K. Scholes, and R. Whittington *Exploring Corporate Strategy: Text and Cases*. Prentice-Hall, 7th ed. Upper-Saddle River, NJ, 2005.
[5] R. Johnson, J2EE Development Frameworks. *Computer, 38(1)*, 107-110, 2005
[6] R. Kaplan, and W. Bruns, *Accounting and Management: A Field Study Perspective*, Harvard Business School Press, Boston, MA, 1987.
[7] K. C. Laudon, and J.P. Laudon *Management Information Systems: Managing the Digital Firm*. Prentice-Hall 9th ed., Upper Saddle River, NJ, 2005.
[8] S. Mueller, JOpenChart Library Tutorial, 2003. http://jopenchart.sourceforge.net/tutorial.html
[9] S. Payette, and T. Staples The Mellon Fedora Project: Digital Library Architecture Meets XML and Web Services. *6th European Conference on Research and Advanced Technology for Digital Libraries. LNCS vol. 2459*. pp 406-421, Springer-Verlag, Berlin-Heidelberg-New York, 2002.
[10] I. Singh, B. Stearns, M. Johnson and the Enterprise Team, *Designing Enterprise Applications with the J2EE Platform*. Addison-Wesley Professional, 2nd ed., 2002.

Supporting Impact Analysis by Program Dependence Graph Based Forward Slicing

Jaakko Korpi[1], Jussi Koskinen[2]
[1] Department of Mathematical Information Technology, University of Jyväskylä,
P.O. Box 35, 40014 Jyväskylä, Finland, jaakko.korpi@cc.jyu.fi
[2] Department of Computer Science and Information Systems, University of Jyväskylä,
P.O. Box 35, 40014 Jyväskylä, Finland, koskinen@cs.jyu.fi

***Abstract*-Since software must evolve to meet the typically changing requirements, source code modifications can not be avoided. Impact analysis is one of the central and relatively demanding tasks of software maintenance. It is constantly needed while aiming at ensuring the correctness of the made modifications. Due to its importance and challenging nature automated support techniques are required. Theoretically, forward slicing is a very suitable technique for that purpose. Therefore, we have implemented a program dependence graph (PDG) based tool, called GRACE, for it. For example, due to the typical rewritings of Visual Basic programs there is a great need to support their impact analysis. However, there were neither earlier scientific studies on slicing Visual Basic nor reported slicers for it. In case of forward slicing there is a need to perform efficient static slicing revealing all the potential effects of considered source code modifications. Use of PDGs helps in achieving this goal. Therefore, this paper focuses on describing automated PDG-based forward slicing for impact analysis support of Visual Basic programs. GRACE contains a parser, a PDG-generator and all other necessary components to support forward slicing. Our experiences on the application of the PDG-based forward slicing has confirmed the feasibility of the approach in this context. GRACE is also compared to other forward slicing tools.**

I. Introduction

Program slicing [1] is useful in many of the central software maintenance task types, including program comprehension, debugging, testing, and impact analysis. We have earlier developed an approach and implemented a slicer called GRACE (GRAph-based sliCEr) providing backward slicing capabilities for supporting debugging and similar tasks [2]. Another variant of slicing is forward slicing [3], which is useful especially in impact analysis [4]. The research problem and contribution of this paper relates to finding out whether and how impact analysis of Visual Basic (VB) programs could be conveniently supported by forward slicing.

Theoretical research on program slicing is extensive but only few forward slicers exist. Moreover, none of them supports VB, even though that there is a great need to support impact analysis of VB programs. Therefore, in this paper, we focus on describing our experimentations on the development of a forward slicer for VB and consequently apply constructive research method. This paper describes needs and requirements of forward slicing, outlines the tool architecture and implementation, and represents a use case, evaluation of the implemented forward slicing and comparison to related works. Therefore, the whole cycle of constructive research approach from setting of requirements to empirical testing has been traversed. The selected technical approach is *program dependence graph* (PDG) *based static forward slicing*.

The paper is organized as follows. Section II characterizes the essential aspects of source code changes and impact analysis. Section III briefly characterizes the supported programming language. Section IV relates the work to the most important theoretical background of forward slicing. Section V presents the implemented forward slicing approach. Section VI presents a use case and evaluation of the implemented forward slicing features. Section VII compares GRACE and other tools with forward slicing capabilities. Finally, Section VIII presents the conclusions.

II. Source Code Changes and Impact Analysis

According to Lehman's first law [5] software systems used in real-world environment must be continually adapted or they become progressively less satisfactory. This is due to the typical changes of user requirements and characteristics of the technical environment. Therefore, the system evolution necessitates *source code changes*. Systems which are implemented with old technology and typically poorly documented are called *legacy systems*. It is often desirable to modernize, rewrite, migrate or otherwise radically update these systems which will cause massive impact analysis tasks.

Lehman's second law [5] states that as the system evolves its complexity increases unless work is done to maintain or reduce it. Since, *complexity* makes maintenance harder and causes costs [6] its increase should be avoided when possible. Moreover, according to Lehman's seventh law [5] the quality of the systems will appear to be declining unless they are rigorously maintained and adapted to operational environment changes. Therefore, there especially is a need to perform the changes on source code in a *disciplined* fashion such that introduction of unwanted side-effects, increases in the system complexity, and deterioration of technical system quality can be minimized.

K. Elleithy (ed.), Advances and Innovations in Systems, Computing Sciences and Software Engineering, 197–202.

Impact analysis means identifying and assessing the effects of making a set of changes to a software system [4]. It should focus on cost-effective minimization of unwanted *side-effects* (ripple-effects) of program changes. In case of non-trivial programs a support tool is a necessity [7]. *Reverse engineering tools* [8] support impact analysis. Moreover, due to the extent and complexity of large scale impact analysis tasks, the applied tools should provide proper and sufficiently detailed ways to satisfy the situation-dependent *information needs* of software maintainers [8].

We feel that potentially important effects of change to the system need to be both *identified* and *understood*. Gaining the understanding requires a systematic *browsing strategy* and versatile *program dependency information* as represented on an appropriate *abstraction level* according to the generally acknowledged virtues of reverse engineering. Also, it would be beneficial if a tool would support identification and understanding of *delocalized program plans* [9]. Delocalization refers here to the dispersed logic in terms of program lines implementing a certain program plan.

A process model for impact analysis is presented in [4]. It includes 1) identification of the change set, 2) identification of the candidate impact objects, 3) their assessment, and 4) determination of whether and how each candidate object should be changed. Phase 2 is best supported by forward slicing. At a more detailed level, the most important information needs in maintenance work has been identified in [8] based on the analysis of earlier empirical studies. The most important ones which directly relate to impact analysis include: 'ripple effect', 'location and uses of identifiers', and 'variable definitions'. These are clearly such that they can be supported based on forward slicing.

III. The Supported Programming Language

We have currently focused on supporting forward slicing of Visual Basic programs. VB's syntax is relatively simple, and therefore, the language is not too ambitious in terms of its complexity as a target of implementing a forward slicer. Most noteworthily, it is also used to build large commercial applications. Visual Basic 6 aims at supporting the syntax of the older versions. The language currently supports structuring based on program blocks, building of graphical user interfaces, event-driven computing, and creation of classes and interfaces. Therefore, it is a procedural language with some object-oriented (OO) features. There is especially a need to support comprehension and impact analysis related to the older VB programs to rewrite them using Visual Basic.NET, which is its OO-variant. There is also a tendency to rewrite older VB programs in other programming languages. VB projects, as specified in Microsoft's standard form, can be compiled into executable program or into DLL. Preprocessing is not needed since the VB-project directly indicates the related source code modules.

IV. Forward Slicing

Program slicing was introduced by Mark Weiser [1]. It is a technique for determining relevant parts of computer programs based on program dependencies. Slicing can be used to assist program comprehension efforts, debugging, impact analysis and other central software engineering activities. In practice, slicing needs to be efficiently automated. Slicing is initiated by a tool user by defining a *slicing criterion*, which typically is an interesting variable occurrence on a certain program line. *Forward slicing* [3] reveals the program parts which are affected due to potential program changes. *Static forward slicing* captures all the potentially affected program parts due to a considered change to the value of the slicing criterion.

There are practical problems in applying especially static forward slicing. The traditional approach of applying iterative solving of data-flow equations is relatively inefficient, as noted e.g. in [10]. The traditional approach of static slicing also requires storing of great amounts of program information. Most of the current implementations also have limitations regarding completeness of analysis in terms of interprocedural analysis, set of supported programming languages, and extensibility. The necessary information can be represented also in graphs, as described e.g. in [11,12]. Among other contributions, PDGs alleviate the efficiency problems of program slicing. The PDG-approach has e.g. been applied in a sophisticated slicing tool called CodeSurfer as described e.g. in [13]. The current slicers, including the PDG-based one, unfortunately mainly include support only for C [10,13,14]. Moreover, there are only few forward slicers. Therefore, there is a need to extend the support to other languages, including VB and to experiment on forward slicing.

V. The Implemented Forward Slicing Approach

This section describes our approach and its implementation for forward slicing.

A. Requirements

The requirements were set especially such that meeting them would enable academic use and feasibility evaluation of the implementation of the forward slicing capabilities also with relatively large applications as input.

1) Necessary backend features for fully analyzing VB: a) Handling of VB project files (including multiple DLLs), b) conventional parsing capabilities, and c) sufficiently efficient PDG-generation.
2) Slicing features: Efficient PDG-based formation of static forward slices.
3) User interface: a) Conventional source code browsing and selection capabilities, b) representation of the forward slice contents to the user as embedded in the browsed program text to enable direct viewing of the context of the slice.
4) Storage capabilities: Storing the produced slice into mass memory to reduce needs for multiple analyses.
5) Extensibility to cover new features and languages based on a clear and modular system architecture.

B. System Architecture

GRACE consists of 118 modules and 12874 LOC. It has been implemented under Linux by using C++ and GCC [15], and runs under Linux. It is aimed at supporting program comprehension and impact analysis. The main components are: 1) parser, 2) PDG-generator, 3) slicer, and 4) user interface. The parser takes as an input VB project files and corresponding source code modules. Handling of multiple separately compiled components (DLLs) is supported. The architecture is modular. Language-independent and dependent classes are separated. GRACE supports dealing with information needed for slicing both procedural and OO-languages, including inheritance relations of types, classes and interfaces.

C. Parser

Both the traditional and PDG-based slicing require as input the information which has traditionally been stored into abstract syntax trees (ASTs) and symbol tables. Here, the parser component produces these structures. The parser component utilizes an *LL*-grammar of VB. Left-recursive descent backtracking parser has been implemented in C++.

D. PDG-Generator

PDG is a directed graph comprising of vertices and edges connecting them. PDG-generation and its traversal follows the data-structures and algorithms as described in [3,11]. The vertices represent program parts and edges program dependencies. Slicing requires control and data dependence information. GRACE holds the information necessary in sets refering to the nodes of the AST. For structured programs the formation of the PDG is relatively straightforward.

1) *Control Flow Graph* (CFG) is created. Creation of the CFG can be achieved based on AST-traversal. CFG serves as the skeleton of the PDG.

2) *Control dependences* are added to the PDG based on the status of traversing control flow paths.

3) *Data dependences* are determined and updated to the PDG based on AST-traversal and by keeping a record of the variables relevant on each statement. Basically, such variables which are used but for which an assigment has not yet been found are relevant in this sense.

4) The formed intraprocedural PDGs are connected to form a *System Dependence Graph* (SDG), which is needed in interprocedural slicing. SDG is formed from the individual PDGs by adding call-edges, in-edges, and out-edges. Also events need to be taken into account since they can be used in VB. They typically activate indeterministically based on user action. Therefore, if an event handler `T1` sets a global variable `x` and an event handler `T2` reads `x`, then data dependence edge is added to the SDG from the out-vertice corresponding to `T1` to the in-edge corresponding to `T2`.

E. Slicer

Slicing is based on straightforward reachability analysis of the formed PDG/SDG [3]. We apply conservative worst-case analysis. Most of the necessary preparatory work for forward slicing has already been done in the previous phases. In this phase PDG is first traversed and nodes reached by following valid dependencies are marked as relevant for the slice to be formed. Then, the relevant PDG-nodes are ordered according to the corresponding source line numbers for later representation to the tool users, based on the included reference to the corresponding AST-node. Proper slicing of event handling requires that while a variable is relevant to forming the slice, then the corresponding event-handler and affected program parts should also be included into the slice. GRACE takes into account also effects of global variables. The formed slice consists of a list of nodes. Each of them includes a reference to the corresponding PDG-node, and start and end position information refering to the source code.

F. User Interface

The graphical user interface has been implemented by using wxWidgets-library [16]. The interface is illustrated in the next section. It enables basic interaction mechanisms, and viewing of the program code, formed slices, and PDGs.

VI. Use Case and Evaluation

We tested GRACE with two programs: MyWS and TanGo [17]. Efficiency and memory consumption were evaluated in an environment with a modern PC with a 64-bit processor.

MyWS is an open-source web-server containing 1495 LOC. It was first successfully parsed and 7930 AST nodes, and 1067 PDG nodes were generated. AST-size was 738 KB and PDG-size 239 KB. Parsing took 0.79 seconds and PDG-generation 0.03 seconds.

TanGo is a larger program containing 18700 LOC, 40 modules, and 456 subroutines. It enables playing Go, which is a board game. TanGo is freely available under GNU GPL license. TanGo was first successfully parsed and 175859 AST nodes and 27046 PDG nodes generated. AST-size was 16367 KB and PDG-size 6646 KB. Parsing took 4.73 seconds and PDG-generation 0.36 seconds. Forward slicing was tested and assessed in case of a representative example. The example slice is here related to the situation in which a maintainer has noticed that handling of warnings as generated by TanGo should possibly be organized better.

While considering whether such changes are possible and desirable based on the possibilities to modify the source code safely with moderate effort, impacts of necessary changes need to be carefully assessed. The process can be started by gaining the general understanding of the relevant parts of the program by browsing its source code for example via GRACE.

Fig. 1 shows the user interface of GRACE. The right-hand-side of the figure shows an editor window, which can be used in browsing the source code. By browsing the source code the maintainer becomes aware that warnings are handled in TanGo in module `frmBoard` which is a user interface component. Part of its subroutine `Timer_Timer` can be seen in Fig. 1. The figure reveals that warnings are organized such that string variables `Warning1` and `Warning2` contain formulas corresponding to the warnings issued during the go games. Therefore, the next task is to check `Warning1`. Its value is set in the subroutine `Timer_Timer`. Browsing the source code reveals that this subroutine is an event handler for a variable `Timer`.

Next task is to find out what program parts would be affected if the value of `Warning1` would be set differently. Therefore, the maintainer has initiated generation of a forward slice by GRACE with `Warning1` as the slicing criterion. Fig. 1 shows the situation. The red line (dark in grayscale) contains the slicing criterion. Generation of the slice by GRACE took less than 0.02 seconds. The slice contains 471 nodes within 3 modules and 8 subroutines. The produced slice is compact (2.5% of the input program) and therefore helps well program comprehension and impact analysis by eliminating the need to study irrelevant lines in detail. This saves maintainer's work time. Automated slicing also guarantees that all relevant lines can be systematically checked out.

The slice contents are shown in GRACE by framing program lines which contain nodes of the slice as shown in Fig. 1. The example screen area contains 5 of the nodes. The contents of the figure reveal that `Warning1` is used by a method call `eval.evaluate(CStr(Warning1))`. It also reveals that `Warning2` is handled similarly. Variables `warn1w`, `warn2w`, `warn1b`, and `warn2b` use the results. Thus, the slice reveals the underlying delocalized program plan which has been applied to achieve handling of warnings.

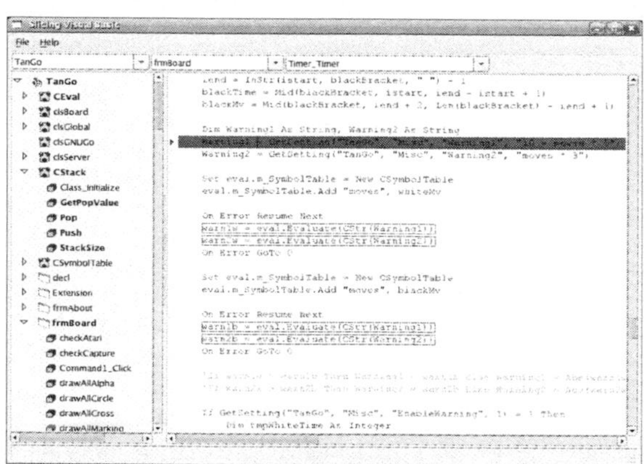

Fig. 1. Interprocedural slice.

Navigation and visualization support mechanisms are also needed and are provided by GRACE. The left-hand-side of Fig. 1 shows a hierarchic view of the names of the projects, classes/modules, and subroutines. Those which contain nodes of the formed slice are written in boldface. By clicking the rows within the view the corresponding source code lines are shown in the editor window and can be browsed further.

Output of the slice contents graphically is supported via Graphviz [18] which is an open source tool producing graph visualizations in AT&T's dot format. It can be started directly from GRACE in an integrated fashion. Fig. 2 shows the automatically formed graph for the example slice. Rectangles are classes/modules, ellipses are subroutines, and arrows are data dependences. The produced slice reveals that large-scale effects of `Warning1` include selection of control flow paths and values of variables in classes `CEval` and `CStack`. Therefore, these classes need to be understood sufficiently well before making final source code modifications, and tested after the modifications. `Warning2` and other relevant variables could be investigated similarly. This way, all the possible direct and indirect effects of considered changes can be investigated systematically and better than in case of elaborate and risky manual analyses.

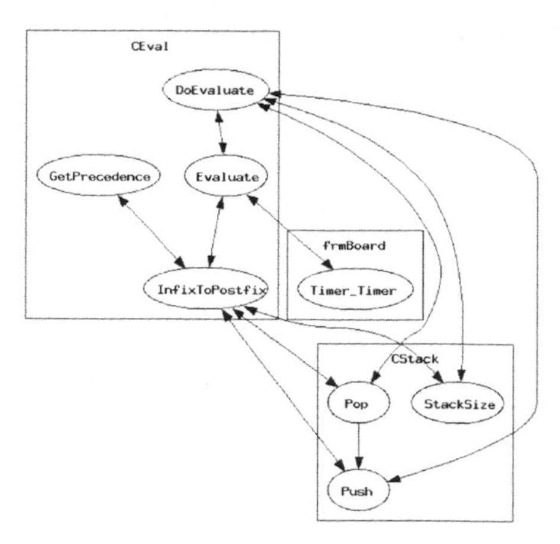

Fig. 2. Dependence graph of the formed slice.

GRACE's parser component supports analysis of VB completely. The slicer component supports complete interprocedural forward slicing. GRACE also supports analysis and slicing of systems which consist of multiple interdependent projects and multiple DLLs. The applied *LL*-parser based analysis is slightly less efficient than the traditional tools such as Bison and Yacc. Therefore, efficiency could be enhanced by optimizing the parser component. However, since there were no special parsing problems and since we do not aim at focusing on incremental analysis, the efficiency of the parser component is less relevant than the efficiency of the slicer component for further development.

Strength of the PDG-based approach is that formation of the slice is mainly reduced to the problem of solving the graph reachability. Formation of the PDG needed in forward slicing was rather straightforward. Formation of the forward slices based on the PDG-information has also been uncomplicated. Traditionally complete interprocedural static slicing has had problems with efficiency as noted in [10]. Worst case forward slices contain the whole program. In GRACE they can be generated for TanGo in less than 6 seconds. Therefore, the speed is sufficient for fluent maintenance work at least in case of programs whose size is less than 20000 LOC.

The system architecture is clear and has been designed taking into account potential needs to extend the support to cover new programming languages by separating language-dependent and -independent data structures and program components. Grammar files of different languages can also be handled in an uniform fashion. McCabe's cyclomatic complexity of GRACE as calculated by CCCC (C and C++ Code Counter) is 21.1/module.

GRACE has fulfilled all the set requirements. Although it is still a prototype its forward slicing capabilities have only minor technical limitations; namely GOTO statements are not yet supported due to their limited significance. This limitation has not affected the conducted feasibility evaluations.

VII. Related Works and Discussion

There are many reverse engineering tools and approaches as compared in [8]. Tools having forward slicing capabilities include those reported in [10,13,14,19]. Main tool features are represented in Table I. Tool aspects which share similarities with GRACE are marked in boldface type. Sufficient efficiency of static forward slicing requires use of PDGs. Currently, only CodeSurfer [13] and GRACE apply PDGs. All these slicers, except CodeSurfer are non-commercial prototypes. The tool reported in Tonella's article [19], and GRACE have been implemented based on open source software. CONSIT [14] has been tested only with small programs. Meaningful support of impact analysis in case of large programs requires taking interprocedural dependencies into account. "Tonella's tool" and CONSIT currently support only intraprocedural analysis.

Table I. Comparison of Forward Slicing Tools

Tool Slicing aspect	SLICE [10]	CodeSurfer [13]	"Tonella's tool" [19]	CONSIT [14]	GRACE
Characterization	Efficient dynamic slicing	**PDG-based slicing**	Decomposition slicing	Conditioned slicing	**PDG-based slicing**
Maturity	**Proto**	Commercial	**Proto**	**Proto**	**Proto**
Analysis B= Backward, F= Forward	**B + F**	**B + F**	**B + F**	**B + F**	**B + F**
Analysis type	Dynamic	**Static** + Dynamic	**Static**	**Static**	**Static**
Technique	Statement-level source code instru-mentation	**PDG-traversal**	Decomposition slicing	Symbolic execution, automated theorem prover	**PDG-traversal**
Implementation tools	Not specified	Not specified	C	Java, Prolog	**C++, GCC**
Slicing criterion level	**Variable**	**Variable**	**Variable**	**Variable**	**Variable**
Analysis scope	**Interproc.**	**Interproc.**	Intraproc.	Intraproc.	**Interproc.**
Supported languages	C	C (C++, alpha vers.)	ANSI C	C subset	**Visual Basic**
Views T= Textual, G= Graphical	T, **Embedded**	**T + G,** **Embedded**	**T + G,** Extracted	T, Extracted	**T + G,** **Embedded**
Linkage	-	-	Variable	-	-
Application focus area(s)	**Deb. & Impact anal.**	**Deb. & Impact anal.**	**Impact analysis**	**Deb. & Impact anal.**	**Deb. & Impact anal.**
Tested input programs	<= 7500 LOC	<= 300000 LOC	<= 2000 LOC	200-500 LOC	**<= 18700 LOC**

All these tools, except GRACE focus on C. VB is supported only by GRACE. Therefore, GRACE fills that research gap on studying application of slicing. Supporting OO-languages is obviously also important due to their increasing practical importance, also in terms of their maintenance. Currently, OO-support is, however, provided only by very few slicers. The best tool in this sense is CodeSurfer which covers C++ at alpha version level.

Special techniques have been proposed concerning specification of the slicing criterion e.g. [14], restriction of the expansion of the slicing analysis, e.g. based on chopping [20] and slice content abstraction as aimed at e.g. by Tonella as presented in [19]. That approach is for forming decomposition slices based on partitioning a program into computations performed on different variables and showing the dependence relations between the computations using concept analysis and by forming so-called concept lattice. Relevant identifiers are grouped enhancing result abstraction level.

Graphical output of the slicing results is important and provided by CodeSurfer, "Tonella's tool", and GRACE. Also the more general approaches as represented in [21,22,23] provide support for impact analysis, although they are not slicing tools. Evaluating impact analysis support in the more general context has been discussed in [24]. Most of the current slicers do not yet include sophisticated features such as query languages, versatile hypertextual navigation, linkages to software documentation, dedicated abstraction mechanisms for representing the produced slicing information, and utilization of the related software metrics information.

Further research directions which are currently under our consideration, related to the further development of GRACE, include support for other programming languages, and application of some of the above mentioned sophisticated slicing techniques to increase flexibility of the slicing.

Supporting OO-languages benefits from the resolution of delayed binding of method invocations, e.g. by applying SDGs and conservative call analysis [25]. For us, support for Java is relevant and promising, since it is currently much applied, and its grammar is relatively simple. Our tool is well extensible to support Java. Special challenges include handling of threads and concurrency [26,27].

VIII. CONCLUSION

There is a clear need to support maintenance and especially impact analysis of programs written in Visual Basic. Program slicing is a theoretically very suitable and promising technique for supporting especially systematic and detailed level impact analysis. There are, however, some problems. 1) Research on slicing has mainly been theoretical; there are only few forward slicing tools. 2) Most of them have practical limitations in terms of the supported programming languages; traditionally, only C is supported. 3) Especially, inefficiency of static slicing has traditionally been a problem. 4) Completeness of the slicing has been rather weak in many of the slicers. 5) Forward slicers tend to be complex due to the complexity of implementing the traditional slicing algorithms.

One way to alleviate these problems and to simplify especially the construction of program slicing capabilities is to apply program dependence graphs (PDGs) as an intermediate form of representing the needed program information. This approach has earlier been successfully applied in CodeSurfer [13]. However, there were no other full-scale implementations of static forward slicing. Moreover, there is no earlier slicing-based support for performing impact analysis of Visual Basic.

Therefore, we have implemented PDG-based static interprocedural forward slicing capabilities for VB into our GRACE-tool. This paper has described the background and need of supporting impact analysis and tool requirements, outlined tool architecture and implementation, represented a use case and tool evaluation focusing on a representative forward slicing example, and compared the tool to other significant forward slicers.

GRACE contains all components which are necessary to enable forward slicing and to represent the results to the tool user via graphical user interface. The components include a VB-parser, and generators of abstract syntax trees, symbol tables and PDGs. The parser of GRACE supports VB fully. This study has confirmed that the selected approach is feasible in terms of implementability and efficiency of the needed analyses. The set objectives in terms of tool requirements have been fulfilled. The slicer also takes into account the effects of event handling features used in VB. By the virtue of PDGs, the tool is modular and extensible to support also other programming languages. On the general level, it can be stated that the development of slicing tools is important both from the theoretical and practical point of views for the improvement of automated software maintenance support and empirical evaluation of slicing variants and application of special techniques to supplement forward slicing. Our main further research avenue is extension of the PDG-based forward slicing support for other programming languages; especially for Java.

REFERENCES

[1] M. Weiser, "Program slicing", in *Proc. 5th Int. Conf. Software Eng.* (ICSE 1981), IEEE, 1981, pp. 439-449.

[2] J. Korpi, and J. Koskinen, "GRACE: Automated slicing for Visual Basic", unpublished.

[3] S. Horwitz, T. Reps, and D. Binkley, "Interprocedural slicing using dependence graphs", *ACM Trans. Program. Lang. Syst.*, vol. 12, issue 1, pp. 26-60, 1990.

[4] J.-P. Queille, J.-F. Voidrot, N. Wilde and M. Munro, "The impact analysis task in software maintenance: A model and a case study", in *Proc. Int. Conf Software Maint.* (ICSM 1994), IEEE, pp. 234-242.

[5] M. Lehman, D. Perry, and J. Ramil, "Implications of evolution metrics on software maintenance", in *Proc. Int. Conf. Software Maint.* (ICSM 1998), IEEE, 1998, pp. 208-217.

[6] R. Banker, S. Datar, C. Kemerer, and D. Zweig, "Software complexity and maintenance costs", *Comm. ACM*, vol. 36, issue 11, pp. 81-94.

[7] S. Black, "Computing ripple effect for software maintenance", *Journal of Software Maint. Evol.*, vol. 13, issue 4, pp. 263-279, 2001.

[8] J. Koskinen, A. Salminen, and J. Paakki, "Hypertext support for the information needs of software maintainers", *Journal of Software Maintenance and Evolution*, vol. 16, issue 3, pp. 187-215, 2004.

[9] S. Letovsky, and E. Soloway, "Delocalized plans and program comprehension", *IEEE Software,* vol. 3, issue 3, pp. 41-49, 1986.

[10] G. Venkatesh, "Experimental results from dynamic slicing of C programs", *ACM Trans. Program. Lang. Syst.*, vol. 17, issue 2, pp. 197-216, 1995.

[11] S. Horwitz, and T. Reps, "The use of program dependence graphs in software engineering", in *Proc. 14th Int. Conf. Softw. Eng.* (ICSE 1992), ACM, 1992, pp. 392-411.

[12] M. Harrold, and B. Malloy, "A unified interprocedural program representation for a maintenance environment", *IEEE Transactions on Software Engineering*, vol. 19, issue 6, pp. 584-593, 1993.

[13] P. Anderson, T. Reps, and T. Teitelbaum, "Design and implementation of a fine-grained software inspection tool", *IEEE Trans. Softw. Eng.*, vol. 29, issue 8, pp. 721-733, 2003.

[14] C. Fox, S. Danicic, M. Harman, and R. Hierons, "CONSIT: A fully automated conditioned program slicer", *Software – Practice and Exper.*, vol. 34, issue 1, pp. 15-46, 2004.

[15] *GCC: GNU Compiler Collection*, http://gcc.gnu.org [13.9.2005]

[16] *wxWidgets*, http://www.wxwidgets.org, [14.12005].

[17] *tangogoclient*. http://www.amourtan.com/ [2006].

[18] *Graphviz - Graph Visualization Software*, http://www.graphviz.org [2006].

[19] P. Tonella, "Using a concept lattice of decomposition slices for program understanding and impact analysis", *IEEE Trans. Software Eng.*, vol. 29, issue 6, pp. 495-509, 2003.

[20] T. Reps, and G. Rosay, "Precise interprocedural chopping", in *Proc. 3rd ACM SIGSOFT Symp. Found. Software Eng., ACM SIGSOFT Software Eng. Notes*, vol. 20, issue 4, 1995, pp. 41-52.

[21] Y.F. Chen, E. Gansner, and E. Koutsofios, "A C++ data model supporting reachability analysis and dead code detection", *IEEE Trans. Software Eng.*, vol. 24, issue 9, pp. 682-694, 1998.

[22] G. Murphy, D. Notkin, and E. Lan, "An empirical study of static call graph extractors", *ACM Trans. Software Eng. Methodology*, vol. 7, issue 2, pp. 158-191, 1998.

[23] M. Hutchins, and K. Gallagher, "Improving visual impact analysis", in *Int. Conf. on Software Maintenance* (ICSM 1998), IEEE, pp. 294-303.

[24] J. Koskinen, and A. Salminen, "Supporting impact analysis in HyperSoft and other maintenance tools", *Proc. IASTED Int. Conf. Softw. Eng.* (SE 2005), Anaheim, USA: ACTA Press, pp. 187-192.

[25] L. Larsen, and M.J. Harrold, "Slicing object-oriented software", in *Proc. 18th Int. Conf. Software Eng.* (ICSE 1996), IEEE, pp. 495-505.

[26] N. Walkinshaw, M. Roper, and M. Wood, "The Java system dependence graph", in *Proc. 3rd IEEE Int. Workshop Source Code Anal. and Manip.* (SCAM 2003), IEEE, pp. 55-64.

[27] J. Zhao, "Slicing concurrent Java programs", in *Proc. 7th Int. Workshop Program Compreh.* (IWPC 1999), IEEE, pp. 126-133.

An Analysis of Several Proposals for Reversible Latches

J. E. Rice
Department of Mathematics & Computer Science
University of Lethbridge
4401 University Drive
Lethbridge, AB, Canada
T1K 3M4

Abstract–**Recent work has begun to investigate the advantages of using reversible logic for the design of circuits. The majority of work, however, has limited itself to combinational logic. Researchers are just now beginning to suggest possibilities for sequential implementations. This paper performs a closer analysis of three latch designs proposed in previous work and suggests advantages and disadvantages of each.**

I. Introduction

Reversible computing has recently been re-introduced as a potential solution to the problem of the ever-growing demand for lower power devices. As stated by Frank [1]

...computers based mainly on reversible logic operations can reuse a fraction of the signal energy that theoretically can approach arbitrarily near to 100% as the quality of the hardware is improved...

However, reversible logic is suffering from two problems. Firstly, there is a lack of technologies with which to build reversible gates. Work is certainly continuing in this area. Secondly, while there is much research into how to design combinational circuits using reversible logic, there is little in the area of sequential reversible logic implementations. There is no limitation inherent to reversible logic preventing the design of sequential circuits; in fact when Tommaso Toffoli first characterized reversible logic in his 1980 work Reversible Computing [2] he stated that "*Using invertible logic gates, it is ideally possible to build a sequential computer with zero internal power dissipation.*"

Table 1. The truth table of a 3x3 reversible function.

Inputs *xyz*	outputs *x'y'z'*
000	000
001	001
010	011
011	010
100	100
101	101
110	111
111	110

Researchers such as Rice [3] and Thapliyal, Srinivas and Zwolinski [4] have begun work in presenting memory elements such as reversible latches and reversible flip-flops. This paper presents an analysis of two reversible SR-latch implementations, and provides a comparison to the traditional SR-latch.

II. Background

A. Reversible Logic

Before discussing sequential reversible logic we first present the basic concepts underlying reversible logic. According to Shende *et al.* [5] a gate is reversible if the (Boolean) function it computes is bijective. This means that a function is reversible if there is a one-to-one and on-to mapping from the inputs to the outputs (and vice versa) of the function. At the very least, a reversible function must have the same number of inputs as it does outputs. For instance, the traditional NOT gate is reversible, but that the traditional AND gate is not.

Table 1 shows the truth table for a 3x3 reversible function. In such a function each output can be thought of as a transformed version of one of the inputs. The symbols and behaviours of the reversible gates used in this paper are shown in Figure 1.

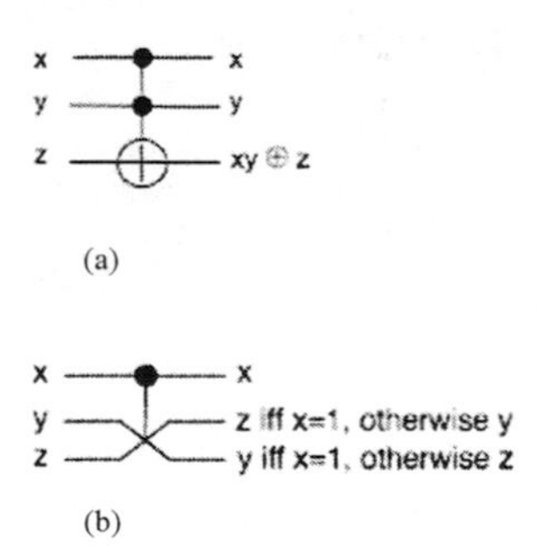

Figure 1. (a) The Toffoli gate and (b) the Fredkin gate.

K. Elleithy (ed.), Advances and Innovations in Systems, Computing Sciences and Software Engineering, 203–206.

Table 2. The next state values for the SR-latch.

inputs $S\,R$	next state $Q+\ \bar{Q}+$
0 0	$Q\ \bar{Q}$ (same as previous state)
0 1	0 1
1 0	1 0
1 1	not permitted

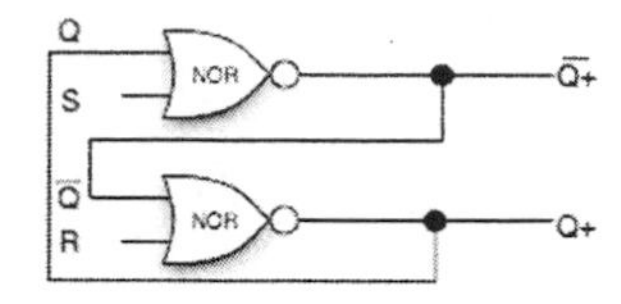

Figure 2.A traditional NOR-based SR=latch.

B. *The SR-Latch*

The primary focus of this paper is the SR-latch. This latch allows the outputs $Q+$ and $\bar{Q}+$ to be "set" to the values 1 and 0, respectively, or "reset" to 0 and 1. The primary inputs to the SR-latch are S (set) and R (reset). The behaviour of the SR-latch is characterized by the truth table given in Table 2. Figure 2 shows a traditional NOR-based structure for this latch.

III. Reversible SR-Latches

Given the need for reversible memory elements in order to build reversible sequential circuits, it seems reasonable to try to mimic the behaviour of the SR-latch using reversible logic gates. At first glance the SR-latch appears to exhibit most of the desirable reversible characteristics, except that there are two output possibilities when the inputs SR are set to 00. Closer examination shows that the SR-latch actually has four inputs: S, R, Q and $\bar{Q}$, where Q and $\bar{Q}$ represent the current state of the latch.

A reversible version of this latch must have at least three inputs; in an ideal situation these would consist of S, R and the non-inverted Q. The truth table for such a latch is shown in Table 3. One can see that an additional output has been added, labeled g for garbage. This output would not be used although is required to maintain the reversibility of the device. The first two rows of the table are illustrative of the process used in constructing such a table. In row 0 in order to maintain the characteristic of the SR-latch $Q+$ must be 0 and $\bar{Q}+$ must be 1. In row 1 the R input is a 1, thus outputs $Q+ = 0$ and $\bar{Q}+ = 1$. However this is a combination we have already used, so to differentiate between rows 0 and 1 we'll arbitrarily assign $g = 0$ in row 0 and $g = 1$ in row 1.

We encounter the same problem in rows 2 and 4, and so again we arbitrarily assign values to the garbage output g. In

Table 3. A truth table for a latch with two primary inputs S and R, and Q representing the current state of the latch.

row number	$Q\,S\,R$	$Q+\ \bar{Q}+\ g$
0	0 0 0	0 1 0
1	0 0 1	0 1 0
2	0 1 0	1 0 0
3	0 1 1	not permitted
4	1 0 0	1 0 1
5	1 0 1	0 1 ?
6	1 1 0	1 0 ?
7	1 1 1	not permitted

row 5, however we encounter problems: the outputs $Q+\ \bar{Q}+ = 01$ are required according to the functionality of the SR-latch, but this combination has been used twice before and so no possible assignment to g can make the outputs unique. The same problem is encountered in row 6, making it clear that a three-input/three-output reversible SR-latch is not possible.

To solve the problem encountered in Table 3 we can add an output. This of course requires that we add an input. The desired behaviour is characterized in Table 4. Note that we have fixed the value of the additional input at 1, as we only need this input to "balance" the added output. The rows containing "XXXX" for the outputs are those where $S = R = 1$, which is not permitted.

The following subsections describe two possible alternatives for reversible SR-latches. Each is based on a similar design, but we describe various differing characteristics.

A. *Fredkin-based SR-Latch*

One of the first researchers to characterize a reversible latch was Picton [6]. He suggested a reversible SR-latch built out of Fredkin gates, as illustrated in Figure 2 (a). The problem with Picton's latch is that it incorporates fan-out, which is not permitted in reversible designs. One solution to this problem is shown in Figure 2 (b).

In this analysis and in all following discussions we assume a delay of 1 for each level of gates and an additional delay of 1 for the long wires propagating the values for $Q+$ and $\bar{Q}+$ back to the inputs. For instance, if the current state of the latch is $Q\ \bar{Q}SR = 0000$ then Table 5 illustrates how each of the values change.

Table 4. The state table for a four-input/four-output reversible latch.

$1\,Q\,S\,R$	$Q+\ \bar{Q}+\ g_1\,g_2$
1 0 0 0	0 1 0 0
1 0 0 1	0 1 0 1
1 0 1 0	1 0 0 0
1 0 1 1	X X X X
1 1 0 1	0 1 1 0
1 1 1 0	1 0 1 0
1 1 1 1	X X X X

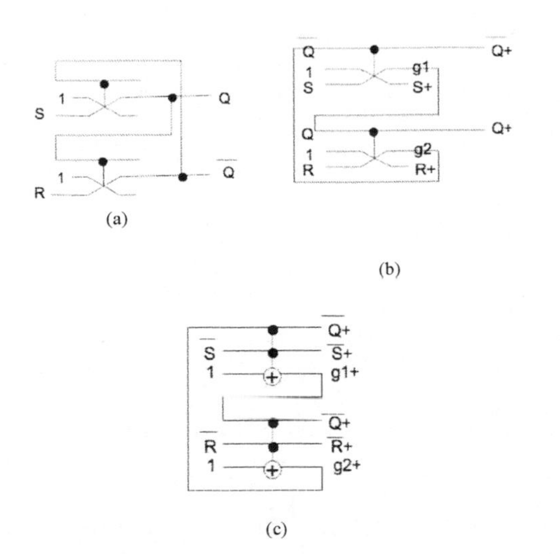

Figure 2. (a) The Fredkin-based reversible SR-latch as suggested by Picton, (b) a modified version of Picton's latch, and (c) A reversible SR-latch based on Toffoli gates.

Table 5 shows that with an initial state of $Q\ \bar{Q}$ = 00 and inputs *SR* = 00 the latch is unstable, oscillating (as does the traditional NOR-based SR-latch) between $Q+\ \bar{Q}+$ = 00 and 11. A similar table can be built starting values of $Q\ \bar{Q}SR$ = 0001. With these values the final state of the latch should be $Q+\ \bar{Q}+$ = 01, and our analysis shows that after 5 time periods the latch is stable at these values. Similar state tables can be derived for all 16 possible initial starting states of the latch, and these tables show that this latch is unstable when any one of the following four starting values for $\bar{Q}QSR$ are used: 0000, 1100, 0100 or 1000

The first two cases are not entirely surprising, since they reflect the behaviour of the NOR-based latch on which this design was modeled. However, the second two cases are dismaying, since this behaviour is in violation of the required characteristics of the SR-latch. One possible fix for this behaviour is to intead add Toffoli gates to Picton's design in Figure 2 (a) in order to provide the fan-out signals. However, this modification was not initially suggested due to the additional delay caused by having two levels of gates internal to the latch and because of the additional requirement of another input.

B. *Toffoli-based SR-Latch*

A second reversible design for the SR-latch was also proposed by Rice [3]. This design is shown in Figure 3 (c). Again, state tables showing the changes in values for the latch can be derived for all 16 input combinations of $\bar{Q}QSR$. This latch is unstable for only two input cases: $\bar{Q}QSR$ = 0000 and 1100. Additionally, this latch responds to inputs such as $\bar{Q}QSR$ = 0100 very quickly, taking only two timesteps to stabilize.

We should note that a similar alternative to the latch shown in Figure 2 (c), again based on Toffoli gates, was also proposed by Rice [3]. Our analysis for this paper showed that this alternative does not have the required behaviour of a SR-latch, and so we have not included this latch in these comparisons.

C. *Comparisons*

During the course of this work we found that the traditional SR-latch became unstable under the input conditions $SRQ\ \bar{Q}$ = 0000 or $SRQ\ \bar{Q}$ = 0011. In both these cases the next state would oscillate between $Q+\ \bar{Q}+$ = 00 and 11. This behaviour is also reflected in the Toffoli and Fredkin-based SR-latches.

Table 5. An illustration of how the internal state of the Fredkin-based SR-latch changes, assuming a delay of 1 for processing at each level of gates and a delay of 1 for each propagation of values from outputs back to inputs.

Time period	$\bar{Q}\,Q\,S\,R$	$\bar{Q}+\ g_1\ S+$	$Q+\ g_2\ R+$
0	0 0 0 0		
1 (after gates process inputs)	0 0 0 0	0 1 0	0 1 0
2 (after propagating values back to inputs)	1 1 0 0	0 1 0	0 1 0
3 (after gates process inputs)	1 1 0 0	1 0 1	1 0 1
4 (after propagating values back to inputs)	0 0 0 0	1 0 1	1 0 1
5 (after gates process inputs)	0 0 0 0	0 1 0	0 1 0

Table 6. The average number of timesteps required for each latch to reach a stable state.

Latch type	Average delay
Traditional SR-latch	3.36
Toffoli-based SR-latch	3.71
Modified Picton SR-latch (no Toffoli gates)	3.83
Picton SR-latch with Toffoli gates	4.57

Table 6 compares the average number of timesteps required for each latch to reach a stable state. We identified a stable state by detecting a repetition of values on the inputs or the outputs. Input combinations resulting in instability were not included in our computations. The values were determined by assuming a delay of one for each level of gates to process their inputs, and a delay of one for signals to be propagated back to the inputs.

IV. CONCLUSION & FUTURE WORK

The purpose of this paper is both to illustrate the feasibility of reversible logic in sequential logic design, and to examine more closely the behaviour of a basic memory element, the SR-latch. We found that one of the proposed latch designs has some flaws, and while correction of these flaws was possible, it resulted in additional delay and the need of an additional input. We would conclude that the better reversible design for an SR-latch is the Toffoli-based design, as shown in Figure 2 (c).

There are many areas of work that may lead from this paper, most notably similar types of analysis for other latches and for flip-flops designed from these latches. Additionally, during the course of this work the authors noted the lack of simulation tools that support reversible gates, and this is most definitely an area worthy of attention. Finally an ongoing goal is to develop a synthesis process that will support reversible logic, and incorporate the sequential elements we are proposing in this work.

REFERENCES

[1] M. P. Frank, "Introduction to Reversible Computing: Motivation, Progress, and Challenges," in Proceedings of the 2nd Conference on Computing Frontiers, 2005, pp. 385–390.

[2] T. Toffoli, Automata, Languages and Programming. Springer Verlag, 1980, chapter: Reversible Computing, pp. 632–644.

[3] J. E. Rice, "A New Look at Reversible Memory Elements," in Proceedings of the International Symposium on Circuits and Systems (ISCAS), 2006, to appear.

[4] H. Thapliyal, M. B. Srinivas, and M. Zwolinski, "A Beginning in the Reversible Logic Synthesis of Sequential Circuits," in Proceedings of Military and Aerospace Programmable Logic Devices (MAPLD) International Conference, 2005, submission 1012.

[5] V. V. Shende, A. K. Prasad, I. L. Markov, and J. P. Hayes, "Reversible Logic Circuit Synthesis," in IEEE/ACM International Conference on Computer Aided Design (ICCAD), 2002, pp. 353–360.

[6] P. Picton, "Multi-Valued Sequential Logic Design using Fredkin Gates," Multiple-Valued Logic, pp. 241–251, 1996.

Implementation of a Spatial Data Structure on a FPGA

J. E. Rice, W. Osborn and J. Schultz

Department of Mathematics & Computer Science
University of Lethbridge
4401 University Drive, Lethbridge, AB, Canada,
T1K 3M4

Abstract - **Many systems exist that store and manipulate data; however, many do no have sufficient support for spatial data Many data structures are proposed that are intended specifically for spatial data; however, software implementations have not performed as well as hoped. This work presents a feasibility study investigating the use of a FPGA for the implementation of a structure to support spatial search and retrieval.**

I. Introduction

Recent advances in the area of data storage have resulted in technology enabling institutions, companies, and individuals to store data in sizes never before envisioned. One area in particular that is leveraging this increase is the area of geographical information systems (GIS). A GIS manages spatial data. Unfortunately many data structures are not appropriate for spatial data. Recent work by Osborn [1] has produced a data structure intended specifically for spatial data. A limitation is that its software implementation is slow. In this work, we address this limitation by investigating the feasibility of implementing the data structure on a reconfigurable chip called a FPGA.

II. Background

A. *FPGAs*

One technique for accelerating computation is a (re)configurable hardware solution. This has been applied to various problems such as image compression [2] and string matching [3], as well as in other bioinformatics applications [5, 6, 7]. Reconfigurable computing utilizes the flexibility and processing power of reconfigurable devices such as Field Programmable Gate Arrays (FPGAs) to achieve an increase in performance. This flexibility allows the development and implementation of a custom hardware circuit as part of a solution. A FPGA consists of many programmable cells, which can be programmed for either I/O or functionality. Cells are comprised of look-up tables (LUTs), which can be programmed for various functions and interconnected in many ways that are determined by the place and route software.

Reconfigurable computing is generally used in a static or dynamic role. In static reconfigurable computing, the device is programmed once for the entire instance of an application. Dynamic reconfigurable computing programs the device many times, producing multiple hardware designs during execution. We are primarily interested in a static solution because reprogramming the FPGA dynamically incurs overhead. However, a dynamic solution is not ruled out at this stage in our investigations.

B. *Spatial Data Representation and Retrieval*

Spatial data is data that exists in multidimensional space. It ranges in complexity from simple points to objects composed of sub-objects, such as points, lines, or arbitrarily-shaped objects. For example, a town is represented with a point, while a province has many towns (*i.e.* points), cities (*i.e.* regions), and roads (*i.e.* linestrings).

Two important issues for spatial data are the efficient retrieval of a specific object (*i.e.* exact match) and the efficient search for subsets of spatial objects (*i.e.* a region search).

Many one-dimensional hierarchical structures are proposed for retrieving spatial data [4]. Most store minimum bounding rectangles (MBRs) of objects and the regions in space that contain objects. Their limitations include overcoverage of empty space, and overlap of the MBRs, which leads to multiple-path searching.

III. The Project

C. *The 2DR-tree*

The 2DR-tree is used to create a two-dimensional hierarchical structure for retrieving objects in two-dimensional space, as opposed to forcing those objects to fit a one-dimensional structure. Using nodes with the same dimensionality as the object space can lead to significant improvement in retrieval performance [1]. The 2DR-tree maps

K. Elleithy (ed.), Advances and Innovations in Systems, Computing Sciences and Software Engineering, 207–210.

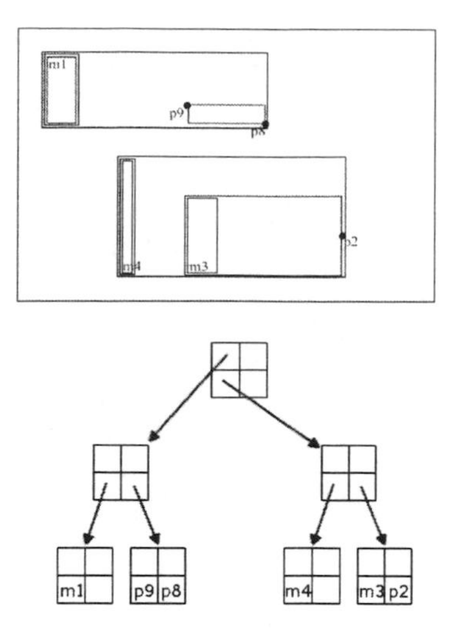

Figure 1. A sample object layout., with the resulting 2DR-tree.

each MBR to an appropriate location in a two-dimensional node, to reduce overlap and overcoverage.

Each node contains individual locations. Each node location holds the MBRs for an object or a region. A MBR consists of two coordinate pairs high *(x,y)* and low *(x,y)*, or *(hx, lx, hy, ly)*. At the leaf level, a MBR in two-dimensional space is placed in an appropriate location of a two-dimensional node. MBRs for regions are placed similarly in non-leaf nodes.

The 2DR-tree differs from other spatial data structures in how the node locations are arranged in a node. Instead of a node being modeled as a flat one-dimensional array, it is modeled as a two-dimensional array. Thus, the *order* of a node is *X*Y*, where *X* is the number of node locations on the x-axis and *Y* is the number of node locations on the y-axis.

Searching is done using a binary search strategy. The main task usually involves finding a split point; however we discuss in the following section the necessity for removing this when implementing the search on a FPGA.

D. *FPGA Implementation*

For this work we used the Virtex-II Pro FPGA Prototyping Station that is provided by the Canadian Microelectronics Corporation (CMC). It is an AMIRIX AP1000 development board that features the Xilinx Virtex-II Pro FPGA. It has 44,000 logic slices and features two embedded IBM PowerPC hard macros and 1.4MB of on-chip RAM. We use the Xilinx ISE and Xilinx EDK software for development and to run simulations.

The basic unit in a 2DR-tree is the node location. With the node location a node of any order can be built. The nodes are then used as the building blocks for a 2DR-tree. It makes

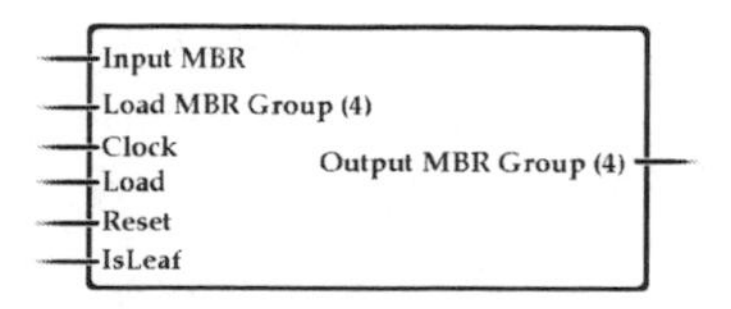

Figure 2. The input/output design for a 2*2 node.

sense for the unit representing the node location to contain most of the functional logic and carry out the bulk of the work (*i.e.* checking two MBRs for overlap). The other calculation performed in software is to find a dividing point for the node during binary search. However, the FPGA can check all nodes simultaneously, and therefore we do not need to find a division; instead all nodes on one level can be tested at once. The node location unit is designed to test the input MBR for overlap with the stored MBR and give the appropriate output.

The output from a node location unit is a MBR. This may be empty if no overlap exists between the input and stored MBRs. If overlap exists, the output depends on whether the node location unit is marked as a leaf-node or not. An *isLeaf* marker is used to identify a leaf-node. A leaf-node will output the stored MBR, otherwise the input MBR is sent as output.

A load signal and a load MBR are used to set the stored MBR value for a node location unit. When the load signal is sent the load MBR value is used to replace the current value in the stored MBR. The initial value of the stored MBR is an empty MBR. The last two incoming signals to a node location unit are the clock and reset signals. The reset signal is used in conjunction with the load signal to re-initialize the stored MBR to the empty MBR, or on its own to signal the node to output an empty MBR.

Using the node location unit as the foundation, a node of any order can be created. The only information stored at the node-level is the status of the node, *i.e.* leaf-node or not. The output from all node location units are passed out of the node unit as a group. The size of this group depends on the node order. It the node has an order of 2*2, then the output group will have 4 MBRs. The load MBRs are also grouped in the same manner. This allows for an entire node to load all of its node locations at once.

The final step is to create, on the FPGA, a 2DR-tree built from the individual nodes. Two issues arise when creating the 2DR-tree. As the height of the tree grows the number of output MBRs increase drastically, as do the number of load MBRs.

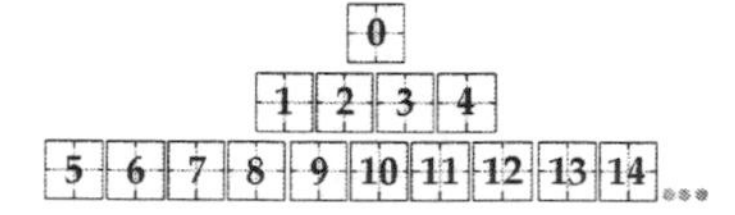

Figure-3. Numbering scheme for the 2DR-tree nodes

To solve these issues two controllers are needed, an input controller and an output controller. The input controller allows a single node to be loaded with a load MBR value, so the overall 2DR-tree on the FPGA only has one load MBR as an input line. Along with the load MBR value the node number is needed. Each node in the tree is numbered starting with the root node as 0 as shown in Figure 3. The reset and load signals for each individual node are also controlled by the input controller.

The output controller allows for a particular node output to be queried, which allows the FPGA 2DR-tree to have only one MBR output line. Each of the output MBRs for individual nodes are fed into the output controller as inputs. The load signal and node number are also provided as inputs to the output controller. If the load signal is set then the output MBR is an empty MBR, otherwise, the output from the node with the given node number is provided. These controllers solve the problem of having too many input and output lines for the FPGA 2DR-tree. It is now possible to create a functional tree to perform binary searches. Using this model a simulation using the Xilinx ISE was created for our testing and analysis purposes.

E. *Results & Analysis*

For the FPGA 2DR-tree performance analysis, we use trees of heights 3 and 4. These heights are chosen due to FPGA space limitations. We can store a height 4 tree by reducing the range of integers in the MBRs, as the sample data did not exceed the range of a 16 bit integer. Further reduction would allow for a deeper tree, but for analysis purposes the two sample tree heights will be adequate.

All of our results are from simulations carried out with the Xilinx design tools, and so the transfer time for sending data from the host system to the FPGA board is not taken into account. Table 1 shows the time required for each of the actions carried out by the FPGA-based 2DR-tree.

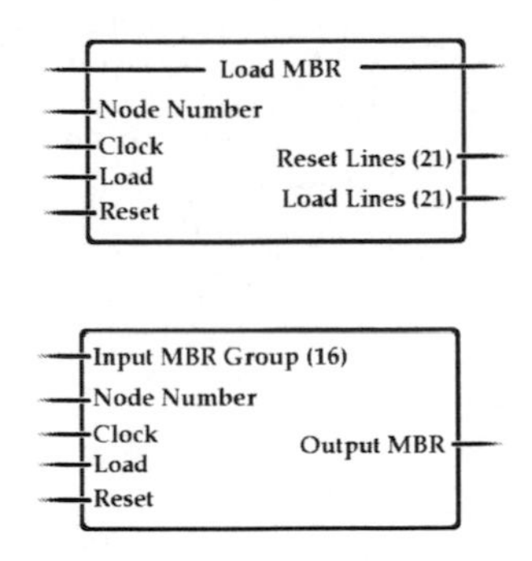

Figure-4. The input controller (top) and the output controller (bottom).

Table 2 shows the time required for the software binary search. The timing results for the software binary search are computed by using the Java code from Osborn's work [1]. The only addition was the code for timing the searches. Because searching takes a fraction of a second, several hundred searches were performed, and the average time was calculated.

Results show that the FPGA implementation will give a tremendous speed increase to the binary search. The concerning factor will be the time it takes for the data to transfer from the host machine to the FPGA board. Currently the transfer is performed via the serial interface, which can be rather slow for large amounts of data. We will investigate data transfer via the PCI Bus, which will produce faster data transfer speeds.

IV. Conclusions & Future Work

This work shows the feasibility of using a FPGA to improve search speeds on a spatial database. Work is continuing in many areas. We will incorporate communication times into our results. We will have the host machine run a Java application that will communicate with the FPGA board. This application will load the FPGA 2DR-tree, run a binary search and query for results. The communication between the application and the FPGA board can be done via the serial port or PCI bus, as investigations allow.

Also, additional logic can be added to support unbalanced trees and self-checking nodes. The current implementation requires that the 2DR-tree be balanced. Extending the logic to allow for leaf-nodes to occur any where in the tree would allow for more realistic sample data to be used. Self-checking nodes will allow for a more robust implementation and will be useful when logic is added for the insert operation. This means a node will have logic to ensure the MBRs within it conform to the required "node validity" presented by Osborn [1]. Finally, for a complete implementation of the 2DR-tree, insert and delete operations need to be incorporated into the design.

Table 1. FPGA 2DR-tree search performance.

Action	Time (Height 3)	Time (Height 4)
Initialize	6 ns	6 ns
Load	264 ns	1,032 ns
Search	48 ns	60 ns
Query Output	216 ns	792 ns
Total Time	534 ns	1,890 ns

Table 2. 2DR-tree software search performance.

Height	Iterations	Avg. Total Time
3	1,000,000	13,422.2 ns
3	5,000,000	13,100.04 ns
4	1,000,000	22,193.8 ns
4	5,000,000	20,684.96 ns

The current work on implementing a 2DR-tree on a FPGA shows much promise. The performance increase is significant. Further research will produce a functional 2DR-tree for binary searches. It can also lead to other speed increases for the 2DR-tree. The ideas presented in this paper can be applied and extended to increase performance for the binary search, insert and delete operations of the 2DR-tree.

References

[1] Osborn, W. *The 2DR-tree: a 2-Dimensional Spatial Access Method.* PhD Thesis, University of Calgary, 2004.

[2] Simpson, A., Hunter, J., Wylie, M., Hu, Y., and Mann, D. *Demonstrating Real-Time JPEG Image Compression-Decompression Using Standard Component IP Cores on a Programmable Logic Based Platform for DSP and Image Processing.* Proceedings of FPL 2001, LNCS 2147, Springer-Verlag, pp. 441-450, 2001.

[3] Lee, H. and Ercal, F. *RMESH Algorithms for Parallel String Matching.* Proceedings of the 3rd International Symposium on Parallel Architectures, Algorithms and Networks (I- SPAN'97), pp. 223-226, 1997.

[4] Gaede, V. and Guenther, O. *Multidimensional Access Methods.* ACM Computing Surveys, 30(2), pp. 170-231, 1998

[5] K. B. Kent, J. E. Rice, S. Van Schaick, and P. A. Evans. *Hardware-Based Implementation of the Common Approximate Substring Algorithm.* In Proceedings of the Euromicro Symposium on Digital System Design: Architectures, Methods and Tools (DSD), pages 314–320, 2005.

[6] J. E. Rice and K. B. Kent. *Systolic Array Techniques for Determining Common Approximate Substrings.* In Proceedings of the International Symposium on Circuits and Systems (ISCAS), 2006. Paper number 1480 (CDROM).

[7] K. B. Kent, R. B. Proudfoot, and Y. Zhao. *Optimizing the Edit-Distance Problem.* In Proceedings of the 17th International Workshop on Rapid System Prototyping (RSP), 2006. to appear.

Security Management: Targets, Essentials and Implementations

Zhao Jing
School of Civil Engineering,
Shijiazhuang Railway Institute, Hebei 050043, China
Zheng Jianwu
Department of Information Engineering,
Shijiazhuang Railway Institute, Hebei 050043, China.
{zhaojing, zhengjw }@sjzri.edu.cn

***Abstract*—We first analyze security targets of implementing security management for nowadays IT infrastructures – information systems created by enterprises for successful business, and detail possible measures for achieving relevant targets. Secondly, we conclude that the essentials of security management are to construct trustworthy network endpoints, and to establish trustworthy communication channel between intending communication parties; then two instances of accomplishing the essentials of security management are exemplified, i.e. trustworthy smart card transaction and trustworthy SOA-Based Web Services. At last, we discuss the main aspects of implementing security management for information systems, precisely, strategic steps, i.e. (1) attestation and negotiation, (2) proposing and implementing application-specific strategies, and (3) considerations for strength and efficiency of security management.**

I. Introduction

With the advent and rapid development of Information Technology (IT), enterprises nowadays increasingly depend on IT to create the flexible and agile IT infrastructures – information systems necessary to be able to be responsive to changing business demands, and therefore to drive business innovation and provide competitive business benefit. Specifically, the enterprises leverage information systems for maximizing business profit, minimizing the cost of product or service, establishing strong relationship with partners and customers, and further earning the enterprises reputation.

When exploiting the possibilities and potentials of the information systems, it is very important and also is required that the enterprises actively complete following tasks.

- IT asset management
- Performance management
- Compliance management
- Security Management

The objectives of implementing security management can be listed as follows.

1) Guarantee the customers' confidence,
2) Build customers' respect to the enterprise or earn the enterprise reputation,
3) Prevent enterprise asset leakage,
4) Fight attacks of varied type, or resist the attackers the ability to gain control of the IT infrastructure.

II. Security Targets and Measures for Achieving Them

For pursuing objectives of security management mentioned above, it is needed first to make clear what lower-level security targets should be achieved while operating information systems, and what possible measures can be taken for achieving them.

A. Security Targets

Data privacy, data integrity, and identity legitimacy are three basic targets, which should be achieved in terms of implementing security management. Furthermore, in nowadays-networked environment, it is highly possible that the adversaries may attack information systems, and following two factors always account for the successful attacks.

1) Vulnerabilities of the IT infrastructure;
2) Tricks and malicious intend of the adversaries.

Therefore, we emphasize the importance of fighting attacks of varied types as well.

B. Measures for Achieving the Targets

Although a wide variety of measures can be implemented to achieve security targets specified above, the most commonly adopted mechanisms are based on the use of cryptographies.

1) Measures for Identity Authentication

Authentication relates to source authentication, i.e. confirmation of the identity of communication source, or peer entity authentication, which assures one entity of the purported identity of the other correspondent. Either secret key [1] [2] or public key cryptography [3] can be implemented for identity authentication.

When utilizing secret key cryptography, authentication can be implemented by demonstrating the knowledge of the cryptographic key. Public key cryptography is always selected for peer identity authentication. Namely, two intending communication parties can authenticate each other with the digital certificates.

2) Measures for Data Confidentiality and Integrity

Both secret key cryptography and public key cryptography can be used to guarantee data privacy, both data stored in servers and data transmitted over communication channels.

Message digest algorithms [4] and private or public cryptography can be combined to effectively detect intentional

K. Elleithy (ed.), Advances and Innovations in Systems, Computing Sciences and Software Engineering, 211–216.

or unintentional data modifications (however, cryptography does not protect message from being altered). When secret key cryptography is used, FIPS-113, Computer Data Authentication [5], specifies a standard technique for calculating a MAC (message authentication code) for integrity verification. Public key cryptography verifies integrity by using of secure hashes (FIPS-180, Secure Hash Standard [6]) and public key signatures (FIPS-186, Digital Signature Standard [7]).

3) Measures for Repudiation and Non-Repudiation

Repudiation: A quality that prevents a third party from being able to prove that a communication between two other parties ever took place. This is a desirable quality if you do not want your communications to be traceable.

Non-repudiation: Non-repudiation is the opposite quality a third party can prove that a communication between two other parties took place. Non-repudiation is desirable if you want to be able to trace your communications and prove that they occurred.

Public key cryptographic systems can be utilized for both repudiation and non-repudiation. For example, RSA signature (or public key signatures, FIPS-186) can provide non-repudiation while RSA encrypted nonce can provide repudiation.

4) Measures for Fighting Potential Attacks

There are so many types of attacks, proper mechanisms should be designed and implemented to fight them. We simply describe measures for addressing Denial-of-Service, Auto-Replay and so on.

1) Against Denial-of-Service: There are network threats not only to authenticity and privacy but also to availability and quality of supplied services. For example, attackers may abuse authentication procedures, which in turn overload the target with heavy verification cost. In order to discourage Denial-of-Service (DoS), cryptographic protocols, which overload attackers as well, can be incorporated in measures for securing IT infrastructure.

2) Anti Counterfeiting: A sequence number combined with the use of authentication can be introduced for distinguishing authentic (or legitimate) communication information from fraudulent information, i.e. attacks launched by adversaries.

   Value of a sequence number should obey following two rules. The first rule is that value of the byte in every legitimate traffic is unique. The second one is that value of the byte in different authentic traffic should vary according to its initial value and value variation principle agreed on by two legitimate communicating parties.

3) Anti-Replay: Mechanism mentioned above for anti-counterfeiting can also be implemented for fighting auto-replay attacks. As an example, IPSec [8] provides optional anti-replay services by introducing sequence number into communication exchanges.

III. Essentials of Accomplishing Security Management

The section above has detailed common lower-level security targets that needed to be fulfilled when implementing security management for nowadays-networked information systems. According to the "Trustworthy" notion presented in [9] and [10], we can conclude that following two things should be accomplished for successful security management.

1) Constructing trustworthy network endpoint.
2) Establishing trustworthy communication channel.

In the sequel, two instances of accomplishing essentials of security management, i.e. trustworthy smart card transaction and trustworthy SOA-Based Web Service, are exemplified.

A. Trustworthy Smart Card Application System

The smart card is an intrinsically secure computing platform, especially, it is ideal for sealed storage and isolated computing. As a result, the smart card is widely utilized for nowadays society.

Fig. 1 illustrates the smart card application system. Objects and links connecting them, framed inside the dashed line, make up of the application terminal.

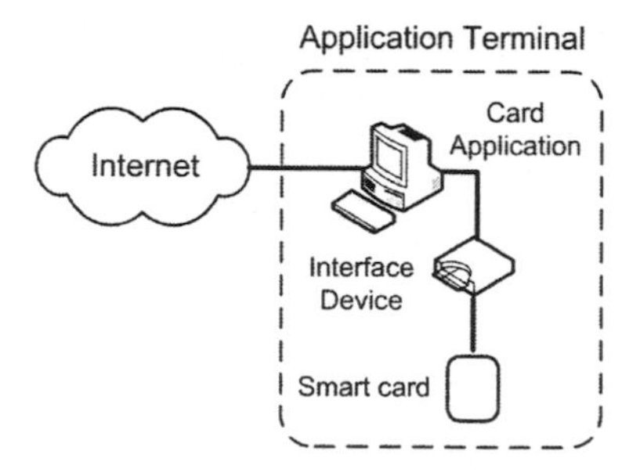

Fig. 1. Smart Card Application System

1) Essentials of Securing Smart Card Transaction

Fig. 2 shows two links of the smart card application system, by which the system is unified as a whole. The card application is the entrance to the Internet, connecting the external world over specific network protocols.

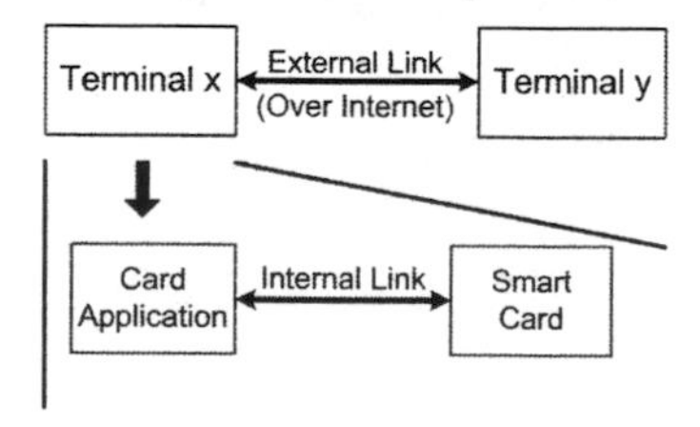

Fig. 2. Two Links of Smart Card Application System

As for these two links shown in Fig. 2, they are equally important in terms of the security of the system, and therefore should be thoroughly safeguarded with proper policies and measures. Essentials of securing smart card application system are as follows.

1) Constructing trustworthy application terminal.

2) Securing traffic through external link (Internet Link).

By implementing SSL/TLS [11], IPSec [8] and so on, traffic between application terminals through the Internet can be secured. The task of constructing trustworthy application terminal can be further divided into following two aspects.

1) Authenticating the opposite identity, i.e. trust authentication between the card application and the smart card, which is discussed in Section 3.1.4.
2) Establishing trustworthy smart card transaction channel between the card application and the smart card, as shown in Fig. 3.

Information exchange between the card application and the smart card of the application terminal is completed in two steps.

a) Card application initiates the transaction and sends command to the smart card.
b) After completing the command received, smart card returns response to the card application.

Therefore, task of establishing trustworthy transaction channel is in essence to propose strategy for trustworthily securing APDU transmission between two legitimate parties.

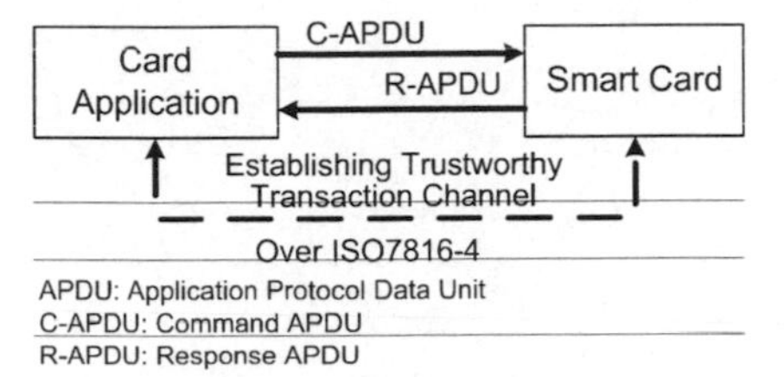

Fig. 3. Anticipated Trustworthy Smart Card Transaction

2) Analysis of Transaction Messages

A specific response corresponds to a specific command, referred to as a command-response pair. An application protocol data unit (APDU) contains either a command data unit (C-APDU) or a response data unit (R-APDU).

Fig. 4 depicts the C-APDU message, consisting of a mandatory header of four bytes, i.e. CLA, INS, $P1$ and $P2$, and a conditional body ([...]) of variable length, i.e. Lc, $IDATA$ and Le.

[CLA] [INS] [P1] [P2] [Lc] [IDATA] [Le]

Fig. 4. Command APDU Message

Fig. 5 depicts the R-APDU message, consisting of a conditional body $ODATA$ of variable length and a mandatory trailer of two bytes. Two mandatory bytes are status bytes $SW1$ and $SW2$, which code the status of the receiving entity after processing command APDU.

[ODATA] [SW1][SW2]

Fig. 5. Response APDU Message

Please refer to [12] for detail descriptions about C-APDU and R-APDU.

3) Strategy for Trustworthy Smart Card Transaction

Because the strategies for securing command APDU and response APDU can be built on the same idea, only strategy, specifically algorithmic steps, for securing command APDU transmission is detailed.

Assuming that the original C-APDU P is expressed as *[CLA] [INS] [P1] [P2] [Lc] [IDATA] [Le]*.

1) If Lc field is absent in P, insert a zero byte to P at Lc position (directly after $P2$). Original message P is updated as *[CLA] [INS] [P1] [P2] [Lc] [IDATA] [Le]*.
2) Call MD5 to hash the message P (MD5 is selected just for exemplification, other proper hash algorithms are also Ok.),
$$H = MD5(P) .$$
H is the output message digest, whose length is exactly 16 bytes.
3) Concatenate output message digest H with $IDATA$ in original message P. Concatenating operation is illustrated as
$$IDATA' = H \,||\, IDATA .$$
4) Concatenate a transaction sequence number byte SN, and $IDATA'$ is modified as
$$IDATA' = SN \,||\, IDATA' .$$
Sequence number for transaction, SN, is introduced for providing anti-counterfeiting service.
5) Call message encryption function $GE()$ to encrypt $IDATA'$ above with secret session key K. Encryption is expressed as
$$C = GE(IDATA', K) .$$
$GE()$ is defined according to symmetric cryptography, such as DES and AES. Where does the session key K come from is discussed in Section 3.1.4.
6) Construct a new command APDU P'. P' is constructed as
P' = *[CLA] [INS] [P1] [P2] [Lc'][C] [Le]*.
Four mandatory bytes are taken directly from original message P. Value of Lc' in P' is different from that of Lc in original message P, Le field in P' is the same as Le field in original P; if Le is absent in P, Le will not be present in P', otherwise, Le will be present in P'.
7) Send P' to smart card instead of original command P.

With the use of symmetric cryptography, message digest and sequence number, implementing the strategy above can therefore guarantee data privacy & integrity, anti-counterfeiting, and structure soundness of APDU message as well, for smart card transaction.

4) Protocol for Trust Negotiation: STS

Authenticating trustworthiness of the card application and the smart card is to convince them that the card application (respectively, smart card) is really the party it claims to be by evidencing notarized document, such as public certificate signed by a proper (publicly recognized) certificate authority.

Symmetric cryptography is introduced in designing strategy for securing APDU transmission, it is required that the card application and the smart card posses an identical session key. Furthermore, sequence number SN is introduced in for the purpose of anti-counterfeiting, it is also required that the two intending communication parties know the byte value of SN for every command-response pair.

W. Diffie's STS (Station to Station) protocol [13] supplies a complete and detail solution to tasks above facing in securing APDU transmission. STS can be implemented for trust negotiation while authenticating identity legitimacy over public communication channel.

B. Trustworthy SOA-Based Web Service

Web Service is one of hot topics nowadays, it is modular software applications built to run over the Internet, and based on open standards for definition, discovery, and interoperability. Service-Oriented Architecture (SOA) is suggested and recommended as its underlying architecture for establishing self-descriptive, reusable, and loosely coupled software components.

The exchange of business information over the Internet – an untrusted public network – gives rise to considerable security concerns, widely believed to be one of the major factors inhibiting the deployment of Web Service.

Fig. 6 illustrates the architecture of SOA-Based Web Service, consisting of three different entities, i.e. UDDI repository, service requestor, and service provider, and three distinct operations, i.e. finding, publishing, and provisioning.

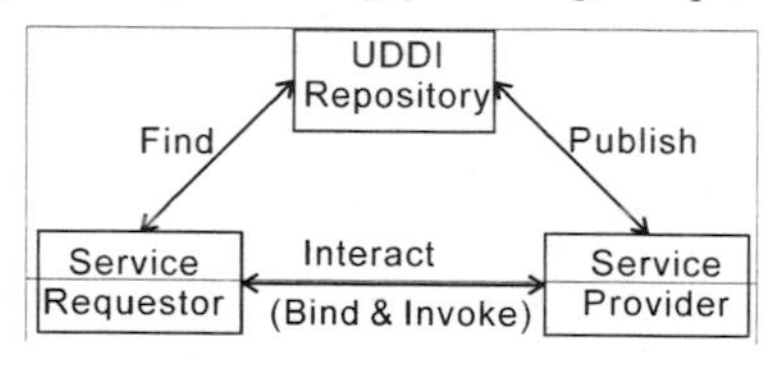

Fig. 6. Architecture of SOA-Based Web Service

1) Analysis of SOAP Service Message

The W3C SOAP recommendation [14] describes how XML messages can be exchanged in SOA, which is XML-based application-level transmission protocol, especially for exchange of information in a decentralized, distributed environment, e.g. Web Services.

In a SOA-Based Web Service, as shown in Fig. 6, SOAP endpoints – service requestors, service providers and intermediaries, send and receive SOAP messages. A SOAP message is an envelope that has child elements of an optional header and a mandatory body containing core service messages. The basic form of a SOAP message is identified as an XML 1.0 document, as illustrated with the XML document below.

```
<?xml version='1.0' ?>
<env:Envelope xmlns:env=
    "http://www.w3.org/ 2003/05/soap-envelope">
<env:Header>
...
</env:Header>
<env:Body>
...
</env:Body>
</env:Envelope>
```

A SOAP message has one element with the local name envelope qualified with the namespace http://www.w3.org/2003/05/soap-envelope; besides qualifying the namespace as a SOAP namespace, the URL identifies the version of SOAP implemented.

2) Essentials of Securing SOA-Based Web Service

When considering to secure service interaction between requestor and provider in SOA. The essential task of securing SOA is to establish trustworthy service channel after finishing trust authentication, between service requestor and service provider, as shown in Fig. 7.

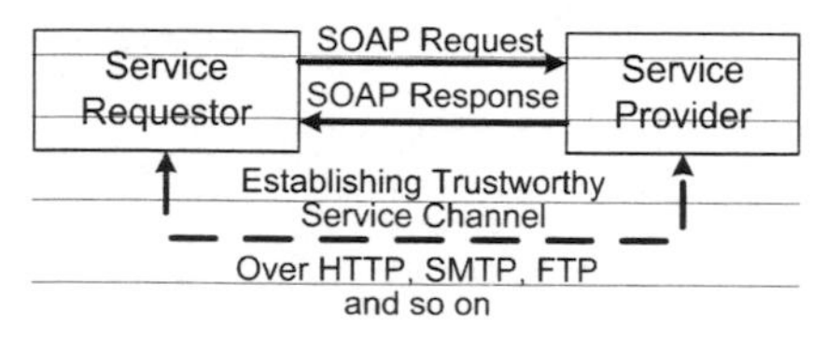

Fig. 7. Anticipated Trustworthy Service Interaction

3) Specification for Trustworthy SOA-Based Service: WS-Security

The WS-Security specification [15] is the standard defined by OASIS (2004) and provides a mechanism for addressing element-level, end-to-end security, specifically for data integrity, confidentiality, and data origin authentication features within a SOAP message. WS-Security makes use of the XML Signature [16] and XML Encryption [17] specifications and defines how to include digital signatures, message digests, and encrypted data in a SOAP message. However, the specification goes beyond these two underlying specifications by tailoring them for manipulating SOAP messages, in analogous to algorithmic steps demonstrated in Section 3.1.3.

The cryptographic mechanisms are utilized by describing how XML Encryption and XML Signature and so on are applied to parts of a SOAP message. That includes processing rules so that a SOAP node can determine the order in which parts of the message have to be validated or decrypted.

4) Specification for Trust Negotiation: WS-SecureConversation

We cannot depend only on WS-Security for completely solving security problems posed by Web Services. WS-Security just aims at providing mechanisms for data privacy & integrity and data origin authentication, it does not provide a session mechanism for key derivation and so forth. WS-SecureConversation (2004) [18] introduces mechanisms to establish and share so-called "security contexts", in analogous to STS protocol demonstrated in Section 3.1.4.

Based on established security contexts or arbitrary already existing shared secret keys, WS-SecureConversation provides mechanisms to derive session keys. How to establish security context is detailed in [18], please refer to it for more detail.

IV. Implementations of Security Management

In this section, main aspects of implementing successful security management for enterprises' information systems are discussed.

A. *Attestation and Negotiation*

Constructing trustworthy network endpoints, as required by accomplishing essentials of security management, is to convince the opposite of the purported identity of the other correspondent.

Moreover, in order to establish trustworthy communication channel between two intending communication parties, it is always required that two parties dynamically negotiate a private session key, some special bytes, e.g. *SN* byte for anti-counterfeiting in strategy for trustworthy smart card transaction, and so on.

In a word, prior to any essential communication taken place, attestation and negotiation must be completed between two intending communication parties. STS protocol is always the underlying algorithm of designing measures for attestation and negotiation.

B. *Application-Specific Strategies for Communication Privacy*

The application-specific strategies are to describe how cryptographic mechanisms are utilized and integrated for achieving specific security targets. In order to successfully implementing security management, it is the most important and complicated to propose proper strategies, which should be carefully tailored toward specific information systems.

1) *What Should be Guaranteed*

People implementing security management should make clear what should be guaranteed and what is not needed prior to proposing the strategies. Data privacy, data integrity, and power to fight attacks are always required.

1) Data privacy: Adversaries cannot read or understand the kernel communication information by intercepting or tampering messages exchanged between two legitimate parties without authority of accessing the proper session key. Legitimate parties possessing right session key can recover any encrypted message to its intelligible representation.
2) Data integrity: To verify data integrity of the received message, the recipient possessing the proper secret session key can recalculate a hash code and compare it with the hash code embedded (e.g. signed by the sender) in the message. If even a single binary bit in the message is altered, the receiving party will generate a different hash code. With data integrity, both parties involved in communicating interaction know that what they are seeing is exactly what the other party sent.
3) Fighting attacks: As in securing smart card transaction, because the initial *SN* byte and the principle for value variation are all trustworthily negotiated by legitimate parties before any essential information exchange, the sender and the receiver can therefore distinguish authentic transaction information from fraudulent information by verifying *SN* byte in the message received. Namely, the value privacy of initial transaction sequence number byte and the privacy of principle deciding the value variation of *SN* lie behind the possibility of anti-counterfeiting.

2) *How to Integrate*

So many cryptographic mechanisms are involved in designing strategies for securing information systems, key to success of security management is how these mechanisms can be combined and integrated.

As exemplified in Section 3.2.3, the WS-Security specification clearly describes how XML encryption and XML signature specifications and so on are applied to SOAP messages. These rules are described using a specific header field, the <wsse:Security> header. This header provides a mechanism for attaching security-related information to a SOAP message, whereas multiple <wsse:Security> headers may exist inside a single SOAP message. Each of these headers is intended for or targeted at a different SOAP intermediary. This enables intermediaries to encrypt or decrypt specific parts of a message before forwarding it or enforces that certain parts of the message must be validated before the message is processed further.

C. *Security Strength and Execution Efficiency: Tradeoff*

1) *Security Strength*

Because of involving a wide variety of cryptographic technologies in designing strategies for securing information systems, including symmetric cryptography, asymmetric cryptography, message digest algorithm, digital signature algorithm, Diffie-Hellman key agreement and exchange algorithm and so forth, not only algorithms should be properly selected, but also related parameters should be properly specified.

Following list gives useful guidance for selecting cryptographic algorithms.

- Authentication Method: RSA Signatures instead of Pre-Shared Keys
- Encryption Algorithm: 128-bit AES-CBC instead of 56-bit DES-CBC
- Hash Algorithm: SHA-1 instead of MD5

- Key Agreement: STS instead of conventional Diffie-Hellman key agreement

2) Efficiency

Cryptographic measures taken in accomplishing security management should not bring noticeably deterious effects to performance of the information system, otherwise, the strategies and related measures are impractical.

As mentioned in Section 2.2, both symmetric cryptography and asymmetric cryptography can be utilized to secure communication. Symmetric cryptography is simpler (such as shorter key length, simpler arithmetic operations, etc.) than asymmetric cryptography; moreover, asymmetric cryptography needs more computational resources (CPU time, memory space, etc.) than that of needed by symmetric cryptography when encrypting or decrypting information. Therefore, symmetric cryptography is preferable to asymmetric cryptography from the viewpoint of efficiency.

Another example, if a Web Service requires each element in SOAP message being encrypted with a 2048-bit RSA algorithm and given the fact that 1000 service invocations may happen during the next 5 min. Security context mentioned in Section 3.2.4 is introduced for dealing with the dilemma, and therefore for guaranteeing service quality.

V. Concluding Remarks

Security is a constantly moving target and should be make its top priority, no enterprise can claim that security measures being taken can make its IT infrastructure proof against all possible future attacks.

However, as outlined in this paper, if the practioners can keep in mind the potential attacks facing the IT infrastructure, the security targets and essentials of achieving them, and implement proper (application-specific) strategies, then successful security management is attainable, and security targets can be achieved.

References

[1] FIPS Publication 46-3, Data Encryption Standard. October 25, 1997.

[2] FIPS Publication 197, Advanced Encryption Standard. November 26,2001.

[3] R. Rivest, A. Shamir, and L. Adleman, A method for obtaining digital signatures and public-key cryptosystems, Communications of the ACM, 1978, 21(2): 120-126.

[4] IETF RFC 1321, The MD5 Message-Digest Algorithm. April 1992.

[5] FIPS-113, Computer Data Authentication. May 30, 1985.

[6] FIPS-180-2, Secure Hash Standard (SHS). August 1, 2002.

[7] FIPS-186-2, Digital Signature Standard (DSS). January 27, 2000.

[8] IETF, IP Security, RFC 2401-2412, http://www.ietf.org/rfc

[9] Trusted Computing Group. https://www.trustedcomputinggroup.org/

[10] Microsoft Corporation, NGSCB. http://www.microsoft.com/resources/ngscb/default.mspx

[11] Netscape, SSL 3.0 specification, http://wp.netscape.com/eng/ssl3/

[12] ISO/IEC 7816, Identification Cards-Integrated circuit(s) cards with contacts, International Organization for Standardization.

[13] W. Diffie, P. C. van Oorschot, and M. J. Wiener, Authentication and authenticated key exchanges, Designs, Codes and Cryptography, 1992, (2): 107-125.

[14] SOAP Version 1.2 Part 1: Messaging framework, http://www.w3.org/TR/2003/REC-soap12-part1-20030624/; June 2003.

[15] WS-Security: SOAP Message Security 1.0, http://docs.oasisopen.org/wss/2004/01/oasis-200401-wss-soap-messages ecurity-1.0.pdf; March 2004.

[16] XML-Signature Syntax and Processing, http://www.w3.org/TR/2002/REC-xmldsig-core-20020212/; February 2002.

[17] XML Encryption Syntax and Processing, http://www.w3.org/TR/2002/REC-xmlenc-core-20021210/; December 2002.

[18] Web Services Secure Conversation Language (WS-Secure-Conversation), http://msdn.microsoft.com/ws/2004/04/ws-secure-conversation/; April 2004.

Application of Fuzzy Set Ordination and Classification to the Study of Plant Communities in Pangquangou Nature Reserve, China

Jin-tun Zhang[1] Dongpin Meng[2]

[1]College of Life Sciences, Beijing Normal University, Xinwaidajie 19, Beijing 100875, China. E-mail: Zhangjt@bnu.edu.cn

[2]Institue of Loess Plateau, Shanxi University, Wucheng Road 36, Taiyuan 030006, China

Abstract: Fuzzy Set Ordination (FSO) and Fuzzy C-means Classification techniques were used to study the relationships between plant communities and environmental factors in Pangquangou Nature Reserve, Shanxi province of China. Pangquangou Nature Reserve, located at N37°20'-38°20', E110°18'-111°18', is a part of Luliang mountain range. Eighty-nine quadrats of 10m x 20m along an elevation gradient were set up and recorded in this area. The results showed that the two methods, FSO and fuzzy C-means classification describe the ecological relations of communities successfully. The results of FSO showed that the distribution of communities is closely related to elevation, water-conditions and humidity, and also related to aspect and slope. Thirteen community types were distinguished by fuzzy C-means classification, and each of them has special characteristics. The combination of FSO and fuzzy C-means classification may be more effective in the studies of community ecology.

INDEX TERMS: **Fuzzy sets, quantitative classification, ordination, plant community, vegetation ecology**

I. INTRODUCTION

Ordination and classification are the most common techniques in vegetation ecology [1-2]. Since 1950s, vegetation ecologists have invented or adopted a wide variety of methods of ordination and classification. Fuzzy set ordination and classification are comparatively new in this field [3-6]. Fuzzy mathematics has been developed by numerous individuals, and is now applied in all the fields of sciences. It is potentially useful in ecology because the description of ecological systems in not always possible in terms of a binary approach. Ecological communities have been shown to vary as their component species respond more or less independently to environmental gradients. Because of this, both overlapping and internal heterogeneities are important characteristics of ecological communities. The techniques based on fuzzy set theory might be more appropriate [2, 5, 7]. The existed applications of fuzzy set techniques in ecology have been proved their efficiency [6, 8-10]. In this work, we used fuzzy set ordination and classification to analyze ecological relations of plant communities in Pangquangou Nature Reserve, China.

II. MATERIALS AND METHODS

Study area

Guandi mountain with the highest peak of 2831 m is located at N37°20'-38°20', E110°18'-111°18', and is a part of Luliang mountain range. The center part of this mountainous area is the National Natural Reserve, Pongquangou Reserve that mainly protect the living environment, *Larix* and *Picea* forests, for a national conservation bird of *Crossoptilon mantchuricum*. The climate of this area is warm temperate and semi humidity with continental characteristics and controlled by seasonal wind. The annual mean temperature varies from 3 °C to 4 °C, the monthly mean temperatures of January and July are –10°C and 16.1°C respectively and the annual accumulative temperature more than 10°C is 2100 °C. The mean precipitation varies from 600mm to 800mm in this mountain. Several soil types, such as mountain cinnamon soil, brown forest soil and mountain meadow soil, can be found in this area. The study area, Shenweigou, is a main valley in the east area. Its elevation varies from 1700 m to 2831 m. Most of this valley is covered by forests, and a small area close to the

K. Elleithy (ed.), *Advances and Innovations in Systems, Computing Sciences and Software Engineering*, 217–222.

mountain top covered by subalpine shrublands and meadows [11].

According to the system of national vegetation regionalization [12], the basal vegetation zone of Pangquangou Nature Reserve is warm temperate deciduous broad-leaved forest. It is rich in plant species in this area. There are 752 species of vascular plants in 345 genuses of 93 families [13]. The main dominant tree species in forest are *Larix principis-rupprechtii*, *Picea wilsonii*, *Picea. meyeri*, *Populus davidiana*, *Betula albosinensis*, *Betula platyphylla*, and *Quercus liaotungensis*. The common shrubs are *Hippophae rhamnoides*, *Ostryopsis davidiana*, *Spiraea* spp., *Caragana jubata*, *Dasiphora fruticosa*, and *Dasiphora glabra*. There are many grass and herbage plants, such as *Carex* spp, *Imperata cylindrical*, *Aconitum cathayensis*, *Sedum aizoon*, *Potentilla supine*, *Sanguisorba officinalis*, *Vicia unijuga*, and *Kobresia bellardii*.

Sampling

Along the elevation gradient of 1700-2700 m in Shenweigou valley, 21 transects separated by 50 meters in altitude were set up, and species data of cover, height, basal area, and individual number from 4-6 quadrats in each transect were recorded. The quadrat size is 10 m × 20 m, in which 4 m × 4 m and 2 m × 2 m small quadrats were used to record shrubs and herbs respectively, and to calculate frequency. There were totally 198 species in 89 quadrats recorded. Elevation, slope, aspect and the depth of litters for each quadrat were also measured and recorded. The elevation for each quadrat was measured by altimeter, the slope and aspect measured by compass meter, and the depth of litters measured by ruler directly.

We used Importance Value of each species as data in community analysis. The importance value was calculated by the formulas [1,2]:

$IV_{\text{Tree and shrub}}$ = Relative cover + Relative frequency + Relative density

IV_{Herbs} = Relative cover + Relative height

Therefore the species data matrix is the importance values of 198 species in 89 quadrats.

Fuzzy set ordination

Ordination is defined here as a mapping from a set of ordered n-tuples to a set of values or symbols. In fuzzy ordination, the ordered n-tuples are the membership values of the elements (quadrats) of the universe in the specified fuzzy sets [5,14]. The procedure of fuzzy set ordination is as follows:

Firstly, we defined the set of quadrats of high elevation (Set ***A***) by scaling the quadrat elevation between 0 and 1 linearly, and assigning the relative elevation to the quadrats as membership values. ***A*** is a fuzzy set whose membership values are in the interval [0, 1]. The notation of ***A*** is

$$A = \mu_A(x) \tag{1}$$

Secondly, we defined the set of low elevation quadrats (Set ***B***) as the complement of the set of high elevation quadrats:

$$B = \mu_{\bar{A}}(x) = 1 - \mu_A(x) \tag{2}$$

Thirdly, we defined the set of quadrats similar to high elevation quadrats (Set ***C***) as follows:

$$\mu_C(x) = \frac{\sum_{x \neq y}\left[R_{xy}(\mu_A(y))\right]}{\sum_{x \neq y}(\mu_A(y))} \tag{3}$$

Where $\mu_C(x)$ is the membership of quadrat x in Set ***C***, and $\mu_A(y)$ is the membership of quadrat y in Set ***A***, R_{xy} is the similarity coefficient between quadrats x and y and is calculated by formula:

$$R_{xy} = \begin{cases} 1 & (\text{when } x = y) \\ 1 - c\sum_{m=1}^{p} | X_{mx} - X_{my} | & (\text{when } x \neq y) \end{cases} \tag{4}$$

Here X_{mx} and X_{my} refer to the importance values of species m in quadrats x and y, P is the number of species and c is a constant.

Fourthly, in the similar manner, the set of quadrats similar to low elevation (Set ***D***) was defined as follows:

$$\mu_D(x) = \frac{\sum_{x \neq y}\left[R_{xy}(\mu_B(y))\right]}{\sum_{x \neq y}(\mu_B(y))} \tag{5}$$

Where $\mu_D(x)$ is the membership of quadrat x in Set ***D***, and

$\mu_B(y)$ is the membership of quadrat y in Set $\boldsymbol{B}$.

Fifthly, we computed the anticommutative difference of fuzzy sets $\boldsymbol{C}$ and $\boldsymbol{D}$ to define the set of quadrats similar to the high elevation quadrats while not similar to low elevation quadrats, which is Set $\boldsymbol{E}$:

$$\mu_E = \mu_{C-D}(x) = \left[1 + (\mu_{\overline{C}}(x))^2 - (\mu_{\overline{D}}(x))^2\right]/2 \qquad (6)$$

Then we use Sets $\boldsymbol{E}$ and $\boldsymbol{A}$ as axes to plot two-dimensional ordination diagram [2,5].

Fuzzy C-means classification

Fuzzy C-means classification is a soft classification technique. It is based on minimizing the within group sum of squares, $J_m(\boldsymbol{U}, \boldsymbol{V}, \boldsymbol{A})$, which is given by

$$J_m(U,V,A) = \sum_{i=1}^{N} \bullet \sum_{j=1}^{C} (U_{ij})^m (dA_{ij})^2 \qquad (7)$$

Where $i = 1, 2, \ldots, N$ = the number of quadrats; $j = 1, 2, \ldots, C$ = the number of clusters; $\boldsymbol{U} = \{U_{ij}\}$ = the matrix of membership values, U_{ij} is the membership of quadrat i in cluster j; $\boldsymbol{V}$ is a matrix of cluster centers; m is fuzziness parameter ($1 \le m < \infty$), usually $m = 2$; $(dA_{ij})^2$ is the diastance index:

$$(dA_{ij})^2 = \| X_i - V_j \|^2 A = (X_i - V_j)^T A (X_i - V_j) \qquad (8)$$

$\boldsymbol{X_i}$ is the vector of attribute measurements in quadrats, usually a vector of ordination scores; V_j is the centre of cluster j, if $\boldsymbol{A}$ is a unit matrix, then

$$(dA_{ij})^2 = \| X_i - V_j \|^2 \qquad (9)$$

The procedure of fuzzy C-means classification is below:

Firstly, we seleceted an ordination method, Detrended Correspondence Analysis (DCA), and carried out ordination analysis for species data. The first DCA axis was used as the basic data X_i.

Secondly, we determined a number of clusters, C.

Thirdly, we assigned the matrix of primary membership values, $\boldsymbol{U_0}$. Any value can be given to a quadrat as its membership value in cluster j, but the sum of memberships for a quadrat must equal to 1: $\sum U_j = 1$.

Fourthly, we calculating the V_j and $(dA_{ij})^2$:

$$V_j = \sum_{i=1}^{N} (U_{ij})^m X_i / \sum_{i=1}^{N} (U_{ij})^m \qquad (10)$$

$(dA_{ij})^2$ was caculted using equations (8) and (9).

Fifthly, we calculated the new membership values:

$$U_{ij} = \left\{ \sum_{k=1}^{C} \left[\frac{(dA_{ij})^2}{(dA_{ik})^2} \right]^{\frac{1}{m-1}} \right\}^{-1} \qquad (11)$$

($i = 1, 2, \cdots, N$; $j = k = 1, 2, \cdots, C$)

Sixthly, based on the new membership values $\boldsymbol{U}$, we go back to the fourth step and calculated the new V_j, $(dA_{ij})^2$ and U_{ij} iteratively, and until the membership values become approximately stable.

At last, we classified quadrats into clusters based on the final $\boldsymbol{U}$. A quadrat belonged to the cluster in which it had the maximum membership value [2, 6].

III. Results

FSO ordination

We selected the elevation as environmental gradient, and calculated fuzzy Sets A, B, C, D and E respectively. The result of quadrats ordination is shown in Fig.1. The first FSO axis is a direct elevation gradient, along which the elevation is increased from the left to the right. The mean temperature is decreased, and the humidity and water-conditions are increased from the left to the right. The second FSO axis is briefly related to water-conditions and humidity, and also correlated to elevation gradient. In addition, slope and aspect are also important to ordination axes. The distribution plant communities (quadrats) in FSO diagram closely related to the FSO axes, i.e. related to environmental gradients. This suggests that FOS describe ecological relations between communities, species and environments successfully.

Fuzzy C-means classification

We used the first DCA axis as the basic information in fuzzy C-means classification. The results shown that 89 quadrats were classified into 13 groups, representing 13 plant communities. The vegetation communities and their brief characteristics are as follows.

I Comm. Valley bottom meadow (found in 2 quadrats). This type is dominated by *Ranunculus japonicus* and *Taraxacum dealbatum*, distributed in hills above 1700-1850 m with slope around 15°, and its soil is subalpine meadow soil. The common species are *Sanguisorba officinalis*, *Thalictrum petaloideum, Taraxacum asiaticum,* and *Carex* spp.

II Comm. *Hippophae rhamnoides, Ostryopsis davidiana* (found in 4 quadrats). Distributed in low hills of 1700-1800 m with slope around 25°, and its soil is brown soi. The common species in the community are *Ribes burejense*, *Rosa bella, Dasiphora glabra, Lonicera chrysantha*, *Artemisia apiacea, Galium verum, Thalictrum squarrosum*, and *Ranunculus japonicus*.

III Comm. *Populus cathayana* (found in 5 quadrats). Distributed in hills about 1800 m with slope around 20°, and its soil is mountain brown soil. The common species are *Betula platyphylla, Evonymus alatus, Spiraea pubescens*, *Cotoneaster acutifolius, Carex lanceolata, Ploygonatum verticilatum, Epilobium hirsutum,* and *Agrimonia pilosa*.

IV Comm. *Pinus tabulaeformis* (found in 12 quadrats). Appeared in altitude 1500-1900 m, slope around 20°, mountain cinnamon soil and brown forest soil. The common species in the community are *Quercus liaotungensis, Betula platyphylla, Salix wallichiana, Ostryopsis davidiana, Rosa xanthina, Vitex negundo* var. *heterophylla, Spiraea pubescens, Potygala tenuifolia, Atractylodes chinensis,* and *Artemisia gmelini*.

V Comm. *Quercus liaotungensis* (found in 11 quadrats). Appeared in altitude 1750-2050 m, slope around 30°, mountain brown soil. The common species in the community are *Betula platyphylla, Picea wilsonii, Crataegus pinnatifida, Rosa xanthina, Corylus mandshurica, Phlomis umbrosa, Carex lanceolata, Convallaria keiskei, Chamaenerion angustifolium*, and *Bupleurum chinense*.

VI Comm. *Populus dividiana, Betula platyphylla* (found in 11 quadrats). It is distributed from 1950 to 2500 m in hills with slope around 20°, and its soil is brown forest soil. The common species in the community are *Quercus liaotungensis, Spiraea pubescens, Hippophae rhamnoides, Carex lanceolata, Lathyrus humilis, Phlomis umbrosa, Galium bungei, Clematis macropetala* and *Maianthemum bifolium*.

VII Comm. *Populus dividiana* (found in 4 quadrats). Distributed from 1950 to 2300 m in hills with slope around 25°, and its soil is brown forest soil. The common species in the community are *Quercus liaotungensis, Lespedeza bicolor, Crataegus pinnatifida, Evonymus alatus, Spiraea pubescens, Carex lanceolata, Convallaria keiskei, Ledebouriella divaricata*, and *Bupleurum chinense*.

VIII Comm. *Larix principis-rupprechti* (found in 14 quadrats). Distributed from 1900 to 2500 m in hills with slope around 20°, and its soil is brown forest soil. The common species in the community are *Betula platyphylla, Picea wilsonii, Quercus liaotungensis, Acer ginnala, Lonicera chrysantha, Lespedeza* sp., *Corylus mandshurica, Carex lanceolata, Lathyrus humilis, Bupleurum chinense,* and *Thalictrum petaloideum*.

IX Comm. *Picea wilsonii* (found in 7 quadrats). Appeared in altitude 1950-2050 m, slope 15-25°, mountain brown forest soil. The common species in the community are *Populus davidiana, Betula platyphylla, Corylus mandshurica, Rosa xanthina, Vicia cracca, Ledebouriella divaricata, Adenophora elata, Carex lanceolata* and *Lathy rushumili*.

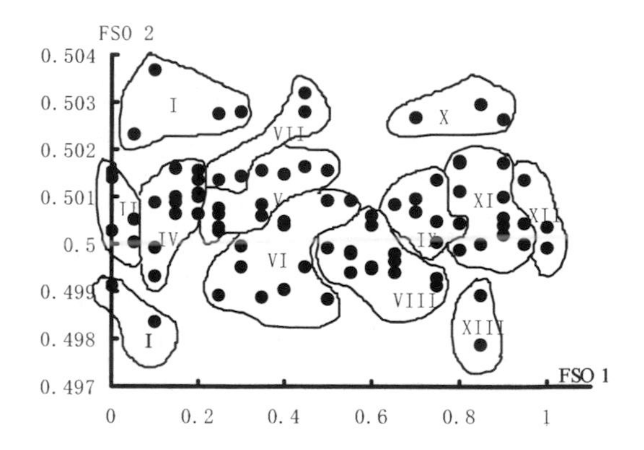

Fig. 1. Fuzzy set ordination diagram of 89 quadrats in Pangquangou Nature Reserve, China. The FSO 1 and FSO 2 refer to the first and second fuzzy set ordination axes, $\mu A(x)$ and $\mu_E(x)$ respectively. I, II, ..., XIII represent community classes produced by fuzzy C-means classification (see the text).

X Comm. *Picea meyeri* (found in 3 quadrats). Distributed from 2000 to 2500 m in altitude with slope 20-25°, brown forest soil. The main species in the community are *Picea wilsonii, Larix principis-rupprechtii, Betula platyphylla, Lonicera hispida, Rosa bella, Lonicera chrysantha, Carex* sp., *Lathyrus humilis, Phlomis umbrosa, Maianthemum bifolium, Cerastium arvense*, and *Vicia unijuga*.

XI Comm. *Potentilla glabra, Spiraea alpina* (found in 11 quadrats). Appeared in altitude 2150-2450 m, slope 20-25°, mountain meadow soil. The common species in the community are *Spiraea pubescens, Dasiphora fruticosa, Carex leorhyncha, Artemisia japonica, Sanguisorba officinalis*, and *Scutellaria scordifolia*.

XII Comm. *Evonymus hamiltonianus* (found in 3 quadrats). Distributed from 2600 to 2800 m close to mountain top, and its soil is mountain meadow soil. The common species are *Dasiphora fruticosa, Trollius chinensis, Saussurea mongolica, Sanguisorba officinalis, Thalictrum petaloideum*, and *Taraxacum dealbatum*.

XIII Comm. Subalpine meadow (found in 2 quadrats). Dominated by *Kobresia bellardii* and *Potentilla nivea*, distributed from 2100 to 2800m with gentle slope, and its soil is subalpine meadow soil. The common species are *Geum aleppicum, Papaver nudicaul , Rhodiola dumulosa, Trollius chinensis*, and *Carex* sp.

IV. DISCUSSION

Fuzzy set ordination raveled the environmental gradients clearly. The first FSO axis is significantly correlated to elevation, humidity and water-conditions, and the second FSO axis related to water-conditions, elevation, aspect and slope. These are the most important factors influencing the spatial patterns of plant communities in this region [11, 13, 15]. Therefore, the results of FSO are reasonable for ecological interpretation [16-18]. The distribution of vegetation in the FSO diagram also supports this point, i.e. communities with low altitude and poor water-conditions distributed in the left part of the FSO diagram, such as Comm. I – IV, while communities with high altitude and rich water-conditions occupied in the right part of the FSO map, such as Comm. X – XIII. The rest communities (Comm. V – IX) occupied in the central part of the map.

The thirteen communities produced by fuzzy C-means classification represent the general vegetation in Pangquangou Nature Reserve [11, 13]. They are almost all secondary natural vegetation. The classification scheme of plant communities is reasonable according to Chinese vegetation classification system [12]. Each type of community occupied a certain area with clear boundaries in FSO space (Fig. 1). This suggests that FSO ordination and Fuzzy C-means classification provide identical ecological relations in this study, and they both are very effective in vegetation analysis [17, 19-21].

Fuzzy set ordination uses information from both vegetation and environment variables and should produce more reasonable results [2, 5, 9]. Fuzzy C-means classification uses information of species composition only [22, 23]. Therefore the combination of fuzzy set ordination and fuzzy C-means classification may be more effective in the studies of community ecology.

ACKNOWLEDGMENT

The study was financial supported by the National Natural Science Foundation of China (No 30070140) and the Teachers' Foundation of Education Ministry of China.

REFERENCES

[1] Greig-Smith, P Quantitative plant ecology. 3rd edition. Blackwell Scientific Publications, London, 1983.

[2] Zhang, J-T. Quantitative ecology. Science Press, Beijing, 2004.

[3] Bezdek J C. Pattern Recognition with Fuzzy Objective Function Algorithms. Plenum Press, New York. 1981.

[4] Zhang, J-T. Application of fuzzy mathematics to vegetation classification of Scrub. *Vitex negundo* var. *heterophylla*. *Acta Phytoecologica et Geobotanica Sinica*,Vol. 9, 306-314, 1985

[5] Robert D W. Ordination on the basis of fuzzy set theory. *Vegetatio*, Vol. 66, 123-131, 1986.

[6] Equihua M. Fuzzy clustering of ecological data. *Journal of Ecology*, Vol. 78, 519-525, 1990.

[7] Kaufmann, A. Introduction to the theory of fuzzy subsets: Vol. 1: foundamental theoretical elements. Academic Press, London. 1975.

[8] Boyce, R. L. Fuzzy set ordination along an elevation gradient on a mountain in Vermont, USA. *Journal of Vegetation Science*, Vol. 9, 191-200, 1998.

[9] Boyce, R. L. and Ellison, P. C. Choosing the best similarity index when performing fuzzy set ordination on binary data. *Journal of Vegetation Science*, Vol. 12, 711-720, 2001.

[10] Sarbu, C. and Zwanziger, H. W. Fuzzy classification and comparison of some Romanian and German mineral waters. *Analytical Letters* Vol. 34, 1541-1552, 2001.

[11] Qiu, Y and Zhang, J-T. The ordination axes clustering based on detrended canonical correspondence analysis ordination and its application to the analysis to the ecological gradients of communities in Bashuigou catchment, Guandi Mountain. *Acta Ecologica Sinica*,Vol. 20, 199-206, 2000.

[12] Wu, Z.Y. Vegetation of China. Science Press, Beijing, 453-615, 1982.

[13] Zhang, J.-T. Succession analysis of plant communities in abandoned croplands in the Eastern Loess Plateau of China. *Journal of Arid Environments*, Vol. 63, 458-474, 2005.

[14] Chen, T-G, Zhang, J-T. Plant species diversity of Shenweigou of Pangquangou Nature Reserve (Shanxi, China) I. Richness, evenness and diversity indices. Chinese *Journal of Applied and Environmental Biology*, Vol. 6, 406-411, 2000.

[15] Zhang, J-T. and Oxley, R. A comparison of three methods of multivariate analysis of upland grasslands in North Wales. *Journal of Vegetation Science*, Vol. 5, 71-76, 1994.

[16] Zhang F. The community characteristics and biomass of *Larix principis-rupprechtii* forest in Pangquangou Nature Reserve. *Journal of Shanxi University* (Nat Sci Ed), Vol. 15, 94-97, 1992.

[17] Maselli, F., Gilabert, M. A. and Conese, C. Integration of high and low resolution NDVI data for monitoring vegetation in Mediterranean environments. *Remote Sensing of Environment*, Vol. 63, 208-218, 1998.

[18] Zhang, J-T. A combination of fuzzy set ordination with detrended correspondence analysis: One way to combine multi-environmental variables with vegetation data.. *Vegetatio*, Vol. 115, 115-122, 1994.

[19] Zhang, J-T. A study on relations of vegetation, climate and soils in Shanxi province, China. *Plant Ecology*, Vol. 162, 23-31, 2002.

[20] Hill, M.O. TWINSPN-A Fortran program for arranging multivariate data in an ordered two-way table by classification of the individuals and atributes. Ithaca: Cornell University, 1979.

[21] Kral, K. Review of current and new supervised classification methods of satellite imagery with increased attention to the knowledge base classification (Case study: Montagne Noir - France) *Ekologia* (Bratislava) Vol. 22 (Suppl. 2), 168-181, 2003.

[22] Zhang, L.J., Liu, C., Davis, C. J., Solomon, D. S., Brann, T. B. and Caldwell, L. E. Fuzzy classification of ecological habitats from FIA data. *Forest Science*, Vol. 50, 117-127, 2004.

[23] Bezdek J C. Numerical taxonomy with fuzzy sets. *Journal of Mathematical Biology*, Vol. 1, 57-71, 1974

On Searchability and LR-Visibility of Polygons

John Z. Zhang
Math and Computer Science
University of Lethbridge
Lethbridge, Alberta
Canada T1K 3M4
zhang@cs.uleth.ca

Abstract – **Imagine that intruders are in a dark polygonal room and move at a finite but unbounded speed, trying to avoid detection. Polygon search problem asks whether a polygon is searchable, i.e., no matter how intruders move, searcher(s) can always detect them. A polygon is LR-visible if there exist two boundary points such that the two polygonal chains divided by them are mutually weakly visible. We explore the relationship between the searchability and LR-visibility of a polygon. Our result can be used as a preprocessing step in designing algorithms related to polygon search.**

I. Introduction

Path planning is a fundamental problem in robotics. At an abstract level, the problem was modeled as *polygon search problem* by Suzuki and Yamashita [17], which is to search for mobile intruders inside a polygonal area by one or more mobile robots or so-called *searchers*. In the model, an enclosed area is delineated by a simple polygon, while searchers and intruders are represented by moving points. A k-searcher is the one who holds k flashlights and whose vision is restricted to the beams from the k flashlights. The light beams from the flashlights can rotate at a bounded speed. Another typical situation is the so-called *two-guard search*, which requires that the two guards be always mutually visible during search and their vision is restricted to the beam between them. In this paper, we focus the polygon search by a boundary 1-searcher, who always stays on the boundary of a polygon. Unless stated otherwise, a 1-searcher is understood as a boundary 1-searcher in the sequel.

Some introductory results regarding searchable polygons by a 1-searcher were obtained in [17]. For instance, if a polygon is searchable by a 1-searcher, it has no three points such that the shortest path between any two of them is invisible to the third one. Park *et al.* [14] discussed the searchability by a 1-searcher, characterizing searchable polygons by identifying a set of forbidden patterns. Based on a case-by-case analysis, Tan [19] established four configurations that make a polygon non-searchable by a non-boundary 1-searcher. Kameda *et al.* [9] proposed a simple characterization. Zhang and Burnett [21] provided another simple proof for the same characterization as the ones proposed in [9], [14]. However, testing those characterizations takes $O(n\log n)$ time, where n is the number of vertices of a polygon. The reason behind this complexity is that one needs to compute the extension points from all the reflex vertices, which puts a lower bound on the complexity. (See Sec. II.A for the definition of extension points.) For other related work, refer to [8], [10], and [15].

Icking *et al.* [7] considered the two-guard *street walk* problem. A *street* is a polygon P with two distinguished vertices s and g, which divide the polygon boundary into two chains, L and R. It is required that L and R be *mutually weakly visible*, i.e., for each point $p \in L$, there exists a point $q \in R$ such that p and q are visible, and vice versa. The two guards, one on each side of the street, start from s, move along the side, and finally meet at g. Any intruder slipping into the street would be, if possible, pushed out through g. Heffernan [6] proposed a linear-time algorithm for checking whether a given street is walkable by two guards. Tseng *et al.* [20] considered whether, for a given polygon, there exists a pair of points, with respect to which the resultant street is walkable. Bhattacharya *et al.* [1] proposed a linear-time algorithm for identifying all those pairs. See [3], [4] for related work.

Room search problem is another variation of the polygon search problem. Instead of two prespecified points, as in the street walk problem, there is only one prespecified boundary point, called *door*, which is required to be always protected, i.e., no intruder shall be able to escape the room through it. Lee *et al.* [11] discussed the problem for a 1-searcher. Park *et al.* [13] investigated the case by two guards. The work in [2], [12], [16], [18] also discussed some other problems in room search.

The main contribution in this paper is to relate the searchability and LR-visibility of a polygon by a 1-searcher. We believe that the result can be used to investigate whether there exists an efficient algorithm (in linear time) for searchability testing on polygons by a 1-searcher. We will delay our further discussions to Sec. II.C after necessary notation.

II. Priliminaries

A. Notation

A polygon P is defined by a clockwise sequence of distinct *vertices* numbered $p_0, p_1, \ldots, p_{n-1}$ ($n \geq 3$), and n *edges*, connecting adjacent vertices. The edge between vertices u and v is denoted by (u,v). The *boundary* of P, denoted by ∂P, consists of all its edges and vertices. The vertices immediately

K. Elleithy (ed.), Advances and Innovations in Systems, Computing Sciences and Software Engineering, 223–228.

preceding and succeeding vertex v in the clockwise direction are denoted by $Prec(v)$ and $Succ(v)$, respectively. For any two points $a, b \in \partial P$, the open (resp. closed) section of ∂P clockwise from a to b is denoted by $\partial P_{cw}(a,b)$ (resp. $\partial P_{cw}[a,b]$). The origin vertex of P is p_0.

A vertex whose interior angle between its two incident edges is more than 180° is called a *reflex vertex*. Consider reflex vertex r. Extend the edge $(Succ(r), r)$ toward the interior of P, and let $B(r) \in \partial P$ denote the *backward extension point*, where this extension leaves P for the first time. We call the polygonal area formed by $\partial P_{cw}[r, B(r)]$ and chord $\overline{rB(r)}$ the *clockwise component* associated with r. Similarly, the extension of $(Prec(r), r)$ determines *forward extension point*, $F(r)$. The *counter-clockwise component* associated with r is bounded by $\partial P_{cw}[F(r), r]$ and chord $\overline{rF(r)}$. A component is *redundant* if it is a superset of another component.

B. 1-Searchable polygons

Two points u and v are said to be *mutually visible* if the line segment $\overline{uv}$ is completely contained inside P. If we consider that $\partial P \subseteq P$, this definition leads to "grazing visibility", allowing $\overline{uv}$ to graze a reflex vertex or a polygon edge. We adopt this visibility definition in this paper.

We formalize the polygon search by a 1-searcher. Let $e(t) \in P$ denote the position of the intruder at time $t \geq 0$. It is assumed that $e:[0,\infty) \rightarrow P$ is a continuous function, representing the unpredictable move path of the intruder. The initial position $e(0)$ and the move path e are unknown to the searcher. Let λ represent a continuous path of the searcher in the form $\lambda:[0,\infty) \rightarrow \partial P$. $\lambda(t)$ denotes the position of the searcher at time $t \geq 0$. The searcher's vision is restricted to the beam emitting from her flashlight, whose direction can be changed continuously with a bounded rotational speed. Let $\theta(t)$ denote the beam head of the flashlight at time t on ∂P, which is the point where the beam intersects ∂P before it leaves the polygon for the first time.

Polygon P is *1-searchable* if we can always find a *search schedule* (λ, θ), such that, for every continuous function $e:[0,\infty) \rightarrow P$, there exists a time $t \in [0,\infty)$ such that $e(t) \in \overline{\lambda(t)\theta(t)}$.

C. LR-visibility and previous results

As defined in [5], a polygon P of n-vertices is *LR-visible* if there exist a pair of points s and t on ∂P such that $\partial P_{cw}(s,t)$ and $\partial P_{cw}(t,s)$ are weakly visible from each other. An LR-visible polygon has some "nice" properties. If a polygon (of n vertices) is LR-visible, we can determine, in $O(n)$ time, all possible pairs of boundary chains (A_i, B_i), $i = 0, 1, \ldots, m$, such that for any $s \in A_i$ and any $t \in B_i$, P is LR-visible with respect to s and t. This is shown in Fig. 1 (adopted from [5]). As a preparatory step, there is another algorithm to calculate all the non-redundant components in linear time. If the polygon is not LR-visible, this algorithm terminates prematurely. These linear-time algorithms have been used as a preprocessing step in various problems related to polygon search, as discussed below.

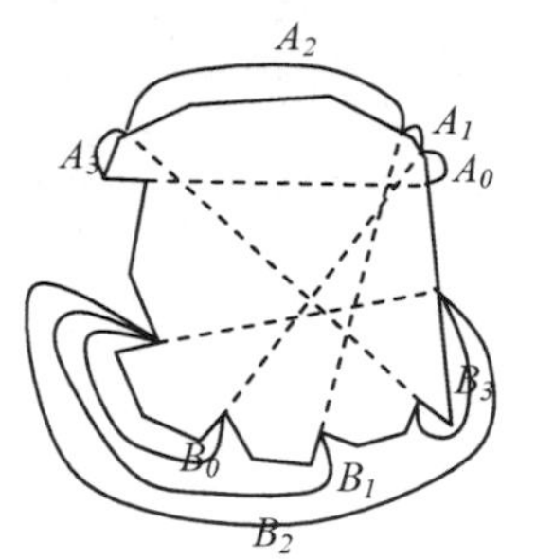

Fig. 1. LR-visibility

The relationship between the LR-visibility of a polygon and its searchability was previously attempted in [17] for a 1-searcher. It was stated that if a simple polygon P is 1-searchable, then there exist a pair of vertices u and v of P such that P is weakly visible from the shortest path between u and v, i.e., every point on the two sides divided by u and v can be seen by at least one point on the shortest path. Then according to [3], P is LR-visible with respect to u and v. However this relationship is not direct, since at that time no characterization was ever obtained for searchable polygons by a 1-searcher and the reasoning for the claim is involved and hard to follow.

LR-visibility of a polygon is part of the definition of the street walk problem, as discussed in Sec. I. However, the properties of an LR-visible polygon were further explored in [1] in designing a linear-time algorithm for identifying all the pairs of points, each of which accommodates a walkable street. This result was a further improvement over the previous $O(n \log n)$ complexity for the same problem [20].

Lee *et al.* [11] and Park *et al.* [13] investigated the room search problem by a 1-searcher and by two guards, respectively. Characterizations were derived for both search models. Bhattacharya *et al.* [2] unified these two search models within the same framework. The LR-visibility of a searchable room was studied, which resulted in a linear-time algorithm for checking whether a room is searchable by a 1-searcher or by two guards.

In this paper, under the light of characterizations of 1-searchable polygons, we attempt to relate the searchability and LR-visibility of 1-searchable polygons. We believe that with our result, we might be able to test whether a polygon is searchable by a 1-searcher in linear time, an improvement over the previous result.

In the sequel, we represent a polygon using a unit circle and only mark relevant reflex vertices, for the sake of clear illustrations.

III. Searchability of a Polygon

We utilize the characterization proposed in [21] for a 1-searchable polygon, since it facilitates our discussions in Sec. IV. The equivalent characterization was discussed in [9], [14] as well, but with different proof approaches.

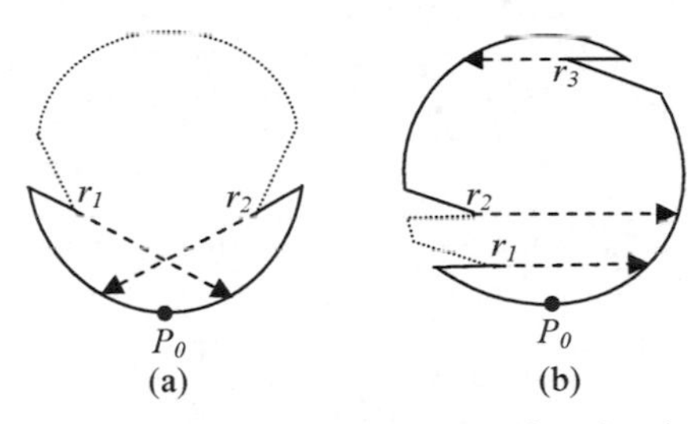

Fig. 2.. Patterns that generate traps on polygon boundary.

A searcher starts her search from somewhere on the boundary of a given polygon. It is obvious that the polygon is not 1-searchable, if there is "no where" to start such a search. As discussed in [21], some patterns, in terms of the reflex vertices in the polygon, are due to this observation. We list in Fig. 2 the patterns that bound some polygonal chains on ∂P. The pattern in Fig. 2 (a) (The extensions from r_1 and r_2 need not intersect.) bounds $\partial P_{cw}(r_1, r_2)$, while the pattern in Fig. 2 (b) (The extensions from r_1 and r_2 can intersect.) bounds $\partial P_{cw}(r_1, r_2)$. In both cases, if a point s is falling in such a polygonal chain, s is said to be *trapped*, which means that the searcher should not start search from it. If every point on ∂P is trapped, ∂P is said *trapped*. Under this situation, it is obvious that the searcher has nowhere to start a search, i.e., the polygon is not 1-searchable.

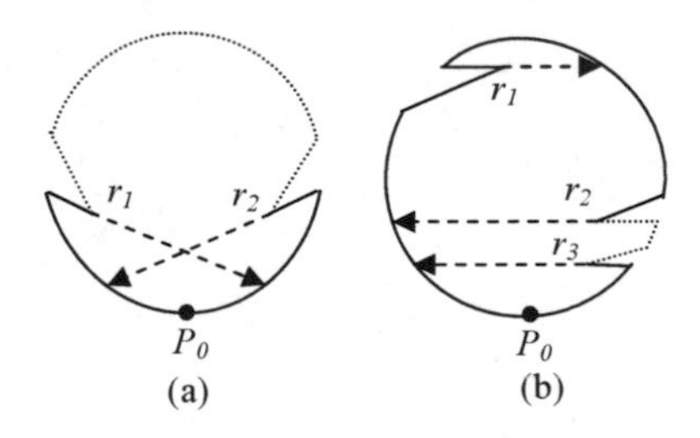

Fig. 3. Patterns that generate unreachables on polygon boundary.

When a search finishes, the beam head and the searcher's position are at the same point t on ∂P, with the whole polygon cleared. We say that t is *reachable*. However, if every point on ∂P is not reachable due to some patterns, the polygon is not searchable by a 1-searcher. We list those patterns in Fig. 3 that make some portions of ∂P unreachable. In Fig. 3 (a), the points falling in $\partial P_{cw}(r_1, r_2)$ are unreachable. The same applies to the ones in $\partial P_{cw}(r_2, r_3)$ in Fig. 3 (b).

In addition, we have two special patterns shown in Fig. 4 ([9], [14], [19], [21]), which are included in neither Fig. 2 nor Fig. 3. Note that the extensions from r_1 and r_3 need not intersect. The following is the major result from [21].

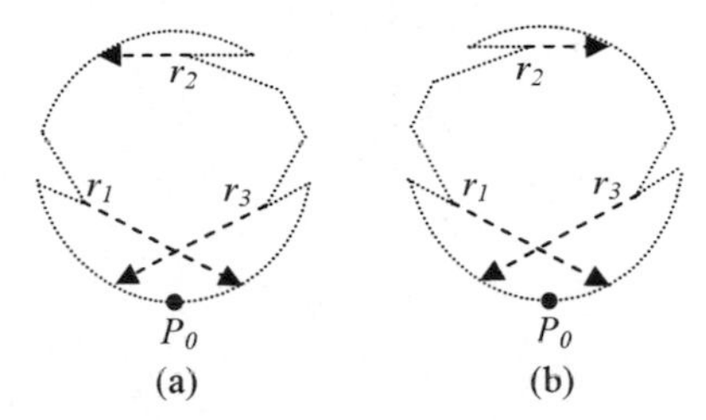

Fig. 4. Two special patterns that make a polygon non 1-searchable.

Theorem III-1: A polygon P is 1-searchable if and only if not every point on ∂P is trapped, not every point on ∂P is unreachable, and the patterns in Fig. 4 are not present. □

IV. Searachability and LR-visibility

In this section, we prove that a searchable polygon by a 1-searcher is also LR-visible.

We first give the following lemma [5].

Lemma IV-1: If a polygon contains three disjoint components, it is not LR-visible. □

See Sec. II.A for the definition of components in a polygon. The following lemma was first proved in [17].

Lemma IV-2: If a polygon contains three disjoint components, it is not 1-searchable. □

By Lemma IV-2, there are only two cases of our further interest in a searchable polygon; either (1) there are no disjoint components or (2) there exist two disjoint components, with all the other components having common polygon boundary with at least one of them.

Let us consider the situation where all the components in a searchable polygon have a common polygon boundary.

Lemma IV-3: If all the components in a searchable polygon share a common polygon boundary, the polygon is LR-visible.

Proof: We can select two distinguished points s and t on the common boundary. We immediately know that the polygon is LR-visible with respect to s and t since every component contains both of them [5]. □

We now only need to consider case (2), as shown in the following series of lemmas.

Lemma IV-4: If a 1-searchable polygon P contains, among all the others, two disjoint components, with one clockwise and the other counter-clockwise, P is LR-visible.

Proof: we will locate two distinguished points s and t on ∂P such that P is LR-visible with respect to them.

Assume that the clockwise component is due to reflex vertex a and the counter-clockwise one is due to reflex vertex

b, as shown in Fig. 5.

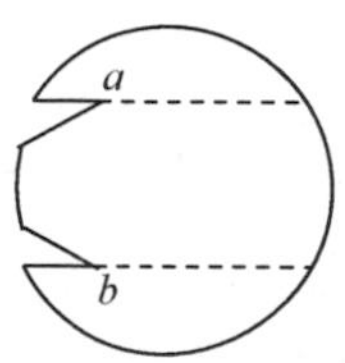

Fig. 5. Two disjoint components. (The clockwise component is due to vertex a while the counter-clockwise component is due to vertex b.)

All the other components must have common polygon boundary with at least one of them, since otherwise the polygon is not 1-searchable.

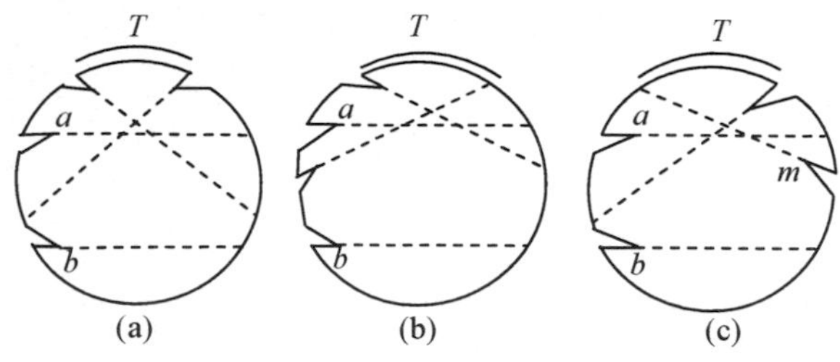

Fig. 6. Definition of T for the components in Fig. 5

We consider the components that only intersect the component due to a. It is easy to see that the common boundary of these components is continuous inside the component. We call it T.

There can be only three situations here, which are shown in Fig. 6. In the figure, we only show the two components that generate the starting point and the ending point of T. The situation shown in Fig. 6 (c) is the special pattern shown in Fig. 4 (a), with $m = r_1$, $b = r_2$, and $a = r_3$. We thus drop it immediately from further discussions.

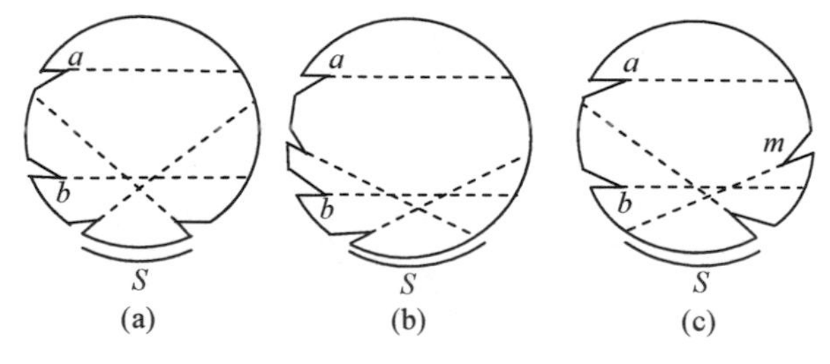

Fig. 7. Defintion of S for the components in Fig. 5

Similarly, we consider the components that only intersect the component due to b. We also have a common intersection of those components, which is within the component (due to b) and is denoted as S. Fig. 7 shows three situations. Again, the situation in Fig. 7 (c) is the special pattern shown in Fig. 4 (b) with $b = r_1$, $a = r_2$, and $m = r_3$, and is thus dropped for further discussions as well.

The only components left are those that intersect (having common boundaries with) both the component due to a and the one due to b. However, we claim that such a component must contain S, T or both, as shown in the following, since otherwise the polygon is not 1-searchable.

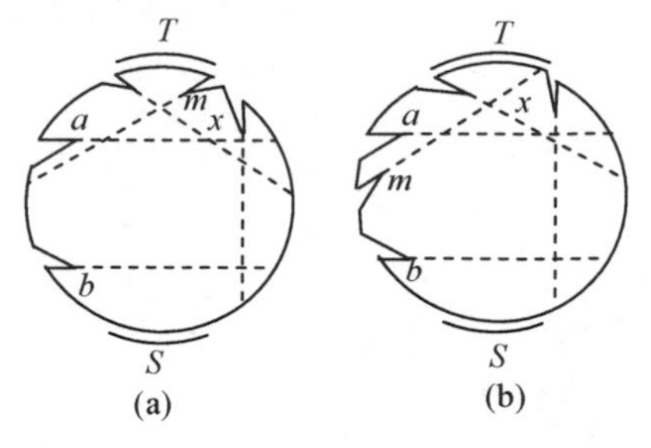

Fig. 8. The situation where the clockwise component due to x does not intersect T or S, and is between T and S clockwise.

We first assume that the component under discussion is due to reflex vertex x and is clockwise. Suppose that it is between T to S in the clockwise direction and does not intersect T or S.

We list all the possible situations in Fig. 8. It is easy to see that all these situations make the polygon non-searchable. For instance, the three components due to the reflex vertices x, m and b in Fig. 8 (a) form the special pattern shown in Fig. 4 (a), while the three components due to the reflex vertices x, m and b in Fig. 8 (b) form the special pattern shown in Fig. 4 (b).

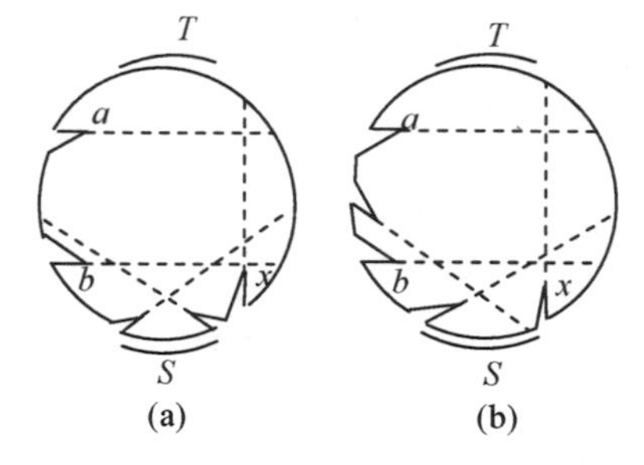

Fig. 9. The situation where the counter-clockwise component due to x does not intersect T or S, and is between T and S clockwise.

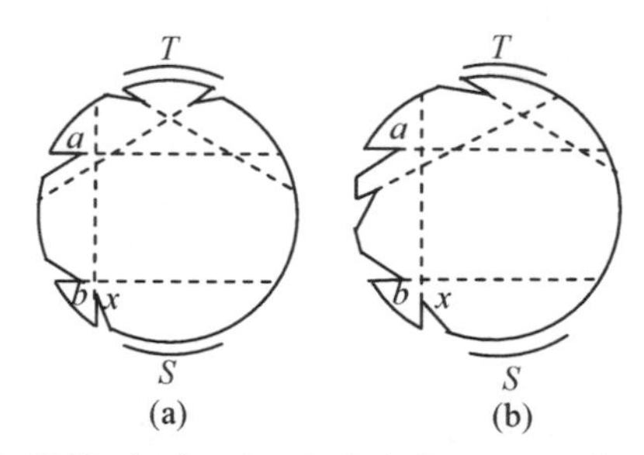

Fig. 10. The situation where the clockwise component due to x does not intersect S or T, and is between S and T clockwise.

Next we assume that the component due to x is counter-clockwise and is between T and S in the clockwise direction. For the same reason as above, the two possible situations, as shown in Fig. 9, make the polygon non 1-searchable.

We then consider the situation where the component due to x is clockwise but is between S and T in the clockwise direction. The possible situations are shown in Fig. 10. Here, we consider Fig. 10 (b) since the same argument also applies to the situation in Fig. 10 (a).

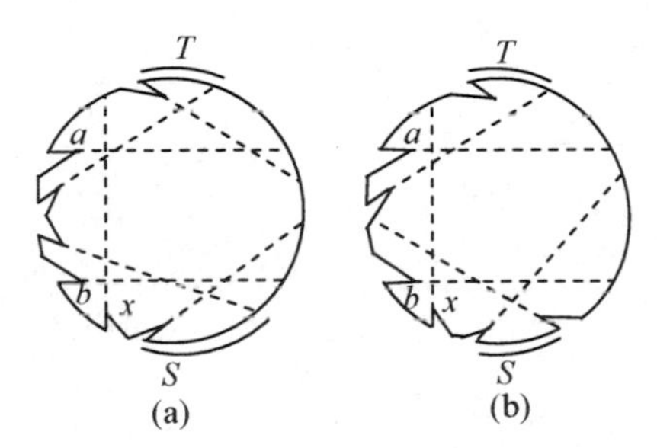

Fig. 11. Further situations for the pattern in Fig. 10.

We show in Fig. 11 the two possible situations for Fig. 10 (b), where we include the components that form S. It is easy to see, using the same reasoning as the one for Fig. 8 and Fig. 9, that in each of these situations in Fig. 11, the polygon is not 1-searchable.

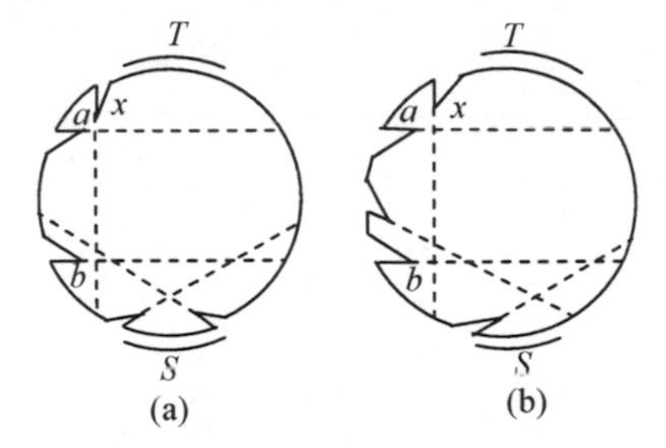

Fig. 12. The situation where the counter-clockwise component due to x does not intersect S or T, and is between S and T clockwise.

The last possibility we need to consider is that the component due to x is counter-clockwise but is between S and T in the clockwise direction. Again we list the two possible situations in Fig. 12. For those patterns, we further consider them, just as what we did above. We list the possible situations in Fig. 13 for the pattern shown in Fig. 12 (b) when the components forming T are included for consideration. It is easy to see that each of the situations makes the polygon non 1-searchable. The pattern in Fig. 12 (a) is dealt with similarly.

We conclude that every such a component must contain S or T. Adjust S and T by taking into account the common boundary of these components. Now select a point $s \in S$ and a point $t \in T$. Every component must contain either s or t, with respect to which P is LR-visible. □

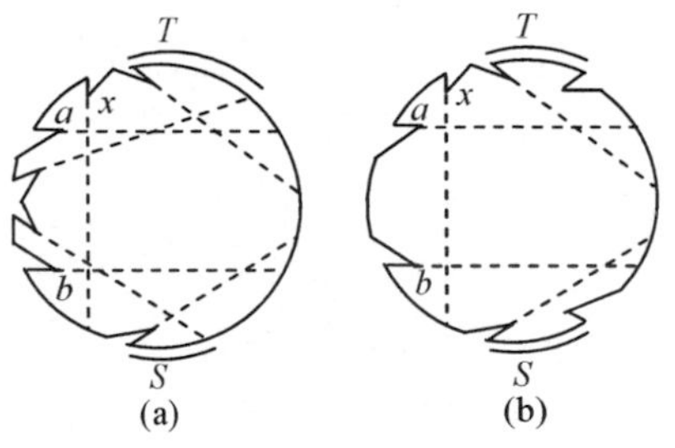

Fig. 13. Further situations for the pattern in Fig. 12 (b)

Lemma IV-5: If a searchable polygon P contains, among all the others, two disjoint components, both clockwise or counter-clockwise, P is LR-visible.

Proof: We consider here the case where the two components are clockwise. The same reasoning applies to the counter-clockwise case.

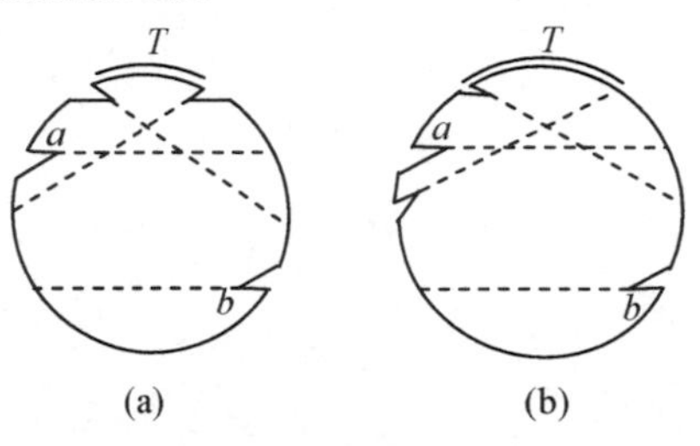

Fig. 14. The situation for T where the components due to a and b are both clockwise.

We list the two situations where T is formed in Fig. 14 when the two disjoint components due to a and b, respectively, are both clockwise. The situation in Fig. 14 (a) involves a counter-clockwise component and was already dealt with in *Lemma IV-4*. The only situation left is shown in Fig. 14 (b).

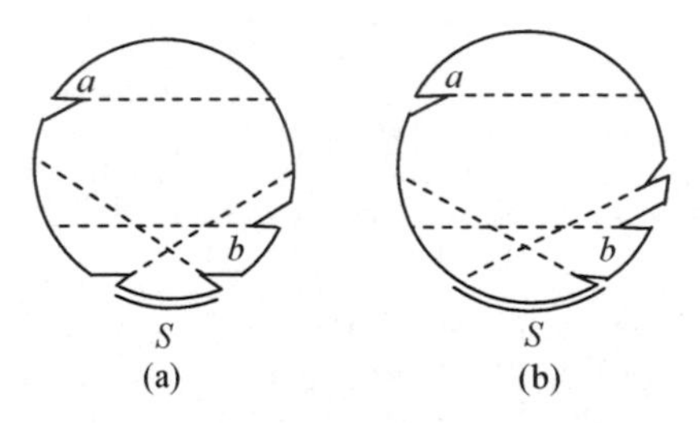

Fig. 15. The situations for S where the components due to a and b are both clockwise.

Similarly, for S we list the two situations in Fig. 15 and only the situation in Fig. 15 (b) needs to be further discussed.

Thus the only combination is between the situation in Fig. 14 (b) and the one in Fig. 15 (b).

We consider a clockwise component between T and S in the clockwise direction, as shown in Fig. 16. It is not 1-searchable since every point on ∂P is unreachable, according to Fig. 3 (b). Note that we need not consider any counter-clockwise components.

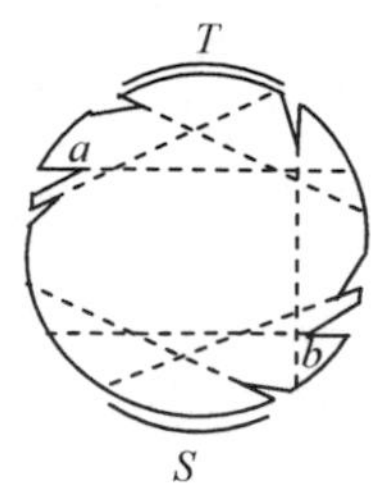

Fig. 16. The situation where there is a clockwise component between T and S clockwise.

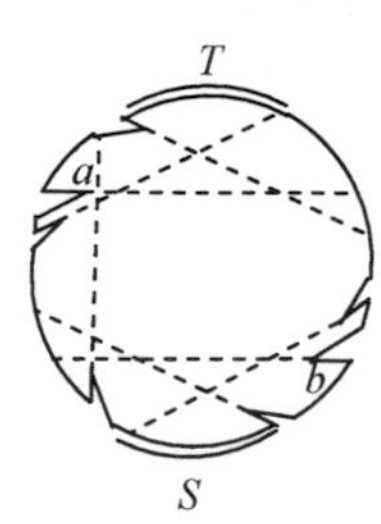

Fig. 17. The situation where there is a clockwise component between S and T clockwise.

The last possibility is shown in Fig. 17, which can also be handled similarly, since every point on ∂P is unreachable.

Summarizing the above, the lemma follows. □

Theorem IV-6: A 1-searchable polygon P is LR-visible.

Proof: It follows directly from *Lemma IV-3*, *Lemma IV-4* and *Lemma IV-5*. □

V. Discussions

The major contribution of the paper is to relate the searchability and the LR-visibility of polygons when a 1-searcher is used. Given a polygon, we first check its LR-visibility. If the answer is negative, we immediately know that the polygon is not 1-searchable. Otherwise, we take advantage of the properties of its LR-visibility and attempt to check in linear time whether it is searchable or not by a 1-searcher. We are working along this direction.

Though there is some subtle difference between a 1-searcher and two guards, we believe that the same conclusion holds for the latter search model as well.

On the other hand, an LR-visible polygon may not be searchable by a 1-searcher. Such an example is shown in Fig. 4, where both polygons are LR-visible but not 1-searchable.

Acknowledgment

The author would like to express their sincere thanks to the anonymous readers from Simon Fraser University, Canada, for their comments during the initial writings of the paper.

References

[1] B. Bhattacharya, A. Mukhopadhyay, and G. Narasimhan, "Optimal algorithms for two-guard walkability of simple polygons," In *Proc. 7th Int'l Workshop on Algorithms and Data Structures, Proc.*, 2001.

[2] B.K. Bhattacharya, J. Z. Zhang, Q. Shi, and T. Kameda, "An optimal solution to room search problem," In *Proc. 18th Canadian Conf. on Computational Geometry*, pages 55-58, 2006.

[3] B. Bhattacharya and S. K. Ghosh, "Characterizing LR-visibility polygons and related problems," In *Proc.10th Canadian Conf. on Computational Geometry*, 1998.

[4] D. Crass, I. Suzuki, and M. Yamashita, "Searching for a mobile intruder in a corridor: the open edge variant of the polygon search problem," *Int'l J. of Computational Geometry and Applications*, 5(4): 397-412, 1995.

[5] G. Das, P. J. Heffernan and G. Narasimhan, "LR-visibility in polygons," *Computational Geometry: Theory and Applications*, 7:37-57, 1997.

[6] P. Heffernan, "An optimal algorithm for the two-guard problem," *Int'l J. of Computational Geometry and Applications*, 6:15-44, 1996.

[7] C. Icking and R. Klein, "The two guards problem," *Int'l J. of Computational Geometry and Applications*, 2(3):257-285, 1992.

[8] T. Kameda, M. Yamashita, and I. Suzuki, "On-line polygon search by a seven-state boundary 1-searcher," *IEEE Trans. on Robotics*, Vol. 22:446–460, 2006.

[9] T. Kameda, J. Z. Zhang, and M. Yamashita, "Simple characterization ofpolygons searchable by 1-searcher," In *Proc. 18th Canadian Conf. on Computational Geometry*, pages 113–116, August 2006.

[10] S. M. LaValle, B. Simov, and G. Slutzki, "An algorithm for searching a polygonal region with a flashlight," *Int'l J. of Computational Geometry and Applications*, 12(1-2):87–113, 2002.

[11] J. Lee, S. M. Park, and K. Y. Chwa, "Searching a polygonal room with one door by a 1-searcher," *Int'l J. of Computational Geometry and Applications*, 10(2):201–220, 2000.

[12] J. H. Lee, S. Y. Shin, and K. Y. Chwa, "Visibility-based pursuitevasions in a polygonal room with a door," In *Proc. ACM Symp. On Computational Geometry*, pages 281–290, 1999.

[13] S. M. Park, J. H. Lee, and K. Y. Chwa, "Characterization of rooms searchable by two guards," In *Proc. Int'l Symp. on Algorithms and Computation*, pages 515–526, 2000.

[14] S. M. Park, J. H. Lee, and K. Y. Chwa, "A characterization of the class of polygons searchable by a 1-searcher," Technical Report CS/TR-2000-160, Korea Advanced Institute of Science and Technology, December 2000.

[15] S. M. Park, J. H. Lee, and K. Y. Chwa, "Visibility-based pursuit-evasion in a polygonal region by a searcher," In *Proc. 28th Int'l Colloquium on Automata, Languages and Programming*, pages 456–468, 2001.

[16] S. M. Park, J.-H. Lee, and K.-Y. Chwa, "Searching a room by two guards," *Int'l J. of Computational Geometry and Applications*, 12(4):339–352, 2002.

[17] I. Suzuki and M. Yamashita, "Searching for a mobile intruder in a polygonal region," *SIAM J. on Computing*, 21(5):863–888, October 1992.

[18] X. Tan, "Efficient algorithms for searching a polygonal room with a door," In *Proc. Japanese Conf. on Discrete and Computational Geometry*, pages 339–350, 2000.

[19] X. Tan, "Searching a simple polygon by a k-searcher," In *Proc. Int'l Symp. on Algorithms and Computation*, pages 503–514, 2000.

[20] L. H. Tseng, P. J. Heffernan, and D. T. Lee, "Two-guard walkability of simple polygons," *Int'l J. of Computational Geometry and Applications*, 8(1):85–116, 1998.

[21] J. Z. Zhang and B. Burnett. "Yet another simple characterization of searchable polygons by 1-searcher," In *Proc. IEEE Int'l Conf. on Robotics and Biomimetics*, December 2006, To appear.

Swarm Intelligence in Cube Selection and Allocation for Multi-Node OLAP Systems

Jorge Loureiro
Dept. de Informática, Escola Superior de Tecnologia
Instituto Politécnico de Viseu
Campus Politécnico de Repeses
3505-510 Viseu, PORTUGAL
Tel. 351 232 480530
jloureiro@di.estv.ipv.pt

Orlando Belo
Departamento de Informática, Escola de Engenharia
Universidade do Minho
Campus de Gualtar
4710-057 Braga, PORTUGAL
Tel. 351 253 604476
obelo@di.uminho.pt

Abstract-**The continuous growth of OLAP users and data impose additional stress on data management and hardware infrastructure. The distribution of multidimensional data through a number of servers allows the increasing of storage and processing power without an exponential increase of financial costs. But this solution adds another dimension to the problem: space. Even in centralized OLAP, cube selection efficiency is complex, but now, we must also know where to materialize subcubes. This paper proposes algorithms that solve the distributed OLAP selection problem under space constraints, considering a query profile, using discrete particle swarm optimization in its normal, cooperative, multi-phase and hybrid genetic versions.**

I. Introduction

Organizations depend heavily on the quality of their information systems. All business actors and, particularly, enterprise decision makers need consolidated, trusty, timely, suitable and handily information to manage uncertainty. Data warehousing (DWing) and satellite technologies and concepts bring the power of information and knowledge into the business management. Among these, OLAP (on-line analytical processing) allows a natural business view perspective and easy data surfing, especially concerning to its inherent multidimensional data analysis, which increases its popularity. The size of OLAP data and queries became huge and the on-line characteristic imposed a great stress on OLAP server and data management. While research has provided a huge set of technologies for the efficient operation of centralized DW/OLAP systems, the architectural style of huge installations has shifted from a centralized to a distributed structure. Among the reasons for such a change (some of them of historical nature) one seems obvious: when a problem becomes simply not manageable, we must divide it. Advantages are obvious: an increase of processing power and storage capacity without an exponential increase of financial costs. This OLAP infrastructure is then composed of several server-nodes interconnected by communication facilities, and then we name it as MultiNode-OLAP (M-OLAP) architecture. But this solution comes at expenses of an additional increase of management complexity. Now, we must deal with an additional dimension: space. Selecting the optimal set of aggregate data structures (views or subcubes), to speed up incoming queries constrained by materializing storage and/or maintenance costs, is now, additionally, constrained by storage capacities per node, maximum global maintenance cost and a query mix per node. The selection must also consider processing power per node, queries redirection, parallelization of tasks, communication costs (when transferring data between nodes) and manage replicates of subcubes (i.e. data of a specific granularity may be stored multiple times to save communication costs and speed up processing). The solutions of the centralized approach must now be migrated to the distributed world. We need 1) distributed cost models, query and maintenance cost formulas, 2) estimation algorithms and 3) to design or adapt general purpose optimizing algorithms, many of them already used for the centralized approach.

Concerning to the first need, the linear cost model and lattice framework proposed in [1] whose dependencies and the ancestor / descendent concept allow the computing of one subcube using others. This framework is extended in [2] to the distributed aggregation lattice, where the communication facilities generate inter-node dependencies. This distributed lattice was adapted in [3],[4] for the M-OLAP architecture considering real communication and processing power parameters, where query and maintenance cost formulas also appear.

Estimation query algorithms for centralized DW/OLAP facilities are straightforward: nothing more than searching the query's ancestor subcube with minimal cost. When the distributed scenario is on concern, the inter-node subcubes' dependencies may be used, and also parallel processing simulation may be interesting [3], although not absolutely necessary: when optimization is the only objective function, a serial approach may be sufficient, as the algorithm needs only a comparative cost value, and not a value that is closer to reality. This could speed-up the execution, once the serial algorithm's complexity is, usually, substantially lower. Maintenance cost estimation algorithms are most complex. Two situations can be distinguished: integral (computing all materialized subcubes from scratch) or incremental. For the distributed scenario, since maintenance cost may be used as constraint, it makes all sense to use the inherent parallelism of M-OLAP architecture. Subcubes or deltas may be processed and integrated as a propagation wave, using updated subcubes (or deltas) to compute others and integrating generated results

K. Elleithy (ed.), Advances and Innovations in Systems, Computing Sciences and Software Engineering, 229–234.

into the destination nodes. Some maintenance cost estimation algorithms and a comparative analysis are presented in [3].

Related to the optimization algorithms, only approximate solutions are intended, as the cube selection problem is recognizably NP-hard [1]. To the centralized approach, many solutions have already been proposed: Greedy [1],[5], evolutionary e.g. [6], random (iterative improvement, simulated annealing and two-phase optimization) [7] and Discrete Particle Swarm Optimization (Di-PSO) [8]. For the distributed scenario, a greedy approach [2] and a genetic and co-evolutionary one have been recently proposed [4].

This research work proposes a study of a Di-PSO approach when applied to the distributed cube selection problem, and it also researches about the performance impact of some algorithm's variants, mainly the inclusion of some genetic hybridizing, cooperative and multi-phase versions with mass extinction and hill climbing.

II. Cost Model and Cost Estimation Algorithms Revised

M-OLAP architecture has several OLAP server nodes (OSN), characterized by their processing power and storage resources, able to answer and redirect user queries. These nodes are interconnected by a possibly heterogeneous communication network, which imposes additional communication costs. The schema is shown in Fig. 1, where we have a 3 nodes' architecture. The communication facilities generates new dependencies beyond the node's border, and then in this distributed lattice we have two kinds of dependencies: the traditional intra-node dependencies (as in [1]) and inter-node dependencies (shown in Fig. 1 as a dash line), as any subcube into one node may be computed using other subcube into a different node.

We're going to use the linear cost model [1], but now immersed into the distributed lattice. Instead of using records, in this paper we're going to use time for the unit of costs, as it matches the purpose of the undertaken optimization – minimizing the answering time to the user's queries – and, on the other hand, time also comes to sight in the maintenance time cost or constraint.

The distributed nature of M-OLAP architecture results in two distinct kinds of costs: 1) intrinsic costs due to scan/aggregation or integration, known from now on as processing costs (Cp), and 2) communication costs (Cc), when transferring data between OSNs. Both are responsible for the query and maintenance costs. In short, we have an M-OLAP Architecture where each node has a processing power (Pp) and storage, where the M distribution of materialized subcubes inhabits (possibly including redundant ones). These nodes are fully interconnected through a general network, where each link is characterized by a Binary Debit (BD) and Latency (La).

Given an M distribution of subcubes and a query set Q, the primary objective function to minimize will be the total answering time (a measure of the time that the user must wait for the answer, related to its productivity and satisfaction). Then, the selection and allocation of distributed M may be stated formally as:

Definition 1: Selection and allocation of distributed M problem. Let $Q=\{q_1, \ldots, q_n\}$ be a set of queries with access frequencies $\{fq_1, \ldots, fq_n\}$, query extension $\{qe_1,\ldots,qe_n\}$; let update frequency and extension be $\{fu_1, \ldots, fu_n\}$ and $\{ue_1,\ldots,ue_n\}$, respectively, and let S_{Ni} be the amount of materializing space by OLAP node i. A solution to the selection and allocation problem is a set of subcubes $M=\{s_1,\ldots,s_n\}$ with a $\sum_j |s_{jN_i}| \leq S_{N_i}$, so that the total costs of answering all queries Q and maintaining M, $Cq(Q, M)+Cm(M)$ are minimal.

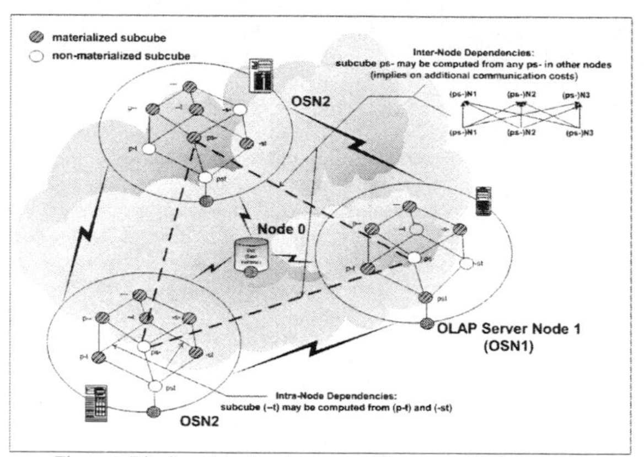

Figure 1. Distributed Lattice. The dashed line shows the inter-node dependence in the same level of granularity relating to the communication interconnections.

Cq and Cm may be computed using the following equations:

$$Cq(Q,M) = \sum_{q_i \in Q} \min(|Anc(S_i,M)| / Pp_n + |S_i|.qe_{q_i}.8.8 / BD + La).fq_{q_i} \quad (1)$$

and

$$Cm(M) = f_u \sum_{S_i \in M} \min|Anc(S_i,M)| / Pp_{on} + (|S_i|.8.8 / BD + La + |S_i| / Pp_{dn}).u_e \quad (2)$$

In these equations:

- Pp is the processing power of node n (in $records.s^{-1}$), that allows to adopt time as the cost referential and differentiate the processing cost of a subcube (or delta), depending on the node where it was processed (Pp_{on}) or integrated (Pp_{dn});
- DB and La are the binary debit and latency of the communication link, only considered if the answering node is different from the query node;
- fq and fu are query and update frequencies;
- qe and ue are the query and update extend, being the first a measure of the size of the decreasing of the produced subcube, when applying restrictions clauses; the second accommodates the impact of the changes into each subcube, if incremental maintenance is on concern;

- finally, $|S_i|.8.8$, is the size (in bits) of subcube S_i - the amount of data to transfer between nodes - supposing that each cell has 8 bytes – the size of a double type number in many representations, that may have to be corrected to an integer number of packets, if the communication link has a minimal packet size.

Concerning to the cost estimation algorithms, we opted by the use of the architecture's parallelism when estimating costs, allowing that multiple queries or subcubes/deltas can be processed using different OSNs and communication links. This implied the control of conflicts and resource allocation with the use of succeeding windows to discretize time and separate each parallel processing where multiple tasks may occur. With these general rules in mind, we will use the Greedy2PRS algorithm to estimate the maintenance costs, proposed in [3], which may be looked for further details about the distributed cost model and, specially, concerning to parallel execution tasks simulation. We also used a query costs estimation algorithm, M-OLAP PQCEA, which is the acronym of Multi-Node OLAP Parallel Query Cost Estimation Algorithm, also used in [4].

III. PARTICLE SWARM ALGORITHMS AND ITS VARIANTS

Life (as its biological basic aspect, or, at a higher level, as the social and cognitive behavior), is a perpetual process of adaptation to the environmental conditions, requiring a continuous demand for solutions in face of succeeding new problems. Particle swarm optimization (PSO) [9] is a computational paradigm based on the phenomenon of collective intelligence exhibited e.g. by swarms of insects or schools of fish, modeling a social and cognitive behavior. PSO paradigm is based on a collection, called a swarm, of fairly-primitive agents, called particles, which can fly on an n-dimensional space, whose position will bring its instant fitness. This position may be altered by the application of an interactive procedure that uses a velocity vector, allowing a progressively best adaptation to its environment. It also assumes that individuals are social by nature, and thus capable of interacting with others within a given neighborhood. For each individual, there are two main types of information available: the first one is its own past experiences (known as individual knowledge), the pbest (particle best) position; the other one is related to the knowledge about its neighbor's performance (referred as cultural transmission), gbest (global best) position. PSO in its original proposal was directed to continuous optimization problems, being latter adapted to discrete ones, named Discrete Particle Swarm Optimization (Di-PSO) [10]; in this work we'll restrict our discussion to the second, given the combinatorial nature of the problem that we intend to solve.

A. Discrete Particle Swarm Algorithm

A simple form of a discrete particle swarm algorithm may be viewed in Algorithm 1. The algorithm has three main pieces: 1) the generation, 2) the iterative part, where each particle is evaluated with the pbest and gbest updating and where particles move to regions of space corresponding to better solutions, and 3) the returning of the gbest position. Although similar to the continuous version, the spatial evolution of particles now is not addictive. The discrete space doesn't allow the addictive continuous relation space-velocity. It is substituted by the introduction of the probabilistic space: the particle's position, in each dimension, will be given by a probability dependent of its velocity vector. Further details, equations and rules may be found in [10] and [8].

ALGORITHM 1. STANDARD DISCRETE PARTICLE SWARM OPTIMIZATION ALGORITHM.

1. *Initialization*: randomly initialize a population of particles (position and velocity) in the n-dimensional space.
2. *Population loop*: **For each** particle, **Do**:
   2.1. *Own goodness evaluation and pbest update*: evaluate the 'goodness' of the particle.
   If its goodness > its best goodness so far, **Then** update *pbest*.
   2.2. *Global goodness evaluation and gbest update*:
   If the goodness of this particle > the goodness that any particle has ever achieved, **Then** update *gbest*.
   2.3. *Evaluate* $v_i^{k+1} = wv_i^k + c1.rnd().(pbest - s_i^k) + c2.rnd().(gbest - s_i^k)$ (3)
   $if \ v_i > v_{max}, \ then \ v_i = v_{max}; \ if \ v_i < -v_{max}, \ then \ v_i = -v_{max}$ (4)
   2.4. *Particle position update*: $if \ rnd() < sig(v_i^{k+1}) \ then \ s_i^{k+1} = 1; \ else \ s_i^{k+1} = 0$ (5)
3. *Cycle*: **Repeat** Step 2 **Until** a given convergence criterion is met, usually a pre-defined number of iterations.

B. Other Variants

Many variations have been proposed to PSO, mainly including: 1) cooperation [11], 2) genetic hybridization [12], [13], 3) mass extinction and 4) multi-phase swarms with hill-climbing [14].

The cooperative version (Di-CPSO) differs mainly from the original version in the number of swarms: several in the first and only one in the second. The original cooperative version was proposed in [11], to the PSO continuous version, where were named as cooperative swarms, due to the cooperation among the swarms. Here, it was migrated to the discrete version. Instead of each particle be allow to fly over all dimensions, it is restricted to the dimensions of only a part of the problem (in this case, the node boundary). But the fitness can only be computed with a global position of a particle. Then, we need a scheme to build a general position, where the particle is included. This is achieved with the denoted "context vector": a virtual global positioning of particles, where each particle's position will be included (into the corresponding dimensions of the node). In the proposed algorithm, some variations were introduced to the scheme proposed in [11]. Not only the best particles (gbest) of each swarm are selected to generate the context vector, but we use instead a probabilistic selection as follows: to each swarm, and with a probability p, the gbest vector is selected, and with (1-p) probability a particle is randomly selected in the same swarm. This scheme was successfully used with genetic co-evolutionary algorithms, in the same problem [4]. Summarizing, we may say that the swarms cooperate in achieving the solution, each being charged of a part of the problem (in this case, the subcubes to materialize into each OSN).

Multi-phase D-PSO (M-DiPSO) proposes the division of the swarm into several groups, each one being in one of possible phases, switching from phase to phase by the use of an adaptive method: phase change occurs if no global best fitness improvement is observed in S recent iterations. The algorithm includes also hill-climbing,.

The rest of the features are implemented as minor variations of these three base versions. The genetic hybridization comes with the possibility of particles having offspring [13] and the selection of particles [12]. Mass extinction is the artificial simulation of mass extinctions which have played a key role in shaping the story of life on Earth. It is performed simply by reinitializing the velocities of all particles at a predefined extinction interval (Ie), a number of iterations.

IV. Proposed Algorithms and Problem Mapping

We have designed and developed three algorithms based on the Di-PSO: 1) M-OLAP Discrete Particle Swarm Optimization (M-OLAP Di-PSO), 2) M-OLAP Discrete Cooperative Particle Swarm Optimization (M-OLAP Di-CPSO) and 3) M-OLAP Multi-Phase Discrete Particle Swarm Optimization Algorithms (M-OLAP M-DiPSO). Some of these have been provided with genetic hybridization (breeding and selection) and mass extinction. Due to space restrictions, we present here only the formal description of the second (see Algorithm 2).

Problem coding in Di-PSO is a simple mapping of each possible subcube in one dimension of the discrete space. Mapping process for the 3 nodes M-OLAP lattice may be seen in Fig. 2. As we have n nodes, this implies a n*subcubes_per_lattice number of dimensions of Di-PSO space. If the particle is at position=1 (state 1) in a given dimension, this means that the corresponding subcube/node is materialized; in its turn, position=0 of the particle implies that the subcube is not materialized.

As we apply a per node space constraint, the particles' move may produce invalid solutions.

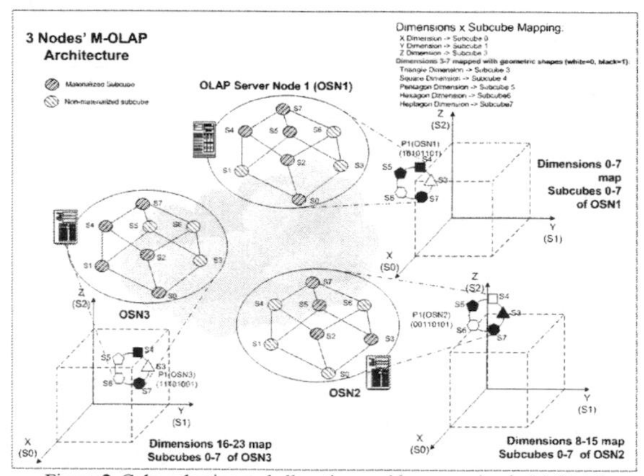

Figure 2. Cube selection and allocation problem mapping in M-OLAP environment in discrete PSO n-dimensional space.

Algorithm 2. Distributed cube selection and allocation algorithm addressing a M-OLAP architecture using a Discrete Cooperative Particle Swarm Optimization.

1. *Initialization* - randomly initialize n swarms (one for each architecture's node) of particles (position and velocity):
   - 1.1. Randomly generate the *velocity* of each particle being its position generated according to the formulas that rule the particle's dynamics.
   - 1.2. *Repairing* those particles that don't satisfy space constraint, changing its position.
   - 1.3. *Generating initial context vector*, taking randomly one particle of each swarm;
   - 1.4. *Initial repairing of the temporary context vector*: computing the maintenance cost of each temporary context vector component (related to one node), rebuilding it if it offends the maintenance constraint, updating the initial context vector with the corrected components.
   - 1.5. *Maintenance cost computing* of the cube distribution proposed by each particle, being repeated 1.1. and 1.2. to the particles that don't obey to the defined constraint, using the initial context vector generated and constrained in 1.3. and 1.4.
   - 1.6. *Initial goodness computing*, updating *pbest* and *gbest*.
   - 1.7. *Generate the context vector* of the initial population.
   - 1.8. *Show the distribution of solutions* of this initial swarm and also of the *pbest* solution.
2. *Populations Loop* - **For Each** swarm and **For Each** particle, **Do**:
   - 2.1. *Using the formulas* that rules the particle's dynamics:
     - 2.1.1. v_i^{k+1} *computing*: to apply equation (3) and rule (4) in algorithm 1.
     - 2.1.2. *position updating*: to apply equation (5) in algorithm 1.
   - 2.2. *Repairing* those particles that don't satisfy space constraint, changing its position.
   - 2.3. *Maintenance cost computing* of the cube distribution proposed by each particle, being repeated 2.1.1. and 2.1.2. to the particles that don't obey to the defined constraint, what means that only the particle's moving that generates valid solutions is allowed.
   - 2.4. *Apply cross operator [13]*, repairing those particles that don't obey to space constraints, just allowing crossings that generate particles that obey to the maintenance of time constraints.
   - 2.5. *Own goodness evaluation and pbest update*: evaluate the 'goodness' of the particle.
     If its goodness > its best goodness so far, **Then** update *pbest*.
   - 2.6. *Global goodness evaluation and gbest update*:
     If the goodness of this particle > the goodness that any particle has ever achieved, **Then** update *gbest*.
   - 2.7. *Selection and cloning* with substitution operator applying.
   - 2.8. *Showing the distribution of solutions* of the actual swarm and also of the *pbest* solution.
3. *Cycle*: **Repeat** Step 2 **Until** a given convergence criterion is met, usually a pre-defined number of iterations.

In this work, instead of applying a fitness penalty that will only avoid the fault particle to become pbest or gbest, we employ a repair method, which randomly dematerialize subcubes generating velocities accordingly, until the particle's proposal is valid. A more complex repair method was tried (that eliminated materialized subcubes which implied the least fitness loss, employing an inverse greedy method), but the performance gain didn't justify the processing cost increase.

V. Experimental Performance Study

For test data, we have used the test set of Benchmark's [15] (TPC-R 2002), selecting the smallest database (1 GB), from which we selected 3 dimensions (customer, product and supplier). To broaden the variety of subcubes, we added additional attributes to each dimension, generating hierarchies, as follows: customer (c-n-r-all); product (p-t-all) and (p-s-all); supplier (s-n-r-all). The 64 subcubes' cube and size may be found in [8]. M-OLAP PQCEA and Greedy2PRS Algorithm (as refereed in section II) were used to compute the query and maintenance cost, supposing qe=1 and ue=0.1 for all OSNs. In this simulation we supposed a three OSNs (with materializing space=10% of total cube space and a processing power of 15E3 records.s^{-1}) plus the base relations (e.g. a DW - considered as node 0 with processing power of 3E4 records.s^{-1}) and a network with BD=1Gbps and delay=[15,50]ms. An

OLAP distributed middleware is in charge of receive user queries and provide for its OSN allocation, and corresponding redirecting. It is supposed that middleware services run into each OSN, on a peer-to-peer approach, although one or many dedicated servers may also exist. To simplify the computations, it is supposed that communication costs between OSNs and any user are equal (using a LAN), and then may be neglected. We generated a random query profile (normalized in order to have a total number equal to the number of subcubes) that was supposed to induce the optimized M (materialized set of selected and allocated subcubes).

Initially we have performed some tests to gain insights about the tuning of some parameters of Di-PSO algorithms, e.g. Vmax, w, c1 and c2 [8]. We also used some information about the values used in other research works. Then, we selected Vmax=10, w=[0.99,1.00], varying linearly with the number of iterations and c1=c2=1.74. We generated the initial velocity vector randomly and the initial position with rule (5) in Algorithm 1. As said in last section, each invalid solution was randomly repaired. As we have three base Di-PSO variants, often we will perform a comparative evaluation, applying each one in turn to the same test. Given the stochastic nature of the algorithms, all presented values are the average of 10 runs.

In the first test we evaluated the impact of the particles' number on the quality of the solution, using, in this case, only the Di-PSO. The results are shown in Fig. 3-plot a). As we can see, a swarm with a higher number of particles achieved good solutions after a reduced number of iterations; but if a number of generations was allowed, the difference between the quality of the solutions of a great and a small swarm vanished. Inspecting the values we observed that a 20 particles' swarm seems to be a good trade-off between quality of final solution and run-time execution, as a low number of particles is favorable in terms of run-time as we can see also analyzing fig. 3, plot b). After these results, we selected a swarm population of 20 particles for all subsequent tests.

Concerning to the comparative quality of Di-PSO, Di-CPSO and Di-MPSO algorithms, as we can see in Fig. 3-plot c), cooperative and multi-phase versions achieve a good solution in a lower number of iterations. Once the interval between evaluations is short, any improvement has a high probability of being captured. But, if a high number of iterations were allowed, normal and cooperative versions would achieve almost the same quality and the multi-phase behaves poorer. A trade-off analysis of quality vs. run-time shows that, even for a low number of iterations (e.g. 100), where Di-CPSO and Di-MPSO performs better, we observed that Di-CPSO spends 89 seconds to achieve a 6507 solution. For a similar solution, Di-PSO uses 63 seconds for a 6479 solution. For Di-MPSO x Di-PSO the plate bends higher to the second: 6,875 in 175 sec. vs. 6,893 in 43 sec.

Next test tries to evaluate if mass extinction will be also beneficial when PSO is applied to this kind of problem. Then we executed M-OLAP Di-PSO algorithm varying Ie. We used Ie from the set [10, 20, 50, 100, 500] and also the no mass extinction option. Fig. 3 – plot d) show the obtained results. As we can see, for low values of Ie, the algorithm has a poorer performance compared to the algorithm with no mass extinction. The best value seems to happen with Ie=100, where mass extinction option surpasses the no mass extinction use.

Fig. 3 - plot e) shows the impact of genetic crossing. Other values of crossing were tried (10, 40 and 60%) but they hurt the quality. As we can see, 20% crossing from Di-PSO seems to be beneficial, and only for the final solution (an observed difference of only 1%). The same isn't true for Di-CPSO. As crossing is, in this case, performed as an intra-node operation, it seems that no further information is gained. Something different happens to Di-PSO: the inter-node crossing seems to be limited interesting. Even for this version, higher crossing values disturb the swarm and damage the quality of the achieved solutions. Those evidences seem to show the very limited impact of crossing for general Di-PSO algorithms applied to this problem.

The last test evaluates the scalability of algorithms, concerning to the number of M-OLAP nodes. We used M-OLAP Di-PSO algorithm. The two others, as similar in terms of complexity (cost estimation algorithms and loop design) must have the same behavior. We used the same 3 nodes M-OLAP architecture and another with 10 nodes. We generate another query profile, also normalized. The plot f) of Fig. 3 shows the observed run-time and corresponding ratio. As we can see, an increase of 10/3=3.3 for the number of nodes implies a run-time increase of 3.8, showing a quasi-linearity, and the support of M-OLAP with many nodes.

VI. CONCLUSIONS AND FUTURE WORK

The executed experimental tests have shown that particle swarm optimization, in its discrete version, allowed the design and building of algorithms that stand as good candidates to the cube selection and allocation problem solution in an M-OLAP environment. As far as we're concerned, this is the first proposal of this kind of algorithm to this problem. The used cost model extends the existing proposals by the explicit inclusion of enlarged characteristics communication costs and processing power of OLAP nodes. A set of proposed variations in PSO continuous and discrete versions were also included in the developed algorithms.

Globally, the tests allow us to conclude that all algorithms achieve good solutions, but multi-phase version seems to perform poorer in this kind of problem (it is neither faster nor better in the final solution). Cooperative version achieves good solutions in fewer iterations than normal version, but in terms of execution time, this is not really true, as the first needs a higher number of fitness evaluations. The quality of final solution is almost the same for both.

We may also say that M-OLAP Di-PSO algorithm is a good solution for this kind of problem once Di-PSO achieves solutions of the same quality as standard genetic algorithms but using a short run-time [16], and scales well in terms of supporting M-OLAP architectures with many nodes.

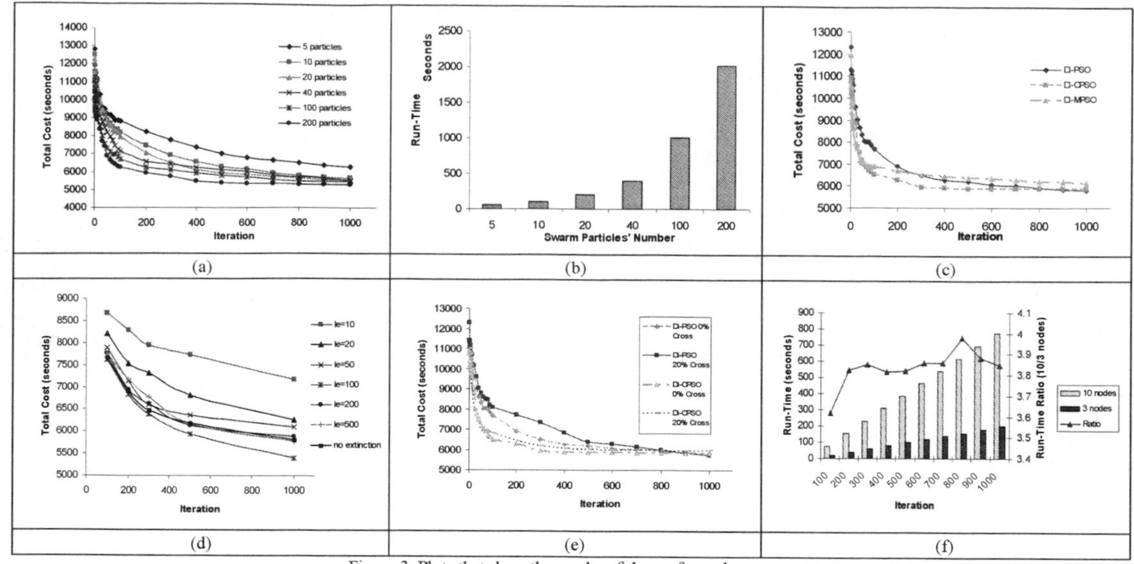

Figure 3. Plots that show the results of the performed tests.

Mass extinction seems to be interesting, since the extinction interval is carefully selected. In this case, a value of 1/10 of the number of iterations improves the quality of achieved solutions. Concerning to genetic swarm operators, our results seem to advise only crossing operator for M-OLAP Di-PSO with a low crossing percent (this case 20%).

For future work, we'll propose the comparative valuation of these algorithms in face of the greedy and genetic ones, but using another cost model [17], introducing non-linearities and coexistence of subcubes under incremental maintenance (typical for the maintenance of great static structures) and by subcube rebuilding (for dynamic or even pro-active cubes selection and allocation). This will be of great interest because it enlarges the application range of all algorithms in new real situations.

ACKNOWLEDGMENT

The work of Jorge Loureiro was supported by a grant from PRODEP III, acção 5.3 – Formação Avançada no Ensino Superior, Concurso N.º 02/PRODEP/2003.

REFERENCES

[1] V. Harinarayan, A. Rajaraman, and J. Ullman, "Implementing Data Cubes Efficiently", in Proceedings of ACM SIGMOD, Montreal, Canada, 1996, pp. 205-216.

[2] A. Bauer, and W. Lehner, "On Solving the View Selection Problem in Distributed Data Warehouse Architectures", in Proceedings of the SSDBM'03, IEEE, 2003, 43-51.

[3] J. Loureiro, and O. Belo, "Evaluating Maintenance Cost Computing Algorithms for Multi-Node OLAP Systems", in Proceedings of the JISBD2006, Sitges, Barcelona, Spain, 3-6 October, 2006.

[4] J. Loureiro, and O. Belo, "An Evolutionary Approach to the Selection and Allocation of Distributed Cubes", in press (IDEAS2006).

[5] W. Liang, H. Wang, and M.E. Orlowska, "Materialized View Selection under the Maintenance Cost Constraint", in Data and Knowledge Engineering, 37(2), 2001, pp. 203-216.

[6] W.-Y. Lin, and I.-C. Kuo, "A Genetic Selection Algorithm for OLAP Data Cubes", in Knowledge and Information Systems, Volume 6, Number 1, Springer-Verlag London Ltd., 2004, pp. 83-102.

[7] P. Kalnis, N. Mamoulis, and D. Papadias "View Selection Using Randomized Search", in Data Knowledge Engineering, vol. 42, number 1, 2002, pp. 89-111.

[8] J. Loureiro, and O. Belo, "A Discrete Particle Swarm Algorithm for OLAP Data Cube Selection", in Proceedings of the 8th International Conference on Enterprise Information Systems (ICEIS 2006), Paphos – Cyprus, 23-27 May 2006, pp. 46-53.

[9] J. Kennedy, and R.C. Eberhart, "Particle Swarm Optimization", in Proc. of IEEE Intl. Conference on Neural Networks (Perth, Australia), IEEE Service Center, Piscataway, NJ (1995) IV, pp. 1942-1948.

[10] J. Kennedy, and R.C. Eberhart, "A Discrete Binary Version of the Particle Swarm Optimization Algorithm", in Proc. of the 1997 Conference on Systems, Man and Cybernetics (SMC'97), 1997, pp. 4104-4109.

[11] F. Van den Bergh, and A.P. Engelbrecht, "A Cooperative Approach to Particle Swarm Optimization", in IEEE Transactions on Evolutionary Computation, Vol. 8, No. 3, June 2004, pp. 225-239.

[12] P. Angeline, "Using Selection to Improve Particle Swarm Optimization", in Proceedings of the ICEC'98, Anchorage, Alaska, USA, May 4-9, 1998.

[13] M. Løvbjerg, T. Rasmussen, and T. Krink, "Hybrid Particle Swarm Optimization with Breeding and Subpopulations", in Proceedings of the 3rd Genetic and Evolutionary Computation Conference (GECCO-2001).

[14] B. Al-Kazemi, and C.K. Mohan, "Multi-Phase Discrete Particle Swarm Optimization", in Proceedings of the FEA 2002, pp. 622-625.

[15] Transaction Processing Performance Council (TPC): TPC Benchmark R (decision support) Standard Specification Revision 2.1.0. tpcr_2.1.0.pdf, available at http://www.tpc.org.

[16] J. Loureiro, ad O. Belo, "Life Inspired Algorithms for the Selection of OLAP Data Cubes", in WSEAS Transactions on Computers, Issue 1, Volume 5, Janeiro 2006, pp. 8-14.

[17] J. Loureiro, and O. Belo, "Establishing more Suitable Distributed Plans for MultiNode-OLAP Systems", in Proceedings of 2006 IEEE International Conference on Systems, Man and Cybernetics (SMC 2006), Taipei, Taiwan, 8-11 October, 2006.

Developing Peer-to-Peer Applications with MDA and JXTA

José Geraldo de Sousa Junior Denivaldo Lopes
Federal University of Maranhão - UFMA
Campus do Bacanga
65080-040 São Luís - MA - Brazil
jgeraldo@dee.ufma.br and dlopes@dee.ufma.br

Abstract- **Recently, Peer-to-Peer (P2P) architecture is being used to explore better the computing power and bandwidth of networks than a client/server architecture. In order to support the creation of P2P applications, some frameworks were proposed such as JXTA. However, large systems using P2P architecture are complex to be developed, maintained and evolved. Model Driven Architecture (MDA) can support the management of the complexity in the software development process through transformations of Platform Independent Models (PIM) into Platform Specific Models (PSM). In this paper, we apply an MDA approach to allow the development of applications based on JXTA. The JXTA implementation in Java is used to demonstrate our approach. We propose a model transformation definition from an UML model to a Java+JXTA model. In order to validate our approach, we present two case studies.**

I. INTRODUCTION

Peer-to-Peer (P2P) architecture provides a better utilization of computing power of the nodes and bandwidth of the networks rather than a client/server architecture. P2P architecture is popular thanks to applications such as Usenet news servers, Napster, OpenNAD, Gnutella and Kazaa. In order to support the creation of P2P applications, some frameworks were proposed as JXTA [8]. JXTA "*is a set of open, generalized peer-to-peer(P2P) protocols that allow any connected device on the network - from cell phone to PDA, from PC to server - to communicate and collaborate as peers*" [8]. JXTA Technology was designed to be portable, thus the protocols can be implemented in any programming language that the developer prefers. Although these benefits, large systems based on a P2P architecture and JXTA platform are complex to be developed, maintained and evolved.

Recently, Object Management Group (OMG) has proposed the use of Model Driven Architecture (MDA) to manage the development, maintenance and evolution of complex distributed systems [14]. One of the most subjects in investigation on MDA is the transformation from Platform Independent Models (PIMs) to Platform Specific Models (PSMs) and the transformation from PSMs to source code [5, 9, 11, 12, 14, 15].

When both the metamodel used to create the PIM and the metamodel used to create the PSM are available, we can define transformations to generate a PSM from a PIM. However, before creating transformation definitions from a PIM to a PSM, we need a PSM pattern. A PSM pattern is necessary to insert the specific platform characteristics in a PSM. In this paper, hereafter, this pattern is named *template*. We show how a pattern or *template* is extracted from the JXTA platform.

MDA has been largely used in Information Systems (IS) [2, 4, 7, 11] in order to automatize the process of generating platform specific models, source code and configuration files. However, the use of MDA to develop software based on P2P platforms is still evolving and little research has applied MDA to support P2P application development [3].

In this paper, we aim to present how we can develop P2P applications according to an MDA approach. For this purpose, we limit our scope to provide a *template* for JXTA platform, transformation definitions from UML to Java+JXTA, and illustrative examples.

This paper is organized in the following way. Section II describes the use of MDA to develop JXTA applications. Section III presents a case study that consists in the parallel/distributed array multiplication using P2P networks. Section IV presents transformation definition from our PIM (that conforms to UML) to a PSM (that conforms to Java+JXTA), including the role of a JXTA *template*. Section V shows another case study in which we apply the same transformation definitions of section IV. Section VI presents a comparison among P2P applications and standalone applications. Finally, in section VII, we conclude this paper presenting final considerations and future research directions.

II. APPLYING AN MDA APPROACH TO DEVELOP P2P APPLICATIONS

Figure 1 presents an MDA approach applied to JXTA platform.

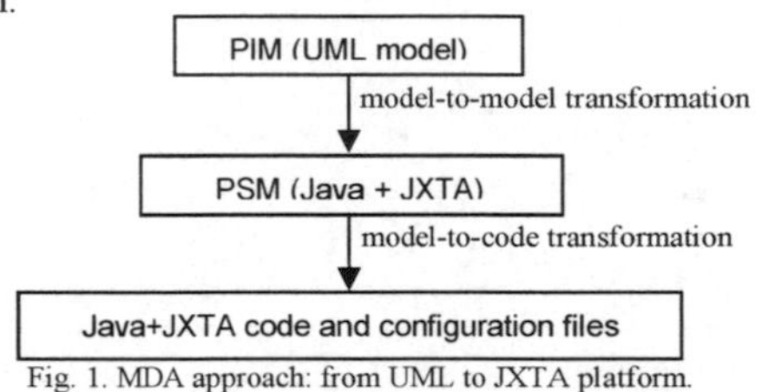

Fig. 1. MDA approach: from UML to JXTA platform.

In this figure, the PIM is described using the UML metamodel (UML version 1.5), and the PSM is described using the Java metamodel[1] and JXTA *template*. In order to realize the model-to-model transformation, before, we need match the UML metamodel and the Java metamodel, and we need create transformation definitions conform to this matching. Afterwards, these transformation definitions must be incremented in order to apply the JXTA template for

[1] In this research, we use the Java metamodel proposed in [10].

K. Elleithy (ed.), Advances and Innovations in Systems, Computing Sciences and Software Engineering, 235–240.

creating the PSM conform to Java metamodel and JXTA *template*. In order to realize the model-to-code transformation, we define a transformation from a PSM (in Java+JXTA) to a Java+JXTA code and configuration files.

In next sections, we illustrate this MDA approach using case studies: array multiplication and creation of the Mandelbrot fractal. These case studies are largely used as benchmark for parallel/distributed solutions [1].

III. CASE STUDY: ARRAY MULTIPLICATION

The parallel/distributed array multiplication using a P2P network can be described in the following way. First, we must read both $A_{i,j}$ and $B_{i,j}$ arrays. Afterwards, we divide the array $A_{i,j}$ in n sub arrays (let $n = \sqrt{m}$, where m is the total of computers). Then, array A is divided in i/n, i.e. $A_{i/n,j}$ sub arrays, and array B is divided in j/n, i.e. $B_{i,j/n}$ sub arrays (Cf. Figure 2). Thus, i and j must be integer numbers that are multiple of n, in order to have sub arrays with the same dimension.

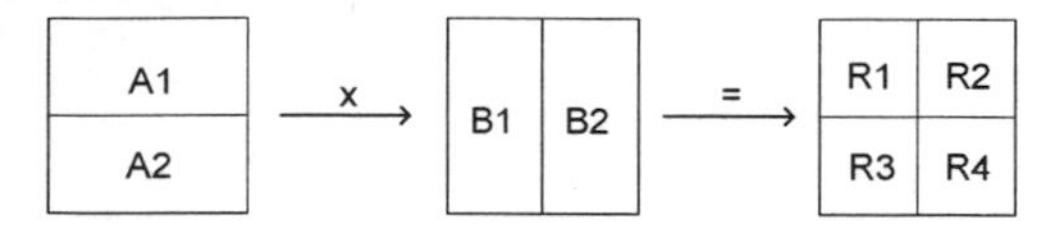

Fig. 2. Array multiplication: overview.

Figure 2 illustrates an example of array multiplication. *A1* and *A2* are sub arrays of *A*, *B1* and *B2* are sub arrays of *B*, and *R1*, *R2*, *R3* and *R4* are sub arrays of the result array. Figure 3 shows a P2P configuration for array multiplication (using $m = 4$ and $n = 2$).

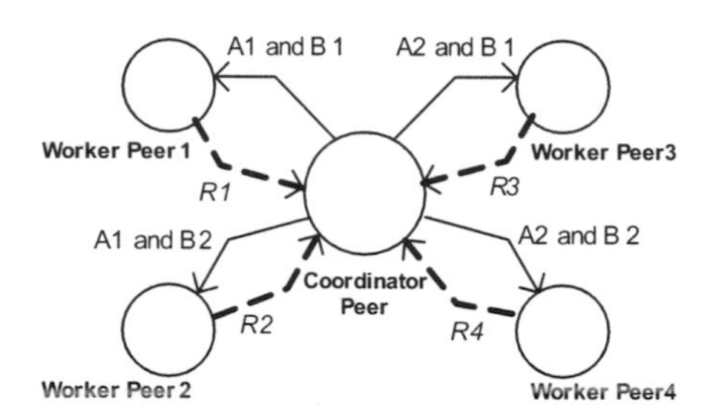

Fig. 3. A P2P configuration to array multiplication.

In order to perform the array multiplication through a P2P architecture, we must have m peers waiting on the network to execute the multiplications of sub arrays. Each sub array A is multiplied for a sub array B, generating a part of the array R.

The *coordinator* is responsible to make the division of the array A and B and to request four peers (named *Worker* peers) on P2P network for doing the sub array multiplications. Once those four peers calculated the sub array multiplications, the final result is constituted with the composition of the results of each worker peer.

A. *UML Model: Array Multiplication*

In Figure 4, we present an use case diagram of our application. In that diagram, we have the agent *Client* that uses the use case *Coordinator* to coordinate *Worker* peer tasks for calculating the multiplication of sub arrays. That representation provides an overview of the system in the analysis view point. *Client* is the user that uses *Coordinator* peer, and *Coordinator* peer has associations with four *Workers*.

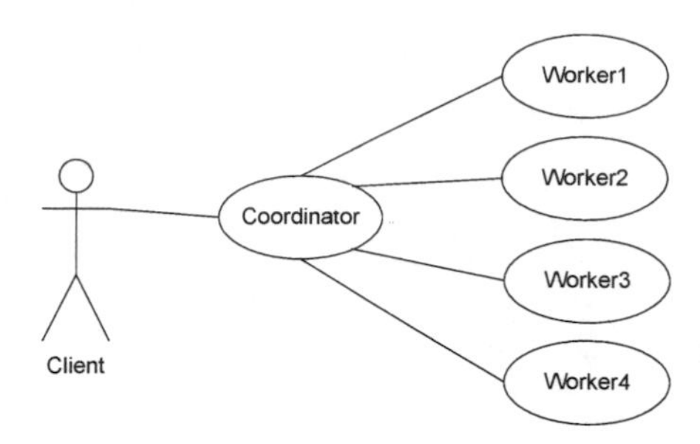

Fig. 4. Use case: array multiplication (fragment).

The class diagram is presented in Figure 5. The class *Coordinator* represents the coordinator peer tasks and the class *Worker* represents worker peer tasks. Each *Coordinator* has association with m *Workers* (let m the total of computers). The PIM of Figure 5 has a class *Operations* that is responsible for some special array operations such as array reading and multiplication.

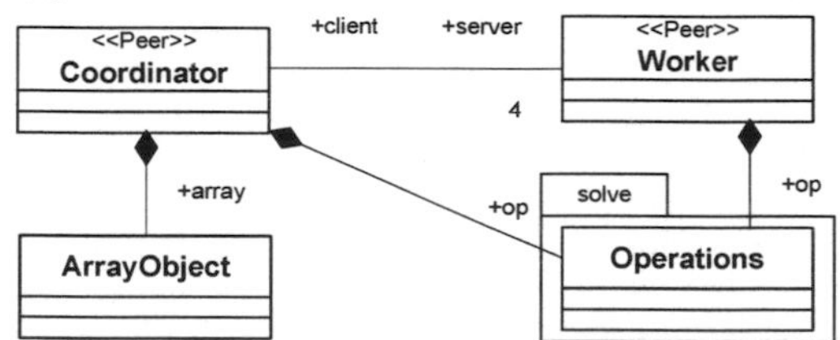

Fig. 5. Class diagram: A PIM for array multiplication (fragment).

IV. TRANSFORMATION FROM UML TO JXTA

In this section, we present a transformation definition from UML to JXTA platform.

Some questions about the mentioned transformation could be done. If the JXTA platform is implemented on Java, then is it only needed the Java metamodel to perform the matching among UML and JXTA? Is it only the Java metamodel sufficient to realize the transformation?

Answering those questions, we can declare that just Java metamodel is not sufficient. A metamodel provides the way as related models can be created. If the transformation is performed just from UML model to Java model, then we obtain a final Java model and not a JXTA model. We need a transformation definition that maps the UML metamodel to Java metamodel and includes a JXTA *template* into the PSM.

In this research, we have used the Java metamodel provided in [10].

A. *A JXTA Template*

In this section, we discuss a JXTA *template* and we show its employing in the transformation definition from a PIM (that conforms to UML) to a PSM (that conforms to Java+JXTA *template*). Figure 6 shows a JXTA *template* for Java applications using UML notation.

In Figure 6, we can think about Java model[2] as following: each *JClass* that implements a *<<Peer>>* must extend a *JClass* named "*PeerJxta*" (e.g. *MyPeerJXTA*). In our approach, the JXTA platform information is present in the *template*. Thus, a transformation definition from UML to Java+JXTA must relate the UML metamodel to a Java metamodel and include the JXTA *template*.

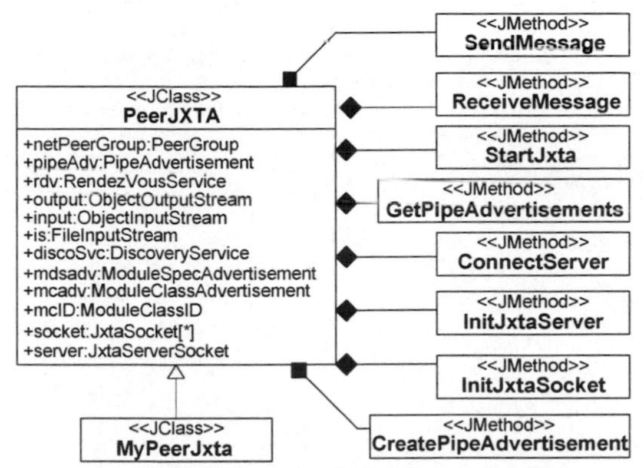

Fig. 6. JXTA *template* for Java (fragment).

PeerJxta class must have all *JMethods* specified in the *template*, as for instance, *startJxta*, *sendMessage* and *receiveMessage*. Pay attention that implementations of these methods are necessary for a class which is a peer, and then we constructed a class *PeerJxta* with all those methods and attributes that are essential for the communication among peers.

B. Transformation Definition: from PIM to PSM

In this section, we provide transformation definitions written in ATL [6] in order to realize a transformation from our PIM in UML to a PSM in Java+JXTA.

Listing 1 presents an ATL transformation rule[3] that transforms UML classes to Java classes. In ATL rule structure, we have filters (see lines 2-3) that are applied to the variable "c". The filter describes basically that, if the expression between parentheses is true, then the rule must be executed. Otherwise, this rule is not executed. For instance, we can say that if an UML class has "*Peer*" stereotype, then this rule is not executed. The transformation rule in Listing 1 transforms a class from UML model which does not have the stereotype "*Peer*" to a common Java class. In Figure 5, this transformation rule is executed for *ArrayObject* and *Operations* UML classes, that are transformed in common Java classes, because they do not have *<<Peer>>* stereotype.

Listing 1. Transformation definition in ATL.

```
1   rule C2JC {
2   from c : UML!Class(c.stereotype->select(e|e.name = 'Peer')-
    >isEmpty())
3   to jc : JAVAM!JClass (
4     name <- c.name,
5     visibility <- if c.visibility = #vk_public then
6       #public
7     else if c.visibility = #vk_private then
8       #private
9     else
10      #protected
11    endif endif,
12    ***(other bindings)  )}
```

2 We use UML graphical notation with profiles to express a Java model.

3 In order to simplify the presentation of this paper, we provide only fragments of transformation definitions, PSM in XMI format and source code.

To create a Java class that is a JXTA peer, we constructed another rule that is exposed in Listing 2. That rule performs the opposite in relation to the previous rule: it selects classes that have *<<Peer>>* stereotype. Note that we used stereotype just for determining what classes are peers and what classes are not.

In Listing 2 and lines 2-3, we have a rule with constraints from *template*. In lines 8-20, new elements of JXTA platform are generated by transformation from classes that have *<<Peer>>* stereotype. A *JClass* with the same name of the UML class and a *JClass* with name *PeerJxta* are generated. When establishes *super* ← *peerC*, it means that *JClass* generated from UML class must extend *PeerJxta*. That class *PeerJxta* contains methods and attributes necessary to P2P communication.

Listing 2. Transformation definition in ATL – applying the template.

```
1   rule C2JXC {
2   from c: UML!Class(c.stereotype->select(e|e.name = 'Peer')-
    >notEmpty())
3   to jxc: JAVAM!JClass (
4     name <- c.name,
5     super <- peerC,
6     ***(other bindings)
7   ),
8   peerC: JAVAM!JClass (
9     name <- 'PeerJxta',
10    visibility <- #public,
11    super <- JAVAM!JClass,
12    ***(other bindings)
13  ),
14  startJ: JAVAM!JMethod (
15    name <- 'startJxta',
16    owner <- peerC,
17    visibility <- #public,
18    ***(other bindings)
19  ),
20   ***(other bindings) }
```

We have, still in Listing 2, *PeerJxta* class that contains *startJxta* method, through *owner* ← *peerC* binding.

The implementation of this rule is conform to the *template* of Figure 6. When these rules generate a JXTA model, they (rules) fill the gaps between UML metamodel and Java metamodel, including the JXTA *template*.

Listing 3 presents the generated PSM (using XMI format). In line 1, the head of an XML file is presented. In line 5, the *JPackage* named "*example*" contains all classes. In line 10, the *JClass* "*Coordinator*" contains *JMethods* as "*run*" (see line 13). In line 25-27, the *JClass* "*Coordinator*" inherits from the *JClass* "*PeerJXTA*" according to *JAVAM.JClassifier.super* that points to *xmi.idref=a36* that is the *xmi.id* of *JClass* "*PeerJXTA*".

Listing 3. PSM in XMI format (fragment).

```
1   <?xml version = '1.0' encoding = 'windows-1252' ?>
2   <XMI xmi.version = '1.2' timestamp = 'Thu Aug 31 10:36:13 BRT
```

```
      2006'>
3      ***
4      <XMI.content>
5        <JAVAM.JPackage xmi.id = 'a1' name = 'example'>
6          <JAVAM.JPackage.jelements>
7            <JAVAM.JPackage xmi.id = 'a2' name = 'coordinator'>
8              <JAVAM.JPackage.jelements>
9                ***
10               <JAVAM.JClass xmi.id = 'a17' name = 'Coordinator'
11     visibility = 'public'  modifier = 'regular' isActive = 'false'>
12                 <JAVAM.JClassifier.jmembers>
13                   <JAVAM.JMethod xmi.id = 'a26' name = 'run' visibility
14     = 'public' modifier = 'regular' isStatic = 'false' isNative = 'false'
       isSynchronized = 'false'>
15                     <JAVAM.JMethod.jparameters>
16                       <JAVAM.JParameter xmi.id = 'a27' name = 'return'
       result = 'true'>
17                         <JAVAM.JParameter.type>
18                           <JAVAM.JPrimitiveType xmi.idref = 'a8'/>
19                         </JAVAM.JParameter.type>
20                       </JAVAM.JParameter>
21                     </JAVAM.JMethod.jparameters>
22                   </JAVAM.JMethod>
23                         ***
24                 </JAVAM.JClassifier.jmembers>
25                 <JAVAM.JClassifier.super>
26                   <JAVAM.JClass xmi.idref = 'a36'/>
27                 </JAVAM.JClassifier.super>
28               </JAVAM.JClass>
29                         ***
30               <JAVAM.JClass xmi.id = 'a36' name = 'PeerJxta'
31     visibility = 'public' modifier = 'regular' isActive = 'true'>
32                 ***
33               </JAVAM.JClass>
34             </JAVAM.JPackage.jelements>
35           </JAVAM.JPackage>
36                       ***
37       </XMI.content>
38     </XMI>
```

C. *Transformation definition: from PSM to source code*

After generating a PSM conform to Java and JXTA *template*, the question is how to generate source code conform to the JXTA platform in Java. At current stage of our research, a complete generation of source code is still in course. However, we have conditions to generate skeletons of classes.

In ATL language, queries have functionalities as navigating on models enabling creation of outputs in textual form. Using this functionality, we can generate source code from a model.

Listing 4. Transformation program: from PSM to source code (fragment).

```
1   query Java2SourceCode =
2   JAVAM!JClassifier.allInstances()->
3   select(e|e.oclIsTypeOf(JAVAM!JClass) or
4   e.oclIsTypeOf(JAVAM!JInterface))->
5   collect(x | x.toString().writeTo
6   ('C:/SourceCode/JXTA/' +
7   x.jpackage.name.replaceAll
8   ('.', '/' ) + '/' + x.name + '.java'));
9
10  helper context JAVAM!JClass def : toString () :
11  String =
12  self.jpackage.toString() +
13  self.visibility() +
14  self.modifierAbstract() +
15  'class ' +
16  self.name +
17  self.getSuperType() +
18  self.getImplements() +
19  ' {\n\n' +
```

Listing 4 shows a query program. It uses a function of *helper* type, which defines *toString* method to Java classes. Other *helpers* are used to generate interfaces, methods and attributes, and so on. *Queries* iterate through packages, classes and interfaces from PSM (Java+JXTA model), extracting information from this PSM, to write them in specific files which contain source code and documents. In lines 1-8, a *query* is defined. This *query* navigates into a PSM conform to Java metamodel, selects all the instances of *JClass* (see line 3) and *JInterfaces* (see line 4), collects them (see lines 5-8) and applies the method *toString* (see line 5). In line 5, the generated strings are written in files through the method *writeTo*. In line 10, the *helper toString* is defined in the context of a *JClass*. The other lines describe the navigation into a *JClass* in order to recuperate information that must be written in a source code.

Listing 5. PeerJxta class generated by query program.

```
1   package example.coordinator;
2   ***
3   public class PeerJxta {
4     public DiscoveryService discoSvc;
5     public PipeAdvertisement pipeAdv;
6     public PeerGroup netPeerGroup;
7     *** (other attributes)
8     public void startJxta( int quant_connections ) {***}
9     public void getPipeAdvs( int num_pipeAdv ) {***}
10    public void sendMessage(Object obj, int order) {***}
11    *** (other methods) }
```

In Listing 5, we have a fragment of generated source code from the class PeerJxta that is mentioned in Figure 6, in Listing 1 and in Listing 2.

V. ANOTHER CASE STUDY: MANDELBROT FRACTAL

A fractal object describes a type of irregular forms, but it follows a repeated standard, usually it is created using an iterative process [1]. As fractals are essentially generated by mathematical functions, we propose a P2P configuration similar to Figure 3, in order to create the Mandelbrot fractal. We divide the fractal image area in blocks and each block is calculated separately in parallel by *Worker* peers.

In order to calculate this fractal, the mathematical function receives the number of maximum colors and the dimension of the fractal figure, and returns a vector of integer values that correspond to pixels. A small number of colors means a small effort to calculate the pixels, i.e. a short time for processing the input parameters. A large number of colors means a big effort to calculate the pixels, i.e. a long time for processing the input parameters. Similarly, the dimension of the fractal figure has impact in the processing time of the mathematical function.

Figure 7 presents the PIM for the creation of Mandelbrot fractal (fragment). *Coordinator* and *FractalCalculator* classes are peers. *MandelPar* class is responsible for presenting the

GUI, that uses *CanvasPar* to coordinate the tasks for generating the pixels and creating the fractal. *CanvasPar* uses four *ThreadCall* to connect with four peers *FractalCalculator*, and uses *DataPar* for operations such as showing coordinates and processing time. *DataPixel* stores the pixels of the mandelbrot fractal. In the communication process, *FractalCalculator* and *ThreadCall* exchange an object instantiated from *DataPixel*.

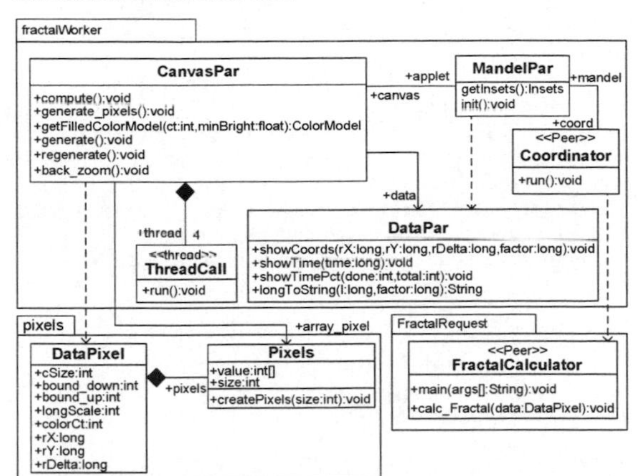

Fig. 7. Class diagram: A PIM for fractal creation (fragment).

This PIM can be transformed in a PSM conforms to Java+JXTA, and, after that, this PSM can be transformed in Java+JXTA code. For this purpose, we can reuse the transformation definitions previously presented in section 4.

In order to simplify the presentation of this case study, we skip some steps and present the skeleton of the generated code. Listing 6 presents a skeleton for the P2P application that creates a Mandelbrot fractal.

Listing 6. Code of FractalCalculator class (fragment).

```
1  package fractalRequest;
2  ***
3  public class FractalCalculator extends PeerJxta {
4    ***
5    public void calc_Fractal( Data_Pixel data ) {***}
6    public void main(String args[] ) {***}
7  *** }
```

Listing 7 presents a fragment of the class *CanvasPar*.

Listing 7. Code of CanvasPar class (fragment).

```
1  package fractalWorker;
2  ***
3  public class CanvasPar{
4    public MandelPar applet;
5    public ThreadCall thread[];
6    public Pixels array_pixel;
7    ***
8    public void generate( ) {***}
9    public void compute( ) {***}
10   public void back_zoom( ) {***}
11   public void regenerate( ) {***}
12 ***}
```

Afterwards, this code can be completed. The execution of this P2P application is illustrated in Figure 8.

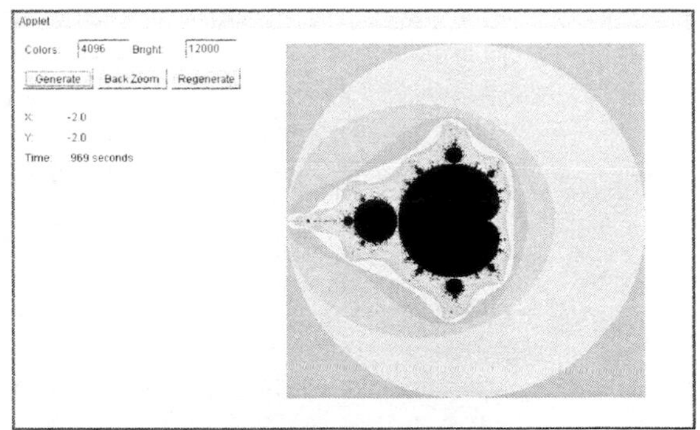

Fig. 8. P2P application for Mandelbrot

VI. RUNNING THE CASE STUDIES

In this section, we present some results obtained with the execution of our case studies, and we establish a comparison among our P2P applications and sequential applications. Thus, we aim to demonstrate the performance of P2P application developed according to an MDA approach.

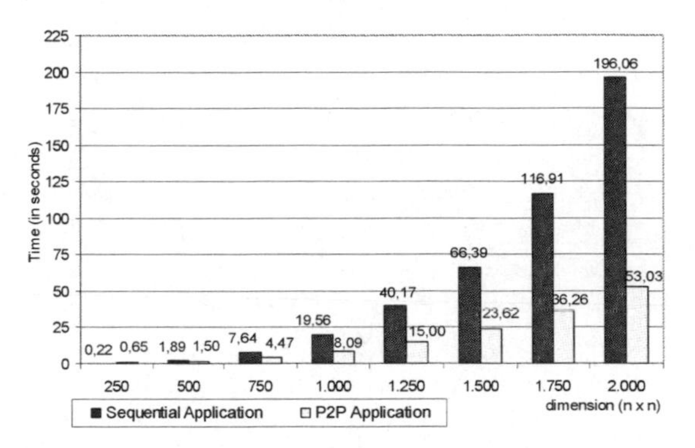

Fig. 9. Array multiplication: times in sequential and P2P application.

In Figure 9, on the horizontal axis, we have the array dimension, and, on the vertical axis, we have the computing time in seconds. In this experiment, we have fixed the number of computers in 4 (four) and varied the array dimension (one of the computers executed the *Coordinator* peer and one *FractalCalculator*). Note that our P2P application has not gain with small dimension arrays (equal or less than 250x250).

Increasing the array dimension, the required processing time consequently increases for both P2P application and sequential application. However, the processing time consumed by our P2P application is less than the processing time consumed by a sequential application.

Figure 10 presents the *speed-up* and *efficiency* using 4 (four) computers. The *speed-up* is a measure that informs how much a parallel algorithm is better than a sequential algorithm.

In the case of arrays with dimension 250x250, we can note that a P2P application is not better than sequential application, because the *speed-up* is not greater than 1. In the array multiplication with dimensions equal or greater than 500x500, the *speed-up* is greater than 1, what means better performance in P2P approach.

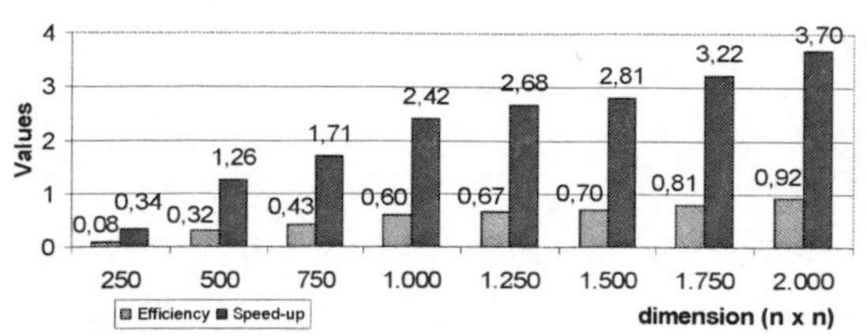

Fig. 10. Array multiplication: speed-up and efficiency of the P2P application.

The *efficiency* relates the *speed-up* and the total of processing units in execution. In our case, the *speed-up* is divided by 4 (four) because we use four computers to solve the problem. For instance, for 2000x2000 dimension, the *efficiency* reaches 92%, what means 92% of the machine processing power were used.

Figure 11 and Figure 12 present the values obtained with the Fractal case study.

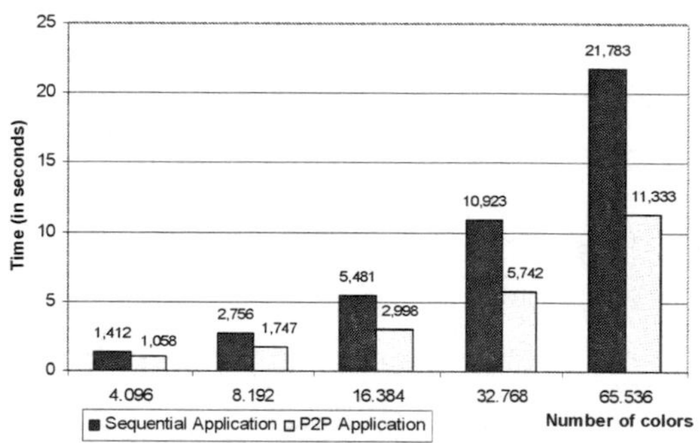

Fig. 11. Fractal Mandelbrot: times in sequential and P2P application.

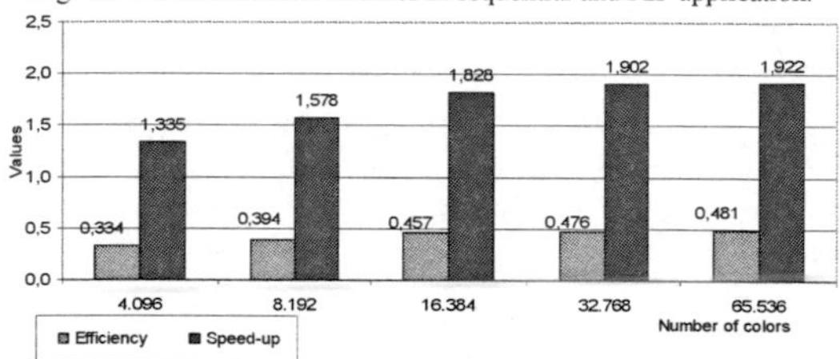

Fig. 12. Fractal Mandelbrot: speed-up and efficiency of the P2P application

Figure 11 shows the processing times in sequential application and P2P application. In this experiment, we have increased the number of colors and fixed the number of computer in 4 (four). The *efficiency* and *speed-up* evaluation are presented in Figure 12. This figure shows that our P2P application is better than a sequential application.

VII. CONCLUSION

In this paper, we presented our approach to develop P2P applications through models. Two case studies were provided to illustrate model driven approach to generate a system on JXTA platform.

Once we created the transformation definitions, we can apply them to different PIMs (that conform to UML metamodel) and obtain the source code for JXTA platform. Thus, the time to develop software applications using an MDA approach is more profited in modeling task than in coding task. Moreover, as the final source code is generated from a model, some prone errors of coding task can be avoided, and best practices can be used in the model and propagated in the final source code.

Researches in MDA have been aimed the improvement of transformation definition. In this way, we provided an important advance in skill about model transformation, since we obtained source code conform to JXTA platform from an independent model (UML model) through transformations. The results obtained in section 6 demonstrate that an MDA approach can be used to develop performant P2P applications.

In future directions, we aim to generate all the source code based on the JXTA platform, including method bodies. For this purpose, we can use action semantics [13].

ACKNOWLEDGMENT

The work described in this paper was financed by ***Fundo Setorial de Tecnologia da Informação (CT-Info), MCT, CNPq (CT-Info/MCT/CNPq)***.

REFERENCES

[1] M. Allen and B. Wilkinson, *Parallel Programming: Techniques and Applications Using Networked Workstations and Parallel Computers*. Prentice-Hall, 2nd edition, 2004.

[2] J. Bézivin, S. Hammoudi, D. Lopes, and F. Jouault, "B2B Applications, BPEL4WS, Web Services and dotNET in the context of MDA," *Knowledge Sharing in the Integrated Enterprise - Interoperability Strategies for the Enterprise Architect*, 183, October 2005.

[3] R. Carroll, C. Fahy, E. Lehtihet, and S. van der Meer, "Applying the P2P Paradigm to Management of Large-Scale Distributed Networks Using a Model Driven Approach." *IEEE/IFIP Network Operations and Management Symposium*, pages 1–14, 2006.

[4] P. Cáceres, V. de Castro, J. M. Vara and E. Marcos, "Model Transformations for Hypertext Modeling on Web Information Systems." *In Proceedings of the 2006 ACM symposium on Applied computing*, pages 1232–1239, 2006.

[5] D. Frankel and J. Parodi, *Using Model-Driven Architecture™ to Develop Web Services*. Technical report, IONA Technologies PLC, April 2002.

[6] Atlas group, LINA, and INRIA. *ATL: Atlas Transformation Language - ATL User Manual version 0.7*, February 2006.

[7] J. Skene and W. Emmerich, "Model Driven Performance Analysis of Enterprise Information Systems," ETAPS 2003 - Electronic Notes in Theoretical Computer Science, 82(6), 2003.

[8] JXTA.org, *JXTA*, April 2006. Available at http://www.jxta.org.

[9] A. Kleppe, J. Warmer and W. Bast. *MDA Explained: The Model Driven Architecture: Practice and Promise*. Addison-Wesley, 1st edition, August 2003.

[10] D. Lopes. "Study and Applications of the MDA Approach in Web Service Platforms." Ph.D. thesis (written in French), University of Nantes, 2005.

[11] O. Patrascoiu, "Mapping EDOC to Web Services using YATL," *In Proceedings of 8th IEEE International Enterprise Distributed Object Computing Conference (EDOC 2004)*, pages 286–297, September 2004.

[12] S. J. Mellor, K. Scott, A. Uhl and D. Weise, *MDA Distilled: Principles of Model-Driven Architecture*. Addison-Wesley, 1st edition, March 2004.

[13] OMG, *Action Semantics for the UML*, OMG ad/2001-08-04, August 2001.

[14] OMG, MDA Guide Version 1.0.1, Document Number: omg/2003-06-01. OMG, June 2003.

[15] QVT-Merge Group, Revised submission for MOF 2.0 Query/Views/Transformations RFP (ad/2005-07-01) version 2.1, July 2005. Available at http://www.omg.org/docs/ad/05-07-01.pdf.

A Case Study to Evaluate Templates & Metadata for Developing Application Families

José Lamas Ríos
ARTech
Av. 18 de Julio 1645/4,
11200 Montevideo-Uruguay
jlr@artech.com.uy

Fernando Machado-Píriz
Universidad Católica del Uruguay
Av. 8 de Octubre 2738
11600 Montevideo, Uruguay
fmachado@ucu.edu.uy

***Abstract*—Automatic code generation of application families emerges as a solid promise to cope with the increasing demand of software in business environments. Using templates and metadata for development of abstract solutions and further automatic generation of the particular cases, helps freeing the developers from the most mechanical and tedious tasks of the implementation phase, allowing them to focus their knowledge in the expression of conceptual solutions.**

In this case study, we adapted the Halstead metrics for object-oriented code, templates, and metadata -in XML format- to measure the effort required to specify and then automatically generate complete applications, in comparison with the effort required to build the same applications entirely by hand. Then we used the same metrics to compare the effort of specifying and generating a second application of the same family, versus the effort required to coding this second application by hand.

***Index Terms*—Automatic Programming, Program transformation, Computer Aided Engineering, Computer-aided design, Computer-aided manufacturing, Automatic code generation, Templates, Metadata, Halstead metric**

I. INTRODUCTION

In the next decade, we can expect an increasing worldwide demand of new software in about an order of magnitude, driven by new forces in the global economy, the ubiquity of software in social infrastructure, new application types like business integration, web services, mobile devices, and smart appliances [3].

The software industry is not in good shape to face this demand, according to project failures, costs and schedules overruns reported by industry analysis like the now infamous series of Chaos Reports, from The Standish Group. In the report of 1994, 31.1% of software projects were cancelled before completion, only 16.2% finished on time and within budget, while the remaining 53% required on average 189% and 222% of their original estimates for cost and time [7]. In the 2004 report, 18% of software projects failed, 29% succeeded, 53% were challenged [8].

These figures are far from being acceptable in a mature and efficient software industry. Undoubtedly, one of the causes of software projects failure rates is that -despite all the undeniable advances in software engineering- software development is still a labor-intensive task.

The way we develop software today has changed dramatically from the old times, when developers used to code entire applications using plain assembly language. Many advances in software engineering have emerged to make software development easier, like object-orientation [5] or design patterns [2], just to name a few. However, the complexity of the applications built nowadays has also increased considerably.

In spite of the advantages of greater conceptual abstractions, there is still a very high requirement of specialized human labor; at the same time, there are serious difficulties for reusing solutions at high scale.

Although we have learned to reuse design decisions, in the form of software architectures or design patterns, for example, the implementation is generally performed by hand, at a much lower level of abstraction.

Automatic code generation has always been one of the solutions proposed in order to speed-up the software development process, especially for application families. Even though automatic code generation is gaining acceptance between software developers, it is far from being widely used. There are at least two reasons for that:

- The first reason is that generated code sometimes needs to be changed by hand, but those changes are usually lost when the code is generated again. If generated code cannot be changed, developers lose the freedom to adapt the code to fit their particular needs.
- The second reason is that code generators are usually black boxes, provided by third parties, and thus out of developer control. These black boxes, and the inherent lack of control, could entail uncertainty and mistrust for most software developers.

In this work, we evaluate automatic application code generation using templates and metadata. Developers can build their own code generators specifying templates, and then use different sets of metadata to build more than one instance in a family of applications. This approach represents a way to smoothly combine different levels of abstraction, taking advantage of the benefits of automatic code generation, but keeping the freedom and control of coding by hand.

The development method that we evaluated uses templates to express high-level solutions and the implementation of known variants at the same time. Then, these templates are combined with metadata for particular cases, using an automated tool that generates the source code of the application.

K. Elleithy (ed.), Advances and Innovations in Systems, Computing Sciences and Software Engineering, 241–246.

This is the approach taken by tools like Velocity[1], CodeSmith[2], or GeneXus Patterns[3].

These kind of automatic code generators, together with domain-specific modeling languages, are joined in the Model-Driven Engineering, a very promising approach proposed to alleviate complexity and effectively express domain concepts [6].

The rest of this paper is organized as follows: in the next section, we describe the approach of using templates and metadata for automatic generation of application families. In section 3, we state the goals of the case study; describe how we gather effort measures from templates and metadata code adapting the Halsted metrics; and we abstract case study results. Finally, in section 4, we summarize our conclusions from this case study.

II. TEMPLATES & METADATA

A template is a partial specification of a source code fragment, where portions that vary in a predictable manner will be substituted by specific data provided elsewhere. In each template, varying portions are specified using meta-programming that includes data input processing and traditional programming constructs -like variables definitions, loops, branches, etc.-, combined with textual portions -in the output programming language- that will be yielded as-is -without change- in the generated output code, once the varying portions are processed.

Meta-programming portions are enclosed within tags "<%" and "%>". The expression used to generate code from meta-programming constructs is enclosed within tags "<%=" and "%>". The Figure 1 shows a template example:

```
<%Property name="fieldName" type="string"%>
<%Property name="fieldType" type="string"%>
<%
string lowerName = fieldName.ToLower();
%>
  private <%=fieldType%> <%=lowerName%>;

  public <%=fieldType%> <%=fieldName%>
  {
    get {return <%=lowerName%>;}
    set {<%=lowerName%> = value;}
  }
```

Figure 1. A simple template example

There is a template compiler[4] and a tool that merges templates with metadata. The result of merging the previous template with the metadata set composed by `fieldName="id"` and `fieldType="int"` is shown in Figure 2.

```
private int id;

public int Id
{
  get {return id;}
  set {id = value;}
}
```

Figure 2. Code generated example

Metadata sets are provided as XML files. Other XML files specify which templates must be merged with which metadata files, and where the resulting source code should be placed (i.e: file name, folder, module); we call these other files the production line specification.

In general, each metadata file is used with many templates. For example, the definition of a business entity *Book* is combined with one template to generate a `Book` class, with another template to generate a `BookManager` class -responsible for inserting, updating, retrieving and deleting `Book` instances in a database-, and with yet another template to generate the a `BooksCollection` class -responsible for containing instances of `Book`-.

Generally speaking, each template represents a family of classes that follow the same pattern. Merging the template with each metadata file produces different classes of the family. Some of these classes might require subtle variations that need to be written by hand, and the integration with the generated code is done by using inheritance, composition, partial classes, or by adding new layers in the application architecture.

III. CASE STUDY

A. Goals

There are many code generation tools using templates and metadata to generate automatically application code[5]. Tools providers claim that this code generation approach has many advantages like quality, consistency, productivity and abstraction, just to name a few. Nevertheless, up to where we know, little research has been done in order to prove or reject these claims.

The first objective of the case study reported here was to demonstrate that the method depicted in the previous section is valid to develop software applications and applications fami-

[1] http://jakarta.apache.org/velocity/

[2] http://www.codesmithtools.com/

[3] http://www.genexus.com/

[4] ARTech Consultores, the publisher of GeneXus, provided the compiler to us.

[5] See, for example, http://www.codegeneration.net for a comprehensive list of code generation tools and approaches.

lies, where family members share architecture, domain and functionality. The second aim was to analyze the behavior of this method in the context of the aforementioned problems and compare it with manual development.

The attribute chosen for these comparisons is the effort required for a developer to specify templates and metadata versus the effort required for manual coding. We choose this attribute because effort is one of the most -if not the most- important driver of productivity.

Industry claims about productivity gains resulting for using templates and metadata for code generation, are generally based on the fact that much less template code and metadata is required to specify and automatically build applications. For example, in one of our experiments, two metadata files and 35 templates files where used to generate 84 source files with 5.133 lines of C# code.

However, differences in the size of the required source code cannot adequately explain claimed improvements in productivity, if the effort required to specify these sources is unknown. In fact, developers not using templates and metadata in every day work might perceive coding this way as more complex and hard to do than manual coding is.

One of the classical measures of effort is one of the Halstead Software Science measures [4]. We decided to use this measure of effort, even when validation of Halstead Software Science measures has been controversial, and different investigations have arrived to contradicting results.

Halstead Software Science measures were defined to be used with no particular programming language, but they are usually applied to procedural languages. We needed to adapt this measure to templates, metadata in XML, and source code in the C# language.

B. Adapting Halstead Effort Measures

Maurice Halstead proposed the Software Science metrics based upon algorithmic complexity theory concepts. Halstead defined four primitives:

- n_1: the number of distinct *operators* appearing in a program;
- n_2: the number of distinct *operands* appearing in a program;
- N_1: the total number of occurrences of the *operators* in a program;
- N_2: the total number of occurrences of the *operands* in a program.

Based on these primitives, Halstead defined many quantitative measures, including "vocabulary", "length", and "mental effort", among others, defined as:

- Vocabulary: $n = n_1 + n_2$
- Length: $N = N_1 + N_2$
- Volume: $V = N \log_2 n$
- Estimated program level[6]: $L_e = (2/n_1)\,(n_2/N_2)$
- Effort = V/L

The effort measure represents the number of mental discriminations -decisions- that a fluent, concentrating programmer should make in implementing an algorithm. The rationale behind this measure is that the difficulty of a program increases as the volume also increases, and decreases as long as the program level increases [1].

Halstead effort primitives, n_1, n_2, N_1, and N_2 can be extracted automatically from source code, after defining what is an *operand* and what is an *operator* in the source code language. In object-oriented code, each method can be seen as a procedure or function in a traditional procedural language, where Halstead metrics have been already applied. Therefore, within each method of a class, we consider *class attributes*, *local variables* and *method arguments* like *operands*, and *messages* or *method calls* like *operators*; we consider other arithmetic and logic operators in the usual way. Additionally, we consider *attributes* of a class like *operands*, since attributes are generally used as variables in traditional procedural languages.

As we have seen in the previous section, each template has portions of meta-programming code -in C#- mixed with portions of output code. The latter usually contain substitution parameters, and in our case, are also written in C#-.

From the template code point of view, we could see it as a procedure body, with some text outputs fragments in it. In order to calculate the Halstead effort metric, we could process all of the template code in the traditional way. In this context, output fragments could be seen as an imaginary call to an "Emit" fictitious procedure, which received the code to be sent to the output as a text string argument. Additionally, substitution parameters could be seen as imaginary calls to the same "Emit" procedure, providing the corresponding parameter.

Measuring the templates in this way would be consistent with the way they are processed by the template compiler itself. It first generates a fixed program, containing an actual "Emit" method, and an empty "Render" one. The template being compiled is used as the body of the "Render" method, but all the output sections and the substitution parameters are converted to the appropriate calls to "Emit". The resulting program is then compiled -using the standard C# compiler-, to produce an executable program. This executable program is executed, calling "Render" to obtain the generated output: the application source code -also in C# in our case-.

However, a measure obtained under these assumptions, although aligned to the processing performed by the compiler, would not exactly match the experience of the template developers; and could not be an appropriate measure of development

[6] Program level is defined as LV=V*, where V* is a constant representing the volume of the most compact representation of a given algorithm; we have used the estimated program level to calculate the effort measure.

effort. For, even if the compiler sees output fragments as meaningless text strings, this is not true at all for the developers; for them, these strings are actual source code too, so the complexity of this code must be taken into account.

It is also reasonable to expect some sort of direct relationship between the effort required to write one of the classes generated by a given template, and the effort required to write the template itself. In other words, a simple class would require a simple template, and a complex class, a complex template.

Unfortunately, we cannot take the shortcut of estimating the effort of the output fragments simply as the effort calculated for one of the generated classes, because the effort required to write the output fragments includes the abstraction of common patterns in the generated class. In fact, that is precisely the difference between writing a template, and writing one of the generated classes by hand.

On the other hand, we cannot process output fragments based on a strict language grammar analysis, the way we could with generated code, because some parts of the syntactic structure of the output fragments might be hidden behind substitution parameters that insert them. For example, if we used a grammar that expected the "class" keyword in order to detect a class declaration, the analysis could fail if the developer of the template wanted to make the declaration vary between "class" and "interface" and used a substitution parameter for that.

Since we could not use a measuring tool based on the complete language grammar, we opted for a loose grammar approach, which along with some assumptions on the way the templates are written, allowed us to obtain the required measures, even in the presence of these limitations. All the templates were then checked to comply with the following five assumptions

- The first assumption was about naming conventions: in template code and in output code, the developer follows usual naming conventions, i.e. camelCase[7] for variable names, including instance variables, and PascalCase for all the other names.
- The second assumption was about uniform structure: the output of every template must be a source code file in which every method is inside a class definition, and every class is inside a namespace definition.
- The third assumption was about the visibility of the braces, i.e.: "{" and "}": templates cannot hide behind substitution parameters the braces used to specify the scope of methods, classes, or namespaces declarations.
- The fourth assumption was about correctness: templates must be valid, i.e., they do not have compilation errors, and the generated code is free of compilation errors too.
- The last assumption was that reserved language keywords should not be used as identifiers in any context, even in those allowed by the language syntax.

C. Data Gathering

We performed the measurement of the effort in two scenarios. The first one was the construction of a business application for an fictitious bookstore. The application uses three logical layers, and allows creating, updating and querying books, book inventory, book orders and invoices to customers and from suppliers. It uses a database management system to store persistent data.

For comparison purposes, we assumed that effort measured in generated code is a rough approximation of manual development effort. In other words, we do not build applications again from the scratch by hand; instead, we consider the generated code as good as the code we can write by ourselves, for the purpose of this case study.

We developed this application using an incremental approach. For each increment, after adding new functionality to the application, we took measures of the effort required so far by the templates and metadata (E_{tm}), and compared it against the effort of the equivalent manual coding of the application (E_m), which we approximated by taking measures on the generated code, as stated above.

In the first increment, the application only had metadata and templates needed to implement functionality required to handle books, genres and authors. In successive increments, we added functionality to support new business entities, as follows:

- Increment 2: customers
- Increment 3: orders, order items
- Increment 4: invoices, invoice items
- Increment 5: suppliers
- Increment 6: purchase orders, purchase order items
- Increment 7: purchase invoices, purchase invoice items

For the second scenario, we used the same templates to build a second application, just by using different metadata files, this time for a car replacement parts store.

This new application reused two of the business entities of the previous one -customers and suppliers- and we added the following:

- Car brand
- Car model
- Car part
- Part order, part order item
- Part invoice, part invoice line
- Part purchase order, part purchase order line
- Part purchase invoice, part purchase invoice line

In this second scenario, we only took measures for the final version.

[7] The names camelCase and PascalCase are used to refer to different ways to mix upper case and lower case characters in source code identifiers built up from one or more words. With camelCase, the first character of all words except the first one apperar in upper case and the rest in lower case, whereas in PascalCase, the first character of all words appear in upper case, and the rest in lower case.

D. Result set

The following table resumes gathered Halstead measures[8]:

TABLE 1. CASE STUDY RESULTS

Scenario		Effort	
		Templates & Metadata (E_{tm})	Manual (E_m)
A	Increment 1	5.483.271	1.537.942
	Increment 2	5.512.880	2.038.793
	Increment 3	5.536.552	2.913.841
	Increment 4	5.563.796	3.853.793
	Increment 5	5.594.344	4.378.169
	Increment 6	5.618.016	5.257.140
	Increment 7	5.645.260	6.202.195
B		149.356	6.259.232
Totals		5.794.616	12.461.427

IV. CONCLUSIONS

The first goal of this work was to validate the claim that using templates and metadata it is possible to create an automated solution to build application families.

After the case study, we can conclude that this is a feasible approach to build application families. In fact, application families can be very broad. For example, although in this work we built only two applications, both typical business applications, both with similar but not the same business entities, the specified solution does not contain any constraint, or anything oriented specifically, to these application domains. Neither there is nothing in the arranged product line nor in the coded templates, specific to these business applications; even more, there is nothing specially designed to support invoices, products or clients.

Thoroughly, if we were forced to enumerate the general characteristics that define the application family we can build with this solution, the answer would be something like "multi-layer applications consuming persistent-storage services from a database management system." Granted, applications are generated for a specific language in a specific platform; and yet, there is still a broad spectrum of applications the solution is capable to generate, for which the applications used in the case study are just two possible cases.

The second goal of the case study was to determine if the approach of using templates and metadata allows reducing development effort. It would be of no benefit to have code generators and product lines, if the effort required to develop and use them were disproportionate compared with the equivalent manual development.

The context in which we evaluated this approach considered a solution built and consumed by the same developer or the same team of developers. When the solution is built in one organization but consumed by third parties, other scales comes into play, and even huge development efforts may be compensated by mass-market use. In our context, we were looking forward for an economy of scale that could be achieved even by small teams –or individuals-, small applications, and small-sized families of applications.

While developing the application, we took measures to compare the evolution of the effort required to develop the application using templates and metadata, with respect to the estimated effort required to build the same application by hand. In advance, we did not expect major savings for small applications; on the contrary, we expected that the effort of developing with templates would be very high.

Indeed, the figures for the first version of the application, when there were only three entities, reveal that the effort of using templates and metadata was three or four times greater than the effort required by manual development -see Table 1, scenario A, increment 1-. The important issue was to watch how these measures evolved -at which rate- while the application functionality grew.

The second increment became very promising because the effort required to add the new functionality using templates and metadata was about 5 per thousand, while the estimated equivalent effort required by manual coding grew more than 32 percent.

This tendency held in following increments. In one side, growing at the rate of new metadata files, of relatively low effort, and in the other side, growing at the rate of the effort estimated for all new generated files -see Figure 3-. In the last increment, the effort for templates and metadata had grown about 3%, while the effort calculated for manual development had grown more than 300%, and was 10% greater than its counterpart.

[8] Halstead effort is a dimensionless measure.

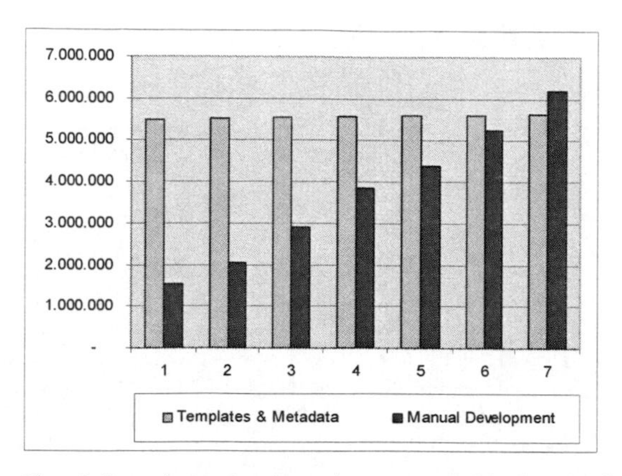

Figure 3. Comparing templates & metadata versus manual development effort for the same application. The y-axis shows dimensionless Halstead effort and the x-axis shows successive iterations

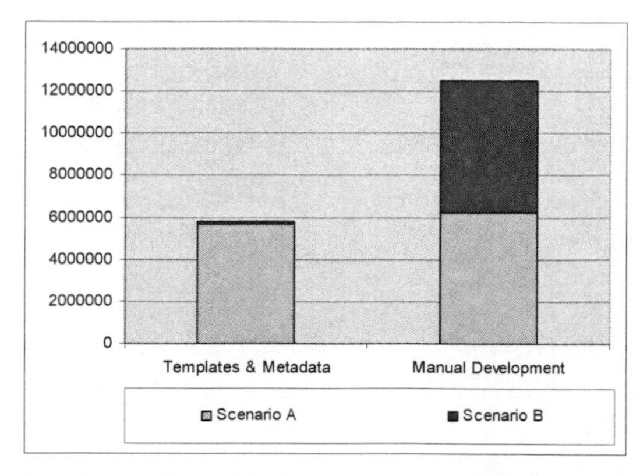

Figure 4. Comparing templates & metadata versus manual development for two different applications. The y-axis shows dimensionless Halstead effort and the x-axis shows evaluated scenarios

It looks like even for a single application, the gains derived from the internal reuse are so high, that significant economies are obtained as soon as the application reaches a relatively small size, and furthermore, the economies keep growing as the size increase.

We developed a second application to evaluate the other possible scenario: external reuse in the context of application families. In this second scenario, we took the templates from the previous one and used them to develop an entirely new and different application, replacing metadata files; i.e. using templates and metadata to build a family of applications.

In this new scenario, results were even better. To build this new application, we only needed to write new metadata files. The effort calculated for the generated application resulted 42 times greater than the effort required by templates and metadata. In this case too, we can expect a constant effort gain as long as application size increases.

If we consider the numbers accumulated from both scenarios, and compare the effort estimated for manual development from both applications, with respect to the effort required by templates and metadata, the ratio is 2:1 -see figure 2-.

Even more interesting is the fact that moving from one to two applications implies a 2.6% effort growing using templates and metadata, while approximately a 100% with manual coding.

In this paper, we analyzed and compared the variations of the effort required to develop applications using templates and metadata versus manual development, in two different dimensions: the application size and the number of applications. In both cases the results from our case study show that it is possible to obtain considerable benefits, even from small applications or few applications, and that those benefits grow as long as the application functionality increases.

REFERENCES

[1] Fitzsimmons, A. and Love, T., A Review And Evaluation Of Software Science, *ACM Computing Surveys*, vol. 10, Mar., 1978.

[2] Gamma, E., Helm, R., Johnson, R., and Vlissides, J.. *Design Patterns: Elements of Reusable Object-Oriented Software*, Reading, MA, USA: Addison-Wesley, 1995.

[3] Greenfield, J., The Case for Software Factories, *Microsoft Architect Journal*, vol. 3, Jul., 2004.

http://msdn.microsoft.com/library/en-us/dnmaj/html/aj3softfac.asp

[4] Halstead, M., "Toward a theoretical basis for estimating programming effort", *ACM/CSC-ER Proceedings of the 1975 annual conference*, 1975.

[5] Meyer, B., *Object Oriented Software Construction, 2nd edition*, Prentice Hall PTR, 1997.

[6] Schmidt, D., Model-Driven Engineering, *IEEE Computer*, vol. 39(2), Feb, 2006.

[7] The Standish Group International, Inc., *The Chaos Report*, 1994.

[8] The Standish Group International, Inc., *The Chaos Report 2004 Q3*, 2004.

Application of Multi-Criteria to Perform an Organizational Measurement Process

Josyleuda Melo Moreira de Oliveira, Karlson B. de Oliveira, Ana Karoline A. de Castro, Plácido R. Pinheiro, Arnaldo D. Belchior
Universidade de Fortaleza (UNIFOR) – Masters Degree in Applied Computer Sciences (ACS)
Av. Washington Soares, 1321 - Bl J Sl 30 - 60.811-341 - Fortaleza – Brasil
{josymm, karlson.oliveira, akcastro}@gmail.com, {placido, belchior}@unifor.br

Abstract - **Software quality has become increasingly important as a crucial factor in keeping organizations competitive. Software process measurement is an essential activity in achieving better quality and guarantees, both in the development process and in the final product. This paper presents the use of multi-criteria in a proposed model for the software measurement process, in order to make it possible to perform organizational planning for measurement, prioritize organizational metrics and define minimal acceptance percentage levels for each metric. This measurement process was based on five well known processes of measurement: CMMI-SW, ISO/IEC 15939, IEEE Std 1061, Six Sigma and PSM.**

Keywords - Metrics. Measurement Software Process. Multi-Criteria

I. INTRODUCTION

"You can't control what you can't measure" [10]. The results of measurements enable one to make decisions based on solid facts, so as to improve processes and products.

The proposed process of software measurement is based on mapping five measurement processes: CMMI-SW [8], ISO/IEC 15939 [15], IEEE Std 1061 [14], Six Sigma [17] and PSM [16]. We used the best practices of each of these processes, including points considered important to facilitate the applicability of the model in question [3].

This paper presents the usage of multi-criteria in a process measurement model proposed to make one of the macro-activities of the process, called organizational planning for measurement feasible. Since the metrics are the key to the entire measurement process, it is necessary to prioritize them and define minimum acceptable percentage values for each organizational metric, in order to facilitate analysis of measurement data.

In Section 2 we expose the proposed measurement process. In Section 3 we show the importance of using multi-criteria in the measurement process. In Section 4 we present the solution and Section 5 displays the conclusions and recommendations for future research.

II. RELATED WORK

As related work, two good papers about Multi Criteria Decision Aid (MCDA) approach used in the software engineer field were used. Both are about prioritization.

- The first one proposes to help test analysts and stakeholders with the selection process of use cases for automation, according to the reality of each organization, considering that they evaluate the quality of the final product. Software Test Automation is a renowned way to enhance the test process, but it is not always the best approach for the project, despite the fact that a crucial matter when applying software tests is to decide the viability of automating them. This decision is not always easy to make, because it involves people expressing their point of view, which can cause a conflict with others' opinions. Moreover, some important criteria may not be considered, causing the automation to be inefficient [23].
- The second is to help to prioritize the criteria in order to answer important questions on User Interface (UI) designer, such as: How to choose the best usability pattern from a list of alternatives productively? Usually, there are many people involved in the project who want to make their opinion have an impact on the final result and such opinions can be controversial since it may involve users' representatives and professionals from the software organization. This fact raises another issue: How can we organize the project participants with different points of views to choose the best usability pattern? [22].

III. ORGANIZATIONAL SOFTWARE MEASUREMENT PROCESS (OSMP)

The proposed measurement process [1] is composed of seven macro-activities (Fig. 1):

i. **Organizational Planning for Measurement**: Involves the activities necessary for starting up the measurement process in a given organization. Information must be gathered in order to understand the characteristics and needs of the organization. Measurement commitments must be established and supported, to then define the scope of the organizational metrics;

ii. **Plan Measurement**: consists of planning the measurement of projects;

iii. **Perform Measurement**; involves gathering and storing project data;

iv. **Analyze Measurement**: is responsible for analyzing the data and communicating the results of measurement;

v. **Monitoring and Control Measurement**: in association with the other macro-activities, checks if the products are being correctly utilized and if the activities are being carried out in accordance with the objectives of the measurement;

vi. **Evaluate Process Measurement**: evaluates the

K. Elleithy (ed.), Advances and Innovations in Systems, Computing Sciences and Software Engineering, 247–252.

measurement process, registering the lessons learned and the process improvements in the organization;

vii. **Organizational Evaluation**: all finished projects are evaluated based on the indicators defined in the macro-activity Organizational Planning for Measurement The frequency is defined early by the organization and with the results in hand, high management can see the main deficiencies in its projects, take corrective actions and consider organizational improvements.

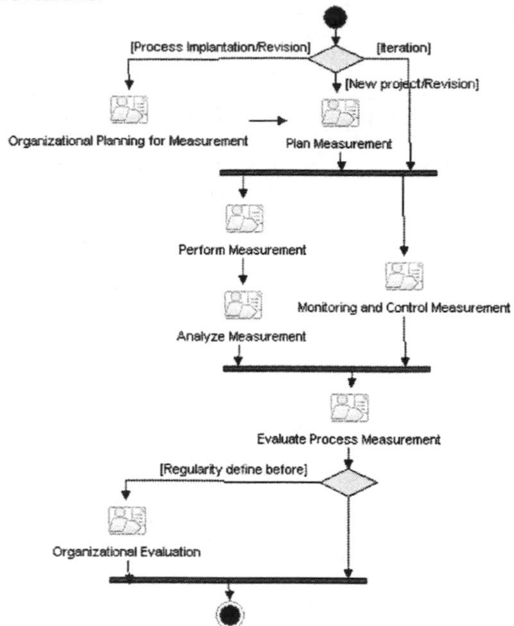

Fig. 1 – Workflow of OSMP.

Each of the mentioned macro-activities contains a set of activities which must be carried out by specific roles. In this context, four roles were identified in the process and subdivided into main roles and collaborative roles. These roles are: Measurement Analyst, Upper Management, Project Manager and Relevant Stakeholders.

A need was also identified to describe the steps to be accomplished in each activity in order to better organize the process and to favor handling by the users, considering the concept of integrating the measurement process to the activities normally performed in the software development process.

Input and output artifacts were identified for each activity. Thus we researched what was available in literature to adapt to the context in question, and new templates were created to be filled out in the course of the process.

When difficulties were perceived in the identification of metrics, we suggested using the GQIM (Goals-Questions-Indicators-Metrics) [18] technique, which is an evolution of GQM (Goals-Questions-Metrics) [5], and based on specialized literature we generated, for this purpose, a document containing a vast set of metrics with the main characteristics of each one.

For the user to understand the work accomplished on the basis of GQIM he must understand a few concepts. Objectives are collected to establish a measurement program in a software organization. Each objective is detailed through a set of questions which express its needs, leading to comprehension of the objective. Many questions have measurable answers that can be expressed through metrics. The metrics are directly related to indicators.

An indicator is a representation of measurement data that provides insight into software development processes and/or software process improvement activities [6]. An indicator is a measure that provides an estimate or evaluation of specified attributes derived from an analysis model, and is the basis for analysis and decision in a measurement process [16].

This paper only details the first macro-activity of the proposed process, which is Organizational Planning for Measurement (Table I), using MCDA (Multi-Criteria Decision Analysis) [4]

TABLE I
Organizational Planning for Measurement

Organizational Planning for Measurement
Activity: Characterize the organizational unit. **Steps:** Search, gather and study information about the organization
Activity: Establish and support measurement commitments. **Steps:** Identify stakeholders; Supply resources; Define responsibilities; Prepare measurement templates; Communicate the measurement process with its objectives and responsibilities to the entire organization; Train personnel involved with the measurement; Define configuration and change procedures; Define tools for data storage.
Activity: Plan metrics selection. **Steps:** Define the objective of the measurement; Define techniques, tools and criteria for selecting the metrics.
Activity: Define organizational metrics. **Steps:** Prepare questions based on the objectives of the measurement; Prepare indicators that answer the questions; Define the necessary metrics to obtain the defined indicators; Disclose the definition of metrics to relevant personnel; Review and improve selected metrics based on feedback from relevant personnel; Categorize the metrics; Prioritize the metrics and define minimum acceptable percentages for each organizational metric.

IV. USING MULTI-CRITERIA IN THE MEASUREMENT PROCESS

In order to prioritize the organizational metrics gathered in the Organizational Planning for Measurement macro-activity, and in order to define priorities in the allocation of tasks, the multi-criteria technique is used due to two main characteristics:

- There is a set of criteria in the process; and
- It pursues solutions that best fit the needs of the actors involved.

Bana and Costa [4] uphold that decision analysis results in better understanding for the environment decider, assuring that the proposed solution is considered adequate within the analyzed context. The decision analysis process is comprised of three phases: (i) structuring, (ii) evaluation and (iii) recommendations. Although they are distinct, these three phases share intrinsic connections. In this paper we approach only the evaluation phase, which consists of a set of three activities: (i) building a quantitative model of values, which includes building an additive value model; (ii) evaluating options, which consists of applying models for a particular set of options; and (iii) analyzing sensitivity and robustness.

This methodology was chosen for having a tool that favors the construction of judgment matrixes. It's an interactive support tool for the construction, on a set S of stimuli or potential actions, of numerical interval scales that quantify the attractiveness of the elements of S in the opinion of the evaluators, based on semantic judgment of the difference in attractiveness between two actions [20].

V. SELECTING OBJECTIVES AND ORGANIZATIONAL METRICS

The proposed measurement process is in use at IVIA (IVIA Comércio e Serviços de Informática Ltda. – IVIA Computer Commerce and Services Ltd.), a software development company with a Research and Development department (R&D), located in Fortaleza, Ceará, founded in March, 1996 and qualified by ISO 9001:2000 in 2003. In this paper we present the selection of objectives and metrics for this organization.

Starting with the application of the GQIM technique [18], and using a field survey carried out by [12] with the purpose of identifying the main problems detected in the development of software, the following measurement objectives were obtained for IVIA (such objectives are adopted as criteria for the multi-criteria technique):

1. Improve the accuracy of delivery date estimates;
2. Improve the quality of products released for use;
3. Reduce Costs and;
4. Increase the value and use of co-workers' potentials.

For each objective a set of questions was prepared (some are listed below), thus obtaining the indicators (Table II) which will be the multi-criteria actions.

1. How accurate is the estimated schedule?
3. How accurate is the effort estimate?
4. How high is the quality of the products before their release for use?
5. How high is the quality of the products after their release for use?
6. Are co-workers assigned to the projects according to their capabilities?

Use of indicators is achieved starting by gathering the metrics presented in Table III.

TABLE II
INDICATORS DEFINED FOR IVIA

Indicator	Calculation / Definition
Customer satisfaction	**Calculation:** average obtained on the customer satisfaction questionnaire (M1).
Accuracy of time estimates	**Calculation:** actual time of the project (M2) / estimated time of the project (M3) [9]
Accuracy of the effort estimates	**Calculation:** actual effort of the project (M4) / estimated effort of the project (M5) [9]
Cost accuracy	**Calculation:** actual cost of the project (M6) / estimated cost of the project (M7)
Defect density	**Calculation:** (number of errors in a project (M8)) / system size (M9). **Definition:** the number of errors of all artifacts are added to the number of modifications of all artifacts in all use cases, and the result is compared to the size of the system (presently measured in lines of code) [18]
Rework index	**Calculation:** rework effort (M10) * 100 / actual effort (M4) [18]
Team experience	**Calculation:** median values obtained by team members (M11) [11]
Deterioration of software	**Calculation:** effort used to correct problems after release of software (M12) / actual effort of entire project (M4). **Definition:** software modifications undertaken after its release for use increase the possibility of inclusion of new problems, reducing the quality of the final product. Thus, the term "software deterioration" was defined as the relationship between the effort spent to correct problems found after release of the system, compared to the actual effort of the entire project [21]
Level of learning	**Calculation:** Average values obtained by team members before the project (M13) / Average values obtained by team members after the project (M9)

TABLE III
METRICS DEFINED AT THE IVIA ORGANIZATION

Metric	Calculation/Definition
M1: Average customer satisfaction	**Calculation:** Sum total of grades on the questionnaire / number of questions. **Definition:** four possible values were established: 1- Not satisfied; 2 – A bit satisfied; 3 – Satisfied; and 4 – Very satisfied
M2: Actual time	**Calculation:** acceptance date – starting date **Definition:** actual time is measured in number of consecutive days between the starting date and the date of acceptance [7]
M3: Estimated time	**Calculation:** forecast acceptance date – forecast start date **Definition:** estimated term corresponds to the number of consecutive days between the forecast start and forecast finish dates [7]
M4: Actual effort	**Calculation:** quantity of man/hours used in the actual term of the project **Definition:** actual effort corresponds to the total working hours of all team members in the achievement of the use case, up until acceptance date [7]
M5: Estimated effort	**Calculation:** quantity of man/hours used in the estimated term of the project. **Definition:** estimated effort corresponds to the quantity of man/hours forecast for the project [7]
M6: Estimated cost	**Calculation:** Estimated cost of the project
M7: Actual cost	**Calculation:** Actual cost of the project
M8: Number of errors	**Calculation:** sum total of errors in use cases produced during the project. **Definition:** error is any new problem detected in an artifact in the act of its approval [19]
M9: System size	**Calculation:** total number of lines of code, except blank lines,, or those containing only comments **Definition:** to favor the measurement of system size we use the total lines of code, except those left blank or containing only comments [7]
M10: Rework effort	**Calculation:** total man/hours spent on rework during the project

	Definition: rework effort corresponds to the sum total of hours spent on rework by all team members throughout entire project [7]
M11: Values obtained by the team members	**Calculation:** values obtained per team member **Definition:** five possible values were established: 0 – no experience; 1 – academic training; 2 – practice on at least three projects; 3 – experienced; 4 – capable of orienting others [11]
M12: Effort spent to correct problems after release of the system	**Calculation:** total number of man/hours spent to correct problems found in the system after user acceptance **Definition:** any work carried out after release of the software, due to problems found, must be counted as effort used to correct problems. This does not include evolutionary maintenance [21]
M13: Average values obtained by team members prior to the project	**Calculation:** Average values obtained per team at the start of the project **Definition:** This evaluation must be accomplished at the start of the project for all the project co-workers. Five possible values were established: 0 – no experience; 1 – academic training; 2 – practice on at least three projects; 3 – experienced; 4 – capable of orienting others [11]
M14: Average values obtained by team members after the project	**Calculation:** Average values obtained per team at the end of the project. **Definition:** This evaluation must be accomplished at the end of the project for all the project co-workers. Five possible values were established: 0 – no experience; 1 – academic training; 2 – practice on at least three projects; 3 – experienced; 4 – capable of orienting others [11]

VI. USE OF MULTI-CRITERIA IN THE ORGANIZATIONAL PLANNING FOR MEASUREMENT MACRO-ACTIVITY

In order to aid judgment the M-MACBETH approach for MCDA [4] was used in this work [2]. Initially a tree was constructed for each task, each with its criteria (Fundamental Point of View – FPV) as defined in the previous section. Fig. 2 shows the tree corresponding to the measurement objectives at the IVIA organization.

Each FPV is comprised of actions that correspond to all possible indicators for each detailed objective. The order of the actions in the judgment matrix will be determined by their respective degrees of importance and by the company's project profile.

Four criteria were defined, corresponding to the objectives detailed in the previous section:

(i) Improve delivery time; (ii) Reduce Costs; (iii) Improve product quality; and (iv) Improve process quality (with this improvement the company will focus on increasing the value and use of co-workers' potentials).

Each criterion identified above will be judged by the actions (Fig. 3) corresponding to the indicator metrics.

After judgment of each criterion of the process, a scale of values is generated which corresponds to the Current Scale field of the judgment matrix represented in Fig. 4. The values are included in the task corresponding to the objective function of the model.

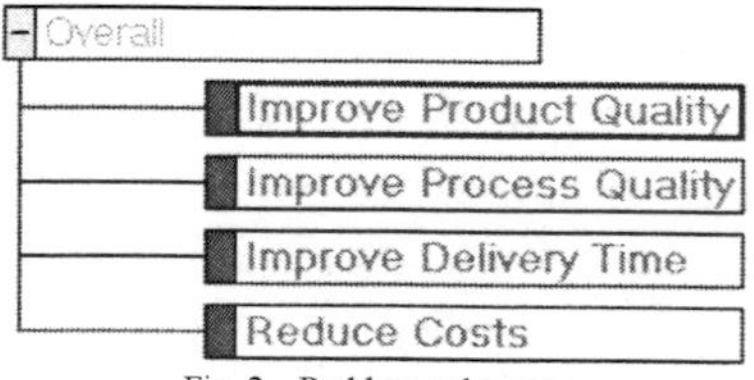

Fig. 2 – Problem value tree

In Figs. 4 and 5 examples of indicators (actions) are evaluated for each of the objectives (criteria). It should be pointed out that the indicators were listed two-by-two for each criterion.

Actions

	Actions	Short	Financial value
1	Customer Satisfaction	CS	0
2	Accuracy of Time Estimates	ATE	0
3	Accuracy of Effort Estimates	AEE	0
4	Cost Accuracy	CA	0
5	Defect Density	DD	0
6	Rework Index	RI	0
7	Team Experience	TE	0
8	Deterioration of Software	DS	0
9	Level of Learning	LL	0

Add | Remove | Properties | Profile | Help

Fig. 3 – Actions of the IVIA organization measurement process

Fig. 6 identifies the relevance of each objective for the IVIA organization, compared in pairs.

Fig. 7 shows the final result obtained from this project, using the Macbeth tool. Starting from this result one can use the rates obtained in the analysis of the metrics of the IVIA organization project. Thus, the decreasing order of importance of each indicator for the organization is shown in the figure, for example, the team experience (TE) is the most relevant indicator for organization and its percentage of importance is of 90.01%. This means that the average of the team experience in the projects must be at least of 90,01%.

Analyzing the deterioration of software (DS), we will see that it is the least important indicator for the organization according to priority order and it has a percentage of 70.01%. This means that the products can present 29.99% of corrections of errors after the delivery of the product to the customer and still will be in the minimum organization expectancy.

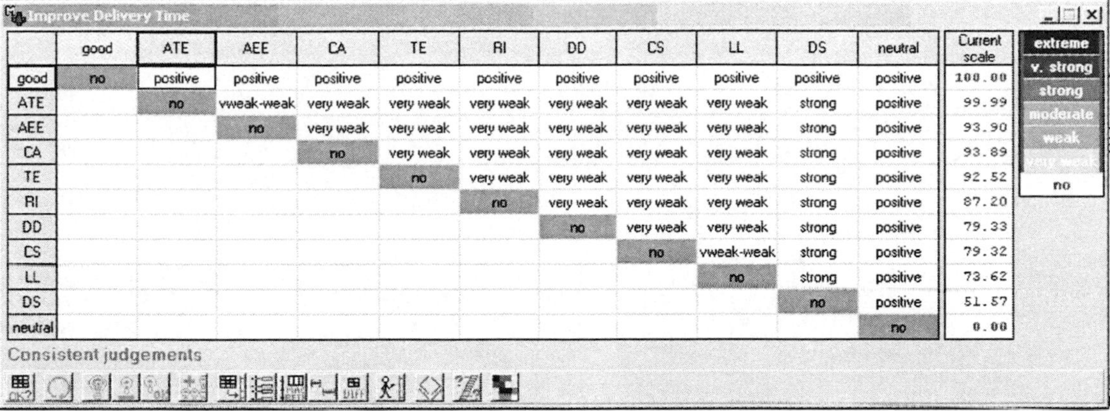

Improve Delivery Time

	good	ATE	AEE	CA	TE	RI	DD	CS	LL	DS	neutral	Current scale
good	no	positive	positive	positive	positive	positive	positive	positive	positive	positive	positive	100.00
ATE		no	vweak-weak	very weak	very weak	very weak	very weak	very weak	very weak	strong	positive	99.99
AEE			no	very weak	very weak	very weak	very weak	very weak	very weak	strong	positive	93.90
CA				no	very weak	very weak	very weak	very weak	very weak	strong	positive	93.89
TE					no	very weak	very weak	very weak	very weak	strong	positive	92.52
RI						no	very weak	very weak	very weak	strong	positive	87.20
DD							no	very weak	very weak	strong	positive	79.33
CS								no	vweak-weak	strong	positive	79.32
LL									no	strong	positive	73.62
DS										no	positive	51.57
neutral											no	0.00

Fig. 4 – Judgment of the indicators to improve accuracy of project estimates

Improve Process Quality

	good	CS	LL	TE	RI	DS	DD	ATE	AEE	CA	neutral	Current scale
good	no	positive	positive	positive	positive	positive	positive	positive	positive	positive	positive	100.00
CS		no	very weak	very weak	very weak	very weak	vweak-weak	vweak-weak	vweak-weak	vweak-mod	positive	99.99
LL			no	very weak	very weak	very weak	vweak-weak	vweak-weak	vweak-weak	vweak-mod	positive	86.67
TE				no	very weak	very weak	vweak-weak	vweak-weak	vweak-weak	vweak-mod	positive	86.66
RI					no	very weak	vweak-weak	vweak-weak	very weak	vweak-mod	positive	86.65
DS						no	vweak-weak	vweak-weak	very weak	vweak-mod	positive	77.76
DD							no	vweak-mod	very weak	moderate	positive	71.65
ATE								no	vweak-strg	vweak-strg	positive	71.64
AEE									no	vweak-weak	positive	60.63
CA										no	positive	32.28
neutral											no	0.00

Consistent judgements

Fig. 5 – Judgment of the indicators to improve process quality

Overall

	[IDT]	[RC]	[IPQ]	[IPCQ]	[all low]	Current scale
[IDT]	no	very weak	weak-mod	vweak-mod	positive	45.45
[RC]		no	vweak-weak	very weak	positive	27.27
[IPQ]			no	very weak	positive	18.19
[IPCQ]				no	positive	9.09
[all low]					no	0.00

Consistent judgements

Fig. 6 – Degrees of relevance between each criterion

These data are so important when the measurement analyst analyzes the collected data of the measurement, because from this order of priority and these minimum or maximum percentages of each indicator it is possible to get a more consistent result to use as a parameter of comparison based on mathematical calculations, not on feeling. This helps a lot in the statistics data analysis.

Actions scores

Actions	Overall	IPQ	IPCQ	IDT	RC
[all high]	100.00	100.00	100.00	100.00	100.00
TE	90.01	99.97	86.66	92.52	80.31
RI	90.01	99.96	86.65	87.20	89.17
ATE	88.40	66.69	71.64	99.99	89.15
AEE	85.18	66.73	60.63	93.90	91.14
DD	85.07	99.98	71.65	79.33	89.16
CA	85.01	66.72	32.28	93.89	99.99
CS	80.02	80.52	99.99	79.32	74.21
LL	70.01	80.31	86.67	73.62	51.57
DS	70.01	80.53	77.76	51.57	91.13
[all low]	0.00	0.00	0.00	0.00	0.00
Scaling constants:	0.18	0.09	0.45	0.27	

Fig. 7 – Final judgment of the IVIA indicators

VII. CONCLUSIONS

This paper presents a proposal for the prioritization and definition of minimum percentages for indicators in a measurement process, using as a means a model of software measurement based on five well known processes of measurement and focused on the Organizational Planning for Measurement macro-activity. The multi-criteria methodology is applied with the objective of enabling the achievement of that macro-activity.

The evaluation of data conducted by the measurement analyst is therefore expected to be made easier, and to attain the diagnosis of software development of a certain organization, this analyst will be able to count on a resource which is no longer based only on feeling.

The software measurement process hereby proposed is being implemented at a company called IVIA. Our main objective is that at the end of the experiments it shall be possible to refine this measurement process even more, so as to contribute to the development of quality software.

REFERENCES

[1] Oliveira, Josyleuda M. M., Oliveira, Karlson B., Belchior, A. D., *Organizational Software Measurement Process*, accepted in Metrikon 2006, Germany - Potsdam, November, 2006

[2] Oliveira, Josyleuda M. M., Oliveira, Karlson B., Belchior, A. D., Castro, Ana Karolina, Pinheiro, Plácido Rogério, *Aplicação de Multicritério para Implementação de um Processo de Medição Organizacional*, accepted in SBPO (Brazilian Symposium of Operational Research), Goiânia – Brazil, September, 2006; in Portuguese.

[3] Oliveira, Josyleuda M. M., Oliveira, Karlson B., Belchior, A. D., Measurement Process: *A Mapping among CMMI-SW ISO-IEC 15939 IEEE Std 1061 Six Sigma and PSM*, accepted in ICSSM (International Conference Service System & Service Management) do IEEE, France, October, 2006

[4] Bana e Costa, C. A, Corte, J. M. D., Vansnick, J. C. *MACBETH. LSE-OR Working Paper*, 56, 2003.

[5] Basili, Victor R. *Software modeling and measurement: the goal/question/metric paradigm*. Motorola SCC Document, Appendix B, 1995.

[6] Baumert, J. H., Mcwhinney, M. S. *Software measures and the capability maturity mode*l. Carnegie Mellon University, Pennsylvania: Software Engineering Institute, 1992.

[7] Carleton, Anita D., Park, Robert E., Goethert, W. B., Florac, William A.; Bailey, Elizabeth K.; Pfleeger, Shari L. *Software measurement for DoD systems: recommendation for initial core measures*. CMU/SEI-92-TR-19, ESC-TR-92-19,Pittsburgh, Software Engineering Institute, Carnegie Mellon University, 2002.

[8] Chrissis, Mary B., Konrad, Mike, Shrum, Sandy. *CMMI: guidelines for process integration and product improvement*. Boston: Addison Wesley, 2003.

[9] Daskalantonakis, M. K. *A practical view of software measurement and implementation experiences within motorola*. In: Applying Software Metrics, IEEE Computer Society Press, p. 168-180, 1992.

[10] Demarco, T. *Controlling software projects*. New York: Yourdon Press, 1982.

[11] Fenton, Norman E., Pfleeger, Shari Lawrence. *Software metrics: a rigorous and practical approach*. 2. ed. PWS Publishing Company, 1997.

[12] Gomes, Augusto G. G. J. *Avaliação de processo de software baseado em medição*. Tese (Mestrado em Informática) – Rio de Janeiro, Brasil, 2001.

[13] Goodwin, P., Wright, G. *Decision analysis for management judgment*. 2. ed. John Wiley & Sons, Chicester, 1998.

[14] IEEE Std 1061-1998. *IEEE standard for a software quality metrics methodology*, 1998.

[15] ISO/IEC 15939:2002, Software – engeneering: *software measurement process*. 2002.

[16] McGarry, John, Card, David, Jones, Cheryl, Layman, Beth, Clark, Elizabeth, Dean, Joseph, Hall, Fred. *Practical software e measurement: objective information for decision makers*. PSM, Addison-Wesley, 2002.

[17] Pande, P. S., Neuman, R. P., Cavanagh, R. R. *The six sigma way*. McGraw-Hill, 2000.

[18] Park, Robert E., Goethert, Wolfhart B.; Florac, William A. *Goal-driver software measurement: a guidebook*. SEI, 1996.

[19] Roberts, M. A. *Experiences in analyzing software inspection data, software engineering process group*. McDonnell Douglas Aerospace, McDonnell Douglas Corporation, St. Louis, MO, 1996.

[20] Souza, G. G. C. de. *Um modelo de multicritério para a produção de um jornal*. 2003. 71f.. Dissertação (Mestrado em Informática Aplicada) – Universidade de Fortaleza, Fortaleza.

[21] Tajima, D., Matsubara, T. *The computer software industry in Japan*. IEEE Computer, 14(5), p. 89-96, 1981.

[22] Sousa, Kênia Soares. *UPi – A Software Development Process Aiming at Usability, Productivity and Integration*, 2005, p. 159-186, Dissertação (Mestrado em Informática Aplicada) – Universidade de Fortaleza, Fortaleza.

[23] PINHEIRO, P. R.; SAMPAIO, Marcia; DONEGAN, Paula et al. *Multicriteria Model for Selection of Automated System Tests*. In: A. Min Tjoa; Li Xu; Sobail C. (Org.). Research and Practical Issues of Enterprise Information Systems. New York, 2006, v. 205, p. 777-782

Institutionalization of an Organizational Measurement Process

Josyleuda Melo Moreira de Oliveira, Karlson B. de Oliveira, Arnaldo D. Belchior
Universidade de Fortaleza (UNIFOR) – Masters Degree in Applied Computer Sciences (ACS)
Av. Washington Soares, 1321 - Bl J Sl 30 - 60.811-341 - Fortaleza – Brasil
{josymm, karlson.oliveira}@gmail.com, belchior@unifor.br

***Abstract* - Software development is a complex activity which demands a series of factors to be controlled. In order for this to be controlled in an effective manner by project management, it is necessary to use software process measurement to identify problems and to consider improvements. This paper presents an organizational software measurement process resulting from the mapping of five relevant software measurement processes: CMMI-SW, ISO/IEC 15939, IEEE Std 1061, Six Sigma, and PSM (Practical Software Measurement). The best practices of each one were used, including relevant keys to facilitate the applicability of a measurement process focused on project management, as well as assuring the software quality.**

Keywords - Measurement process, CMMI-SW, ISO/IEC 15939, IEEE Std 1061, Six Sigma, PSM, Goal/Questions/Metrics (GQM), Goal/Questions/Indicators/Metrics (GQIM).

I. INTRODUCTION

Computer systems are more and more common in many human activities. This is true in people's daily tasks, as well as in the competitiveness of organizations, which are dependent on the quality of the software developed or acquired for their use. This has generated a need for managers to monitor their projects continuously and keep precise, predictable and repeatable control. For this, it is of great importance that organizations have an institutionalized software development process that will give them more refined management control.

However since "you can't control what you can't measure" [6], the measurements carried out through the software life cycle are the most efficient way to execute such control. With the measurement results, it is possible to make decisions based on concrete data and to improve software processes and products. These aspects can be covered by a software measurement process to be implemented in the organization. However, in many organizations, measurement is still seen as expendable or is inconsistent and unplanned.

This paper presents a measurement process resulting from the mapping of five relevant software measurement processes: CMMI-SW [3] [4], ISO/IEC 15939 [9], IEEE Std 1061 [8], Six Sigma [14] [16] and PSM (Practical Software Measurement) [1] [11]. The best practices of each one were used, including relevant keys to facilitate the applicability of a measurement process focused on project management, as well as assuring the software quality.

The Organizational Software Measurement Process (OSMP) has a set of activities which must be executed in an organization in order to institutionalize a measurement process and the usage during the software project life cycle, aiming for better legibility, understanding and learning. It is used on an iterative and incremental life cycle. A website support tool was created, which details the process, process workflow, macroactivities, activities, roles and templates to facilitate the navigability and the use of the considered process.

This paper is organized in the following way: Section 2 accounts for related work. Section 3 is the core of the paper: here the OSMP is discussed. Section 4 illustrates the tool that supports the proposed process. Section 5 illustrates conclusions and sketches possible future developments.

II. RELATED WORK

We studied five software measurement processes and mapped them to investigate the similarities and the gaps among these approaches, finding the strongest point of each one (best practices) [12]. Thus, was born the Organizational Software Measurement Process (OSMP). The following measurement processes were used:

- **CMMI-SW:** handles measurement on two levels of maturity: in level 2, in the process area "Measurement and Analysis", and in level 4, in the process area "Organizational Process Performance" (OPP) and "Quantitative Projects Management" (QPM), where strong statistical data resulting from the measurement carried out in level 2 are used, seeking to identify possible improvements in the organizational processes [4].
- **ISO/IEC 15939:** the intention of ISO/IEC 15939 (2002) (Software Engineering - Software Measurement Process) is to collect, analyze, and report data relating the products developed and processes within the organizational unit, to support effective management of the processes, and to objectively demonstrate the quality of the products [9].
- **IEEE Std 1061:** provides a framework for software quality metrics and a methodology for software quality metrics for establishing quality requirements and identifying, implementing, analyzing and validating process and product software quality metrics [8].
- **Six Sigma:** is a structured problem-solving methodology widely used in business. It is more used for bigger organizations, due to the fact that it is a very costly process. However, as it uses methods and tools it becomes simpler to understand and to implement [14]. Six Sigma presents two models: DMAIC (define, Measure, Analyze, Improve, Control) and DCOV (define, characterize, Optimize, Verify).

K. Elleithy (ed.), Advances and Innovations in Systems, Computing Sciences and Software Engineering, 253–258.

• **PSM** (Practical Software Measurement) [1] teaches a method to select and apply software measures that directly support their project needs and address project-specific issues.

III. ORGANIZATIONAL SOFTWARE MEASUREMENT PROCESS (OSMP)

The principal workflow of measurement process model proposed (Fig. 1) contains seven macroactivities to achieve its objectives [12]: (i) Organizational Planning for measurement; (ii) Plan measurement; (iii) Perform measurement; (iv) Analyze measurement; (v) Monitoring and control Measurement; (vi) Evaluate process measurement; and (vii) Organizational Evaluation.

The macroactivity Organizational Planning for measurement must be executed on organizational measurement process implantation and be repeated when there are organizational measurement process changes. The macroactivity Plan Measurement must be always executed when a new project is beginning or when there are plan measurement changes and the project isn't finished yet. The macroactivities of items (iii), (iv) and (v) must be worked at least on each iteration when an iterative incremental life cycle is used. The macroactivity Evaluate Process Measurement must be executed at the end of the project to verify process effectiveness, to capture lessons learned and point out improvements. The Organizational Evaluation macroactivity must be performed in a regularity defined in the Planning of the Organizational Measurement macroactivity

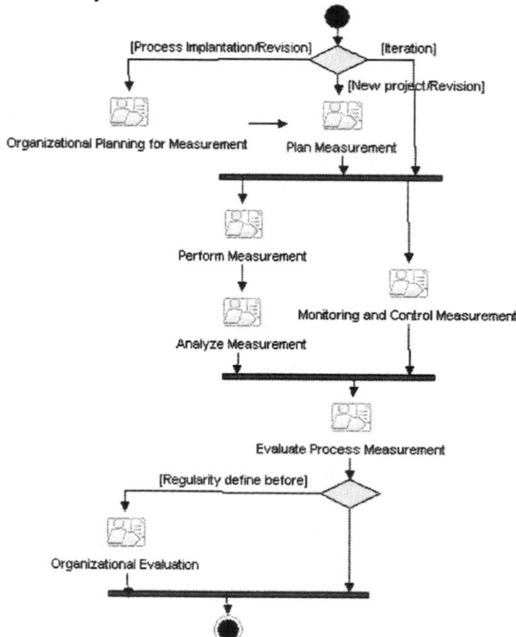

Fig. 1 – Workflow of OSMP.

Each one of these macroactivities contains a set of activities to be executed by specific roles. These roles are: (i) Measurement Analyst: Responsible for knowing the functioning of the measurement process well; (ii) Senior manager: responsible for business in the organization, he must know the objectives and needs of the organization very well; (iii) Project Manager: responsible for project management such as planning, problem solving and customer contact and; (iv) Relevant Stakeholders: Other people involved in the process such as the customers and the software development team.

These roles are classified as essential (the activity is carried out by the main actors) and contributing (a secondary role in which the activities are carried out by the collaborators, and in the absence of the essential role the activities are carried out by the contributing role).

For each proposed activity the source was identified (CMMI-SW, ISO/IEC 15939, IEEE Std 1061, Six Sigma and PSM) if there was any. It was also necessary to describe the steps to be executed for each activity in detail to structure the process adequately and to facilitate use. In/Out artifacts were generated for each activity. Some of these artifacts were taken from literature and tailored to the context of the considered process. Also some new templates were created to be filled in during the process. It shows some tips, tools and techniques to facilitate applicability and indicate who must work in the execution of each activity.

A *Macroactivity Organizational Planning for Measurement*

It involves the necessary activities to institutionalize the measurement process in the organization. Information must be elicited to understand the organizational characteristics and needs. Measurement commitments must be established and supported in order to define the target of the organizational metrics and indicators. It is formed by the following activities and all macroactivities have a workflow showing activities, role and artifacts (Fig. 2).

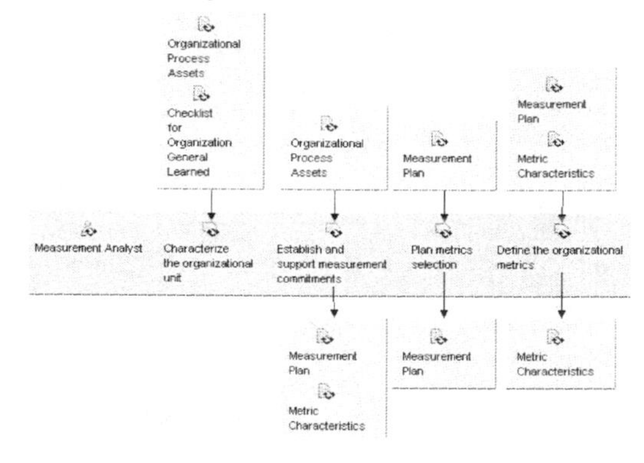

Fig. 2 – Workflow of Macroactivity Organizational Planning for Measurement

Activity: Characterize the organizational unit

The measurement analyst needs to have a general view of the organization. How it works, its goals, main needs, benefits and difficulties. All of this is to facilitate definition of the future measurement goals and process specialization.

At this moment, the organization's mission must be established, describing the organization's statement of purpose; what it is doing and why. The statement of vision describes what the organization aspires to, the organization's special task, and what specific motivation binds the organization's stakeholders together[10].

Tools and techniques: Brainstorming, Jade and cause-and-effect diagram to learn how the organization works. Table I shows a summarized information of what was done for each activity.

TABLE I
CHARACTERIZE THE ORGANIZATIONAL UNIT

Purpose: Understand how organization works.	
Source: ISO/IEC 15939, Six Sigma and IEEE Std 1061 [12]	
Steps: (i) Search, collect and understand information about the organization	
Input Artifacts: Organizational process assets; Checklist for Organization General Learned	Output Artifacts: -
Essential role: Measurement Analyst	Contributing role: Senior manager

Activity: Establish and support measurement commitments

It is necessary to prepare the organizational environment with correct infra-structure in order to achieve the success of the measurement process. Thus, the organization must specialize the OSMP, tailoring it to its context, adapting or creating templates such as metric plan template that will be used as a guide in all of the organization projects. Later, the specialized process must be communicated to the organization staff.

All the software development staff is identified to define who will work in the organizational measurement process, to get their commitment and provide appropriate training.

Tips: (i) Discover how much the organization wants to invest in measurement to define the scope in accordance with the reality of the company; (ii) It is necessary that the whole organization understand how important the measurement process is, and that the main point is to measure process and product, not people, in order to get the complete commitment of the team; (iii) It is very important to use a storage tool which can vary from an electronic spread sheet to a data base.

Activity:Plan metrics selection

The organizational measurement goal must be defined based on the characterize the organizational unit results activity. Here one must define how to carry out selection of the organizational metrics, defining techniques, tools and criteria.

Tools and techniques: (i) GQIM [15] or GQM [2] is used. However, this doesn't hinder the organization from using other techniques that are more adequate; (ii) Use techniques such as Brainstorming, Jade and Diagram cause-and-effect to get the organizational goals; (iii) The multi-criteria technique [13] was used to prioritize and generate minimum acceptable values for each metric in accordance with the profile of the organization; (iv) Use PSM categories [11] to improve metrics structure and presentation.

Activity: Define the organizational metrics.

Here, organizational metrics are selected based on the Plan metrics selection activity, and they must be revised for each iteration, verified if the organization goals are to be achieved by metrics and indicators.

B *Macroactivity Plan Measurement*

It consists of the planning of measurement and creating measurement plan for the projects, adapting when the project demands. The measurement analyst instantiates the specialized process for the projects. Adjustments can be made, but must be allowed in the artifact Criteria to adapt process. It carries out the following activities:

Activity: Define metrics

The project metrics are defined. It is necessary to use the organizational metrics listed in the artifact Criteria to adapt process (OSMP). Metrics can be included depending on the project needs.

It is necessary to identify who will be part of the project measurement team in order to get their commitment. If there is a change in the organizational metrics set used in the project measurement instance, then it is necessary to prioritize the metric again.

Tips: The measurement analyst must be in accordance with the project manager to identify possible project difficulties to be faced, so the measurement can be used as an activity of support to the project management, such as for example, helping to detect and monitor the project risks.

Activity: Plan data collection and storage

It is responsible for planning how data is collected and stored. It is very important because the resulting data are used in data analysis. Thus, any data problem will impact on analysis results and information could be wrong and the organization could be harmed.

Tips: The level of complexity of the revision of the collected data will depend on the structure provided by the organization, varying from a simple revision up to a complete audit. However, the verification of whether the data is complete, the integrity and the conformity of the data is indispensable in this revision.

Activity: Plan data analyses

The success of this activity is to select tools and criteria correctly, because tools can help in the future data analyses and the analysis criteria defines how each metric is going to be analyzed. In the revision of the analysis results it is necessary to assure that the data are correctly analyzed. Selected criteria and tools can vary for each metric depending on its nature and complexity.

Tools and techniques: Here are suggested some analysis techniques that can be used [18][7]: Cause-and-Effect Diagram, Histogram, Pareto Diagram, Bar Chart, Scatter Diagram and Run Chart.

Activity:Plan the communication of the results.

Disclosing the measurement results is important because the staff needs to know the organization's situation and thus collaborate in improvements.

Tool and techniques: Results can be communicated by several means, for example e-mail, folder, and poster, as described in the measurement plan.

Tips: It is best that the result is disclosed first to upper management, and after approval it is disclosed to everybody. The result can be controlled by change management.

C *Macroactivity Perform Measurement*

All the activities are related to the data collection and storage, and must be executed as described in the project measurement plan (in section Plan data collection and storage). It carries out the following activities:

Activity: Collect data

Here data collection is accomplished.

Activity: Store data

All collected data will be brought together and put in a unified format to help to perform the next activity.

D *Macroactivity Analyze Measurement*

It is responsible for all the measurement data analysis, helps the organization carry out the evaluation of the collected data and is able to detect problems and risks in the projects. Must be accomplished according to the project metrics plan (in section Plan data analyses). It carries out the following activities:

Activity: Analysis of the consolidated data

The analysis of the consolidated data is carried out and the data results are interpreted, according with the indicators previously defined for the project. Simple improvements do not need to wait until the execution of the macroactivity Evaluate Measurement to be done, they can be accomplished here.

Activity: Communication of the results.

All processes shown in the literature and used as bases for this measurement process show how important it is to disclose the data analysis results to relevant stakeholders, mainly to the project manager to help him make decisions in the project, as for example action for risks mitigation based on concrete diagnostics supplied by the result of the measurement. This activity must be executed as indicated in the metric plan (section planning of the communication of results).

E *Macroactivity Monitoring and Control Measurement*

It is related to the other macroactivities of the process. It verifies whether the measurement artifacts are being used correctly and if the activities are being executed as planned.

The execution of this macroactivity figures in the monitoring of all the measurement process and can result in the need for creating a new metric or changing the measurement plan, in order to continue to take care of the objectives of the measurement for the project. It has the following activity:

Activity: Monitor and control.

Revise all the measurement process activities to verify if each activity is done as planned or defined, and if the artifacts were produced. This activity must be carried out during the measurement process life cycle and handled by the project manager.

Tips: This activity could be added to the responsibilities of the quality assurance team.

F *Macroactivity Evaluate Process Measurement*

The evaluation of the measurement process is carried out in this macroactivity, registering lessons learned and considering process improvements. It carries out the following activities.

Activity: Process evaluation

It evaluates the measurement process by three perspectives [11]: (i) Performance: evaluate input and output artifacts; (ii) Conformity: compare the actual measurement process with the proposed implementation described; and (iii) maturity: Compare the process maturity with an external benchmark process maturity.

Activity: Register lessons learned

It was necessary to create this specific activity in order to disclose the lessons learned from the measurement for future projects and avoid further similar mistakes.

Activity: Consider improvements

All tasks related to measurement process improvement must be carried out. Improvements should be developed based on data analysis results, measurement evaluation and lessons learned, achieving the main goal which is not just to measure the process, but improve it [5].

G *Macroactivity Organizational Evaluation*

All the projects finished are evaluated based on the indicators defined in the macroactivity Organizational Planning for Measurement The frequency is previously defined by the organization, and with the results in hand upper management can see the main deficiencies in its projects and take corrective actions and consider organizational improvements.

Activity: Evaluate the Organization based on the Measurement

The Organizational evaluation will be carried out based on the measurement. Thus, it is necessary to integrate the resulting data of all the projects finished in the previously defined period in the macroactivity Organizational Planning for Measurement, later analyzing the organizational indicators and finally considering organizational improvements based on the results.

IV. WEBSITE TOOL

A tool was created to support the OSMP. It is a website which details the whole process such as process workflow, macroactivities, activities, roles and templates. It helps all the staff understand and use the process, especially the quality team, because it shows all the steps necessary for introducing the process. Moreover, the tool should support the successive phases of the measurement process and help maintain the information, because all data will be in the same place and their access is optimized. Thus, the use of the measurement process becomes very easy [19].

EPF-Composer (Eclipse Process Framework Composer) was used for the development of the website (the process itself, with its elements and diagrams) and TextPad 4.7.3 for the web pages customization.

V. CASE STUDY

The case study of this work it was carried out at IVIA, with 152 employees and ISO 9001:2000 certified since 2003, it is a software development company with a research and development area (R&D).

The measurement process of this organization was defined and institutionalized from the proposed Organizational Software Measurement Process (OSMP), passing through for the following stages: (i) Process specialization for the organization; (ii) Process instantiation in three pilot projects; (iii) Institutionalization of the process and (iv) Evaluation of the Organization based on the Measurement. .

A *Process Specialization for the Organization*

At the beginning, the OSMP was presented to SEPG (Software Engineering Process Group) and to the software director of the company in order to identify the applicability of the process to the company reality. They concluded that the process was completely adherent to its measurement needs however, it was necessary to specialize the OSMP in accordance with the organization standards.

At the specialization, the company decided not to create the measurement analyst role, although recognizing its importance. These role responsibilities were incorporated by the quality analyst role responsibilities, a role previously defined in the company project management process.

Some proposed artifacts proposed by the OSMP had been replaced by existing corresponding artifacts already in the processes repository of the company. The following specific measurement artifacts had been elaborated: Measurement Plan, Metrics Definition and Measurement Tool (electronic spread sheet).

Some meetings with the SEPG, the software director and project managers had been carried out during the definition of the process in order to get information about projects management and to find difficulties provided by the lack of concrete data in the decision making process.

The next step was to elaborate the flows of the specialized OSMP. In this case the organization opted for using the original flows of the OSMP, and generating a document detailing the described procedures for each activity. The specialized OSMP was submitted to the approval of the company and was approved without significant changes and authorized for its instantiation in a pilot project.

B *Process Implantation in a Pilot Projects*

Three pilot projects were chosen for the instantiation of the specialized OSMP. After that, the project manager was guided in his new responsibility in the execution of the measurement process.

A mentoring for the use of the specialized OSMP was carried out for the pilot project. The SEPG (or quality analysts) periodically monitored the measurement process usage and provided help to solve identified inconsistencies.

The measurement process instantiation in the pilot project was a highly important stage for the accomplishment of the needed adjustments and stabilization of this process in the organizational context.

Lessons learned in pilot projects have been incorporated to the specialized process as a result of the use of this measurement process. The use of the measurement process in pilot projects has been considered a success, and the company SEPG team gave endorsement for the institutionalization of the specialized OSMP for the new projects of the company.

C *Process Institutionalization*

The specialized measurement process knowledge was disseminated to the main people involved in this process through planned training, in order to assure they had the knowledge and the necessary abilities to play their roles.

With the main people involved entirely trained and a stable process, the institutionalization of the specialized OSMP in the whole organization occurred in a gradual way. The project teams felt motivated to execute the process, therefore they understood its importance and they could objectively follow the result of its work efforts.

D *Evaluate the Organization based on the Measurement*

For the organizational analysis, data collected before the measurement process were used, that were in the organization Balanced ScoreCard (BSC) and data were collected during the measurement process. Thus, for all indicators defined in the process data analysis was carried out, using its priority order as well as the percentage defined with reference for analysis of each one of them [13].

As an example of the analysis carried out for each indicator, we can consider the effort estimation (Fig. 3), and thus verify that almost all projects (projects A – G) estimated effort before using the quality processes. They exceeded the values defined as maximum variation for effort (15% - horizontal parallel line to the X axis on the chart) in the macroactivity, Organizational Planning for Measurement, using a multicriteria technique [13].

Although, after the use of quality process as project management and management of the requirements (projects H and I), lower values were observed for effort variation. After OSPM implantation, the organization was able to identify the improvement in the projects estimated effort and evidencing the importance of the use of the quality processes in the software development.

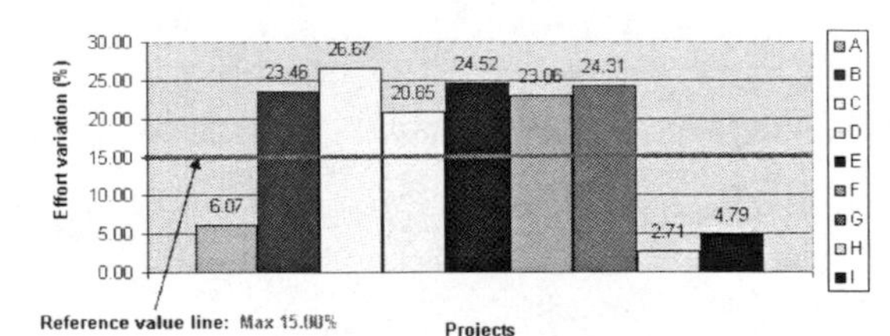

Fig. 3 – IVIA Projects, estimated effort

VI. CONCLUSIONS AND FUTURE WORK

This paper presents the Organization Software Measurement Process (OSMP) based on five relevant measurement processes. A Website was created to support the OSMP which details the whole process, such as process workflow, macroactivities, activities, roles and templates.

From the specialization of the OSMP for the organization to the instantiation in software projects, some improvements were implemented in the specialized process in order to make it more functional and efficient in the organizational context.

The process was known better after the specialized OSPM Website and training as to its use was provided. These actions motivated the collaborators to produce improvement suggestions.

It was important to highlight that the objective of the measurement is to measure processes, not people. With this, one could count on the project team support and contribution. This can be considered as a main point of the success of the specialized OSMP, since the activities of the measurement had been incorporated into the activities of the development team, mainly activities related to the data collection and storage.

The measurement process is now an important tool of support for project managers to monitor project activities. Moreover, upper management can see the current situation of the organization software development, being able to make more realistic decisions and to carry out improvements in the organization.

REFERENCES

[1] Bailey, Elizabeth, Card, David, Dean, Joseph, Hall, Fred, Jones, Cheryl, Layman, Beth and Mcgarry, John, *Practical Software and Systems Measurement*, DoD and US army, 2003.

[2] Basili, Victor R. 1992, *Software modeling and measurement: The Goal/Question/Metric Paradigm*, Motorola SCC Document, Appendix B, in (Solingen, 1995).

[3] Baumert, J. H., Mcwhinney, M. S., *Software Measures and the Capability Maturity Model*. – Carnegie Mellon University, Pennsylvania: Software Engineering Institute, 1992.

[4] Chrissis, Mary B., Konrad, Mike, Shrum, Sandy. *CMMI: Guidelines for Process Integration and Product Improvement*. – Boston: Addison Wesley, 2003.

[5] Daskalantonakis, M.K., 1992, *A Practical View of Software Measurement and Implementation Experiences Within Motorola*, In: Applying Software Metrics, IEEE Computer Society Press, pp. 168-180.

[6] Demarco, T., *Controlling Software Projects*, Yourdon Press, New York, 1982.

[7] Kan, Stephen H., *Metrics and Models in Software Quality Engineering*, Second Edition, Addison Wesley, 2002

[8] IEEE Std 1061-1998, *IEEE Standard for a Software Quality Metrics Methodology*, 1998.

[9] ISO/IEC 15939:2002, *ISO/IEC 15939:2002 - Software Engineering - Software measurement process*, 2002.

[10] Goethert, Wolfhart, Fisher, Matt, *Deriving Enterprise-Based Measures Using the Balanced Scorecard and Goal-Driven Measurement Techniques*, Software Engineering Measurement and Analysis Initiative, Technical Note CMU/SEI-2003-TN-024, October 2003

[11] McGarry, John, Card, David, Jones, Cheryl, Layman, Beth, Clark, Elizabeth, Dean, Joseph, Hall, Fred, *Practical Software Measurement – Objective Information for Decision Makers*, PSM, Addison-Wesley, 2002

[12] Oliveira, Josyleuda M. M., Oliveira, Karlson B., Belchior, A. D., *Measurement Process: A Mapping among CMMI-SW ISO-IEC 15939 IEEE Std 1061 Six Sigma and PSM*, ICSSM (International Conference Service System & Service Management) IEEE, France, October, 2006

[13] Oliveira, Josyleuda M. M., Oliveira, Karlson B., Belchior, A. D., Castro, Ana Karolina, Pinheiro, Plácido Rogério, *Aplicação de Multicritério para Implementação de um Processo de Medição Organizacional*, SBPO (Brazilian Symposium of Operational Research), Goiânia – Brazil, September, 2006; in Portuguese.

[14] Pande, P. S., Neuman, R. P., Cavanagh, R. R., *The Six Sigma Way*, McGraw-Hill, 2000.

[15] Park, Robert E., Goethert, Wolfhart B.; Florac, William A., *Goal-Driver Software Measurement - A Guidebook*, SEI, 1996.

[16] Pyzdek, Thomas, *The Six Sigma Project Planner*, McGraw-Hill, 2003.

[17] RUP 2003. Rational Software Corporation, *Rational Unified Process*, Version 2003.06.00.65, CD-ROM, Rational Software, Cupertino, California, 2003.

[18] Florac, William A., Carleton, Anita D., *Measuring the Software Process – Statistical Process Control for Software Process Improvement*, SEI, Addison – Wesley, 1999

[19] Oliveira, Josyleuda M. M., Oliveira, Karlson B., Belchior, A. D., *Organizational Software Measurement Process*, accepted in Metrikon 2006, Germany - Potsdam, November, 200

Decomposition of Head Related Impulse Responses by Selection of Conjugate Pole Pairs

Kenneth John Faller II[1], Armando Barreto[1], Navarun Gupta[2] and Naphtali Rishe[3]

Electrical and Computer Engineering Department[1] and School of Computer and Information Sciences[3]
Florida International University
Miami, FL 33174 USA

Department of Electrical and Computer Engineering[2]
University of Bridgeport
Bridgeport, CT 06604 USA

Abstract - **Currently, to obtain maximum fidelity 3D audio, an intended listener is required to undergo time consuming measurements using highly specialized and expensive equipment. Customizable Head-Related Impulse Responses (HRIRs) would remove this limitation. This paper reports our progress in the first stage of the development of customizable HRIRs. Our approach is to develop compact functional models that could be equivalent to empirically measured HRIRs but require a much smaller number of parameters, which could eventually be derived from the anatomical characteristics of a prospective listener. For this first step, HRIRs must be decomposed into multiple delayed and scaled damped sinusoids which, in turn, reveal the parameters (delay and magnitude) necessary to create an instance of the structural model equivalent to the HRIR under analysis. Previously this type of HRIR decomposition has been accomplished through an exhaustive search of the model parameters. A new method that approaches the decomposition simultaneously in the frequency (Z) and time domains is reported here.**

I. Introduction

The emergence of inexpensive and powerful computers has expanded virtual spatial audio to many areas. Virtual spatial audio is the use of digital signal processing (DSP) techniques to assign an artificial sense of directionality to digital sound signals.

Currently, there are two approaches to virtual spatial audio: multi-channel and two-channel approaches. The multi-channel approach uses multiple (more than two) speakers placed around the listener at strategically defined locations (e.g., Dolby 5.1 array) to physically reproduce the directionality of sounds generated around the listener. This approach produces emulated spatial sounds in a limited listening region which are then perceived by the listener, much like he/she would perceive naturally occurring sounds. However, this relies on the proper positioning of the speakers around the listener, which limits the use of the approach to stationary uses such as a home theater system.

The two-channel approach uses DSP techniques to create binaural sound pairs (Left ear signal, Right ear signal) for virtual spatial audio digitally so that they can be delivered to the listener through stereo headphones. It is known that sound signals are altered by the physical environment (e.g., floor, ceiling, walls, listener's torso, listener's head, and listener's outer ear) as they travel from the source to the eardrums of the listener. The two-channel approach strives to replicate this process synthetically, so that the listener can locate the virtual spatial audio source, at the location being emulated. The synthetic transformation is performed by application of special digital filters that are characterized by their impulse responses, called head-related impulse responses (HRIRs). Every position around the listener will have a specific HRIR, for each ear, associated with it. Convolving a sound signal with each HRIR for a desired location modifies the signal in a way that is similar to modifications the environment would have produced on the signal.

Logically, HRIRs depend on the anatomical features (outer ear, head, and torso) of the listener. As a result, HRIRs for each different location differ from person to person. Ideally, the HRIRs of each prospective listener would have to be measured empirically, at numerous source locations, in order to achieve highly convincing virtual spatial audio. However, this requires specialized personnel and expensive equipment that includes a small, wide bandwidth speaker and miniature microphones placed in the ear canal of the subject (Fig. 1)

Fig. 1 Empirical HRIR measurement at FIU.

K. Elleithy (ed.), *Advances and Innovations in Systems, Computing Sciences and Software Engineering*, 259–264.

Since it is not possible to provide access to this measurement process for every potential user of virtual spatial audio, commercial developers have resorted to the use of "generic" HRIR pairs obtained experimentally from a mannequin of "average anatomical dimensions" (e.g., MIT's measurements of a KEMAR Dummy-Head Microphone [1]) or using a limited number of subjects to represent the general population (e.g., the CIPIC Database [2]). These databases include HRIR pairs for many different positions around the listener, defined in terms of their azimuth (θ), elevation (ϕ) and distance (r) in a spherical coordinate system (Fig. 2)

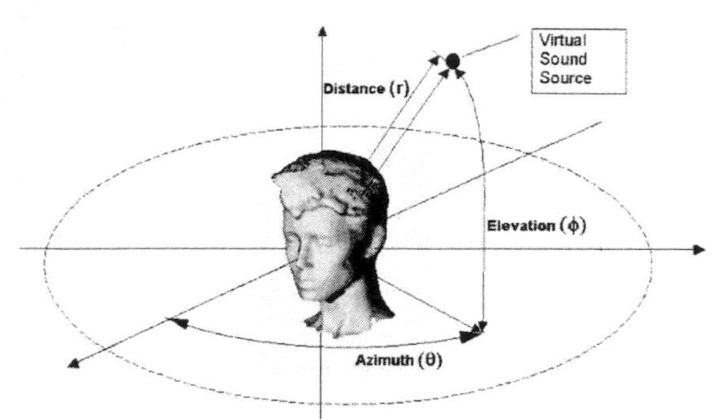

Fig. 2 Diagram of spherical coordinate system

Unfortunately, this type of "generic" HRIRs yield only an approximate sense of source location in many users, lacking the high spatialization fidelity of individual HRIRs [3].

The overall purpose of our research is to develop customizable HRIRs from a generic dynamic model. The generic model can be customized using physical measurements of the listener to provide similar spatialization fidelity as measured HRIRs. The current representation of HRIRs prohibits customization using geometric characteristics of the intended listener. Therefore, we believe that decomposition of HRIRs into partial components will allow their re-generation from a reduced number of parameters that are related to the geometry of each intended listener. Efficient HRIR customization could have significant practical impact because it would extend the benefits of high-fidelity audio spatialization to the overall computer user population.

II. Methodology

The following subsections describe the methodology used in this paper.

A. Structural Pinna Model

Brown and Duda in [4] proposed a "structural" model for binaural sound synthesis. In this approach, effects of the head, shoulders and pinna (outer ear) are "cascaded" to create a transfer function that contains all the spectral cues necessary to generate synthesized binaural sound. However, they did not provide a method to define the parameters of their pinna sub-model. A customizable functional model developed by Algazi models a listener's head with only 3 simple anatomical measurements [5].

In [6] we proposed a pinna model in which the sound entering the ear canal is the summation of signals with different delays. The delays are a result of waves bouncing off of the geometrical structures of the pinna, into the ear canal. The effect of the pinna cavities is modeled with a resonator. Therefore, the HRIRs were broken down into one direct wave and three delayed waves. Recent research by our group has achieved improvements in the decomposition of HRIRs augmenting the model with an additional delayed wave. A block diagram of this augmented model is shown in Figure 3.

In this model, the parallel paths represent the multiple signals entering the ear canal. Each indirect signal will arrive at the ear canal after a delay, τ_i, with respect to the direct wave. Additionally, the indirect signal will also have less energy, represented by a magnitude factor ρ_i , in comparison with the direct wave. The pinna model shown in Figure 3 only requires 11 parameters (the resonator is represented by two parameters), and could be "cascaded" with Algazi's functional head model to represent a complete HRIR.

An efficient method must be found to obtain the parameters in the model of Figure 3 from HRIRs obtained empirically as long sequences of impulse response samples. This will enable the development of databases of these parameters (at numerous azimuths and elevations) for subjects whose relevant anatomical characteristics will also be measured. Our expectation is that once the data set is large enough, empirical relationships can be developed between the model parameters and the anatomical features. At that point the geometric characteristics of a new intended user could be measured and "converted" to parameter values to instantiate the model at a desired location.

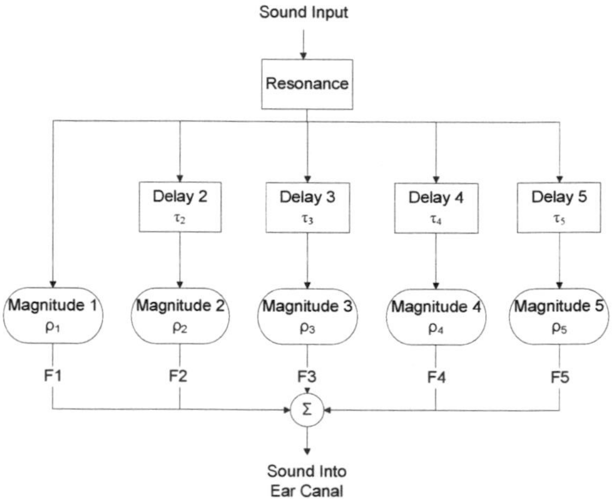

Fig.3: Block diagram of the pinna model

B. Iterative Decomposition Method

The impulse response of the model shown in Figure 3 will be the superposition of a damped sinusoidal (i.e., the impulse response of the resonator) with other damped sinusoids that appear delayed and scaled. Decomposition of a measured HRIR into this kind of sinusoidal components will reveal the delays and scaling factors that should be used in the model to create an instance that will have a close approximation of the HRIR being decomposed as impulse response.

Time-domain methods for this decomposition have been suggested before. In [6-8], two of the these methods, based on the Prony and Steiglitz-McBride (STMCB) signal modeling methods, were compared for decomposition of HRIRs. All these methods sought to apply second-order signal modeling to windowed sections of the HRIR that could be assumed to contain only a single damped sinusoid, which Prony and STMCB approximate with reasonable accuracy [9-11]. A full description of those methods can be found in [6-8].

A major drawback of this approach, however, is that the window sizes are not initially known. To discover the appropriate window sizes, a program was written to iterate through all possibilities. The windows were gradually widened starting from 2 to 10 sampling intervals for each window (for a total of five windows). In each tentative window the signal would be approximated using one of the modeling methods (second-order Prony or STMCB) and each possible sequence of second-order approximations (considered at the appropriate delays) would be summed together resulting in a candidate HRIR. All the possible resulting candidate HRIRs would be temporarily stored and eventually compared to the original measured HRIR using Equations 1 and 2 to assess their individual similarity or "fit" to the original HRIR. The candidate HRIR with the highest fit at the end of this process would be kept as the "reconstructed" HRIR that represents the most accurate decomposition achievable for that original (measured) HRIR. Analysis of the results from this process showed that, in general, it approximates the original HRIR with relatively high accuracy. Figure 4 shows the components extracted from a measured HRIR by this process.

$$\text{Error} = \text{Original HRIR} - \text{Reconstructed HRIR}, \quad (1)$$

$$\text{Fit} = [1 - \{\text{MS(Error)}/\text{MS(Original HRIR)}\}]. \quad (2)$$

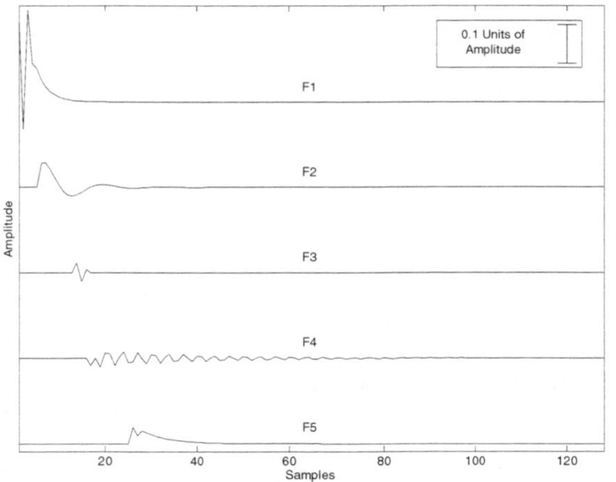

Fig.4: Five damped sinusoidal components obtained from a measured HRIR

Although this iterative process resulted in high fits for most of the HRIRs explored (about 96% average fit), the iterative search approach is extremely computationally intensive, even with just the 5 windows processed in those studies. In fact, the tree-diagram needed to track all possible width combinations of 5 sequential windows has 9x9x9x9x9=59,049 leaf nodes and the addition of any subsequent windows with this approach will multiply the number of leaf nodes by 9, per additional window. To truly select the best of all possible alternatives, all the branches of the tree need to be explored and the reconstructed HRIR defined at each leaf node compared with the measured HRIR to assess its fit. It became clear that increasing the number of windows of analysis (which may be necessary to model late components in the HRIRs) would be impractical using the iterative search method.

Another drawback is that when the delay between sinusoids is small (less than 5 samples), the second-order STMCB or Prony sequential methods alone tend to inaccurately reconstruct the signal. To verify this, a single damped sinusoidal (x) was created and tested with the iterative method using Prony and STMCB. A short window containing only the first three samples from x was processed by STMCB and Prony in an attempt to approximate the original signal. The results of STMCB (xs) and Prony (xp) are shown in Figure 5. The approximations xs and xp appear to capture the details of the beginning part of x but fail to approximate the rest of it. This will lead to inaccurate approximation of the parameters for the pinna model. These drawbacks have prompted us to develop a new, faster and potentially more accurate method of HRIR decomposition into sequential damped sinusoids.

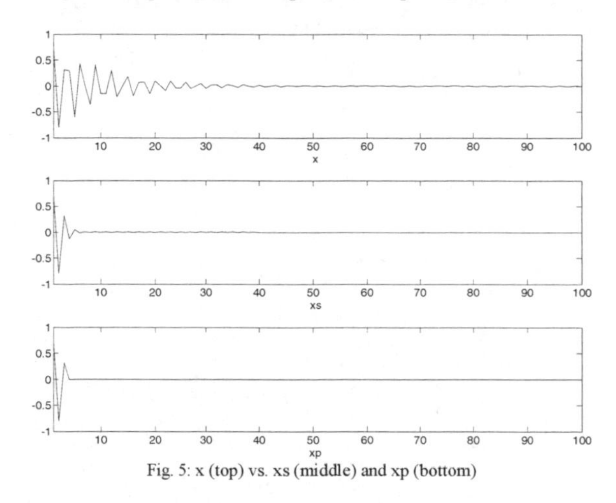

Fig. 5: x (top) vs. xs (middle) and xp (bottom)

C. Pole-Decomposition Method

In previous HRIR decomposition approaches the goal was always to isolate a segment of the HRIR that could be considered constituted by a single damped sinusoid. Under that assumption a second-order modeling approach (Prony or STMCB) was used to deal with every window along the HRIR. However, the correct demarcation of the boundaries for these second-order windows was crucial to the accuracy of the process and, therefore, all probable window widths, for each of the sequential windows, had to be considered. This resulted in a search tree with a branching factor that remained high (e.g., 9) from the root node all the way to the leaf nodes.

In the new decomposition approach the end of the analysis windows does not need to be defined in advance. Instead a higher-order approximation is used on the complete remnant of the HRIR (at any point during the decomposition) to pre-define multiple damped sinusoids contained in the HRIR remnant, and then they are individually isolated according to their pole signature in the Z-domain, and pursued as candidates for the second-order representation of the particular HRIR segment in question.

In general, a single damped sinusoidal component sequence will be represented by a conjugate pair of poles within the unit circle and a zero at the origin of the Z-plane (Figure 6) [12]. Hence, a damped sinusoid in the Z-domain can be described with the following general equation:

$$X(z) = \frac{k \cdot z}{(z - p_1)(z - p_2)} \qquad (3)$$

where k is a scalar and p_1 and p_2 are complex poles. According to Equation 3, if the scalar k and the poles are known then, using the inverse Z-transform, it is possible to characterize the corresponding time domain sequence as a specific damped sinusoid.

In this new approach, instead of iterating through all possible window width combinations, an attempt is made to identify multiple delayed and damped sinusoids in the complete HRIR remnant available. Each of the viable damped sinusoids will be separated according to their conjugate pole signatures in the Z-domain. Then each damped sinusoid will be investigated as the approximation of that particular segment. The end of the segment is not pre-determined, but instead will be defined by the time index at which the reminder of the previous HRIR remnant minus the second-order approximation being investigated surpasses a predetermined threshold. That point will be considered the time at which a new damped sinusoidal contribution starts. The origin for analysis will be shifted to that point and the process will be repeated, except that using a modeling order which is two less than the previous modeling order used.

This also results in a tree-search approach. However, the branching factor of this search tree starts at the amount of damped sinusoids being extracted from the whole HRIR but decreases by one in every subsequent stage of the decomposition, which makes the number of leaf nodes much smaller than for the previous algorithm. For example, if 5 damped sinusoids will be extracted, only 5x4x3x2x1 = 5! = 120 leaf nodes will exist. An experiment using simulated damped sinusoids was performed in order to verify this pole decomposition method. Three damped sinusoids with different magnitudes and delays where created and summed together, to be analyzed by the pole decomposition method.

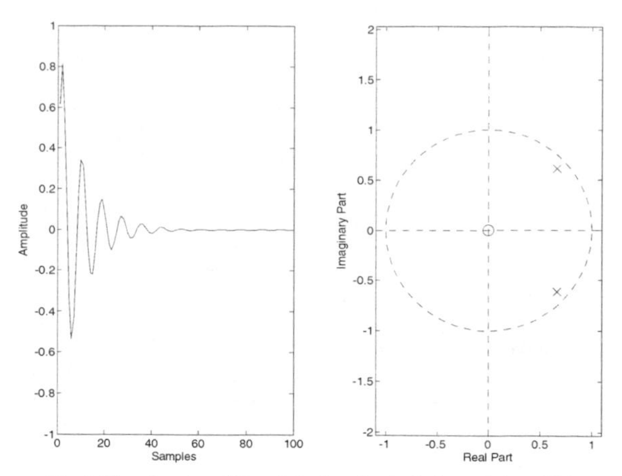

Fig. 6: Time domain and Zero-Pole plot of a single damped sinusoidal

The damped sinusoids used in this example were created using equation 4 where N is the length of the signal, n = 0,…,N-1, d_i is the negative damping factor and ω_d is the digital frequency. Once the three sinusoids (x1, x2 and x3) were created, the desired delays (τ_2 and τ_3) were applied to the last two sinusoids respectively, resulting in x2s and x3s. Finally, the sinusoids were summed point-to-point to produce the test signal (x). In this example N=100, τ_2=3, τ_3=6, ω_d=0.711, d_1=-0.1, d_2=-0.125 and d_3=-0.15. The three signals (x1, x2s and x3s) and the resulting signal (x) are shown in Figure 7.

$$x_i(n) = e^{d_i * n} \cdot \sin(\omega_d \cdot \pi \cdot n) \qquad (4)$$

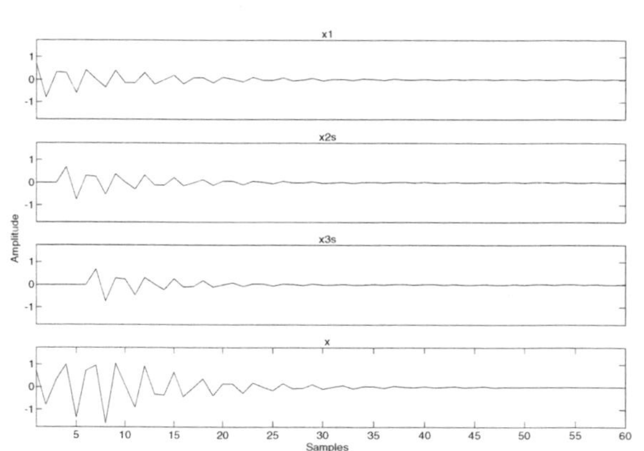

Fig. 7: Plot of the three damped sinusoids (x1, x2s with delay τ_2 and x3s with delay τ_3) and the sum of them (x)

The process starts by applying a sixth order STMCB approximation process to the complete x. Sixth order is used initially because the decomposition of x into three second – order signals (damped sinusoids) is sought. The results from the sixth-order STMCB approximation will have the pole structure shown in Figure 8. As seen in the figure, there are two complex conjugate pairs of poles. Each of these will be investigated as a candidate to represent the first sinusoidal present in x (there could be up to three branches at the initial node of this search tree, if all the poles were complex).

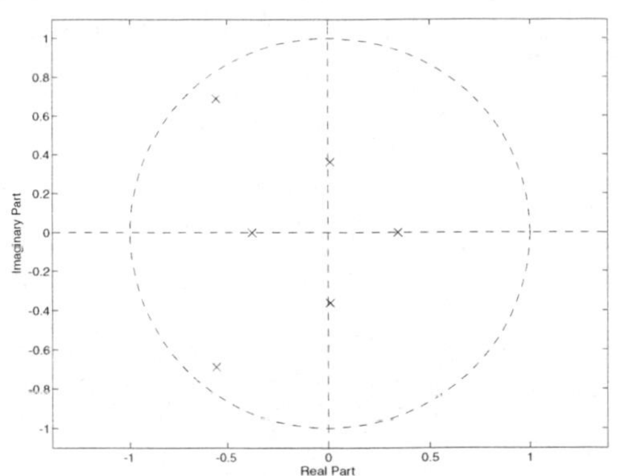

Fig. 8: Poles obtained from the sixth-order STMCB approximation of the complete sequence x

The investigation of each of these alternatives involves its subtraction from x to define a residue sequence, as shown in Figure 9, which will then be thresholded. The threshold level used for this segmentation was set at 25% of the signal peak, in this synthetic example. A slightly different threshold to process real HRIRs was found as described in the following section. This is the only adjustable parameter in our method and the rationale for the value recommended is also presented in the following section of the paper.

The time at which the residual surpasses this threshold will be considered the onset of the next damped sinusoidal, i.e., the estimate of τ_2. As in the previous method, the decomposition process will continue on to a second stage after re-establishing the origin of analysis at the estimated τ_2. The assumption made in every subsequent decomposition stage is that there should be one less damped sinusoidal present in the new remnant (since one has just been removed in the previous stage). As such, a fourth-order STMCB approximation will be applied in the second decomposition stage, yielding 4 poles, which will then be used to synthesize up to two candidates for the second damped sinusoid extracted from x. The same pattern of steps will be applied through all subsequent stages of the decomposition, until the stage in which a second-order STMCB approximation will be applied to the last remnant to identify the last damped sinusoid in it.

After M stages of decomposition there will be M! leaf nodes in the search tree, each representing a set of M delayed and scaled damped sinusoids that, when added together, form candidate approximations to the original signal x. The fit of each of those M! candidate approximations with respect to x will be evaluated (Equations 1 and 2) and the candidate with the highest fit will be selected as the final decomposition of x. In our example, the winning candidate approximation had a 99.99% fit with the original x, and the individual damped sinusoids obtained through each stage of decomposition also matched x1, x2s and x3s very closely.

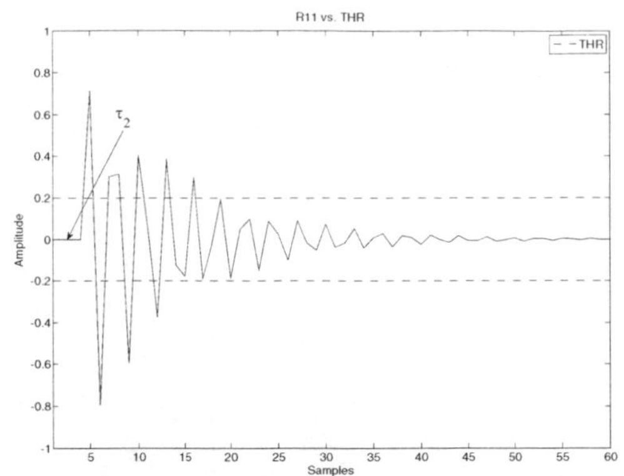

Fig. 9: Plot of the first remnant with threshold lines (THR)

III. POLE DECOMPOSITION OF MEASURED HRIRS

The method described in Section II-C was applied to the decomposition of 14 actual HRIRs, recorded from 14 subjects using the AuSIM HeadZap system at Florida International University (Fig. 1). The goal in each case was to obtain M = 5 damped sinusoidal components. Therefore, the order of the first STMCB approximation process was 10. The procedure was identical as the one explained for the decomposition of the synthetic sequence x, in Section II-C, with the exception that an empirically defined threshold level (18% of the signal peak value) was applied to each reduced remnant of the HRIR.

The empirical determination of the best threshold level to use in decomposing actual HRIR signals was performed by plotting the average fit for the reconstructed HRIR as the threshold used changed in increments of 0.005 for HRIRs measured from 14 subjects and corresponding to sound sources at +/-90° azimuth (i.e., directly lateral from the ear measured) and elevations from -36° to 54° at increments of 18°. For example, Figure 10 shows this plot for ϕ = -36°. As can be seen in this plot, there is a curvature that has a maximum at a threshold value of about 0.18. Similar observations were made for other elevations. Thus 18% was selected as the recommended threshold.

HRIRs from 14 subjects for an elevation of 0° and azimuths from -150° to 180° at increments of 30° (along the horizontal plane) were decomposed using the old and the new algorithms. The results for each ear are displayed in Table I.

TABLE I: HRIR DECOMPOSITION RESULTS

METHOD:	Average Fit Left Ear	Average Fit Right Ear
Exhaustive, variable window width (old)	97.57%	97.57%
Pole decomposition w/ Threshold (new)	89.40%	88.15%

While the goodness of fit achieved by both methods is similar, the pole decomposition method has been found to be much faster than the old method, as detailed in the next section. Figures 11 to 13 show the highest, average and lowest fits for HRIRs using the "new" method, respectively.

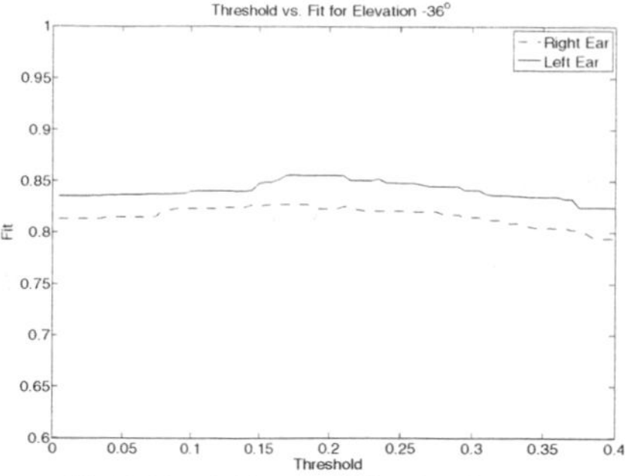

Fig.10: Threshold versus fit for elevation -36°

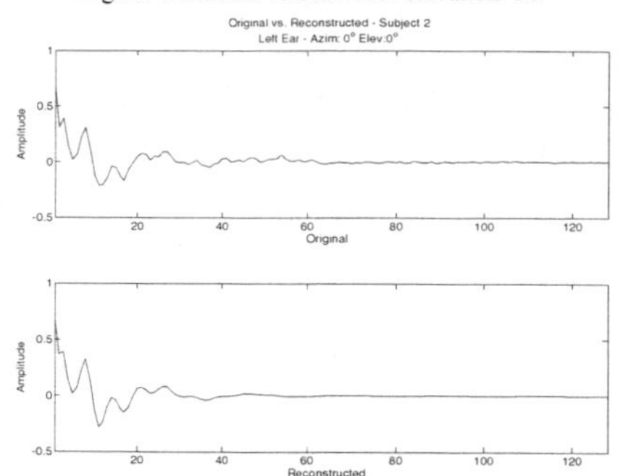

Fig.11: Original (top) vs. reconstructed HRIRs for the left ear of subject 2 for azimuth 0° and elevation 0° - Highest Fit example

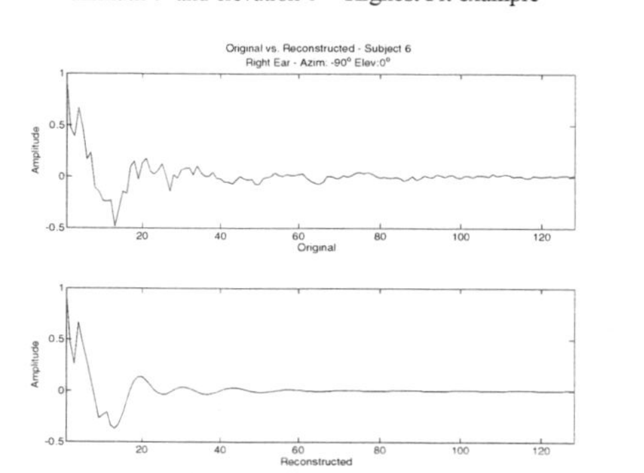

Fig.12: Original (top) vs. reconstructed HRIRs for the right ear of subject 6 for azimuth -90° and elevation 0° - Average Fit example

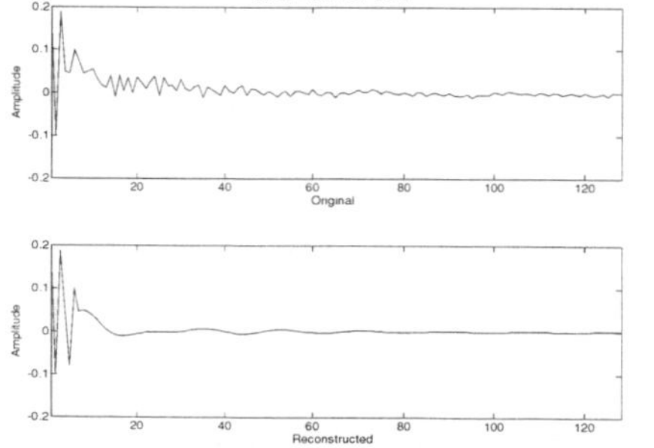

Fig.13: Original (top) vs. reconstructed HRIRs for the left ear of subject 6 for azimuth -120° and elevation 0° - Lowest Fit example

IV. Conclusion

The results shown in Table I indicate that the "old" method achieved a slightly higher average fit, but exhibited several drawbacks. First, the average calculation time was found to be about 100 times longer for the "old" method when a test set of 14 HRIRs were decomposed by both approaches (429 s to 4.2 s). Secondly, when the delay is small (less than 5 samples wide), the second-order STMCB sequential method alone tends to inaccurately reconstruct the signal.

Therefore, according to the observations indicated above, it seems that the new HRIR decomposition method, based on the separation of damped sinusoids according to their pole pair signature in the Z-domain, may be a more practical approach to the creation of a large database of decomposed HRIRs (particularly if more than 5 components will be sought), which is a pre-requisite to the establishment of relationships between model parameters and measurable anatomic characteristics of the subjects.

Acknowledgment

This work was sponsored by NSF grants IIS-0308155, CNS-0520811, HRD-0317692 and CNS-0426125.

References

[1] B. Gardner, K. Martin, and Massachusetts Institute of Technology. Media Laboratory. Vision and Modeling Group., *HRFT measurements of a KEMAR dummy-head microphone*. Cambridge, Mass.: Vision and Modeling Group, Media Laboratory Technical R4port 280, Massachusetts Institute of Technology, 1994.

[2] V. Algazi, R. Duda, D. Thompson, and C. Avendano, "The Cipic HRTF database," in *2001 IEEE Workshop on Applications of Signal Processing to Audio and Acoustics* New Paltz, NY, 2001, pp. 99-102.

[3] E. M. Wenzel, M. Arruda, D. J. Kistler, and F. L. Wightman, "Localization Using Nonindividualized Head-Related Transfer-Functions," *Journal of the Acoustical Society of America*, vol. 94, pp. 111-123, 1993.

[4] C. P. Brown and R. O. Duda, "A structural model for binaural sound synthesis," *IEEE Transactions on Speech and Audio Processing*, vol. 6, pp. 476-488, 1998.

[5] V. R. Algazi, C. Avendano, and R. O. Duda, "Estimation of a spherical-head model from anthropometry," *Journal of the Audio Engineering Society*, vol. 49, pp. 472-479, 2001.

[6] A. Barreto and N. Gupta, "Dynamic Modeling of the Pinna for Audio Spatialization," *WSEAS Transactions on Acoustics and Music*, vol. 1, pp. 77-82, January 2004.

[7] K. J. Faller II, A. Barreto, N. Gupta, and N. Rishe, "Decomposition and Modeling of Head-Related Impulse Responses for Customized Spatial Audio," *WSEAS Transactions on Signal Processing*, vol. 1, pp. 354-361, 2005.

[8] K. J. Faller II, A. Barreto, N. Gupta, and N. Rishe, "Performance Comparison of Two Identification Methods for Analysis of Head Related Impulse Responses," in *Advances in Systems, Computing Sciences and Software Engineering*, T. Sobh and K. Elleithy, Eds. Netherlands: Springer, 2006, pp. 131-136.

[9] L. Ljung, *System Identification - Theory For the User*. Englewood Cliffs, NJ: Prentice-Hall, 1987.

[10] T. W. Parks and C. S. Burrus, "Digital filter design," Wiley-Interscience, 1987, pp. 226-228.

[11] K. Steiglitz and L. McBride, "A technique for the identification of linear systems," *IEEE Transactions on Automatic Control*, vol. 10, pp. 461-464, 1965.

[12] L. P. Charles and H. T. Nagle, *Digital control system analysis and design (3rd ed.)*: Prentice-Hall, Inc., 1995.

GIS Customization for Integrated Management of Spatially Related Diachronic Data

K. D. Papadimitriou, T. Roustanis
Dep. of Rural and Surveying Engineering
Polytechnic School
Aristotle University of Thessaloniki
54622, Thessaloniki, GREECE

***Abstract*-This study presents the development of an interface for the management of diachronic spatial data that describe the evolution of an area.**

Subsequent data that represent specific spatial characteristics for various time periods are organized and processed in a customized GIS environment.

Vector and raster data (old scanned maps, air photos and satellite imagery) are related based on their spatial and temporal properties and they are archived adequately.

Part of the data set contains digital documentation in form of digital photos, audio and video files, thus the customization includes multimedia playback for selected geographic features that are described by these means.

As a case study is used an extended area in Northern Greece that includes various archaeological sites along Egnatia road (nearby ancient Via Egnatia).

I. INTRODUCTION

The majority of scientific researches that imply field work (such as ecology, archaeology and geology) involve large amount of heterogeneous data sets from multiple sources that may contain both spatial information (raster, vector, tabular data and meta data) as well as non-spatial [1] such are digital documentation (reports, photos, sound recordings and video) for subsequent time periods.

Specialist 's needs for appropriate management and representation of both geographic data and multimedia files within the same interface imposes the combination of different components in customized software environments [2], [3], [4].

This paper demonstrates the capabilities of a customized GIS environment that is used for the management of geographic and multimedia data that describe an area of interest through time. The customization is based on eight developed forms which provide geographic data archiving and representation based on the spatial and temporal properties of features [5] as well as multimedia playback for spatially referenced documentation data [6]. The development was performed with the combination of GIS and COM tools, taking advantage of GeoMedia Professional and Microsoft Visual Basic. The result of this collaboration offers an integrated interface for:

a) the massive import of imagery,
b) the linking of selected geographic features to multimedia files,
c) the automated attribution of metadata to raster files,
d) the discrimination among vector data which represents features that "are" and "are not" recognizable over selected imagery,
e) the automatic display of geographic data at any selected area,
f) the selective display of available data over the working view,
g) the selective display of spatially related vector data over selected raster data and
h) the automated playback of linked multimedia in GIS environment.

The developed functions are presented using a data set that describes the archaeological sites along Egnatia road (ancient Via Egnatia) at the prefecture of Seres at Northern Greece.

The majority of GIS applications in the field of archaeology are dedicated to prediction models for site locations [7] rather than an interface for digital documentation [8], [9].

The aim of this interface is to facilitate GIS users from various disciplines, who cope with diachronic spatial data and the relative digital documentation, and who require special functions (apart from GIS functionality) that are not offered from official software distributions.

II. MATERIALS

As a case study is used the surrounding area of Egnatia road at the prefecture of Seres in Northern Greece (Fig. 1). Contemporary Egnatia road is a major motorway that follows the ancient Via Egnatia, build by Romans, that links several archaeological sites from early Hellenistic through late Byzantine years. As a consequence it has a special historic and archaeological interest.

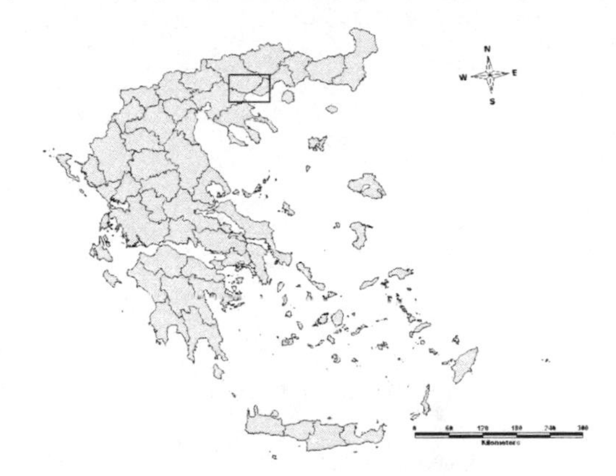

Fig. 1. Map overview showing the study area.

A. Data

The historical documentation of the study area consists an heterogeneous data-set that contains three major categories of data types:

a) Raster Data, including aerial photos, aged from 1945 to 1996, at scales from 1:6000 to 1:42000, satellite images, scanned historical maps, aged from 1901 to 1945, at scales 1:20000 to 1:200000, scanned topographic maps (scale 1:50000), scanned topographic diagrams, aged from 1925 to 1987 (scale 1:5000) and scanned geological maps.

b) Vector Data describing administrative boundaries, the coast line, urban limits, buildings (structures), soil data, coverage data and archaeological traces.

c) Multimedia documentation from field work (findings, excavation details, oral reports etc) containing digital photos, audio and video files [10].

B. Software Component Objects

The interface combines GIS tools for the analysis of the spatial relations

K. Elleithy (ed.), Advances and Innovations in Systems, Computing Sciences and Software Engineering, 265–268.

between geographic features and their representation, DB management tools for the archiving and data retrieval and multimedia tools for playback. The combination is obtained using the appropriate COM tools (Fig. 2) using the programming language Microsoft Visual Basic.

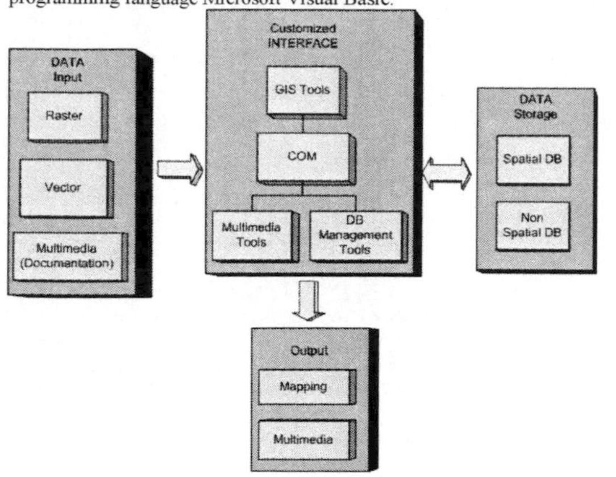

Fig. 2. Architecture of customization and schema of collaborating components.

The graphic environment that is used to host the developed interface is GeoMedia Professional and all the geographic operations are executed using its component objects [11]. The management of data is carried out using the SQL Server which can cope with large amount of data, while MS Media Player is controlled for multimedia playback in GeoMedia 's environment.

The following component objects were used from Geomedia Professional:

a) GMMapview object (Fig. 3.a) which is used for Map View display and provides the methods and properties to control the display, retrieve information as well as extend automation functions for a Map View display. It defines the cartographic environment (interface) of the application.

b) LegendEntries object (Fig. 3.b) is a collection of legend entry objects that comprise a legend (features that are displayed on the Map View). It is used to define the overlay sequence of cartographic layers and the way of representation.

c) EventServer object (Fig. 3.c) which listens to a single or multiple Map Views for window, keyboard, and mouse events occurring in the active Map View. It recognizes events occurring in a Map View and processes them according to an EventControl object. It serves the interaction between the user and the application.

d) EventControl object (Fig. 3.d) associates and disassociates an EventServer object (as described) with one or more GMMapView objects. The EventControl object also provides exposed automation for responding to events occurring in an associated Map View. The EventControl object handles the actual communication between a Map View and the occurring events.

e) SmartLocateService object (Fig. 3.e) which contains methods and properties for locating objects within specified geometric regions. These regions can be within or intersecting a closed geometric object (such as a rectangle). They can also be within or intersecting a circle with a radius that is a specified number of pixels from a locate point. This service defines the spatial criteria for the location – selection of displayed geographic features.

f) LocatedObjectsCollection object (Fig. 3.f) contains all objects found by a locate method for the SmartLocateService object. This collection can contain geometry objects (objects having a located geometric element) or raster objects. The collection of located (selected) objects is the subset of displayed features that will be used later for geoprocessing.

g) OriginatingPipe object (Fig. 3.g) defines a database query criterion and produces a recordset according to that criterion and provides is used to specify the database table from which to extract data, an SQL-equivalent WHERE clause to limit the retrieved records, a spatial criteria to set geometric extents and relationships to limit the retrieved records and to generate an output recordset according to the query requirements. This output recordset has been defined spatially by the LocatedObjectsCollection object and extracted through the OriginatingPipe object.

h) Geographic Data Objects – GDO (Fig. 3.h) are programmable database objects based on the Microsoft Data Access Object (DAO) model. The Microsoft Component Object Model (COM) design provides the automation for data access and data update. GeoMedia combines DAO concepts and COM with geographic aspects of geodetic coordinate systems, geometry, and spatial filtering.

Additionally, two component objects from Microsoft Visual Basic was used:

i) Scripting.FileSystemObject (Fig. 3.i) that allows the creation, delete and gain of information about folders and files and

j) Windows Media Player ActiveX control object (Fig. 3.j) which adds Windows Media Player functionality to other programs (it brings multimedia playback capabilities in the environment of GeoMedia).

The following schema (Fig. 3) demonstrates the collaboration between the component objects that has been used.

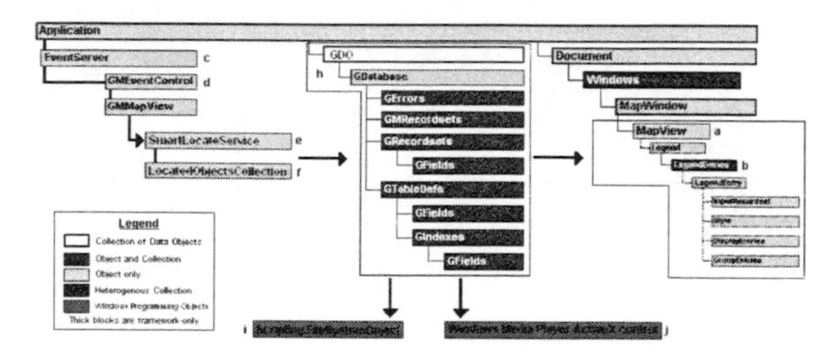

Fig. 3. Schema of the collaboration between the component objects that has been used in the customized environment.

III. RESULTING FUNCTIONS

Three main categories of functions (tools) are defined; Input, Storage and Management and Display, which provide the following operations:

A. Input

This category contains three forms for massive import of imagery (Fig. 4.a), vector digitizing (Fig .4.b) and the linking of selected geographic features to multimedia files (Fig. 4.c).

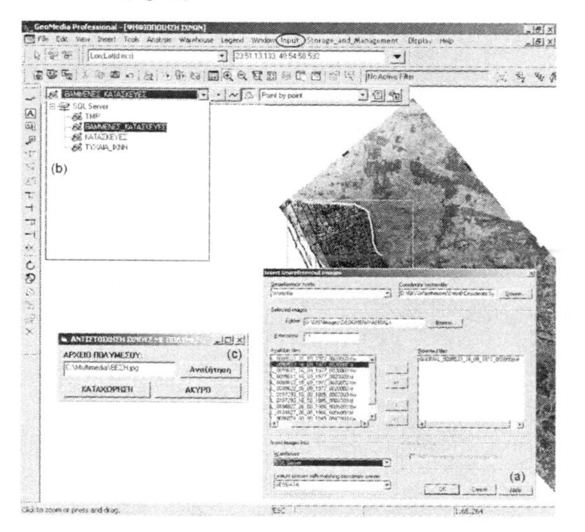

Fig. 4. The first category of functions is used for data input and contains two GeoMedia commands (a,b) and a form for multimedia documentation associating media files (photos, audio, video) to vector data (c).

Massive input of raster data and vector digitizing are provided from original GeoMedia tools and the link between geographic features and multimedia files is done using a developed form.

B. Storage and Management

The second category is dedicated to the update of descriptive properties

(Fig. 5.a), the discrimination among vector data which represents features that "are" and "are not" recognizable over selected imagery (Fig. 5.b) and the automated attribution of metadata to raster files.

Those forms are used to update the descriptive properties of input data (raster or vector), linking attribute 's values with related table indexes of the data-base while the attribution of raster metadata is done automatically, based on a predefined manner for the file names (raster files without appropriate name require that the user inserts the attribute values manually).

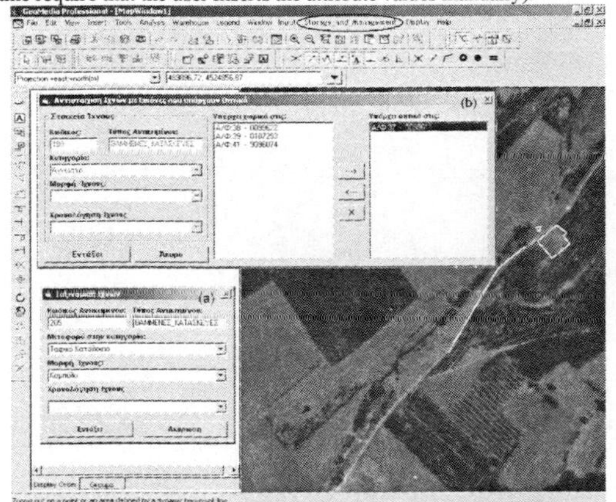

Fig. 5. The developed forms of the second category of functions are used for the management of descriptive properties (a) and for the definition of optical and spatial relationship between selected vectors and images (b).

File names of raster data to be imported, has to conform with the following format in order to automate the attribution of metadata (Fig. 6):

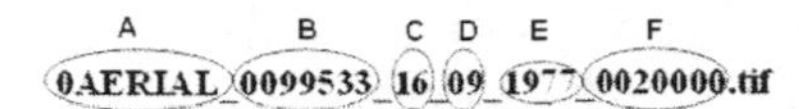

Fig. 6. File name format for the automated attribution of metadata to raster files (a: Image Type ID, b:Image ID, c: Day, d: Month, e: Year, f: Scale).

Image type ID (Fig. 6. a) describes the type of image (scanned map, airphoto, satellite image), Image ID (Fig. 6. b) is a code number for each image, Day (Fig. 6. c), Month (Fig. 6. d) and Year (Fig. 6. e) describe the date and Scale (Fig. 6. f) refers to the denominator of the scale.

C. Display

This set of functions is used for displaying data interactively and include forms for the automatic display of geographic data in any selected area (Fig. 7.a), the selective display of available data over the working view (Fig. 7.b), the selective display of spatially related vector data over selected raster data (Fig. 7.c) and the automated playback of linked multimedia files in GeoMedia' s environment (Fig. 7.d).

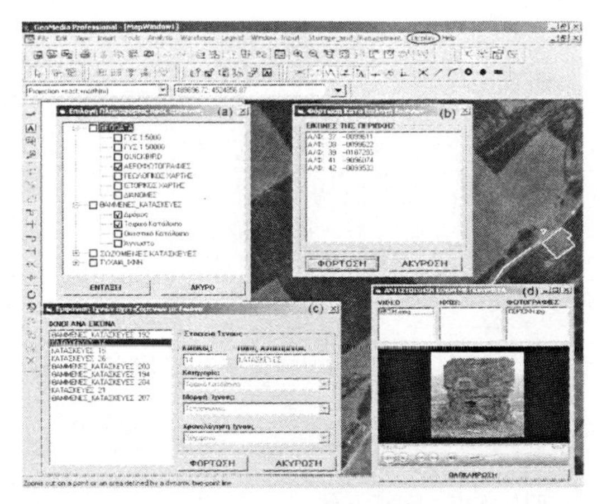

Fig. 7. The display operations, in the third category of functions, are controlled by a) a selection form, b) a form for displaying desired image based on selected position, c) a form for displaying vector data related to selected image and d) a window for automated multimedia playback.

The functions of the above three categories are organized respectively in three new menu items at GeoMedia 's menu bar (Fig. 5,6,7) as well as the according toolbox, from which the described operations can be executed.

IV. CONCLUSIONS

The majority of raster data is used for representing the differences over subsequent time periods (evolution of the area) In the same time, archaeological traces (vector data) although spatially related with overlapping raster data, they are not visible (or recognizable) on every image (e.g. a new archaeological finding appears on the post- excavation air photos but not on the old historical maps).In such cases the developed interface allows this discrimination and lets the user define relationships that are based on the recognition of geographic features (vector data) on an image (which are already spatially related). Through the above new kind of relationship, geographic features without temporal properties are related to time-stamped imagery. This is a side-way for their attribution with time and could be used as a bridge for spatio-temporal analysis (of data without temporal properties). On the other hand raster data, through this kind of relationship with vector data, can participate indirectly in both spatial and non-spatial queries.

During extended fieldwork (such as archaeology implies), a large amount of digital documentation is produced that in many cases contains much more details than tabular data may describe. In such cases multimedia association and playback provides a more detailed description of examined features. Attributing geographic features with multimedia files (additionally to standard information) facilitates their description and understanding from researchers and inexperienced users (viewers).

The developed functions could be useful in various GIS applications that are used for spatio-temporal documentation and data management or diachronic analysis, such as forestry and vegetation monitoring, landscape and urban change detection, where combinations of raster and vector data are taking place.

Further development includes automatic feature recognition and extraction (from raster data) as well as customization using open source software.

To conclude, the presented customization brings together common operations from different software (GIS, DB management, multimedia) that provide an integrated interface for storage, management, retrieval and representation of geographic and spatially related data from various time periods.

ACKNOWLEDGEMENT

We appreciate the contribution of D. Kaimaris (Phd Student at the Department of Rural and Surveying Engineering) who gave us part of the data he is using in his thesis study and demonstrated us practical matters on the management of spatial data for archaeological documentation.

REFERENCES

[1] Y. Zhou, "Multi-source data management in soil erosion monitoring system at regional scale", Geoscience and Remote Sensing Symposium, Proceedings, vol. 5, pp. 2875 – 2877, IGARSS 2004.

[2] L. Bordoni, A. Colagrossi," A Multimedia Personalized Fruition of Figurative Artistic Heritage by a GIs-Based Methodology", Multimedia Computing and Systems, IEEE International Conference, Vol. 2 pp.184 – 188, July 1999

[3] S. Fabrikant, "Spatialized Browsing in Large Data Archives", Invited seminar. Queen's Centre for Knowledge-Based Enterprises (KBE), Kingston, Canada, Mar. 26, 2003.

[4] B. Ducke, "Open-Source-GIS-Technologie, Stand und Perspektive", presented in the Berlin Head Office of the German Archaeological Institute (DAI) during the conference "GIS in der Archaologie", (http://www.arc-team.com/archeos/), 2006.

[5] T. Roustanis, "Application Development and Database Model to support Spatiotemporal analysis on Common Utility Network Damages through Data Warehouse technology", Msc Thesis, Dep. of Rural and Surveying Engineering, Aristotle University of Thessaloniki, Greece, 2004

[6] K.D. Papadimitriou, "Location Based Information System for Decision Support in Emergencies", Ph.D. dissertation, Dep. of Rural and Surveying Engineering, Aristotle University of Thessaloniki, Greece, 2004

[7] K.L. Wescott, R.J. Brandon, "Practical Applications of GIS for Archaeologists", Taylor & Francis, Great Britain, 2000.

[8] C. Güney, R.N. Çelik, "Multimedia Supported GIS Application for the Documentation of Historical Structures", Survey Review, Vol 37, No 287, pp. 66-83, 2003.

[9] Bakourou, E., Tsioukas, V., Katzougraki, I., Stylianidis, E., Papadimitriou, K., Patias, P., "The promotion of cultural heritage through internet using advanced audio-visual information: The venetian castles of Peloponnisos", IAPRS&SIS, Vol. XXXIV, Part 5, Com V, Corfu, Greece, pp. 298-301, 2002.

[10] D. Kaimaris, "Photogrammetric processing of Digital Images in service of Archaeological research, tracing Via Egnatia from Amfipolis to Philippoi", Ph.D. dissertation, Dep. of Rural and Surveying Engineering, Aristotle University of Thessaloniki, Greece,2006.

[11] Intergraph Corporation, "Geomedia Professional Object Reference", 2005.

BlogDisc: A System for Automatic Discovery and Accumulation of Persian Blogs

Kyumars Sheykh Esmaili
Web Intelligence Laboratory
Sharif University of Technology
Tehran, Iran
shesmail@ce.sharif.edu

Hassan Abolhassani
Web Intelligence Laboratory
Sharif University of Technology
Tehran, Iran
abolhassani@sharif.edu

Zeinab Abbassi
Web Intelligence Laboratory
Sharif University of Technology
Tehran, Iran
z_abbasi@ce.sharif.edu

Abstract — **One of the important elements of the new generation of the Web is the emergence of blogs. Currently a considerable number of users are creating content using blogs. Although Persian blogs have a short history, they have improved significantly during this short period. Because of fundamental differences between Persian and other languages, limited work has been done to analyze Persian blogs. In this work, a system named BlogDisc for automatic discovery and accumulation of Persian blogs is developed. This system uses content as well as link structure of the blogs. As an important part of this research, we propose an algorithm to recognize blogs that are not hosted on special blog hosts.**

Keywords: Blogsphere, Weblog, Blog Discovery, Web Graph

I. Introduction

Nowadays, Blogs (also know as Weblogs) are known as an important element of the Web. Using them, users can produce content on the Web very easily. Informally, a blog is a personal page maintained by a person and is updated based on its author's thoughts regularly and in the order of time. Normally a blog also has links to other blogs. On the other hand the content of a blog can be highly varied: links, explanations about other blogs, photos and also diaries and news. A more formal definition for a blog is given in [4]. According to that definition, a blog is a web page which has dated posts sorted in reverse chronological order. Some other characteristics of them are as follows: containing the author's profile in its margins, using blogging tools and templates, personalized content, being read by special users, and membership in some active communities.

The first blog appeared in 1996 but they became popular in 1999 when blogger[1] and other easy to use tools were announced. In the year 2002 an article in the Newsweek[2] magazine estimated that there are half million blogs. A couple of electronic magazines' articles in year 2000 brought many attentions to them [5]. As reported by the Technorati[3] site currently there exist more than 55 million blogs. They build a huge source of information and hence their analysis becomes very important. Generally two reasons for systematic study of the blog space can be mentioned:

- **Social aspects**: blogs have structural differences from normal web pages because they display a hierarchy of messages, like news groups, but each is written by a single person. However, there are more important structural differences: blogsphere culture is highly concentrated on small community interactions between a limited number of bloggers. Members of such informal societies put links to other blogs in theirs and often they leave some comments on each other's posts. Often the thread of responses is very high in a short period but then suddenly it stops. Naturally we can study the structure of such highly dynamic societies and model them.
- **Technical aspects**: popular studies on the Web and its graph are based on snapshots and static pictures of it that are obtained by crawling. Those studies are concerned with what happens during long periods of time. Some of them are responding to this concern by generating and analyzing of sequences of snapshots [3,4,9]. Developing tools and methods to analyze such sequences demands time-consuming efforts. On the other hand, blogsphere has a technical advantage: for ordinary web pages if a new crawling is done it is impossible to find exact time for changes but since each entry in a blog has date and time associated with it, it is possible to provide a more exact image of blogsphere evolutions [5].

Because of special characteristics of Persian language (right to left direction, encoding, fonts, etc.) creation of Persian blogs requires sophisticated tools. Therefore before the appearance of specific hosts for Persian language, the number of Persian blogs was too low. Fortunately at the present time, there are a considerable number of hosts for Persian bloggers. Nowadays Persian speakers show a lot of

[1] http://blogger.com
[2] http://www.msnbc.msn.com/id/3037881/site/newsweek/
[3] http://www.technorati.com/

K. Elleithy (ed.), Advances and Innovations in Systems, Computing Sciences and Software Engineering, 269–273.

interest in blogging such that currently Persian is among 10 most active languages in blogging[4,5].

This research aims at discovering, gathering and analyzing Persian blogs. As an important part of this research, a system for automatic discovery and gathering of Persian blogs named BlogDisc is introduced that will be explained in this paper.

The rest of this paper is organized as follows. In Section 2, a review of the related work in automatic discovery of blogs and in particular Persian blogs are discussed. Then in Section 3 a general categorization for Persian blogs is introduced which has an important role in our algorithm. In Section 4, the main algorithm of BlogDisc is introduced. The implementation of this algorithm and some of the results are discussed in Section 5.

II. Related Work

A. *Persian blogsphere Analysis*

Authors in [7] have gathered Persian blogs on a specific blog host – PersianBlog which is the largest and oldest host for Persian blogs. After crawling they have found 106699 blogs and 215765 links between them. Therefore, the average number of links in a blog is 2.022 but the variance is high. Near 45% of them are singular blogs having no outlinks. Also around 48% of them (51535 blogs) construct a huge single connected component having 208231 links which means there is 4.04 links per page in it. All of the links from Persian blog to outside it is about 87359 links covering a wide variety of destination pages. Blogs constitute a social network because the ratio of inlinks to outlinks is about 2.46. The authors compiled the gathered blogs in a test collection containing following information:

- list of all crawled blogs
- list of links between nodes in these blogs
- list of connected components
- ranks of blogs in the largest component (based on number of in-links, HITS and PageRank scores)

Also in [8] based on the same test collection, recommender techniques in Persian blogsphere have been evaluated.

B. *Automatic recognition and gathering of blogs*

In automatic blog discovery the only reported works arerelated to Japanese blogs [4,6]. In the papers, authors named the sites that are not hosted in special blog hosts and also does not use special tools for blogging as "web diaries". In fact these two papers explain the differences between blogs and also provide techniques for automatic recognition of "web diaries" for Japanese blogs. In their view, a blog is a personal web page which has order for dated entries and contains its author thoughts and beliefs. The main approach introduced in the papers for automatic recognition of "web diaries" uses following modules:

- Crawling module: it tries to collect candidate blog pages from the Web. To do so it: 1) crawls the Web, 2) uses information of Blogs directories and their ping servers, and 2) uses the links between the candidate blog and pages which are recognized as blog by recognizer module.
- Recognizer module: it tries to select only pages that are blogs. Such recognition is based on the characteristics of blogs. A page is recognized as a blog if and only if a sequence of entries that are written for a given date can be extracted from it. Also entries should satisfy following conditions:
    - They should have a date and the date should appear in top of them
    - Dates in their entries should be consistent (e.g. 01-Jan-2003 and 2003/1/1 are considered to be inconsistent) and were ordered in ascending or descending order.
    - Post date should be in a chronological order

III. Categorization of Persian Blogs

It is possible to categorize Persian blogs to the following subcategories:

A. *First group*

This group includes the blogs that are hosted on Persian blog hosts; they are well-defined blogs which it means we are sure that they are blogs and ordinary web pages. The positive point of this group is that blog URLs have a specific pattern which facilitates discovery and recognition of them. The other positive point is that they can be used as starting points of crawling to find blogs in group 2 and 3.

The main Persian blog hosts which provide blogging tools are

- Persianblog[6]
- Blogfa[7]
- Mihanblog[8]
- Blogsky[9]
- Parsiblog[10]

---

[4] http://www.timesonline.co.uk/article/0,,1068-1957461,00.html
[5] http://technology.guardian.co.uk/online/weblogs/story/0,,1377538,00.html
[6] http://www.persianblog.com/
[7] http://www.blogfa.com/
[8] http://www.mihanblog.com/
[9] http://www.blogsky.com/
[10] http://www.parsiblog.com/

Among these hosts, the Persianblog is the most popular among Iranian bloggers.

B. *Second group*

Blogs in this group are hosted on non Persian hosts but are updated in Persian. After a preliminary scanningww of the list of crawled blogs in this research it seems that Blogspot is the only major non Persian site which is used by Iranian Persian speakers for blogging. This group also constitutes well-defined blogs since by recognition of usage of Persian language in a page in this domain we are sure to have a Persian blog.

C. *Third group*

The blogs in this group are similar to ordinary web pages in public domains. Therefore to recognize these blogs, we cannot rely on their URL pattern, but we should use their content as well as their link structure. Although, the number of such is fewer than blogs in the first and second group, these blogs are important. In fact, those bloggers that started writing in the first or the second group after gaining popularity and experience move to this group (a blog with personal address and independent). According to Webstats4u[10] statistics a high percent of mostly read Persian blogs are in this group. Therefore discovery of them is of high importance.

IV. PROPOSED METHOD FOR PERSIAN BLOG DISCOVERY

As mentioned above some of the Persian blogs are developed outside special hosts. To discover such blogs we propose a new algorithm in this section. It outputs a value between 0 and 1 which we name BV, showing the probability of a page to be a blog. Therefore for well-defined blogs (those in groups one and two), this value is 1. For the blogs in the third group, this value is determined by an analysis of the content of blog and afterwards by the link structure.

A. *Recognition by content*

According to the definition of a blog mentioned before, the most important characteristic of a blog is that it should have some entries sorted in reverse chronological order. This characteristic is the main clue to discover a blog. As the discovery of blogs in third group is the target of this algorithm we cannot assume a specific format for date and time. Many of these blogs use some special services but those services also let the users use their desirable format and hence we need to process the html source of blogs for discovery.

Since a page having a sequence of dates in reverse order is not necessarily a blog to improve the correctness of the recognition probability of algorithm we need to use other characteristics of blogs. We use following points in this regards:

- Blogs are normally have archives
- Most of them have facilities to receive comments for each post
- Some of them use facilities provided by blogrolling site
- Most of the blogs not hosted in special hosts use specific templates. For example Movable Type[11] and WordPress[12] are popular templates
- Some of them use special stat softwares.

More formally, in our algorithm, we first set BV to 0 for not well-defined blogs and increase it based on the above features as below:

If a page does not have a sequence of dates in reverse order it is not a blog, otherwise it is a blog with probability of 0.25. In addition if it has any of the following characteristics its probability to be a blog is increased by 0.05 percent (we assign these values heuristically after evaluating some arbitrary values):

- Uses standard templates for blogs
- Uses stat software
- Has archive
- Has a system for commenting
- Uses blogrolling

Therefore if a page has all the above characteristics it is a blog with probability 0.5.

B. *Recognition by link structure*

In addition to blog content analysis it is possible to use the structure of graph obtained during crawling. In fact normally blogs has links to other blogs but it is rare to have a web page to have links to some blogs. More specifically, if a site has many output links and most of them are links to other blogs then with high probability it is a blog. On the other hand, if a site has many outlinks but almost none of them are links to blogs then with high probability it is not a blog. Also if a site has many in-links and they are from some blogs then it is more probable that it would be a blog. Therefore, after calculating BVs - between 0 and 0.5- from previous phase we can update them as follows:

1) If outdegree of a page in greater than 5, and half of its outlinks are pages with BV of greater than or equal to 0.5 we add μ[13] to their BV.

2) If the number of outlinks is between 3 and 5, and only one of them has BV less than 0.5, then we add μ to their BV.

3) If the number of outlinks is less than 3 only in the case that all the pages represented by outlinks have BV greater than 0.5, we add μ to the BV of the site.

4) If the number of in-links are higher than 50 and more than 80% of them are from blogs, then we add $\mu/20$ to the BV. It should be noted that in our experiments, we did not

[11] http://www.movabletype.org/

[12] http://wordpress.org/

13 μ is a real small value (in our experiments we used 0.05)

use this part because our crawl is not very broad and therefore does not cover a lot of pages having link to a page in consideration.

C. Analysis of algorithm convergence

In this section we analyze the run-time convergence of the algorithm for the link analysis. We consider a case where all the steps in algorithm are applied. In each iteration of algorithm either there is no change in BV values, which results in termination of the algorithm, or some values are changed and totally at least $\mu/20$ is added. However the sum of values at most is equal to the number of vertices, i.e. N. This simply shows that when the step 4 of algorithm is not executed it finishes after $20N/\mu$ iterations.

V. IMPLEMENTATION AND RESULTS

A. Crawler

We have implemented a special crawler. Each crawler needs some pages to start its operations. In our case, it is better to select pages from blogs having high number of links. For this purpose we used directories available in Persian blog hosts. They usually have directories with a considerable number of links. In addition they have links to recently updated blogs, more visited blogs, most active blogs, popular blogs and recent items. Such lists are updated frequently and we can use them as starting point for crawl.

This project only aims at discovering blogs and we need to use an optimized crawler for this purpose. It means that we should not gather non-blog pages. On the other hand since all blogs are not hosted in special hosts we cant limit the crawling process. The crawler is implemented as a focused crawler [1]. If current page is a well-defined blog, crawl is continued, otherwise, its internal links are analyzed and if none of them are pointing to a well-defined blog it is much probable that it is not a blog. But since we are not sure about our guess we keep it and we stop continuing crawl process from this page. Operation of focused crawler is showing in the figure 1. With that crawler more than 40000 pages are extracted from the Web.

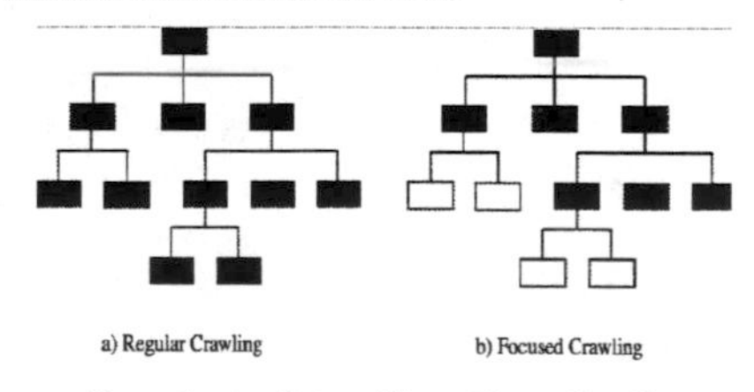

Figure 1 – An Abstract View of Focus Crawling

B. Link structure and their analysis

Characteristics of the graph constructed from the crawler are summarized in table 1. In this table, marked pages represent pages visited by our crawler but not necessarily collected.

Also with the definition of strongly connected components (a component which every node can be reached from other nodes considering links as directed edges) we have:

- The largest strongly connected component has 5439 nodes
- The second strongly connected component has 4000 and the next one has 154 nodes.
- The number of strongly connected components with at least 2 nodes is 51.

Such results agree with the specification of blog graph reported in [7].

Property	No.
Nodes	40643
Outdegree=0	1392
AverageOutdegree	34.35
MaxOutdegree	1238
Marked Pages	361243
Indegree=0	4579
MaxIndegree	861

Table 1 – Characteristics of the Crawled Graph

C. BlogDisc algorithm results

Every page in the link structure is not necessarily a blog. To find blogs, we apply the algorithm explained in Section 4. Therefore we give an initial value to each page according to its content and then upgrade it with the links from it to other pages. This is an iterative process which as explained before finally converges. The obtained results are summarized below:

From the total accumulated pages in the crawling phase, 65.36% are blog hosts in special blog hosts or are well-defined blogs (BV=1). After the content analysis phase, the rests are:

- 15.5% (5.36% of total crawled pages) of them are not recognized as blogs (BV=0)
- The rest (29.27% of total crawled pages) are recognized as weak blogs ($0<BV<0.5$)

After link structure analysis:

- 63% of weak blogs (18.44% of total crawled pages) are recognized as blogs (BV=1)
- The rest (7.08% of total crawled pages) are remained as weak blogs ($0< BV <1$)

Aforementioned results of the algorithm are shown in the Figure 2. It can be deduced from this figure that combining content and hyperlink analyses of blogs drastically improve precision of blog discovery process.

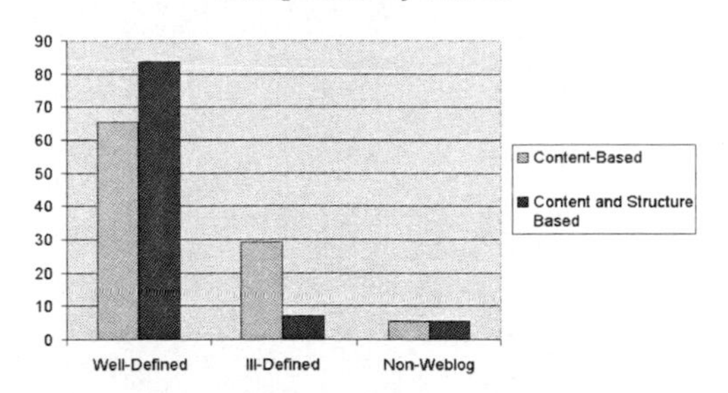

Figure 2- Comparing Performance of Content and Content-Structure Based Blog Discovery Methods

VI. CONCLUSION

Blogs are valuable resources in the Web and their analysis is important from social as well as technical points of view. One of the important parts of such analysis is the automatic recognition of blogs and gathering of them. In this article, a combined content and link analysis methods for automatic discovery of Persian blogs is explained and the result of the experiments shows its effectiveness.

Since our proposed method doesn't use any specific characteristics of Persian language, it can be applied to blogs in other languages as well.

REFERENCES

[1] S. Chakrabarti, M. van den Berg, and B. E. Dom. "Focused crawling: a new approach to topic-specific web resource discovery". In Proceedings of the Eighth International World Wide Web Conference, Toronto, Canada, May 1999.

[2] K. Balog and M. de Rijke, "Decomposing Bloggers' Moods. Towards a Time Series Analysis of Moods in the Blogsphere", In Proceedings of WWW2006 Workshop on Blogging Ecosystem, Edinburgh, Scotland, Apr. 2006.

[3] S. Nakajima, J. Tatemura, Y. Hino, Y. Hara, and K. Tanaka, "Discovering Important Bloggers Based on a Blog Thread Analysis", In Proceedings of WWW2005 Workshop on Blogging Ecosystem, Chiba , Japan, May 2006.

[4] T. Nanno, "Automatic Collection and Monitoring of Japanese Blogs", WWW 2004 Workshop on the Blogging Ecosystem: Aggregation, Analysis and Dynamics, New York, May 18th 2004

[5] R. Kumar, J. Novak, P. Raghavan, and A. Tomkins. On the bursty evolution of blogsphere. In Proc. Of the 12th International World Wide Web Conference, pages 568-576, 2003.

[6] T. Nanno, T. Fujiki, Y. Suzuki, M. Okumura, "Automatically collecting, monitoring, and mining japanese blogs". WWW (Alternate Track Papers & Posters) 2004: 320-321

[7] K. Sheykh Esmaili, M. Jamali, M. Neshati, and H. Abolhassani, Y. Soltan-Zadeh, "Experiments on Persian Blogs", WWW2006 Workshop on Blogging Ecosystem, Edinburgh, Scotland, Apr. 2006.

[8] K. Sheykh Esmaili, M. Neshati, M. Jamali, H. Abolhassani, J. Habibi, "Comparing Performance of Recommendation Techniques in the Blogsphere", Proceedings of ECAI2006 Workshop on Recommender Systems, Trento, Italy, August 2006.

[9] R. Kumar, J. Novak, P. Raghavan, and A. Tomkins, "Structure and Evolution of Blogsphere", Communications of the ACM, Volume 47, Issue 12 (December 2004).

[10] Webstats4u ,http://www.Webstats4u.com.

Fuzzy Semantic Similarity Between Ontological Concepts

Ling Song, Jun Ma, Hui Liu, Li Lian, Dongmei Zhang
School of Computer Science & Technology, Shandong University,
Jinan, 250061, P.R.C

Abstract- **The main focus of this paper concerns the measuring similarity in a content-based information retrieval and intelligent question-answering environment. While the measure of semantic similarity between concepts based on hierarchy in ontology is well studied, the measure of semantic similarity in an arbitrary ontology is still an open problem. In this paper we define a fuzzy semantic similarity measure based on information theory that exploits both the hierarchical and non-hierarchical structure in ontology. Our work can be generalized the following: firstly each concept is defined as a semantic extended fuzzy set along its semantic paths; secondly the semantic similarity between two concepts is computed with two semantic extended fuzzy sets instead of two concepts themselves. Our fuzzy measure considers some factors synthetically such as ontological semantic relation density, semantic relation depth and different semantic relations, which can affect the value of similarity. Compared with existed measures, this fuzzy similarity measure based on shared information content could reflect latent semantic relation of concepts better than ever.**

I. INTRODUCTION

Query processing plays an important role in several application fields such as intelligent question answering, search engineer and information retrieval. The purpose of similarity measures in connection with querying is to look for similar rather than for exactly matching values. For example, in information retrieval, user provides keywords to search; actually, the search operation may need better and closer concepts that are similar to the keywords on a semantically level rather than keywords. Ontology is a specification of a conceptualization of a knowledge domain, which is a controlled vocabulary that describes concepts and the relations between them in a formal way, and has a grammar for using the concepts to express something meaningful within a specified domain of interest. So determining semantic similarity between ontological concepts becomes more and more important [1-3].

The central question addressed in this paper is how to derive semantic similarity between ontological concepts that measure degree of similarity proportional to how much the concepts x and y share or how close they are in generalized ontology. Several papers have reported measurements aiming only at hierarchical structure (hypernym/ hyponym) relation. In this paper we propose an approach not only deal with hierarchical structure relation but also non-hierarchical structure relation. Semantic similarity between ontological concepts is generalized as follows: each concept is regarded as a fuzzy set along its semantic paths. The semantic similarity between two concepts can be determined by computing two semantic extended fuzzy sets. This similarity measure can mine latent semantic relation and describe conceptual semantic similarity more precisely and more closely to human's intuition.

The organization of this paper is as follows: Section II introduces a formal model for ontology. Section III gives brief overview of several relevant similarity measures and analyze their properties and existing limitations. This is necessary for the work of Section IV, in which we define an information-theoretic measure of semantic similarity that exploits both the hierarchical and non-hierarchical structure in ontology, which is a weak fuzzy similarity measure. Finally, Section V concludes this paper.

II. ONTOLOGICAL REPRESENTATIONS

In this section we introduce basic definitions of ontology. Upon these we build our fuzzy similarity framework. The purpose of the ontology is to formally express a shared understanding of information. In order to compare and measure semantic similarity between concepts in ontology, one may consider semiotic levels. Researchers have define a frame system as the underlying model for the ontology [4-6], thereby we work on a slightly revised excerpt of the OKBC (Open Knowledge Base Connectivity) knowledge models:

Definition 1. (Core Ontology) An Core ontology is a sign system $O := (L;F;G;C;Root;H;R)$, which consists of:

A Lexicon: The lexicon consists of a set of terms (lexical entries) for concepts, L^C, and a set of terms for relations, L^R. Their union is the lexicon $L := L^C \cup L^R$.

Two Reference Function: The reference functions F, G, with $F: 2^{L^C} \rightarrow 2^C$ and $G: 2^{L^R} \rightarrow 2^R$. F and G link sets of lexical entries $\{L_i\} \subset L$ to the set of concepts and relations they refer to, respectively. In general, one lexical entry may refer to several concepts or relations and one concept or relation may be refered to by several lexical entries. Their inverses are F^{-1} and G^{-1}.

A set of concepts C (classes in OKBC). About each $c \in C$ exists at least one statement in the ontology.

A particular top concept $ROOT$. $ROOT$ is not in C, but in the taxonomy, it is above every other concept.

Concept Hierarchy structure H: Concepts are taxonomically related by the acyclic, transitive relation H, $H \subset C \times (C \cup \{ROOT\})$. $H(c_1; c_2)$ means that c_1 is a subconcept of c_2. It holds that $\forall c \in C : H(c, ROOT)$.

Concept non-hierarchical structure R: Concepts are semantic related by relations R, $R \subset C \times C$, $R(c_1,c_2)$ means that c_1 has a kind of semantic relation with c_2.

III. PREVIOUS WORK

Many measuring similarity between ontological concepts have been proposed in the past researches [4,7-12]. These

K. Elleithy (ed.), Advances and Innovations in Systems, Computing Sciences and Software Engineering, 275–280.

measures can be divided into three main approaches. The first is based on a feature-theoretical approach; the second uses distance between the two concepts within the link structure; the third is based on shared information content of the most specific common ancestor of the two concepts. The definition of general objects' similarity is given:

Definition 2. (Similarity) A similarity measure is a real-valued function $sim(x,y): U \times U \rightarrow [0,1]$ on a set U measuring the degree of similarity between x and y. where U is the set of universe.

A. Feature-Theoretical Approach

Intuition tells us that the similarity between object A and B is related to both common and different characteristics. The more commonality they share, the more similar they are. The more differences they have, the less similar they are. Tversky[7-8] developed a feature-theoretical approach to the analysis of similarity relations. See equation (1):

$$S(A,B) = f(A \cap B) / [f(A \cap B) + \alpha * f(A - B) + \beta * f(B - A)] \quad (1)$$

Where the features shared by A and B, $A \cap B$; the features of A that are not shared by B, $A - B$; the features of B that are not shared by A, $B - A$.

B. Edge-based Approach

Using the simple edge count distance in measuring conceptual distance is the underlying assumption that edges or links between concepts represent uniform distances. The shorter the path from one node to another, the more similar they are. To overcome the limitation of uniform distances, it is necessary to consider that the edge connecting the two nodes should be weighted[12].

Sussna[13] think that the weight between two nodes c_1 and c_2 is calculated as follows:

$$wt(c_1,c_2) = \frac{wt(c_1 \rightarrow_r c_2) + wt(c_2 \rightarrow_{r'} c_1)}{2d} \quad (2)$$

given $$wt(x \rightarrow_r y) = \max_r - \frac{\max_r - \min_r}{n_r(x)} \quad (3)$$

where r is a relation of type r, □r' is its reverse, d is the depth of the deeper one of the two, max_r and min_r are the maximum and minimum weights possible for a specific relation type r respectively, and $n_r(x)$ is the number of relations of type r leaving node x.

Wu and Palmer[14] proposed a measure for semantic similarity :

$$sim_{Wu\&Palmer}(c_1,c_2) = \frac{2 \times N_3}{N_1 + N_2 + 2 \times N_3} \quad (4)$$

Where N_1 and N_2 are the length (in number of nodes) of the path from c_1 and c_2 to their most specific common superconcept c_3, and N_3 is the length of the path from c_3 to the root .

In Fig.1, $sim_{Wu\&Palmer}(hill, coast) = \frac{2 \times N_3}{N_1 + N_2 + 2 \times N_3} = \frac{2 \times 3}{2+2+2 \times 3} = 0.6$

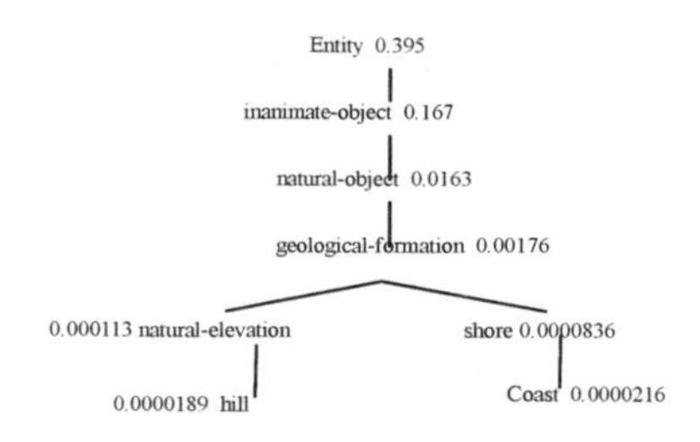

Fig 1. A Fragment of WordNet[15]

C. Information Theoretic Models

Conceptual similarity between two concepts c_1 and c_2 may be judged by the degree to which they share information. The more information they share, the more similar they are. Following the notation in information theory, the information content (IC) of a concept c can be quantified as follows [14]:

$$IC(c) = -\log p(c) \quad (5)$$

Where $P(c)$ is the probability of encountering an instance of concept c. Formally,

$$Freq(c) = \sum_{n \in words(c)} count(n) \quad (6)$$

Where $words(c)$ is the set of words subsumed by concept c. Concept probabilities were computed as:

$$p(c) = \frac{Freq(c)}{N} \quad (7)$$

Where N was the total number of words observed.

Therefore the amount of information contained in x_1□c_1 and x_2□c_2 is

$$-\log p(c_1) - \log p(c_2) \quad (8)$$

If concept c_3 is the most specific concept that subsumes both c_1 and c_2, the similarity of two concepts can be formally defined as [12]:

$$sim_R(c_1,c_2) = \max_{c \in Sup(c_1,c_2)} [IC(c)] = \max_{c \in Sup(c_1,c_2)} [-\log p(c)] = -\log p(c_3) \quad (9)$$

Where $Sup(c_1,c_2)$ is the set of concepts that subsume both c_1 and c_2 .To maximize the value, the similarity value is the information content value of the node whose IC value is the largest among those super classes. While the node c_3 is the "lowest upper bound" among those that subsume both c_1 and c_2. In Fig. 1, attached data is corresponding probability to a concept.

$sim_R(hill,coast) = -log(P(geological\text{-}formation)) = 2.75$

Another similarity approach is [3,16]:

$$sim_L(c_1,c_2) = \frac{2 \times \log p(c_3)}{\log p(c_1) + \log p(c_2)} \quad (10)$$

In Fig. 1,

$$sim_L(hill, shore) = \frac{2 \times \log P(Geological-formation)}{\log P(hill) + \log P(shore)} = 0.63$$

The previous three similarity approaches have a common property: equivalence relation.

Formally, For U is a universal set of all sets defined on a given domain D [16]:

Definition 3. (A Equivalence relation) An equivalence relation is a mapping, $sim : U\times U \to [0,1]$, such that for $x, y, z \in U$,

(a) Reflexivity: $sim(x, x) = 1$

(b)Symmetry: $sim(x, y) = sim(y, x)$

(c)Transitivity:$sim(x,y)=1, sim(y,z)=1 \Rightarrow sim(x,z)=1$.

It is well known that it is the most restrictive similarity relation for crisp data. But we do not think such an equivalence relation could be suitable for semantic similarity between concepts in ontology. For a counter example, with concept inclusions relation, intuitions imply dogs have strong similarity with animals, but animals are only to some degree similar to dogs. That is, the similarity should be anti-symmetry. In the next section we will give a new definition of similarity and a new similarity measure.

IV. OUR WORK: A FUZZY SIMILARITY MEASURE

Firstly we give a definition of Anti-Symmetry Similarity:

Definition 4. (Anti-Symmetry Similarity) A similarity measure is a real-valued function $sim(x,y):U\times U\rightarrow[0,1]$ on a set U measuring the degree that y is similar to x, and $sim(y,x):U\times U\rightarrow[0,1]$ on a set U measuring the degree that x is similar to y, where U is the set of universe.

In order to find a reasonable similarity measure to deal with complex situation for arbitrary ontology, we introduce fuzzy sets and property of similarity relations.

A. *Fuzzy Sets and Weak Fuzzy Similarity Relation*

The theory of Fuzzy Sets proposed by Zadeh[17] has achieved a great success in various fields. Semantic similarity measure may be generalized to a fuzzy semantic similarity measure if the weights for the relation link are replaced by membership degrees indicating the strength of the relationships between the parent and child concepts.

In fact, we think that the degree of similarity between two concepts in ontology is neither necessarily symmetric nor necessarily transitive. A weak fuzzy similarity relation as a generalization of a fuzzy similarity relation is proposed [18].

Definition 5. (A weak fuzzy similarity relation) A weak fuzzy similarity relation is a mapping, $sim : U\times U \to [0,1]$, such that for $x, y, z \in U$,

(a) Reflexivity: $sim(x, x) = 1$

(b)Conditional symmetry: *if* $sim(x, y) > 0$ *then* $sim(y, x)>0$

(c)Conditional transitivity: *if* $sim(x, y) \geqslant sim(y, x)>0$ and $sim(y,z) \geqslant sim(z,y)>0$ then $sim(x,z) \geqslant sim(z,x)$

Only the weak fuzzy similarity relation in Definition 5 can be suitable for similarity between two ontological concepts. There are three reasons: Firstly, conceptual similarity has property of reflexivity; secondly, ontological structure such as hyponym/hypernym make concepts have property of conditional symmetry, that is, if it exists some kind of semantic relation between concept x and concept y, then it also exists some kind of semantic relation between concept y and concept, as semantic relation is bi-directional in ontology; lastly, it satisfies conditional transitivity, Intuitively, in hierarchical structure ontology, the similarity between hyponym concept to its immediate hypernym concept is larger than the similarity between hypernym concept to its immediate hyponym concept, after transitive of this kind of semantic relation, the similarity between hyponym concept to its indirect hypernym concept is still larger than the similarity between hypernym concept to its indirect hyponym concept. Next we'll propose a new approach that satisfies weak fuzzy similarity relation to measure conceptual similarity.

B. *A Fuzzy Similarity Measure*

The extension of hierarchy ontology to an arbitrary ontology raises two questions. First, how to determine the ancestral concepts that are relative closely to a concept? Second, how to determine shared information of a pair of concepts in ontology? Our work is defining an extended semantic fuzzy set that has close semantic relation to a concept.

An ontology graph $G=(V, E)$ is a graph of nodes representing concepts. Each edge represents a kind of semantic relation, that is, if it exists a kind of semantic relation between two concepts, then it exists an edge between two nodes that represent these two concepts. V is set of nodes and E is set of edges. E is divided to two sub-sets, which are a set of hierarchy component and a set of non-hierarchy component respectively. Fig. 2 is a fragment of an abstract ontology. The solid edges are hierarchy structure (hypernym/hyponym relation) and broken are non-hierarchy structure (other semantic relations). All semantic relations are bi-directional. In order to represent semantic relations explicitly, adjacency matrixes are introduced and corresponding operations are defined. This is necessary for the work to compute conceptual similarity of arbitrary ontology.

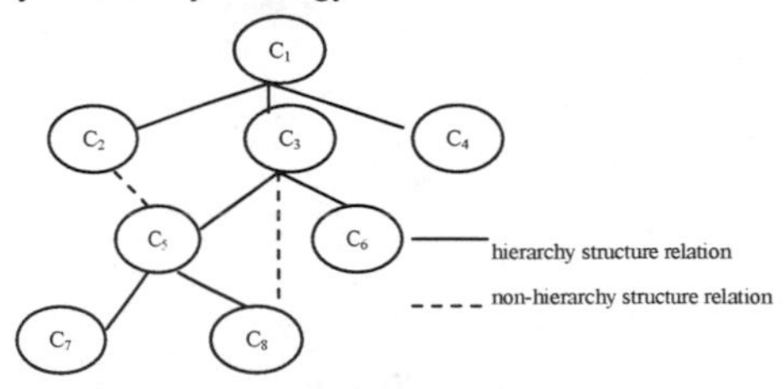

Fig. 2. a fragment of abstract ontology

To distinguish the role of different semantic relations, it is useful to assign weights and vary these weights according to the edge's semantic type. In order to make the implicit membership relations explicit, we represent the graph structure by means of adjacency matrices and apply a number of operations to them. Adjacency matrix T' is used to express hierarchical structure of an ontology. Assuming that α_1 and α_2 are expressed immediate Inclusion relation and IS-A relation respectively. T_{ij}' is used to express weights between concept i and concept j. Adjacency matrix T' is defined:

$$T' = \begin{cases} 1 & \text{if } i=j \\ \alpha_1 & \text{if } i\neq j \text{ and } (i,j)\in Inclusion \\ \alpha_2 & \text{if } i\neq j \text{ and } (i,j)\in IS\text{-}A \\ 0 & \text{otherwise} \end{cases}$$

Let α_1=0.4 ,α_2=0.9, then in fig. 2,

$$T' = \begin{bmatrix} 1 & 0.4 & 0.4 & 0.4 & 0 & 0 & 0 & 0 \\ 0.9 & 1 & 0 & 0 & 0 & 0 & 0 & 0 \\ 0.9 & 0 & 1 & 0 & 0.4 & 0.4 & 0 & 0 \\ 0.9 & 0 & 0 & 1 & 0 & 0.4 & 0 & 0 \\ 0 & 0 & 0.9 & 0 & 1 & 0 & 0.4 & 0.4 \\ 0 & 0 & 0.9 & 0.9 & 0 & 1 & 0 & 0 \\ 0 & 0 & 0 & 0 & 0.9 & 0 & 1 & 0 \\ 0 & 0 & 0 & 0 & 0.9 & 0 & 0 & 1 \end{bmatrix}$$

The method that T' unifies definition weights of adjacency edge couldn't consider density of edges. Here we borrow the original thought of equation (3) to verify weights and reflect our intention. Assuming $max_{inclusion}=0.4$ and $min_{Inclusion}=0.3$ are the maximum and minimum weights for an inclusion relation respectively, and $n_{Inclusion}$ (x) is the number of inclusion relations leaving node x. In fig. 2, $n_{Inculde}(C_1)=3$ $n_{Include}(C_5)=2$, combined with equation (3):

$$wt(C_1, C_2)= \sigma_{\max}-(\sigma_{\max}-\sigma_{\min})/3=0.37$$

$$wt(C_5, C_7)= \sigma_{\max}-(\sigma_{\max}-\sigma_{\min})/2=0.35$$

After verifying weights matrix T' becomes matrix T:

$$T = \begin{bmatrix} 1 & 0.37 & 0.37 & 0.37 & 0 & 0 & 0 & 0 \\ 0.8 & 1 & 0 & 0 & 0 & 0 & 0 & 0 \\ 0.8 & 0 & 1 & 0 & 0.35 & 0.35 & 0 & 0 \\ 0.8 & 0 & 0 & 1 & 0 & 0.3 & 0 & 0 \\ 0 & 0 & 0.8 & 0 & 1 & 0 & 0.35 & 0.35 \\ 0 & 0 & 0.85 & 0.85 & 0 & 1 & 0 & 0 \\ 0 & 0 & 0 & 0 & 0.8 & 0 & 1 & 0 \\ 0 & 0 & 0 & 0 & 0.8 & 0 & 0 & 1 \end{bmatrix}$$

We use additional adjacency matrix S to represent the non-hierarchical components of ontology. To represent different semantic relations, an explicit way is introduce different similarity factors for different semantic relations. Assume that we have k different semantic relations R_1, R_2, ... ,R_k and let β_1, β_2, ... β_k be the corresponding similarity factors(weights). Note that semantic relation in S is conditional symmetry. A non-hierarchical semantic relation adjacency matrix S is defined:

$$S_{ij}=\begin{cases} 1 & \text{if } i=j \\ \beta_i & \text{if } i\neq j \text{ and } (i,j)\in R_i \\ 0 & \text{otherwise} \end{cases}$$

In fig. 6, let $S_{52}=0.5$, $S_{25}=0.2$, $S_{83}=0.5$, $S_{38}=0.2$, then

$$S = \begin{bmatrix} 1 & 0 & 0 & 0 & 0 & 0 & 0 & 0 \\ 0 & 1 & 0 & 0 & 0.2 & 0 & 0 & 0 \\ 0 & 0 & 1 & 0 & 0 & 0 & 0 & 0.2 \\ 0 & 0 & 0 & 1 & 0 & 0 & 0 & 0 \\ 0 & 0.5 & 0 & 0 & 1 & 0 & 0 & 0 \\ 0 & 0 & 0 & 0 & 0 & 1 & 0 & 0 \\ 0 & 0 & 0 & 0 & 0 & 0 & 1 & 0 \\ 0 & 0 & 0.5 & 0 & 0 & 0 & 0 & 1 \end{bmatrix}$$

Adjacency matrix G is used to represent all semantic relations in graph G, let $G=T \vee S$, in fig. 2,

$$G = \begin{bmatrix} 1 & 0.37 & 0.37 & 0.37 & 0 & 0 & 0 & 0 \\ 0.8 & 1 & 0 & 0 & 0.2 & 0 & 0 & 0 \\ 0.8 & 0 & 1 & 0 & 0.35 & 0.35 & 0 & 0.2 \\ 0.8 & 0 & 0 & 1 & 0 & 0.3 & 0 & 0 \\ 0 & 0.5 & 0.8 & 0 & 1 & 0 & 0.35 & 0.35 \\ 0 & 0 & 0.85 & 0.85 & 0 & 1 & 0 & 0 \\ 0 & 0 & 0 & 0 & 0.8 & 0 & 1 & 0 \\ 0 & 0 & 0.5 & 0 & 0.8 & 0 & 0 & 1 \end{bmatrix}$$

To represent transitivity of semantic relations, we define matrix composition operation $\otimes$ on matrices as follows:

$$[A\otimes B]_{ij}=\max(A_{ik}\cdot B_{kj}) \qquad (11)$$

Where A and B are two adjacency matrices respectively. A_{ik} is the immediate weight that the ith concept is similar to the jth concept in a kind of semantic relation, and B_{kj} is the immediate weight that the jth concept is similar to the kth concept in another kind of semantic relation.

We know that conceptual similarity in hierarchical structure of ontology satisfies conditional symmetry. Does such conditional symmetry still exist after transitivity of semantic relation? We can obtain the following theorem.

Theorem 1. If A is a adjacency matrix that represents hierarchical structure of ontology, then $[A\otimes A]_{ij}=max(A_{ik}\cdot A_{kj})$ still satisfies conditional transitivity.

Proof: Without loss of generality, assume that $A_{ik}\geqslant A_{ki}>0$ and $A_{kj}\geqslant A_{jk}>0$, according to definition of matrix composition operation: $A_{ij} = [A\otimes A]_{ij}=max(A_{ik}\cdot A_{kj})$, $A_{ji} = [A\otimes A]_{ji}=max(A_{jk}\cdot A_{ki})$, we have known that $max(A_{ik}\cdot A_{kj}) \geqslant max(A_{jk}\cdot A_{ki})$, so $A_{ij} \geqslant A_{ji}$.

Let $T^{(0)}=T$, $T^{(r+1)}=T^{(0)} \otimes T^{(r)}$, We define the closure of T, denoted T^+ as follows:

$$T^+ = \lim_{r\to\infty} T^{(r)} \qquad (12)$$

Note that the computation of the closure T^+ converges in a number of steps that is bounded by maximum depth of the ontological hierarchical structure. In fig. 2,

$$T^+ = \begin{bmatrix} 1 & 0.37 & 0.37 & 0.37 & 0.13 & 0.13 & 0.045 & 0.045 \\ 0.8 & 1 & 0.3 & 0.3 & 0.1 & 0.1 & 0.036 & 0.036 \\ 0.8 & 0.3 & 1 & 0.3 & 0.35 & 0.35 & 0.12 & 0.12 \\ 0.8 & 0.3 & 0.3 & 1 & 0.1 & 0.3 & 0.036 & 0.036 \\ 0.64 & 0.24 & 0.8 & 0.24 & 1 & 0.28 & 0.35 & 0.35 \\ 0.13 & 0.25 & 0.85 & 0.85 & 0.3 & 1 & 0.1 & 0.1 \\ 0.5 & 0.19 & 0.64 & 0.19 & 0.8 & 0.2 & 1 & 0.28 \\ 0.5 & 0.19 & 0.64 & 0.19 & 0.8 & 0.2 & 0.28 & 1 \end{bmatrix}$$

Transitivity closure T^+ guarantee semantic transitivity of hierarchical structure in actually, however, semantic relations of non-hierarchical structure are not always transitivity, so we define semantic relation matrices W: $W=T^+\otimes G\otimes T^+$, W reflects similarity of ontological concepts from the aspect of semantic transitivity, which means that we only consider one layer non-hierarchical relations compose with transitivity closure. In fig. 2,

$$W=\begin{bmatrix} 1 & 0.37 & 0.37 & 0.37 & 0.13 & 0.13 & 0.045 & 0.074 \\ 0.8 & 1 & 0.3 & 0.3 & 0.2 & 0.1 & 0.07 & 0.07 \\ 0.8 & 0.3 & 1 & 0.3 & 0.35 & 0.35 & 0.12 & 0.2 \\ 0.8 & 0.3 & 0.3 & 1 & 0.1 & 0.3 & 0.036 & 0.036 \\ 0.64 & 0.5 & 0.8 & 0.24 & 1 & 0.3 & 0.35 & 0.35 \\ 0.68 & 0.25 & 0.85 & 0.85 & 0.3 & 1 & 0.1 & 0.17 \\ 0.5 & 0.4 & 0.64 & 0.19 & 0.8 & 0.8 & 1 & 0.28 \\ 0.5 & 0.4 & 0.64 & 0.19 & 0.8 & 0.8 & 0.28 & 1 \end{bmatrix}$$

In fig. 2, the degree that c_3 is similar $_{to}$ c_5 is: W_{35}=0.35, the degree that c_5 is similar to c_7 is: W_{57}=0.35, however, intuition tells us the degree that c_5 is similar to c_7 is larger than the degree that c_5 is similar to c_3, as c_5 and c_7 locate in the deeper layer, the different is smaller. So although a semantic relation matrix solves both semantic transitivity problem and edge density problem, it didn't solve the problem that ontological depth how to effect conceptual similarity. In section II, we have reviewed the information theory models, which consider information content of ancestral nodes. If we consider information content of the ancestors, then the information content in the deeper layer is more, and the similarity of a pair of concepts that is located in the deeper layer is higher naturally.

Computation of similarity raises two important problems: The first is how to define ancestral nodes that is related close to a concept; the second is how to determine shared information of a pair of concepts. The first problem is analyzed in the following: 1) Only consider ancestral nodes to a concept in hierarchical structure; 2) Consider all nodes that have any semantic relations with a node;

The first case only considers the most strongest semantic relation: IS-A relation, that is, all concepts in the path from a concept node to the root are considered as having close semantic relations with the concept. However in the second case, all semantic relations are bi-directional and semantic relations are various, which will increases computation complexity. Therefore, in our work, except considering the ancestral concepts to a concept in hierarchical structure, we choose concepts that have a layer non-hierarchical semantic relation with these ancestral concepts.

Definition 6 ((Extended semantic Set) For a concept *c*, we define a extended semantic set, which includes all ancestral concepts in hierarchical structure and all concepts that have a layer non-hierarchical semantic relation with these ancestral concepts.

For example, in fig.3, for C_8, the extended semantic set is defined: {C_5, C_3, C_2, C_1}, for C_6, the extended semantic set is defined: { C_2, C_3, C_4, C_1}

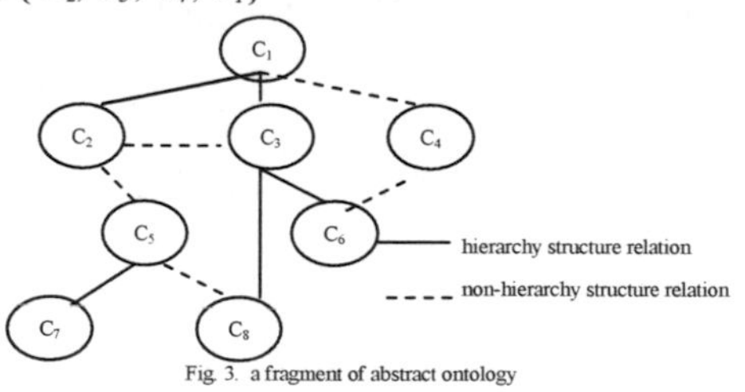

Fig. 3. a fragment of abstract ontology

Definition 7. (Extended Semantic Fuzzy Set): Given a concept *c*, the extended semantic set is $S=\{c_1', c_2', \ldots ROOT\}$, the degree that every concept *c'* in *S* is similar to *c* is regarded as fuzzy relation between concepts, so extended semantic fuzzy set *c+* can be described by:

$c+=\{1/c, sim(c,c_1')/c_1', sim(c,c_2)/c_2', \ldots, sim(c,ROOT)/ROOT\}$

Where $sim(c,c_i')$ is the degree that *c'* is similar to *c*, here we use directly the elements in semantic relation matrix *W* to represent similarity.

For example, in fig.3, $c_8+=\{1/c_8, 0.35/c_5, 0.2/c_3, 0.07/c_2, 0.074/c_1\}$

With Definition 7 any concept can be extend to be a semantic fuzzy set. Now it's time to determine shared information of a pair of concepts. According to information theoretic models, we should retain the greatest possible shared information of the concepts being compared.

Definition 8. (Fuzzy similarity): Let μ_X and μ_Y be two membership functions over a given domain *D* for two labels *X* and *Y* of a universe *U*. A fuzzy similarity relation is a mapping, $R: U \times U \to [0,1]$, defined by:

$$R(X,Y)=\alpha\frac{|X\cap Y|}{|X|}+(1-\alpha)\frac{|X\cap Y|}{|y|}=\alpha\frac{\sum_{d\in D}\min\{\mu_X(d),\mu_Y(d)\}}{\sum_{d\in D}\mu_X(d)}+(1-\alpha)\frac{\sum_{d\in D}\min\{\mu_X(d),\mu_Y(d)\}}{\sum_{d\in D}\mu_Y(d)} \quad (13)$$

Where *R(X,Y)* is the degree that *Y* is similar to *X* and $|X|=\sum_{d\in D}\mu_X(d)$ and $|Y|=\sum_{d\in D}\mu_Y(d)$ are regarded as cardinality of X, and α is a parameter, $\alpha\in[0,1]$.

Equation (13) is based on information theory. Especially, semantic path and cardinality of $|X\cap Y|=\sum_{d\in D}min\{\mu_X(d),\mu_Y(d)\}$ are actually based on shared information content.

What is more, it satisfies properties of reflexivity, conditional symmetry and conditional transitivity, which is a weak fuzzy similarity relation.

Compared with two existed similarity measures, in the following we use fuzzy similarity (equation (13)) to compute conceptual similarity in fig.3, let α =2/3.

	$sim_{Wu\&Palmer}$	sim_{Henrik}	Fuzzy similarity R
Similarity between C_7 and C_3	0.5	0.81	0.136
Similarity between C_3 and C_7	0.5	0.16	0.130

Table 1 Similarity comparing table 1 of fig.3

R and sim_{Henrik} satisfy conditional symmetry, that is, the degree that specific concept is similar to general concept is larger than the degree that general concept is similar to specific concept. But $sim_{Wu\&Palme}$ thinks that similarity is symmetry.

	$sim_{Wu\&Palmer}$	sim_{Henrik}	Fuzzy similarity R
Similarity between C_4 and C_3	0	0.36	0.27
Similarity between C_6 and C_5	0.5	0.36	0.28
Similarity between C_8 and C_7	0.67	0.36	0.37

Table 2 Similarity comparing table 2 of fig.3

R and simWu&Palme solve the similarity between for siblings on low levels are higher. But simHenrik can't solve.

	$sim_{Wu\&Palmer}$	sim_{Hennik}	Fuzzy similarity R
Similarity between C_3 and C_6	0.67	0.4	0.30
Similarity between C_4 and C_6	0.67	0.4	0.26

Table 3 Similarity comparing table 3 of fig.3

The density of Inclusion relation from c_3 is larger than that from c_4, so similarity is higher. Only fuzzy similarity R can reflect this characteristic.

From above evaluation, our approach about semantic similarity could compute conceptual similarity of arbitrary ontology, which considers factors such as ontological semantic relation density, semantic relation depth and different semantic relations. Compared with existed measures, this fuzzy similarity measure based on shared information content could reflect latent semantic relation of concepts better than ever.

V. CONCLUSION

The key issue is how the ontology influences the similarity if we want to compute conceptual similarity in ontology. In this paper we propose a fuzzy similarity measure that exploits both the hierarchical and non-hierarchical structure in ontology. Our measure considers some factors synthetically such as ontological semantic relation density, semantic relation depth and different semantic relations, which can affect the value of similarity. Compared with existed measures, this fuzzy similarity measure based on shared information content could reflect latent semantic relation of concepts better than ever.

REFERENCES

[1] Ullas Nambiar and Subbarao Kambhampati, "Mining approximate functional dependencies and concept similarities to answer imprecise queries," *Seventh InternationalWorkshop on theWeb and Databases*, Paris, France,2004, pp.73-78.

[2] Ishwinder Kaur and Anthony J. Hornof, "A comparison of LSA, wordNet and PMI-IR for predicting user click behavior," *Conference on Human Factors in Computing Systems*, Portland, Oregon, USA, 2005, pp.51 – 60.

[3] Valerie Cross. "Fuzzy semantic distance measures between ontological concepts," *Fuzzy Information. 04*, IEEE Annual Meeting of the Volume 2, Issue , 27-30 June 2004 Page(s): 635 - 640 Vol.2

[4] Alexander Maedchel and Steffen Staab. "Measuring similarity between ontologies," *Proceedings of the 13th International Conference on Knowledge Engineering and Knowledge Management*, Springer-Verlag, London, UK,2002, pp. 251 – 263.

[5] Vinay K. Chaudhri, Adam Farquhar Richard Fikes, Peter D. Karp and James P. Rice. OKBC: "A progammatic foundation for knowledge base interoperability," *Proceedings of the fifteenth national/tenth conference on Artificial intelligence/Innovative applications of artificial intelligence*, Madison, Wisconsin, United States, 1998, pp.600-607.

[6] Andreas Hotho, Alexander Maedche and Steffen Staab, "Ontology-based text document clustering," http://citeseer.ist.psu.edu/585623.html.

[7] Amos Tversky, "Features of Similarity," *Psychological Review*, 1977, 84(4): pp.327-352.

[8] Amos Tversky and Itamar Gati, "Studies of similarity," http://ruccs.rutgers.edu/forums/seminar1_fall03/Lila2.pdf.

[9] Philip Resnik. "Semantic similarity in a taxonomy: an information-based measure and its application to problems of ambiguity in natural language," *Journal of Articial Intelligence Research*, 1999, 11: pp95-130.

[10] M. Andrea Rodrı´guez and Max J. Egenhofer, "Determining semantic similarity among entity classes from different ontologies," *IEEE Transactions on Knowledge and Data Engineering*.2003,15(2): pp442 – 456.

[11] Peter Haase, Mark Hefke and Nenad Stojanovic, "Similarity for Ontologies - a comprehensive framework," http://citeseer.ist.psu.edu/ehrig04similarity.html.

[12] Jay J. Jiang and David W. Conrath, "Semantic similarity based on corpus statistics and lexical taxonomy," *In Proceedings of International Conference Research on Computational Linguistics (ROCLING X)*, Taiwan, 1997.

[13] Michael Sussna, "Word sense disambiguation for free-text indexing using a massive semantic network," *Proceedings of the Second International Conference on Information and Knowledge Management*, Washington, D.C., United States, 1993, pp.67 - 74.

[14] Wu, Z. and Palmer, M., "Verb semantics and lexical selection," *In Proceedings of the 32nd Annual Meeting of the Associations for Computational Linguistics*, Las Cruces, New Mexico, 1994, pp. 133–138.

[15] Dekang Lin, "An information-theoretic definition of similarity," *Proceedings of the Fifteenth International Conference on Machine Learning*. Morgan Kaufmann Publishers Inc, San Francisco, CA, USA,1998. pp.296 – 304.

[16] Rolly Intan, "Rarity-based similarity relations in a generalized fuzzy information system," *Proceedings of the 2004 IEEE Conference on Cybernetics and Intelligent Systems*, Singapore, December 1-3, 2004, pp.462-467.

[17] L. A. Zadeh, "Similarity relations and fuzzy orderings," *Information Science*, 1970, 3(2): 177-200.

[18] Rolly Intan and Masao Muhidono, "A proposal of fuzzy thesaurus generated by fuzzy covering," F*uzzy Information Processing Society-22nd International Conference of the North American*, 2003, pp.167- 172.

Research on Distributed Cache Mechanism in Decision Support System

LIU Hui[1,2] JI Xiu-hua[1,2]

(1. Computer Science & Technology Department, Shandong Economic University, Shandong Jinan 250014)
(2. Computer Science & Technology Department, Shandong University, Shandong Jinan 250014)
liuh_lh@sdie.edu.cn

Abstract-**With the development of Internet technology, the business decision makers put forward higher requirements with the performance in Decision Support System (DSS). In order to improve the query response -time in DSS, this paper proposes a DSS architecture with distributed cache mechanism, gives the working flow of the system, and introduces an admission & replacement algorithm. Experiments prove that the system performance is favorable.**

Keywords-**Decision Support System, distributed cache, On_Line Analytical Processing, Data Warehouse**

I. INTRODUCTION

In the global competition environment, it is more and more important to establish the effective Decision Support System (DSS). Data Warehouse (DW) and On_Line Analytical Processing (OLAP) are two essential parts of DSS, they convenient the business decision makers to process, analyze and understand the commercial questions effectively [1]. In DW there is historical unification data of business database, which supports the precise Ad-hoc(Dynamic) inquires and issues the useful information. OLAP is the on-line data accessing and analysis aiming at the specific questions.

With the development of Internet and 3W technology, the demand to DSS becomes higher, specially there are more and more Ad-hoc (Dynamic) inquires. Because the data in warehouse are all historical data, and the frequency of data renew is not so high, we can save the inquiry result in caches, and make the result as the inquiry gist in next time. Therefore we can enhance DSS inquiry response-time by applying the cache mechanism to DSS in the system, for the purpose of improving the DSS performance. [2] has proposed one cache manager names WATCHMAN, the result it inquired each time is saved in manager in the form of view, meanwhile the corresponding inquiry field. Only when there is the same inquiry field in cache manager, the new inquiry can be explained, otherwise the inquiry will stay in DW, in this way the inquiry cost is increased. [3] designed the standard form for the inquiry sentence and proposed the decomposition inquiry algorithm, but because all inquiry results are still saved in only one cache manager, the overload of cache manager has affected the inquiry response-time. The distributed cache mechanism has been proposed in the operating system, [4] proposed one kind of agent server frame, in this frame each cache manager is called a node, this node either respond the entire inquiry or cannot respond.

This article designs many cache managers based on the distributed cache mechanism, makes it possible for many cache managers to complete one inquiry task together by integrating the decomposition inquiry algorithm in [3]. The research indicates that this kind of structure can reduce the load of buffer manager and the inquiry cost, ultimately enhance the inquiry efficiency.

This article introduces the distributed cache mechanism in the second part, and proposes the corresponding architecture and working flow, especially the admission & replacement algorithm in the cache mechanism, then we analyze the system performance in the third part.

II. THE DISTRIBUTED CACHE MECHANISM IN DSS

A. System structure

Supported by the national natural science foundation of China under grant No. 60403036 and 60573114, the science project of Shandong Economic University under grant No. 01610786.

K. Elleithy (ed.), *Advances and Innovations in Systems, Computing Sciences and Software Engineering*, 281–284.

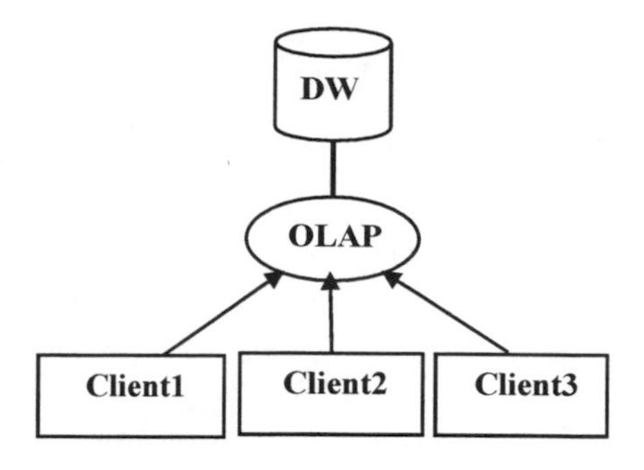

Fig1:General DSS Structure

Fig1 is the structure of many DSS, in this way the entire cache exists in OLAP Server. This kind of concentrated cache mechanism causes OLAP Server overload, and increases the inquiry response-time. At the same time, on account of the data increment in DW and the inquiry frequency augment in the client side, the cache space cannot satisfy the actual needs. Also there are some structures in which the caches exist in the client side, while the client not corresponding mutually, these structures cannot realize the share of the client resource, on the contrary, they cause the resource waste and the inquiry response-time increasing. Fig2 brings forth a new kind of DSS cache structure this article proposed.

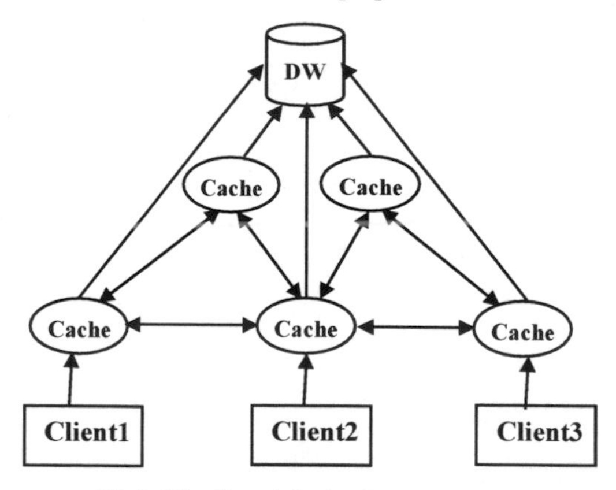

Fig2: Distributed Cache Structure in DSS

Fig2 includes three layers named DW layer, OLAP cache servers layer(namely the data is placed in OLAP server), and client side layer respectively. In Fig2 each cache delegates one OLAP cache server, which is called a node, Cache1, Cache2 and Cache3 are called local OLAP cache server, while Cache4 and Cache5 are called long-distance OLAP cache server. The method that increases the number of the cache server can reduce every cache's load, and the communication between the caches can raise the using efficiency of the resource. In actual facts, the number of DW maybe more than one, then we only need to extend the DW layer.

B. System's working flow

While Client1(For example in Fig2) sends inquiry request names 'Query', it obtains the result through caches and views saved in DW. This working flow includes user request, buffer examination, cooperate works and service response. Describing workflow as follows:

(1)Query is decomposed several sub-inquiry Q_i which has same granularity according to split algorithm in [3] (which is omitted here), supposing Q_{All} as set of sub-inquiries.

(2)Client1 first inspects the local Cache1, supposing Q_{local} as the sub-inquiry set which can be replied in the local buffer, and Q_{miss} as the sub inquiry set that cannot.

(3)Cache1 sends inquiry request to all nodes of next level, if neighbor Cache *i* can reply some sub-inquiry or its set, Cache *i* gives the answer instead of sending request to its neighbor nodes. If Cache *i* cannot, then it continues to send request to its neighbor until the progressive layer required by algorithm.

(4)Cache1 accepts the neighbors' response in period of appointed time. Supposing Q_{cache} as the sub-inquiry set that can be replied in the neighbor buffer. If the view in buffer of neighbor Cache i can reply the sub-inquiry, but these view have been signed cleared or it's data hasn't been the newest, then Cache i reverts this kind of sub-inquiry set to Cache1, these cleared views are marked as $Q_{evicted}$. Supposing Q_{DW} as sub-inquiry set hasn't been replied: $Q_{DW} = Q_{miss} - (Q_{cache} - Q_{evicted})$. Cache1 obtains the answer through the data warehouse.

(5)Cache1 synthesizes the answer of these sub-inquiries and returns to the user. At the same time Q_{All} is delivered to the buffer control module to carry out the admission & replacement algorithm.

In step 3, maybe the of computing inquiry result of Cache *i* is not the lowest, but the method continues to search other neighbors can increase the network cost, in usual time the network cost may be much bigger than the computation cost.

C. Admission & replacement algorithm

The buffer views in Cache is important to the inquiry response-time, so the algorithm is: The system first inspects the buffer whether it has had the corresponding view, if not, then it inspects that the buffer whether has the enough space to save the new view, if still not, then we save the new inquiry result to replace the content which has the smallest weight.

Algorithm:

if Q_{query} has been in the buffer,
then $W(Q_{query}):=G(Q_{query}, P)$;
else
{ // buffer Q_{query}
if the buffer has enough space, then
{ buffer Q_{query};
$W(Q_{query}):=G(Q_{query}, P)$;
}
else
{
Sort the buffer content of sub-inquiry by des according to the weight W;
victims:=Φ; next:=0;
if

$$FreeCacheSpace + \sum_{vi \in victims} size(vi) < size(Q_{query})$$

then
{//Supposing Q_{tag} is the set has the smallest weight, put Q_{tag} in the cleared queue $Evict(Q_i)$;
victims:=victims□$Evict(Q_i)$;
next++;
}

$$victims := \sum_{vi \in victims} W(vi)$$

if $W_{victims}$□$G(Q_{query})$,
then
{
Evict victims from cahce;
Insert Qquery;
$W(Q_{query}):=G(Q_{query}, P)$;
}
}
}
end if

In algorithm the weight is defined as:
$W(Q_{Query}) := G(Q_{query}, P)$ and

$$G(Q_{query}, P) = \frac{f(Q_{query}) * cost(Q_{query})}{size(Q_{query})}$$

this refers to the income value of view replies the query in node P, Q_{query} refers to the view replies the query, $f(Q_{query})$ expresses the frequency the buffer view replies the query, $cost(Q_{query})$ refers the memory view cost (including computation cost and network transmission cost), $size(Q_{query})$ expresses the space the view takes to reply the query.

III. SYSTEM PERFORMANCE ANALYSIS

A. Analysis basis

This article uses the Detailed Cost Saving Ratio(DCSR)[6] to express the cost of admission & replacement algorithm above:

$DCSR=(\Sigma_i wcost(q_i) - \Sigma_i cost(q_i))/\Sigma_i wcost(q_i)$

Here $wcost(q_i)$ refers to the whole cost to reply q_i in the worst situation (namely the system doesn't use buffer mechanism), $cost(q_i)$ refers to the actual cost to reply q_i in DSS. The bigger DCSR is, the lower the cost to reply q_i is, the better the system performance is.

B. Performance analysis

This article use an example of empirical datum, supposing the size of buffer space is equally. First we concentrate all buffer space in one server (dashed line), secondly we disperse the buffer space to many servers (solid line).

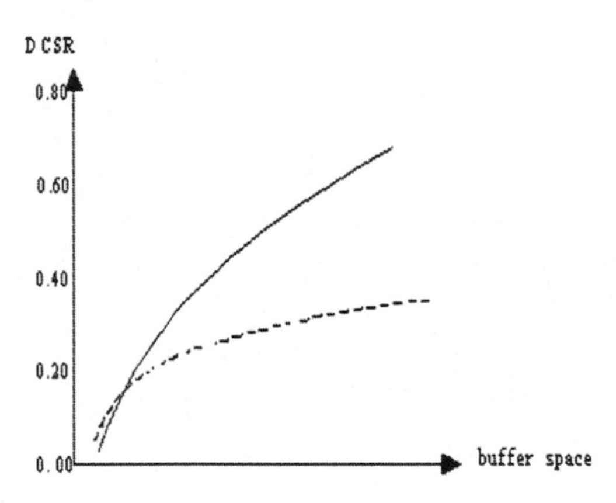

Fig3:Central buffer (dashed line) VS Distributed buffer (solid line)

In Fig3 the dashed line expresses the DCSR when the central buffer carries on the inquiry, correspondingly the solid line expresses the DCSR of the distributed buffer. We can see, when the buffer space is not very big, the central buffer's processing result is good somewhat, the reason is that the distributed buffer DSS includes the work of inquiry transmission, decomposition, final outcome synthesis and so on, which waste some network costs. But along with buffer space increasing, DCSR of the distributed buffer DSS begins to increase, obviously the cost is lower compared to central buffer DSS.

According to above analysis, distributed buffer DSS is super than some traditional central buffer system, this is because: The inquiry decomposition synthesis technology has decided whether the sub-inquiry is answered by DW or retrieved from Cache directly. The buffer mechanism makes the buffer storage cooperating each other, which removes the nonessential repetition.

IV. CONCLUSION AND FUTURE WORK

In order to enhance the performance of DSS and make the system to satisfy the demand of data quantity and inquiry complication, this article has proposed one kind of decision support system using distributed buffer mechanism. The system establishes many buffer managers to be able to complete an inquiry together, which benefits all users.

Research indicates the performance efficiency this system obtains is super than general DSS, but it also needs to be improved, for example the dynamic connection between buffers instead of static can make the system more flexible, how to assign the server space to buffer the data from different DW and so on. In the future work, we will continue to improve the system, and will apply this to the actual decision support system.

REFERENCES

[1]S.Santani, M.K.Mohania, V.Kumar, Y.Kambayashi: *Recent Advances and Research Problems in Data Warehousing*. ER Workshops 1998.

[2]P.Scheuermann,J.Shim,R.Vingralek: *WATCHMAN:A data warehouse intelligent cache manager.* In Proceedings of the International Conference on Very Large Database, 1996.

[3]Junho S.,Peter S.,Radek V.: *Dynamic Caching of Query Results for Decision Support Systems.* Proc. of the 11th International Conference on Scientific and Statistical Database Management, July 1999, IEEE Computer Society.

[4]Kalnis P.,Papadias D.: *Proxy-Server Architectures for OLAP*. Proc. of ACM-SIGMOD,2001.

[5]Deshpande P.,Ramasamy K.,Shukla A.,Naughton J.F.: *Caching Multidimensional Queries Using Chunks.* Proc.ACM-SIGMOD, 1998.

[6] Kotidis Y.,Roussopoulos N.,Dynamar: *A Dynamic View Management System for Data Warehouses.* Proc. ACM-SIGMOD, 1999.

AUTHOR DETAIL

LIU Hui was born in May, 1978. She is now an instructor in Department of Computer Science and Technology, Shandong Economic University, China. She received her B.S. and M.E. degree in Computer Science from Shandong University, in 2001, 2004 respectively. She is now a Ph.D. candidate in Department of Computer Science and Technology, Shandong University. Her research interests include information management, analysis and design of algorithm.

Ji Xiuhua was born in October, 1964, She received a BS and a ME in radio electronics from the Shandong University in 1985 and 1988, respectively. She is an associate professor of the computer science department at the Shandong Economic University. She is currently a Ph.D. candidate in computer science and technology. Her research interests include computer graphics and image processing.

Research on Grid-based and Problem-oriented Open Decision Support System

Liu Xia [1, 2]
[1] Department of Control Science and Engineering, Huazhong University of Science and Technology, Wuhan, 430074
[2] School of Business of Zhengzhou University, Zhengzhou, 450001
lx@zzu.edu.cn

Xueguang Chen [1]
[1] Department of Control Science and Engineering, Huazhong University of Science and Technology, Wuhan, 430074
xgchen9@mail.hust.edu.cn

Zhiwu Wang [1,3]
[1] Department of Control Science and Engineering, Huazhong University of Science and Technology, Wuhan, 430074
[3] Zhengzhou Economic Management Institute, Zhengzhou, 450052
zhiwuwang@mail.hust.edu.cn

Qiaoyun Ma[4]
4 Information & Management Science College of Henan Agricultural University, Zhengzhou,450002
maqiaoyun@163.com

***Abstract* - The characteristics of grid technology is suitable for constructing the platform of DSS, and can solve the problems of the development of DSS in distributed and dynamic decision-making environments. Open Grid Service Architecture (OGSA) is a new type grid architecture, which can support service creating, keeping and applying. With the inspiration of virtual enterprise building idea, this paper puts forward the architecture model of Grid-based and Problem-oriented Open Decision Support System (GPODSS) and discusses its operational process.**

I. INTRODUCTION

With the development of the society and technology, decision-making and decision environment are getting more complex, the needed resources for decision-making are distributed widely. Decision Support System (DSS) is facing great challenges practically and theoretically. At present, the deficiencies of DSS are as follows:

1) The structure of DSS can't adapt the development of distributed decision;

2) DSS can't satisfy the requirements of dynamical change of decision problems;

3) Lack of efficient mechanisms of trading and market-oriented decision resource sharing;

4) The complexity of system makes it difficult to provide transparent DSS services.

Grid technology can integrate various distributed resources such as computers, database and storage facilities, provide a high-efficiency calculation and service environment [1], which is widely distributed and transparent to users. Foster believes that, the nature of the grid is the resource-sharing in virtual organization [2]. With the inspiration of virtual enterprise building idea, Grid technology can bring new idea on how to solve those problems on DSS.

DSS begins with solving specific problem, and then turns it into a structural problem by various means [3]. Therefore, we can see that DSS has the characteristic of problem directing, which makes DSS relies on specific problems. But in reality, decision problem varies from each other, and this calls for the change of DSS to meet the requirements.

DSS is problem0oriented, and requires its proper response to different problems. Virtual organization, as a loosely-coupled system, is market-oriented, can response to the changes rapidly. Therefore, the idea of virtual enterprise building inspires us to build a new type DSS system, which can adapt to the changing market. OGSA is the newest grid architecture [4], which defines the "Grid Service" by service focusing. Besides its flexible, simple storing structure, quick formation, it fits for the demand of building an adoptable, quick and dynamic DSS. This article provides a way to create DSS through the creation of virtual organization [5] on the grid circumstance of OGSA, and study the problem-oriented virtual decision support organization under the grid environment.

II. THE MODEL OF GPODSS

A. Grid-based and Problem-oriented Open Decision Support System (GPODSS)

Grid-based and Problem oriented Open Decision Support System (GPODSS) is a new type of DSS based on grid platform, which makes full use of the good characteristics of grid technology (computational grid, data grid, information grid, knowledge grid, service grid, ect). To make it adapt the open, dynamic and heterogeneous Internet environment and have a quick response to a decision under the grid environment, it draws lessons from the idea of virtual enterprise building, makes structural and operational improvement of the traditional DSS.

B. The architecture of GPODSS

GPODSS is based on OGSA, which is guided by problems, driven by tasks and supported by negotiation. Based on OGSA, this article puts forward a model of GPODSS which creates a system focusing on service, provides decision resource

K. Elleithy (ed.), Advances and Innovations in Systems, Computing Sciences and Software Engineering, 285–288.

management and decision service. The core of this model is decision service electronic market, [6]. In this framework, while drawing lessons from "service oriented architecture" [7], it regards the decision service as the basic unit to organize and manage various decision resources, and it also provides all kinds of efficient decision service.

Model structure is showed as figure 1.

It is consisted of four parts: virtual decision organization, a decision support agent set, decision service electronic market environment and distributed resource, among which decision service electronic market is the core of the model.

1. Virtual Decision Organization (VDO)

VDO which contains decision makers (decision service demander, final decision makers) and decision participants (decision supporting service supplies) is the main part to complete decision-making tasks. The aim of decision support organization is to solve a specific problem (can be seen as a virtual organization in a grid). Also it is the joint between a decision maker and GPODSS.

2. The set of decision support agent

Decision support agent set, an assembly of decision-making agents, is the agency of virtual organization and decision service electronic market. Its main function is to provide problem description, problem analysis, and decision-making service organization formation, according to the decision maker. It is also the central part on providing decision support to virtual decision organization in GPODSS. It can realize the description and solution of a problem in the traditional DSS, but also realize

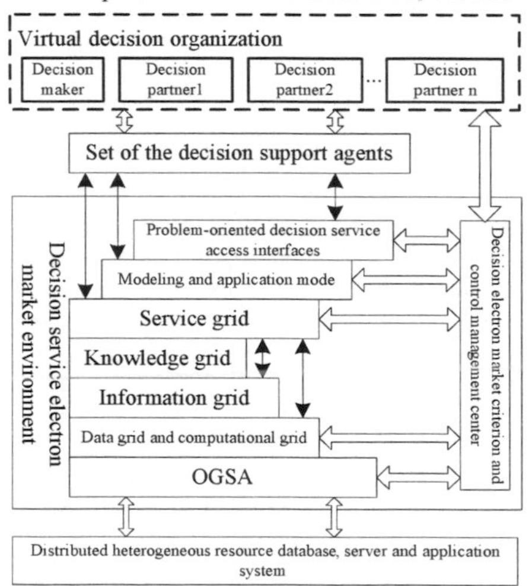

Fig. 1 GPODSS architecture model

the formation of virtual decision organization, the decomposition of a problem and the distribution of decision tasks.

3. Decision Service Electronic Market

Based on OGSA, GPODSS is the core functional unit of decision service electronic market. OGSA defines the Grid service, and transforms the decision services in GPODSS into grid services.

Data grid and computational grid technology support data access and model calculating in the GPODSS decision making process [8]. Information grid processes the results and data that comes from computational grid and data grid, and forms homogeneous decision supporting information, then provide that information to knowledge grid or service grid, knowledge grid further provides high quality, close-to-answer information to decision task using "knowledge mining". Service grid is the key part and main functional unit in the platform of decision service electronic market. It not only provides all kinds of service supported by service market, also provides decision service for decision main role (decision maker and decision participant) on how to deal with decision problem.

Modeling and application template output decision model and applied mode in accordance with different requests from the decision service agents and problem oriented decision visiting joint, giving proper decision service to them.

Problem oriented decision service interface, the important functional unit of decision service electronic market platform, is the key part to realize its universal, interaction, adaptability.

Decision electronic market normalization and control management center is the central controlling unit of decision service electronic market environment. It can make the various subsystem in current OGSA platform satisfied the uniform standard, and easy to control these subsystems. In this way, it is beneficial for co-communication, interconnection and co-operation in GPODSS.

III. The Process of Decision Support Operation in GPODSS

The operational process of GPODSS begins with the decision problem put forwarded by decision makers, ends with the solution of the problem (make a decision). According to the GPODSS architecture that has been given before, here is the process of decision supporting operation (showed as figure 2).

The decision supporting process of GPODSS is described in the form of virtual decision organization. Its basic idea is: to make a quick and correct decision for a specific decision problem, using the pattern of virtual decision supporting organization and various kinds of service provided by grid. It integrates the various decision resources and decision supporting ability of various places dynamically, just like the construction and operation of virtual organization. Then it evaluates and synthesizes the decision results, and fulfils the decision task quickly and effectively.

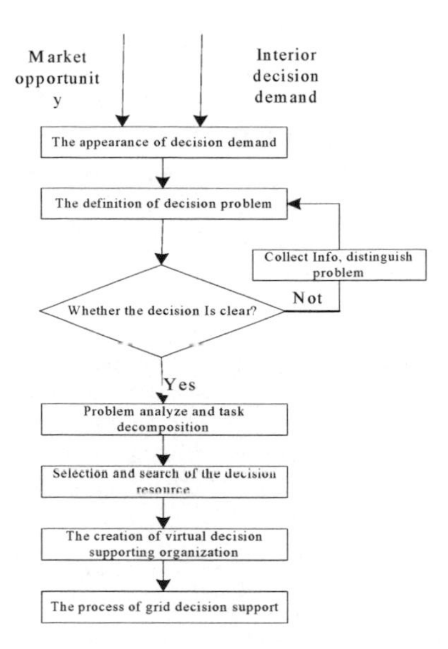

Fig. 2 The decision support processed for virtual decision organization

A. The appearance of decision-making demands

The demands often come from the market and the interior of the organization. These problems are always dynamically appearing in the system stratagem and development, differing from the routine problem in the operation layer.

B. The definition of the decision-making problem

The definition of the decision-making problem is to clarify the object and the nature of the problem, to set the goal and to analyze the restriction etc. Once the problem is clearly defined, the operation of system will go to the next phase quickly. Otherwise, the decision makers and technicians are required to collect more information, distinguish it till they have a full comprehension. The definition of the decision-making problem is realized by the interaction among decision supporting agent module, decision makers and decision service electronic market.

C. The analysis of the decision and task decomposing

After clarifying the definition of a decision-making problem, we new concern is to see whether the decision makers have the ability to solve the problem (needed decision resource and time span included). If they can't solve it, they will send the task to the capable decision making role in the decision service electron market for help.

A strategic and development concerned decision-making is always of complexity. For this kind of problem, the decision maker should adopt the divide and conquer strategy, decomposing the complex problem into independent sub-problems and solving them respectively. Here, we call these independent sub-problems as decision tasks. The decomposition of a decision-making problem is realized by the interactions between decision makers and decision service agents.

D. Select decision partners and asking for decision service

1. Select decision partners

Take a to-be-solved problem as a system, through the analysis and disintegration of the problem, the sub-problems or sub-modules are transformed into comparatively independent systems. After this, the decision service agents put them into the decision service electronic market of GPODSS to select the capable decision-making partners.

1) Decision makers (decision-making needed) need decision-making partners, and submit the requests to the decision supporting agents of GPODSS.

2) The decision supporting agents make the requests into bids and register them on UDDI (Universal Description, Discovery and Integration) center in the decision service electronic market platform.

3) Decision service suppliers check the requests through the UDDI center and bids in the electronic market.

4) The decision maker selects the proper decision partners via agent by partnership evaluation and selection strategy, and then negotiates and makes trade with them.

5) After the trade is made, the decision partners will accept the task and participate in the decision support process.

The whole process forms a task-driven, virtual decision supporting organization, directed by solving the decision-making problem for the decision makers and decision partners.

2. Asking for decision service

The decision supporting process is realized through virtual decision organization. Decision service for decision makers and decision partners is to search for the solution of the problem via the grid service electronic market and decision organization, under the grid environment of OGSA.

Decision service is a grid service which has clear decision supporting function. It is also the basic unit of organizing and managing decision resources. It belongs to problem directed, task-driven. It is temporarily created for the task and can be regarded as an encapsulation consisted of two parts:

1) All the resources such as data, information, model, and knowledge of computing task.

2) All the software, hardware and protocol needed for accomplishing data process, model calculation, knowledge reasoning and the interaction between the decision makers and decision partners.

Asking for a decision service is similar to the process of looking for a decision partner by a decision supporting agent in the grid decision service electronic market.

E. Distributed decision support under grid environment

Distributed decision support under grid environment is: under the coordination and control of the decision makers, decision partners quickly response to the decision-making process via agents and platform of GPODSS.

1. Cooperate with modeling

The foundation of decision supporting service is to build a proper decision model for sub-problems. As required by the decision-making and the decision makers, the decision partners who are specialized in the specific fields create a systematical and applicable operation model. The process is realized by the service provided by decision service electronic market, and interaction between the decision service agents and the decision market.

2. Decision management process

Decision management is the activity of the decision makers who have to schedule and harmonize the whole decision process under grid environment. It includes:

1) Process coordination: orderly dispatch the solving process of the timely restricted sub-problems; provide management method to ensure the solving of dispatched sub problems.

2) Resource management: As requested by the decision partners, building an interior resource sharing service on GPODSS for the sub-problems (data, model, knowledge, and calculate resources) can make the system operation effectively, especially in the situation of resources conflict.

F. Decision making synthesis, evaluation and accomplishment

As the solutions of sub-problems are completed by different decision partners, the decision makers should evaluate and synthesize them, and then provide a final decision.

Decision making synthesis is a dynamic process. The process of support provided by decision partners is the process of integrations with decision makers under grid environment. The decision makers have to constantly response to the results given by the decision partners. Decision makers rely on their past experience or personal preference to evaluate whether to accept or to modify them. If they don't satisfy with the results, decision-making partners must rebuilt or make amendment on the model. So synthesis, evaluation and decision support is a dynamic and perfecting process.

When the decision maker is satisfied with the results, the whole decision-making period will be over. Virtual decision organization is disjointed.

IV. CONCLUSION

GPODSS puts the idea of grid technology and virtual enterprises into the process and the creation of decision support system. It takes full consideration of the problem-oriented DSS. To adapt the DSS and to response to decision problem quickly, it takes advantage of OGSA framework and service-oriented, and provide decision support via decision service electronic market to decision maker. As it absorbs the flexible, dynamic swift characteristic of virtual enterprise, combined with capable exterior decision body, it can realize the sharing in the virtual organization and accomplish the decision support process.

ACKNOWLEDGMENTS

Great thanks for the support from SRFDP Grant # 20040487076.

REFERENCES

[1] I. Foster, C. Kesselman. The Grid 2: Blueprint for a New Computing Infrastructure. CA: Morgan Kaufmann, San Fransisco, 2003.

[2] I. Foster, C. Kesselman, S. Tuecke. The Anatomy of the Grid: Enabling Scalable virtual Organizations. International Journal supercomputer applications .2001, vol. 15, no. (3), pp: 200-222.

[3] X. G. Chen, J. P. Wang, J. Hu, Q. Fei. A Discussion on the Leading Difference between EIS and Traditional DSS. Journal of Huazhong University of Science and technology. Wuhan, China., vol. 26, no.9, pp: 24-26, September 1998

[4] I. Foster, C. Kesselman, J. Nick, S. Tuecke. The Physiology of the Grid: An Open Grid Services Architecture for Distributed Systems Integration. http://www.globuse.org/research/paper/ogsa.pdf, June 2002. [EB/OL]

[5] M. Weng, J. B. Liang, D. F. Su. Creation of the Grid Virtual Organization Based on OGSA. Journal of Computer Engineering & Science, Changsha, China, vol. 28, no. 2, pp:140-142, 2006.

[6] J. Y. Chi, X. G. Chen. A Model of Grid Based Decision Support System. Computer Science, vol. 33, no. 3, pp. 121-124, 2004.

[7] X. L. Chai, Y. Q. Liang. Web Services of Technology, Framework and Application. CA: Publishing House of Electron Industry, pp: 12-13 January 2003.

[8] G. W. Yang. Three-layer Framework of Grid: Data, Information and Knowledge. Computer education, Beijing, China, pp: 27-28 ,July 2004.

Development and Analysis of Defect Tolerant Bipartite Mapping Techniques for Programmable Cross-points in Nanofabric Architecture

Mandar Vijay Joshi and Waleed Al-Assadi

Department of Electrical & Computer Engineering,
University of Missouri-Rolla, Rolla, Missouri- 65401
Email: {mvjvx8, waleed} @umr.edu

Abstract

Chemically Assembled Electronic Nanotechnology (CAEN) using bottom-up approach for digital circuit design has imposed new dimensions for miniaturization of electronic devices. Crossbar structures or Nanofabrics using silicon nanowires and carbon nanotubes are the proposed building blocks for CAEN, sizing less than 20 nm, allowing at least 10^{10} gates/cm^2. Along with the decrease in size, defect rates in the above architectures increase rapidly, demanding for an entirely different paradigm for increasing yields, viz. *greater defect tolerance*, because the defect rates can be as high as 13% or more. In this paper, we propose a non-probabilistic approach for defect tolerance and evaluate it in terms of its coverage for different sizes of fabric and different defect rates.

1. Introduction

Moore's Law, governed by advances in Photolithography techniques, is expected to be revised in coming seven to twelve years [1, 2], because of limits enforced by molecular sizes.
Silicon Nanowires and Carbon Nanotubes exhibit a reasonable promise as a technology substitute to CMOS VLSI. Research by Phaedon Avouris, IBM Research Division [3] demonstrates that these novel structures have excellent operating characteristics as compared to lithographic scale silicon devices, and therefore can be used to construct Nano-diodes and FETs. Recent developments have shown how to build Programmable Logic Arrays (also called the Nano-PLA's) using bottom up synthesis, using Silicon Nanowires [4].

A system involving two planes, one of CNT's or Si Nanowires and the other of p-type Si Nanowires, can be made to behave as a programmable-Crosspoint array [5]. Such an assembly is called "Nanofabric" (Goldstein, et. al.) or Crossbar (Andre et. al.) [3, 6]. Every Crosspoint can be individually addressed and programmed . Since these devices show *"hysteresis"* in behavior, in response to the voltage differential applied, they are programmed at a higher voltage, and read at a lower voltage differential. A programming technique for Nano-PLA is proposed by Researchers in Caltech. These architectures exhibit ultra high density with respect to the size of digital circuits that can be designed using them.

The increase in density is accompanied by substantial increase in the defects, introducing a new challenge, i.e. the ability to tolerate defects to as to compensate for yield and cost [8]. The Cross points in the Nanofabric are the programmable structures, with which logic elements can be implemented. These cross points can have defects due to following reasons [1, 2];

Breaks in Nanowires: It has been observed that the breaks in Nanowires increase with increase in length. There may be some breaks during the fabrication of nanowires, on account of limitation of the fabrication techniques, and axial stress. Therefore, their lengths should, nominally, not exceed 10s of microns.
Crosspoint defects: These defects are characterized by inability of a crosspoint to be programmed "closed" or programmed "open". The defect that a crosspoint cannot be programmed "open" is extremely unlikely [1], and therefore we neglect its occurrence in this paper.

K. Elleithy (ed.), Advances and Innovations in Systems, Computing Sciences and Software Engineering, 289–294.

The distribution of the above defects is statistical in nature, and it demands for precise modeling of defects in order to establish a programming methodology that enables the designer to tolerate defects and optimize the available programmable resources. Such a methodology will, therefore, differ for different defect rates. In this work, we propose a general mathematical defect model for a Nanofabric, and use the model to evaluate two different approaches to improve defect tolerance, mainly in terms of yield. We focus on nanofabric with higher defect rates than 10%, assumed in the previous methodologies and algorithms proposed.

Section 2 gives a brief introduction to the previous research carried out in terms of fabrication of Nano-scale wires, defects in NWs, defect mapping techniques and defect tolerating algorithms. It is followed by mathematical modeling of the defective nanofabric and formulation of defect tolerance methodology discussed in section 3. Section 4 proposes the new approach to improve yield and time complexity. This approach is quantitatively evaluated in the sections to follow.

2. Related Work

After the discovery of Carbon Nanotubes (CNT's) in 1991, a significant amount of research has been carried out to study their electronic properties, and check feasibility as a technology substitute for lithographic CMOS VLSI. Carbon Nanotube FET's were fabricated at IBM Research Division, New York [3].
Synthesis of Silicon Nanowires was carried out at Department of Chemistry, Harvard University. Architecture of NanoPLA was proposed by Caltech, and a suitable addressing technique, called "Stochastic Addressing" was developed for Nano PLA's and memories using Nanofabric architectures [7, 9].

Probabilistic Design Paradigm used by Margarida Jacome et. al. in proposes architectures suitable for a Nanofabric in the form of design flows that are the behavioral primitives of the target component[10, 11]. By introducing redundancy in the structural equivalents of the same, a desired accuracy in mapping can be obtained, i.e. false negatives can be minimized. This can also be used as a technique to obtain the defect map.
Development of Teramac Computer by Culbertson et. al uses the concept of introduction of redundancy in terms of using more FPGA's than actually needed [10]. This is helpful in achieving the desired performance and avoiding faulty sites in the FPGA's that have a large number of defects. This reduces cost by a great extent.
Greedy Heuristic Algorithms proposed by DeHon use probability of a match between the rows of Matrix of Boolean functions and the Defect Matrix, as a tool for mapping. A sorting methodology is followed that enables the rows with minimum probability of a match, to find the maximum candidates for a match from the opposite matrix. Since the probability of finding a match largely depends on the number of ON inputs in a given Boolean function, sorting process just has to sort the elements in decreasing order of the number of ON inputs in each element [1, 2]. It can be summarized that we need some mechanism to address the issue of mapping between the two matrices in order to tolerate the defect rate, increase yield thereby reducing the cost. This can be done by enforcing certain defect tolerating algorithms, and also by introducing redundancy with respect to programmable resources, The former approach is dealt with in [1, 2] and it is seen that the problem comes down to a classical example of a *"Bipartite Mapping Problem"*.

3. Formulation of the Problem

The following figure illustrates an example of the function matrix "F" and the defect Matrix "W". The solid circles in W represent the defective crosspoint. Therefore, while mapping, care has to be taken that any of the ON inputs should NOT coincide with the defect site.
If a i th row in matrix F (considered as one Boolean function) can be mapped successfully at jth row in Matrix W, it is said that F(i) has an edge with W(j). The algorithm demands for maximum number of actual matches from the given edges.

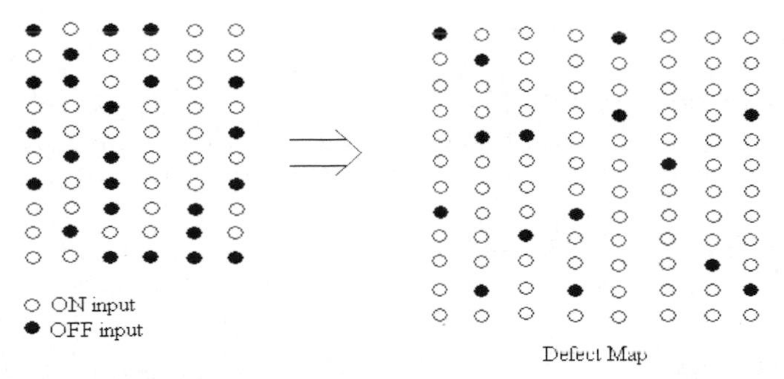

Figure 1: Pictorial Illustration of Matrix F and Matrix W respectively. (All the small circles represent Crosspoints).

3.1 Mathematical Representation of Edge detection [1]:

Using the analysis in [1], we have:

If f_i represents i^{th} row in the matrix F,

$f_i = (I_{i,0}, I_{i,1}, ..., I_{i,N-1})$,
Where:
$I_{i,j} = \{$
0 if input j is an OFF-input of f_i
1 if input j is an ON-input of f_i
}
Similarly the defect configuration of each nanowire Wi can be defined as below:

$W_i = (J_{i,0}, J_{i,1}, ..., J_{i,N-1})$, where:
$J_{i,j} =$
0 if crosspoint corresponding to input j is non-programmable
1 if crosspoint corresponding to input j is programmable

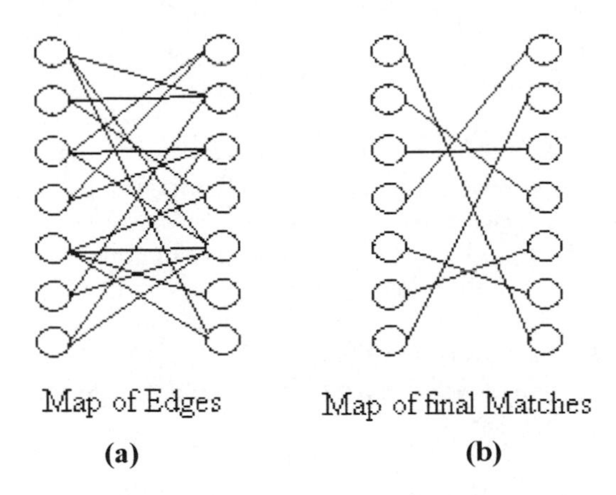

Figure 2: An illustration of ideal bipartite mapping.

Now we define a bipartite graph G (F, W, E), where F and W are the set of nodes as defined above. For every OR function f_i in F and nanowire W_j in W, (f_i, w_j) Є E if and only if:

$$\forall_{(1 \le k \le N)} (Ii,k \le Jj,k)$$

3.2 Mathematical Representation of Match Detection based on [1]:

In a bipartite graph G (V1, V2, E), the set M Є E is a matching from V1 to V2 if and only if the following conditions hold:
There is exactly one v Є V2, such that. (u, v) Є M, and There is at most one u Є V1, such that (u, v) Є M.

The probability of a Boolean function of having an edge with an element in W decreases with increase in the number of ON inputs. Therefore, more is the number of on inputs; less is the number of edges it finds.

E.g. If n_i represents the number of ON inputs in i^{th} row, the expected number of edges the row will find is given by:

$$E(D(f_i)) = \{W\} \cdot P^{ni} \ldots\ldots (I)$$

Where, {W} represents the total number of rows (elements) in matrix W,
E represents expectation of "D" that's edge density, and
P represents the probability of occurrence of an ON input. [1]
Throughout, it has been assumed that the defect rate in the Nanofabric does not exceed 10%, and therefore, the mapping process is primarily governed by the number of ON inputs in an element in F.
But, as the defect rate goes on increasing, the impact of result (I) becomes weaker, as explained in the following section.

Let the defect rate be'd'.
Let the number of columns in the matrices be "r".
Therefore, the expected number of defective cross points in each row in matrix W is,

K= r.d

4 Proposal of New Approach

The level of defect rates assumed in [1, 2] are less than or equal to 10%, and it is observed that the probabilistic sorting gives a sufficiently high coverage, i.e. number of matches.

It is known that the defect rates can be much higher than considered above, for which the Greedy Heuristic Algorithms are not tested. We, therefore propose two approaches to improve coverage rate at higher defect rates (up to 20%)

4.1 Double Variable Redundancy

We propose to establish a new defect-mapping scheme targeting a nanofabric with a greater defect rate. The algorithms discussed above, assume that there exists a single vertical nanowire per input variable in the Nano-PLA, and therefore, occurrence of a defect at the site of an ON input results in skipping of the whole

Evaluation of combinational logic outputs is performed by using AND-OR, OR-AND, NAND-NAND or NOR-NOR methods. If we consider AND-OR implementation, each single product is called "Minterm", and the output is the logical sum of several Minterms. A PLA has many outputs, and therefore it needs an "AND-Array". To evaluate individual Minterms, it needs "OR array". A single vertical Nanowire can be dedicated for a single minterm, or a single product. The presence of even a single defect at a Crosspoint to be programmed on such a wire would make the entire Minterm faulty, and therefore it would result in an error at the output. Although occurrence of defects is less than 15%, the overall probability of error becomes very high, with increase in the number of Crosspoints to be configured. We, therefore dedicate TWO wires per Minterm and TWO wires per product term. It follows that we have a set of four Crosspoints, any of which being programmable, makes the PLA work. We call this set as a Programmable "Quad" in the further discussion. A defective quad is therefore the one that has ALL FOUR Crosspoints defective. The redundancy now established is called as *"Double Variable Redundancy" (DVR)*.

On account of the redundant Crosspoints available in the PLA, the reliability of every Quad is increased by a great extent. Figure3 and Figure 4 illustrate the DVR principle.

4.2 Time Complexity

In this section we illustrate and compare the time complexity involved in the programming of PLA's based on DVR, and the time complexity that exists in the programming technique using Greedy Heuristic Algorithm.

Summarizing the results in [1], we obtain the time complexity for Greedy Heuristic Algorithm as follows:

$$T_c = O(|F| \log(|F|))\, c + O(|F| \cdot P_J^{-cm} \cdot cm)$$

Where,
T_c= Time complexity

|F| = size of the array of Boolean functions
cm= maximum number of Crosspoints to be Programmed in a minterm
P_J= Probability that the given junction is Programmable.

Now,

The time complexity involved in the configuration of a DVR based PLA is given as:

T_c= Total number of Crosspoints to bc Programmed

Therefore,

$T_c = O(r*c*P_{on})$

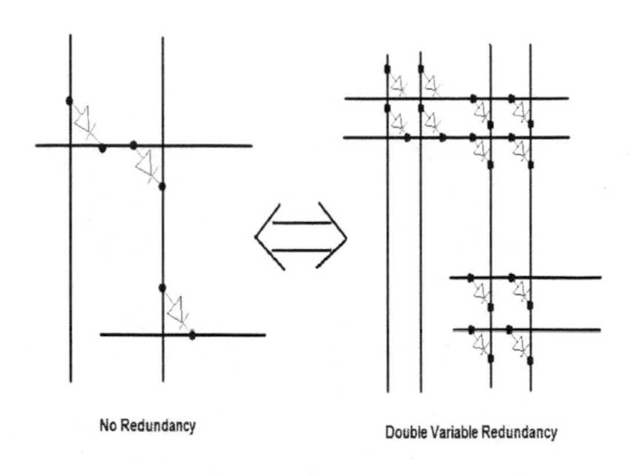

Figure 3: Illustration of DVR

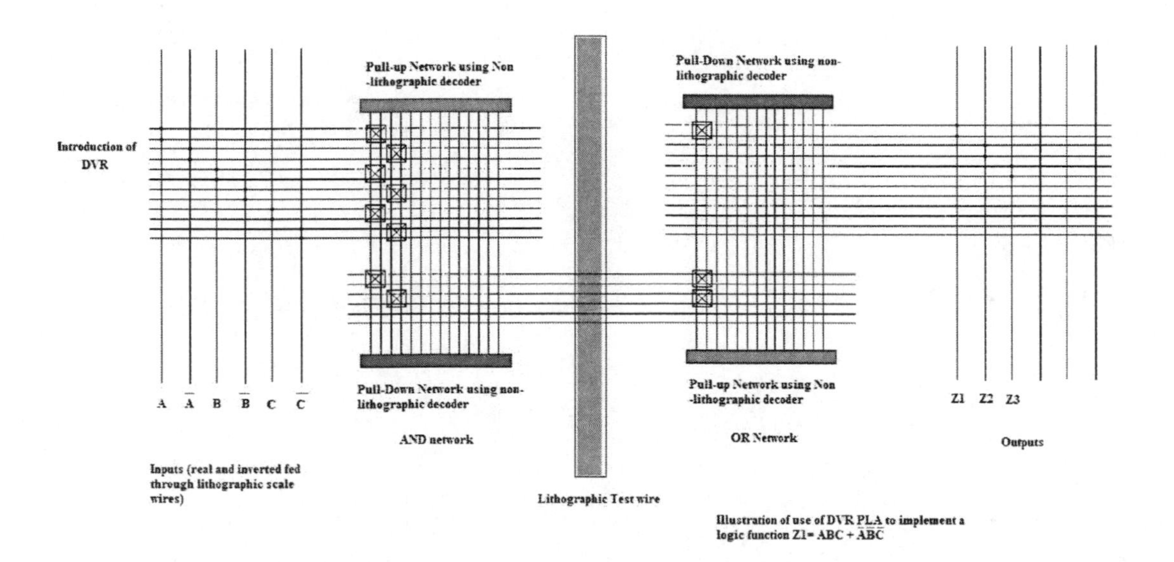

Figure 4: ***The DVR based proposed PLA***

5.1Results:

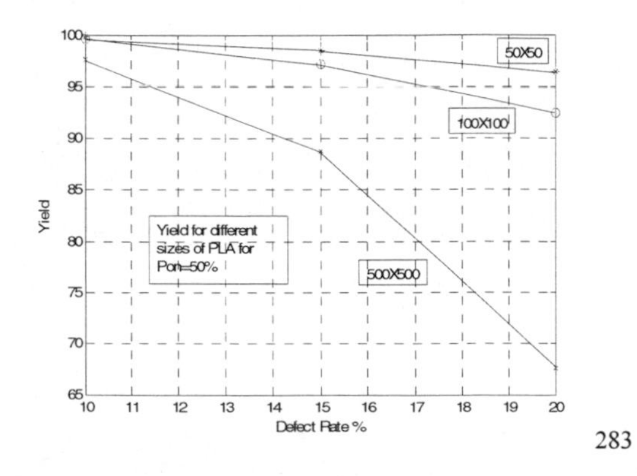

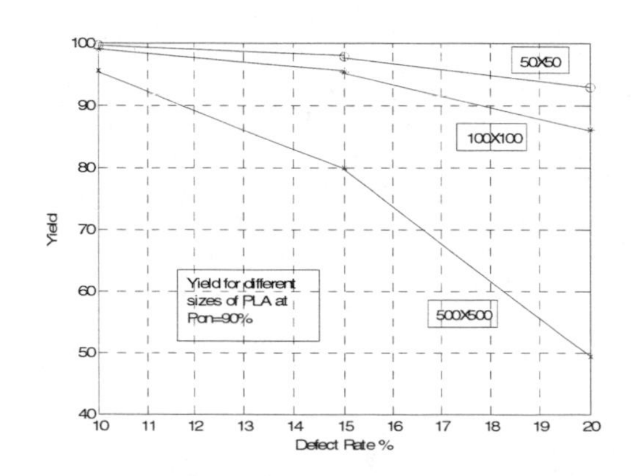

(a)

(b)

Figure 5: ***The relation between Yield and Defect Rates for different PLA sizes. It also establishes the relation between Yield and the Probability Pon. For (a) Pon= 50%, (b) Pon=70%***
(c) Comparison between time complexities of DVR and Greedy Heuristic Approach.

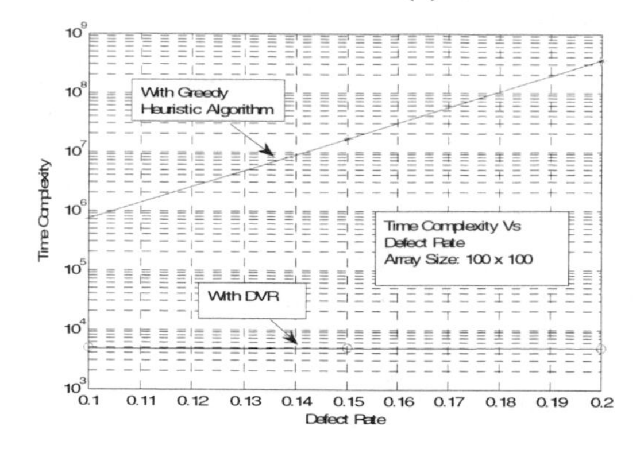

(c)

6. Conclusions

The DVR approach gives excellent improvement with respect to number of mappings and time complexity. It is less area efficient, as the Redundancy introduced demands for quadruple the number of cross-points.

The development of the DVR approach may demand for change in the PLA architecture in order to enable the introduction of an input variable via different, but predefined, nanowires.

References

[1] H. Naeimi, *A. DeHon, "A Greedy Algorithm for Tolerating Defective Cross points in Nano PLA Design",* M.S. Thesis, Calif. Inst. Of Technology, 2005

[2] S. Zhang, *"Cost-driven Optimization of Repair Strategies for Tolerating Defective Cross Points in Nano fabric"* M.S. Thesis, University of Missouri-Rolla, 2006

[3] P. Avouris, *"Molecular Electronics with Carbon Nanotubes"*, Accounts of Chemical Research, Vol. 35, Nov.12, 2002.

[4] A. De Hon *"Nanowire Based Sub lithographic Programmable Logic Arrays"*, ACM 1-58113-829-6/04/0002, 2002

[5] A. De-Hon, *"Nanowire-Based Programmable Architectures"*. ACM Journal on Emerging Technologies in Computing Systems, Vol. 1, No. 2, July 2005.

[6] M. Mishra and S. C. Goldstein *"Defect Tolerance at the end of the Roadmap"*. ITC , Paper 46.3 p.p.1201-1210, 2003

[7] A. De Hon *"Stochastic Assembly of Sub lithographic Nanoscale Interfaces",* IEEE transactions on Nanotechnology 2(3)-:165-174, 2003

[8] M. Jacome, C. He, G. de Veciana, S. Bijansky, *"Defect Tolerant Probabilistic Paradigm for Nanotechnologies"*,in Proc. Design Automation Conference, 2004.

[9] Yi Cui, J. Lauhon, M. S. Gudiksen, J. Wang, C. M. Lieber, *"Diameter controlled synthesis of single layer-crystal silicon nanowires",* American Institute of Physics, 10.1063/1.1363692, 2001.

[10] W. B. Culbertson, R. Amerson, R. J. Carter, P. Kuekes, G. Snider, *"Defect Tolerance on the Teramac Custom Computer"* , Proceedings of 1997 IEEE symposium, on FPGA's for Custom Computing Machines, April 16-18 1997.

[11] M. Jacome, C. He,G. de Veciana, S. Bijansky, *"Scalable Defect Mapping and configuration of memory based Nanofabrics"*, High-Level Design Validation and Test Workshop, 2005. IEEE 30 Nov.-2 Dec. 2005 p.p.11-18

Nash Equilibrium Approach to Dynamic Power Control in DS-CDMA Systems

J. Qasimi M and M. Tahernezhadi
Department of Electrical Engineering
Northern Illinois University, DeKalb, IL

ABSTRACT

This papers aims at the power control aspect of Resource allocation for wireless data via employing microeconomics concepts of utility and pricing in relation to the non-cooperative game theory and Nash Equilibrium. Specifically, an efficient algorithm based on stochastic gradient formulation is proposed to adaptively converge to arrive at optimal power set with higher utilities with the pricing factor is a parameter. Both single cell and multi-cell are presented in this paper. Comparative numerical and graphical results provided attest to practical usefulness of the proposed power control algorithm.

Index terms: Signal to Interference Ratio, Quality of Service, Frame Success Ratio, Code Division Multiple Access (CDMA), and Bit Error Rate.

1. Introduction

Resource allocation is concerned with relegation of scarce resources to individual users in a group. Most prominent scarce resources in wireless networks are bandwidth, power and capacity. Resource allocation in CDMA systems has taken center stage in recent years given tremendous growth in demand for wireless services. Among the various radio resources, power resource is significant due to its impact on interference and battery consumption. In every wireless network, a user would aim to achieve high reception quality as measured by signal to interference ratio (SIR), while trying to spend least amount of power to attain that quality. Therefore, achieving high SIR and low battery consumption (transmit power) is the goal of every wireless terminal. There exists a tradeoff between high SIR and low transmit power. Finding the efficient balance between these two commodities is the primary area of focus in the power control branch of the resource allocation. Game theoretic approaches related to microeconomics have been considered for problem formulation of power control in wireless networks, where quality of service (QoS) of a wireless user is considered as the utility of that user. The distributed power control is said to be a 'non-cooperative power control game' when each user tries to maximize its utility without cooperating with other users.

Herhold and Rave [1] tried to introduce relaying in CDMA networks for power control. Their simulation results demonstrated that relaying is effective only for limited loaded networks and low data rates. Vasilios Siris [2] proposed an algorithm for elastic resource allocation in CDMA networks considering continuous data rates as opposed to discrete data rates commonly used in CDMA systems utilizing discrete spreading factors.

Narayan and Goodman [3] proposed a novel algorithm for CDMA network power control by pricing to obtain Pareto improvement of the non-cooperative power control game. Pareto improvement refers to obtaining improvements in user utilities relative to the case with no pricing. Their proposed formulation resulted in increased individual utilities and reduced powers. Specifically, their formulation considers a centralized power control mechanism where the base station would inform the terminal to either increase or decrease its power level to obtain a fixed target SIR. As there is a price to be paid for an increase in transmit power, each terminal would save its power because of its associated price, leading to higher individual utilities. However, this algorithm enjoys high computational complexity since the base station has to undergo an exhaustive search to find the optimal power set which leads to maximization of the individual utilities. Moreover, the base station has no control over the pricing factor and would remain static for a given rate of congestion.

An improved pricing model is proposed in this paper to overcome the above-stated shortcomings. The pricing factor is cast in terms of the sum of all the transmitted powers as opposed to individual power as done in [3] . Owing to a proposed stochastic gradient-based formulation, this in turn results in faster convergence of the transmitter powers for a given pricing factor. Also, in this approach, since the base station is given control over the pricing factor, the base station can adjust the price for each cell according to its congestion statistics. As such, the base station can either decide to increase its revenue by taxing the terminals or ease them depending on the congestion. Improved utilities are attained for a single cell environment with multiple terminals. Subsequently, the proposed formulation for a single is extended to a multi-cell environment.

In summary, salient contributions in this paper relative to previous work are as follows:

- Pricing factor is made dynamic and relative to the congestion traffic.
- A simplified stochastic gradient based search algorithm to find optimal power set is proposed.
- Utilities higher than those from previous formulations are achieved.

This paper is organized as follows. Section 2 introduces the concepts of SIR, utility functions and QoS. Section 3 discusses the non-cooperative power control game. It is then followed by section 4 where the non-

K. Elleithy (ed.), Advances and Innovations in Systems, Computing Sciences and Software Engineering, 295–300.

cooperative power control game with the pricing concept is introduced. The formulations presented in Sections 2, 3 and 4 closely follow [6]. Section 5 presents the proposed stochastic gradient based search algorithm for a single cell network. Section 6 extends the single cell network formulation to multi-cell networks. Simulations and results are presented in section 7. Finally, Section 8 presents conclusions.

2. SIR and the Utility Function

In a multiple access wireless communication system, each user's transmission is a source of interference to other users. It is desirable for any user to achieve a high SIR while at the same time spending low power. SIR can be increased if the user expends more energy and spends more transmitter power. On the other hand, the SIR decreases if there is an increase in power of other users, thereby increasing the interference for that user [12].

The Utility function in microeconomics is usually referred to as the satisfaction of a user in a group of users. A utility function describes the preference relation between the elements of an individual's set of actions. It is defined as the number of information bits received successfully per joule of energy expended. An optimum power control algorithm for wireless voice systems maximizes the number of conversations that can simultaneously achieve a certain Quality of Service (QoS) objective with a minimum acceptable SIR. However, this approach is not appropriate for the efficient operation of a wireless data system [3, 4] given that QoS objective for data signals varies from the QoS objective for telephony signals. For example, a voice user is indifferent to the changes in the SIR so long as it remains above a certain target SIR leading to a utility of step-function characteristic.

In a data system, error-free communication has high priority. When a data system detects an error in transmission, the data in error has to be retransmitted. We consider a cellular system where each user transmits L information bits in frames (packets) of M > L bits at a rate R bits/second using p Watts of power. Let P_c denote the probability of correct reception of a frame at the receiver or frame success rate (FSR). FSR is a function of the terminal's SIR (γ) at the base station and depends on the system's properties such as modulation, radio propagation channel and receiver structure. If the number of transmissions necessary to receive a packet correctly is denoted by K, then the probability density function of the random variable K can be expressed as

$$\begin{aligned} P_k &= P_c(1-P_c)^{k-1} && \text{for } k = 1,2,\ldots,K \\ &= 0 && \text{otherwise} \end{aligned}$$

(1)

where the transmissions are assumed to be statistically independent. The expected value of K is $E[K] = 1/P_c$. The total transmission time required for successful reception is KM/R seconds and the total energy expended is pKM/R joules with expected value pM/RP_c. Therefore, in order to transmit an L bit packets successfully, a terminal expends pM/RP_c Joules of energy. Finally, the utility function is given as

$$u = \frac{L}{pM/RP_c} = \frac{LRP_c}{Mp} \text{ bits / Joule}$$

(2)

Assuming perfect error detection and no error correction, one has

$$P_c = (1-P_e)^M$$

(3)

where P_e is the bit error rate (BER). The BER decreases monotonically with SIR (γ). Consequently, P_c increases monotonically with SIR. The FSR can be approximated by an efficiency function that closely follows the behavior of the probability of correct reception while producing $P_c = 0$, when p = 0 or $(\gamma = 0)$. Accordingly, the efficiency function is defined by:

$$f(\gamma) = (1 - 2P_e(\gamma))^M \quad (4)$$

Replacing the FSR by the above efficiency function in the utility function to yield

$$u = \frac{LRf(\gamma)}{Mp} \text{ bits / Joule} \quad (5)$$

3. Non-cooperative Power Control Game

Consider a multi-user game with N users where N = {1, 2, 3... N}, and let the power vector **p**= $p_1, p_2 \ldots, p_N$ denote the outcome of the power set from the strategic game theory, where p_i denotes the i-th user's transmit power. The resulting utility level for the jth user is denoted by $u_j(\mathbf{p})$. The utility of user j can be expressed as

$$u_j(p_j, \mathbf{p}_{-j}) = \frac{LR}{Mp_j} f(\gamma_j) \text{ bits / Joule}$$

(6)

Where $\mathbf{p}_{-j}$ denotes the strategy space formed by all the terminal's strategy sets except the jth user. And, γ_j is the SIR of user j defined as,

$$\gamma_j = \frac{W}{R} \frac{h_j p_j}{\sum_{j \neq i} h_j p_j + \sigma^2}$$

(7)

where, W is the available spread spectrum bandwidth [Hz], R is the bit rate, σ^2 is the AWGN power at the receiver [Watt], and $\{h_j\}$ is the set of path gains from the mobile to the base station. We assume that the strategy space P_j of each user is a compact, convex set with minimum and maximum power constraints denoted by $\underline{p_j}$ and $\overline{p}_j$, respectively. For NPG, we let $\underline{p_j} = 0$ for all j which results in the strategy space $P_j = [0, \underline{p_j}]$.

Note that (6) demonstrates the strategic interdependence between users, where the utility of each user depends not

only on its own power level but also on the choice of other players' strategies.
In the power control game, each user maximizes its own utility in a distributed fashion. Formally, the non-cooperative power control game (NPG) is expressed as

$$(\textbf{NPG}) \qquad \max_{p_j} \; u_j(p_j, p_{-j}) \; for \; all \; j \in N$$
$$s.t. \quad p_j \in P_j, \quad for \; all \quad j \in N$$

(8)

Where u_j is given in (6) and $P_j = [0, \bar{p}_j]$ is the strategy space of user j. Since utility is a function of the transmit powers of all users in the system, there exists a strong interdependence between the course of action each user takes in order to maximize its utility. The question then is to predict an operating point for the system. Indeed, it is possible to characterize a set of powers where the users are satisfied with the utility they receive given the power selections of other users. The equal SIR equilibrium power vector can also be calculated using the SIR balancing algorithm discussed in [13]. We need to explore the existence of such a set of powers that will constitute equilibrium of power choices for the users.

In [5], it is clearly shown that the Nash Equilibrium is inefficient for the power control in CDMA and there exists a power vector lower than that obtained through Nash equilibrium [6]. The benefits of pricing and the efficiency obtained through pricing the terminals are also presented in [6].

4. Non-cooperative Power Control with Pricing

In order to improve the equilibrium utilities of NPG in the Pareto sense, we resort to usage-based pricing schemes. Through pricing, we can increase system performance by implicitly inducing cooperation and yet maintaining the non-cooperative nature of the resulting power control solution. Within the context of a resource allocation problem for a wireless system, the resource being shared is the radio environment and the resource usage is determined by terminals transmit power. If all the terminals agree to reduce their transmit powers at the NPG equilibrium by the same marginal amount, one will see increased utilities for all terminals. Therefore, we conclude efficiency in power control can be promoted by a usage-based pricing strategy where each user pays a price proportional to it's transmit power [10], [11]. The fixed point equation in [9] offeres a similar QoS as a function of pricing but with a significant delay. Consequently, we develop a non-cooperative game with pricing. Let $G_c = [N, \{P_j\}, \{u_j^c(\cdot)\}]$ denote an N-player non-cooperative power control game with pricing (NPGP). Utilities for NPGP are

$$u_j^c(\mathbf{p}) = u_j(\mathbf{p}) - c_j(p_j, \mathbf{p}_{-j}) \qquad (9)$$

Where $c_j : P \rightarrow R_+$ is the pricing function for terminal $j \in N$. The multi-objective optimization problem that NPGP solves can be expressed as,

$$\max_{p_j} u_j^c(p_j, \mathbf{p}_{-j}) = u_j(\mathbf{p}) - c_j(p_j, \mathbf{p}_{-j}), \; \forall \, j \in N$$
$$s.t. \; p_j \in P_j, \; for \; all \; j \in N \qquad (10)$$

The above formulation does not assume any particular form for the pricing function, $c_j(\cdot)$. However, as discussed in the previous sections, we impose a price that increases monotonically with the transmit power of the user. Particularly, we restrict our attention to pricing schemes of the form

$$c_j(p_j, \mathbf{p}_{-j}) = c\beta \, p_j \qquad (11)$$

Where c and β are positive scalars. The pricing factor, c, can be considered to have units bits/sec/Watt^2 so that it is consistent with the units of the net utility u_j^c in bits/Joule. The pricing factor, c, needs to be adapted such that user self-interest leads to best possible improvement in overall network performance. The NPGP with such adaptive pricing is as follows:

$$NPGP \quad \max_{p_j} u_j^c(\mathbf{p}) - c\beta \, p_j, \quad for \; all \; j \in N$$
$$s.t. \, p_j \in P_j, \; for \; all \; j \in N$$

(12)

Notice that the NPGP is practically the same game as the NPG with different payoff functions. We seek a Nash equilibrium point that solves NPGP, if one exists. In game $G = [N, \{P_j\}, \{u_j(\cdot)\}]$, each utility function is quasi-concave in its own strategy. We established that in a game with such utility functions there exists a unique equilibrium. The NPGP, however, does not have quasi-concave utility functions. Analytical techniques used to prove Nash existence under strong assumptions of convexity and differentiability is no longer applicable.

5. Improved Pricing in Power Control (Single Cell Network)

The Utility after pricing as described in the above non-cooperative game is as

$$u_{c_j}(n) = u_j(n) - cp_j(n)$$

(13)

The sum of the utilities with this pricing would essentially be the sum of the individual utilities, given as

$$U_{sum}(n) = \sum_{j=1}^{N} u_{c_j}(n)$$

(14)

Using (13) in (14) we obtain,

$$U_{sum}(n) = \sum_{j=1}^{N} (u_j(n) - cp_j(n)) \qquad (15)$$

In an attempt to formulate a stochastic gradient search algorithm for maximization of (15), the gradient of the above sum with respect to c is sought:

$$\nabla_c U_{sum}(n) = -\sum_{j=1}^{N} p_j(n)$$

(16)

Using a stochastic gradient approach, the update equation for the pricing factor is then given as

$$c(n+1) = c(n) + \alpha \sum_{j=1}^{N} p_j(n)$$

(17)

where α is a step size parameter. This formulation provides a rationalized search process for the pricing factor leading to maximization of sum of utilities. Effectively, it also leads to a faster search process for the least power that can be transmitted by a terminal without sacrificing its SIR.

To summarize the process of finding the optimum powers at the terminal, an algorithm for this search is stated below.

Algorithm *(at the terminal)*:

Step 1: At instant n=0, set the initial powers of all terminals to the powers obtained at equilibrium ($\tilde{p}$), i.e. the powers obtained with the equilibrium SIR $\tilde{\gamma}$.

Step 2: For each terminal for all $j \in N$ compute the best power that gives the maximum utility: $r_j(n) = \arg\max_{p_j \in P} u_j^c\left(p_j(n), \mathbf{p}_{-j}(n-1)\right)$, where $r_j(n)$ is the set of best transmit powers at instance 'n' in response to the interference vector $\mathbf{p}_{-j}(n-1)$.

Step 3: In this game, more than one transmitted power might constitute a best response to a given interference vector. Hence, it is important to assign the least power to the terminal among the possibilities as given by Step 2. Assign the transmit power to that terminal given as $p_j(n) = \min(r_j(n))$.

The equilibrium SIR was found in this case to be $\tilde{\gamma}$=12.38. Once the equilibrium is obtained, the NPGP is played again after incrementing the pricing factor, c, by a positive value found in (16). The algorithm discussed below returns a set of powers at equilibrium with this incremented pricing factor. If the sum of utilities at this new pricing factor improves with that of the old one, then the process is repeated. This continues until an increase in the pricing factor causes the sum of utilities to be lower than that with the previous pricing factor. This value of pricing factor is declared to be the best pricing factor for that system.

Algorithm (*at the network*): To obtain the best pricing factor

Step 1: Initialize the pricing factor to c=0 and declare it to all terminals.

Step 2: Increment the value of c using (17). Declare this new value of pricing factor to all terminals and find the new $u_j^{c+\Delta c}$'s for all terminals.

Step 3: If $U_{sum}^{c} \leq U_{sum}^{c+\Delta c}$ then repeat step 2, else stop. This final value of c is the best pricing factor for that system.

6. Improved Pricing in Power Control For Multi Cell Networks

In the previous section, the power control algorithm for a single cell network was presented. The same analysis is now extended to the multi-cell networks in a multi-cell environment; in a multi-cell environment, the interference is not only due to terminals within its cell but also from terminals from other cells as well. S.V.Hanley [7] considered the combined problem of transmit power control and the base station assignment in a wireless system where multiple cells are involved. The goal was to determine a transmit power and a base station assignment for each user such that the target SIRs were provided to all terminals. In this approach, the total transmit power was minimized over the set of powers and the base station assignments are subject to minimum SIR constraints. Hence, the resource allocation problem in a multi cell is now a two-dimensional problem where utility is to be maximized over both transmit power and base station assignment. The formulation is same as the single cell network, but the maximization now is two-dimensional.

A wireless CDMA system is considered with *K* cells and *N* users spread randomly throughout the area. The base stations are denoted by *K*= {1, 2, 3, . . . K). d_{ij} is the distance in meters of terminal *j* from base station *i*, with a path gain h_{ij}. It is assumed that the terminals are stationary and therefore the path gains do not change. Each terminal is assumed to transmit data at the rate of *R* bits/second over a spectrum of *W* Hz. Noise is assumed to be additive white Gaussian (AWGN) with noise power of σ^2 watts at the receiver. And the SIR attainted for the jth terminal at base station *i* is denoted as γ_{ij}. This SIR is given as

$$\gamma_{ij} = \frac{W}{R} \frac{h_{ij} p_j}{\sum_{k=1, k \neq j}^{N} h_{ik} p_k + \sigma^2} \quad (18)$$

In a multi cell environment, each base station receives different power levels from its terminals resulting in different SIR levels and therefore different QoS levels at each base station. It is also assumed that there is no base station diversity, i.e. each terminal is connected to a single base station at any given time, although all the terminals interfere with each other all the time. The Utility of terminal *j* at its assigned base station a_j is,

$$u_{a_j j}(\mathbf{p}) = \frac{LR}{Mp_j} f(\gamma_{a_j j})\ bits\ /\ Joule \quad (19)$$

where $f(\cdot)$ is the efficiency function that approximates the probability of successful reception, *L* is the number of information bits in a packet of size *M* bits, **p** denotes the power vector $\mathbf{p} = (p_1, p_2 \ldots, p_N)$ and $a_j \in K$ is the base station assigned for user *j*. The base station assignment, denoted by the vector $\mathbf{a} = (a_1, a_2 \ldots, a_N)$, can be arbitrary. Notice that there are a total of K^N different base station assignments possible. The SIR of each terminal from each of the base station is obtained and the base station which gives the maximum SIR is assigned to that terminal. After the base station assignment, the following algorithm is implemented.

Algorithm: The algorithm for multi-cell network is implemented the same way as described for single cell networks. The only difference being that the algorithm is implemented for all the base stations.
Considering a multi cell power control game with pricing under fixed base assignment strategy, the sequence of steps is as follows:
Step 1: For cell $K=1$, at instant $n=0$, set the initial powers of all terminals to the powers obtained at equilibrium ($\widetilde{p}$), i.e. the powers obtained with the equilibrium SIR $\widetilde{\gamma}$
Step 2: Initialize the pricing factor to c=0 and declare it to all terminals for that base station K.
Step 3: For each terminal in cell K, for all $j \in N_k$ compute the best power that gives the maximum utility,
$r_j(n) = \arg\max_{p_j \in P} u_j^c(p_j(n), \mathbf{p}_{-j}(n-1))$, where $r_j(n)$ is the set of best transmit powers at instance 'n' in response to the interference vector $\mathbf{p}_{-j}(n-1)$
Step 4: Assign the transmit power to that terminal given as
$p_j(n) = \min(r_j(n))$ for all $j \in N_k$
Step 5: Increment the value of 'c' as given in (17). Declare this new value of pricing factor to all terminals in base station *K* and find the new $u_j^{c+\Delta c}$'s for all terminals.
Step 6: If $U_{sum}^{c} \leq U_{sum}^{c+\Delta c}$ then repeat step 2, else stop. This final value of c is the best pricing factor for that system.
Step 7: Repeat steps 1 to 6 for the next base station

7. Numerical Results and Simulations

A. *Single Cell Network*

The algorithms described in section 5 is implemented for a single cell network, with a single CMDA base station having nine terminals with a fixed frame size of 80 bits per frame and no forward error correction. The terminals are at arbitrary distances d_j = [310, 460, 570, 660, 740, 810, 880, 940, 1000] meters from the base station, where *j* represents the terminal. A simple path loss model is assumed and the path losses are obtained by using $h_j = \frac{K}{d_j^4}$, where K is a constant of value 0.097 and d_j is the distance of jth terminal from the base station. The efficiency function in this numerical example approximates the FSR (Pc) for non-coherent FSK and is taken to be

$$f(\gamma_j) = \left(1 - e^{-0.5 j}\right)^M$$

(20)

The network parameters considered throughout the single cell network are as follows:
Number of Terminals, N: 9
Total number of bits per frame, M: 80
Number of information bits per frame, L: 64
Spread spectrum bandwidth, W: 10^6
Bit rate, R: 10^4 bits/second
Noise power at the receiver, $\sigma^2 = 5 \times 10^{-15}$ Watts
The equilibrium powers at NPG are obtained by solving the $\gamma_j = \widetilde{\gamma}$ for all *j*. The Utility maximizing SIR, $\widetilde{\gamma}$, is obtained by solving the First Order Necessary Optimality Condition (FONOC) given as,

$$f'(\gamma_j)\gamma_j - f(\gamma_j) = 0, \quad j = 1, \ldots, N$$

(21)

This Utility maximizing SIR was found to be $\widetilde{\gamma}$ =12.38 from (21). The feasibility condition for the existence of equal SIR equilibrium for the CDMA system is then obtained as N < 9.05. After the equilibrium with NPG (Zero pricing) is obtained, the algorithm is implemented again with pricing. The pricing factor 'c' is set to a zero to begin the iterations. After the first iteration, the pricing factor 'c' was increased by an incremental value given by (17). The set of powers for this new pricing factor were determined and the utilities were observed to increase than those in the previous instance. The pricing factor is thereby incremented by the sum of transmitted powers obtained in the previous iteration as described in the algorithm in Section 5, and the procedure is repeated. This is continued until further increase in the pricing factor would result in the utility levels worse than that of the previous iteration. Using the parameters and the efficiency function above, the algorithms at the network and at the terminal are implemented. The value of pricing factor 'c' was initially set as 0. The transmitted powers and the utilities were calculated for all the nine terminals. The iteration were repeated incrementing the pricing factor as given by (17) until the Sum of Utilities falls below the value in its previous iteration($U_{sum}^{c} \leq U_{sum}^{c+\Delta c}$). The control factor 'α' was introduced in (17) to make the system more dynamic. This is the used as a multiplication factor for the pricing in the algorithm.

Table I shows the dynamic variation of the pricing factor with change in α . This variation was not described in any prior work, but is clearly demonstrated in this paper. Increasing or decreasing this value of α , can increase or decrease the revenue of the system respectively. The significance of α is that it is a controlling factor for the overall revenue (pricing factor) of the system.

B. Multi Cell Network

The approach for single cell network is extended to a multi cell network network. However, in a multi-cell environment, each terminal experiences interference from terminals in other cells in addition to the ones from its own cell. When a base station establishes a connection with a terminal, we say that the terminal is assigned to the base station. The problem is now two dimensional: assigning base station and assigning transmit powers to each terminal. As mentioned earlier in Section 6, there are now K^N combinations of base station assignments. The SIRs for each terminal from all the base stations is first calculated. The base station yielding the best SIR(nearest to the terminal in most cases) is assigned to that terminal. Once the base station assignment is completed, the algorithm described in Section 6 is implemented with four base stations. The assumptions made with the single cell network are still valid in the multi cell network (fixed frame size, no forward error correction, e.t.c). The Euclidean distances were arbitrarily assumed and used similarly as they were used in the single cell networks. Similar path loss model was sought and the same design parameters were used for the simulations of the multi cell network.

7. Conclusions

This paper presented an efficient and simple power control algorithm based on utility maximizing with a fixed SIR that is encountered in wireless voice communications systems. Exploiting the fact the the power control game with self-optimizing terminals using Nash equilibrium is inefficient [5], it was shown that given the current congestion statistics, there exists a power vector which has values lower than those attained through Nash equilibrium leading to maximized utility. Also the obtained powers were also lower than those determined through linear pricing used by Mandayam and Goodman [3]. It can be observed that while each terminal tries to achieve higher individual utility, the overall utility (sum of utilities) increases itself. In a single cell network, is evident from section 7(A) that the terminals closer to the base station operate with higher SIR than terminals farther away. One drawback with this approach is that this function is highly non-linear and varies drastically with minor changes in the parameters. However, it was observed that given the fixed set of parameters (bandwidth, chip rate, noise power, etc), the convergence is fast and dynamic. The pricing factor can be controlled by the base station and was made dynamic with the congestion statistics. The results showed that while users with better channels received better QoS, they also made proportionally higher contributions to the network revenue.

REFERENCES

[1] Patrick Herhold, Wolfgang Rave, and Gerhard Fettweis, "Relaying in CDMA networks: Path loss Reduction and Transmit Power Savings," *In VTC2003, Korea, spring 2003.*

[2] Vasilios A Siris, "Resource control for elastic traffic in CDMA networks," *MOBICOM 2002: 193-204.*

[3] D. J. Goodman and N. B. Mandayam, "Power control for wireless data," *IEEE Personal Communications., 7:48–54, April 2000.*

[4] V. Shah, N. B. Mandayam, and D. J. Goodman, "Power control for wireless data based on utility and pricing," *In Proceedings of the PIMRC, pages 1427-1432, 1998.*

[5] P. Dubey, "Inefficiency of Nash equilibria," *Mathematics of Operations Research, 11(1):1-8, February 1986.*

[6] Cem U Sarayadar, "Pricing and power control in wireless data networks," *Dissertation Rutgers, The state university of New Jersey, January 2001*

[7] S. V. Hanly, "An algorithm of combined cell-site selection and power control to maximize cellular spread spectrum capacity," *IEEE Journal on Selected Areas in Communications, 13(7):1332-1340, September 1995.*

[8] Y. A. Korilis and A. Orda, "Incentive compatible pricing strategies for QoS routing," *In Proceedings of the IEEE INFOCOM, volume 2, pages 891-899, 1999.*

[9] M. L. Honig and K. Steiglitz, "Usage-based pricing of packet data generated by
A heterogeneous user population," *In Proceedings of the IEEE INFOCOM, pages 867-874, 1995.*

[10] F. P. Kelly, "Charging and rate control for elastic traffic," *European Transactions on Telecommunications, 8:33-37, 1997.*

[11] F. P. Kelly, A. K. Maulloo, and D. K. H. Tan, "Rate control in communication networks: shadow prices, proportional fairness and stability," *Journal of the Operational Research Society, 49:237-253, 1998.*

[12] R. D. Yates, "A framework for uplink power control in cellular radio systems," *IEEE Journal on Selected Areas in Communications, 13(7):1341-1347, 1995.*

[13] J. Zander, "Performance of optimum transmitter power control in cellular radio systems," *IEEE Transactions on Vehicular Technology, 41(1):57-62, February 1992.*

α	Sum of Utilities	Optimum Pricing Factor 'c'
1	3.03×10^7	729
50	3.13×10^7	2.2×10^4
100	3.14×10^7	3.9×10^4

Table I. Variation of pricing factor with alpha

Natural Language Processing of Mathematical Texts in mArachna

Marie Blanke[1], Sabina Jeschke[1], Nicole Natho[1], Ruedi Seiler[1], and Marc Wilke[1]
[1]*Technische Universität Berlin, Institut für Mathematik*
Straße des 17. Juni 136
10623 Berlin, Germany
{blanke, sabina.jeschke, natho, seiler, wilke}@math.tu-berlin.de

***Abstract*-mArachna is a technical framework designed for the extraction of mathematical knowledge from natural language texts. mArachna avoids the problems typically encountered in automated-reasoning based approaches through the use of natural language processing techniques taking advantage of the strict formalized language characterizing mathematical texts. Mathematical texts possess a strict internal structuring and can be separated into text elements (entities) such as definitions, theorems etc. These entities are the principal carriers of mathematical information. In addition, Entities show a characteristic coupling between the presented information and their internal linguistic structure, well suited for natural language processing techniques. Taking advantage of this structure, mArachna extracts mathematical relations from texts and integrates them into a knowledge base. Identifying sub elements within new elements of information with already stored mathematical concepts defines the structure of the knowledge base. As a result, mArachna generates an ontology of the analyzed mathematical texts. In response to user queries, parts of the knowledge base are visualized using OWL. In particular, mArachna aims to provide an overview of single fields of mathematics, as well as showing intra-field relations between mathematical objects and concepts. The following paper gives an overview of the theoretical basis and the technologies applied within the mArachna framework.**

I. Background

Information and knowledge are central concepts in today's society. Numerous publications, books and the World Wide Web create an "info glut" that is not easy to manage. Furthermore, manual information processing is a very time-consuming process. A reasonable approach to this problem is the development of mechanisms for the automated extraction of knowledge from natural language texts. Such an automated extraction requires methods for natural language processing and for the classification and visualization of knowledge. Artificial intelligence and psychology provide key impulses for the development of knowledge classification mechanisms such as semantic networks, associative networks and knowledge maps. These mechanisms provide an effective way to organize knowledge, a prerequisite for modeling knowledge. In the context of this work, *knowledge* is defined as relations between propositions and terms. Hence, knowledge classification mechanisms help to answer the question of how to make knowledge accessible for and processable by a computer.

II. Basic Concepts

mArachna is a system for automatically extracting knowledge from mathematical natural language texts. mArachna aims to create a knowledge base for mathematical knowledge, which, in combination with an intelligent retrieval interface, could serve as an "Intelligent Mathematical Encyclopedia".

To achieve this goal, mArachna follows a minimalistic approach based mainly on computer-linguistic techniques for analyzing natural language texts. It should be noted that mArachna is **not an automated reasoning system**: the system does not provide any interpretation of the actual mathematical content by the computer, i.e. it does not attempt to "understand" mathematics. mArachna performs its analysis solely on the linguistic structure of the text.

Mathematical texts show a distinctive structure, both on the linguistic level and in the presentation of knowledge chosen by the author. This structure is characterized by typical text elements, such as definitions, theorems and proofs. In the following, we will refer to these text elements as entities. In mathematical textbooks, entities are commonly used to describe mathematical objects and concepts. These entities form a complex network of relationships that can be described by an ontology. Since, following the ideas of Hilbert [8] and Bourbaki [1], mathematics can be described starting with a small set of axioms (in our case set theory) and propositional logic[1], it can be said that mathematics possess an inherent structure, or, in other words, an inherent ontology. The network of mathematical terms and their relations as created by mArachna should closely recreate that structure inherent to mathematics itself. As a result, we expect mArachna to be able to integrate mathematical entities from very different sources, such as different mathematical

[1] This approach is valid within the context of mArachna, despite Gödel's Incompleteness Theorem [7], since we map existing and proven mathematical knowledge into a knowledge base and do not want to prove new theorems or check the consistency of the theorems we store.

K. Elleithy (ed.), *Advances and Innovations in Systems, Computing Sciences and Software Engineering*, 301–305.

textbooks, independent of the upper ontology preferred and used by the authors of those sources.

As a result of the described approach it follows that the concept of mArachna depends on the assumption, that textbooks and papers are both mathematically correct and follow a sensible upper ontology, i.e. possess an inherent structure. The upper ontology of the sources is mapped by both the structure of the knowledge base and additional annotation of the stored information.

III. Linguistic Approach

Entities are the principal carriers of information in mathematical texts. They are analyzed using natural language processing techniques, based on a linguistic classification scheme [3, 4]. This scheme defines four levels: relations between different types of entities are described on the ***entity level***.

On the ***internal entity structure level*** specifies the internal structure of an entity (i.e. the assumptions and proposition of a theorem). Characteristic sentence structures, which are commonly found in mathematical texts, are described on the ***sentence level***. On the ***word and symbol level*** at the bottom single symbols and words and their relations between each other are schematized [2].

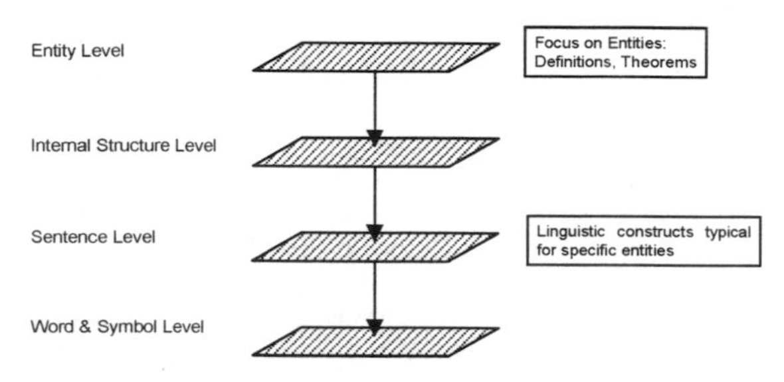

Fig. 1: Linguistic Structure of mathematical texts.

Mathematical information is extracted from a text using the structures and linguistic relations as defined by this classification scheme (see Fig. 1). The information is integrated into a knowledge base. This knowledge base consists of one or more directed graphs representing terms, concepts and their relations between each other. It is based on an ontology of the language of mathematics encoded with the web ontology language OWL [5]. The linguistic analysis of entities yields triples consisting of two nodes and one relation. Nodes represent mathematical terms and propositions, with the relation describing how they are connected to each other. It should be noted that triples themselves can be used as nodes in other triples, allowing the representation of more complex interrelations. In this context, different types of relations describe different types of linguistic phrases or key words in mathematical texts (e.g. two nodes corresponding to two propositions A and B, connected by the relation "is equivalent to"). These triples are then integrated into the knowledge base. This process closely maps the actual language structure, resulting in a very fine-grained knowledge base.

IV. Technical Aspects of mArachna

mArachna (Fig. 2) provides a flexible interface based on TEI (Text Encoding Initiative [9]) as the generic input format, with all mathematic symbols being encoded in MathML (Presentation) [10]. Currently, mArachna provides a LaTeX/TEI-converter; using Java based XML and MathML parsers. LaTeX was chosen since it is the standard format for publication in mathematics and the natural sciences.

The TEI representation of the text is analyzed to extract the entities and their relations to each other, as well as the information provided on their internal structure level. Following this preliminary analysis, mArachna extracts single sentences from the entities, annotating them with additional information concerning their parent entity and their role within the internal structure of that entity. Mathematical formulae are generally separated from the surrounding natural language text for further processing (simple formulae are replaced with corresponding natural language text). This preliminary analysis is implemented in Java using rule-based string comparison.

In the next step, the extracted natural language sentences are analyzed using computer-linguistic methods. The natural language analysis is implemented in the TRALE-System [12] using a Head-Driven Phrase Structure Grammar. TRALE is a PROLOG adaptation of ALE for the German language [11]. The underlying dictionary and grammar have been extended to include the specifics of mathematical language. Complex formulae have to be processed separately from the natural language analysis. This process is, however, not yet implemented.

The natural language analysis performed by TRALE provides detailed syntactic and even some limited semantic information about the sentences. At this point, the separately processed formulae will have to be reintegrated into the natural language text for further semantic analysis to generate numerous of the above-described triples of {subject, predicate, object} (see ch. III.) to be stored as OWL documents.

The semantic analysis is based on simple rules coded in Java. These rules map typical mathematical language constructs onto the corresponding basic mathematical concepts (e.g. proposition, assumption, definition of a term etc). The resulting triples are annotated with additional information. This includes both the context within the original document as well as their classification within the context of the final OWL documents generated by mArachna. For each element of the triples it has to be decided if they represent OWL classes or instances of OWL classes. This distinction poses a significant challenge for the automated analysis, currently forcing mArachna to use OWL Full.

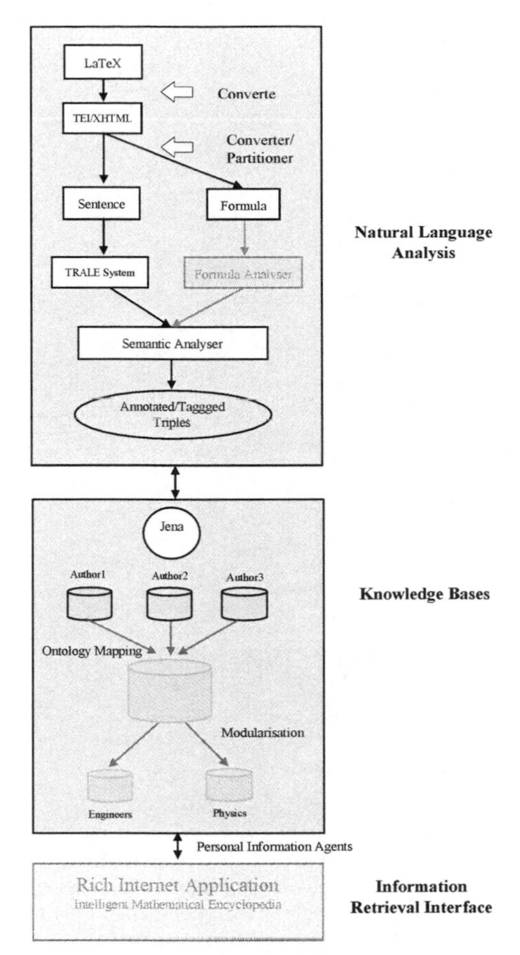

Fig.2: Technical Concept of mArachna

The conversion of the generated, annotated triples into OWL documents for storage within the knowledge base is performed within the Java-based Jena framework [13]. Jena has the additional advantage that it provides APIs for persistent storage and the SPARQL query engine.

The knowledge extracted from each analyzed text generates a separate, independent knowledge base. This separation is necessary to retain the idiosyncrasies and preferences of each author. Initially, these knowledge bases contain basic mathematical knowledge (see ch. 2) comprising axiomatic set theory and first order logic.

To avoid inconsistencies within the knowledge base, new information is added using a semi-automated approach: information is integrated into the knowledge base if and only if there are no conflicts with existing entries. This means that new nodes have to be connected to existing ones; duplications or contradictions are inadmissible. In case of conflicts, the user may provide additional information or delete the conflicting entries manually. This user-intervention is performed through Jena. This dual model of information management is based on well-known models of human knowledge processing: humans will be able to integrate new knowledge into their world view only if it can be linked to existing knowledge. Insufficient or incorrect prior knowledge may lead to misinterpretations of new information. As a consequence, incorrect knowledge may be deleted or corrected under certain conditions (see ch. 1). It should be noted that, given the nature of the sources, human intervention into the automated integration is not the rule; most of the knowledge presented in a textbook follows the rules of consistency required for the automated integration.

V. LINGUISTIC ANALYSIS: AN EXAMPLE

Take as an example (Fig. 3) the definition of a group [6][2].

Let M be a set, M ≠ , and

: M M → M

a, b ↦ a b

a map.

The tuple M, * is called a *group* if the following holds:

1. The map is associative: (a b) c = a (b c), (a, b, c M).
2. For each a, b M exists a x M satisfying a x = a.
3. For each a, b M exists a y M satisfying y b = b.

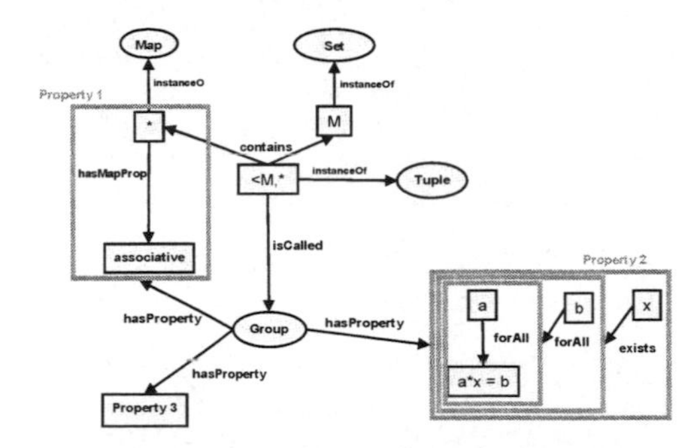

Fig. 3: Extract of mArachna's ontology of the above definition.

Based on the knowledge of the use of certain keywords in mathematical texts, the linguistic analysis has identified "Let M ... a map" as the assumption, "The tuple ... if the following holds:" as the proposition and the following three points as the properties of a definition. The morphological, syntactical and semantic analysis results in the representation as triples of predicate, relation and object. For example, the phrase "Let M be a set" is mapped to the triple (is_a, set, M). Finally, identifying identical predicates and objects of different triples with each other generates the network of relations.

The relations between entities are described using OWL. The representation in OWL provides an overview of the

[2] The definition given might look somewhat unusual but is equivalent to the more common definition of a group as a set with an associative map and an inverse for each element of the set as well as a neutral element [6].

mathematical knowledge that is generated by extracting relevant information from the knowledge base, with special attention given to the underlying field ontology.

VI. Evaluation of the Basic Concept

At the moment mArachna exists as a prototypical implementation to analyze text written in German (it will be extended to analyze texts written in English). For selected text elements, the prototype demonstrates the feasibility of a semi-automated approach to semantic extraction as it has been described in this paper. The semantic extraction leads to information snippets of the analyzed mathematical text. This information can be integrated into the discussed knowledge base.

Future evaluation will include analyzing complete textbooks on Linear Algebra and merging the resulting knowledge bases to test the validity of the ontology mapping approach. The results should be compared with a "standard" mathematical encyclopedia, particularly concerning the generated (or, in the case of the standard encyclopedia, used) ontologies for this field of mathematics.

VII. Related Works

mArachna consists of several separate subprojects. The results of the linguistic analysis of mathematical language [15] are based on TRALE [12]. The structuring of mathematical language and the modeling of the resulting ontology is similar to the work performed in [16, 17]. An example for a different approach, based on automated reasoning would be DIALOG [14].

VIII. Knowledge Representation and Knowledge Management in mArachna: an Outlook

A. Summary

Mathematical language, in particular within the entities, places a strict emphasis on the transfer of knowledge. Thus, the knowledge base consists of mathematical knowledge without any need of interpretation. The acquisition process (semantic extraction) follows strict rules based on the structure of entities. The same is true for the organization and storage of mathematical content in the knowledge base. But still, some problems remain with the knowledge base with regard to merging or extracting knowledge.

B. Knowledge Management and Ontology Mapping

The notations and phrasing used can vary significantly between different authors, based on personal preferences, didactical goals etc. As a result, each text is stored in its own corresponding knowledge base. Future developments of mArachna will have to implement a sensible ontology mapping to create one, unified knowledge base from the separate smaller bases associated with one text each (see Fig. 4). In reverse, strategies and tools for the modularization of this unified knowledge base into smaller bases, specifically adapted to the needs and preferences of specific target user groups, have to be conceived and implemented. These "modularized" knowledge bases in turn could serve as the foundation for an intelligent retrieval system based on the idea of PIAs (personal information agents). For example, in education it is advisable to use different versions of an entity for elementary school and high school. Modularized knowledge bases in different notations would support the harmonization of the teaching material within a series of courses.

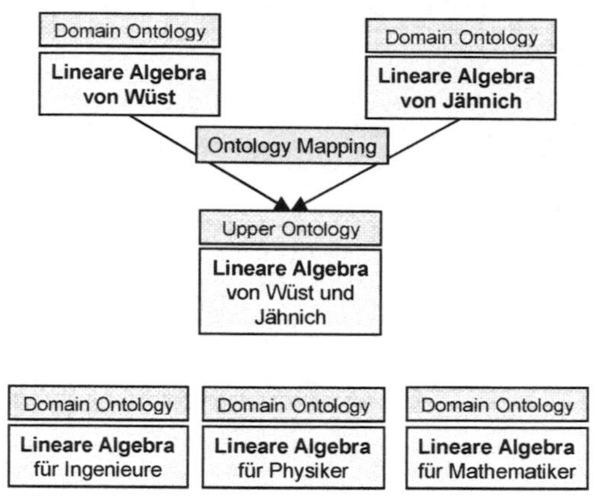

Fig. 4: Ontology mapping in mArachna

C. Retrieval

The aim of the future design of a retrieval interface will be the support of users in learning and understanding mathematics. The interface should provide different tools for selecting information based on personal preferences (more axiomatic oriented, more example oriented). In addition it would be desirable to integrate the administration of different roles into a generic interface (students, teacher and administrator).

D. Processing of Mathematical Formulae

As formulae form a major portion of mathematical texts and constitute a primary source of information in these texts, it is desirable to be able to include their content in the analysis and representation created by mArachna. Currently, mArachna is not capable of this important feature yet. However, we are investigating an approach to rectify this deficiency. We propose using a syntactical analysis similar to those used in computer algebra systems in combination with contextual grammars (e.g. Montague grammars) to correlate the information given in a formula with information already provided in the surrounding natural language text. Using this approach should enable mArachna to integrate formulae and their informational content in the network created by the analysis of the natural language text. It should be pointed out that we do not aim for machine-based understanding of the formulae, as automatic reasoning systems would require. Instead, formulae are to be treated as a different representation of mathematical knowledge, to be integrated into the knowledge base in a similar manner to that used for the natural language text. However, the analysis proposed here can be used as a first step in a further process leading to viable input for such reasoning systems, providing additional assistance in building the knowledge base.

REFERENCES

[1] Bourbaki, N. (1974). *Die Architektur der Mathematik. Mathematiker über die Mathematik*, 1974, Springer, Berlin, Heidelberg, New York.

[2] Grottke, S., Jeschke, S, Natho, N., Seiler, R. (2005). *mArachna: A Classification Scheme for Semantic Retrieval in eLearning Environments in Mathematics*. Proceedings of the 3rd International Conference on Multimedia and ICTs in Education, June 7-10, 2005, Caceres/Spain, 2005, ISBN 609-5994-5

[3] Jeschke, S. (2004). *Mathematik in Virtuellen Wissensräumen - IuK-Strukturen und IT-Technologien in Lehre und Forschung*. PhD Thesis, Technische Universität Berlin, Berlin.

[4] Natho, N. (2005). *mArachna: Eine semantische Analyse der mathematischen Sprache für ein computergestütztes Information Retrieval*. PhD Thesis, Technische Universität Berlin, Berlin.

[5] W3Cb. Web Ontology Language (OWL). http://www.w3.org/2004/OWL/

[6] Wüst. *Mathematik für Physiker und Mathematiker*, Bd.1. Wiley-VCH.2005

[7] K. Gödel. *Über formal unentscheidbare Sätze der Principia Mathematica und verwandte Systeme I*, Monatsheft f. Mathematik und Physik, 1931-1932, p. 147f

[8] D. Hilbert, 1928, *Die Grundlagen der Mathematik*, Abhandlungen aus dem mathematischen Seminar der Hamburgischen Universität 6,p. 65-85

[9] http://www.tei-c.org

[10] http://www.w3.org/Math

[11] Stefan Müller, *Deutsche Syntax deklarativ: Head-Driven Phrase Structure Grammar für das Deutsche*, Linguistische Arbeiten, No. 394, Tübingen: Max Niemeyer Verlag, 1999

[12] Stefan Müller, *TRALE*, http://www.cl.uni-bremen.de/Software/Trale/index.html

[13] *Jena – A Semantic Web Framework for Java*, http://jena.sourceforge.net

[14] *DIALOG*, http://www.ags.uni-sb.de/~dialog/

[15] J. Baur, *Syntax und Semantik mathematischer Texte*, Master Thesis, Saarbrücken 1999

[16] *MBase*, http://www.mathweb.org/mbase/

[17] A. Franke, M. Kohlhase, *MBase: Representing Knowledge and Context for the Integration of Mathematical Software Systems*, JSymComputation, 23 (4): 365-402, 2001

Humanization of E-services: Human Interaction Metaphor in Design of E-services

Mart Murdvee
PE Consult Ltd. / Tallinn University of Technology (Estonia)
mart.murdvee@pekonsult.ee

Abstract: **A possible way of introducing better e-services is to regard an e-service not as a package of functions but as a person or people offering the service. It finally means emulation of a particular structure of human functioning, human interaction and communication by an instrument such as a computer or a mobile phone etc in the most human-like way possible. It means using human schematas and scripts in programming where the e-service provider is a "personality" who has "his" social role in human-computer interaction. That requires the use of psychological principles that are common in human goal-orientated behaviour, human-human interaction and natural intercourse between people. There is a need for research on human-human interaction in service and/or other situations, and introduction of the findings in the technical solutions of e-services.**

The following is more an ideological and idealistic than academic approach to the issue of human-computer interaction. The principal aim of the article is to underline the importance of psychological knowledge of human behaviour, interaction and communication, and show how such knowledge could be applied to designing e-services.

The aim is also to provoke discussion and show the need to study human-human interaction more and use the knowledge in designing e-services, programs and gadgets.

Development of e-services (the e-services are thought here in very broad sense – it can be a program what runs some gadget – mobile phone, MP-player, or a program run on computer, or a web-site – everything what does something for people in computational way), be it web sites, mobile phones, all kinds of other intelligent gadgets etc, is all about developing more functions and not so much effort is put into making such services easy to use. For example, mobile phones have a multitude of functions (and more are introduced every day) but most people do not use the majority of them or hardly any of them at all. People do not use those functions not because they do not need them, but because using them is difficult or people do not understand the possibilities they offer. The same applies to computer programs.

Metaphors in Desing

Many different metaphors are used in the design of human-computer interfaces. We see newsletter or book metaphors in the design of homepages and software written for the purpose of disseminating information. The information collecting sites use the metaphors of questionnaires. The metaphors of TV or radio sets are used in programs and on web sites providing audio-visual services, i.e. television and radio programmes. Different computerized services use different metaphors; sometimes it is an exact imitation of the equipment both in terms of the appearance and the level of interaction – the person has to move an adjuster or click on buttons etc. The metaphors for services are selected according to the familiarity of the action for the customer, the similarity of the action to the customer's habitual action in the past or to the customs of the customer. The more similar and customary the action, the easier it is to use the service and the more satisfied the customer is.

The use of a human interaction metaphor in design of human-computer or human-gadget interfaces is rare although human-human interaction is the most common type of interaction in our everyday life. But it is easy to do; for example, we can use the concept "secretary" when we design a phone book and a scheduler for the mobile phone and then it is easy to look upon them as a complex set of functions and interactions.

Two Ways of Getting Help and Developing E-services

When the man is confronted with a problem beyond his natural abilities, he can act in two ways – either use instruments or other people. The first option is to use an instrument to extend his abilities, e.g. a hammer when he intends to hit something harder than is possible with his fist; a car when he wants to travel faster and farther than he can on foot; a phone when the others are farther than his voice can reach etc. The second option is to get help from other people who have knowledge, abilities and other qualities to do things the man wants to be done but is unable to do himself, things the man wants to be done but the others can do better or by saving time and other resources. So, people ask relatives, friends or hired specialists to achieve specific goals; we employ servicemen and attendants such as secretaries, librarians, doctors, workers; all in all, we use human services.

We constantly ask two related questions, sometimes without realising it – what things, instruments and what kind of people we need in our lives. We need different instruments depending on our needs, goals, abilities and situations we are in. It is the same with people – we need a friend and an entertainer as a teenager, a secretary in our business life, a trainer-instructor in spare time, a nurse or a helpmate in older age etc. – and all this

K. Elleithy (ed.), Advances and Innovations in Systems, Computing Sciences and Software Engineering, 307–311.

depends on our needs, goals, situations, personal resources, age etc.

When creating or developing e-services we can also ask people two questions: one about instruments and the other about the persons they need; we can also combine the two questions and ask what kind of a person with which instrument they need in various situations. Then we can monitor and research the functions of the person needed. We can research ways of natural (or common) communication and interaction between the person and the person needed. The third step can be creating gadgets or e-services which emulate the behaviour and functioning of the people needed - we can introduce e-attendants, e-people. It is possible to introduce new services and new instruments and gadgets when we start not by asking what services people need but by asking who they need.

Example: "e-nurse".

Older people need a nurse and a helpmate. What does the nurse do, what are the nurse's responsibilities? The nurse monitors the person's health by measuring certain parameters such as temperature, blood pressure, pulse rate, sugar level etc. The nurse draws conclusions based on the monitored parameters by using diagnostic reference levels and the tendencies evident in those parameters. The nurse can inform the doctor about the patient's health and be the mediator of the doctor's advice and instructions. The nurse monitors the taking of medication, conducting the necessary procedures and reminds the person of this. The nurse is there to call an ambulance. The nurse monitors the patient's behaviour and, if necessary, calls for help. All these functions can be performed by equipment making use of the existing technology – e.g. a scheduler to remind us to take medicine, sensors detecting movement, the SMS or the Internet for communicating with doctors etc. The e-nurse's communication style must be as caring and informative as in case of the human nurse. Actually, the nurse-patient interaction requires specific research – the dialogues, what is said and what is replied in various situations of human-human interaction etc. The findings enable us to establish the best and most effective way of interacting, and, thus, we can program menu structures and the wording of questions and responses to be understandable and natural for most elderly people. As a result, the e-nurse offered to elderly people should not only be a piece of equipment with various medical and communicational functions but a piece of equipment with various medical and communicational functions imitating the human-nurse and the patient-nurse interaction. The emphasis, the accentuating of the interactive function and the emulation of human interaction are the primary issues.

Now we can ask whether it is possible to have the e-nanny, e- pal, e-entertainer, e-secretary, e-trainer, e- librarian and other e-attendants.

Social Role

"Humanization" of e-services and producing e-attendants can be viewed in the light of the theory of social role known in psychology. The social role is a particular behaviour others expect from the person in social interactions. The social role is also a set of norms concerning how the person has to behave in particular social interaction situations [6].

In the light of the theory we can, on the one hand, find out what the customer's expectations are for mobile phones or specific programs, what the customer expects from interaction with the equipment and what the expected, preferable normal interaction is like.

By applying the theory of social role upon "humanising" e-services we can establish

- What kind of professional people are needed?
- What kind of role expectations are there?
- What are the behavioural and communicative norms of the roles?
- How are the roles performed?
- How are interactions built up? What kind of interaction structures are there?

In designing, the technological possibilities available for e-services (mobile phones, computer programs, web sites etc) must be analyzed with the purpose of establishing how to make the service function in the most human-like way and how to emulate the possible roles in e-services.

Differences in Use of Instrument and Human Service. Scripts.

There is a difference between using equipment and services of a human being. If we use a hammer, we have to know how to hold the hammer and a nail, how to hit the nail with the hammer etc. In case of the services provided by people, the main point is communication between the customer and the service provider. If we need a nail to be driven in a wall and have hired a person to do it, we just tell them: "Could you drive a nail in the wall here?"; we interact with the other person and define a goal. The interaction can continue: "More to the right, please!" or "More to the left, please!"; we can also reset the goal or cancel the action altogether.

The main differences between employing equipment and services of other people are as follows:

- In case of man provided services, the user does not need to know how their goal is achieved (naturally, sometimes it is necessary to avoid big mistakes); suffice the customer knows that the service provider can meet their goals;
- the customer can communicate their goal to the service provider:
- the customer can, if necessary, control the process of goal-achievement
- the customer can exploit the achieved goal.

From the point of view of the service provider, communication with the customer must include information about the following:

- availability of the service (possible goals of the customer) or provision of the service – the customer has certain objectives/needs and the service provider has to show that a service meeting such objectives/needs is available;

- capability of providing the service;
- the status of the function or the goal-achieving process, i.e. the current situation regarding provision of the service;
- possible ways of managing the goal-achieving process;
- reporting on the achieved goal and/or the goal itself.

The situation of using human services can be described using concept of script. Script is a coherent sequence of event expected by the individual either as a participant or as an observer. Scripts answer the questions: "What can we do together?", "How shall I proceed?" and "What do I do next?" [7].

The main parameters and interaction structures, scripts of human goal-orientated behaviour in situations where co-operation occurs should be studied under circumstances that are similar to the e-services or e-persons to be developed.

LOGIC OF GADGETS AND LOGIC OF HUMAN INTERACTIONS. SCHEMATAS.

Design and construction has developed towards making instruments and gadgets. This tradition has worked very well. And using instruments is becoming easier – we can solve a lot of our problems just by pushing a button.

The gadget-centred designing process usually proceeds from the possibilities gadgets offer and only then human actions and human-gadget interaction are considered - the designer knows the possibilities of a gadget and humans are taught to use it. The person is forced to follow the inner logic of the gadget. The use of gadgets mostly means providing the user with various buttons, combinations of buttons or their sequences. The more there are possibilities, the more buttons and sequences of button-pushing are provided. Of course, buttons are grouped, mostly in a logical way. But, unfortunately, it is usually the logic of functions, not the logic of human thinking and action. The same applies to program menus and program user interfaces.

This problem may discussed using concept of schema. Schema is a conceptual framework that individuals use to make sense out of stored and processed information [3]. The concept of schema is also known in programming. The main problem here is mismatch of schematas of people and of programming.

The impact of the programming language and functional principles of the computer have to be emphasized in designing of programs and gadgets. The designer must realise that the functioning schematas of computer programs differs from the schematas of human and human action. The most obvious and well-known discrepancy between the programming schema and human schema is the "Turn off Computer" option in Microsoft Windows programs – to turn off the computer you have to push the "Start" button. In terms of program schema and for the programmer, it is entirely logical – the "Start" button initiates various program sequences, including turning off the computer, but in terms of common sense and human schema, this is confusing.

People are used to acting and communicating in a certain way and try to act in seemingly similar situations in a customary way according to their past experience – they generalize their behaviour. Actually, the paradox of the active user - "Users never read manuals but start using the software immediately" [2] - is not a paradox at all; this is the normal way of human action – generalization [4] - to apply previous experience in new situations. In fact, the existence of the paradox is an indication of the programs and gadgets being designed without consideration to natural human behaviour, the existing human schematas and scripts and of the fact that designers are aware of it. Why expect the user to read the manual; why expect them to learn? A good designer should expect the people to be able to use the program or the gadget based on experience they already have.

This means that the designer should be familiar with natural habits and experience of the people who start using the program or the gadget. The knowledge can be gained by monitoring and researching people's behaviour in specific or similar situations, in situations where people provide services to others and by studying the structures of human mind. Such knowledge would help us design human-centred gadgets and e-services.

E-CLERK

E-services will exist and do exist parallel to the services provided by human beings and, one way or another, people remain part of the system. The main point is that the consumer of e-services is a human being with their experience and customs. It is only natural for the man to expect the same or even better communication and intercourse from e-services as from services provided by people.

Introduction of various Internet services is actually translation of services provided by human beings into the program language and the Internet environment. The translation process can be described as replacing the human clerk with the e-clerk.

The principal criterion of such a translation should be that, for the customer and the organization providing the e-service, the service in the Internet environment should be of similar quality to the service provided by people. It means that e-services should:

- retain the properties and qualities of services provided by human beings;
- have the same or similar input and output than services provided by human beings, especially in terms of communication and intercourse;
- have an inner arrangement, i.e. control, information exchange and storage, imitating that of services provided by human beings.

Providing e-services, we can emulate the functioning of whole existing organizations. The functional unit there is the e-clerk and they can be united in organizations. The e-clerk is the program unit with particular input-output and functions that emulates the behaviour of the human clerk. We can introduce:

- e-security for watching entrances;
- e-record keeper for recording information and entering it in databases;
- e-archivist for finding the necessary information from databases;
- e-secretary for organising the information and passing it on to the appropriate recipient;
- e-consultant for consulting the client about what to do to achieve the desired goal;
- e-superior for checking and supervising the system and reporting to the human operator on the system and errors in it etc.

Job descriptions of real clerks and organizational charts of real organizations can be used for developing e-clerks and e-organizations. Also, existing application forms and interaction experience of real clerks regarding what customers do and ask, where they make mistakes and need help, effective wording of questions and replies etc., can be drawn from upon designing e-clerks.

In theory, the client-centred input and output must imitate the best ways of human communication and intercourse, and the structure of electronic action should be similar to the actions of the effective and pleasant human clerk.

Some principles:

- The more directly the e-clerk is connected to the client, the more human and familiar its functioning should be.
- The less the client is used to all kinds of e-services, incl. computers etc., the more human and familiar the functioning of e-services should be.

The use of the above principles concerning emulation of human clerks and organizations in developing e-services improves the parallel functioning of human and e-organizations, services provided by people and e-services because the structure of actions and acting persons are similar in both cases. Therefore, co-operation/co-functioning of human and e-services is better if human clerks understand the e-organization better.

In principle, we can work out a program of various ready-made e-clerks and it is possible to introduce them in a working organization by customizing input-output, setting functions, working criteria etc.

Human services contain extensive experience of best ways and administrative management of services. The organizations and their functioning principles have been developed over an extended period of time. The experience should be used for development of services in the new environment.

"Personality" of Gadget

There is always a tendency to treat instruments like persons - give them names and talk to them. The more advanced in communications the instruments are the more people tend to perceive them as persons. Take, for example, the Tamagotchi toy that was developed to be a person with needs and ability to communicate the needs to its owner and consider its popularity.

The instrument can have "a personality" of its own. Or the people are interpreting the complex behaviour of machines in a human-like way, the most understandable way for people. "A personality" of machines can be built unintentionally not with purpose ("a stupid machine" – the machine that behaves in a way not understandable to us). But we can intentionally build machines that can be "personalities" with specific characteristics of "a person". For example, gadgets can be built to use various styles of interaction expressed by a variety of wording and styles of interaction. We can have a gadget, for example, with a businesslike, motherly or pally "personality" or gadget with communication style of some character from cartoon on demand. The customer can be given the possibility to choose the "personality" of the gadget so that it is more suitable as a partner for the personality of their own. It is also possible to show the virtual person on the screen as a picture and let the virtual person communicate also in a non-verbal way by means of facial expressions and gestures that are illustrations and symbols, to represent "the personality" and improve communication. In summary, the human-like interaction and "personality" of a gadget makes using the gadget more pleasant and effective.

Cultural Origin of Gadget

We must not forget that human interaction is culture dependent. The ways of acting are learned by being in various situations and from other people. The ways of acting are based on culture and determined by the norms of a specific culture. This can be seen in verbal and non-verbal behaviour [6].

Accepting this, the designer of human-machine interaction such as program menus, wording etc., should realise that what is "normal" in one culture (his own culture), can be wrong or not suitable in another. Therefore, comparative research into interaction norms of various cultures should be done. Since there are likely to be differences, we can come up, for example, with neutral wording that does not insult the norms of all cultures or employ separate wording for each culture to satisfy the customer.

The same can be true about the structure of interaction. Even if the functions of interacting people are the same, the culture can alter structures of interaction. We can ask, for example, whether the secretary in Japan acts the same way as the secretary in the USA, or are there differences? If there are differences, the designer of human-machine interaction has to correct the interaction structure of the gadget that is emulating the behaviour of a secretary, and make it more acceptable in the culture the gadget is used in.

Conclusion

Using the term "natural" when speaking about behaviour, interaction and communication etc, is understandable but misleading. There is no universal "natural" human behaviour. Each particular type of behaviour is determined by the goal, situation and acquired ways of acting. Therefore, the human-computer interaction can also be considered natural even when this way of interaction contradicts the schematas and scripts of everyday human action. That is why people can work with programs and use gadgets – they learn and adapt. But this learning is costly. It takes time and requires resources. Purchased but not learned and employed functions are stored in megabytes on computer discs and in the memories of various gadgets, and people are nonetheless less effective in their actions. Maybe now it is no longer the time to force people to learn how to use computers but teach computers how to act like human beings by starting with human-centred designing of human-computer interaction.

One possible option of making e-services better and easier to use is to regard e-services not as packages of functions but as person(s) providing services. It means emulation of a particular structure of human functioning, human interaction and communication by an instrument such as a computer or a mobile phone etc in the most human-like way possible. It means using human schematas and scripts in programming where the e-service provider is a "personality" who has "his" social role in human-computer interaction. That requires the use of psychological principles that are common in human goal-orientated behaviour, human-human interaction and natural intercourse between people.

There is a need for more extensive research on human-human interaction in service and/or other situations, and introduction of the findings in the technical solutions of e-services.

CRITERIA OF EXCELLENCE FOR E-SERVICE

What would an ideal e-service be like? The program or gadget would have to act as the best human service provider so that the naive person, knowing nothing about the service and getting no external help, would want to say afterwards: "Thank you!"

REFERENCES

[1] Anderson, R. C. (1977) "The notion of schemata and the educational enterprise: General discussion of the conference." In Anderson, Richard C., R. J. Spiro, and W. E. Montague (editors). 1984.Schooling and the acquisition of knowledge. Hillsdale, NJ: Lawrence Erlbaum.

[2] Carroll, J.M., Rosson, M.B. Paradox of the Active User. In Interfacing Thought: Cognitive Aspects of Human-Computer Interaction, 1987, edited by John M. Carroll, Cambridge, MA, MIT Press, pp. 80-111.

[3] Crooks, R.L. (1991). Psychology: Science; Behavior and Life. 2nd ed. Holt, Rinrhart and Wilson, Inc.

[4] generalization. (2006). In Encyclopædia Britannica. Retrieved October 12, 2006, from Encyclopædia Britannica Online: http://www.britannica.com/eb/article-9036382

[5] Neil McBride, N., Elbeltagi, I. (2004). Service-Oriented Human Computer Interaction and Scripting. Idea Group Inc. www.idea-group.com/downloads/excerpts/sarmento.pdf

[6] Ridgeway, C. L. (2006). Linking Social Structure and Interpersonal Behavior: A Theoretical Perspective on Cultural Schemas and Social Relations. Social Psychology Quarterly; Vol. 69 Issue 1, p 5-16.

[7] Treuholm, S., Jensen, A. (1992). Interpersonal Communication. Waldsworth Publishing Company.

[8] Turner, R. H. (1978). The Role and the Person. The American Journal of Sociology, Vol. 84, No. 1, pp. 1-23.

Introducing The (POSSDI) Process

The Process of Optimizing the Selection of The Scanned Document Images

Mohammad A. ALGhalayini [1] and Abad Shah [2]

[1]King Saud University
Vice Rectorate for Studies, Development, and Followup
Riyadh, Saudi Arabia
[2]Departement of Computer Science and Engineering
University of Engineering and Technology
Lahore, Pakistan

Abstract

Today, many institutions and organizations are facing serious problem due to the tremendously increasing size of documents, and this problem is further triggering the storage and retrieval problems due to the continuously growing space and efficiency requirements. This problem is becoming more complex with time and the increase in the size and number of documents in an organization. Therefore, there is a growing demand to address this problem. This demand and challenge can be met by developing a process to enable specialized document imaging people to select the most suitable image type and scanning resolution to use when there is a need for storing documents images. This process, if applied, attempts to solve the problem of the image storage type and size to some extent. In this paper, we present a process to optimize the selection of the scanned image type and resolution to use prior to acquire the document image which we want to store and hence to retrieve; therefore, we optimize the document image storage size and retrieval time.

Introduction

To approach the problem of document storage and retrieval, we have to examine the amount of circling documents in the various institution offices, the size of the daily flowing documents (incoming and outgoing), the need for storing and retrieving those flowing documents, and the hardware and software infrastructure in the institution.

In this paper we will focus on selecting a close to optimal scanning size for the (Black and White) and the (256 Gray Shades) scanned document images by applying the **(POSSDI)** Process which is **The Process of Optimizing the Selection of The Scanned Document Images.**

Introducing The (POSSDI) Process Through Examining The Uncolored Scanned Document Images

Our first consideration yielded that most of the documents being used in most of the governmental institutions are A4 documents and in most cases there is a need to store an image of the document in (256 gray shades) or in (black and white). However; since in some instances there is a need for saving color documents as well, we may consider applying the same **(POSSDI)** process. We selected a sample A4 document shown below as **Figure (1-1).** The scanning of the sample A4 document was done and resulted document images were saved in 2 different output types, the 256 gray shades and the Black and white.

In each of the 2 output types above, we scanned the A4 sample document in 10 different resolutions (dpi)[1], 75 dpi, 100 dpi,150 dpi, 200 dpi, 250 dpi, 300 dpi, 350 dpi, 400 dpi, 450 dpi, and 500 dpi

The scanned images were stored in 21 different image type as follows :

1 - FlashPic *.pcx
2 - Graphic Interchange Format *.gif

[1] Dots per inch (dpi) is the resolution measurement unit.

K. Elleithy (ed.), Advances and Innovations in Systems, Computing Sciences and Software Engineering, 313–324.

3 - JPEG *.jpg
4 - PaperPort Browser-Viewable *.htm
5 - PaperPort Image *.max
6 - PaperPort Self-Viewing *.exe
7 - PC PaintBrush *.pcx
8 - PCX Multi-page *.dcx
9 - PDF Image *.pdf
10 - Portable Network Graphics *.png
11 - TIFF *.tif
12 - TIFF - Class F *.tif
13 - TIFF - Group 4 *.tif
14 - TIFF - LZW *.tif
15 - TIFF - Uncompressed *.tif
16 - TIFF Multi-page *.tif
17 - TIFF Multi-page - Class F *.tif
18 - TIFF Multi-page - Group 4 *.tif
19 - TIFF Multi-page - LZW *.tif
20 - TIFF Multi-page - Uncompressed *.tif
21 - Windows Bitmap *.bmp

Finally, the scanned images were evaluated quality-wise by a group of 10 average users and each scanned image was given a quality mark out of 10 points. Our goal is to obtain a close to optimal scanning size and type while not sacrificing the quality of the scanned image by selecting a suitable scanning resolution as well. We shall call this process the **(POSSDI)** Process which is **"The Process of Optimizing the Selection of The Scanned Document Images"** which aims to enable for selecting a balanced document images in size and quality for archiving purposes.

In this paper, will apply this process and analyze it's results for the (256 Gray Scale) and the (Black and White) scanned document images.

1 – 1 Optimizing The Selection Of The (256 Gray Scale) Scanned Document Images

In the beginning of our research study, we scanned a sample A4 document page, since most flowing documents are A4 size, shown in **Figure (1–1)** using PaperPort Document Management software[2] which allows scanning and saving the scanned pages using multiple image types and resolutions as shown in **Figure (1-2)**.

From observing the figure, we can see that different image types and resolutions are stored with different sizes based on the scanning resolution and the output image type.

Applying the (POSSDI) Process

It could be appropriate in our first **(POSSDI)** process application to assume that an A4 (256 Gray Scale) page should not exceed the size of 300 KB when stored. Therefore, our **first step of the (POSSDI) Process** would be to eliminate all the scanned pages that consumed more than 300 KB in size. The result of this step is shown in **Figure (1–3)** below.

It may be necessary at this point to mention that the scanned pages were observed by 10 different average users for quality evaluation and were given a readable quality mark out of 10 points[3]. The table of the (256 Gray Shades) quality evaluation is shown below as **Table (1–1)**. It was agreed on that the scanned images with qualities less than 6.5 is considered low, and the quality more than 8 is considered high[4]; therefore, in our second step of our **(POSSDI)** process we will eliminate all the scanned pages that were low in quality (less than 6.5 marks) and also the high quality as well (more than 8.0 points). The result of the **second step of the (POSSDI) process** is shown in **Table (1-2)** below.

Figure (1–4) below shows our **third step of the (POSSDI) process** in the approach of optimizing the selection based on the storage size and the quality of the stored image. It shows only the images that consumed 300 KB in size or less and in the same time have acceptable quality according to our quality assumption stated previously.

Analyzing Figure (1–4) To Obtain The Best Storing Size And Scanning Resolution For Each Considered Image Type

As we can observe from **Figure (1–4)**, some image types are completely unacceptable in our study, according to our assumption, since they failed to have any acceptable image occurrence[5] (in the occurrence table) either in quality[6] or in size or in both. These image types are: PaperPort Self-Viewing *.exe, PC PaintBrush *.pcx, PCX Multi-page *.dcx, TIFF - Uncompressed *.tif,, and Windows Bitmap *.bmp . In addition, we will ignore the type that have (multi) part since they are actually duplicates of their single page size and quality but were available in our document management software. These multi page image types are : TIFF Multi-page *.tif, TIFF Multi-page - Class F *.tif, TIFF Multi-page - Group 4 *.tif, TIFF Multi-page - LZW *.tif, TIFF Multi-page - Uncompressed *.tif .

In addition, we eliminated totally the columns for the 75 dpi scanning resolution since it had no acceptable

[2] PaperPort is a trade name of ScanSoft Co. The version used in the research study is PaperPort ver. 9.0 professional.

[3] An evaluation session was held and a group of 10 users voted for a quality mark out of 10 points for each scanned document image.

[4] Most of the stored images with quality 8.5 up to 10.0 were consuming a high storage space and hence they were considered unacceptable and eliminated.

[5] The acceptable image storing size for the (256 Gray Scale) is less than or equal to 300 KB as assumed.

[6] The acceptable image storing quality for the (256 Gray Scale) is 6.5, 7.0, 7.5, 8.0 as assumed.

occurrences for any acceptable image qualities (or sizes) and for the 500 dpi scanning resolution column for the same reason as well. Now we are left with fewer acceptable image types and scanning resolutions as shown in **Figure (1–5).**

Refining the Selection for Further Optimization

By observing **Figure (1–5)**, we may keep our accepted images sizes and quality interval **or** narrow it further more if we wish as the **final step of the (POSSDI) Process**. This decision depends on many factors such as the quantity of flowing documents to be scanned and stored, the IT infrastructure in the institution, the storage media used, the hardware specifications., the network speed, and need for a reliable and quality stored document images to be retrieved in the future. To see the effect of narrowing the acceptance interval, we assume that for the (256 Gray Scale) scanned images, the acceptable may not exceed 200 KB instead of 300 KB and the accepted quality interval is [7.0 , 7.5] marks only instead of [6.5 , 7.0 ,7.5 , 8.0]. The resulting figure is shown below as **Figure (1–5B).**

Analyzing Figure (1–5B)

It is possible now for any document archiving specialist to observe **Figure (1–5B)** and make an easier selection of which image type to use and which resolution to use when scanning and storing the scanned images. As we can notice that the images with the type (gif) are accepted when scanned with a resolution 150 dpi and they consume 80 KB in storage space. Also the images with type (jpg) are accepted when scanned with resolution 150 dpi and they consume 160 KB of storage space when stored. Further more, the images of type (max), (pdf), and (png) are accepted when scanned using 100 dpi resolution and they occupy 112 KB, 128 KB, and 192 KB respectively. Finally the images of type (tif class F) are accepted when scanned with resolutions 300 dpi and 350 dpi and they consume 96 KB and 144 KB respectively; were the images of type (tif group 4) are accepted when scanned with resolutions 150 dpi , 200 dpi and 250 dpi and they consume 32 KB and 48 KB and 80 KB respectively.

Which resolution to use when more than one scanning resolution is accepted is not a major question since the quality for all of them is accepted according to our study assumption, again it is a matter of the occupied size on disk. The size selected may very much make a difference in speeding up the process of scanning, storing, and retrieving the scanned document. In addition, we recommend to store the image in such a type that is standard and commonly used among most, if not all, users to avoid installation of a specific image viewer of such a type on all users pc's. For example, if we select the image type (max) which is the PaperPort software default image type, we need to install the PaperPort viewer on all users machines to enable them to view the stored document images; however, saving the scanned image with type (gif) will allow all users to view the image normally, since it is supported by Windows Operating System, and saves a good chunk of space per image if compared with the 112 KB (max) type since the (gif) image consumes only 80 KB. The 32 KB difference in storage space makes it clear why we should go for a lower storage size when we are dealing with thousands of documents to be scanned and have their images stored daily.

In addition, when we observed the need for storing document images in many institutions, we found that in most cases the flowing documents may contain more than one single page, and it was necessary most of the time to store the whole document including the accompanying pages for later retrieval; therefore, it is necessary to consider which type allows saving multiple pages as a single object, the question which makes **Figure (1–5B)** very helpful to decide upon.

From **Figure (1–5B)** we can see that the (gif), (jpg), and the (png) image types may be excluded since there is no multi page image type available for these types, also the (max) image type was excluded previously since it requires a special image viewer. Hence, we are left with the image types (pdf) , (tif Class F) , and (tif group 4) to choose from. These types occupied 128 KB when a scanning resolution of 100 dpi was used for the (pdf) image type, 96 KB and 144 KB when scanning resolution of 300 dpi and 350 dpi was used respectively for the (tif class F) image type, and 32 KB, 48 KB, and 80 KB when scanning resolution of 150 dpi, 200 dpi, and 250 dpi was used respectively for the (tif group 4) image type. Any of these three image type is perfectly acceptable to use, the recommended image type is obviously the (tif group 4) type since it occupies the smallest storage space, but if the storage space is not a crucial issue to consider, then the standard Acrobat (pdf) type is a good image type selection since it is a well known image type and it occupied a relatively acceptable storage space for the good quality it shows in addition to the flexibility of the Acrobat viewer provided for the user as far as zooming, printing, and browsing the saved document pages images.

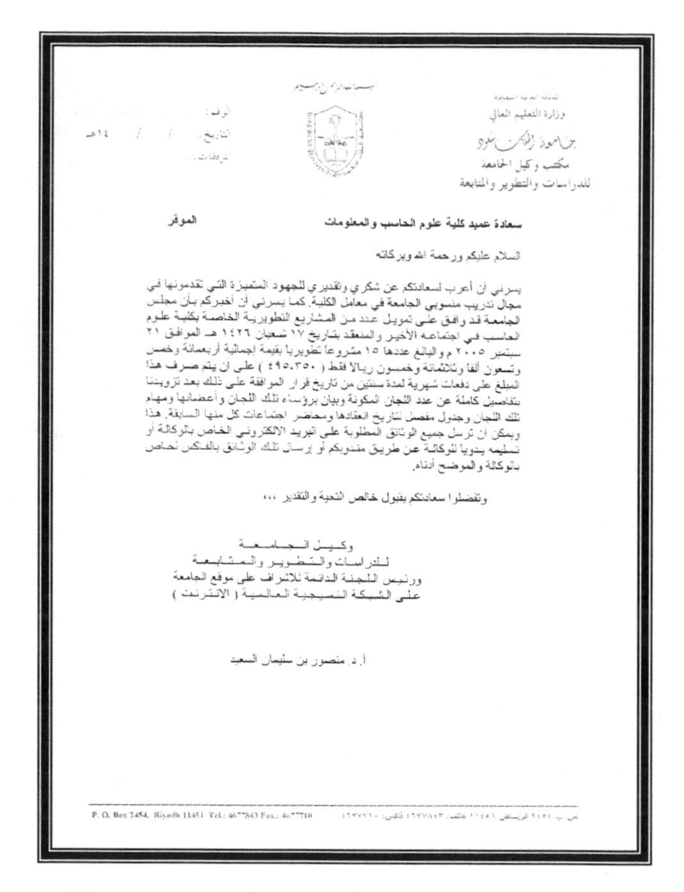

وزارة التعليم العالي
جامعة الملك سعود
مكتب وكيل الجامعة
للدراسات والتطوير والمتابعة

الرقم:
التاريخ: / / ١٤هـ
المرفقات:

سعادة عميد كلية علوم الحاسب والمعلومات الموقر

السلام عليكم ورحمة الله وبركاته

يسرني أن أعرب لسعادتكم عن شكري وتقديري للجهود المتميزة التي تقدمونها في مجال تدريب منسوبي الجامعة في معامل الكلية. كما يسرني أن أخبركم بأن مجلس الجامعة قد وافق على تمويل عدد من المشاريع التطويرية الخاصة بكلية علوم الحاسب في اجتماعه الأخير والمنعقد بتاريخ ١٧ شعبان ١٤٢٦ هـ الموافق ٢١ سبتمبر ٢٠٠٥ م والبالغ عددها ١٥ مشروعا تطويريا بقيمة إجمالية أربعمائة وخمس وتسعون ألفا وثلاثمائة وخمسون ريالا فقط (٤٩٥،٣٥٠) على أن يتم صرف هذا المبلغ على دفعات شهرية لمدة سنتين من تاريخ قرار الموافقة على ذلك بعد تزويدنا بتفاصيل كاملة عن عدد اللجان المكونة وبيان برؤساء تلك اللجان وأعضائها ومهام تلك اللجان وجدول مفصل لتاريخ انعقادها ومحاضر اجتماعات كل منها السابقة. هذا ويمكن أن ترسل جميع الوثائق المطلوبة على البريد الالكتروني الخاص بالوكالة أو تسليمه يدويا للوكالة عن طريق مندوبكم أو إرسال تلك الوثائق بالفاكس الخاص بالوكالة والموضح أدناه.

وتفضلوا سعادتكم بقبول خالص التحية والتقدير ،،،

وكيل الجامعة
للدراسات والتطوير والمتابعة
ورئيس اللجنة الدائمة للاشراف على موقع الجامعة
على الشبكة النسيجية العالمية (الانترنت)

أ.د. منصور بن سليمان السعيد

Figure (1–1) Showing the Sample Full A4 Document

Full A4 Scanned Pages Images Types vs Storage Sizes in KiloBytes (256 Gray Shades)

No.	Scanning Resolution	75	100	150	200	250	300	350	400	450	500
	Image Type	**dpi**	**dpi**	**dpi**	**dpi**	**dpi**	**dpi**	**dpi**	**dpi**	**dpi**	**dpi**
1	FlashPic *.fpx	112	176	304	480	688	896	1,168	1,472	1,760	2,096
2	Graphic Interchange Format *.gif	32	48	80	112	176	208	272	352	432	496
3	JPEG *.jpg	64	96	160	240	352	448	576	720	864	1,008
4	PaperPort Browser-Viewable *.htm	96	160	272	416	592	784	1,040	1,280	1,552	1,840
5	PaperPort Image *.max	80	112	192	304	432	560	752	928	1,120	1,328
6	PaperPort Self-Viewing *.exe	432	464	544	656	784	912	1,104	1,280	1,488	1,696
7	PC PaintBrush *.pcx	224	400	720	1,024	1,808	1,808	2,848	3,552	4,816	5,760
8	PCX Multi-page *.dcx	224	400	720	1,024	1,808	1,808	2,848	3,552	4,816	5,760
9	PDF Image *.pdf	80	128	224	336	480	608	800	976	1,168	1,376
10	Portable Network Graphics *.png	112	192	352	512	896	912	1,408	1,776	2,352	2,832
11	TIFF *.tif	128	208	384	544	944	976	1,504	1,904	2,544	3,056
12	TIFF - Class F *.tif	16	32	48	48	96	96	144	192	256	304
13	TIFF - Group 4 *.tif	16	32	32	48	80	80	128	176	240	288
14	TIFF - LZW *.tif	128	208	384	544	944	976	1,504	1,904	2,544	3,056
15	TIFF - Uncompressed *.tif	544	976	2,176	3,856	6,000	8,640	11,760	15,360	19,440	24,016
16	TIFF Multi-page *.tif	128	208	384	544	944	976	1,504	1,904	2,544	3,056
17	TIFF Multi-page - Class F *.tif	16	32	48	48	96	96	144	192	256	304
18	TIFF Multi-page - Group 4 *.tif	16	32	32	48	80	80	128	176	240	288
19	TIFF Multi-page - LZW *.tif	128	208	384	544	944	976	1,504	1,904	2,544	3,056
20	TIFF Multi-page - Uncompressed *.tif	544	976	2,176	3,856	6,000	8,640	11,760	15,360	19,440	24,016
21	Windows Bitmap *.bmp	544	976	2,160	3,840	6,000	8,640	11,744	15,344	19,424	23,968
	Minimum Storage Size	**16**	**32**	**32**	**48**	**80**	**80**	**128**	**176**	**240**	**288**
	Maximum Storage Size	**544**	**976**	**2176**	**3856**	**6000**	**8640**	**11760**	**15360**	**19440**	**24016**

Figure (1–2) Showing the sizes of scanned images in 21 image types and 10 different resolutions.

Full A4 Scanned Pages Images Types vs Storage Sizes in KiloBytes (256 Gray Shades)

No.	**Scanning Resolution** **Image Type**	**75 dpi**	**100 dpi**	**150 dpi**	**200 dpi**	**250 dpi**	**300 dpi**	**350 dpi**	**400 dpi**	**450 dpi**	**500 dpi**
1	FlashPic *.fpx	112	176								
2	Graphic Interchange Format *.gif	32	48	80	112	176	208	272			
3	JPEG *.jpg	64	96	160	240						
4	PaperPort Browser-Viewable *.htm	96	160	272							
5	PaperPort Image *.max	80	112	192							
6	PaperPort Self-Viewing *.exe										
7	PC PaintBrush *.pcx	224									
8	PCX Multi-page *.dcx	224									
9	PDF Image *.pdf	80	128	224							
10	Portable Network Graphics *.png	112	192								
11	TIFF *.tif	128	208								
12	TIFF - Class F *.tif	16	32	48	48	96	96	144	192	256	
13	TIFF - Group 4 *.tif	16	32	32	48	80	80	128	176	240	288
14	TIFF - LZW *.tif	128	208								
15	TIFF - Uncompressed *.tif										
16	TIFF Multi-page *.tif	128	208								
17	TIFF Multi-page - Class F *.tif	16	32	48	48	96	96	144	192	256	
18	TIFF Multi-page - Group 4 *.tif	16	32	32	48	80	80	128	176	240	288
19	TIFF Multi-page - LZW *.tif	128	208								
20	TIFF Multi-page - Uncompressed *.tif										
21	Windows Bitmap *.bmp										
	Minimum Storage Size	16	32	32	48	80	80	128	176	240	288
	Maximum Storage Size	544	976	2176	3856	6000	8640	11760	15360	19440	24016

Figure (1-3) Showing the result after eliminating all the scanned images with sizes above 300 KB

Full A4 Scanned Pages Images Quality Evaluation Sheet (256 Gray Shades)

No.	**Scanning Resolution** **Image Type**	**75 dpi**	**100 dpi**	**150 dpi**	**200 dpi**	**250 dpi**	**300 dpi**	**350 dpi**	**400 dpi**	**450 dpi**	**500 dpi**
1	FlashPic *.fpx	5.0	6.5	7.5	8.0	8.5	8.5	8.5	8.5	8.5	9.0
2	Graphic Interchange Format *.gif	5.5	6.5	8.0	8.5	8.5	8.5	8.0	8.0	8.0	8.0
3	JPEG *.jpg	5.5	6.5	7.5	8.0	8.0	8.0	9.0	9.0	9.0	9.0
4	PaperPort Browser-Viewable *.htm	4.5	6.5	7.5	8.0	8.5	9.0	9.0	9.0	9.0	9.0
5	PaperPort Image *.max	4.0	7.0	8.0	8.5	8.5	8.5	8.5	8.5	8.5	9.0
6	PaperPort Self-Viewing *.exe	4.0	6.5	7.5	8.5	8.5	9.0	9.0	9.0	8.5	8.5
7	PC PaintBrush *.pcx	6.0	7.0	7.5	7.5	7.5	7.5	8.5	8.5	8.5	8.5
8	PCX Multi-page *.dcx	5.0	6.5	8.5	8.5	9.0	9.0	9.0	9.0	9.0	8.5
9	PDF Image *.pdf	6.0	7.0	8.0	8.5	8.5	9.0	9.0	9.0	9.0	9.0
10	Portable Network Graphics *.png	6.0	7.5	7.5	7.5	7.5	8.0	8.5	8.5	8.5	8.5
11	TIFF *.tif	5.5	6.5	7.5	7.5	8.0	8.0	8.5	8.5	8.0	8.0
12	TIFF - Class F *.tif	1.5	3.0	6.0	6.5	6.5	7.0	7.0	6.5	6.5	7.0
13	TIFF - Group 4 *.tif	2.0	5.5	6.5	6.5	7.0	7.5	7.0	6.5	6.5	6.5
14	TIFF - LZW *.tif	5.0	6.5	7.0	7.5	7.5	8.0	8.0	8.0	8.0	8.0
15	TIFF - Uncompressed *.tif	6.0	6.5	6.5	7.0	7.5	7.5	7.5	7.5	7.5	8.0
16	TIFF Multi-page *.tif	5.5	6.5	7.5	7.5	8.0	8.0	8.5	8.5	8.0	8.0
17	TIFF Multi-page - Class F *.tif	1.5	3.0	6.0	6.5	6.5	7.0	7.0	6.5	6.5	7.0
18	TIFF Multi-page - Group 4 *.tif	2.0	5.5	6.5	6.5	7.0	7.5	7.0	6.5	6.5	6.5
19	TIFF Multi-page - LZW *.tif	5.0	6.5	7.0	7.5	7.5	8.0	8.0	8.0	8.0	8.0
20	TIFF Multi-page - Uncompressed *.tif	6.0	6.5	6.5	7.0	7.5	7.5	7.5	7.5	7.5	8.0
21	Windows Bitmap *.bmp	5.5	7.0	8.0	8.5	8.5	8.5	9.0	9.0	9.0	8.5
	Minimum Storage Quality	1.5	3.0	6.0	6.5	6.5	7.0	7.0	6.5	6.5	6.5
	Maximum Storage Quality	6.0	7.5	8.5	8.5	9.0	9.0	9.0	9.0	9.0	9.0
	Average Quality	4.6	6.2	7.3	7.6	7.8	8.1	8.2	8.1	8.0	8.1

Table (1–1) Showing the quality of scanned images in 21 image types and 10 different resolutions.

Full A4 Scanned Pages Images Quality Evaluation Sheet (256 Gray Shades)

No.	Scanning Resolution Image Type	75 dpi	100 dpi	150 dpi	200 dpi	250 dpi	300 dpi	350 dpi	400 dpi	450 dpi	500 dpi
1	FlashPic *.fpx		6.5	7.5	8.0						
2	Graphic Interchange Format *.gif		6.5	8.0				8.0	8.0	8.0	8.0
3	JPEG *.jpg		6.5	7.5	8.0	8.0	8.0				
4	PaperPort Browser-Viewable *.htm		6.5	7.5	8.0						
5	PaperPort Image *.max		7.0	8.0							
6	PaperPort Self-Viewing *.exe		6.5	7.5							
7	PC PaintBrush *.pcx		7.0	7.5	7.5	7.5	7.5				
8	PCX Multi-page *.dcx		6.5								
9	PDF Image *.pdf		7.0	8.0							
10	Portable Network Graphics *.png		7.5	7.5	7.5	7.5	8.0				
11	TIFF *.tif		6.5	7.5	7.5	8.0	8.0			8.0	8.0
12	TIFF - Class F *.tif			6.0	6.5	6.5	7.0	7.0	6.5	6.5	7.0
13	TIFF - Group 4 *.tif			6.5	6.5	7.0	7.5	7.0	6.5	6.5	6.5
14	TIFF - LZW *.tif		6.5	7.0	7.5	7.5	8.0	8.0	8.0	8.0	8.0
15	TIFF - Uncompressed *.tif		6.5	6.5	7.0	7.5	7.5	7.5	7.5	7.5	8.0
16	TIFF Multi-page *.tif		6.5	7.5	7.5	8.0	8.0			8.0	8.0
17	TIFF Multi-page - Class F *.tif			6.0	6.5	6.5	7.0	7.0	6.5	6.5	7.0
18	TIFF Multi-page - Group 4 *.tif			6.5	6.5	7.0	7.5	7.0	6.5	6.5	6.5
19	TIFF Multi-page - LZW *.tif		6.5	7.0	7.5	7.5	8.0	8.0	8.0	8.0	8.0
20	TIFF Multi-page - Uncompressed *.tif		6.5	6.5	7.0	7.5	7.5	7.5	7.5	7.5	8.0
21	Windows Bitmap *.bmp		7.0	8.0							
	Minimum Storage Quality	1.5	3.0	6.0	6.5	6.5	7.0	7.0	6.5	6.5	6.5
	Maximum Storage Quality	6.0	7.5	8.5	8.5	9.0	9.0	9.0	9.0	9.0	9.0
	Average Quality	4.6	6.2	7.3	7.6	7.8	8.1	8.2	8.1	8.0	8.1

Table (1–2) Showing the quality of scanned images in 21 image types and 10 different resolutions
After eliminating the low quality images (less than 6.5) and the high quality images (more than 8).

Full A4 Scanned Pages Images Types vs Storage Sizes in KiloBytes (256 Gray Shades)

No.	Scanning Resolution Image Type	75 dpi	100 dpi	150 dpi	200 dpi	250 dpi	300 dpi	350 dpi	400 dpi	450 dpi	500 dpi
1	FlashPic *.fpx		176								
2	Graphic Interchange Format *.gif		48	80							
3	JPEG *.jpg		96	160	240						
4	PaperPort Browser-Viewable *.htm		160								
5	PaperPort Image *.max		112	192							
6	PaperPort Self-Viewing *.exe										
7	PC PaintBrush *.pcx										
8	PCX Multi-page *.dcx										
9	PDF Image *.pdf		128	224							
10	Portable Network Graphics *.png		192								
11	TIFF *.tif		208								
12	TIFF - Class F *.tif			48	48	96	96	144	192		
13	TIFF - Group 4 *.tif			32	48	80	80	128	176	240	
14	TIFF - LZW *.tif		208								
15	TIFF - Uncompressed *.tif										
16	TIFF Multi-page *.tif		208								
17	TIFF Multi-page - Class F *.tif			48	48	96	96	144	192		
18	TIFF Multi-page - Group 4 *.tif			32	48	80	80	128	176	240	
19	TIFF Multi-page - LZW *.tif		208								
20	TIFF Multi-page - Uncompressed *.tif										
21	Windows Bitmap *.bmp										
	Minimum Storage Size	16	32	32	48	80	80	128	176	240	288
	Maximum Storage Size	544	976	2176	3856	6000	8640	11760	15360	19440	24016

Figure (1-4) Showing the sizes of scanned images in 21 image types and 10 different resolutions
After eliminating the unacceptable quality images and eliminating the images with sizes more than 300 KB.

Full A4 Scanned Pages Accepted Images Types vs Storage Sizes in KiloBytes (256 Gray Shades)

No	Scanning Resolution	100	150	200	250	300	350	400	450
	Image Type	**dpi**	**dpi**	**dpi**	**dpi**	**dpi**	**dpi**	**dpi**	**dpi**
1	FlashPic *.fpx	176							
2	Graphic Interchange Format *.gif	48	80						
3	JPEG *.jpg	96	160	240					
4	PaperPort Browser-Viewable *.htm	160							
5	PaperPort Image *.max	112	192						
6	PDF Image *.pdf	128	224						
7	Portable Network Graphics *.png	192							
8	TIFF *.tif	208							
9	TIFF - Class F *.tif		48	48	96	96	144	192	
10	TIFF - Group 4 *.tif		32	48	80	80	128	176	240
11	TIFF - LZW *.tif	208							
	Minimum Storage Size	32	32	48	80	80	128	176	240
	Maximum Storage Size	976	2176	3856	6000	8640	11760	15360	19440

Figure (1–5) Showing the acceptable document Images in size and in scanning resolutions.

Full A4 Scanned Pages Accepted Images Types vs Storage Sizes in KiloBytes (256 Gray Shades)

No	Scanning Resolution	100	150	200	250	300	350
	Image Type	**dpi**	**dpi**	**dpi**	**dpi**	**dpi**	**dpi**
1	Graphic Interchange Format *.gif		80				
2	JPEG *.jpg		160				
3	PaperPort Image *.max	112					
4	PDF Image *.pdf	128					
5	Portable Network Graphics *.png	192					
6	TIFF - Class F *.tif					96	144
7	TIFF - Group 4 *.tif		32	48	80		
	Minimum Storage Size	32	32	48	80	80	128
	Maximum Storage Size	976	2176	3856	6000	8640	11760

Figure (1–5B) Showing the sizes of scanned images in 7 image types and 6 different resolutions after eliminating the unacceptable quality images (outside [7.0 , 7.5]) and eliminating the images with sizes more than 200 KB.

1 – 2 Optimizing The Selection Of The (Black And White) Scanned Document Images

The scanned A4 sample document page shown previously in **Figure (1–1)** was also stored as (Black and White) output images for all the selected scanning resolutions and image types[7]. The sizes of these scanned document images are as shown in **Figure (1-6).**

From observing **Figure (1–6)**, we can see that different image types and resolutions are stored with different sizes based on the resolution and the image type.

Applying the (POSSDI) Process

It could be appropriate in our first analysis that we assume that an A4 (Black and White) page should not exceed the size of 150 KB when stored. Therefore, our **first step of applying the (POSSDI) process** would be to eliminate all the scanned pages that consumed more than 150 KB in size. The result of this step is shown in **Figure (1–7)** below.

It may be necessary at this point to mention that the scanned pages were also observed by 10 different average users for quality evaluation and were given a readable quality mark out of 10 points as we proceeded with our (256 Gray Scale) scanned images analysis. The table of the (Black and White) quality evaluation is shown below as **Table (1–3).** It was agreed on that the scanned images with quality less than 6.5 is considered low, and the quality more than 8 is considered high[8]; therefore, in our **second step of applying the (POSSDI) process**, we will eliminate all the scanned pages that were low in quality (less than 6.5 marks) and also the high quality (more than 8.0 points). The result of the second step is shown in **Table (1-4)** below.

Figure (1–8) below shows our **third step of applying the (POSSDI) process** which is approach of optimizing the selection based on the storage size

[7] Scanning was done using PaperPort Document Management software which allows scanning and saving the scanned page using multiple image types and different scanning resolutions.

[8] Most of the stored images with quality 8.5 up to 10.0 were consuming a high storage space and hence they were considered unacceptable and eliminated.

and the quality of the stored image. It shows only the images that consumed 150 KB in size or less and in the same time have acceptable quality according to our quality assumption stated previously.

Analyzing Figure (1–8) to obtain the best storing size and scanning resolution for each considered image type

As we can observe from **Figure (1–8),** that some image types are completely unacceptable in our study, according to our assumption, since they failed to have any acceptable image occurrence[9] (in the occurrence table) either in quality[10] or in size or in both. These image types are: FlashPic *.fpx, JPEG *.jpg, PaperPort Self-Viewing *.exe, TIFF - Uncompressed *.tif, and Windows Bitmap *.bmp .

Also we will ignore the type that have (multi) part , as we did with the 256 grey shades scanned images, since they are actually duplicates of their single page size and quality but were available in our document management software.

In addition, we eliminated totally the columns for the 75 dpi and the 100 dpi scanning resolution since they had no acceptable occurrences for any acceptable image qualities (or sizes) and for the 500 dpi scanning resolution column for the same reason as well.

Now we are left with fewer acceptable image types and scanning resolutions as shown in **Figure (1–9).**

Refining the Selection for Further Optimization

By observing **Figure (1–9)**, we may keep our accepted images sizes and quality interval **or** narrow it further more if we wish as the **final step of the (POSSDI) Process**. To see the effect of narrowing the acceptance interval, we assume that for the (Black and White) scanned images the acceptable may not exceed 100 KB instead of 150 KB and the accepted quality interval is [7.0 , 7.5] marks only instead of [6.5 , 7.0 ,7.5 , 8.0]. The resulting figure is shown below as **Figure (1–9B)**.

Analyzing Figure (1–9B)

It is possible now for any document archiving specialist to observe **Figure (1–9B)** and make an easier selection of which image type to use and which resolution to use when scanning and storing the scanned images. As we can notice that images with the type (gif) are accepted when scanned with a resolution 300 dpi and they consume 96 KB in storage space. Also the images with type (htm) are accepted when scanned with resolution 200 dpi and 250 dpi and they consume 48 KB and 64 KB of storage space respectively when stored. Further more the images of type (max) are accepted when scanned using 250 dpi resolution and they occupy 32 KB, where the images of type (pdf) are accepted when scanned using 300 dpi resolution and they occupy 48 KB. In addition, the images of type (png) are accepted when scanned using 300 dpi, 350 dpi, and 400 dpi scanning resolutions and they occupy 80 KB, 96 KB, and 96 Kb respectively. Finally the images of type (tif) are accepted when scanned with resolutions 200 dpi, 250 dpi, 300 dpi, and 350 dpi and they consume 32 KB, 32 KB, 48 KB, and 48 KB respectively, where the images of type (tif class F) are accepted when scanned with resolutions 250 dpi, 300 dpi, 350 dpi and 400 dpi and they consume 48 KB, 64 KB,64 KB, and 80 KB respectively; were the images of type (tif group 4) are accepted when scanned with resolutions 200 dpi , 250 dpi, 300 dpi, and 350 dpi and they consume 32 KB, 32 KB, 48 KB, and 48 KB respectively. Finally the images of type (tif LZW) are accepted when scanned with resolutions 200 dpi , 250 dpi, 300 dpi, and 350 dpi and they consume 48 KB, 64 KB, 80 KB, and 96 KB respectively.

From **Figure (1–9B)** we can see that the (gif), (htm), and the (png) image types may be excluded since there is no multi page image type available for these types, also the (max) image type was excluded previously since it requires a special image viewer. Hence, we are left with the image types (pdf), (tif), (tiff Class F), (tif group 4), and (tif LZW) to choose from.

These types occupied 48 KB when a scanning resolution of 300 dpi was used for the (pdf) image type, 32 KB, 32 KB, 48 KB and 48 KB when scanning resolution of 200 dpi, 250 dpi, 300 dpi, and 350 dpi was used respectively for the (tif) image type, and 48 KB, 64 KB, 64 KB and 80 KB when scanning resolution of 250 dpi, 300 dpi, 350 dpi, and 400 dpi was used respectively for the (tif Class 4) image type. Where the image types (tif group 4) consumed 32 KB, 32 KB, 48 KB, and 48 KB when scanning resolution of 200 dpi, 250 dpi, 300 dpi, and 350 dpi was used respectively. Finally, the image types (tif LZW) consumed 48 KB, 64 KB, 80 KB, and 96 KB when scanning resolution of 200 dpi, 250 dpi, 300 dpi, and 350 dpi was used respectively.

---

[9] The acceptable image storing size for the (Black and White) is less than or equal to 150 KB as assumed.

[10] The acceptable image storing quality for the (Black and White) is 6.5, 7.0, 7.5, 8.0 as assumed.

Full A4 Scanned Pages Images Types vs Storage Sizes in KiloBytes (Black and White)

No.	Scanning Resolution	75	100	150	200	250	300	350	400	450	500
	Image Type	dpi	dpi	dpi	dpi	dpi	dpi	dpi	dpi	dpi	dpi
1	FlashPic *.fpx	112	176	336	528	720	976	1,232	1,520	1,792	2,112
2	Graphic Interchange Format *.gif	16	16	32	64	80	96	112	144	160	192
3	JPEG *.jpg	48	96	176	288	384	496	624	752	864	1,008
4	PaperPort Browser-Viewable *.htm	32	32	32	48	64	64	80	80	96	96
5	PaperPort Image *.max	16	16	32	32	32	48	48	64	64	64
6	PaperPort Self-Viewing *.exe	368	368	384	384	400	400	400	416	416	416
7	PC PaintBrush *.pcx	16	32	64	112	144	192	224	272	320	384
8	PCX Multi-page *.dcx	16	32	64	112	144	192	224	272	320	384
9	PDF Image *.pdf	16	16	32	32	32	48	48	48	48	64
10	Portable Network Graphics *.png	16	16	32	48	48	80	96	96	96	112
11	TIFF *.tif	16	16	16	32	32	48	48	48	64	64
12	TIFF - Class F *.tif	16	16	32	32	48	64	64	80	80	96
13	TIFF - Group 4 *.tif	16	16	16	32	32	48	48	48	64	64
14	TIFF - LZW *.tif	16	16	32	48	64	80	96	112	128	144
15	TIFF - Uncompressed *.tif	80	128	272	496	752	1,088	1,472	1,920	2,448	3,008
16	TIFF Multi-page *.tif	16	16	16	32	32	48	48	48	64	64
17	TIFF Multi-page - Class F *.tif	16	16	32	32	48	64	64	80	80	96
18	TIFF Multi-page - Group 4 *.tif	16	16	16	32	32	48	48	48	64	64
19	TIFF Multi-page - LZW *.tif	16	16	32	48	64	80	96	112	128	144
20	TIFF Multi-page - Uncompressed *.tif	80	128	272	496	752	1,088	1,472	1,920	2,448	3,008
21	Windows Bitmap *.bmp	80	128	288	496	768	1,104	1,488	1,936	2,448	3,024
	Minimum Storage Size	16	16	16	32	32	48	48	48	48	64
	Maximum Storage Size	368	368	384	528	768	1104	1488	1936	2448	3024

Figure (1–6) Showing the sizes of scanned images in 21 image types and 10 different resolutions.

Full A4 Scanned Pages Images Types vs Storage Sizes in KiloBytes (Black and White)

No.	Scanning Resolution	75	100	150	200	250	300	350	400	450	500
	Image Type	dpi	dpi	dpi	dpi	dpi	dpi	dpi	dpi	dpi	dpi
1	FlashPic *.fpx	112									
2	Graphic Interchange Format *.gif	16	16	32	64	80	96	112	144		
3	JPEG *.jpg	48	96								
4	PaperPort Browser-Viewable *.htm	32	32	32	48	64	64	80	80	96	96
5	PaperPort Image *.max	16	16	32	32	32	48	48	64	64	64
6	PaperPort Self-Viewing *.exe										
7	PC PaintBrush *.pcx	16	32	64	112	144					
8	PCX Multi-page *.dcx	16	32	64	112	144					
9	PDF Image *.pdf	16	16	32	32	32	48	48	48	48	64
10	Portable Network Graphics *.png	16	16	32	48	48	80	96	96	96	112
11	TIFF *.tif	16	16	16	32	32	48	48	48	64	64
12	TIFF - Class F *.tif	16	16	32	32	48	64	64	80	80	96
13	TIFF - Group 4 *.tif	16	16	16	32	32	48	48	48	64	64
14	TIFF - LZW *.tif	16	16	32	48	64	80	96	112	128	144
15	TIFF - Uncompressed *.tif	80	128								
16	TIFF Multi-page *.tif	16	16	16	32	32	48	48	48	64	64
17	TIFF Multi-page - Class F *.tif	16	16	32	32	48	64	64	80	80	96
18	TIFF Multi-page - Group 4 *.tif	16	16	16	32	32	48	48	48	64	64
19	TIFF Multi-page - LZW *.tif	16	16	32	48	64	80	96	112	128	144
20	TIFF Multi-page - Uncompressed *.tif	80	128								
21	Windows Bitmap *.bmp	80	128								
	Minimum Storage Size	16	16	16	32	32	48	48	48	48	64
	Maximum Storage Size	368	368	384	528	768	1104	1488	1936	2448	3024

Figure (1–7) Showing the result after eliminating all the scanned images with sizes above 150 KB

Full A4 Scanned Pages Images Quality Evaluation Sheet (Black and White)

No.	Scanning Resolution / Image Type	75 dpi	100 dpi	150 dpi	200 dpi	250 dpi	300 dpi	350 dpi	400 dpi	450 dpi	500 dpi
1	FlashPic *.fpx	1.0	2.0	4.0	6.5	7.0	8.0	8.5	8.5	8.5	8.5
2	Graphic Interchange Format *.gif	1.0	5.0	6.0	6.5	6.5	7.0	7.5	8.0	8.0	8.5
3	JPEG *.jpg	1.5	5.0	6.0	7.0	7.0	7.0	7.5	7.5	8.0	8.5
4	PaperPort Browser-Viewable *.htm	1.0	2.0	5.0	7.0	7.5	8.0	8.5	8.5	8.5	8.5
5	PaperPort Image *.max	1.5	2.5	5.0	6.5	7.5	8.0	8.5	8.5	8.5	8.5
6	PaperPort Self-Viewing *.exe	1.0	2.0	5.5	6.5	7.0	7.5	8.0	8.0	8.5	8.5
7	PC PaintBrush *.pcx	2.0	5.0	6.5	7.0	7.0	7.0	7.0	7.5	8.0	8.0
8	PCX Multi-page *.dcx	1.5	2.5	5.5	6.5	7.0	7.5	8.0	8.0	8.5	8.5
9	PDF Image *.pdf	1.0	3.5	5.0	6.0	6.5	7.0	8.0	8.5	8.5	8.5
10	Portable Network Graphics *.png	2.0	5.0	6.0	6.5	6.5	7.0	7.0	7.5	7.5	8.0
11	TIFF *.tif	2.0	5.0	6.5	7.0	7.0	7.5	7.5	8.0	8.0	8.5
12	TIFF - Class F *.tif	1.5	5.0	6.5	6.5	7.0	7.0	7.5	7.5	8.0	8.0
13	TIFF - Group 4 *.tif	1.5	5.0	6.5	7.0	7.0	7.0	7.5	8.0	8.0	8.0
14	TIFF - LZW *.tif	1.5	4.5	6.5	7.0	7.0	7.0	7.5	8.0	8.0	8.5
15	TIFF - Uncompressed *.tif	1.0	5.0	6.5	7.0	7.0	7.0	7.5	7.5	8.0	8.0
16	TIFF Multi-page *.tif	2.0	5.0	6.5	7.0	7.0	7.5	7.5	8.0	8.0	8.5
17	TIFF Multi-page - Class F *.tif	1.5	5.0	6.5	6.5	7.0	7.0	7.5	7.5	8.0	8.0
18	TIFF Multi-page - Group 4 *.tif	1.5	5.0	6.5	7.0	7.0	7.0	7.5	8.0	8.0	8.0
19	TIFF Multi-page - LZW *.tif	1.5	4.5	6.5	7.0	7.0	7.0	7.5	8.0	8.0	8.5
20	TIFF Multi-page - Uncompressed *.tif	1.0	5.0	6.5	7.0	7.0	7.0	7.5	7.5	8.0	8.0
21	Windows Bitmap *.bmp	1.5	2.5	5.0	6.5	7.0	8.0	8.5	8.5	8.5	8.5
	Minimum Storage Quality	1.0	2.0	4.0	6.0	6.5	7.0	7.0	7.5	7.5	8.0
	Maximum Storage Qality	2.0	5.0	6.5	7.0	7.5	8.0	8.5	8.5	8.5	8.5
	Average Quality	1.4	4.1	5.9	6.7	7.0	7.3	7.7	8.0	8.1	8.3

Table (1–3) Showing the quality of scanned images in 21 image types and 10 different resolutions.

Full A4 Scanned Pages Images Quality Evaluation Sheet (Black and White)

No.	**Scanning Resolution / Image Type**	**75 dpi**	**100 dpi**	**150 dpi**	**200 dpi**	**250 dpi**	**300 dpi**	**350 dpi**	**400 dpi**	**450 dpi**	**500 dpi**
1	FlashPic *.fpx				6.5	7.0	8.0				
2	Graphic Interchange Format *.gif				6.5	6.5	7.0	7.5	8.0	8.0	
3	JPEG *.jpg				7.0	7.0	7.0	7.5	7.5	8.0	
4	PaperPort Browser-Viewable *.htm				7.0	7.5	8.0				
5	PaperPort Image *.max				6.5	7.5	8.0				
6	PaperPort Self-Viewing *.exe				6.5	7.0	7.5	8.0	8.0		
7	PC PaintBrush *.pcx			6.5	7.0	7.0	7.0	7.0	7.5	8.0	
8	PCX Multi-page *.dcx				6.5	7.0	7.5	8.0	8.0		
9	PDF Image *.pdf					6.5	7.0	8.0			
10	Portable Network Graphics *.png				6.5	6.5	7.0	7.0	7.5	7.5	
11	TIFF *.tif			6.5	7.0	7.0	7.5	7.5	8.0	8.0	
12	TIFF - Class F *.tif			6.5	6.5	7.0	7.0	7.5	7.5	8.0	
13	TIFF - Group 4 *.tif			6.5	7.0	7.0	7.0	7.5	8.0	8.0	
14	TIFF - LZW *.tif			6.5	7.0	7.0	7.0	7.5	8.0	8.0	
15	TIFF - Uncompressed *.tif			6.5	7.0	7.0	7.0	7.5	7.5	8.0	
16	TIFF Multi-page *.tif			6.5	7.0	7.0	7.5	7.5	8.0	8.0	
17	TIFF Multi-page - Class F *.tif			6.5	6.5	7.0	7.0	7.5	7.5	8.0	
18	TIFF Multi-page - Group 4 *.tif			6.5	7.0	7.0	7.0	7.5	8.0	8.0	
19	TIFF Multi-page - LZW *.tif			6.5	7.0	7.0	7.0	7.5	8.0	8.0	
20	TIFF Multi-page - Uncompressed *.tif			6.5	7.0	7.0	7.0	7.5	7.5	8.0	
21	Windows Bitmap *.bmp				6.5	7.0	8.0				
	Minimum Storage Quality	1.0	2.0	4.0	6.0	6.5	7.0	7.0	7.5	7.5	8.5
	Maximum Storage Qality	2.0	5.0	6.5	7.0	7.5	8.0	8.5	8.5	8.5	8.5
	Average Quality	1.4	4.1	5.9	6.7	7.0	7.3	7.7	8.0	8.1	8.3

Table (1–4) Showing the quality of scanned images in 21 image types and 10 different resolutions after eliminating the low quality images (less than 6.5) and the high quality images (more than 8).

Full A4 Scanned Pages Images Types vs Storage Sizes in KiloBytes (Black and White)

No.	Scanning Resolution	75	100	150	200	250	300	350	400	450	500
	Image Type	dpi	dpi	dpi	dpi	dpi	dpi	dpi	dpi	dpi	dpi
1	FlashPic *.fpx										
2	Graphic Interchange Format *.gif				64	80	96	112	144		
3	JPEG *.jpg										
4	PaperPort Browser-Viewable *.htm				48	64	64				
5	PaperPort Image *.max				32	32	48				
6	PaperPort Self-Viewing *.exe										
7	PC PaintBrush *.pcx			64	112	144					
8	PCX Multi-page *.dcx				112	144					
9	PDF Image *.pdf					32	48	48			
10	Portable Network Graphics *.png				48	48	80	96	96	96	
11	TIFF *.tif			16	32	32	48	48	48	64	
12	TIFF - Class F *.tif			32	32	48	64	64	80	80	
13	TIFF - Group 4 *.tif			16	32	32	48	48	48	64	
14	TIFF - LZW *.tif			32	48	64	80	96	112	128	
15	TIFF - Uncompressed *.tif										
16	TIFF Multi-page *.tif			16	32	32	48	48	48	64	
17	TIFF Multi-page - Class F *.tif			32	32	48	64	64	80	80	
18	TIFF Multi-page - Group 4 *.tif			16	32	32	48	48	48	64	
19	TIFF Multi-page - LZW *.tif			32	48	64	80	96	112	128	
20	TIFF Multi-page - Uncompressed *.tif										
21	Windows Bitmap *.bmp										
	Minimum Storage Size	16	16	16	32	32	48	48	48	48	64
	Maximum Storage Size	368	368	384	528	768	1104	1488	1936	2448	3024

Figure (1-8) Showing the sizes of scanned images in 21 image types and 10 different resolutions after eliminating the unacceptable quality images and eliminating the images with sizes more than 150 KB.

Full A4 Scanned Pages Images Types vs Storage Sizes in KiloBytes (Black and White)

No.	Scanning Resolution	150	200	250	300	350	400	450
	Image Type	dpi	dpi	dpi	dpi	dpi	dpi	dpi
1	Graphic Interchange Format *.gif		64	80	96	112	144	
2	PaperPort Browser-Viewable *.htm		48	64	64			
3	PaperPort Image *.max		32	32	48			
4	PC PaintBrush *.pcx	64	112	144				
5	PCX Multi-page *.dcx		112	144				
6	PDF Image *.pdf			32	48	48		
7	Portable Network Graphics *.png		48	48	80	96	96	96
8	TIFF *.tif	16	32	32	48	48	48	64
9	TIFF - Class F *.tif	32	32	48	64	64	80	80
10	TIFF - Group 4 *.tif	16	32	32	48	48	48	64
11	TIFF - LZW *.tif	32	48	64	80	96	112	128
	Minimum Storage Size	16	32	32	48	48	48	48
	Maximum Storage Size	384	528	768	1104	1488	1936	2448

Figure (1–9) Showing the acceptable document Images in size (150 KB or less) and in scanning resolutions (150 dpi – 450 dpi).

Full A4 Scanned Pages Accepted Images Types vs Storage Sizes in KiloBytes (Black and White)

No.	Scanning Resolution	200	250	300	350	400
	Image Type	dpi	dpi	dpi	dpi	dpi
1	Graphic Interchange Format *.gif			96		
2	PaperPort Browser-Viewable *.htm	48	64			
3	PaperPort Image *.max		32			
4	PDF Image *.pdf			48		
5	Portable Network Graphics *.png			80	96	96
6	TIFF *.tif	32	32	48	48	
7	TIFF - Class F *.tif		48	64	64	80
8	TIFF - Group 4 *.tif	32	32	48	48	
9	TIFF - LZW *.tif	48	64	80	96	
	Minimum Storage Size	32	32	48	48	48
	Maximum Storage Size	528	768	1104	1488	1936

Figure (1–9B) Showing the sizes of scanned images in 9 image types and 5 different resolutions after eliminating the unacceptable quality images (outside [7.0 , 7.5]) and eliminating the images with sizes more than 100 KB.

Any of these five image types are perfectly acceptable to use, the recommended image types are obviously the (tif) and the (tif group 4) types since they occupy the smallest storage space, but if the storage space is not a crucial issue to consider, then the standard Acrobat (pdf) type is a good image type selection since it is well know image type and it occupied a relatively acceptable storage space (48 KB) for the good quality it shows (300 dpi), in addition to the flexibility of the Acrobat viewer gives as far as zooming and printing, and browsing the saved document pages images.

Conclusion and Summary

Our goal in this research paper was to obtain a close to optimal scanning size and type while not sacrificing the quality of the scanned image by selecting a suitable scanning resolution and an acceptable quality for the scanned document image as well. We introduced the **(POSSDI)** Process which is "**The Process of Optimizing the Selection of The Scanned Document Images**". The process aims to enable the archiving systems users for selecting a balanced document images in size, type, and quality for archiving purposes. It was shown practically that by applying this process, we were able to reach an acceptable image type and size to store and hence to achieve efficiently. The yielded schedules out of applying the **(POSSDI)** process makes it easy for all documents achieving specialists to refer to and decide what type and resolution to select prior to storing their documents images. Same **(POSSDI)** process steps were applied to different color scanned document images and yielded the expected results.

References

1 - L. Alshuler and R. Dolin. (Ed.) Version 3 Standard: Clinical Document Architecture Release 1.0. Canada HL7. November 6, 2000.

2 - Frei, H.P. Information retrieval - from academic research to practical applications. In: Proceedings of the 5th Annual Symposium on Document Analysis and Information Retrieval, Las Vegas, April 1996.

3 - Clifton, H., Garcia-Molina, H., Hagmann, R.: ''The Design of a Document Database'', Proc. of the ACM

4 - ISO/IEC, Information Technology - Text and office systems - Document Filing and Retrieval (DFR) -

5 - Part 1 and Part 2, International Standard 10166, 1991

6 - Sonnenberger, G. and Frei, H. Design of a reusable IR framework. In: SIGIR'95: Proceedings of 18th ACM-SIGIR Conference on Research and Development in Information Retrieval, Seattle, 1995. (New York: ACM, 1995). 49-57.

7 - Fuhr, N. Toward data abstraction in networked information systems. Information Processing and Management 5(2) (1999) 101-119.

8 - Jacobson, I., Griss, M. and Jonsson, P. Software reuse: Architecture, process and organization for business success. (New York: ACM Press, 1997).

9 - Davenport, T.H. Information ecology: Mastering the information and knowledge environment. (Oxford: Oxford University Press, 1997).

10 - Diaz, I., Velasco, M., Llorens, J. and Martinez, V. Semi-automatic construction of a thesaurus applying domain analysis techniques. International Forum on Information and Documentation, 23(2) (1998) 11-19.

11 - Bocij, P. et al., Business information systems: Technology, development and management. (London: Financial Times Management, 1999).

12 - Kaszkiel, M. and Zobel, J. Passage retrieval revisited. In: SIGIR'97: Proceedings of 20th ACM-SIGIR Conference on Research and Development in Information Retrieval Philadelphia, 1997. (New York: ACM, 1997).

13 - Richard Casey, " Document Image Analysis" , available at:

http://cslu.cse.ogi.edu/HLTsurvey/ch2node4.html

14 - H Bunke And P S P Wang, "Handbook Of Character Recognition And Document Image Analysis", available at

http://www.worldscibooks.com/compsci/2757.html

15 - M Simone, " Document Image Analysis And Recognition" , available at :

http://www.dsi.unifi.it/~simone/DIAR/

16 - "Document Image Analysis", avalable at :

http://elib.cs.berkeley.edu/dia.html

Infrastructure for Bangla Information Retrieval in the Context of ICT for Development

Nafid Haque, M. Hammad Ali, Matin Saad Abdullah, Mumit Khan
BRAC University
66 Mohakhali, Dhaka-1212
Bangladesh.
Email: nafid99@yahoo.com, hammad2099@yahoo.com, mabdullah@bracuniversity.net, mumit@bracuniversity.ac.bd

Abstract

In this paper, we talk about developing a search engine and information retrieval system for Bangla. Current work done in this area assumes the use of a particular type of encoding or the availability of particular facilities for the user. We wanted to come up with an implementation that did not require any special features or optimizations in the user end, and would perform just as well in all situations. For this purpose, we picked two case studies to work on in our effort to finding a suitable solution to the problem. While working on these cases, we encountered several problems and had to find our way around these problems. We had to pick and choose from a set of software packages for the one that would best serve our needs. We also had to take into consideration user convenience in using our system, for which we had to keep in mind the diverse demographics of people that might have need for such a system. Finally, we came up with the system, with all the desired features. Some possible future developments also came into mind in the course of our work, which are also mentioned in this paper.

Introduction

Over the last decade or so, there has been a huge increase in the use of computers for storing information and ensuring access to this information. Computers have become our preferred method of interaction, work and most importantly, storage of data. The small lifetime of paper documents, and the difficulty of extracting important information from such documents has made it necessary for important information to be stored in digital form. As the amount of data stored electronically increases day by day, so does the need for efficiently searching through this vast collection of data. A huge body of data will serve us no purpose if at times of need we cannot find the desired information without having to sequentially browse through all of it. Not so long ago, almost all computer data was exclusively in English. In recent years, though, there has been a rise in the use of other languages with computers. Since the learning of the English language proved to be a bottleneck in the process of familiarizing people with a computerized system, there has been a greater frequency of using one's native language with computers. Bangla is one such language. However, the plethora of text corpus in Bangla is not represented accurately if we only take into consideration the number of websites being presented in Bangla. Nevertheless, newspapers and magazines are being increasingly attentive towards creating and maintaining their own websites. From that there is an automatic rise in the need for a search engine that can take a Bangla query string, consult the indexed database and produce results. This is particularly true for newspaper websites, which have extremely dynamic content but at the same time also need to ensure easy access to their archives. Apart from the Bangla documents available over the Internet there are many other documents in Bangla that are stored locally for certain purposes and are used to get the information required at the time of need. The most common example of a considerable amount of Bangla data residing in local machines would be the data repository that certain organizations like D.Net [8] want to make available to people in the rural areas of the country. Ensuring this access to information is absolutely integral to the development of a country like Bangladesh. People in the rural areas are often in need of information regarding issues such as the legal system of Bangladesh, or the symptoms of certain diseases, or the best practices regarding the harvest of crops. All of these data can be made available in portable media that can be sent out to information kiosks in remote parts of the country. With this comes the need for efficiently searching through all of this data, based on a set of relevant keywords. All of this data would not be of any use if certain important information could only be retrieved through systematically going through the entire *corpus*. The time and manpower required for such an operation makes the solution prohibitively

K. Elleithy (ed.), Advances and Innovations in Systems, Computing Sciences and Software Engineering, 325–330.

expensive and ineffective. Towards this end, what we are trying is not to coin a revolutionary new technology, something that the world has never seen before. We are more interested in *harnessing available technology for the sake of national development*. A lot has been done to ensure human rights in a diverse number of sectors, ranging from education and medical facilities to the rights of practicing one's own religion and philosophy in life. However, little has been done to ensure the rights of access to information for people from all walks of life. This is the era of information and communications technology. Just like before this there have been upsurges in the fields of industries, electricity or even the advent of computers as a tool, this is the era of the upsurge in Information and Communications Technology for Development. In today's world, the most influential men are those who have the most convenient access to the largest body of information. As stated in the WSIS Geneva 2003 Declaration, "our common desire and commitment to build a people-centered, inclusive and development-oriented Information Society, where everyone can create, access, utilize and share information and knowledge, enabling individuals, communities and peoples to achieve their full potential in promoting their sustainable development and improving their quality of life, premised on the purposes and principles of the Charter of the United Nations and respecting fully and upholding the Universal Declaration of Human Rights" [15].

Computers today have become a common household tool in developed countries. However, the technology made possible by the use of computers is not equally accessible in all areas of a developing country like Bangladesh. Digital divide has become a matter of great concern. In many countries, the situation is such that in the cities people are enjoying the facilities provided by the latest and the best in communications infrastructure, whereas in other parts of the country people do not even have telephone facilities, let alone access to the internet. For instance, in a developing country like Bangladesh, digital divide is a very serious problem. The Internet is a novel concept in most parts of Bangladesh. Ensuring Internet access all over the country is a faraway dream, something that might yet take decades. Even in the areas that do have internet facilities, the quality of service varies widely. What we want is an alternative, one that can be achieved in a much shorter time but perform just as well. So *in this paper we propose an infrastructure to extract information from a collection of Bangla text with all the modern features of an online search engine*. The proposed solution will be able to extract information from any online sites as well as data that are stored locally on any portable media, and will not have any dependencies that would make it impractical for use in remote areas of a developing country like Bangladesh.

Background Study

Extracting information from a collection of data is not a new concept. Many proprietary and open-source search engines are capable of this task. The search engines that support Unicode can easily be used for searching for a query string in any language supported by Unicode. One of the leading online search engines today is Google which has recently come up with an application called Google Desktop [1]. Google Desktop indexes the data stored on a local machine and allows an interface to search through the data with all the features of its online search engine. There are many online search engines other than Google, like Yahoo [2] and MSN [3]. However, since these leading search engines are all proprietary products of their respective developers, they cannot be customized to meet specific user needs. Apart from these proprietary search engines there are many open-source engines freely available that can be customized to meet specific user needs. Lucene [4] is one of the most popular and mature open-source search engines. Nutch [5] is a useful environment built upon Lucene that gives the user the full advantage of a search engine. Vicaya [6] is a search engine built on Nutch and Lucene. Vicaya can itself reside on any portable media like a CD and search through its contents without any prior installation on the machine. Our paper proposes a system similar to Vicaya but with additional features to make searching in Bangla easier for the user.

Methodology

The reason behind focusing not just on search engines for the web but also on an Information Retrieval system for the local machine stemmed from the fact that there is a greater demand for searching Bangla text from a local machine or portable media than from the Internet. This is due to the lack of availability of Internet facilities in remote areas of the country. *In remote areas of the country people are in need of specific information. Keeping this in mind many Government and Non-government organizations have taken initiatives to setup information centers where a person can come for their required information.* The organizations working with this concept wish to setup these information centers as close as possible to those who need the information. However, Internet facilities cannot be provided to these centers. Thus arises the need for providing all the information in some portable media like CD/DVD ROM. Obviously, then comes the need to search through the corpus at regular intervals. At present, all the available information retrieval/search facilities from a local media are based on the concept of raw string matching – probably the simplest and most inefficient method of searching. The problem with string matching is that it finds all and only the words that are spelled exactly like the query word. That is, it does not allow for the concept of fuzzy searches, words that are spelled very similarly, or words that sound phonetically similar and are very often confused for each other. Another limitation, although not very common, could be the fact that such searching techniques do not use the power of reverse indexing, something that has become very common in the arena of web search engines. This means that each time a query word is typed in, the entire corpus is searched all over again to find each and every instance of the word. As can be readily seen, this can be a great obstacle towards efficient retrieval of data.

So, we needed a state-of-the-art implementation that would work for the Internet as well as for the local machine. Crawling through the data at regular intervals would be a must for websites with contents that change on a regular basis. Thus the index file should be updated along with the website so that the new data can be searched too. For data repositories in a local machine, crawling need only be a one-time thing. We can collect all the data, index it and provide the indexed database in the same media along with the actual collection of data. For these two purposes, we chose two entities as a case study: Prothom Alo as the website and projects managed by an organization called D.Net for the local machine information retrieval system.

Case 1:
We chose Prothom Alo [7] since it is the largest circulating Bangla newspaper at present and regularly manages a website. Though being the largest online daily, Prothom Alo was found to be using a proprietary encoding format rather than Unicode. Thus there is a need for a system that would identify the type of encoding being used and if the encoding is not Unicode, then the system will convert it to Unicode (text only). The crawler can then create the index database of the available content. Also we found that the main subscribers/users of the online version of the newspaper are the Non-Resident Bangladeshi (NRB) citizens. These people mostly prefer writing Bangla using the English alphabets. Thus a mechanism allowing users to write their Bangla search query in English would be an added advantage in a search engine package.

Case 2:
For the part concerning information retrieval from a local machine, we chose the D.Net [8] projects Pallitathya [9] and Abolombon [10]. The objective of these projects is to ensure access to information to people in remote areas of the country, where Internet is not an option. Details about these projects are given below:

Abolombon is a project geared towards creating awareness among rural people about governance and human rights issues. The project has several dimensions: lack of awareness about their rights, lack of awareness related to the role and obligations of government institutions, lack of availability of information related to legal support, inadequate legal references for legal aid, among others [10]. The project has developed substantive digital legal content in simple Bangla. This information has also been made available through a web portal for open access to target people [10]. The legal contents are also available off line through CDs in project areas. D.Net delivers the legal contents through rural information centers in remote areas. A D.Net employee trained with the use of computers acts as the intermediary and searches for precise data from these CDs for people to come to these centers looking for help and advice. This intermediary would be the end user of our system, the person typing in the query words, obtaining the results and making it useful for the person in need of it.

Thus, with the vision of building a fully-functional Bangla text search engine with all modern features, we have chosen Lucene and Nutch to be the base of our work [4, 5]. Our aim is to customize the open-source search engine Nutch so that it can search for Bangla text from anywhere, be it portable media or the Internet. The implementation would also be able to do this regardless of the encoding method used. The system we are proposing will have three different parts integrated into a common package. The three parts can be referred to as content extraction, indexing and user-friendly input methods. The following sections will highlight each individual part of the package.

1. Content Extraction

From case 1 we have found that even the largest daily newspaper available online is not using Unicode as their encoding format. As the search engine only supports Unicode, there is a need to convert any other encoding format to Unicode. We have used a simple Java program that crawls through a website and tries to identify the encoding being used by analyzing the Meta Tags, Font types etc. After identifying the type of encoding, the extractor program then converts only the available text into its corresponding Unicode format leaving out any graphics. Thus, given an URL, it would try to find out the encoding used for the site. If it is using Unicode then a crawl can be performed directly. However, if the site is using some other encoding format, then it tries to find out more about the type of encoding used through a set of heuristics developed from stochastic modeling. Once the format is known, it can then use this knowledge to extract all the contents and convert them to its equivalent Unicode encoding. Nutch can then simply crawl through this set of files and create the index. Searching can then proceed on the indexed Unicode database.

2. Indexing

Initially, we focused on the search engine API Lucene. Lucene is an open source project that provides the basic functionalities of a search engine without providing the user interface [4]. It is written in Java and suitable for full-text search, especially in cross-platform scenarios. However, since Lucene is just an API it does not provide the framework for user input. So we moved onto Nutch, a package built upon Lucene. In addition to all the functionalities of Lucene, Nutch also provides an easy-to-use interface for user interaction [5]. Since it is written in Java, we would also have Unicode support. Specifically designed to crawl through Unicode text data over the Internet, Nutch was our best choice for our case study on the Prothom Alo website. Of course, to exactly meet our needs we had to make certain changes to the default configurations for Nutch. Following we give a quick overview of what changes we had to make based on our needs.

First of all, we had to create a file containing the set of URLs that we wanted Nutch to crawl. In our case, this was the URL to the Prothom Alo site. We kept the set of websites to be crawled limited to just the one due to bandwidth limitations, and also because we were still at a prototype stage and testing on one domain would be enough to gauge how our solution is performing. Following this step, we had to modify some of the parameters in the Nutch configuration files according to the site that we wanted to crawl, the settings of the Internet connection being used and other issues such as the depth to which we wanted the crawl to run. Once all this has been done, we can just run the crawl and the database would be created. Then this database could be used to search through the contents of the site. More details on how these configurations are to be tweaked can be found from the Nutch tutorial [12] and the Nutch Wiki [11]. For the searching, we would also need to have a Tomcat server running on the machine [13]. Some changes had to be made to ensure that the Tomcat server could support UTF-8 characters [14], which it does not by default. With all of these ready, we could crawl any site, create the database and then search through that database. So we essentially had an information retrieval system for the web, for the Bangla language. If a Bangla site uses Unicode, we could just crawl through it and get the database ready, limited only by the available bandwidth. If the site does not use Unicode, then we would have the added step of using the content extractor to grab the text content, converting it to Unicode and then crawling through that corpus. Nevertheless, the process of crawling and then searching through the database remains the same.

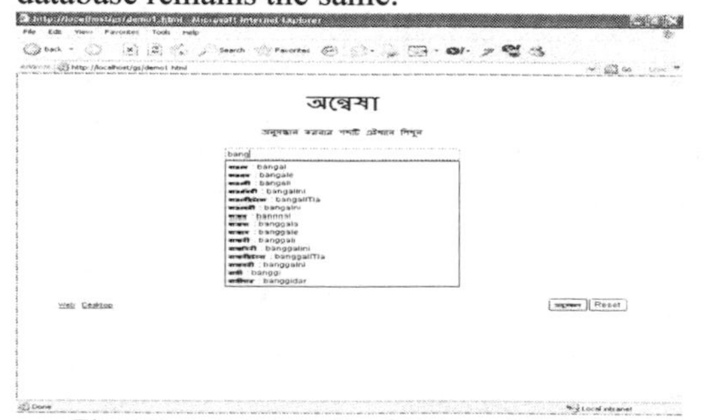

Fig. 1 Prototype Interface for our system

Fig. 2 Search results being displayed

3. User-friendly input methods
Taking the search query as an input needs a good consideration to make the entire project a success. Though it has been quiet a long time that Bangla is being used in daily computing, yet there is lack of a national standard keyboard layout. A wide range of Bangla keyboard layouts are available and thus the users of Bangla keyboards are not focused to a single keyboard layout. Also from recent trends it can be seen that people prefer to write Bangla phonetically using the English alphabets. Keeping all this in mind we have designed our system so that it can be of help for users regardless of the keyboard layout they use. As the whole system is based on Unicode, users can use any Unicode supported Bangla keyboard to enter their search string in to the system. In case users do not have a Unicode supported Bangla keyboard, the system will provide the user an option to enter their search strings in a transliterated form (writing Bangla using English alphabets.) A JavaScript code has been developed and integrated with the system so that the user can enter their Bangla text in a transliterated form. Here the user enters the Bangla query string in English alphabets and the JavaScript code converts that to its Bangla equivalent. The script code is also capable of suggesting the user with near matches of their input words. It can even suggest words from the index database created by Nutch so that the user can further refine his query string for the nearest match. This will be a great help in a situation where the user does not know the exact spelling of the word, a common scenario in the case of searching for names of individuals, among other things.

With the three above stated segments for the proposed system, the entire system can be a very strong tool for information retrieval from a corpus of any size. All the contents need not be on the Internet but can reside on any portable storage medium (e.g. CD/DVD ROM). In case of searching data from a portable media, the contents need to be indexed first and then along with a live version of the system can be put on the storage medium preferred. Then, the storage medium (e.g. CD/DVD ROM) can be carried to any place and the data in it can be retrieved with a query string. In case the system is to be used for searching data from the Internet, the system can be fed with a list of URLs at the beginning. Then the system will crawl all the contents of the given URLs and create its index file. Later the system can be used to search for information within this set of URLs. As the contents to be grabbed from the Internet may change periodically, thus a need for re-indexing may arise. In that case the system needs to be configured so that the given URLs can be crawled at regular intervals.

Future Work
At present, the content extractor that we are using only works for text content. In future, we would like to hone this implementation further so that it can work for other, more diverse types of content. This is important, because more often than not websites carry a lot of non-text information that could be just as important to the user. We want the user to be able to see all of this content while using our implementation and not just stay confined to seeing text. So we plan on working further with the content extractor in relation to our system. In addition, we want to enhance the system further, and go from an *information retrieval* system to an *information extraction* system. Such a system would have uses in the field of text summarization, text categorization using the presence of keywords or information extraction from a newspaper article or any such information portal.

Conclusion
In this paper, we have tried to suggest a system that would make it possible to search through a large collection of data in Bangla within a feasible amount of time and expense, regardless of the type of encoding being used and the availability of other facilities. Instead of focusing exclusively on searching for the web, we paid equal attention to information retrieval from portable media. We did not try to devise something completely original, something that no one has ever thought of before. *On the contrary, we were trying to suggest ways in which we could capitalize on open-source*

technology to ensure elimination of digital divide in different parts of Bangladesh. We wanted to suggest an infrastructure that would make sure that access to important information within limited time becomes a right and not a privilege. The main incentive behind our work was not the intention to come up with a system that no one has thought of or worked on before. Rather, it was trying to contribute something to the mission planned by the United Nations and the International Telecommunications Union in their endeavor to ensure fair access to information for everyone regardless of their geographical location or other background information, as expressed in the Geneva Declaration of 2003 [15]. This is the biggest reason why we did not focus solely on web search, and just try to hone it further. In a country like Bangladesh, the majority does not have access to the Internet. Furthermore, there does not seem to be any probability of ensuring Internet access in all parts of the country even within the next decade or so. So we tried to find a way to do the best we can *to harness the power of Information and Communications Technology for development*, even within the limitations that are imposed on us in the context of this country. We wanted to make sure that what we implemented would be of use not only to the city-dwellers wanting to browse through today's newspaper, but also the peasant in a remote village wanting to know about the best practices regarding his livelihood. We are still far away from the day when people from such diverse backgrounds will achieve equal rights in every walk of life, but we can always hope that our work will ensure equal rights to them at least in their pursuit of information. *Our work may not be groundbreaking in terms of innovation or pioneering in a field, but we trust that it will prove to be very important for the sake of national development. We believe it was our responsibility to use our abilities, and available technology, to do something for the betterment of the nation.* We sincerely hope that the infrastructure suggested in this paper can go far towards this end.

Acknowledgment

This work has been partially supported through PAN Localization Project (www.PANL10n.net) grant from the International Development Research Center, Ottawa, Canada, administered through Center for Research in Urdu Language Processing, National University of computer and Emerging Sciences, Pakistan.

References

1. Google Desktop, available online at http://desktop.google.com/about.html
2. Yahoo! Search, available online at http://search.yahoo.com
3. MSN Search, available online at http://search.msn.com
4. Erik Hatcher and Otis Gospodnetic, 'Lucene in Action', April 2006.
5. The Official Nutch Website – http://www.lucene.apache.org/nutch
6. Vicaya, available online at http://vicaya.sourceforge.net
7. Prothom Alo, the largest online daily newspaper in Bangla, available online at www.prothom-alo.net
8. D.Net – Development Research Network, www.dnet-bangladesh.org
9. Pallitathya, a research program of D.Net on understanding information needs from a village perceptive, http://www.pallitathya.org/
10. Abolombon, a program of D.Net designed to improve access to legal information on governance and human rights issues, http://www.abolombon.org/
11. The Nutch wiki, available online at http://wiki.apache.org/nutch/
12. The Nutch tutorial for Version 0.7.x, available online at http://www.lucene.apache.org/nutch/tutorial.html
13. A step by step guideline on how to configure and use Tomcat, available online at http://www.coreservlets.com/Apache-Tomcat-Tutorial/
14. Weblog on enabling Tomcat to support UTF-8 Encoding - http://rollerweblogger.org/page/roller/20040415
15. FAQ on the World Summit Information Society at http://www.itu.int/wsis/

AN IMPROVED WATERMARKING EXTRACTION ALGORITHM

Ning Chen, Jie Zhu
School of Electronic, Information and Electrical Engineering
Shanghai Jiao Tong University
Shanghai, China, 200030
(chenning_750210@163.com)

***Abstract*-Echo hiding is one of the prevailing techniques in audio watermarking due to its good perceptual quality. However, the detection ratio of this method is relatively low and its robustness against many common signal-processing operations is not satisfactory. In this paper, an improved watermarking extraction algorithm, which is based on auto-power-cepstrum, is proposed. Computer simulation results prove that the new method achieves higher detection ratio when compared with conventional auto-complex-cepstrum based algorithm and its robustness against various signal processing manipulations, such as Mp3 compression, re-sampling, cropping, re-quantization, filtering, amplitude amplifying, noise addition and time delay, is great.**

I. INTRODUCTION

Digital watermarking, a method of data hiding, has been proposed as a means for owner identification of digital data. Unlike cryptographic techniques, for example encryption, digital watermarking does not restrict or regulate access to the host signal, but ensures the embedded data remain inviolate and recoverable [1]. Recently, digital audio watermarking technique has found many other applications, such as copyright protection, integrity verification and carrying side information of audio clips.

To be effective in the protection of the ownership of intellectual property, the embedded watermarks should be imperceptible, which means they bring no subjective distortion to original audio signal and robust, which means they are able to survive manipulation and common signal processing, such as amplitude amplifying, Mp3 compression, re-sampling, filtering, time-delay and noise addition [2]. On the other hand, watermark detection should unambiguously identify the owner, which can be achieved by blind detection.

As an important audio watermarking technique, echo hiding is widely used due to good perceptual quality. It was first proposed by W. Bender et al [1]. Bender proposed using two delay kernels to represent the binary 0 and 1 in watermark that is to be embedded. In decoding system, the auto-complex-cepstrum (the autocorrelation of complex cepstrum) is employed to estimate echo delay to extract watermarks. References [3]-[9] adopt the same decoding method. But the experimental results show that the detection ratio of the auto-complex-cepstrum based decoding scheme is relatively low and its robustness against many common signal processing manipulations is not satisfactory.

In this paper, the autocorrelation of power cepstrum (auto-power-cepstrum), which is found most efficient in recognizing wavelet arrival times and amplitudes [10], is proposed to estimate the echo delay to extract watermark. Computer simulation results indicate that this algorithm achieves high detection ratio even in the presence of various common signal-processing manipulations. In the next two sections, the echo data hiding technique is briefly reviewed and the auto-power-cepstrum based echo detection is described. Experimental results are given in Section Ⅳ in terms of bit accuracy rate (BAR) of extracted watermark data and the similarity (Sim) between original watermark and the extracted watermark in the presence of various attacks. The conclusion and future work is presented in Section Ⅴ.

II. ECHO DATA HIDING

K. Elleithy (ed.), Advances and Innovations in Systems, Computing Sciences and Software Engineering, 331–336.

The embedding process of audio watermarking system based on echo hiding can be represented as a system that has two possible system functions (shown in Fig. 1). The delay between the original signal and the echo is dependent on which system function we use. Specifically, if binary 1 is embedded, the original signal is echoed with delay d_1 and if binary 0 is embedded, the original signal is echoed with delay d_0 (shown in Fig. 2). When the delay d_0 (or d_1) is relatively small, say 2ms, the human auditory system cannot perceive clear echo [11]. In order to embed more than one bit, the original signal is segmented into smaller portions. Each individual portion can then be embedded with the desired bit by considering each as an independent signal. The final encoded signal is the recombination of all independently echoed signal portions.

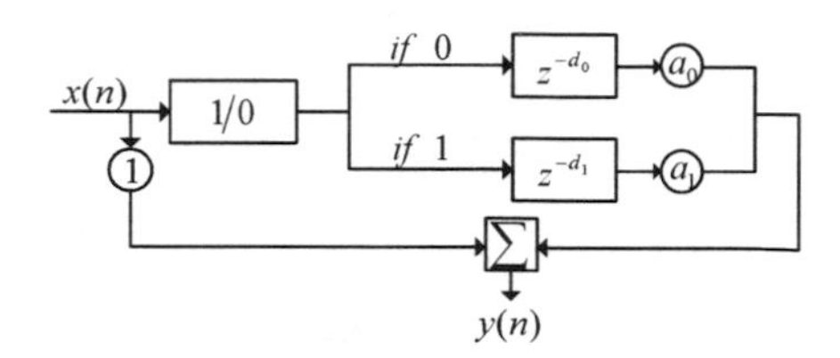

Fig. 1. Diagram of Echo Hiding

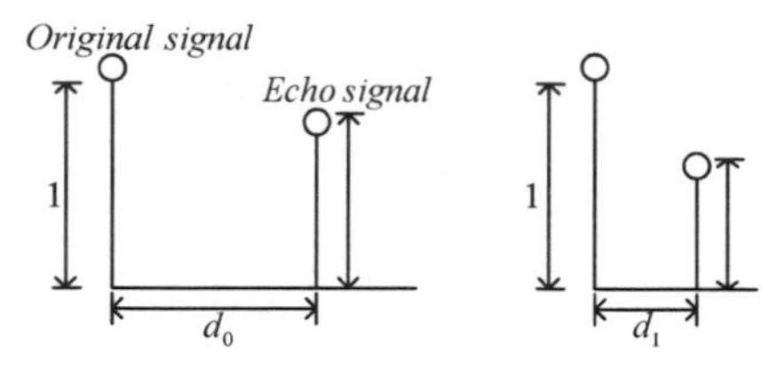

Fig. 2. Echo Kernels

Extraction of the embedded information involves the detection of spacing between the original signal and the echo (namely, the echo arrival time). In references [3]-[9], complex cepstrum technique is applied in this stage; waveform, which contains echo of delay d_1 (d_0), will exhibit peak in the location of d_1 (d_0) in its complex cepstrum. As the spike corresponding to echo delay in complex cepstrum can be easily corrupted by complex cepstrum of the original signal or noise, autocorrelation is applied to complex cepstrum to estimate echo delay. But computer experimental results show that the detection ratio of this extracting scheme is relatively low and its robustness against many common signal processing manipulations is not satisfactory, as is the result of the inherent phase unwrapping [12] and aliasing problems [13] during the calculation of complex cepstrum.

III. ECHO DELAY ESTIMATION BASED ON AUTO-POWER-CEPSTRUM

The power cepstrum was first described by Bogert et al [14]. It is the power spectrum of power spectrum of a signal. The block diagram of power cepstrum analysis is shown in Fig. 3. If smoothing window function (e.g., hamming window) is applied immediately before or after the logarithm operation and zeros are added at the end of the signal, aliasing problem can be avoided.

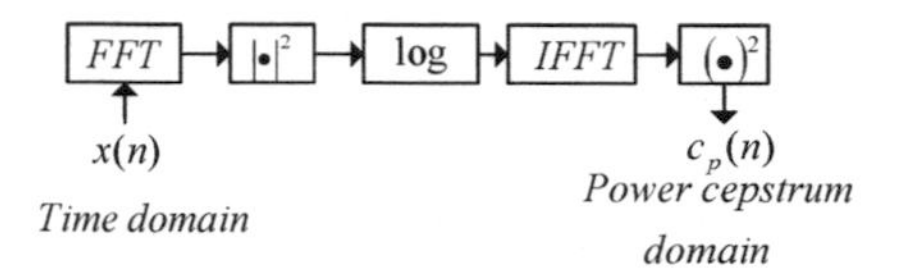

Fig. 3. Block Diagram of Power Cepstrum Analysis

To explain why power cepstrum analysis can be used for detecting echo, an example of a single additive echo is given; the formula for the total signal can be written as follows,

$$\tilde{x}(t) = x(t) + a \cdot x(t - t_0) \tag{1}$$

where t_0 is the echo delay and a is the amplitude of the echo. The power spectral density of $\tilde{x}(t)$ can be represented as

$$P_{\tilde{x}}(\omega) = P_x(\omega) \cdot (1 + a^2 + 2a\cos\omega t_0) \tag{2}$$

Thus, from Eq. (2) we can get that the spectral density of a signal with an echo has the form of an envelope that modulates the periodic function of frequency. The envelope is the spectrum of the original signal and the periodic function of frequency is the spectrum contribution of the echo. By taking the logarithm of the spectrum, this product is converted to the sum of two components, specifically

$$\log P_{\tilde{x}}(\omega) = \log P_x(\omega) + \log(1 + a^2 + 2a\cos\omega t_0) \tag{3}$$

If $a << 1$, a^2 can be neglected in above expression and

remembering that

$$\log(1+x)=\sum_{m=1}^{+\infty}\frac{1}{m}(-1)^{m+1}x^{m} \qquad (-1<x\leq 1) \qquad (4)$$

$\log P_{\hat{x}}(\omega)$ can be well-approximated by the follow expression:

$$\log P_{\hat{x}}(\omega)\cong \log P_{x}(\omega)+2a\cos\omega t_0 \qquad (5)$$

Thus, the logarithm of the power spectrum is viewed as a waveform that has an additive periodic component whose fundamental frequency is the echo delay t_0 [15]. In conventional analysis of time waveforms, such periodic components show up as sharp peaks in the corresponding Fourier spectrum. Therefore, the spectrum of the log spectrum would likewise shows a peak when the original time waveform contains an echo. This is the reason why power cepstrum can be used to estimate the echo arrival time.

Usually, the following equation is applied to calculate the power cepstrum of signal $x(n)$.

$$c_p(n)=(IFFT(\log|FFT(x(n))|^2))^2 \qquad (6)$$

From Eq. (6), it can be seen that in power cepstrum the phase information is lost. As a result, it does not have phase unwrapping problem, which is sensitive to noise.

In this paper, the auto-power-cepstrum is proposed to extract the watermark, which is embedded with echo hiding technique, from the watermarked audio signal. The calculation of the autocorrelation of power cepstrum using DFT is as follows,

$$r_{c_p}(\tau)=\sum_{n=0}^{N-1}c_p(n)c_p(n+\tau)=\frac{1}{N}\sum_{k=0}^{N-1}|C_p(k)|^2 e^{j\frac{2\pi}{N}k\tau} \qquad (7)$$

where N is the length of power cepstrum $c_p(n)$ and $C_p(k)$ is the discrete Fourier transform (DFT) of $c_p(n)$.

The auto-power-cepstrum and the auto-complex-cepstrum of echoed audio signal are shown in Fig. 4. There appear peaks corresponding to the echo delay ($t_0=100$ *units*) in both waveform, but the peak in auto-power-cepstrum is much more prominent than that in the auto-complex-cepstrum, which indicates that the former is superior to the later in echo detection.

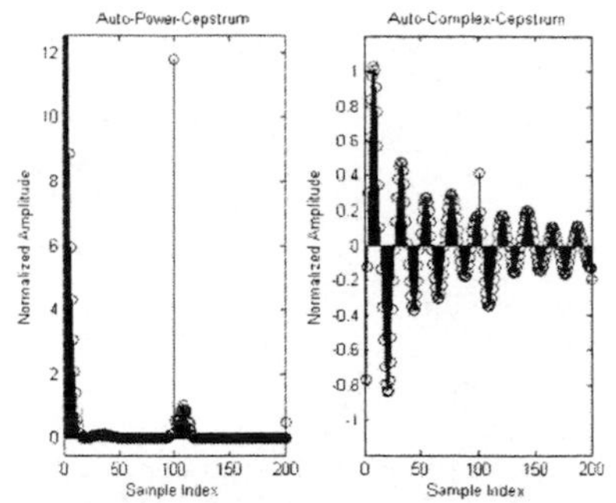

Fig. 4. The auto-power-cepstrum and auto-complex-cepstrum of echoed audio signal when $a=0.4$, $t_0=100$ units .

IV. EXPERIMENTAL RESULTS

In order to test the perceptual quality of our method and its robustness to common signal processing manipulations, a set of experiments were conducted. A piece of pop music audio signal was used as host signal, which was sampled at 44.1 kHz, represented by 16 bits per sample, and 22.32 seconds in length. The watermarking algorithm was implemented in MATLAB 7.1 under Windows XP. In the experiment, a binary random sequence was embedded into the host signal using echo hiding technique. Then auto-power-cepstrum and auto-complex-cepstrum based algorithms are used to extract the watermark from the watermarked audio signal respectively.

A. *Performance analysis*

In this study, reliability was measured as:

1) Bit accuracy rate (BAR) of extracted watermark data: for embedded and extracted watermark sequence of length B bits, the BAR (in percent) is given by the expression:

$$BAR=\frac{100}{B}\sum_{n=0}^{B-1}\begin{cases}1, & \tilde{w}(n)=w(n)\\ 0, & \tilde{w}(n)\neq w(n)\end{cases} \qquad (8)$$

Fig. 5 shows the bit accuracy rate of auto-power-cepstrum based detection algorithm and auto-complex-cepstrum based one as a function of echo decay rate.

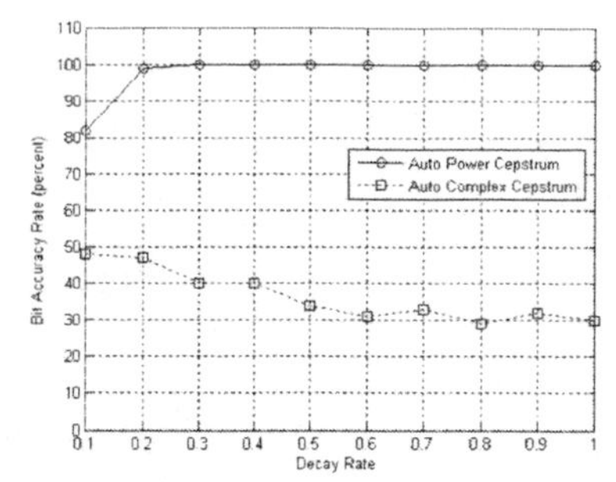

Fig. 5. Bit accuracy rate as a function of echo decay rate

2) Similarity (SIM) measure given in Eq. (9):

$$Sim(W,\tilde{W}) = W \cdot \tilde{W} / \sqrt{W \cdot W} \quad (9)$$

Where W denotes the original watermark and $\tilde{W}$ denotes the extracted watermark. The comparison of the SIM of two detection algorithms at different echo decay rate is shown as Fig. 6.

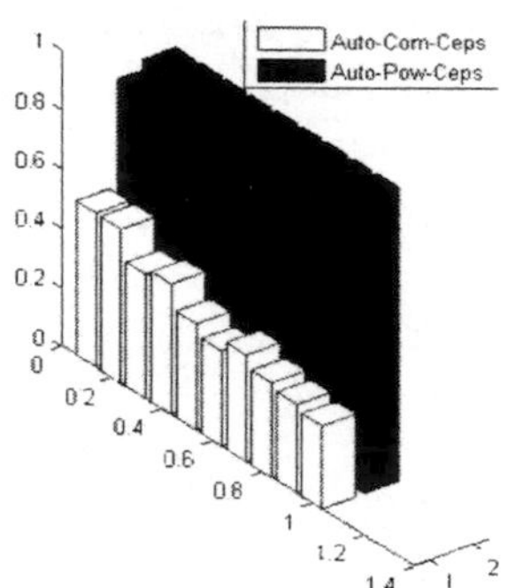

Fig. 6. SIM for different echo decay rate

It is clear that the auto-power-cepstrum based algorithm has better reliability than the auto-complex-cepstrum based one.

B. *Perceptual quality*

Perceptual quality refers to the imperceptibility of embedded watermark data within the host signal. In this study, the SNR of the watermarked signal versus the host signal was used as quality measure:

$$SNR = 10 \cdot \log_{10} \left\{ \frac{\sum_{n=0}^{N-1} x^2(n)}{\sum_{n=0}^{N-1} \left[\tilde{x}(n) - x(n)\right]^2} \right\} \quad (10)$$

Fig.7 shows the SNR of watermarked audio signal as a function of echo decay rate.

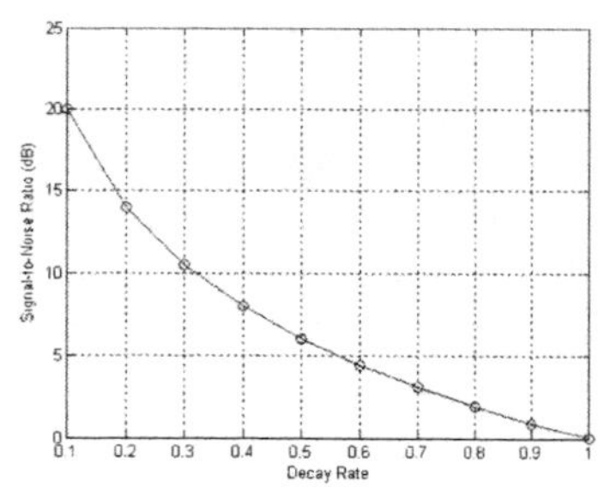

Fig. 7. SNR as a function of echo decay rate

C. *Robustness against common signal processing manipulations*

1) Robustness to loss compression:

When decay rate was 0.4, the watermarked audio signal underwent ISO/MPEG-1 Audio Layer III encoding/decoding at various bit rates, such as 48Kbps, 56Kbps, 64Kbps, 80Kbps, 96Kbps, 112Kbps and 128Kbps. And the bit accuracy rate of the two detection algorithm as a function of compression rate is shown in Fig. 8.

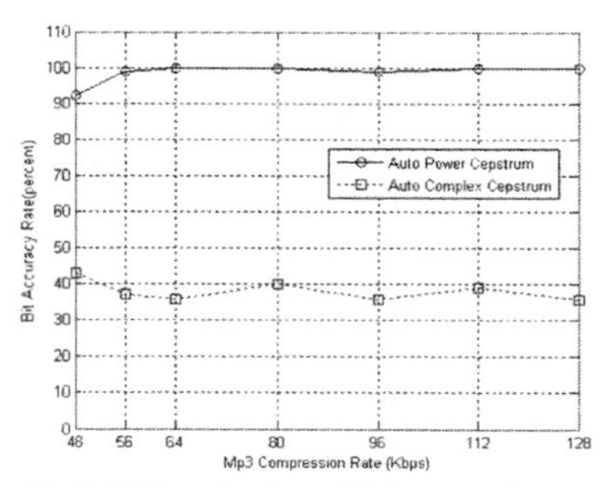

Fig. 8. BAR as a function of Mp3 compression rate

Results show that the watermark detection based on auto-power-cepstrum is performed with a 100% success when the compression rate is larger than 64Kbps. And even when the compression at 48 Kbps is applied, it achieves 92 percent

BAR. It is obvious that when compared with auto-complex-cepstrum based algorithm, auto-power-cepstrum based one has greater robustness against Mp3 compression.

2) Robustness to filtering:

The robustness of the new algorithm against low-pass filtering was also investigated. Fig. 9 depicts the bit accuracy rate of two algorithms as a function of cutoff frequency of low-pass filtering.

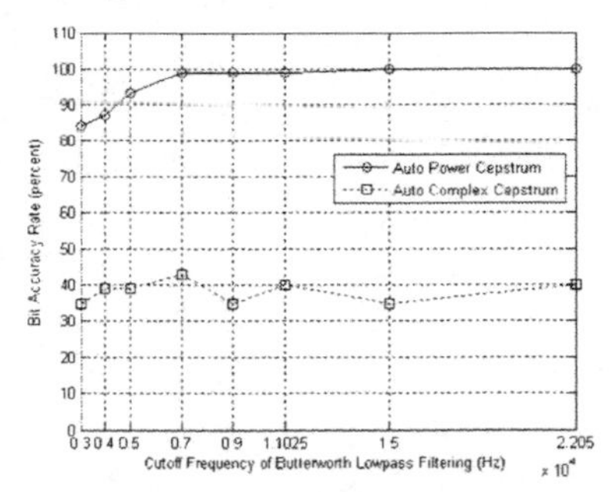

Fig. 9. BAR as a function of cutoff frequency of low-pass filtering

The auto-power-cepstrum based algorithm proves to be robust against low-pass filtering, more specifically filtering with a 23th order Butterworth low-pass filter with cutoff frequency of larger than 7 KHz. The solid curve of Fig. 9 displays the BAR of auto-power-cepstrum based algorithm after filtering, while the dashed one indicates the corresponding BAR of auto-complex-cepstrum based algorithm. It is shown that the former is superior to the later in resisting low-pass filtering.

3) Robustness to re-sampling, re-quantization, and cropping:

Watermarked audio signal with original sampling rate 44100 Hz was sub-sampled down to 22050 Hz and 11025 Hz and interpolated back to their initial frequency. Although the above procedure caused noticeable distortion in relation to the original signal, the BAR of auto-power-cepstrum based algorithm remained 99 percent. While the BAR of auto-complex-cepstrum based algorithm was just 39 percent and 38 percent respectively.

The original 16-bit watermarked audio signal was re-quantized down to 8 bits/sample and backward. Despite the resulting loss of information, the auto-power-cepstrum based algorithm achieved 100 percent success. The new method had greater robustness against re-quantization than auto-complex-cepstrum based one had, which achieved a BAR of 39 percent.

Robustness to cropping has also been tested. Fig. 10 depicts the BAR of the two algorithms when cropping of different rates is applied to the watermarked audio signal.

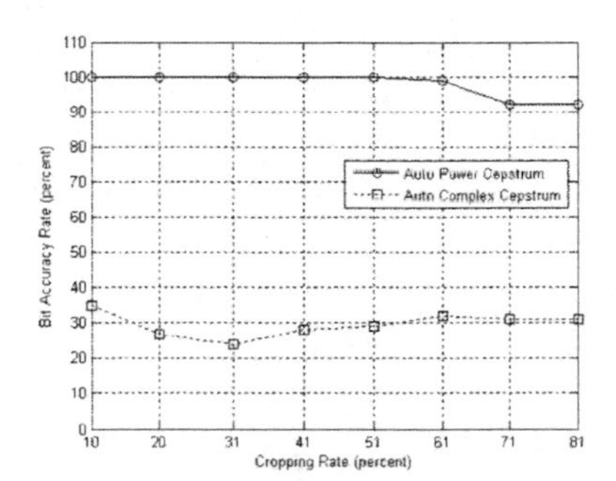

Fig. 10. BAR as a function of cropping rate

The results verify that the auto-power-cepstrum based algorithm is inherently robust to cropping attack and even when the cropping rate is larger than 70 percent it can achieve a BAR above 90 percent. While the BAR of auto-complex-cepstrum based algorithm is always below 35 percent.

4) Robustness to amplitude amplifying, white noise addition, and time-delay:

When the amplitude of the watermarked audio signal was amplified by different rates, the BAR of two watermark extraction methods was shown in Fig. 11.

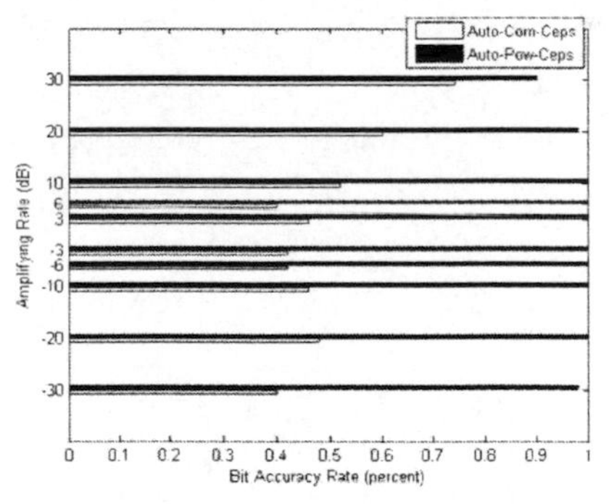

Fig. 11. BAR as a function of amplifying rate

When white noise with a constant level of 36 dB was added to the watermarked audio signal, the Sim and BAR was 1 and 96 for auto-power-cepstrum based algorithm and 0.3 and 28 for auto-complex-cepstrum based one.

When a time-delay of 500 ms was applied to the watermarked audio signal, the Sim and BAR was 1and 100 for the new method and 0.44 and 41 for the old one.

The above experimental results show that the auto-power-cepstrum based algorithm has higher robustness against amplitude amplifying, white noise addition and time delay attacks than auto-complex-cepstrum based one has.

V. CONCLUSION AND FUTURE WORK

Echo hiding is one of the most prevailing audio watermarking methods for copyright protection. However, its relatively low detection ratio and poor robustness against common signal processing manipulations restrict its practical application. By using auto-power-cepstrum to estimate the echo delay in the watermarked audio signal to extract watermarks, a significant improvement in detection ratio is achieved. Further more, its robustness to common signal processing operations of Mp3 compression, re-sampling, cropping, re-quantization, low-pass filtering, amplitude amplifying, noise addition and time-delay is enhanced greatly.

In order to further improve the performance of the algorithm, further research will focus on the cepstral domain modification of audio signal for data embedding and watermark generating using binary chaotic sequence, by which the perceptual quality and robustness will be enhanced further.

REFERENCES

[1] W. Bender, D. Gruhl, N. Morimoto, A.Lu, "Techniques for data hiding," IBM Systems Journal, Vol 35, Nos 3 & 4, pp. 313-336, 1996.

[2] Michael Arnold, "Audio Watermarking: Features, applications and algorithms," 2000 IEEE International Conference on Multimedia and Expo.

[3] Chou, S.A., Hsieh, S.F., "An echo-hiding watermarking technique based on bilateral symmetric time spread kernel," IEEE International Conference on Acoustics, Speech and Signal Processing, Volume 3, Page(s): III-1100 - III-1103, May,2006.

[4] Wen-chih Wu, Chen, O.T.-C., "Analysis-by-synthesis echo hiding scheme using mirrored kernels," IEEE International Conference on Acoustics, Speech and Signal Processing, Volume 2, Page(s):II-325 - II-328, May, 2006.

[5] B. Ko, R. Nishimura, Y. Suzuki, "Time-spread echo method for digital audio watermarking using PN sequences," IEEE, International Conference on Acoustics, Speech, and Signal Processing, vol.2, pp.2001-2004, 2002.

[6] Hyoung Joong Kim and Yong Hee Choi, "A Novel Echo-Hiding Scheme With Backward and Forward Kernels," IEEE Transactions on circuits and systems for video technology, vol.13, No.8, August, 2003.

[7] Fabien A.P. Petitcolas, Ross J. Anderson and Markus G. Kuhn, "Information hiding-a survey," Proceedings of the IEEE, special issue on protection of multimedia content, 87(7):1062-1078, July 1999.

[8] Say Wei Foo, Theng Hee Yeo and Dong Yan Huang, "An adaptive audio watermarking system," Proceedings of IEEE Region 10 International Conference on Electrical and Electronic Technology, Page(s): 509-513 vol.2, Aug, 2001.

[9] Daniel Gruhl, Walter Bender, "Echo hiding," http://citeseer.ist.psu.edu/gruhl96echo.html.

[10] R. C. Kemerait, "Signal detection and extraction by cepstrum techniques," Ph.D. dissertation, Univ. Florida, Gainesville, 1971.

[11] Hyen O Oh, Jong Won Seok, Jin Woo Hong, Dae Hee Youn, "New echo embedding technique for robust and imperceptible audio watermarking," Proceedings of IEEE international conference on Acoustics, Speech and Signal Processing, Vol.3 pp. 1341-1344, 2001.

[12] Bednar, J. Watt, T., Joplin Ave., Tulsa, OK, "Calculating the complex cepstrum without unwrapping or integration," Acoustics, Speech, and Signal Processing, 1985.

[13] T.K. Bysted, "Aliasing in the complex cepstrum of linear-phase signals," International Conference on Information, Communications and Signal Processing, Singapore, 9-12, September, 1997.

[14] B.P. Bogert, M.J.R. Healy, and J.W. Tukey, "The quefrency analysis of time series for echo: cepstrum, pseudo- autoconvariance, cross-cepstrum, and saphe cracking," in Time Series Analysis, M. Rosenblatt, Ed., ch.15, pp.209-243, 1963.

[15] Oppenheim, A.V., Schafer, R.W., "From frequency to quefrency: a history of the cepstrum," IEEE Signal Processing Magazine, Volume 21, Issue 5, Page(s):95 – 106, Sept, 2004.

Building Knowledge Components to Enhance Frequently Asked Question

Noreen Izza Arshad [1], Savita K. Sugathan[2], Mohamed Imran M. Ariff [3], Siti Salwa A. Aziz[4]
[1,2]*{noreenizza, savitasugathan}@petronas.com.my* , [3]*moham588@perak.uitm.edu.my*, [4]*apachigurl84@yahoo.com*

Abstract

A web page that adopts knowledge components concepts is able to help its users in delivering or obtaining information and knowledge through websites itself. In this study, the adoption of knowledge components is implemented through an agent and it is compared to other types of FAQ. A well-structured agent and its knowledge components will benefit the websites users and motivates them to leverage on usage of the FAQ. This research also includes the development of knowledge components, agent and the result of observation is done to view user's perception. Program D works as the interpreter is part of the methodology, which is created using open source Artificial Intelligence Mark-up Language (AIML) and Program D. The system implementation includes Knowledge Warehouse, which stores and organized the knowledge components. The goal of this study is to develop the knowledge components and it is implemented through agent. The outcome of this study is measured by comparing the agent with knowledge components with link type FAQ and top down FAQ. These three types of FAQ serve the same purpose, which is to cater frequent normal questions and ad-hoc queries from website users. It is hope that a well-structured knowledge component could enhance the usage of Internet FAQ where people could benefit an appropriate and relevant answer upon having any enquiries.

Keywords
Knowledge Components (KC), Frequently Ask Question (FAQ), Knowledge Warehouse (KW).

1.0 Introduction

It is important for website developers to embrace the methods and trends to ease the website users in fulfilling their needs and thirst for knowledge and information. Although sometimes a website seem to have a perfect content, but somehow website users are characterized by a different attitude with expectations that sites will provide all the information they need [3]. This is where Frequently Ask Question (FAQ) becomes vital. A good FAQ will be able to reduce cost in answering phone, fax, e-mails by a person and ad hoc queries.

A survey was conducted for "frequent internet users" from September 4 2006 until September 14 2006, with some test scripts which are available at http://FreeOnlineSurveys.com/rendersurvey.asp?sid=e5y9avvm90hr320219857. The objective of this survey was to get online users' feedback regarding Traditional FAQ and Agent-based FAQ.

Many users claimed that they always get questions regarding various disciplines and FAQ is supposed to serve the purposes of answering these questions. They stressed out that FAQ is supposed to be used to clarify terms and concepts that are frequently encountered and often misunderstood.

Anyway, FAQ does not change the core interaction between websites users and a websites' owner, but it is an assistant to gain information and knowledge in a convenient manner. Perhaps it is a cost effective way rather than having a help desk staff waiting at the concierge answering the same normal questions for many times. FAQ is perceives as an opportunity that can be enhanced since it is a great assistant to a web page and to eliminate answering same question repeatedly.

From the survey conducted, it shows that the normal online FAQ is not being fully leveraged by the internet user. This is an element which could be improved and it is an opportunity for web developers. Fig. 1 shows that only 44.83% of the internet user actually uses FAQ. Internet has been a widely used medium in information delivery, hence believed with a better FAQ it would be a great help to users.

K. Elleithy (ed.), Advances and Innovations in Systems, Computing Sciences and Software Engineering, 337–343.

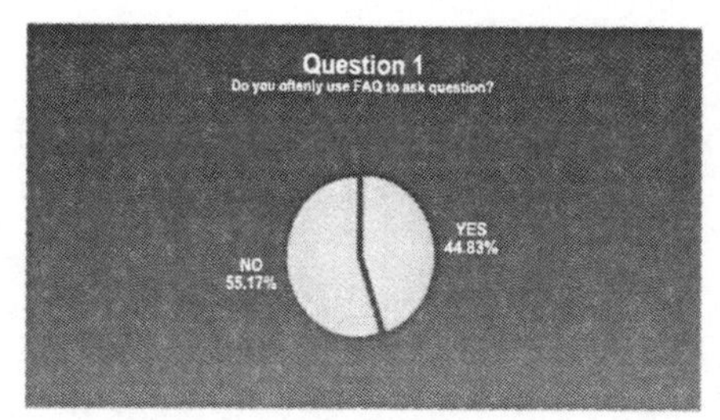

Figure 1: FAQ usage

Thus, the aim of this project is to develop the knowledge components, which is represented by the agent as an alternative to traditional approaches of FAQ. Besides, elements of human-computer interaction and knowledge warehouse are incorporated in building the knowledge components and agent. An excellent FAQ will benefit the researchers who uses website to seek information and knowledge in a fair duration of time. As for the website developers, it is a tool for them to deliver clear and satisfactory information rather than needing to be contacted via phone, e-mail or faxes.

As mentioned, the objective of this study is to develop the knowledge components of FAQ and build it together with the agents and present it online. The outcome should be measured in terms of to what extend does an agent-based FAQ enhance the usability of websites. Knowledge Warehouse (KW), which stores and organizes the knowledge components, is another core of this study and implementation as it is the brain that stores the FAQ of the agent. The agent acts like a mediator, which is integrated online. This implementation is a mean of conveying information such as education, medication, entertainment and more.

The reason of utilizing knowledge components and storing them into knowledge warehouse is due to the term 'knowledge' itself. Reference [4] mentioned that knowledge consist of work procedures and processes, precedents, details and conceptual relationships between topics in a domain. Reference [1] also mentioned that interpretation of knowledge is that knowledge is generated when information is combined with context and experience. The components of knowledge are then arranged in the knowledge warehouse. Reference [4] mentioned that a major economic benefit of any database lies in storing information once, in a form that is accessible by other system that needs it. Knowledge warehouse is a repository, which promotes sharing and reusability. On the other hand, agent-based is perceived to be acknowledging by the world to be one of the technologies that help, assist and ease human activities [5]. This is the primary reason of connecting the knowledge components and representing it through agents.

2.0 System Architecture

System implementation is shown in Fig. 2 where Program D is used to represent Java AIML interpreter since this project interface is a web based application. Program D requires a web application server to runs. Apache Tomcat 5.5 is installed and serves as java servlet container that is used in the official reference implementation for the Java Servlet and JavaServer Pages technologies. Apache Tomcat also acts as a deployer for Program D where developer will interact through Tomcat Web Application Manager.

Knowledge components in Fig. 2 are referring to all the AIML files that are created and organize in Program D. Program D that contains all the Knowledge Component (KC), which is AIML files that acts as an AIML interpreter and will process all the loaded KC (AIML files) in its directory. Knowledge components are where the agent knowledge resides.

Referring to Fig. 2, a client with specific task and goals will send request/input in question form to the agent through the web page. Program D as the AIML interpreter will search and retrieve for the matched response to the request/input. After crawling through the KC and found the matched response, then the agent will response back to the client's question.

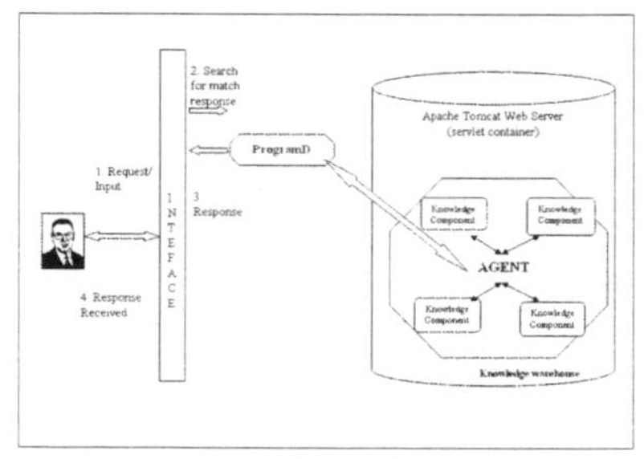

Figure 2: System framework

3.0 Agent Development

In developing the agent, a few important elements are considered. The agent must be able to understand messages/questions and retrieve answers from the appropriate knowledge components. The communication should takes place between the agent and user. Fig. 3 is showing simple agent architecture for the agent implementation.

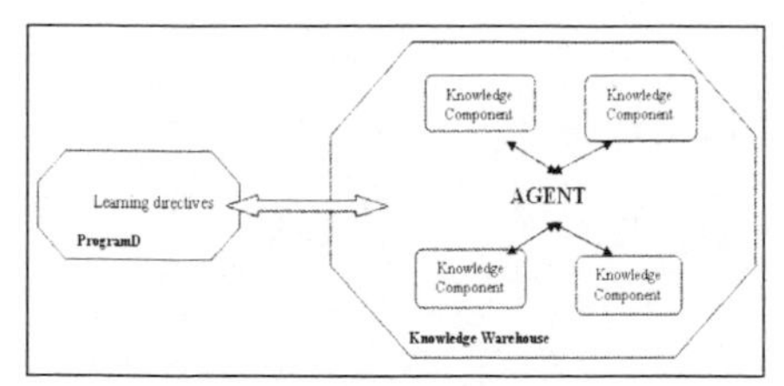

Figure 3: Simple agent architecture

A knowledge specialized agent is developed which have its own sets of knowledge. Each of the agents could have its specialized knowledge and sets of knowledge components. The KC could be shared and reused by any specialized agents as long as it is retrieving from the respective knowledge warehouse. As for this study, agent is developed with the knowledge components on hotel reservation. The researchers' implements by taking the advantage of *agentbot*; an agent with fixed rules and a finite, controlled set of things to say.

4.0 Knowledge Warehouse and Components Development

Agent which is created acts like a communicator between user and has the capability to search for the answer from one of the KCs. However, the answers given are limited to knowledge stored in the KC. The KC is storing knowledge and is maintained in a knowledge warehouse. Knowledge warehouse consists of knowledge components (KCs) that are defined as the smallest level in which knowledge can be decomposed. The agent's knowledge is developed, stored, organized, and processed in a location or a knowledge warehouse named "hotelReservation" that has the directory of "...var\ProgramD\resources\ hotelReservation". KW "hotelReservation" consists of AIML files as the knowledge components. One of the KC and KW develop in our study is shown in Fig. 4 and Fig. 5.

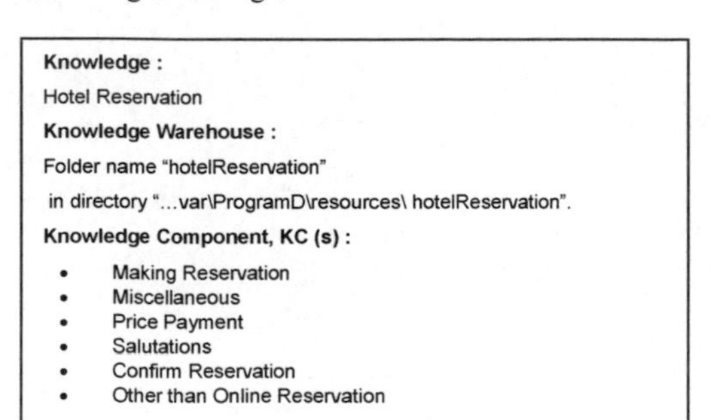

Knowledge :

Hotel Reservation

Knowledge Warehouse :

Folder name "hotelReservation"

in directory "...var\ProgramD\resources\ hotelReservation".

Knowledge Component, KC (s) :

- Making Reservation
- Miscellaneous
- Price Payment
- Salutations
- Confirm Reservation
- Other than Online Reservation

Figure 4: Part of knowledge warehouse

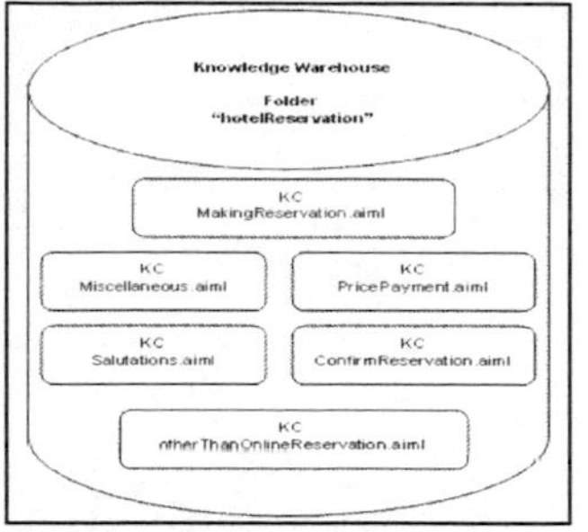

Figure 5: Knowledge warehouse architecture

Fig. 4 and Fig. 5 are rationalizing the basic structure of AIML concept, which creates the KC. While, Fig. 6 is the human-agent conversation which is in natural human language as seen in the online interface. On the other hand, Fig. 7 is the converted KC from natural language to (Artificial Intelligence Markup Language) AIML. AIML is utilized in developing the KC for agent's brains simply because AIML has been designed for ease of implementation and ease of use by newcomers. Since the agent is presented online, AIML is one of the best languages to consider as it is interoperable with XML and XML derivatives such as XHTML

Human: Hello, my name is Doubly Aimless
Robot: Hello Mr. Aimless or is it Dr. Aimless?
Human: I am a profound doubter of Alice
Robot: Why do you doubt Alice?
Human: Because, for me, AIML technology can do little beyond being a platform to display witticisms
Robot: Ok.

Figure 6: Human-robot conversation

```
<?xml version="1.0" encoding="UTF-8"?>
<aiml version="1.0">
        <category>
        <pattern>
        Hello, my name is Doubly Aimless </pattern>
        <template>
        Hello Mr. Aimless or is it Dr. Aimless?
</template>
        </category>
        <category>
        <pattern> I am a profound doubter of Alice
</pattern>
        <template> Why do you doubt Alice?
</template>
        </category>
        <category>
        <pattern> Because, for me, AIML technology
can do
         little beyond being a platform to display
witticisms
        </pattern>
        <template> Ok. </template>
        </category>
  </aiml>
```

Figure 7: Converted human-robot conversation to AIML

4.1 *Using learning directives to retrieve the KC*

As mentioned, the KC is developed using AIML and stored as .aiml formats. Each of this KC may hold a single topic or various topics. Good naming conventions would be a great way to recognize the functions of each KC. Based from Fig. 8, in Program D, <learn> ... </learn> learn directive is use to call the KC or the .aiml components. With this learn directive, Program D knows where to locate and load/retrieve the knowledge components or .aiml files. Fig. 8 also shows how learn directive is loaded and the KC or .aiml file is added. Learn directive function is processed immediately upon startup of Program D. To specify a new .aiml file to load, open the bots.xml in the ProgramD\conf directory.

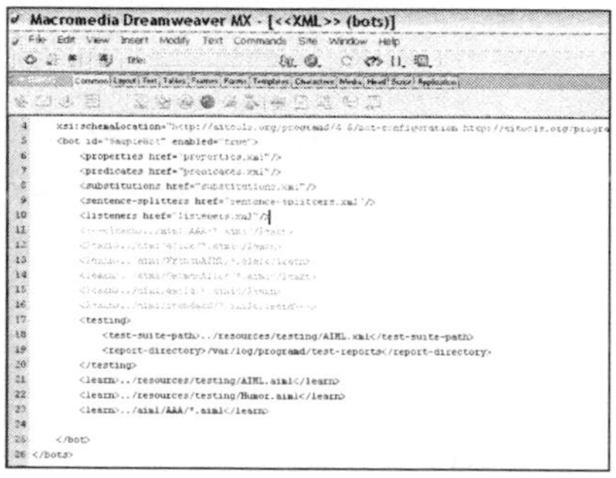

Figure 8: Adding .aiml file using learn directives

Learning directives in program D is an important step since it retrieves the KC. As shown in Figure 8 in line 23 <learn>../aiml/**AAA /*.aiml**</learn> is retrieving KC within a directory reached by going one level up (to the main PROGRAMD directory), and then into a directory called aiml, and into a subdirectory called AAA. Meaning more than one AIML files (KC) in AAA subdirectory will be loaded. As shown in Fig. 8 in line 22, <learn> ./resources/testing/**Humor.aiml** </learn> is retrieving the specified KC which is humor component.

5.0 Results and Discussions

In measuring the KC and its effectiveness, a survey is conducted which involves three types of FAQ. The reason of choosing FAQ is because the KC is implemented for hotel reservation FAQ and the agent is an agent, which answers FAQ regarding hotel reservation. Hence, the measurement of result outcome, three different FAQ is compared; (I) Link type question-to-answer, (II) Top down question-to-answer and (III) Knowledge Component Agent based FAQ.

The first comparison item is the link type question-to-answer which is a link type of FAQ. When user clicks on the question it will be directed to the answers. The second comparison item is the top down question-to-answer, which is a FAQ displaying all questions and answers from top to bottom. User will need to go to the list to look for intended question and the answer displayed together. The last item is the knowledge component agent that being implemented in this study, which serves as an agent-based FAQ. User needs to type in the intended question and an agent will gives the answer subsequently. These three types of FAQ was developed, which stored with hotel reservation components in order to conduct the survey as well as measuring the outcome of agent-based FAQ in comparison to others (I and II).

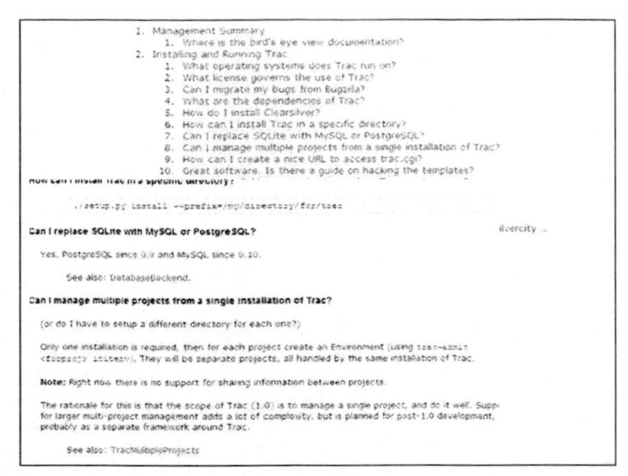

(I) Link type question-to-answer FAQ

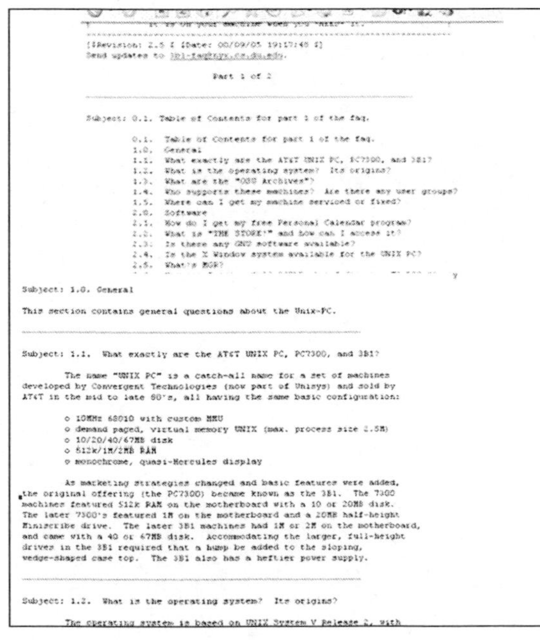

(II) Top-down FAQ

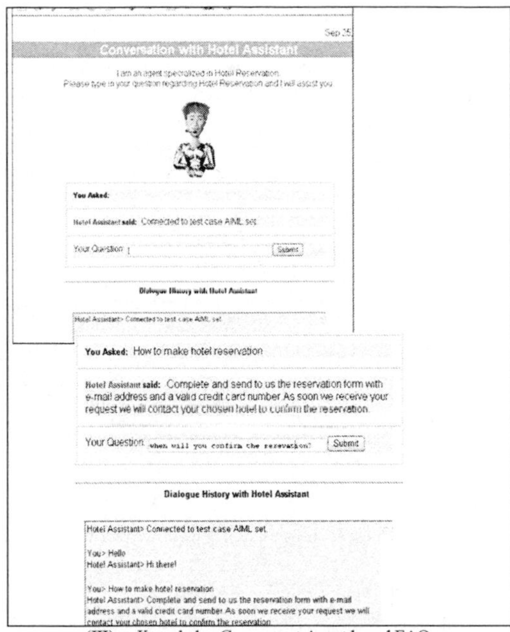

(III) Knowledge Component Agent based FAQ.

The result of comparison, which is based on user survey conducted in October 2006, with frequent Internet users are shown in Figure 9. The survey conducted has shown a result where user prefers to use KC and agent based FAQ compared to others due to some factors. We further explore why (I) Link type question-to-answer FAQ and (II) Top-down FAQ is not favored by users.

Based on the testing result, users mentioned that FAQ (I) and (II) are too lengthy which takes a lot of their time to read and scanning through for the appropriate answer. User mostly prefers a very answer-focused result, less text and words. Easy to get answer and efficient response is what users are searching in an efficient FAQ.

On the other hand, most of the users feel it is convenient to use the (III) Knowledge Component Agent based FAQ. This type of FAQ is favored since it allows user to type-in their intended questions and directly retrieves the relevant answer. The knowledge warehouse components are designed to stores precise information. Thus, the setup of the knowledge components eliminates the user from retrieving unnecessary information. The design of page which is presented in a single page manage to deliver response in a way which not confusing.

Many users like the idea of having (III) FAQ since it is promoting communicative environment for user to ask questions and receive responses. The relevant and focus answer manage to overcome the time consuming problem of in FAQ (I) and (II).

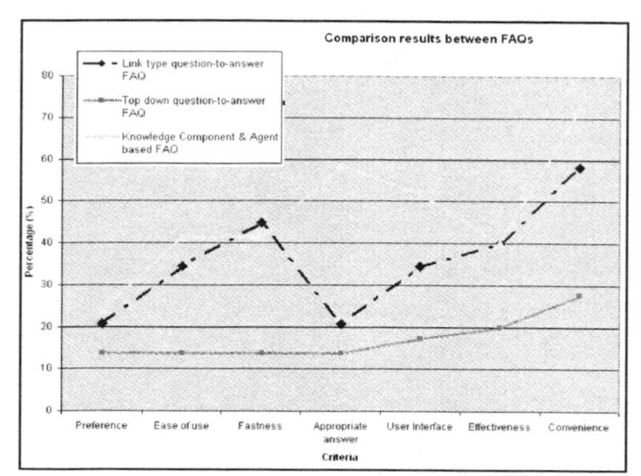

Figure 9: Comparison results between FAQ

Referring to Fig. 9, in terms of duration of time taken to get feedback, it seems like many user prefer to use the link type question to answer FAQ, yet fastness does not mean appropriate answer is retrieved. Most of the user prefers KC and agent based FAQ which caters appropriate and suitable answer which is an important element of any FAQ.

After the survey in Fig. 9 was conducted, we have conducted another type of testing. From Fig. 9, we concluded that knowledge component agent based FAQ is the prominent FAQ. Hence, further testing was conducted to support this study. From Fig. 10, the result of testing is shown based on the functionality of the knowledge component agent based FAQ created. Most of the user agreed that this type of FAQ is easy to use, user friendly and clear. When the agent is answering the inquiries from users, the agent is able to provide a specific and exact answer. Since the result is relevant, this has created a motivation for user to use FAQ as a point of reference. This is a way where customer services could be improved and benefited by user and the organization itself. Yet, the result of functionality testing shows that number of irrelevant answer/response from the agent seems to be high whenever a certain conversation takes place. When further investigation was carry out, it was found that the studied knowledge warehouse is still limited to limited knowledge. This limit the agents in answering questions only to what is stored in the knowledge components. Researchers also realized that knowledge components should be added to robust sets of knowledge for a better relevant of answers. Yet, this initial research is acceptable since questions related to hotel reservation are well answered. Those which are not answered are questions about hotel yet not specific to reservations.

User/Tester	Criteria					
	Easy to use	Communicative	Easier to get specific answer	Attract in using FAQ more often	Improve customer service	Number of irrelevant answer/number of question asked
User 1	Y	Y	N	Y	Y	14/21
User 2	Y	Y	Y	Y	Y	11/12
User 3	Y	Y	Y	Y	Y	9/13
User 4	Y	Y	N	Y	Y	9/20
User 5	Y	Y	Y	Y	Y	5/15
User 6	Y	Y	Y	N	Y	6/11
User 7	Y	Y	Y	Y	Y	14/17

Figure 10: Knowledge component agent based FAQ testing results

This study is considered as at a very initial phase, where future enhancement is a very excellent opportunity to make knowledge component agent based FAQ widely used. Thus, a set of survey was created to allow gathering of recommendations based on user perception shown in Fig. 11. Yet, to avoid too much category of suggestions at this point of time, we catered two categories which are (1) adding voice feature (2) agent based hotel reservations on behalf of user. These two categories are the components that will be incorporated and focused in next phase of researchers' study. Open comments from user are gathered as to explore further room for improvements.

User/Tester	Criteria		
	Incorporate with voice	Agent make reservation on behalf of user	Comments
User 1	Preferred	Most Preferred	
User 2	Not Preferred	Preferred	This approach should be implemented on lengthy FAQ
User 3	Moderate	Moderate	
User 4	Not Preferred	Most Preferred	Interesting an easy to use Interface should be more interesting
User 5	Not Preferred	Moderate	Make it have more knowledgble and understand users' question
User 6	Most Preferred	Preferred	Enhance as is function with direct answer Make it more interactive
User 7	Not Preferred	Most Preferred	Make it able to answer more questions

Figure 11: Knowledge component agent based FAQ future recommendation.

6.0 Conclusion

Therefore, this study of Knowledge Component Agent based FAQ is hoped to be a great online assistance to answer ad-hoc questions and clarify ambiguous inquiries. In comparison of Link type question-to-answer FAQ and Top-down FAQ, Knowledge Component Agent based FAQ is designed to cater user needs especially in retrieving relevant answers. It is a tool which motivates people to use FAQ; which is less time consuming yet interactive. The knowledge components and knowledge warehouse are the components which are considered core parts of this study have contributed important elements to the Agent based FAQ. Hence, robust sets of knowledge components will make the agent an efficient and beneficial one.

For future enhancement, user which involves in the survey suggested for the agent to have voice while communicating with user. This is mimicking the real person at the helpdesk counter or phone line. Furthermore, this study is based on knowledge components and single-agent. Yet, the knowledge components can later expand its capability by communicating and interacting with multi-agents.

References

[1] Huang, K., Lee, Y.W., Wang, R.Y. (1999). Quality Information and Knowledge. Upper Saddle River, NJ: Prentice Hall PTR. P209.

[2] Klein, G.A. (1992). Using Knowledge Engineering to Preserve Corporate Memory. *The psychology of expertise*. In: R.R. Hoffman. Springer-Verlag New York. p170-187.

[3] Maswera, T., Dawson, R. and Edwards, J. (2005). Assessing the Levels of Knowledge Transfer Within e-Commerce Websites of Tourist Organisations in Africa. In: Remenyi, D. Proceedings of 6th European Conference on Knowledge Management, Ireland September 8-9, 2005. Reading: Academic Conference Limited. P293-300.

[4] Yacci, M. (1999). The Knowledge Warehouse: Reusing Knowledge Components. *Performance Improvement Quarterly*. 12(3).

[5] *What Is The Future Of Chatbots. United Kingdom.* Available: URL http://www.icogno.com/what_is_future_of_chatbots.html. Last accessed 16 October 2006.

Semantic Representation of User's Mental Trust Model

Omer Mahmood, John D Haynes
School of Information Technology
Charles Darwin University
Darwin, NT, 0909, Australia

Abstract- It is believed that trust will be the primary mental force in the electronic environment as it is in the current physical environment. At the core trust is impacted by users' propensity to trust (internal mental state), reliance on the trustee and external direct and indirect factors.

Decentralization of publication is one of the great advantages of the internet infrastructure. Web 2.0 applications aim to promote and assist online users to publish and contribute freely for Collective Intelligence. They resolve around the notion that people can add contents for Collective Intelligence and enterprises can also use the contents to reduce costs and increase profits through Social Network Analysis.

This paper proposes a conceptual mental trust recognition and evaluation model and meta-document structure to represent, distribute and store users trust evaluations. The proposed document design is based on decentralized information structure that semantically represents contributed contents and the contributor. The contents are represented and distributed by using Atom, Resource Description Framework (RDF) and RDF Schema. The proposed meta-document structure uses RDF Schema to semantically represent users' internal (inner) online trust evaluation model within Web 2.0 environment. It can be used as a blueprint to develop new vocabularies for any e-domain. However, trust recognition model is selected due to its importance in electronic environment.

Keywords Inner Trust, Semantic Web, Electronic Trust Evaluation, Information Management, Perception, Conception, Mental Trust Model

1. Introduction

"Trust is the subjective probability [1] by which an individual, A, expects that another individual, B, performs a given action on which its welfare (interests) depends" [2]. However recently it has been argued by reference [3] that the above definition of trust is not comprehensive as it only refers to one dimension of trust (predictability), while ignoring the "competence" dimension; it does not represent the meaning of "I trust B" where there is also the decision to rely on B; and it doesn't explain what such an evaluation is made of and based on [3]. Reference [3] identified subjective probability and user expectation about the profitable behavior as necessary components of mental trust. They pointed that one's trust in another is relative to a certain goal. Although trust is composite of subjective probabilities, usefulness of the other party and the users' dependence on achieving the desired goals impacts the users' final decision. Several models have been proposed to model users' trust in electronic environment [4,5,6,7]. However in this paper we have proposed a new mental trust evaluation model. The proposed trust evaluation and recognition model is adopted from the model proposed by reference [8].

Web 2.0 is not all about new technologies. Rather it is more related to how the technology can be used to link the physical world with the electronic one by adding semantic information with the contents. Web 2.0 is not only a set of technologies; it also has properties which aim for social integration, user-contributed content, user-generated metadata and decentralized and participatory products and processes [9]. Within Web 2.0, Social Network Analysis (SNA) is considered to be of high impact and Collective Intelligence has been rated as transformational [10]. Collective Intelligence is an approach used to produce intelligent contents such as documents and decisions that result from individuals working together with no centralized governing body. SNA applies new ways to deploy and engineer web applications, so that enterprises can increase revenues or save costs.

Trust has been recognized as one of the main factors affecting electronic commerce. According to *WISTA International E-Commerce Survey* (2000) [11], trust (26%) is the most important barrier to electronic commerce in 27 surveyed countries. Reference [11] recognized "trust as significant stumbling block in electronic commerce development due to the fact that electronic commerce is global and its international reach means that participants must deal with unknown or anonymous individuals and companies". Reference [11] also identified payment security (25%), trust in infrastructure (17%) and information privacy (15%) as the most important trust related issues for acceptance of electronic commerce. The survey established the impact of trust in electronic commerce (strongly 42%, moderately 35%).

Because of the above, this paper concentrates on achieving Collective Intelligence in the domain of online trust

K. Elleithy (ed.), Advances and Innovations in Systems, Computing Sciences and Software Engineering, 345–350.

by semantically representing the proposed model of mental trust evaluation in electronic environment. For this purpose, we have proposed a conceptual Web 2.0 application architecture that uses Atom [12], RDF [13] and RDF Schema (RDFS) [14]. A unique blend of these technologies, proposed here, will assist online users to make decisions while committing electronic transactions on the basis of semantic representation of previously contributed mental trust evaluations.

2. Four Beliefs of Trust

Trust is recognized as being a composite of the following four beliefs [8] which also represent internal mental state of the trustor at the time of trust evaluation and recognition:

A. Competence

User's estimation on agent B's ability to deliver the expected results

B. Disposition

User's estimation that agent B will actually deliver the expected results. Disposition is based on temperament/nature of agent B

C. Dependence

User's estimation of reliance on agent B's ability to achieve the desired goal i.e. the user believes that her/his interest depends on another individual

D. Fulfillment

User's belief that user's goal cannot be achieved without the results produced by agent B

Reference [8] stated that the above four beliefs actually represent users' internal mental state and users' internal (inner) trust.

3. Internal and External Trust

In comparison with internal trust which only focuses on trustor, external trust takes into account the environment conditions. An agent 'A' decision to rely on agent 'B' for some service primarily depends on A's personal estimation of B's competence, ability, honesty and reliability i.e. user's internal trust. However in order to rely on agent B, agent A must also trust the environment i.e. agent A must have a positive estimation of the external conditions that might influence B's actions or ability. Such external conditions determine user's external trust (estimated subjective probability) of having appropriate conditions. In an electronic environment, external trust is more vital than internal trust in determining user's final decision, as unfavorable external conditions may change the agent A's perception to believe that agent B will not be able to achieve the desired task. In relation to above, favorable external conditions may create extremely constructive conditions which may increase the agent A's perceived internal trust level and induce the agent to complete online transaction.

Combining internal and external trust would produce a conceptual model presented in fig 1.

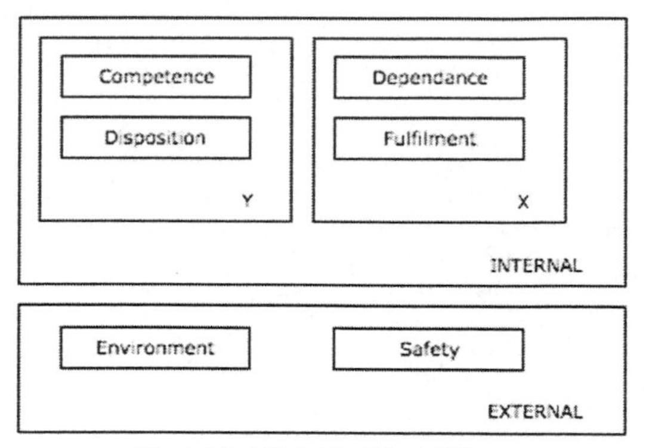

Fig. 1. Mental Model of Trust. [15]

4. Proposed Mental Trust Model

Reference [16] stated that users' "internal trust is developed before external trust", which is represented by the four beliefs of trust. In this paper we propose that inner (internal) mental trust is actually a composite of user's propensity to trust (innate ability to trust) and user's reliance on the trustee.

For the propensity to trust we further subcategorize into conscious and unconscious propensity to trust. The conscious propensity to trust consists of 'Competence' belief and it is influenced by user's past experiences. Whereas, the unconscious propensity to trust is based on 'Disposition' belief and it is influenced by the user's current internal mental state and as reference [17] points out elements of this current mental state can be examined by considering the "conception" of the user.

In the existing literature on mental models of trust, perception is the key element for entry into considerations of the mental model of the user. Apart from Haynes, more specifically reference [17] the user's conception is not mentioned. But the conscious perception of the user which concentrates on online objects and information presented to the user for comparison with past experiences is only part of the mental model. Conception, a measure of unconscious mental processes, as distinct from conscious perception, must be considered, and can be provided for in the design of trust based websites. Such provision can take the form of feedback to the user which consequently can provide the user with greater confidence in the trustee. Accordingly conscious action is based on the user's perception, whereas the unconscious state is the ground from which the user's conception arises.

Thus the distinction is made that perception as a viewing element of consciousness provides the user with the capacity to compare what is currently in view as objects, with what was experienced in the past and to make a conscious decision concerning, in this case trust, accordingly [17]. On the other hand, conception which happens unconsciously and invokes the imagination, allows the user to emotionally feel through the parameters of a decision, based on the user's deeper personal recognition of what perception has, in the absence of conception, decided [17]. User perception and user conception is a three step process: first the user perceives, then the user conceives, then the action of the consequent decision takes place again consciously through user perception. Thus a user could perceive sufficient elements of trust to continue with a transaction yet not proceed in a trusting way because of their conception (a deeper personal feeling) that something is still wrong.

What implications does this distinction between conscious perception and unconscious conception have for the design of trust models in relation to mental models?

Firstly, it makes designers aware that there is a difference between user conscious perception and user unconscious conception. Furthermore the relationship between perception (as conscious) leading to a competence belief and conception (as unconscious) leading to a disposition belief which together influence propensity is reflected in Figure 2.

Secondly, in previous models, the unconscious state of the user has not been considered specifically in relation to a feedback mechanism to the user. Therefore, what we propose is a feedback to the user to so that the user, who may have feelings that they cannot identify which consequently negatively moderate against the building of trust, can begin to feel more trusting.

This feedback mechanism may take the form of questions composed independently of the transaction that the user (as trustor) can put to the trustee or if that is not possible or not desirable then the questions can be put to other bodies of people who originally had similar unconscious feelings. Or it may take the form of FAQs now essentially built up to address the unconscious conception of the user with suggested ways of following up a concern.

In any case the contributing point we make in relation to mental models is that unconscious concerns of the user can be treated with appropriate feedback, and we believe that since the unconscious conception of the user can often be a decisive contributing factor shifting the user to a state of trust it needs to be effectively addressed.

Reliance, the act of relying, is when an individual depends on another individual for support, help or assistance. The propensity (innate) trust is different from reliance, as reliance on trustee is primarily influence by the users' dependence on trustee to achieve the desired goals. Reliance on trustee is composed of 'Dependence' and 'Fulfillment' beliefs. In addition to propensity to trust (internal mental state) and reliance on the trustee, it is proposed here that users trust in electronic environment is influenced by external factors. Where, the external trust is composed of direct and indirect communication medium dimensions. The direct aspects are represented by the properties of communication channel at any given time and the indirect aspect consists of culture and user's context.

The proposed mental trust model is represented in Fig. 2.

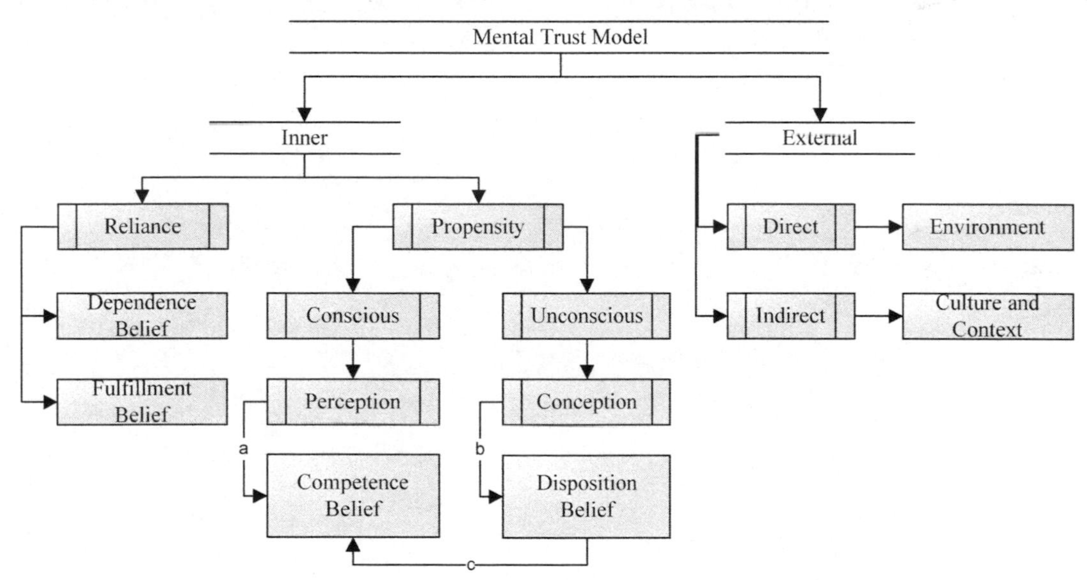

Fig. 2. Proposed Mental Model of Trust.

By using the above model online users can evaluate electronic merchants on basis of four beliefs and external - direct and indirect factors. External direct factors which impact user's trust in communication channel are the use of appropriate security technologies such as SSL and presence of trust seals such as Visa, VeriSign and Master. The indirect external factors are not explicitly recognized by the user such as the impact of user's culture [18], web interface [19] and user's context at the time of transaction.

The proposed separation of internal and external trust will enable the online user, who utilizes the contributed information and contents, to clearly distinguish between user's trust in the trustee and user's trust in environment. Further categorization of internal trust, will enable the online users to identify the user's actual trust in e-merchant and user's internal mental ability (state) to trust.

The proposed categorization targets to enable the future Web 2.0 applications to perform SNA to estimate the impact of context and culture on trust and identify (if any) link between reliance and propensity.

5. Semantic Representation of Proposed Model

The meta-document storage and linkage architecture, proposed in this paper, enables the online users to store and share their mental trust evaluations (states) so that the contributed information can be later used for Collective Intelligence and SNA. The proposed application architecture is divided into two modules, discussed in detail below.

A. Content Distribution

We have selected Atom Syndication Format 1.0 for information sharing, which is an XML language used for web feeds. The two obvious choices were RSS and Atom, however Atom 1.0 is selected on the basis of reasoning that is outlined in Table 1.

TABLE 1: COMPARING ATOM 1.0 WITH RSS

Atom 1.0	RSS
Atom is defined within XML Namespace [20].	RSS is not defined within XML Namespace.
Atom has standardized autodiscovery. Atom has registered IANA MIME type 'application/atom+xml'. Atom syndication format is published as an IETF standard in RFC 4287 [21].	RSS uses many non standard variants of autodiscovery. RSS 2.0 feeds are often sent as 'application/rss+xml', although it is not a registered MIME-type. RSS has multiple incompatible and widely-adopted versions of RSS.
Atom uses the Atom Publishing Protocol (APP). APP is a simple HTTP-based protocol for creating and updating Web resources. APP is IEFT draft protocol.	MetaWeblog and Blogger are the two popular protocols widely used with RSS. However there are interoperability and feature shortcomings issues with them.

B. Proposed Meta-document Structure

RDF is the language of Web 2.0; therefore RDF and RDFS are used to store each user's inner and external metal trust evaluations. To semantically represent the proposed trust model, a new RDF Scheme "Mental Trust Model RDFS (MTMRS)" is engineered, which uses Dublin Core defined elements, IANA, W3C and ISO standards. The standards used in MTMRS include ISO8601:2004 for date format, ISO639-2 for language definition, DCMI Vocabulary for genre of the resource and IANA media types for service output format. Find below the conceptual diagram of MTMRS (Fig. 3).

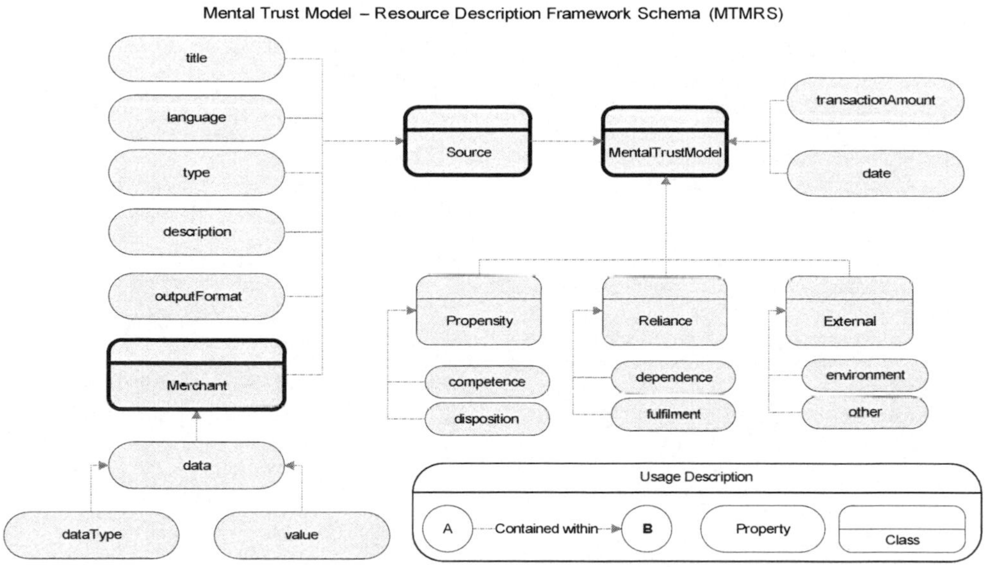

Mental Trust Model – Resource Description Framework Schema (MTMRS)

Fig 3: MTMRS Conceptual Diagram.

6. Connecting Evaluations With Contributor

To feed the contents of user mental trust evaluation model (MTEM) and to represent and link contents with contributor, we have decided to use Atom 1.0. The proposed structure of Atom and MTEM documents is as follows:

A. Atom1.0

Feed information provides description of the Atom document, including a reference to the author web page. Within the Atom document it is proposed to use 'author' element to specify the name, email and URL to author home page. 'Updated' element is used to specify the last modification date and time of the Atom document. Like RSS, any change to Atom will result in automatic notification to subscribed users.

The users online trust metal evaluation model is semantically represented in MTEM document. We propose to use 'entry' element in Atom document to provide link(s) to user MTEM document(s). Within entry, the type of the entry is specified by using 'category element'. For example "MTMRS" is specified as the value of 'term attribute' and 'scheme attribute' points to MTMRS (Schema) document. After specifying the category of entry, 'link' element is used to provide a reference to MTEM document. This structure will enable the SNA Web applications to identify and retrieve semantic details of MTEM document, thus enabling them to process MTEM documents autonomously. Besides above the last modification date is specified by the 'updated element' and the publication/creation date and time of the corresponding MTEM document is specified by 'published element'.

B. Mental Trust Evaluation Model Document

The MTEM document is divided into description and evaluation sections. The description section is represented by using RDF 'description' element. Within description element only Dublin Core elements are used to represent and provide information about the author of MTEM document.

For each trust evaluation, MTMRS 'evaluation' element is used. The structure of 'evaluation' element is presented in fig. 3.

The Fig 4 presents the links between Atom, MTEM and MTMRS documents.

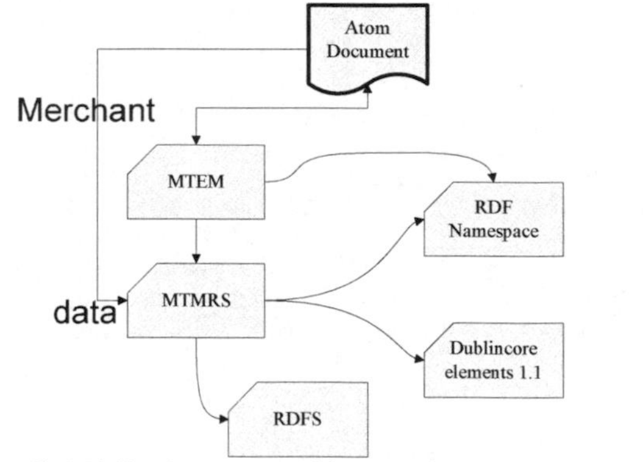

Fig 4: Linking the components of the proposed system through Atom

7. Conclusion

In this paper we have proposed a conceptual user mental trust recognition and evaluation model. Our model proposes that inner mental trust is actually a composite of user's propensity to trust and user's reliance on trustee.

The propensity to trust is further subcategorized into conscious and unconscious propensity to trust. The conscious propensity to trust consists of 'Competence' belief and it is influenced by the user's past experiences compared to the user's perception of objects that are currently in view. On the other hand, the unconscious propensity to trust is based on 'Disposition' belief and is influenced by the user's current internal mental state which happens unconsciously and invokes the imagination that allows the user to deeply feel a decision. The user conceives this outcome (decision), hence we have referred to this process as conception

In previous models, the unconscious state of the user has not been considered specifically in relation to a feedback mechanism to the user.

The contributing point we make in relation to mental models is that unconscious concerns of the user (which often negatively moderate against the building of trust) can be treated with appropriate feedback. We believe that the unconscious conception of the user can often be a decisive contributing factor shifting the user to a state of trust. Thus effective content sharing and feedback based on this model will positively affect the unconscious concerns of the user.

To effectively store, process and share trust evaluation contents and to represent the user mental model a meta-document structure is proposed. The proposed document structure is based on decentralized information storage, access and retrieval. The architecture uses Web 2.0 technologies such as, Atom, RDF and RDFS for data storage, representation, processing and sharing. For semantic presentation of the mental trust model we have proposed a RDF Vocabulary (MTMRS).

By using the proposed model online users can evaluate electronic merchants on basis of four beliefs and external - direct and indirect factors. The proposed separation of internal and external trust, targets to enable the online user, who utilizes the contributed information and contents, to clearly distinguish between user's trust in the trustee and user's trust in environment. Further categorization of internal trust enables the online users to identify the user's actual trust in e-merchant and user's internal mental ability (state) to trust. The proposed categorization of mental trust model targets to enable the future Web 2.0 applications to perform SNA to estimate the impact of context and culture on online trust and identify (if any) link between reliance on trustee and propensity to trust in electronic environment.

REFERENCES

[1] Anscombe F.J. and Aumann R.J. (1963) "A definition of subjective probability", Annals of Mathematical Statistics 34, pp. 199-205.

[2] Gambetta, D. (1990) "Can We Trust? " In D. Gambetta, editor, Trust: Making and Breaking Cooperative Relations, Basil Blackwell. Oxford, pp. 213 – 237.

[3] Castelfranchi, C. and Falcone, R. (1998) "Principles of Trust for MAS: Cognitive Anatomy, Social Importance and Quantification". In Third International Conference on Multi Agent Systems (ICMAS'98), pp. 72 – 79

[4] Mahmood, O (2006). "Online Initial Trust Evaluation and Comparison in Electronic Environment". Proceedings of the IADIS International Conference e-Commerce 2006, Barcelona, Spain, ISBN: 972-8924-23-2 (Unpublished)

[5] Ang, L.; Dubelaar, C. & Lee, B.-C. (2001) "To Trust or Not to Trust? A Model of Internet Trust From the Customer's Point of View" Proceedings of the 14th Bled Electronic Commerce Conference, Bled, Slovenia, 25-26 June 2001, pp 40 - 52

[6] Egger, F.N. (2000). "Trust Me, I'm an Online Vendor": Towards a Model of Trust for E-Commerce System Design*. In: G. Szwillus & T. Turner (Eds.), CHI2000 Extended Abstracts: Conference on Human Factors in Computing Systems, The Hague (The Netherlands), April 1-6, 2000: 101-102

[7] Mahmood, O. (2006) "Modelling Trust Recognition and Evaluation in Electronic Environment". International Journal of Networking and Virtual Organisations (IJNVO). Special Issue on "Trust for Virtual Organisations and Virtual Teams". Inderscience Publishers Ltd. ISSN (Online): 1741-5225, ISSN (Print): 1470-9503 (In Press)

[8] Castelfranchi, C. and Falcone, R. (1999) "The Dynamics of Trust: from Beliefs to Action", Autonomous Agents '99 Workshop on "Deception, Fraud and Trust in Agent Societes", Seattle, USA, May 1, pp. 41-54

[9] Gartner (2006a) "Gartner Says Web 2.0 Offers Many Opportunities for Growth, But Few Enterprises Will Immediately Adopt All Aspects Necessary for Significant Business Impact" Gartner.com, Retrieved from http://www.gartner.com/press_releases/asset_152253_11.html on 18th September 2006

[10] Gartner (2006b) "Gartner's 2006 Emerging Technologies Hype Cycle Highlights Key Technology Themes" Gartner.com, Retrieved from http://www.gartner.com/it/page.jsp?id=495475 on 18th September 2006

[11] World information technology and services alliance 2000) World information technology and services alliance (WISTA) International survey of E-Commerce 2000 Retrieved from http://www.witsa.org/papers/EComSurv.pdf on 26 August 2006

[12] Nottingham, M & Sayre, R. (2005). The Atom Syndication Format. Internet Society Taskforce (IEFT) FRC 4287 Retrieved from http://tools.ietf.org/html/rfc4287 on 19 September 2006

[13] W3C (2004). "Resource Description Framework Specification" W3C.org. Retrieved from http://www.w3.org/RDF/ on 16 August 2006

[14] W3C (2004). "RDF Vocabulary Description Language 1.0: RDF Schema" W3C.org. Retrieved from http://www.w3.org/TR/rdf-schema/ on 19 August 2006

[15] T3 Group (2006). "Internal and external trust", T3 Group. Adapted from http://www.istc.cnr.it/T3/ trust/pages/internalexternal.html on 18 October 2006

[16] Burt, D (2001). "Institutional Trust". In Proceeding of 86th Annual International Conference (2001). Institute of Supply Management. Retrieved from http://www.ism.ws/pubs/proceedings/ConfProceedingsIndex.cfm?&navItem Number=13123 on 6 October 2006

[17] Haynes J D and Mahfouz A, (2002) "Internet Interface Design: E-commerce and the User", Chapter 3 in book, *Internet Management Issues: A Global Perspective*, Edited by Haynes, J D, Idea Group Publishers, Hershey, USA.

[18] Karvonen, K (2001): Designing Trust for a Universal Audience: A Multicultural Study on the Formation of Trust in the Internet in the Nordic Countries , in the Proceedings of the First International Conference on Universal Access in HCI (UAHCI'2001), August 5-10, 2001, New Orleans, LA, USA

[19] Kubilus, N. J. (2002) Designing an e-commerce site for users. ACM Crossroads. 2002,8. Retrieved on 28 March 2006 from http://www.acm.org/crossroads/xrds7-1/ecuser.html

[20] W3C (2005). Atom Syndication Format, Namespace. W3C, December 2005, Retrieved from http://www.w3.org/2005/Atom on 21 September 2006

[21] The Internet Society (2005). The Atom Syndication Format, Internet Official Protocol Standards, RFC 4287, December 2005. Retrieved from http://tools.ietf.org/html/rfc4287 on 19 September 2006

Access Concurrents Sessions Based on Quorums

Ousmane THIARE
Department of Computer Science
LICP EA2175
University of Cergy-Pontoise-France
E-mail: othiare@dept-info.u-cergy.fr

Mohamed NAIMI
Department of Computer Science
LICP EA2175
University of Cergy-Pontoise-France
E-mail: naimi@dept-info.u-cergy.fr

Mourad GUEROUI
PRISM Lab
45, Avenue des Etats-Unis
78000 Versailles-France
E-mail: mogue@prism.uvsq.fr

***Abstract*- This paper presents a quorum-based distributed algorithm for the group mutual exclusion. In the group mutual exclusion problem, multiples processes can enter a critical section simultaneously if they belong to the same group. This algorithm assumes that only one session can be opened at any time, several processes can access to the same session, and any requested session can be opened in a finite time. The message complexity of this algorithm is $O(\sqrt{n})$ for the finite projective plane of order 2 (Fano plane) and $O(2\sqrt{n}-1)$ for a grid where n is the total number of processes.**

I. INTRODUCTION

Group mutual exclusion is a natural generalization of the classical mutual exclusion problem [2][3] that was recently identified and solved by Joung [1]. In mutual exclusion, we have a collection of processes that alternate repeatedly between two sections of code: a possibly nondeterminating *noncritical section* (*NCS*) and a terminating *critical section* (*CS*). The problem is to provide a synchronisation algorithm in the form of a *waiting* and an *exit* protocol to be executed, respectively, immediately before and after the *CS*.

In the classical *group mutual exclusion*, a process wishing to enter the CS requests a "session". Processes that request different sessions cannot be in the CS simultaneously, but processes that have requested the same session can. Sessions represent ressources each of which can be used simultaneously by an arbitrary number of processes, but no two of which can be used simultaneously.

An exemple of group mutual exclusion, described by Joung, is a CD "juke box" shared by multiple processes [1]: Any number of processes can simultaneously access the currently loaded CD, but processes wishing to access a CD other than the currently loaded one must wait. In this case, the sessions are the CDs in the juke box.

Solutions for the group mutual exclusion problem under share-memory model can be found in [5][6][18][8]. In this paper, we investigate the group mutual exclusion problem under message-passing model. For message-passing model, solutions to group mutual exclusion have been proposed for ring networks [9][10] and tree networks [11]. Typically, solutions for ring and tree networks incur high synchronization delay and have high waiting time. For a fully connected network, group mutual exclusion algorithms based on modification of the Ricart and Agrawala's algorithm for mutual exclusion [13] have been proposed in [12]. These algorithms have high message complexity and high message overhead of $O(n)$ and $O(n^2)$, respectively.

The quorum-based mutual exclusion algorithm by Maekawa [19] also been modified to derive two quorum-based algorithms for group mutual exclusion [14]. These algorithms use a special type of quorum system called the group quorum system. In a group quorum system, any two quorums belonging to the same group need not intersect. The maximum number of pair-wise disjoint quorums offered by a group quorum system is called the degree of the quorum system. In [17], Joung introduce a group quorum system called the *surficial quorum system*, which has a degree of $\sqrt{\frac{2n}{m(m-1)}}$, where m is the number of groups. When used with Maekawa's algorithm, the surficial quorum system can only allow up to the degree number of process of the same group to execute concurrently.

The rest of paper is organized as follows. Section II gives the problem definition. Section III decribes the computational model assumed in this paper and we then present our quorum-based algorithm for group mutual exclusion. In section IV, we show correctness of the proposed algorithm. We present our conclusion in section V.

II. DEFINITIONS

A.. System Model

We assume a distributed system comprising of a set of n processes $V = \{P_1, P_2, ..., P_n\}$ and a set of communication channels $E \subseteq V \times V$. Let $G = \{x_1, x_2, ..., x_m\}$ be a set of sessions. The distributed system is asynchronous, i.e, there is no common global clock or shared memory. Channels are reliable and first-in-first-out (*FIFO*). Message delays are finite but may be underbounded.

B. The Group Mutual Exclusion Problem

We consider a system of n asynchronous processes $P_1, ..., P_n$, each of which cycles through the following three states, with *NCS* being the initial state:

K. Elleithy (ed.), Advances and Innovations in Systems, Computing Sciences and Software Engineering, 351–356.

- *NCS*: the process is outside *CS* and does not wish to enter *CS*.
- *Wait*: the process wishes to enter *CS*, but has not yet entered *CS*.
- *CS*: the process is in *CS*.

In this section, we present a quorum system for group mutual exclusion. We begin with the definition of group quorums systems.

Definition 1. Let $P = \{1,...,n\}$ be a set of nodes[1]. An **m-group quorum system** $C = \{C_1,...,C_m\}$ over $\mathcal{P}$ consists of m sets, where each $C_i \subseteq 2^P$ is a set of subsets of $\mathcal{P}$ satisfaying the following properties:

- **intersection:**
  $\forall 1 \le i, j \le m, i \ne j, \forall Q_1 \in C_i, \forall Q_2 \in C_j :$
  $Q_1 \cap Q_2 \ne \varnothing.$
- **minimality:**
  $\forall 1 \le i \le m, \forall Q_1, Q_2 \in C_i, Q_1 \ne Q_2 : Q_1 \not\subset_1 Q_2.$

We call each C_i a **cartel** and each $Q \in C_i$ a **quorum**.

Intuitively, C_i can be used to solve group mutual exclusion as follows: Each process *P*, when attemping to enter *CS*, must acquire a quorum $Q \in C_j$ it has choosen by obtaining permission from every member of the quorum. Upon exiting *CS*, process *P* sends a release to the members of the quorum. Suppose a quorum member gives permission to only one process at a time. Then, by the intersection property, no two processes requesting different groups can be in *CS* simultaneously. The minimality property is used rather to enhance efficiency.

In addition, the quora in the system satisfy the following for extra conditions:

- $\forall 1 \le i, j \le m, |C_i| = |C_j|.$
- $\forall 1 \le i, j \le m, \forall Q_1 \in C_i, \forall Q_2 \in C_j : |Q_1| = |Q_2|.$
- $\forall p, q \in P : |n_p| = |n_q|$, where n_p is the multiset $\{Q | \exists 1 \le i \le m : Q \in C_i \wedge p \in Q\}$, similar for n_q.

  In other words, $|n_p|$ is the number of quora involving p.

Intuitively, the first condition ensures that each group has an equal chance in competing for *CS*. The second condition ensures that the number of messages needed per entry to *CS* is independant of the quorum a process chooses. The third condition means that each node share the same responsability in the system. As argued by Maekawa [4], these three conditions are desirable for an algorithm to be truly distributed. The last condition simply minimizes the number of nodes that must be common to any two quora of different cartels, thereby reducing the size of a quorum.

[1] The terms processes and nodes will be used interchangeably throughout the paper.

C. Maekawa's Algorithm

Maekawa's algorithm implements mutual exclusion by using a coterie that satisfies the aforementioned properties. Lamport's logical clock [16] is used to assign a timestamp to every request for critical section. A request with a smaller timestamp has a *higher priority* than a request with a larger timestamp. (ties are broken using process identifiers). Maekawa's algorithm works as follows:

1. When a process wishes to enter critical section, it selects a quorum and send a REQUEST message to all the quorumm members. It enters the critical section once it has successfully locked all its quorum members. On exiting the critical section, the process unlocks all its quorum members by sending a RELEASE message.
2. A node, on receiving a REQUEST message, checks to see whether it has already been locked to the requesting process by sending a LOCKED message to it. Otherwise, the node uses timestamps to determine whether the process currently holding a lock on it--hereafter referred to as the *locking process*--should be preempted. In case the node decides not to preempt the locking process, it sends a FAILED message to the requesting process. Otherwise, it sends an INQUIRE message to the locking process.
3. A process, on receiving an INQUIRE message from a quorum node, unlocks the quorum node by sending a RELINQUISH messageas and when it realizes that it will not be able to successfully lock all its quorum members. This is ascertained when a FAILED message is received from one of the quorum members.
4. A node, on receiving a RELINQUISH or RELEASE message, grants the lock to the process whose request has the highest priority among all the pending requests, if any.

Maekawa [4] prove that the message complexity of above algorithm is $O(q)$, where *q* is the maximum size of a quorum. Further, its (best-case) synchronization delay and waiting time are both two message hops. (When analyzing the synchronization delay of a quorum-based algorithm derived from the Maekawa'a algorithm, we ignore the delay incurred due to deadlock resolution and only analyze the *best-case* synchronization delay. This is consistent with the practice used by other researchers [16][14]).

III. PRINCIPLE OF THE ALGORITHM

In this section, we propose a distributed algorithm for the group mutual exclusion algorithm based on quorum. This algorithm is classified as Maekawa-type algorithm [4][14][15]: send a request for permissions to some processes to enter a critical section, and releases permissions to exit a critical section. Different from the traditional mutual exclusion, request for the same group (session) must be granted simultaneously in case of group mutual exclusion. On the other hand, even if some processes are continuously making request for releasing the same session x, a process P_i requesting for $x'(\neq x)$ must be granted eventually. In addition deadlock must not be happen.

To avoid starvation and deadlock, totally ordered priority is defined for each request by a timestamp, which is a pair logical clock value on request and process identifier [16][14]. Permissions given to some processes may be preempted by a request with higher priority. For simplicity, we omit explicit code for handling logical clocks. The timestamp for current request is maintained by variable H_i. An ordering $<$ on timestamps are defined as:

$\langle P_i, H_i \rangle < \langle P_j, H_j \rangle$ iff $H_i < H_j$ or $(H_i = H_j) \wedge (P_i, P_j)$.

The outline of our algorithm is as follows. Each process is in one of the three states: *Wait* (request a session), *CS* (executing in a critical section), and *NCS* (otherwise).

1. A process P_i whishing to access to the session x sends a request message to a process P_j in a quorum already built. It waits for a grant message from P_j in the quorum. Then, P_i can access to session x with its group members if they are not requested for another session $x'(\neq x)$.
2. When P_i release a session, it sends a release message to $Pred_i$.
3. Suppose that when a request message of P_i arrives at P_j, P_j sends an *OK*() message to some P_k. If the priority of P_i is higher than that of P_k,i.e, $\langle P_i, H_i \rangle < \langle P_j, H_j \rangle$, P_j sends an inquire message to preempt the *OK*() message sent to P_k. If P_k is in Waiting state, *OK*() message from P_j is preempted, and P_k sends a relinquish message to P_j. If P_k is in *CS* or *NCS* state, the message is ignored because P_j eventually receives a release message from P_k.
4. When a process P_j receives a request message for session x for a process P_i. There are four several cases to study.
   a) no proces swas granted by P_j: P_j sends an *OK*() message to P_i, and the current session number becomes x.
   b) some process was granted by P_j: P_j is granted, if $x = x'$ holds and there is no requests in WS whose priority is less than that of P_j.
   c) some process was granted by P_j, their priority is less than that of P_i, and $x \neq x_j$. Here deadlock can occur and we use Maekawa [4].
   d) otherwise: the request by P_i is deferred, and stored in WS.
5. When a process P_j receives a release message for session x from a process P_i. If there is no more process granted by P_i, request for another session is granted.

A. Variables of the algorithm

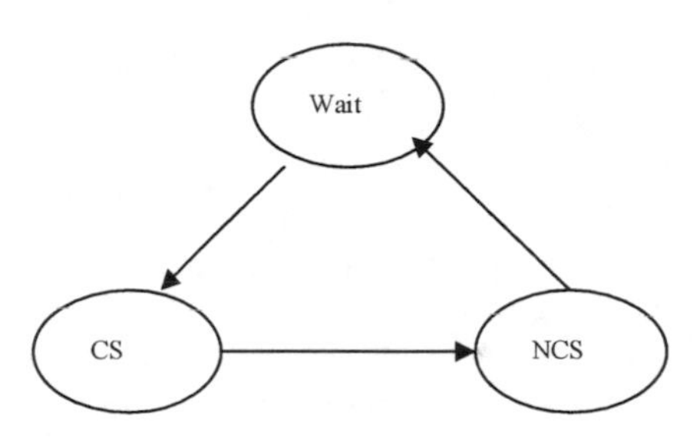

Fig. 1. States Process

The variables used in the algorithm for process P_i are listed below.

status: indicates whether a process P_i is the *Wait*, *CS*, or *NCS* section. Initially, *status=NCS*.
H_i: a timestamp for current request. Initially $H_i=0$.
x_i: maintains the current session number.
$g(P_i)$: denotes the current group on which P_i belongs.
$Pred_i$: gives the authorization to access the session.
nOK: the number of *OK*() message received by a process.
Nrel: denotes the number of *REL*() message received by a process.
WS: a set of waiting request. Initially *WS*=$\emptyset$.
OKset: a set of pairs of processes which have received *OK*() message and timestamps. Initially *OKset*=$\emptyset$.
Head(): returns the first node in the set queue.
REQ(P_i ,x, H_i): message sent by process P_i to request a session before entering the *CS*.
OK(): message for granting a process.

$REL(P_i, x, H_i)$: signify that process P_i with timestamp H_i has closed the session x.

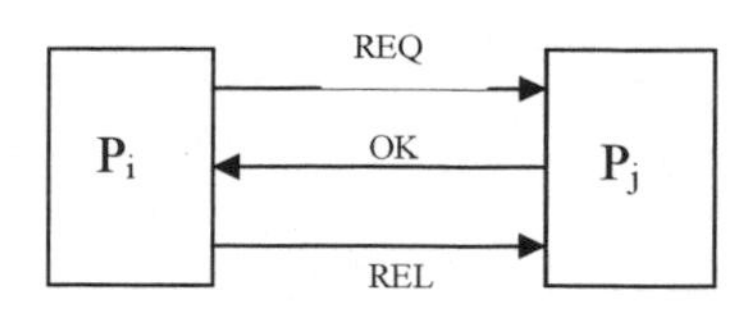

Fig. 2. Messages exchanged by Processes

B. Algorithm description

The distributed algorithm is based on the following rules: rules of application processes and, rules of management processes.

Rules of application processes

When a process P_i wants to enter *CS*

1. $status \leftarrow Wait$
2. $group \leftarrow g(P_i)$
3. $H_i \leftarrow H_i + 1$
4. **Send** $REQ(P_i, x, H_i)$ **to all** q in g
5. $nOK \leftarrow |g(P_i)|$ -1

When a process P_i receives $OK(P_j, x, H_j)$ *from* P_j

1. $H_i \leftarrow max(H_i, H_j)$
2. $nOK \leftarrow nOK-1$
3. **If** $(nOK=0)$ **then** $status=CS$
4. **For all** $P_k \in g(P_i)$ **Send** $\Theta(x)$ *to* P_k
5. $Nrel \leftarrow |g(P_i)|$ -1

When a process P_i exits *CS*

1. $status \leftarrow CS$
2. **If** ($Pred_i$ $=nil$) **then**
3. **If** ($Nrel=0$) **then**
4. **Send** *OK()* **to** *Head(WS)*
5. **Else**
6. **Send** *REL()* **to** $Pred_i$
7. $Pred_i \leftarrow nil$

When a process P_i receives $\Theta(x)$ from P_k

1. $Pred_i \leftarrow P_k$ /* P_i can acceed to session x

Rules of management processes

When $REQ(P_i, x, H_i)$ is received at process P_j from P_i

1. **Do**
2. **If** $(OKset=0)$ **then**
3. $OKset \leftarrow \{(P_i, x, H_i)\}$
4. $x_j \leftarrow x$
5. **Send** *OK()* **to** P_i
6. **Else If** $(OKset \neq \emptyset) \wedge (x = x_j)$ **then**
7. **If** $(WS \neq \emptyset)$ **then**
8. $(P'_k, H'_k) \leftarrow \min(WS)$
9. **If** $(WS \neq \emptyset) \wedge (P'_k, H'_k) < (P_j, H_j)$ **then**
10. $WS \leftarrow WS \cup \{(P_i x, H_i)\}$
11. **Else**
12. $OKset \leftarrow OKset \cup \{(P_i, x, H_i)\}$
13. **Send** *OK()* **to** P_i
14. **Else** /* $(OKset \neq \emptyset) \wedge (x \neq x_j)$
15. $(P''_k, H''_k) \leftarrow \min(OKset)$
16. **If** $(P''_k, H''_k) < (P_i, H_i)$ **then**
17. $WS \leftarrow WS \cup \{(P_i x, H_i)\}$
18. **Else** /* $(OKset \neq \emptyset) \wedge (x \neq x_j) \wedge (P_i, H_i) < (P''_k, H''_k)$
19. See Maekawa [4].
20. **End If**
21. **End If**
22. $OKset \leftarrow \emptyset$ $x_j \leftarrow x$
23. **For all** $(P_l, x_l, H_l) \in WS$ **s.t.** $x_l = x_j$ **do**
24. $OKset \leftarrow OKset \cup (P_l, x_l, H_l)$
25. $WS \leftarrow WS \setminus \{(P_l x_l, H_l)\}$
26. **Send** *OK()* **to** P_l
27. **End If**
28. **End If**
29. **Od**

When $REL(P_i, x, H_i)$ is received at process P_j from P_i

1. **Do**
2. $OKset \leftarrow OKset \setminus \{(P_i, x\ H_i)\}$
3. **If** $(OKset = \emptyset) \wedge (WS \neq \emptyset)$ **then**
4. $(P'_k, x'_k, H'_k) \leftarrow \min(WS)$
5. $x_j \leftarrow x'_k$
6. **For all** $(P_l, x_l, H_l) \in WS$ **s.t.** $x_l = x_j$ **do**
7. $OKset\ OKset \cup \{(P_l, x_l, H_l)\}$
8. $WS \leftarrow WS \setminus \{(P_l x_l, H_l)\}$
9. **Send** *OKset()* **to** P_l
10. **od**
11. **End If**
12. **Od**

C. Example

Let us consider the finite projective plane of 7 elements, which consists of a subset from V such that every subset has exactly 3 elements, every element is contained in exactly 3 subsets, and every two subsets intersect in exactly one element. Processes are labelled as $1,2,...,n$. The groups are:
$g(P_1)=\{1,2,3\}$, $g(P_2)=\{2,4,6\}$, $g(P_3)=\{3,5,6\}$, $g(P_4)=\{4,1,5\}$
$g(P_5)=\{5,2,7\}$, $g(P_6)=\{6,1,7\}$, $g(P_7)=\{7,4,3\}$.
We assume that we have two sessions x_1 and x_2. Now we illustrate our algorithm by the following scenario:

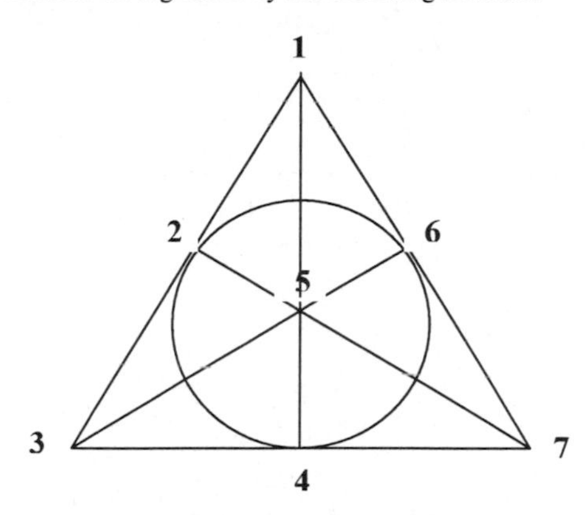

Fig. 3. The finite projective plane of order 2 (Fano plane)

E_1: The processes 1, 2 and 3 $\in g(P_1)$.
E_2: 1 sends a $REQ(1,x_1,H_1)$) message to processes 2 and 3.
E_3: 1 waits an *OK*() message from all its group members i.e. 2 and 3.
E_4: 2 and 3 receive message $REQ(P_1,x,H_1)$) from process 1.
E_5: If 2 and 3 have not requested anothter session x_2, they can send *OK*() message to process 1.
E_6: Process 1 receives all *OK*() message (*nOK*=$|g(P_1)|$-1) from its group members.
E_7: 1 sends a $\Theta(x_1)$ message to processes 2 and 3 for accessing the session x_1.
We can also use another example for a grid quorum on n=9 elements. A quorum is the union of a full row and one elements from each row below the full row.

1	2	3
4	5	6
7	8	9

Fig. 4. A grid quorum of n=9 elements

IV. Proof of correctness

A. Mutual Exclusion

Lemma 1. Let $x \in \{1,2,...,m\}$ be any session and suppose that each request is for session x_i. Then, each process requesting the session x_i is eventtually enters their critical sections.

Proof 2. A process $P_k \in g(P_k)$ for session x_k sends an *OK*() message to any process $P_i \in g(P_i)$ requesting session x_i if $x_k = x_i$ and there is no requs for other sessions.

Lemme 2. At most one session is opened at a time.

Proof 3. If two processes P_i and P_j are in *CS*, we have two cases:

i) P_i and P_j have requested the same session.

ii) P_i and P_j have requested two different session x_i and x_j.

If P_i's priority is higher than P_j's priority i.e.

$\langle P_i, H_i \rangle < \langle P_j, H_j \rangle$ iff $H_i < H_j$ or $(H_i = H_j) \wedge (P_i, P_j)$. (1)

If P_j's priority is higher than P_i's priority i.e.

$\langle P_j, H_j \rangle < \langle P_i, H_i \rangle$ iff $H_j < H_i$ or $(H_j = H_i) \wedge (P_j, P_i)$. (2)

By (1) and (2), we have ($P_i = P_j$) and P_i and P_j must belong to the same session.

Because the non-empty intersection property, we have $g(P_i) \cap g(P_j) \neq \varnothing$ and hence P_i and P_j cannot be granted simultaneously. This is a contradiction with the specification of the algorithm that allow only one session to be opened at any instance.

Theorem 4. Group mutual exclusion algorithm problem is satisfied by the proposed algorithm.

Proof 5. The theorem directly follows from Lemma 1 and Lemma 2.

B. Starvation Freedom

The starvation of process P_i occurs when other preceding requests are continuously locking for waiting a member of

$g(P_i)$. Assume that starvations occurs. Consider $P_i \in g(P_i)$ be the highest priority request which cannot ever enter *CS*. From the assumption, P_i is the highest priority request that does not enter *CS* from some time H_i. Let *H* be the time at witch request from P_i arrives at every member of $g(P_i)$. Let us consider the system state after time $T = \max(H, H_i)$. Each process $P_k \in g(P_i)$ $(k \neq i)$ must try to send *OK*() message to P_i because P_i has the highest priority. If P_k has not sent *OK*() message to any process, obviously it sends *OK*() message to P_i. Otherwise P_k can send *OK*() message to another process, say P_j. If P_j has not entered *CS*, P_k will be able to send *OK*() message to P_i. If P_j has entered *CS*, it eventually exits *CS*. After T, P_k does not send *OK*() message to any other requests. Therefore, P_k will be able to send *OK*() message to P_i and no starvation occurs.

V. Conclusion

In this paper we have presented a quorum-based distributed algorithm for group mutual exclusion. In the proposed algorithm a process can enter CS only with its members group. This algorithm is optimal in terms of the number of messages used to create group mutual exclusion algorithm.

References

[1] Y.-J. Joung. ''Asynchronous Group Mutual Exclusion Algorithm extended abstract),'' Proc. *17th Ann. ACM Symp. Principles of Distributed Computing*, pp. 51-60, June. 1998.

[2]. E. Dijkstra. ''Solution of a problem in concurrent programming control, '' *Communication of the ACM*, 8:569, 1965.

[3] L. Lamport. ''The mutual exclusion problem: Parts I \& II,'' *Journal of the ACM*, 33(2):313-348, 1986.

[4] M. Maekawa., ''A $\sqrt{n}$ Algorithm for Mutual Exclusion in Decentralized Systems,'' *ACM Trans. Computer Systems*, Vol. 3, no. 2, pp. 145-159, May. 1985.

[5] Y.-J. Joung. ''Asynchronous Group Mutual Exclusion Algorithm,'' *Distributed Computing(DC)*, 13(4):189-206, 2000.

[6] P.Keane and M. Moir. ''A Simple Local-Spin Group Mutual Exclusion, '' *In ACM Symposium on Principles of Distributed Computing(PODC)*, pages 23-32, 1999.

[7] I. Suziki and T. Kasami. ''A Distributed Mutual Exclusion Algorithm,'' *ACM Transactions on Computer Systems*, 3(4):344-349, 1985.

[8] V. Hadzilacos. ''A Note on Group Mutual Exclusion,'' *In Proceedings of the 20th ACM Symposium on Principles and Distribute Computing(PODC)*, Aug. 2001.

[9] S. Cantarell, A. K. Datta, F. Petit, and V. Villain. ''Token Based Group Mutual Exclusion for Asynchronous Rings,'' *In Proceedings of the IEEE Conference on Distributed Computing Systems(ICDCS)*, pages 691-694, 2001.

[10] K.-P. Wu and Y.-J. Joung. ''Asynchronous Group Mutual Exclusion in Ring Networks,'' *IEEE Proceedings--Computers and Digital Techniques*, 147(1):1-8, 2000.

[11] J. Beauquier, S. Cantarell, A. K. Datta, and F. Petit. ''Group Mutual Exclusion in Tree Networks,'' *Journal of Information Science and Engineering*, 19(3):415-432, May 2003.

[12] Y.-J. Joung. ''The Congenial Talking Philosophers Problem in Computer Networks,'' *Distributed Computing(DC)*, pages 155-175, 2002.

[13] G. Ricart and A. K. Agrawala. ''An Optimal Algorithm for Mutual Exclusuion in Computer Networks,'' *Communications of the ACM (CACM)*, 24(1):9-17, Jan. 1981.

[14] B. A. Sanders. ''The information structure of mutual exclusion algorithms,'' *ACM Transactiond on Computer Systems*, 5(3):284-299, August 1987.

[15] M. Singhal. ''A class of deadlock-free maekawa-type algorithms for mutual exclusion in distributed systems, '' *Distributed Computing*, pages 131-138, April 1991.

[16] L. Lamport. ''Time, clocks, and the ordering of events in a distributed system,'' *Communications of the ACM*, 21(7)558-565, July 1978.

[17] Y.-J. Joung. ''Quorum-Based Algorithms for Group Mutual Exclusion,'' *In Proceedings of DISC, volume 2180 of LNCS*, pages 16-32. Springer-Verlag, 2001.

[18] K. Alagarsamy and K. Vidyasankar. ''Elegant Solutions for Group Mutual Exclusion Problem, '' *Technical report, Department of Computer Science, Memorial University of Newfoundland, St. John's, Newfoundland, Canada, 1999.*

[19] M. Maekawa. ''Algorithm for Mutual Exclusion in Decentralized System., '' *ACM Transactions on Computer Systems*, 3(2):145-159, May 1985.

A Dynamic Fuzzy Model for Processing Lung Sounds

P.A. Mastorocostas, D.N. Varsamis, C.A. Mastorocostas, C.S. Hilas
Department of Informatics and Communications
Technological Educational Institute of Serres
62124, Serres, GREECE

Abstract— **This paper presents a dynamic fuzzy filter, with internal feedback, that performs the task of separation of lung sounds, obtained from patients with pulmonary pathology. The filter is a novel generalized TSK fuzzy model, where the consequent parts of the fuzzy rules are Block-Diagonal Recurrent Neural Networks. Extensive experimental results, regarding the lung sound category of coarse crackles, are given, and a performance comparison with a series of other fuzzy and neural filters is conducted, underlining the separation capabilities of the proposed filter.**

I. Introduction

Pathological discontinuous adventitious sounds (DAS) are strongly related to pulmonary dysfunction. Their clinical use for the interpretation of respiratory malfunction depends on their efficient and objective separation from vesicular sounds (VS). In order to achieve this kind of separation, the nonstationarity of DAS must be taken into account. A number of nonlinear processing methods have been used in the past, with the wavelet transform-based stationary-nonstatonary (WTST-NST) filter [1] providing the most accurate separation results. However, this method could not be easily implemented in real-time analysis of lung sounds.

During the last years computational intelligence models, such as neural and fuzzy-neural networks have been proposed, providing encouraging results. In particular, The Orthogonal Least Squares-based Fuzzy Filter has been suggested in [2] for real-time separation of lung sounds. A recurrent neural filter has been reported in [3], based on the Block-Diagonal Recurrent Neural Network (BDRNN) [4], which is a simplified form of the fully recurrent network, with no interlinks among neurons in the hidden layer. As shown in [3], due to its internal dynamics, the BDRNN filter is capable of performing efficient real-time separation of lung sounds.

It is common knowledge that the fuzzy neural networks combine **the benefits of fuzzy systems and those of the neural networks. In particular, fuzzy inference provides an efficient way of handling imprecision and uncertainty while neural learning permits determining the model parameters. Stem from this fact and the modeling capabilities of the BDRNN mentioned above, in this work a recurrent fuzzy neural network is proposed, for real-time separation of DAS from VS. The novelty of the proposed model lies in the consequent parts of the fuzzy rules, which are small block-diagonal recurrent neural networks, thus introducing dynamics to the overall network.**

Extensive experimental results are given and a comparative analysis with previous works is conducted, highlighting the efficiency of the proposed filter.

II. The Dynamic Fuzzy Inference System and the Learning Algorithm

A. The Dynamic Block-Diagonal Fuzzy Neural Network

The suggested dynamic block-diagonal fuzzy neural network (DBD-FNN) comprises r Takagi-Sugeno-Kang rules [5] of the following form:

$$IF\ \boldsymbol{u}(k)\ is\ \mathsf{A}^{(l)}\ THEN\ g_l(k) = BDRNN_l\left(\boldsymbol{u}(k)\right),\ l = 1,...,r \quad (1)$$

where $\mathsf{A}^{(l)}$ is the fuzzy region in the premise part and the sub-model $BDRNN_l$ is a block-diagonal recurrent neural network that implements the consequent part of the l-th rule. For the sake of simplicity, a multiple–input – single–output model is used. Based on the structural characteristics of the TSK model, it can be divided into three major parts: the premise, the consequent and the defuzzification part.

At each time instant k, the premise part is fed with the process variables $u_1(k),...,u_m(k)$, which are used for defining the fuzzy operating regions. The firing strengths of the rules are calculated by the following static function:

$$\mu_l(k) = \prod_{j=1}^{m} \exp[-\frac{1}{2}\cdot\frac{(u_j(k)-m_{lj}(k))^2}{(\sigma_{lj}(k))^2}],\ \ l = 1,...,r \quad (2)$$

where $\boldsymbol{m}_l = \{m_{lj}, j = 1,...,m\}$ and $\boldsymbol{\sigma}_l = \{\sigma_{lj}, j = 1,...,m\}$, $l = 1,...,r$, are the premise part parameters.

The consequent parts of the model are dynamic, including the r sub-models of the rules. Each sub-model $BDRNN_l$ is a block-diagonal recurrent neural network, which is a two-layer network, with the output layer being static and the hidden layer being dynamic. The hidden layer consists of pairs of neurons (blocks); there are feedback connections between the neurons of each pair, introducing dynamics to the network.

K. Elleithy (ed.), Advances and Innovations in Systems, Computing Sciences and Software Engineering, 357–362.

The operation of the BDRNN with m inputs, r outputs and N neurons at the hidden layer is described by the following set of state equations:

$$x_{2i-1}^{(l)}(k) = f_a[\sum_{j=1}^{m} b_{2i-1,j}^{(l)} \cdot u_j(k) + w_{1,i}^{(l)} \cdot x_{2i-1}^{(l)}(k-1) + w_{2,i}^{(l)} \cdot x_{2i}^{(l)}(k-1)] \quad (3a)$$

$$x_{2i}^{(l)}(k) = f_a[\sum_{j=1}^{m} b_{2i,j}^{(l)} \cdot u_j(k) - w_{2,i}^{(l)} \cdot x_{2i-1}^{(l)}(k-1) + w_{1,i}^{(l)} \cdot x_{2i}^{(l)}(k-1)] \quad (3b)$$

$$g_j(k) = f_b[\sum_{j=1}^{N} c_{lj} \cdot x_j^{(l)}(k)] = f_b[\sum_{j=1}^{N/2} c_{l,2j-1} \cdot x_{2j-1}^{(l)}(k) + \sum_{j=1}^{N/2} c_{l,2j} \cdot x_{2j}^{(l)}(k)] \quad (4)$$

$l = 1,\ldots,r$, $i = 1,\ldots, N/2$

where

- f_a and f_b are the neuron activation functions of the hidden and the output layer, respectively. In the following, the activation functions are both chosen to be the sigmoid function $f(z) = \frac{1-e^{-a\cdot z}}{1+e^{-a\cdot z}}$.
- $\mathbf{x}^{(l)}(k) = [x_i^{(l)}(k)]$ is a N-element vector, comprising the outputs of the hidden layer of the *l-th* fuzzy rule. In particular, $x_i^{(l)}(k)$ is the output of the *i-th* hidden neuron at time *k*.
- $B = \left[b_{ij}^{(l)}\right]$ and $C = \left[c_{lj}\right]$ are $r \times N \times m$ and $r \times N$ input and output weight matrices, respectively.
- $w_{1,i}^{(l)}, w_{2,i}^{(l)}$ are the rules' feedback weights, that form the block diagonal feedback matrices, $W^{(l)}$. The *scaled orthogonal* form is employed in this work, [4], where the feedback matrices are described by the following formula:

$$W^{(l)} = \begin{bmatrix} w_{1,i}^{(l)} & w_{2,i}^{(l)} \\ -w_{2,i}^{(l)} & w_{1,i}^{(l)} \end{bmatrix} \quad i = 1,2,\ldots, N/2 \quad (5)$$

The output of the model at time *k*, *y(k)*, is determined using the weighted average defuzzification method:

$$y(k) = \frac{\sum_{l=1}^{r} \mu_l(k) \cdot g_l(k)}{\sum_{l=1}^{r} \mu_l(k)} \quad (6)$$

The configuration of the proposed BDRNN consequent part is presented in Fig. 1, where, for the sake of simplicity, a single–input–single–output BDRNN with two blocks of neurons is shown.

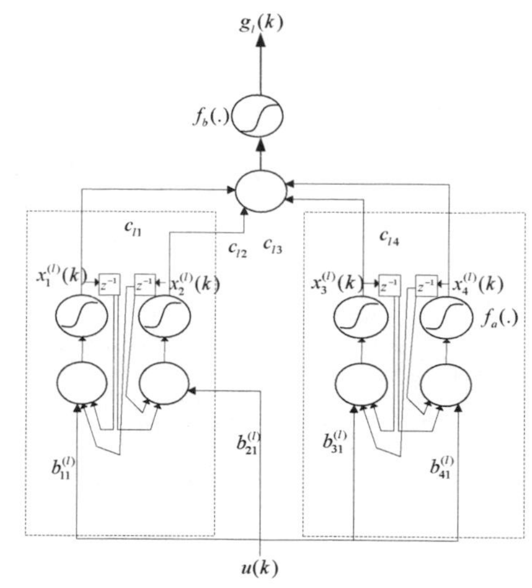

Fig. 1. Configuration of the consequent part of the fuzzy rules

B. *The Training Algorithm*

The DBD-FNN is trained by use of the D-FUNCOM algorithm, which was developed in [6] for fuzzy models, whose consequent parts are recurrent neural networks. Only minor modifications are made, such that the method takes into consideration the special features of the BDRNN, requiring calculation of the error gradients for the feedback weights and extraction of the gradients for the premise part weights. Since the scope of the paper is to highlight the model's operation rather than the selected training algorithm, a brief derivation of the error gradients is given in the sequel, while the complete update formulas are fully described in [6]:

- Let us consider a training data set of k_f input–output pairs. The Mean Squared Error is selected as the error measure, where $y_d(k)$ is the desired output:

$$E = \frac{1}{k_f} \cdot \sum_{k=1}^{k_f} [y(k) - y_d(k)]^2 \quad (7)$$

Since the premise and defuzzification parts are static, the gradients of E with respect to the weights of the premise part are derived by use of partial derivatives. For batch learning, $m_{ij}(1) = \ldots = m_{ij}(k_f)$ and $\sigma_{ij}(1) = \sigma_{ij}(2) = \sigma_{ij}(k_f)$. Thus, the error gradients of the premise part weights are given by the following formulas:

$$\frac{\partial E}{\partial m_{ij}} = \sum_{k=1}^{k_f} \{ \frac{2}{k_f} \cdot (y(k) - y_d(k)) \cdot \frac{g_i(k) - y(k)}{\sum_{l=1}^{r} \mu_l(k)} \cdot \mu_i(k) \cdot \frac{z_j(k) - m_{ij}(k)}{(\sigma_{ij}(k))^2} \} \quad (8a)$$

$$\frac{\partial E}{\partial \sigma_{ij}} = \sum_{k=1}^{k_f} \{ \frac{2}{k_f} \cdot (y(k) - y_d(k)) \cdot \frac{g_i(k) - y(k)}{\sum_{l=1}^{r} \mu_l(k)} \cdot \mu_i(k) \cdot \frac{z_j(k) - m_{ij}(k)}{(\sigma_{ij}(k))^3} \} \quad (8b)$$

• The gradients of E with respect to the weights of the consequent part should be calculated using ordered derivatives, since there exist temporal relations through the feedback connections. The calculation of the error gradients is based on the use of Lagrange multipliers, as shown below:

$$\frac{\partial^+ E}{\partial b_{2i-1,j}^{(l)}} = \sum_{k=1}^{k_f} \{ \lambda_{x,2i-1}^{(l)}(k) \cdot u_j(k) \cdot f_a'(k,l,2i-1) \} \quad (9a)$$

$$\frac{\partial^+ E}{\partial b_{2i,j}^{(l)}} = \sum_{k=1}^{k_f} \{ \lambda_{x,2i}^{(l)}(k) \cdot u_j(k) \cdot f_a'(k,l,2i) \} \quad (9b)$$

$$\frac{\partial^+ E}{\partial c_{l,2i-1}} = \sum_{k=1}^{k_f} \{ \lambda_{g,l}(k) \cdot x_{2i-1}^{(l)}(k) \cdot f_b'(k,l) \} \quad (10a)$$

$$\frac{\partial^+ E}{\partial c_{l,2i}} = \sum_{k=1}^{k_f} \{ \lambda_{g,l}(k) \cdot x_{2i}^{(l)}(k) \cdot f_b'(k,l) \} \quad (10b)$$

$$\frac{\partial^+ E}{\partial w_{1,i}^{(l)}} = \sum_{k=1}^{k_f} \{ \lambda_{x,2i-1}^{(l)}(k) \cdot x_{2i-1}^{(l)}(k-1) \cdot f_a'(k,l,2i-1) + \lambda_{x,2i}^{(l)}(k) \cdot x_{2i}^{(l)}(k-1) \cdot f_a'(k,l,2i) \} \quad (11a)$$

$$\frac{\partial^+ E}{\partial w_{2,i}^{(l)}} = \sum_{k=1}^{k_f} \{ \lambda_{x,2i-1}^{(l)}(k) \cdot x_{2i}^{(l)}(k-1) \cdot f_a'(k,l,2i-1) - \lambda_{x,2i}^{(l)}(k) \cdot x_{2i-1}^{(l)}(k-1) \cdot f_a'(k,l,2i) \} \quad (11b)$$

$$\lambda_{g,l}(k) = \frac{2}{k_f} \cdot (y(k) - y_d(k)) \cdot \frac{\mu_l(k)}{\sum_{j=1}^{r} \mu_j(k)} \quad (12a)$$

$$\lambda_{x,2i-1}^{(l)}(k) = \lambda_{g,l}(k) \cdot c_{l,2i-1} f_b'(k,l) + \lambda_{x,2i-1}^{(l)}(k+1) \cdot w_{1,i}^{(l)} \cdot f_a'(k+1,l,2i-1) - \lambda_{x,2i}^{(l)}(k+1) \cdot w_{2,i}^{(l)} \cdot f_a'(k+1,l,2i) \quad (12b)$$

$$\lambda_{x,2i}^{(l)}(k) = \lambda_{g,l}(k) \cdot c_{l,2i} f_b'(k,l) + \lambda_{x,2i-1}^{(l)}(k+1) \cdot w_{2,i}^{(l)} \cdot f_a'(k+1,l,2i-1) + \lambda_{x,2i}^{(l)}(k+1) \cdot w_{1,i}^{(l)} \cdot f_a'(k+1,l,2i) \quad (12c)$$

$$\lambda_{x,2i-1}^{(l)}(k_f) = \lambda_{g,l}(k_f) \cdot c_{l,2i-1} f_b'(k_f,l) \quad (13a)$$

$$\lambda_{x,2i}^{(l)}(k_f) = \lambda_{g,l}(k_f) \cdot c_{l,2i} f_b'(k_f,l) \quad (13b)$$

where $f_a'(k+1,l,i)$ and $f_b'(k,l)$ are the derivatives of $x_{il}(k)$ and $g_l(k)$, with respect to their arguments. Equations (12b) and (12c) are backward difference equations that can be solved for $k = k_f - 1, \ldots, 1$ using the boundary conditions in (13a) and (13b), respectively, for $k = k_f$.

III. THE DBD-FNN FILTER

The DBD-FNN is for the estimation of the nonstationary part of the input signal. The network operates in parallel mode and is fed with the input signal $u(k)$, which is the normalized zero-mean recorded lung sound. As a result, the output of the filter is an estimation of the DAS.

The same pre-classified lung sound signals used in [1]–[3], i.e. ten cases, are used as model generation sets. The lung sounds are divided to three categories [1]: the coarse crackles (CC), the fine crackles (FC) and the squawks (SQ). The case of coarse crackles is examined in the present work. The sounds have been drawn from an international sound database, [7]. The data set has been obtained by digitizing sections of 15sec of the signals from the lung sounds databases by a 12-Bit Analog-to-Digital (A/D) converter at a sampling rate of 2.5kHz, divided into successive records of 1024 or 2048 samples each, with zero mean value and normalized. Then, all these records have been processed by the WTST-NST filter in order to obtain an accurate estimation of their stationary and non-stationary parts. Therefore, the nonstationary output of the WTST-NST filter are considered to be the desired ones.

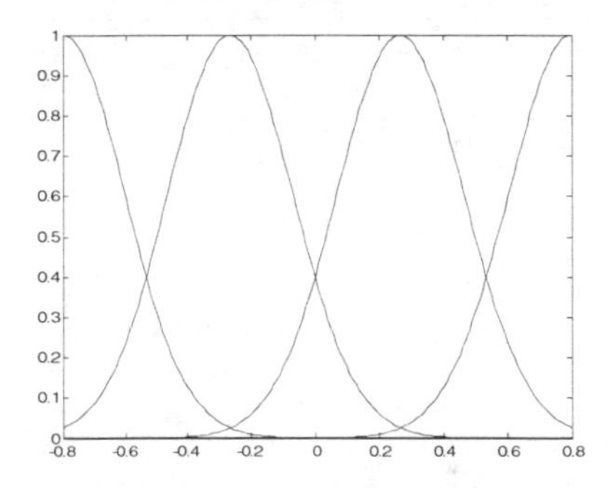

(a) Initial partition of the input space

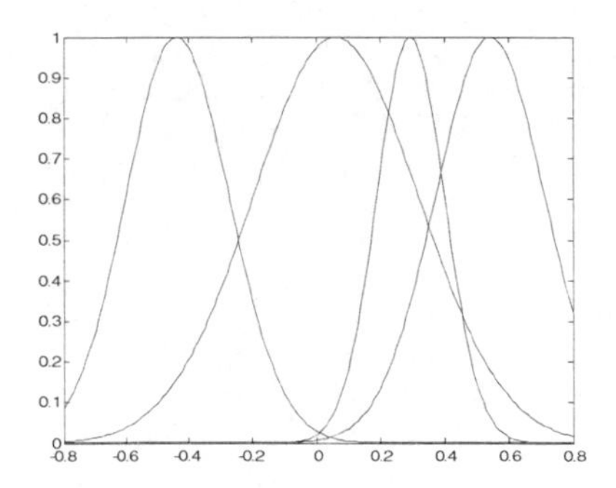

(b) Final partition of the input space

Fig. 2

Several DBD-FNNs with different structural characteristics are examined. Additionally, various combinations of the learning parameters are tested. For each case, 100 trials are conducted with random initial weight values and the results are averaged. Selection of the model and the parameter combination is based on the criteria of *(a)* effective separation of DAS from VS and *(b)* a moderate complexity of the resulting model. The selected structural characteristics are given in Table I. Training lasts for 500 epochs and the selected learning parameters are given in Table II. The initial and final membership functions are shown in Fig. 2. It can be noticed that the final model efficiently covers the input space, while the membership functions do not overlap considerably, thus preserving the local modeling approach of the Takagi-Sugeno-Kang fuzzy systems.

TABLE I
CHARACTERISTICS OF THE DBD-FNN STRUCTURE

Parameter	Value
Number of rules	4
Number of blocks	2
Coefficient of the sigmoid function, a	2
Overlapping coefficient between initial membership functions, [0,1]	0.4

TABLE II
D-FUNCOM LEARNING PARAMETERS

Premise part				
n^+	n^+	Δ_{min}	Δ_0	ξ
1.05	0.8	1E-4	0.02	0.9
Consequent part				
n^+	n^+	Δ_{min}	Δ_0	ξ
1.1	0.6	1E-4	0.1	0.9

IV. RESULTS AND DISCUSSION

A. *Experimental Results*

In this subsection, the results obtained using the DBD-FNN on the case of coarse crackles are presented. The processed records were selected so that the main structural morphology of the DAS would be clearly encountered. The efficiency of the filter is tested using for evaluation the same cases that were used in [1]–[3].

The results obtained using the DBD-FNN filter on the case of coarse crackles are presented in Figs 2 and 3, where "noisy" recorded vesicular sounds and separated DAS are depicted in parts (a) and (b), respectively. The position of waves identified visually (by a physician) as crackles were marked with arrowheads, in order to be compared with nonstationary signals separated automatically by the filter. In this way, the performance of the proposed filter in separating the nonstationary parts of breath sounds was verified, according to the evaluation procedure for the WTST-NST, OLS-FF and BDRNN models.

A case of coarse crackles (C1-2048 samples), recorded from two patients with chronic bronchitis, is displayed in Fig. 3a. Comparing the nonstationary part of Fig. 3a with Fig. 3b, it is clear that in Fig. 3b all DAS sounds are easily identified; their various morphologies and locations are clearly distinguished without any distortion. The efficiency of the proposed filter is highlighted in Fig. 4 as well. A time section of 0.015sec, corresponding to a coarse crackle, is presented. It is clear that the model has effectively detected the existence and the shape of the crackle.

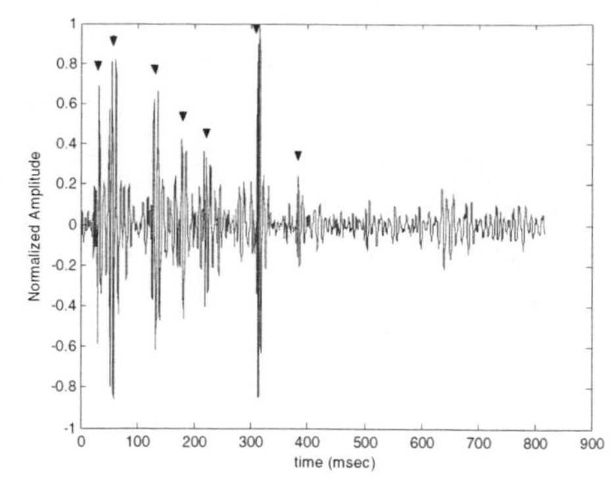

(a) A time section of 0.8192sec of coarse crackles recorded from a patient with chronic bronchitis (case C1), considered as input. The arrowheads indicate waves, which have been visually identified as coarse crackles.

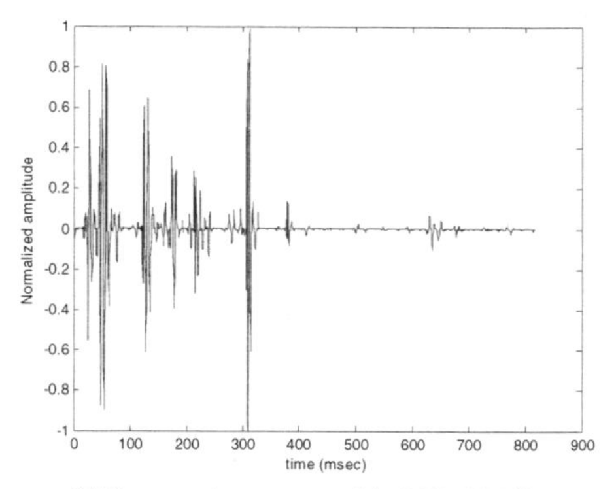

(b) The nonstationary output of the DBD-FNN filter.
Fig. 3.

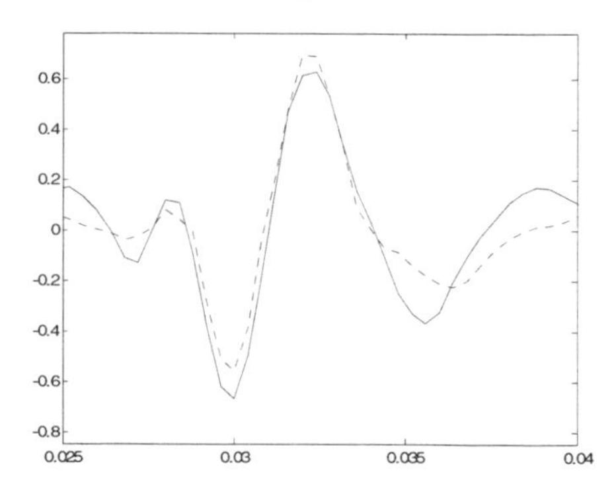

Fig. 4. A section of 0.015sec that contains a coarse crackle together with the estimation of the nonstationary part by the DBD-FNN filter.

B. Performance Evaluation

The evaluation of the performance of the DBD-FNN filter is based on qualitative and quantitative measures introduced in [1]:

• Auditory inspection of the DBD-FNN's stationary output. The effect of the DBD-FNN on input breath sounds was tested in a qualitative manner by listening to its stationary outputs after digital-to-analog (D/A) conversion.

• The rate of detectability: $D_R = \frac{N_E}{N_R} \cdot 100\%$

where N_E is the number of estimated DAS and N_R is the number of visually recognized DAS by a physician (considered as the true number of DAS in the input signal).

• The root mean squared error:

$RMSE = \sqrt{\frac{1}{k_f}\sum_{k=1}^{k_f}[y(k) - y_d(k)]^2}$, where $y(k)$ is the DBD-FNN output of the *k-th* sample, $y_d(k)$ is the respective actual output and k_f is the number of samples. It should be noted that the values of the RMSE do not always represent good separation results since they do not focus on the particular signal details a physician is interested in. They are only intended to provide an indication of the quality of achieving the desired input-output relationships, *given* the evaluation by the first two criteria.

The results of the quantitative evaluation are presented in Table III for each case. These results show that the proposed filter produces a very efficient separation, since the average rate of detectability is 98.18%. Additionally, during the qualitative testing of the DBD-FNN filter, by listening to its stationary outputs after Digital-to-Analog (D/A) conversion, the sounds were practically not heard, confirming a high quality separation performance by the DBD-FNN filter.

TABLE III
PERFORMANCE OF THE DBD-FNN FILTER

Case	*N*	N_E/N_R	D_R(%)	*RMSE*
C1	2048	7/7	100	0.04122
C2	2048	7/7	100	0.05307
C3	1024	8/8	100	0.06056
C4	1024	8/8	100	0.09622
C5	1024	10/11	90.90	0.07863

• *N*: Number of samples
• N_R: Number of visually recognized DAS by a physician
• N_E: Number of estimated DAS using the DBD-FNN filter
• D_R: Rate of detectability of the DBD-FNN filter
• ***RMSE***: Root Mean Squared Error of the estimated output of the DBD-FNN filter

It should be mentioned that, since the proposed filter requires only four neurons in the hidden layer of the consequent parts of the fuzzy rules, the operations of the DBD-FNN (multiplications, additions and look-up table operations) can be delivered within the sampling period (0.4ms), ensuring its real-time operation and, consequently, improving the procedure of clinical screening of DAS.

C. Comparison with previous works

The DBD-FNN filter's performance is evaluated with regard to the WTST-NST [1], the OLS-FF [2] and the DBRNN [3]. All four models are applied to the same 16 cases of patients and the results are shown in Table IV.

TABLE IV
PERFORMANCE EVALUATION OF DIFFERENT ALGORITHMS

	WTST-NST	OLS-FF	BDRNN	DBD-FNN
Average D_R **(%)**	*97.5*	*96.36*	*98.18*	*98.18*
Average RMSE	Not defined	*0.0711*	*0.0638*	*0.0630*

As shown in Table IV, the DBD-FNN filter exhibits a superior separation performance compared to the WTST-NST and the OLS-FF, with respect to the average rate of detectability. Moreover, it outperforms the static fuzzy filter OLS-FF with respect to the RMSE, leading to the conclusion that the recurrent model tracks effectively the dynamics of the nonstationary signal. Compared to the other recurrent model, the DBD-FNN exhibits similar results but requires 75% less training time than the BDRNN does. Furthermore, as discussed in the previous subsection, the proposed filter satisfies the real-time implementation issue, a fact that constitutes a clear advantage over the WTST-NST model since it improves the procedure of clinical screening of DAS.

V. CONCLUSION

A new recurrent fuzzy filter has been implemented for real-time separation of lung sounds. The filter model is based on the classic TSK fuzzy model, with the consequent parts of the fuzzy rules consist of block-diagonal recurrent neural networks. The scheme has been evaluated on pre-classified DAS, i.e., coarse crackles, selected from an international lung sound database. From the experiments and the comparative analysis with other separation schemes, it is concluded that the DBD-FNN filter performs very efficiently in separating the DAS from VS despite the differences in their structural character, and is capable of performing real-time separation. Hence, the proposed filter can be used as an objective method for real-time DAS analysis.

ACKNOWLEDGMENT

The Project is co-funded by the European Social Fund and National Resources – (EPEAEK–II) ARXIMHDHS.

REFERENCES

[1] L.J. Hadjileontiadis and S.M. Panas, "Separation of Discontinuous Adventitious Sounds from Vesicular Sounds Using a Wavelet-Based Filter," *IEEE Trans. Biomedical Eng.*, vol. 44, pp. 1269-1281, 1997.

[2] P.A. Mastorocostas, Y.A. Tolias, J.B. Theocharis, L.J. Hadjileontiadis, and S.M. Panas, "An Orthogonal Least Squares-Based Fuzzy Filter for Real-Time Analysis of Lung Sounds," *IEEE Trans. Biomedical Eng.*, vol. 47, pp. 1165-1176, 2000.

[3] P.A. Mastorocostas and J.B. Theocharis, "A Stable Learning Method for Block-Diagonal Recurrent Neural Networks: Application to the Analysis of Lung Sounds," *IEEE Trans. Syst., Man, and Cybern. – Part B*, vol. 36, pp. 242-254, 2006.

[4] S.C. Sivakumar, W. Robertson, and W.J. Phillips, "On-Line Stabilization of Block-Diagonal Recurrent Neural Networks," *IEEE Trans. Neural Networks*, vol. 10, pp. 167-175, 1999.

[5] T. Takagi and M. Sugeno, "Fuzzy Identification of Systems and Its applications to modeling and Control," *IEEE Trans. Syst., Man, and Cybern.*, vol. 15, pp. 116-132, 1985.

[6] P.A. Mastorocostas and J.B. Theocharis, "A Recurrent Fuzzy Neural Model for Dynamic System Identification," *IEEE Trans. Syst., Man, and Cybern. – Part B.*, vol. 32, pp. 176-190, 2002.

[7] S.S. Kraman, *Lung Sounds: An Introduction to the Interpretation of the Auscultatory Finding*, Northbrook, IL: Amer. College of Chest Physicians, 1993, audio tape.

A Formal Specification in JML of Java Security Package

Poonam Agarwal*, Carlos E. Rubio-Medrano, Yoonsik Cheon and Patricia. J Teller
The University of Texas at El Paso,
500 W. University Avenue, El Paso, TX 79968 U.S.A.
{pdagarwal, cerubio, ycheon, pteller}@utep.edu

Abstract-**The Java security package allows a programmer to add security features to Java applications. Although the package provides a complex application programming interface (API), its informal description, e.g., Javadoc comments, is often ambiguous or imprecise. Nonetheless, the security of an application can be compromised if the package is used without a concrete understanding of the precise behavior of the API classes and interfaces, which can be attained via formal specification. In this paper, we present our experiences in formally specifying the Java security package in JML, a formal behavior interface specification language for Java. We illustrate portions of our JML specifications and discuss the lessons that we learned, from this specification effort, about specification patterns and the effectiveness of JML. Our specifications are not only a precise document for the API but also provide a foundation for formally reasoning and verifying the security aspects of applications. We believe that our specification techniques and patterns can be used to specify other Java packages and frameworks.**

I. INTRODUCTION

Java makes it possible to write secure applications by providing a security model based on a sandbox, an execution environment in which a program runs and the program's execution is confined within certain bounds [7]. The classes in the java.security package as well as those in the security extension allow security features to be added to an application [9]. It is necessary to precisely understand the behaviors of these classes and interfaces because a small misuse of an API can significantly compromise the security of an application. However, an informal document, such as Javadoc comments, often is inadequate for precisely specifying the behavior of the API because of the inherent nature of a natural language, i.e., its ambiguity and impreciseness. Even though the source code can provide precise information, in it essential features are tangled with implementation decisions and details.

A formal specification may complement the informal document, since it can document essential features in a concise and precise manner. A formal behavioral interface specification language, such as JML [5], is designed to precisely document both the syntactic interface and the behavior of program modules. A formal behavioral interface specification provides not only a precise document describing the API, but also a formal foundation for rigorously proving properties about the implementation [1] [8].

In this paper, we present a specification case study of the Java security package using JML. This work is currently being done under the Milaap Project, the research goal of which is to unify and integrate several different verification methods. During this study, we specified a significant number of core classes of the security package. In this paper we present some of the interesting classes and discuss the lessons that we learned from this specification effort. The case study allowed us to identify several specification patterns that facilitate writing JML specifications. It also permitted us to critically evaluate the effectiveness of JML as an API documentation language, which led us to a JML wish list. We expect our specifications to be a valuable document to the users of the Java security package, and we believe that our specification techniques and patterns are applicable to the specification of other Java packages and frameworks.

Our work is not the first to formally specify Java packages. The JML distributions are shipped with specifications of several JDK classes, such as collection classes (refer to the JML website at www.jmlspecs.org). Our main contribution is not only the final specifications, but also identification of reusable specification patterns and a critical evaluation of JML. Poll et al. specified in JML all the classes of the Java Card API [8], and their specification made many of the implicit assumptions underlying the implementation explicit; this agrees with our findings. Their specifications are written in a lightweight style, while most of our specifications are written in a heavyweight style. Catano and Huisman found a similar result in that even lightweight use of formal specifications contributed greatly to the improvement of the quality of applications [2].

The remainder of the paper is organized as follows. The Java security package and JML are described in Section II. Sections III and IV focus on the two main aspects of the security package: protection mechanism and cryptographic architecture. A discussion of the results and conclusions of our specification case study conclude the paper.

II. BACKGROUND

A. *The Java Security Package*

The Java security package (java.security) provides a framework for the Java security architecture, which can be broadly classified into two different aspects of security: protection mechanism and cryptographic architecture (see Fig. 1).

* The work of authors was supported in part by the NSF, CNS-0509299.

K. Elleithy (ed.), Advances and Innovations in Systems, Computing Sciences and Software Engineering, 363–368.

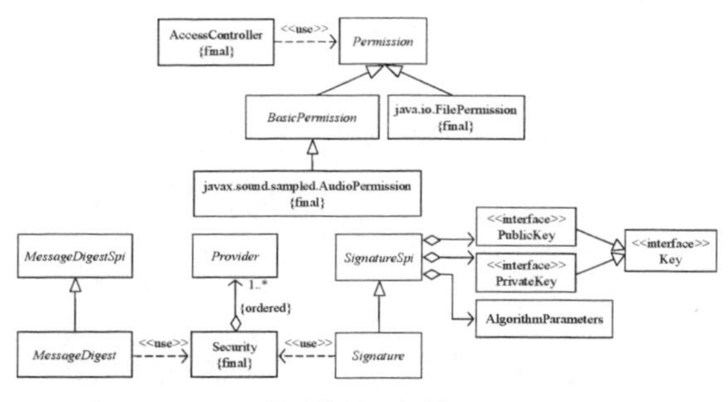

Fig. 1. Security-related classes

The protection mechanism deals with access control and prevention of unauthorized modifications. It is built upon concepts such as code sources (locations from which Java classes are obtained), permissions (requests to perform particular operations), policies (all the specific permissions that should be granted to specific code sources), and protection domains (particular code sources and permissions granted to those code sources) [7]. The class AccessController determines based on the security policy in effect, whether an access to a resource should be granted or denied. Any section of code that performs a security-sensitive operation should consult the access controller to verify if the operation is allowed. An operation or access is represented by permission, and different types of permissions form a class hierarchy rooted at the abstract class Permission.

The Java Cryptography Architecture (JCA) deals with authentication and supports algorithm and implementation independence [7]. Some of the JCA core classes are Security (handling all installed providers and security properties), Provider (interface to the concrete implementation of the cryptographic services), engine classes such as MessageDigest and Signature, and key and parameter classes such as Key and AlgorithmParameters. The code below shows a typical use of the MessageDigest engine class that provides an interface to a message digest (hash) algorithm, such as MD5 and SHA. A message digest is a secure one way hash function that takes arbitrary-sized data and returns a fixed-length hash value.

```
MessageDigest md = MessageDigest.getInstance("MD5","SUN");
md.update("Hi, JCA!".getBytes());
byte[] hash = md.digest();
```

Every engine class implements factory methods, named *getInstance*, each of which returns an instance of the specified algorithm from the optionally specified provider. To find an appropriate algorithm implementation, the factory methods ask the Security class, which, in turn, consults the available providers to check whether they can supply the desired algorithm. The example code gets an instance of Sun's implementation of the MD5 algorithm. The *update* method adds a sequence of bytes to the internal buffer, and the *digest* method computes the message digest of the bytes stored in the internal buffer.

B. JML

The Java Modeling Language (JML) [5] is a formal interface specification language for Java and describes both the syntactic interface and the behavior of Java program modules. The syntactic interface of a Java class or interface, commonly called an application programming interface (API), consists of the signatures of its methods and the names and types of its fields. The behavior of a program module is specified by writing, among other things, pre and postconditions of the methods exported by the module. The assertions in pre and postconditions are written using a subset of Java expressions and are annotated in the source code. The pre and postconditions are viewed as a contract between the client and the implementer of the module. The client must guarantee that, before calling a method *m* exported by the module, *m*'s precondition holds, and the implementer must guarantee that, after such a call, *m*'s postcondition holds. A method specification can consist of, among other things, a precondition (specified in the *requires* clause), a frame axiom (specified in the *assignable* clause), and a normal or exceptional postcondition (specified in the *ensures* or *signals* clause).

III. ACCESS CONTROL

The access control mechanism is built upon concepts such as code sources, permissions, policies, and protection domains [7]. A permission—an encapsulation of a request to perform a particular operation—is a key concept underlying the access control mechanism and provides an interesting aspect on formalization, as there exist several types of permissions organized into a class hierarchy, with some sharing certain common properties. In this section, we specify representative methods of the permission classes.

A. Permission Classes

A permission, a specific action that code is allowed to perform, consists of three components: *type, name,* and *actions*. The *type,* which specifies the type of the permission, is represented by a particular Java class that implements the permission. The *name* is based on permission type, e.g., the name of a file permission is a file or directory name, and a few permissions require no name. The *actions* also vary based on the permission type, and many permissions have no actions associated with them. The *actions* specifies what may be done to the target of the permission, e.g., a file permission may specify that a particular file can be read, written, deleted, executed, or some combination of these. In this paper we specify two representative classes and their superclasses, one for name-based permissions and the other for the name and action-based permissions. We focus only on one method in permission classes, *implies*, that determines whether one permission implies another permission. The *implies* method is one of the most important methods of permission classes because it is the primary method used by the access controller to make access decisions. We illustrate the specification technique that we used to factor out common or general properties into the superclass and leave the specifics to subclasses.

The abstract class Permission is the ancestor of all permission classes (see Fig. 2). In JML, specifications are typically annotated in source code as special comments, i.e., //@ and a pair of /*@ and @*/, and the specification of a method precedes its declaration. All permissions have a name—the interpretation of which depends on the subclass—and several abstract methods (e.g., *implies*) that define the semantics of the particular subclass. The JML modifier *spec_public* states that the field is treated as public for specification purpose; e.g., it can be used in public specifications. The specification of the *implies* method illustrates a specification pattern that we use to leave the specification of subclass-specific properties to subclasses. The keyword *normal_behavior* specifies the behavior of the method when it terminates normally, i.e., without throwing an exception. The *pure* modifier states that the method has no side-effect; only pure methods can appear in JML assertions. The postcondition states that the method should return true if and only if (a) the given permission, p, is of the same type as the receiver, and (b) jmlImpliesPerm(p) holds. Property (a) is common to all permission classes and, thus, is specified in this class. However, determining whether the receiver implies the argument permission is subclass-specific because different subclasses may have different names or actions, and their interpretations may be different. Thus, it cannot be completely specified in the class Permission. Our approach is to delegate the specification responsibility to subclasses by introducing a model method, i.e., *jmlImpliesPerm*, as shown in property (b). A model method is a specification-only method and can be used only in assertions. Each (concrete) subclass is responsible for overriding this model method to specify the subclass-specific precise meaning of the *implies* method (see BasicPermission and FilePermission below).

```
public abstract class Permission implements Guard, java.io.Serializable {
  private /*@ spec_public @*/ String name;

  /*@ public normal_behavior
    @ assignable this.name;
    @ ensures this.name == name; @*/
  public Permission(String name);

  /*@ public normal_behavior
    @ ensures \result <==> getClass() == p.getClass() && jmlImpliePerm(p);
    @*/
  public abstract /*@ pure @*/ boolean implies(Permission p);

  //@ public model pure boolean jmlImpliesPerm(non_null Permission p);
  // ...
}
```

Fig. 2. Partial specification of class Permission

The class BasicPermission is an abstract subclass and implements permissions that have a name (or target) string but do not support actions. It implements hierarchical property names, i.e., names with dots. An asterisk may appear by itself, or if immediately preceded by a ".", it may appear at the end of the name, to signify a wildcard match (see the *implies* method below). The BasicPermission class does not introduce any new state component. However, it constrains the inherited state by requiring the name to be non-null, as stated in the invariant clause below.

```
//@ public invariant name != null;

/*@ public normal_behavior
  @ assignable this.name;
  @ ensures this.name == name; @*/
public BasicPermission(/*@ non_null @*/ String name);
```

The BasicPermission class provides a simple wildcarding capability, e.g., "*" implies permission for any target and "x.*" implies permission for any target that begins with "x.". This is implemented by the *implies* method, the behavior of which is specified below indirectly by overriding the inherited model method *jmlImpliesPerm*. The *ensures_redundantly* clause specifies facts that can be inferred from other assertions, such as those inherited from the superclass, and the *old* clause introduces a short name for an expression.

```
/*@ also public normal_behavior
  @ ensures_redundantly \result ==> (p instanceof BasicPermission)
  @     && jmlImpliesPerm(p); @*/
public /*@ pure @*/ boolean implies(/*@ non_null @*/ Permission p);

/*@ also public normal_behavior
  @ old String n1 = name;
  @ old String n2 = p.name;
  @ ensures \result <==> n1.equals(n2) || "*".equals(n1) ||
  @     (n1.endsWith(".*") && n1.length() < n2.length() &&
  @      n1.regionMatches(0, n2, 0, n1.length()-1));
  @ public pure model boolean jmlImpliesPerm
  @     (non_null Permission p) { /*...*/ } @*/
```

An example of a concrete class that implements a simple named permission is the class AudioPermission defined in the package javax.sound.sampled. It represents access rights to the audio system resources, and the permission string (or name) can be play, record, etc. The class only defines a constructor; all the permission methods are inherited from the superclasses.

The class FilePermission is a final, concrete subclass of Permission and consists of a pathname and a set of actions. An action represents access that can be granted to a pathname, and a possible value is "read", "write", "execute", or "delete". The pathname is modeled by the *name* field inherited from Permission, and the actions are modeled by the model field *actions* of type JMLValueSet (see below).

```
//@ public model non_null JMLValueSet actions;
//@ public invariant actions.isSubset(FILE_ACTIONS);
```

The JML model class JMLValueSet, from the package org.jmlspecs.model, defines immutable sets that use equals for a membership test. The *invariant* clause states that actions can contain only valid action names; FILE ACTIONS, not shown here, is a specification-only constant denoting the set of all file actions. In the implementation, the actions are represented as a bit mask (see below). The JML *represents* clause specifies an abstraction function that maps program values such as a bit mask to abstract values such as a JMLValueSet; the model method *toSet* that does this mapping is not shown in this paper. That is, the value of the model field *actions* is given by the program field *mask*. The *in* clause states that *mask* belongs to the data group of *actions*, meaning that a method that can change the value of *actions* also can change the value of *mask* [6]. Note that the *represents* clause is private and, thus, is for the implementer, not for the client of this class.

```
private transient int mask; //@ in actions;
//@ private represents actions <- toSet(mask);
```

The constructor establishes the invariant about actions by requiring the action string, *acts*, to be a sequence of comma-separated file actions (see below). This is indirectly asserted by stating that the value toSet(acts) should be a subset of FILE_ACTIONS. The overloaded model method *toSet* converts a sequence of comma-separated strings to a JMLValueSet. As shown below, a model method also may have the method body. The constructor is allowed to mutate the model field *actions* and, thus, is allowed to change the program field *mask* to initialize it.

```
/*@ public normal_behavior
  @ requires toSet(acts).isSubset( FILE_ACTIONS);
  @ assignable name, actions;
  @ ensures name == path && actions.equals(toSet(acts)); @*/
public FilePermission(/*@ non_null @*/ String path,
                      /*@ non_null @*/ String acts);

/*@ public model pure JMLValueSet toSet(non_null String acts) {
  @  JMLValueSet r = new JMLValueSet();
  @  StringTokenizer tok = new StringTokenizer(acts, ",");
  @  while (tok.hasMoreTokens())
  @    r = r.insert(new JMLString(tok.nextToken().trim().toLowerCase()));
  @  return r;
  @ } @*/
```

The specification of the overriding *implies* method is shown below. As in the class BasicPermission, its precise behavior is specified by overriding the model method *jmlImpliesPerm*.

```
/*@ also public normal_behavior
  @ ensures_redundantly \result <==>
  @   p.getClass() == FilePermission.class && jmlImpliesPerm(p); @*/
public /*@ pure @*/ boolean implies(/*@ non_null @*/ Permission p);

/*@ also public normal_behavior
  @ ensures \result <==> actions.isSuperset(((FilePermision) p).actions)
  @   && jmlImpliesPath(p.name);
  @ public model pure boolean jmlImpliesPerm(non_null Permission p);
  @*/
```

The *jmlImpliesPerm* method defines the *implies* relation of file permissions, based on both actions and file path. Its postcondition asserts that the return value be true if and only if the receiver's action set includes all the actions of the argument permission, p, and jmlImpliePath(p.name) is true. A new model method *jmlImpliesPath*, not shown in this paper, formulates the *implies* relation on file path names, e.g., both "/tmp/*"and "/-" imply "/tmp/t.txt".

IV. CRYPTOGRAPHY

The Java Cryptographic Architecture (JCA) includes APIs for message digests, digital signatures, and key and certificate management. JCA is based on the provider architecture to support multiple and interoperable cryptography implementations. In this section we describe key JCA classes such as Security, MessageDigest, and Signature.

A. Security Class

This final class manages the installed providers and the security properties, i.e., it centralizes all security properties and common security methods. For this, it provides a set of static methods, including methods for adding new providers or properties, retrieving existing providers or properties, and removing existing providers. All of these are typical methods for managing a collection of data. The class also defines a method named *getAlgorithms* that returns a set of strings containing the names of all available algorithms or types for the specified cryptographic service, e.g., message digest and signature. In this section, we specify the *getAlgorithms* method, but first let us define the abstract model of the class Security.

```
/*@ public static model non_null JMLObjectSequence jmlPrs;
  @ public static invariant !jmlPrs.has(null)
  @    && (\forall Object o; jmlPrs.has(o); o instanceof Provider)
  @    && jmlPrs.size() == jmlPrs.toSet().size(); @*/
private static /*@ non_null @*/ Provider[] providers; //@ in jmlPrs;
/*@ private static represents jmlPrs <-
  @    JMLObjectSequence.convertFrom(providers); @*/
```

This specification shows an interesting pattern. For our purpose, it is sufficient and even advantageous to model the class as a sequence of providers, but the implementation uses an array of providers, a private field named *providers*. We may use this array as our abstract model, but we opted for sequences by introducing a model field *jmlPrs*. The resulting specification is more abstract and maintainable. For example, a change of representation has only one affect, i.e., it affects the *represents* clause; the rest of the specification, such as pre and postconditions, remains the same. In addition, sequences are a lot easier to manipulate in pre and postconditions. As stated in the invariant, *jmlPrs* contains only instances of Providers, no null, and no duplicates.

The specification of *getAlgorithms* is given below. It looks a bit complicated due to nested quantifiers, but it precisely documents that the returned value is a set of strings, the elements of which are collected from the keys of the properties (maps) of all providers. The elements are built from the keys that contain no blanks and the prefixes of which match case-insensitively the given service name (n) by discarding the matching prefix and the next character ("."). For example, if the key is "MessageDigest.MD5" and the service name is "MessageDigest", then "MD5" is added to the result.

```
/*@ public normal_behavior
  @ requires n != null && n.length() > 0 && !n.endsWith(".");
  @ ensures (\forall Object o; \result.contains(o); o instanceof String) &&
  @    (\forall String s; \result.contains(s) <==>
  @       (\exists Provider p; jmlPrs.has(p);
  @          (\exists String k; algKeys(p, n).contains(k);
  @             s.equals(k.toUpperCase().substring(n.length()+1))))); @*/
public static /*@ pure non_null @*/ Set getAlgorithms(String n);
```

Given a provider and a service name, the model method *algKeys* returns the set of keys of the provider that contain no blanks and the prefixes of which match case insensitively the given service name.

B. Engine Classes

The engine classes provide different cryptographic services, such as message digests and digital signatures. All engine classes are abstract and define several factory methods named *getInstance* to create objects that implement specific cryptographic algorithms. All factory methods take an algorithm name, and some also take an optional provider name or object. If the provider is not specified, an algorithm available from the default provider is returned. In addition to factory methods, each engine class defines a set of service-specific API methods.

The MessageDigest engine class provides an interface to a secure one-way hash function that takes arbitrary-sized data and returns a fixed-length hash value, called a *message digest*. We specify all the methods (i.e., *getInstance*, *update*, and *digest*) used in the example in Section II.A. For these methods, it is sufficient to model a message digest object as a sequence of bytes to be digested, as shown below; i.e., we ignore other implementation fields.

```
//@ public model non_null JMLValueSequence data; initially data.isEmpty();
```

The JML model class JMLValueSequence, imported from the org.jmlspecs.models package, defines an immutable sequence of values. Initially, the model field data is empty, i.e., there is nothing to digest.

The class MessageDigest defines several factory methods to create new instances. Specified below is the one that takes an algorithm name (e.g., SHA and MD5) and a provider name.

```
/*@ public normal_behavior
  @  requires Security.getProvider(p) != null && (* alg available from p *);
  @  ensures \fresh(\result) && \result.data.isEmpty() &&
  @     \result.getAlgorithm().equals(alg) &&
  @     \result.getProvider().getName().equals(p);
  @ also public exceptional_behavior
  @  requires Security.getProvider(p) != null && (* no such alg from p *);
  @  signals (NoSuchAlgorithmException e1);
  @ also public exceptional_behavior
  @  requires Security.getProvider(p) != null;
  @  signals (NoSuchProviderException e2); @*/
public static /*@ pure @*/ MessageDigest getInstance(
  /*@ non_null @*/ String alg, /*@ non_null @*/ String p)
  throws NoSuchAlgorithmException, NoSuchProviderException;
```

This specification uses informal descriptions in its preconditions. In JML, one can escape from formality by enclosing plain English or other applicable languages in a pair of (* and *). If the named algorithm is available from the named provider, this method should return a new object of type MessageDigest; otherwise, it should throw an exception. The *\fresh* expression in the *ensures* clause asserts that the object is newly created, i.e., it does not exist in the pre-state but does exist in the post-state. The specification also shows that one can mix formal and informal descriptions in JML assertions.

One uses *update* methods to append data to a message digest object and then calls *digest* methods to actually compute a message digest of the accumulated data. There are several forms of *update* and *digest* methods. The ones used in Section II.A are specified below.

```
/*@ public normal_behavior
  @ assignable data;
  @ ensures data.equals(\old(data.concat(toSeq(d)))); @*/
public void update(/*@ non_null @*/ byte[] d);

/*@ public static model pure JMLValueSequence toSeq(non_null byte[] b) {
  @  JMLValueSequence r = new JMLValueSequence();
  @  for (int i = 0; i < b.length; i++)
  @     r = r.insertBack(new JMLByte(b[i]));
  @  return r;
  @ } @*/

/*@ public normal_behavior
  @ assignable data;
  @ ensures data.isEmpty() && (* \result is hash of \old(data) *); @*/
public /*@ non_null @*/ byte[] digest();
```

The *update* method is specified in terms of the model method *toSeq*, which converts an array of bytes to a sequence of bytes. The postcondition states that the argument bytes are appended at the end; the expression *\old(e)* denotes the value of *e* evaluated in the pre-state. The digest method has a side-effect in that it also trashes the accumulated data.

The Signature engine class provides an interface to compute and verify digital signatures. It uses a private key to sign or produce a new digital signature and a public key to verify a signature; both keys are implemented as interfaces, i.e., PrivateKey and PublicKey. The API of this class is similar to that of MessageDigest, however, an interesting aspect of this class from a specification perspective is that the same object may be used for both signing data and verifying signed data. This means that methods should be called in a particular order. For example, to sign data, one must first call one of the *initSign* methods that initialize the object by supplying a private key; then call the *update* methods, possibly several times, to append data; and finally call one of the *sign* methods. A similar sequence of method calls (i.e., *initVerify*, *update*, and *verify*) are required to verify signed data. But how can this protocol property, i.e., ordering dependency among method calls, be specified? Since JML does not allow a protocol property to be expressed explicitly, we encode it as a finite state machine in pre and postconditions. For example, the following gives partial specifications of the *initSign* and *sign* methods.

```
protected /*@ spec_public @*/ int state;
//@ initially state == UNINITIALIZED;

//@ assignable state;
//@ ensures state == SIGN;
public final void initSign(PrivateKey k) throws InvalidKeyException;

//@ requires state == SIGN;
public final byte[] sign() throws SignatureException;
```

The field *state* represents the current protocol state and the allowed transitions are specified by manipulating it in pre and postconditions of the methods involved. For example, the *initSign* method sets the state's value to the constant SIGN, and the *sign* method requires that it be called in a state where the *state*'s value is SIGN.

V. DISCUSSION

An interface specification of a module may be written at different abstraction levels for different users, e.g., a highly abstract specification for the module's client and a more implementation-oriented specification for the module's implementer. JML facilitates writing specifications at different abstraction levels and, through specification visibility and model elements, mixing them in a single file. In our case study, we attempted to produce client-oriented specifications by defining abstract models for classes. Occasionally, however, we made connections to the representations through private abstraction functions (e.g., the Security class in Section IV.A).

Writing an interface specification is an iterative and incremental process, often starting from the definition of an abstract model to the specification of each method. We found that a concise mathematical notation such as Z, which makes it easy to see commonalities and variations of different classes, especially when a complex class hierarchy is involved, is an excellent tool to sketch out abstract models. We also used UML class diagrams to identify abstract models: First, we reverse engineered a class diagram from source code to highlight various relationships among classes, and then extended or decorated it with model elements such as model fields and classes. Since JML facilitates incremental development of specifications, often we started with an informal/lightweight specification, which evolved into a formal/heavyweight one. Although JML also allows multiple specifications of a module to be in separate files, we didn't explore this refinement feature in our incremental development.

Another feature of JML that we used heavily in our case study is informal descriptions. This feature allowed us to tune the level of formality and mix formal and informal text in our specifications. We prefer a simple English description when a formalization does not add much or is practically impossible to completely formalize (e.g., the *digest* method in Section IV.B).

Additionally, we identified several features missing in JML. For example, we could not separate and specify cleanly the ordering of dependencies among methods [4]. We had to mix the so-called protocol properties with functional properties in pre and postconditions by encoding them as finite state machines (e.g., the Signature class in Section IV.B). Writing such specifications is laborious and error-prone, and, since the protocol properties have to be inferred, the specifications become hard to understand. Another missing feature may be a more succinct and expressive notation to manipulate mathematical structures such as sets and sequences, e.g., something like Haskell's list comprehension notation. Although JML provides so-called model classes for such mathematical structures, we found that the use of these classes often becomes cumbersome and verbose, e.g., we had to convert values back and forth between Java types and JML types used by model classes.

Finally, we mention several JML specification techniques or patterns that we used recurrently in this case study. The most frequently used pattern is "abstract with model fields", which produces an abstract, client-oriented specification that is easy to maintain and allows runtime assertion checking. The approach is to introduce model fields to define the abstract model of a class and to write pre and postconditions in terms of the model fields (see Section IV.A). If the representations of the model fields are known, private abstraction functions may be specified to make pre and postconditions checkable at runtime [3]. Another pattern illustrated in this paper is "delegate to model methods", which is used to specify method overrides (see Section III.A). The idea is for a superclass to delegate to a subclass the responsibility of specifying an overridden method. The approach is to introduce a model method to specify the behavior of an overridden method and to let a subclass to override the model method to provide a subclass-specific behavior specification. A related pattern, "know your class" also allows the specification of class-specific behavior by using the *getClass* method in assertions.

VI. CONCLUSION

We documented formally in JML a significant portion of the java.security package and its extensions. The starting point of our specifications was informal API descriptions obtained from the Javadoc comments of the source code. We quickly found that in many places the descriptions were incomplete or ambiguous. Thus, we resorted to the source code and experimented with it to resolve ambiguity and to determine the precise behavior of the API. Given our specification, we expect other Java programmers to be able to avoid this "code experimentation". We also expect that our specifications will facilitate the use of the Java security API and increase the reliability of the code on which it is based.

We are in the process of verifying both our specifications and the source code using various JML tools. All our specifications will be soon available from our project website at opuntia.cs.utep.edu/milaap.

REFERENCES

[1] L. Burdy, Y. Cheon, D. R. Cok, M. D. Ernst, J. R. Kiniry, G. T. Leavens, K. R. M. Leino, and E. Poll. An overview of JML tools and applications. *International Journal on Software Tools for Technology Transfer*, 7(3):212–232, June 2005.

[2] N. Catano and M. Huisman. Formal specification of Gemplus's electronic purse case study. In L. H. Eriksson and P. A. Lindsay, editors, *FME 2002*, volume LNCS 2391, pages 272–289. Springer-Verlag, 2002.

[3] Y. Cheon, G. T. Leavens, M. Sitaraman, and S. Edwards. Model variables: Cleanly supporting abstraction in design by contract. *Software—Practice and Experience*, 35(6):583–599, May 2005.

[4] Y. Cheon and A. Perumendla. Specifying and checking method call sequences of Java programs. *Software Quality Journal*, 2006. To appear.

[5] G. T. Leavens, A. L. Baker, and C. Ruby. JML: A notation for detailed design. In H. Kilov, B. Rumpe, and I. Simmonds, editors, *Behavioral Specifications of Businesses and Systems*, pages 175–188. Kluwer Academic Publishers, Boston, 1999.

[6] K. R. M. Leino, A. Poetzsch-Heffter, and Y. Zhou. Using data groups to specify and check side effects. In *ACM PLDI 2002*, volume 37(5) of ACM SIGPLAN Notices, pages 246–257, June 2002.

[7] S. Oaks. *Java Security*. O'Reilly, second edition, 2001.

[8] E. Poll, J. van den Berg, and B. Jacobs. Formal specification of the Java Card API in JML: the APDU class. *Computer Networks*, 36(4):407–421, 2001.

[9] Sun Microsystems, Inc. *Java 2 platform API specification*. Available online from http://java.sun.com (Date retrieved: April 2, 2006).

Enterprise Integration Strategy of Interoperability

Raymond, Cheng-Yi Wu
IBM Software Group
P.O.Box 3077, 7 CBP1, Singapore 486048
Faculty of Information Technology
University of Technology, Sydney
P.O. Box 123, Broadway, NSW 2007, Australia

Jie Lu
Faculty of Information Technology
University of Technology, Sydney
P.O. Box 123
Broadway, NSW 2007, Australia

Abstract

In this new computing age of high complexity, a common weakness in the interoperability between business and IT leaves IT far behind the direction business is taking; poor business responsiveness and IT governance makes it even harder to achieve the enterprise goal. To cope with this common issue, we introduce the enterprise interoperability to integrate the metadata between business, service and information layers, this create visibility of vertical alignment within enterprise architecture and use metadata configuration to construct the mapping between each layer.

Introduction

The main purpose of this paper is to conduct a research in the enterprise architecture issue as a means of addressing the vertical gap with new knowledge contributions in the vertical integration of the interconnectivity of cross-layers include direction, business, process, service and information.

We identify this overall tactical issue that emits from the vertical gap as the "invisible enterprise issue" and these are the findings in our early analysis stage. Realization require multi-dimensional aspects from business innovation, service strategy, solution architecture, implementation and integration [1].

Two key processes were used to achieve this objective. The first process relates to how the enterprise goal may be achieved by impacting business process and its service components level, which we call "business semantic", the reason componentization needed to be set first because visibility and traceability [2]. The second process follows one level below the first one and is called "information metadata." It constitutes the repository of metadata storing information and its process of moving information between service and data; it thus supports the upper layers by repository synchronization. The micro process integration was initiated by the domain context analysis process. H. Lee & J. Lee pointed out, once sets of use case scenarios are formulated, the next step will be the extraction of verbs and nouns for identifying domain objects [3]. The three different types of repositories (metadata, services and semantics) collaborate each other, the fact that all of the organizations share the same service means they are all managed under the same rules which were developed using the top-down approach [4], [5].

Discussion

First Area - Technical Environment Setup
First, the concept of "centric master data and single view of information" to consolidate all fragmentation into master data which called "single source of truth". Before the built, the metadata was categorized into two groups - system and transaction, the transaction metadata was related to those business transaction while system metadata was related to those system configuration and DDL (data definition language) which supported parameterization in metadata computing. Metadata computing was followed by the built of metadata repository, the reason of the introduction of recursive metadata iteration and cursor computing was an intention to build the metadata engine in communicating with other repositories in Java/XML/SQL formats [6]. One of the examples was by allocating summarized "course" information in metadata repository as driving force in extracting associated students' study plan gained the query response time from several seconds down to one nanosecond.

Second Area - Business Process Strategy
There are two steps to achieve "componentization", as we were all aware that traditional silo implementation aimed end-to-end business process, it supported project architecture but also caused many problems of fragmentation in enterprise architecture. The concept of componentization takes the identified business component(s) into decomposition process and produce common elements, the process use litmus test for its boundary decision, when we take the componentization concept into realization, the key of this phase was using DCRP reuse methodology [7] in achieving this goal and was the important creative work in this phase. The DCRP reuse methodology was intended to address the vertical integration concerns which were currently labor-intensive development and maintenance issue inherent in multiple instances of the enterprise

K. Elleithy (ed.), Advances and Innovations in Systems, Computing Sciences and Software Engineering, 369–374.

architecture. Interoperability capability in a vertical direction allows mediation computing entities such as messaging carrier and class mapping (i.e. XML and JDBC) choreography between service and information to become easier and leading toward a seamless vertical integration [8]. The integration need an interlingua as mediation solution, the constraints and dependency of interoperability between two applications and tendency that the integration has been moving downward as the high complexity combination of application, process and database integration enforce the mapping between theories, and these require separate mediation in integration.

Third Area - Enterprise Integration

The realization of enterprise Integration initiatives started from the outcomes of previous two areas in deriving a "loosely coupled" mediation (which leveraged to service oriented architecture's principles) in creating configurable linkages between business and IT. In this final stage, the resolutions and approaches were leveraged to those common methodologies and architecture practices (such as Zachman, RUP, SOA and MDA, to proceed the verification and validation of the approaches we had. This is evident from the nature of business direction, virtualization and interoperability perspectives [9]. *Desmond DSouza define vertical integration in his "*Model-Driven Architecture and Integration Opportunities and Challenges" said, vertical Integration came from different levels of abstraction of the same system, from physical data models, logical data models, networks, applications specifications, component assemblies, business process models, and business goals and strategies [10].

Enterprise Interaction Strategy

The semantic interpretation from business to service, and from service to technical need decomposition process to work on those selected business components, so each business component can be decomposed into business use cases for further breakdown. Service components derived from business use case are the mediation between business and technical element which integrate all business semantic, interfaces and technical configuration into a logical component, initially all the service components are in atomic level [11] [12]. S. Melnik (1999) proposed a universal interface which avoids common models and languages, the Generic Interoperability Framework developed to facilitate integration of heterogeneous information systems. [13]. Based on this common framework, we can build the component of both functional and non-functional by using industry standard. DSouza specified the standard and common component as followed - To manage models of large scale, a coherent modeling architecture is required, with standard ways to describe shared concepts, rules, patterns, frameworks, mappings, and generators. [14]. M. Schmalz argued that database schema and application code can be efficiently derived from various types of schema representations, particularly the relational model, and supports comparison of a wide variety of schema and code constructs [15]. According to L. Yu et al., An obstacle to software reuse is the large number of major modifications that frequently have to be made as a consequence of dependencies within the reused software components. Component base implementation is the foundation in coping with this bottleneck [16]. Therefore the determination of Common Services and Composite Services is a use cases driven parsing process, the purpose of the parsing iteration process is trying to find out the common areas and dependency between services. The determination of common component and service aggregation, as mentioned, is through the parsing process using use case driven model.

C. Foste et al. indicated Metadata has a key role to play in allowing systems to be built and automated. A management system composed of multiple distributed components implements a service such as content validation, transcoding, search or generation; and relies on metadata information to achieve success [17].

In building up the enterprise interoperability between layers, componentization will be the foundation for integrating business and service components to derive the architecture of common component and service aggregation, store the intelligence of mapping and routing information in repository in metadata form, this strengthen the foundation of enterprise vertical integration. However, the tactical issue in the ontology skeleton is that currently there are no principles to direct each layer when working on the above-mentioned three issues using vertical integration. In other words, when we are looking for the principle of "separation of concerns" in vertical integration, to maximize the output from equation –

$$\tilde{\mathbf{O}}_{Output} = \bullet\ \tilde{O}_n = \bullet\ (\mathbf{d}_n \times \mathbf{b}_n) \times (\mathbf{p}_n \times \mathbf{s}_n) \times \mathbf{i}_n$$

Each $\tilde{\mathbf{O}}_n$ needs to be a concatenation from an assembly of "atomic" level components;
each component can be the assembly of a semantic sub-context ($\mathbf{d}_n$), a business sub-component ($\mathbf{b}_n$), a decomposed sub-process ($\mathbf{p}_n$), a fine-grained service component ($\mathbf{s}_n$), or a piece of metadata ($\mathbf{i}_n$). The routing of $\tilde{\mathbf{O}}_n$ forms a map for approaching a business context. Before approaching the resolution, we need to have an insight into the ontology principles in order to determine the framework. The reason the SOA approach is required in this foundation and should be extended to the business layer is because we need to create a broad "loosely-coupled" and "separation of concern" ontology environment covering all business, process and service layers.

Ontology skeleton in Integration strategy
Referring to Figure 1 – the concept of the ontology approach is based on the driving force of the business layer which takes the business context and translates it into a semantic message. The semantic message stored in the repository further directs the decomposition and input of the element message to the service component which we call "business semantic injection." Referring to Figure 2 - the business component in each cell denotes a unique and independent business entity with an individual function. Figure 3 – what we are looking for is a driving force in creating a common foundation of "semantic injection" downward to a service component so that each service component holds an associated business semantic element from micro process. For an overview of how the business context and semantic message can be transmitted to the component, please refer to Figures 4, 5 and 6. Figure 4 shows that the business component and its element decomposition are accomplished by means of a parallel process; one is "logical" and is a class object element while the other is a "physical" component which carries the message.

Metadata Strategy
The database partition shown in Figure 7 is an efficient way to achieve a "separation of concern." By storing the partition information in a metadata repository with linkage to service metadata, the system integrity and performance will be improved dramatically. Each pending record can be centralized in 1% of the database for further service until completed. By this means each query hit-ratio is nearly 100% and greatly improves SLA (service level agreement) efficiency. Thus interoperability can be attained within the meta-layer to impact the individual entity effectively [18]. In each strategic implementation, the interoperability is built by using semantic metadata between each layers. Figure 8 shows how to top down and bottom up approaches, how the information and service form common components and aggregated components in creating the vertical coherence of enterprise integration.

Conclusion

The concept of "enterprise vertical integration" covered many aspects of technology and knowledge which we already discussed, the issue was from misalignment of business and IT, so the business context and semantics cannot effectively impact IT development and cause all development, change and maintenance very difficult. The lack of interconnectivity between context, semantics, service and data is the tactical issue with the clear implication that enterprise architecture needs a robust micro-mechanism of semantic messaging and metadata to coordinate across layers. The alignment can reuse the mapping between business service and information component from those successful experience, however in most of the project scenario, only part of the empirical transformation process from business to service and to physical element can be reused, and leave a large portion of the alignment require starting from business decomposition. In our current research, an insight in how to start from business component to derive the service component will be described in a separate research paper.

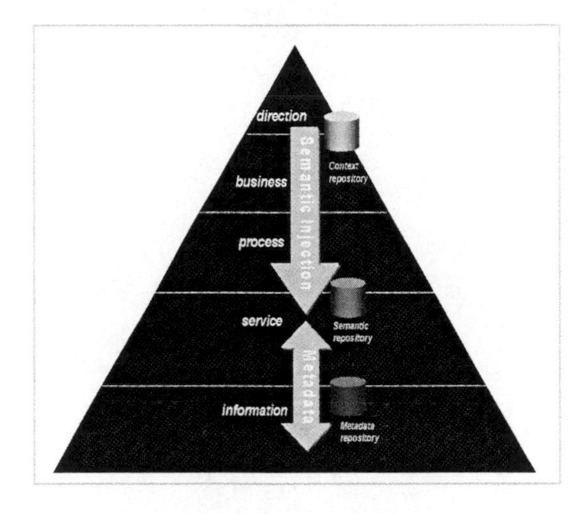

Figure 1 – Conceptual ontology interoperability between layers by using "semantic injection" and "metadata"

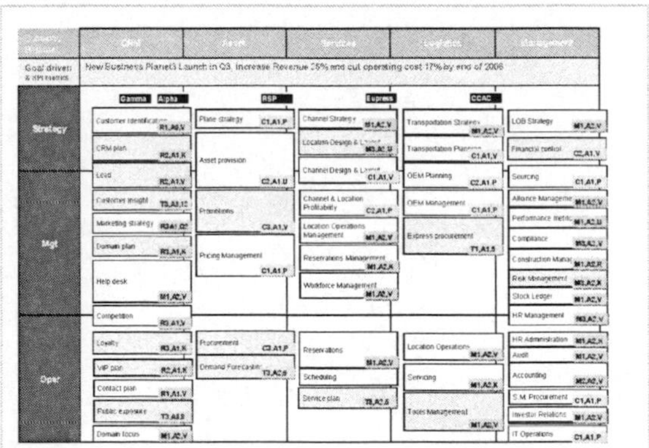

Figure 2 - Mapping between context and semantic notation

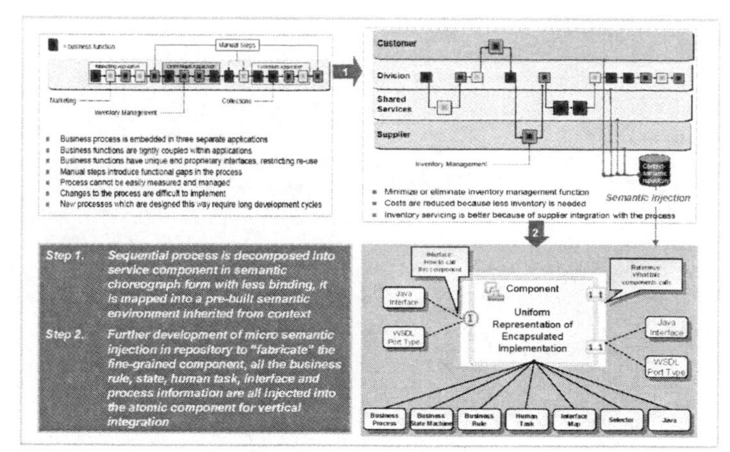

Figure 3 – Through business component modeling, each business element is identified as a strategic direction element to provision a specific business goal

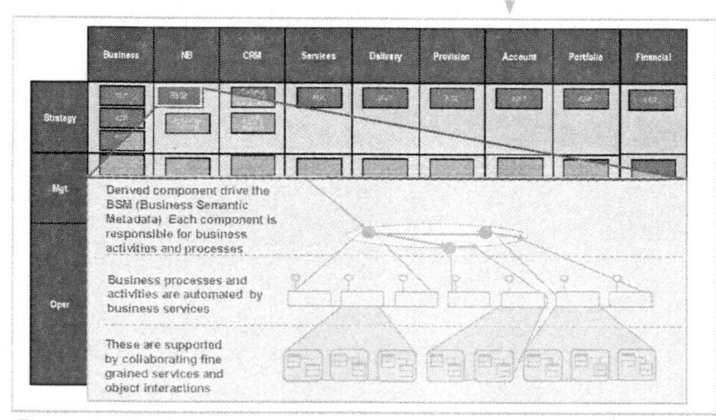

Figure 4 – Business semantic injection into object model

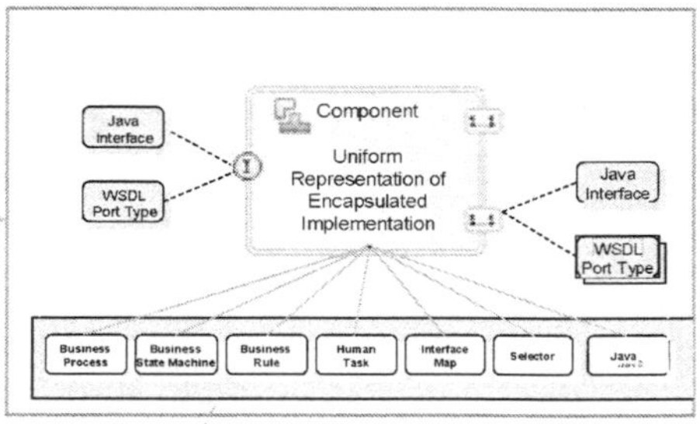

Figure 5 – Service component used as a dynamic micro repository holding business semantic

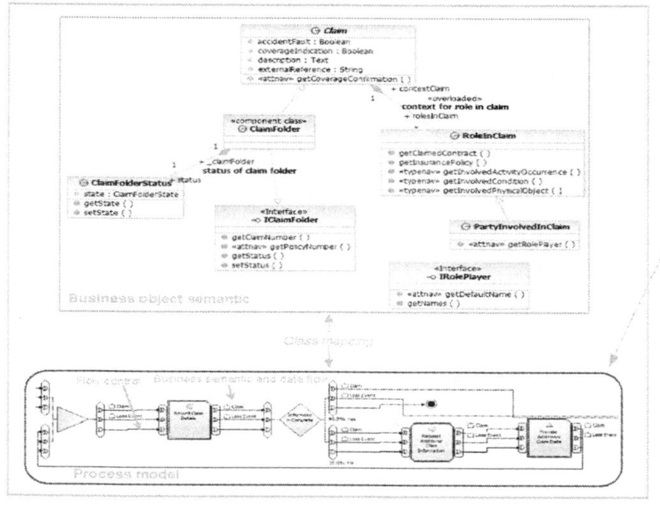

Figure 6 – Service component inherited semantic micro properties and assembly into process by using business rules

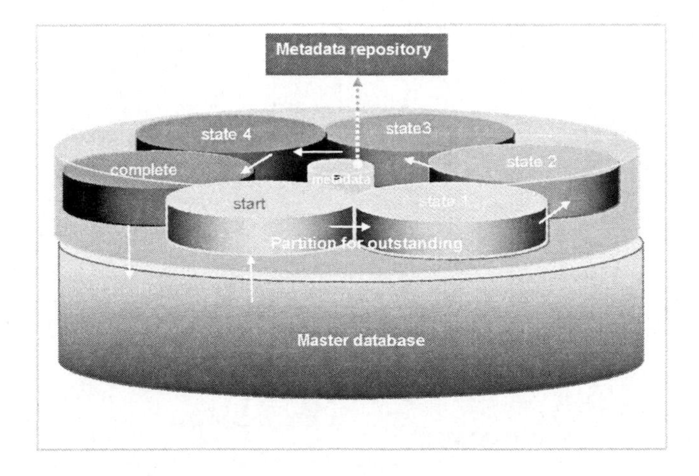

Figure 7 - Database partition in metadata environment

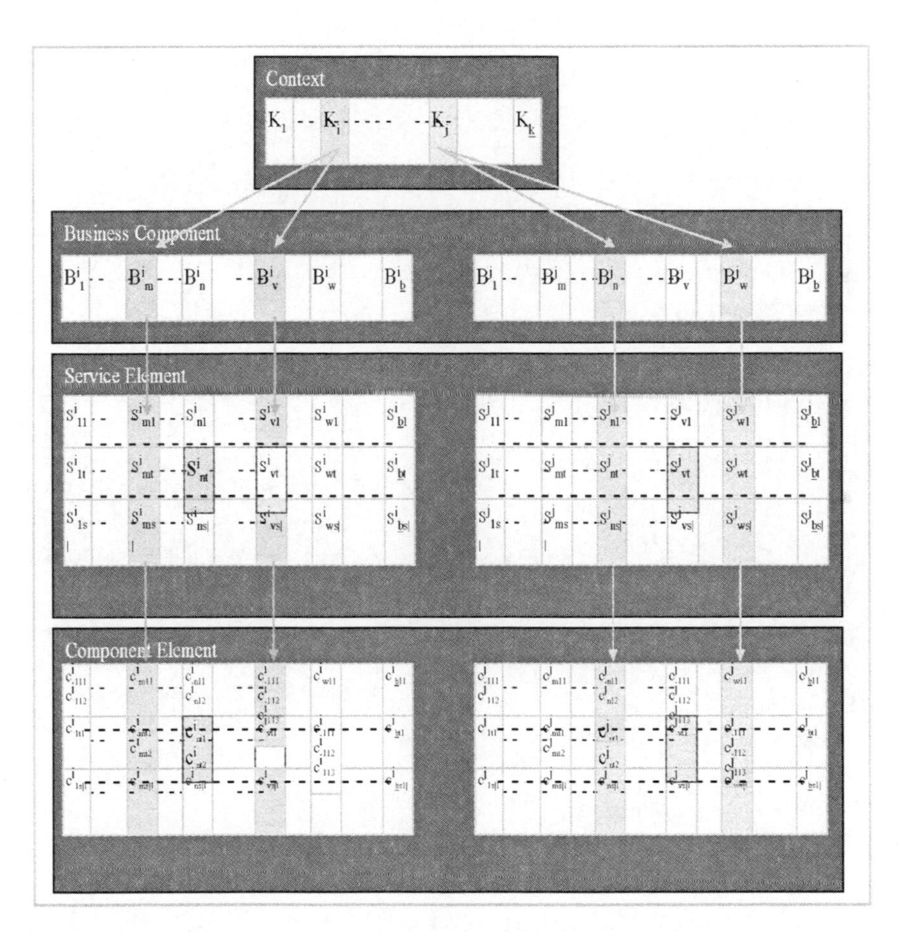

Figure 8 - Metadata in component element provisioning the component services for upper layers of services and business

Reference

1. W. Yao, (2005), "Interoperability: an indispensable propellant of e-China", International Journal of Services and Standards 2005 - Vol. 1, No.4 pp. 446 – 452
2. M. Lytras, (2006), "The Semantic Electronic Government: knowledge management for citizen relationship and new assessment scenarios", Electronic Government, an International Journal 2006 - Vol. 3, No.1 pp. 5 – 17
3. Heeseok Lee and Jae Lee (2006), "Analyzing Business Domain: A Methodology and Repository System", Korea Advanced Institute of Science and Technology
4. C. Wagner, K. Cheung, R. Ip, S. Bottcher, (2006), "Building Semantic Webs for e-government with Wiki technology", Electronic Government, an International Journal 2006 - Vol. 3, No.1 pp. 36 - 55
5. G. Moynihan, R. Batson, (2006), "Development of a matrix methodology for database reengineering and improvement: an e-government case study", Electronic Government, an International Journal 2006 - Vol. 3, No.2 pp. 190 - 203
6. Tannenbaum. Metadata Solutions: (2001), "Using metamodels, repositories, xml, and enterprise portals to generate information on demand", Addison Wesley Professional, Boston, MA
7. Dublin Core. (1997), "Dublin Core Metadata Element Set", Version 1.1: Reference Description, http://purl.org/dc/documents/rec-dces-19990702.htm
8. C. Liu (2005) "Using e-governmental indicators to build virtual value chain", Electronic Government, an International Journal 2005 - Vol. 2, No.3 pp. 277 – 291
9. Batini, R. Grosso, G. Longobardi, (2006), "Design of repositories of conceptual schemas for large-scale e-government projects", Electronic Government, an International Journal 2006 Volume 3 - Issue 3 pp 306 - 328
10. ntti Vehviläinen, Olli Alm and Eero Hyvönen: (2006), "Combining Case-Based Reasoning and Semantic Indexing in a Question-Answer Service", June 20, 2006. Poster paper, 1st Asian Semantic Web Conference
11. M. Staab, N. Stojanovic, R. Studer, and Y. Sure, (2001), "Semantic Portal - The SEAL approach. to appear in: Creating the Semantic Web.", MIT Press
12. O. Valkeapää and E. Hyvönen, (2006), "Semantic Annotation with Browser-based Annotation Tool", 1st Asian Semantic Web Conference
13. Sergey Melnik, (1999), "Generic Interoperability Framework", Distributed and Parallel Databases Journal 1999
14. DSouza,(2001)," Model-Driven Architecture and Integration Opportunities and Challenges", Kinetium
15. M. Schmalz, 2003, "EITH – A Unifying Representation for Database Schema and Application Code in Enterprise Knowledge Extraction" , 22nd International Conference on Conceptual Modeling
16. L. Yu, (2005), "Categorization of Common Coupling in Kernel-Based Software," 43rd ACM Southeast Conference, 2005
17. C. Foster, P. Stentiford, (1999), "Metadata - The Key to Content Management Services, Proceedings of the Meta-Data", The third IEEE Meta-Data conference

A Method for Consistent Modeling of Zachman Framework Cells

S. Shervin Ostadzadeh
Computer Engineering Department,
Faculty of Engineering,
Science & Research Branch of
Islamic Azad University, Tehran, Iran

Fereidoon Shams Aliee
Computer Engineering Department
Faculty of Electrical & Computer Eng.,
Shahid Beheshti University,
Tehran, Iran

S. Arash Ostadzadeh
Computer Engineering Department,
Faculty of Engineering,
Islamic Azad University of Mashhad,
Mashhad, Iran

Abstract - **Enterprise Architecture has been in center of attention in late 90s as a comprehensive and leading solution regarding the development and maintenance of information systems. An enterprise is considered a set of elaborate physical and logical processes in which information flow plays a crucial role. The term Enterprise Architecture encompasses a collection of different views within the enterprise which constitute a comprehensive overview when put together. Such an overview can not be organized regardless of incorporating a logical structure called Enterprise Architecture Framework. Among various proposed frameworks, the Zachman Framework (ZF) is one of the most prominent ways of conceptualization. The main problem faced in using ZF is the lack of coherent and consistent models for its cells. Several distinctive solutions have been proposed in order to eliminate the problem, however achieving no success in thoroughly covering all the cells in ZF. In this paper, we proposed an integrated language based on Model Driven Architecture (MDA) in order to obtain compatible models for all cells in ZF. The proposed method was examined in practice, revealing its advantages and the efficiency gained in comparison to previously studied techniques.**

Keywords: Zachman Framework, Enterprise Architecture Framework, Model Driven Architecture, Software Architecture.

I. INTRODUCTION

It goes without saying that nowadays utilizing information and communication technologies in enterprises is one of the most challenging tasks. Considering the fact that an opportunity can turn into a threat if misused, proper practice of the innovative technology in developing information systems can result in noticeable improvement of procedures within an enterprise. Changes are inevitable within enterprises as time proceeds, which accordingly bring up changes in relevant information systems. Managements are hesitant to adopt the regular changes due to the huge development and maintenance costs of information systems; as a result, outdated systems form a barrier toward the enterprise evolvement. Enterprise Architecture is the novel promising concept to address this problem, intended to unify an enterprise and the underlying information technology by employing a structured framework and methodology.

The common way to comprehend procedures in an enterprise is to provide views of components within that enterprise, which is called architecture. Architecture, such as Data Architecture represents only a single view of an enterprise, but Enterprise Architecture refers to a collection of architectures which are assembled to form a comprehensive view of an enterprise. Organizing such great amounts of information requires a framework. Various enterprise architecture frameworks have been proposed; among them are Zachman Framework, FEAF, TEAF, and C4ISR.

ZF is widely accepted as the main framework in EA [1,7]. Compared to other proposed frameworks, it has evident advantages to list: (1) using well-defined perspectives, (2) using comprehensive abstracts, (3) normality, and (4) extensive usage in practice [1]. They were the motivations for ZF adoption in our work, nevertheless; there are challenges to overcome, among them is the absence of an integrated language to model cells in the framework.

ZF does not recommend any specific tool or model for a particular cell, i.e. there is no identifiable technique to address the cells in the Zachman matrix which itself resembles an obstacle. In order to elegantly define and implement the concepts of EA, ZF needs a consistent modeling approach to describe its cells. Validity, sufficiency, necessity, integrity, authenticity, fitness and suitability of models are achieved through such a modeling method [1]. We aim to resolve the problem by proposing an approach based on MDA in order to model all cells in ZF.

The challenge we referred to is also addressed in other researches. A complete overview is given in [7]. Applying UML to ZF seems to be the best solution proposed up to now. Unfortunately, UML is not mature enough to support all aspects of an EA [2,8]; as a result, a lot of cells in ZF remain unmapped. Some other solutions practice the use of nonstandard symbols which leave the initial problem intact.

The rest of this paper is organized as follows. In Section 2, we introduce some basic concepts and principles. Next, we plot an MDA overview in section 3. We discuss our proposed approach in section 4, and present a case study in section 5. Section 6 contains a comparison between the proposed method and other methods. Finally, we make conclusions and suggest some comments for future works.

II. BASIC CONCEPTS

In this section we briefly introduce some basic concepts and principles. We believe these remarks can help readers to clearly understand what we mean by the concepts that are used in this article.

K. Elleithy (ed.), Advances and Innovations in Systems, Computing Sciences and Software Engineering, 375–380.

A. Architecture

Architecture has emerged as a crucial part of design process. Generically, architecture is the description of a set of components and the relationships between them. In computer science, there are software architectures, hardware architectures, network architectures, information architectures, and enterprise (IT) architectures.

B. Enterprise

An enterprise consists of people, information, and technologies; performs business functions; has a defined organizational structure that is commonly distributed in multiple locations; responds to internal and external events; has a purpose for its activities; provides specific services and products to its customers [2]. An IT-related enterprise is an enterprise in which IT plays an important role in its activities. In this paper, we refer to an IT-related enterprise as an enterprise.

C. Enterprise Architecture

Enterprise Architecture is a comprehensive view of an enterprise. EA shows the primary components of an enterprise and depicts how these components interact with or relate to each other. EA typically encompasses an overview of the entire information system in an enterprise; including the software, hardware, and information architectures. In this sense, the EA is a meta-architecture. As regards, EA contains different views of an enterprise; including work, function, process, and information. It is at the highest level in the architecture pyramid. Refer to [2] for more details.

D. Enterprise Architecture Framework

A framework is a classification schema that defines a set of categories into which various things can be arranged. An enterprise architecture framework is a way of organizing and classifying the types of information that must be created and used for enterprise architecture. John A. Zachman, the creator of Zachman Framework, states that "The Framework for Enterprise Architecture is a two dimensional classification scheme for descriptive representations of an Enterprise [9]". As some examples, we can refer to Zachman Framework, FEAF, TEAF, and C4ISR.

E. Zachman Framework

In 1987, an IBM researcher, named John A. Zachman, proposed a framework for Information System Architecture, which is now called Zachman Framework [10]. ZF is a two dimensional information matrix consisting of 6 rows and 6 columns.

The rows describe the perspectives of various stakeholders. These rows starting from the top include: Planner (Scope), Owner (Enterprise Model), Designer (System Model), Builder (Technology Model), Contractor (Detail Representation), and Functioning Enterprise.

The columns describe various abstractions that define each perspective. These abstractions are based on six questions that one usually asks when s/he wants to understand a thing. The columns include: Data (What is it made of?), Function (How does it work?), Network (Where are the elements?), People (Who does what work?), Time (When do things happen?), and Motivation (Why do things happen?). To find cell definitions of ZF refer to [11].

III. Model Driven Architecture

The Object Management Group (OMG) was formed to help reduce complexity, lower costs, and hasten the introduction of new software applications. The OMG is accomplishing this goal through the introduction of the Model Driven Architecture architectural framework with supporting detailed specifications. These specifications will lead the industry towards interoperable, reusable, and portable software components.

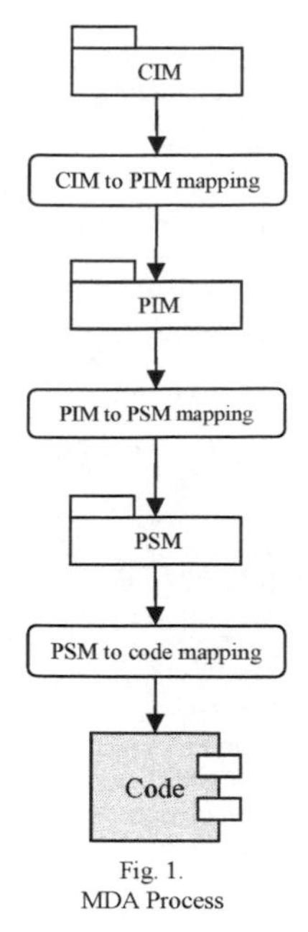

Fig. 1.
MDA Process

A. Models in the MDA

The MDA separates certain key models of a system, and brings a consistent structure to these models [14]:

- ***Computation Independent Model (CIM):*** A computation independent model is a view of a system from the computation independent viewpoint. A CIM does not show details of the structure of systems. A CIM is sometimes called a domain model and a vocabulary that is familiar to practitioners of the domain in question is used in its specification.
- ***Platform Independent Model (PIM):*** A platform independent model is a view of a system from the platform independent viewpoint. A PIM exhibits a specified degree of platform independence so as to be suitable for use with a number of different platforms of similar type.
- ***Platform Specific Model (PSM):*** A platform specific model is a view of a system from the platform specific viewpoint. A PSM combines the specifications in the PIM with the details that specify how that system uses a particular type of platform.

Fig. 1 demonstrates how these models will be created in an MDA development process.

B. Model Standardization in the MDA

OMG has adopted a number of technologies, which together enable the Model Driven Architecture. These include MOF, UML, CWM, XMI, and profiles (such as the profiles for EDOC, EJB …). Refer to [19] for more information.

- ***MOF (Meta Object Facility):*** MOF is an extensible model driven integration framework for defining, manipulating, and integrating metadata and data in a platform independent manner. MOF-based standards are in use for integrating tools, applications, and data. All MDA models are based on MOF. We discuss about this issue later in more details.

- ***UML (Unified Modeling Language):*** A specification defining a graphical language for visualizing, specifying, constructing, and documenting the artifacts of distributed object systems. The current official version for MDA is UML 2.0. Note that UML 1.x does not support all aspects of MDA. Refer to [16] for more information concerning UML 2.0.
- ***OCL (Object Constraint Language):*** The OCL is an addition to the UML specification which provides a way of expressing constraints and logics on models.
- ***CWM (Common Warehouse Meta-model):*** The CWM standardizes a complete and comprehensive meta-model that enables data mining across database boundaries at an enterprise and goes well beyond. It forms MDA mapping to database schemas. The CWM does for data modeling what UML does for application modeling.
- ***BPDM (Business Process Definition Meta-model):*** This Meta-model is developed for business processes. Such a meta-model is independent of specific process definition languages and would allow MOF-compliant models to interface with languages like WSBPEL and notations like BPMN.
- ***BSBR (Business Semantics for Business Rules):*** This meta-model is developed for capturing business rules in business terms, and the definitions and semantics of those terms in business vocabularies. In fact, there will be two specifications: a more generic standard for business rules, and a more specific one for production rules that are actually used by rule engines. To find more information about this meta-model refer to [15].
- ***WSM (Web Services Meta-model):*** This meta-model is developed in order to facilitate the development of MOF-compliant web service models.
- ***UML Profiles:*** UML Profiles tailor the language to particular areas of computing (such as Enterprise Distributed Object Computing) or particular platforms (such as EJB, .Net, or CORBA).

IV. CONSISTENT MODELS FOR ZF

As mentioned earlier, one of the difficulties we are facing to use ZF is the problem of incoherent models. ZF expresses what information must be created for each cell of the framework; however, it doesn't indicate how this information must be created and which model(s) must be applied to present that information. In addition, there is not a set of coherent and consistent models across the entire cells. We are not intended to address these as ZF weak points, since ZF is just a framework, not a methodology. Anyhow, an architect who uses ZF has to overcome these problems.

In this section, we investigate the use of MDA in Zachman Framework in order to suggest a practical model for each cell. This can improve the usability of ZF for the developers who intend to put it in practice.

A. The Problem Space

We mentioned that ZF contains six rows, each representing one stakeholder perspective. The question is: "shall we propose a method for the entire rows?" We answer the question in this section.

The first row of ZF specifies the architecture boundary. J.A. Zachman indicates that this row is a list of important things to the enterprise [11], and not a model. It seems that using natural language is the best way to describe this row. So, the first row does not exist in our problem space. Second till fifth rows model businesses, systems, technologies, and detailed-representations, respectively, and they exist in our problem space. The sixth row is not a model at all. It represents the actual deployed or running elements, data, and people of the organization. It is not a perspective, as such, but the "real world" in all its complexity. This row is usually omitted from ZF, and does not exist in our problem scope.

B. Acceptability

An accepted modeling approach shall be:

- Integrable: We shall integrate all models together.
- Wide covering: We shall apply an approach to model as more cells as possible.
- Be standard: We shall not use strange and uncommon modeling elements, since they are not standard.
- Easily understandable
- Up-to-date

C. The Proposed Method

Fig. 2 depicts the initial scheme with respect to the mapping between the rows in ZF and different model types in MDA. In the primary plan, CIM is employed for the second row. As a result, all the models related to the cells of the second row (owner), to be proposed later; have to be described in the CIM level. In the third row, we use PIM, therefore all the models for the designer row cells have to be in PIM level. PSM is recommended for the fourth and fifth rows and all the cells constituting the builder and contractor rows will be designed in the PSM level.

CWM is the best option for Data column modeling; however it should be noted that using simpler models is recommended for the higher rows in the framework. We use

<table>
<tr><th>*Zachman Framework*</th><th>Data</th><th>Function</th><th>Network</th><th>People</th><th>Time</th><th>Motivation</th></tr>
<tr><td>Planner</td><td colspan="6"></td></tr>
<tr><td>Owner</td><td colspan="6">CIM</td></tr>
<tr><td>Designer</td><td colspan="6">PIM</td></tr>
<tr><td>Builder</td><td colspan="6" rowspan="2">PSM</td></tr>
<tr><td>Contractor</td></tr>
<tr><td>Functioning Enterprise</td><td colspan="6">Code</td></tr>
</table>

Fig 2. Applying MDA models to Zachman Framework rows

Business profile for the owner/data cell. Relations between data are also demonstrated using Specialization/ Generalization, Aggregation, and Inclusion relationships. In order to model the remaining rows we adopt CWM. Since the stakeholders of these rows are familiar with formal methods of software modeling, it is expected that no problem will occur using CWM. However, it should be noted that the CWM models for the third row have to be in the PIM level and for the fourth and fifth rows they should be in the PSM level.

We prefer employing BPDM to model the Function column. As with the Data column, we try to use simpler models for the higher rows in the framework. We utilize Business Use Case diagram for the owner/function cell. Each process in the enterprise is modeled with a Use Case and organized with Aggregation and Inclusion relationships. Activity diagram is also used to present the workflow of processes. We utilize BPDM for the designer/function cell. Since builder/function cell is considered as design of applications, we employ UML 2.x diagrams to model this cell. Finally, for the contractor/function cell we refer to PSM-specific profiles. Considering chosen technologies, it's possible to use WSM and CORBA, EJB, and .Net profiles.

The owner/network cell is modeled with the Organization Unit stereotype contained in UML Packages. Relations between organization sections are depicted using the Dependency relationships and their essences are clarified with extra comments. To model the designer/network cell, we employ the EDOC profile. It enables us to model different parts of organizational computations which are distributed within an enterprise. Deployment diagram is used for the builder/network cell. This gives us the opportunity to model hardware resources and their relations. We also utilize WSM and PSM-specific profiles (CORBA, EJB, and .Net) for the contractor/network cell.

In order to model the People column in the second row, we employ the Use Case diagrams along with the Business profile. Organization structure is depicted with nested Packages and Organization Unit stereotype. Worker, Case Worker, Internal Worker, External Actor or Work Unit symbols are used for organizational roles. Communication diagram is also employed to show the workflow. We use BPMD for the designer/people cell. Interaction Overview diagram included in UML 2.x is utilized to model the builder/people cell. Unfortunately, modeling the contractor/people cell causes a problem since no particular diagram in MDA is suitable for defining the out-of-context characteristic of the workflow. However, utilizing the method proposed in the builder/people cell, and with the help of extra comments, it's possible to model this cell to some extent.

Time column can be modeled with Timing diagram and Scheduling profile which are presented in UML 2.x. In the past, proposed solutions often utilized Sequence diagram to model this column [7], however it is not possible to correctly demonstrate the duration of an event with this diagram. Using the Timing diagram allows us to specify the exact start and end times and the duration for each event. In the third row, it is possible to employ the State diagram along with the Timing diagram to show different states.

OCL is often recommended for the Motivation column modeling. However, it should be noted that using this language causes trouble in higher rows. The OCL language is a formal language to specify rules and constraints. Its thorough understanding requires familiarity with some concepts of formal specifications in mathematics. This requirement leaves it vague for the managements who are probably unfamiliar with formal concepts. Besides, OCL is designed for stating rules and constraints upon objects, nevertheless the abstraction level related to the higher rows of the framework is beyond the concept of an object and using objects to model these levels is inappropriate [20]. We use Business Motivation Model (BMM) diagrams related to BSBR, to model the owner/motivation cell. By defining goals, viewpoints, missions, objects, facts, strategies, and so on along with their relations, BMM diagrams try to model the concepts governing the processes within an enterprise. To model the builder/motivation cell, Production Representation Rule (PRR) is employed. This meta-model, let us specify the rules which are defined in the second row in the form of CIM as a presentation of production rules in PIM level. OCL is used to state the rules supervising current objects for the fourth and fifth rows.

Fig. 3 summarizes our proposed method. All models can be integrated, since they are MOF-compliant. This characteristic guarantees that all models used in an MDA system can communicate with every other MOF-compliant model.

Zachman Framework	Data	Function	Network	People	Time	Motivation
Planner	Out of problem space					
Owner	Business profile	Business Use Case	Organization Unit stereotype	Business profile/use case	Timing diagram	BSBR BMM
Designer	CWM	BPDM	EDOC	BPDM	Timing & State diagrams	PRR
Builder	CWM	UML 2.x	Deployment diagram	Interaction Overview diag.	Timing diagram	OCL
Contractor	CWM	CORBA, EJB, .Net, WSM	CORBA, EJB, .Net, WSM	Interaction Overview diag.	Timing diagram	OCL

Fig 3. Consistence models for Zachman Framework cells

V. Case Study

The solutions provided in the field of Software Engineering are less likely to be proved using formal methods. Instead, we usually analyze them through case studies. Certainly, a case study is an initial test in a sample environment and the actual performance of a method can only be evaluated after years of experiments in real world.

Here are the steps we used in our case study:

1. Choosing an enterprise and checking its need for architecture.
2. Understanding the enterprise (architecture domain)
3. Modeling the first row of Zachman framework utilizing the information from the previous step.
4. Modeling the remaining rows in the Zachman framework using the proposed method.
5. Fixing weak points in the modeled cells.

The enterprise that we chose for our case study is one of the *Functional Architectures* of *FAVA* project [3] in the I.R.I presidential office. FAVA architectures consist of two main parts: *Fundamental Architectures* and *Functional Architectures*.

One of the *Functional Architectures* is the *Trace/Action Functional Architecture* on which this case study is based. FAVA project is designed using CHAM [5]. We do not intend to specify the details of CHAM, however it is worthwhile to mention that in CHAM, the architecture design is done in two phases: *Macro Architecture* design and *Detailed Architecture* design.

In the macro architecture design phase we have specified the architecture essence. This phase for the *Trace/Action Functional Architecture* is presented in [6].

In the detailed architecture design phase we specify the architecture details and the implementation methods. In this phase we had to use one of the enterprise architecture frameworks, so we used Zachman framework based on our proposed method. You can find the detailed architecture design of the Trace/Action Functional Architecture in [6].

In this case study we planned to answer the following key questions:

- Are all the cells modeled in one paradigm?
- Are the Zachman framework cells created using simple models based on E/R relationship?
- Is this architecture capable of utilizing different technologies?
- Does the proposed modeling method satisfy the architect needs to express his/her ideas?

VI. The Proposed Method vs. Other Methods

We defined some characteristics in sections 4 (B) and 5. To compare the proposed approach with other approaches we evaluate these characteristics on five-scale criterion: (0) Null - The approach does not have the characteristic at all. (1) Unknown - The status of the characteristic is unknown in the relevant approach. (2) Weak - The approach has the characteristic in an initial way. (3) Average - The approach has the characteristic. (4) Strong - The approach completely supports the characteristic. Table 1 shows the result of this comparison.

TABLE 1. The Proposed Approach Rank

	[7]	[17]	[18]	Ours
Integrable	3	2	2	4
Wide covering	2	4	2	3
Being standard	3	3	0	4
Easily understandable	4	3	3	2
Being up-to-date	3	3	3	4
One paradigm	4	0	3	4
E/R	4	4	4	4
Technology	0	3	2	4
Idea expressibility	1	2	1	3

VII. Conclusions

In this paper a method for consistent modeling of Zachman framework cells was proposed. Compared to other frameworks, ZF has some evident advantages. These advantages have caused its extensive usage as the basic framework of enterprise architecture in the present time. However, an architect should face the lack of consistent and coherent models for its cells. In this paper, we proposed a novel method for modeling the ZF cells. The presented method adopts the latest standards of OMG modeling (MDA). We inspected the method in practice by employing it in a case study. The results were promising compared to previously proposed methods, indicating that it is well suited to diminish the modeling problems one is facing using the ZF, to a great extent. This research demonstrated how the models in ZF cells can be made coherent in order to avoid inconsistencies within the framework. In general, our method can increase the success rate of information system projects within enterprises.

In our proposed solution, the forth and fifth rows of People column don't have well defined models. In future works, one is expected to suggest some MOF-compliant models for these cells. Considering the fact that those models are MOF-compliant, they can be integrated with our proposed solution, achieving full support of the Zachman Framework.

References

[1] S. S. Ostadzadeh, *A Unified Modeling Approach for Zachman Framework based on MDA*, MSc thesis, Science & Research branch of Islamic Azad University, Tehran, 2006.

[2] S. S. Ostadzadeh, "MDA role in Enterprise Architecture", Technical Report, Science & Research branch of Islamic Azad University, Tehran, 2005.

[3] A. Majidi, "FAVA: Strategic Plan & Architecture", Technical Report, I.R.I Presidential Office, ITC, Tehran, 2004.

[4] A. Majidi, "eGovernment: Strategic Plan & Architecture", Technical Report, Institute for Strategic Researches in Information Technology (IRIT), I.R.I Presidential Office, ITC, Tehran, 2005.

[5] A. Majidi, "CHAM: National Framework & Methodology for Macro/Micro Systems", Technical Report, Institute for Strategic Researches in Information Technology (IRIT), I.R.I Presidential Office, ITC, Tehran, 2005.

[6] S. S. Ostadzadeh and A. Majidi, "Track/Action Functional Architecture", Technical Report, Institute for Strategic Researches in Information Technology (IRIT), I.R.I Presidential Office, ITC, Tehran, 2005.
[7] A. Fatholahi, *An Investigation into Applying UML to Zachman Framework*, MSc thesis, Shahid Beheshti University, Tehran, 2004.
[8] D.S. Frankel, *Model Driven Architecture: Applying MDA to Enterprise Computing*, OMG Press, Wiley Publishing, 2003.
[9] J.A. Zachman, *The Zachman Framework: A Primer for Enterprise Engineering and Manufacturing*, 2003.
[10] J.A. Zachman, "A Framework for Information Systems Architecture", IBM Systems Journal, Vol. 26, No. 3, 1987.
[11] J.A. Zachman, "The Framework for Enterprise Architecture – Cell Definitions", ZIFA, 2003.
[12] G. Booch, B. Brown, S. Iyengar, J. Rumbaugh, and B. Selic, "An MDA Manifesto", MDA Journal, May 2004.
[13] D. D'Souza, "Model-Driven Architecture and Integration", Kinetium, March 2002.
[14] MDA Guide, OMG document, 2003. http://www.omg.org/
[15] S.J. Mellor, K. Scott, A. Uhl, and D. Weise, *MDA Distilled: Principles of Model-Driven Architecture*, Addison Wesley, 2004.
[16] D. Pilone and N. Pitman, *UML 2.0 in a Nutshell*, O'Reilly, 2005.
[17] System Architect Manuals, *Building Enterprise Architecture: The Popkin Process*, Popkin Company, 2004.
[18] S. S. Ostadzadeh, "Development of Zachman Framework using the Rational Unified Process and the Unified Modeling Language", Technical Report, Science & Research branch of Islamic Azad University, Tehran, 2004.
[19] Object Management Group Document, http://www.omg.org/technology/documents/
[20] F. Shams Aliee, *Modeling the Behavior of Processes Using Collaborating Objects*, PhD Thesis, University of Manchester, Manchester, May 1996.

Beyond User Ranking: Expanding the Definition of Reputation in Grid Computing

Said Elnaffar
College of IT, UAE University, UAE
elnaffar@uaeu.ac.ae

Abstract-**Shopping around for a good service provider in a Grid Computing environment is no less challenging than the traditional shopping around in non-virtual marketplace. A client may consult a service broker for providers that can meet specific QoS requirements (e.g., CPU speed), and the broker may return a list of candidate providers that satisfy the client's demands. If this computing platform is backed up by some reputation system, the list of providers is then sorted based on some reputation criterion, which is commonly the user rating. We argue in this paper that judging the reputation of a provider based on user rating is not sufficient. The reputation should additionally reflect how trustworthy that provider has been with respect to complying with the finalized SLA (using a metric called *conformance*) and how consistent it has been with respect to honouring its compliance levels (using a metric called *fidelity*). Accordingly, we perceive the reputation as a vector of three dimensions: user rating, conformance, and fidelity. In this paper, we define these metrics, explain how to compute them formally, and how to use them in the reputation-enabled framework that we describe.**

I. INTRODUCTION

Grid computing [5] is a service model in which providers provision computing resources, services and infrastructure management to clients as needed, and charge them per use. As a consequence of the rapid growth of such on-demand computing applications and the profusion of service providers, clients eventually have to select the provider that they can rely on in order to achieve their business objectives [2]. To that end, quality of service (QoS) metrics [1] are typically used to differentiate and rank service providers.

QoS-enabled Grids are typically associated with service level agreements (SLA) that, as binding contracts between the service provider and the service requestor, guarantee application performance quantitatively.

Throughput, Response time, and availability are among the many SLA metrics that can be negotiated and investigated when it comes to assessing the QoS a provider can offer to a client [1]. They comprise the main criteria of compelling a client to favour one provider over others. Unfortunately such set of metrics lack an attribute that measures how often a particular service provider was able to comply with the SLA in the past. Many researchers define reputation (or trustworthiness) of a service provider as the average ranking assigned by the clients who interact with that provider [8]. If a service provider is known to offer certain qualities over a period of time irrespective of its limitations, then it is assumed to have a good reputation. A reputation system is defined as a secure informative system responsible for maintaining a dynamic and adaptive reputation metric for its community. Grid players, such as service providers, brokers, and clients, continuously interact with the reputation system to establish a community ranking mechanism that co-operatively help them make future decisions based on the overall community experiences. However, we contend that designing a reputation system that solely relies on the temporal perspective of humans (i.e., the clients) can expose the system to dishonest ratings caused by the following types of users:

- *Emotional Reactors*: those clients who give non-subjective, inaccurate ratings influenced by some personal (could be temporal) issues with the service provider.
- *Bad Mouthers*: clients who unfaithfully exaggerate by giving negative ratings to service providers.
- *Ballot Stuffers*: those clients who unfaithfully exaggerate by giving positive ratings to service providers.

Moreover, client views may not necessarily be able to capture the degree of inconsistency (variance) in the service provider's compliance levels over an extended period of time. These views are predominantly influenced by the most recent experience. Such vulnerability gives devious providers the opportunity to manipulate user perceptions psychologically. For examples, and in order to improve their reputation, providers may choose to boost their image in a season that witness soaring competition from their rivals by honouring the compliance levels promised to be delivered to users. As a result, users tend to express their satisfaction through that short period of time, which might be temporary, by giving high rating scores right after that pleasant experience. After that critical season elapses, service provider might opt to return to its relaxed policy with respect to adhering to the SLA. Such behaviour leaves the compliance profile of a given provider with superficial

K. Elleithy (ed.), *Advances and Innovations in Systems, Computing Sciences and Software Engineering*, 381–386.

positive and negative ratings over various points of time, lacking in-depth analysis.

In order to reflect providers' degree of compliance and their consistency at providing such compliance, we may need to aggregate some technical performance measures over time. This analysis can be done for two types of provider profiles: *local* profile and *global* profile. Each client locally maintains a private profile for each provider it deals with. This local profile reflects the direct personal experience that client had with a specific provider. The global profile, on the other hand, reflects the aggregate experience of the client community with a specific provider. Therefore, it is typically stored at a central global entity such as a Grid broker. Retaining the two profiles has manifold advantages. For example:

- In case of unavailability of the global profile, clients can still select providers based on their privately retained local profiles
- A service provider selection algorithm can take into account both profiles
- A client has the choice of adopting a selection criterion that assigns different weights to the two profiles. Suspicious clients may choose to give more weight to their personal experience, under the belief that other's ratings cannot be trusted.

The traditional QoS metrics such as throughput and response time are no longer adequate to effectively assess the goodness of provider as seen by the client [4]. Rather, the reputation of a provider should be factored in. However, as we explained earlier, banking on user ranking solely to represent the reputation of a provider has its own shortcomings. We believe that the overall reputation of a provider should be a combination of the following metrics:

- *User Ranking*: reflects the user view of the quality of service received from a provider. It is usually the ultimate means for capturing non-measurable aspects of quality. Therefore, it is deemed a qualitative metric as it is based on the human perception.
- *QoS Conformance*: measures the discrepancies between the projected QoS, as outlined in the SLA, and the eventually delivered QoS. Therefore, it is deemed a system-based metric as it is based on quantitative computations.
- *Fidelity*: is a new notion we present in this work to measure the creditability of a provider by assessing how consistently it abides by the QoS conformance.

The rest of the paper is structured as follows. In Section II, we give background knowledge pertaining to the commonly used QoS metrics in Grid computing and describe related work in reputation systems. In Section III, we formally define how to compute QoS conformance, and introduce the new *fidelity* metric along with its formulation. In Section IV, we describe a reputation-enabled framework in which these concepts fit together. The last section summarizes and concludes our paper.

II. RELATED WORK

QoS is one of the chief aspects in the Grid [1],[2], [3], [5], [6], [7]. Buco [2] advocates that a successful utility computing provider must be able not only to satisfy its customers' demand for high service-quality standards, but also to fulfill its service-quality commitments based upon business objectives (e.g., cost-effectively minimizing the exposed business impact of service level violations). To this end, he presents the design rationale of a business-objectives based utility computing SLA management system. Menasce et al. [1][2] explore some of the relevant issues (e.g., definition of quantitative QoS metrics, and the relationship between resource allocation and SLAs) pertinent to designing grid applications that deliver appropriate QoS. Typically, when a discussion of QoS attributes is raised, the following QoS metrics are considered:

- *Latency*: it is the time elapsed between the moment of receiving the provider a request and the moment of commencing its processing. This time is also called the *waiting time*.
- *Throughput*: is the number of interactions that a service provider can process per time unit.
- *Availability*: it is the percentage of time that a service provider is up and operational through an observed period of time.
- *Response Time*: it is the time elapsed between sending a request by a client and receiving the results from the service provider.
- *Reliability*: this is a metric that represents the success rate (or probability) of processing a client request correctly within a maximum expected time frame.
- *Cost*: it is the monetary fee as set by the service provider.
- *Reputation*: it is an index of trustworthiness that is traditionally derived from the users' assessment for a particular service provider.

In general, reputation indicates how much users have confidence in a provider based on their perceptions. Therefore, the reputation system is one of the important tools that clients should consult as they shop around for a provider. Depending on the reputation algorithm, the reputation system works on aggregating user ratings and distributes (or makes) the results accessible to prospective clients [8]. E-Bay [11] and Amazon [12] are examples of e-business that essentially count on reputation systems that are based on aggregating the numerical ratings.

We believe that reputation systems are vital to Grid environments in order to increase reliability, utilization, and popularity. Reputation serves as an important metric to avert the usage of under provisioned and malicious resources with the help of community feedback; it provides the ability to simplify the selection process while focusing first on qualitative concerns (e.g., what storage or CPU power are available at a particular resource).

Reputation systems have been the focus of several researchers [9], [10], [13], [14], [15]. The work in [9] suggests a reputation management framework for Grids to facilitate a distributed and efficient mechanism for resource selection. The framework uses an algorithm that combines the two known concepts of EigenTrust [13] and global trust integration [14]. In [15] the authors present a trust model for Grid systems and show how the model can be used to incorporate the security implications into scheduling algorithms.

A quantitative comparison of reputation systems in the Grid environment is conducted in [10]. This study shows that using a reputation system to guide service selection can significantly improve client satisfaction with minimal overhead. We noticed that the majority of the studied systems rely primarily on basic statistical computation (e.g., averaging) on user rating. What we advocate in this paper is to maintain system-based time series statistics that help infer how providers are likely to behave in the future, leading to improving the quality of selection. Among these statistics are the *Conformance* and *fidelity*, as explained next.

III. Reputation = (User Ranking, Conformance, Fidelity)

As shown in the previous section, the majority of reputation systems count ultimately on ratings given by the end users. Such interpretation of trustworthiness remains merely a user perception that fails to express the real performance quality of a provider or its service. Confining reputation to user ranking creates a vulnerable metric that is sensitive to manipulative providers (e.g., *bad mouthers* and *ballot stuffer*) and to devious practices by service providers such as delivering extremely high quality of service at specific times of the year in order to manipulate the perception of the end users hoping that they forget the provider's past shortcomings. Reputation should rather be an indication of the *truthfulness* of a service provider to deliver the promised performance along with user's opinions. Therefore, we express reputation as a three dimension vector comprising *User Ranking*, *Conformance*, and *Fidelity*. User ranking gives users the opportunity to express their opinion. Conformance gauges the compliance of a provider with the agreed SLA. Fidelity reflects how much clients can trust a specific provider with respect to delivering *consistent* conformance levels.

A. Local vs. Global Reputation Profiles

For each provider, the three metrics introduced above can be computed for two reputation profiles: *local* and *global*. The local profile is internally maintained by each client to reflect its sole view and experience with certain service provider. The global profile, on the other hand, is a publicly shared profile, typically maintained by the Grid broker, that reflects the collective view and experience of all clients with that provider. Relying on the global profile solely denies the client from factoring in its personal past experience (good or bad) with a particular provider. Furthermore, the local profile can be the last resort in case the global profile is inaccessible for some reason.

Having the two reputation profiles entails having two versions for each metric. *lRank*, *lConformance*, and *lFidelity* to denote the reputation components maintained in the local profile. Similarly, *gRank*, *gConformance*, and *gFidelity* denote the reputation components maintained in the global profile.

To obtain reliable statistics, reputation components (user ranking, conformance, fidelity) should be calculated based on observing a number, n, of client-provider interactions that occurred over a time window (*assessment window*). The client and the Grid broker can control the size of this window, which can be as big as days, weeks, months, or even year. The subsequent sections explain how to compute the three reputation components formally.

B. User Ranking

When a transaction is completed, each user is given the opportunity to rate the provider with regard to the quality of the transaction. Depending on the ranking system and the aggregation algorithm [10], a rating can be, for example, either +1 for satisfied, -1 for dissatisfied, or 0 for neutral; a user's feedback score is the sum of these individual ratings. Other aggregation algorithms can produce a normalized ranking metric as follows:

$$lRank = \frac{r-s}{s+r} \qquad (1)$$

Where r and s represent the number of positive and negative ratings assigned to the provider respectively over a given assessment window.

The *gRank* can be computed as follows:

$$gRank = \frac{p-v}{v+p} \qquad (2)$$

Where p and v respectively represent the number of clients that voted positively and negatively through the assessment window of the last n clients interacted with the service provider.

C. Conformance

Conformance is the concept that indicates the degree by which a service provider complies by the SLA. Bhoj et al. [16] view a compliance as a contract template and a system dictionary used to verify a contract. The specific data of a client are used to fill out the contract template. Such data get verified against the system measurements that were collected while forming the contract. The contract is evaluated and the compliance results are produced as customized reports for each client. What we propose in this paper is to quantitatively express the discrepancy between what the provider promised to deliver in its SLA and the actual quality attained.

To compute the conformance, the reputation system should progressively record the actually delivered attribute values for each interaction between the client and the provider. The conformance of each quality attribute (e.g., response time) in the SLA is computed as the average of the normalized difference between the projected and actual values of the attribute. The normalized difference could be positive or negative depending on whether the agreed upon value was greater than or lesser than the delivered value, respectively. A positive average indicates a positive compliance which means the agreed values have been delivered without violations. A negative average indicates a negative compliance and that the provider failed to deliver the agreed values. An average value of zero is ideal compliance indicating the delivered values being exactly equal to the agreed values.

Formally, the conformance of a provider is computed from the conformance values of all SLA attributes. At interaction (or service invocation) number I between the client and the provider, there are two vectors of attributes values:

$$P = (p_{1i}, p_{2i}, \dots, p_{qi})$$
$$A = (a_{1i}, a_{2i}, \dots, a_{qi})$$

Vector P represents the projected values of all q attributes negotiated in the SLA. Vector A represents the actually delivered values of these attributes. I denotes the interaction number observed within the assessment window of the most recent n interactions ($1 \le i \le n$).

The normalized difference, c_{xi}, between any pair a_{xi} and p_{xi} ($1 \le x \le q$) can be positive or negative and it denotes the compliance of the provider after executing interaction I with respect to attribute x:

$$c_{xi} = \frac{a_{xi} - p_{xi}}{p_{xi}} \tag{3}$$

This leads to producing the conformance vector ($c_{1i}, c_{2i}, \dots, c_{qi}$). However, since all dimensions are normalized and we are interested in obtaining the scalar value, C_i, that reflects the collective conformance of the provider upon the completion of interaction i, we can define

$$C_i = \frac{1}{q} . \sum_{x=1}^{q} Cxi \tag{4}$$

To compute *lConformance* over an assessment window of n interactions, we can write:

$$lConformance = \frac{1}{n} . \sum_{i=1}^{n} Ci \tag{5}$$

To compute *gConformance*, the broker views the assessment window as the last m clients that interacted with a certain provider. Therefore, we can write:

$$gConformance = \frac{1}{m} . \sum_{i=1}^{m} Ci \tag{6}$$

where C_i is the conformance value reported by client i resulting from its last interaction with the provider.

D. Fidelity

Conformance is useful to assess the overall compliance of a provider; however, it does not show the degree of adherence to that compliance. The role of the *fidelity* metric proposed in this paper is to assess the degree of consistency in the compliance levels of a service provider, adding a new dimension to the assessment of the reputation of providers.

The notion of fidelity is based on computing the variance in the conformance levels. The lower the variance is, the more successful the provider is in delivering consistent performance levels. Fidelity is computed over a range of conformance measures obtained from past interactions with the service provider. Doing so gives an insight into providers' historical performance by progressively assessing their conformance levels over a range of past interactions (assessment window). In other words, fidelity is deemed a measure of truthfulness and verity.

Since fidelity represents the degree of variation in the conformance values achieved by a service provider, we can use the statistical standard deviation (σ) to define fidelity. The standard deviation can assess the degree of spread in the conformance values of a service provider, which indicates how efficiently and how often the guaranteed levels of quality are met. However, and since we are in need for a normalized metric that enables ranking providers among themselves objectively, we further normalize the standard deviation by the mean (μ) producing what is statistically known as the *coefficient of variance* (COV), which is basically σ/μ.

Hence, the local fidelity can be formulated as:

$$lFidelity = \frac{\sigma}{\mu} = \frac{\sqrt{\frac{1}{n}\sum_{i=1}^{n}(C_i - \mu)^2}}{\mu} \quad (7)$$

Where n is the number of the conformance measures spawn from the last n interactions (i.e., size of the assessment window as set by the client). C_i is the conformance level measured upon the completion of the ith interaction. μ is the mean conformance, that is

$$\mu = \frac{\sum_{i=1}^{n} C_i}{n} \quad (8)$$

Similarly, the global fidelity can be computed as follows

$$gFidelity = \frac{\sqrt{\frac{1}{m}\sum_{t=1}^{m}(C_t - \psi)^2}}{\psi} \quad (9)$$

Where m is the number of the conformance measures collected from the last m clients that interacted with the provider. Therefore, m represents the assessment window size as set by the Grid broker. C_t is the conformance obtained from the tth client upon the completion of its interaction. ψ is the mean conformance, that is

$$\psi = \frac{\sum_{t=1}^{m} C_t}{m} \quad (10)$$

E. *Reputation (Rep) of Providers*

Since we define reputation as a combination of user ranking, conformance, and fidelity, the local reputation profile of a provider can be represented by the vector *lRep=(lRank, lConformance, lFidelity)* that is maintained at the client side. Likewise, the global reputation profile of that provider can be represented by the vector *gRep=(gRank, gConformance, gFidelity)* that is maintained at the broker side. The overall reputation (*Rep*) of the provider as seen by the client can be expressed as the weighted sum of the local and global reputation vectors:

$$Rep = W_l * lRep + W_g * gRep \quad (11)$$

Such that

$$W_l + W_g = 1$$

Where W_l and W_g represent the significance (or trust) weights of the local reputation and the global reputation, respectively, as arbitrarily assigned by each client. Therefore, we can write

$$Rep = (Rank, Conformance, Fidelity) \quad (12)$$

where

$$Rank = W_l * lRank + W_g * gRank$$
$$Conformance = W_l * lConformance + W_g * gConformance$$
$$Fidelity = W_l * lFidelity + W_g * gFidelity$$

IV. A FRAMEWORK FOR REPUTATION-BASED GRIDS

In light of the new definition of reputation, Fig 1 depicts the reputation-based framework that we suggest for the Grid. As we can see, we have the three typical main players in this framework: clients, Grid brokers, and service providers. A provider publishes its services by registering them, along with their associated QoS specifications, with a broker. The

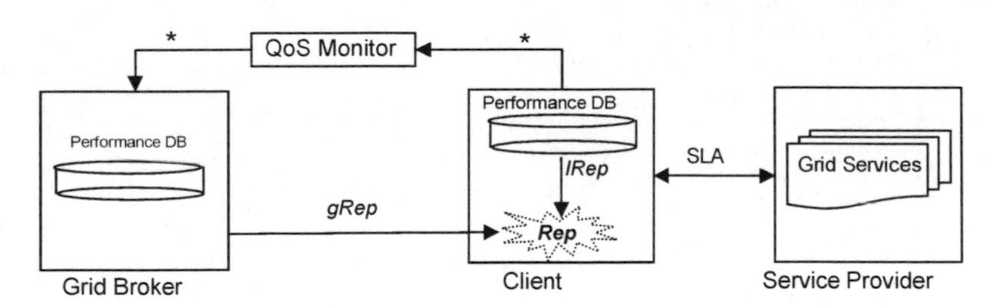

Fig 1. Reputation-Based Framework for the Grid

broker maintains a database of historical performance measures (e.g., user ratings and conformances) for each service provider. The broker uses this database to compute the global reputation (*gRep*) of a provider, as explained in the previous section. Likewise, each client maintains a database that stores a time series of performance measures experienced with providers that client dealt with. This internal database is referenced when the client needs to compute the local reputation (*lRep*) of a given provider.

Upon a request from a client, the broker returns a list of candidate providers. The client, based on the *Rep* vector, selects its prospective provider and negotiates the relevant QoS attributes with it. A copy of the finalized SLA is sent to the broker in order to compute reputation components (e.g., *gConformance*).

The execution of client-provider interactions gets monitored by an independent module *QoS Monitor* [17] in order to intercept the eventually delivered QoS attribute values and relay them to the performance database of the broker. It is important to have such independent monitoring entity in order to prevent clients from tampering with real performance readings attainted by service providers. In addition to the finally attained performance data, each client sends its own personal rating to the broker in order to compute the global ranking (*gRank*), which is required to complete the computation of the global reputation (*gRep*) of a provider.

V. CONCLUSION

Finding a service provider that can satisfy the client's computing requirements, such as CPU power or storage capacity, is the first step towards forming the initial set of candidate providers. The client has yet to rank these candidates according to their past compliance with the agreed SLA. This compliance and providing consistent conformance with the promised QoS attributes constitute provider's reputation. We contended in this paper that counting exclusively on the typical ratings given by users is not a reliable means of assessing the goodness of a provider due to numerous pitfalls. Rather, we view reputation as a function of three combined qualities: user ranking, conformance, and fidelity. We explained how to produce local and global reputation vectors and described the general framework in which the new concepts fit together in order to establish a reputation-based computing platform.

The next item in our research agenda is to quantitatively compare our reputation system with others. We intend to build a generic Grid simulation framework in which we can seamlessly plug into it any arbitrary reputation policy that we intend to evaluate.

ACKNOWLEDGMENT

We are grateful to the research support we got from the Scientific Research Council of the UAE University through research grant #02-06-9-11/06.

REFERENCES

[1] D. Menascé and E. Casalicchio, Quality of Service Aspects and Metrics in Grid Computing, *Proc. 2004 Computer Measurement Group Conference*, Las Vegas, NV, December 5-10, 2004.

[2] D. Menascé & E. Casalicchio, QoS in Grid Computing, *IEEE Internet Computing 8*(4), July/August 2004.

[3] M. J. Buco, R. N. Chang, L. Z. Luan, C. Ward, J. L. Wolf & P. S. Yu, Utility Computing SLA Management Based Upon Business Objectives, *IBM Systems Journal 43*(1), 2004, pp. 159-178.

[4] A. Bouch, A. Kuchinsky & N. Bhatti, Quality is in the Eye of the Beholder: Meeting Users' Requirements for Internet Quality of Service. *Proc. SIGCHI on Human Factors in Computing Systems*, April 2000.

[5] I. Foster & C. Kesselman, *The Grid: Blueprint for a New Computing Infrastructure* (2nd ed., Morgan Kaufmann, 2004).

[6] A. Leff, J. Rayfield & D. Dias, Service-Level Agreements and Commercial Grids, *IEEE Internet Computing 7*(4), 2003, pp. 44-50.

[7] J. Nabrzyski, J. Schopf & J. Weglarz, eds., *Grid Resource Management - State of the Art and Future Trends*, (Kluwer Academic Publishers, 2004).

[8] P. Resnick, R. Zeckhauser, E. Friedman, and K. Kuwabara, Reputation Systems, *Communications of the ACM 43*(12), December 2000.

[9] K. Alunkal, I. Veljkovic, G. von Laszewski & K. Amin, Reputation-Based Grid Resource Selection, *Workshop on Adaptive Grid Middleware (AGridM 2003)*, New Orleans, LA, USA, Sept. 28, 2003.

[10] J. Sonnek & J. Weissman, A Quantitative Comparison of Reputation Systems in the Grid, *Proceedings of the Sixth ACM/IEEE International Workshop on Grid Computing*, 2005.

[11] Ebay Web Page [online]. Available: http://www.ebay.com.

[12] Amazon Web Page [online]. Available: www.amazon.com.

[13] S. Kamvar, M. Schlosser & H. Garcia-Molina, The Eigentrust Algorithm for Reputation Management in P2P Networks, *12th International World Wide Web Conference 2003*, Budapest, Hungary, ACM Press, May 2003.

[14] F. Azzedin & M. Maheswaran, Evolving and Managing Trust in Grid Computing Systems, *IEEE Canadian Conference on Electrical Computer Engineering, IEEE Computer Society Press*, May 2002.

[15] F. Azzedin and M. Maheswaran, Integrating Trust into Grid Resource Management Systems, *Proceedings of the International Conference on Parallel Processing*, pages 47-54, 2002.

[16] P. Bhoj, S. Singhal & S. Chutani, SLA Management in Federated Environments, *HP Labs Technical Report*, December, 1998.

[17] A. Keller & H. Ludwig, Defining and Monitoring Service Level Agreements for Dynamic e-Business, *Proceedings of the 16th USENIX System Administration Conference (LISA '02)*, November 2002.

A Comparative Study for Email Classification

Seongwook Youn and Dennis McLeod
University of Southern California,
Los Angeles, CA 90089
USA

Abstract *- Email has become one of the fastest and most economical forms of communication. However, the increase of email users have resulted in the dramatic increase of spam emails during the past few years. In this paper, email data was classified using four different classifiers (Neural Network, SVM classifier, Naïve Bayesian Classifier, and J48 classifier). The experiment was performed based on different data size and different feature size. The final classification result should be '1' if it is finally spam, otherwise, it should be '0'. This paper shows that simple J48 classifier which make a binary tree, could be efficient for the dataset which could be classified as binary tree.*

I. INTRODUCTION

Email has been an efficient and popular communication mechanism as the number of Internet users increase. Therefore, email management is an important and growing problem for individuals and organizations because it is prone to misuse. The blind posting of unsolicited email messages, known as spam, is an example of misuse. Spam is commonly defined as the sending of unsolicited bulk email - that is, email that was not asked for by multiple recipients. A further common definition of a spam restricts it to unsolicited commercial email, a definition that does not consider non-commercial solicitations such as political or religious pitches, even if unsolicited, as spam. Email was by far the most common form of spamming on the internet.

Text classification including email classification presents challenges because of large and various number of features in the dataset and large number of documents. Applicability in these datasets with existing classification techniques was limited because the large number of features make most documets undistinguishable.

In many document datasets, only a small percentage of the total features may be useful in classifying documents, and using all the features may adversely affect performance. The quality of training dataset decides the performance of both the text classification algorithms and feature selection algorithms. An ideal training document dataset for each particular category will include all the important terms and their possible distribution in the category.

The classification algorithms such as Neural Network (NN), Support Vector Machine (SVM), and Naïve Bayesian (NB) are currently used in various datasets and showing a good classification result.

The problem of spam filtering is not a new one and there are already a dozen different approaches to the problem that have been implemented. The problem was more specific to areas like Artificial intelligence and Machine Learning. Several implementations had various trade-offs, difference performance metrics, and different classification efficiencies. The techniques such as decision tree (J48), Naive Bayesian classifiers, Neural Networks, Support Vector Machine, etc had various classification efficiencies. The remainder of the paper is organized as follows: Section 2 describes existing related works; Section 3 introduces four spam classification methods used in the experiment; Section 4 discusses the experimental results; Section 5 concludes the paper with possible directions for future work.

II. RELATED WORKS

[17] compared a cross-experiment between 14 classification methods, including decision tree, Naïve Bayesian, Neural Network, linear squares fit, Rocchio. KNN is one of top performers, and it performs well in scaling up to very large and noisy classification problems.

[14] showed that bringing in other kinds of features, which are spam-specific features in their work, could improve the classification results. [11] showed a good performance reducing the classification error by discovering temporal relations in an email sequence in the form of temporal sequence patterns and embedding the discovered information into content-based learning methods. [13] showed that the work on spam filtering using feature selection based on heuristics.

Aproaches to filtering junk email are considered [2, 5, 14]. [6] and [7] showed approaches to filtering emails involve the deployment of data mining techniques. [3] proposed a model based on the Neural Network to classify personal emails and the use of Principal Component Analysis (PCA) as a preprocessor of NN to reduce the data in terms of both dimensionality as well as size. [1] compared the performance of the Naïve Bayesian filter to

K. Elleithy (ed.), Advances and Innovations in Systems, Computing Sciences and Software Engineering, 387–391.

an alternative memory based learning approach on spam filtering.

[15] and [18] developed a algorithm to reduce the feature space without sacrificing remarkable classification accuracy, but the effectiveness was based on the quality of the training dataset.

In the classification experiment for spam mail filtering, J48 showed better result than NB, NN, or SVM classifier.

III. SPAM CLASSIFICATION METHODS

Generally, the main tool for email management is text classification. A classifier is a system that classifies texts into the discrete sets of predefined categories. For the email classification, incoming messages will be classified as spam or legitimate using classification methods.

A. Neural Network (NN)

Classification method using a NN was used for email filtering long time ago. Generally, the classification procedure using the NN consists of three steps, data pre-processing, data training, and testing. The data pre-processing refers to the feature selection. Feature selection is the way of selecting a set of features which is more informative in the task while removing irrelevant or redundant features. For the text domain, feature selection process will be formulated into the problem of identifying the most relevant word features within a set of text documents for a given text learning task. For the data training, the selected features from the data pre-processing step were fed into the NN, and an email classifier was generated through the NN. For the testing, the email classifier was used to verify the efficiency of NN. In the experiment, an error BP (Back Propagation) algorithm was used.

B. Support Vector Machines (SVM) Classifier

SVMs are a relatively new learning process influenced highly by advances in statistical learning theory. SVMs have led to a growing number of applications in image classification and handwriting recognition. Before the discovery of SVMs, machines were not very successful in learning and generalization tasks, with many problems being impossible to solve. SVMs are very effective in a wide range of bioinformatic problems. SVMs learn by example. Each example consists of a m number of data points(x1,……xm) followed by a label, which in the two class classification we will consider later, will be +1 or -1. -1 representing one state and 1 representing another. The two classes are then separated by an optimum hyperplane, illustrated in figure 1, minimizing the distance between the closest +1 and -1 points, which are known as support vectors. The right hand side of the separating hyperplane represents the +1 class and the left hand side represents the -1 class.

This classification divides two separate classes, which are generated from training examples. The overall aim is to generalize well to test data. This is obtained by introducing a separating hyperplane, which must maximize the margin () between the two classes, this is known as the optimum separating hyperplane

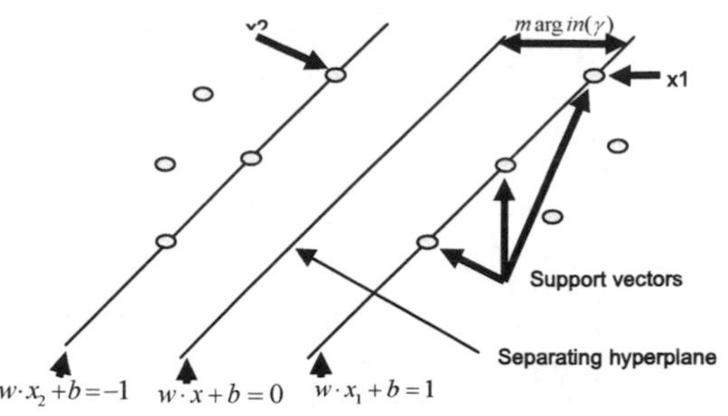

Let's consider the above classification task with data points x_i, i=1.....m, with corresponding labels $y_i = \pm 1$, with the following decision function:

$$f(x) = sign(w \cdot x + b)$$

By considering the support vectors x1 and x2, defining a canonical hyperplane, maximizing the margin, adding Lagrange multipliers, which are maximized with respect to α:

$$W(\alpha) = \sum_{i=1}^{m} \alpha_i - \sum_{i,j=1}^{m} \alpha_i \alpha_j y_i y_j (x_i \cdot x_j)$$

$$(\sum_{i=1}^{m} \alpha_i y_i = 0, \alpha_i \geq 0)$$

C. Naïve Bayesian (NB) Classifier

Naïve Bayesian classifier is based on Bayes' theorem and the theorem of total probability. The probability that a document d with vector $\vec{x} = < x_1, ..., x_n >$ belongs to category c is

$$P(C = c \mid \vec{X} = \vec{x}) = \frac{P(C = c) \cdot P(\vec{X} = \vec{x} \mid C = c)}{\prod_{k \in \{spam, legit\}} P(C = k) \cdot P(\vec{X} = \vec{x} \mid C = k)}$$

However, the possible values of $\vec{X}$ are too many and there are also data sparseness problems. Hence, Naïve Bayesian classifier assumes that $X_1, ... X_n$ are conditionally independent given the category C. Therefore, in practice, the probability that a document d with vector $\vec{x} = < x_1, ..., x_n >$ belongs to category c is

$$P(C = c \mid \vec{X} = \vec{x}) = \frac{P(C = c) \cdot \prod_{i=1}^{n} P(X_i = x_i \mid C = c)}{\prod_{k \in \{spam, legit\}} P(C = k) \cdot \prod_{i=1}^{n} P(X_i = x_i \mid C = k)}$$

$P(X_i \mid C)$ and P(C) are easy to obtain from the frequencies of the training dataset. So far, a lot of researches showed that the Naïve Bayesian classifier is surprisingly effective.

D. J48 Classifier

J48 classifier is a simple C4.5 decision tree for classification. It creates a binary tree.

IV. RESULTS

In this section, four classification methods (Neural Network, Support Vector Machine classifier, Naïve Bayesian classifier, and J48 classifier) were evaluated the effects based on different datasets and different features. Finally, the best classification method was obtained from the training dataset. 4500 emails were used as a training dataset. 38.1% of dataset were spam ad 61.9% were legitimate email. To evaluate the classifiers on training dataset, we defined an accuracy measure as follows.

$$Accuracy(\%) = \frac{Correctly\_Classified\_Emails}{Total\_Emails} * 100$$

Also, Precision and Recall were used as the metrics for evaluating the performance of each email classification approach.

A. Effect of dataset on performance

An experiment measuring the performance against the size of dataset was conducted using dataset of different sizes listed in Fig.1. The experiment was performed with 55 features from TF/IDF. For example, in case of 1000 dataset, Accuracy was 95.80% using J48 classifier.

Data Size	NN	SVM	Naïve Bayesian	J48
1000	93.50%	92.70%	97.20%	95.80%
2000	97.15%	95.00%	98.15%	98.25%
3000	94.17%	92.40%	97.83%	97.27%
4000	89.60%	91.93%	97.75%	97.63%
4500	93.40%	90.87%	96.47%	97.56%

With 55 features

Fig. 1. Classification result based on data size

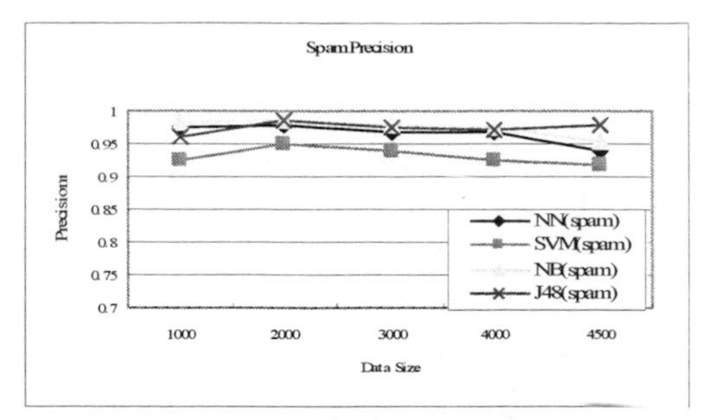

Fig. 2. Spam precision based on data size

A few observations can be made from this experiment. As shown in Fig. 1, the average of correct classification rate for both J48 and NB was over 95%. Dataset size was not an important factor in measuring precision and recall. The results show that the performance of classification was not stable. For four different classification methods, precision of spam mail was shown in Fig. 2, likewise, precision of legitimate mail was shown in Fig. 3.

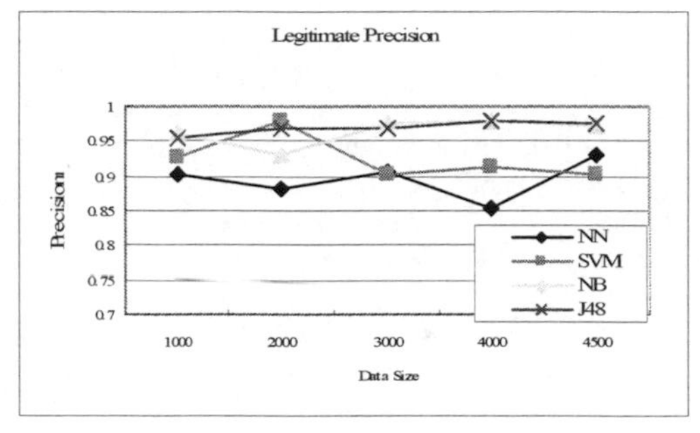

Fig. 3. Legitimate precision based on data size

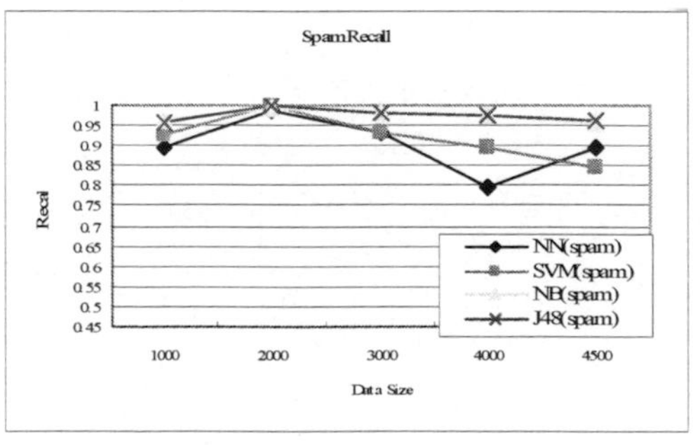

Fig. 4. Spam recall based on data size

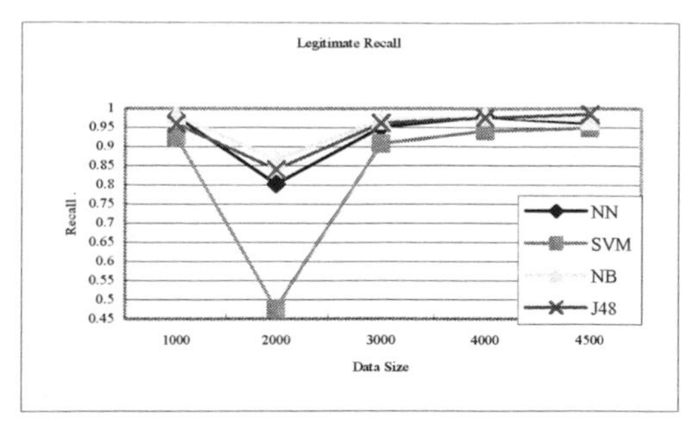

Fig. 5. Legitimate recall based on data size

As shown in Fig. 2, 3, 4, and 5, the precision and recall curves of J48 and NB classification were better than the ones of NN and SVM. Also, the average precision and recall for both J48 and NB was over 95%. In Fig. 5, legitimate recall values were sharply decreased at the data size 2000. The increase of spam mail in the training dataset between 1000 and 2000 result in a sharp decrease of legitimate recall values for all classifiers

B. Effect of feature size on performance

The other experiment measuring the performance against the size of dataset was conducted using different features listed in Fig. 6. 4500 email dataset was used for the experiment. For example, in case of 10 features, Accuracy was 94.84% using J48 classifier. The most frequent words in spam mail were selected as features. Generally, the result of classification was increased for all classification methods according the feature size increased.

Feature Size	NN	SVM	Naïve Bayesian	J48
10	83.60%	81.91%	92.42%	94.84%
20	89.87%	85.73%	95.60%	96.91%
30	93.31%	88.87%	95.64%	97.56%
40	92.13%	89.93%	97.49%	97.13%
50	93.18%	90.27%	96.84%	97.67%
55	93.10%	90.84%	97.64%	97.56%

Fig. 6. Classification result based on feature size

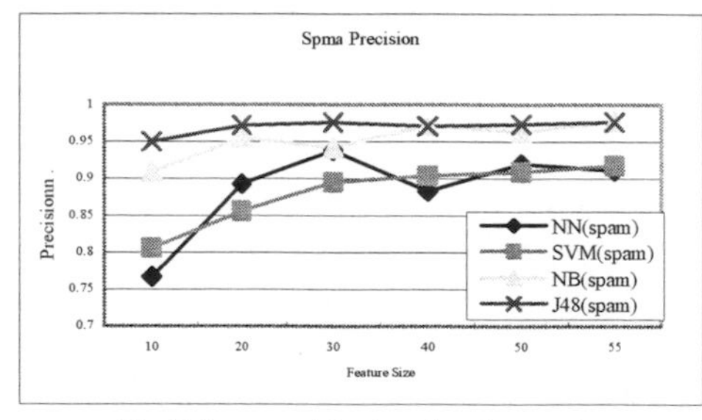

Fig. 7. Spam precision based on feature size

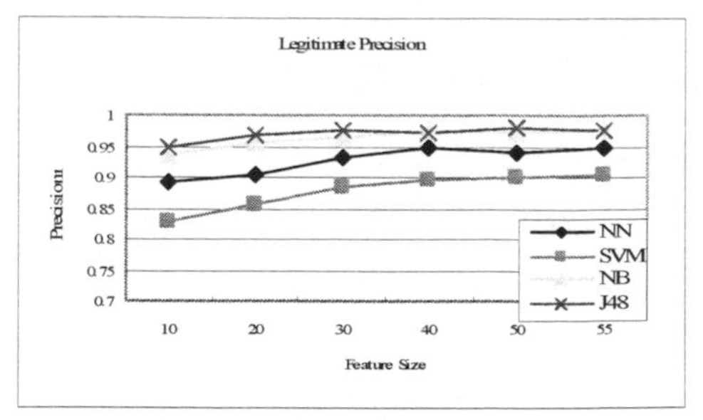

Fig. 8. Legitimate precision based on feature size

As shown in Fig. 7, 8, 9, and 10, good classification result order in the experiment was J48, NB, NN, and SVM for all cases (spam precision, legitimate precision, spam recall, and legitimate recall). The overall precision and recall for email classification increase and become stable according to the increase of the number of feature. Gradually, the accuracy increase and finally saturated with the increased feature size. As shown in Fig. 7 and 8, J48 classifier provided the precision over 95% for every feature size irrespective of spam or legitimate. Also, J48 classifier supported over 97% of classification accuracy for more than 30 feature size. For the recall, J48 and NB showed better result than NN and SVM for both spam and legitimate mail, but J48 was a little bit better than NB.

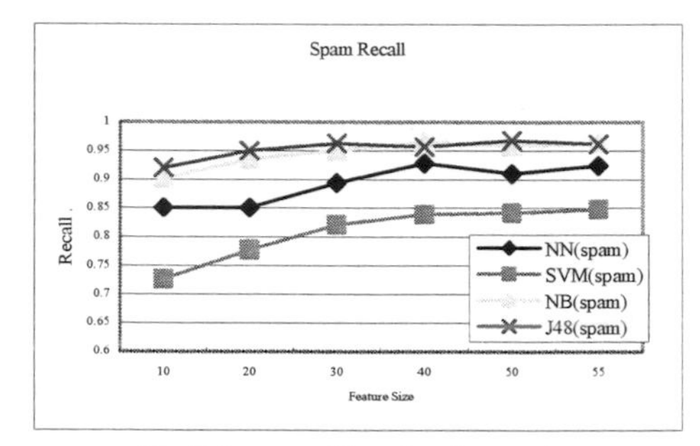

Fig. 9. Spam recall based on feature size

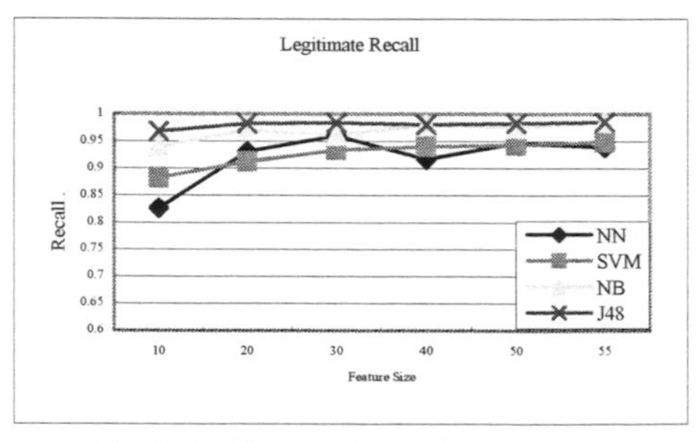

Fig. 10. Legitimate recall based on feature size

V. Conclustion and Future Work

In this paper, four classifiers including Neural Network, SVM, Naïve Bayesian, and J48 were tested to filter spams from the dataset of emails. All the emails were classified as spam (1) or not (0). That was the characteristic of the dataset of email for spam filtering. J48 is very simple classifier to make a decision tree, but it gave the efficient result in the experiment. Naïve Bayesian classifier also showed good result, but Neural Network and SVM didn't show good result compared with J48 or Naïve Bayesian classifier. Neural Network and SVM were not appropriate for the dataset to make a binary decision. From this experiment, we can find it that a simple J48 classifier can provide better classification result for spam mail filtering. In the near future, we plan to incorporate other techniques like different ways of feature selection, classification using ontology. Also, classified result could be used in Semantic Web by creating a modularized ontology based on classified result. There are many different mining and classification algorithms, and parameter settings in each algorithm. Experimental results in this paper are based on the default settings. Extensive experiments with different settings are applicable in WEKA. Moreover, different algorithms which are not included in WEKA can be tested. Also, experiments with various feature selection techniques should be compared.
Furthermore, we plan to create an adaptive ontology as a spam filter based on classification result. Then, this ontology will be evolved and customized based on user's report when a user requests spam report. By creating a spam filter in the form of ontology, a filter will be user customized, scalable, and modularized, so it can be embedded to many other systems. This ontology also may be used to block porn web site or filter out spam emails on the Semantic Web.

Acknowledgement

This research has been funded in part by the Integrated Media Systems Center, a National Science Foundation Engineering Research Center, Cooperative Agreement No. EEC-9529152.

References

[1] I. Androutsopoulos, G. Paliouras, V. Karkaletsis, G. Sakkis, C. Spyropoulos, and P. Stamatopoulos, "Learning to Filter Spam E-Mail: A Comparison of a Naive Bayesian and a Memory-Based Approach," CoRR cs.CL/0009009, 2000.
[2] W. Cohen, "Learning rules that classify e-mail," In Proc. of the AAAI Spring Symposium on Machine Learning in Information Access, 1996.
[3] B. Cui, A. Mondal, J. Shen, G. Cong, and K. Tan, "On Effective E-mail Classification via Neural Networks," In Proc. of DEXA, 2005, pp. 85-94.
[4] E. Crawford, I. Koprinska, and J. Patrick, "Phrases and Feature Selection in E-Mail Classification," In symposium of ADCS, 2004, pp. 59-62.
[5] Y. Diao, H. Lu, and D. Wu, "A comparative study of classification based personal e-mail filtering," In Proc. of fourth PAKDD, 2000.
[6] T. Fawcett, "in vivo spam filtering: A challenge problem for data mining," In Proc. of ninth KDD Explorations vol.5 no.2, 2003.
[7] K. Gee, "Using latent semantic indexing to filter spam," In Proc. of eighteenth ACM Symposium on Applied Computing, Data Mining Track, 2003.
[8] Z. Gyöngyi, H. Garcia-Molina, and J. Pedersen, "Combating Web Spam with TrustRank," In VLDB, 2004, pp. 576-587.
[9] T. Joachims, "A Probabilistic Analysis of the Rocchio Algorithm with TFIDF for Text Categorization," In ICML, 1997, pp. 143-151.
[10] T. Joachims, "Structured Output Prediction with Support Vector Machines," SSPR/SPR, 2006, pp. 1-7
[11] S. Kiritchenko, S. Matwin, and S. Abu-Hakima, "Email Classification with Temporal Features," Intelligent Information Systems 2004, pp. 523-533.
[12] S. Martin, B. Nelson, A. Sewani, K. Chen, and A. Joseph, "Analyzing Behavioral Features for Email Classification," CEAS, 2005.
[13] T. Meyer, and B. Whateley, "SpamBayes: Effective open-source, Bayesian based, email classification system," In Proc. of first Conference of Email and Anti-Spam, 2004.
[14] M. Sahami, S. Dumais, D. Heckerman, and E. Horvitz, "A Bayesian Approach to Filtering Junk E-Mail," In Proc. of the AAAI Workshop on Learning for Text Categorization, 1998.
[15] S. Shankar and G. Karypis, "Weight adjustment schemes for a centroid based classifier," Computer Science Technical Report TR00-035, 2000.
[16] I. Stuart, S. Cha, and C. Tappert, "A Neural Network Classifier for Junk E-Mail," in Document Analysis Systems, 2004, pp. 442-450.
[17] Y. Yang, "An Evaluation of Statistical Approaches to Text Categorization," Journal of Information Retrieval, Vol 1, No. 1/2, 1999, pp. 67-88.
[18] Y. Yang and J. Pedersen, "A Comparative Study on Feature Selection in Text Categorization," In ICML, 1997, pp. 412-420.
[19] S. Youn and D. McLeod, "Ontology Development Tools for Ontology-Based Knowledge Management," In Encyclopedia of E-Commerce, E-Government and Mobile Commerce. Idea Group Inc, 2006.

Noise Reduction for VoIP Speech Codecs Using Modified Wiener Filter

Seung Ho Han[1], Sangbae Jeong[1], Heesik Yang[1], Jinsul Kim[2], Won Ryu[2], and Minsoo Hahn[1]

[1]Speech and Audio Information Laboratory, Information and Communications University,
103-6, Munji-dong, Yuseong-gu, Daejeon, 305-732, Republic of Korea
{space0128, Sangbae, sheik, mshahn}@icu.ac.kr
[2]Convergence Network Inter-working Technology Team, BcN Service Research Group,
Broadband Convergence Network Research Division, Electronics and Telecommunications
Research Institute, 161, Gajeong-dong, Yuseong-gu, Daejeon, 305-350, Republic of Korea
{jsetri, wlyu}@etri.re.kr

Abstract- **Noise reduction is essential to achieve an acceptable QoS in VoIP systems. This paper proposes a Wiener filter-based noise reduction scheme optimized to the estimated SNR at each frequency bin as a logistic function is used. The proposed noise reduction scheme would be applied as pre-processing before speech encoding. For various noisy conditions, the PESQ evaluation is performed to evaluate the performance of the proposed method. In this paper, G.711, G.723.1, and G.729A are used as test VoIP speech codecs. The PESQ results show that the performance of our proposed noise reduction scheme outperforms those of the noise suppression one in the IS-127 EVRC and the noise reduction one in the ETSI standard for the advanced distributed speech recognition front-end.**

I. INTRODUCTION

A noise reduction technique is indispensable to achieve acceptable speech quality since the presence of background noise degrades the speech quality. Many approaches including spectral subtraction schemes and adaptive filtering techniques including Kalman and Wiener filtering have been proposed to suppress the background noise [1]-[9].

In VoIP systems, a background noise is one of the important factors degrading the QoS. A single microphone-based noise reduction scheme is required because voice signals are generally acquired by one microphone. Since the end-to-end delay is also the significant QoS factor, the noise reduction scheme should not induce intolerable delays. This paper chooses a Wiener filter-based noise reduction scheme because it satisfies the above constraints and shows a basically high performance as one of the popular approaches. Our Wiener filter is optimized to the estimated SNR at each frequency bin. It is achieved by the use of a logistic function. The ITU-T P.862 PESQ (Perceptual Evaluation of Speech Quality) is used as an objective performance measure [10]. Various noise conditions such as white Gaussian, office, babble, and car noises are considered.

The performance of the proposed method is compared with that of the noise reduction methods in the IS-127 EVRC (Enhanced Variable Rate Codec) [11] because it includes the noise reduction block while VoIP speech codecs have no noise reduction one. And, the performance is compared with that of noise reduction method in the ETSI (European Telecommunications Standards Institute) standard for the distributed speech recognition front-end [12] because it adopts the standardized noise reduction method based on the Wiener filter. Our proposed noise reduction scheme is applied as pre-processing of VoIP speech codecs such as G.711, G.723.1, and G.729A.

The paper is organized as follows. The derivation of the Wiener filtering algorithm is given in Section II and our Wiener filter-based noise reduction is proposed in Section III. In Section IV, our experiments are described and the performance evaluation results are shown. Finally, in Section V, we draw the conclusions including further studies.

II. WIENER FILTERING ALGORITHM

For a noncausal IIR (Infinite Impulse Response) Wiener filter [8], a clean speech signal $d(n)$, a background additive noise $v(n)$, and an observed signal $x(n)$ can be expressed as

$$x(n) = d(n) + v(n)\,. \tag{1}$$

It is assumed that $d(n)$ and $v(n)$ are jointly wide-sense stationary. A Wiener filter is designed to minimize the mean square error

$$\xi = E\left[\left(d(n) - \hat{d}(n)\right)^2\right] = E\left[e^2(n)\right] \tag{2}$$

where $\hat{d}(n)$ is the output of the Wiener filter expressed as

$$\hat{d}(n) \quad \sum_{l=-\infty}^{\infty} w(l)x(n-l) \tag{3}$$

where $w(l)$ is the unit sample response of the Wiener filter. In order to minimize ξ, the derivative of ξ with respect to $w(k)$ should be equal to zero for each k. Thus, we have

$$E\left[d(n)x(n-k)\right] = \sum_{l=-\infty}^{\infty} w(l)E\left[x(n-l)x(n-k)\right]\,. \tag{4}$$

K. Elleithy (ed.), Advances and Innovations in Systems, Computing Sciences and Software Engineering, 393–397.

The equation can be rewritten as

$$r_{dx} = \sum_{l=-\infty}^{\infty} w(l) r_x(k-l). \tag{5}$$

where $r_x(k)$ is the autocorrelation of $x(n)$ and $r_{dx}(k)$ is the cross-correlation between $x(n)$ and $d(n)$. Because it is assumed that $d(n)$ and $v(n)$ are uncorrelated, the autocorrelation and the cross-correlation can be respectively given as

$$r_x(k) = r_d(k) + r_v(k), \tag{6}$$

$$r_{dx}(k) = r_d(k). \tag{7}$$

Thus, (5) can be expressed in the frequency domain as

$$P_d(e^{j\omega}) = W(e^{j\omega})\left[P_d(e^{j\omega}) + P_v(e^{j\omega})\right]. \tag{8}$$

and the frequency response of the Wiener filter becomes

$$W(e^{j\omega}) = \frac{P_d(e^{j\omega})}{P_v(e^{j\omega}) + P_d(e^{j\omega})} = \frac{\zeta(e^{j\omega})}{1+\zeta(e^{j\omega})}. \tag{9}$$

in which $\zeta(e^{j\omega})$ is considered as the SNR (Signal-to-Noise Ratio) defined by

$$\zeta(e^{j\omega}) = \frac{P_d(e^{j\omega})}{P_v(e^{j\omega})}. \tag{10}$$

III. Proposed Wiener Filter

Since a noncausal IIR filter is unrealizable in practice, we propose a causal FIR (Finite Impulse Response) Wiener filter. The proposed noise reduction procedure is shown in Fig. 1.

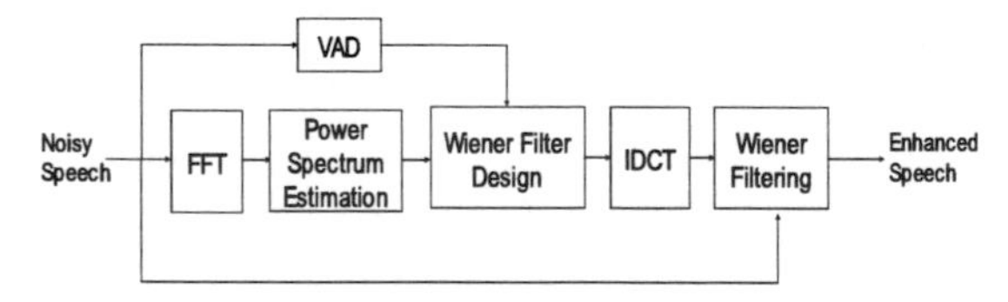

Figure 1. Proposed noise reduction procedure

Noise reduction is processed frame-by-frame. The length of the processing frame is 80 samples or 10 msec. For the power spectrum calculation of the processing frame, total 100 samples, i.e., the current 80 and the past 20 samples, are used. The past samples are initialized to zero in the first frame. There is no buffering delay to analyze the power spectrum. For the power spectrum analysis, the signal is windowed by the 100 sample-length asymmetric window $h(n)$ whose center is located at the 70th sample. The length and the center of the asymmetric window are empirically chosen to make the algorithm produce the best performance. The asymmetric window shown in Fig. 2 is expressed as (11).

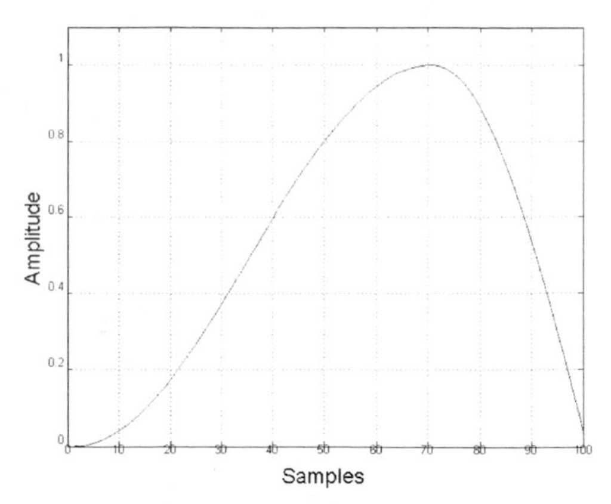

Figure 2. Asymmetric window for power spectrum analysis

$$h(n) = \begin{cases} 0.5 - 0.5\cos\left(\dfrac{2\pi n}{139}\right) & ;0 \le n < 70 \\ \cos\left(\dfrac{2\pi(n-70)}{119}\right) & ;70 \le n < 100 \end{cases} \tag{11}$$

The signal power spectrum is computed for this windowed signal using the 256-FFT. Based on the VAD decision, the noise power spectrum is updated only for non-speech intervals in the Wiener filter design. For speech intervals, the last noise power spectrum is reused. And the speech power spectrum is estimated by the difference between the noise power spectrum and speech power spectrum. In our proposed Wiener filter, the frequency response is expressed as

$$W(k) = \frac{\zeta^{\alpha}(k)}{1+\zeta^{\alpha}(k)}, \qquad 0 < \alpha \le 1. \tag{12}$$

and $\zeta(k)$ is defined as

$$\zeta(k) = \frac{P_d(k)}{P_v(k)}. \tag{13}$$

where k is the frequency bin, $\zeta(k)$, $P_d(k)$, and $P_v(k)$ are the SNR, the speech power spectrum, and the noise power spectrum, respectively. Therefore, filtering can be controlled by the parameter α. As α increases, $\zeta^{\alpha}(k)$ also increases for $\zeta(k)$ greater than one, while $\zeta^{\alpha}(k)$ decreases for $\zeta(k)$ less than one. Therefore, the signal is more strongly filtered out to reduce the noise for small $\zeta^{\alpha}(k)$. On the other hand, the signal is more weakly filtered with little attenuation for large $\zeta^{\alpha}(k)$.

To analyze the effect of α, we evaluate the performances for α values from 0.1 to 1. The performance is evaluated not for the coded speech but for the original speech in white Gaussian conditions. The other experimental conditions are explained in Section IV. As α is increased up to 0.7, the performance is improved. The performance becomes worse after that. Table I shows the PESQ scores according to the

SNR when α is set to 0.6 and 0.7, from now on, will be referred as case A and case B, respectively.

The performance of the case B is worse for low SNRs but better for high SNRs compared with case A. Thus, we can adaptively select the optimal α according to the estimated SNR by a logistic function. The logistic function is trained to decide the optimal α for the estimated SNR at each frequency bin. The logistic function used in this paper can be expressed as

$$p(SNR) = Min + \frac{2(Max - Min)}{1 + e^{(|n-1|/\beta)}}. \tag{14}$$

Because the shape of the logistic function changes with the variation of β, if the maximum and the minimum values of the logistic function are fixed, we should find the appropriate β. Fig. 3 shows the shape of the logistic function when β, *Max*, and *Min* are 1.65, 0.65 and 0.6, respectively.

The appropriate β value is decided by the simple gradient search algorithm [9]. At the first iteration, for the initial β value, the corresponding α as the output of the logistic function is calculated with the estimated SNR as the input of logistic function at each frequency bin for each frame. The average spectral distortion J for all frames is measured with the log spectral Euclidean distance defined as

$$J = \frac{1}{N}\sum_{i=0}^{N-1}\frac{1}{L}\sqrt{\sum_{k=0}^{L-1}\left[\log\left|X_{ref_i}(k)\right| - \log\left\{\left|X_{in_i}(k)\right|W_i(k)\right\}\right]^2}. \tag{15}$$

where i is the frame index, N is the total number of the frames, k is the frequency bin index, L is the total number of the frequency bins, $|X_{ref_i}(k)|$ is the spectrum of the clean reference signal, and $|X_{in_i}(k)|W_i(k)$ is the noise-reduced signal spectrum after filtering with the designed Wiener filter. At the second iteration, β is updated by the simple gradient search procedure. The average spectral distortion is measured with the new logistic function defined by the updated β. Until the termination condition is satisfied, the iteration is repeated. Finally, β is decided after the final iteration.

The designed Wiener filter coefficients in the frequency domain are transformed into the time-domain ones by the IDCT (Inverse Discrete Cosine Transform). Finally, the noise is suppressed by the convolution sum between the impulse response of the proposed Wiener filter and the noisy speech. Because the proposed Wiener filter is a causal filter, the algorithmic delay is unavoidable. However, the delay is almost the half of the filter length and almost negligible compared to the total VoIP delay.

When our proposed algorithm is applied, the highest performance is achieved for all SNR conditions except the 0 dB case, an inconsiderable condition in a practical sense as shown in Table I. Thus, the proposed Wiener filter is believed to reduce the noises effectively.

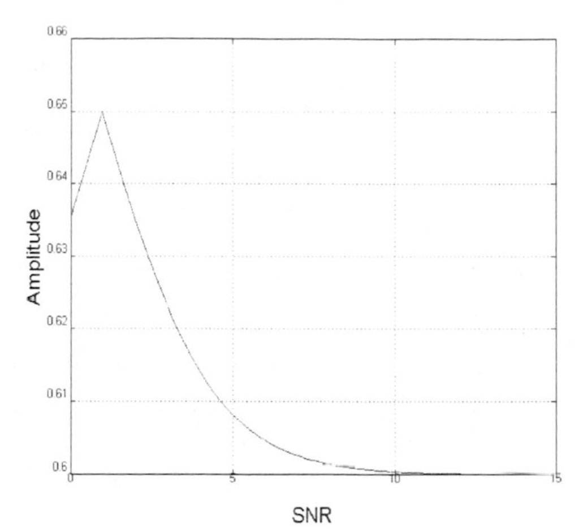

Figure 3. Shape of the logistic function

TABLE I
COMPARISON OF PESQ SCORES

SNR(dB)	20	15	10	5	0
Case A	3.25	2.94	2.61	2.23	1.76
Case B	3.24	2.90	2.60	2.26	1.81
Proposed	3.26	2.94	2.63	2.26	1.80

IV. EXPERIMENTS AND RESULTS

To evaluate the performance, the PESQ is measured as the objective speech quality assessment. After comparing the original signal with its degraded signal, the PESQ gives us the reliable estimation of the subjective measurement as an MOS-like value from -0.5 to 4.5.

In our experiments, one hundred mono spoken sentences sampled at 8 kHz with 16 bit resolution are used as the clean speech. The duration of each utterance is almost 10 seconds. The utterances are spoken by 2 males and 2 females. As the open test, 40 spoken sentences are used to train the logistic function and the others, to evaluate the performance. In the training process, the maximum and the minimum values of the logistic function are fixed as 0.65 and 0.6, respectively. After the training process, β is determined to be 1.65. Thus, in the Wiener filter design, α is decided by the output of the logistic function characterized by these parameter values when the estimated SNR is the input of the logistic function at each frequency bin for each frame. The number of the proposed Wiener filter coefficients is 65. Therefore, the algorithmic delay becomes 4 msec.

Zero mean white Gaussian noises are generated and office, babble, and car noises are recorded with a Sony digital audio tape recorder. Office noises are collected in a room where many computers and an air-conditioner are working. Babble noises are collected in a cafeteria where people make noises. Car noises are collected in a driving car at 70 km/h speed with the windows closed but the air-conditioner turned on. The noise signals are added to the clean speech ones to produce noisy ones with the SNR of 0, 5, 10, 15, and 20 dB. Therefore, total 800 noisy spoken sentences are trained because there are 5 SNR levels, 40 speech utterances, and 4 types of noises.

Noise is reduced as pre-processing before encoding the speech in a codec. Final decoded speech is evaluated by the PESQ. Popular VoIP speech codecs such as G.711, G.723.1, and G.729A are tested.

Our results are compared with those of the noise suppression in the IS-127 EVRC and the noise reduction in the ETSI standard as mentioned. The ETSI noise reduction scheme generates 40 msec buffering delay for the power spectrum analysis while there is no buffering delay in the EVRC noise suppression scheme.

Table II, III, and IV show the PESQ results in G.711, G.723.1, and G.729A, respectively. In most noisy conditions, the performance of the proposed method is higher than those of the EVRC noise suppression method and those of ETSI noise reduction method.

TABLE II
PESQ RESULTS IN G.711

Noise type	SNR (dB)	None	EVRC	ETSI	Proposed
White Gaussian noise	0	1.42	1.39	1.69	1.83
	5	1.61	1.83	2.09	2.28
	10	1.89	2.32	2.48	2.64
	15	2.20	2.73	2.84	2.96
	20	2.52	3.07	3.17	3.27
Office noise	0	1.44	1.64	1.64	1.73
	5	1.76	2.02	2.04	2.12
	10	2.11	2.42	2.44	2.52
	15	2.44	2.78	2.80	2.88
	20	2.77	3.13	3.18	3.24
Babble noise	0	1.51	1.67	1.73	1.75
	5	1.79	2.04	2.11	2.12
	10	2.11	2.45	2.48	2.48
	15	2.40	2.76	2.76	2.81
	20	2.73	3.11	3.14	3.18
Car noise	0	2.28	2.57	2.55	2.63
	5	2.64	2.90	2.92	3.01
	10	2.97	3.22	3.26	3.30
	15	3.28	3.52	3.60	3.65
	20	3.61	3.84	3.85	3.89

Table V shows the average PESQ gains for all types of noises. We can observe that our proposed method is superior to the others. Significant PESQ gains are achieved when the proposed method is compared with the EVRC noise suppression one. Furthermore, the proposed method is slightly better than the ETSI noise reduction one. Also, our proposed method is effective for the noise reduction of the SNR from 5 dB to 15 dB. In G.711 based on a waveform-based coding technique, higher PESQ gains are achieved by our method.

Table VI shows the average PESQ gains for all test speech codecs. The proposed method is most effective to reduce the white Gaussian noises. The performance of the proposed method is slightly worse than that of ETSI noise reduction one while being better than that of EVRC noise suppression one.

TABLE III
PESQ RESULTS IN G.723.1 (6.3KBPS)

Noise type	SNR (dB)	None	EVRC	ETSI	Proposed
White Gaussian noise	0	1.44	1.40	1.78	1.86
	5	1.70	1.90	2.20	2.27
	10	2.03	2.36	2.55	2.58
	15	2.34	2.70	2.85	2.84
	20	2.64	2.95	3.07	3.09
Office noise	0	1.54	1.68	1.72	1.77
	5	1.91	2.07	2.12	2.14
	10	2.28	2.46	2.51	2.51
	15	2.58	2.77	2.82	2.83
	20	2.87	3.05	3.09	3.11
Babble noise	0	1.59	1.72	1.79	1.76
	5	1.91	2.11	2.19	2.14
	10	2.27	2.48	2.52	2.46
	15	2.55	2.77	2.79	2.79
	20	2.84	3.04	3.07	3.06
Car noise	0	2.22	2.36	2.41	2.54
	5	2.55	2.68	2.75	2.90
	10	2.84	2.97	3.02	3.12
	15	3.09	3.21	3.26	3.34
	20	3.29	3.40	3.40	3.45

TABLE IV
PESQ RESULTS IN G.729A

Noise type	SNR (dB)	None	EVRC	ETSI	Proposed
White Gaussian noise	0	1.43	1.40	1.79	1.85
	5	1.71	1.93	2.23	2.27
	10	2.06	2.40	2.59	2.58
	15	2.40	2.74	2.89	2.86
	20	2.71	2.98	3.13	3.14
Office noise	0	1.51	1.67	1.67	1.70
	5	1.92	2.08	2.09	2.09
	10	2.31	2.48	2.51	2.52
	15	2.63	2.80	2.84	2.84
	20	2.93	3.09	3.13	3.14
Babble noise	0	1.59	1.74	1.77	1.74
	5	1.94	2.13	2.19	2.13
	10	2.30	2.51	2.54	2.46
	15	2.60	2.80	2.83	2.82
	20	2.90	3.08	3.11	3.09
Car noise	0	2.23	2.40	2.37	2.51
	5	2.57	2.74	2.73	2.89
	10	2.87	3.06	3.03	3.16
	15	3.15	3.32	3.30	3.39
	20	3.36	3.49	3.46	3.51

TABLE V
AVERAGE PESQ GAINS FOR ALL TYPES OF NOISE

Codec	SNR (dB)	EVRC	ETSI	Proposed
G.711	0	0.16	0.24	0.32
	5	0.25	0.35	0.43
	10	0.33	0.40	0.47
	15	0.37	0.42	0.50
	20	0.38	0.43	0.49
G.723.1 (6.3 kbps)	0	0.09	0.23	0.28
	5	0.17	0.29	0.34
	10	0.22	0.30	0.32
	15	0.22	0.29	0.31
	20	0.20	0.25	0.27
G.729A	0	0.11	0.21	0.26
	5	0.19	0.28	0.31
	10	0.23	0.28	0.29
	15	0.22	0.27	0.28
	20	0.19	0.23	0.25

TABLE VI
PESQ GAINS FOR ALL TEST VOIP SPEECH CODECS

Noise type	SNR (dB)	EVRC	ETSI	Proposed
White Gaussian noise	0	-0.03	0.32	0.42
	5	0.21	0.50	0.60
	10	0.37	0.55	0.61
	15	0.41	0.55	0.57
	20	0.38	0.50	0.54
Office noise	0	0.17	0.18	0.24
	5	0.19	0.22	0.25
	10	0.22	0.25	0.28
	15	0.23	0.27	0.30
	20	0.23	0.28	0.31
Babble noise	0	0.15	0.20	0.19
	5	0.21	0.28	0.25
	10	0.25	0.29	0.24
	15	0.26	0.28	0.29
	20	0.25	0.28	0.29
Car noise	0	0.20	0.20	0.32
	5	0.19	0.21	0.35
	10	0.19	0.20.	0.30
	15	0.18	0.21	0.29
	20	0.16	0.15	0.20

V. CONCLUSIONS

This paper proposes a new Wiener filtering scheme optimized to the estimated noisy signal SNR at each frequency bin by the logistic function to reduce additive noises. The proposed noise reduction scheme is applied before encoding as pre-processing of VoIP speech codecs. For all noisy conditions, average PESQ gains of 0.44, 0.30, and 0.28 are achieved for G.711, G.723.1, and G.729A, respectively, by the proposed method. The PESQ results show that the performance of the proposed method outperforms those of the EVRC noise suppression one and the ETSI noise reduction one. Thus, our proposed noise reduction can be effectively used to reduce additive background noises in VoIP applications.

The other speech distortion measures such as the Kullback-Liebler distance and the Bark spectral distance can be considered as future studies to find more appropriate distortion measure incorporating with human perceptual characteristics. And, a filter coefficient smoothing method can also be considered to prevent an abrupt filter coefficient change.

REFERENCES

[1] S. F. Boll, "Suppression of acoustic noise in speech using spectral subtraction," *IEEE Trans. Acoust., Speech, Signal Process.*, vol. 27, no. 2, pp. 113-120, Apr. 1979.

[2] R. J. McAulay and M. L. Malpass, "Speech enhancement using a soft decision noise suppression filter," *IEEE Trans. Acoust., Speech, Signal Process.*, vol. 28, no. 2, pp. 137-145, Apr. 1980.

[3] Y. Ephraim and D. Malah, "Speech enhancement using a minimum mean-square error short-time spectral amplitude estimator," *IEEE trans. Acoust., Speech, Signal Process.*, vol. 32, no. 6, pp. 1109-1121, Dec. 1984.

[4] Y. Ephraim and D. Malah, "Speech enhancement using a minimum mean-square error log-spectral amplitude estimator," *IEEE trans. on Acoustics, Speech, and Signal Processing*, vol. 33, no. 2, pp. 443-445, April 1985.

[5] K. K. Paliwal and A. Basu, "A speech enhancement method based on Kalman filtering," *Proc. ICASSP'87*, pp. 177-180, 1987.

[6] S. Gannot, D. Burshtein, and E. Weinstin, "Iterative and sequential Kalman filter-based speech enhancement algorithms," *IEEE Trans. Speech Audio Process.*, vol. 6, no. 4, pp. 373-385, Jul. 1998.

[7] R. Martin, "Noise power spectral density estimation based on optimal smoothing and minimum statistics," *IEEE Trans. Speech Audio Processing*, vol. 9, no. 5, pp. 504-512, Jul. 2001.

[8] M. H. Hayes, *Statistical Digital Signal Processing and Modeling*, John Wiley&Sons, 1996.

[9] B. Widrow and S. D. Stearns, *Adaptive Signal Processing*, Englewood Cliffs, NJ 07632: Prentice Hall, 1985.

[10] Perceptual evaluation of speech quality (PESQ), an objective method for end-to-end speech quality assessment of narrow-band telephone networks and speech codec, *ITU-T Recommend. P.862*, Feb. 2001.

[11] Enhanced variable rate codec, *Speech service option 3 for wide-band spectrum digital systems*, 1996.

[12] Speech processing, transmission and quality aspects (STQ); distributed speech recognition; extended advanced front-end feature extraction algorithm; compression algorithms; back-end speech reconstruction algorithm, *ETSI ES 202 212*, Nov. 2005.

A Formal Framework for "Living" Cooperative Information Systems

Shiping Yang, Martin Wirsing
Institut für Informatik, Ludwig-Maximilians-Universität München,
Oettingenstr. 67, München 80538, Germany
{yangs, mwirsing}@pst.informatik.uni-muenchen.de

***Abstract* - This paper constructs a high-level Abstract State Machine (ASM) model of our conceptual software architecture for "living" cooperative information systems founded in living systems theory. For practical execution, we use AsmL, the Abstract state machine Language developed at Microsoft Research and integrated with Visual Studio, to refine the ASM model to an executable system model for evaluation.**

I. Introduction

Today's business environment is becoming more complex, organizations are facing increased competition, global challenges, and market shifts together with rapid technological developments. A successful business is constantly adapting to change, and therefore so should the information systems, which must support rather than hinder this changeability. The problem then is not how to build information systems which share goals with their organizational environment, human users, and other existing systems at some time point. Rather, the problem is how to build information systems which continue to share goals with their organizational environment, human users, and other existing systems as they all evolve. Therefore, what is required for "living" businesses in a changing world are "living" cooperative information systems (CIS), which should be more flexible to respond quickly to the needs of business as they evolve.

However, the way in which to design, build, integrate, and maintain information systems that are flexible, reusable, resilient, and scalable is now becoming well understood but not well supported. In [1], we presented a conceptual software architecture (see Fig. 1) founded in Living Systems Theory (LST) to build "living" CIS for virtual organizations (VOs), where both the organizational environment of a system and the system itself are described in terms of the same concept: peer. This paper uses the Abstract State Machine (ASM) paradigm to formalize the informal software architecture for all peers.

The remaining of this paper is organized as follows. Section II gives a brief overview of the informal software architecture for all peers introduced in [1]. In section III, this architecture is firstly described in the form of an UML (Unified Modeling Language) class diagram, and then is formalized as a high level ASM model. Section IV presents the Graphical User Interface (GUI) of the refined ASM model using the Abstract state machine Language (AsmL). Based on this executable system prototype, the result of our case studies introduced in [1] is discussed. Finally, section V concludes this paper and outlines our topics for further research.

II. The Software Architecture for All Peers

For the purpose of understanding our formal specification, this section gives a brief overview of the informal software architecture for all peers (see Fig. 1) introduced in [1]. This architecture is built atop the internet infrastructure, and consists of 6 critical subsystems: connector, request manager, resource manager, grid, resources, and user. These subsystems carry out the essential services of a peer, and are integrated together to form an actively self-regulating, developing, unitary system with purposes and goals.

A. Connector

Connectors are interfaces through which peers can interact with their environment. Any communication with a peer must go through the connector. The responsibilities of connectors are to check and package messages destined for other peers in the environment, and to receipt and interpret messages from other peers and from a peer's internal components, e.g. request manager, resources. The connector subsystem can be paralleled with, respectively, the input transducer subsystem, decoder subsystem, encoder subsystem, output transducer subsystem, or the combination of these subsystems defined in LST [2, 3].

B. Grid

The grid subsystem provides services such as connectivity, event, policy-based control, resource reservation etc. It enables the "plug-and-play" of kinds of resources, and the exchange of data and coordination between plug-in resources. It thus allows custom configuration of services for a peer, in most cases even while the peer is running. The grid subsystem can be paralleled with the channel-net subsystem defined in LST [2, 3].

C. Request Manager

The request manager subsystem receives the parsed request from the underlying connector and, based on the content, invokes services provided by resources within this peer to handle the request. It provides services such as workflow,

K. Elleithy (ed.), Advances and Innovations in Systems, Computing Sciences and Software Engineering, 399–404.

scheduling, and brokering. Request manager functions as the associator subsystem defined in LST [2, 3].

D. Resource Manager

The resource manager subsystem is a peer's repositories for knowledge about itself and resources in its environment, e.g. the services provided by resources, the status of resources, and the groups of resources. It automatically discovers (locates) all the resources "plugging" into the environment, once the resources are identified. The resource manager subsystem can be paralleled with the memory subsystem defined in LST [2, 3].

E. Resources

For us, a resource is a provider of a service, such as software, hardware, and even human. Either type of resource is represented by a peer service. Note that the role that is assigned to a person in social organizations can also be represented as a (composite) service. Resources can represent entities owned or managed by a peer. They implement the local, resource-specific operations that occur on specific resources (whether physical or logical) and the interfaces that interact with users. For example, a printer may be a physical resource; a distributed file system is a logical resource. Resource subsystem can be paralleled with the internal transducer subsystem defined in LST [2, 3].

F. User

Frequently a peer will be under the control of a human operator, the user subsystem, and will interact with the network on the basis of the user's direction. User can be any individual or collection of individuals who receive messages, associate them with past experience or knowledge, and then choose a course of action that may alter the behavior or state of the system or its components. The user subsystem can be paralleled with the decider subsystem defined in LST [2, 3].

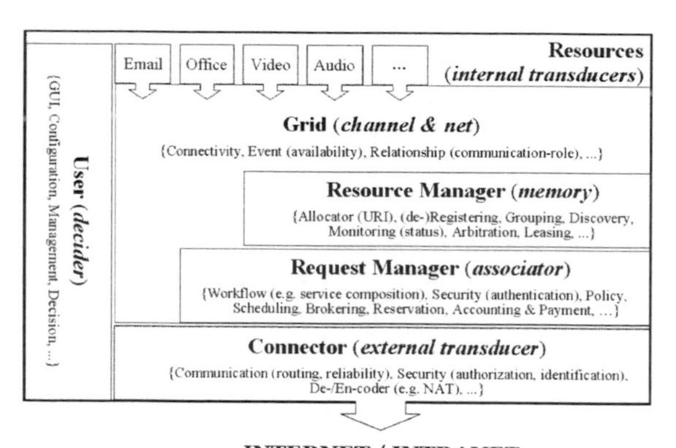

Fig. 1: The general software architecture for all peers

III. Abstract State Machine Models

A reasonable choice for the construction of an abstract CIS model is a distributed real-time ASM consisting of an arbitrary number of concurrently operating and asynchronously communicating components. ASM, formerly known as Evolving Algebras, have been introduced to bridge the gap between formal models of computation and practical specification methods. The result is a formal method for transparent design and specification of discrete dynamic systems. ASM combines advantages of informal methods (understandability, executability) with the advantages of formal methods (precision and applicability of mathematical methods and results). For detailed information on ASM, please see e.g., [4], [5], [6].

Based on the architecture defined in the previous section, we can now define the high-level ASM models for each subsystem of a peer, which behaviors can be defined by the composition of its subsystems' ASM models. Note that we introduce an additional GUI (see section IV) that allows for user-controlled interaction between the user subsystem and the other subsystems.

A. Overview

Intuitively, a component either represents a peer or some fraction of the underlying communication network. At the component level, peers are further decomposed, where each individual component splits into a collection of synchronously operating functional units, i.e. the subsystems. This decomposition is such that each of the resulting units provides specific services. Accordingly, we model peers as parallel compositions of synchronously operating ASMs.

Given above consideration, an overview of the various component types and the relations between them is presented in the form of an UML class diagram in Fig. 2.

- *Agents*

We formulate dynamic properties of the peer in terms of component interactions. Components operate autonomously and so that we can identify them with ASM *agents* in the distributed ASM. Agents come as elements of a dynamically growing and shrinking universe (or *domain*) *AGENT*, where we associate with each agent a universal unique identifier from a dynamic domain *UUID*, and a *program* defining its behavior. A program consists of guarded update *rules* specifying state transitions through local updates on global states. We distinguish different types of agents according to different types of programs as represented by a static universe *PROGRAM*.

```
domain AGENT
static domain PROGRAM
domain UUID
uuid : AGENT -> UUID
```

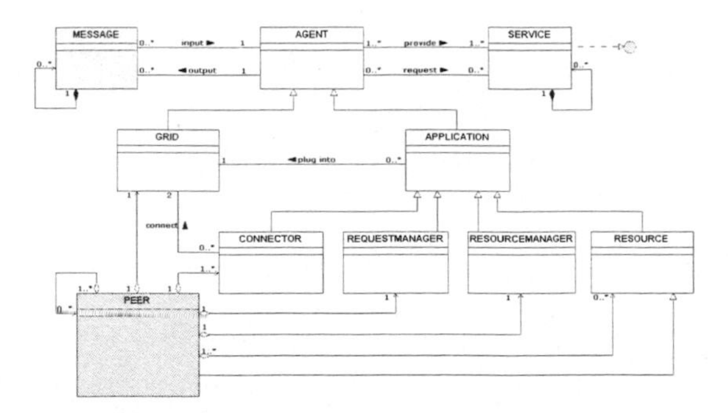

Fig. 2: UML class diagram of "living" CIS

In any given state of an ASM run, the behavior of an agent is well defined as stated by a unary dynamic function *program*. Being dynamic, this function allows introducing new agents at run time.

```
program: AGENT -> PROGRAM
forall a ∈ AGENT : program(a) ∈ PROGRAM
```

In our systems, there are two basic types of *AGENT*: *GRID*, and *APPLICATION*, which includes the other four subsystems of a peer: connector, request manager, resource manager, and resource. Each application is plugged into a specific grid of a peer. The operation of sending a message as well as the delivery of a message both require some form of direct interaction between the application and its local grid, which is uniquely determined by a peer. Abstractly, this relation is expressed using a few unary dynamic functions e.g. *grid* defined on applications.

```
AGENT ≡ GRID ∪ APPLICATION
APPLICATION ≡ CONNECTOR ∪ REQUESTMANAGER ∪
              RESOURCEMANAGER ∪ RESOURCE
grid : APPLICATION -> GRID
```

In LST [3], Miller identified 8 distinct levels for all living systems: cells, organs, organisms, groups, organizations, communities, societies, supranational systems. In our "living" CIS for VOs, we will be concerned primarily with the subset of human systems and with the middle levels for all peers in the context of VOs: individuals, groups, organizations, communities, and societies. Similarly, every peer belongs to one of the 5 defined levels and is composed of components that themselves are peers on the next lower level. These organizational levels are also characterized by the fact that each level has the same types of components but different specializations. Note that peer is also a type of resource providing services.

```
RESOURCE ≡ SOFTWARE ∪ HARDWARE ∪ PEER
```

```
PEER ≡ INDIVIDUAL ∪ GROUP ∪ ORGANIZATION ∪
       COMMUNITY ∪ SOCIETY
```

- *Service*

We identify the services associated with an agent using a static function *SERVICEs* defined on agents. Whether a particular peer can access a specific service or not is defined by a predicate *canAccess*.

```
domain SERVICE
SERVICEs : AGENT -> Set of SERVICE
canAccess : RESOURCE × SERVICE -> BOOL initially
                                        false
svcStatus : SERVICE -> {Initialized, Ready, Running,
                     Suspended, Terminated, Completed}
provideService : AGENT × SERVICE -> BOOL
executeService : SERVICE × REQUEST -> RESULT
```

Every service has specific properties or *attributes* that may be used to check whether one service is compatible with another.

```
domain ATTRIBUTE
attribute : SERVICE -> ATTRIBUTE
compatible : ATTRIBUTE × ATTRIBUTE -> BOOL initially
                                            false
```

- *Messages*

Messages are uniformly represented as elements of a dynamic domain *MESSAGE*. Each message is of a certain type from the static domain *MSGTYPE*. The message type determines a specific message subject from the dynamic domain *SUBJECT*, and how a message is transmitted using different protocols such as UDP or TCP, though we do not make this distinction explicit.

```
domain MESSAGE initially empty
domain MSGTYPE ≡ {Advertise, Request, Response, ...}
domain SUBJECT ≡ {Peer, Resource, Service, ...}
```

A message uniquely identifies a sender, a receiver, a message type, a message subject, and the actual message content. The content can be any finite representation of information to be transferred from a sender to a receiver.

```
sender : MESSAGE -> AGENT
receiver : MESSAGE -> AGENT
type : MESSAGE -> MSGTYPE
subject : MESSAGE -> SUBJECT
content : MESSAGE -> DATA
```

To limit the maximum number of peers that a message can pass on its way from the sender to a destination, a time-to-live (*ttl*) is assigned when the message is created. Each connector decrements the *ttl* by one until the message eventually reaches its final destination or will be discarded.

```
ttl : MESSAGE -> {0,1,2,3,4} initially 4
```

In addition, an agent has a local mailbox for storing messages until these messages will be processed.

```
mailbox : AGENT -> Set of MESSAGE initially empty
```

The operation of delivering a message to the mailbox of a given agent is defined below. All subsystems and the peer itself are treated uniformly. They are both agents that have a mailbox and the operation performed on this mailbox (e.g., inserting a copy of some message) does not depend on the particular type of agent.

```
DeliverMessageTo(ag:AGENT, msg:MESSAGE) =
  extend MESSAGE with m
    sender(m) := sender(msg)
    receiver(m) := receiver(msg)
    type(m) := type(msg)
    subject(m) := subject(msg)
    content(m) := content(msg)
    ttl(m) := ttl(msg) - 1
    mailbox(ag) := mailbox(ag) ∪ {m}
```

The following *output* operation is defined for subsystems such as connector, request manager, resource manager, and resources, to send a message over a local grid. Here *me* refers to the agent sending the message. The *extend* operation below creates a new message object. The effect of the send operation is that the message is put into the *mailbox* of the grid subsystem.

```
output(rcv: AGENT, mtp: MSGTYPE, sbj: SUBJECT,
      cont: DATA) =
  extend MESSAGE with m
    sender(m) := me
    receiver(m) := rcv
    subject(m) := sbj
    type(m) := mtp
    content(m) := cont
    mailbox(grid(me)) := mailbox(grid(me)) ∪ {m}
```

The following *reply* operation is defined for applications to send a response message, which identifier is the same one as that of the request message.

```
reply(msg: MESSAGE, cont: DATA) =
  extend MESSAGE with m
    id(m) := id(msg)
    sender(m) := uuid(me)
    receiver(m) := sender(msg)
    subject(m) := subject(msg)
    type(m) := Respond
    content(m) := cont
    mailbox(grid(me)) := mailbox(grid(me)) ∪ {m}
```

B. Initial States

An initial state reflects the particular system configuration under consideration. As such it identifies some finite collection of a priori given agents, one for each connector, each grid, each request manager, each resource manager, and each resource. Depending on the type of an agent, it executes the relevant program *runCONNECTOR*, *runGRID*, *runREQUESTMANAGER*, *runRESOURCEMANAGER*, *runRESOURCE*, or *runPEER*.

In every state of an ASM run, including the initial state, the following property holds:

```
∀a ∈ AGENT : program(a) =
  runCONNECTOR, if a ∈ CONNECTOR
  runGRID, if a ∈ GRID
  runREQUESTMANAGER, if a ∈ REQUESTMANAGER
  runRESOURCEMANAGER, if a ∈ RESOURCEMANAGER
  runRESOURCE, if a ∈ RESOURCE
  runPEER, if a ∈ PEER
```

C. Grid Model

Based on above specification, we would like to use the grid subsystem as an example to explain the formalization of our system, which is presented in details in [8].

As discussed in section II, the grid subsystem handles the delivery of messages to destinations within a local peer. That is, given the destination *UUID* of a message in conjunction with a local group, a (possibly empty) set of related destinations on this group must be identified. We therefore introduce a dynamic mapping of UUIDs called *addressTable*, which is a mapping from UUIDs of multicast groups to the related group members that are plugged into a grid by a connector. Therefore, we identify the resources owned by a peer using a static function *RESOURCEs*, and associated connectors, request manager and resource manager with a peer using a static function *CONNECTORs*, *requestManager* and *resourceManager* defined on a grid, respectively.

```
addressTable : GRID -> Map of UUID to CONNECTOR
RESOURCEs : GRID -> Set of CONNECTOR initially empty
CONNECTORs : GRID -> Set of CONNECTOR initially empty
requestManager : GRID -> REQUESTMANAGER
resourceManager : GRID -> RESOURCEMANAGER
```

The behaviors of the grid subsystem can thus be defined as the following program:

```
runGRID ≡
  forall msg ∈ mailbox(me): readyToDeliver(msg) do
    if receiver(msg) = broadcast then
      if subject(msg) ∈ {Addressing, Discovery} then
        deliverMessageTo(resourceManager(me), msg)
      else
        forall r ∈ RESOURCEs: status(r)≠InActive do
          deliverMessageTo(r, msg)
    else
      if receiver(msg) = me then
        ... //deliver message based on proper rules
      else
        let members = RESOURCEs(me) ∪
                      {resourceManager(me),
                       requestManager(me)}
        choose m ∈ members: receiver(msg) = m and
                            status(m) ≠ InActive) do
          deliverMessageTo(m, msg)
        ifnone
          if ttl(msg) > 0 then
            choose con ∈ CONNECTORs(me):
                receiver(msg) ∈ routingTable(con) do
              deliverMessageTo(con, msg)
    mailbox(me) := mailbox(me) - {msg}
```

D. Peer Model

In our systems, individuals, groups, organizations, communities, and societies are all peers, each of which is composed of the 6 defined subsystems: connector, grid, request manager, resource manager, resources, and user (the external world). The behaviors of every peer are thus defined as follows:

```
runPEER ≡
  if status(me) ≠ InActive then
    runCONNECTOR()
    runGRID()
    runREQUESTMANAGER()
    runRESOURCEMANAGER()
    runRESOURCE()
```

IV. An Executable System Prototype

The ASM model defined in the section III can be refined into an executable AsmL program, which is written in an object-oriented style. Our prototype implementation utilizes the language interoperability features of Microsoft Visual Studio .NET platform, which integrates AsmL into the C# project system. Moreover, a GUI is created using Visual C# .NET. The interoperability between these components is made possible through the common language runtime that is integrated with the .NET platform [7]. Based on our executable system prototype, this section presents the result of the case studies introduced in [1].

A. Graphical User Interface

It is important to have a GUI that allows us to visualize the state in a way that is close to the intuitive understanding, and that allows us to interact with the AsmL model. Fig. 3 shows a snapshot of one peer GUI in a setup of the case study (see Fig. 4) for the Beijing Organizing Committee for the Games of the XXIX Olympiad (BOCOG) with 12 separate peers, one for Society (bottom), 4 for organizations, 7 for individuals (see Fig. 5).

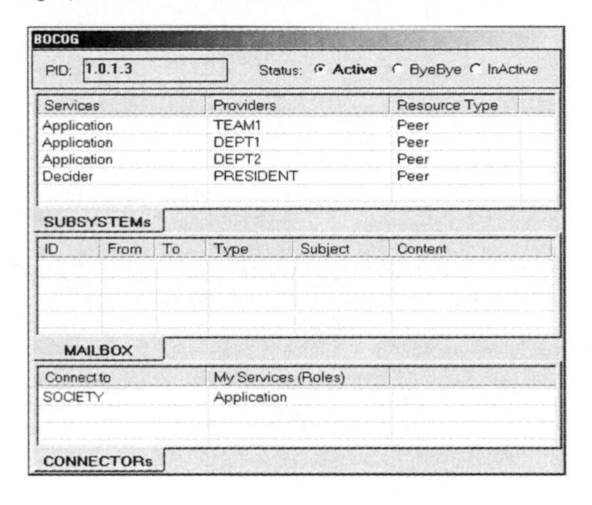

Fig. 3: Snapshot of the peer GUI

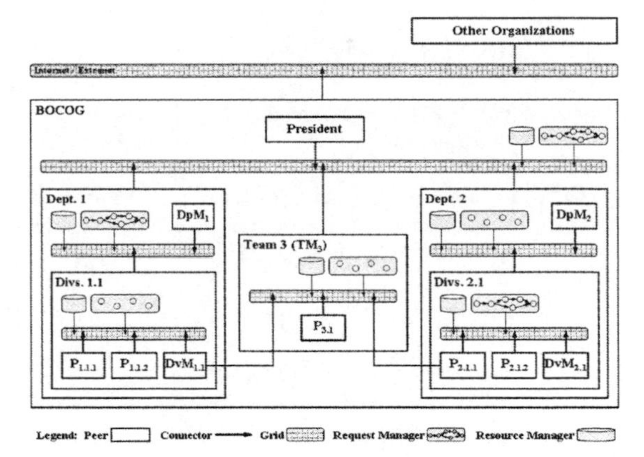

Fig. 4: A case study for BOCOG

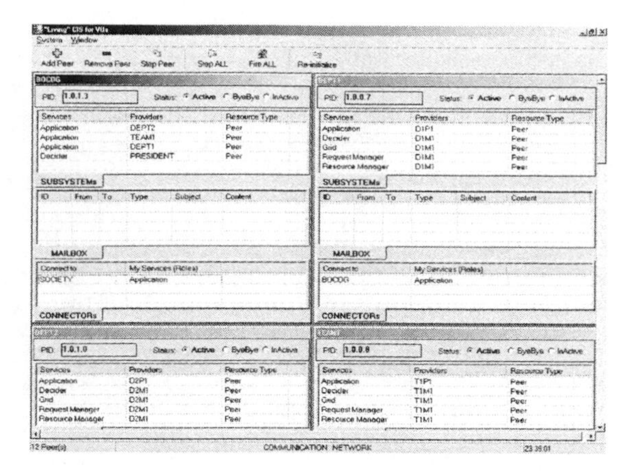

Fig. 5: Snapshot of the System Prototype GUI

B. Discussion

The runs of the high-level executable AsmL model are controlled and inspected at various levels of detail as required for e.g. simulation and conformance testing, by means of a GUI. The case studies demonstrate the advantages of our approach that supports changing organization, dynamic business processes, and decentralized ownerships of resources [1].

The resulting formalization of our AsmL model captures all the significant aspects of a peer: extensibility, autonomy, communication, and dynamics. The abstract operational view of ASMs thereby allows a seamless integration of control and data-oriented aspects of behavior specifications. Moreover, we combine synchronous and asynchronous execution paradigms, associated with the peer programs and the communication network respectively, in one common model. In that sense, there are no conceptual difficulties to include the missing details as they all rely on the same architectural model.

V. Conclusion and Future Work

This paper constructs a high level ASM model of the general software architecture for all peers introduced in [1]. This ASM model serves as a conceptual framework for dealing with system behavior at a more abstract level. Conceptually, the ASM model is designed to meet the fundamental requirements on technical descriptions of our "living" CIS for VOs [1], namely:

- *Extensible*. System should be open and can be constructed from parts and survive over time;
- *Autonomous*. System should support decentralized ownership of resources, and different levels of freedom and perspectives.
- *Accessible*. System should enable seamless connections among people, software agents, and various kinds of IT systems.
- *Dynamic*. System (and its model) should accurately represent the organization at all times.

This high level model with a formal semantics can be realized in multiple implementations, specialized as necessary for particular platforms. Note that our system architecture is based on the high level abstraction of the underlying communication network. That is, it does not cover other non-functional requirements such as security, performance, and usability, especially under the real operation environment. Nevertheless, these form a good base for our following future work.

Therefore, the main direction for our future research is obviously the development of a distributed usable software system to further evaluate our architectural approach for building "living" CIS in a lab environment, as well as in a real business environment if possible, where the system should furthermore provide mechanisms for security and robustness to enable real-world applications, beyond pilot implementations.

References

[1] S. Yang, M. Wirsing: Towards Living Cooperative Information Systems for Virtual Organizations Using Living Systems Theory. In *Proc. of the 1st International Joint e-Conferences on Computer, Information, and Systems Sciences, and Engineering (CISSE'05)*, December 10-20, 2005

[2] J. G. Miller: *Living Systems*. McGraw-Hill, 1978, ISBN 0-07-042015-7

[3] J. G. Miller: *Living Systems*. Colorado: University Press of Colorado, 1995

[4] E. Börger, R. F. Stärk: *Abstract State Machines: A Method for High-Level System Design and Analysis*. Springer, 2003

[5] Y. Gurevich: Evolving Algebras 1993: Lipari Guide. In E. Börger (eds.), *Specification and Validation Methods*, pp.9–36. Oxford University Press, Oxford, UK, 1995

[6] Y. Gurevich: *May 1997 Draft of the ASM Guide*, Technical Report CSE-TR-336-97, EECS Department, University of Michigan, 1997, http://www.eecs.umich.edu/gasm/papers/guide97.html

[7] Microsoft Corporation: *AsmL: The Abstract State Machine Language*, Foundations of Software Engineering -- Microsoft Research, 2002, http://research.microsoft.com/fse/asml/

[8] S. Yang: *Towards "Living" Cooperative Information Systems For Virtual Organizations: Based On Peers, Founded In Living Systems Theory*, unpublished PhD Thesis, University of Munich, Germany, 2006

Crime Data Mining

Shyam Varan Nath
Principal Consultant, Oracle Corporation
Shyam.Nath@Oracle.com
+1(954) 609 2402

Abstract

Solving crimes is a complex task and requires a lot of experience. Data mining can be used to model crime detection problems. The idea here is to try to capture years of human experience into computer models via data mining. Crimes are a social nuisance and cost our society dearly in several ways. Any research that can help in solving crimes faster will pay for itself. According to Los Angeles Police Department, about 10% of the criminals commit about 50% of the crimes. Here we look at use of clustering algorithm for a data mining approach to help detect the crimes patterns and speed up the process of solving crime. We will look at k-means clustering with some enhancements to aid in the process of identification of crime patterns. We applied these techniques to real crime data from a sheriff's office and validated our results. We also used semi-supervised learning technique here for knowledge discovery from the crime records and to help increase the predictive accuracy. Our major contribution is the development of a weighting scheme for attributes, to deal with limitations of various out of the box clustering tools and techniques. This easy to implement data mining framework works with the geo-spatial plot of crime and helps to improve the productivity of the detectives and other law enforcement officers. It can also be applied for counter terrorism for homeland security.

Keywords: *Crime-patterns, clustering, data mining, k-means, law-enforcement, semi-supervised learning*

1. Introduction

Historically solving crimes has been the prerogative of the criminal justice and law enforcement specialists. With the increasing use of the computerized systems to track crimes, computer data analysts have started helping the law enforcement officers and detectives to speed up the process of solving crimes. Here we will take an inter-disciplinary approach between computer science and criminal justice to develop a data mining paradigm that can help solve crimes faster. More specifically, we will use clustering based models to help in identification of crime patterns[1].

We will discuss some terminology that is used in criminal justice and police departments and compare and contrast them relative to data mining systems. Suspect (also called subject or target someimes) refers to the person that is believed to have committed the crime. The suspect may be identified or unidentified. The suspect is not a convict until proved guilty. The victim is the person who suffers due to the crime. Most of the time the victim is identifiable and in most cases is the person reporting the crime. Additionally, the crime may have some witnesses. There are other words commonly used such as homicides that refer to manslaughter or killing someone. Within homicides there may be categories like infanticide, eldercide, killing intimates and killing law enforcement officers. For the purposes of our modeling, we will not need to get into the depths of criminal justice but will confine ourselves to the major kinds of crimes for crime patterns.

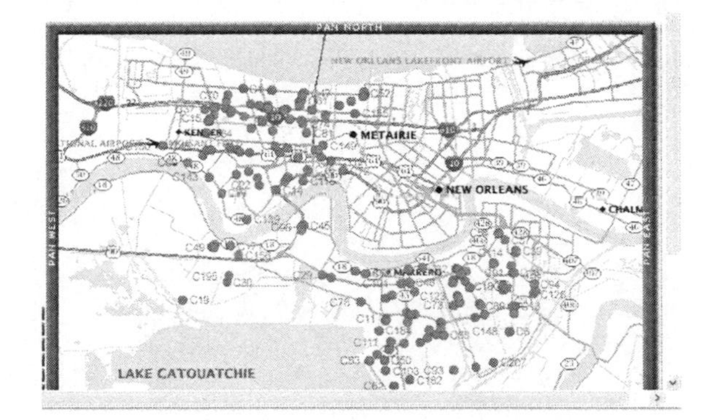

Fig 1 Geo-spatial plot of crimes, each red dot represents a crime incident.

Cluster (of crime) has a special meaning and refers to a geographically close group of crime, i.e. a lot of crimes in a given geographical region. Such clusters can be visually represented using a geo-spatial plot of the crime overlayed on the map of the police jurisdiction. The densely populated group of crime is used to visually locate the 'hot-spots' of crime. Fig 1 shows the red dots as crime incidents and a close group of red dots represent a cluster or hot spot of crime. It is the geo-spatial plot of clusters of crime. However, when we talk of clustering

K. Elleithy (ed.), Advances and Innovations in Systems, Computing Sciences and Software Engineering, 405–409.

from a data-mining standpoint, we refer to similar kinds of crime in the given geography of interest. Such clusters are useful in identifying a crime pattern or a crime spree. Some well-known examples of crime patterns are the DC sniper, a serial-rapist or a serial killer. These crimes may involve single suspect or may be committed by a group of suspects.

2. Crime Reporting Systems

The data for crime often presents an interesting dilemma. While some data is kept confidential, some becomes public information. Data about the prisoners can often be viewed in the county or sheriff's sites. However, data about crimes related to narcotics or juvenile cases is usually more restricted. Similarly, the information about the sex offenders is made public to warn others in the area, but the identity of the victim is often prevented. Thus as a data miner, the analyst has to deal with all these public versus private data issues so that data mining modeling process does not infringe on these legal boundaries.

Most sheriffs' office and police departments use electronic systems for crime reporting that have replaced the traditional paper-based crime reports. These systems are generically called Record Management System of RMS. The crime reports out of these systems have the following kinds of information categories namely - type of crime, date/time, location etc. Then there is information about the suspect (identified or unidentified), victim and the witness. Additionally, there is the narrative or description of the crime and Modus Operandi (MO) that is usually in the text form. The police officers or detectives use free text to record most of their observations that cannot be included in checkbox kind of pre-determined questions. While the first two categories of information are usually stored in the computer databases as numeric, character or date fields of table, the last one is often stored as free text.

The challenge in data mining crime data often comes from the free text field that is also called the unstructured data. While free text fields can give the newspaper columnist, a great story line, converting them into data mining attributes is not always an easy job. To overcome this issue we use text searching/mining tools. Commercial databases like Oracle database contain the text engine and help to faciliate such tasks. Since crime suspects often use different names or aliases, text search is also useful in name searches with nicknames or partial names or other slight variations of names. We will look at how to arrive at the significant attributes for the data mining models.

3. Data Mining and Crime Patterns

We will look at how to convert crime information into a data-mining problem [2][5], such that it can help the detectives in solving crimes faster. We have seen that in crime terminology a cluster is a group of crimes in a geographical region or a hot spot of crime. Whereas, in data mining terminology a cluster is group of similar data points – a possible crime pattern. Thus appropriate clusters or a subset of the cluster will have a one-to-one correspondence to crime patterns.

Thus clustering algorithms in data mining are equivalent to the task of identifying groups of records that are similar between themselves but different from the rest of the data. In our case some of these clusters will useful for identifying a crime spree committed by one or same group of suspects. Given this information, the next challenge is to find the variables providing the best clustering. These clusters will then be presented to the detectives to drill down using their domain expertise. The automated detection of crime patterns, allows the detectives to focus on crime sprees first and solving one of these crimes results in solving the whole "spree" or in some cases if the groups of incidents are suspected to be one spree, the complete evidence can be built from the different bits of information from each of the crime incidents. For instance, one crime site reveals that suspect has black hair, the next incident/witness reveals that suspect is middle aged and third one reveals there is tattoo on left arm, all together it will give a much more complete picture than any one of those alone. Without a suspected crime pattern, the detective is less likely to build the complete picture from bits of information from different crime incidents. Today most of it is manually done with the help of multiple spreadsheet reports that the detectives usually get from the computer data analysts and their own crime logs.

We choose to use clustering technique over any supervised technique such as classification, since crimes vary in nature widely and crime database often contains several unsolved crimes. Therefore, classification technique that will rely on the existing and known solved crimes, will not give good predictive quality for future crimes. Also nature of crimes change over time, such as Internet based cyber crimes or crimes using cell-phones were uncommon not too long ago. Thus, in order to be able to detect newer and unknown patterns in future, clustering techniques work better.

4. Clustering Techniques in Use

We will look at some of our contributions to this area of study. We will show a simple clustering example here. Let us take an oversimplified case of crime record. A crime data analyst or detective will use a report based on this data sorted in different orders, usually the first sort will be on the most important characteristic based on the detective's experience.

Crime Type	Suspect Race	Suspect Sex	Suspect Age gr	Victim age gr	Weapon
Robbery	X	M	Middle	Elderly	Knife
Robbery	Y	M	Young	Middle	Bat
Robbery	X	M	?	Elderly	Knife
Robbery	X	F	Middle	Young	Piston

Table 1 Simple Crime Example

We look at table 1 with a simple example of crime list. The type of crime is robbery and it will be the most important attribute. The rows 1 and 3 show a simple crime pattern where the suspect description matches, same race=X, same sex=M and similar age group. The rows 2 and 4 cannot be same if the information about the suspect's race and sex is correct. The victim profile is similar too in 1 and 3 and so is the weapon used. Since major characteristics of the suspect cannot change between the different crime incidents, these will be our major data mining attributes. The victims of crimes may be different but typically criminally target a similar type of people or property for crime and often use the same or similar weapon. The aim here is that we can use data mining to detect much more complex patterns since in real life there are many attributes or factors for crime and often there is partial information available about the crime. In a general case it will not be easy for a computer data analyst or detective to identify these patterns by simple querying. Thus clustering technique using data mining comes in handy to deal with enormous amounts of data and dealing with noisy or missing data about the crime incidents.

We used k-means clustering technique here, as it is one of the most widely used data mining clustering technique. Next, the most important part was to prepare the data for this analysis. The real crime data was obtained from a Sherriff's office, under non-disclosure agreements from the crime reporting system. The operational data was converted into denormalised data using the extraction and transformation. Then, some checks were run to look at the quality of data such as missing data, outliers and multiple abbreviations for same word such as blank, unknown, or unk all meant the same for missing age of the person. If these are not coded as one value, clustering will create these as multiple groups for same logical value. The next task was to identify the significant attributes for the clustering. This process involved talking to domain experts such as the crime detectives, the crime data analysts and iteratively running the attribute importance algorithm to arrive at the set of attributes for the clustering the given crime types. We refer to this as the semi-supervised or expert-based paradigm of problem solving. Based on the nature of crime the different attributes become important such as the age group of victim is important for homicide, for burglary the same may not be as important since the burglar may not care about the age of the owner of the house.

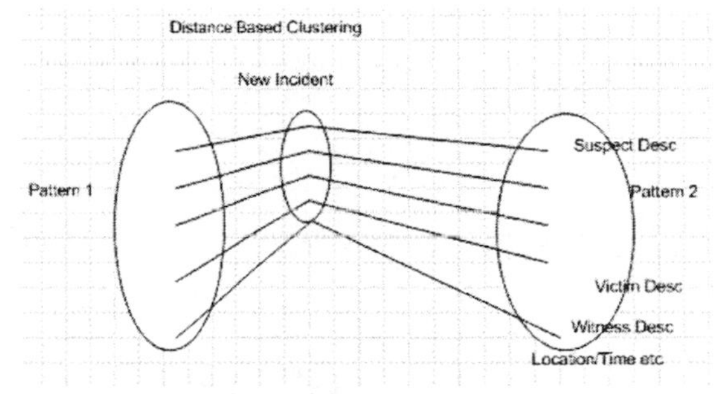

Fig 2 a Clustering without weights

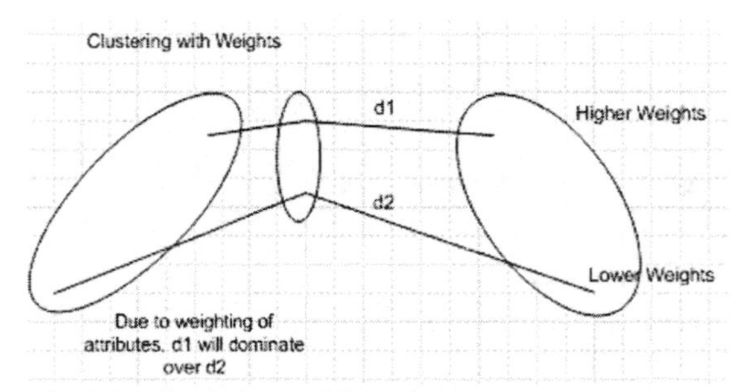

Fig 2 b Clustering with weights

To take care of the different attributes for different crimes types, we introduced the concept of weighing the attributes. Fig 2a shows distance based clustering without weights. Most out of box clustering algorithms will use this kind of clustering. However, out main contribution here is the use of weights shown in Fig 2 b. This allows placing different weights on different attributes dynamically based on the crime types being clustered. This also allows us to weigh the categorical attributes unlike just the numerical attributes that can be easily scaled for weighting them. Using the integral weights, the categorical attributes can be replicated as redundant columns to increase the effective weight of that variable or feature. We have not seen the use of weights for clustering elsewhere in the literature review, as upon normalization all attributes assume equal importance in clustering algorithm. However, we have introduced this weighting technique here in light of our semi-supervised or expert based methodology. We can think of weights as skewing the different dimensions over which the distances between the clusters are measured. As a result, some dimensions become more important than the other dimensions. The important dimensions represent the attributes like suspect identity related attributes. Based on our weighted clustering attributes, we cluster the dataset for crime patterns and then present the results to the detective or the domain expert along with the statistics of the important attributes. These weights will changes based on major crime time. For instance, in homicide the chactertisics of the victim will be important but in burgalry these may not be much important as in most cases the burgler is not likely to confront the owner of the home. However, in burglary, the nature of the property stolen may be important.

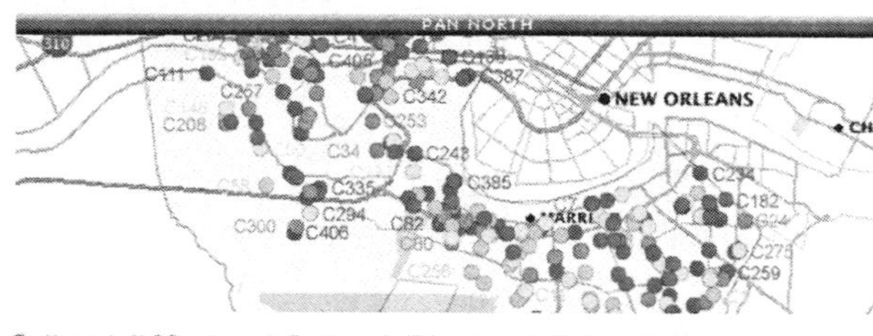

Pattern 1 (129 crimes)	Pattern 2 (79 crimes)	Pattern 3 (29 crimes)
Suspects point of entry	Suspects point of entry	Suspects point of entry
Victims Race	Victims Race	Suspects count (number)
Suspects count (number)	Number of days old	Victims Race
Number of days old		Number of days old

Pattern 5 (50 crimes)	Pattern 6 (9 crimes)	Pattern 7 (13 crimes)
Suspects race	Suspects city	Suspects Sex
Suspects Average height	Suspects point of entry	Suspects point of entry
Suspects Average Weight	Suspects Average Age	Suspects city
Suspects Average Age	Suspects count (number)	Suspects Average height
Suspects point of entry	Number of days old	Suspects Average Weight

Figure 3 Plot of crime clusters with legend for significant attributes for that crime pattern

The Fig 3 shows the color coded clusters along wit h the legends to list the significant attributes of crime that characterise that cluster. The detective looks at the clusters, smallest clusters first and then gives the expert recommendations. This iterative process helps to determine the significant attributes and the weights for different crime types. Based on this information from the domain expert, namely the detective, future crime patterns can be detected. First the future or unsolved crimes can be clustered based on the significant attributes and the result is given to detectives for inspection. Since, this clustering exercise, groups hundreds of crimes into some small groups or related crimes, it makes the job of the detective much easier to locate the crime patterns.

The other approach is to use a small set of new crime data and score it against the existing clusters using tracers or known crime incidents injected into the new data set and then compare the new clusters relative to the tracers. This process of using tracers is analogous to use of radioactive tracers to locate something that is otherwise hard to find.

5. Results of Crime Pattern Analysis

The proposed system is used along with the geo spatial plot. The crime analyst may choose a time range and one or more types of crime from certain geography and display the result graphically. From this set, the user may select either the entire set or a region of interest. The resulting set of data becomes the input source for the data mining processing. These records are clustered based on the predetermined attributes and the weights. The resulting, clusters have the possible crime patterns. These resulting clusters are plotted on the geo-spatial plot.

We show the results in the figure below. The different clusters or the crime patterns are color-coded. For each group, the legend provides the total number of crimes incidents included in the group along with the significant attributes that characterize the group. This information is useful for the detective to look at when inspecting the predicted crime clusters.

We validated our results for the detected crime patterns by looking the court dispositions on these crime incidents as to whether the charges on the suspects were accepted or rejected. So to recap the starting point is the crime incident data (some of these crimes already had the court dispositions/ rulings available in the system), which the measured in terms of the significant attributes or features or crime variables such as the demographics of the crime, the suspect, the victim etc. No information related to the court ruling was used in the clustering process. The detailed results are not presented here due to confidentiality of the nature of the data.

Subsequently, we cluster the crimes based on our weighing technique, to come up with crime groups (clusters in data mining terminology), which contain the possible crime patterns of crime sprees. The geo-spatial plot of these crime patterns along with the significant attributes to quantify these groups is presented to the detectives who now have a much easier task to identify the crime sprees than from the list of hundreds of crime incidents in unrelated orders or some predetermined sort order. In our case, we looked at the crime patterns, as shown in same colors below and looked at the court dispositions to verify that some of the data mining clusters or patterns were indeed crime spree by the same culprit(s).

6. Conclusions and Future Direction

We looked at the use of data mining for identifying crime patterns crime pattern using the clustering techniques. Our contribution here was to formulate crime pattern detection as machine learning task and to thereby use data mining to support police detectives in solving crimes. We identified the significant attributes; using expert based semi-supervised learning method and developed the scheme for weighting the significant attributes. Our modeling technique was able to identify the crime patterns from a large number of crimes making the job for crime detectives easier.

Some of the limitations of our study includes that crime pattern analysis can only help the detective, not replace them. Also data mining is sensitive to quality of input data that may be inaccurate, have missing information, be data entry error prone etc. Also mapping real data to data mining attributes is not always an easy task and often requires skilled data miner and crime data analyst with good domain knowledge. They need to work closely with a detective in the initial phases.

As a future extension of this study we will create models for predicting the crime hot-spots [3] that will help in the deployment of police at most likely places of crime for any given window of time, to allow most effective utilization of police resources. We also plan to look into developing social link networks to link criminals, suspects, gangs and study their interrelationships. Additionally the ability to search suspect description in regional, FBI databases [4], to traffic violation databases from different states etc. to aid the crime pattern detection or more specifically counter terrorsim measures will also add value to this crime detection paradigm.

Another area that needs further work is the ability to find unsoved crimes that were possiblity done by the suspect arrested for a new crime. This will help to charge all the crimes committed by the criminal and that may result in a higher class of resultant crime such as a number of petty thefts may accumulate to felony. This will help to punish the repeat offenders and keep them off the street for longer time.

7. References

[1] Hsinchun Chen, Wingyan Chung, Yi Qin, Michael Chau, Jennifer Jie Xu, Gang Wang, Rong Zheng, Homa Atabakhsh, "Crime Data Mining: An Overview and Case Studies", AI Lab, University of Arizona, proceedings National Conference on Digital Government Research, 2003, available at: http://ai.bpa.arizona.edu/

[2] Hsinchun Chen, Wingyan Chung, Yi Qin, Michael Chau, Jennifer Jie Xu, Gang Wang, Rong Zheng, Homa Atabakhsh, "Crime Data Mining: A General Framework and Some Examples", IEEE Computer Society April 2004.

[3] C McCue, "Using Data Mining to Predict and Prevent Violent Crimes", available at: http://www.spss.com/dirvideo/richmond.htm?source=dmpage&zone=rtsidebar

[4] Whitepaper, "Oracle's Integration Hub For Justice And Public Safety", Oracle Corp. 2004, available at: http://www.oracle.com/industries/government/IntegrationHub_Justice.pdf

[5] John Zeleznikow, "Using data mining techniques to detect criminal networks" presented at Australian Institute of Criminology April 2005, available at: http://www.aic.gov.au/conferences/occasional/2005-04-zeleznikow.html

[6] Shyam Varan Nath, "Crime Patterns Detection Using Data Mining Framework", Knowledge Discovery, Data Mining, and Machine Learning Proceedings, University of Hildesheim, Germany, Oct 2006

About the Author: Shyam Varan Nath is a Principal Analytical Consultant, specializing in Data Mining with Oracle Corporation. Parts of the above work were done while working on Oracle Protect for Law Enforcement. He is an Oracle Certified Professional and has worked in the industry since 1991. He has an undergraduate degree in EE from Indian Institute of Technology (IIT), Kanpur India and has obtained an MBA and MS (Comp Sc) from Florida Atlantic University, Boca Raton, FL. He is currently a part-time doctoral student in Computer Science, specializing in data mining. Shyam is also the President of the Business Intelligence, Warehousing and Analytics Special Interest Group (BIWA SIG) (http://OracleBIWA.org).

Combinatorial Hill Climbing Using Micro-Genetic Algorithms

Spyros A. Kazarlis
Technological Educational Institute of Serres,
Dept. Of Informatics & Communications
Terma Magnesias St., 621 24, Serres, GREECE
kazarlis@teiser.gr

Abstract- **This paper introduces a new hill-climbing operator, (MGAC), for GA optimization of combinatorial problems, and proposes two implementation techniques for it. The MGAC operator uses a small size second-level GA with a small population that evolves for a few generations and serves as the engine for finding better solutions in the neighborhood of the ones produced by the main GA. The two implementations are tested on a Power Systems' problem called the Unit Commitment Problem, and compared with three other methods: a GA with classic hill-climbers, Lagrangian-Relaxation, and Dynamic Programming. The results show the superiority of the proposed MGAC operator.**

I. INTRODUCTION

In order to boost the convergence speed of GAs towards the exact optimum, many hybrid genetic schemes have been proposed in the literature that combine GAs with hill climbing or local search techniques [1], [6], [8], [9], [10], [11], [12], [14], to solve both continuous variable and combinatorial problems. Memetic Algorithms [13] are probably the most well known paradigm of such schemes. Most GA-hill climbing hybrids are implemented for continuous variable problems. In such problems the hill climbers used are often designed to perform independent steps along each axis in the space (one variable at a time) searching for better solutions. One such operator is the PhenoMute (PM) hill climbing operator suggested in [2], [11], [12]. Such operators cannot follow the potential "ridges" created in the search space of difficult constrained optimization problems, as the direction of the "ridge" usually does not coincide with that of a single axis.

This problem has led the author together with other researchers in introducing the Micro GA hill climbing operator (MGA) [4], [5]. The MGA operator uses a small second-level population, or a Micro GA [3], [7], that evolves for a small number of generations and acts in a small neighborhood around the best solution produced by the main GA at each generation. The MGA operator is capable of genetically evolving paths of arbitrary direction leading to better solutions and following potential ridges in the search space regardless of their direction, width, or even discontinuities. Although proven effective the MGA operator was designed only for continuous variable problems.

In this paper the combinatorial version of the MGA operator is presented, called the Micro GA combinatorial hill climbing operator, or MGAC for brevity. The major difference in combinatorial problems is that the definition of the "neighborhood" of a solution is not very easily conceived, as every small perturbation to a combinatorial problem solution may be considered to reside within the "neighborhood" of the solution. Thus, before introducing the new MGAC operator, a definition of the combinatorial "neighborhood" is given.

Moreover, two specific implementations of MGAC are proposed, namely MGAC-ARM and MGAC-CNS. The first one (MGAC-ARM) genetically searches the space of possible perturbations within a single neighborhood at a time, but the specific neighborhood that it searches is selected among all possible neighborhoods using an Adaptive Ranking Multi-neighborhood scheme, that promotes "fertile" neighborhoods. The second one (MGAC-CNS) uses the Micro GA to search the space of possible neighborhoods (Combinatorial Neighborhood Space) and each neighborhood is evaluated by checking the quality of a number of random chosen sample perturbations within the neighborhood.

The organization of the paper is as follows: in section II the definition of the combinatorial neighborhood is given. In section III the MGAC-ARM implementation is described in detail, while the MGAC-CNS implementation is described in section IV. Section V presents the test problem on which the new operators are tested, together with the simulation results. Conclusions of this work are presented in section VI.

II. DEFINITION OF THE COMBINATORIAL NEIGHBORHOOD

In order to give a definition of the combinatorial "neighborhood" that is essential to develop the MGAC operator, a few principles must be considered: a) the neighborhood must be small compared to the whole solution, b) the neighborhood must be allocated "around" the original solution, and c) perturbations within the neighborhood may result in small alterations of the whole solution. With these requirements in mind the definition of the combinatorial "neighborhood" can de formed as follows:

Without loss of generality we can assume that every combinatorial problem can be encoded using the binary alphabet. This means that every possible solution can be represented as a binary string of some length, depending on the specific problem. Lets define this length as ℓ. Then the search space SS is composed of 2^{ℓ} different solutions. The neighborhood of every solution $S \in SS$ can be defined as a

The Project is co-funded by the European Social Fund and National Resources - (EPEAEK-II) ARXIMHDHS.

K. Elleithy (ed.), *Advances and Innovations in Systems, Computing Sciences and Software Engineering*, 411–416.

subset N_S of SS, that is composed of solutions produced from solution S, by allowing n of the ℓ bits of solution S to change and keeping the rest ℓ-n bits constant. Thus, the number of solutions that reside within the neighborhood N_S is 2^n. This definition covers the requirements described earlier in this chapter as a) by keeping n small, the resulting neighborhood search space can be small compared to the original problem's search space, b) the neighborhood is around the original solution as ℓ-n bits of the solution are kept unchanged, and the rest are allowed to perturb, and c) by selecting the n bits in such a way that they are semantically close (their positions affect the same or similar regions of the decoded real solution), it can be ensured that perturbations within the neighborhood will result in alterations of certain portions of the whole solution.

From the above definition it is clear that there isn't only one single neighborhood that can be defined for every solution S. In fact the number of different neighborhoods (NN) is ℓCn, where xCy is the combination of x elements taken as groups of y elements and is given by :

$$NN = xCy = \frac{x!}{y!(x-y)!} \tag{1}$$

In a real implementation, though, this number may be reduced when considering the third requirement (requirement c). According to this requirement, not all combinations may be considered as neighborhoods, because the selected bits of each neighborhood must be semantically close.

III. Implementation 1 (MGAC-ARM): Micro-GA Searches the Neighborhood Selected by a Ranking Algorithm

According to this variant, the MGAC operator searches one neighborhood at a time, using a Micro GA. In other words the MGAC operator genetically scans the space of all possible perturbations of the specific n bits of a single neighborhood ($n \ll \ell$). The neighborhood searched by the MGAC operator should not remain the same during the main GA's run, but it should be possible for the MGAC to examine a large number of neighborhoods. Moreover, if the MGAC operator succeeds in improving the best-so-far solution by examining a specific neighborhood NEi, (i=1..NN), it is wise to insist on examining this neighborhood, as it is possible to come up with even better solutions in the future.

With the above considerations in mind, an adaptive ranked based multi-neighborhood scheme (ARM) has been developed for this MGAC variation. This scheme works as follows :

1. Before the beginning of the GA evolution, all the possible neighborhoods are calculated, and each one is assigned a neighborhood identification number and a rank. At the beginning all ranks are set equal to 1, which means that at the beginning all neighborhoods have equal probability of being selected by the MGAC operator for examination.
2. Every time the MGAC is invoked, it selects one neighborhood out of NN, with probability proportional to the neighborhood's rank (roulette wheel selection).
3. If the MGAC finds a better solution by examining the selected neighborhood, it increases its rank by 2, with a maximum of 10.
4. If it fails to find a better solution, it decreases its rank by 1, with a minimum of 1.

In the long run, neighborhoods that produce better solutions consistently, are ranked better than the ones that don't produce better solutions or the ones that managed to produce a better solution once, but proved to be unproductive later on.

The MGAC operator itself works as follows (also see Figure 1):

1. When invoked, the MGAC is fed with the best-so-far solution S_{best} of the main GA.
2. Then, it selects a neighborhood NEi with probability proportional to the neighborhoods rank, via roulette wheel selection.
3. All the bits of S_{best} that do not belong to NEi are kept unchanged. Those that belong to NEi are allowed to change.
4. It forms a population of 5 solutions that are bit strings of length n, (where n is the number of bits allowed to change), randomly generated at the beginning.
5. It evaluates each of the 5 solutions by injecting the bits of each solution to the corresponding bits of solution S_{best}.

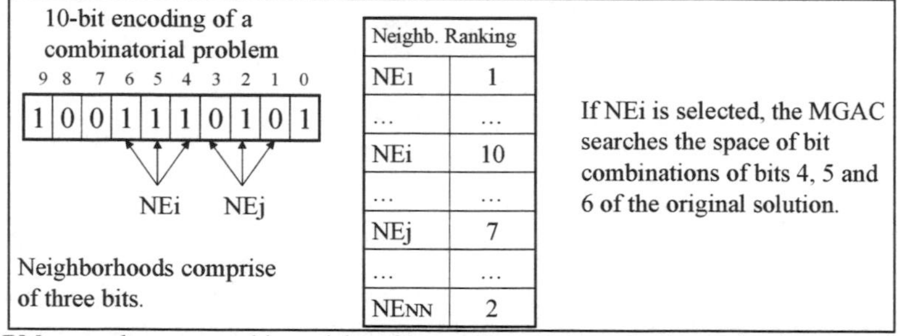

Figure. 1. MGAC-ARM example on a combinatorial problem with a 10-bit encoding and neighborhoods of 3 bits. Possible neighborhoods are 10C3, i.e. 10!/(3!(7!))=120. Neighborhoods are ranked depending on whether they produce better solutions or not. Neighborhoods are selected with probability proportional to their ranks.

6. Then solutions (among the 5) are selected in pairs, with probability proportional to their fitness, to mate recombine and produce 5 new solutions (the next generation).
7. The steps 5-6 are repeated for 7 generations.
8. If the best solution produced has better fitness than the original solution S_{best}, then the rank of neighborhood NEi is increased by 2 (max 10). Otherwise, it is decreased by 1 (min 1).
9. Finally, if the best solution produced is better than S_{best}, it replaces it in the main GA's population.

From the above algorithmic description it is evident that With the MGAC-ARM scheme the following targets are achieved:

1. The neighborhood of the best-so-far solution is genetically searched by a Micro GA
2. All neighborhoods have the opportunity to be searched by the operator
3. "Fertile" neighborhoods are searched more often than "sterile" ones

However, the MGAC-ARM scheme has a scaling-up problem: when dealing with large scale combinatorial problems, the number NN of possible neighborhoods (sets of n out of ℓ bits) can be extremely large, and make it practically impossible for the algorithm to a-priori calculate and enumerate all possible neighborhoods.

IV. IMPLEMENTATION 2 (MGAC-CNS): MICRO-GA SEARCHES THE NEIGHBORHOOD SPACE AND EACH NEIGHBORHOOD IS EVALUATED BY SAMPLES

This scheme tries to solve the scaling-up problem of the MGAC-ARM variation. Instead of using the Micro GA to search a single neighborhood at a time (as in MGAC-ARM), that is selected among all possible neighborhoods via a ranking strategy, the MGAC-CNS variation uses the Micro GA to search the space of possible neighborhoods, in order to discover regions of "fertile" neighborhoods, the perturbations of which may give better solutions.

In order for the MGAC-CNS to work, a representation method is needed, to encode all possible neighborhoods in a string of symbols. For example, in a combinatorial problem with a 10-bit encoding and 3-bit neighborhoods, as in the previous section (Fig.1), one might adopt an integer encoding to represent possible neighborhoods. Each MGAC-CNS solution could be a vector S consisting of three (3) integers S_i, i=1..3, each of which can take values in the range 0..9, representing a bit position in the 10-bit solution of the main problem. Thus, the solution S=(4,5,6) coincides with neighborhood NEi of Figure 1. Of course care should be taken during the reproduction phase so that there are no duplicate values in every produced solution S (e.g. S=(2,2,6) is invalid).

Every time it is called, the MGAC-CNS operator initially produces a population of 5 such vectors at random. Then, it evolves this population, using common integer crossover and mutation operators, for 7 generations. Every produced vector S, that represents a specific 3-element set of bit positions (0..9), must be evaluated in order for the genetic evolution to work. Thus the fitness function of the MGAC-CNS, has to evaluate a whole neighborhood of 2^3, or in general 2^n solutions.

The most proper thing to do this might be the exhaustive search method. However, this technique will consume 2^n fitness evaluations, for every evaluated solution (neighborhood) of the MGAC-CNS operator, or $5x7x2^n$ evaluations every time the operator is invoked.

In order to overcome this, we have used a neighborhood evaluation function that samples the specific neighborhood's solutions, by evaluating a small number m of bit combinations within the neighborhood under evaluation. In the simulations performed in this work we have used m=2. The bit combinations evaluated for each neighborhood are taken at random. After the evaluation of the samples, their fitness values are averaged to produce the final quality of the neighborhood under examination.

The above process is summarized in the following (see also Figure 2):

1. For the Nth time the MGAC-CNS is invoked:
2. Produce P-1 random parent vectors (neighborhoods) S_i, i=1..P-1, $S_i=(S_{i1}, S_{i2}, \ldots, S_{in})$ (P is the no of genotypes in the MGAC-CNS population, e.g. P=5, n is the bits per neighborhood, e.g. n=3).
3. Inject the best neighborhood S_{best} of the previous (N-1) MGAC-CNS run as the Pth parent genotype.
4. Evaluate each S_i neighborhood:

4-1. Randomly produce m (e.g. m=2) samples of bit combinations for this neighborhood (e.g. 101, 011)

4-2. Inject the bits of each sample to the corresponding bits of the best-so-far main GA solution.

4-3. Calculate Fitness Fj (j=1..m) of the resulting genotype, using the main GA's fitness function.

4-4. Average Fj (j=1..m), to calculate the fitness of neighborhood S_i.

5. Mate and reproduce neighborhoods S_i, i=1..P, of the parent population, using common crossover and mutation operators, and produce the generation of offspring neighborhoods.
6. Continue for G generations (e.g. G=7)
7. If the MGAC-CNS finds a solution better than the best-so-far solution of the main GA, then the MGAC-CNS solution replaces the corresponding main GA solution.

One drawback of MGAC-CNS though is the fact that each neighborhood produced, is evaluated by a limited number of samples. This is adopted for practical computational reasons, because the exhaustive search of the neighborhood may consume quite a large number of fitness evaluations.

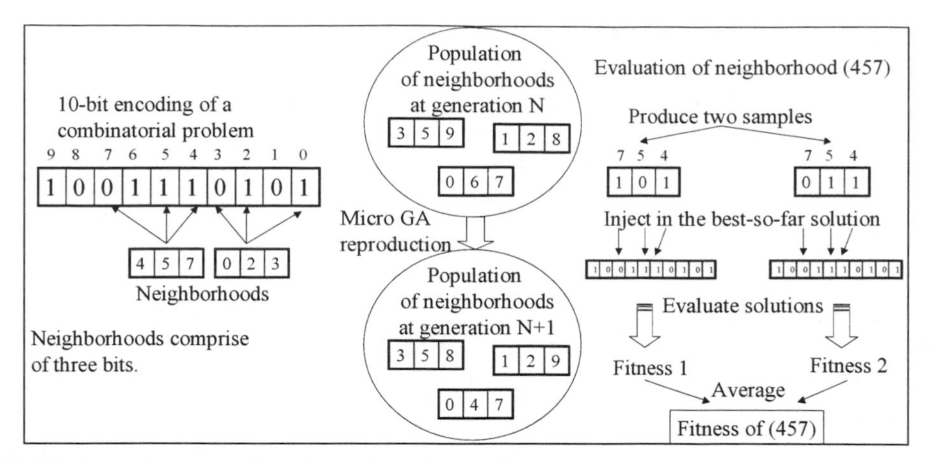

Figure. 2. MGAC-CNS example on a combinatorial problem with a 10-bit encoding and neighborhoods of 3 bits. Micro GA searches the space of all possible neighborhoods. Each produced neighborhood is evaluated by producing two random samples (bit combinations), injecting them into the best-so-far genotype of the main GA, evaluate the two solutions and average the fitness values.

V. Simulation Results on the Unit Commitment Problem

In order to test the efficiency of the two MGAC operator variations, a difficult constrained combinatorial problem was selected, the Unit Commitment problem (UC).

The UC problem comes from the field of Power Systems and it is in fact a time scheduling problem. It involves determining the start-up and shut down schedules of thermal units to be used to meet forecasted demand over a future short term (24-168 hour) period. The objective is to minimize total production costs while observing a large set of operating constraints.

The total costs consist of a) Fuel costs, b) Start-up costs and c) Shut-down costs. Fuel costs are calculated using unit heat rate and fuel price information usually as :

$$FC = a + b \cdot P + c \cdot P^2 \quad (2)$$

where P is the power output of a unit, and a, b, c are fuel cost coefficients. The power outputs of the committed units for every hour of the schedule are easily calculated by the λ-iteration algorithm [2], [6].

Start-up costs are expressed as a function of the number of hours the unit has been down. Here we have used the following formula:

$$SUC = \begin{cases} HSC, if(down \leq cold\_start\_hour: \\ CSC, if(down > cold\_start\_hour: \end{cases} \quad (3)$$

where HSC is the "Hot Start Cost" value and CSC is the "Cold Start Cost" value for the specific unit, "down" is the number of hours the unit has been down, and "cold_start_hours" is different for each unit. Shut-down costs are defined as a fixed dollar amount for each unit per shut-down, and is taken equal to 0 in this work. Thus the cost (objective) function to be minimized can be formulated as :

$$O(s) = \sum_{i=1}^{U}\sum_{j=1}^{H} FC_{ij}(s) + \sum_{i=1}^{U}\sum_{j=1}^{SU(i)} SUC_{ij}(s) \quad (4)$$

where s is a specific solution, U is the number of units, H is the number of scheduling hours, FC_{ij} is the fuel cost of unit i at hour j, SU(i) is the number of start-ups of unit i during the schedule, and SUC_{ij} is the start-up cost of unit i at start-up j.

The constraints taken into account in this work, are: (a) System power balance (demand + losses + exports), (b) System reserve requirements, (c) Unit initial conditions, (d) Unit high and low MW limits (economic, operating), (e) Unit minimum-up time, (f) Unit minimum-down time.

For testing the MGAC operator we have used an instance of the UC problem presented in [6]. This instance includes 10 thermal power production units and a scheduling horizon of 24 hours. The unit data and the forecasted demand for the schedule are omitted for brevity and can be found in [6] .

For the application of GAs to the UC problem a simple binary alphabet was chosen to encode a solution. With the assumption that at every hour a certain unit can be either ON or OFF, an H-bit string is needed to describe the operation schedule of a single unit. In such a string, a '1' at a certain location indicates that the unit is ON at this particular hour, while a '0' indicates that the unit is OFF.

The main GA used a population of 50 genotypes, Roulette Wheel parent selection, multi-point crossover and bit mutation with adaptive application probabilities. Also some advanced operators were used just like in [6]. Additionally, the problem constraints were handled by adding penalty terms to the objective function, for every constraint violation, and applying the Varying Fitness Function technique [6], [12] for the gradual application of penalties through the GA run, just like in [6].

TABLE I. SIMULATION RESULTS OF 3 METHODS, ON THE UC PROBLEM, REPORTED IN [6].

units	Dynamic Programming	Lagrangian Relaxation	Simple GA					
	best solution	best solution	gener. limit	Mean No of Evaluat	success %	Best solution	worst solution	Differ. %
10	565825	565825	500	35,000	60%	565825	570032	0.74
20	-	1130660	1000	70,000	75%	1126243	1132059	0.51
40	-	2258503	2000	140,000	90%	2251911	2259706	0.34
60	-	3394066	3000	210,000	100%	3376625	3384252	0.22
80	-	4526022	4000	280,000	100%	4504933	4510129	0.11
100	-	5657277	5000	350,000	100%	5627437	5637914	0.19

TABLE II. SIMULATION RESULTS OF THE TWO MGAC VARIANTS ON THE UC PROBLEM.

units	GA with the MGAC-ARM operator						GA with the MGAC-CNS operator					
	gener limit	success %	Best solution	worst solution	Differ. %	Average solution	gener. limit	success %	best solution	worst solution	Differ. %	average solution
10	350	60%	565825	566764	0,166	566066	320	40%	565825	566977	0.203	566129
20	700	100%	1126242	1130521	0,379	1128534	640	80%	1126028	1131951	0.526	1129558
40	1400	100%	2250921	2255540	0.205	2253883	1280	100%	2251090	2256468	0.238	2254899
60	-	-	-	-	-	-	1920	100%	3376429	3386609	0.301	3381095
80	-	-	-	-	-	-	2560	100%	4501905	4510403	0.188	4506909
100	-	-	-	-	-	-	3200	100%	5623547	5634774	0.199	5630006

(The "Differ.%" figures are the percent difference between the best and worst solutions. A GA run is considered successful when it outperforms the corresponding Lagrangian Relaxation solution. All costs are in US$.)

The simulations included test runs for 10, 20, 40, 60, 80 and 100 unit problems. For the 20-unit problem the units were duplicated and the demand values were doubled. The data were scaled appropriately for the larger problems. The two GA-MGAC variations are compared to three other algorithms implemented and tested in [6]: a Dynamic Programming algorithm, a Lagrangian Relaxation algorithm, and a "simple" GA without the MGAC operator. In order to have a fair comparison, the runs are conducted on the basis of 'equal number of evaluations' for all GA algorithms. On the 10 unit problem, simple GA runs consume an average of 35,000 evaluations 25,000 (population 50 X 500 generations) of which are consumed by the simple GA and the rest 10,000 by custom hill-climbers. In order for the MGAC-ARM scheme to consume the same total number of evaluations, the generation limit was reduced to 350 generations (Table II), and for the MGAC-CNS to 320 generations.

For every problem and method, 20 independent runs have been performed as in [6]. The results are shown in Table II. Dynamic Programming was run only for the 10-unit problem, as for the larger problems the memory and cpu-time resources needed were prohibitive. The GA-MGAC-ARM variation was applied to problems of up to 40 units, as the number of possible neighborhoods that have to be enumerated and ranked rises to extremely high values for larger problems (13,160,160 neighborhoods for the 40-unit problem and 109,230,240 for the 60-unit one. The numbers represent not all possible neighborhoods, but all valid ones. Valid neighborhoods consist of 5units X 5hours, where the hour bits are contiguous and the same for all units).

From the results presented in Table II the following conclusions can be drawn:

For the 10-unit problem the two GA-MGAC schemes find the same solution as the other 3 algorithms with a success rate of 60% for the ARM variant, and 40% for the CNS variant, compared to 60% of the simple GA. Moreover, the worst solution produced by the two GA-MGAC schemes is much closer to the optimum than the corresponding worst solution of the simple GA, despite the reduced generation limits. This explains the low percent difference of 0.166% (ARM) and 0,203 (CNS). Thus, for the 10 unit problem ARM seems to slightly outperform the simple GA, in terms of average solution quality, while CNS seems to be slightly worse in terms of the success rate of finding the exact optimum, and slightly better in average solution quality than the simple GA. This could be due to the reduced generation limit of CNS.

For the 20-unit problem the GA-MGAC-ARM finds a solution slightly better than that of the simple GA, yet exhibiting better success rate and average solution quality, while the CNS variation discovers a new up-to-now best solution with a cost of 1126028, and exhibits similar performance compared to the simple GA. All of the 20 ARM runs produce solutions better than the best solution produced by the Lagrangian Relaxation method. Additionally, the worst solution is much closer to the best one, than the corresponding worst solution of the simple GA, and this leads to a smaller percent difference value of 0.379 (simple GA: 0.51). Here the ARM variant seems to clearly outperform the simple GA, while CNS is slightly worse and more or less even with simple GA, except for the new best solution it discovers.

For the 40-unit problem both GA-MGAC variants discover new up-to-now solutions with a cost of 2250921 (ARM) and 2251090 (CNS). Again, all 20 runs of both variants produce solutions better than that of the Lagrangian Relaxation method. Moreover, the worst solution of both variants is again much closer to the best one, than that of the simple GA, resulting again in a smaller percent difference value of 0.205 (ARM) and 0.238 (CNS). In this problem the two variants exhibit similar performances, but clearly outperform the simple GA in terms of the best and average solution quality.

For the 60, 80, and 100-unit problems the CNS variant seems to outperform the simple GA considering the best solutions found, exhibiting thus better in-depth search efficiency, but is slightly worse in terms of average solution quality (% difference). This could be due to the reduced generation limit of the GA-MGAC-CNS algorithm, that is mandatory in order to achieve the same total number of fitness evaluations.

From the above it is evident that both MGAC operator variations really enhance the GA performance on a difficult constrained combinatorial problem, like the UC problem presented. They are capable of discovering even better solutions than the best known ones for the specific instances of the UC problem used in this work. They also seem to enhance the robustness of GAs in consistently finding solutions close to the optimum. Also it is evident that the ARM variant is most suitable for relatively small problem scales (e.g. up to 40X24=960 bits problems), where it has a clear advantage over the simple GA, while the CNS variation is most suitable for large scale problems, where the ARM variation is not applicable. In large scale problems the CNS variation also seems to have a clear advantage over the simple GA, especially in discovering solution close to the global optimum.

VI. CONCLUSIONS

In this paper, a new hill climbing operator for genetic optimization of combinatorial problems has been presented, the Micro GA Combinatorial hill climbing operator (MGAC). Two variants of MGAC have been presented: the ARM and the CNS variants.

Both variants try to search combinatorial neighborhoods or subsets of the symbol strings that encode complete solutions of the main GA, keeping a part of the best-so-far solution constant, and allowing a subset (neighborhood) of it to change.

The performance of the two operator variations has been demonstrated by their application on a power systems scheduling problem, the Unit Commitment problem. The simulation results have shown that the MGAC operator is able to boost the performance of the main GA in difficult combinatorial problems, by improving its search efficiency and robustness.

More effort is also needed in applying the two MGAC operator variations to more combinatorial problems, in order to test if their performance and robustness can be generalized.

REFERENCES

[1] D. H. Ackley, "Stochastic Iterated Genetic Hill Climbing," Ph.D. Thesis, Department of Computer Sciences, Carnegie Mellon University, Pittsburgh, PA, 1987.

[2] A. Bakirtzis, V. Petridis, S. Kazarlis, "A Genetic Algorithm Solution to the Economic Dispatch Problem," *IEE Proceedings - Generation, Transmission, Distribution*, Vol. 141, No. 4, July 1994, pp. 377-382.

[3] G. Dozier, J. Brown and D. Bahler, "Solving Small and Large Scale Constraint Satisfaction Problems using a Heuristic-based Microgenetic Algorithm," in *Proc. of the 1st IEEE Int. Conf. on Evolutionary Computation*, vol. 1, Piscataway, NJ: IEEE Press, pp. 306-311, 1994.

[4] S.Kazarlis, S.Papadakis, J.Theocharis and V.Petridis, "Micro-Genetic Algorithms as Generalized Hill Climbing Operators for GA Optimization," *IEEE Transactions on Evolutionary Computation*, Vol. 5, No. 3, June 2001, pp. 204-217.

[5] S. A. Kazarlis, "Micro-Genetic Algorithms As Generalized Hill-Climbing Operators for GA Optimization of Combinatorial Problems – Application to Power Systems Scheduling", *Proceedings of the the 4th Conference on Technology and Automation, October 2002*, Thessaloniki, Greece, pp. 300-305.

[6] S. A. Kazarlis, A. G. Bakirtzis, and V. Petridis, "A Genetic Algorithm Solution to the Unit Commitment Problem," *IEEE Trans. on Power Systems*, vol. 11, no. 1, pp. 83-92, Feb 1996.

[7] K. Krishnakumar, "Micro-genetic algorithms for stationary and non-stationary function optimization," in *SPIE Proceedings: Intelligent Control and Adaptive Systems*, pp. 289-296, 1989.

[8] J. A. Miller, W. D. Potter, R. V. Gandham, and C. N. Lapena, "An Evaluation of Local Improvement Operators for Genetic Algorithms," *IEEE Trans. on Syst., Man, Cybern.*, vol. 23, No. 5, pp. 1340-1351, 1993.

[9] M. Mitchell, J. Holland, and S. Forrest, "When will a genetic algorithm outperform a hill-climbing?," in J. D. Cowen, G. Tesauro, and J. Alspector, editors, *Advances in Neural Information Processing Systems 6*, San Mateo, CA, 1994, Morgan Kaufmann.

[10] H. Muhlenbein, "How Genetic Algorithms really work: I. Mutation and hill-climbing," in R. Manner and B. Maderick, editors, *Parallel Problem Solving from Nature 2*, pp. 15-25, Elsevier, 1992.

[11] V. Petridis and S. Kazarlis, "Varying Quality Function in Genetic Algorithms and the Cutting Problem," in *Proceedings of the 1st IEEE Conference on Evolutionary Computation* (vol. 1). Piscataway, NJ: IEEE Press, 1994, pp. 166-169.

[12] V. Petridis, S. Kazarlis, and A. Bakirtzis, "Varying Fitness Functions in Genetic Algorithm Constrained Optimization: The Cutting Stock and Unit Commitment Problems," *IEEE Transactions on Systems, Man, and Cybernetics*, Vol. 28, Part B, No. 5, October 1998, pp. 629-640..

[13] N. J. Radcliffe and P. D. Surry, "Formal Memetic Algorithms," in *Proceedings of the 1st AISB Workshop on Evolutionary Computing* (AISB '94), T. C. Fogarty, Editor, Springer-Verlag, pp. 1-16, 1994.

[14] J-M. Renders and H. Bersini, "Hybridizing Genetic Algorithms with Hill-Climbing Methods for Global Optimization: Two Possible Ways," in *Proc. of the 1st IEEE Int. Conf. on Evol. Computation*, vol. 1, Piscataway, NJ: IEEE Press, 1994, pp. 312-317.

Alternate Paradigm for Navigating the WWW Through Zoomable User Interface

Sumbul Khawaja
ISRA University
P.O. Box 313 Hala Road Hyderabad
sumbul@isra.edu.pk
+92 - 022- 2030181

Prof. Dr. Asadullah Shah
ISRA University
P.O. Box 313 Hala Road Hyderabad
asadullah@isra.edu.pk
+92 - 022- 2030181

Kamran Khowaja
ISRA University
P.O. Box 313 Hala Road Hyderabad
kamran@isra.edu.pk
+92 - 022- 2030181

Abstract - web browsing has become extremely important in every field of life whether it is education, business or entertainment. With a simple mouse click, user navigates through a number of web pages. This immediacy of traversing information links make it difficult to maintain an intuitive sense of where one is, and how one got there. A zooming browser is designed in java to explore alternate paradigm for navigating the www. Instead of having a single page visible at a time, multiple pages and the links between them are depicted on a large zoomable information surface. Links are shown in hierarchy so that user can see the relationship of web pages with their parent and child nodes. Browser also maintains the history of links traversed.

I. Introduction

It is a truism of modern life that there is much more information available than one can readily and effectively access. To make the navigation effective in WWW there should be a way to keep track of all the information visited from start till end of surfing. This can be done by providing user an area where user can have all the history of navigation along with the hierarchy of navigation. An alternate interface for internet browsing is proposed through a Zoomable User Interface (ZUI) that is an alternative to traditional window and icon-based approaches to interface design.

ZUIs have been discussed since 1993 [1]. No definition of ZUI has been generally agreed upon. It is considered that two main characteristics of ZUIs are (a) that information objects are organized in space and scale, and (b) that users interact directly with the information space, mainly through panning and zooming [2]. In ZUIs, space and scale are the fundamental means of organizing information [1][3]. The appearance of information objects are based on the scale at which they are shown.

The second main characteristic of ZUIs is that the information space is directly visible and can be manipulated through panning and zooming. Panning changes the area of the information space that is visible, zooming changes the scale at which the information space is viewed. Usually, panning and zooming are controlled with the mouse or the keyboard, so that change in the input device is linearly related to how much it is panned or zoomed.

However, these characteristics do not define zoomable interfaces in general. ZUIs are combined with other interfaces techniques, such as transparent overviews [4], some overview detail interfaces are extended with animated zooming [5] and some effort has been put into extending zoomable user interfaces with navigation mechanisms that can implement direct zoom and pan.

II. Architecture

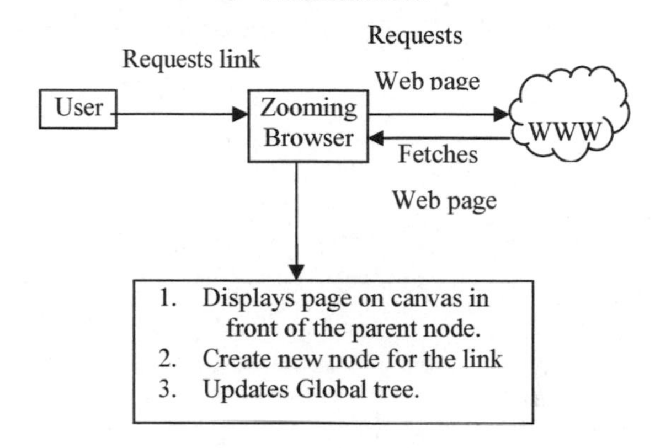

III. Related Technology

The proposed work is the motivation of Pad++ [6]. In 1989, a system called Pad was developed and presented at an NSF workshop. Pad integrated zooming into a single program that ran on inexpensive hardware. However to use Pad as a system for building general user interfaces requires a higher level structure called a Pad. Object to interpret events and control these display items so they behave as a single application. In Pad an object consists of a region together with a package of code and data which respond to event messages. An object's behavior is specified by the application developer. In order to make itself seen, each object manages a collection of display items, creating, modifying, and deleting them [1].

Display is complicated by the fact that objects may be continually creating and destroying display items. Before the display is created, first there is need to give each object an opportunity to know at what magnification it will be called upon to appear, since this will probably influence what display items it chooses to show. The Pad system is written in three layers, a real-time display layer written in C++, a Scheme interpreter providing an interface to the C++ layer, and a collection of Scheme code implementing the Pad application interface. It currently runs under X Windows and MS-DOS. The X Windows version has been compiled and run on SunOS, AIX and Linux. Pad++ was built, a direct but substantially more sophisticated successor to Pad. [7].

K. Elleithy (ed.), Advances and Innovations in Systems, Computing Sciences and Software Engineering, 417–420.

which describes the implementation, and prototype applications. Pad++, in addition, introduces an informational physics strategy for interface design and briefly compares it with metaphor-based design strategies [6].

IV. FEATURES

In order to overcome navigation problems faced in Pad++ and give users an environment where they can effectively navigate through the web information, a zooming browser is developed. Although there are currently a handful of commercial browsing tools available, the tools that clearly dominates the market are Microsoft IE and Netscape Navigator. Therefore, to have the greatest potential impact on daily WWW users, a Zooming Browser is designed having similar features as above two.

Consequently, the model in this proposal has envisioned for using Zoomable Browser begins same as IE. Fig. 1 shows the Zooming Browser.

The browser generates a hierarchical structure of all the visited web pages so at any time user can view from where user started the journey of WWW as shown in Fig. 2. At any instance of time user can go to any of the page from the hierarchy of web pages and can have zoomable view of the web page. This is currently experimenting with more flexible mechanisms for tree layout and interaction. It also includes exploring alternative visualization and better methods to manage interacting with large dynamic trees.

A. *Meaningful Spatial Structure*

One of the Zooming Browser potential advantages over previous browsing tools is the ability to spatially organize data in two dimensions at different magnifications. This spatial layout may provide the users with an additional attribute or memory pathway with which to recall the web content.

B. *Sense of Semantic Zooming*

Zooming Browser's spatial layout indicates the semantic difference between two links by their separation in the virtual representation space. Pages are organized in a manner that they can portray virtual separation between them in case of navigation through them.

C. *Improved Overview Support*

Overviews are intrinsic in the nature of ZUIs. It is always possible in Zooming Browser to zoom out so that all web pages are in view. The overview visualization capability exists at different magnifications for the web pages without any additional effort or input from the user.

D. *Inherently Hierarchical*

Hierarchies are a natural format for organizing data. Zooming Browser presents hierarchies in a format that more closely approximates a 2D representation of a tree. Alternatively, it allow for visually distinguishing hierarchy levels by placing them at varying levels of scale or magnification.

V. EXAMPLE

A user begins by writing URL in the address bar in much the same manner as in IE. The web page for the given URL is brought into the canvas. Once the URL given in address bar is brought into the browser, its link is being stored and added in hierarchy as shown in Fig. 3. On this simple page displayed on canvas, user can perform zooming or panning. Fig. 4 shows panning applied on the canvas. Fig. 5 and 6 show the applications of zooming i.e. zoom-out and zoom-in on canvas.

Browsing proposed here is fundamentally hierarchical. Pages are displayed in hierarchical manner in canvas that is the sub links are opened in front of the main page and the further sub links are opened in front of the parent page and so on and so forth. Like Fig. 7 shows this hierarchy of web pages when displayed on canvas. It can also be viewed from the Fig. 7 that along with displaying pages on canvas global tree is also showing the hierarchy of links traversed.

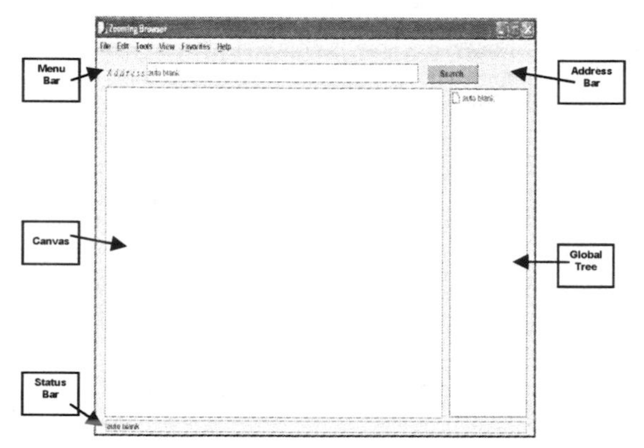

Fig. 1. Zooming Browser

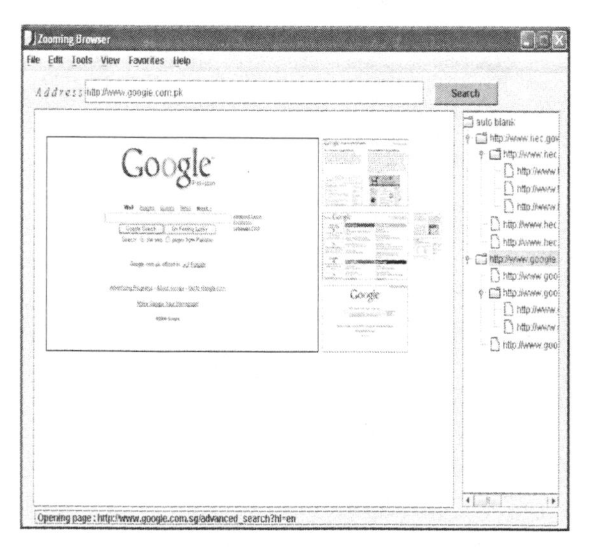

Fig. 2. WWW journey

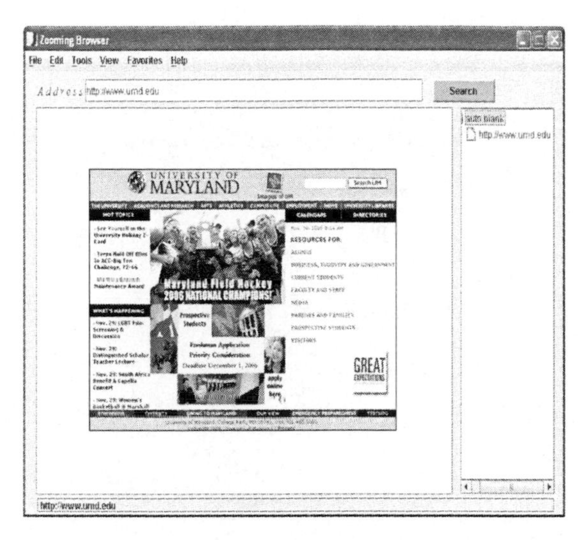

Fig. 3: Displaying URL in Canvas

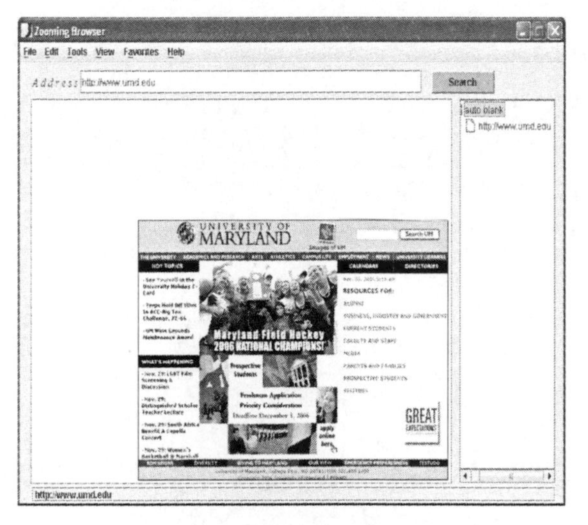

Fig. 4: Applying Panning

VI. CONCLUSIONS

Navigating the WWW presents a struggle between focus and context. The challenge is that how to best support incidental and intentional access while organizing useful information so that it can be effectively retrieved again.

Zoomable Browser provides an extensive graphical workspace where dynamic objects can be placed at any position and at any scale. It also allows WWW pages to remain visible at varying scales while they are not specifically being visited, so the viewer may examine many pages at once. This browser is also exploring a tree layout that permits users to dynamically add to and reorganize a tree of web pages. That selection animates the page to occupy a larger section of the display. Zoomable Browser proposed here, accomplishes the above task.

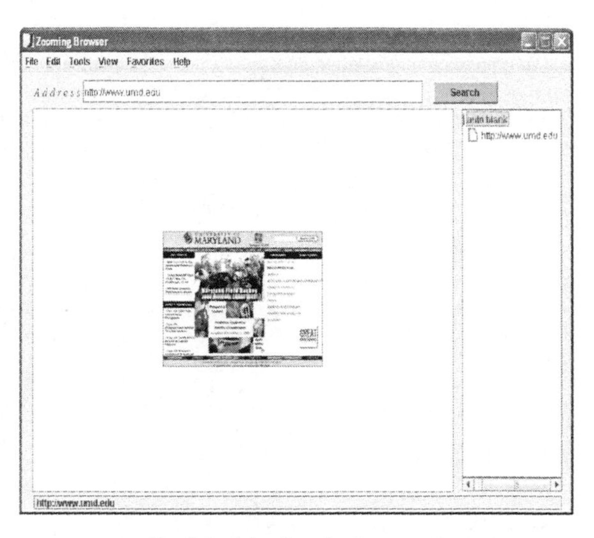

Fig. 5: Applying Zooming (zoom-out)

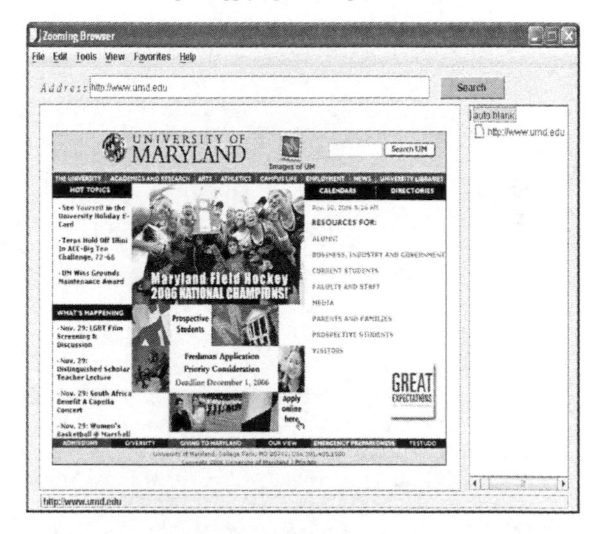

Fig. 6: Applying Zooming (zoom-in)

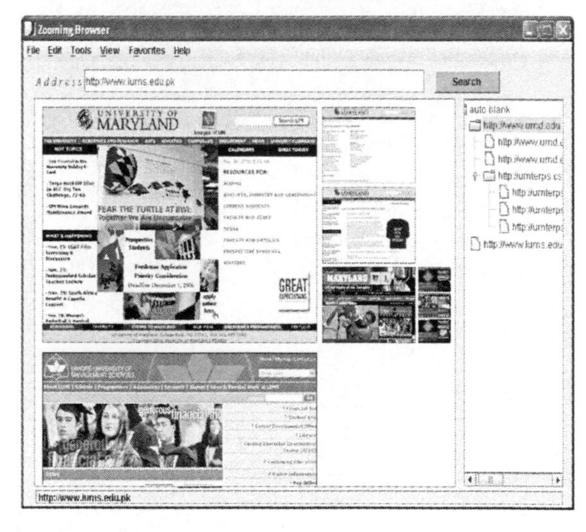

Fig. 7: sub-links view

VII. RECOMMENDATIONS

The browsing technique suggested in this proposal is showing successful web browsing in a zoomable environment when there are HTML pages. No support is currently given for dynamic web navigation. Proposed ZUI browsing paradigm can be applied for dynamic web pages. Further work can be done by implementing support for dynamic web pages navigation, ActiveX controls, applets and DOM object. Security concerns needed to secure web navigation can be implemented when the user privacy is main issue in web sites. Secure Socket Layer (SSL) can be implemented for security reasons.

VIII. REFERENCES

[1] K. Perlin, & D. Fox, "Pad: An alternative approach to the computer interface," proc. of the Computer Graphics (SIGGRAPH '93) 57–64, ACM (1993), New York.

[2] K. Hornbaek, B. B. Bederson, and C. Plaisant, "Navigation Patterns and Usability of Overview+Detail and Zoomable User Interfaces for Maps," proc. of the ACM Transactions on Computer-Human Interaction, 2001, Vol. 9, No. 4, pp. 362-389.

[3] G. W. Furnas, & B. B. Bederson, "Space-Scale Diagrams: Understanding Multiscale Interfaces," proc. of ACM Conference on Human Factors in Computing Systems (CHI '95, Denver, CO, May 7-11) Katz, I. R., Mach, R., Marks, L., Rosson, M. B., and Nielsen, J. Eds. ACM Press, 1995, New York, NY, 234-241.

[4] S. Pook, E. Lecolinet, G. Vaysseix, & E. Barillot, "Context and Interaction in Zoomable User Interfaces," proc. of the 5th International Working Conference on Advanced Visual Interfaces (AVI 2000, Palermo, Italy, May 23-26). ACM Press, 2000, New York, NY, 227-231.

[5] Ghosh, Partha And Shneiderman, "Ben Zoom-Only vs. Overview-Detail pair: A study in Browsing Techniques as Applied to Patient Histories," University of Maryland Technical Report No. CS-TR-4028, 1999, ftp://ftp.cs.umd.edu/pub/hcil/Reports-Abstracts-Bibliography/99-12html/99-12.html.

[6] B. B. Bederson, & J. D Hollan, "Pad++: A zooming graphical interface for exploring alternate interface physics," proc. of User Interface Software and Technology (UIST '94), ACM, 1994, New York, 17–26.

[7] B. B. Bederson, J. D. Hollan, J. Stewart, D. Rogers, A. Druin, D. Vick, I. Ring, E. Grose, C. Forsythe, "A zooming web browser, human factors and web development," 1997.

[8] B. B. Bederson, J. Meyer, & L. Good, "Jazz: An Extensible Zoomable User Interface Graphics ToolKit in Java," proc. of the 13th Annual ACM Symposium on User Interface Software and Technology (UIST'00, San Diego, CA, Nov. 6-8). ACM Press, New York, 2000, 171-180.

[9] G. W. Furnas, & X. Zhang, "MuSE: a multiscale editor," proc. of the 11th Annual ACM Symposium on User Interface Software and Technology (UIST '98, San Francisco, CA, Nov. 1-4). ACM Press, 1998, New York, NY, 107-116.

[10] A. U. Frank, And S. Timpf, "Multiple Representations for Cartographic Objects in a Multi-Scale Tree-An Intelligent Graphical Zoom. Computers & Graphics," 1994, 18, 6, 823-829.

[11] Guo, Huo, Zhang, Weiwei, And Jing Wu "The Effect of Zooming Speed in a Zoomable User Interface," Report from Student HCI Online Research Experiments, 2000.

[12] T. Igarishi, & K. Hinckley, "Speed-dependent automatic zooming for browsing large documents," proc. of the 13th Annual ACM Symposium on User Interface Software and Technology (UIST 2000, San Diego, CA, Nov. 5-8). ACM Press, 2000, New York, NY, 139-148.

[13] D. Schaffer, Z. Zuo, S. Greenberg, L. Bartram, J. Dill, S. Dubs, and M. Roseman, "Navigating Hierarchically Clustered Networks through Fisheye and Full-Zoom Methods," ACM Trans. on Computer-Human Interaction, 1996, 3, 2, 162-188.

[14] C. Ware, "Information Visualization: Perception for Design," San Francisco, CA: Morgan Kaufmann Publishers, 2000.

[15] A. Woodruff, J. Landay, and M. Stonebreaker, "Constant Information Density in Zoomable Interfaces," proc. of the 4th International Working Conference on Advanced Visual Interfaces, 1998 AVI '98, L'Aquila, Italy, Maya 24-27. 110-119.

[16] A. Woodruff, J. Landay, & M. Stonebreaker, "Goal-Directed Zoom," In Summary of the ACM Conference on Human Factors in Computing Systems (CHI '98, Los Angeles, CA, Apr. 18-23). Karat, C.-M., Lund, A., Coutaz, J., and Karat, J. Eds. ACM Press, 1998, New York, NY, 305-306.

A Verifiable Multi-Authority E-Voting Scheme for Real World Environment

T. Taghavi
Communication and Computer Research Laboratory,
Faculty of Engineering,
Ferdowsi University of Mashhad
to_ta70@stu-mail.um.ac.ir

M. Kahani
Computer Engineering Department,
Facultyof Engineering,
Ferdowsi University of Mashhad
kahani@um.ac.ir

A. G. Bafghi
Computer Engineering Department,
Faculty of Engineering,
Ferdowsi University of Mashhad
ghaemib@um.ac.ir

Abstract- In this Paper, we proposed a verifiable multi-authority e-voting scheme which satisfies all the requirements of large scale general elections. We used blind signature for voters' anonymity and threshold cryptosystem for guarantee fairness of the voting process. Our scheme supports all types of election easily without increasing complexity of the scheme. Moreover, our scheme allows open objection that means a voter can complain in each stage while his privacy remains secret. Furthermore, the simplicity and low complexity of computation of the protocol makes it practical for general use.

Keywords- Blind Signature, Digital signature, Threshold Cryptosystem, Open Objection

I. Introduction

Voting can be time consuming, inconvenient as well as expensive, especially when voters and administrators are geographically distributed. With the rapid expansion of the Internet, electronic voting appears to be a less expensive alternative to the conventional paper voting. It also reduces the chances of errors in the voting process. However, in order for electronic voting to replace the conventional mechanisms, it must provide the whole range of features that the conventional voting systems have. Further, due to the inherent lack of security in the Internet, electronic voting systems need to be carefully designed; otherwise these systems become more susceptible to fraud than conventional systems.

Electronic voting has been intensively studied for over the last twenty years. Up to now, many electronic voting schemes have been proposed, and their security as well as their effectiveness has been improved. However, no complete solution has been found in either theoretical or practical domains [1].

In this paper, we propose an effective and secure electronic voting protocol that is suitable for large scale voting over the Internet.

For an electronic voting scheme to be usable in practice, it has to satisfy some requirements. These requirements can be classified into the following three categories as follows:

1. *Basic Requirements:* unreusability, eligibility, privacy, completeness, soundness and fairness.
2. *Extended Requirements:* individual verifiability, universal verifiability, receipt-freeness and open objection.
3. *Practical Requirements:* flexibility, mobility and scalable.

Basic requirements are satisfied in most electronic voting systems and their implementation is relatively easy. But extended requirements are hard to implement and in many cases, they require a large amount of computation and communication.

A. Classification of Schemes

Electronic voting schemes can be classified by their approaches into the following three categories:

1. Schemes using mix-net; eg. [2]
2. Schemes using homomorphic encryption; eg. [3]
3. Schemes using blind signature; eg. [1]

Voting schemes based on mix-net are generally not efficient because they require huge amount of computation for multiple mixers (mixing and proving correctness of their jobs)

The idea of using homomorphic encryption in electronic voting is to sum the encrypted votes, and then decrypt the sum without decrypting individual votes. So, concealing the voter's identity is not required. Voting schemes based on homomorphic encryption use zero-knowledge proof techniques to prove the validity of the ballot. The communication complexity in these schemes is quite high. In addition, these schemes is only suitable for *yes-no* voting. More choices can be added by doing several simultaneous *yes/no* polls, but each added choice then adds to the complexity of the scheme.

In voting schemes based on blind signature technique, each individual vote is decrypted. Therefore, voters must send their votes anonymously to ensure privacy. These schemes are simple, efficient, and flexible, but providing receipt-freeness in these schemes is rather hard, because the voter's blind factor can be used as a receipt of his vote. These schemes support all kinds of voting and since ballots are decrypted, invalid ballots can be detected, and we don't need zero knowledge to construct proofs for verifying correctness of the ballots. This approach is considered to be the most suitable and promising for large scale elections. Since the communication and computation overhead is fairly small even if the number of voters is large. We use blind signature technique in our proposed scheme.

B. Outline of the Paper

In the next section, some of the related work is discussed. In session 3, our administering agents are presented and we explain our protocol briefly without details. The details of the

K. Elleithy (ed.), Advances and Innovations in Systems, Computing Sciences and Software Engineering, 421–426.

protocol are explained in session 4, which is followed by the analysis of the proposed voting protocol and conclusion.

II. Related Works

The first actual voting protocol employing blind signatures appeared in 1992 [4] and the first implementation in 1997 [5]. A somewhat improved version was offered by He & Su in 1998 [6]. The basic idea underlying all of these schemes is to employ two logical authorities, a registrar and a tallier. The registrar validates eligible voters and provides them with anonymized certified voting tags, using blind signatures. The voters then cast their (blindly signed) ballots over anonymous communication channels to the tallier.

In [4] the encrypted ballot that is cast by the voter contains the voter's signature that allows the voter to identify its ballot in the published list. Thus any entity that can verify the voter's signature is able to link a voter to a cast ballot and anonymity of the voter is not ensured.

Reference [7] proposed a voting scheme, which he himself later showed to lack the postulated receipt-freeness; a repaired version by the same author, making use of blind signatures, appears in [8]. Although theoretically sound, this scheme suffers from the fact that it depends on untappable channels or voting booths. These cannot be implemented over the Internet, making the schemes not practical for a real world remote electronic election.

In [9] voters trying to vote twice will be traced. The scheme requires existence of the anonymous channel supporting replays (recipient of the anonymous message can send a replay to the anonymous sender).
The eligibility is achieved if the authority is honest. If a voter complaints, his privacy can be compromised (at least the authority will get to know his vote).

In [1] voter cast his or her ballot anonymously, by exchanging untraceable yet authentic messages. It is suitable for large scale voting over the Internet; however it can't provide receipt-freeness and fairness.

III. Model of Our Electronic Voting

In this section, a brief overview the proposed voting scheme is presented.

A. Administering Agents

1. A Certificate Keys Authority – *CKA*. He certificates the public keys of voters and authorities.
2. A Voter Registration Authority – *VRA*. He verifies the identities and eligibilities of voters and then issues Register Certificate (*RC*) to voters in the registration stage.
3. A Voter Certifying Authority – *VCA*. He prepares blank ballots and distributes one to each voter also certifies a ballot that is cast, has been cast by a registered voter and that voter has cast one and only one ballot.
4. A Vote Compiler – *VC*. Each filled ballot cast by a voter is delivered to the vote compiler. After creating decryption key, the vote compiler tallies the votes and announces all the relevant statistics pertaining to this voting process.
5. N talliers $-T_j$ (j = 1... N). They cooperate with each other and create private key of the system that has been shared between them before starting the election. This key will use to decrypt the votes.
6. A *judge* – An independent authority for handling the complaints. Each voter that has complaint about his vote sends his complaint to this authority.

B. Notations

In this session we define a set of notations, and then use them to describe our protocols. The notations are defined as follow:

1. X – The identity of an agent
2. X_e – agent X's public key
3. X_d – agent X's private key
4. h (M) – a digest of the message M
5. $[m, X_e]$- an entity m encrypted with X's public key, where X is the recipient of m
6. $[m, X_d]$– an entity m signed with X's private key, where X is the originator of m
7. VS_e – public key of the voting system
8. S–secret key of the voting system

C. Trap-door Bit Commitments

In a trap-door bit commitment scheme, where a voter V has committed to a message M, it is possible for V to open M in many different ways. We use trap-door bit commitment that [7] used to achieve receipt freeness.

In this section we describe the trap-door bit-commitment that we will use in vote casting stage. But there is a weakness in this approach that we are currently working on it to solve efficiently.

Several parameters p,q,g,h are generated and published before the election by the voting system. Where p and q are prime, $q \mid p-1$, g and h are in Z_p^* and

q = *order (g)* = *order (h)* $\quad g^q \equiv h^q \equiv 1 (\bmod p), g \neq h \neq 1$

Here ω such that $h = g^{\omega} \bmod p$ is not known to any party.

These prime numbers p and q are different from prime numbers that are used to generate pair keys of voting system.

Voter randomly generates $\alpha \in Z_q$ and calculates $G = g^{\alpha} \bmod p$. Trap-door bit-commitment is defined as follow

$$\beta = BC(v,k) = g^v G^k \bmod p$$

Where v is voter's vote and k is a random number. Voter can open this bit commitment in many ways, $(v,k),(v',k')$ etc., using α such that $v + \alpha k \equiv v' + \alpha k' (\bmod q)$

The trap door bit-commitment is essential for satisfying receipt freeness. If the value of α is generated by voter as specified, then the scheme satisfies the receipt- freeness. However, if α is generate by a coercer and he forces voter to use $G = g^{\alpha} \bmod p$ for voter's bit-commitment, then voter cannot open $\beta = BC(v,k)$ in more than one way, since voter does not know α. Hence, the voting scheme is not receipt free and coercer can coerce voter. To prevent this attack, voting authority must be sure that voter knows α that is used to create bit-commitment. On the other hand the authority must not know α because he can trace the voter through α. This is still an open area in this research and we will work on it in our future works.

D. Overview of the Proposed Voting Protocol

Before the voting period, voters have to register with Voter Registration Authority (*VRA*) to be an eligible voter. The authority then issues a certificate for each such registered voter. Then during the election voter sends his certificate to the Voter Certifying Authority (*VCA*) to obtain a blank ballot. If he is an eligible voter, *VCA* sends a blank ballot to him. Each blank ballot has a unique serial number. Voter uses this serial number to create voter mark, and then he blinds and signs it. After that voter sends blinded voter mark to the *VCA* to obtain *VCA*'s signature on it. If *VCA* hasn't already received a blinded voter mark from that voter, he signs blinded voter mark for him and then sends signed blinded voter mark to voter. Voter un-blinds it and obtain *VCA*'s signature on it. In vote casting stage voter sends his signed voter mark and his vote to the Vote Compiler(*VC*) through an anonymous channel. Vote casting in our protocol is a process similar to uploading to a site-that is the identity of the voter is not provided in the message. At the deadline of the voting talliers cooperate with each other to create secret key of the voting system and then send this secret key to the *VC* to decrypt the votes.

IV. Proposed Multi-Authority Electronic Voting Scheme

In this session we have described our protocol completely.

A. Pre-System Set up

1. Each entity goes to the election site and download key generation applet. After that he generates a pair of keys (such as RSA public key and private key).
2. Each entity (voters and authorities) must go to the Certificate Keys Authority (*CKA*) and registers his public key and receives his public key certificate and *CKA*'s public key. So each entity has his own public key that is certified by *CKA*.

B. System Set up

- *Sharing secret key of voting system:* N talliers execute the key generation protocol of $(t; N)$-threshold encryption scheme and as a result each tallier T_i possesses his share S_i of a secret S. Any cooperation of more than t talliers can decrypt an encrypted ballot. [10]
- *Publishing authorities' public keys:* public key of the *VRA*, *VCA* and *VC* that contains *CKA*'s signature, are published on the site.
- *Publishing the list of candidates:* The *VRA* publishes the list of L candidates, their certificates and their advertisement on the public site. Candidate's certificate has *VRA*'s signature and ensures that this candidate has already registered as a candidate and he is eligible.

C. Voter Registration

Voter Registration Authority => *VRA*

Voter Identification =
Request, [request, V_d], voter's public key certificate

1. $V \rightarrow VRA$: Voter Identification
2. $VRA \rightarrow V$: [{request, name of the e-voting}, VRA_d]
= voter Register Certificate (RC)

In any election, an individual must register to be an eligible voter. This is done before the voting period. The voter must register with Voter Registration Authority. Voter registration is done as follows:

1. Voter sends his request, signed request and his public key certificate to VRA. (Request at least consists of voter's *ID*). *VRA* makes sure that the voter himself sends this request because of the voter's signature on it.
2. Registration Authority checks the users with the National Registration Database[1] to determine the eligibility of a voter and his precinct. If a voter is eligible and hasn't registered before, The authority issues a certificate for each such registered voter that contains the voter's request and name of the e-voting that are signed with *VRA*'s private key. If voter doesn't have the right, *VRA* gives him an error message.

Note: *VRA* makes sure that the voter himself has sent the request because of the voter's signature on it. If voter doesn't sign his request, anyone that knows his public key certificate and his *ID* can send this request and gives register certificate. So when original voter wants to register, *VRA* sends an error message to him because his *ID* has been already existed in *VRA*'s database. However counterfeit voter won't be able to cast vote instead of original voter because in voter certification stage he must sign voter mark. But he can prevent original voter to vote.

D. Voter Certification

Voter Certifying Authority => *VCA*

1. $V \rightarrow VCA$: RC, voter's public key certificate
2. $VCA \rightarrow V$: [{y, [h (y), VCA_d], N}, V_e]
N is a random nonce; y is a ballot serial number
3. $V \rightarrow VCA$: [{m×[r,VCA_e],[h(m×[r,VCA_e]),V_d],V_{ID}, N},VCA_e]
m is a voter mark generated by the voter
r is a random number that is blind factor
4. $VCA \rightarrow V$: [[{m× [r,VCA_e]},VCA_d],V_e]

1. The voter sends his Register Certificate and his public key certificate to *VCA*.
2. *VCA* first makes sure that the voter is a registered voter by verifying the *VRA*'s signature on the voter register certificate, and then checks whether he or she has received blank ballot before. if he hasn't received blank ballot before *VCA* sends a blank ballot to the voter that is encrypted with the voter's public key. The blank ballot is a message of two fields (i) the ballot serial number field, y and (ii) *VCA* signed digest of the ballot serial number, [h(y), VCA_d]. *VCA* generates a unique serial number, y, for every voter then

[1] This database contains information about all people who lived in the country such as name, family, father's name, date of birth, user ID, etc.

creates a list of ballot serial numbers and voter register certificates. This is the list of the blank ballots issued and will be published by *VCA* at the end of the voting. So everyone can check that blank ballots issued to the registered voters.

In this stage *VCA* sends a random nonce with the blank ballot to the voter to ensure that the voter himself responds the message. In the next stage that voter sends his blinded voter mark to the *VCA*, he adds this nonce to his message. We use nonce to prevent replay attacks.

3. When the voter receives the message, he makes sure that *VCA* sends this blank ballot because of *VCA*'s signature on the blank ballot also he makes sure that the blank ballot has not been tampered with during transit, including that nobody has put an identifying mark within the blank ballot .The voter then retrieves the serial number, *y*, from the received message. Using the serial number, *y*, the voter creates a voter mark, *m*, as follows: The voter pads *y* with a fixed length random number to obtain a number *x*. The voter then computes a hard to invert permutation, *m*, of *x*. The value, *m*, is the voter mark [1]. Note that since the serial number *y* is unique for every voter the voter mark is unique to every voter. However from the voter mark it is not possible to obtain the serial number *y* and hence impossible to identify the voter.

   The voter then blinds the voter mark with a random number, *r*, to get the blinded voter mark $m\times [r, VCA_e]$. The voter also computes a digest of the blinded voter mark and signs the digest. The voter encrypts the blinded voter mark, the signed digest of the blinded voter mark, his *ID* and the random nonce that *VCA* had sent to him in the previous stage, with *VCA*'s public key.
4. *VCA* first makes sure that the voter is a registered voter and has get a blank ballot from *VCA,* by searching voter's *ID* in his database. (All registered voters that have given blank ballot are in *VCA*'s database). *VCA* also makes sure that the voter has not submitted earlier, another blinded voter mark to sign. Since the vote cast by the voter later on will be accompanied by the voter mark, this step effectively ensures that the voter casts one and only one vote. By verifying the voter's signature on the digest of blinded voter mark and verifies random nonce, *VCA* makes sure that the voter himself sends blinded voter mark also *VCA* makes sure that the voter mark has not been tampered with in transit.

   VCA verifies voter's signature on his blinded voter mark then signs his blinded voter mark. *VCA* encrypts signed blinded voter mark, with voter's public key and sends it to the voter.

Objection: Voter verifies *VCA*'s signature on his blinded voter mark if it isn't *OK*, he can complain to the *judge*. In this case voter must send the signed blinded voter mark that receives from *VCA* and there are *VCA*'s signatures on it, to the *judge,* on the other hand *VCA* must send blinded voter mark that receives from voter and there is voter's signature on it, to the *judge*. The *judge* verifies the *VCA*'s signatures on them. If the *VCA*'s signature isn't *OK*, *judge* forces *VCA* that re-sings the blinded voter mark for that voter.

$V \rightarrow judge$:

$[\{[\{m\times [r, VCA_e]\}, VCA_d], \text{public key certificate}, V_{ID}\}, judge_e]$

$VCA \rightarrow judge$:

$[\{m\times[r, VCA_e], [h\,(m\times[r, VCA_e]), V_d], V_{ID}\}, judge_e]$

Note that here blinded voter mark is sent to the *judge* therefore *judge* doesn't know actual voter mark and so the privacy of voter remains secret.

E. Vote Casting

Vote Compiler => *VC*

1. $V \rightarrow public - site$

$[\{[\{vote, \beta, k, G, [m, VCA_d]\}, VS_e], \beta, [m, VCA_d]\}, VC_e]$

Reminde : β is trap door bit-commitment and $G = g^{\alpha} \bmod p$

2. $public - site \rightarrow V$

$[\{[\{vote, \beta, k, G, [m, VCA_d]\}, VS_e], \beta, [m, VCA_d]\}, VC_e]$

3. $VC \rightarrow public - site$

$[h([\{vote, \beta, k, G, [m, VCA_d]\}, VS_e]), VC_d], [h(\beta, [m, VCA_d]), VC_d]$

4. $public - site \rightarrow VC$

$[h([\{vote, \beta, k, G, [m, VCA_d]\}, VS_e]), VC_d], [h(\beta, [m, VCA_d]), VC_d]$

Vote casting in our protocol is a process similar to uploading to a site – that is the identity of the voter is not provided in the message – unlike the other steps. At best, an IP address be traced back but cannot be linked with a voter. That way we ensure the anonymity of the voter. The voter "un-blinds" the signed blinded voter mark and obtains *VCA's* signature on it.

1. The voter now prepares a fixed length message of a pre-determined format (the format is announced to all voters prior to voting initiation) and indicates his or her vote as the message's content. The voter appends his signed voter mark, to this message (signed by *VCA*). Recall that it is not possible to recover the serial number from the voter mark, so it is not possible to identify the voter by the voter mark. Also voter appends his G, trap door bit-commitment and the random number used in the trap door, to his vote and then encrypts them with public key of the voting system afterwards, encrypts this message, signed voter mark and trap door bit-commitment (*β*) with *VC*'s public key and, after waiting for a random amount of time, uploads the same onto a publicly up loadable site announced to voters before. The mechanism used to upload does not associate, in any manner, the voter's identity with the uploaded material – for example an anonymous/guest ftp mechanism or proxy mechanism.
2. Periodically, *VC* downloads cast votes from this site. *VC* checks that it has not received this voter mark before.
3. If *VC* has not received this voter mark before, he signs the digest of the submitted vote and the digest of the voter mark and *β* then uploads them to the public place.

   submitted vote means: $[\{vote, \beta, k, G, [m, VCA_d]\}, VS_e]$
4. Sometime later the voter retrieves these signed from the public place. This guarantees that *VC* has received the voter's vote. Voter verifies the *VC*'s signature on the hash value of his submitted vote and makes sure that his vote hasn't been tampered.

Note: $[h(\beta, [m, VCA_d]), VC_d]$ will use for universal verifying and complaining. We discuss more about this in next stages.

Objection: If submitted vote has been tampered, voter can complain. The voter sends his submitted vote and signed digest of it with *VC*'s private key to the *judge* through anonymous channel (public sit in our protocol).

$V \rightarrow judge: [\{[\{vote, \beta, k, G, [m, VCA_d]\}, VS_e],$

$[h([\{vote, \beta, k, G, [m, VCA_d]\}, VS_e]), VC_d]\}, judge_e]$

Note that since submitted vote exists on the public site the voter can't change it before sending to the judge

F. Vote Counting

At the deadline of voting, N talliers jointly execute the $(t; N)$-threshold decryption protocol to obtain secret key of the voting system, any subset of t talliers can construct the secret key.

1. Talliers construct the secret key.
2. Alternatively each tallier that participates in constructing the secret key, signs it with his private key.
3. Signed secret key is encrypted with Vote Compiler's public key.
4. Sequence of talliers that sing the secret key is encrypted with Vote Compiler's public key.
5. Encrypted secret key and encrypted sequence send to the Vote Compiler.

$T \rightarrow VC: \quad [\{d_1, d_2, d_3 \ldots d_t \ldots d_n\}, VC_e]$

$$[[\ldots[\ldots[[secret - key, T_{d_1}], T_{d_2}] \ldots, T_{d_t}] \ldots, T_{d_n}], VC_e]$$

T_{d_1} : First tallier's private key T_{d_n} : Nth tallier's private key

When *VC* receives the message, he first decrypts it and then verifies tallier's signature on it. To verify tallier's signature, *VC* must know the sequence of the talliers that sign the secret key. This is the reason that we send sequence of talliers that sing the secret key together with signed secret key.

VC makes sure that at least t talliers participate in decryption process. When *VC* obtains secret key, he decrypts the votes.

G. Publishing

At the end of voting each entity publishes the following data:

a. VRA publishes:

i. list of registered voters (voter certificate)

b. VCA publishes:

i. number of blank ballots (N_{bb})

ii. serial numbers of ballots together with *RC*

iii. blinded voter marks received from the voter, $m \times [r, VCA_e]$

iv. their digests signed by the voters $[h(m \times [r, VCA_e]), V_d]$

v. the corresponding blinded voter marks signed by *VCA*

c. VC publishes:

i. Signed voter marks and their corresponding β in random order. $[m, VCA_d], G, \beta$

ii. List of voter marks that haven't valid vote

iii. List of correct votes in random order.

iv. Non-interactive modification of zero-knowledge proof, σ to prove that the list of valid votes contains only correct open values of the list of β (bit-commitment votes) without revealing the linkage between β and vote [7].

H. Objections

After publishing the results, some voters may have a complaint about their published vote. Here we describe that in each condition the voter how can send his complaint to the judge that his privacy remain secret.

Condition 1: when *VC* decrypts a vote he check that whether β is the correct bit-commitment of the submitted vote or not. If it's not *OK* he doesn't count that vote and at the end of counting he publishes corresponding voter mark with error message instead of β. Also a voter may cast empty ballot or ballot with invalid content, for example the name that is in the vote, isn't the name of candidates that are published before. In these cases *VC* publishes corresponding voter mark with error message instead of β and that vote isn't counted.

When the results are published if a voter mark of the voter contains an error message instead of β he can complain to the *judge*. He sends following message to the *judge* through an anonymous channel (in our protocol through public sit):

$[h([\{vote, \beta, k, G, [m, VCA_d]\}, VS_e]), VC_d], judge_e]$

Condition 2: After publishing the result if the voter's vote has been changed, the voter can protest. He sends his signed voter mark and β that there is *VC*'s signature on them and the signed voter mark and β that are published, to the *judge* through public sit.

$[\{[h(\beta, [m, VCA_d]), VC_d], published\ \beta, [m, VCA_d]\}, judge_e]$

Condition 3: After publishing the result if the voter's vote has been deleted, the voter can protest. He sends his signed voter mark and β that there is *VC*'s signature on them to the *judge* through public sit. $[[h(\beta, [m, VCA_d]), VC_d], judge_e]$

V. Analysis of the Voting Protocol

A. Basic Requirements

Unreusability: When *VCA* signs a blinded voter mark it makes sure that it does not sign two blinded voter mark from the same voter. Also each voter mark corresponds to one and only one voter and each voter mark is unique. Because the seed that is used to generate the voter mark is the unique serial number, the voter mark guarantees that two cast votes are not erroneously attributed to the same voter.

Eligibility: *VCA* makes sure that ineligible voters do not get a blank ballot because *VCA* first makes sure that the voter is a registered voter by verifying the *VRA* signature on the voter certificate. Also when *VCA* wants to sign a blinded voter mark he checks that voter's *ID* has already existed in his database as a valid voter.

Completeness: When the vote compiler, *VC* receives a vote with the signed voter mark; it signs the digest of it. *VC's* signature on the digest ensures that *VC* cannot claim later that it did not receive a valid vote. When the votes are officially published, a voter will be able to identify his or her vote by the voter mark. These two together guarantees every vote that cast to the *VC* is counted in the final tally.

Soundness: The content of the vote may be invalid. For example the ballot is empty or voter writes name of person who isn't a candidate, etc. Since all votes are decrypted at the end of e-voting so invalid votes are detected and they won't be counted in the final tally.

Fairness: In this protocol we use threshold encryption to encrypt the votes. At the deadline of vote casting talliers cooperate with each other and construct the secret key so before vote counting stage *VC* doesn't know the secret key. So during the voting *VC* won't be able to decrypt the votes.

Privacy: At every stage, till the vote is cast, the voter makes sure that none of the agents has put an identification mark on his vote. When a voter casts a vote the only thing that can possibly be identified with the ballot is the IP address of the server that the voter used to cast the vote (the server can be thought to be like an open work station accessible everyone that wants to vote). This does not reveal the voter's identity. Although the voter mark generated by the voter contains the unique serial number, y, it is computationally infeasible to compute y from the voter mark. Also when the filled ballot is transferred to *VC*, it is encrypted in transit. Thus only the voter knows about his vote till such time as the vote is cast.

B. Extended Requirements

Individual verifiability: When the votes are published a voter can identify his/her vote by the voter mark. If the identified vote does not match the vote that the voter actually cast, voter can send a complaint to the judge without revealing his privacy. We discuss about this situation in session 4.8.

Universal verifiability: *VRA* publishes certificates of registered voters. *VCA* publishes a list of ballot serial numbers and voter register certificates. So everyone can check that blank ballots issued to the registered voters and everyone can check that if The number of blank ballots is greater than or equal to the number of received votes but less than or equal to the number of registered voters.

VCA publishes blinded voter marks that there is voter's signature on them. So everyone makes sure that voter marks are created by eligible voters also *VCA* publishes signed blinded voter marks with his private key so everyone can check that *VCA* signs all valid voter marks. When voters send their vote through public site, *VC* signs their voter mark and their bit-commitment and put it on the site. $[h(\beta, [m, VCA_d]), VC_d]$ is used for universal verifying. For each β and signed voter mark that published at the end of e-voting,, corresponding signed of them that have *VC*'s signature should be existed on the public site. So everyone makes sure that all counted votes are confirmed by *VC* and send through public site. Also this message ensures that *VC* cannot claim later that it did not receive a vote from that voter mark

Receipt-freeness: The proposed election scheme provides receipt-freeness by using of trap door bit-commitment. So voter can open this bit commitment in many ways. The most important thing in this way is that voter knows α (random number used in create G) to be able to open his bit-commitment in many ways. When VC publishes the result, he publishes signed voter marks and corresponding bit-commitment index by the voter marks. On the other hand *VC* publishes List of correct votes in random order. And then by using of Non-interactive modification of zero-knowledge proof, σ he proves that the list of valid votes contains only correct open values of the list of β (bit-commitment votes) without revealing the linkage between β and vote [7].

If voter himself creates α and he knows it our protocol is receipt-free but if voter doesn't know α he can't open his vote in many ways and our protocol isn't receipt-free. We are working to solve this weakness.

Open Objection: In each stage we explained that how voter can complain while his privacy remains secret.

C. Practical requirements

Flexibility: In this protocol we use blind signature and anonymous channel (public site) so we can use different type of voting because all individual votes are decrypted

Mobility: In this protocol, voters use internet to cast their votes so anywhere that a voter can access to the internet, can cast his vote.

Scalable: our scheme is good for large scale election with a lot of voters and candidates because our scheme doesn't involve complex and high computational overheads.

VI. Conclusion

In this paper we have proposed an efficient electronic voting scheme that is suitable for large scale voting over the Internet. The protocol satisfies the core properties and extended requirement of secure voting systems. Receipt-freeness in our protocol isn't achieved completely and we want to satisfy it in our future works. We use the Internet for electronic voting so voters can participate in voting in any place they like over the Internet. Then electronic voting system can play an important role to increase the participation rate in voting.

Acknowledgment

The first author would like to thank FUM Communication and Computer Research Laboratory for their in-kind support and encouragement during this research.

References

[1] Indrajit Ray and Indrakshi Ray. "An Anonymous Electronic Voting Protocol for Voting Over The Internet".2001

[2] M. Abe, "Universally verifiable mix-net with verification work independent of the number of mix- servers", Advances in Cryptology – Eurocrypt'98

[3] B. Lee and K. Kim. "Receipt-free Voting Scheme with a Tamper-Resistant Randomizer".2002

[4] A. Fujioka, T. Okamoto, and K. Ohta. "A practical secret voting scheme for large scale elections." Advaced in Cryptology - AUSCRYPT'92, 1992.

[5] L. F. Cranor and R. K. Cytron, "Sensus: A security-conscious electronic polling system for the Internet". Proceedings of the Hawaii International Conference on System Science, 1997.

[6] Q. He and Z. Su, "A new practical secret voting scheme for large scale election". Information Security Conference, 1998

[7] T. Okamoto, "An Electronic Voting Scheme", proc. Of IFIP'96, Advanced IT Tools, 1996.

[8] T. Okamoto. "Receipt-free electronic voting scheme for large scale election", Proc. of Workshop on Security Protocols'97.

[9] M.J. Radwin. An untraceable, universally verifiable voting scheme. 1995.

[10] Adi Shamir, "How to Share a Secret", Massachusetts Institute of Technology, 1979.

Stochastic Simulation as an Effective Cell Analysis Tool

Tommaso Mazza
t.mazza@unicz.it
University "Magna Græcia" of Catanzaro
Viale Europa, Campus of Germaneto
88100, Catanzaro ITALY

Abstract – **Stochastic Simulation is today a powerful tool to foresee possible dynamics of strict subsets of the real world. In recent years, it has been successfully employed in simulating cell dynamics with the aim of discovering exogenic quantities of chemicals able to deflect typical diseased simulation paths in healthy ones. This paper gives a large overview of the stochastic simulation environment and offers an example of a possible use of it on a pathway triggered by DNA damage.**

I. Introduction

Often when the knowledge of the probability of a sure event is fundamental, but there are too many conditioning variables to exactly get an analytical solution, one resorts to *simulated sampling methods*. Replacing the analytic appraisal with the empiric observation of a phenomenon and mining implicit information not analytically noticeable from this, to reproduce a generic mechanism the stochastic simulation is largely used. As an example, the observed frequency of a sure event constitutes an appraisal of the probability of that event (providing that the sampling has been simulated for a consisting number of times). Then, with the term *simulation* one intends the capability of replicate either an existing or a virtual reality by means of opportune models with the aim of studying either the effects of possible interventions or those events in some way foreseeable.

In the past, models have been widely used in decisional process. Those said "scale models", are able to faithfully represent a local reality but, in the same time, they exhibit elevate rigidity and high realization costs. Other models very often used are the "analytic models" in which the reality to be observed is represented by means of logic/mathematic variables and relations. The "linear programming" problems belong to this class. In many cases, they allow to determine one or some optimal (or suboptimal) solutions for a given problem. Moreover, the more complex the problem is and the more difficult and onerous the resolution of this one will be.

In this context, take place the *computational models* so named because of the use of calculators not only as a computational unit but also as a way to represent parts of the model by means of information objects (data structures, programs, etc…) take place. Therefore, the correspondence between reality and model is not based on dimensional reduction but on a functional relation. Respect to a direct, very expensive and usually not practicable experimentation, a computer aided simulation has the advantage of versatility, rapidity of realization and low costs. By means of the simulation, it is possible to try quickly policies and planning chooses and to model systems also of big dimensions and biggest complexity exploring in the time theirs behaviours and trends.

Just biological and chemical systems are very complex and intricate for manual inspection and recent predictive techniques to figure out possible evolutions of interesting strict subsets of a more exhaustive reality are emerging. One will talk about techniques of discrete and stochastic simulation that will be the object of this dissertation.

In this context, the next section will open a wide window on the basis of the simulation; the section 3 will go deeply into stochastic methods applied in biology and medicine successfully up to now. The section 4 will explicate the modelling approach adopted here; the section V will show a real biological use case while the section VI will be the conclusion of this dissertation.

II. Simulation: Basic Concepts

Simulation is a discipline acting to develop a high level of understanding about interactions happening inside a system. After some consideration regarding a meaningful way to put System, Model, and Simulation in an appropriate perspective, one arrives at the following distinction:

- **System.** A system is an entity which maintains its existence through the mutual interaction of its parts. A system exists and operates in time and space.
- **Model.** A model is a simplified representation of a system over some time period or spatial extent intended to promote understanding of the real system.
- **Simulation.** A simulation is the manipulation of a model in such a way that it operates on time and/or

K. Elleithy (ed.), Advances and Innovations in Systems, Computing Sciences and Software Engineering, 427–432.

space to compress it, thus enabling one to perceive the interactions that would not be otherwise apparent because of their separation in time or space. This compression also provides a perspective on what happens within the system, which, because of the complexity of the system, would probably otherwise not be evident.

In general, to simulate a system it's needed to build a model and the first step to do this is to analyze the system one wants to face and solve. Therefore, the component, their inter-relations and activities it's needed to discover.

Entities and **resources** are the basic compounds in a system. Entities are individual objects of which keeping track during the simulation, while resources are common elements which generally don't require to be individually modelled and of which the information kept is just the number of the resources. The same element could be considered as an entity or a resource according with the aims of the simulation. Entities are structured in hierarchical *classes* characterized both by *attributes* more o less shared between the other classes and by *operations* that entities can carry out. In the time, entities can book many *states,* where a state is the situation in a given instant in which the system or an entity can be found. Entities pass from a state to another with the firing of an **event**, namely of an action that causes a significant change in the system. An event marks the beginning and the end of an **activity** that is responsible of changing some characteristics of the system. Sometimes it's worth grouping a series of events (and then of states) into an events collection named **process**. In this context, simulation acts as an effective tool to describe interrelations between entities and furnishes *time-course-processes* according to wished probability distribution functions.

But upstream it's useful to carefully describe the simulation architecture taking care of its characterizing elements:

- **Entity, Event and Activity.** Just seen.
- **Current State of the System.** The system is fully characterized by any between a finite set of instantaneous states: $S = \{s, s', ...\}$. Activities in the system perform actions $Ac = \{\alpha, \alpha', ...\}$ which transform the state of the system. A *run* or process is thus a sequence of interleaved system states and activities: $r\colon e_0|\alpha_0 \rightarrow e_1|\alpha_1 \rightarrow e_2|\alpha_2 \rightarrow ... \; e_{u-1}|\alpha_{u-1} \rightarrow e_u$.
- **Random Number Generator.** Random number generators use iterative deterministic algorithms to produce a sequence X_i of pseudo-random numbers that approximate a truly random sequence. For a survey look at [3] and [4].
- **Time of Simulation.** Time is an important concept in any performance model. It is distinctly different than the CPU time used in executing the model or the "real world" time of the person running the model. Simulation time starts at zero and then advances unevenly, jumping between times at which the state of the model changes. It is impossible to make move time backwards during a simulation run.
- **The controller.** It is the system component that handles the sequence of the events and the evolutions of the states in the time. In particular it takes care of the advancement of the simulation time. Generally, this function is provided by each kind of simulation architecture, even if it changes frequently in functioning modalities. To diverse approaches of the simulation, many implementations exist about the controller. The most famous are: *activity-based* and *process-based* controllers.
  - **Activity-based:** the system activities are decomposed in elementary ones. These activities correspond to the events that lead to state changing in the system. The role of the controller will be that to handle the list of activities in such a way they will be correctly carried out. Each activity is characterized by the condition for which it becomes and by the action consequentially fired: *event[condition]* $\rightarrow$ *action.*
  - **Process-based:** In this approach, all events of the life-cycle of an entity, with its relative operations, are grouped in a process. A process can be *active* or *waiting* on a condition and, for each entity/process, the controller must keep a list with two information: (i) *future events* and (ii) *current events*. This method is very natural and used even if it requires care in the inter-process relations modelling.
- **Events List.** Many algorithms which can be used to schedule events in a general purpose discrete simulation system are considered. Proper management of the event list is critically important to achieve execution efficiency in discrete event modelling. For such systems, the use of a poor event list algorithm can result in a sure unbearable inefficiency. Conversely, replacement of a poor algorithm with a good algorithm can highly reduce execution times. Look [1] for a survey and [2] for a tutorial.
- **Statistic Results.** Generally, according to a desired probability distribution function, simulation techniques give single trajectories between all possible. But generating many trajectories, one may estimate any parameter of interest by calculating the value of the parameter for each trajectory and observing the statistics of those calculated values. For example, to find the average number of chemical A present at time t, one can runs many trajectories (hundreds or thousands) and plot a histogram of the values of the number of chemical A at time t.

III. STOCHASTIC SIMULATION: USE IN BIOLOGY

A stochastic process [5] is one whose behaviour is non-deterministic in that the next state of the environment is partially but not fully determined by the previous state of the environment. Then, it is a sequence of measurable functions, that is, a random variable X defined on a probability space (Ω, S, Pr) with values in a space of functions F. The space F in turn consists of functions $I \rightarrow D$. Thus a stochastic process can also be regarded as an indexed collection of random variables $\{X_i\}$, where the index i ranges through an index set I, defined on the probability space (Ω, S, Pr) and taking values on the same codomain D (often the real numbers R). *This view of a stochastic process as an indexed collection of random variables is the most common one.*

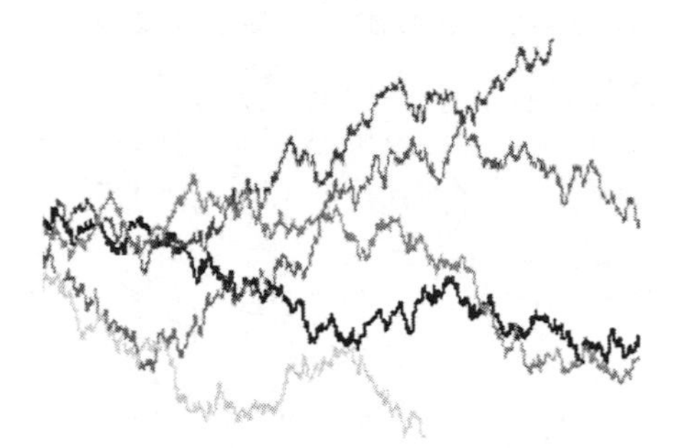

Fig. 1. Five Stochastic Processes.

In a continuous stochastic process the index set is continuous (usually space or time), resulting in an uncountable infinite number of random variables. Each point in the sample space Ω corresponds to a particular value for each of the random variables and the resulting function (mapping a point in the index set to the value of the random variable attached to it) is known as a realisation of the stochastic process. In the case the index family is a real (finite or infinite) interval, the resulting function is called a sample path. A particular stochastic process is determined by specifying the joint probability distributions of the various random variables.

In chemistry before and systems-biology then, great steps have been made after the determination of the two celebre versions of the algorithm of simulation still now frequently used. The Gillespie algorithms [6] generate a statistically correct trajectory (possible solution) of a stochastic equation. They were developed and published by Dan Gillespie in 1977 to simulate chemical or biochemical systems of reactions efficiently and accurately using limited computational power. As computers have become faster, the algorithm has been used to simulate increasingly complex systems. The algorithm is particularly useful to simulate reactions within cells where the number of reagents typically number in the tens of molecules (or less). Mathematically, it is a variety of a Dynamic Monte Carlo method and similar to the Kinetic Monte Carlo methods. It is used heavily in Computational systems biology.

Traditional continuous and deterministic biochemical rate equations do not accurately predict cellular reactions since they rely on bulk reactions that require the interactions of millions of molecules. They are typically modelled as a set of coupled ordinary differential equations. In contrast, the Gillespie algorithm allows a discrete and stochastic simulation of a system with few reactants because every reaction is explicitly simulated. When simulated, a Gillespie realization represents a random walk that exactly represents the distribution of the Master equation.

The physical basis of the algorithm is the collision of molecules within a reaction vessel. It is assumed that collisions are frequent, but collisions with the proper orientation and energy are infrequent. Therefore, all reactions within the Gillespie framework must involve at most two molecules. Reactions involving three molecules are assumed to be extremely rare and are modelled as a sequence of binary reactions. It is also assumed that the reaction environment is well mixed.

Below is a summary of the steps to run the Gillespie algorithm:

0	**Initialization:** Initialize the number of molecules in the system, reactions constants, and random number generators.
1	**Monte Carlo Step:** Generate random numbers to determine the next reaction to occur as well as the time interval.
2	**Update:** Increase the time step by the randomly generated time in Step 1. Update the molecule count based on the reaction that occurred.
3	**Iterate:** Go back to Step 1 unless the number of reactants is zero or the simulation time has been exceeded.

Table. 1. Computational Steps for a stochastic simulation algorithm

The algorithm is computationally expensive and thus many modifications and adaptations exist, including the Gibson & Bruck [8] method, tau-leaping [7], as well as hybrid techniques where abundant reactants are modelled with deterministic behaviours. Adapted techniques generally compromise the exactitude of the theory behind the algorithm as it connects to the Master equation, but offer reasonable realizations for greatly improved timescales.

IV. UNDERSTANDING P-SYSTEMS

Membrane Computing is an area of computer science aiming to abstract computing ideas and models from the structure and the functioning of living cells, as well as from the way the cells are organized in tissues or higher order structures. In short, it deals with distributed and parallel computing models, processing multisets of symbol-objects in a localized manner (evolution rules and evolving objects are encapsulated into compartments delimited by membranes), with an essential role played by the communication among compartments (with the environment as well). [10]

The essential ingredient of a P system is its membrane structure [13], [14], which can be a hierarchical arrangement of membranes. The intuition behind the notion of a membrane is that from biology, of a three-dimensional vesicle, but the concept itself is generalized/idealized to interpreting a membrane as a separator of two regions, a finite "inside" and an infinite "outside", also providing the possibility of a selective communication among the two regions. *The variety of suggestions from biology and the range of possibilities to define the architecture and the functioning of a membrane-based-multiset-processing device are practically endless.*

Thus, membrane computing is not a theory related to a specific model, but it is a framework to devise compartmentalized models. Although the domain is rather young, not only many types of P systems have been proposed, but the flexibility and the versatility of P systems seem to be, in principle, unlimited.

The compartments of a cell contains multisets of substances (ions, small molecules, macromolecules) swimming in an aqueous solution. A multiset can be represented in many ways, but the most compact one is in the form of a string. For instance, if the objects *a, b, c* are present in, respectively, 5, 2, 6 copies each, we can represent this multiset by the string $a^5b^2c^6$; of course, all permutations of this string represent the same multiset. They are processed by means of rewriting-like rules [15]. This means rules of the form $u \rightarrow v$, where u and v are multisets of objects (represented by strings). Therefore, if many copies of the same chemical compound exist, so when it is impossible to distinguish between them, rules and the objects are chosen in a non-deterministic manner. This is also related to the idea of parallelism. Biochemistry is not only non-deterministic, but it is also parallel. If two chemicals can react, then the reaction does not take place for only two molecules of the two chemicals, but, in principle, for all molecules. This is the suggestion supporting the maximal parallelism used in many classes of P systems: in each step, all rules which can be applied must be applied to all possible objects. Here we close with the observation that membrane computing deals with models which are intrinsically discrete and evolves through rewriting-like rules.

Systems based on multiset-rewriting rules as above are usually called transition P systems. Of course, when presenting a P system it is needed to specify:

1. **the alphabet of objects** (an usual finite non-empty alphabet of abstract symbols identifying the objects),
2. **the membrane structure** (it can be represented in many ways, but the most used one is by a string of labelled matching parentheses),
3. **the multisets of objects** present in each region of the system (represented in the most compact way by strings of symbol-objects),
4. **the sets of evolution rules** associated with each region, as well as the indication about the way the output is defined – see below.

Formally, a transition P system (of degree m) is a construct of the form:

$$\Pi = (\mathbf{O}, \mathbf{C}, \mu, \mathbf{w_1}, \mathbf{w_2}, \ldots, \mathbf{w_m}, \mathbf{R_1}, \mathbf{R_2}, \ldots, \mathbf{R_m}, \mathbf{i_o}) \quad (1)$$

where:

1. Π is the (finite and non-empty) alphabet of objects,
2. $C \subset O$ is the set of catalysts,
3. μ is a membrane structure, consisting of m membranes, labelled with 1, 2, …, m; one says that the membrane structure, and hence the system, is of degree m,
4. $w_1, w_2, \ldots, w_m$ are strings over O representing the multisets of objects present in the regions 1, 2, …,m of the membrane structure,
5. $R_1, R_2, \ldots, R_m$ are finite sets of evolution rules associated with the regions 1, 2, …,m of the membrane structure,
6. i_o is either one of the labels 1, 2, …,m, and then the respective region is the output region of the system, or it is 0, and then the result of a computation is collected in the environment of the system.

The rules are of the form $u \rightarrow v$ or $u \rightarrow v\delta$, with $u \in O^+$ and $v \in (O \times Tar)^*$, where Tar = {here, in, out}. The rules can be cooperative (with u arbitrary), non-cooperative (with $u \in O - C$), or catalytic (of the form $ca \rightarrow cv$ or $ca \rightarrow cv\delta$, with $a \in O - C$, $c \in C$, $v \in ((O - C)\times Tar)^*$); note that the catalysts never evolve and never change the region, they only help the other objects to evolve.

V. MODELING A REAL BIOLOGICAL CASE

In a recent work [11] it has been analyzed the interruption, after a DNA damage, of the cell division cycle due to the degradation of Cdc25A, which is a phosphatase crucial in the mitosis process. This is even more interesting if one considers that up to now the arrest induced by DNA damage has been ascribed only to the transcription factor and tumour suppressor protein p53. Surprisingly though, transient inhibition of Cdk2 (the kinase whose complex is activated by Cdc25A) in response to DNA damage occurs even in cells lacking p53 or p21, which is an inhibitory protein transcriptionally regulated by p53.

Such a p21 is an important effector of the mitosis arrest and plays a critical role in the well-documented p53 function. It has also important implications for understanding cell cycle checkpoints and the mechanism(s) through which p53 inhibits human neoplasia. Given the importance of checkpoints for preventions of genetic diseases including cancer, we explore these alternatives mechanisms of mitosis arrest by identifying a signalling pathway p53-independent that causes the cell division arrest after DNA damage. We formalize such a biological reality as a P-System [10], with the aim of obtaining a model to reproduce and observe the evolution of the system. The fluctuation of the key elements concentrations is crucial to capture typical healthy states of the system. The goal is to deflect whichever diseased path to healthy paths modulating, in a few times, the considered concentrations.

We will examine the role of Cdc25 family members of which at least Cdc25A is essential both for the entry into S phase at the checkpoint control of the G1-S transition, and for the cell cycle arrest in response to a DNA damage. In particular, we analyze the degradation of phosphorylated Cdc25A by ubiquitin mediation, which inhibits the activation of the complex cdk2-cyclinE and provokes the G1 arrest. Such a degradation takes place in the cytosol (which is the fluid portion of the cytoplasm, exclusive of organelles and membranes) and is mediated by the 'endopeptidase activity' of '26S proteasome', causing the dissociation of phosphoCdc25A in its constituting aminoacids (see Fig. 2).

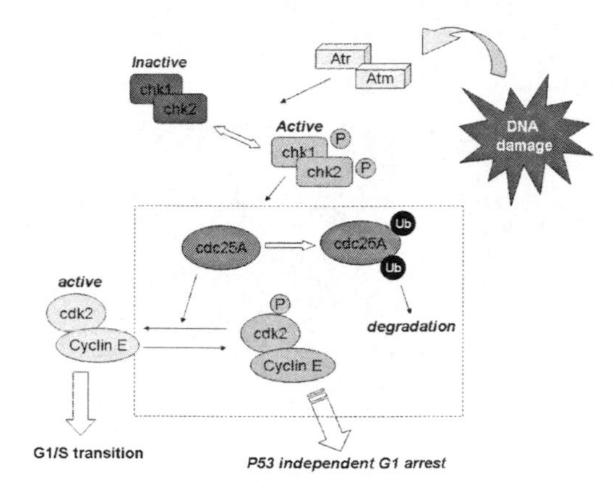

Fig. 2. p53 independent G1 arrest.in response to stress.

This mechanism is intriguing because we could learn how to module the quantity of Cdc25A in order to arrest the proliferation of tumored cells (that have DNA damaged).

The model relating to this system has been written according to the syntax of **CytoSim** [12], a stochastic simulator of biological processes that involves compartments and membranes with peripheral and integral proteins. It is available online as a java applet at the address: http://www.msr-unitn.unitn.it/Rpty_Soft_Sim.php. By means of it, it is possible to write rules (see table 2) as chemical equations and specify velocity constants for each rule.

<table>
<tr><td>J0</td><td>Cdc25A 10→ pCdc25A</td></tr>
<tr><td>J1</td><td>|| + pCdc25A → pCdc25ACytosol + ||</td></tr>
<tr><td>J2</td><td>pCdc25ACytosol (0.3*Cdc25A) → ubiquitinatedpCdc25ACytosol</td></tr>
<tr><td>J3</td><td>ubiquitinatedpCdc25ACytosol + proteasome26s (0.3*Cdc25A)→ proteasome26s</td></tr>
<tr><td>J4</td><td>pCdc25ACytosol 10→ Cdc25ACytosol</td></tr>
<tr><td>J5</td><td>Cdc25ACytosol + || 10 → || + Cdc25A</td></tr>
</table>

Table. 2. Interaction-Rules and Reaction Rates

The model also encompasses compartments, i.e. rules becoming to a compartment can be explicitly specified other than beginning quantities of species insides it (see Table. 3).

C1	cytosol[J1, J2, J3, J4, J5, 1 proteasome26s]
C2	nucleoplasm[cytosol, J0, 100000 Cdc25A]

Table. 3. Compartments and Species

By means of this description, simulations have been carried out on the previous model. It is assumed that initially, the only present species is Cdc25A in a number of 100000 molecules. One will observe the course of the concentrations of Cdc25A (phosphorilated and not) in the cytoplasm. It's well known that the phosphorilated form is in charge of the arrest of the cell cycle and the other of the transition in the G1/s. With low degradation and low activity of the proteosome, it is expected that both quantities oscillate. When ubiquitin and the proteosome activity increase, a decadent course in the time until the consumption of all Cdc25A it is expected. The aim of the simulation is to calculate the boundaries of the proteosome activity (in terms of the kinetic constant of the J2 rule) about the degradation of Cdc25A in the cytoplasm, over which the trend of degradation does not significantly change.

The outcomes of the simulation show how the previous expectations are revealed true.

S0	In the beginning, Cdc25A (phosphorilated and not) equally grow (Fig 3a);
S1	When the proteosome activity is low, concentrations of both species oscillate (Fig. 3b);
S2	Degradation times go down when the proteosome activity increases (k[J2] = 1 in Fig. 3c - k[J2] = 5 in Fig. 3d);
S3	Degradation times arrest when k[J2]=k[J4] therefore when k[J2] = 10 (Fig. 3e). In fact, for k[J2] = 100 the degradation time remain unchanged (Fig. 3f).

Table. 4. Evaluation of modification of k [J2]

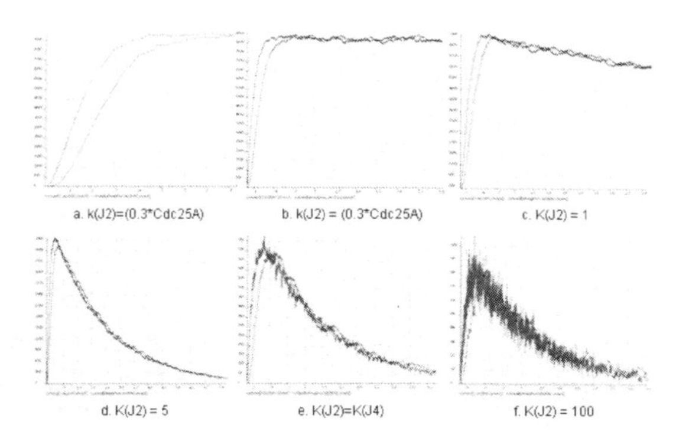

Fig. 3. Outcomes of the simulations.

VI. CONCLUSIONS

Encouraging results scientists are acquiring in the cell machineries understanding. But today, too many chemical species and too many relations exist between them. Therefore, manual procedures to discover inter-relations and inter-actions become tremendously slow and discouraging. In this context biology and informatics must walk together to capitalize their own potentialities. The simulation is just one of the infinite ways in which these two worlds can meet each other. Biology produces well-founded models and Informatics produces well-founded observation of that biological reality. One deals with continuous changes of information between both sides, of hard collaboration and sometimes of misunderstanding, but surely this is the only right way to reach the full knowledge.

REFERENCES

[1] Jean G. Vaucher and Pierre Duval, "A comparison of simulation event list algorithms", *Commun. ACM*, Vol. 18, No. 14, pp. 223-230, (1975).

[2] James O. Henriksen, "Event list management - a tutorial", *WSC '83: Proceedings of the 15th conference on Winter Simulation*, IEEE Press, pp. 543-551, Piscataway, NJ, USA (1983).

[3] H. Niederreiter, "Random number generation and quasi-Monte Carlo methods", Society for Industrial and Applied Mathematics, Philadelphia, PA, USA, (1992).

[4] M. Donald MacLaren and G. Marmaglia, *"Uniform Random Number Generators"*, J. ACM, Vol. 12, No. 1, ACM Press, New York, NY, USA, (1965).

[5] Papoulis, Athanasios & Pillai, S. Unnikrishna. *"Probability, Random Variables and Stochastic Processes"*. McGraw-Hill Science/Engineering/Math, (2001).

[6] Daniel T. Gillespie, "Exact Stochastic Simulation of Coupled Chemical Reactions". The Journal of Physical Chemistry, Vol. 81, No. 25, pp. 2340-2361 (1977).

[7] M. Rathinam, L. R. Petzold, Y. Cao, and Daniel T. Gillespie, *"Stiffness in stochastic chemically reacting systems: The implicit tau-leaping method"*. Journal of Chemical Physics, Volume 119, Issue 24, pp. 12784-12794 (2003).

[8] M. A. Gibson and J. Bruck, *"Efficient Exact Stochastic Simulation of Chemical Systems with Many Species and Many Channels"*. J. Phys. Chem. A 104: 1876-1889 (2000).

[9] Daniel T. Gillespie, *"A rigorous derivation of the chemical master equation"*, Physica A, Volume 188, Issue 1-3, p. 404-425 (1992).

[10] G. Paun. From cells to computers: Computing with membranes (P systems). Biosystems, 59(3):139-158, (2001).

[11] G. Franco, P. H. Guzzi, V. Manca, T. Mazza, *"Mitotic Oscillators as MP Graphs"*, Proceedings of The Seventh Workshop on Membrane Computing (WMC7), July 17-21, 2006, Leiden, The Netherlands. LNCS, to appear.

[12] Matteo Cavaliere, Sean Sedwards, *"Modelling Cellular Processes using Membrane Systems with Peripheral and Integral Proteins"*, Computational Methods in Systems Biology, *Lecture Notes in Computer Science series, vol.* 4210/2006, pp 108-126.

[13] I.I. Ardelean: *"The Relevance of membranes for P Systems"*. Fundamenta Informaticae, 49, 1–3 (2002), 35–43.

[14] Gh. Paun: *"Computing with Membranes – A Variant"*. International Journal of Foundations of Computer Science, 11, 1 (2000), 167–1 82.

[15] Y. Suzuki, H. Tanaka, "Abstract Rewriting Systems on Multisets, and Its Application for Modelling Complex Behaviours". In Proceedin gs of the Brainstorming Week on Membrane Computing, pp. 313–331. Tarragona, February 2003. Rovira I Virgili University, Tarragona, 2003.

Bond Graph Causality Assignment and Evolutionary Multi-Objective Optimization

Tony Wong, Gilles Cormier
Department of automated manufacturing engineering
École de technologie supérieure, University of Québec
1100 Notre-Dame West, Montréal, Québec, H3C 1K3, Canada

Abstract—**Causality assignment is an important task in physical modeling by bond graphs. Traditional causality assignment algorithms have specific aims and particular purposes. However they may fail if a bond graph has loops or contains junction causality violations. Some of the assignment algorithms focuses on the generation of differential algebraic equations to take into account junction violations caused by nonlinear multi-port devices and is not suitable for general bond graphs. In this paper, we present a formulation of the causality assignment problem as a constrained multi-objective optimization problem. Previous solution techniques to this problem include multi-objective Branch-and-Bound and Pareto archived evolution strategy – both are highly complex and time-consuming algorithms. A new solution technique called gSEMO (global Simple Evolutionary Multi-objective Optimizer) is now used to solve the causality assignment problem with very promising results.**

Keywords: **Causality assignment, constrained optimization, evolutionary algorithms, multi-objective optimization, physical modeling**

I. Introduction

In physical modeling and simulation, system equations' formulation is generally based on the input-output relationships of all interconnected components. If the set of relationships is invariant then the task of equation formulation is trivial. However, topological invariance is seldom possible especially in cases where cross-domain applications are involved. For the latter cases, one has to resolve the proper input-output relationships before deriving system equations.

One approach to determine the input and output sets is to model system components as abstract multi-port devices. Then interconnect each device according to system topology and devices' intrinsic characteristics. Finally, perform a systematic analysis of the resulting device graph to yield useful information on the input and output sets. This approach corresponds to the use of bond graph as a framework to physical modeling and simulation. In bond graph formalism, the determination of the input and output sets is defined as a causality assignment procedure [1]. As it turns out, causality assignment not only determines the input and output sets but also detects the presence of algebraic loops and dependent storage elements. The knowledge of algebraic loops can provide better insight in the computer simulation of a given system. Furthermore, the occurrence of dependent storage elements may indicate violations of principles of conservation of energy (i.e. nonlinear cross-domain system modeling or unintentional modeling errors).

This paper is organized as follows. In section II, we present an overview of the bond graph modeling technique with emphasis on the CAP (Causality Assignment Problem). The operating principles of gSEMO (global Simple Evolutionary Multiobjective Optimizer), a multi-objective optimization algorithm used to solve the CAP, are given in section III. Section IV details the mapping of the CAP into a constrained MOOP (Multi-Objective Optimization Problem) and its solution by the gSEMO. Finally, section V conatins some interesting results of this work.

II. Bond Graph Modeling

This section gives an overview of the bond graph formalism. It is based on the works by Van Dijk [1] and Karnopp et al. [2]. The goal of bond graphs is to represent a dynamic system by means of basic multi-port devices (or simply multi-ports) and their interconnections by bonds. These basic multi-ports exchange and modulate power through their bonds. There exist two power variables and two corresponding energy variables on each connected bond. They are: i) effort variable $e(t)$; ii) flow variable $i(t)$; iii) momentum $\phi(t)$; iv) displacement variable $q(t)$. The effort variable is the time derivative of some momentum and conversely, the momentum is the time integral of an effort,

$$e(t) = d\phi(t)/dt, \quad \phi(t) = \phi_0 + \int e(t)dt. \tag{1.1}$$

The same relationship applies to the flow variable and the displacement variable,

$$i(t) = dq(t)/dt, \quad q(t) = q_0 + \int i(t)dt. \tag{1.2}$$

Graphically a bond is a directed line connected to two ports sharing the bond variables. The direction of the line shows the power flow between ports and is called the power half arrow. The direction of the power flow is chosen by convention and does not need to reflect the true polarity of the flow. Fig. 1 illustrates the power flow between two multi-ports m_1 and m_2.

$$m_1 \;\xrightarrow[i(t)]{e(t)}\; m_2$$

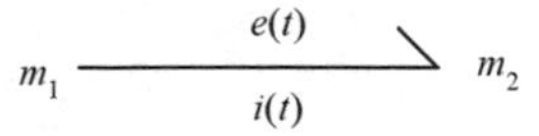

Fig. 1. A bond connecting two multi-ports m_1 and m_2. The half arrow indicates that power flows from m_1 to m_2.

K. Elleithy (ed.), Advances and Innovations in Systems, Computing Sciences and Software Engineering, 433–438.

A. *Multi-ports*

The basic multi-ports of a bond graph are elements that are source, dissipator, storage and modulator of power [2]. Table I enumerates the set of basic multi-ports. Note that in bond graph modeling it is not necessary to have a bijective mapping between the basic multi-ports and the system components. In fact, most of the basic multi-ports can represent subsystems comprising a number of system components. The important factor to consider is that each system component or subsystem obeys the constitutive laws of a given multi-port. In Table I, the second column presents the usual power flow convention of the multi-ports. The third column presents the mandatory, constrained, preferred and indifferent computational causality of the multi-ports. The fourth column presents the constitutive laws of the basic multi-ports.

TABLE I
BOND GRAPH'S BASIC MULTI-PORTS ELEMENTS

Multi-port Name	Power flow	Computational Causality	Constitutive laws
Effort source	S_e ⇀	S_e ⇀|	$e(t)=E(t)$
Flow source	S_f ⇀	S_f|⇀	$i(t)=I(t)$
Resistor	R ↼	R ↼| R|↼	$e(t)=f_R i(t)$, $i(t)=f_R^{-1}e(t)$.
Capacitor	C ↼	C ↼| C|↼	$e(t)=f_C^{-1}\int i(t)dt$, $i(t)=f_C\, de(t)/dt$.
Inertia	L ↼	L|↼ L ↼|	$i(t)=f_L^{-1}\int e(t)dt$, $e(t)=f_L\, di(t)/dt$.
Transfor-mer	$\xrightarrow{1}$ $\overset{m}{TF}$ $\xrightarrow{2}$	|$\xrightarrow{1}$ $\overset{m}{TF}$|$\xrightarrow{2}$ $\xrightarrow{1}$| $\overset{m}{TF}$ $\xrightarrow{2}$|	$e_1=m\,e_2$, $i_2=m\,i_1$. $i_1=m^{-1}i_2$, $e_2=m^{-1}e_1$.
Gyrator	$\xrightarrow{1}$ $\overset{r}{GY}$ $\xrightarrow{2}$	|$\xrightarrow{1}$ $\overset{r}{GY}$ $\xrightarrow{2}$| $\xrightarrow{1}$| $\overset{r}{GY}$|$\xrightarrow{2}$	$e_1=r\,i_2$, $e_2=r\,i_1$. $i_1=r^{-1}e_2$, $i_2=r^{-1}e_1$.
0-junction	1, n, n-1 → 0	1, n, n-1 → 0	$e_1=\ldots=e_{n-1}=e_n$, $i_1+\cdots+i_{n-1}+i_n=0$.
1-junction	1, n, n-1 → 1	1, n, n-1 → 1	$i_1=\ldots=i_{n-1}=i_n$, $e_1+\cdots+e_{n-1}+e_n$ $=0$.

The small stroke, called the causal stroke, at one end of a bond indicates the direction of travel of the effort variable information. The reaction to the effort information is the presence of a flow variable traveling in the opposite direction. Thus, the causal stroke represents the flow causality at one end of a bond. The opposite end of a bond must have complementary causality. This constitutes the fundamental causal constraint of a bond graph. Fig. 2 shows the presence of a causal stroke which determines the direction of travel of $e(t)$ and $i(t)$.

The 1-port sources S_e and S_f represent the interaction of a system with its environment. For example, in electrical and electronics systems, they may represent voltage and current sources.

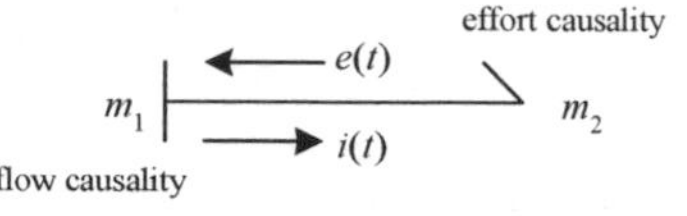

Fig. 2. The fundamental causal constraint: Causality must be complementary at both ends of a bond.

The 1-port resistive element R, capacitor element C and inertia element L behave analogously to their electrical counterpart. The 2-port transformer and gyrator are power continuous elements (no power storage and no power dissipation). The n-port junction elements are also power continuous. The constitutive laws of a 0-junction are analogous to the Kirchoff's current law. The constitutive laws of a 1-junction are analogous to the Kirchoff's voltage law.

B. *Causality Assignment*

In bond graphs, the inputs and outputs are characterized by the effort causality and flow causality. Thus, causality assignment is a process by which the bond variables ($e(t)$ and $i(t)$) are partitioned into input and output sets. It establishes the cause and effect relationships between the factors of power. There exist four types of causal constraints in bond graphs:

- *Mandatory causality*
  The constitutive laws allow only one of the two port variables to be the output. Sources S_e and S_f have mandatory causality. Multi-port elements R, L, C, TF and GY can also have mandatory causality if their constitutive laws cannot be inverted.
- *Preferred causality*
  For the storage elements C and L, there can be time derivative causality or time integral causality. The preferred causality here refers to the integral causality of these elements.
- *Constrained causality*
  For TF, GY, 0- and 1-junction there are relations between the causality of the different ports of the element. The relations are causal constraints because the causality of a particular port imposes the causality of the other ports.
- *Indifferent causality*
  Indifferent causality means there is no causal constraints. The linear resistor element R exhibits indifferent causality since both power variables $e(t)$, $i(t)$ can be made member of the input and output sets.

Most traditional causality assignment procedures use a local constraint propagation scheme to label bond causality. From a starting point, usually one of the source elements, bond causality is assigned sequentially, according to the multi-port connected to the bond, until all element ports are labeled. These causality assignment procedures must also satisfy the four causality types and the fundamental causal constraint.

The SCAP [1] (Sequential Causality Assignment Procedure) is an example of causal labeling by local propagation. However, this widely used procedure may produce incorrect causality assignment if the bond graph contains loops [2]. The modified form of SCAP, called MSCAP, can resolve incorrect causality assignment and guarantee optimal labeling by introducing the so-called basis-variable order and some graph-theoretic considerations [2]. The MSCAP is still a local propagation scheme and is actually composed of two different sub-procedures. In order to decide which of the sub-procedures is applicable, it requires some non trivial preprocessing to determine: i) if the bond graph has one or more loops along some subgraphs comprising 0-, 1-junctions and transformer multi-ports; ii) if the loop gain of all causal cycles is different than +1. However, MSCAP may still fail if the bond graph has unsatisfiable causal constraints.

Bond graphs with unsatisfiable causal constraints are usually found in nonlinear applications. Most often their constitutive laws are not amendable to the invert operation. This results in mandatory causality that may lead to junction causality violations. The RCAP [3] (Relaxed Causality Assignment Procedure) is a method that assigns causality consistent to the nonlinear constitutive laws of the elements. The causality constraints for 0-, 1-junctions are not strictly enforced. However, the constitutive laws for the junctions are still maintained and are used to generate the so-called algebraic constraint equations [4]. The resulting system equations are known as DAEs (Differential Algebraic Equations). It is obvious that RCAP is useful when there are constraint violations caused by nonlinear elements. There exists also a modified form of RCAP and is designed to generate more efficient DAEs [1].

In summary, the traditional SCAP may produce incorrect assignment due to bond graph loops. The modified SCAP requires non-trivial preprocessing and may fail in presence of junction causality violations. RCAP and its modified form are mainly concerned with the generation of DAEs for nonlinear systems. They cannot detect unintentional modeling errors. In the following sections, we present an alternate formulation of the CAP that does not require a priori knowledge of the bond graph topology or its intent. This alternate formulation considers the CAP as a constrained MOOP (Multi-Objective Optimization Problem) which is to be solved by a suitable multi-objective evolutionary optimizer.

III. MULTI-OBJECTIVE OPTIMIZATION

A MOOP has a number of objective functions which are to be minimized or maximized. In multi-objective optimization with conflicting objectives there is no single optimal solution. Instead a set of solutions exist which are all optimal with respect to some objectives [5]. This solution set arises because of trade-offs between conflicting objectives.

In this work, we apply an algorithm called gSEMO [6] to solve the MOOP representing the CAP (Causality Assignment Problem). The gSEMO is based on the dominance concept and the resulting solution set contains non-dominated solutions. The set of non-dominated solutions is known as the Pareto-optimal set if it is the non-dominated set of the entire feasible search space. Assuming a minimization problem, the dominance relation is defined as follows [9]:

Definition 1. A vector $\boldsymbol{u} = [u_1, \ldots, u_K]$ is said to dominate a vector $\boldsymbol{v} = [v_1, \ldots, v_K]$, denoted by $\boldsymbol{u} \succ \boldsymbol{v}$, if the following conditions are true:

The vector $\boldsymbol{u}$ is not worse than $\boldsymbol{v}$,

$$u_i \leq v_i, \quad \forall i \in \{1, \ldots, K\}. \tag{2.1}$$

The vector $\boldsymbol{u}$ is strictly better than $\boldsymbol{v}$ in at least one element,

$$\exists i \in \{1, \ldots K\} : u_i < v_i. \tag{2.2}$$

Note that the dominance relation is non reflexive, non symmetric but is transitive. All Pareto-based MOEA (Multi-Objective Evolutionary Algorithm) will attempt to find non-dominated solutions that are close to or are members of the Pareto-optimal set.

A. global Simple Evolutionary Multiobjective Optimizer (gSEMO)

This technique was devised by Giel and Lehre as a simple yet efficient evolutionary algorithm for the theoretical study of multi-objective optimizers [6], [7]. The basic gSEMO is a population-based evolutionary algorithm equipped with a single uniform mutation operator. It is in fact a variant of SEMO and FEMO (Fair Evolutionary Multiobjective Optimizer) algorithms proposed by Laumanns et al. [8]. gSEMO, SEMO and FEMO are designed to enable analytical computation of its expected runtimes. Results for several pseudo-Boolean and discrete problems can be found in [7], [8], [13], [14], [15]. The following steps show the general operating principles of the gSEMO algorithm.

1. Generate uniformly a random solution vector $\boldsymbol{x} \in \{0, 1\}^n$.
2. Evaluate objectives' value $f(\boldsymbol{x})$.
3. $\mathrm{P} \leftarrow \boldsymbol{x}$.
4. Select $\boldsymbol{x} \in \mathrm{P}$ uniformly at random.
5. Mutate $\boldsymbol{x}$ to create $\boldsymbol{x}'$ by changing the value of each pseudo-Boolean variable of $\boldsymbol{x}$ with probability $1/n$.
6. Evaluate objectives' value $f(\boldsymbol{x}')$.
7. if $\neg\exists\, z \in \mathrm{P}$ such that $z \succ \boldsymbol{x}' \vee f(\boldsymbol{z}) = f(\boldsymbol{x}')$ then $\mathrm{P} \leftarrow (\mathrm{P} \setminus \{z \in \mathrm{P} \mid \boldsymbol{x}' \succ z\}) \cup \{\boldsymbol{x}'\}$.
8. Repeat Steps 4 – 7 for G iterations.

The gSEMO maintains a population which is a collection of the best solutions found during the search. Initially the population is a singleton (step 3). The uniform mutation operator simply negate, with probability $1/n$, each of the n pseudo-Boolean variables of a candidate solution $\boldsymbol{x}$. If the mutate solution $\boldsymbol{x}'$ is not dominated by a solution in P, then $\boldsymbol{x}'$ is added to P and all solutions in P dominated by $\boldsymbol{x}'$ are removed. These simple steps are repeated for a prescribed number of iterations.

In summary, the gSEMO is a population-based meta-heuristic designed for the search of non-dominated solutions in an iterative manner. It requires a single user-defined control parameter – the maximum number of iterations G.

IV. Problem Mapping

This section presents the mapping of the CAP into a constrained MOOP. The first step is to define the proper bond graph representation. Then we proceed to define the proper candidate solution representation. The other steps of the problem mapping procedure are given in the following subsections.

We define a bond graph as $BG = <G, I>$. G is a directed and labeled graph and I a function identifying the multi-port type. We represent the graph G as a triplet $G = (M, B, \lambda_m)$, where $M = \{m_1, \ldots, m_{\|M\|}\}$ is the set of multi-ports, $B = \{b_1, \ldots, b_{\|B\|}\}$ is the set of bonds and $\| M \|$ denotes the cardinality of the set M. For each bond $b \in B$, there is a ordered couple (m_i, m_j) joined by b which defines the power direction of the bond. While λ_m is the set of bonds incident to a multi-port $m \in M$. We define $\lambda^+ : B \rightarrow M$, a function that returns the starting multi-port of a bond. Similarly, we define $\lambda^- : B \rightarrow M$ as a function returning the ending multi-port of a bond. We thus have

$$\lambda_m = \{b \mid b \in B, \lambda^+(b) = m\} \cup \{b \mid b \in B, \lambda^-(b) = m\}. \quad (3)$$

Finally, the identifying function I of the bond graph is simply $I : M \rightarrow T$, where $T \in$ {SF, SF, R, L, TF, GY, 0-junction, 1-junction} are the basic multi-port element labels.

A. Candidate Solution Vector Representation

The natural candidate solution representation is a pseudo-Boolean $1 \times \|B\|$ vector. Each element of the vector represents a bond within the bond graph. Since every bond has two causal labels (one for each end) and are complementary because of the fundamental causal constraint, a binary-valued representation (0-value representing flow causality and 1-value representing effort causality or vice versa) is adequate to capture all causality assignment of a bond graph. More formally, let $L_b^m \in \{0,1\}$ be the value of an element, of a solution vector, corresponding to bond b connected to multi-port m. The fundamental causal constraint states that,

$$L_b^{m_i} = 1 - L_b^{m_j}, \quad \begin{matrix} m_i = \lambda^+(b), m_j = \lambda^-(b), \\ \forall b \in B. \end{matrix} \quad (4)$$

Thus a candidate solution is a vector where each element represents the causality at one end of a bond. The causality at the other end of a bond can easily be obtained by (4). The gSEMO maintains a population of non-dominated solution vectors and it is defined by

$$\begin{matrix} \mathbf{X} = \left[x_{i,j}\right], \\ x_{i,j} \in \{0,1\}, \end{matrix} \quad i = 1,2,\ldots,N \quad j = 1,2,\ldots,\|B\|; \quad (5)$$

where N is the population size. This pseudo-Boolean encoding ensures that all candidate solutions always satisfy the fundamental constraint (4).

B. Objective Functions Evaluation

The main objective of the CAP is to satisfy, if possible, all mandatory, preferred and constrained causality of bond graph element ports while maintaining the fundamental causal constraint of the bond graph. In order to solve this problem by gSEMO, it is necessary to have a measure that indicates the assignment goodness for a candidate solution vector. In this context, the evaluation of a candidate solution vector is simply the direct counting of the number of causal violations for each multi-port of the bond graph. Recall that a solution vector has binary-valued elements. Using the convention: a 0-value indicating flow causality and a 1-value indicating effort causality, it is trivial to derive the following counting schemes.

Multi-port S_f, L

$$f_{S_f,L} = \sum_{b \in \lambda_m} L_b^m, \quad \forall m \in \{M_{S_f}\} \cup \{M_L\}, \quad (6.1)$$

where $M_{S_f} = \{p \mid I(p) = \mathrm{SF}, \forall p \in M\}$ is the set of flow sources and $M_L = \{p \mid I(p) = \mathrm{L}, \forall p \in M\}$ is the set of inertia elements in the bond graph.

Multi-port S_e, C

$$f_{S_e,C} = \sum_{b \in \lambda_m} \left(1 - L_b^m\right), \quad \forall m \in \{M_{S_e}\} \cup \{M_C\}. \quad (6.2)$$

Multi-port TF

$$f_{TF} = \sum_{b_i, b_j \in \lambda_m, i \neq j} \left|L_{b_i}^m + L_{b_j}^m - 1\right|, \quad \forall m \in \{M_{TF}\}, \quad (6.3)$$

where $|\cdot|$ denotes the absolute value operator.

Multi-port GY

$$f_{GY} = \sum_{b_i, b_j \in \lambda_m, i \neq j} \left|L_{b_i}^m - L_{b_j}^m\right|, \quad \forall m \in \{M_{GY}\}. \quad (6.4)$$

Multi-port 0-junction

$$f_{0\text{-junction}} = \left| \|\lambda_m\| - \sum_{\forall b \in \lambda_m} L_b^m - 1 \right|, \quad \forall m \in \{M_{0\text{-junction}}\}. \quad (6.5)$$

where $\|\cdot\|$ denotes the set cardinality.

Multi-port 1-junction

$$f_{1\text{-junction}} = \left| \sum_{\forall b \in \lambda_m} L_b^m - 1 \right|, \quad \forall m \in \{M_{1\text{-junction}}\}. \quad (6.6)$$

Note that these counting schemes do not involve the linear resistor elements because they have indifferent causality. For completeness sake, the counting scheme for nonlinear resistors is given by (6.7).

Multi-port nonlinear R

$$f_{R_x} = \sum_{\forall b \in \lambda_m} \left(L_b^m - 2xL_b^m + x\right), \quad \forall m \in \{M_{R_x}\}, \quad (6.7)$$

where $x \in \{0, 1\}$ and represents the mandatory causality imposed by the nonlinear constitutive laws.

C. Constrained Multi-Objective Optimization

The causality assignment problem can be expressed as a constrained MOOP given by:

$$\begin{aligned} &\text{minimize} && f_k(\mathbf{x}), && k = 1,\ldots,K; \\ &s.t. && g_j(\mathbf{x}) \le 0, && j = 1,\ldots,J; \\ & && x_i \in \{0,1\}, && i = 1,\ldots,\|B\|. \end{aligned} \tag{7}$$

In this work, the K objective functions are the number of causal violations for the inertia, capacitor, 0- and 1-junction multi-ports. The J inequality constraints are the number of causal violations for the sources, TF, GY and nonlinear R multi-ports. All decision variables are binary-valued and their number is equal to that of the bonds. Thus, we have 4 objective functions represented by (6.1), (6.2), (6.5) and (6.6) and five inequality constraints represented by (6.1), (6.2), (6.3), (6.4) and (6.7). In this forumaltion, we try to enforce mandatory, constrained causality while minimizing non preferred causality and junction violations.

The standard gSEMO is an unconstrained multi-objective evolutionary algorithm. In order to solve (7), we need to adopt a suitable constraint handling technique. A number of constraint handling techniques are given in [10] and [11]. The penalty approach is the most straightforward *external* technique for the gSEMO. In this approach all inequality constraints are normalized (necessary when constraints are incommensurable). For each candidate solution vector we compute the number of constraint violations using

$$\omega_j(\mathbf{x}) = \begin{cases} g_j(\mathbf{x}), & \text{if } g_j(\mathbf{x}) > 0; \\ 0, & \text{otherwise.} \end{cases} \tag{8}$$

Then we sum all constraint violations from (8)

$$\Omega(\mathbf{x}) = \sum_{j=1}^{J} \omega_j(\mathbf{x}) \tag{9}$$

and add the resulting value of (9) to each of the K objective function value. Thus, we have

$$F_k(\mathbf{x}) = f_k(\mathbf{x}) + \alpha_k \Omega(\mathbf{x}), \quad k = 1,\ldots,K \tag{10}$$

where α_k are constants such that both terms on the right-hand side of (10) have the same magnitude. In this work, $\alpha_k = 1$, $\forall k$. For our gSEMO implementation, instead of direct evaluation of the objective functions, we will use (8) – (10) to take into account constraint violations.

V. Results

This section presents some results using the gSEMO. As an example, this evolutionary optimizer is used to solve a bond graph inspired by the hedgetrimmer system modeling study of [12]. Fig. 3 shows the bond graph representation of this system. Our hedgetrimmer model has 21 bonds, four nonlinear multi-ports. The NL1 box represents a subsystem that translates rotary displacement into linear displacement. The NL2 box converts the linear displacement into a nonlinear load to simulate the cutting action on branches and leaves.

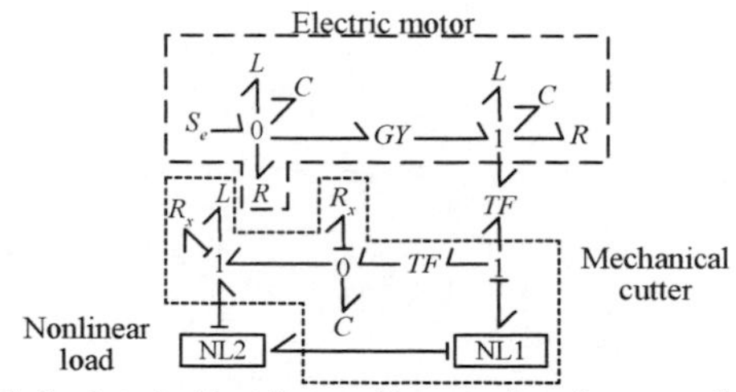

Fig. 3. Bond graph with nonlinear multi-ports and mandatory causality.

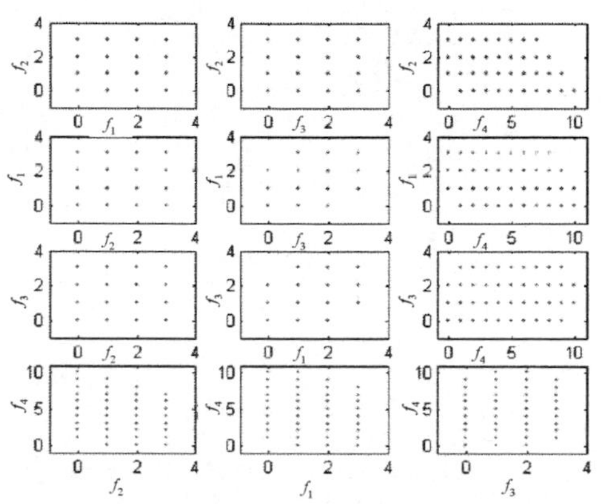
Fig. 4. Objective space of the causality assignment problem.

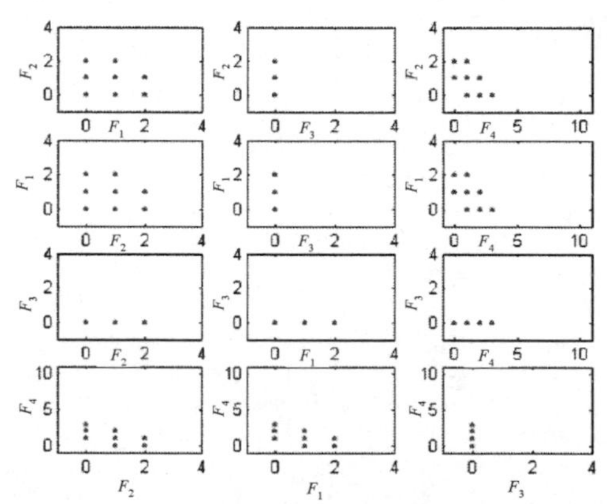
Fig. 5. Objective space of the non-dominated solutions obtained by gSEMO.

TABLE II
NON-DOMINATED SET OBJECTIVE FUNCTION VALUES

F_1	F_2	F_3	F_4	F_1	F_2	F_3	F_4
0	0	0	3	1	2	0	0
1	1	0	1	2	0	0	1
1	0	0	2	0	1	0	2
0	2	0	1	2	1	0	0

In Fig. 4, the pair-wise XY plots show the discrete objective space of the CAP. We applied the gSEMO with G equal to 1000 iterations. The resulting non-dominated solutions are plotted in Fig. 5 using their objective values. The same data are given in Table II in numerical form. F_1 to F_4 represent respectively the number of inertia, capacitor, 0- and 1-junction multi-ports causal violations.

One way to select a unique solution from the non-dominated set, is to define an ideal vector f^* representing our preference in the objective space [16]. Using an appropriate distance metric we then select the closest objective vector and its corresponding solution vector as the preferred solution of the problem. If we define $f^* = [0\ 0\ 0\ 0]^T$ and using the L_2-norm, the objective vector $[1\ 1\ 0\ 1]^T$ is the selected one. This vector corresponds to 1 non-preferred causality for C element, 1 non-preferred causality for L element and 1 junction violation. However, if we define $f^* = [2\ 2\ 0\ 0]^T$, the selected objective vector becomes $[1\ 2\ 0\ 0]^T$. That is, a causality assignment with three non preferred causalities and no junction violation.

VI. References

[1] J. Van Dijk, "On the role of bond graph causality in modeling mechatronics systems," Ph.D. dissertation, Univ. of Twente, Enschede, The Netherlands, 1994.

[2] D. C. Karnopp, D. L. Margolis, and R. C. Rosenberg, *System Dynamics: A Unified Approach*, 2nd ed., New York: John Wiley & Sons, 1990.

[3] J. Van Dijk and P. C. Breedveld, "Relaxed causality: a bond graph oriented perspective on dae-modeling," *Proc. International Conf. on Bond Graph Modeling and Simulation*, pp. 225-231, 1995.

[4] B. J. Joseph and H. R. Martens, "The method of relaxed causality in the bond graph analysis of nonlinear systems," *ASME Trans. Journal of Dynamic Systems, Measurement and Control*, no. 96, pp. 95-99, 1974.

[5] A. Ben-Tal, "Characterization of Pareto and lexicographic optimal solutions," in *Multiple Objective Decision Making – Methods and Applications*, vol. 177, *Lecture Notes on Economics and Mathematical Systems*. Berlin: Springer-Verlag, pp. 1-11, 1980.

[6] O. Giel, P. K. Lehre, "On the effect of populations in evolutionary multi-objective optimization," *Proc. of the 8th Annual Conference on Genetic and Evolutionary Computation*, pp. 651-658, 2006.

[7] O. Giel. "Expected runtimes of a simple multi-objective evolutionary algorithm," *Proc. of the 2003 Congress on Evolutionary Computation*, pp. 1918-1925, 2003.

[8] M. Laumanns, L. Thiele, E. Zitzler, "Running time analysis of multiobjective evolutionary algorithms on Pseudo-Boolean Functions," *IEEE Transactions on Evolutionary Computation*, 8(2), pp. 170-182, 2004.

[9] C. M. Fonseca and P. J. Fleming, "Multiobjective optimization and multiple constraint handling with evolutionary algorithms – Part I: A unified formulation," *IEEE Transactions on Systems, Man and Cybernatics*, vol. 28, no. 1, pp. 26-37, 1998.

[10] C. A Coello, "Constraint-handling using an evolutionary multiobjective optimization technique," *Civil Engineering Systems*, Gordon and Breach Science Publishers, vol. 17, pp. 319-346, 2000.

[11] C. A. Coello, "Constraint handling through a multiobjective optimization technique," in Annie S. Wu, editor, *Proc. of the Genetic and Evolutionary Computation Conference*, pp. 117-118, 1999.

[12] M. A. Atherton, R. A. Bates, "Bond graph analysis in robust engineering design," *Quality and Reliability Engineering International*, vol. 16, pp. 325-335, 2000.

[13] M. Laumanns, L. Thiele, E. Zitzler, "Running time analysis of evolutionary algorithms on a simplified multiobjective knapsack problem," *Natural Computing*, vol. 3, pp. 37-51, 2004.

[14] F. Neumann, "Expected runtimes of a simple evolutionary algorithm for the multi-objective minimum spanning tree problem," *Proc. of the 8th Conference on Parallel Problem Solving from Nature*, pp. 80-89, 2004.

[15] R. Kumar, N. Banerjee, "Running time analysis of a multiobjective evolutionary algorithm on simple and hard problems," in Alden H. et al. (editors), *Foundations of Genetic Algorithms. 8th International Workshop*, pp. 112--131, 2005.

[16] K. Deb, *Multi-Objective Optimization using Evolutionary Algorithms*, Chichester, UK: Wiley, 2001.

Multi-criteria Scheduling of Soft Real-time Tasks on Uniform Multiprocessors Using Fuzzy Inference

Vahid Salmani[†] Mahmoud Naghibzadeh[†] Mohsen Kahani[†] Sedigheh Khajouie Nejad[‡]
[†]Department of Computer Engineering, Ferdowsi University of Mashhad, Iran
[‡]Department of Computer Engineering, Islamic Azad University of Mashhad, Iran
salmani@ferdowesi.um.ac.ir naghib@ferdowesi.um.ac.ir kahani@ferdowesi.um.ac.ir khajouie_nejad@mshdiau.ac.ir

Abstract- **Scheduling algorithms play an important role in design of real-time systems. Due to high processing power and low price of multiprocessors, real-time scheduling in such systems is more interesting, yet more complicated. Uniform multiprocessor platforms consist of different processors with different speed or processing capacity. In such systems the same piece of code may require different amount of time to execute upon different processing units. It has been proved that there in no optimal online scheduler for uniform parallel machines. In this paper a new fuzzy-based algorithm for scheduling soft real-time tasks on uniform multiprocessors is presented. The performance of this algorithm is then compared with that of EDF algorithm. It is shown than our proposed approach has supremacy over EDF in some aspects, since it usually results in higher success ratio, better utilizes the processors and makes a more balanced schedule.**

I. INTRODUCTION

In real-time systems each task has a deadline before or at which it should be completed using a scheduling algorithm. A scheduling algorithm is a set of rules that determines which task should be executed in any given instance. Due to the tasks' criticality scheduling algorithms should be timely and predictable.

One of the research fields in the area of real-time systems to which more attention has been recently paid are *multiprocessor* real-time platforms that include several processors on which jobs can get executed. As multiprocessor systems are applied in real-time applications, scheduling of real-time tasks in such systems is of much significance. Two important types of multiprocessor systems are *identical* and *uniform* parallel machines. In the former the processing power of all processors is the same, whereas, each processor might have a different processing power in the latter case [1]. In this paper we have concentrated our research on the second type i.e. uniform multiprocessors. It has been shown that in general there is no optimal scheduling algorithm for multiprocessors [2].

Although many scheduling algorithms focus on timing constraints, there are other implicit constraints in the environment, such as uncertainty and lack of complete knowledge about the environment, dynamicity in the world, bounded validity time of information and other resource constraints. In real world situations, it would often be more realistic to find viable compromises between these parameters. For many problems, it makes sense to partially satisfy objectives. The satisfaction degree can then be used as a parameter for making a decision. One especially straightforward method to achieve this is the modeling of these parameters through fuzzy logic [3].

This paper proposes a multi-criteria algorithm for scheduling soft real-time tasks using fuzzy inference. Contrary to many traditional scheduling algorithms which use only one parameter for scheduling tasks, our novel policy uses several parameters for calculating the tasks' priority.

The rest of this paper is organized as follows. In part two, the related works are studied, and in part three we survey the scheduling policies in multiprocessor environments. A brief review of fuzzy inference is presented in the fourth section. Then, we present our new approach in part five. Real-time system model and our research framework details are presented in section six. Section seven includes performance evaluation of the proposed algorithm and experimental results based on simulation, and finally we conclude in part eight.

II. RELATED WORK

In this part the scheduling algorithms which are served as the basis of our new approach are studied. This approach uses three criteria, namely *deadline*, *laxity* and *fairness* each of which is corresponded to one of the following algorithms.

A. Earliest Deadline First Algorithm

The earliest deadline first [4] (also known as *nearest deadline first*) is a dynamic priority algorithm which uses the deadline of a task as its priority. The task with the earliest deadline has the highest priority, while the lowest priority belongs to the task with the latest deadline. This algorithm has been proved to be optimal on uniprocessors [5]. However, it is not optimal on multiprocessor platforms [6].

B. Least Laxity First Algorithm

The least laxity first [7] (also known as *minimum laxity first*) assigns higher priority to a task with the least laxity. The *laxity* of a real-time task T_i at time t, $L_i(t)$, is defined as in (1):

$$L_i(t) = D_i(t) - E_i(t) \qquad (1)$$

where $D_i(t)$ is the deadline by which the task must be completed and $E_i(t)$ is the amount of computation remaining to be performed. In other words, laxity is a measure of the flexibility available for scheduling a task. A laxity of $L_i(t)$ means if the task T_i is delayed at most by $L_i(t)$ time units, it will still meet its deadline.

A task with zero laxity must be scheduled right away and executed without preemption or it will fail to meet its deadline. The negative laxity indicates that the task will

K. Elleithy (ed.), *Advances and Innovations in Systems, Computing Sciences and Software Engineering*, 439–444.

miss the deadline, no matter when it is picked up for execution.

Similar to EDF, LLF is optimal on uniprocessors and has schedulability bound of 100% [6] .However, it has been proved that LLF is not optimal on multiprocessors [6].

A major problem with LLF algorithm is that it is impractical to implement [8] because *laxity ties* result in the frequent context switches among the corresponding tasks. This will cause the system performance to remarkably degrade. A laxity tie occurs when two or more tasks have the same laxities.

C. Pfair Algorithm

Pfair algorithm [9] is the only known approach for optimally scheduling periodic tasks on identical multiprocessors. Under Pfair scheduling, each task is executed at an approximately uniform rate by breaking it into a series of quantum-length *subtasks*. Time is then subdivided into a sequence of (potentially overlapping) subintervals of approximately equal lengths, called *windows*. To satisfy the *Pfairness* rate constraint, each subtask must execute within its associated window. Different subtasks of a task are allowed to execute on different processors.

This work proved that the problem of optimally scheduling periodic tasks on identical multiprocessors could be solved *on-line* in polynomial time by using Pfair scheduling algorithms. Nevertheless, due to the quantum-based nature of Pfair scheduling, the frequency of preemptions and migrations is a potential concern. As a result, Pfair schedules are likely to contain a large number of job preemptions and context-switches. For some applications, this is not an issue; for others, however, the overhead resulting from excessive preemptions may prove unacceptable.

III. MULTIPROCESSOR SCHEDULING

Two important parameters affecting the performance of multiprocessor scheduling algorithms are *preemption* and *migration*. A task is said to be preempted if it is not completed, but its execution is not continued on the processor it was just running on. If a task whose execution has been preempted, resumes its running on another processor, a migration has occurred. Based on the two aforementioned factors, there are two types of scheduling policies in multiprocessor environments named *global* and *partition* scheduling.

Global scheduling algorithms put all the arrived tasks with non-zero remaining execution time into a queue that is common among the processors. In a system with *m* processors, in every moment, *m* tasks having the highest priorities should be executing considering preemptions and migrations, if necessary [10].

Partition scheduling algorithms divide the task set into partitions (i.e. subsets) such that all the tasks within a partition are assigned to a processor. In this policy task migrations are not allowed [10].

Many efforts have been done to compare the two mentioned policies whose result show the partitioning is a batter approach for use in hard real-time systems, whereas, in case of soft real-time systems global is the preferred policy [11][12]. As a result, global policy has been applied in our proposed approach.

IV. FUZZY INFERENCE

A Fuzzy Inference System (FIS) tries to derive answers from a knowledgebase by using a fuzzy inference engine. The inference engine, which is considered to be the brain of the expert systems, provides the methodologies for reasoning around the information in the knowledgebase and formulating the results. A FIS is conceptually very simple. It consists of an *input* stage, a *processing* stage, and an *output* stage. The input stage maps the inputs to the appropriate membership functions and truth values. The processing stage invokes each appropriate rule and generates a result for each. It then combines the results of the rules. Finally, the output stage converts the combined result back into a specific output value [13].

The processing stage, which is called the inference engine, is based on a collection of logic rules in the form of *IF-THEN* statements, where the *IF* part is called the *antecedent* and the *THEN* part is called the *consequent*. A typical FIS has dozens of rules. These rules are stored in a knowledgebase. An example of fuzzy IF-THEN rules is "IF *deadline* IS *near* THEN *priority* IS *high*" in which *deadline* and *priority* are linguistics variables and *near* and *high* are linguistics terms. The five steps toward a fuzzy inference are as follows:

1) *fuzzifying inputs*
2) *applying fuzzy operators*
3) *applying implication methods*
4) *aggregating outputs*
5) *defuzzifying results*

Fuzzifying inputs is the act of determining the degree to which they belong to each of the appropriate fuzzy sets via membership functions. Once the inputs have been fuzzified, the degree to which each part of the antecedent has been satisfied for each rule is known. If the antecedent of a given rule has more than one part, the fuzzy operator is applied to obtain one value that represents the result of the antecedent for that rule. The implication function then modifies that output fuzzy set to the degree specified by the antecedent. Since decisions are based on the testing of all of the rules in the FIS, the results from each rule must be combined in order to make the final decision.

Aggregation is the process by which the fuzzy sets that represent the outputs of each rule are combined into a single fuzzy set. The input for the defuzzification process is the aggregated output fuzzy set and the output is then a single crisp value [13]. This can be summarized as follows: mapping input characteristics to input membership functions, input membership function to rules, rules to a set of output characteristics, output characteristics to output membership functions, and then output membership function to a single crisp valued output.

There are two common inference methods [13]. The first one is called *Mamdani*'s fuzzy inference method [14] and the second one is Takagi-Sugeno-Kang, or simply *Sugeno*, method of fuzzy inference [15]. These two methods are the same in many aspects such as the procedure of fuzzifying inputs and fuzzy operators.

The main difference between Mamdani and Sugeno methods is that the Sugeno's output membership functions are either linear or constant but Mamdani's inference expects the output membership functions to be fuzzy sets.

Sugeno's method has three advantages. Firstly, it is computationally efficient, which is an essential benefit to real-time systems. Secondly, it works well with optimization and adaptive techniques. The third advantage of Sugeno type inference is that it is well-suited to mathematical analysis [3]. Owing to the aforementioned benefits of Sugeno method, it has been applied in our proposed approach.

V. PROPOSED APPROACH

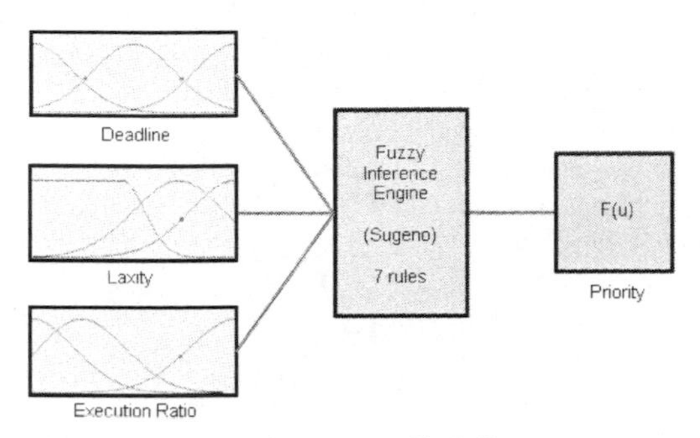

Figure 1. Inference system block diagram

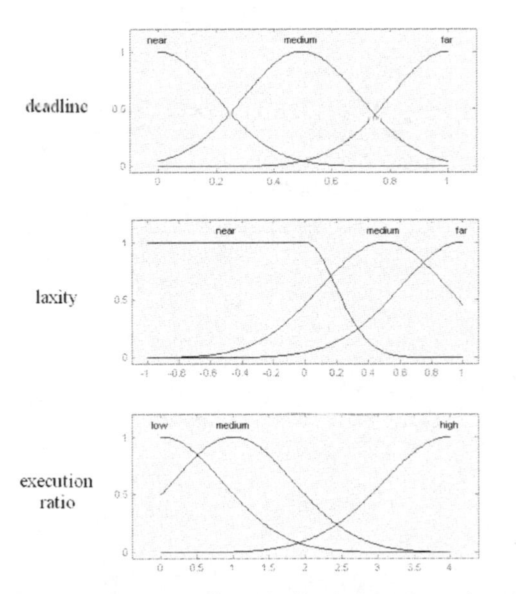

Figure 2. Fuzzy sets corresponding to deadline, laxity, and execution ratio

The block diagram of our fuzzy system is presented in Fig. 1. In the proposed model, the input stage consists of three linguistic variables, namely deadline, laxity and execution ratio. The values of these input variables are obtained from the corresponding characteristics of tasks. Since the mentioned characteristics may vary a lot from task to task, they must be normalized. Therefore, we normalize the deadline to a number between 0 and 1. Similarly, the laxity is normalized to a number between -1 and 1. In order to obtain the value of the third parameter for the task T_i we apply the formula (2):

$$Execution\_ratio_i = \frac{(WCET_i - RET_i) \times P_i}{(current\_time - RT_i) \times WCET_i} \quad (2)$$

in which $WCET_i$, RET_i, P_i and RT_i are worst case execution time, remaining execution time, request interval, and request time of the task T_i, respectively. An execution ration of 1 means the corresponding task has already been executed as much time as it is supposed to, according to Pfair algorithm. If it is less than 1, it means the task has already been executed less than what it is expected, and vise versa.

The input variables are mapped into the fuzzy sets as illustrated in Fig2. The shape of the membership function for each linguistic term is determined by an expert. There are some techniques for adjusting membership functions [16] which are not considered in this work.

Our proposed system has 7 rules which are shown in Fig. 3. In an FIS, the number of rules has a direct effect on the time complexity of the inference process. Therefore, having fewer rules may result in a better system performance.

1. IF *deadline* IS *near* THEN *priority* IS *very high*
2. IF *laxity* IS *near* THEN *priority* IS *very high*
3. IF *execution ratio* IS *low* THEN *priority* IS *high*
4. IF *deadline* IS *near* AND *laxity* IS *near* THEN *priority* IS *urgent*
5. IF *deadline* IS *near* AND *laxity* IS *far* AND *execution ratio* IS *high* THEN *priority* IS *medium*
6. IF *deadline* IS *medium* AND *laxity* IS *far* AND *execution ratio* IS *high* THEN *priority* IS *low*
7. IF *deadline* IS *far* AND *laxity* IS *far* AND *execution ratio* IS *high* THEN *priority* IS *very low*

Figure 3. Fuzzy rule-base

A. *Highest Fuzzy Priority First (HFPF) Algorithm*

We use the aforementioned FIS to calculate the dynamic priority of tasks. Consequently, the following algorithm is performed at every scheduling event:

1) For each task T feed its corresponding deadline, laxity, and execution ratio to fuzzy inference engine. Then consider the output as the priority of task T.

2) Select n tasks having highest calculated priorities.

3) Assign the task with the highest priority to the fastest processor and so on.

4) Loop (until the next scheduling event)

A. Execute n selected tasks on their designated processors.

B. Update the system status.

5) End loop

In this scheme, a scheduling event occurs whenever a task finishes its execution or a new request arrives.

VI. SYSTEM MODEL

This research is concentrated on uniform multiprocessors in *soft* real-time environments. The algorithms being investigated are in the class of *dynamic best effort*, i.e. no feasibility check is performed before a task's execution, and there is no guarantee that tasks will not miss their deadlines. These algorithms are *on-line* and use up-to-date information for the scheduling activities during the systems execution.

We have focused on *periodic* tasks and each task's deadline is equal to its period. The reason for this choice is that it has been proved that a periodic task model is useful for modeling and analysis of real-time systems [17]. Moreover, load factor measurement is easier and more accurate for periodic tasks. All tasks are *synchronous* i.e. their first request arrive simultaneously at the time zero. Such systems are common and applicable [18].

Tasks are *preemptable* and in each scheduling event a dispatcher decides which task to be performed next. In addition, a task is not allowed to run concurrently (on more than one processor at a time). Tasks must declare their characteristics and requirements such as interval, deadline and Worst Case Execution Time (WCET) at their arrival. The intervals and execution times are correct multiples of one time slice. The actual execution time of each task is equal to its WCET.

Scheduling algorithms must prevent simultaneous access to resources and shared devices. We assume the tasks are independent and do not need to do I/O operations. Therefore, the concurrency control matters have not been considered.

Tasks are assigned to the processors in a FCFS manner. This means that we allocate a task to the first processor which is found to be idle. The algorithms are work-concerning i.e. the processors are not idle as long as there are some tasks to be executed.

VII. Experimental Results

In this section, we study the performance of our proposed algorithm (HFPF) based on simulation. In [19] Baruah et al. came to the conclusion that despite its non-optimality, EDF is an appropriate algorithm to use for online scheduling on uniform multiprocessors. Consequently, we compare the performance of HFPF with that of global EDF algorithm. This evaluation can be conducted in two major ways [20].

1) Examining the effect of a specific parameter (e.g. load factor) on a variety of performance metrics as dependent variables.

2) Investigating the effect of different parameters as independent variables on a specific performance metric (e.g. success ratio).

Since presenting the result of the above methods together requires a lot of space, we choose the former approach. The latter method can be used in a separate paper. Therefore, the load factor is considered as main parameter and its effect on the performance metrics below as dependent variables is presented [21]:

1) success ratio

2) response time

3) preemptions and migrations

4) load balance

5) CPU utilization

In the experiments that we present later, in order to minimize the influence of exceptional states, each experiment was repeated 100 times and the results were averaged out. The simulation time is equal to a *meta period* which is equivalent to the smallest common multiple of all tasks' periods. It should be considered that presented results are in fact the average of the obtained values from all processors. In the following experiments, the values of the parameters are considered as below, unless mentioned otherwise.

1) The number of tasks in the system is a random number with a uniform distribution between 30 and 100.

2) The period of tasks is a random number with a uniform distribution between 10 and 10000.

3) The WCET of each task is a random number between 1 and 40% of its period.

4) Tasks are preemptable.

5) The number of processors is a random number with a uniform distribution between 4 and 8.

6) The processing capacity of each processor is at most 4 times as much as that of the slowest one.

7) The overall processing capacity of the system is equal to 10.

8) The Domino effect is omitted.

9) The load factor fluctuates between 0 and 20 in order to evaluate the performance of algorithms in both non-overloaded and overloaded conditions.

Load factor of task T_i is defined as the ratio of its WCET (E_i) to its request period (P_i). For n periodic tasks, load factor is equal to:

$$L = \sum_{i=1}^{n} \frac{E_i}{P_i}. \tag{3}$$

In multiprocessor environments, the overall load factor is the sum of all processors' load factor.

A. Success Ratio

Success ratio is defined as the ratio of the jobs that have been successfully completed to the jobs that arrived to the system [12]. As illustrated in Fig. 4, both algorithms show a near optimal performance in non-overloaded situations. Nonetheless, in overloaded conditions, the performance of the both methods dramatically descends. However, HFPF shows a slightly better performance.

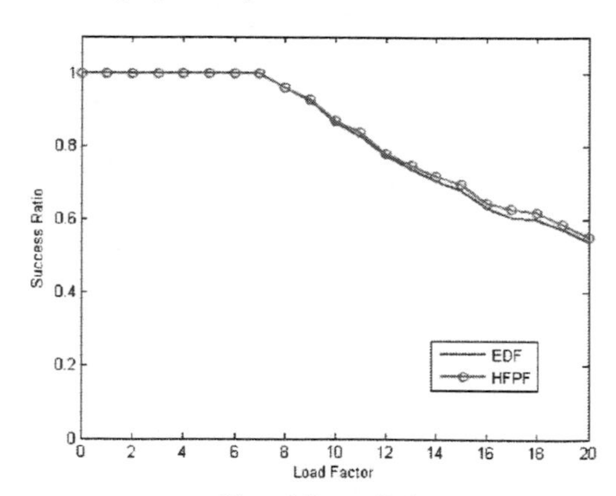

Figure 4. Success Ratio

B. Response Time

Response time is defined as the time between arriving a request and completion of its processing. Obviously it is not influenced by the tasks which fail to meet their deadlines.

A point to be considered is that due to presence of the tasks with different periods, the absolute numeric values are meaningless. Therefore, we use *response ratio* instead of response time and define it as the ratio of a task's response time to its period. In calculation of these values, scheduling overhead and communication time between processors has been ignored.

Fig. 5 depicts the observed response ratio. The two algorithms in non-overloaded conditions have close performances. In overloaded conditions, however, EDF algorithm shows a better performance.

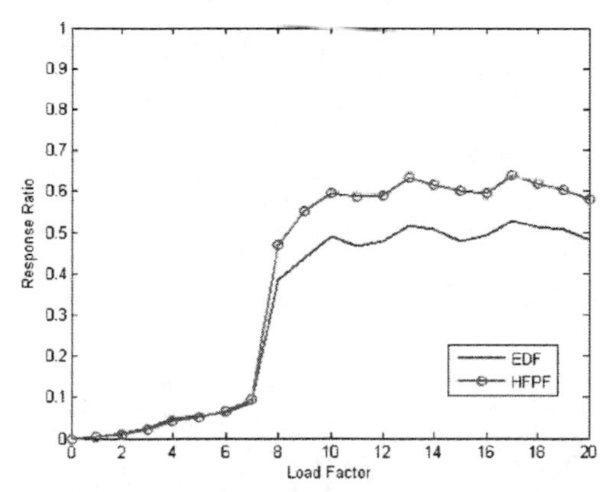

Figure 5. Response Ratio

C. *Preemptions and Migrations*

One of the most significant factors influencing scheduling overload is the number of produced preemptions and migrations, and our aim here is to measure their values for each of the compared algorithms. In this case, due to different number of tasks in diverse conditions, applying absolute numeral values is meaningless. As a result, we use *preemption ratio* and *migration ratio*, instead of preemption and migration, and define them as the ratio of the total number of preemptions to the total number of arrived requests that take the chance to be executed and the ratio of the total number of migrations to the total number of arrived requests that are picked up for execution, respectively.

Fig. 6 illustrates the produced (inter-processor) preemption ratio in this experiment indicating that the supremacy is with HFPF algorithm.

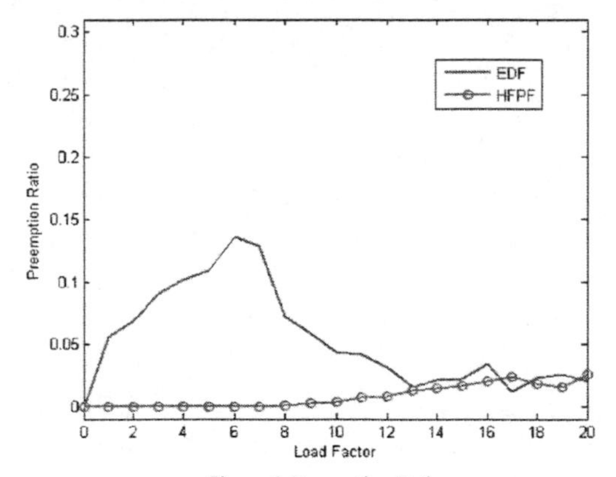

Figure 6. Preemption Ratio

In Fig. 7, the produced migration ratio has been drawn. Unlike the previous diagram, HFPF algorithm produces more migrations than EDF does.

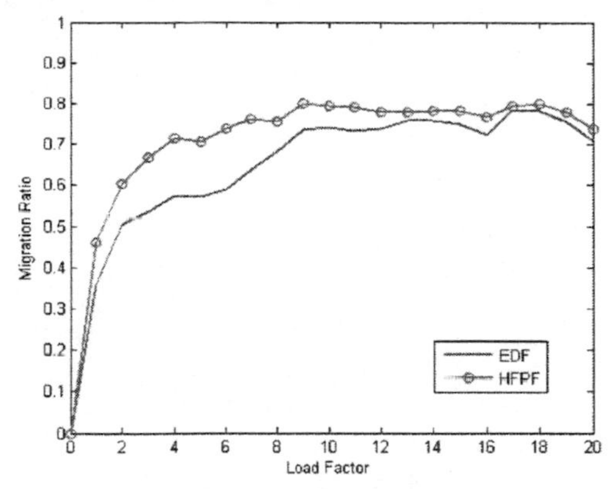

Figure 7. Migration Ratio

If the architecture of the multiprocessor machine is considered to be *idealized*, meaning that the cost of preemption equals the cost of a migration, then the *overhead ratio* is defined as summation of the preemption ratio and migration ratio. Fig. 8 shows the observed overhead ratio or, in other words, the sum of the two previous diagram's values. Since the tasks' priorities are influenced by three parameters in HFPF, they experience more fluctuation which results in more context switches. Therefore, the HFPF algorithm imposes a slightly more overhead on the system.

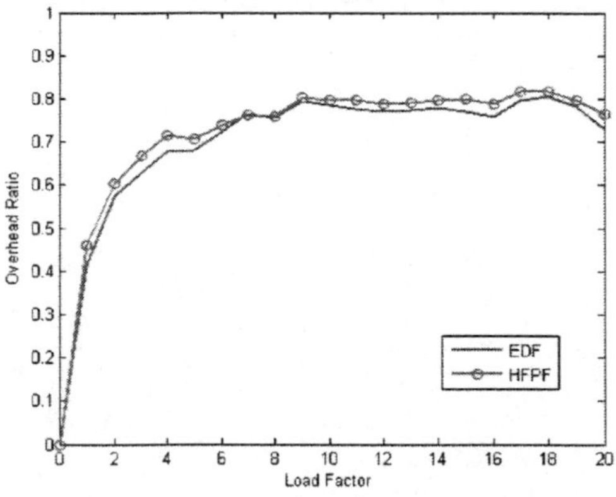

Figure 8. Overhead Ratio

D. *Load Balance*

Load balance means steady distribution of load among processors in such a way that minimizes the load difference. Regular load balance among processors not only decreases the response time, but also increases system's reliability which is very significant in real-time systems. Another advantage of a balanced system is the minimized total power consumption. The length of schedule in balanced case is also minimized. We apply the formula (4) for defining the system's load balance [21]:

$$1-\frac{\sum_{i=1}^{p}\left|U-U_i\right|}{p\times U} \quad (4)$$

in which the p is the number of processors. U is the average CPU utilization and U_i is equal to the ith processor's utilization. Fig. 9 illustrates the load balance for both algorithms. Apparently, HFPF algorithm results in a higher balance.

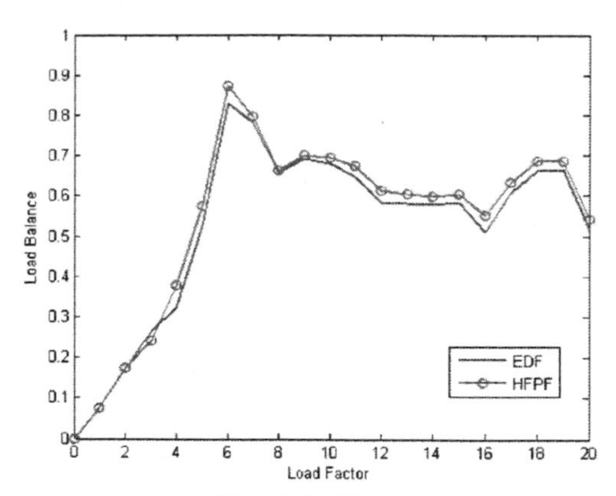

Figure 9. Load Balance

E. CPU Utilization

CPU utilization is the percentage of CPU time in which, the CPU has not been idle with respect to the time passed. Therefore, it does not include the times in which CPU has had idle processing or has been processing the jobs which have ultimately been missed.

In Fig. 10, the CPU utilization of the algorithms has been illustrated. Both algorithms have approximately the same performance in low load factor conditions and use the maximum possible CPU resources. However, in overloaded conditions the HFPF algorithm shows a slightly better CPU utilization.

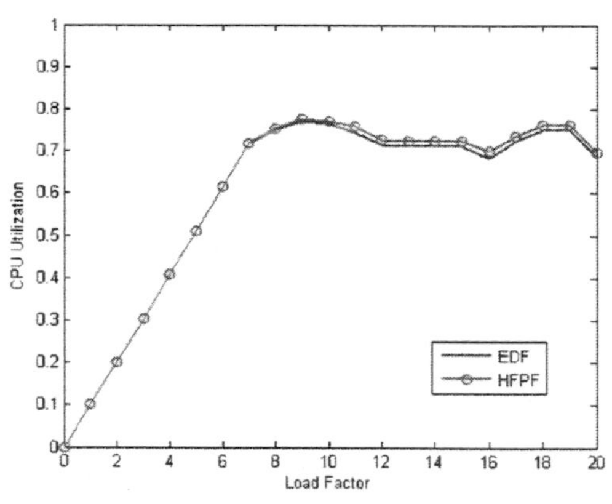

Figure 10. CPU Utilization

VIII. CONCLUSION

In this paper the use of fuzzy logic for scheduling soft real-time tasks on uniform multiprocessors was studied. We proposed a novel multi-criteria scheduling algorithm, called HFPF, and compared its performance with that of EDF algorithm in the global sense. The presented results show that HFPF has supremacy over EDF in some aspects such as success ratio, CPU utilization and Load balance. However, it produces higher number of context switches and thus imposes more overhead on the system. Finally, since identical machines are special cases of uniform ones, our results can be applied to identical multiprocessors as well.

REFERENCES

[1] J. Goossens, and P. Richard, Overview of real-time scheduling problems, *ninth international workshop on project management and Scheduling*, Nancy, France, April 2004, pp. 13-22.

[2] K. Mok, and M. L. Dertouzos, Multiprocessor scheduling in a hard real-time environment, *the 7th IEEE Texas Conference on Computing Systems*, November 1978, pp. 5-12.

[3] M. Sabeghi, M. Naghibzadeh, T. Taghavi, Scheduling Non-Preemptive Periodic Tasks in Soft Real-Time Systems Using Fuzzy Inference, *the 9th IEEE International Symposium on Object and Component-Oriented Real-Time Distributed Computing*, April 2006.

[4] L. Liu, and J. W. Layland, Scheduling algorithms for multiprogramming in a hard real-time environment, JACM, Vol. 20, No. 1, Jan. 1973, pp. 46-61.

[5] M. L. Dertouzos, Control robotics: the procedural control of physical processes, *IFIP Congress*, Stockholm, Sweden, August 5-10, 1974, pp. 807-813.

[6] M. Dertouzos, and A. Mok, Multiprocessor on-line scheduling of hard-real-time tasks, IEEE Trans. Software Engineering, 15, Dec. 1989, pp. 1497-1506.

[7] A. Mok. "Fundamental Design Problems of Distributed Systems for Hard Real-time Environments", *PhD thesis*, Massachusetts Institute of Technology, Cambridge, MA, 1983.

[8] S. H. Oh, and S. M. Yang, A Modified Least-Laxity-First Scheduling Algorithm for Real-Time Tasks, *the 5th International Workshop on Real-Time Computing Systems and Applications (RTCSA '98)*, Hiroshima, Japan, October 1998, pp. 31-36.

[9] S. Baruah, N. Cohen, C.G. Plaxton, and D. Varvel. Proportionate progress: A notion of fairness in resource allocation. *Algorithmica*, 15:600–625, 1996.

[10] S. Funk, J. Goossens, and S. Baruah, On-line scheduling on uniform multiprocessors, *the 22nd IEEE Real-Time Systems Symposium (RTSS 2001)*, London, UK, December 2001, pp. 183-192.

[11] S. Lauzac, R. Melhem, and D. Mosse, Comparison of Global and Partitioning Schemes for Scheduling Rate Monotonic Tasks on a Multiprocessor, *10th EUROMICRO Workshop on Real-Time Systems*, Berlin, June 17-19, 1998, pp. 188-195.

[12] B. Andersson, and J. Jonsson, Fixed-priority preemptive multiprocessor scheduling: to partition or not to partition, *7th International Conference on Real-Time Computing Systems and Applications (RTCSA'2000)*, Cheju Island, South Korea, December 12-14, 2000, pp. 337-346.

[13] L. Wang, A course in fuzzy systems and control, Prentice Hall, August 1996.

[14] E.H. Mamdani, S. Assilian, An experiment in linguistic synthesis with a fuzzy logic controller, International Journal of Man-Machine Studies, Vol. 7, No. 1, pp1-13, 1975.

[15] M. Sugeno, Industrial applications of fuzzy control, Elsevier Science Inc., New York, NY, 1985.

[16] J. Jang, ANFIS: Adaptive-Network-based Fuzzy Inference Systems, IEEE Transactions on Systems, Man, and Cybernetics, Vol. 23(3), pp 665-685, May 1993.

[17] J. Goossens, S. Baruah, and S. Funk, Real-time scheduling on multiprocessor, *the 10th International Conference on Real-Time System*, 2002.

[18] J. Goossens, S. Funk, and S. Baruah, EDF scheduling on multiprocessor platforms: some (perhaps) counterintuitive observations, the International Conference on Real-Time Computing Systems and Applications (RTCSA 2002), Tokyo, Japan, March 18-20, 2002, pp. 321-329.

[19] S. Baruah, S. Funk, and J. Goossens, Robustness Results Concerning EDF Scheduling upon Uniform Multiprocessors, IEEE Trans. Computers, VOL. 52, NO. 9, Sep 2003, pp. 1185-1195.

[20] V. Salmani, M. Naghibzadeh, A. Habibi, H. Deldari, Quantitative Comparison of Job-level Dynamic Scheduling Policies in Parallel Real-time Systems, *IEEE TENCON 2006*, Hong Kong, November 14-17, 2006.

[21] V. Salmani, M. Naghibzadeh, A. Taherinia, M. Bahekmat, and S. K. Nejad, "Efficiency assessment of job-level dynamic scheduling algorithms on identical multiprocessors", *WSEAS Trans. Computers*, vol. 5, pp. 2948-2955, December 2006, in press.

A Finite Element Program Based on Object-Oriented Framework for Spatial Trusses

Vedat TOĞAN, Serkan BEKİROĞLU
Karadeniz Technical University, Department of Civil Engineering, 61080, Trabzon, Turkey

Abstract- **Spatial truss structures are very popular in architecture and civil engineering. These types of structures have single structural elements with small size. Due to this, spatial trusses can be easily manufactured, transported and assembled in practice. The aim of this work is to develop a Java software package for linear simulation of spatial truss structures using the finite element method. In this program, in contrast to the node-oriented description of element quantities the element-oriented matrix notation is used and it is possible to visualize the model as well as the simulation results. The functionality of the software is demonstrated at hand of some application examples. The results and visualizations of the numerical examples are confirmed that presented finite element analysis program based on object-oriented methodology for spatial truss can be used effectively.**

I. Introduction

Physical and engineering problems can be described by differential equations based analytical models or integral expressions based numerical models. The analytical models are less preferred than the numerical one due to increasing computational effort in complex structures solutions. Thereby, the bottleneck leads to use numerical models. One of the well-known numerical methods to assemble numerical one is the Finite Element Method (FEM) so that the FEM is recently conducted in wide range of engineering problems. Typical finite element programs consist of several hundred thousand lines of procedural code, usually written in FORTRAN [1, 2]. The codes contain many complex data structures, which are accessed throughout operations in the code. The access, especially global access, decreases the flexibility of the system. Further, it is difficult to modify the existing codes and to extend the codes to adapt them for new uses, models, and solution procedures [1]. On the other hand, the object oriented programming (OOP) paradigm with its characteristics of abstraction, inheritance, modularity, and encapsulation of data and operations provides a highly flexible and modular programming environment for analysis and design of complex software systems [3] so that modular programming allows simple and easy addition of subsequent code parts .

This paper presents Java based object-oriented architecture for linear simulation of spatial truss structures using the finite element method. The goal of the system architecture is to provide a flexible and extendible set of objects that facilitate finite element modeling and analysis. The flexibility is achieved by the separation of the finite element analysis tasks into distinct objects. This results in an implementation that is manageable, extendible, easily modified, and provides finite element developers [4-8] with a system that can be adapted to meet future requirements [4].

II. Object-oriented programming

The basis of object-oriented philosophy depends on some definition such as abstraction, class, object, message, encapsulation, polymorphism and so on. Roughly speaking, the abstraction manages to branch distinct data structures and so makes programming acquire flexibility. Other flexibility facilities are classes, objects and messages, which are the main blocks of object-oriented programming, so that a group of objects with the same character is called a class and it is the key form that comprises great code or software. Furthermore, to invoke an operation of an object means to send a message to this object so that objects communicate through sending and receiving messages [9]. By the way, classes and objects encapsulate data and data operations in such a way that encapsulating the data and operations isolates them each other and promotes reuse of them. The ability of isolation of encapsulation makes the software maintenance so easy. Other vital aspect of OOP is polymorphism which allows the same operation behaving differently in distinct classes and thus allows object of one class to be used in place of that of another related class.

C++, Java, C#, and Perl are well known as OOP languages recently. Although C++ is the premier language for systems and applications [10] and is the most widely used programming language due to its runtime efficiency and hybrid properties that may allow both procedural and OOP [11], However, Java is used as programming language in the present study since some advantages of Java over C++ are that Java eliminates manual memory allocation, multiple inheritance, pointer arithmetic, and defining header file.

The key concept as beginning OOP is the separation of tasks into distinct classes of objects. It describes the purpose of the abstraction and the inherent qualities of the class, but the description is not covered in any programming languages so that this is called Object Oriented Analysis (OOA).

III. Object-oriented analysis

Object-Oriented analysis can be said to draw concepts of classes that are parts of the whole code in a way that classes

K. Elleithy (ed.), Advances and Innovations in Systems, Computing Sciences and Software Engineering, 445–450.

have distinct responsibilities. In this sense, a Java software package for the linear simulation of spatial truss structures using the finite element method is identified as follows;

Structure class is responsible for managing nodes and elements, and assembling and solving the linear system equations.

Node class's objects are responsible for recalling positions of nodes, managing **forces** and applying **boundary conditions**, keeping track of the degree of freedom (dof) of nodes, and storing the displacement.

Element class is responsible for computing the element stiffness matrix, k, and the response of the spatial truss. To make these, the class must know the related nodes comprising itself, the equation numbers of element dofs, and the area of cross-section, A, and the Young modulus, E.

Finally, **Visualizer** class is responsible for creating of the graphical objects that represent the structure and its response results.

Having identified the classes, next step is to specify relation of the classes along with attributes, methods and the relation is called Object Oriented Design (OOD). Several modeling tools are wrapped under the heading of the UML, which stands for Unified Modeling Language. In this sense, general overview of the classes, mentioned above, of the object design based on UML notation is given in Fig. 1.

UML class notation corresponds to a box divided into three parts: class name, attributes, and operations. The top-level class is the **Structure**, which is associated with a class library "inf.jlinalg". Now that the time seems ripe to express, another useful phonemena in OOP is association that assists to reduce algorithm by allowing to use preceding classes or class library. For instance, in this study in order to idendify vector and matrix and also make corresponding operations of those in the algorithm, the class libraries "inf.jlinalg"and "inf.v3d" [12] are used. Further, the class Vector3D, an existing class taken from inf.jlinalg, is included in UML class diagram shown in Fig. 1, which illustrates association at class level.

In Fig. 1, class names are shown only. Names and attributes of the some methods of the classes and constructor of the classes are not presented due to fixed place.

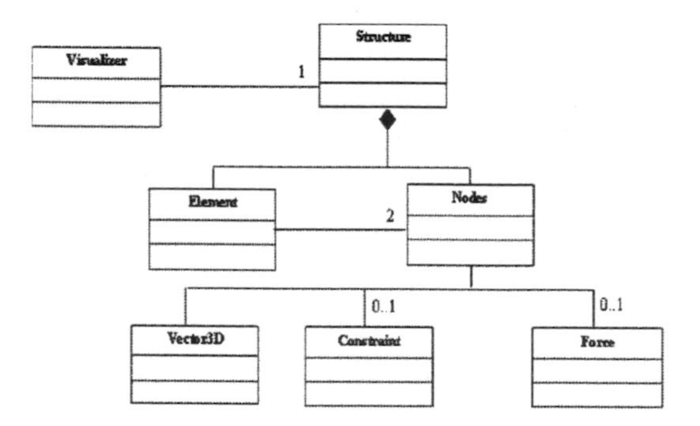

Fig. 1. UML class diagram

IV. Element-Oriented Matrix Notation for Spatial Truss Element

Finite truss element is presented in depth [13, 14, and 15]. Generally, the element quantities are considered by the node-oriented description, which is mostly known, and the element-oriented description. In the node-oriented description, the element stiffness matrix and the element vector of external loads are described in terms of the nodal virtual strain-displacement operator, $\mathbf{B}^i$, and the nodal matrix of shape functions, $\mathbf{N}^i$, [12]. However, if all element degrees of freedom are contained in the element displacement vectors in local and global coordinates, the element load vector and the element stiffness matrix requires the transformation matrix, **T**, [15, 16] in the element-oriented description.

After the description of all elements quantities, they are grouped into an ensemble. The assembly of finite elements is based on the identity of the sum of the element quantities. This induces a relation between the element displacements and the structural displacements.

The relation between structural and element degrees of freedom (dof) are manifold [12]. The construction of such a relation based on the element-oriented notation is illustrated through the assembly of truss elements in Fig. 2.

First of all, the structural dof are numbered as can be seen in Fig. 2a. The main rule of numbering is that the degrees of freedom corresponding to boundary are not numbered during this process if boundary conditions are zero (fixed). After numbering in elements, the element nodes are connected with structural nodes (Fig. 2b). The numbering is also shown in Fig. 6.

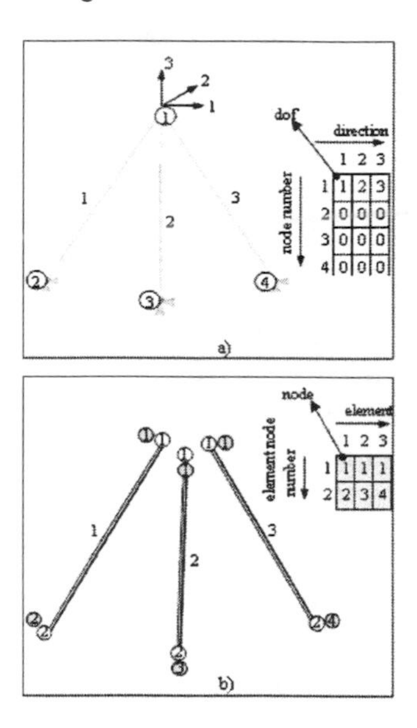

Fig. 2. a) Numbering of structural nodes and structural degrees of freedom b) Relation between structural nodes and elements

V. NUMERICAL EXAMPLES

The functionality of the software presented in the current work is demonstrated at hand of two application examples. The first example is solved for showing how the classes specified in Fig. 1 are used to create a spatial truss structure model. A shed structure is modeled and solved as the second example. The visualizations of the models as well as the analysis results are illustrated in the related Fig.

A. A Simple Spatial Truss Model

A simple spatial truss model seen in Fig. 3 contains 4 nodes and 6 elements and the length of the model's elements is 15 $x10^3$ mm, the area of cross section 14.835 $x10^3$ mm^2, and the modulus of elasticity 210 $x10^3$ N/mm^2. The model is subjected to one load acting on the node 1. The load composes of a vector, [0, -20, -100] $x10^3$ N. The code of test program, the listing of given input data for the model, and the results of the structural analysis of the model are presented in Figure 4, 6, 5, respectively. Further, Fig. 3 also shows the response of model under the load case.

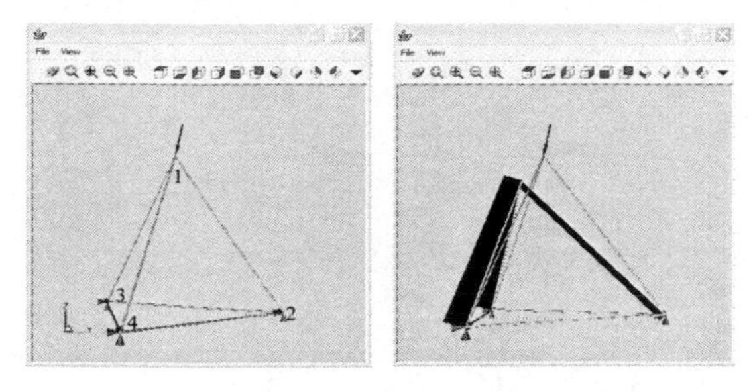

Fig. 3. Visualization of the model and the analysis results

B. A Shed Spatial Truss Model

A shed structure illustrated in Fig. 7 covers a square field with size of 4 10^4 x 4 10^4 mm and has a height of 5359 mm. The structure is designed in a way that its sub-parts are reproduced along direction Z per 5000 mm and are connected by lateral elements on the lower arch level in that direction. In a sub-part each node on lower arch connects to 4 nodes on upper arches with diagonals. So, it consists of 409 nodes and 1512 truss elements, whose cross-sectional areas are 15 $x10^3$ mm^2 and elasticity modulus 210 $x10^3$ N/ mm^2.

It is subjected to a load scheme that is applied to all nodes of upper arches and load value 15000 N are the same along direction Z and decreases 1000 N per each node level in direction Y so that in the two nodes near symmetry axis the value is 5000 N. Under the load scheme, Fig. 8 and 9 show that distribution of internal force on upper chord in the front side and of displacements along three orthogonal axes.

```
public class SimpleModel {

      public static Structure createStructure(){
            Structure struct=new Structure();
            double a=0.0140492;  //m²
            double e=2.1e11;   //N/m²
            Constraint c=new Constraint(false, false, false);
            Force f=new Force(0, -20e3, -100e3); //N
            Node n1=struct.addNode(0.0, 0.0, 12.24745);   //create
            Node n2=struct.addNode(0.0, 8.660254, 0.0 );  //nodes
            Node n3=struct.addNode(-7.50, -4.330127, 0.0);
            Node n4=struct.addNode(7.50, -4.330127, 0.0);
            n1.setForce(f);       //Apply Boundary Conditions
            n2.setConstraint(c);
            n3.setConstraint(c);
            n4.setConstraint(c);
            struct.addElement(e,a,0,1);   //create elements
            struct.addElement(e,a,0,2);
            struct.addElement(e,a,0,3);
            struct.addElement(e,a,1,2);
            struct.addElement(e,a,2,3);
            struct.addElement(e,a,3,1);
            return struct;  // return the new structure
      }
      public static void main(String[] args) {
            //create objects
            Viewer viewer=new Viewer();
            Structure struct=createStructure();
            Visualizer viz=new Visualizer(struct);
            //perform analysis
            struct.solve();
            //print the structure and the results
            struct.printStructure();
            System.out.println();
            struct.printResults();
            //draw
            viz.drawElements();
            viz.drawConstraints();
            viz.drawForces();
            viz.drawDisplacedStructure();
            viz.drawElementForces();
            //show viewer
            viewer.setVisible(true);
      }
}
```

Fig. 4. Program of simple model

```
Listing analysis results
*******************************************************
Displacements
node        u1              u2              u3
1     0.00000E00   -2.033669E-04   -2.542086E-04
2     0.00000E00     0.00000E00      0.00000E00
3     0.00000E00     0.00000E00      0.00000E00
4     0.00000E00     0.00000E00      0.00000E00
*******************************************************
Element Force
elem                  force
1                -1.773082E04
2                -5.237183E04
3                -5.237183E04
4                 0.00000E00
5                 0.00000E00
6                 0.00000E00
```

Fig. 5. Output of test program

```
Number of equation: 3
Element degrees of freedom:
    0    1    2   -1   -1   -1
    0    1    2   -1   -1   -1
    0    1    2   -1   -1   -1
   -1   -1   -1   -1   -1   -1
   -1   -1   -1   -1   -1   -1
   -1   -1   -1   -1   -1   -1
Listing structure
*******************************************************
Nodes
idx              x              y              z
1       0.00000E00     0.00000E00    1.224745E01
2       0.00000E00    8.660254E00     0.00000E00
3      -7.50000E00   -4.330127E00     0.00000E00
4       7.50000E00   -4.330127E00     0.00000E00
*******************************************************
Constraints
node        u1             u2             u3
2         fixed          fixed          fixed
3         fixed          fixed          fixed
4         fixed          fixed          fixed
*******************************************************
Forces
node            f1             f2             f3
1       0.00000E00    -2.00000E04    -1.00000E05
*******************************************************
Elements
idx              E              A              L
1       2.10000E11    1.40492E-02     1.50000E01
2       2.10000E11    1.40492E-02     1.50000E01
3       2.10000E11    1.40492E-02     1.50000E01
4       2.10000E11    1.40492E-02     1.50000E01
5       2.10000E11    1.40492E-02     1.50000E01
6       2.10000E11    1.40492E-02     1.50000E01
```

Fig. 6. Listing output of test program for simple spatial truss

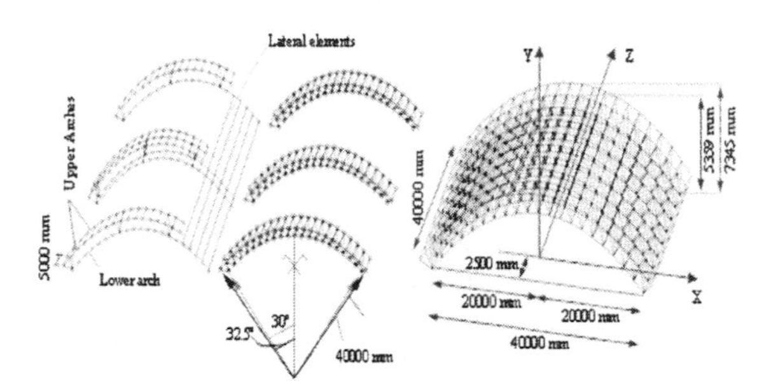

Fig. 7. A shed spatial truss system

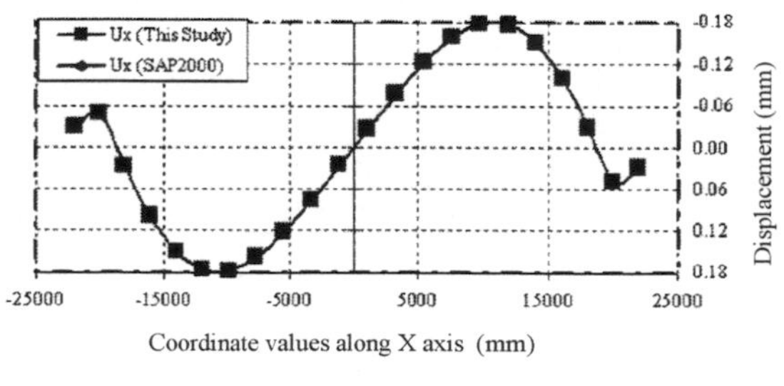

a)

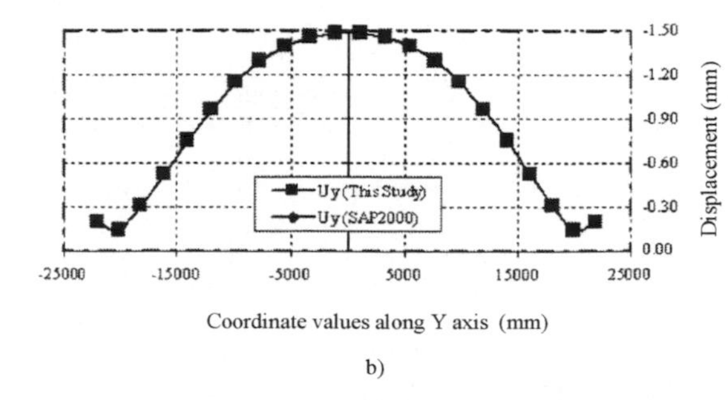

b)

Fig. 8. Displacement value along a) X axis, and b) Y axis.

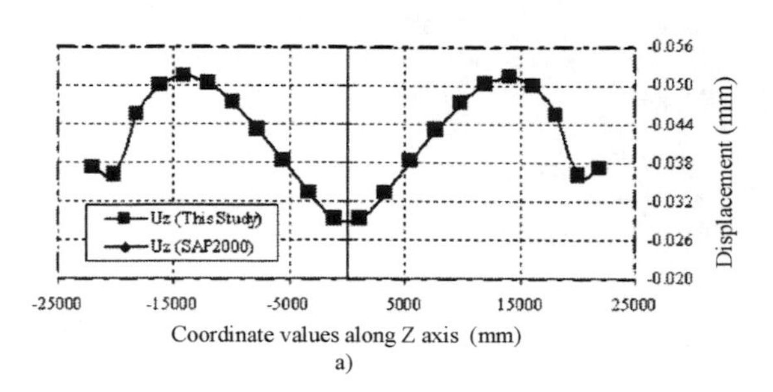

a)

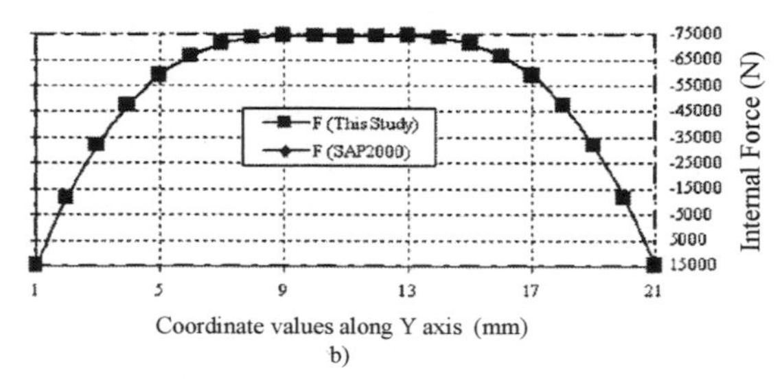

b)

Fig. 9. a) Displacement value along Z axis, and b) Internal Force value

From SAP2000 and this study the obtained the results of the displacements and internal force follow same path in satisfactory way. Similarly, Fig. 10 and 11 highlight the same achievement with the above graphs but on bottom chord in the front side.

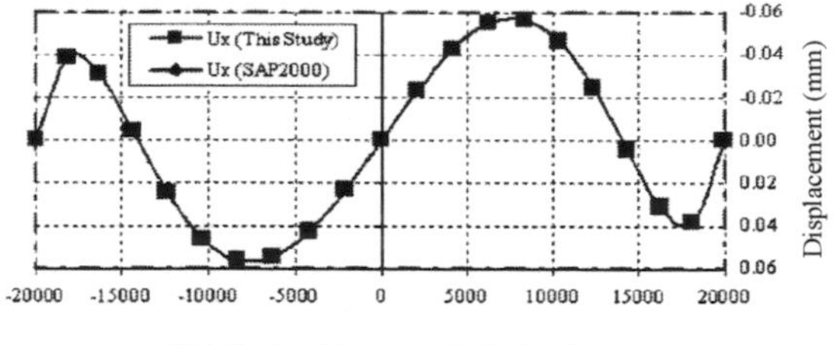

a)

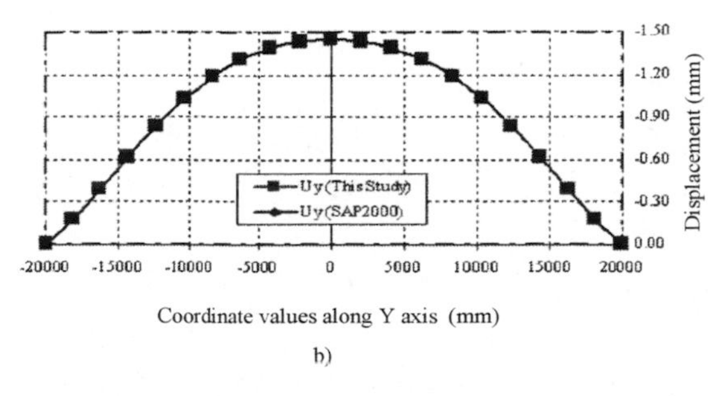

b)

Fig. 10. Displacement value along a) X axis, and b) Y axis.

The shed structure gives maximum displacement in the two nodes of upper surface near symmetry axis in front and back sides of the structure and the value is 1.4812 mm. Further, maximum internal force is observed in the bottom members of the lower arches in front and back sides of the structure and its value is 176.5299 10^3 N. These results were also verified by SAP2000. The maximum values for current program and SAP2000 are seen in Fig. 12.

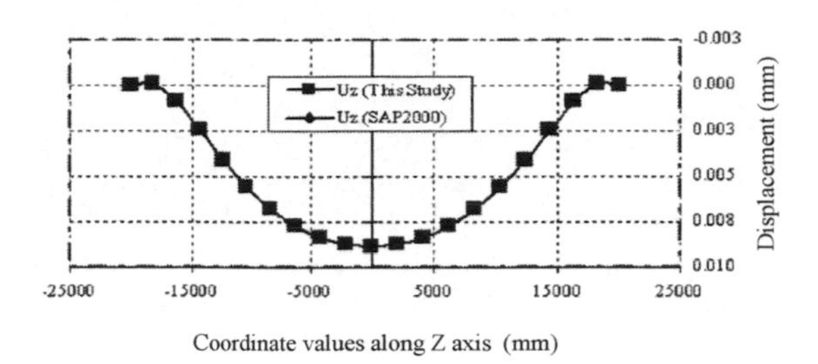

a)

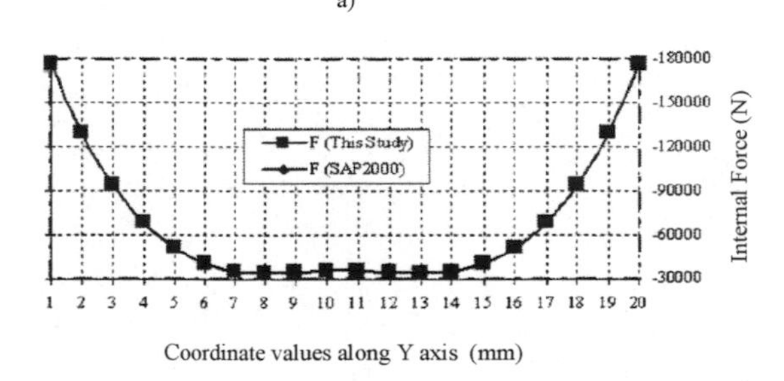

b)

Figure 11. a) Displacement value along Z axis, and b) Internal Force value

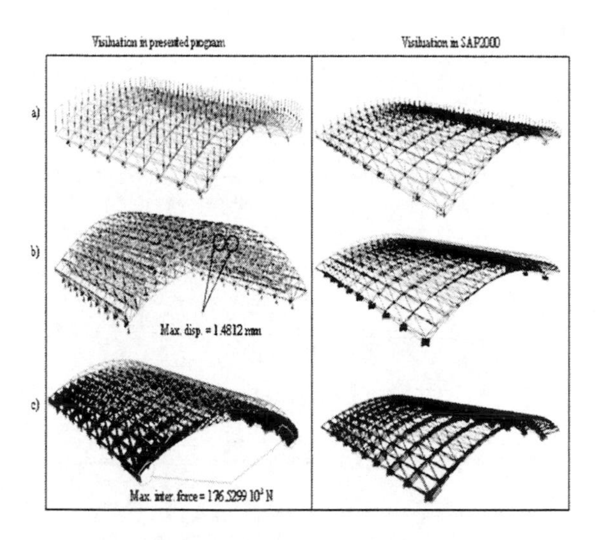

Figure 12. Visualization of a) external forces, b) deformed and undeformed shape of the shed structure, and c) internal forces of elements in the presented program and SAP2000, respectively.

VI. CONCLUSIONS

Most software using object-oriented programming manner were coded in C++ programming language. However, in the current work JAVA was preferred. Besides, proposed program is capable of visualizing of the structure response and its own under the given load condition. To show the efficiency, applicability and reliability of the coded software, the listing of the analysis results and the response of the first example were presented and last example was also solved a commercial finite element program, SAP2000.

The results and visualizations of the numerical examples confirm that presented finite element analysis program based on object-oriented methodology for spatial truss can be used effectively.

ACKNOWLEDGMENT

The authors thank Dr. M. Baitsch and Dr. D. Kuhl for their valuable support and encouragement.

REFERENCES

[1] G.C. Archer, "Object-oriented finite element analysis," Ph.D. Thesis, University of California at Berkeley, USA, 1996.

[2] G.G. Yu, "Object-oriented models for numerical and finite element analysis," Ph.D. Thesis, The Ohio State University, USA, 1994.

[3] G. Booch, "Object-oriented design with applications," Benjamin-Cummings, 1991.

[4] G.C. Archer, G. Fenves, and C. Thewalt, "A new object-oriented finite element analysis program architecture," Computers and Structures, Vol. 70, pp. 63-75, 1999.

[5] P.R.B. Devloo, and J.S.R.A. Filho, "On the development of finite element program based on the object oriented programming philosophy," Numer Meth Eng'92 , First European Conference on Numerical Methods in Engineering, Hirsch, O.C. Zienkiewicz, and E. Oñate, Elsevier; 1992. pp. 39–42.

[6] J. Cui, and L. Han, "Object-oriented FE software development," First Int Conf Eng Comput Comput Simul, Changsha, 1995. pp. 18–23.
[7] B. W. R. Forde, R. O. Foschi, and S. F. Stiemer, "Object-oriented finite element analysis," *Computers and Structures*, Vol. 34, pp. 355-374, 1990.
[8] Y.U. Li-chao, and A. V. Kumar, "An object-oriented modular framework for implementing the finite element method," *Computers and Structures*, Vol. 70, pp. 919–928, 2001.
[9] J. Mackerle, "Object-oriented programming in FEM and BEM-a bibliography (1990-2003)," Advances in Engineering Software, Vol. 35, pp. 325-336, 2004.
[10] C.S. Horstmann, and G. Cornell, "Core Java-Volume 1.," Sun Microsystems, Inc., USA, 1997.
[11] M. Sönmez, "Object-oriented finite element programming for nonlinear analysis of steel frames," Sixth International Congress on Advances in Civil Engineering, Bogazici University, Istanbul, 2004, pp. 763-772.
[12] M. Baitsch, and D. Kuhl, "Object-oriented modeling and implementation of structural analysis software," Seminar notes, Ruhr University Bochum, Bochum, Germany, 2005.
[13] K. J. Bathe, Finite Element Procedures. Prentice Hall, Upper Saddle River, New Jersy, USA, 1996.
[14] R. D. Cook, D. S. Malkus, and M. E. Plesha, Concepts and Applications of Finite Element Analysis, Third edition, John Wiley and Sons, New York, USA, 1989.
[15] M. A. Crisfield, Non-linear Finite Element Analysis of Solids and Structures-Volume 1, John Wiley and Sons, Chichester, England, 1997.
[16] D. Kuhl, and G. Meschke, Finite element Methods for linear structural mechanics, Lecture notes. Ruhr University Bochum, Institute for Structural Mechanics, Bochum, Germany, 2004.

Design for Test Techniques for Asynchronous NULL Conventional Logic (NCL) Circuits

Venkat Satagopan, Bonita Bhaskaran, Waleed K. Al-Assadi, Scott C. Smith, and Sindhu Kakarla
Department of Electrical and Computer Engineering, University of Missouri - Rolla
1870 Miner Circle, Rolla, MO 65409
Email: {venkat, bonita, waleed, smithsco, sk9qd} @umr.edu

Abstract

Conventional ATPG algorithms would fail when applied to asynchronous circuits due to the absence of a global clock and presence of more state holding elements that synchronize the control and data paths, leading to poor fault coverage. This paper presents three DFT implementations for the asynchronous NULL Conventional Logic (NCL) paradigm, with the following salient features: 1) testing with commercial DFT tools is shown to be feasible; 2) this yields a high test coverage; and 3) minimal area overhead is required. The first technique incorporates XOR gates for inserting test points; the second method uses a scan latch scheme for improving observability; and in the third scheme, scan latches are inserted in the internal gate feedback paths. The approaches have been automated, which is essential for large systems; and are fully compatible with industry standard tools.

1. Introduction

The digital world has been dominated by the growth of synchronous techniques for nearly four decades due only to their ease of design. Also, CAD tools for synchronous designs have become more advanced and sophisticated allowing total automation of several stages of the design process. However, with clock speeds nearing the GHz range and CMOS technology reaching the deep submicron range, serious concerns have been raised over the suitability of synchronous designs for next-generation devices due to clock synchronization, power consumption, and noise issues [1].

Designers are looking at asynchronous circuits as a potential solution to these problems as they are modular and do not require clock synchronization. Some of the possible benefits of asynchronous techniques include low power, less EMI, less noise, increased robustness, and design-reuse [2-4]. Asynchronous circuits fall into two main categories: delay-insensitive and bounded-delay models [5]. Paradigms, like NCL, assume delays in both logic elements and interconnect to be unbounded, although they assume that wire forks are isochronic [6]. NCL circuits often outperform other self-timed methods since they target a wider range of logical operators as opposed to others targeting standard, restricted sets [2].

In order for an asynchronous ASIC to be successfully implemented on a silicon wafer, the design flow needs to be automated and include test capabilities. The test methods are aimed at early detection of defects and improving reliability/yield. Design-For-Test (DFT) implementations of chips help make them more viable for being tested easily using prototype testers. Asynchronous NCL designs present a challenging test case to the tester/DFT CAD tools. Testability can be strengthened by making design modifications that are dormant under normal circuit operation, and only come into play during test mode.

This paper is organized as follows: Section 2 overviews the NCL paradigm; Section 3 reviews the challenges faced in testing NCL circuits and the previous work; Section 4 details the proposed DFT implementations and the developed automated tool; Section 5 compares the fault coverage results of the proposed techniques as applied to single and two-stage NCL pipelines; Section 6 provides conclusions.

2. NCL Circuits

NULL Convention Logic (NCL) provides an asynchronous design methodology by incorporating data and control information into one mixed path. NCL is a self-timed logic paradigm in which control is inherent in each datum, so there is no need for worse-case delay analysis and control path delay matching [7].

NCL utilizes threshold gates with hysteresis state-holding capability to achieve delay-insensitivity [2]. Such an operator consists of a **set** condition and a **reset** condition that the environment must ensure are not both satisfied at the same time. If neither condition is satisfied then the operator maintains its current state. NCL uses symbolic completeness of expression to achieve self-timed behavior [8]. Traditional Boolean

K. Elleithy (ed.), Advances and Innovations in Systems, Computing Sciences and Software Engineering, 451–456.

logic is not symbolically complete; the output of a Boolean gate is only valid when referenced with time. NCL eliminates this problem of time-reference by employing dual-rail or quad-rail signals. A dual-rail signal, D, consists of two mutually exclusive wires, D^0 and D^1, which may assume any value from the set {DATA0, DATA1, NULL} while a quad-rail signal, Q, consists of four mutually exclusive wires, Q^0, Q^1, Q^2, and Q^3, which may assume any value from the set {DATA0, DATA1, DATA2, DATA3, NULL}.

All NCL systems must satisfy the following two criteria to be delay-insensitive [2]:

- Input-Completeness – (i) the outputs of a circuit may not transition from NULL to DATA until all inputs have transitioned from NULL to DATA. (ii) The outputs of a circuit may not transition from DATA to NULL until all inputs have transitioned from DATA to NULL. According to Seitz's "weak conditions" of delay-insensitive signaling, an output can transition without a complete input set as long as all outputs cannot transition before all inputs arrive [9].
- Observability – ensures that every gate transition is observable at one or more circuit outputs, which means that any gate that transitions will cause at least one output to transition.

Each stage in a pipelined NCL system consists of three components: combinational logic, registration, and completion logic – all built from the threshold gates. In an NCL system, the DATA wavefront and NULL wavefront are applied alternatively [2]. The NCL registers interact with one another using handshaking signals and are responsible for ensuring that successive DATA wavefronts are separated by a NULL wavefront. When all outputs of a combinational circuit are DATA, request for NULL (*rfn* or logic 0), is generated on the K_o output of the register; and vice versa, when all outputs of a combinational circuit are NULL, request for DATA (*rfd* or logic 1), is generated on the K_o output of the register.

3. Testing of NCL Designs

DFT methods collectively refer to the design practices used to modify the existing designs in order to make them easily testable using Automatic Test Pattern Generator (ATPG) [10]. In a VLSI design flow, commercial DFT CAD tool suites are used to simulate the ATPG programs of the testers. Failures in VLSI circuits can be modeled at different levels of abstraction; DFT CAD tools target the design at the gate-level. The tools achieve a good correlation between actual failures at the Physical Design level and the stuck-at-fault models at the gate-level.

Testing asynchronous circuits has been a major challenge [11]. In order to compete with its synchronous counterparts, asynchronous schemes must be capable of producing VLSI circuits that are at least as readily testable as synchronous circuits. NCL uses a delay-insensitive, self-timed paradigm to achieve synchronization by means of handshaking, leading to the presence of many feedback paths, which in turn pose a serious problem for the test pattern generation programs of the tools. Conventional Boolean ATPG libraries cannot be used along with the NCL circuits comprised of threshold gates. The creation of the NCL ATPG library is the first step to validate the tests using the DFT tools. Analysis of the NCL fault coverage results from the tools revealed two important causes for fault degradation – 1) untestable faults in the feedback paths 2) unobserved faults in paths propagating through many logic levels. Such untestable or unobservable faults occur due to poor controllability and observability [12]. Controllability of a net is defined as the ease with which test vectors can be determined to excite the fault location. Observability is the ease with which the excitation of the fault can be observed at a primary output node or a latch. Dedicated schemes need to be applied to the problematic paths of NCL systems to tackle these critical issues.

Ongoing research activities aim to bring about a good improvement in the testability of NCL circuits. Previous DFT methods for asynchronous delay-insensitive circuits use insertion of scan latches. This is a well-established technique for synchronous circuits, wherein the latches and flip-flops in the design are converted into scan chains to better control their states and observe individual outputs by a simple shift operation. The work by Kang et al. [13] proposed a new scan design with low overhead for asynchronous micropipeline circuits to efficiently detect stuck-at and delay faults. A partial-scan technique for targeting delay faults for clockless systems was demonstrated in [14]. Blunno et al. [1] have successfully demonstrated automated synthesis of micropipelines from behavioral HDL. The work in [15] suggests techniques for designing and synthesizing NCL circuits using conventional CAD tools. The work by Sorensen et al [11] uses commercial DFT CAD tools for testing the NCL designs. The methods proposed in [11] are for acyclic and cyclic NCL pipelines, wherein the fault sites in the NCL design are mapped onto an equivalent Boolean design. A cyclic pipeline has a feedback loop in its data path as well as the completion feedback loops, and is more complex to test.

4. Proposed DFT for NCL Systems

The commercial DFT CAD tools require the presence of a clock for sequential logic and cannot support feedback paths in combinational logic because, trying to control these faults could lead to hazardous race conditions. The use of a self-testing mechanism like a BIST would prove to be beneficial, but would lead to a large area overhead. Since the tools cannot be directly applied to NCL systems, three different structured-DFT schemes, which use minimal additional logic to modify a design, make it readily testable with commercial design tools, are discussed below.

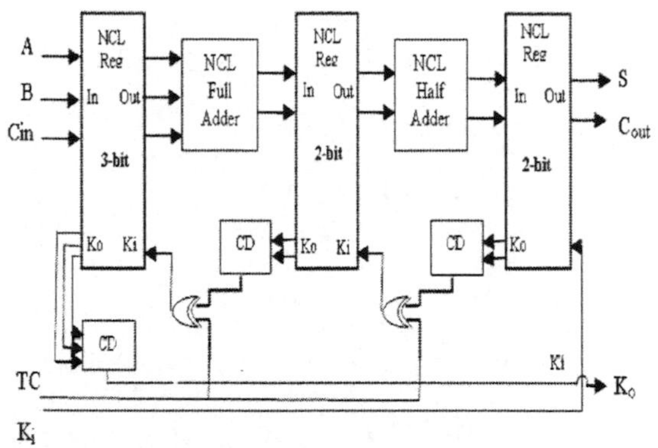

Figure 1. Test Point Insertion

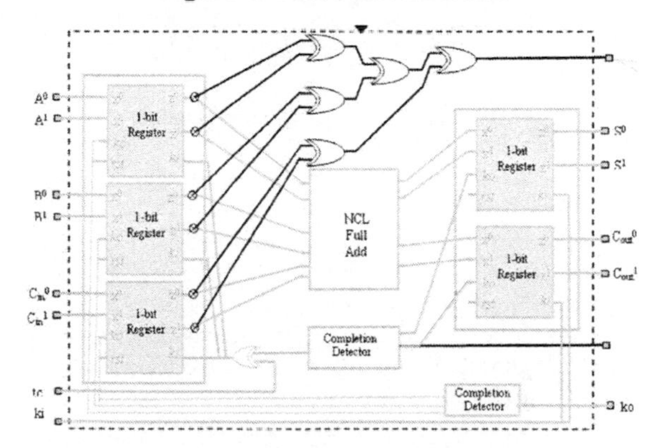

Figure 2. Balanced XOR Tree Solution

4.1. Non-Scan Technique

This two-fold approach [3] caters to solve both controllability and observability issues by 1) inserting test control pin to break feedback loops and 2) adding balanced gate structures. Feedback significantly degrades stuck-at fault coverage by increasing the complexity of the test to that of testing a sequential circuit, which needs two vectors $<t_1, t_2>$, where t_1 is the initialization vector and t_2 is the test vector. Breaking these loops ensures that the complexity of the test is reduced to that of a combinational circuit. Inserting test control pin allow the tools to probe and control the feedback paths effortlessly. The scheme proposed here simply uses an XOR gate to do so. Probability of occurrence of a '1' and '0' at the output of an XOR gate is equal, thus making fault propagation easy. As an example, let us consider a two-stage pipelined adder – a Full Adder followed by a Half Adder, as shown in Figure 1. Note that this is an extremely simplified system, used only to demonstrate the testing methodology.

For every registration stage, there is a feedback path from the succeeding stage, which comes from the Completion Detector's (CD) output. An XOR gate is inserted in this path, whose other input is controlled by an external test signal, *TC*, which is a primary input (PI), whose value is set to '0' in the normal functional mode. The XOR gate would therefore not interfere in the normal operation of the system, and provides a means for the tester to excite the required faults on the feedback net. While controllability issues are resolved by breaking the feedback paths, fault sites in long paths cannot easily propagate to the output due to the topology of the design. An easy solution for this problem would be to identify the unobservable nodes in the design and make them primary outputs. Although observability is greatly enhanced by doing this, design complexity would increase, due to the addition of one primary output pin for each observation point added. Clearly, this would have several undesirable effects - increase in cost, long wire connections leading to signal coupling (bridging faults), and EMI problems of ground bounce and SSN. Instead, a prudent solution strives to limit the number of added observation points (OPs) at the primary output pins to a single one. This can be accomplished by consolidating the OPs by the inclusion of a Balanced XOR tree structure, as shown in Figure 2 for a slice of the pipelined adder. The inherent characteristic of this structure is that all the input fault locations have the same probability of occurrence and are equally likely to occur on the primary output pin. Apart from reducing the number of primary output pins and the routing of long wires, this method also helps in shortening the length of a signal path in the design from an input edge, thus making it more easily observable. In terms of cost, the resulting design would additionally have a single test pin, one output pin, and a few XOR gates. Weighted Random Pattern Test

(WRPT) can then be effectively since there is less probability of having random resistance faults [16].

4.2. Scan Technique

The size of the balanced XOR tree grows at an exponential rate with increasing number of unobserved faults. Furthermore, in a large SOC, one or few specific modules are asynchronous to achieve benefits in terms of power, speed, etc. In such mixed-mode systems, an efficient testing strategy would be to integrate the control of the asynchronous modules through the synchronous ones [17]. This implies that the DFT implementations of such systems should incorporate the system clock. As a solution to these issues, a clocked DFT implementation for NCL is discussed here using observation latches [4]. The latches are generic components inserted in long paths that suffer from poor observability and can be stitched into the scan chains of the synchronous modules during the test phase. They serve as internal probe points for the design and improve the fault propagation on these nodes. The trade-off between probability of fault propagation (fault coverage) and additional area overhead due to the added scan latches dictates the number of internal nodes that could be combined. Due to this trade-off, four internal nodes or potential observation points (POPs) were mapped to a single latch as illustrated in Figures 3 and 4. The number of nodes that are combined could be reduced if higher fault coverage is desired. The functionality of the NCL design is not affected since the latches are enabled only during test mode. Any number of POPs could be observed in this fashion. These observation latches form a scan chain with the PI, *scan_in*, tied to the scan input of the first latch, and the scan output of the last latch tied to the PO, *scan_out*. In test mode, faults on any of these nodes could be activated by applying suitable test patterns. An applied test clock would cause the latches to capture and propagate the fault values through the scan chain to the single PO.

4.3. Algorithmic DFT Tool

Automation of the schemes presented in Sections 4.1 and 4.2 was achieved by developing an algorithmic DFT tool. NCL designs can be modified to conform to existing DFT tools without any loss in functionality [11]. Any NCL threshold gate can be mapped to its equivalent Boolean form by using gates from the DFT tool's library. For example, the set condition for a TH22 gate is defined as Z=A•B. Using the equivalence principle, a TH22 gate is replaced by a 2-input AND gate for the purpose of analyzing faults and generating test patterns. An NCL Test library has been created in this fashion for all NCL gates. This mapping forms the basis for the approaches detailed in Sections 4.1 and 4.2 [3, 4].

Repeated tests on NCL pipelines helped to evolve design strategies for the automated DFT tool [3,4]. These strategies allow the automated tool to identify nodes in NCL designs with poor controllability and observability. Fault nodes were marked as Potential Observation Points (POPs) and combined to a PO using the techniques described in Sections 4.1 and 4.2. This process is automated and is illustrated in Figure 5. The algorithmic DFT tool is explained in great detail in [4] and its outline is described here. The tool uses PERL scripts to implement the automation.

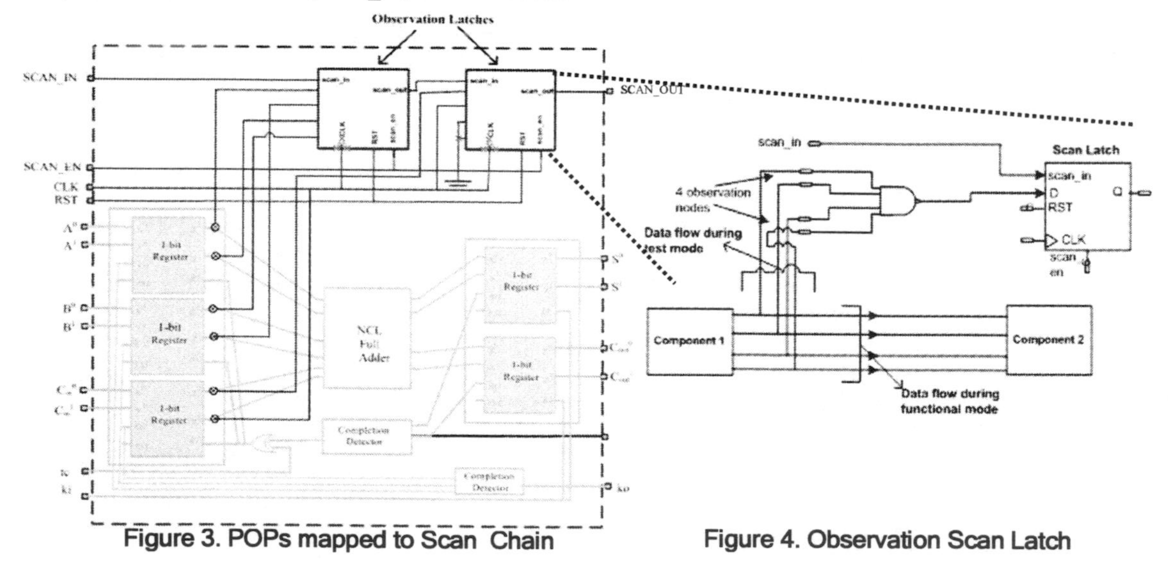

Figure 3. POPs mapped to Scan Chain

Figure 4. Observation Scan Latch

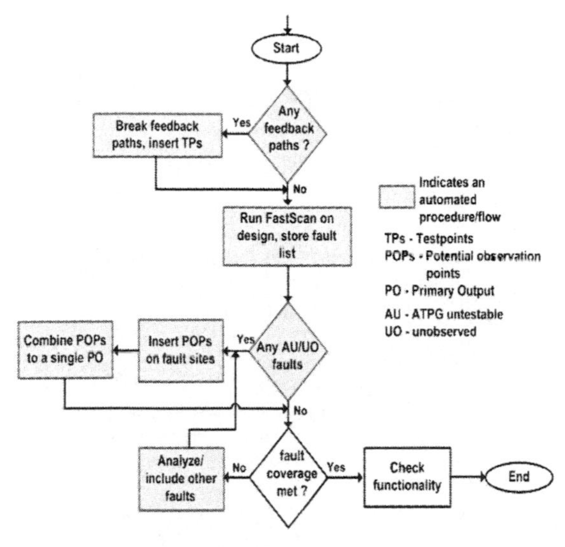

Figure 5. DFT Tool Flow

The developed PERL script requires a structural VHDL netlist of the NCL pipeline as its input. The program executes the following steps to convert the original NCL design to its DFT equivalent. The command line invocations for the different scripts are shown in italics.

1) The first step is to break the feedback paths using the *break_feedback* script.

> *break_feedback design_name.vhd*

2) The structural VHDL design is now flattened using Leonardo Spectrum. This reduces the complexity of the script used to identify the fault nodes.

3) A command file, *fs.do,* invokes FastScan on the design and stores the fault list in the fault file, *design_name.fault.*

4) Script *find_faults* then searches the fault file for specific fault types (i.e. UnObserved - UO, ATPG Untestable - AU) and stores the fault location of these fault types in a data file, *faults.dat.*

> *find_faults design_name.faults*

5) Now, POPs can be identified and combined into a PO. The script, *map_faults*, performs this function.

> *map_faults design_name.vhd*

Step 5 reads *faults.dat* to identify the POPs, processes the input VHDL design file to incorporate the POPs, automatically inserts generic Balanced XOR trees or observation latches to combine the POPs onto a PO, and finally outputs a VHDL file, *output_dft.vhd*, which has all the added DFT features incorporated within. Steps 3 through 5 may be repeated as necessary. Once the target fault coverage is achieved, functional verification is performed as the final step.

4.4. Modeling of Gate Internal Feedback

Any NCL gate can be represented by the following Boolean expression: $Z = F + G \bullet Z^{-}$, where F represents the set condition, G represents the reset condition and Z^{-} is the previous value of the output Z. NCL gates used in the non-scan and the scan latch insertion techniques were modeled using only the set condition and the gate internal feedback (GIF) paths were ignored. This model fails to accurately represent the state holding functionality of NCL gates. Here we present a new DFT model for NCL gates, which takes into account the GIF paths. These paths are in turn broken by scan insertion to improve testability such that the clock is active high (logic 1) during functional mode. To demonstrate, consider the TH23 gate, a threshold gate in which two of the three inputs should be asserted to have logic 1. The internal feedback path is broken by scan insertion as shown in Figure 6. The NCL Test library explained in Section 4.3 was rebuilt using the new gate models. Fault coverage of NCL pipelines improved significantly when the new gate models were incorporated with the non-scan and the scan insertion techniques. The automation scheme presented in Section 4.3 was modified to incorporate the new NCL gate model library.

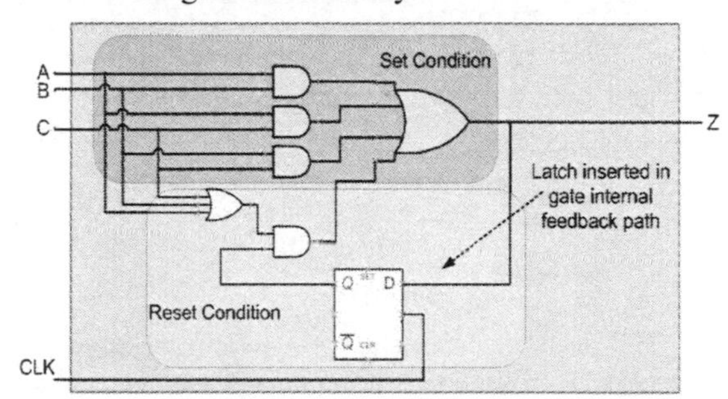

Figure 6. DFT Library Model of TH23 Gate

5. Results and Analysis

Feedback in asynchronous designs is a challenge to any DFT tool. We demonstrate the effectiveness of the proposed DFT approaches using pipelined adder designs that have up to six feedback paths in the handshaking circuits. The results using the three DFT approaches are compared in Table 1. For a given design, the number of untestable faults (UT), the fault coverage percentage (FC) and CPU time are specified. Simulations were run on a 900 MHz Sun SPARC machine. For designs without any DFT, the number of untestable faults increases with the order of the pipeline as evident from Table 1. The non-scan approach is not suitable for higher order pipelines due to limitations in the propagation of faults [4]. However, this method allows the designer to keep the design purely asynchronous in test mode.

For embedded asynchronous systems, the Scan Latch and GIF techniques are more suitable since they allow the overall circuit to be tested by a synchronous tester. Scan insertion was applied to the two-stage adder using 10 scan latches in a single scan chain. Higher coverage can be obtained by increasing the number of scan latches and scan chains. The GIF method produces the best fault coverage with minimal number of UTs for the adder pipelines. The effectiveness of the GIF technique is currently being evaluated for a ten stage Multiply-and-Accumulate design, which includes data feedback paths, as well as the feedback control loops.

Table 1. Testability Analysis of NCL Pipelines

Design Approach	# of UT	FC (%)	CPU (sec.)
Single Stage Adder			
No DFT methods	60	52.19	0.8
Automated-(XOR Tree)	15	98.32	1.8
Automated-(Scan Latch)	15	88	3.5
Automated (GIF)	0	100	3.5
Two stage Adder			
No DFT methods	143	26.92	0.8
Automated-(XOR Tree)	86	97.63	1.8
Automated-(Scan Latch)	86	82.33	3.5
Automated (GIF)	1	99.9	2.3

6. Conclusions

This work demonstrates a framework to effectively improve testability of NCL designs with minimal hardware overhead. The methodology proposed is automated based on algorithms developed and uses industry-standard DFT tools. Results indicate a significant improvement in overall fault coverage with reasonable computation time. Most important feature of the proposed method is its suitability to either a pure asynchronous system or an embedded asynchronous system. Non-scan techniques help to improve fault coverage while retaining the asynchronous nature of NCL designs. Scan techniques provide an extremely useful way to take advantage of scan-based techniques. We provide a platform for enhancing testability of NCL designs which would reduce costs and design time. The tool is currently being evaluated on different sized benchmark circuits and on circuits available in the NCL literature.

7. References

[1] I. Blunno, L. Lavagno, "Automated synthesis of micro-pipelines from behavioral Verilog HDL", *Proc. International Symposium on Advanced Research in Asynchronous Circuits and Systems*, IEEE, Apr 2000.

[2] S.C. Smith, R.F. DeMara, J.S. Yuan, D. Ferguson, and D. Lamb, "Optimization of NULL Convention Self-Timed Circuits," *Integration, the VLSI Journal*, Vol. 37/3, pp. 135-165, August 2004.

[3] B. Bhaskaran, V. Satagopan, W.K. Al-Assadi, S. C. Smith, "Implementation of Design for Tests for NCL Designs", *CDES* Jun, 2005.

[4] V. Satagopan, B. Bhaskaran, W.K. Al-Assadi, S.C. Smith, "Automation in Design for Test for Asynchronous Null Conventional Logic (NCL) Circuits", *NASA Symposium on VLSI Design*, Oct 2005.

[5] Roig, "Formal Verification and Testing of Asynchronous Circuits", *Ph.D. Dissertation*, Universitat Politecnica de Catalunya, May 1997.

[6] C.H. Berkel, M. Rem, and R. Saeijs, "VLSI Programming," *1988 IEEE International Conference on Computer Design: VLSI in Computers and Processors*, pp. 152-156, 1998.

[7] S.H. Unger, *Asynchronous Sequential Switching Circuits,* Wiley, New York, 1969.

[8] K. M. Fant and S. A. Brandt, "NULL Convention Logic: A Complete and Consistent Logic for Asynchronous Digital Circuit Synthesis," *International Conference on Application Specific Systems, Architectures, and Processors*, pp. 261-273, 1996.

[9] C. L. Seitz, "System Timing," in *Introduction to VLSI Systems*, Addison-Wesley, pp. 218-262, 1980.

[10] M.L. Bushnell, V.D. Agrawal, *Essentials of Electronic Testing for Digital, Memory, and Mixed-Signal VLSI Circuits*, Kluwer Academic Publishers, Nov 2000.

[11] A. Kondratyev, L. Sorensen, A. Streich, "Testing of Asynchronous Designs by Inappropriate Means. Synchronous approach." *IEEE* 2002.

[12] W.K. Al-Assadi et. al, "Faulty Behavior of Storage Elements and its Effects on Sequential Circuits", *IEEE Transactions on VLSI Systems,* Dec 1993.

[13] Y.S. Kang, K.H. Huh and S. Kang, "New Scan Design Of Asynchronous Sequential Circuits", Dept. of Electrical Eng., Yonsei University.

[14] M. Kishinevsky, A. Kondratyev, L. Lavagno, A. Taubin, "Partial-Scan Delay Fault Testing of Asynchronous Circuits", *IEEE,* Nov 1998.

[15] M. Ligthart, K. Fant, R. Smith, A. Taubin, A. Kondratyev, "Asynchronous design using commercial HDL synthesis tools", *Proc. International Symposium on Advanced Research in Asynchronous Circuits and Systems*, IEEE, Apr 2000.

[16] E.B. Eichelberger, E.Lindbloom et al, *Structured Logic Testing*, Prentice-Hall, Inc. 1991.

[17] S. Banerjee, S.T. Chakradhar, R.K. Roy, "Synchronous test generation model for asynchronous circuits", *9th International Conference on VLSI Design,* IEEE, Jan 1996.

Ant Colony Based Algorithm for Stable Marriage Problem

Ngo Anh Vien, Nguyen Hoang Viet, Hyun Kim, SeungGwan Lee[1] and TaeChoong Chung[*]

Artificial Intelligence Lab, Dept. of Computer Engineering
School of Electronics and Information, Kyunghee University
(Corresponding author*)
School of General Education, Kyunghee University[1]
SoChonDong, GiHungGu, YongInSi, GyongGiDo 446-701, South Korea

Abstract-**This paper introduces ant colony system (ACS), a distributed algorithm that is applied to the Stable Marriage Problem (SM). The stable marriage problem is an extensively-studied combinatorial problem with many practical applications. It is well known that at least one stable matching exists for every stable marriage instance. However, the classical Gale-Shapley [2] algorithm produces a marriage that greatly favors the men at the expense of the women, or vice versa. In our proposed ACS, a set of cooperating agents called ants cooperate to find stable matchings such as stable matching with man-optimal, woman-optimal, egalitarian stable matching, sex-fair stable matching. So this ACS is a novel method to solve Stable Marriage Problem. Our simulation results show the effectiveness of the proposed ACS.**

I. Introduction

An instance of the classical Stable Marriage Problem (SM) [1] comprises n men and n women, and each person has a preference list (PL) in which they rank all members of the opposite sex in strict order. A matching M is a bijection between the men and women. A man m_i and a woman w_j form a blocking pair for M if m_i prefers w_j to his partner in M and w_j prefers m_i to her partner in M. A matching that admits no blocking pair is said to be stable, otherwise the matching is unstable. SM arises in important practical applications, such as the student-project allocation problem (SPA) [4] and the works of Roth [3] which are applications to the National Resident Matching Program (NRMP) and related labor markets have given rise to interesting theoretical questions, resulting a better understanding of the area.

As a method for the solution, Gale-Shapley (GS) [2] algorithm is well known. They showed that every instance contains a stable matching, and gave an $O(N^2)$-time algorithm to find one. The stable matching found by this algorithm is extremely among many (for the worst case, in exponential order) stable matchings in that each member in the man group gets the best partner over all stable matchings, then each woman gets the worst partner. We say Gale-Shapley algorithm can search the men-optimal (or women-pessimal) stable matching.

The natural metaphor on which ant algorithms are based is that of ant colonies. Real ants are capable of finding the shortest path from a food source to their nest [5], [6] without using visual cues [7] by exploiting pheromone information. While walking, ants deposit pheromone on the ground and follow, in probability, pheromone previously deposited by other ants. This idea has first been applied to the traveling salesman problem (TSP) [8], [9] where an ant located in a city chooses the next city according to the strength of the artificial trails.

In this paper we propose an ant colony-based algorithm to find stable matchings depending on different criteria [1] such as stable matching with man-optimal, woman-optimal, egalitarian and sex-fair. Each ant starts at a random man and chooses the woman for this man by using pheromone on this man-woman pair. Then the ant moves to the next man and the choosing process is iterated similarly. Once all ants have completed their matching tours (when all men of the matching tour have their own distinct woman) a global pheromone updating rule is applied. The ACS algorithms we propose are very general; they can be used, with minor modifications, to find many different stable matchings based on various criteria.

The remainder of this paper is organized as follows: In Section 2, the preliminaries for this problem is reviewed. The proposed ant colony based algorithm is described in Section 3. In Section 4, experimental results are presented with different criteria. Suggestions for future development are given in Section 5. Finally, conclusions are presented in Section 6.

II. Preliminaries

In this section, we survey one example and then formally define some concepts for the problem: stable matchings with man-optimal, woman-optimal, egalitarian and sex-fair.

Example 1:

m_1	5	3	1	4	2	6	8	7	w_1	4	7	3	8	1	5	2	6
m_2	8	2	4	5	3	7	1	6	w_2	5	3	4	2	1	8	6	7
m_3	5	8	1	4	2	3	6	7	w_3	6	8	2	4	3	7	5	1
m_4	8	4	3	2	5	6	1	7	w_4	5	6	8	3	4	7	1	2
m_5	6	5	4	8	1	7	2	3	w_5	1	3	5	2	8	6	4	7
m_6	7	4	2	5	6	8	1	3	w_6	8	6	2	5	1	7	4	3
m_7	8	5	6	3	7	2	1	4	w_7	2	5	8	3	6	4	7	1
m_8	4	7	1	3	5	8	2	6	w_8	7	5	4	1	6	2	8	3

Fig. 1. Preference List PL

$$M_0 = \{(1,5),(2,2),(3,1),(4,3),(5,6),(6,7),(7,8),(8,4)\}$$
$$M_1 = \{(1,5),(2,3),(3,1),(4,2),(5,6),(6,7),(7,8),(8,4)\}$$
$$M_2 = \{(1,5),(2,2),(3,1),(4,3),(5,6),(6,4),(7,8),(8,7)\}$$
$$M_3 = \{(1,5),(2,3),(3,2),(4,1),(5,6),(6,7),(7,8),(8,4)\}$$
$$M_4 = \{(1,5),(2,3),(3,1),(4,2),(5,6),(6,4),(7,8),(8,7)\}$$

K. Elleithy (ed.), *Advances and Innovations in Systems, Computing Sciences and Software Engineering*, 457–461.

M_5 = {(1,5),(2,2),(3,1),(4,3),(5,4),(6,6),(7,8),(8,7)}
M_6 = {(1,5),(2,3),(3,2),(4,1),(5,6),(6,4),(7,8),(8,7)}
M_7 = {(1,5),(2,7),(3,1),(4,2),(5,6),(6,4),(7,8),(8,3)}
M_8 = {(1,5),(2,3),(3,1),(4,2),(5,4),(6,6),(7,8),(8,7)}
M_9 = {(1,5),(2,7),(3,2),(4,1),(5,6),(6,4),(7,8),(8,3)}
M_{10} = {(1,5),(2,7),(3,1),(4,2),(5,4),(6,6),(7,8),(8,3)}
M_{11} = {(1,5),(2,3),(3,2),(4,1),(5,4),(6,6),(7,8),(8,7)}
M_{12} = {(1,5),(2,7),(3,1),(4,2),(5,4),(6,3),(7,8),(8,6)}
M_{13} = {(1,5),(2,7),(3,2),(4,1),(5,4),(6,6),(7,8),(8,3)}
M_{14} = {(1,5),(2,7),(3,2),(4,1),(5,4),(6,3),(7,8),(8,6)}

Fig. 2. All Stable Matchings

TABLE 1
EVALUATIONS

	sm	sw	sm+sw	|sm-sw|
M_0	13	25	38	12
M_1	17	23	40	6
M_2	15	22	37	7
M_3	22	20	42	2
M_4	19	20	39	1
M_5	20	19	39	1
M_6	24	17	41	7
M_7	22	17	39	5
M_8	21	17	41	7
M_9	21	17	41	13
M_{10}	27	14	41	13
M_{11}	29	14	43	15
M_{12}	34	12	46	22
M_{13}	32	11	43	21
M_{14}	39	9	48	30

Fig.1 shows a marriage instance, PLs. All stable matchings of the marriage instance in Fig.1 are shown in Fig.2. Table 1 shows evaluations of each stable matching. The column indicated by *sm (sw)* represents the sum of the rank of every man's (woman's) partner in his (her) preference for each matching. M_0 is man-optimal since it minimizes *sm* and maximizes *sw*. On the contrary, M_{14} is woman-optimal. M_2 is the solution of the egalitarian stable matching since it minimizes $sm + sw$. M_4 and M_5 are the sex-fair stable matchings since they minimize $|sw - sm|$.

We define the man score, $sm(M_i)$ (the ith stable matching in the set of all stable matchings *M*), and the woman score, $sw(M_i)$, in the stable matching M_i as follows:

$$sm(M_i) = \sum_{(m,w)\in M_i} mr(m,w)$$

$$sw(M_i) = \sum_{(m,w)\in M_i} wr(m,w)$$

where $mr(m,w)$ means the rank of *w* in *m*'s PL and $wr(m,w)$ the rank of m in the *w*'s PL. Now we define the man-optimal stable matching, woman-optimal stable matching, egalitarian stable matching and sex-fair stable matching.

Definition 1: The stable matching M_i is man-optimal when it gives

max $sw(M_i)$ over all $M_i \in M$

Definition 2: The stable matching M_i is woman-optimal when it gives

max $sm(M_i)$ over all $M_i \in M$.

Definition 3: The stable matching M_i is egalitarian when it gives

min $(sm(M_i)+sw(M_i))$ over all $M_i \in M$.

Definition 4: The stable matching M_i is sex-fair when it gives

min $|sm(M_i)-sw(M_i)|$ over all $M_i \in M$.

III. ANT COLONY SYSTEM and STABLE MARRIGAGE PROBLEM

Ant system has been applied to combinatorial optimization problems such as the traveling salesman problem (TSP) [10], [11], [12], and the quadratic assignment problem [13]. The main idea is that of having a set of agents, called *ants*, search in parallel for good solutions to the TSP, and cooperate through pheromone-mediated indirect and global communication. Informally, each ant constructs a TSP solution in an iterative way: it adds new cities to a partial solution by exploiting both information gained from past experience and a greedy heuristic. Memory takes the form of pheromone deposited by ants on TSP edges, while heuristic information is simply given by the edge's length.

In this section, we propose ant colony-based algorithm description for SM problem. The ant colony-based algorithm for SM problem is stated in Fig. 3: in addition to the cost measure $\delta(i,j)$, each pair (i,j) also has a desirability measure $\tau(i,j)$, called pheromone, which is updated at run time by artificial ants (ants for short). Informally, ant system works as follows: Each ant, starting from a random man, it prefers to choose a woman making a pair which has a high amount of pheromone according to a probabilistic state transition rule, and then the ant moves to choose another woman for the next man. Once all ants have completed their matching tours (when all men of the tour have their own distinct women, the ant makes a new matching), a global pheromone updating rule is applied: A fraction of the pheromone evaporates on all pairs (pairs that are not refreshed become less desirable), and then each ant deposits an amount of pheromone on pairs which belong to its matching tour in proportion to how good its matching tour was. We will design the cost functions that depend on different criteria: man-optimal, woman-optimal, egalitarian stable matching and sex-fair (in other words, pairs which belong to many good matching tours are the pairs which receive greater amount of pheromone). The process is then iterated.

Each criteria of the stable matching finding (man-optimal, woman-optimal, egalitarian stable matching, sex-fair stable matching) will have different global cost function and cost measure on pairs. The pheromone updating rules are designed so that they tend to give more pheromone to pairs which should be visited by ants. The common state transition rule used by ant system, called a random-proportional rule, is given by (1), which gives the probability with which ant k in the ith man chooses the woman j. The ACS algorithm is reported in Fig. 4.

```
Stable Marriage Problem (SMP)
  Let m = {1, ... , n}, w = {1, ... , n}
be a set of men and women respectively, A
= {(i,j) : i∈M,j∈W} be the pair set,
δ(i,j)and τ(i,j) be a cost measure and
pheromone associated respectively with
pair (i,j) ∈A.
  The SMP is the problem of finding stable
matching man-woman. With each stable
criteria, we have different cost measure
δ(i,j), pheromone τ(i,j) and different
cost function of the matching.
  We see that: if δ(i,j)≠ δ(j,i), then
the SMP is asymmetric.
```

Fig. 3: The SM Problem

```
Initialize
Loop /* at this level each loop is called
an iteration */
        Each ant is positioned on the
        random man.
        Loop /* at this level each loop is
        called a step */
        Each ant applies a state transition
        rule to incrementally build a
        matching.
        Until all ants have built a
        complete matching. A global
        pheromone updating rule is applied.
Until End_condition
```

Fig. 4. The ACS algorithm

$$p_k(i,j) = \begin{cases} \dfrac{[\tau(i,j)].[\eta(i,j)]^{\beta}}{\sum_{u \in J_k(i)} [\tau(i,u)].[\eta(i,u)]^{\beta}} & j \in J_k(i) \\ 0 & otherwise \end{cases} \quad (1)$$

where τ(i,j) is the pheromone and η(i,j) is heuristic value on pair (i,j), $J_k(i)$ is the set of women that remains to be proposed by ant k positioned at the ith man (to make the solution feasible), and β is a parameter which determines the relative importance of pheromone versus η (β>0).

The result of simulations shows that this random choosing rule by the probability makes the evolution speed slow. Therefore we adjust the choosing rule. We combine the deterministic choice and random choice. So the state transition rule is as follow: an ant at the ith man chooses the woman by applying the rule given by (2):

$$w = \begin{cases} \arg\max_{u \in J(i)} \{[\tau(i,u)].[\eta(i,u)]\} & q \le q_o \\ W & otherwise \end{cases} \quad (2)$$

where q is a random number uniformly distributed in [0, 1], q_0 is a parameter ($0 \le q_0 \le 1$), and W is a random variable selected according to the probability distribution given in (1).

In ant system, the global updating rule is implemented as follows. Once all ants have built their matchings, pheromone is updated on all pairs according to

$$\tau(i,j) \leftarrow (1-\alpha).\tau(i,j) + \alpha.\Delta\tau(i,j) \quad (3)$$

where

$$\Delta\tau(i,j) = \begin{cases} F & \text{if (i,j)} \in \text{ global best matching done by ant k} \\ 0 & otherwise \end{cases}$$

Let us define the cost function F of one matching M:

$$F = f(M) + S \quad (4)$$

$f(M)$: The number of stable pairs in the matching. So $0 \le f(M) \le n$.

S: We will define this variable depending on different criteria of the stable matching.

$0 < \alpha < 1$ is a pheromone decay parameter, F is the cost of the matching performed by ant k. Pheromone updating is intended to allocate a greater amount of pheromone to larger cost. The pheromone updating formula was meant to simulate the change in the amount of pheromone due to both the addition of new pheromone deposited by ants on the visited pairs, and to pheromone evaporation.

In the next parts, we are going to analyze the global cost function S and heuristic values η of each matching criterion.

A. Men-optimal stable matching

According to definition 1, if a stable matching M has maximum sw then it is man-optimal stable matching. So we choose heuristic value η(i,j) = wr(i,j) (the ranking of the ith man in the jth woman's PL). In (2) we multiply the pheromone on pair (i,j) by the corresponding heuristic value η(i,j). In this way we chose pairs which have greater ranking of men in the women's PL and a greater amount of pheromone. We define the variable S of the cost function for man-optimal stable matching as follows

$$S = \frac{\prod_{i=1}^{n-1}(f(M)-i)}{(n-1)!} sw \quad (5)$$

sw: the sum of the rank of every woman's partner in her PL.

If $f(M) < n$ (the matching M is not stable) then $S = 0$, so according to (4) we have $F = f(M)$.

And if $f(M) = n$ (if M is a stable matching) then $S = sw$, so according to (4) we have $F = f(M) + sw$.

Definition 1 tells that if the matching M is a stable matching which has maximum sw, M is the man-optimal stable matching. Hence when we maximize the function F with S defined in (5), we will obtain the man-optimal stable matching.

B. Women-Optimal Stable Matching

According to definition 2, if a stable matching M has maximum *sm* then it is woman-optimal stable matching. So we set heuristic value $\eta(i,j) = mr(i,j)$ (the ranking of the jth woman in the ith man's PL). In this way we chose pairs which have greater ranking of women in the women's PL and which have a greater amount of pheromone.

The variable S of the cost function of one matching M is set like that: we will change *sw* in (5) to *sm*.

$$S = \frac{\prod_{i=1}^{n-1}(f(M)-i)}{(n-1)!} sm \tag{6}$$

sm represents the sum of the rank of every man's partner.

We also have $F = f(M)$ with $f(M) < n$ (because $S = 0$ with all $f(M) < n$). And if $f(M) = n$ (if M is a stable matching) then $S = sm$, so $F = n + sm$.

Based on definition 2, if *sm* is maximum and $f(M) = n$ M is stable matching) then the solution is woman-optimal stable matching. So if we maximize the function F, we will obtain the woman-optimal stable matching.

C. Egalitarian Stable Matching

In definition 3, if a stable matching M has minimum $(sm+sw)$ then it is egalitarian stable matching. Heuristic value $\eta(i,j) = (mr(i,j)+wr(i,j))^{-1}$, the inverse of the sum between the ranking of the jth woman in the ith man's PL and the ranking of the ith man in the jth woman's PL. In this way we chose pairs which have smaller sum of the rankings and which have a greater amount of pheromone.

The variable S of the cost function of one matching M: It is a little different with (5) and (6).

$$S = \frac{\prod_{i=1}^{n-1}(f(M)-i)}{(n-1)!}(sm+sw)^{-1} \tag{7}$$

$F = f(M)$ with $f(M) < n$ (because $S = 0$ with all $f(M) < n$). And $S = sm$ if $f(M) = n$ (if M is a stable matching), so $F = n + sm$.

Therefore if $sm+sw$ is minimum and $f(M) = n$, then $F = n + (sw+sm)^{-1}$, then the solution is egalitarian stable matching. So when we maximize the function F, we will obtain the egalitarian stable matching.

D. Sex-Fair Stable Matching

In definition 4, if a stable matching M has minimum $|sm-sw|$ then it is sex-fair stable matching. We have $0 \le |mr(i,j) - wr(i,j)| \le n-1$. So heuristic value is set as following:

$$\eta(i,j) = \begin{cases} |mr(i,j)-wr(i,j)|^{-1} & mr(i,j) \ne wr(i,j) \\ 2 & otherwise \end{cases}$$

In this way we chose pairs which have smaller deviation between the rankings and which have a greater amount of pheromone.

The variable S of the cost function of one matching M: we will change $(sw+sm)$ in (7) to $|sw-sm|$.

$$S = \frac{\prod_{i=1}^{n-1}(f(M)-i)}{(n-1)!}|sm-sw|^{-1} \tag{8}$$

We have: $0 \le |sm-sw| \le n-1$. Therefore if $|sm-sw| = 0$ then we set S as follow:

$$S = 2.\frac{\prod_{i=1}^{n-1}(f(M)-i)}{(n-1)!} \tag{9}$$

$F = f(M)$ with $f(M) < n$ (because $S = 0$ with all $f(M) < n$). So if $|sm-sw|$ is minimum and $f(M) = n$, then $F = n + |sw-sm|^{-1}$ (or $F = n + 2$ if $sw = sm$), it means that the solution is sex-fair stable matching. So when we maximize the function F, we will obtain the sex-fair stable matching.

IV. Experiment Results

To evaluate the performance of the proposed algorithm, we performed simulations with the problem size n (the number of men) from 3 to 20. The basic settings for our performance are $\alpha = 0.1$, $\beta = 2$, $q_0 = 0.9$. With $n=20$, the number of ants $m=15$, the simulations were stopped after 3500 iterations. We implemented the simulation 20 times, each set ants' positions randomly. We also made Gale-Shapley algorithm simulation, and did the comparison. The results with man-optimal and woman-optimal stable matching finding are the same as Gale-Shapley algorithm.

With sex-fair finding experiment, we did 90 experiments with n from 6 to 20 (6 for each run). The results were 90% percents of 90 instances having deviation between woman score and man below 3 and 94% percents having the deviation below 4. The experiments with egalitarian stable matching finding were evaluated as follows: with the same instance we found simultaneously man-optimal, woman-optimal and egalitarian stable matching. We obtained the maximum woman score sw^*, maximum man score sm^* and egalitarian score se^* $(sw+sm)$ respectively. The egalitarian score se^* were approximately the same as $(sw^*+sm^*)/2$. This result can tell that this stable matching is egalitarian or nearly egalitarian. Fig.5 shows the comparison between se^* and $(sw^*+sm^*)/2$ with simulations running n from 3 to 20.

In the table 2, we have a comparison between ACS based algorithm, Gale-Shapley algorithm, and genetic algorithm (GA) [14]. In this experiment, we set $n=15$, $m=10$.

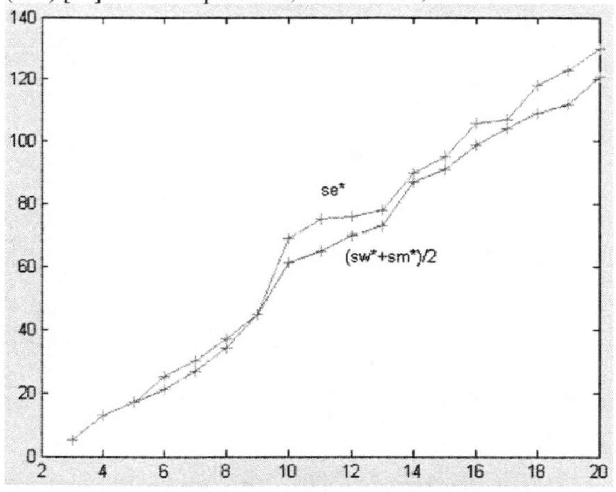

Fig. 5. Evaluated result of the egalitarian stable matching

TABLE 2
A COMPARISON BETWEEN ALGORITHMS

	ACS	Gale-Shapley	GA
Men-optimum (*sw*)	77	77	77
Women-optimum (*sm*)	82	82	82
Egalitarian (*sm+sw*)	103	-	107
Sex-fair |*sm-sw*|	4	-	8

V. SUGGESTIONS FOR FUTURE DEVELOPMENT

We only solved the well-known stable marriage problem which has the same number of men and women. But our proposed approach can be generalized to wide range of matching problems including roommate problems, variable group sizes including singles, couples, and larger groups, preference list ties, and others.

VI. CONCLUSIONS

In this paper, we proposed an ant colony system based algorithm for finding optimum stable matchings depending on different criteria such as stable matching with man-optimal, woman-optimal, egalitarian stable matching and sex-fair stable matching. It can find the man-optimal or woman-optimal stable matching the same as Gale-Shapley can. Moreover, the proposed algorithm can find the sex-fair and egalitarian effectively. These results could be considered optimum. By experimental evaluation, we show the effectiveness of the proposed ant colony system based algorithm. This is a novel approach, and this is the first time ant colony system was applied to the stable marriage problem.

REFERENCES

[1] D.Gusfield and R.W.Irving, *The Stable Marriage Problem: Structure and Algorithms*, The MIT Press, 1989.

[2] D.Gale and L.S.Shapley, College admissions and the stability of marriage, *American Mathematical Monthly*, Vol.69, pages 9-15, 1962.

[3] A.E.Roth. The Evolution of the Labor Market for Medical Interns and Residents: A Case Study in Game Theory, *Journal of Political Economy*, pages 991-1016, 1984.

[4] Abraham, David J and Irving, Robert W and Manlove David F (2003), The Student-Project Allocation Problem, Proceedings of ISAAC 2003: *The 14th Annual International Symposium on Algorithms and Computation*. 15-17 December, 2003 Lecture Notes in Computer Science Vol.2906, pages 474-484, Kyoto, Japan.

[5] R. Beckers, J.L. Deneubourg, and S. Goss, Trails and U-turns in The Selection of The Shortest Path by The Ant, Lasius Niger, *Journal of Theoretical Biology*, vol. 159, pp. 397–415, 1992.

[6] S. Goss, S. Aron, J.L. Deneubourg, and J.M. Pasteels, *Self-organized Shortcuts in The Argentine Ant*, Naturwissenschaften, vol. 76, pp. 579–581, 1989.

[7] B. Hölldobler and E.O. Wilson, *The ants*. Springer Verlag, Berlin, 1990.

[8] Dorigo M, Gambardella L M, Ant Colonies for The Travelling Salesman Problem, *BioSystems*, 1997, pp. 73-81.

[9] M Dorigo, and L. Ganbardella. Ant Colony System: A Cooperative Learning Approach To The Traveling Salesman Problem, *IEEE transactions on Evolutionary Computing*, 1(1):53-66, 1997.

[10] A. Colorni, M. Dorigo, and V. Maniezzo, Distributed Optimization by Ant Colonies, *Proceedings of ECAL91 - European Conference on Artificial Life*, Paris, France, 1991, F. Varela and P. Bourgine (Eds.), Elsevier Publishing, pp. 134–142.

[11] A. Colorni, M. Dorigo, and V. Maniezzo, An Investigation of Some Properties of An Ant Algorithm. Proceedings of the Parallel Problem Solving from *Nature Conference* (PPSN 92), 1992, R. Männer and B. Manderick (Eds.), Elsevier Publishing, pp. 509–520.

[12] M. Dorigo, V. Maniezzo, and A.Colorni. The Ant System: Optimization by A Colony of Cooperating Agents. *IEEE Transactions on Systems, Man, and Cybernetics*–Part B, vol. 26, No. 2, pp. 29–41, 1996.

[13] V. Maniezzo, A.Colorni, and M.Dorigo, The Ant System Applied To The Quadratic Assignment Problem, *Tech. Rep. IRIDIA/94-28*, 1994, Université Libre de Bruxelles, Belgium.

[14] Ngo Anh Vien, TaeChoong Chung, Multiobjective Fitness Functions for Stable Marriage Problem using Genetic Algorithm, *SICE-ICASE International Joint Conference* 2006 in Pusan, Korea.

Q-Learning Based Univector Field Navigation Method for Mobile Robots

Ngo Anh Vien, Nguyen Hoang Viet, HyunJeong Park, SeungGwan Lee[1] and TaeChoong Chung**

Artificial Intelligence Lab, Dept. of Computer Engineering
School of Electronics and Information, Kyunghee University
(Corresponding author**)
School of General Education, Kyunghee University[1]
SoChonDong, GiHungGu, YongInSi, GyongGiDo 446-701, South Korea

***Abstract*-In this paper, the Q-Learning based univector field method is proposed for mobile robot to accomplish the obstacle avoidance and the robot orientation at the target position. Univector field method guarantees the desired posture of the robot at the target position. But it does not navigate the robot to avoid obstacles. To solve this problem, modified univector field is used and trained by Q-learning. When the robot following the field to get the desired posture collides with obstacles, univector fields at collision positions are modified according to the reinforcement of Q-learning algorithm. With this proposed navigation method, robot navigation task in a dynamically changing environment becomes easier by using double action Q-learning [8] to train univector field instead of ordinary Q-learning. Computer simulations and experimental results are carried out for an obstacle avoidance mobile robot to demonstrate the effectiveness of the proposed scheme.**

***Index Terms*—Reinforcement learning, Q-learning, double action Q-learning, navigation, mobile robots, univector field.**

I. Introduction

Generally, the mobile robot navigation with obstacle avoidance is one of the key issues to be looked into for successful applications of autonomous mobile robots. This approach involves three tasks: mapping and modeling the environment, path planning and selection, and path following [1]. In robot navigation problem, path planning and path following were really complicated and they attracted much study. The path following task is composed of trajectory planning, which makes the generated path a time parameterized line, and generating tracking control. In this task, tracking problem is the key issue to be studied and there are various approaches such as sliding mode control and feedback linearization control [3]–[5]. In the path planning task, a path generation algorithm is developed which does not cross the obstacles and connects the destination with the starting point. On the other hand, in unified navigation such as potential field method [2], [6], [7], these two tasks are unified in one task.

The potential field method of avoiding obstacle consists of evaluating a repulsive force for each obstacle. That evaluation is made taking into account the distance to the obstacle and the relative velocity between the robot and the obstacles. An attractive that tends to drive the robot to its target is also calculated. Each of these forces has the direction of the object gave rise to it. However, when a constant velocity can not be maintained and the obstacle is too big, it makes robot oscillate and the direction at the final point can not be guaranteed [6, 7]. To solve this problem, univector field method [10, 11] is used to guarantee the final posture of the robot.

In dynamical environment, the velocity of both obstacles and robot change rapidly, so efficiency of trajectories and short navigation time must to be ensured. These are important issues of soccer robots also [9]. In addition, the posture of a soccer robot must be ensured also because robot posture (position and orientation) is of the utmost importance for dribbling and kicking action. Recently research interested in robot soccer is being focused on the application of fuzzy logic, evolutionary computation, reinforcement learning, unified navigation method and so on [10]–[13]. In [11], the author aimed to obtain the suboptimal univector field for navigation hence a function approximator and its learning algorithm were used and based on evolutionary programming (EP) which takes into consideration the kinematic properties of the robot. Two kinds of univector fields are trained. One is concerned about the final posture of the robot and hence deals with the field for the desired orientation. This takes the robot to a desired posture. The other is the field for obstacle avoidance. Combining these two well-trained fields, the complete field for an environment with obstacles can be generated. But this proposed method has been applied only to static obstacles and EP is an offline learning strategy.

In this paper, we proposed an improved univector field navigation method which used Q-learning to train the univector field vector, called Q-learning based univector field (QUF). Using the QUF method the robot can navigate rapidly to the desired posture (position and orientation) without colliding with obstacles. With this proposed method, we can use double action Q-learning [8] to navigate the robot in a dynamically changing environment that suits to robot soccer game. In this kind of game, the dynamic environment comprises of many obstacles and moving targets.

In Section II, the kinematic properties of differential-drive mobile robots are discussed. In Section III, the definition of univector field and the concept of univector field navigation are described. The Q-learning algorithm based univector field is also presented. Sections IV reports simulations and experimental results respectively. Concluding remarks follow in Section V.

K. Elleithy (ed.), Advances and Innovations in Systems, Computing Sciences and Software Engineering, 463–468.

II. Control of Mobile Robots

In this paper, a nonslipping and pure rolling differential-drive mobile robot is considered [14]. It is assumed that the posture (position and orientation) of the robot is known at each instant. The experimental setup is described in Section IV. The mechanical structure of the mobile robot is shown in Fig. 1(a), where L is the base width of the robot and R is the radius of the wheel.

The kinematics of the robot can be described using Fig. 1(b). Posture p_s and position p of the robot are defined as

$$p_s = \begin{bmatrix} x_c \\ y_c \\ \theta_c \end{bmatrix}, \quad p = \begin{bmatrix} x_c \\ y_c \end{bmatrix} \tag{1}$$

where (x_c, y_c) is the position of the center of robot, and θ_c is the heading angle of the robot with respect to absolute coordinates (x, y). Velocity vector s is defined as follows:

$$s = \begin{bmatrix} v \\ \omega \end{bmatrix} = \begin{bmatrix} \dfrac{V_R + V_L}{2} \\ \dfrac{V_R - V_L}{L} \end{bmatrix} = \begin{bmatrix} \dfrac{1}{2} & \dfrac{1}{2} \\ \dfrac{1}{-L} & \dfrac{1}{L} \end{bmatrix} \begin{bmatrix} V_R \\ V_L \end{bmatrix} \tag{2}$$

where v is the translational velocity of the center of robot and ω is the angular velocity with respect to the center of robot. Equation (2) gives the relation between the velocity vector and the velocities of two wheels, V_L and V_R, where V_L is the left wheel velocity and V_R is the right wheel velocity.

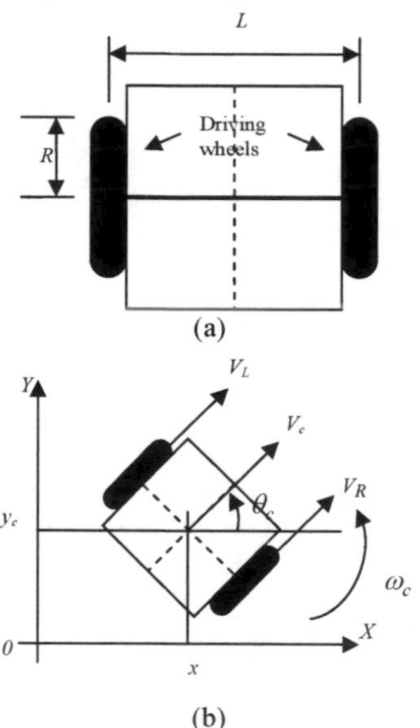

(b)

Fig. 1. Shape of differential-drive mobile robot and its modeling. (a) Shape of the robot. (b) Robot modeling.

The robot kinematics associated with the Jacobian matrix and velocity vector is defined as

$$p_s = \begin{bmatrix} \cos\theta_c & 0 \\ \sin\theta_c & 0 \\ 0 & 1 \end{bmatrix} \begin{bmatrix} v \\ \omega \end{bmatrix} = J(\theta_c)s \tag{3}$$

To get the robot position and orientation, (3) should satisfy the following nonholonomic constraint:

$$x_c \sin\theta_c - y_c \cos\theta_c = 0 \tag{4}$$

which is equivalent to $dx_c / dy_c = \tan\theta_c$, meaning that the moving direction at every instant is the same as the heading angle of robots. It implies pure rolling and nonslipping as assumed.

III. Univector Field Navigation Method

A. Univector Field Navigation

Fig. 2 shows the univector field, where the tiny circles with small dash attached to it denote the robot heading direction. The tiny circle is meant to represent the robot position and the straight line attached to it represents its heading directions [13]. A vector field (x, y) at position p is defined as $F(p)$ or $F(x, y)$, all vector fields constitute a univector field space $N(p)$ for robot control. It is assumed that the magnitude of the vector field is unity and is the same at all points [11]. The angle ϕ(p) of the vector at a robot position p is generated by:

$$\phi(p) = \angle \overrightarrow{pg} - n\alpha \quad \text{with } \alpha = \angle \overrightarrow{pr} - \angle \overrightarrow{pg} \tag{4}$$

where n is a positive constant. The larger n is, the smaller the ϕ(p) is at the same robot position. Thus, if n increases, the univector field spreads out to a larger area, making the path to be traversed by the robot in reaching its goal larger. The shape of the field and the turning motion of the robots changes according to the parameter n and the length of the line gr. The univector field method is based on (4), through which the vector field at all points can be obtained. In Fig. 2, g represents the target position of the robots. A dummy point r is used for deriving the vector field. The dummy point r is selected heuristically close to the goal point g. In practical robot soccer applications, the point g will be the position of the ball.

The following relationships are used to reduce the error in angle between the robots and the field vectors:

$$\begin{aligned} \omega &= K_P\theta_e + K_D\dot{\theta}_e, \\ \theta_e &= \angle F(p) - \theta_c, \\ \dot{\theta} &= \frac{d\theta_e}{dt} \end{aligned} \tag{5}$$

where $F(p)$ is the vector field at position p with unit magnitude, θ_e is the error in angle between the robot heading and the vector field direction, $\dot{\theta}_e$ is the derivative of θ_e, K_P is the proportional feedback gain, and K_D is the time derivative feedback gain.

The translational velocity υ is constant. If $\upsilon = 0$, the robot's heading angle will be towards the direction of $F(p)$ without any changes in position. As indicated by (5), the robot motion is controlled through its right and left wheel velocities which are functions of time:

$$V_R = V_C + K_P^{'}\theta_e + K_D^{'}\dot{\theta}_e,$$
$$V_L = V_C - K_P^{'}\theta_e - K_D^{'}\dot{\theta}_e \qquad (6)$$

According to (2), $K_P^{'} = \frac{L}{2}K_P$ and $K_D^{'} = \frac{L}{2}K_D$, V_C is the constant robot center velocity. The robot's vector field will be oriented towards the target position and the associated angle of the robot motion is as shown in Fig.2.

To exploit the univector field $N(p)$ for robot control with better performance, the field has to be modified. For this purpose, a Q-Learning based Univector Field navigation method was developed. In (4): The larger n is, the smaller the ϕ(p) is at the same robot position. Thus, if n changes, the univector field will change accordingly. Hence, this paper aims to train the constant n in order to modify univector field $N(p)$ by using Q-learning. For this reason, we use the position of the robot as states in Q-learning, and use actions of choosing n for each position p as action space of each state.

The shaded region in Fig. 2 is the expanded region. In this region the vector field's angles change very little in comparison with the changing of constant n.

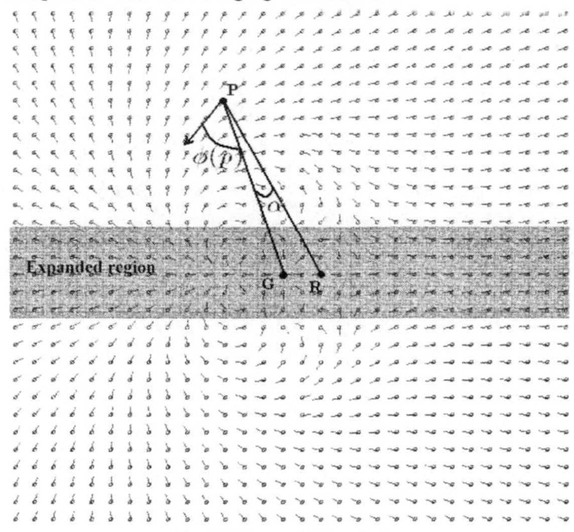

Fig. 2. Univector field method.

Consequently, it makes the robots difficult to avoid big obstacles within this narrow area. Due to this drawback, this narrow region is expanded as shown in Fig.2. The width of the expanded region is chosen to be based on the biggest obstacle's size in the environment. Therefore, the modified univector field has different n action space from the vector field outside the modified area. The robot follows the Q-learning univector field and then enters into the expanded region.

B. Q-Learning Based Univector Field

The Q-learning was implemented to train constant n_p (the constant corresponds to each univector field $N(p)$) to perform obstacle avoidance. Thus, QUF can guarantee the desired posture at the target's position with collision-free.

As Q-learning [15] needs no model of environment, it is easy to implement and it is an off-policy algorithm. Q-learning has been used in many application areas. In the QUF method, the value of taking action n in state p (position of the robot) is called the action value as depicted in Fig. 3. The robot can learn the action value using the selected action in the state without any model of the environment. The updating equation of the Q value in the Q-learning is as follows:

$$Q(p,n_p) \leftarrow Q(p,n_p) + \alpha(r + \gamma \max_{n_{p'}} Q(p',n_{p'}) - Q \qquad (7)$$

where Q is the action value, $p^{'}$ is the next state after taking action n_p, α is the learning rate, r is the reward value and γ is the discount rate.

The robot selects the optimal action in each state based on Q values. After the action is selected, the robot receives the reward and the Q-value of the state-action pair is updated.

The action n space: is a set of discrete numbers in range $\{-n_k<-n_{k-1}<...-n_1<n_1<...<n_{k-1}<n_k\}$, with all $i=1...k$, $n_i > 0$, $n_i = n_{i+1} - \Delta$, where $\Delta>0$ is a smooth decrement constant. For this reason, we have the range of modified ϕ(p) :

$$\phi(p) = \angle \vec{pg} - n\alpha$$
$$\begin{cases} \angle \vec{pg} - n_k \,|\alpha| \le \phi(p) \\ \angle \vec{pg} + n_k \,|\alpha| \ge \phi(p) \end{cases} \qquad (8)$$

It means that modified angle $\theta(p)$ of univector field will be modified to avoid obstacles. With specific problem, there are different kinds of obstacles (big or small), different speeds of obstacles, or with different setting of dummy point r, then Δ and k will be set differently.

With state position p in expanded region (α is very small): action space is $\{m \times n_i\}$ where m is amplification constant.

Initialize $Q(p, n_p)$ arbitrarily
Repeat (for each episode)
Initialize p
Repeat (for each step of episode):
Choose n_p from p using policy derived from Q
Take action n_p
Observe the new state p' and r

$$Q(p,n_p) \leftarrow Q(p,n_p) + \alpha(r + \gamma \max_{n_{p'}} Q(p',n_{p'}) - Q(p,n_p))$$

$p \leftarrow p'$
until p is terminal

Fig.3 The Q-learning based univector field

But when $\alpha = 0$, to avoid there is no affect of n on $\theta(p)$, the univector field is changed to: $\phi(p) = \angle \overrightarrow{pg} - n_i$.

In the next section, simulations will show the effectiveness of the QUF method. When applying the QUF method, univector fields for getting the desired posture at the target are trained to avoid obstacles. So QUF method is unified navigation method which unifies path planning and path following step together. The applied Q-learning in the QUF method is ordinary reinforcement learning based on Markov Decision Process (MDP) [15]. And the obstacles in the QUF method were assumed static. But our approach is open to apply modified Q-learning algorithm in a dynamically changing environment.

Initialize $Q(p, n_p, a)$ arbitrarily
(a is action taken by the obstacle)
Repeat (for each episode)
Initialize p
Repeat (for each step of episode):
Get a by predicting the action that will be taken by the obstacle
Choose n_p from p using policy derived from Q
Take action n_p
Observe the new state p' and r
Determine the action a that have been taken by the environment

$$Q(p,n_p,a) \leftarrow Q(p,n_p,a) + \alpha\left[r + \gamma \frac{\sum \max_{a'} \max_{n_{p'}} Q(p',n_{p'},a')}{count(a')} - Q(p,n_p,a)\right]$$

$p \leftarrow p'$
until p is terminal

Fig.4 The double action Q-learning based univector field

In a dynamically changing environment, the robot must face with both moving and static objects, so the robot should have the ability to deal with both of them. For this reason, double action Q-learning method [8] for obstacle avoidance in a dynamically changing environment was applied to train constant n as depicted in Fig. 4.

This method explored the possibility of improving the ordinary Reinforcement Learning (RL) approach by considering not only the action of the robot but also the predicted action taken by the environment (or objects in the environment). First of all, a state defines the relationship between two objects in the environment. Secondly, the environment can change by itself. It changes whether the robot acts or not. If we can predict the action that will be taken by the environment accurately, the function Q will be very useful to determine our best action in the next time step. Fortunately, it is possible to predict the motion of the environment through sensory inputs from which the robot can take appropriate action to maximize its reward.

In our multi-obstacle environment, we treat each obstacle as an independent object; Q-learning can be applied on each pair of object-agent relationship. Now we must combine Q-values from different blocks. In [8], the author took the sum of all Q-values from all blocks. Hence, we also use this approach to combine the Q-values from different object-agent relationships. That is:

$$Q(p,n_p) = \sum_i q_i(p,n_p,a^i) \qquad (9)$$

where $Q(p, n_p)$ is the resultant Q-value for the entire environment and $q_i(p,n_p,a^i)$ is the Q-value of the particular type of object when the agent faces that object along. The action a^i is determined through the prediction module. $\max_{n_p} Q(p,n_p)$ is then chosen as the final action n_p.

IV. Simulation and Experiment Results

To illustrate the efficiency of the proposed Q-learning based univector field for posture acquisition and obstacle avoidance, computer simulations were carried out on an Intel Pentium IV PC. For each individual, the simulation was carried out 20 times with uniformly distributed random starting postures. Throughout simulations, Δ=5 and k=4 so action n space {-20, -15, -10, -5, 5, 10, 15, 20} is used to train the univector field, if collision occurs, the robot are brought back to the original position and given a punishment of -10. The proposed method is applied to static obstacles for designated final posture and final posture in dynamically changing environment.

A. Q-Learning Based Univector Field for Final Posture and Static Obstacle Avoidance

Figs. 5 and 6 show the results for rectangular obstacle avoidance with the desired posture at the target position. To avoid the obstacles, univector field is trained to get optimal univector angle. And to obtain the desired posture at the target position, the robot must converge to the desired point with its heading angle to the desired orientation. It was assumed that the final position is the center of the field (0, 0) and the final orientation is to the left (π rad) as shown in Fig. 5. The final position is in the upper-left and the final orientation is to the right ($\pi / 4$ rad) as shown in Figs. 6. Each simulation was stopped when the robot reached an allowable error bound of the target point. Fig. 6 shows simulation results in an environment with multiple obstacles. The vector field in the figures is the initial vector field when the robot starts learning.

B. Q-Learning Based Univector Field for Final Posture in Dynamically Changing Environment

The testing environment consists of numbers of obstacles and the robot itself. The state indicating the robot-object relationship is represented by the x and y coordinates of the object with the robot placing in the origin. The obstacles [8] can choose one of 5 actions: up, down, left, right, rest. For evaluation purpose, we focus on the double action Q-learning based univector field method and thus assume that the prediction and action determination of obstacles were accurately. A goal is given to the robot and obstacles are randomly placed in between so that the robot needs to handle the problem of obstacle avoidance in its navigation.

Fig. 7 depicts a screen capture of the simulator when the robot (white square) was moving among 10 obstacles (blue square) in between the starting point (upper left), goal (lower right) and the final orientation is to the right (0 rad).

Fig. 8, we can see that the number of collisions also starts to decrease at episode 10 and there are only a few collisions after episode 300. The reason is that the robot is learning in the first 300 episodes and it has nearly learned all the Q-values for all the cases that would cause collision after episode 300. Therefore, the number of collisions decreases naturally.

The proposed method has been applied only to static target in dynamically changing environment. Hence, in its present form it cannot be applied to robot soccer. Therefore subsequent research will investigate the possibility of modifying the univector field to suit a dynamic environment with moving target.

V. CONCLUSION

A novel learning navigation method for final posture acquisition and obstacle avoidance was developed for fast moving mobile robots. The method was the result of introducing a Q-learning based univector field into the unified navigation method. To obtain the trained vector field, reinforcement learning is used to train the coefficient in the univector angle's formula. The new method has taken into account of the obstacle avoidance action with univector field. With the new method, we can apply other moving obstacle avoidance methods to the unified navigation method. The simulations results show that the proposed method successfully navigates fast moving mobile robots through collision-free trajectories and results in desired final posture of the robots. Especially, we can use double action Q-learning based univector field method to navigate the robot with the desired final posture of the robots in dynamically changing environment.

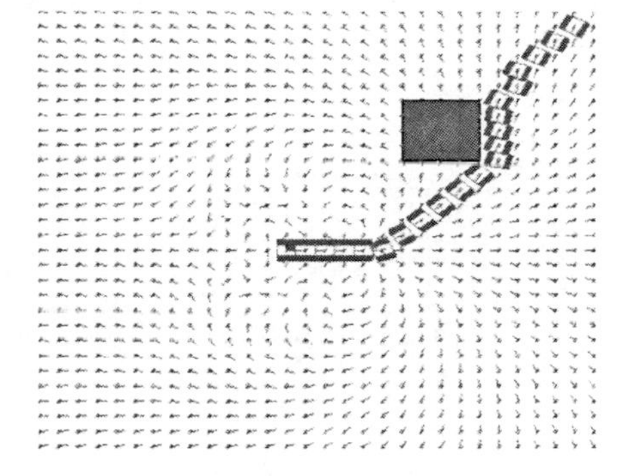

Fig. 5: Simulation with a static obstacle.

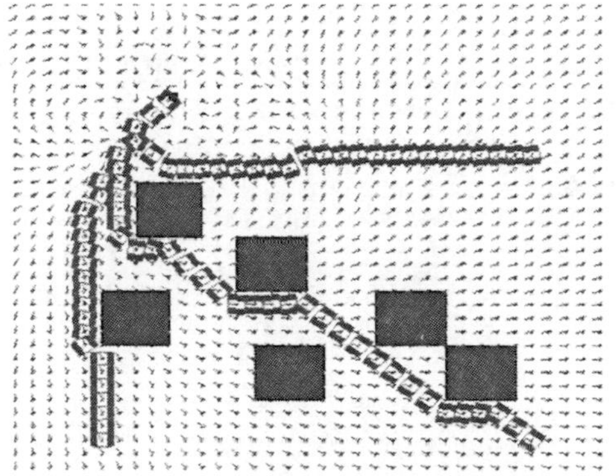

Fig. 6: Simulation results of multiple obstacles

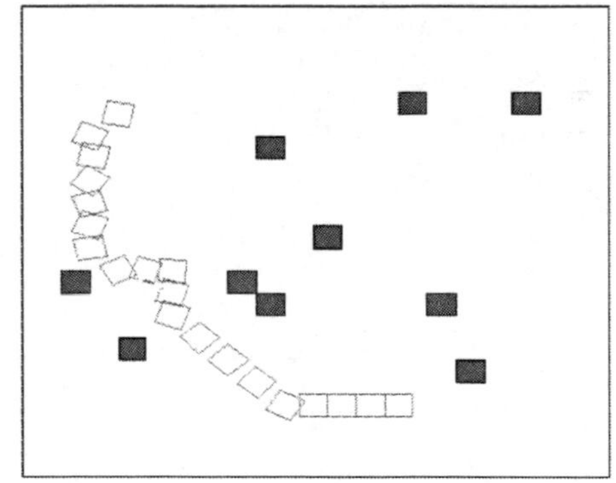

Fig. 7: Screen capture of the simulator.

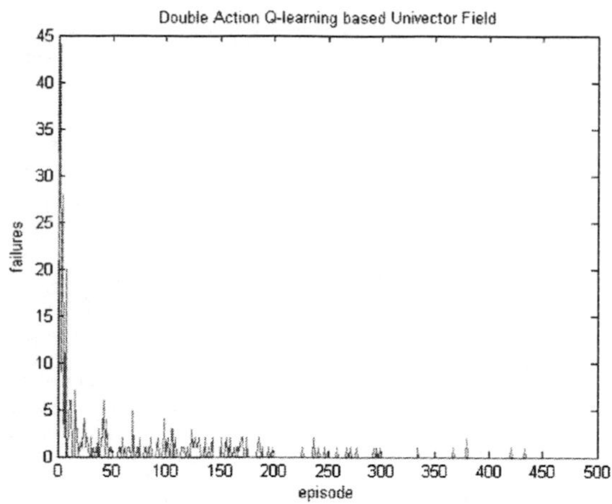

Fig.8: Sum of reward

REFERENCES

[1] P. J. McKerrow, *Introduction to Robotics*, Reading, MA: Addison-Wesley, 1991, pp. 431–435.

[2] E. Rimon, "Exact robot navigation using artificial potential functions," *IEEE Trans. Robot. Automat.*, vol. 8, pp. 501–518, Oct. 1992.

[3] J.-M. Yang and J.-H. Kim, "Control of nonholonomic mobile robots," *IEEE Control Syst. Mag.*, vol. 19, pp. 15–23, 73, Apr. 1999.

[4] C.-Y. Su and Y. Stepaneko, "Robust motion/force control of mechanical systems with classical nonholonomic constraints," *IEEE Trans. Automat. Contr.*, vol. 39, pp. 609–614, Mar. 1994.

[5] B. d'Andrea-Novel, "Control of nonholonomic wheeled mobile robots by state feedback linearization," *Int. J. Robot. Res.*, vol. 14, no. 6, pp. 543–559, Dec. 1995.

[6] J. Borenstein and Y. Koren, "Real-time obstacle avoidance for fast mobile robots," *IEEE Trans. Syst., Man, Cybern.*, vol. 20, pp. 1179–1187, Apr. 1989.

[7]. J. Borenstein and Y. Koren, "The vector field histogram—Fast obstacle avoidance for mobile robots," *IEEE Trans. Syst., Man, Cybern.*, vol. 7, pp. 278–288, Mar. 1991.

[8] Ngai, D.C.K.; Yung, N.H.C, **"**Double action Q-learning for obstacle avoidance in a dynamically changing environment**"**, *Intelligent Vehicles Symposium*, 2005 Proceedings IEEE 6-8 June 2005 Page(s):211 – 216.

[9] H.-S. Shim *et al.*, "Designing distributed control architecture for cooperative multi-agent system and its real-time application to soccer robot,"*J. Robot. Autonom. Syst.*, vol. 21, no. 2, pp. 149–165, Sept. 1997.

[10] J.-H. Kim, K.-C. Kim, D.-H. Kim, Y.-J. Kim, P. Vadakkepat, Evolutionary Programming-Based Univector Field Navigation Method for Fast Mobile Robots, *IEEE Trans. Syst., Man, Cyber*, vol. 31, no. 3,pp. 450-458, June-2001.

[11] Y.-J. Kim, D.-H. Kim, J.-H. Kim, Evolutionary programming-based vector field method for fast mobile robot navigation, in: *Proceedings of the Second Asia–Pacific Conference on Simulations, Evolutions and Learning*, 1998.

[12] W.-G. Han *et al.*, "GA based on-line path planning of mobile robots playing soccer games," in *Proc. IEEE 40th Midwest Symp. Circuit Syst.*, Sacramento, CA, Sept. 1998, pp. 522–525.

[13] J.-H. Kim *et al.*, "Path planning and role selection mechanism for soccer robots," in *Proc. IEEE Int. Conf. Robot. Automat.*, vol. 4, Leuven, Belgium, 1998, pp. 3216–3221.

[14] G. Camoin *et al.*, "Structural properties and classification of kinetic and dynamic models of wheeled mobile robots," *IEEE Trans. Robot. Automat.*, vol. 12, pp. 47–62, Feb. 1996.

[15] R. S. Sutton and A. G. Barto, *Reinforcement Learning: An Introduction*, Bradford Books/MIT Press, 1998.

Statistical Modeling of Crosstalk Noise in Domino CMOS Logic Circuits

Vipin Sharma and Waleed K. Al-Assadi, *Senior Member, IEEE*
Department of Electrical & Computer Engineering, University of Missouri – Rolla
1870 Miner Circle, Rolla, MO 65409 U.S.A
Email: {vsmq7, waleed}@umr.edu

Abstract- Domino logic circuits have been aggressively explored for vulnerabilities due to crosstalk noise. In these circuits, statistical modeling of crosstalk noise seems to be a promising approach due to factors like: large unpredictability in crosstalk noise with technology trends pushing process variations to their extreme end and reducing feature sizes ensuing unevenness in device geometries. We present here a general model for crosstalk noise with cross-coupling capacitive variance and MOS devices' channel width variation effects and progressively refine it to get the most accurate circuit analysis model for deriving the crosstalk distribution. The statistical model derived is validated with 1000 runs of Monte Carlo simulations.

Keywords: Domino circuits, noise, statistical model, process variations.

I. Introduction

Domino logic circuits belong to the dynamic logic circuits family offering the designers the advantages of higher speeds, smaller area, and lower power dissipation then their equivalent static circuits. Technology trends are pushing the problem of inherent reduced noise immunity in domino logics on the extreme end [1]. Factors like aggressive device scaling, one dimensional interconnects scaling, and fabrication of analog and digital circuits on the same substrate and process variations are making the problem worse. As technology is advancing, the effects on crosstalk noise due to process variations is gaining more and more attention recently as it is becoming increasingly difficult to control reducing feature sizes [2]. Since coupling capacitances are highly sensitive to these process variations, capacitive cross-talks is the problem which demands a thorough analysis especially if recent technologies trends are continued to be followed in future.

This work is supported by Missouri Research Board Under Grant No. R5003188.

The existing literature offers various statistical models for different types of target areas. The work in [3] considers process variations specifically in estimating the crosstalk noise. Aggarwal et al. in [4] present a statistical approach of crosstalk noise between two lines. The authors in [5] develop analytical closed form expressions for modeling of crosstalk. The works in references [6] and [7] develop statistical model for MOS devices and delay – fault testing respectively to model process variations. Crosstalk noise is analyzed by authors in [8] and [9] specifically for domino CMOS logic circuits.

The works cited above model crosstalk noise based on essential factors; however, as process variations command a statistical approach, accounting for the ones which have significant effect on the derived model on the crosstalk noise improves the model for judging the final circuit parameters against initial design specifications. Predicting the variance in noise as precisely as variance parameters, requires the victim node to be sampled at a specific instant. A general circuit analysis model is needed which is not only accurate in predicting the final distributions, but also efficient in terms of simplicity and computational complexity. The work presented in this paper addresses these concerns and supports the results with several Monte Carlo simulation runs. Since inputs to output coupling is aggressively explored in the literature the analysis in this paper excludes the possibilities of coupling through gate inputs. Although process variations transform into variations in delay, process specific device parameters, etc., we limit our statistical analysis on capacitive cross couplings and device channel widths as dominant parameters.

The rest of the paper is organized as follows: Section II presents a general capacitive cross-coupling circuit analysis. Section III discusses an appropriate sampling instant of the victim node at which the crosstalk noise voltage should be measured. Section IV presents a statistical model for the output node of dynamic gate. Section V validates the model derived with Monte Carlo simulation results.

K. Elleithy (ed.), Advances and Innovations in Systems, Computing Sciences and Software Engineering, 469–474.

II. Capturing Geometrical Variations

Crosstalk noise is an undesirable electrical interaction between two or more signal nets due to capacitive, resistive or inductive coupling. It is generally modeled as cross-coupling between the victim and aggressor nets. Victim nets are the signal lines receiving undesirable cross-coupling effects from nearby nets, called aggressors. This section identifies several possible cross-coupling effects in a domino CMOS logic gate between the aggressor and potential victims and progressively refines the model for gaining efficiency in analysis while keeping the model as accurate as possible for precise predictions by the distribution functions derived later.

A. Potential Victims

Figure 1 shows potential cross-couplings effects and their associated victim nodes in a general structure of domino gate with I_1-I_n as ‘n’ number of primary inputs. The figure shows five victim nodes labeled as node 2 (dynamic node of the gate), node 3 (output of the gate), node 4 (static node external to the gate), node 5 (static node external to the gate but coupled to dynamic node) and node 6 (the clock node). Node 1 is assumed to be a nearby net acting as an aggressor to all other nodes. Each of these nodes is coupled to the aggressor node (line 1 or node 1). We assume the aggressor is affected by process variations resulting into corresponding capacitive variations. All these capacitances have potential to impact the crosstalk nose at the dynamic gate output. Determining the dominant coupling is critical and is explored in the next subsection.

B. Dominant Cross-Couplings

We consider all the couplings through dynamic node to the output and directly to the output node as crosstalk analysis through inputs is explored in depth in the existing literature. For capacitive coupling C_{16} (aggressor–clock node) following four points are worth noting: gate-drain coupling capacitance of pre-charge transistor M1 should be large enough to couple noise to the dynamic node.

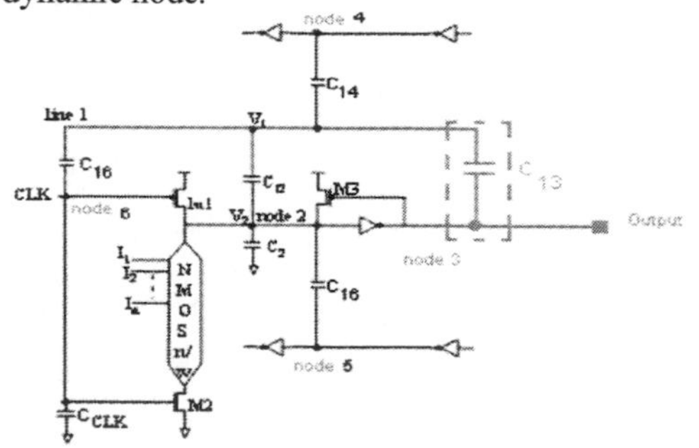

Fig. 1. Potential victims

To couple the noise from gate of M1 to the output node the noise has to couple through dynamic node. The gate-drain capacitance of transistor M1 varies with the driving strength [1]. Also the CLK node being a global net has a very large capacitance to the ground. Considering these factors and noting that driving strength of precharge transistor is least in evaluation mode (hence the gate-drain capacitance), we can safely infer that the amount of cross-coupling from aggressor node to CLK node through the dynamic node to the output node is minimal and can safely be ignored.

The presence of node 5 and node 6 in the Figure 1 on the final crosstalk noise model is also examined with several Monte Carlo simulation results. The simulation result suggests that these nodes can safely be ignored without much affecting the accuracy of the model prediction. For rest of the analysis part, we have the circuit modeled wherein only two coupling capacitances are present viz., C_{12} and C_{13}. Based on Monte Carlo simulation results we obtained, considering C_{12} and C_{13} varying jointly with the same values results into better estimate of variance in crosstalk noise voltage with the corresponding variance in the capacitances. We finally have the circuit model for crosstalk analysis with two capacitive cross-couplings which vary jointly with the same distribution.

III. Appropriate Sampling Instant

The prediction of the crosstalk noise voltage as a function of maximum dependence on the variant parameter depends heavily on chosen sampling instant for the output victim node. To exclude the dependence of the statistical model on the aggressor signal characteristics and to better estimate the effect of crosstalk noise for the fan outs of the victim node we consider the re-settling time of the victim node as the sampling time for the crosstalk nose voltage. We define the re-settling region of the crosstalk affected node as the region in which the victim node recovers from the crosstalk noise pulse on the aggressor. This concept was also partly well treated in [1] which construes that even the properties of the crosstalk affected node in the re-settling region are essential for proceeding with the crosstalk noise analysis. The aim remains to derive the maximum dependence function of the crosstalk noise voltage on the variance parameters. This in fact is a good approach towards a more general statistical model of crosstalk noise. We illustrate the concept of having appropriate samplings instants with the help of a simple RC circuit charging phase with all charging waveforms overlapped (Figure 2).

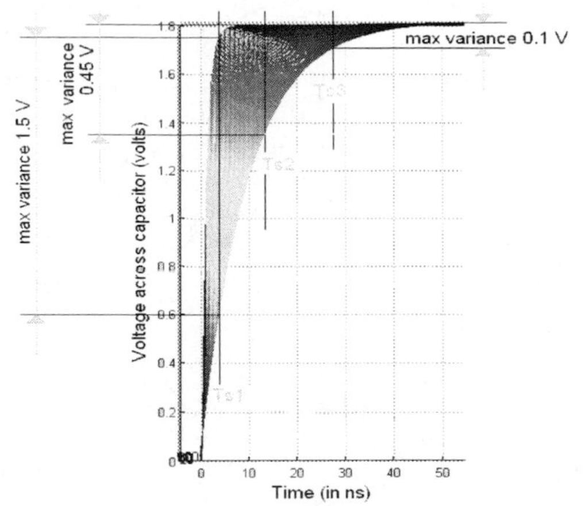

Fig. 2. Appropriate sampling instants

The aim is to find the appropriate sampling instant of the victim node. If T_{s1} is selected as the time instant for sampling the crosstalk noise voltage at the victim node, realistic information regarding the maximum dependence of crosstalk noise voltage on the randomly varying parameters is obtained. The crosstalk noise voltage at the victim node if expressed as a function of the capacitance variance, say $f_C(t)$ and if $f_{CMAX}(t)$ and $f_{CMIN}(t)$ are $+3\sigma$ and -3σ points respectively of the source distributions of capacitance variance, the appropriate sampling instant is given by the following expression:

$$F(t_S) = |f_{CMAX}(t_S) - f_{CMIN}(t_S)| \quad (1)$$

Where t_S is varied from 0 to the maximum time required by any of the function f_{CMAX} or f_{CMIN} to resettle up to 90% of the final value. The appropriate sample instant, t_{SMAX} is the value of t_S where $F(t_S)$ is maximum which can easily be derived using the properties of first and second order differentiations.

IV. STATISTICAL MODEL

For crosstalk noise in domino gates is derived we model the non linearity of MOS devices (mostly channel ON resistances) involved in domino logics with varying passive parameters (resistances). On the basis of results obtained with Monte Carlo runs on an example domino logic circuitry (AND gate), a circuit model (Figure 6) resembling the behavior of active devices under process variations is presented upon which the crosstalk noise voltage on the victim node is derived.

Figure 3 shows the results from 100 runs of Monte Carlo simulation runs. The dynamic node voltage is in the first row and the second row is the output voltage.

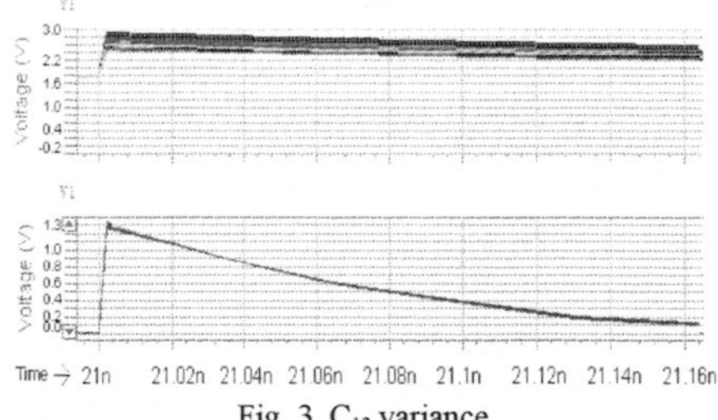

Fig. 3. C_{12} variance

C_{12} varies with 20% variance in $\pm 1\sigma$ points following Gaussian distribution. As can be seen from the figure, simply varying C_{12} and keeping C_{13} constant, results into very minimal variations in the crosstalk noise voltage at the o/p node. We thus ignore the randomness of C_{12} while still take its presence into account while considering the final equivalent circuit for analysis.

We also investigate the effect of NMOS/PMOS transistor width variance due to process variances on crosstalk noise distributions. Figure 4 shows the Monte Carlo simulation results wherein the widths of the keeper transistor and of the NMOS transistor in the output static inverter were varied individually. With different driving strengths of these devices the victim node 3 is restored to its original value with different speeds resulting into crosstalk re-settling region variance. We therefore include the variance of channel widths in our analysis.

A. *Special Treatment to C_{12}*

Previous analysis concluded that C_{12} variance is unimportant but the presence of C_{12} is important in the statistical model. This statement is best illustrated with 100 Monte Carlo runs of the same domino AND gate. Figure 5 shows two results. First the case wherein C_{12} was varied (20% variance with Gaussian distribution) and the second case shows simulation runs wherein C_{12} is removed from the circuitry. We perform this second simulation in order to emphasize the point that even with unimportance of C_{12} variance its presence in statistical model is critical.

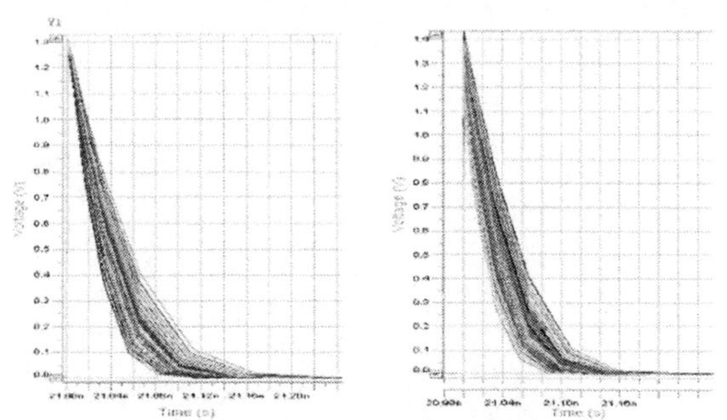

NMOS in Static Inverter Keeper Device

Fig. 4. Crosstalk with channel width variance

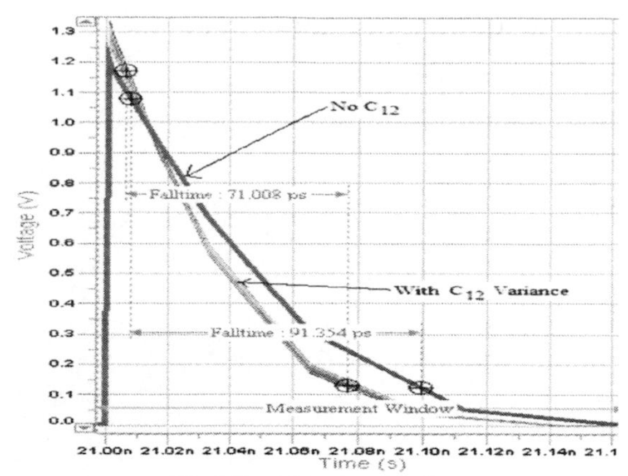

Fig. 5. Special treatment to C_{12}

Figure 5 shows that if C_{12} is removed two differences are introduced: The curve resettles from a smaller value and it has a large fall time associated with its resettling phase. With C_{12} presence, the large peak amplitude is due to the parasitic capacitances and comparatively small resettling time results as the static inverter starts restoring the crosstalk affected output node only by noticing the presence of rising crosstalk voltage at its input through C_{12}. This explains the peculiar behavior (presence is critical but variance is not) of C_{12} with crosstalk distributions on output node.

B. Equivalent Circuit Model

We now present an equivalent yet simple circuitry to accurately model variant parameters in terms of passive components. Figure 6 shows our equivalent circuit model which captures dominant process variations viz., channel width variances in keeper and static CMOS inverter and the coupling capacitance C_{13}. $R_{WS} = f_1(W_{NStatic})$, $R_{WK} = f_2(W_{Keeper})$ and $R_{C12} = f_3(C_{12})$ are mapping functions identifying the dependence of modeled parameters on their real variant counterparts. R_{WS} models the channel width variances in static inverter's NMOS transistor while R_{WK} models the variance in channel width of keeper transistor. R_{C12} models the presence of C_{12} capacitance.

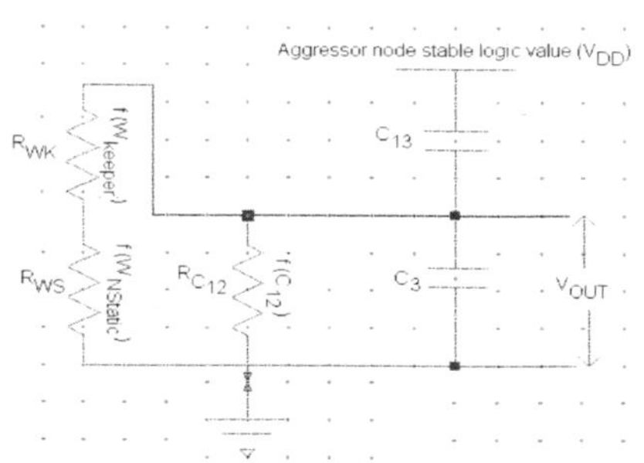

Fig. 6. Equivalent Circuit Model

We model the presence of C_{12} with a resistor since the sole function of C_{12} presence in our equivalent circuitry is to capture the effect of reduced fall time in the re-settling region. The model shows the case wherein the aggressor node has switched to a maximum value of V_{DD} and the static inverter and keeper transistor are acting together to restore the victim to its original value. Functions f_1, f_2 and f_3 are derived empirically with several statistical simulation results in the next sub-section.

C. Empirical Derivation of f_1 and f_2

We restrict our analysis on the empirical derivation of function f_1 as f_2 is similar. The derivation consists of series of Monte Carlo simulations with an example circuitry of domino AND gate. The results of Figure 7 validate our assumption of mapping channel width variances to passive resistive parameters.

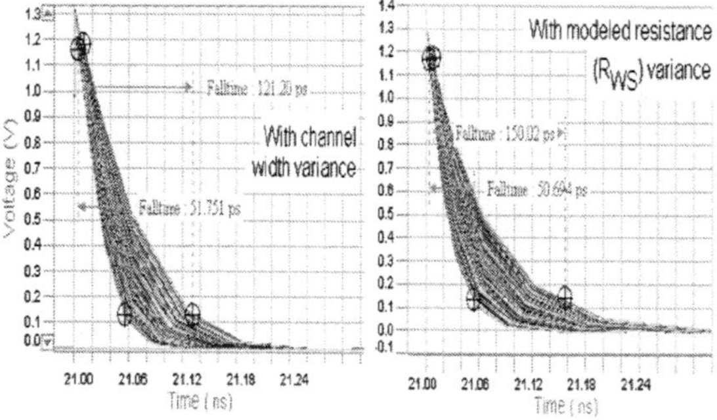

a) Crosstalk noise

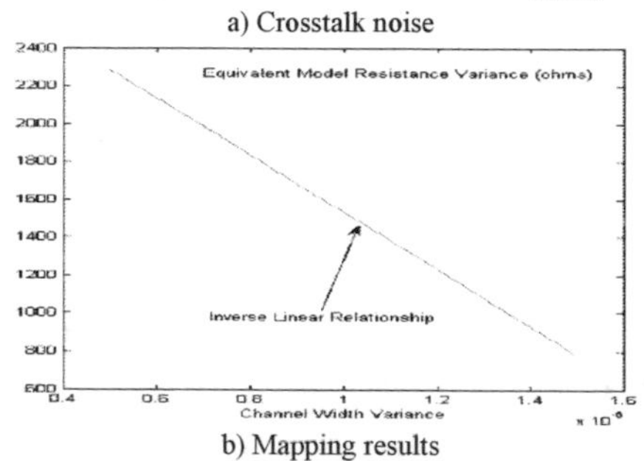

b) Mapping results

Fig. 7. Mapping the modeled parameters to their corresponding variants

To map the same crosstalk re-settling distributions to a resistive parameter, an RC circuit was simulated with uniform variances in values of R. The characteristics of uniform variance in the resistive parameter were varied until a similar re-settling region was obtained. The mapping results are shown in Figure 7 (b). An inverse linear relationship was observed between the actual width variances and its modeled equivalent. This gives us a simple mapping equation as follows:

$$R_{WS} = f_1(W_{NStatic}) = K_1 - K_2 W_{NStatic} \quad (2)$$

$$R_{WK} = f_2(W_{Keeper}) = K_3 - K_4 W_{Keeper} \quad (3)$$

Where, parameters K_1-K_4 are constants. K_1 and K_3 specify the maximum values of R_{WS} and R_{WK} respectively obtained such that the fall times in the re-settlings regions match with the falltimes in the re-settlings region of the actual circuit. However, constants K_2 and K_4 depend on several parameters viz., the process technology, device strength, etc. and can be derived empirically.

C. Empirical Derivation of f_3

Presence of C_{12} in our equivalent circuit model shows a somewhat complex influence on the distribution of crosstalk noise. Simulation results in Figure 8 capture the presence of C_{12} on the fall time of re-settling region. It shows how different values of C_{12} change the falltimes in the re-settling region. C_{12} attains all possible values in the range, and creates complex effects on the distribution of crosstalk noise. Deriving closed form expressions to capture this effect is not only a difficult and computationally intensive process, but it is also not required.

A realistic view on the possible range of values of C_{12} is taken and we therefore restrict our approximation in the region shown as "Range of practical interest" in Figure 8. This assumption is perfectly valid, although determining the range of interests of C_{12} will differ with different process technologies. Also it is worth noting the fact that increasing C_{12} is causing reduction in fall time which is contrary to the 10-90% fall time equation: t_{FALL}=2.2*RC. This paradox finally results into a model, which is simple yet includes all behavior of active devices for the crosstalk distribution. The range of practical interest of C_{12} can be approximated by the following expression where K_5 is a constant derived empirically:

$$R_{C_{12}} = \frac{K_5}{C_{12}^{\ 2}} \tag{4}$$

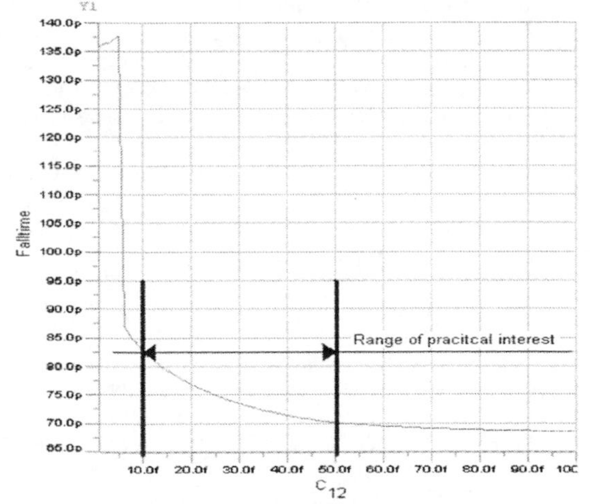

Fig. 8. Complexity of C_{12} in the model

D. Crosstalk Statistical Model

The final model is derived using an expression for output voltage as a function of the resistive and capacitive parameters of our equivalent circuit in Figure 6:

$$V_{xtalk} = \left[\frac{V_{DD}C_{13}}{C_T}\right]\left[e^{-\frac{t}{RC_T}}\right] \tag{5}$$

Where, R a random variable is equivalent resistance of R_{WS}, R_{WK} and R_{C12} and C_T, again a random variable is sum of C_{13} and C_3. 't' is simply the sampling time, which if substituted by t_{SMAX} gives the information of maximum dependence of crosstalk noise voltage on process variations. We conclude two important results from equation 5.

The distribution of maximum crosstalk noise can be derived using the first term of equation 5 where

$$R_1 = \frac{V_{DD}C_{13}}{C_T}$$

forms a random sample from the same distribution as of C_{13}. Thus the maximum crosstalk noise model in the domino circuit resembles a simple model of aggressor and victims coupled with the ratio of C_{13} / (C_T) capacitance.

The second term of equation 5 represents another random variable

$$R_2 = e^{-\frac{t}{RC_T}}$$

which is a complex exponential function of three random variables viz., channel widths of keeper and static inverter MOS devices and the coupling capacitance C_{13}, representing the re-settling region crosstalk noise distribution.

If $V_{MAXXTALK}$ represents maximum crosstalk noise, its cumulative probability density function (CDF) can be derived using transformation methods [10]:

$$F_{V_{MAXXTALK}} = F_{C_{13}}\left(\frac{V_{MAXXTALK}C_3}{V_{DD} - V_{MAXXTALK}}\right) \tag{6}$$

Where, F_{C13} is the CDF of the source distribution i.e. of C_{13}. The CDF of the crosstalk noise distribution at the victim output node can be derived in the same way as for the maximum crosstalk noise and is given by:

$$F_{V_{RSXTALK}} = F_{RC_T}\frac{t_{SMAX}}{\ln\left(\frac{1}{V_{RSXTALK}}\right)} \tag{7}$$

Where, F_{RCT} is the CDF of RC_T term. We note that since R and C_T appear in the same position of the crosstalk noise distribution expression 5, the effect of variance in these parameters on the crosstalk noise distribution is exactly same.

V. MODEL VALIDATION

The circuit used for model validation is a dynamic AND gate with two inputs. C_{13} follows a Gaussian distribution with 30 fF of nominal (mean) value and 20% $\pm 1\sigma$ variation. The probability distribution function of the maximum crosstalk noise statistical model derived from the cumulative distribution function (equation 6) as given below:

$$f_{V_{MAXXTALK}} = \left(\frac{(V_{DD} - V_{MAXXTALK})^2}{C_3 V_{DD}} \right) f_{C_{13}} \left(\frac{V_{MAXXTALK} C_3}{V_{DD} - V_{MAXXTALK}} \right) \quad (8)$$

Where f_{C13} is the probability distribution function of C_{13} capacitance and $f_{VMAXXTALK}$ is the density function of the maximum crosstalk function. Figure 9 gives the PDF obtained from 1000 runs of Monte Carlo simulations.

Figure 10 shows results obtained for crosstalk noise in re-settling region. C_{13}, and channel widths followed Gaussian distribution with 20% $\pm 1\sigma$ variance.
The PDF obtained from the derived models and from Monte Carlo runs are in good agreement with a maximum percentage error within 10% in mean and variance parameters.

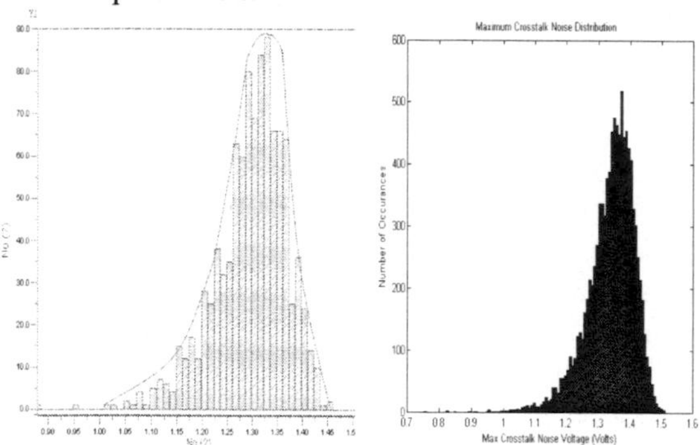

Monte Carlo Model prediction

Fig. 9. Maximum Crosstalk PDF Comparison

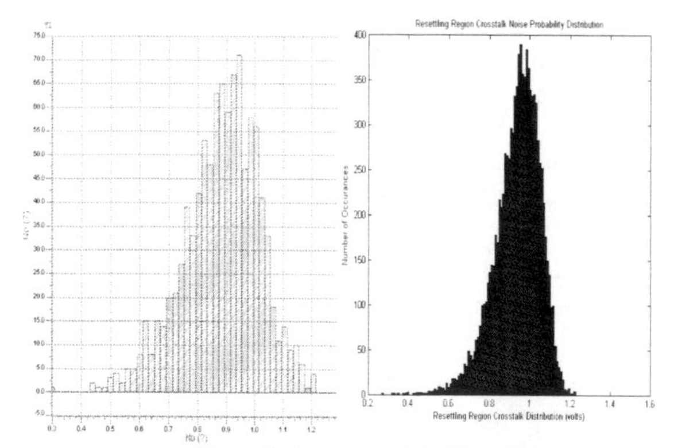

Monte Carlo Model prediction

Fig. 10. Resettling Crosstalk PDF Comparison

VI. CONCLUSION

The proposed approach of statistical modeling of crosstalk noise in domino CMOS logic circuits considers all potential cross talk couplings and progressively refines the equivalent circuit analysis model. As opposed to complex closed form analytical expressions, it captures the geometric variations in terms of simple passive electrical parameters. The model can be applied to any domino gate independent of the logic function implements. We can infer that estimating maximum crosstalk noise is simply a problem of identifying parasitic elements and active devices closed form expressions are not required. Along with the maximum crosstalk noise, the derived model also predicts the crosstalk resettling region distribution. This information in the resettling region can be extremely useful as appropriate safety measures can be taken for gate fanouts with the knowledge of crosstalk noise voltage distributions at the output node.

REFERENCES

[1] Waleed K. Al-Assadi, Vipin Sharma and Pavankumar Chandrashekar, "Crosstalk at Dynamic Node of Domino Logic Circuits", *Proc. of the 2006 International Conference of Computer Design*, pp. 57-63, June 26-29.

[2] D. Boning and S. Nassif, "Models of Process Variations in Device and Interconnect", *Design of High Performance Microprocessor Circuits*, IEEE Press, 2000.

[3] Maurizio Martina and Guido Masera, "A Statistical Model for Estimating the effect of Process Variations on Crosstalk Noise", *SLIP'04*, Feb. 14–15, 2004.

[4] Mridul Agarwal, Kanak Agarwal, Dennis Sylvester and David Blaauw, "Statistical Modeling of Cross-Coupling Effects in VLSI Interconnects", *ASP-DAC* 2005.

[5] O. Nakagawa, M. Sylvester, G. McBride and Young Oh, "On-Chip Cross Talk Noise Model for Deep-Submicrometer ULSI Interconnect", *Article 4, 1998 Hewlett-Packard Company*.

[6] Zhang, Liou, McMacken, K. Stiles, J. Thompson and P. Layman, "An Efficient and Practical MOS Statistical Model for Digital Applications", *IEEE International Symposium on Circuits and Systems*, May 28-31, 2000.

[7] Park and M.R. Mercer, "A Statistical Model for Delay-Fault Testing", Feb.89 - *IEEE Design & Test of Computers*.

[8] Ki-Wook Kim, Sung-Mo Kang, "Crosstalk Noise Minimization in Domino Logic Design", *IEEE Trans. On Computer-Aided Design of Integrated Circuits and Systems*, vol.20, no.9, September 2001.

[9] R. Kundu, R.D. Blanton, "Identification of crosstalk switch failures in domino CMOS circuits," *Proc. Intl. Test Conference*, pp. 502-509, October 2000.

[10] Athanasios Papoulis and S. Unnikrishna Pillai, "Probability, Random Variables and Stochastic Processes", 4th *edition, Tata McGraw-Hil.*

A Decision Making Model for Dual Interactive Information Retrieval

Vitaliy Vitsentiy
National ICT Australia[1]
Queensland University of Technology
v.vitsentiy@qut.edu.au

***Abstract*-A new task in Interactive Information Retrieval (IIR) is considered – optimization of information retrieval taking into account impact on quality of interaction with the user. Dual IIR (DIIR) is defined. An integer programming model for DIIR is given.**

I. Introduction

There is a world tendency of increasing informatization of society. Huge information volumes, which are joined now in computer networks, complicate the task of Information Retrieval (IR), while people more often need it. We encounter IR when we use internet web search engines, on-line store search, library systems, etc. IR concerns satisfaction of peoples' direct information needs, in such cases a person is an active participant of IR. However often information retrieval systems can not cooperate with users optimally. Therefore there is a need of improving Interactive Information Retrieval (IIR) systems.

IRSs are used for search of indistinct information. The user has some information need, which they translate into some query language that usually can not describe the information need precisely. From the point of view of the IRS there is an uncertainty what is searched by the user. This is caused by: 1) information need is not precise; 2) information need is not represented precise enough by the query; and 3) information need changes. So, IRSs make decisions about which documents to select based on the user query in situation where there is not enough information for unambiguous decisions. IRS usually retrieves not one document but a portion of documents, among which the user may find the necessary document. Often information need is not satisfied at once and the search process goes on (see Fig. 1). Thus IIR may be considered as a problem of multistage decision making under uncertainty. Such problems are usually solved by stochastic programming methods [1].

From control theory point of view, IRS is a stochastic control system (see Fig. 2), i.e. a control system that acts under conditions of uncertainty. There is no uncertainty in behavior of the controlled object – document database, giving information on what to retrieve but information coming from the user is uncertain. The control system is not fully automated because the feedback is provided by a human.

[1] National ICT Australia is funded by the Australian Government's *Backing Australia's Ability* initiative, in part through the Australian Research Council and the Queensland Government.

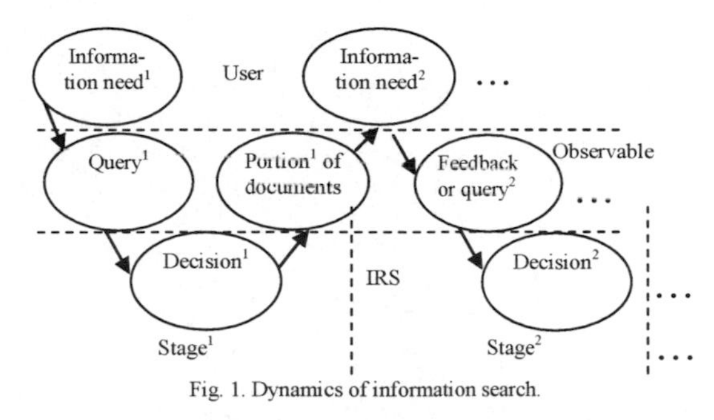

Fig. 1. Dynamics of information search.

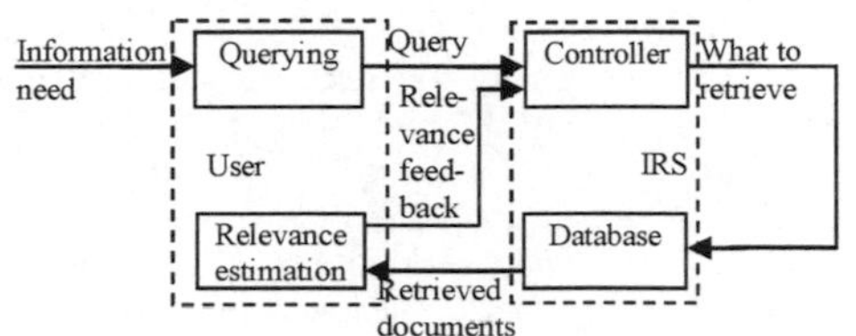

Fig. 2. IRS as a control system.

Let us consider the query coming from the user as an uncertain sample and topics of documents or individual documents as patterns to which the IRS tries to assign the query. Then such decision making during IR can be regarded as pattern recognition. But the feedback in IR is not usually instructional as it is in pattern recognition where feedback tells what pattern is correct for the sample but evaluative where feedback tells how much the pattern is correct for the sample. Such kind of pattern recognition with evaluative feedback better corresponds to intelligent behavior.

II. Dual Interactive Information Retrieval

Dual IIR (DIIR) is information retrieval with feedback after each stage of retrieval when at selection of documents the IRS takes into account not only their relevance but also possible effect of the retrieved portion of documents on user feedback for optimization of the whole user search session. The term "dual" is taken by analogy with "dual control" from adaptive control theory. The dual features are: 1) direction: DIIR System retrieves more relevant documents in the current portion; 2) probing: DIIR System encourages receiving better user feedback. The need for duality is caused: 1) user information need is uncertain and the uncertainty should be

K. Elleithy (ed.), Advances and Innovations in Systems, Computing Sciences and Software Engineering, 475–480.

reduced; 2) the feedback is evaluative, not instructional and therefore it does not allow determining whether there are more relevant documents among the not retrieved documents than the retrieved documents. The expediency of the development of the methods is based on the hypothesis: if user feedback is encouraged to be better especially on early stages of search, even by retrieving documents with somewhat lower relevance, then IRS using feedback can estimate relevance of documents more precisely and this leads to better relevance of all documents in the whole user search session.

Retrieving documents in DIIR is not simple selection of documents with the largest relevance but is an optimization problem. Because the optimal value is determined by: 1) probing may lead to lower relevance of documents in the current portion but a better feedback will allow to determine relevance better and retrieve more relevant documents on the next stages; 2) an increase of quantity of retrieved similar documents may lead not only to increase of relevance of the current portion but also to decrease of feedback quality and accordingly to decrease of relevance on the next stages.

Not dual but a traditional IIR is a kind of greedy approach [2] to optimization problems from the theory of algorithms. Greedy algorithms take decisions on the basis of information at hand without worrying about the effect these decisions may have in the future. But not all of optimization problems can be solved precisely with greedy algorithms, for example knapsack problem can be solved with a greedy algorithm but fractional knapsack problem can not be solved well and usually dynamic programming is used instead [2].

The problem of DIIR includes into the IR decision making also the task similar to active learning [3]. So, DIIR is a problem of adaptive dual control [4], where the problem of optimal balance between control and estimation is studied.

Note, that solution to the problem of DIIR also solves the problem of determining the value of additional information at decision making under uncertainty [5]. Feedback plays the role of additional information. The value of additional information is computed as difference between effect of decision with and without additional information.

III. An Example

A. Numerical Example

Let relevance of every pair of documents topics is estimated as in TABLE I.

TABLE I
RELEVANCE ESTIMATION

Topic \ Topic	1	2	...
1	0.9	0.0	...
2	0.0	1.0	...

Suppose that for some user query, probability distribution that a given topic is the topic of the information need is estimated as in TABLE II.

TABLE II
PROBABILITY DISTRIBUTION ESTIMATION

Topic	1	2	...
Probability	0.5	0.5	0

How many documents of each of the topics to retrieve in two portions of 2 documents each?

Expected relevance of the first topic = 0.9 * 0.5 + 0.0 * 0.5 = 0.45. Expected relevance of the second topic = 0.0 * 0.5 + 1.0 * 0.5 = 0.5. Since expected relevance of the second topic is greater, the greedy solution is:

TABLE III
GREEDY SOLUTION

Portion \ Topic	1	2
1	0	2
2	0	2

The solution has expected relevance in two portions = 2 * 0.5 + 2 * 0.5 = 2.

Note, that even if the decision is recomputed after the first stage, the information about the relevance of the first topic will not be known.

Consider another, non-greedy solution, first portion:

TABLE IV
NON-GREEDY SOLUTION, FIRST PORTION

Portion \ Topic	1	2
1	1	1

The solution has expected relevance in the first portion = 0.9 * 0.5 + 1.0 * 0.5 = 0.95. This value is smaller than the expected relevance of the first portion of the greedy solution.

In the case if feedback after the first stage shows that the first scenario occurred, the second portion of the non-greedy solution is:

TABLE V
NON-GREEDY SOLUTION, SECOND PORTION, FIRST CASE

Portion \ Topic	1	2
2	2	0

The expected relevance of this portion = 2 * 0.9 = 1.8.

In the case if feedback after the first stage shows that the second scenario occurred, the second portion of the non-greedy solution is:

TABLE VI
NON-GREEDY SOLUTION, SECOND PORTION, SECOND CASE

Portion \ Topic	1	2
2	0	2

The expected relevance of this portion = 2 * 1.0 = 2.0.

Thus the expected relevance of the second portion = 1.8 * 0.5 + 2.0 * 0.5 = 1.9. Total expected relevance of the first and the second portion of the non-greedy solution = 0.95 + 1.9 = 2.85. So, the non-greedy solution has greater expected relevance than the greedy one.

B. Qualitative Example

Let the user has an information need regarding natural resources of energy, i.e. wind energy, sun energy, water energy, etc. that don't pollute the environment and are ecologically safe. For information search the user uses query: natural energy. According to the traditional approach to IIR IRS retrieves documents which it estimates as the most relevant to the query.

But it is likely that: 1) many wrong documents are estimated as relevant. For example, documents about energy resources from natural sources like gas and oil are not relevant because their usage pollute the environment; 2) the documents estimated as relevant may belong to one topic. Then the user does not get documents that belong to other topics. For example, the user may get documents about more general concepts of natural energy and does not get documents about usage of wind energy, sun energy, water energy etc.

The combination of the mentioned two cases when the IRS is wrong and also it does not suggest new directions for search to the user is possible too. In the both cases, users do not get documents which they can use as an example for an advice to the IRS about which documents should be added to the set of relevant documents.

IV. Related Works

A. Information Retrieval

The considered direction of research is new, even terminology is different in different authors.

Term "active information retrieval", used in [6, 7], is taken by analogy with term "active learning" from the theory of machine learning; in [8, 9], authors talk about control and planning of IIR; in [10, 11], term "active feedback" is used. But both terms "active IR" and "active feedback" do not describe correctly the underlying notion. For conviction note that feedback is information which is provided by the user, we can not say active interactive IR and it is not clear what the difference is between active and interactive IR. It is rather to talk about control of IIR or about combination of IR with active learning or about non greedy IIR or about dual IIR.

A. Spink showed that users search is multitasking [12]. This may be an argument for semantically broad information retrieval.

Microsoft Cambridge team tried to design an active feedback method and showed their results at TREC 2003 [13]. But they got worse performance.

In [6, 7] retrieval of documents is combined with presenting the user with clusters that relate to their information need for getting feedback. The clusters are chosen according to active learning paradigm.

Maximal Marginal Relevance (MMR) (1) is used [14] as a measure of relevance. The authors use MMR for retrieval of passages and for IR but without consideration of effect of the method on feedback. However it is expedient to research MMR in iterative IR with feedback as an approximate method of DIIR.

$$MMR = \arg\max_{D_i \in R \setminus S} \left(\lambda Sim_1(D_i, Q) - (1-\lambda) \max_{D_j \in S} Sim_2(D_i, D_j) \right) \quad (1)$$

where R – ranked list of documents; S – subset of already selected documents in R, Sim_1, Sim_2 – similarity metrics; λ – constant.

I. Campbell [15] considered cognitive and philosophical aspects of information need and its dynamics: 1) characteristics of information need (developing, multiple, tangential, embedded, threaded); 2) model of user knowledge (see Fig. 3); 3) uncertainty profiles – scenarios of information need uncertainty development and age of evidence.

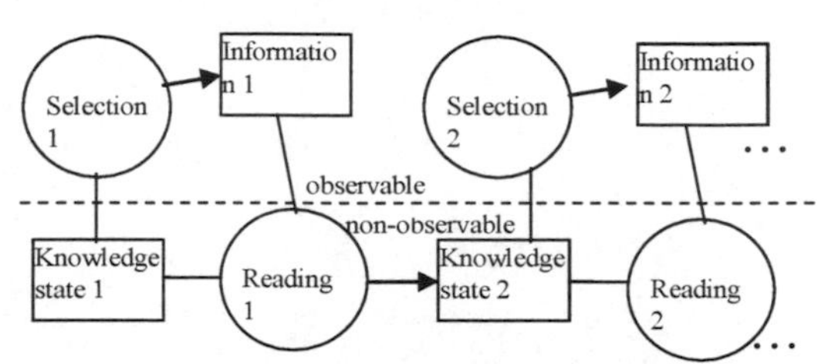

Fig. 3. Iterative update of user knowledge state.

Based on the conducted analysis I. Campbell proposes ostensive model developed on the basis of the probabilistic model. The author gives an example of using the model and shows that the model better incorporates ostensive evidence (time when the document was assessed), giving the terms which documents were assessed as relevant more recently in the session more value.

In [10, 11], the problem DIIR is interpreted within statistical decision theory as a problem of risk minimization. In [6, 7], the problem is interpreted within Bayesian learning methods as a problem of maximization of informativness. In [9], reinforcement learning is used. In general, usually theoretical apparatus of active learning is used [16, 3], especially in works on image retrieval [17, 18, 19].

There is not an acknowledged solution to the problem of DIIR yet. Often authors just consider the problem of active learning without taking into account the problem of combination of active learning with on-line performance.

B. Adaptive Dual Control

Dual control problem was first formulated by Fel'dbaum [4]. A review of solutions to the problem can be found in [20, 21].

It is possible to show that dual control problem is a problem of multistage stochastic dynamic programming [22], though in stochastic programming theory, models with endogenous uncertainty, where the underlying stochastic process depends on the optimization decisions, are not studied well [23]. The task is to find control function u that minimizes some N-stage Performance Index J:

$$J_N = \sum_{k=0}^{N-1} E\left\{c_k\left(s(k+1), u(k)\right)\right\}, \tag{2}$$

where $E\{\cdot\}$ – mathematical expectation over all random variables; c_k – function that reflects how s and u are required to behave; $s(k)$ – some system variable s to be controlled at time k; $u(k)$ – control input at time k.

Optimal value of the performance index, J^* is equal:

$$J^* = \min_{u(0)} E\left\{\min_{u(1)} E\left\{\min_{u(2)} \ldots \right.\right. \tag{3}$$
$$\left.\left. \ldots \min_{u(N-2)} E\left\{\min_{u(N-1)} E\left\{J_N \mid I^{N-1}\right\} \mid I^{N-2}\right\} \ldots \mid I^1\right\} \mid I^0\right\},$$

where $I^k = \{Y^k, U^{k-1}\}$; $Y^k = \{y(i)\}_{i=0}^{k}$; $U^k = \{u(i)\}_{i=0}^{k}$; $y(k)$ – feedback at time k.

For additive type of J as (2), a discrete-time stochastic programming Bellman equation (4) with terminal condition (5) is equivalent to (3).

$$J_k^* = \min_{u(k)} E\left\{c_k\left(s(k+1), u(k)\right) + J_{k+1}^* \mid I^k\right\} \tag{4}$$

$$J_N^* = E\left\{c_{N-1}\left(s(N), u(N-1)\right) \mid I^N\right\} \tag{5}$$

A solution of (4), (5) according to dynamic programming can be found by starting at $k = N-1$ by computing (4) for every possible value of I^{N-1} and working backwards. Because of the "curse of dimensionality" it is difficult to solve in practice.

In practice, instead of exact solution of the dual control problem, approximate suboptimal dual methods are used. There are two directions of accomplishing approximation: 1) implicit methods try to find an approximate way to solve equations (4), (5); 2) explicit methods try to modify performance index J so that the correspondent control u is easier to find analytically or numerically and it still has dual features.

1) Implicit methods.

One of approaches is to find an approximation for a posteriori conditional probability distribution $P\left(s(k) \mid I^k\right)$ used in computation of $E\{\cdot\}$. In [24], the probability distribution is approximated by Gaussian sums. In [25, 26], recursive methods for generating the probability distribution are developed.

Wide-sense adaptive dual control is proposed in [27]. A nominal trajectory $\{s_0(k+1), \ldots, s_0(N)\}$ is computed that corresponds to some nominal control sequence $\{u_0(k), \ldots, u_0(N-1)\}$. The nominal control sequence is a non-dual solution to the control problem. Then using the nominal trajectory in (2), $u(k)$ is found from optimization of (2).

M-measurement feedback control policy [20] assumes feedback for future m steps and then no feedback after the time $m + k$.

Monte-Carlo methods were used for approximation of integration thus reducing computational requirements in the dynamic programming statement of the problem [28]. Neural networks were used for approximation of nonlinear functions of the dual controlled system [29].

2) Explicit methods.

One of approaches is to alter the performance index by addition a term f that encourages probing:

$$J = \min_{u(k)} E\left\{c_k\left(s(k+1), u(k)\right) + \lambda f \mid I^k\right\} \tag{6}$$

where λ – is a constant parameter.

In [30], $f = -\frac{|P(k+1)|}{|P(k+2)|}$, where P is covariance matrix of the estimated parameters. In [31], $f = -\left(y(k+1) - \hat{y}(k+1)\right)^2$, where $\hat{y}(k)$ is computed using the estimated parameters.

In Bicriterial method [32], first a non-dual solution is found. Then a term f similar to that used in [31] is minimized such that u remains inside some domain that is symmetrically distributed around the found non-dual solution. The size of the domain is kept proportional to the uncertainty.

Another explicit approach involves direct modification of the control u, for example by adding a white noise [33] when uncertainty exceeds some limit.

V. Model

We consider information retrieval to be conducted on topics, and not a particular document is searched but a set of documents semantically close to the topic of information need. The documents are clustered on topics. Uncertainty of the information need is described as probability distribution that a given topic is the topic of information need. Relevance of every pair of topics and approximate number of stages are known.

There is a requirement for the model to be easily computable because it should be computed in real time. Therefore the following assumptions are introduced to simplify the task: 1) the probability distribution does not change; 2) feedback completely reduces uncertainty in the relevance of the topic of the retrieved document but not in the probability distribution; 3) uncertainty in the user's relevance assessment of a topic given the topic of the information need is not taken into account.

So, there is only one random variable – topic of the information need, with corresponding vector of outcomes – relevance of the retrieved documents. But the outcomes do not change the probability distribution but only completely reduce the uncertainty in the relevance of the retrieved topics.

Note, these assumptions are only for the task of finding documents to retrieve, another tasks may not conform to all of them.

The probability distribution corresponding to the query and to the feedback is the input to the model. Quantity of documents to retrieve for every topic in the portions of the stages ahead is the output. After the decision the documents in the current portion are retrieved, feedback is obtained, probability distribution corrected according to the feedback and the model is recomputed again for the next stages and so on till the last stage. Recomputation of the model is necessary because of the assumptions: the probability distribution may be different on the next portions, feedback on different documents topics may not correspond to the same information need topic.

Used notation: x, y, g – number of documents to retrieve; a – relevance; P – probability; r – relevant topic; n – number of documents in the topic; m – number of documents in one portion; T – number of stages; I – number of topics of documents; J – number of topics of a query; $\mathbf{N}$ – set of natural numbers. Indexes: t, τ – stage; i – document topic; j – query topic.

A greedy model is to find a number of documents to retrieve for each topic such that expected relevance is maximized and constraints on number of documents in one portion and in every topic are satisfied:

$$\max \sum_t \sum_i x_{ti} \sum_j P(r=j) \cdot a_{ji} \tag{7}$$

s.t.

$$\sum_i x_{ti} = m \quad (t=1,...,T), \tag{8}$$

$$\sum_t x_{ti} \le n_i \quad (i=1,...,I), \tag{9}$$

$$x_{ti} \in \mathbf{N} \quad (t=1,...,T; i=1,...,I). \tag{10}$$

A solution of model (7)-(10) can be found by sorting documents over their expected relevance and retrieving the top of them but such that the constraints are satisfied.

A recourse multistage stochastic programming model $Q(t, j)$ is:

$$\max \sum_{i|\neg\exists x_{\tau i}>0 \wedge \tau<t} x_{ti} \sum_k P(r=k) \cdot a_{ki} + \tag{11}$$
$$+ \sum_{i|\exists x_{\tau i}>0 \wedge \tau<t} y_{jti} \cdot a_{ji} + \sum_k P(r=k) \cdot Q(t+1,k)$$

s.t.

$$\sum_i x_{ti} + y_{jti} = m \quad (j=1,...,J; t=1,...,T), \tag{12}$$

$$\sum_\tau^t x_{\tau i} + y_{j\tau i} \le n_i \quad (i=1,...,I; j=1,...,J), \tag{13}$$

$$x_{ti} \in \mathbf{N} \quad (i=1,...,I; t=1,...,T), \tag{14}$$

$$y_{jti} \in \mathbf{N} \quad (i=1,...,I; j=1,...,J; t=1,...,T). \tag{15}$$

where the objective function of $Q(T, j)$ is :

$$\sum_{i|\neg\exists x_{\tau i}>0 \wedge \tau<T} x_{Ti} \sum_k P(r=k) \cdot a_{ki} + \sum_{i|\exists x_{\tau i}>0 \wedge \tau<T} y_{jTi} \cdot a_{ji} \tag{16}$$

We used another variable, y, in the equations to denote the topics, some of documents of which have been already retrieved and therefore which relevance is not random but known. The set of such topics is not fixed and depends on the previous decisions. Thus, the summation in (11) is conditioned on the previous decisions. This complicates the model.

Deterministic equivalent of the model:

$$\max \sum_t \left(\sum_{i|\neg\exists x_{\tau i}>0 \wedge \tau<t} x_{ti} \sum_j P(r=j) \cdot a_{ji} + \right. \tag{17}$$
$$\left. + \sum_{i|\exists x_{\tau i}>0 \wedge \tau<t} \sum_j y_{jti} \cdot P(r=j) \cdot a_{ji} \right)$$

s.t.

$$\sum_i x_{ti} + y_{jti} = m \quad (t=1,...,T; j=1,...,J), \tag{18}$$

$$\sum_t x_{ti} + y_{jti} \le n_i \quad (i=1,...,I; j=1,...,J), \tag{19}$$

$$x_{ti} \in \mathbf{N} \quad (t=1,...,T; i=1,...,I), \tag{20}$$

$$y_{jti} \in \mathbf{N} \quad (t=1,...,T; i=1,...,I; j=1,...,J). \tag{21}$$

Model (17)-(21) is not recursive now but still has complex objective function.

Lemma 1. If an optimal solution $\{x^*_{ti}, y^*_{jti}\}$ $(j=1,...,J; t=1,...,T; i=1,...,I)$ to (17)-(21) has a pair (x^*_{ti}, y^*_{jti}), such that y^*_{jti} and x^*_{ti} are both greater zero than another optimal solution can be found by assigning zero to one of y^*_{jti} or x^*_{ti} and appropriately increasing the other.

Proof. Let solution $\{x^*_{ti}, y^*_{jti}\}$ is equal $\{x'_{ti}, y'_{jti}\}$. If $\neg\exists x_{\tau i} \mid x_{\tau i} > 0 \wedge \tau < t$, let $y'_{jti} = 0$, $x'_{ti} = x^*_{ti} + y^*_{jti}$; otherwise let $x'_{ti} = 0$, $y'_{jti} = y^*_{jti} + x^*_{ti}$. Then $\{x'_{ti}, y'_{jti}\}$ is feasible and gives greater or equal value of the objective function. The lemma is proved.

Consider the following model:

$$\max \sum_j \sum_i a_{ji} \cdot P(r=j) \sum_t g_{jti} \tag{22}$$

s.t.

$$\sum_i g_{jti} = m \quad (t=1,...,T; j=1,...,J), \tag{23}$$

$$\sum_t g_{jti} \le n_i \quad (i=1,...,I; j=1,...,J), \tag{24}$$

$$g_{jti} = g_{j+1,ti} \quad if \quad \neg\exists g_{k\tau i} \mid g_{k\tau i} > 0 \wedge \tau < t \tag{25}$$
$$(j=1,...,J-1; t=1,...,T; i=1,...,I),$$

$$g_{jti} \in \mathbf{N} \quad (t=1,...,T; j=1,...,J; i=1,...,I). \tag{26}$$

Theorem 1. Model (17)-(21) is equivalent to (22)-(26).

Proof. Let $g_{jti} = x_{ti} + y_{jti}$. (17) is equivalent to (22) by constraint (25) and Lemma 1. (20) and (21) are equivalent to (26) by Lemma 1. So, the objective function and the constraints are equivalent. The theorem is proved.

Model (22)-(26) has $I \times T$ fewer variables than (17)-(21).

Theorem 2. Constraint (25) is equivalent to (27).

$$g_{tji} \leq \frac{1}{J}\sum_{k} g_{tki} + m \cdot \sum_{\tau}^{t-1} g_{\tau ji} \qquad (27)$$

$$\left(j = 1,\ldots,J; t = 1,\ldots,T; i = 1,\ldots,I\right)$$

Proof. Consider two cases: 1) $\exists x_{\tau i} \mid x_{\tau i} > 0 \wedge \tau < t$ is false. Then (25) is $g_{jti} = \frac{1}{J}\sum_{k} g_{kti}$. $\sum_{\tau}^{t-1} g_{j\tau i} = 0$, so (27) is $g_{jti} \leq \frac{1}{J}\sum_{k} g_{kti}$. It can be satisfied only if $g_{jti} = \frac{1}{J}\sum_{k} g_{kti}$. 2) $\exists x_{\tau i} \mid x_{\tau i} > 0 \wedge \tau < t$ is true. Then (25) is always satisfied. $\sum_{\tau}^{t-1} g_{j\tau i} \geq 1$. Since always $g_{jti} \leq m$ then $g_{jti} \leq m \cdot \sum_{\tau}^{t-1} g_{j\tau i}$. So, (27) is always satisfied. The theorem is proved.

So, the model is now linear that should reduce the computational time.

VI. Conclusion

DIIR is defined. A stochastic non-linear non-differentiable integer programming model with decision dependent uncertainty reduction that supports DIIR is given. An equivalent linear integer programming model of smaller dimensionality is derived. A positive effect from usage of DIIR should be expected in the situation of uncertainty in the user information need, long search interaction process, broad semantic range of search.

References

[1] *Stochastic Programming*, ed. by A. Ruszczynski, and A. Shapiro, Elsevier, 2003.

[2] T.H. Cormen, C.E. Leiserson, R.L. Rivest, and C. Stein, *Introduction to Algorithms*, 2nd ed., MIT Press, 2001.

[3] S. Tong, *Active learning: theory and applications*, PhD thesis, Stanford University, 2001.

[4] A.A. Fel'dbaum, "Dual Control Theory. I-IV" *Automation Remote Control*, 21, 22, pp. 874-880, 1033-1039, 1-12, 109-121, 1960-1961.

[5] V. Vitsentiy, "A method of evaluating the expediency of additional information at decision making under uncertainty", *Transactions of TANE*, Ternopil, 10, 91-97, 2000.

[6] T. Jaakkola, and H. Siegelmann, "Active Information Retrieval", *Advances in Neural Information Processing Systems*, 777-784, 2001.

[7] O. Loureiro, and H. Siegelmann, "Introducing an Active Cluster-Based Information Retrieval Paradigm", *Journal of the American Society for Information Science and Technology*, 56(10), 1024-1030, 2005.

[8] V. Vitsentiy, "Improvement of human-machine interaction with applications to information retrieval system", *First International IEEE Symposium on Intelligent Systems*, 363-368, 2002.

[9] V. Vitsentiy, A. Spink, and A. Sachenko, "Planning of interactive information retrieval by means of reinforcement learning", *IEEE International Workshop on Intelligent Data Acquisition and Advanced Computing Systems: Technology and Applications*, 396-399, 2003.

[10] X. Shen, and C. Zhai, "Active feedback in ad hoc information retrieval", *Annual ACM SIGIR Conference on Research and Development in Information Retrieval*, 59-66, 2005.

[11] X. Shen, and C. Zhai, "Active Feedback – UIUC TREC2003 HARD experiments", *Twelfth Text Retrieval Conference*, 662-666, 2003.

[12] A. Spink, M. Park, and B.J. Jansen, "Information Task Switching and Multitasking Web Search", *ASIS&T Annual Meeting*, 213-217, 2004.

[13] S.E. Robertson, H. Zaragoza, and M. Taylor, "Microsoft Cambridge at TREC-12: HARD track", *Text REtrieval Conference*, 418-425, 2003.

[14] J. Carbonell, and J. Goldstein, "The Use of MMR, Diversity-Based Reranking for Reordering Documents and Producing Summaries", *Annual ACM SIGIR Conference on Research and Development in Information Retrieval*, 335-336, 1998.

[15] I. Campbell, *The ostensive model of developing information-needs*, PhD thesis, University of Glasgow, 2000.

[16] D.A. Cohn, Z. Ghahramani, and M.I. Jordan, "Active Learning with Statistical Models", *Neural Information Processing Systems*, 705-712, 1994.

[17] E. Chang, S. Tong, K. Goh, and C. Chang, "Support Vector Machine Concept-Dependent Active Learning for Image Retrieval", *IEEE Transactions on Multimedia*, in press.

[18] C.K. Dagli, S. Rajaram, and T.S. Huang, "Combining Diversity-Based Active Learning with Disriminant Analysis in Image Retrieval", *Third International Conference on Information Technology and Applications*, vol. 2, 173-178, 2005.

[19] T. Qin, T. Liu, X. Zhang, W. Ma, and H. Zhang, "Subspace Clustering and Label Propagation for Active Feedback in Image Retrieval", *11th International Multimedia Modeling Conference*, 172-179, 2005.

[20] H. Unbehauen, "Adaptive dual control systems: a survey", *IEEE Symposium on Adaptive Systems for Signal Processing, Communication and Control*, 171-180, 2000.

[21] B. Wittenmark, "Adaptive dual control methods: an overview", *5th IFAC symposium on Adaptive Systems in Control and Signal Processing*, 67-72, 1995.

[22] Y. Bar-Shalom, "Stochastic Dynamic Programming: Caution and Probing", *IEEE Transactions on Automatic Control*, vol. 26, no 5, 1184-1195, 1981.

[23] V. Goel, and I.E. Grossmann, "A Class of stochastic programs with decision dependent uncertainty", *Mathematical Programming*, 108, (2-3), 355-394, 2006.

[24] D.L. Alspach, "Dual control based on approximate a posteriori density functions", *IEEE Transactions on Automatic Control*, 17, 689-693, 1972.

[25] M. Aoki, *Optimization of Stochastic Systems: Topics in Discrete Time Dynamics*, Academic Press, San Diego, 1989.

[26] A.A. Fel'dbaum, *Optimal Control Systems*, Academic Press, New York, 1965.

[27] E. Tse, Y. Bar-Shalom, and L. Meier, "Wide-sense adaptive control for nonlinear stochastic systems", *IEEE Transactions on Automatic Control*, 18(2), 98-108, 1973.

[28] A.M. Thompson, and W.R. Cluett, "Stochastic iterative dynamic programming: a Monte Carlo approach to dual control", *Automatica*, 41, 767-778, 2005.

[29] S. Fabri, and V. Kadirkamathan, "Dual adaptive control of nonlinear stochastic systems using neural networks", *Automatica*, 34(2), 245-253, 1998.

[30] G.C. Goodwin, and R.L. Payne, *Dynamic system identification: Experiment design and data analysis*, Academic Press, New York, 1977.

[31] R. Milito, C.S. Padilla, R.A. Padilla, and D. Cadorin, "An innovations approach to dual control", *IEEE transactions on Automatic Control*, 27(1), 133-137, 1982.

[32] N.M. Filatov, U. Keuchel, and H. Unbehauen, "Dual control for an unstable mechanical plant", *IEEE Control Systems Magazine*, 16(4), 31-37, 1996.

[33] O.L.R. Jacobs, and J.W. Patchell, "Caution and probing in stochastic control", *International Journal of Control*, 16, 189-199, 1972.

BUSINESS RULES APPLYING TO CREDIT MANAGEMENT

Vladimir Avdejenkov
Vilniaus Gedimino technikos universitetas
Lithuania Saulėtekio al. 11, Vilnius, LT-10223
vladimir@isl.vtu.lt

Olegas Vasilecas
Vilniaus Gedimino Technikos Universitetas
Lithuania Saulėtekio al. 11, Vilnius, LT-10223
olegas@fm.vtu.lt

Report annotation: this article analyses business rules, their application to data analysis and decision making as an aid for business participants. The main focus is on application of business rules, as an example, debtor management problem arising to many marketing companies has been examined. Also, this article scrutinizes principles of business rules running, along with opportunities of their application and implemented practical employment of these rules.

INTRODUCTION

Worldwide business trends respectively adjust attitude towards development of information systems. Lately, in order to ensure effective functioning of companies in the constantly changing environment, it is required to accordingly change the logics of business systems and functioning of business. Thus, the accustomed information system model becomes insufficiently effective, and in such environment it is the best choice to apply rules-oriented information systems [1], [2]. The main advantage of business rules is that they can relatively easily be changed without changing the information system itself.

Information systems users often need a large amount of data to be analyzed before making important decisions. Uppermost, it takes a lot of time, plus, human factor makes an impact. There is a potential to assign data analysis and decision making to the system itself. This paper discusses the potential to analyze the data and to aid system users when making decisions with the help of business rules.

BUSINESS RULES AND BUSINESS RULES MANAGEMENT

Business rules are guidelines that affect business, manage it, determine its development, internal and external threats [3]. Business rules also include the restrictions or instructions indented for business participants. Denotation where smoking is permitted, orders and guidelines for certain situations may also be treated as business rules. Business rules manage company policies, pricelist and structure – that is, the entire company functioning. A couple of examples:

- There are two sorts of goods – standard goods and special goods.
- Alcohol may not be sold to those under 18.

These two informal propositions are business rules; however, in order to be used in information systems, they have to be formalized.

Business rules management systems are used to make rules implementation faster and more effective. Such systems employ elaborate algorithms to optimize computer resources by integrating business rules management system to the business system, each rule is recorded as an independent proposition describing a certain aspect of business, and such being the case, business rules work as an autonomous item, separated from application logics.

Thus, there are two main reasons why implementation of business rules management systems is expedient:

- The detachment of business rules from applications allows changing business logics without programmers and without changing the application.
- The time of application creation decreases as well as the maintenance and development costs are reduced significantly. Studies show 15 to 20 percent of annual IT budget spent on adjustments, testing and installation.
- Business processes taking place in the system become more comprehensible for business people.

BUSINESS RULES OF DEBTOR MANAGEMENT

Every company selling products or providing services does it without prepayment. In other words, the company provides its clients with a short-term credit that allows clients to pay up later. Companies practicing suchlike relationships with customers face the task to effectively manage such clients as debtors. Tracking of their credit level and timeliness of payments have to be considered when blocking the supply of goods, applying special discounts or calculating penalties. Normally, customer credit conditions are often changed, given a large turnover of goods; situation with debtors can rapidly

K. Elleithy (ed.), Advances and Innovations in Systems, Computing Sciences and Software Engineering, 481–483.

change as well, therefore, in order to manage the company effectively, it is very important to automate this process.

Usually most frequently employed business management systems (Scala, Navision, Axapta, etc.) have a limited number of functions for running an automated debtor management process. Mostly the customer card parameters are used to limit the supply of goods. The parameter is activated either manually or by launching a special procedure referring to the customer's credit information and his maximum credit allowed. In order to manage the debtors effectively, this is clearly not enough; therefore, debtor management is topical for many companies. It is suggested to employ business rules, implementing them to the company's main information system.

We will try to classify the business rules of the analyzed objective field. On the basis of ECA rules concept, in order to describe the business rules, *Events*, *Conditions* and *Actions* must be defined. First of all, we will describe possible events the system can react to:

- Input of new orders
- Customer payments
- Deliveries of goods to buyers

Another stage is the determination of conditions that need to be checked following the event:

- Maximum credit allowed.
- Client's debt at a particular time.
- Client's credit data.

It is only left to define actions that we can take in this case:

- Blocking deliveries of goods to clients.
- Sending notifications to clients.
- Encouraging debtors.

Possible events, conditions and actions are necessary to define the business rules of application domain. For a better understanding of application domain, we will present several examples of debtor management business rules:

- Prohibit shippings if client's debt overruns the credit allowed.
- Sending reminders of payments to the buyers.
- Providing a client with a complementary 3% discount if terms of lending are not violated during the past six months.

Implementing the rules described above, data analysis related work for information system users can be facilitated. A research on business rules and business rules management systems proved new active DBMS rules – triggers – to be an expedient solution in tackling this problem [4]. That is because majority of contemporary business management systems in order to secure data employ contemporary active database management systems (MS SQL Server, Oracle, InterBase) that support use of triggers.

Triggers are identical to business rules: their business logic is separated from the information system logic; therefore the triggers are stored separately. All these reasons determine the use of this particular architecture, tackling the described problem of facilitation of data analysis and solution-making for system users. One part of Fig. 1 displays a business management system with its logic; another part demonstrates business rules entered as triggers that implement their business logic that is independent from business management system.

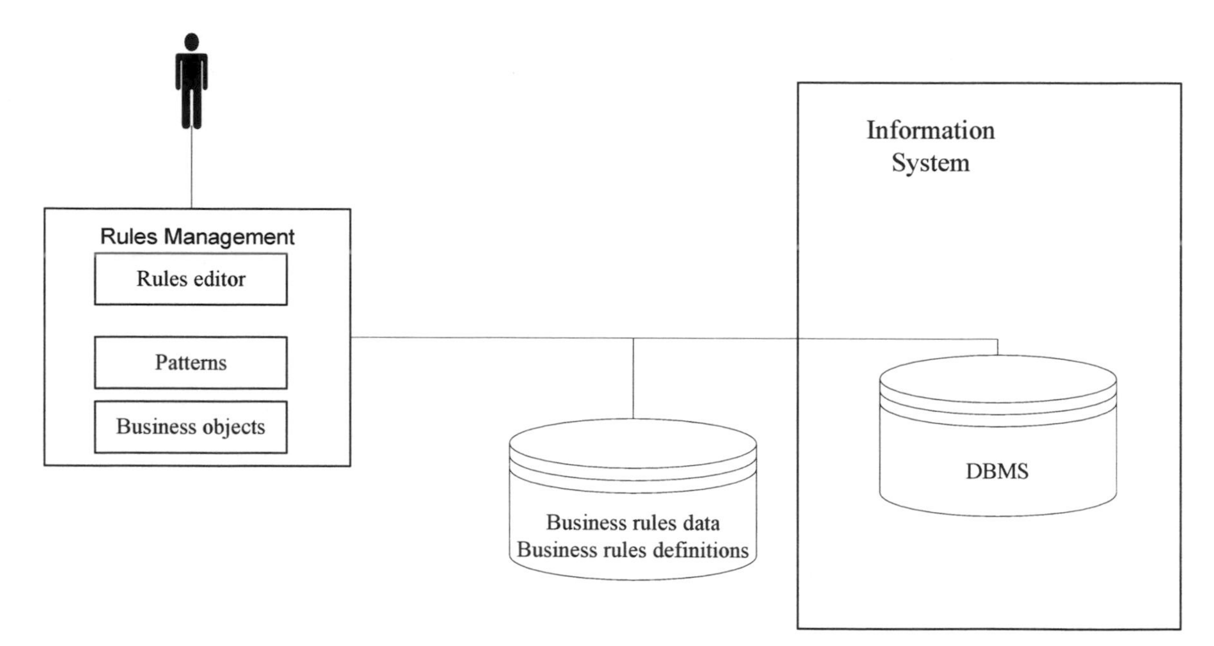

Fig. 1. Integration of business rules to DBMS

BUSINESS RULES INSTALLATION METHODOLOGY

RDBMS triggers are active DBMS elements that become active in certain conditions and insure a wide scope of functionalities from consistency support till complex business rules implementation.[5], [6].

A trigger can be treated as a ECA rule: *Event*, *Condition* and *Action*. Triggers react to certain events, conditions are checked and depending on the results of the condition check, an action is performed. Three variants of event are possible: insertion of a line, deletion of a line, or an update of the line. If a stated event takes place, a condition is checked – verification of entered, deleted data, or data from a different chart, etc. Depending on the condition check results, an action is performed – that is practically any type of an SQL language command (cancellation of transaction, changing data both in a basic table and in any other table, or sending messages) [7].

To conclude, the problem described above can be solved using DBMS triggers. Let‘s scrutinize one of the rules of the business sphere of interest. Let‘s take this rule: ‚Prohibit new sales orders entries for clients having overrun the credit‘. In this business rule, an event is an attempt to enter a new order, a condition is comparing the maximum credit allowed with a client‘s debt at that particular time, and an action performed is either permission or prohibition of a new order entry.

Such triggers involve sales order, client description and debt charts. A trigger will react to the new order entry and depending on the comparison of the client card with the client debt either prohibits or permits the new order entry. Application of this method allows documenting all business rules defined above. We believe that a thorough representation of a suchlike trigger in this article could be inexpedient. Examples presented are used in a particular system.

In order to make business rules adjustment apprehensible to the ordinary business people (not programmers); possibilities are explored to use patterns [8] and Unified Modeling Language [9] for business rules modeling.

CONCLUSION

The article analyses business rules, their application to data analysis and decision making as an aid for business participants. As an example, debtor management problem arising to many trading companies is explored.

With specific examples presented, the principles of business rules running are scrutinized along with opportunities of their application and implemented practical employment of these rules. Visual UML tools are suggested for business rules modeling, as well as the employment of patterns is expedient for tackling a specific problem. DBMS triggers are offered to be used in business rules realization.

Research shows the practical value of this method, however, imperfect methods of business rules interpretation for the eventual users offer great opportunities for the development of business rules interpretation tools.

REFERENCES

[1] John Medike, Feng-Wei Chen, Margie Mago. Creating an Intelligent and Flexible Solution with BPM, Business Rules, and Business Intelligence. http://www.106.ibm.com/developerworks/db2/library/techarticle/0310mago/0310mago.html

[2] Barbara von Hale. Business Rules Applied - Buldning Better Systems Using the Business Rules Approach. John Wiley & Sons, Inc., New York ISBN 9955-09-335-8

[3] Object Management Group, 2002, Business Rules in Models, Request For Information, 2002, September, URL: http://cgi.omg.org/cgi-bin/doc?ad/2002-9-13.

[4] J. Reinert, N. Ritter. Applying ECA Rules in DB-based Design Environments Tagungsband CAD'98 "*Tele-CAD - Produktentwicklung in Netzen*" , März 1998, Informatik Xpress 9, pp. 188-201

[5] Jeffery Garbus, David Pascuzzi, Alvin Chang. Database Design on SQL Server. ISBN 5-272-00003-X

[6] I. Valatkaitė, O. Vasilecas. 2002, Verslo taisyklių modeliavimas koncepciniais grafais ir jų realizavimas naudojant aktyvių duomenų bazių trigerius, *Lietuvos matematikos rinkinys*, T.42, spec. nr., p. 289-293.

[7] O. Vasilecas. ECA taisyklių realizacija aktyviose duomenų bazių valdymo sistemose. *Lietuvos mokslas ir pramonė. Konferencijos "Informacinės technologijos'2002" pranešimų medžiaga*, Kaunas, Technologija, 2002, p. 71-77.

[8] Gunnar Övergaard, Karin Palmkvist. *Use Cases Patterns and Blueprints*. ISBN 0-13-145134-

[9] V.Avdejenkov, I.Valatkaitė, O.Vasilecas. Verslo taisyklių modeliavimas UML ir jų realizavimas reliacinėje DBVS, *Informacinės technologijos 2003*, KTU, ISBN 9955-09-335-8.

Information System in Atomic Collision Physics

V.M. Cvjetković
Faculty of Science
Radoja Domanovića 12
34000 Kragujevac, Serbia
B. M. Marinković
Institute of Physics
Pregrevica 129
11000 Belgrade, Serbia
D. Šević
Institute of Physics
Pregrevica 129
11000 Belgrade, Serbia

***Abstract* - Fundamental aspects of scientific research in the field of atomic physics are discussed in this paper from the point of view of information system that would cover the most important phases of research. Such information system should encompass the complexity of scientific research trying to incorporate data scattered in various books, articles, research centers, databases, etc. We started from scratch with principal analysis of basic research processes and data that represent needs and condensed research experience. Particular problem of search for data is specially discussed and the main idea for new proposed approach is described. We developed a prototype of information system to be used by researchers in various research phases. Search for data is based on the web, as it is the standard way for easy data access.**

I. INTRODUCTION

Organizing of any research implies several stages of activity. It starts with defining the process of interest, getting information about published data concerning the chosen processes, planning the research by choosing the methods and parameters, then performing research, obtaining and processing data and finally presenting and publishing results in appropriate form. Each of these stages could be time consuming and there is no doubt that specific research could last for years.

In this paper we have developed the logical model of Information System (IS) in the field of Atomic Collision Physics (ACP). The model should reflect all stages of activities. It is also implemented as a web application based on developed logical model. The aim of such IS is to facilitate the search for published data in a specific field of ACP, to make possible the critical evaluation of published data and used methods of research, as well as to keep track of own research. During the same period of time, the thematic of research could be changed and/or increased in sense of new results or new methodology invented.

Research activity has complex structure and its stages overlap not only in time domain, but also in domain of concepts. That often leads the researchers to repeat the same activity that is usually unnecessary. In planning own research, the researchers have at their disposal only the extensive databases of whole publications that can usually be browsed by authors and keywords. However, the analysis of research procedures is left to researchers. That analysis in its essence includes the reading of the whole text i.e., expressed in the informatics terms, the selection of research categories by "free text search". Also, there are available specific databases that include numeric or graphics data but without any knowledge considering the methodology, preparation, parameters describing how data were obtained, and these are the key for evaluation of existing research. Without such performed analysis, these numerical results could hardly be compared or evaluated. There is no IS available that would comprise the research process in whole.

Large number of data bases exist on the Internet nowadays, that cover the field of atomic and molecular physics. Atomic and molecular databases can be divided in two main groups, numerical and bibliographical databases. Numerical databases are specific and comprise for instance Fundamental Physical Constants – NIST [1], spectroscopic data, data for collision processes, etc. Some representatives of these databases are NIST Atomic Spectroscopic Data [2], NIST Molecular Spectroscopic Data [3], TOPbase [4], Center for Astrophysics, Atomic and Molecular Physics Division [5] for spectroscopic data, and NIFS [6], NIST [7], IAEA ALADDIN [8], Atomic Data for Astrophysics, Univ. Kentucky, USA [9] for collisional processes data.

Bibliographical databases are somewhat less numbered, and there are also two basic kinds - spectroscopic and databases for collision processes. Bibliographical databases for spectroscopy are for instance NIST [10], STARK [11], while collision are IAEA [12], GAPHYOR [13], ORNL CFADC [14] etc.

There are also systems that act as "search engines" for databases, collecting data from several databases. Typical representative is GENIE [15] collecting data from nine different databases throughout the world, among which are some of the above mentioned. The other representative is DANSE [16] that was developed for the ICAMDATA [17] 2002, Gatlinburg, Tennessee. Most of the features in DANSE are still in the

K. Elleithy (ed.), Advances and Innovations in Systems, Computing Sciences and Software Engineering, 485–490.

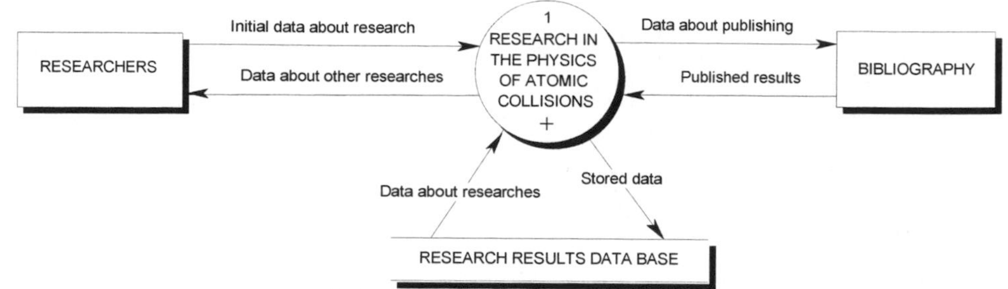

Fig. 1. Context diagram of the function RPAC

preparation level. There are systems that offer online computing for calculation of some important specific atomic features like AAEXCITE [18] and RATES [19]. All mentioned examples are undoubtedly very important and contain large number of valuable data at disposal for scientists committing various researches thru ought the world. On the other side, these databases are specific for some area and offer search for data based on criteria that are most frequently used, not necessarily optimal for any research. Also, these databases are no complete, and in general they lack customization. Therefore, the support for research is only partial, not to mention that they are just "one way" and generally do not offer the easy input or storing of some specific data of interest for current ongoing research or for the whole field of activity.

The proposed answer to all the mentioned could be the creation of Information System (IS) designed to support specific research tasks, but at the same time to be based on ground wide enough to provide common foundation for research in the area of atomic and molecular physics. The significance of existing databases will be primarily to serve as the source for data of interest for the given institution.

The basic idea for IS that is developed here would be to serve as a tool for researcher, which implies that it should comprise as much of research activities in all stages, as possible. As the IS was made from scratch, the starting point for designing such IS are detailed and elaborated process model and data model. Process model is largely based on, and expresses the knowledge and the long term experience of researchers in the ACP field. Process model also served as a guideline for the data model. During data modeling, very important concept or idea emerged and became the unique characteristic of this IS. We call it the Expert Decomposition of the Article (EDA). It expresses the semantic meaning of the article text being analyzed, using a number of selected, universal, most important notions that characterize the research in the area of ACP. EDA enables the IS to selectively store the most important data that characterize the research described in the article, and these include results of numerical and graphical type, important parameters, preparation, methodology, used particles, bibliographical and other data. In that way, the IS contains the most necessary data that characterize the research, without the need to contain the full text of the article, as it can be obtained by well known means. This IS also enables very selective and efficient data retrieval such as particular numerical or graphical results, that are obtained on the basis of expert characterization of results. By forming the complex search condition consisting of mentioned notions obtained by EDA, researcher can quickly obtain particular research results with corresponding bibliographical data. Whether the search yields results or not, researcher spends minimum time on the search through numerous scientific articles. Testing of the partial implementation of the IS that contains some decomposed scientific articles, confirms the basic expectations, and thus justifies the fundamental structure of the IS.

In the following text, process and data models are presented in order to describe the specific "science research" system and to give the basis for EDA. After that, EDA its use and possibilities with some implementation details are discussed.

II. Process Model

Process model defines those processes from the research area of atomic physics that are to be modeled and supported by the appropriate informational technology. Generally, all activities performed by researchers that contribute to scientific research are described as Research in the Physics of Atomic Collisions (RPAC). In the process model, it is represented by the corresponding RPAC function. All other activities are derived from that function by the process of hierarchical decomposition. On the highest hierarchical level - context level, process model can be represented with Data Flow Diagram (DFD) as shown in Fig 1. Researchers perform the RPAC function, and publish the research results in some of the bibliographical forms. For obtaining necessary data for other researches in this area, researchers have the opportunity to search the existing database. Database also serves as storage for own research results, and research results of other researchers. Researchers supply the RPAC function with initial data for the research that they plan, and from the RPAC function they get data about other earlier published researches. Data flows between RPAC function and bibliography as the element from its surrounding, are established in both directions, so that RPAC gives data for publishing, and also accepts published results from the earlier performed researches. To make the search for data more effective, RPAC stores new and published data in Research Results Database.

RPAC function is a complex one, as it encompasses all research activities. For more detailed analysis, it has to be decomposed on a number of hierarchical levels. The structure for the next level can be obtained by following reasoning.

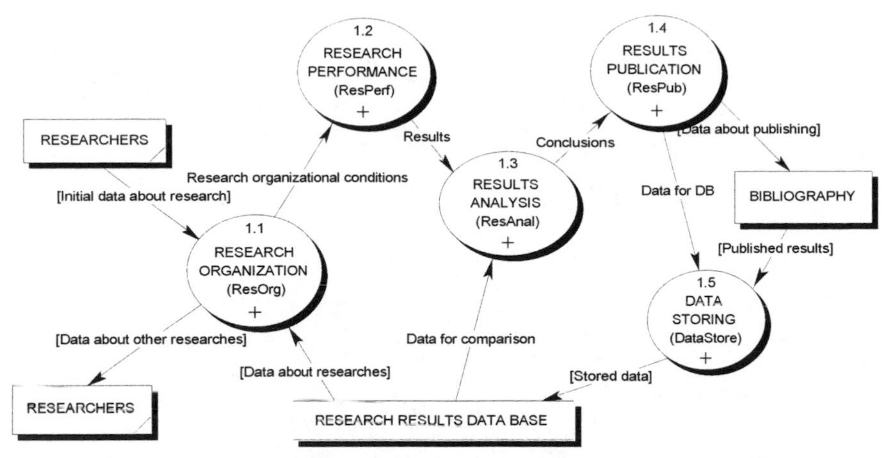

Fig 2 DFD of the decomposed RPAC function

Researchers, on the base of the set aim define criteria for searching and make insight into available data from the earlier performed researches. In this procedure, which can be iterative, initial conditions of the planned research are prepared, i.e. preparation for computer aided research, and for its organization is performed. After the organizing process is over, the next step is performing of the research in which the measurement of the observed phenomenon is made, i.e. some results are obtained. By the analysis of these results certain conclusions are drawn, that are the basis for their publishing. In the phase of publishing there are two important activities. On one side relevant data and information are published in some of the bibliographical forms, and on the other preparation of data is made for their processing and storing in the information system database. Processing and storing in the database is also enabled for data that were published earlier. Accordingly, RPAC function can be decomposed on sub processes as shown in corresponding DFD on Fig. 2. This data flow diagram makes the first decomposition level.

III DATA MODEL

Data model implemented here is relational and it is developed up to the level of recognition of all the entities It defines identifiable and descriptive attributes of entities, relationships between entities and properties of these relationships, in details enough for implementation.

DFD's that describe the process model are defined both on the context level and on the following two or three levels of decomposition (not discussed here in detail). While the processes are the basis for design of the process logical and physical model that has to enable data transformation from input to output data flows, the very data flows are the basis for defining the data logical and physical model. In the following text on the basis of data flows analysis, the structure of the relational model for support of the RPAC function is given. This is achieved by using the diagrams of the parts of whole model with symbolic according to Integration Definition for Information Modeling (IDEF1X) [20] standards, and which relates to development of the data logical model. For that purpose, method of logical design for data modeling and database design - "Entity-Relationship Modeling" ("ER" modeling) was used. Analysis of the data flows in DFD's of the function RPAC data model, led to independent entities and their characteristics. As a result, the following entities are considered in data model: Author, Laboratory, Country, Editor, Journal, Publication, Particle, Preparation, Process, Interaction Method, Method, Parameter, Research, Quantity, Result, Table Result, and Graph.

Data model consists of two main sub models, the first describing entities and their relationships important for publishing, and the other describing entities and their relationships important for experiment or research. Entity relationship diagram for the whole data model is shown in Fig. 3.

Experiment is the central entity of the data model. It gathers around and connects all the other entities. That model characteristic corresponds to the fact that experiment is the central part of research in the ACP.

Upper left part of the data model in Fig. 3 includes entities and relationships that are important for publishing. Publication can contain one or more experiments. Experiment can be understood here in wider context, and can include both theoretical and experimental research.

Particle is subject of the research in the physics of atomic collisions. In experiment particle can appear in two roles, as a target and as a projectile. Every particle can be in both roles. Particle can be categorized to chemical element – atom or compound – molecule, that can be ionized or not. Chemical element compound describe some common physical and chemical characteristics for atoms and molecules. Besides, particle can be electron or photon, that are simple elementary particles with very well known attributes, so there is no need for photon or electron entities. It is enough just to evident photon or electron. Some characteristics like energy, polarization can be specified as parameters or with Preparation entity.

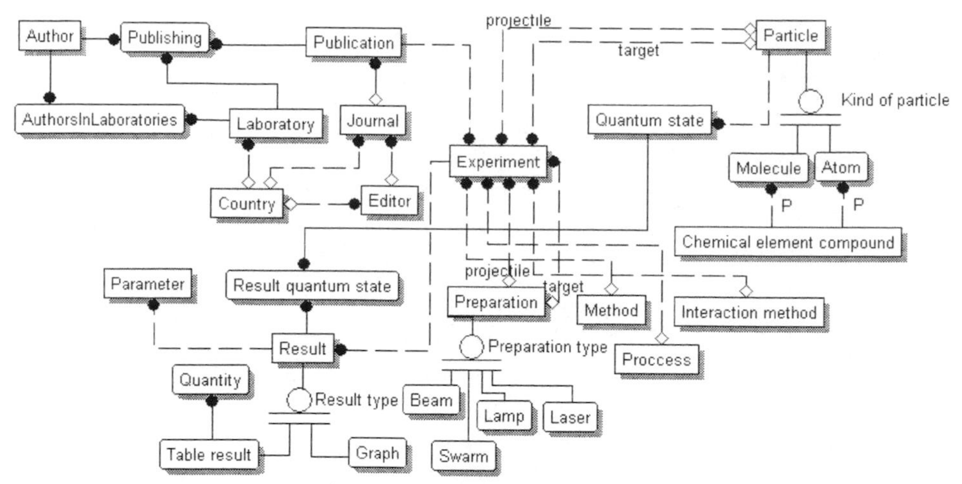

Fig. 3. Relational data model diagram

Preparation is the entity used to describe the form in which the particle is during the experiment performance. Particle can be prepared in the form of light beam (laser, lamp), particle beam, swarm and gas cell, independently from whether it has the role of projectile or target in the experiment. Not all combinations of particles and preparations are allowed depending on whether the particle is photon (laser or lamp) or not (beam and swarm). Particles in atomic physics – atoms and molecules are characterized by quantum states, as it is usual way of describing particles that are governed by the laws of quantum physics. For each particle there can be number of quantum states that also depend on degree of particle ionization.

Result is outcome of every experiment and its final aim. Results can be displayed in many forms: tabular, graphical, descriptive texts and combined.

Process is the kind of interaction between particles with the target role and particles with projectile role during the experiment. It can be elastic scattering, excitation, ionization, absorption, decomposition, etc.

Interaction method describes interaction between target particle and projectile during the experiment.

Method is the way the research of some process in physics of atomic collisions is made.

Parameter is characteristic that can be used for additional description of some research. The need for use of the Parameter entity appears if some specific property of the research being performed has to be defined. Using parameters, it is possible to enter in database quite arbitrary data that are necessary to further characterize the research, in addition to model entities.

IV IS MAIN CHARACTERISTICS

When searching for very specific data like experimental results from some research, that can not be retrieved by keyword search, even when data are in some kind of electronic form, one has to know in advance where to look for, exact references, titles, authors, etc. If the fact whether some result is published or not, or where it is published is not known to researcher, than there is high probability that the necessary data if exist, will remain hidden somewhere in the huge mass of published literature. In the case of experimental results in atomic physics, search for results is complex one for the simple reason that the given result is connected or depends on a number of characteristics - parameters that characterize it. The intention is to create a system for search and retrieval on the basis of given conditions i.e. experimental parameters. Output of the search for experimental result should contain result value or corresponding graph, values of parameters and full reference of article in which the result is published. In that way, it is possible to search for particular result starting from experimental conditions and parameters that characterize result, without knowing whether it exists or not, and where it is published. If it exists, i.e. if it is entered in data base, complete reference will be obtained as a result of search together with quantified values.

Search is based here on fragmentation – decomposition of text into many important categories that characterize the field of investigation, in this case atomic collisions. Of course, principles are of universal kind, and are directly applicable to other areas. Main categories for text fragmentation of the article or book contents are in principle the same as independent entities that arose in the data model of information system. Those entities are for instance Author, Publication, Laboratory, Projectile Particle, Target Particle, Process, Method, Interaction Method, Parameter, Variable Quantity, Measured Quantity, and others already mentioned in the previous section. Entity named publication contains data about texts and documents that were published and fragmented. For the time being, the necessary work on text fragmentation is human, more precisely the expert, as the document fragmentation must be done by people very well acquainted with the research area that the given document that is analyzed belongs to. Data resulting from the text fragmentation are stored in the tables of the database that ensures minimum redundancy of data and that the fragmentation of the given text has to be done only once, but that the results of fragmentation can be used unlimited number of times by various people.

Fragmentation categories could be viewed as dimensions of text, and data for each category resulting from fragmentation as particular coordinate values, thus positioning the analyzed text into some kind of point in multidimensional space in which the document resides. With such concepts introduced, we could say that the search is to be performed by giving one or more "coordinate" values, and checking the set of "points" thus obtained.

Second very important specificity of such approach is to realize that no text or document has to be in the underlying database, but just the data resulting from the text decomposition or fragmentation. Search would yield just the one or more particular results from articles or books that were already fragmented - analyzed. The useful information obtained in such way besides results, is the complete reference to text, enabling researcher to find it in a conventional way (library or publisher's documentation), not the text itself. This information system is not to compete with the library, but is rather complementary one.

V EXAMPLES OF IMPLEMENTATION

Access to data – information retrieval, in this IS is based on the WEB server. Certain access, review and search are also possible from the interfaces of programs for data entry. Data search can be performed in many ways and specially allows specific search for experiments results, that is one of the most important characteristic of IS structure and implementation. Currently implemented search consists of "Search for publications" and "Search for results". Those two kinds of the search are complementary from the aspect of starting data, but both enable access to particular results in publications.

Search for publications can in principal start from various groups of data – categories that can be selected. Possible groups of data are authors, particles, processes, methods, and journals – independent entities in data model. As this kind of search resembles classic search with keywords, (although it is not) attention will be paid to search for results.

The other, complementary kind of search starts from data that describe particular results in the research. Fig. 4 gives the web page with overview of quantities that define particular experiment result. Combo box on the top gives the list of various result types, with DCS value selected. Differential cross section (DCS) is one of the basic quantities that are measured in collision type experiments. Evidenced particles in database - atoms and molecules can be selected from their combo boxes. Depending on chosen particle – atom or molecule, and degree of ionization, corresponding quantum states appear in atom quantum states or molecule quantum states combo box. Quantum states characterize the state of excited target particle right after the collision. Pressing the

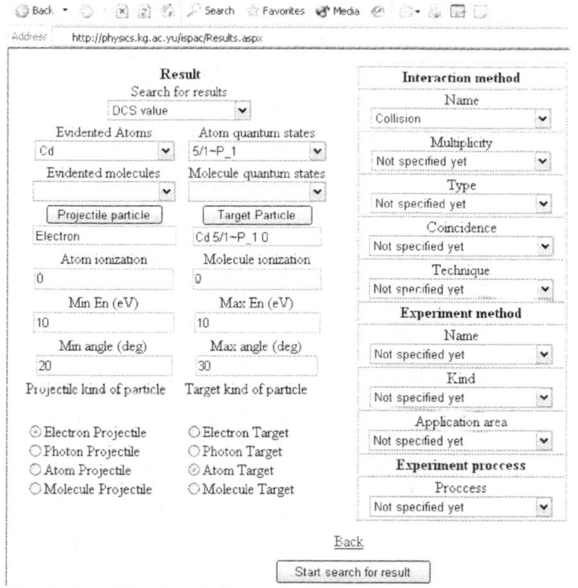

Fig. 4 Quantities that define experiment results

buttons named projectile particle and target particle, selected particles, quantum states and ionization are entered in text boxes below mentioned buttons. Values in these text boxes are used for results search. Text boxes with minimum and maximum energy define energy interval for projectile particles that will be used in the search. Similarly, minimum and maximum angle define the angle interval for scattered particles that will be used for search. Two groups of option buttons below are used for specifying type of particle for projectile and target. Selection in Fig. 4 is electron as projectile and atom as target. On the right there are three groups of combo boxes for specifying the attributes for interaction method, experiment method and experiment process. In the case when "Not specified yet" value is selected, corresponding attribute does not limit the search. If it is important to limit the search with some attribute, then any other value but "Not specified yet" has to be selected from corresponding combo box. After all desired adjustments are set, search starts with pressing the "Start search for result" button.

Information system is fully implemented using Microsoft technology. Microsoft Access 2002 is used for database and programs for data entry. Web server is Internet information server 5.1, which executes Active Server Pages – ASP.NET for all web pages. Contents of web pages are dynamically generated, depending on search conditions. For access to database from web server ADO.NET is used.

VI CONCLUSION

Information system in the physics of atomic collisions presented in this paper was developed with the aim to enable fast and simple access to various data, which are necessary to every researcher. It has double role, on one side using this information system researcher can form own (local) information system, and on the other side, using the query for search, selective access to information from the bibliography is enabled As these both sides are just parts of the same information system, it is easy to set up connection between them.

Besides, as the decomposition of articles was performed on large number of attributes, the search is not performed on "free text". That enables much faster data selection process. It is important to emphasize that data systematization in RPAC enables its use in other areas of physics or chemistry, and whereas specific data are needed (cross sections, rates, etc.).

Information system model was built so that it follows the procedures and processes that researcher goes through during consideration, organization and realization of his experiment. Process model presented in this paper is in fact the logical decomposition of the usual procedures of every researcher that is performed not just in physics of atomic collisions but also in majority of other areas in natural sciences. Even though here presented information system is connected by its attributes for physics of atomic collisions, its logical structure is easy adaptable to other scientific areas.

Implemented IS is based on given process and logical models, and has WEB access for data search. Described process model is not fully implemented, as it includes complex activities in research that are performed by competent scientists exclusively. Part of process model that is implemented includes activities for data input, logical consistency check of data, various kinds of local and WEB data search. Data input enables evidence of data for planned research, current research or data from any kind of reference. The main feature of data search is that it is possible to obtain particular results with full reference to article they belong to, on the basis of various selected experiment parameters. That kind of search gives researcher the unique opportunity to check whether one or more experimental results described with its parameters exist, and if it is the case, to see the value, and reference data. This kind of IS enforce various kinds of standardization, such as for given research model, article decomposition, data organization, search for data, used verbal expressions for entities, attributes and their values.

References

[1] The NIST Reference on Constants, Units, and Uncertainty, http://physics.nist.gov/cuu/index.html

[2] Energy levels, wavelengths and transition probabilities of atoms and ions, http://physics.nist.gov/cgi-bin/AtData/main_asd

[3] Wavenumber tables for calibration of Infrared spectrometers and frequencies for interstellar molecular microwave transitions, http://physics.nist.gov/PhysRefData/contents-mol.html

[4] Wavelength, energy level, oscillator strength, opaicty, and photoionization cross sections, http://astro.u-strasbg.fr/OP.html

[5] Atomic and molecular data for astronomy and aeronomy. Wavelength, energy levels, http://cfa-www.harvard.edu/amdata/ampdata/amdata.html

[6] Collisional excitation, ionization, recombination, charge transfer, sputtering, and backscattering data, http://dbshino.nifs.ac.jp/

[7] Electron-Impact Ionization Cross Section Database for Molecules, http://physics.nist.gov/PhysRefData/Ionization/Xsection.html

[8] Collsional data, H Neutral Beam Data, Particle-Surface Interaction Data, and Data for Elementary Processes in H-He Plasmas, http://www-amdis.iaea.org/ALADDIN/

[9] Photoionization, recombination, collisional ionization, autoionization, charge transfer, auger processes, energy levels, wavelengths, transition probabilities stark broadening, and opacities, http://www.pa.uky.edu/~verner/atom.html

[10] Atomic Transition Probability Bibliographyics Database, http://physics.nist.gov/PhysRefData/Fvalbib/html/reffrm0.html

[11] Obserbatoire de Paris-Section de Meudon: Bibliography on Atomic Line Shapes and Shifts, http://www.obspm.fr/estark

[12] AMBDAS, http://www-amdis.iaea.org/AMBDAS/

[13] Various processes for atoms, ions and molecules (structure, phtonic collisions, electron collisions, atomic and moleuclar collisions, http://gaphyor.lpgp.u-psud.fr/

[14] Oak Ridge National Laboratory's - Controlled Fusion Atomic Data Center, http://www-cfadc.phy.ornl.gov/bibliography/search.html

[15] GENIE - A General Internet Search Engine for Atomic Data, http://www-amdis.iaea.org/GENIE/

[16] DANSE - Atomic and Molecular Bibliographic Data Search Engine, http://www-amdis.iaea.org/DANSE/

[17] ICAMDATA –International Conference on Atomic and Molecular Data, http://physics.nist.gov/Divisions/Div842/Icamdata/Homepage/icamdata.html

[18] Electron Impact Cross Section Calculations Using The Average Approximation, http://www-amdis.iaea.org/AAEXCITE

[19] Effective ionization and recombination rate coefficients, http://www-amdis.iaea.org/RATES

[20] Processing Standards Publication 184 "Integration Definition for Information Modeling (IDEF1X)", (1993) December 21. NIST, U.S.A

Incremental Learning of Trust while Reacting and Planning

W. Froelich
Institute of Computer Science,
Silesian University ,
Ul.Będzińska 39
Sosnowiec, Poland

M.Kisiel-Dorohinicki, E. Nawarecki
Institute of Computer Science
AGH University of Science and Technology
Al. Mickiewicza 30
Kraków, Poland

Abstract - The general idea of the proposed approach is to integrate simple reactive intelligence acquired by experimentation together with planning and learning processes. The autonomous agent [1] can be considered as a representative of an intelligent entity located in the real world. It is expected that it should express rational behavior and possess the ability to learn relevant to it's goals. The main objective of this paper is to construct a cognitive model of an agent that is capable of rational behavior in a dynamical environment. The concepts, like the goal, reactivity, and planning are investigated in the context of an agent that undertakes decisions and actions in completely or partly unknown environment. We have also proposed the integration of reactive and planning decision selection mechanisms by applying the concept of trust to its decisions on the basis of reinforcement. When designing our agent, we applied the bottom up approach, aiming to present some of the relevant research in this area. The primary advantage of this approach is shown by the improved performance of the agent during the execution of the given task. The effectiveness of the proposed solution has been initially tested in a simulated environment (evasive maneuver problem).

I. INTRODUCTION

The rational behavior of an autonomous system located in a dynamical environment is a research topic that can be investigated from many angles. There are some fundamental questions on how such a system can be modeled and how it can express rationality relevant to it's goals. For instance, the actual agent's state at a discrete point in time can be represented in the simplest way by the tuple:

$$ag = \Re s, d, W >, \qquad (1)$$

where: $s = [s_1,...,s_n] \in S$ denotes the agent's observation vector, S is an n-dimensional observation space (in general it can be continuous), $d = [d_1,...,d_m] \in D$ denotes the decision vector and D is a finite set of decisions available to the agent. After the decision process is completed, as a result an action or to the chain of actions is found to have taken place in the environment. The agent *ag* is equipped with the internal model W of the environment in which it is situated. In the simplest case the model W can be a decision table storing observation-decision pairs which are called reflexes [1]. In more apparent solutions there is the need for some form of generalization to achieve a more compact and universal model. Some of the proposed solutions include applying rules, decision tables, neural networks or functions which approximate the decision table. We assume that the agent's decision process is considered only in discrete points in time. In each time step an agent perceives the environment, recognizes vector s, then it selects decision d. The mapping π: $S \rightarrow D$ is taken here to denote agent's decision policy. We define a utility function u: $\pi \rightarrow R$ for the entire policy. The consequence of this is that the way the goal has been attained influences the evaluation of agent's performance measured by the utility. The concrete construction of $u(\pi)$ depends naturally on a particular task and its implementation. The goal of the agent Ag is to find and follow such an acceptable policy π^* that the utility $u(\pi^*) \geq u_t$, where $u_t \in R$ denotes certain utility threshold.

II. REACTING AND PLANNING

The formulated problem can be considered as the search within a plan space or within a state space. In the first case the agent tries to obtain the acceptable policy by constructing and testing the entire possible observation-decision chains. Among many examples, these methods apply neural networks, genetic programming or the evolutionary algorithms with reinforcement learning (EARL). Different types of classifier systems [9][10] are well known. In general, systems that search in plan space do not involve the analysis of a single decision step with respect to a single observation (reactive step). Therefore we have decided to focus our attention on state space searching systems. In such systems the decision selection methods try to construct the policy gradually, usually using step by step considerations. In our opinion, this solution is also closely connected to biological systems observed in nature. The state space search methods can be investigated in the following three situations.

1) In the first case, the environment is deterministic, fully observable and known to the agent, which means, it has a complete model W. In particular, on the basis of W, the agent attempts to compute the next expected environmental state. Having a complete knowledge of the environment, the problem of finding the policy π^* is reduced to the search in W. In particular, when the utility of the policy depends only on the situation observed at the end of the episode, the agent can use for example A* search or backwards planning algorithms

K. Elleithy (ed.), Advances and Innovations in Systems, Computing Sciences and Software Engineering, 491–495.

[2] to find the appropriate path to the desired situation. It's possible to apply symbolic computations, where maybe the most known solution is the STRIPS planning system [2].

2) In second case, the agent is situated in a partially known environment that can have stochastic properties or be partially observable. In such a case the model W is partial; it's not possible to compute every state of the environment but the agent can obtain the evaluation of possible decisions (actions) in the form of a reinforcement signal. After each action has taken place or at the end of the episode, the reinforcement indicates the effectiveness of the performed actions. It is assumed that in stochastic environments the transitional function between states and the expected value of the reinforcement it is taken into account. In such a situation, it is possible to model the system by Markov's decision process (MDP) and apply a dynamic programming method. Note that the agent can use memory M to avoid aliasing states in non-Markovian environments. The agent also applies memory to store the actual evaluation of the states (value function) for the entire policy. It should be noted here that in the described situation there is no need for an agent to perform experiments (actions) in the real environment to find the acceptable policy (π*).

3) The third case seems to be the most problematical for an agent. The world model is completely unknown. The agent needs to experiment and learn in the environment to at least partially constitute the model W. Among the solutions that are known to exist, the agent can apply TD-learning algorithms to learn the value function.

The detailed description of the existing solutions is not discussed in this paper. Depending on how much knowledge about the environment is available, the situations described above are considered in most cases separately.

There are very few published suggestions to integrate above mentioned solutions. One of the most interesting is the Dyna system [3][4] and it's multi-agent extension M-Dyna-Q [5]. In the following chapters we set out another solution to this problem.

III. MODELLING THE COGNITIVE AGENT

In the first step of the analysis we propose the following extension of a simple agent (1):

$$Ag = \Re s, d, W, M, G>, \quad (2)$$

where M is the memory for storing the history of the agent's activities and G is a set it's goals. Observations, decisions or state variables of the agent are continuously stored in memory M. In particular it can assume the form of a time window containing full or limited behavior chain, of observation-decision pairs $<s_t, d_t>$, where t is the time index. It can be stated that the history of agent's operations, which constitutes the temporal context of the agent, can in some cases significantly influence the decision process, especially in non-Markov environments.

The set of agent's goals can be acquired a-priori from the domain expert or it can constitute them on the basis of experience and reinforcement from the environment. The goals in G can be described in a language on different level of abstractions. The language in which the goals in G are represented should be consistent with the meaning of information obtained from the observation vector s. This should enable autonomous recognition (observation) of satisfactory situations after attaining at least one of it's goals. For instance, where the agent is purely reactive the goals can be represented as a set of the desired observations of vector s.

Let us look closer now at the agent's decision process. For the sake of simplicity, we will consider the one step decision process. In the case of a complex decision problem, we assume that the agent enters the repetitive reactive or planning mode. An agent can also follow the elaborated plan with the possibility of periodical checking and adaptation of it. The case where one agent's decision corresponds to the entire chain of actions is not considered here - in our opinion without loss of generality of the proposed model.

First we describe the reactive decision selection mechanism, which has been shown in Fig. 1. From the cognitive point of view one can distinguish some elements within an agent model that belong to a perception field or to an unconscious field. The notions of cognitive fields are introduced here to better understand the nature of and some of the differences between reaction based and planning based mechanisms of the decision selection process. In the reactive mode, the agent perceives only the indirect association between sensory input and the decision vector. In the planning mode, the agent simulates different possibilities of future occurrences and thus perceives and controls (in an imaginary way) possible intermediate associations that should lead from sensory observations to the decisions and a goal. The unconscious association is thus shifted to a fully controlled perception field. While operating in the reactive mode, the agent perceives the content of s, memory M, decision vector d and the associations among them. Here we denote the decision with a subscript r for the reactive type deselection mechanism: $<s, M> \rightarrow d_r$. As Fig. 1 outlines, one of the possible and most widely used methods of such a decision selection is to apply a weighted association mechanism. The probability of choosing decision d_i is proportional to its weight w_i. The weights $<w_1, w_2, \dots, w_n>$ of possible decisions can be found for example by applying TD algorithm and by being computed on the basis of value function. The reactive decision can be chosen for instance by applying non-deterministic proportional selection algorithm (roulette wheel). One of the obvious advantages of the reactive mechanism is a relatively fast processing in relation to planning.

For the purposes of our analysis we assume that at the initial state the agent knows nothing about the environment, W and M are empty. The agent must gradually explore the environment to achieve goals.

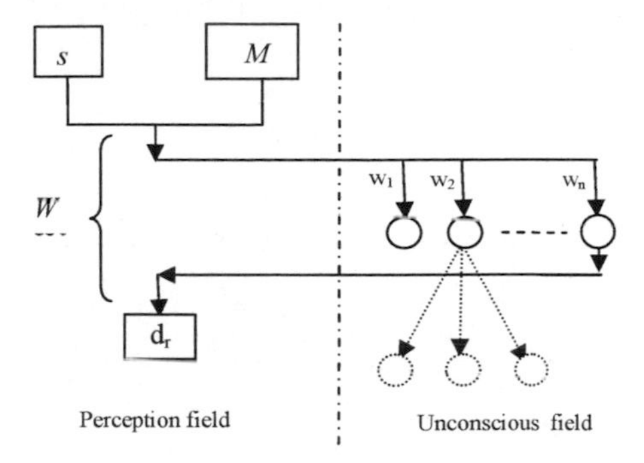

Fig. 1. Reactive decision selection mechanism

It stores its experiences in W for instance in a form of the set of pairs (associations): $<s, M> \rightarrow d_r$. Such traces lead the agent to the construction of its policy. The content of W can be approximated to save the agent resources. As the experienced associations are only the estimations of the uncertain environment, the agent always has a restricted level of confidence in them. The more experience it has, the more its world model W is more completed and closer to the "objective" state of the world.

The complexity of the environment on one hand and limited computational resources of the agent on the other imply the need of symbolical computations. Every agent that that is planning performs simulations of future states using its world model. The simulation plays a key role as an imaginary experience [5]. To store the intermediate states of the simulations we have equipped our agent with the appropriate set I in which the agent stores statements expressed in a planning language dedicated to describing the states of the system on different levels of abstraction. Thus the final representation of our agent assumes the following form:

$$Ag = \Re s, d, W, M, G, I > \quad (3)$$

In most cases of planning systems the representation language of W is dedicated to the planning algorithms and operates on a higher abstraction level (e.g. rules). In I the planning process reconstructs simulated experiences on the basis of heretofore learned W. The problem is that our reactive agent usually stores experiences on the basis of simple associations between observations and decisions. One possibility is to switch the planning language to W; the second is to apply the appropriate transformation algorithm. The problem is not discussed here. For our purposes we assume only that the representation language of I must be consistent with W.

The decision selection mechanism based on planning is described in Fig.2. The entire process of building the associations $<s, M> \rightarrow d$ is under the control of planning algorithm - thus being located within an agent conscious perception field.

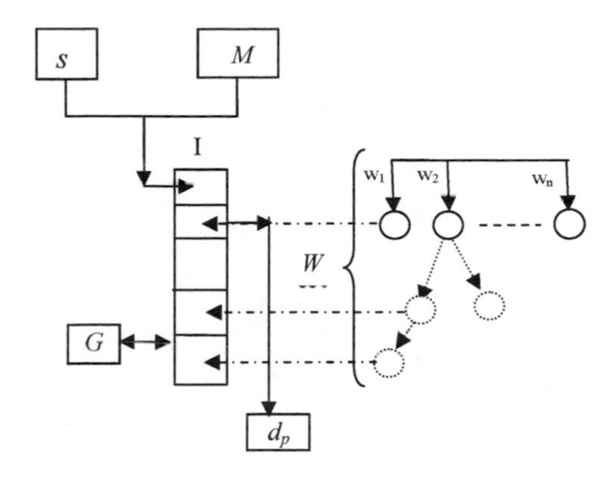

Fig. 2. Decision selection mechanism based on planning.

The subscript p denotes the planned version (in opposite to reactive d_r) of the actual decision. Associations between the world states are stored and retrieved from the model W. The decision selection algorithm searches the model W for the appropriate path between the actual observed (initial) state to the goal (terminal) state. With subscript p we denote the decision computed by the planning algorithm. The multi-step search process is usually complex and takes some time to deliver the acceptable chain of associations *from s* to one of the goals in G.

In practical solutions it is not always obvious which of the above described approaches the agent should employ. As previously mentioned, one of the proposed approaches to integration of reactive and plan based systems is the system called Dyna [3][4]. In brief, the idea is to use planning as the simulated experience for additional adjustment of the value function. The number of updating steps towards the goal is a parameter of the system. The proposed solution integrates methods based on value function with planning. The confidence of the knowledge acquired in W is to a certain extent measured by the number of simulation steps that are used to adjust the value function. The advantage of quick response times of reactive mechanism is retained and depends only on the number of look-ahead steps of the planning part of the algorithm.

In many situations the quality of the agent's policy can depend on the appropriate decision: to what extent the world model W is complete and can be applied to planning. On the other hand any plan in a stochastic environment is uncertain; hesitancy and periodical verification are expected.

For these reasons, we would like to put forward the concept of trust for the effective integration of the reactive and plan based decision selection methods.

IV. LEARNING TRUST

We propose to complement the agent's memory by two state variables: $rt_t \in R$ and $pt_t \in R$, which represent the agent's trust to reactive and plan based decisions respectively. The memory includes also the constant t_c, which plays a role of time constraint for the planning module. The complete scheme of the decision mechanism has been sketched on Fig.3.

The algorithm of an agent life cycle is represented as follows:

1. Observe the environment s;
2. If one of the goals in G has been reached – stop.
3. Execute in parallel:
    a. propose a reactive decision d_r and compute the function $rtf(d_r)$,
    b. propose the plan based decision d_p and compute the function $ptf(d_p,I)$ on the basis of the achieved plan stored in I, continuously check the computation time, if it exceeds t_c ,interrupt the computations in step 2, assume immediately d_r as final decision and go to the point 1.
4. Check the following conditions:
    a. if $rtf(d_r) > rt_t$ then choose d_r as final decision,
    b. if $ptf(d_p, I) > pt_t$ then choose d_p as final decision,
    c. if both of $rtf(d_r) > rt_t$ and $ptf(d_p, I) > pt_t$ choose the final decision corresponding to the greater value,
    d. if $rtf(d_r) \leq rt_t$ and $ptf(d_p, I) \leq pt_t$ then generate random exploratory decision.

The values of thresholds rt_t and pt_t can be set a-priori by the domain expert or learned i.e by evolution [6][7]. The question, how to compute the trust functions $rtf(d_r)$ and $ptf(d_p, I)$ can be an interesting challenge for the future research. The objective is to achieve the expected increase in the effectiveness of the agent. Let us notice that when the agent computes the function for the reactive type of decision, it does not have access to any anticipation of the following next observation of environment. It can't estimate the effects of the decision only on the basis of the value function stored in W. Therefore the trust to the decision d_r can be evaluated on the basis of the corresponding value function or on the basis of the separately obtained (e.g. from domain expert) lookup table with <d_r, cf> pairs, where cf denotes confidence factor for every decision.

The computation of $ptf(d_p, I)$ is expected to be more complex. It is possible to take into account the evaluation of the trust to the decision on the basis of the expected future observations.

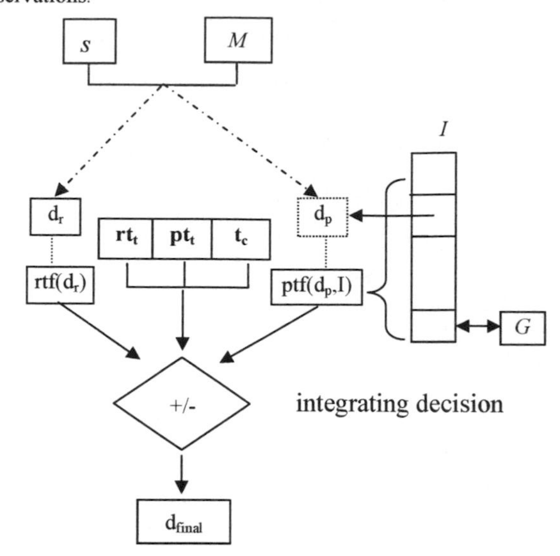

Fig. 3. Application of trust measure to the decision selection.

The agent can compute this on the basis of the plan that has been worked out in I. One of the possibilities is to apply the analysis of the confidence factors for the situations described in the plan. The minimum, maximum, mean, standard deviation or other values can be considered. Notice, that such joint indicator of trust cannot be computed before the entire plan is constructed.

V. EXPERIMENTAL RESULTS

The above presented idea has been verified in the simulated environment. The values of thresholds rt_t and pt_t have been involved in the agent genome [6]. We have applied a dedicated reinforcement learning algorithm for the reactive part of the system and a forward search algorithm for the planning module.

The simulation has been implemented for the evasive maneuvers problem [8] (pursue and evasion game) in a two-dimensional space for a single pursuer (rocket) and a single evasive object (plane). A brief introduction to the problem is as follows. The rocket is launched towards the flying plane, whose goal is to hit the plane. The rocket is led automatically (in a deterministic way) at the target. The plane should be capable of evading the racket by changing the parameters of flight, i.e. the direction and speed. The plane is equipped with measuring devices that measure the parameters of the approaching rocket flight: the distance, mutual flight angle, and the rocket's speed. On the basis of the radar readings the agent controlling the plane should be capable of learning how to change the direction and speed of the plane to avoid it being hit. The agent learning task has been divided into learning episodes.

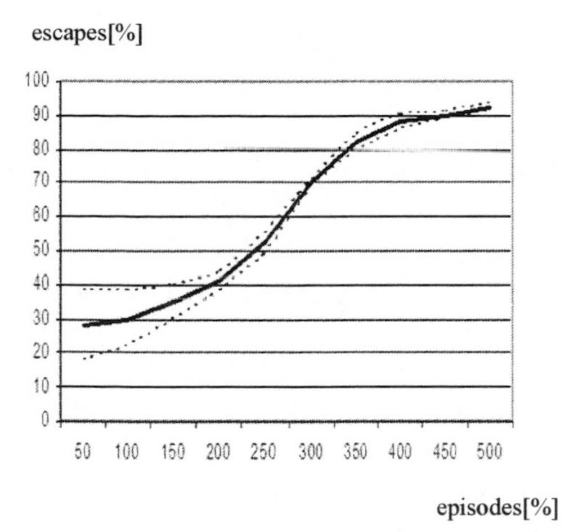

Fig. 4. The effectiveness of the agent.

One of the initial conditions of each episode was the mutual location of the objects: the rocket and the plane. The episode ends when the plane is hit or manages to avoid being destroyed. The reference point for executed experiments was the case of the agent using the random strategy of controlling the plane's escape (consisting in random changes of flight direction and speed). In this case, the number of instances where the plane avoided hit did not exceed 40% of all learning episodes. Further, the tests were carried out in conjunction with the agent's decision process based on the suggested model and a hybrid learning method proposed in [6]. The effectiveness for the best agent led to more than 90% among all learning episodes avoiding hit (fig. 4).

VI. Concluding Remarks

The main idea of the approach was based on the agent paradigm in order to represent interactions between an intelligent entity and its environment. First of all, we have shown some of the aspects of reactive and planned based decision selection processes. It has been also suggested that the ambivalence between them can be resolved to a certain extent by applying trust indicators and learning. The computational experiments have shown that the method of trust indicators adjustment based on multiagent evolutionary technique can lead to an acceptable solution. Based on the presented realization and performed experiments, it may be said that the suggested idea fulfills the assumed expectations. It is also foreseen to have great universality within the chosen range of applications. Further improvements of the suggested approach can be foreseen for the near future.

References

[1] M. Wooldridge, editor. An Introduction to Multiagent Systems. John Wiley & Sons, 2002

[2] S.Russel,P.Norvig, Artificial Intelligence. A modern Approach.Second Edition, Prentice Hall, 2003.

[3] R.S. Sutton. Dyna, an integrated architecture for learning, plan ning, and reacting. SIGART Bulletin, 2:160–163, 1991.

[4] R.S. Sutton. Planning by incremental dynamic programming. In Proceedings of the Eigtth International Workshop on Machine Learning, pages 353–357, 1991.

[5] G. Weiß, An architectural framework for integrated multiagent planning, reacting, and learning. In Proceedings of the 7th International Workshop on Agent Theories, Architectures, and Languages (ATAL). Lecture Notes in Computer Science, Volume 1986. Springer-Verlag. 2001.

[6] W. Froelich, Evolutionary multi-agent model for knowledge acquisition. In Inteligent Information Processing and Web Mining (IIPWM'05), Advances In Soft Computing. Springer, 2005.

[7] M. Kisiel-Dorohinicki, G. Dobrowolski, and E. Nawarecki. Agent populations as computational intelligence. In L. Rutkowski and J. Kacprzyk, editors, Neural Networks and Soft Computing, Advances in Soft Computing. Physica-Verlag, 2003.

[8] J. Grefenstette, C. Ramsey, and A. Schultz. Learning sequential decision rules using simulation models and competition. Machine Learning, 5(4), 1990.

[9] J.Holland:"Adaptation in Natural and Artificial Sytems", The University of Mitchigan Press, 1975

[10] W.Stolzman: "Antizipative Classifier Systems", PhD Dissertation,Universitaet Osnabrueck, Shaker Verlag, 1997

Simulation of Free Feather Behavior

Xiaoming Wei [1] , Feng Qiu [2] and Arie Kaufman [2]
[1] Computer Science Department, Iona College
[2] Center for Visual Computing and Department of Computer Science
Stony Brook, New York, U.S.A.

Abstract

We present a general framework for simulating the behaviors of free feather like objects inside a dynamic changing flow field. Free feathers demonstrate beautiful dynamics, as they float, flutter, and twirl in response to lift and drag forces created by its motion relative to the flow. To simulate its movement in 2D, we adopt the thin strip model to account for the effect of gravity, lift and inertial drag. To achieve 3D animations, we implement two methods. For the first approach, we extend the thin strip model, use either flow primitive or noise functions to construct a time-varying flow field and extract external forces to update the thin strip computation. For the second approach, we implement a physically based simulation of the flow field and adopt the momentum-exchange method to evaluate the body force on the feather. As a result, the natural flutter, tumble, gyration dynamics emerge and vortices are created all in response to local surface-flow interactions without the imposition of the thin strip model.

1 Introduction

The modeling of complex scenes including realistic clouds, plants, furry animals, fabrics, liquids, and objects having all manner of albedo, reflectance, and translucence have always been a challenging and exciting work in computer graphics. The ubiquity of these simulations is now driving the research frontier into the domain of dynamics. The task is to bring the graphics alive so that trees are blown in the wind, flags flap, fluids flow, and hair flies. In this paper, we concentrate our work on the simulation of free feather liked objects immersed in a flow field.

Free feather dancing in the wind illustrates beautiful and complex dynamics. In response to lift and drag forces they float and flutter. The dynamics of a falling feather is non-trivial not only because the forces of lift and drag couple the downward and side-to-side motions, but also because of the production and influence of vortices in the surrounding flow. In this paper, we present approaches to simulate 2D and 3D feather dynamics in a flow field. A thin strip model is used for 2D simulation. Since the model itself is defined in a motionless medium, to capture the effects of a gusty wind, we simply add appropriate terms to account for the external forces along ***X*** and ***Y*** coordinates. While in 3D, we propose two methods. The first one is to use an extended thin strip model with a random generated and time-varying flow field. External forces are obtained by considering the change in the flow velocity at the mass center of the feather. The additional ***Z*** coordinate of the feather velocity is updated with a simple function as the feather flies away. To further improve the simulation quality, especially to be able to simulate a free falling feather in a quiescent flow field, we propose the second approach to use physically based simulation of flow field combined with the momentum-exchange method. The feather itself is imposed as a real boundary object in the flow field. Its linear and angular velocity are updated based on the external forces. As its position and orientation change, the boundary condition of the flow field changed accordingly. The complex dynamics of the feather are created directly by the local surface-flow interaction. Vortices are created as the free feather falls in the otherwise quiescent flow. Between these two approaches, the first one gives us a fast performance speed, while the second one results in a more accurate animation.

Over the past decade, researchers have used a variety of approaches for modeling the interaction of wind with leaves, trees, grass, and human hair [6, 11, 12, 13, 14, 22]. Specifically, Wejchert and Haumann [22] developed an aerodynamic model for simulating the detachment and falling of leaves from trees in a wind

K. Elleithy (ed.), Advances and Innovations in Systems, Computing Sciences and Software Engineering, 497–502.

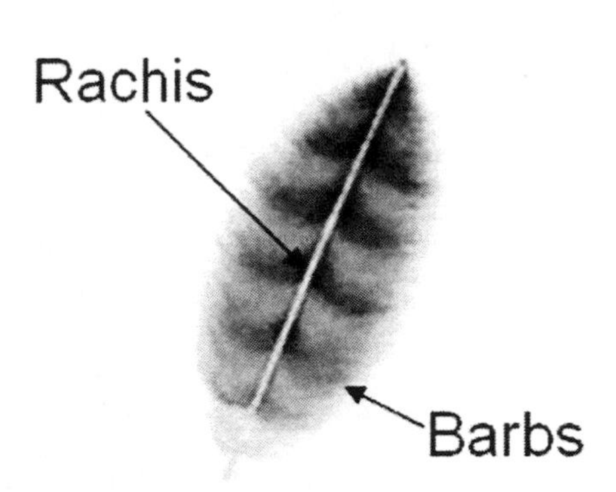

Figure 1: A close-up view of the 3D texture rendered feather.

field built from four basic flow primitives. Shinya and Fournier [13] animated trees and grass blown by complex flow fields by modeling the wind field in Fourier space and then converting to a time-varying force field. Using similar methods, Stam [14] animated trees in turbulent winds. More recently, Perbet and Cani [11] simulated the movement of grass in a wind field which was generated by combining procedural primitives and an additional time-varying stochastic component. In this paper, we concentrate our work on simulating the motion of free, light-weight feather. However, our thin strip model can easily be adapted to the leaves simulation. The flow model and force evaluation method could also be used with grass and tree models.

The remainder of the paper is organized as follows. The modeling of feather is described in Section 2. In Sections 3 and 4, we present the models used for 2D and 3D simulations and demonstrate the results.

2 Feather

A feather is a branching structure developed on a spine called the *rachis*. From the rachis, a number of *barbs* originate. The collection of the barbs on each side of the feather comprises two *vanes* [3]. Following Streit and Heidrich's work [16], we model the shapes of the rachis and the contour of the left and right vanes with three different Bezier curves. In order to construct each individual barb, four control points are used: *P0*, *P1*, *P2*, and *P3*. Points *P0* and *P3* are located on the rachis and the vane contour. Points *P1* and *P2* are generated by perturbing the positions of two points lying on the line between *P0* and *P3* using a noise function. Finally, a texture is mapped onto the barbs to give the feather the desired color and

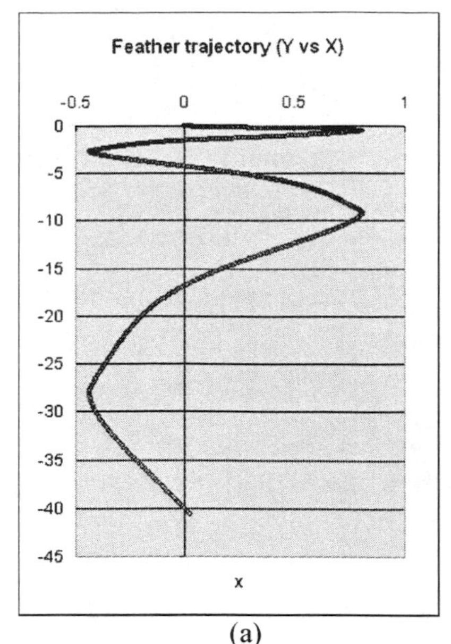

(a)

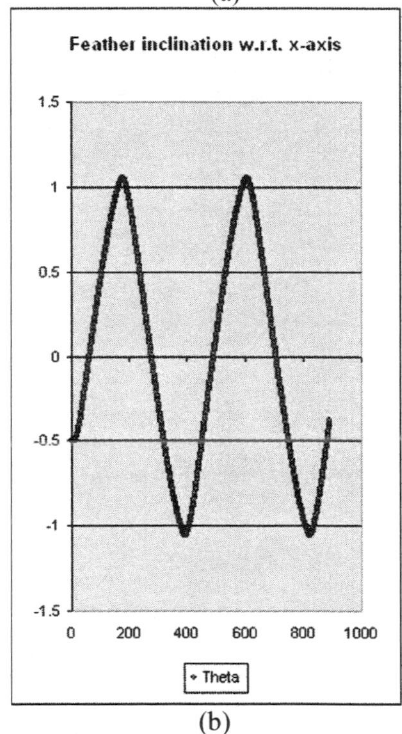

(b)

Figure 2: 2D Feather trajectory on X and Y plane, shown in (a) and the change of θ angle along simulation time t, shown in (b), with an initial parameter setting of $x = 0, y = 0$, $v_x = 0$, $v_y = 0.001$, $dt = 0.015$, $A_\perp = 4.1$, $A_\parallel = 0.88$, $A_\omega = 0.067$ and $Fr = 0.45$.

quality. The bottom fluffy part of the feather is modeled using a 3D noise function. To further improve the rendering and to achieve a nice fuzzy effect of the feather, we voxelize the whole feather model into a volume texture and render it on Graphics Processing Unit (GPU) in real time [21], as shown in Figure 1.

3 2D Feather Dynamics

To simulate the dynamics of the feather in 2D is not a easy task. Even for the simplified system of a thin, flat strip, the effort to construct faithful physical models continues to challenge researchers (see [1] and references therein). To capture the essential flutter dynamics of a feather, we use the thin strip model of Belmonte et al., which includes the effects of gravity, lift, and inertial drag. The scaled equations of motion in this 2D model are given by:

$$\dot{v}_x = \frac{V^2}{Fr}[A_\perp \sin\phi \sin\theta - A_\parallel \cos\phi \cos\theta \mp 4\pi |\sin\phi| \sin\gamma]$$

$$\dot{v}_y = \frac{V^2}{Fr}[A_\perp \sin\phi \cos\theta + A_\parallel \cos\phi \sin\theta \mp 4\pi |\sin\phi| \cos\gamma] - \frac{1}{Fr}$$

$$\dot{\omega} = \pm A_\omega \frac{12\omega^2}{Fr^2} - 6\pi V^2 \sin(2\phi)$$

$$\dot{\theta} = \omega$$

Here, $\mathbf{V} = (v_x, v_y)$ represents the velocity of the feather relative to the flow with the **y** axis pointing vertically upward. The angle θ is that of the rachis with respect to the positive **x** axis, and γ is the angle similarly defined by V. The angle $\phi = \gamma - \theta$. The constants $A_\parallel$ and $A_\perp$ represent the parallel and perpendicular drag coefficients, respectively. Finally, the dimensionless quantity Fr is the *Froude* number which represents the ratio of downward and side-to-side motion time scales. For an explanation of the sign convention we refer the reader to Tanabe and Kaneko [17].

To define the Froude number, we use the following equation:

$$Fr = \left(\frac{M}{\rho L^2 w}\right)^{\frac{1}{2}}$$

where, M is the mass of the strip, L and w are its length and width and ρ is the density of the flow field. To solve the equations, we employ standard fourth-order Runge-Kutta integration within a reference frame that follows any rotation of the feather about the x-axis. Such rotations may arise due to unbalanced torques exerted on the feather by the wind. We model this rotational dynamics in terms of a damped rotator. In Figure 2(a), we show the 2D feather trajectory based on an initial feather speed of $v_x = 0$, $v_y = 0.001$ and the parallel and perpendicular drag coefficients $A_\perp = 4.1$, $A_\parallel = 0.88$ and $A_\omega = 0.067$. In Figure 2(b), we demonstrate the change of q, which is the angle between the rachis and the positive **x** axis, with the initial q to be -0.5 and Fr number to be 0.45. For a 2D thin strip in a flow, two distinct types of motion are observed -fluttering, a kind of rocking motion, and tumbling, a kind of rolling motion. Which of these actually occurs depends on the Froude number. Belmonte et al. observed that for a 2D thin strip, flutter (tumble) occurs for low (high) values of Fr with the transition occurring at $Fr = 0.67$.

4 3D Feather Dynamics

One way to achieve 3D feather dynamics is to extend the 2D thin strip model. We first use flow primitives and noise functions to construct a time-varying wind field. Since the thin strip model defined above assumes that the strip falls through a motionless medium, to capture the effects of a gusty wind, we simply add a term proportional to the appropriate component of the wind force to the 2D Equations. This force is obtained from the flow field by considering the change in the flow velocity at the center of mass of the feather over consecutive time steps. An additional linear equation is implemented to update the v_z component of the feather as the feather is flying away. Snapshots from the feather simulation are shown in Figures 3. Although we did not include deformation dynamics in our current feather model, the feather is seen to flutter and dance in wind as it slowly falls under the influence of gravity. Using the above method, we could not reproduce the vortices created in the flow by the action of the feather. In particular, the model could not simulate a feather in a static flow field since the feather is not presented as a true moving boundary to the flow. To achieve this effect, the underlying flow field needs to be modeled in a physically correct way. In recent years, the application of Computational Fluid Dynamics (CFD) for modeling fluid behavior has led to significant developments in the graphics and visualization field. Especially the impressive work by Foster and Metaxas [5] for simulating the turbulent rotational motion of gas; the semi-Lagrangian advection schemes proposed by Stam [15] and the most recent Fedkiw et al.'s work to simulate smoke [4]. Another interesting approach is

(a) (b) (c) (d)

Figure 3: A sequence of snapshots taken from the feather simulation based on the extension of thin strip model.

(a) (b) (c) (d)

Figure 4: A sequence of snapshots taken from the feather animation based on the Lattice Boltzmann simulation of flow field and the momentum exchange method.

based on the Lattice Boltzmann Model (LBM) [18, 19, 20, 21].Recently, In this work, we have chosen to pursue this alternate approach because it provides good physical accuracy, particularly with complex boundary geometries [2, 9, 10], while being simple to implement and amenable to hardware acceleration.

The surface of the feather is represented as a triangle mesh defined by the shape of the rachis and barbs. The triangle mesh is then defined as boundary conditions inside the flow field. The local forces of interaction between the feather and the underlying flow field are computed separately. For LBM, it is computed by updating particle net force, which is then used to advance its linear and angular velocity. As the feather changes its position and orientation, the boundary condition of the LBM computation is updated accordingly. As a result, the natural flutter, tumble, gyration dynamics emerge and vortices are created all in response to local surface-flow interactions without the imposition of additional physics. For a detailed explanation of LBM and its hardware implementation on GPU we refer the reader to Wei et al. [18, 19, 20, 21] and Li et al. [7, 8]. Snapshots from the feather animation are shown in Figure 4. In lattice units, the feather measures 20 $\leftarrow$2.7 in size with a mass of 22

7g). The size and the mass of the 3D feather

(a) (b)

Figure 5: Snapshots of the feather as it falls through an otherwise quiescent flow field. The streamlines, originating from a vertical plane passing through the feather, illustrate the effect of the feather on the flow. Red (green) streamlines indicate flow to-ward (away from) the viewer.

Section 3, flutter occurs if the Froude number is less than 0.67. In our simulation, in addition to flutter, we observe a third type of motion which we call gyration. In gyration, the feather performs a helicopter-like rotation as it falls. Figures 5(a) and (b) show the streamlines in a horizontal plane containing the feather during its fall. The action of the feather on the flow produces weak vortices. It is the interplay between the feather motion and flow field motion that automatically gives rise to the flutter and gyration of the feather as it falls. Figure 5(a) is taken 20 steps after the feather starts to fall; while figure 5(b) is taken 100 simulation steps later and it gives a much closer view of the feather.

5 Conclusion and Future Work

In this paper, we propose the use of a thin strip model to simulate feather behavior in 2D. For 3D simulations, we present two approaches. One extends the thin strip model, construct a time-varying flow field and use the external forces to update the thin strip computation. The other implements a physically-based simulation of the flow field and adopts the momentum-exchange method to evaluate the body force on the feather. Both of these two approaches give satisfying results. The first one has a faster performance speed, while the latter generates a more realistic animation. As we observe feather in real life, we see its weight, size and shape all affect its behavior.

In our current system, in addition to flutter, we have observed a third type of motion which we call gyration, where, the feather performs a helicopter-like rotation as it falls. For our future work, we would like to study the effect of parameters such as mass, length of the feather and the property of the underlying flow, understand the transition between various motions and eventually establish an interactive animation system that allows users to control these parameters to get the desirable fluttering, tumbling and gyration effect.

References

[1] A. Belmonte, H. Eisenberg, and E. Moses. From flutter to tumble: Inertial drag and froude simi-larity in falling paper. *Physical Review Letters*, 81(2):345–348, 1998.

[2] S. Chen and G. D. Doolean. Lattice Boltzmann method for fluid flows. *Annu. Rev. Fluid Mech.*, 30:329–364, 1998.

[3] Y. Chen, Y. Xu, B. Guo, and H. Shum. Modeling and rendering of realistic feathers. *Proceedings of SIGGRAPH*, Computer Graphics Proc., Annual Conf. Series:630–636, 2002.

[4] R. Fedkiw, J. Stam, and H. W. Jensen. Vi-sual simulation of smoke. *Proceedings of SIG-GRAPH*, Computer Graphics Proc., Annual Con-ference Series:15–22, 2001.

[5] N. Foster and D. Metaxas. Modeling the mo-tion of a hot, turbulent gas. *Proceedings of SIG-GRAPH*, Computer Graphics Proc., Annual Con-ference Series:181–188, 1997.

[6] S. Hadap and N. Magnenat-Thalmann. Modelling dynamic hair as a continuum. *Computer Graphics Forum*, 20(3):329–338, 2001.

[7] W. Li, Z. Fan, X. Wei, and A. Kaufman. Gpu-based flow simulation with complex boundaries. *Chapter 47: GPU Gems II for Graphics and Compute-Intensive Programming*, March 2005.

[8] W. Li, X. Wei, and A. Kaufman. Implementing lattice Boltzmann computation on graphics hard-ware. *The Visual Computer*, 19(7-8):444–456, December 2003.

[9] R. Mei, L. S. Luo, and W. Shyy. An accurate curved boundary treatment in the lattice Boltzmann method. *Journal of Comp. Phys.*, 155:307–330, June 1999.

[10] R. Mei, W. Shyy, D. Yu, and L. S. Luo. Lattice Boltzmann method for 3-d flows with curved boundary. *Journal of Comp. Phys.*, 161:680–699, March 2000.

[11] F. Perbet and M. P. Cani. Animating prairies in real-time. *ACM Symposium on Interactive 3D Graphics*, pages 103–110, 2001.

[12] L. M. Reissel and D. K. Pai. Modelling stochastic dynamical systems for interactive simulation. *Computer Graphics Forum*, 20(3):339–348, 2001.

[13] M. Shinya and A. Fournier. Stochastic motion - motion under the influence of wind. *Computer Graphics Forum*, 11(3):C119–128, September 1992.

[14] J. Stam. Stochastic dynamics: Simulating the effects of turbulence on flexible structures. *Computer Graphics Forum*, 16(3):159–164, 1997.

[15] J. Stam. Stable fluids. *Proceedings of SIGGRAPH*, Computer Graphics Proc., Annual Conference Series:121–128, 1999.

[16] L. Streit and W. Heidrich. A biologically parameterized feather model. *Eurographics*, 21(3):565–573, September 2002.

[17] Y. Tanabe and K. Kaneko. Behavior of a falling paper. *Physical Review Letters*, 73(10), 1994.

[18] X. Wei, W. Li, K. Mueller, and A. Kaufman. Simulating fire with textured splats. *IEEE Visualization*, pages 227–234, October 2002.

[19] X. Wei, W. Li, K. Mueller, and A. Kaufman. The lattice Boltzmann method for gaseous phenomena. *IEEE Transaction on Visualization and Computer Graphics*, 10(2):164–176, March/April, 2004.

[20] X. Wei, Y. Zhao, Z. Fan, W. Li, S. Yoakum-Stover, and A. Kaufman. Blowing in the wind. *ACM SIGGRAPH/Eurographics Symposium on Computer Animation*, pages 75–85, July 2003.

[21] X. Wei, Y. Zhao, Z. Fan, W. Li, S. Yoakum-Stover, and A. Kaufman. Lattice-based flow field modeling. *IEEE Transaction on Visualization and Computer Graphics*, 10(6):719–729, November/December 2004.

[22] J. Wejchert and D.Haumann. Animation aerodynamics. *Proceedings of SIGGRAPH 1991*, Computer Graphics Proceedings, Annual Conference Series: 19–22, 1991.

Evolutionary Music Composer integrating Formal Grammar

Yaser M.A.Khalifa, Jasmin Begovic, Badar Khan, Airrion Wisdom and M. Basel Al-Mourad*
Electrical and Computer Engineering Department
State University of New York
New Paltz, NY 12561, USA
*School of Computing & Information Technology
University of Wolverhampton
Wolverhampton, WV1 1SB, UK

Abstract— **In this paper, an autonomous music composition tool is developed using Genetic Algorithms. The production is enhanced by integrating simple formal grammar rules. A formal grammar is a collection of either or both descriptive or prescriptive rules for analyzing or generating sequences of symbols. In music, these symbols are musical parameters such as notes and their attributes. The composition is conducted in two Stages. The first Stage generates and identifies musically sound patterns (motifs). In the second Stage, methods to combine different generated motifs and their transpositions are applied. These combinations are evaluated and as a result, musically fit phrases are generated. Four musical phrases are generated at the end of each program run. The generated music pieces will be translated into Guido Music Notation (GMN) and have alternate representation in Musical Instrument Digital Interface (MIDI). The Autonomous Evolutionary Music Composer (AEMC) was able to create interesting pieces of music that were both innovative and musically sound.**

I. Introduction

While the concept of using genetic algorithms to compose music has been attempted in the past, due to the nature of music as a creative activity, there is still a need for further work in this area. In [1], Gartland-Johnes and Colpey provide an excellent review of the application of Genetic Algorithms in musical composition. Miranda, in [2] discusses different approaches to using evolutionary computation in music. Most systems listed in literature need a tutor, or an evaluator. The development of autonomous unsupervised music composers is still very limited, but yet has lots of potential. In particular, in identifying chromosome structures and fitness functions that have not been studied in other attempts. The concept of using pattern extraction techniques to extract primary patterns, or motives, in established pieces of music has not been extensively explored in the literature. This is somewhat surprising, since composers have made use of motives for composition for centuries. The problem of composing music based on a library of motives is, however, near or perhaps slightly beyond the frontier of current capabilities of artificial Intelligence (AI) technology. Thus, this area of research spearheads a new direction in automated composition.

The work presented in this paper is an attempt in that direction. It presents an autonomous music composition system. The system composes musical pieces based on a library of evolving motifs. The critique of the generated pieces is based on two evaluation functions: intervals and ratios, that each describes an aspect of the musical notes and/or system as explained in the following sections.

II. Genetic Algorithms Implementation

GA are a stochastic combinatorial optimization technique [8]. It is based on the evolutionary improvement in a population using selection and reproduction, based on fitness that is found in nature. The GA operates on a population of chromosomes,

K. Elleithy (ed.), Advances and Innovations in Systems, Computing Sciences and Software Engineering, 503–508.

where each chromosome consists of a number of genes, and each gene represents one of the parameters to be optimized. The composition of music is performed in two Stages. In Stage I, a set of motifs is generated. In Stage II, motifs and their transpositions are combined to form two music phrases, A and B. At the end of Stage II, phrase $A^{\#}$ is generated by sharing each note of the phrase. At the end, a combination of $ABA^{\#}A$ is produced, which is one of the common combinations in music composition theory.

III. Music Background

In this section, some basic fundamentals of music composition are given. Because the piano has a good visual explanation of music, it will be used for illustration, however these concepts can be transposed to any musical instrument including the human voice.

We begin by analyzing the most basic set of notes called the C major scale, which consists entirely of all the white notes. We will dissect what major scales are, how they composed and further our discussion to how they form what are called chords, or simultaneously depressed single notes.

Music regardless of the instrument has a maximum 12 different distinct pitches or tones which are called keys. A pitch is simply a frequency of sound; within these pitches a multitude of combinations can be formed to produce "music". However, how can we be assured that a specific combination will be musically pleasing to the ear? Of course the term "musically pleasing" is subjective to the listener, but there must be some fundamental principle underlying the organization of the combination in question.

There is an *interval* that exists between to consecutive pitches, the term musical interval refers to a *step* up or down in musical pitch. This is determined by the ratios of the frequencies involved. "...an octave is a music interval defined by the ratio 2:1 regardless of the starting frequency. From 100 Hz to 200 Hz is an octave, as is the interval from 2000 Hz to 4000 Hz." In music we refer to the interval between two consecutive notes as a *half step*, with two consecutive half steps becoming a *whole step*. This convention is the building block of our major scale.

A scale is a set of musical notes that provides the blueprint of our musical piece. Because our starting point is the musical note C, this major scale will be entitled as such. The major scale consists of a specific sequence of whole steps of and half steps, that being W W H W W W H, where W is a whole step and H is a half step. A typical musical convention is to number the different notes of the scale corresponding to their sequential order, usually called *roots*. Using the sequence of the major scale, our C major scale consists of the notes C D E F G A B, returning to note C completing what is known as an *octave*, or a consecutive sequence of 8 major scale notes. Numeric values are now assigned where C corresponds to value 1, D would be 2, E being 3 and so on. The next C in terms of octaves would restart the count therefore the last value or root would be 7 corresponding to note B.

We build on these scales by combining selected roots simultaneously to form what are known as *chords*. Chords can be any collection of notes, this leads to almost endless possibilities in music, however for our purposes we implement the *C major chord* and use its sequence of notes. A major chord consists of the 1st, 3rd, and 5th root of the major scale, this would mean that we utilize notes C E and G.

IV. Stage I

In Stage I, motifs are generated. A table of the 16 best motifs is constructed that is used in Stage II. These motifs will be used both in their current, and transposed locations to generate musical phrases in Stage II. Fig .1 shows the chromosome structure in Stage I. Each chromosome will contain 16 genes, allowing a maximum of 16 notes per motif. Each motif is limited to a four-quarter-note duration.

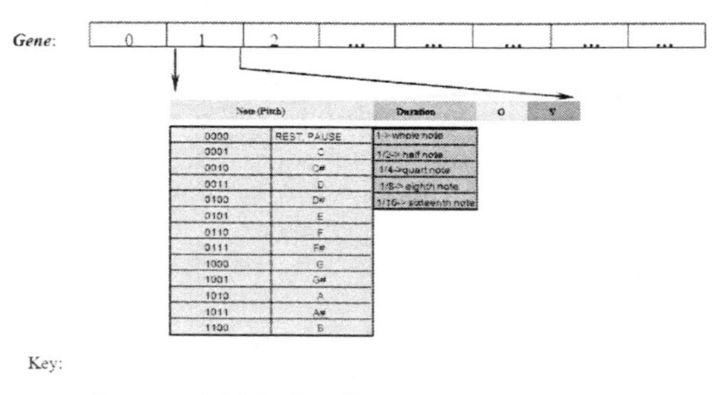

Key:

Note : Limited to 13 possible outcomes
pause, C, C#, D, D#, E, F, F#, G, G#, A, A#, and B.
Duration : Controls how long a note will be held. Limited to five time measures
1/16. 1/8, ¼, ½, and a whole notes.
O : Octave, Limited to two octaves
low and high.
V : Velocity piano (soft) and forte (loud)

Fig .1. Chromosome and Gene Structure for Stage I

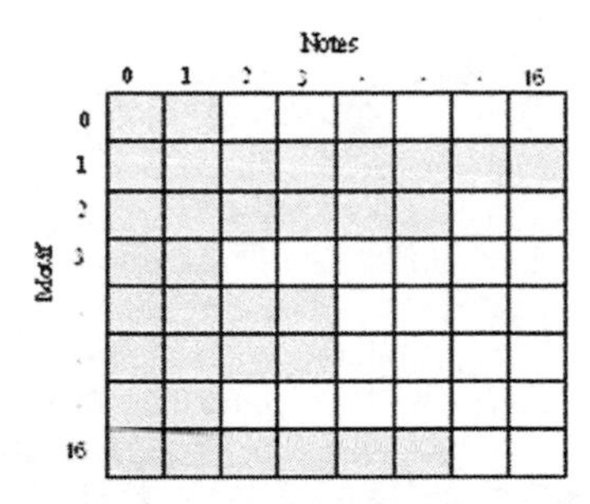

Fig .2. Motif Look-up Table Generated in Stage I

At the end of Stage I, a table of the top 16 motifs is constructed (Fig .2). Each row in this look-up table represents a motif. The columns represent the different notes in the motif. Although all motifs generated are one whole note in duration, they could be composed of either one, two, four, six, or eight notes. However, single note motifs are highly discouraged.

A. Formal Grammar Evaluation Function

As previously mentioned the C major chord consists of the 1st 3rd and 5th root of the scale. This will correspond to values 1, 5 and 8. We can also do an *inversion* of the major chord by using the same notes C E G, this time however starting at note E leaving the inversion to be E G C. If we consider the next octave up and assign the value 13 to the repeating C, the inversion E G C will correspond to values 5, 8, and 13.

These two versions of the major chords give us two production rules of which we are assured will be musically pleasing. The production rules will take the difference of the values assigning a good fitness if the production rule is met. Our first production rule will be the difference or skip of 4 and 3 (5 - 1 = 4 and 8 - 5 = 3), describing our first version of the major chord C E G. The second rule will be the difference or skip of 3 and 5 (8 – 5 = 3 and 13 – 8 = 5), describing the inversion of the major chord E G C.

V. STAGE I EVALUATION FUNCTIONS

A. Intervals Evaluation Function

Within a melody line there are acceptable and unacceptable jumps between notes. Any jump between two successive notes can be measured as a positive or negative slope. Certain slopes are acceptable, while others are not. The following types of slopes are adopted:

Step: a difference of 1 or 2 half steps. This is an acceptable transition.
Skip: a difference of 3 or 4 half steps. This is an acceptable transition.
Acceptable Leap: a difference of 5, 6, or 7 half steps. This transition must be resolved properly with a third note, i.e. the third note is a step or a skip from the second note.
Unacceptable Leap: a difference greater than 7 half steps. This is unacceptable.

As observed from the information above, leaps can be unacceptable in music theory. We model this in GA using penalties within the interval fitness function. Certain resolutions between notes are pleasant to hear, but are not necessary for a "good" melody. These resolutions therefore receive a bonus. Dealing with steps in the chromatic scale, we can define these bonus resolutions as the 12-to-13 and the 6-to-5 resolutions. The 12-to-13 is a much stronger resolution, and therefore receives a larger weight. It was experimentally suggested that the 12-to-13 resolution have double the bonus of the 6-to-5 one, and that the bonus does not exceed 10% of the total fitness. Thus the bonuses are calculated by dividing the number of occurrences of each of the two bonus resolutions by the number of allowed resolutions (15 resolutions among 16 different possible note selections), see equations (1) and (2).

$$\text{12-to-13 bonus} = (\#\text{occurances}/15) * 0.34 \quad (1)$$

$$\text{6-to-5 bonus} = (\#\text{occurances}/15) * 0.34 \quad (2)$$

The total interval fitness:

$$\text{Interval Fitness} = \frac{1}{total\_error(1 - total\_bonus)} \quad (3)$$

VI. STAGE II

In Stage II, motifs from the look-up table constructed in Stage I are combined to form two phrases, A and B. Each phrase is eight measures, and each measure is a four quarter-note duration motif, Fig 3.

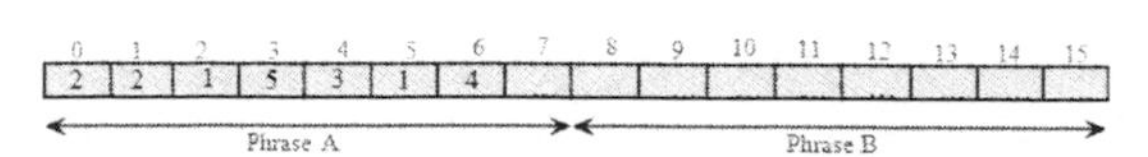

Fig .3. Chromosome Structure for Stage II

VII. Stage II Evaluation Functions

In Stage II, two evaluation functions are implemented: intervals, and ratio. The intervals evaluation function described in the previous section is used to evaluate interval relationships between connecting notes among motifs, i.e. between the last note in a motif and the first note in the following motif. The same rules, as described above in Stage I, are used in Stage II. Other evaluation functions are described below.

A. Ratios Evaluation Function

The basic idea for the ratios section of the fitness function is that a good melody contains a specific ideal ratio of notes, and any deviation from that ideal results in a penalty. There are three categories of notes; the Tonal Centers that make up the chords within a key, the Color Notes which are the remaining notes within a key, and Chromatic Notes which are all notes outside a key. Each type of note is given a different weight based on how much a deviation in that portion of the ratio would affect sound quality. The ideal ratios sought were: Tonal Centers make up 60% of the melody; Color Notes make up 35% of the melody; and Chromatic Notes make up 5% of the melody. Although these ratios choices could be quite controversial, they are a starting point. Ongoing research is looking into making these ratios editable by the user, or music style dependent.

VIII. Results

A. Analysis of Motif Selection

The four motifs in Fig 4 all resulted from a single running of the program. They were handpicked from the final 16 motifs selected by the program as the most fit. It can be observed that each motif has an identical rhythm consisting of four eighth-notes, one quarter-note, and two more eighth notes. Summing the durations of the notes yields the correct four quarter-note duration indicated by the time signature $\begin{pmatrix} 4 \\ 4 \end{pmatrix}$ at the beginning of each motif.

Using the intervals evaluation algorithm as a reference, we can see why these motifs were chosen to be the elite of the population. Examining motif *a*, the first three notes are all $F^{\#}$'s, indicating that no penalty will be assigned (a step size of 0). The next note is a $G^{\#}$ (2 half-steps away from $F^{\#}$). This transition is classified as a step and no penalty is assigned. The following notes are $F^{\#}$, $G^{\#}$, and E (a difference of 2, 2, and 3 half-steps, respectively). These transitions are also acceptable; therefore the intervals evaluation function would not assign any penalty to the motif. When zero error is assigned to a motif, a high fitness value will result. Similar analysis of motifs *b*, *c*, and *d* yield the same result.

So what is the musical difference between the motifs? Since the notes in each motif are slightly different, the musical 'feel' of each motif will vary. Compare motifs *a* and *d* for example. Motif *a* contains four $F^{\#}$'s. They are arranged in such a way that the first beat and a half of the measure are all $F^{\#}$'s, and also the 3rd downbeat (the quarter-note). This repeatedly drives the sound of the $F^{\#}$ into the listener, resulting in an unconscious comparison of this note to every other note in the measure. This in turn will make dissonant notes sound more dissonant, and resolving notes sound more resolved. In the case of motif *d*, the $F^{\#}$'s are arranged in a manner that accents the steady background rhythm of the measure (the repetitive rhythm that your foot taps to when you listen to music). This does not accent the sound of the $F^{\#}$ as much, but rather accents the other rhythms of the measure that occur between the $F^{\#}$'s. A more 'primal' feel will result, as opposed to the more 'melodic' feel of motif *a*.

By analyzing the sequence of notes in the musical piece generated in Figure 5, it is seen that a direct correlation to our Formal Grammar production rule one. The skip from the B to D is a skip of 3 and C to E is a skip of 4 which meets our first production rule. We do see a skip of 2 that being D to C, a skip of 0 in C to C, and a skip of 1 C to B, but we must consider that formal grammar is not our only fitness criteria in the autonomous music composer.

(a)

(b)

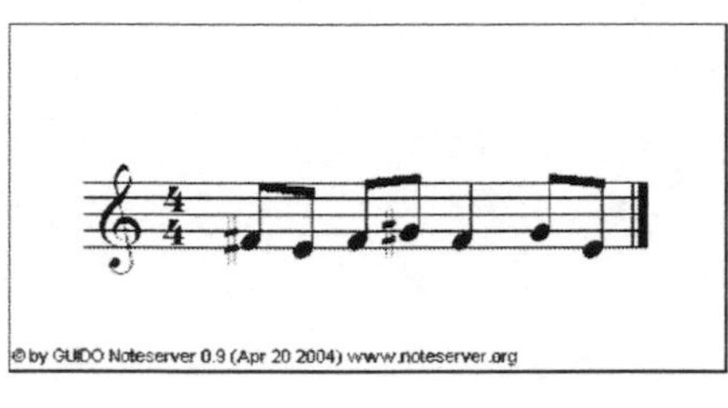

(c)

(d)

Fig .4. Sample Motif Generated in Stage I of the Evolutionary Music Composer

Fig .5. Sample Motifs Generated in Stage I of the Evolutionary Music Composer

Our second production rule is visibly met in Figure 6. The note sequence seen above is C E C E G D and repeat. The skip from E to G is a skip of 3 and G to D is a skip of 5 satisfying our second production rule. We can in this Figure however that we combine both production rules, because the skip from C to E is a skip of 4, therefore this musical piece followed our Formal Grammar rules entirely!

Fig .6. Sample motifs generated in Stage II of the Evolutionary Music Composer

IX. Discussion and Future work

New techniques in evaluating combinations of motives are needed. The evaluation of motive combination should take into consideration the overall musical piece rather than the note transition resolutions of the first and last notes in the motif only. One approach that will be further investigated is the application of formal grammars. In a multi-objective optimization problem such as music composition, different evaluation functions are applied and contribute to the fitness measure of a generated piece. The main functions that have been implemented are intervals, and ratios. They have been equally considered in evaluating the evolutionary generated music so far. Different weighing methods for various evaluation functions is expected to effect the quality of the resulting music. These could also be affected by types of music sought, e.g. classical, Jazz, Blues, etc. In Stage II of the project, methods that use weighted combinations of different fitness functions, or composition rules, will be explored.

REFERENCES

[1] Gartland-Jones, A. Copley, P.: What Aspects of Musical Creativity are Sympathetic to Evolutionary Modeling, Contemporary Music Review Special Issue: Evolutionary Models of Music, Vol. 22, No. 3, 2003, pages 43-55.

[2] Burton, A.R. and Vladimirova, T.: Generation of Musical Sequences with Genetic Techniques, Computer Music Journal, Vol. 23, No. 4, 1999, pp 59-73.

[3] Miranda, E.R.: At the Crossroads of Evolutionary Computation and Music: Self-Programming Synthesizers, Swarm Orchestra and Origins of Melody, Evolutionary Computation, Vol. 12, No. 2, 2004, pp. 137-158.

[4] Miranda, E.R., Composing Music Using Computers, Focal Press, Oxford, UK, 2001.

[5] Khalifa, Y.M.A., Shi, H., Abreu, G., Bonny, S., and Ziender, J.: "Autonomous Evolutionary Music Composer", presented at the EvoMUSART 2005, held in Lausanne, March 2005.

[6] Horowitz, D.: Generating Rhythems with Genetic Algorithms. Proc. of the 12th National Conference on Artificial Intelligence, AAAI Press, 1994.

[7] Marques, M., Oliveira, V., Vieira, S. and Rosa A.C.: Music Composition Using Genetic Evolutionary Algorithms, Proc. of the Congress of Evolutionary Computation, Vol. 1, 2000.

[8] Pazos, A., and Del Riego, A.: Genetic Music Compositor. Proc. of the Congress of Evolutionary Computation, Vol. 2, 1999.

[9] Goldberg, D.E., Genetic Algorithms in Search, Optimization and Machine Learning, Addison-Wesley, Reading , USA, 1989.

A New Algorithm and Asymptotical Properties for the Deadlock Detection Problem for Computer Systems with Reusable Resource Types

Youming Li and Robert Cook
Department of Computer Sciences
Georgia Southern University
Statesboro, GA 30460
Email: yming@georgiasouthern.edu, bobcook@georgiasouthern.edu

***Abstract*-We study the classical problem of deadlock detection for systems with *n* processes and *d* reusable resource types, where $d<<n$. We present a novel algorithm for the problem. The algorithm enjoys two properties. First, its cost is $n/log(n)$ times sm aller than that of the well-known Dijkstra's algorithm, when $d=O(log(n))$. Secondly, its data structures are simple and easy to maintain. In particular, the algorithm employs no graph or tree based data structures. We also derive a linear-time algorithm when *d* and the resource requests are bounded by constants. The linear-time algorithm is asymptotically optimal. The algorithms are applicable to improving the Banker's algorithm for deadlock avoidance.**

Categories and Subject Descriptors:
D.4.1 Operating Systems: Process Management;
General Terms: Deadlock, algorithms, performance

I. Introduction

In a uniprocessor computer system, deadlock detection for a set of processes is a problem that has been studied by many researchers [1, 3, 6, 7, 8, 9, 12, 15]. The more recent references [4, 10] contain practical implementations for deadlock detection. In this model, each process in the set requests additional resources while holding certain amount of resources; a process will not be able to run until the request is granted. A set of processes is in a deadlock state if there is no schedule of execution by which every process in that set can satisfy its resource requests.

In general, deadlock detection in the optimal way has been proven to be an NP-complete problem [5]. Similar results hold for deadlock avoidance [15]. However, if all the resources are reusable, algorithms of polynomial running time exist. An example of a polynomial time deadlock detection algorithm is formulated in [1]. The algorithm, which we will call Dijkstra's algorithm in this paper, has cost $\Theta(dn^2)$. Here n is the number of processes, d is the number of resource types and the cost is the number of arithmetic and logic operations used by the algorithm.

The algorithm is essentially the same as the safe state detection in the Banker's algorithm proposed by Dijsktra [3]; also see [6] and Operating Systems textbooks such as [13, 14]. Thus the problems of deadlock detection and Banker's algorithms for systems with reusable resources are both tractable.

The algorithm has a good aspect in that its cost has linear dependence on d. However, the quadratic dependence on n could make the algorithm far from optimal, especially for systems with large number of processes. One of our main aims in this paper is to provide a measurement for the discrepancy between the complexity of the deadlock detection problem and the cost of the Dijkstra's algorithm for systems with large number of processes. The cost of algorithms and complexity are measured in the worst case. We do this by designing a new algorithm solving the deadlock detection problem.

This paper is organized as follows. In Section II, relevant notations and concepts are given. In Section III, we present an algorithm for solving the deadlock detection problem and we estimate the cost of the algorithm. In Section IV, we discuss the asymptotic behavior. In Section V, we discuss applications to the Banker's algorithm.

II. Preliminaries

Throughout the paper, we use the following notation. Let P_i, $i = 1, \dots, n$ be n processes, and T_j, $j = 1, \dots, d$ be the resource types. Let $H_i = (h_{i1}, \dots, h_{id})$ be the resource vector being held by the process P_i. Let $R_i = (r_{i1}, \dots, r_{id})$ be the resource vector being requested by P_i. Let $V = (v_1, \dots, v_d)$ be the free resource vector. The arithmetic operations and comparison operations for two vectors are defined component-by-component in an obvious way.

The *system state information* consists of the request matrix $R = [r_{ij}]$ $(1 \le i \le n,\ 1 \le j \le d) = (R_1, \dots, R_n)^t$, the resource allocation matrix $H = [h_{ij}]$ $(1 \le i \le n,\ 1 \le j \le d) = (H_1, \dots, H_n)^t$, and the free resource vector V.

Definition: A system is deadlock free if there exists $s \in S_n$, the permutation group on the set $\{1, \dots, n\}$, such that

$$R_{s(i)} \le V + \sum_{1 \le k \le i-1} H_{s(k)} \text{ for all i: } 1 \le i \le n.$$

The system is in a deadlock state or simply deadlocked if it is not deadlock free. In other words, no sequence of execution will result in every process having their resource requests satisfied.

K. Elleithy (ed.), Advances and Innovations in Systems, Computing Sciences and Software Engineering, 509–512.

Definition: Let $A = A(d, n)$ be an algorithm that solves the deadlock detection problem.
(i). We define $cost(A)$ to be the number of arithmetic and logical operations used by A.
(ii). The complexity $comp(d, n)$ of the problem of deadlock detection is the minimal cost of all algorithms that solve the deadlock detection problem.
(iii). A is called almost optimal if $cost(A) = \Theta(comp(d, n))$ as n goes to infinity.
In this paper, we will focus on a reasonable assumption that d is a function of n: $d = d(n)$. One special case is that d is simply a constant. We stress that in this case $comp(d, n)$ is essentially a function of n.

III AN ALGORITHM AND AN UPPER BOUND

In this section, we present an algorithm J that solves the deadlock detection problem with cost $O(dn(log(n)+d))$, which is thus an upper bound of the complexity of the deadlock detection problem. One feature of the algorithm is that it does not employ any graph based data structures, which can be costly and complicated for system implementations employing deadlock detection.

A. Algorithm J

Input: A system of n processes with the following system state information: resource request matrix R, resource allocation matrix H, and resource available vector V. All entries in the matrixes and vector are general nonnegative integers.

Output: TRUE if the system is deadlocked, FALSE otherwise.

Data Structures: For technical reasons, we add two dummy processes P_0 and P_{n+1} with resource requests to be initialized to be 0 and infinity respectively.
An integer b, initialized to be 0, an integer matrix $[t[i, j]]$ ($0 \leq i \leq n+1$, $1 \leq j \leq d$), an array $Mark[n]$ of the Boolean type, all elements are initialized to be FALSE.

Begin Algorithm J
Step 1: For every j from 1 to d, $t[0, j] := 0$, and find $\sigma_j \in Sn$ such that $r_{\sigma j(1), j} \leq \ldots \leq r_{\sigma j(n), j}$.
Step 2: For every j and k from 1 to d, compute $\eta_{j\ k}$ as the composite of inverse of σk and σj.
Step 3: A loop starts with condition "as long as there exists a q such that $t[b, q] < n$". The body of the loop consists of the following actions:
Step 3a: For every j from 1 to d using binary search find $t[b+1, j]$ so that $r_{\sigma j(t(b+1,j)), j} \leq v_j$ and $r_{\sigma j(t(b+1,j+1)), j} > v_j$.
Step 3b: If $t[b+1, j]$ equals $t[b, j]$ for all j, then return TRUE.
Step 3c: For every j from 1 to d and every p from $t[b, j] + 1$ to $t[b+1,j]$ if $Mark[\sigma j(p)]$ is FALSE and $\eta_{k,j}(p) \leq t[b+1 ,k]$ for all k, then increment V by $H[\sigma j(p)]$ and set $Mark[\sigma j(p)]$ to be TRUE.
Step 3d: Increment b by 1.
Step 4: Return FALSE
End Algorithm J

The algorithm works as follows. First, for each resource type, we sort the request array by the processes, in nondecreasing order. We need permutations σ_j to keep track of the processes after sorting. Obviously, we can use any sorting algorithm. The permutation $\eta_{j\ k}$ establishes a map between the processes of two sorted columns of the resource types k and j.

Secondly, for each resource type, we find the last position in the (sorted) request array so that the request of the type by the process at that position can be satisfied.

Thirdly, for every resource type, the algorithm checks which processes before or at the last position can satisfy their requests. In order to identify a process that satisfy requests for all the types, the algorithm uses permutation maps to check whether the same process ID falls in the corresponding position range for each resource type. If such a process is found, the algorithm deallocates all resources allocated to the process, and finds the next and all such processes, and continues with the next iteration. A key point for the efficiency of the algorithm (see the cost analysis below) is that in the next turn, it suffices to only make "cross" checks for limited position ranges (Step 3c), this is due to the facts that the arrays are sorted and available resource vector is increasing.

B. Correctness of the Algorithm

To formally verify the correctness of the algorithm, we define the following sets of process IDs:

$$K_b := \bigcup_{1 \leq j \leq d} \{ \sigma j(t[b, j] + 1), \ldots , \sigma j(n) \},$$
$$L_b := \bigcap_{1 \leq j \leq d} \{ \sigma j(1), \ldots , \sigma j(t[b + 1, j]) \}.$$

By saying that iteration I_b is executed normally, we mean it is not broken by the statement return in Step 3b. We have the following two properties:
(i). K_b is the set of the IDs of the processes that have not satisfied their requests right after the iteration I_{b-1}, if it is executed normally, and
(ii) L_b is the set of the IDs of the processes that have satisfied their requests right after the iteration I_b, if it is executed normally.

The correctness of algorithm J follows from the above two properties. In fact, suppose that in the iteration I_b we have $t[b+1, j] = t[b, j]$ for all j from 1 to d. We consider two cases: $b = 0$ and $b > 0$.

If $b = 0$ then $t[1, j] = 0$ for all j from 1 to d. This means that for each resource T_j, the sorted array r_{ij} ($0 \leq i \leq n+1$) only has one element ($r_{0j} = 0$) less than or equal to v_j, by the way $t[1, j]$ is chosen. Thus only the dummy process P_0 satisfies its requests. The system is deadlocked, and L correctly detects it (Step 3b). Now we consider the case of $b > 0$. Let any z be in K_b, which is the set of the IDs of the processes that have not been marked as TRUE right after the iteration I_{b-1}. Notice that K_b is not empty, since for some j, $t[b, j] < n$, when I_{b-1} is executed (the Boolean condition for the loop}). Thus $\sigma j(t[b, j]+1)$ belongs to K_b. Then there exists j such that z belongs to the set $\{ \sigma j(t[b, j]+1), \ldots , \sigma j(n) \}$. Thus $z = \sigma j(q)$ for some q between $t[b, j] + 1$ and n. Since $t[b+1, j] = t[b, j]$, $r_{z,j} = r_{\sigma j(q),j}$ must be greater than the current v_j by the way $t[b+1, j]$ was chosen. Thus process Pz cannot satisfy its requests; the system is deadlocked since z was arbitrarily chosen. In this case L correctly returns TRUE. On the other hand, suppose that we can increase some $t[b, j]$, or equivalently $t[b+1, j] > t[b, j]$, for some j in each iteration I_b. Then eventually, we have $L_k = \{1, \ldots , n\}$ for some iteration I_k and thus the system is deadlock free, and L correctly detects it by returning FALSE (Step 4). Also observe that for $b > 0$ if we can increase some $t[b, j]$ in I_b, then the code from Step 3c must have been executed during the iteration I_{b-1}, since it is the only way to increase the free resource vector (hence it is the only way to increase $t[b, j]$ for some j, since $\sigma j(t[b, j])$ is the largest position at which the request is less than or equal to v_j.). In other words, at least one process satisfies its requests during iteration I_{b-1}. Moreover, if $b = n$, then all processes have satisfied their requests right after I_{b-1} by induction. This implies that $t[b, q] = n$ for all q from 1 to d, or equivalently, the condition for the loop is false. We thus have the following conclusion, which will be used later.

Conclusion: The number of the iterations of the loop in algorithm L is at most $n + 1$.

Remark: The matrix $[t[i, j]]$ can be replaced with two vector variables, one of them is for the storing the last positions for the columns, and the other one is for storing the updated positions. Thus the variable b is not needed either. The only reason to use the matrix and b is for the clarity of the proof and cost analysis.

C. Cost Analysis

We now turn to the cost of the algorithm.

The cost for Step 1 can be controlled within $O(dnlog(n))$. The cost for Step 2 can be done within $O(d^2n)$, since it takes linear time to compute a permutation and inverse permutation using arrays. Now consider Step 3. Let A be the number of iterations. Then by the above Conclusion $A \leq n+1$. The cost for Step 3a is $O(dlog(n))$ since a binary search takes $O(log(n))$ operations. The cost for Step 3b is $O(d)$. Thus the total cost for the A iterations of Steps 3a and 3b is $O(dnlog(n))$.

The cost for Step 3c is $O(d \;\; \Sigma_{1 \leq j \leq d} Z(j, b))$ where

$$Z(j, b) = t[b+1, j] - t[b, j] + \;\; \Sigma_{t[b,j]+1 \leq p \leq t[b+1,j]} G(j, p, b).$$

Here $G(j, p, b)$ is the number of the executions of the "then" statement in Step 3c (resource deallocation and marking TRUE/done). Hence the total cost for the A iterations of Step 3c is at most proportional to

$$d\Sigma_{0 \leq b \leq A-1}\Sigma_{1 \leq j \leq d} Z(j, b)$$
$$= d\Sigma_{1 \leq j \leq d}\Sigma_{0 < b \leq A-1} Z(j, b)$$
$$= O(d^2n) + d\Sigma_{b,j,p} G(j, p, b) = O(d^2n) + dn$$
$$= O(d^2n).$$

We used the fact that $\Sigma_{b,j,p} G(j,p,b) \leq n+1$, which is due to the fact that each process has at most one chance to be deallocated and marked "TRUE". Thus the total cost of the loop is $O(dn(log(n)+d))$. The total cost of the algorithm is the sum of costs for the three steps; it is $O(dn(log(n)+d))$.

IV ASYMPTOTIC PROPERTIES

We are interested in the behavior regarding deadlock detection for systems with large numbers of processes. Similar foci have been seen in other areas, e.g., [11, 16]. In many situations, it is reasonable to assume that d, the number of resource types, is relatively small comparing to n, the number of processes. The main results are readily obtained, and are summarized in the following proposition.

Proposition. If $d = O(log(n))$ then the complexity of the problem of deadlock detection for systems consisting of n processes and d reusable resource types satisfies

$$comp(d, n) = O(nlog^2(n)).$$

Moreover if $d = O(1)$ then

$$comp(d, n) = O(nlog(n)).$$

The Dijkstra algorithm for deadlock detection has cost proportional to dn^2. Intuitively, even in the average case, its cost is roughly $d*n/2 + d*(n-1)/2 + \ldots + 1 = \Theta(dn^2)$.

Thus, asymptotically, our algorithm is $n/log(n)$ times less costly than Dijkstra's algorithm, for any fixed d, or for that matter, as long as $d = O(log(n))$. Savings can be more dramatic if we assume that d is a constant, and that the request vectors are bounded by a constant vector. This will be discussed shortly.

The algorithm we presented detects deadlock for systems with arbitrary numbers of processes and resource types, and arbitrary request vectors and allocation vectors. In application, these assumptions can be too general to get a more efficient approach. Now we assume the number of resource types d is a constant and the request vectors are bounded by a constant vector. A special case of this setting is that the constant vector consists of 1s, which roughly corresponds to the case that there are fixed number of resources types and that every resource has only one unit in total, which is typical for critical section or locking applications.

As one may expect, such systems can be checked for deadlock in linear time. In fact, a linear time algorithm can be easily obtained from algorithm J. The major step is to use a counting sort (see e.g., [2]) in Step 1 in algorithm J. Here we shall not give a detailed account of the algorithm. However, we point out that we can significantly simplify the data structures used in algorithm J, which makes implementation even more efficient. Namely, we do not need the dummy processes, or the matrix $[t[i, j]]$. We however need two vector variables. By doing so, the variable b can be removed as well; see the previous Remark. Secondly, we do not need the permutations η_{ij}. They are computed on the fly.

V CONCLUSION AND PERSPECTIVE

We have presented a new algorithm for deadlock detection for systems with reusable resource types. We stress that our analysis has been focused on the asymptotic properties, for systems with large numbers of processes.

The novel algorithm we designed is $n/log(n)$ faster than the classic Dijkstra algorithm. The algorithm utilizes simple data structures, and thus can be easily incorporated into implementations. For systems with bounded requests and a fixed number of resource types, we derived from the algorithm a linear-running-time algorithm, which is obviously almost optimal.

As mentioned in the Introduction, safe state detection in the Banker's algorithm for deadlock avoidance is mathematically equivalent to the problem of deadlock detection. Therefore, the cost for Banker's algorithm can be controlled in $O(n^2log^2(n))$ when $d = O(log(n))$ and algorithm J is used in safe state detection, or $O(n^2)$ if d is a constant and the linear-time algorithm is used in safe state detection.

REFERENCE

[1] E.G. Coffman, M.J.Elphick, and A. Shoshani, "System Deadlocks," *Computing Surveys*, 3.2, 67-78, 1971.

[2] T.H. Cormen, C.E. Leiserson, and R.L. Rivest, *Introduction to Algorithms,* The MIT Press, Cambridge, MA, 1998.

[3] E.W. Dijkstra, "Cooperating Sequential Processes," *Technical Report*, Technological University, Eindhoven, the Netherlands, 43-112, 1965.

[4] D. Engler, and K. Ashcraft, "RacerX: Effective, Static Detection of Race Conditions and Deadlocks," SOSP 2003, Bolton Landing, New York, 2003.

[5] E.M. Gold, "Deadlock Prediction: Easy and Difficult Cases," *SIAM Journal of Computing*, Vol. 7, 320-336,1978.

[6] A.N. Habermann, "Prevention of System Deadlocks"," *Communications of the ACM*, 12.7, 373-377, 385, 1969.

[7] R.C. Holt (1971), "Comments on Prevention of System Deadlocks," *Communications of the ACM*, 14.1, 179-196, 1971.

[8] R.C. Holt, "Some Deadlock Properties of Computer Systems," *Computing Surveys*, 4.3, 179-196, 1972.

[9] A.J. Jammel, and H.G. Stiegler, "On Expected Costs of Deadlock Detection," *Information Processing Letters 11,* 229-231, 1980.

[10] T. Li, C.S. Ellis, A.R. Lebeek, and D.J. Sorin, "Pulse: A Dynamic Deadlock Detection Mechanism Using Speculative Execution," *Proceedings of the 2005 USENIX Annual Technical Conference*, Anaheim, California, April 10-15, 2005.

[11] P.B. Miltersen, M. Paterson, and J. Tarui, "The Asymptotic Complexity of Merging Networks," *Journal of the ACM*, 43.1, 147-165, 1996.

[12] T. Minura , "Testing Deadlock-Freedom of Computer Systems," *Journal of the ACM*, 27.2, 270-280, 1980.

[13] A. Silberschatz, P.B. Galvin, and G. Gagne, *Operating System Concepts*, John Wiley \& Sons, Inc., NY, 2002.

[14] W. Stallings, *Operating Systems, Internals and Design Principles,* Prentice-Hall, NJ, 1997.

[15] Y. Suguyama, T. Araki, J. Okui, and T. Kasami, "Complexity of the Deadlock Avoidance Problem," *Trans. Inst. of Electron. Comm. Eng.* Japan J60-D, 4, 251-258, 1977.

[16] J.F. Traub, G.W. Wasilkowski and H. Wozniakowski, *Information-Based Complexity,* Academic Press, Inc., 1988.

On Path Selection for Multipath Connection*

Yu Cai
School of Technology
Michigan Technological University
Houghton, MI 49931
cai@mtu.edu

C. Edward Chow
Department of Computer Science
University of Colorado at Colorado Springs
Colorado Springs, CO 80933
chow@cs.uccs.edu

Abstract

Multipath connection, which utilizes the multiple paths between network hosts in parallel, has been used to improve the network performance, security and reliability. Path selection is a critical decision in a multipath connection network. Different selection results in significantly different result. In this paper, we present several heuristic algorithms including genetic algorithm to solve the path selection problem in a multipath connection environment. The reasons to choose genetic algorithm are because of its flexibility and extensibility when the context of problem changes. We define two objective functions and two constrains in this problem. The performance results of the proposed algorithms on the simulated network topology as well as a real-world network topology are presented. It is observed that genetic algorithm can produce satisfactory results within reasonable execution time.

1. Introduction

Multipath connection, which utilizes the potential multiple paths between network hosts and aggregates the available bandwidth of these paths, is used to improve network performance, security and reliability. In a multipath system, the traffic from a source is spread over multiple paths and transmitted in parallel through the network. Therefore, multipath connection not only can increase the network aggregate bandwidth, but also cope well with network congestion, link breakage and potential attacks.

In [1], [2], [3], the author proposed a new multipath connection approach named Proxy Server based Multipath Connection (PSMC), in which the multiple paths are set up via a set of intermediate connection relay proxy servers. The proxy servers examine the incoming packets and forward them to the appropriate destination. PSMC is an application layer overlay network.

One of the key issues in a multipath system is path selection, or proxy server selection in a PSMC network (We will mix the usage of path selection and proxy server selection in this paper). There might be a large number of proxy servers available, we need to select the "optimal" subset of proxy servers from them to achieve the maximum aggregate bandwidth. Different server selections result in significantly different performance [16]. Therefore, server selection is a critical decision in PSMC.

When the network paths are disjointed, the network reliability is improved, the available throughput increases, the traffic along the paths are load-balanced and least likely to be correlated.

There are two types of disjoint: link disjoint if no common links between paths, and node disjoint if no common nodes between paths besides the end host nodes. In general a link-disjoint paths algorithm can be extended to a node-disjoint algorithm with node splitting [4, 5]. In this paper, we focus on finding link-disjoint paths.

In the real world scenarios, there may not exists disjoint paths between two given end hosts, because the paths are likely to share some common links on the edge of the Internet. We define a disjoint function between path i and j as follows [13].

Disjoint = 1 - (total number of shared links) / (total number of links)

We can also use bandwidth or latency instead of the number of links in the disjoint function definition. The disjoint of path set is defined as the sum of the disjoint function of each pair of paths in the set.

The first objective function in PSMC path selection is to maximize the aggregate bandwidth, then maximize the disjoint function of paths.

The second objective function is to maximize the linear combination of bandwidth and disjoint function, (bandwidth + α*disjoint function), here α is a parameter. It makes sense because sometimes it is hard to tell which is a better selection when we separate the bandwidth and disjoint function. For example, given a path pair of 50Mb/s and 0.98 disjoint (smaller bandwidth larger disjoint), and a path pair of 60Mb/s and 0.7 disjoint (larger bandwidth smaller disjoint), it is hard to tell which path pair is better. By using the second objective function then we can tell the difference.

There are also some constrains on the path selection. The experimental results in PSMC show that the bandwidth distribution among the selected multiple paths has significant impact on the overall system performance [3]. A large-capacity link and a small-capacity link may have worse aggregate bandwidth than two moderate links. The experimental results suggest that if the bandwidth of a path

* This research work was supported in part by a NISSC AFOSR Grant award under agreement number F49620-03-1-0207.

K. Elleithy (ed.), Advances and Innovations in Systems, Computing Sciences and Software Engineering, 513–518.

is smaller than 1/10 of the average of the bandwidth, then this path is treated as "bad" path and should be eliminated.

Another constrain is the number of paths selected. The experimental results in PSMC show that the total number of paths should be smaller than 10 [3]. The bandwidth gain over 10 paths is limited, or even become negative.

With the objective functions and constrains, the path selection in a multipath system is NP-Complete. Heuristic algorithm is a feasible solution. In this paper we propose to use genetic algorithm and a greedy algorithm to solve this problem.

This paper is structured as follows. Section 2 reviews the related works. Section 3 defines the problems and presents the algorithms. Section 4 discusses the performance results. Section 5 is the conclusion.

2. Related Works

The problem of finding disjoint paths in a network has been given much attention in the literature. Various methods have been devised to find a pair of shortest link-disjoint paths with minimal total length [5, 6, 7, 8, 9]. In [7], Suurballe proposes an algorithm to find K node-disjoint paths with minimal total length using the path augmentation method. The path augmentation method is originally used to find a maximum flow in a network [14]. In [5], the authors improved Suurballe's algorithm such that pairs of link-disjoint paths from one source node to n destination nodes could be efficiently obtained in a single Dijkstra-like computation. In general, this type of problems can be solved in polynomial time [6].

However, similar problems with additional multiple constrains become NP-Complete [4, 10, 11]. For example, if requiring the maximal length of the two disjoint paths to be minimized, then the problem becomes NP-Complete [11]. Heuristic algorithms based on matrix calculation like [12] have been proposed.

An optimal algorithm for finding K-best paths between a pair of nodes is given by Lee and Wu in [15], where they transfer the K-best paths problem into a maximum network flow and minimum cost network flow algorithm via some modifications to the original graph. Distributed algorithms for the link/node-disjoint paths algorithms can be found in [8].

Another related area is server selection problem or server placement problem. In [22], the author proposes a solution by setting the graph equal to a tree topology. The algorithm was originally designed for web proxy cache placement. However, the computation time of this algorithm is high, and it is not always true to map network topology to a tree topology. In [23], the author proposes a greedy algorithm by placing replicas on the tree iteratively in a greedy fashion without replacing already placed replicas.

3. NP-hardness

In this section we study the NP-hardness of path selection problem in multipath connection.

1) **Max-flow problem**. The problem of finding maximum aggregate bandwidth in a given network is close to the classic maximum flow problem [14] and can be solved in polynomial time. The basic idea is to use augmenting path and labeling scheme.

2) **Best k-path problem**. The best k-path problem refers to the problem of finding k disjoint path whose aggregate bandwidth is maximum. This problem is close to the K-best paths problem in [15] and can be solved in polynomial time. The basic idea is to transfer the K-best path problem into a maximum network flow problem.

3) **Max-flow, min-size disjoint problem**. We study the problem of finding a set of disjoint paths, whose aggregate bandwidth is maximum, and the set size is minimum. This can be solved in polynomial time based on problem1 and 2.

Instance: A graph G=(V, E), with capacity $BW_{i,j}$ associated with each edge $e_{i,j}$, a source node s, a destination node d,

Question: is there a set of node-disjoint paths $\{p_1, p_2, \ldots p_k\}$ (set size is K) from s to d such that $\sum_{p_i \in S} BW(p_i)$ is maximum and K is minimum.

Theorem 1: The max-flow, min-size disjoint problem problem can be solved in polynomial time.

Proof: First, we can use the solution of problem 1 to find the maximum value of $X = \sum_{p_i \in S} BW(p_i)$ for all possible path set in at most $O(n^2)$ time (n is the number of vertex in G), assuming $\{p_1, p_2, \ldots p_m\}$ is a possible solution.

Second, we use the solution of problem 2 to find the maximum value of best (m-1), (m-2), (m-3) … 2 paths. Assuming $\{q_1, q_2, \ldots q_h\}$ is a possible solution for best h paths (h=m-1, m-2, …, 2), and Y is the maximum aggregate bandwidth of $\{q_1, q_2, \ldots q_h\}$.

If $Y<Z$, then omit this solution because it can not reach the bandwidth upper limit.

If $Y=Z$, then $\{q_1, q_2, \ldots q_h\}$ is a better solution with smaller set size h. We repeat the process until we can not find a smaller size.

The complexity of best k path algorithm is at most $O(n^2)$.

Therefore, this problem can be solved in $O(n^2)$ time.

4) **Max-flow, min-slowest path disjoint problem**. In multipath environment, path latency is also an important factor. It is critical for some real time multimedia applications.

When selecting multiple paths, we usually want to find a set of disjoint paths to achieve maximum aggregate bandwidth, at the same time, the path with longest latency in this set is minimum among all possible path sets.

Instance: A graph G=(V, E), with capacity $BW_{i,j}$ and latency $L_{i,j}$ associated with each edge $e_{i,j}$, a source node s, a destination node d, path latency is $L(P) = \sum_{e_{i,j} \in p} L(l_{i,j})$.

Question: is there a set of node-disjoint paths $\{p_1, p_2, \ldots p_k\}$ from s to d such that $\sum_{p_i \in S} BW(p_i)$ is maximum, and the maximum path latency $L(P_m)$ (m=1...k) is minimum. Or given a non-negative number X, is the latency of the slowest path $L(P_m)$ is less than or equal to X?

Theorem 2: The max-flow, min-slowest path disjoint problem is NP-Complete.

Proof: We show the problem is NP-complete by giving a transformation from a well known NP-complete problem – the Partition problem [24].

The partition problem: given a finite set A and a size $s(a) \in Z^+$ for each $a \in A$, is there a subset $A' \subseteq A$ such that

$$\sum_{a \in A'} s(a) = \sum_{a \in A - A'} s(a)$$

It is easy to see that this problem is NP, since a non-deterministic algorithm can get a set of disjoint paths with maximum aggregate bandwidth and check if the length of the longer path is less than or equal to X.

Let's construct an instance of the problem with a graph G=(V,E), with the following characteristic:

1) (3n+2) nodes, s is source node, d is destination node
2) $V = \{v_i, i=1\ldots2n\} \cup \{u_i, i=1\ldots n\} \cup \{s, d\}$
3) The edges are:
   a) (s, v_1), (s, u_1), (v_{2n}, d), (u_n, d)
   b) (v_i, v_{i+1}), i=1...2n-1
   c) (u_i, u_{i+1}), i=1...n-1
   d) (v_{2i}, u_{i+1}), i=1...n-1
   e) (u_i, v_{2i+1}), i=1...n-1
4) The bandwidth on all edges $BW_{i,j} = 1$
5) The latency is as follows:
   a) $L(v_{2i-1}, v_{2i}) = s(a_i)$, i=1...n
   b) The Latency of every other edge is 0 (very small compared with $s(a_i)$)
6) The $X = 1/2 * \sum_{ai \in A} s(ai)$

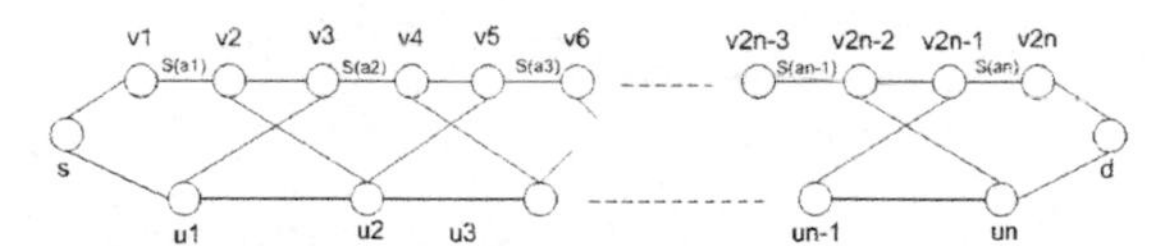

Figure 1: Max-flow, min-slowest path disjoint problem

It is easy to see that in Figure 1 there are at most two disjoint paths from s to d, and the maximum aggregate bandwidth is 2. We need to show that instance of the original problem will have two disjoint paths from s to d of length at most X, if and only if elements of the instance of the Partition problem can be divided into two groups, such that the sums of these two groups are equal.

First, suppose that the set A can be divided into A' and A-A' such that $\sum_{a \in A'} s(a) = \sum_{a \in A - A'} s(a)$. In this case we need to show that we can construct two disjoint paths from s to d that the longest latency is at most $1/2 * \sum_{ai \in A} s(ai)$.

Suppose $A' = \{a_{f(1)}, a_{f(2)} \ldots a_{f(m)}\}$, $f(1) < f(2) < \ldots < f(m)$.

The two paths can be constructed as follows.

Path 1: $s \rightarrow u_1 \rightarrow \ldots \rightarrow u_{f(1)-1} \rightarrow v_{2f(1)-1} \rightarrow v_{2f(1)} \rightarrow u_{f(1)+1} \rightarrow \ldots \rightarrow u_{f(2)-1} \rightarrow v_{2f(2)-1} \rightarrow v_{2f(2)} \rightarrow u_{f(2)+1} \rightarrow \ldots \rightarrow u_{f(m)-1} \rightarrow v_{2f(m)-1} \rightarrow v_{2f(m)} \rightarrow u_{f(m)+1} \rightarrow \ldots \rightarrow u_n \rightarrow d$.

Path 2: $s \rightarrow v_1 \rightarrow \ldots \rightarrow v_{2f(1)-2} \rightarrow u_{f(1)} \rightarrow v_{2f(1)+1} \rightarrow v_{2f(1)+2} \rightarrow \ldots \rightarrow v_{2f(2)-2} \rightarrow u_{f(2)} \rightarrow v_{2f(2)+1} \rightarrow v_{2f(2)+2} \rightarrow \ldots \rightarrow v_{2f(m)-2} \rightarrow u_{f(m)} \rightarrow v_{2f(m)+1} \rightarrow v_{2f(m)+2} \rightarrow \ldots \rightarrow v_{2n} \rightarrow d$.

Figure 2 is an instance of the path selection, here $A' = \{a_1, a_3 \ldots a_n\}$. The edges with bold lines are the selected edges.

We can verify that the path latency is X and the aggregate bandwidth is 2.

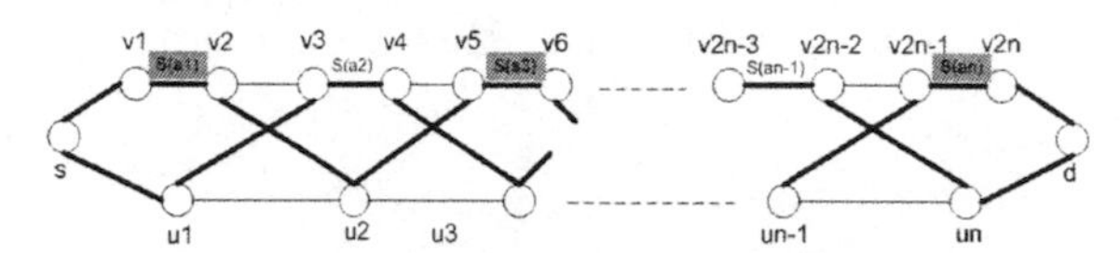

Figure 2: An instance of path selection

Second, now suppose that there are two disjoint path P1 and P2 from s to d, the latency of slower path P2 is at most X and the aggregate bandwidth is 2.

It is easy to see that for edge (v_{2i-1}, v_{2i}) with latency $s(a_i)$, i=1...n, it must be part of P1 or P2, and can not be in P1 and P2 at the same time. So L(P1) + L(P2) =X.

We also have L(P1) <= L(P2) and L(P2) <=1/2*X, therefore, L(P1) = L(P2) = 1/2*X.

Assuming P1 contains edges with latency of $a_{f(1)}, a_{f(2)} \ldots a_{f(m)}$, $f(1) < f(2) < \ldots < f(m)$, then we can set $A' = \{a_{f(1)}, a_{f(2)} \ldots a_{f(m)}\}$, and the sums of two groups A' and A-A' are equal 1/2*X.

Therefore, we prove the theorem.

5) **Max-flow, min-jointness problem**. In the real world scenarios, there may not exists disjoint paths between two given end hosts, because the paths are likely to share some common links on the edge of the Internet. Another scenario is as follows: two end hosts are in China and US respectively, since there are limited gateways connection between China and International network [25], therefore, the multiple paths between China and US are likely to share some common links in the middle.

We define a jointness and disjoint function as follows [13].

Jointness = (total number of shared links) / (total number of links)

Disjoint = 1 - (total number of shared links) / (total number of links) = 1 - jointness

We can also use bandwidth or latency instead of the number of links in the function definition.

We study the problem of selecting a set of paths to achieve maximum aggregate bandwidth with minimum jointness.

Instance: A graph G=(V, E), with capacity $BW_{i,j}$ associated with each edge $e_{i,j}$, a source node s, a destination node d.

Question: is there a set of paths $\{p_1, p_2, \ldots p_k\}$ from s to d such that $\sum_{p_i \in S} BW(p_i)$ is maximum, and the jointness of $\{p_1, p_2, \ldots p_k\}$ is minimum.

Theorem 3: The max-flow, min-jointness problem is NP-Complete.

To proof theorem 3, similar methods in the proof of theorem 2 can be used. Due to the page limitation, the proof is omitted.

3. Path selection

3.1 Network Model

By using networking measurement tools, like traceroute[19], pathchar[17], cprobe[18], we can obtain the IPs of the routers along a path and estimate the bandwidth or latency on each network link. Therefore we can get the network topology between two end hosts. Extensive works have been done in this field [13, 17, 18, 19].

For simplicity, in this paper we assume the network topology and bandwidth on network links are known; the network forwarding route and return route are the same. We model the problem as followings.

Let G = (V, E) be a graph which models the network topology. V represents the set of nodes including proxy server nodes, end host nodes, and the routers in between; E represents the set of edges or link segments that connect the nodes in the network.

Let P be the set of proxy servers. For a proxy server p_i, $BW(p_i)$ represents the available bandwidth of the indirect route via p_i. For an edge e, BW(e) represents the available bandwidth on edge e. Assume the route via p_i consists of a set of edges $e_{i\_1},\ldots,e_{i\_mi}$, then $BW(p_i) = \min\{BW(e_{i\_1})\ldots BW(e_{i\_mi})\}$

Given a sender s, a receiver r and a set of proxy servers p in G, let S be the subset of proxy servers that we selected, $S \subseteq \{p_1,\ldots p_n\}$. The problem is to find a subset S to maximize the objective functions and meet constrains.

There are two types of objective functions:

a) Maximize $\sum_{p_i \in S} BW(p_i)$ then $\sum_{p_{i,j} \in S} Disjoint(i, j)$

b) Maximize $\sum_{p_{i,j} \in S} (\alpha * BW(p_i) + Disjoint(i, j))$, here α is a parameter.

There are two constrains:

a) $BW(p_i) > T * \sum_{p_i \in S} BW(p_i) / Sizeof(S)$, here T is a parameter set to be 1/10.

b) $Sizeof(s) < W$, here W is a parameter set to be 10.

The proposed problem above is a NP-complete problem[4].

3.2) Genetic Algorithm

We proposed to use genetic algorithm to solve this problem. The reasons why to choose genetic algorithm are as follows.

a) It provides more flexibility and extensibility on this problem. If the objective functions and constrains are later changed, we can easily modify the fitness function in genetic algorithm to accommodate such changes. Other heuristic algorithms may require more significant modification under such circumstances.

b) It provides better scalability. The execution time of genetic algorithm scales well with regard to the network size.

c) It provides more controls for the end users. It can easily produce multiple outputs and give end users the opportunity for further selection.

The disadvantages of genetic algorithm are as follows.

a) It is a heuristic algorithm and can not always give the optimal answer.

b) The execution time might be long for a small scale network.

We implement a fix-length genetic algorithm in which the length of chromosomes is fixed, and a variable-length genetic algorithm in which the length of chromosomes can change.

The genetic algorithm works as below.

1) Assign sequential server number, node number and path number to denote each proxy server, node and path. Assign the initial bandwidth to each path.
2) Initialize the first generation of chromosomes by filling server number in chromosome.
3) Crossover and mutation at certain probability. Make sure no duplicated server in chromosome, and the length of chromosome is less than the given upper limit. Several different crossover and mutation methods have been combined together for better performance [20].

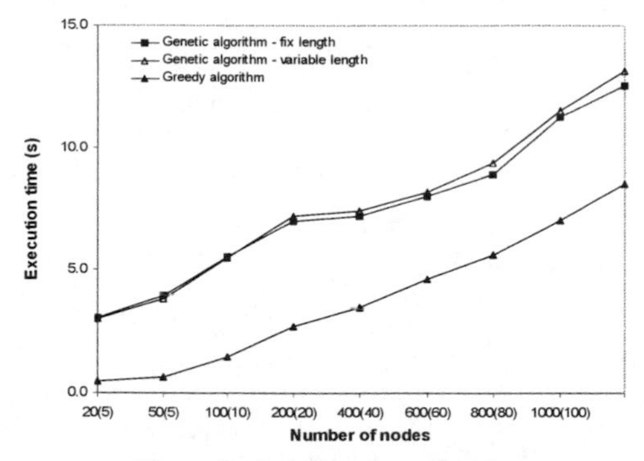

Figure 3: algorithm execution time

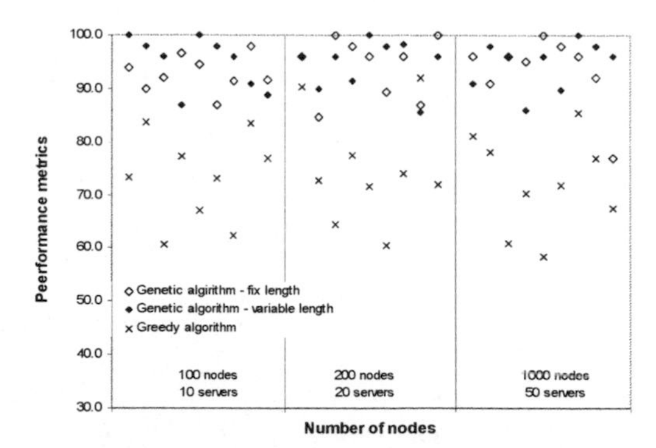

Figure 4: algorithm running results

4) Fitness function. For a given chromosome, use the objective function as fitness function, and check constrains.

5) Run certain generations, and output the result.

3.3) Greedy Algorithm

We also proposed another heuristic algorithm to solve the path selection problem as follows.

1) Initialize the data set.

2) Pick the path with the maximum bandwidth and check if it is disjoint from other selected paths, and all constrains are met.

3) If yes, select this path, otherwise select the path with the next maximum bandwidth.

4) Loop until no path is available to select.

This is a greedy algorithm. As we can see from the algorithm, the execution time will be small, but it may not always yield good selection result. Our experimental results prove this.

4. Results Analysis

We tested the proposed algorithms on simulated network topologies as well as a real-world network topology.

GT-ITM [21], which is one of the most commonly used internet topology models, is used to generate network topologies of various sizes for evaluating the performance of the proposed algorithms. We create transit-stub graphs with 20-100 nodes. We randomly pick 10% nodes as proxy servers, and two nodes as end host nodes. We randomly generate background traffic whose average is 60% of the total network bandwidth.

Figure 3 shows the algorithm execution time vs. the simulated network size. The x axis notation 20(10) means there are 20 nodes plus 10 proxy server nodes. It is observed that the execution time of both algorithms increases close to liner when the size of network increases. The greedy algorithm has lower execution time, but as we can see from Figure 3, the running result is not satisfactory.

Figure 4 shows the algorithm running results. The notation on the chart "100 nodes, 10 servers" means there are 100 nodes in the network plus 10 proxy server nodes. We can observe from the chart that the genetic algorithm can yield satisfactory result (80-100), which are %close to the optimal result (100). There is no significant difference between fix-length genetic algorithm and variable-length algorithm. It is also observed that the greedy algorithm can not yield good result (60-80).

Figure 5 shows the real network topology from a node at University of Colorado at Colorado Springs (UCCS) to the selected Redhat mirror servers. This topology can be viewed as half of a PSMC network, with Redhat mirror sites selected as proxy server. We use the topology and perform some tests on our algorithm. Table 1 shows the running result. It is observed that the execution time of all algorithms are in acceptable range, but the running result of greedy algorithm is far below the optimal result (62% of optimal), while genetic algorithm always yield acceptable result.

Table 1: running results on a real-world topology

Algorithm	Execution time (s)	Running result / Optimal result
Genetic algorithm – fix length	7.1	90%
Genetic algorithm – variable length	7.2	88%
Greedy algorithm	3.1	62%

5. Conclusion

Multipath connection is a topic gaining interest. Path selection is a critical decision in a multipath connection network. In this paper, we present genetic algorithm to solve the path selection problem in a multipath connection environment. Genetic algorithm has better flexibility and extensibility when the context of problem changes. From the performance result, it is observed that genetic algorithm can produce satisfactory results within reasonable execution time.

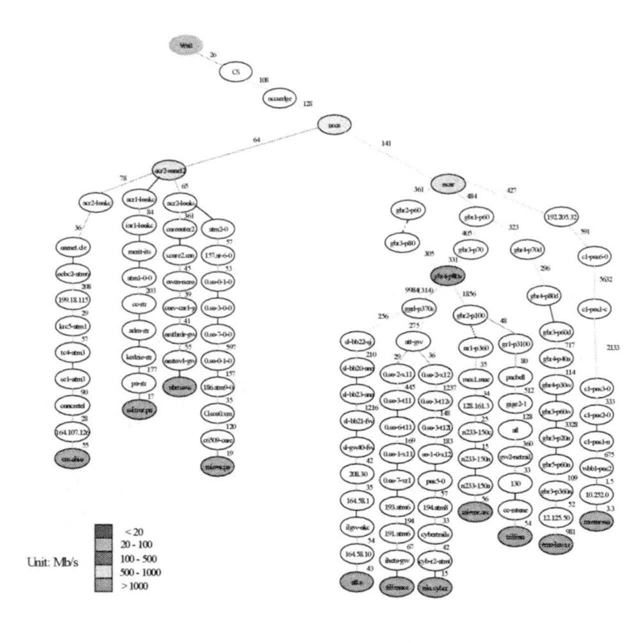

Figure 5: network topology from a node at UCCS to the selected Redhat mirror servers

References

[1] Edward Chow, Yu Cai, David Wilkinson, and Ganesh Godavari, "Secure Collective Defense System", In Proceedings of GlobeCom 2004.

[2] David Wilkinson, Edward Chow and Yu Cai, "Enhanced Secure Dynamic DNS Update with Indirect Route", In Proceedings of the IEEE Workshop on Information Assurance, 2004.

[3] Yu Cai, Edward Chow, and Frank Watson, "On Proxy Server based Multipath Connection", technical report, http://cs.uccs.edu/~scold/psmc.pdf, submitted to GlobeCom 2005.

[4] Guo, Y., F. A. Kuipers and P. Van Mieghem, "Link-Disjoint Paths for Reliable QoS Routing", International Journal of Communication Systems, vol. 16, pp. 779-798, 2003.

[5] J.W. Suurballe and R.E. Tarjan, "A Quick Method for Finding Shortest Pairs of Disjoint Paths", Networks, Vol. 14, pp. 325-333, 1984.

[6] R. Bhandari, "Optimal Diverse Routing in Telecommunication Fiber Networks", Proc. IEEE INFOCOM 1994, Toronto, Ontario, Canada, Vol.3, pp.1498-1508, June 1994.

[7] J.W. Suurballe, "Disjoint Paths in a Network", Networks, Vol. 4, pp. 125-145, 1974.

[8] R. Ogier and N. Shacham, "A distributed algorithm for finding shortest pairs of disjoint paths," in Proceedings of IEEE INFOCOM '89

[9] Deepinder Sidhu , Raj Nair , Shukri Abdallah, "Finding disjoint paths in networks", Proceedings of the conference on Communications architecture & protocols, p.43-51, 1991, Zurich, Switzerland

[10] Z. Wang and J. Crowcroft, "QoS Routing for supporting Multimedia Applications", IEEE J. Selected Areas in Communications, Vol. 14, No.7, pp. 1228-1234, September 1996.

[11] C-L Li, S.T. McCormick, D. Simchi-Levi, "The complexity of finding two disjoint paths with min-max objective function", Discrete Applied Mathematics, Vol. 26, No. 1, pp. 105-115, January 1990.

[12] E. Oki and N. Yamanaka, "A recursive matrix calculation method for disjoint path search with hop link number constraints", IEICE Trans. Commun., Vol. E78-B, No.5, pp. 769-774, May 1995.

[13] M. Zhang, et. al., "A Transport Layer Approach for Improving End-to-End Performance and Robustness Using Redundant Paths", In Proc. of the USENIX 2004 Annual Technical Conference. 2004.

[14] C.H. Papadimitriou and K. Steiglitz, "Combinatorial Optimization . Algorithms and Complexity", Prentice-Hall, Inc., Englewood Cliffs, New Jersey, 1982.

[15] S.W. Lee and C. S. Wu, ."A k-best paths algorithm for highly reliable communication networks"., IEICE Trans. on Commun., Vol. E82-B, No.4, pp.586-580, April 1999.

[16] Pablo Rodriguez, Andreas Kirpal, Ernst W. Biersack, "Parallel-Access for Mirror Sites in the Internet", Proceeding of Infocom, 2000

[17] pathchar, http://www.caida.org/tools/utilities/others/pathchar/

[18] cprobe, http://cs-people.bu.edu/carter/tools/Tools.html

[19] traceroute, http://www.traceroute.org/

[20] John R. Koza, "Genetic Programming", MIT Press, 1992.

[21] Ellen W. Zegura, "GT-ITM: Georgia Tech Internetwork Topology Models", http://www.cc.gatech.edu/projects/gtitm/

[22] B. Li, M. J. Golin, G. F. Ialiano, and X. Deng, "On the optimal placement of web proxies in the internet," Proc. of IEEE INFOCOM, Mar. 1999.

[23] P. Krishnan, D. Raz, and Y. Shavitt, "The cache location problem," ACM/IEEE Transactions on Networking, vol. 8, no. 5, Oct. 2000.

[24] M. Gary, D. Johnson, "Computers and intractability, a guide to the theory of NP-completeness", W.H. Freeman Press 1979

[25] Internet in China, http://austlii.edu.au/~graham/hkitlaw/Choy_and_Cullen.html

Some Results on the Sinc Signal with Applications to Intersymbol Interference in Baseband Communication Systems

Zouhir Bahri
Electrical Engineering Dept., University of Bahrain
P.O. Box 32038, Isa Town, Bahrain

***Abstract*- Some useful results related to the Sinc signal are presented. They are derived using Fourier Series Decomposition and Parseval's Identity. A simple convergence analysis is provided. These results should be useful in many practical situations involving band-limited or time-limited signals. This is illustrated by examples dealing with bandwidth requirements in Baseband Data Communication Systems in the presence of Additive Noise, Intersymbol Interference, and Timing Problems. Using simple trigonometric identities, a larger generalized set of new infinite series results related to the sinc signal is also obtained.**

I. INTRODUCTION

Whenever a signal is windowed in time or frequency, it gets either multiplied or convolved with a *Sinc* function depending on which domain it is represented in. This basic result makes the Sinc quite frequently encountered in a broad range of applications involving finite bandwidth systems (eg. Intersymbol Interference (ISI) in digital communication systems [1]) or finite duration signals (eg. edge effects due to segmentation in speech signals [2]). The Sinc signal is also found in the Power Spectral Densities (PSD) of practically all digital modulation techniques (e.g., ASK, BPSK, QPSK, OQPSK, FSK, MSK) [1]. It is also found in digital systems, either in Analog-to-Digital (A/D) conversion with Zero-Order Hold, or in perfect Digital-to-Analog conversion, as it ideally interpolates the samples to produce the original analog signal[3]. The Sinc signal has different definitions in the literature. In our work, we use

$$\mathrm{sinc}(x) = \frac{\sin(\pi\, x)}{\pi\, x}, \tag{1}$$

and it is shown in Fig. 1. It also finds importance in the sense that in the limiting case, it tends towards the impulse function [3] , i.e.,

$$\lim_{\gamma \to \infty} \gamma\ \mathrm{sinc}(\gamma\ \mathrm{x}) = \delta(\mathrm{x}). \tag{2}$$

It is the intent of this paper to present some infinite series results related to the Sinc signal. Such results should provide more tools in dealing with a variety of practical applications. Even though one of our results (namely (13) below) may be found elsewhere [4], we use a different way of obtaining it and propose an alternative approach to establishing a more generalized set of identities.

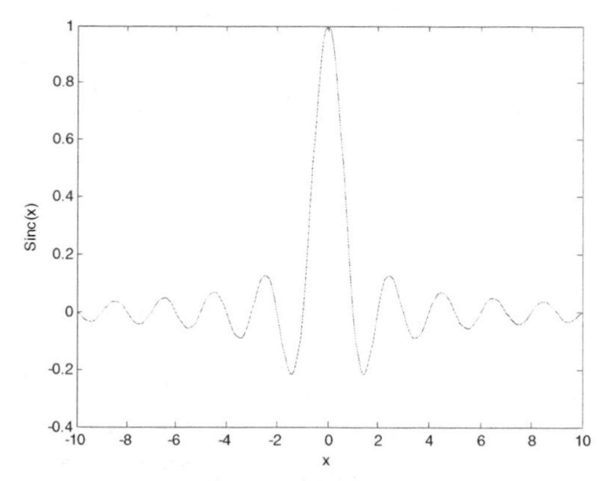

Figure 1 : The Sinc Signal

The remainder of this paper is divided as follows. In Section 2, we provide the derivations of the results related to the Sinc function. Section 3 presents a simple convergence analysis partly to justify a step in the derivations. In Section 4, we illustrate the usefulness of the derived results with the help of two examples dealing with bandwidth requirements in Baseband Data Communication Systems in the presence of Additive Noise, Intersymbol Interference (ISI), and timing problems. Finally we summarize our work in Section 5.

II. DERIVATIONS

In deriving the proposed results, we shall make use of Fourier Series decomposition of a periodic waveform along with Parseval's Identity[3]. Fig. 2 shows a periodic waveform of period T that consists of a train of pulses each of width $2\alpha T$ ($0 < \alpha < 1/2$). It is easy to check that the Fourier Series decomposition of that signal is

$$x(t) = 2A\alpha + 4A\alpha \begin{bmatrix} sinc(2\alpha)cos(\omega_0 t) + sinc(4\alpha)cos(2\omega_0 t) + \\ sinc(6\alpha)cos(3\omega_0 t) + ... \end{bmatrix}, \tag{3}$$

where $\omega_0 = 2\pi/T$. By equating the average power of both sides of (3) (Parseval's Identity[3]), we get

$$2A^2\alpha = 4A^2\alpha^2 + 8A^2\alpha^2 \sum_{n=1}^{\infty} sinc^2(2n\alpha). \tag{4}$$

K. Elleithy (ed.), *Advances and Innovations in Systems, Computing Sciences and Software Engineering*, 519–524.

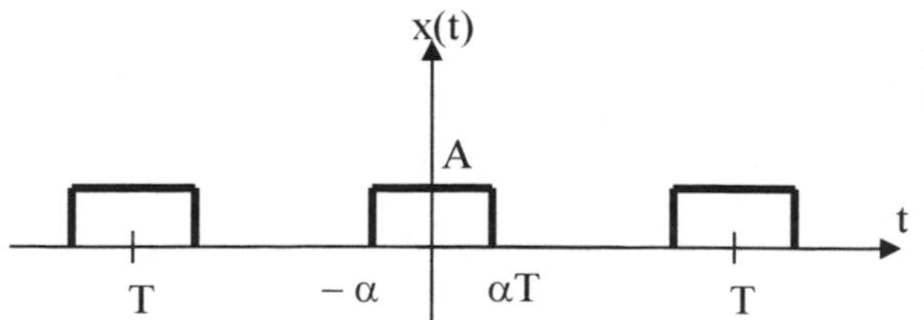

Figure 2. Periodic Waveform used in the Fourier Series Decomposition.

A simple manipulation of (4) leads to the result

$$\sum_{n=1}^{\infty} sinc^2(n\alpha) = \frac{1}{2}\left(\frac{1-\alpha}{\alpha}\right), \quad 0<\alpha\le 1 \quad (5)$$

$\Leftrightarrow$

$$\sum_{n=0}^{\infty} sinc^2(n\alpha) = \frac{1}{2}\left(\frac{1+\alpha}{\alpha}\right), \quad 0<\alpha\le 1 \quad (6)$$

$\Leftrightarrow$

$$\sum_{n=-\infty}^{\infty} sinc^2(n\alpha) = \frac{1}{\alpha}, \quad 0<\alpha\le 1, \quad (7)$$

where we have used the fact that sinc(0)=1 and sinc(-x)=sinc(x) in (6) and (7). (The limiting case α=1 in (5)-(7) follows by straightforward inspection). Now it is easy to check that

$$\frac{d\, sinc^2(n\alpha)}{d\alpha} = \frac{2}{\alpha}\left[sinc(2n\alpha) - sinc^2(n\alpha)\right]. \quad (8)$$

Hence, taking the derivative of both sides of (5), we get

$$\sum_{n=1}^{\infty} \frac{2}{\alpha}\left[sinc(2n\alpha) - sinc^2(n\alpha)\right] = -\frac{1}{2\alpha^2}, \quad 0<\alpha<1. \quad (9)$$

Using the result of (5) in (9), we get

$$\sum_{n=1}^{\infty} sinc(2n\alpha) = \frac{1}{2}\left(\frac{1-2\alpha}{2\alpha}\right), \quad 0<\alpha<1 \quad (10)$$

Hence, results very similar to (5)-(7) follow from (10), namely

$$\sum_{n=1}^{\infty} sinc(n\alpha) = \frac{1}{2}\left(\frac{1-\alpha}{\alpha}\right), \quad 0<\alpha<2 \quad (11)$$

$\Leftrightarrow$

$$\sum_{n=0}^{\infty} sinc(n\alpha) = \frac{1}{2}\left(\frac{1+\alpha}{\alpha}\right), \quad 0<\alpha<2 \quad (12)$$

$\Leftrightarrow$

$$\sum_{n=-\infty}^{\infty} sinc(n\alpha) = \frac{1}{\alpha}, \quad 0<\alpha<2, \quad (13)$$

Using simple trigonometric identities, more infinite series results involving the sinc (hence sines and cosines) can be easily obtained For example,

$$\sum_{n=1}^{\infty} n^2 sinc^4(n\alpha) = \frac{1}{4\pi^2\alpha^3}, \quad 0<\alpha\le\frac{1}{2}, \quad (14)$$

and more generally,

$$\sum_{n=1}^{\infty} n^{2p} sinc^{2p+2}(n\alpha) = \sum_{n=1}^{\infty} n^{2p} sinc^{2p+1}(n\alpha)$$
$$= \frac{1}{4\pi^{2p}\alpha^{2p+1}}\left[\frac{3}{4}\frac{5}{6}\frac{7}{8}\cdots\frac{2p-1}{2p}\right], \quad 0<\alpha<\frac{2}{2p+2}, \quad p=2,3,\ldots \quad (15)$$

Some more examples along with their generalizations are provided in Table 1. The derivations use straightforward trigonometric manipulations along with deductions and are omitted for the sake of conciseness.

III. CONVERENCE ANALYSIS

In this section we investigate the convergence of the series results given in (5)-(7) and (11)-(13). It is not difficult to observe that all these series results have a *non-uniform* convergence profile over the indicated interval of α. This follows from the fact that they all diverge for $\alpha = 0$. Hence, the closer α approaches zero, the slower the convergence to the indicated sums. This, however, does not invalidate the fact that we have differentiated both sides of (5). Uniform convergence is *sufficient* but *not necessary*. The earlier manipulation that led to (9) is justified by the fact that the series in (5) converges uniformly over *compact sets* of the indicated interval (this can be easily verified using Weierstrass -M test[5]). This is also justified by the fact that the series in (11) converges *in the mean* over *compact sets* of the indicated interval (this can be tested by directly applying the definition[5]).

Rather than dwell in the mathematical aspect of the problem, we now turn our attention to a simple numerical investigation of the convergence issue by evaluating the relative error resulting from taking partial sums of N terms from the series. For the summation in (11), this is given by

$$e_N(\alpha) = \left|\frac{\frac{1}{2}\left(\frac{1-\alpha}{\alpha}\right) - \sum_{n=1}^{N} sinc(n\alpha)}{\frac{1}{2}\left(\frac{1-\alpha}{\alpha}\right)}\right| = \left|\frac{2\alpha \sum_{n=N+1}^{\infty} sinc(n\alpha)}{(1-\alpha)}\right|. \quad (16)$$

A similar equation is obtained for the $sinc^2$ (.) series result in (5). We have numerically computed (16) for various values of α and plotted the convergence error as a function of N for some values of α (0.02, 0.9, and 1.9). This is shown in Fig. 3. As expected, we got slower convergence for values of α closer to zero. Hence, it is seen that the series corresponding to

TABLE1
Some more infinite series results with their generalizations

$\sum_{n=1}^{\infty}\frac{\cos(n\pi\alpha)}{(n\pi\alpha)^2}=\frac{1}{6\alpha^2}\left(\frac{3}{2}\alpha^2-3\alpha+1\right),\quad 0<\alpha\le 2$	$\sum_{n=1}^{\infty}\frac{\cos(n\pi\alpha)\sin(n\pi\alpha)}{n\pi\alpha}=\frac{1}{4\alpha}(1-2\alpha),\ 0<\alpha<1$
$\sum_{n=1}^{\infty}\frac{\cos^3(n\pi\alpha)}{(n\pi\alpha)^2}=\frac{1}{6\alpha^2}\left(\frac{9}{2}\alpha^2-\frac{9}{2}\alpha+1\right),\quad 0<\alpha\le\frac{2}{3}$	$\sum_{n=1}^{\infty}\frac{\sin^3(n\pi\alpha)}{n\pi\alpha}=\frac{1}{4\alpha},\quad 0<\alpha<\frac{2}{3}$
$\sum_{n=1}^{\infty}\frac{\cos^4(n\pi\alpha)}{(n\pi\alpha)^2}=\frac{1}{6\alpha^2}\left(6\alpha^2-\frac{9}{2}\alpha+1\right),\quad 0<\alpha\le\frac{1}{2}$	$\sum_{n=1}^{\infty}\frac{\sin^4(n\pi\alpha)}{(n\pi\alpha)^2}=\frac{1}{4\alpha},\quad 0<\alpha\le\frac{1}{2}$
$\sum_{n=1}^{\infty}\frac{\cos^5(n\pi\alpha)}{(n\pi\alpha)^2}=\frac{1}{6\alpha^2}\left(\frac{15}{2}\alpha^2-\frac{45}{8}\alpha+1\right),\quad 0<\alpha\le\frac{2}{5}$	$\sum_{n=1}^{\infty}\frac{\sin^5(n\pi\alpha)}{n\pi\alpha}=\frac{3}{16\alpha},\quad 0<\alpha<\frac{2}{5}$
$\sum_{n=1}^{\infty}\frac{\cos^6(n\pi\alpha)}{(n\pi\alpha)^2}=\frac{1}{6\alpha^2}\left(9\alpha^2-\frac{45}{8}\alpha+1\right),\quad 0<\alpha\le\frac{1}{3}$	$\sum_{n=1}^{\infty}\frac{\sin^6(n\pi\alpha)}{(n\pi\alpha)^2}=\frac{3}{16\alpha},\quad 0<\alpha\le\frac{1}{3}$
$\sum_{n=1}^{\infty}\frac{\cos^7(n\pi\alpha)}{(n\pi\alpha)^2}=\frac{1}{6\alpha^2}\left(\frac{21}{2}\alpha^2-\frac{105}{16}\alpha+1\right),\quad 0<\alpha\le\frac{2}{7}$	$\sum_{n=1}^{\infty}\frac{\sin^7(n\pi\alpha)}{n\pi\alpha}=\frac{5}{32\alpha},\quad 0<\alpha<\frac{2}{7}$
$\sum_{n=1}^{\infty}\frac{\cos^{2p-1}(n\pi\alpha)}{(n\pi\alpha)^2}=\frac{1}{6\alpha^2}\left[\frac{3}{2}(2p-1)\alpha^2-(\frac{3}{2}\frac{5}{4}\frac{7}{6}\ldots\frac{2p-1}{2p-2})3\alpha+1\right],$ $0<\alpha\le\frac{2}{2p-1},\quad p=2,3,\ldots$	$\sum_{n=1}^{\infty}\frac{\sin^{2p+1}(n\pi\alpha)}{n\pi\alpha}=\frac{1}{4\alpha}\left[\frac{3}{4}\frac{5}{6}\frac{7}{8}\ldots\frac{2p-1}{2p}\right],$ $0<\alpha<\frac{2}{2p+1},\quad p=2,3,\ldots$
$\sum_{n=1}^{\infty}\frac{\cos^{2p}(n\pi\alpha)}{(n\pi\alpha)^2}=\frac{1}{6\alpha^2}\left[\frac{3}{2}(2p)\alpha^2-(\frac{3}{2}\frac{5}{4}\frac{7}{6}\ldots\frac{2p-1}{2p-2})3\alpha+1\right],$ $0<\alpha\le\frac{2}{2p},\quad p=2,3,\ldots$	$\sum_{n=1}^{\infty}\frac{\sin^{2p+2}(n\pi\alpha)}{(n\pi\alpha)^2}=\frac{1}{4\alpha}\left[\frac{3}{4}\frac{5}{6}\frac{7}{8}\ldots\frac{2p-1}{2p}\right],$ $0<\alpha\le\frac{2}{2p+2},\quad p=2,3,\ldots$

$\alpha = 0.02$ converges with the slowest rate and that corresponding to $\alpha = 1.9$ converges with the fastest. (Interestingly, the error profile also seems to have a sinc shape). A similar convergence assessment was carried out for the series result relative to the $sinc^2(.)$ given by (5) for $\alpha = 0.01, \alpha = 0.1,$ and $\alpha = 0.8$. This is shown in Figure 4. Even though the error in this case shows a smoother and steeper profile, the same observation can be made regarding the effect of α on the convergence rate.

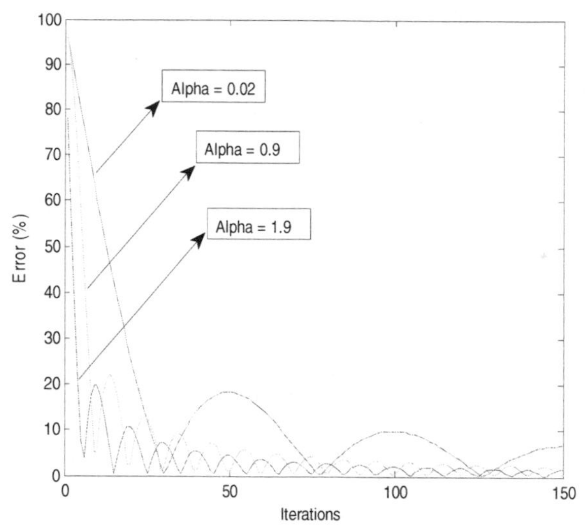

Figure 3: Relative convergence error (in %) versus the number of terms in the partial sum for the *sinc*(.) function.

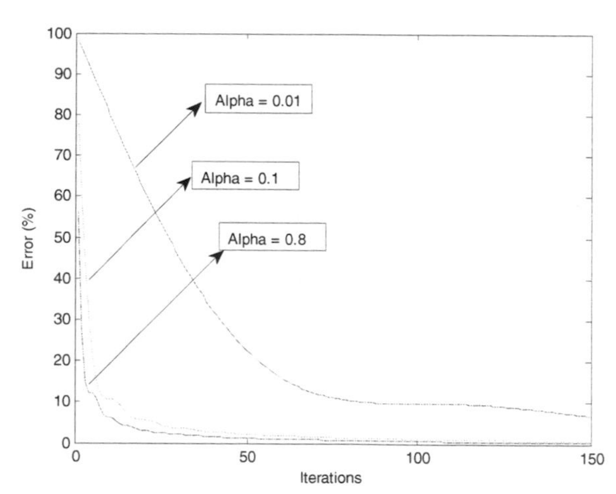

Figure 4: Relative convergence error (in %) versus the number of terms in the partial sum for the $sinc^2(.)$ function

IV. EXAMPLES

We now provide two examples related to the performance of baseband data communication systems in the presence of additive noise, Intersymbol Interference (ISI) and timing problems. In the first example we address the following question: If the channel's bandwidth in such systems is reduced by an ε factor from its absolute minimum, how much degradation in performance does this incur? The second example provides a quantitative analysis of the effect of pulse shaping in baseband data communication systems. In both examples, we shall make use of the series results we developed earlier.

A. Example 1

We now assume binary pulses p_i are transmitted with a rate of 1/T Hz through a channel of bandwidth B Hz and with normalized impulse response so that $h_{ch}(0)=1$. We further assume the pulses p_i to take on values +V (if 1 is being trasmitted) and -V (if 0 is being transmitted) with equal probabilities. In the absence of noise, the output signal at the receiver is given by (assuming the pulse widths to be negligible)

$$y(t) = \sum_{i=-\infty}^{\infty} p_i h_{ch}(t - iT). \qquad (17)$$

If $y(t)$ is sampled at multiple of T (without loss of generality, we consider the output at the origin , i.e., at the sample t = 0), we get

$$y(0) = p_0 + \sum_{\substack{i=-\infty \\ i\neq 0}}^{\infty} p_i h_{ch}(iT). \qquad (18)$$

The second term in (18) represents the Intersymbol Interference (ISI). It is known[1] that the smallest channel bandwidth acceptable is given by

$$B = \frac{1}{2T}, \qquad (19)$$

and corresponds to the ideal case where the channel behaves as an ideal LPF. In this case, the channel's impulse response is given by

$$h_{ch}(t) = sinc(2Bt). \qquad (20)$$

It is easily seen from the fact that $sinc(n)=0$ for all $n \neq 0$ that the ISI term in (18) disappears. Now assume that additive white Gaussian noise with zero mean and Power Spectral Density (PSD) of $N_o/2$ (W/Hz) gets added to the pulses. Hence, (18) (assuming (19) and (20) satisfied) yields

$$y(0) = p_o + n(0), \tag{21}$$

where $n(0)$ represents now a Gaussian random variable with zero mean and variance

$$\sigma_n^2 = BN_o. \tag{22}$$

With this problem setup, we now address the same question mentioned earlier, namely, suppose the channel's bandwidth is reduced from its minimum limit by an ε factor ($0 < \varepsilon < 1$), how much degradation in performance will this incur? In other words, we are now modifying (19) as

$$B = \frac{1}{2T}(1-\varepsilon) \tag{23}$$

Equation (23) is also equivalent to saying that timing errors occurred at both the transmitter and receiver. To investigate the above question, we define as our performance index the standard Signal-to-Noise Ratio (SNR) at the receiver's output as

$$SNR = \frac{E\{p_o\}^2}{noise\ variance}, \tag{24}$$

where $E\{.\}$ denotes the expected value. Hence, using (23) and (20) in (18) (with additive noise), we get

$$y(0) = p_0 + \sum_{\substack{i=-\infty \\ i\neq 0}}^{\infty} p_i sinc(i(1-\varepsilon)) + n(0) \tag{25}$$

The last two terms in RHS of (25) can be regarded as noise corrupting the output, hence we can treat

$$\tilde{n}(0) = \sum_{\substack{i=-\infty \\ i\neq 0}}^{\infty} p_i sinc(i(1-\varepsilon)) + n(0) \tag{26}$$

as the noise sample affecting the output. It is easily seen that $\tilde{n}(0)$ is zero mean and with variance equal to

$$E\{\tilde{n}^2(0)\} = \sigma_n^2 + \sum_{\substack{i=-\infty \\ i\neq 0}}^{\infty} \sum_{\substack{j=-\infty \\ j\neq 0}}^{\infty} E\{p_i p_j\} sinc(i(1-\varepsilon)) sinc(j(1-\varepsilon)). \tag{27}$$

The cross-terms in (27) are zero assuming that $n(0)$ and p_i are independent. Assuming further that the pulses are uncorrelated, we get

$$E\{p_i p_j\} = V^2 \delta[i-j]. \tag{28}$$

Using (22), (23), and (28), (27) is rewritten as

$$E\{\tilde{n}^2(0)\} = \frac{N_o}{2T}(1-\varepsilon) + 2V^2 \sum_{i=1}^{\infty} sinc^2(i(1-\varepsilon)). \tag{29}$$

Using the series result derived earlier in (5), Equation (29) is simplified as

$$E\{\tilde{n}^2(0)\} = \frac{N_o}{2T}(1-\varepsilon) + V^2 \frac{\varepsilon}{1-\varepsilon}. \tag{30}$$

Thus, referring to (24), the modified SNR (due to this reduction in BW) now becomes

$$SNR_{new} = \frac{V^2}{\frac{N_o}{2T}(1-\varepsilon) + V^2 \frac{\varepsilon}{1-\varepsilon}}. \tag{31}$$

The degradation in SNR is therefore given by

$$\Gamma = \frac{SNR_{new}}{SNR_{old}} = \frac{1}{1-\varepsilon + \frac{2TV^2\varepsilon}{N_o(1-\varepsilon)}} = \frac{1}{1-\varepsilon + \frac{\varepsilon}{1-\varepsilon} SNR_{old}}, \tag{32}$$

where as defined in (24) and using (19) and (22), SNR_{old} is given by

$$SNR_{old} = \frac{2TV^2}{N_o}. \tag{33}$$

We have numerically evaluated (32) for various values of SNR_{old}. The results are shown in Fig. 5. Here, we plot the degradation in SNR (in dB) as a function of the percentage reduction in BW for three cases: SNR_{old} = 33 dB ($T=10^{-7}$, $V=10$, and $N_o=10^{-8}$), SNR_{old} = 23 dB ($T=10^{-7}$, $V=10$, and $N_o=10^{-7}$), and SNR_{old} = 13 dB ($T=10^{-7}$, $V=10$, and $N_o=10^{-6}$). These values were chosen to be close to practical baseband data communication systems. As can be seen from Fig. 5., the degradation in SNR becomes more noticeable for higher SNR_{old} values. This is expected since for lower SNR_{old} a reduction in BW will, in addition to introducing ISI, reduce some of the now stronger additive noise, hence alleviating the SNR degradation. Furthermore, and also as expected, the effect of ISI is quite deleterious on the system's performance (for example, for the case of SNR_{old} =33 dB, a 1 % reduction in BW leads to a loss of about 13 dB in SNR).

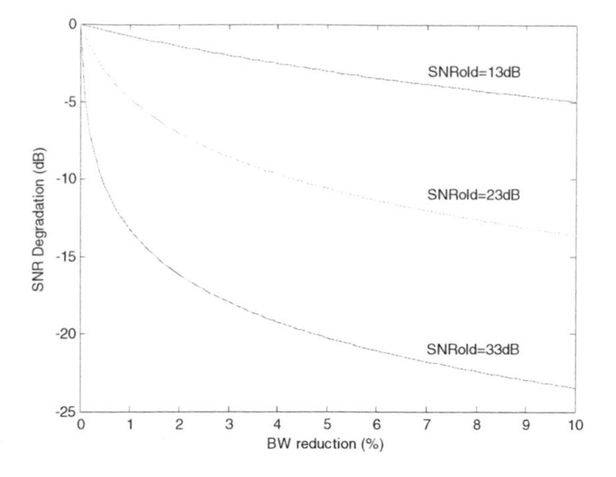

Figure 5. Degradation in SNR (in dB) versus BW reduction (in %).

B. Example 2

In this second example, we make use of the same formulation as in the previous one, except that now we assume a pulse shaping filter is added at the transmitter so that the resulting frequency response of the channel (whose bandwidth is now allowed to exceed 1/2T) gets a *raised cosine* shape[1]. Hence,

$$B = \frac{1}{2T}(1 + r) \quad , \tag{34}$$

where r is the *roll-off* factor. With this pulse shaping, the ISI term is (ideally) removed and the SNR is increased (hence pulse detection is improved). To quantitatively assess the effect of the pulse shaping filter on the system's performance, we now address the following question: Suppose pulse shaping is removed so that the edge tapering of the channel's frequency response is eliminated, how much degradation in SNR does this incur?

With the edge tapering removed, the channel's frequency response is now assumed to be that of an ideal LPF with bandwidth

$$B = \frac{1}{2T}(1 - r) \cdot \tag{35}$$

Equation (35) is identical to (23) with r (roll-off factor) replacing ε (BW reduction factor). Hence, the results obtained in example 1 exactly apply in this case.

V. CONCLUSION

In this paper, we presented several new infinite series results (along with their generalizations) related to the sinc signal. These are derived using Fourier Series Decomposition of periodic signals and Parseval's identity along with some simple trigonometry. Such results should provide more tools in dealing with numerous practical problems that inherently involve the Sinc signal (eg. band-limited or time-limited signals). The two examples we provided, even though academic and with predictable findings, helped to illustrate this point by quantitatively analyzing the performance of data communication systems in the presence of additive noise, Intersymbol Interference (ISI), and timing problems. A simple convergence analysis was provided partly to justify a step in our derivations.

REFERENCES

[1] Proakis, J. G., *Digital Communications*. New York: McGraw-Hill, 1983.

[2] L. R. Rabiner, and R. W. Schafer, *Digital Processing of Speech Signals*. Englewood, NJ: Prentice-Hall, 1978.

[3] A. V. Oppenheim, A. S. Willsky, and I. T. Young, *Signals and System*. Englewood, NJ: Prentice-Hall, 1983.

[4] K. Knopp, *Theory and Application of Infinite Series*. London: London & Glasgow, 1951.

[5] G. A. Korn, and T. M. Korn, *Mathematical Handbook for Scientists and Engineers*. New York: McGraw-Hill, 1968.

Multi-Focus Image Fusion Using Energy Coefficient Matrix

Adnan Mujahid Khan*, Mudassir Fayyaz*, Asif M. Gillani*
*Faculty of Computer Science & Engineering, GIK Institute Topi, Swabi, Pakistan.

***Abstract:* An image fusion algorithm based on Energy Coefficient Matrix (ECM) is presented in this paper. Energy coefficient matrix is computed and proved to indicate the clarity of a pixel at a particular physical location. Based on this clarity value, a fusion decision map has been created. This decision map provides information about which pixel to choose at a particular physical location in wavelet domain. Comparison of the scheme proposed in this paper has been done with two other well known techniques based on Discrete Wavelet Transform (DWT) and Discrete Wavelet Frame Transform (DWFT). Experimental results have shown that the performance of the algorithm proposed in this paper is superior to that of DWT and DWFT-based one.**

I. INTRODUCTION

Limited depth of field is a big problem that optical lenses normally suffer. According to lens formula, only object(s) at a particular depth will truly be in focus. While all the other objects that are at a different depth will be out of focus. However, from image processing knowledge, we know that anything that is out of focus will be blurred. The more the difference of depth as compared to the actual depth of field, the more will be the blur in the resultant image [12]. However, in various image processing applications, we need sharp images, so that various image processing tasks such as edge detection, image segmentation and stereo matching could be performed on the image. Since, these operations give better results on focused images as compared to blurred images, so it is often required to have an everywhere- in-focus image.

Image fusion provides an elegant solution to this problem. In general, the problem that image fusion tries to solve is to combine the information from several images taken from the same scene in order to achieve a new fused image which contains the best focus information coming from the original images [2]. Depending on the stage at which the fusion mechanism takes place; image fusion strategies can be broadly divided into three categories, namely, pixel level *(data level)*, feature level and decision level [1]. Pixel level *(data level)* image fusion works directly on the sensor's output. In this case, fusion decision is based on the pixel's strength in representing the clarity of image at a particular pixel location. In feature level fusion, salient features of the image are extracted from the source images. Fusion decision is based on the fusion of features not pixels, as in the case of pixel level fusion. In decision level fusion, which is a further level of abstraction in image fusion, first, understanding (interpretation) of the image takes place, than fusion of interpretations, instead of pixels or features, takes place.

The simplest image fusion algorithm just takes the pixel-by-pixel average of the source images. This, however, often leads to undesirable side effects such as reduced contrast. Due to great interest of researchers in area of image fusion, various alternatives based on multi-scale transforms (MST) have been proposed. The basic idea behind multi-scale transform based fusion is to perform a multi-resolution decomposition on each source image. Next step is the integration of these decompositions into a composite representation. Finally, the fused image is reconstructed by performing an inverse multi-resolution transform. Several examples of these multi-scale transforms include Gaussian pyramid, Laplacian Pyramid [3], Ratio of low pass pyramid, gradient pyramid [6], and morphological pyramid [5]. In the following few paragraphs, we explain these pyramid-based approaches in few words.

An image pyramid consists of a set of low pass or band pass copies of an image, each copy representing pattern information of different scale. In pyramid based transforms, usually, every level is a factor two smaller than its predecessor. An image pyramid contains all the information necessary to reconstruct the original image.

The Gaussian pyramid is a sequence of images in which each member of the sequence is a low pass filtered version of its predecessor [9].

Laplacian pyramid of an image composed of band pass filtered copies of its predecessor. Band pass copies can be obtained by calculating the difference of low pass images at successive scales of Gaussian pyramid [10].

A comparatively recent approach is one that is based on Discrete Wavelet Transform (DWT). The advantage of scheme based on DWT is, that it gives promising results as compared to the multi-scale (pyramid-based) techniques stated earlier. There are many arguments in favor of DWT based schemes. For example, Wavelet representation provides directional information while pyramids do not. Also, the wavelet basis functions can be chosen to be orthogonal and so, unlike the pyramid-based methods. Strong arguments in favor of Multiresolution decomposition can be found in psycho-visual research, which provides evidence that human visual system processes images in multiscale way. And at last, DWT based schemes provide frequency as well as time domain information at the same time. In DWT based scheme, however, the multi-resolution decomposition and consequently fusion results are shift variant because of an underlying down-sampling operation. So, a slight movement of object or change in camera position can drastically deteriorate the performance of the system. A good remedy is to use discrete wavelet frame transform (DWFT). Although this slightly increases the computation cost, but, however gives a better multi-scale representation. This paper uses DWFT for the computation of energy coefficient matrix (ECM).

K. Elleithy (ed.), Advances and Innovations in Systems, Computing Sciences and Software Engineering, 525–529.

As coefficients with large magnitude correspond to important features in the image, therefore, choose max rule is usually employed to fuse the multi-resolution decomposition [7]. Choose max rule, simply compares the corresponding wavelet coefficients at each decomposition levels, and chooses the one with larger magnitude. However, this simple rule may not produce optimal results [2]. This paper aims to present a better, alternate to choose max rule.

The rest of this paper is organized as follows. Section II describes the proposed fusion scheme. Section III describes evaluation criterion. Experimental results are given in Section IV, while in Section V, some concluding remarks are given.

II. PROPOSED SCHEME

The proposed algorithm deals with the fusion of registered images; however, it gives promising results on slightly un-registered images. For registration of images Global [8] or local [3] motion estimation techniques can be employed. Following are the algorithmic steps involved in the proposed scheme.

2.1 Decomposition

Decompose the source images by Discrete Wavelet Frame Transform (DWFT) to k levels resulting in a total of 3k detail sub-bands and one approximation sub-band. Since, the approximation sub-band is a low-pass *(smooth)* version of the original image and contains very small edge information *(because low-pass filter reduces the signal strength)*, therefore will not help in deciding the clarity of the source image. Thus, only the detail sub-bands will be used in constructing the ECM. We denote the wavelet coefficient for image A (or B) at position *(i,j)* of the b_1b_2 detail sub-band (where b_1b_2 can be either *HL, LH, HH*) at decomposition level *k* by $C^{(A)}_{b_1b_2,k}$ (or $C^{(B)}_{b_1b_2,k}$).

2.2 Activity Level Computation.

The activity level of a multi-scale decomposition (MSD) coefficient reflects the local energy in the space, spanned by the term in the expansion corresponding to this coefficient [11]. For Each pixel location, compute the Activity level at each pixel location by using Coefficient Based Activity *(CBA)* method. Note that this will be defined for every pixel location at all levels because all detail sub-bands of DWFT are of the same size as the input image. The activity in this particular case has been the absolute value of the coefficient. Denote the resulting activity levels for A and B by $A^{(A)}_{b_1b_2,k}$ and $A^{(B)}_{b_1b_2,k}$, respectively.

2.3 ECM Computation

Basic idea behind ECM computation is to exploit the fact that a sharp image has greater Energy as compared to a blur image. This is because of filtration process, which when applied on a sharp signal (to make it smooth), makes the signal lose some of its peaks and as a result loss in signal strength (Energy). Since blurred image is smooth version of a sharp image, so its energy will be less as compared to the energy of a sharp image.

ECM Computation is further divided into 3 steps. They are stated as below.

1. Form a composite matrix that contains accumulative information regarding all the detail bands of the decomposition tree.
$$DE_X(n,m) = \sum A^{(X)}_{b_1b_2,k}(i,j) \quad (1)$$
Where 'X' is the corresponding image name (A or B), *k* is the decomposition level and *(i,j)* is the position within the composite matrix. This step is necessary for the next step.
2. Compute local energy E_A and E_B of high frequency coefficients from the two multi-focus images using the following relation over composite matrix.
$$E_X(n,m) = \sum [DE_X(n+n', m+m')]^2 \quad (2)$$
where, 'X' is the corresponding image name, n' and m' are variables of a small window.
Now clearly, regions which are clear, will have better signal representation and of course high energy as compared to the regions with blurred focus, having smooth (un-sharp) signal representation and less energy.
3. Compute ECM using following ratio.
$$E_F(n,m) = \frac{E_A(n,m)}{E_A(n,m) + E_B(n,m)} \quad (3)$$

This ECM (given by eq.3) represents the regional clarity difference between two source images in a local region centered at $E_F(n,m)$. In areas where image A has clear focus as compared to image B, the value of E_A (n, m) will be much more than E_B (n, m), So, the values of E_F (n, m) will be close to 1, contrarily, in the areas where image B is clearer as compared to image A, the values E_F (n, m) will be closer to 0. In nutshell, ECM will contain values in the rage [0, 1].

2.4 Coefficient Combination

Based on the values in ECM E_F (i, j) at position (i, j) and decomposition level b_1b_2, wavelet co-efficient of the fused image F (i, j) are chosen as stated below.

$$F_{b_1b_2}(i,j) = \begin{Bmatrix} C^{(A)}_{b_1b_2}(i,j) & \text{if} & E_F(i,j) >= 0.5 \\ C^{(B)}_{b_1b_2}(i,j) & & E_F(i,j) < 0.5 \end{Bmatrix} \quad (4)$$

Where b_1b_2 can be LL, *HL, LH, HH* of the fused and source images and *(i, j)* is the corresponding coordinate position in transform domain.

2.5 Consistency Verification

Perform consistency verification on the binary decision map obtained in step 4. Consistency Verification step ensures that fused coefficient does not come from different source images for most of its neighbors. This operation can be either performed using a majority filter or using 'fill' and 'clean' morphological operations. In current implementation, majority filter as explained by Li et al. [2] has been used. It is an optional step but results show that this step produces more accurate results. In results table, *WBV* scheme is one, in which majority filter has been used to remove the isolated points in the binary decision map.

2.6 Reconstruction

Reconstruct the fused image using inverse discrete wavelet frame transform (IDWFT).

III. EVALUATION CRITERIA

Two criteria are used for comparison of results. The basic criterion is the root-mean-squared error (RMSE). Mathematically, it is stated as follows.

$$RMSE = \sqrt{\frac{1}{MN}\sum_{i=1}^{M}\sum_{j=1}^{N}(R(i,j)-F(i,j))^2} \quad (5)$$

where, F is the fused image and R is the reference image. Note that, the size of reference and fused image should be same (i.e. M x N). Here, note that value of RMSE closer to 0 shows superiority of any scheme. And similarly, a larger RMSE value shows that the fused image is not closely matched with the reference image

The other criterion for performance assessment of fusion scheme is Spatial Frequency (SF). Spatial Frequency is directly related to image clarity. It is used to measure the overall activity of an image. For an N x M image Z, whose gray value at pixel position (i,j) is denoted by Z(i, j), its SF is defined as :

$$SZ = \sqrt{RZ^2 + CZ^2} \quad (6)$$

where, RZ and CZ are the row frequency and column frequency respectively.

$$RZ = \sqrt{\frac{1}{NM}\sum_{n=1}^{N}\sum_{m=1}^{M}(Z(n,m)-Z(n,m-1))^2} \quad (7)$$

$$CZ = \sqrt{\frac{1}{NM}\sum_{m=1}^{M}\sum_{n=1}^{N}(Z(n,m)-Z(n-1,m))^2} \quad (8)$$

where, M x N is the dimension of the fused image. Greater the value of spatial frequency, greater will be the clarity of image and vice versa.

IV. EXPERIMENTS

We perform wavelet-based image fusion on four sets of 256 gray level images: *balloon (of size 480 x 640), Pepsi (of size 512 x 512), lab (of size 480 x 640) and clock (of size 512 x 512).* From among these set of images, all the images except the *clock,* are registered. For all of these images, reference image is available. If blurred images are not available, then any reference image can be blurred using some image processing software like Adobe Photoshop.

A number of wavelet based fusion schemes, with different combination of wavelet transforms (DWFT and DWT), coefficient combining methods (CM and ECM), and consistency verification methods (NV and WBV) are compared. For both wavelet transforms, we use bi-orthogonal wavelet *(biorth3.7)* and three levels of decomposition. For ECM computation, window size is set to 3 x 3.

Table 1 and 2 show the experimental results based on two sets of multi-focus images. Compared with wavelet based method, we find that RMSE values obtained by our proposed fusion method are lesser than those commonly known techniques. Even subjective analysis verifies the truthfulness of the objective results.

Similarly, in Table 2, the spatial frequency values of fused image is more then the individual spatial frequency values. This proves the fact that fused image contains more information as compared to the source images. On the basis of experimental results, we propose fusion scheme based on validation procedure (WBV) to be best for fusion of multi-focus images. Although the validation procedure will slightly increase the execution time of the algorithm, but in applications where accuracy is critical, WBV scheme is recommended.

V. CONCLUSION

A pixel-based multi-focus image fusion method based on multi-resolution wavelet decomposition has been proposed in this paper. The advantage of this scheme is that the multi-directional and multi-scale information of wavelet decomposition without down sampling are fully used to extract normalized feature image. Edge information (*horizontal, vertical and diagonal*) from these normalized feature images are accumulated in a single image. Then, the Energy of these feature images are computed with-in a local region. Since Energy of a region represents the strength of underlying signal and the strength of a smooth signal is always less as compared to the strength of a sharp signal. Therefore, ECM of both images will represent the relative strength of images in representing the image clarity *(sharpness)* at some particular location (i, j). Experimental results show that the proposed method can keep the edge information of two spatially registered images to utmost extent and also achieve better fusion performance than other known wavelet transform method.

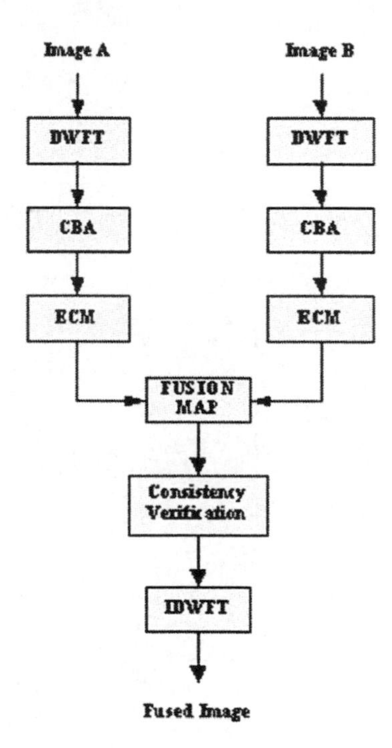

FIG. 1. SCHEMATIC DIAGRAM OF PROPOSED SCHEME

(a)

(b)

(c)

FIG. 2 (a) IMAGE WITH RIGHT HALF IN BLUR (b) IMAGE WITH LEFT HALF IN BLUR (c) FUSED IMAGE USING PROPOSED SCHEME.

Evaluation Criteria		BALLOON	PEPSI	LAB	CLOCK
RMSE	NV	1.91	3.1	3.29	6.7
	WBV	**1.38**	**2.9**	**3.09**	**4.78**
RMSE DWFT(CM+AVG)		2.1	3.74	4.4	5.6
RMSE DWT(CM+AVG)		2.3	3.82	4.85	5.8

TABLE 1 EXPERIMENTAL RESULTS SHOW COMPARISON OF RMSE VALUES OF PROPOSED SCHEME AGAINST OTHER KNOWN TECHNIQUES. BEST RESULTS ARE GIVEN IN BOLD. CLEARLY, THE PROPOSED SCHEME OUT-PERFORMS THE TRADITIONAL SCHEMES BOTH OBJECTIVELY AND SUBJECTIVELY.

Evaluation Criteria	BALLOON	PEPSI	LAB	CLOCK
SPATIAL FREQUENCY (IMAGE 1)	14.73	9.21	10.9	8.9
SPATIAL FREQUENCY (IMAGE 2)	19	13	8.63	6.75
SPATIAL FREQUENCY (FUSED)	**21**	**14**	**13.9**	**10.99**

TABLE 2 EXPERIMENTAL RESULTS SHOW COMPARISON OF SPATIAL FREQUENCY VALUES OF PROPOSED SCHEME AGAINST THE SPATIAL FREQUENCY VALUES OF SOURCE IMAGES 1 & 2. BEST RESULTS ARE GIVEN IN BOLD. CLEARLY, THE PROPOSED SCHEME PERFORMS REASONABLY WELL.

VI. REFERENCES

[1] P. K. Varshney, "Multisensor data fusion," *Electronics and Communication Engineering Journal*, pp. 245-253, Dec. 1997.

[2] S. LI, J. T. Kwok, I. W. Tsang, Y. Wang, "Fusing Images with different focus Using Support Vector Machines," *IEEE Transactions on Neural Network*, Vol. 15, No. 6, pp. 1555-1561, Nov. 2004.

[3] L. Bogoni and M. Hansen, "Pattern Selective Color image Fusion," *Pattern Recognition*, Vol. 34, pp. 1515-1526, 2001.

[4] A. Toet, L.J. van Ruyven, and J. M. Valeton, "Merging thernal and visual images by a contrast pyramid," *Optical Engineering*, Vol. 28, No. 7, pp. 789-792, July 1989.

[5] G. K. Matsopoulos, S. Marshall. And J. N. H. Brunt, "Multi-resolution morphological fusion of MR and CT images of the human brain," *Proceedings of IEE: Vision, Image and Signal Processing*, Vol. 12, No.9, pp.1834-1841, 1995.

[6] P. T. Burt and R. J. Kolezynski, "Enhanced image capture through fusion," *in Proceedings of International Conference on Computer Vision*, pp. 173-182, 1993.

[7] H. Li, B. S. Manjunath, and S. K. Mitra, "Multisensor image fusion using the wavelet transform," *Graphical Models and Image Processing*, Vol. 57, No. 3, pp. 235-245, May 1995.

[8] G. W. M. Hansen, K. Dana, and P. Burt, "Real Time scene stabilization and motion construction," in *Proceedings of Second IEEE International Workshop on Applications of Computer Vison*, 1994, pp. 54-62.

[9] H. Olkkonen and P. Pesola," Gaussian Pyramid Wavelet Transfoorm for Multi-resolution Analysis of Images," *Graphical Models and Image Processing*, vol. 58, pp.394-398, 1996.

[10] P. Burt, E. Adelson,"Laplacian pyramid as a compact image code," *IEEE Transactions on Communications*, Vol. 31, No. 4, 1983.

[11] Z. Zhang and R. S. Blum, "A Categorization of multiscale-decomposition-based image fusion schemes with a performance study for a digital camera application," *Proceedings of IEEE*, Vol. 87, pp. 1315-1326, Aug, 1999.

[12] Y. Wu, Chongyang Liu and Guisheng Liao, "Multi-focus image fusion based on SOFM Neural Networks and Evolution Strategies," *ICNC, LNCS 3612, pp. 1-10, 2005.*

Measuring Machine Intelligence of an Agent-Based Distributed Sensor Network System

Anish Anthony and Thomas C. Jannett
Department of Electrical and Computer Engineering
The University of Alabama at Birmingham
Birmingham, AL 35294-4461 USA
anish@uab.edu

***Abstract* – A measure of machine intelligence facilitates comparing alternatives having different complexity. In this paper, a method for measuring the machine intelligence quotient (MIQ) of human-machine cooperative systems is adapted and applied to measure the MIQ of an agent-based distributed sensor network system. Results comparing the MIQ of different agent-based scenarios are presented for the distributed sensor network application. The MIQ comparison is contrasted with the average sensor network field life, a key performance indicator, achieved with each scenario in Monte Carlo simulations.**

INTRODUCTION

What may seem intelligent to one person may not seem intelligent to another. One reason for this is that the term intelligence has a broad meaning [1]. Numerous definitions of intelligence and numerous criteria for measuring intelligence have been proposed. One of the simplest defines intelligence as that which produces successful behavior [2]. Intelligence requires the ability to sense the environment, make decisions, and control action. Advanced forms of intelligence provide the capacity to act successfully under a large variety of circumstances so as to survive, prosper, and reproduce in a complex and often hostile environment. In guidelines to be followed for developing an intelligence metric, it is asserted that the intelligence metric developed should be application specific, dynamic and adaptive, reflect generalization capabilities, and be able to evaluate interrelations and interactions among multiple systems [3]. Survivability and competence are mentioned as additional criteria for measuring intelligence [4]. Survivability is the ability of a system to cope with diversity in the environment, as well as internal faults (hardware and software) and competence is the ability of a system to successfully perform tasks. Expressiveness and perceptiveness have been suggested as important criteria for measuring intelligence of machines [5]. Expressiveness is a measure of the output richness of an electromechanical system. Perceptiveness is a measure of the fidelity of an electromechanical system's effective mapping from environmental change to output.

In this paper, an agent-based distributed sensor network system is considered as a machine whose intelligence is to be measured. This measured intelligence is one means of evaluating and comparing different scenarios that use multiple agents and different functionalities for the agents. The distributed sensor network application is one in which a large number of sensor nodes will be placed on the ocean floor near the mouth of a port or bay. The sensors measure the range and bearing of vessels that are moving in and out of a port or bay. Target detection, classification, and tracking are of interest. The network will collect information about the target location, fuse this data and then send the fused data to higher levels in the system hierarchy. Individual sensor nodes in the network have limited computational capacity, battery power and thus limited lifetime. A node's batteries may be depleted through normal network operation, leaving the node useless and possibly weakening the network. With reduced resources, the network may no longer be able to deliver the required level of performance. Thus, the field life of the sensor network is an important criterion for designing wireless sensor networks [6].

Agents have many possible advantages in distributed sensor networks. The use of agents for the sole purpose of selecting optimal sensors and fusion algorithms for achieving the performance goals of the system has been suggested [7]. A multi-resolution integration algorithm for data fusion in a mobile-agent distributed sensor network is presented in [8]. In [9, 10] the authors have detailed a case study of using agents for a fire-tracking application. An agent based adaptive urban traffic control system has been proposed in [11]. Following the concepts of using agents in distributed sensor network applications, this paper focuses on comparing different agent based scenarios in order to achieve a goal of increasing the field life of the sensor network. Agents will be used for efficient data fusion and target tracking. At the same time, the agents will use intelligent control and coordination algorithms for reconfiguration of the system and its resources so as to efficiently manage the power consumption in the system, resulting in increased field life of the sensor network.

A measure of the intelligence quotient (IQ) can be used to compare different systems. Considering different agent-based scenarios as different systems, the intelligence of these different agent-based scenarios can be compared. A practical and systematic method of measuring machine intelligence quotient (MIQ) presented in [12] is reviewed in the following and is utilized in this paper.

This paper is organized as follows. First, measuring machine intelligence for human-machine cooperative systems

This work was supported in part by the Office of Naval Research under Grant N00014-03-1-0751

K. Elleithy (ed.), Advances and Innovations in Systems, Computing Sciences and Software Engineering, 531–535.

using an intelligence task graph (ITG) is reviewed. Next, an agent-based distributed sensor network application, for which measuring machine intelligence is of interest, is discussed. A modified scheme to measure the machine intelligence of an agent-based distributed sensor network is presented. Finally, results comparing the machine intelligence of different agent-based scenarios are presented.

MEASURING MACHINE INTELLIGENCE OF A HUMAN-MACHINE COOPERATIVE SYSTEM USING INTELLIGENT TASK GRAPH – A REVIEW

As outlined in [12], in a human-machine cooperative system MIQ is expressed as

$$MIQ = CIQ - HIQ \tag{1}$$

where control intelligence quotient (CIQ) is the sum of all task intelligent costs and the human intelligence quotient (HIQ) is the intelligence quantity required from the humans to control the plant.

$$CIQ = \sum_{i=1}^{n} a_{i1} r_i + \sum_{i=1}^{n} a_{i2} r_i \tag{2}$$

$$HIQ = \sum_{i=1}^{n} a_{i2} r_i + c_{mh} \sum_{i=1}^{n} \sum_{j=1}^{n} a_{i1} a_{j2} f_{ij} + c_{hm} \sum_{i=1}^{n} \sum_{j=1}^{n} a_{i2} a_{j1} f_{ij} \tag{3}$$

The symbols are defined as follows.

- n: Number of tasks
- Task set, T: The set of n tasks required to control events
  $$T = \{T_1, T_2, T_3, \ldots, T_n\} \tag{4}$$
- Task intelligence cost, r_i: The intelligence required to execute T_i is r_i
  $$r = \{r_1, r_2, r_3, \ldots, r_n\} \tag{5}$$
- Data transfer matrix, F: f_{ij} is the quantity of data transferred from T_i to T_j
  $$F = \begin{bmatrix} 0 & f_{12} & f_{13} & \cdots & f_{1n} \\ f_{21} & 0 & f_{22} & \cdots & f_{2n} \\ \vdots & & \cdots & & \vdots \\ f_{n1} & f_{n2} & f_{n3} & \cdots & 0 \end{bmatrix} \tag{6}$$
- Interface complexity, c_{hm} and c_{mh}: c_{hm} is the complexity of transferring data from human to machine and c_{mh} from machine to human.
- Task allocation matrix, A: The elements of matrix A can have only binary values 0 or 1. $a_{i1} = 1$, $a_{i2} = 0$ and $a_{i3} = 0$ if the machine performs T_i. $a_{i1} = 0$, $a_{i2} = 1$ and $a_{i3} = 0$ if the human performs T_i. If T_i can be assigned neither to machine nor human then $a_{i1} = 0$, $a_{i2} = 0$ and $a_{i3} = 1$. Thus, $a_{i1} + a_{i2} + a_{i3} = 1$ for $\forall i, 1 \le i \le n$

$$A = \begin{bmatrix} a_{11} & a_{12} & a_{13} \\ a_{21} & a_{22} & a_{23} \\ \vdots & \vdots & \vdots \\ a_{n1} & a_{n2} & a_{n3} \end{bmatrix} \tag{7}$$

AGENT-BASED DISTRIBUTED SENSOR NETWORK SYSTEM

Fig. 1 shows the field layout of the sensor network. The sensor nodes (small circles) report measured range and bearing of submarine targets to cluster nodes (black dots) that perform local data fusion. The active master node (diamond) then gathers the data from the cluster nodes and performs global data fusion. When one master node fails, one of the redundant master nodes (big circles) becomes active providing information about the target to the command center. The agents are located on the different cluster nodes and the master node. The agents in the system are used for intelligent data fusion and tracking and for control and coordination.

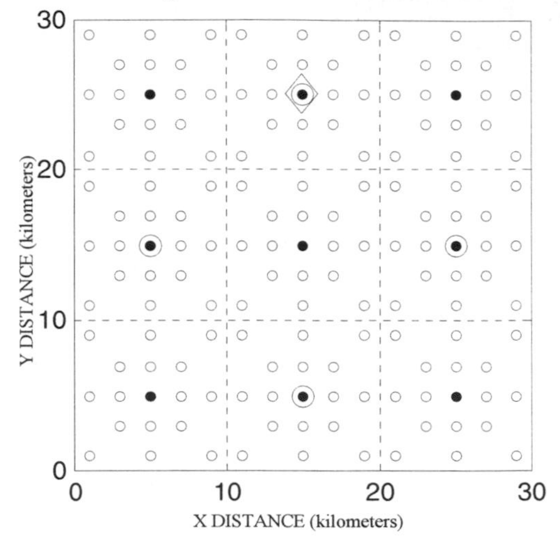

Fig. 1. Field layout of the sensor network

A distributed target tracking algorithm based on the Kalman filter [13] has been adapted and is encapsulated into the agent-based system to achieve the goals of data fusion and target tracking. Agents provide intelligence by deciding when to track, which tracking algorithms to use, how frequently to run the tracking algorithms, and if there are communication delays in the network, whether to skip the old data or use the delayed data and generate a new target track. If there is a slow moving target then the agents can run the tracking algorithms infrequently and save power. Some tracking algorithms may have better resolution but may be more computationally intensive. Thus, the agents can choose the tracking algorithms efficiently to increase the field life of the system while maintaining the tracking performance.

Intelligent agents can also be used in the sensor network system to implement control and coordination strategies in order to maximize the useful life of the field. Agents for control and coordination can gather information regarding the battery life of the nodes in the network and modify the communication routes through the network. Changing the communication routes can help to decrease communication delays in the system as well as manage the power consumption of the nodes that are being used for communication. Also, agents can selectively turn on and off nodes in the system. Thus, nodes that are not in the vicinity of the target can be switched off thereby decreasing the power consumption of the node.

SIMULATION SCENARIOS

Using multiple agents and different functionalities for the agents, numerous agent-based scenarios can be developed, some of which are described below. The goal of the agents in the network is to maximize the field life of the network. Table 1 shows the number of cluster agents (*nca*), number of master agents (*nma*), and their functionalities in the system.

- Scenario 1: Running simulations without agents
- Scenario 2: One master node, one master agent in periodic tracking mode. In this scenario the master agent runs the tracking algorithm periodically (Timed Tracking (TT)). The periodicity of tracking is determined by a fuzzy logic algorithm within the agent, depending on the estimated velocity of the target, and the battery life remaining on the master node.
- Scenario 3: Four master nodes, four master agents running the tracking algorithm periodically and switching the master node in proximity to the target-switching scheme. Four master nodes are present in the system but only one is active at a time. When the battery of the current master node is about to die the master agent at that node transfers the control to a master node in the nearest proximity to the target (Control and Coordination (TT, CC)).
- Scenario 4: One master node, one master agent in covariance tracking mode (Covariance Tracking (CT)). In covariance tracking mode, the master agent checks the target position estimate covariance computed by the Kalman filter tracking algorithm. If the covariance is below a certain level, the master agent stops data fusion process. Kalman filter time updates give an estimate of the target until the covariance of the Kalman filter increases due to lack of measurements. Once the covariance increases above a certain level the fusion process is started again.
- Scenario 5: Four master nodes, four master agents in covariance tracking mode and proximity to the target switching scheme (CT, CC).
- Scenario 6: This scenario incorporates scenario 5 (CT, CC) but in addition has cluster agents on each of the 9 cluster nodes. The master agent instructs the clusters that are far away from the estimated target position to turn off. The cluster agents in turn will instruct the sensors to turn on or off (CC), trying to minimize the battery consumption of the different nodes in the system.
- Scenario 7: No master agents are present in the system, but a cluster agent is present on each cluster in the system. The cluster agents run the target estimate fusion algorithm periodically (TT). The periodicity of fusion is determined by a fuzzy logic algorithm within the agent, depending on the velocity of the target, and the battery life remaining on the cluster nodes.
- Scenario 8: Four master nodes, four master agents used in a switching scheme based on proximity to the target (CC). Also, the master agent has the ability to turn the clusters on or off. Cluster agents are present on all the cluster nodes and perform periodic fusion of target estimates (TT).
- Scenario 9: This scenario incorporates scenario 8 (CC), and in addition the cluster agents have the ability to turn the sensors within each cluster on and off (TT, CC).

TABLE 1
NUMBER OF MASTER AGENTS (*nma*) AND CLUSTER AGENTS (*nca*) IN TIMED TRACKING (TT), COVARIANCE TRACKING (CT), OR CONTROL AND COORDINATION (CC) MODE

	Scenarios								
	1	2	3	4	5	6	7	8	9
nma	0	1	4	1	4	4	0	4	4
nca	0	0	0	0	0	9	9	9	9
MA mode		TT	TT,CC	CT	CT, CC	CT,CC		CC	CC
CA mode						CC	TT	TT	TT,CC

MEASURING MACHINE INTELLIGENCE OF AN AGENT-BASED DISTRIBUTED SENSOR NETWORK SYSTEM

The entire distributed sensor network is decomposed into a number of intelligent machines (IMs) working together to achieve goals in the system. Agents in the system provide intelligence for the system. The goal of these agents is to increase the field life of the network. Fig. 2 shows such a distributed sensor network system that is broken up into different IMs, each encompassing an agent. Each IM in the system is assumed to perform 8 tasks (Fig. 3).

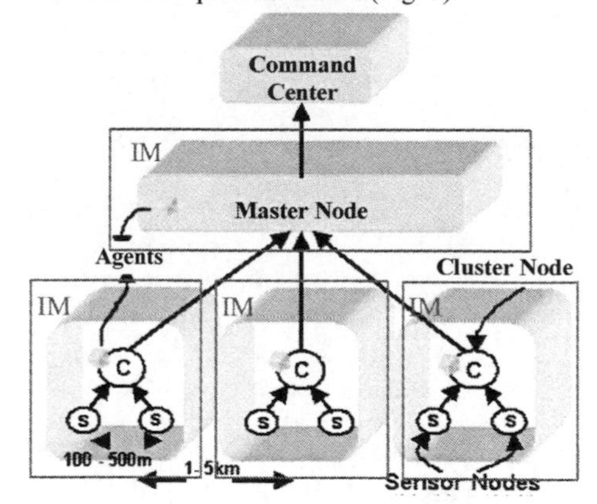

Fig. 2. Agent-based distributed sensor network decomposed into number of intelligent machines.

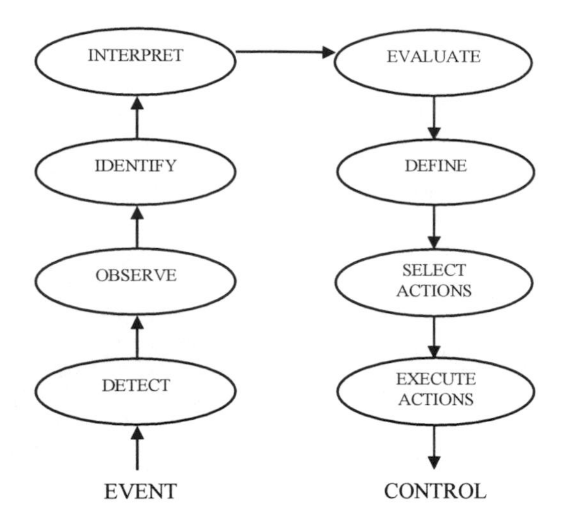

Fig. 3. Tasks performed by each intelligent machine in the agent-based distributed sensor network

Since all the tasks in the system are done by the IM's, the human intelligence quotient (HIQ) can be discarded from (1). Thus,

$$MIQ = CIQ \quad (8)$$

The machine intelligence quotient (MIQ) of the total system can then be calculated as the sum of the MIQ's of each intelligent machine in the system.

$$MIQ(total) = \sum_{i=1}^{nma} MIQ_i(ma) + \sum_{i=1}^{nca} MIQ_i(ca) \quad (9)$$

$MIQ_i(ma)$ is the MIQ contributed by the master agents
$MIQ_i(ca)$ is the MIQ contributed by the cluster agents
nma is the number of master agents in the system
nca is the number of cluster agents in the system

From (2) and (8),

$$MIQ(ma) = CIQ(ma) = \sum_{i=1}^{n} a_{i1}r_i + \sum_{i=1}^{n} a_{i2}r_i \quad (10)$$

$$MIQ(ca) = CIQ(ca) = \sum_{i=1}^{n} a_{i1}r_i + \sum_{i=1}^{n} a_{i2}r_i \quad (11)$$

Since all the tasks in the distributed sensor network system are performed by the agents in the system without human interaction $a_{i2} = 0$ and $a_{i3} = 0$. Thus, the task allocation matrix (7) becomes

$$A = \begin{bmatrix} a_{11} \\ a_{21} \\ a_{31} \\ a_{41} \\ a_{51} \\ a_{61} \\ a_{71} \\ a_{81} \end{bmatrix} = \begin{bmatrix} 1 \\ 1 \\ 1 \\ 1 \\ 1 \\ 1 \\ 1 \\ 1 \end{bmatrix} \quad (12)$$

Assume the task intelligence cost is measured as

$$r = \{r_1, r_2, r_3, r_4, r_5, r_6, r_7, r_8\} = \{5, 8, 12, 18, 19, 15, 11, 9\} \quad (13)$$

From (10) - (13)

$$\begin{aligned} MIQ(ma) = MIQ(ca) &= \sum_{i=1}^{8} a_{i1}.r_i \\ &= r_1 + r_2 + r_3 + r_4 + r_5 + r_6 + r_7 + r_8 \\ &= 5 + 8 + 12 + 18 + 19 + 15 + 11 + 9 \\ &= 97 \end{aligned} \quad (14)$$

From (14), the MIQ contributed by each IM in the system is 97. Thus,

$$MIQ(total) = \sum_{i=1}^{nma} 97 + \sum_{i=1}^{nca} 97 \quad (15)$$

Results and Conclusion

The MIQ's of different agent-based scenarios evaluated by substituting values of *nma* and *nca* from Table 1 in (15) are presented in Table 2. It is clear from inspection of (15) and Table 2 that the number of intelligent agents in the system is an important factor in determining the MIQ for the total agent-based sensor network system. Scenarios 6, 8, and 9 had the highest number of agents and the highest MIQ.

The goal of the agents in the different scenarios was to increase the field life of the network. The field life is defined as the life of the field before a sensor, a cluster, or all the masters in the system run out of battery life. For each scenario, 30 trials were done using Monte Carlo simulations of a target moving through the field. The average field life achieved in simulations of the different scenarios is shown in Fig. 4. The highest field life was achieved for scenarios 8 and 9.

From Table 2 and Fig. 4 it is observed that high field life performance was correlated with high MIQ, except for scenario 6. Scenarios 8 and 9 had the highest MIQ and also achieved the highest field-life performance. However, a low field-life performance was achieved with scenario 6 even though it had had a high MIQ.

Thus, higher machine intelligence does not guarantee an improved system performance. MIQ represents a capability to perform, but the simulation testing presents specific test cases that are used to measure performance. A review of the results indicated that although scenario 6 had a high MIQ, the simulation parameters were not such that the algorithms available in scenario 6 could be effectively employed in minimizing battery consumption to improve performance.

TABLE 2
MIQ OF DIFFERENT AGENT-BASED SCENARIOS

	Scenario								
	1	2	3	4	5	6	7	8	9
MIQ	0	97	388	97	388	1261	873	1261	1261

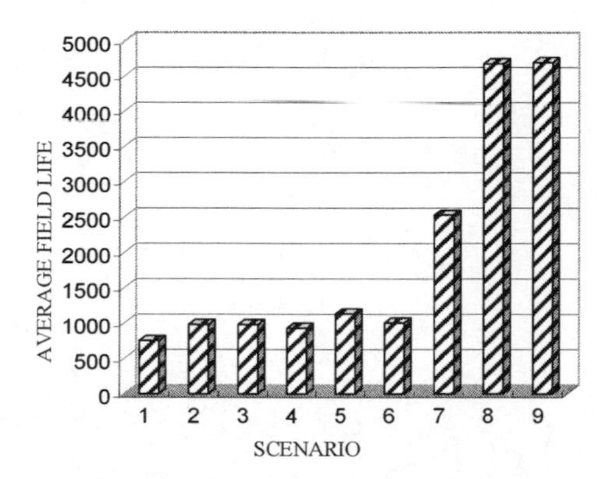

Fig. 4. Average field life achieved in Monte Carlo simulations for each agent scenario (N=30)

ACKNOWLEDGMENT

Parts of this effort were sponsored by the Department of the Navy, Office of Naval Research. Any opinions, findings, and conclusions or recommendations expressed in this material are those of the author(s) and do not necessarily reflect the views of the Office of Naval Research.

REFERENCES

[1] R. R. Brooks, "An intelligence metric for emergent distributed behaviors," in *NIST Workshop on Performance Metrics for Intelligent Systems*, Gaithersburg, MD, August 2000.

[2] J. Albus, "Outline for a theory of intelligence," *IEEE Transactions on Systems, Man, and Cybernetics*, vol. 21, pp. 473-509, 1991.

[3] R. Gao and L. H. Tsoukalas, "Performance metrics for intelligent systems: An engineering perspective," in *NIST Workshop on Performance Metrics for Intelligent Systems*, Gaithersburg, MD, August 2002.

[4] R. Simmons, "Survivability and Competence as Measures of Intelligent Systems," in *NIST Workshop on Performance Metrics for Intelligent Systems*, Gaithersburg, MD, August 2002.

[5] I. R. Nourbakhsh, "Two measures for the "intelligence" of human-interactive robots in contests and in the real world: expressiveness and perceptiveness," in *NIST Workshop on Performance Metrics for Intelligent Systems*, Gaithersburg, MD, August 2002.

[6] V. A. Kottapalli, A. S. Kiremidjian, J. P. Lynch, E. Carryer, T. W. Kenny, K. H. Law, and Y. Lei, "Two-tiered wireless sensor network architecture for structural health monitoring," in *Proc. the SPIE 10th Annual International Symposium on Smart Structures and Materials*, San Diego, CA, August 2003, pp. 8-19.

[7] R. S. Bowyer and R. E. Bogner, "Cooperative behaviour in multi-sensor systems," *Proceedings of the 6th International Conference on Neural Information Processing*, vol. 2, pp. 807-812, November 1999.

[8] H. Q. X. Wang, S. S. Iyengar, and K. Chakrabarty, "Multi-resolution data integration using mobile agents in distributed sensor networks," *IEEE Transactions in Systems, Man, and Cybernetics*, vol. 31, pp. 383-391, 2001.

[9] "Intelligent mobile agents in wireless sensor networks." http://www.cs.wustl.edu/mobilab/projects/agilla/, May19, 2005.

[10] C. L. Fok, G. C. Roman, and C. Lu, "Mobile agent middleware for sensor networks: An application case study," in *Proceedings of the 4th International Conference on Information Processing in Sensor Networks (IPSN'05*, 2005.

[11] F. Y. Wang, "Agent-based control for networked traffic management systems," *Intelligent Systems, IEEE*, vol. 20, pp. 92-96, September,October 2005.

[12] H.-J. Park, B. K. Kim, and K. Y. Lim, "Measuring the machine intelligence quotient (MIQ) of human-machine cooperative systems," *IEEE Transactions on Systems, Man, and Cybernetics – Part A:Systems and Humans*, vol. 31, pp. 89-96, March 2001.

[13] R. Kasthurirangan, "Comparison of architectures for multi-sensor data fusion in deployable autonomous distributed systems," *Masters Thesis in Electrical Engineering*, The University of Alabama at Birmingham, 2003.

Image Processing for the Measurement of Flow Rate of Silo Discharge

Cédric DEGOUET, Blaise NSOM, Eric LOLIVE and André GROHENS
Université de Bretagne Occidentale -. LIME / IUT de Brest. BP 93169. Rue de Kergoat. 29231
BREST Cedex 3 - France blaise.nsom@univ-brest.fr

Abstract : In this work, silo discharge was viewed as a complex fluid flow, in order to perfect a new technique for the measurement of flow rate. Flow rate was investigated using a non intrusive method measuring the evolution of the free surface profile during the discharge flow. This method consisted of recording via a CCD sensor, the evolution of the free surface by laser planes, and then obtaining by processing the free surface position and shape over time.

I. INTRODUCTION

In food technology, a major concern is silo operation. Today, reliable methods and systems are available to silos designers and operators ensuring operation free of major technological risks, such as silo failure or explosion, grain deterioration, etc. This rule book (Reimbert and Reimbert [1], [2]; Jenike [3]; Brown and Nielsen [4]), in force since the 50's is useful for the engineer and does not require any more scientific research. Nonetheless, possible costly defects subsist during silo emptying, such as the formation of steady arches, capable of blocking the operation. These defects are caused by physico-chemical reactions which are in turn linked to grain properties and storage conditions, such as consolidation, relative humidity and temperature (Teunou and Fitzpatrick [5]).

The flowing material being considered to be a bulk quantity and familiar descriptive elements, such as the velocity field or the flow rate, can be described (Nedderman [6]; Tüzün et al. [7]). In fact, the flow rate depends on the flow mode (mass or funnel flow), on the particle's internal friction angle, on the particle's size and the hopper's opening. If α denotes the slope of the (conical) hopper walls, ϕ the dead zone angle with the horizontal, and ϕ_c its complementary $\left(\phi_c = \frac{\pi}{2} - \phi\right)$, the flow rate is given by the generalized formula of Beverloo (Beverloo et al. [8]; Bideau and Ammi [9]; Rose and Tanaka [10])

$$Q = C \cdot \rho \cdot g \cdot D^{1/2} \cdot \left(D - kd\right)^{5/2} \cdot F\left(\alpha, \phi_c\right) \quad (1)$$

with

$$F\left(\alpha, \phi_c\right) = \left[\frac{tg\alpha}{tg\phi_c}\right]^{-0.35} \quad \text{for } \alpha < \phi_c \quad (2)$$

$$F\left(\alpha, \phi_c\right) = 1 \quad \text{for } \alpha > \phi_c \quad (3)$$

where C and k are empirical constants whose dependency on the grains' internal friction coefficient is not yet clearly determined; ρ the grain density under flow, g the gravity, d the grain size and D the hopper opening's diameter. For spherical particles, the previous Beverloo correlation is valid if $D > 6d$. For smaller values of D, the flow rate becomes irregular and is not reproducible; a stable arch forms and blocks the flow for very small openings. The limit $\alpha = \phi_c$ corresponds to the critical angle defining the two flow modes: mass flow for $\alpha < \phi_c$ and funnel flow for $\alpha > \phi_c$.

A common property of Newtonian fluids is to develop a horizontal free surface, while the discharge flow of granular materials generally develop a non horizontal free surface. This property is used to characterize its profile and derive an innovative technique based on the use of a laser beam and derive the flow rate and of discharge flows of soya, colza and rye seeds.

II. PROBLEM STATEMENT

The Brest Chamber of Commerce, who manages the silos at the Brest Commercial Port, provided us with samples soya, colza and rye seeds, to characterize their flowability and the emptying flows of a slender model small. Soya seeds are light yellow hard beads with a d_{50} equal to 6.1×10^{-3} m, while Colza seeds are black hard beads with a d_{50} equal to 2.3×10^{-3} m, and rye seeds have a rounded cylindrical shape, and are brown-green in colour with a d_{50} equal to 3.5×10^{-3} m. The granulometry of each of these materials was characterized using appropriate sieves.

The different grains were then characterized using a Hitachi S-3200N scanning electron microscope (SEM), which uses a beam of highly energetic electrons to examine objects on a very fine scale, and produces high resolution images of a sample surface. The tests were carried out at ambient temperature under a high vacuum (6.0×10^{-4} Pa) obtained by means of a diffusion pump. For the analysis, the grain samples were laid on 2.5×10^{-2} m diameter cylindrical container using a double side adhesive copper strip. Then, the container is metalized using gold, under primary vacuum. This step is necessary for non metallic samples to avoid a superficial accumulation of electric charges from the incident electronic beam, emitted by the microscope tungsten filament gun. The resulting images of the grains studied have a characteristic three-dimensional aspect and are useful for judging the surface structure of the sample. Soya and colza seeds have a smooth and regular surface while rye seeds have a rough irregular surface.

The flowability of each of these granular solids was studied in a recent paper using a modified testing shear box ; thanks to this equipment, the limit stress could be studied up to the breaking point by shearing a sample with its natural consolidation under a known normal stress following an imposed breaking plane (Degouet et al. [11]). Moreover, in the same paper, the specific bulk characteristics and flow properties of these granular solids, that is, densities, compacity, porosity and critical angles were measured. The summary table below lists the resulting grain characteristics :

K. Elleithy (ed.), Advances and Innovations in Systems, Computing Sciences and Software Engineering, 537–542.

Materials		Soya	Colza	Rye
Real density (kg/L)		1177±18	1053±14	1212±19
Apparent density (kg/L)		682±4	641±4	741±4
Porosity		0.42±0.01	0.39±0.01	0.39±0.01
Angle of internal friction (°)		28.8	28.3	25.2
Critical angles (°) ±0.5°	Stop angle	30.3	32.0	29.2
	Start angle	34.2	35.6	33.6

Table 1: Characteristics of the grains studied

Flowability, the outcome of the physico-chemical properties, was then derived and the summary table below lists the resulting flowability indexes :

Materials	soya	colza	rye
Flowability index	0.012	0.026	0.111

Table 2: Flowability indexes of the studied grains

III. THE MODEL SILO

A silo consists of a cell in which the grain material is stored and an exit element which can be either a flat plate or a converging hopper. In order to use optical facilities, the silo is completely transparent (plexiglas) with a slender aspect ratio to reproduce the main flow configurations encountered in the industrial environment. For the same reason, the following three exit elements were built: a flat plate, a single hopper and a double hopper in the form of a convergent four-sided pyramid. They are equipped with an appropriate lightweight lock gate which can be instantaneously locked. Nielsen [12] gives the scale effects in silo flows.

Figure 1: The small-scale silo provided with a flat bottom(*0.8m*x*0.4m*x*0.25m*)

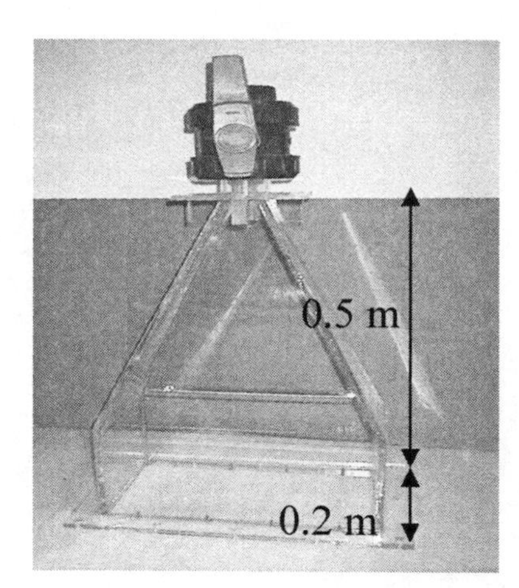

Figure 2: The single hopper

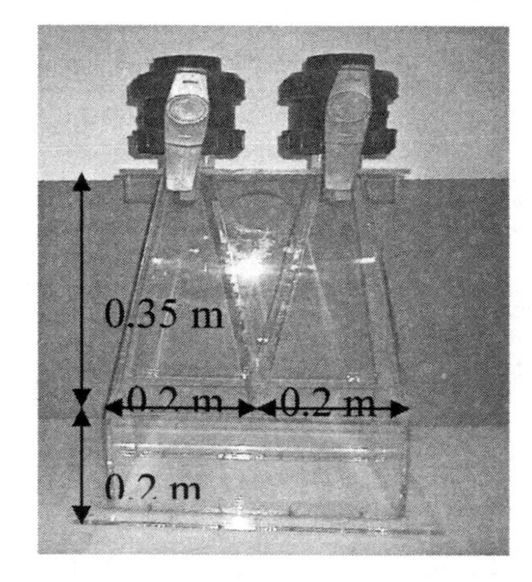

Figure 3: The double hopper.

Facilities

There are two main flow modes during silo emptying: mass flow and funnel flow. In mass flow, all the ensiled medium is in motion during the flow and the free surface shape inside the cell does not change. In funnel flow, a convergent flow channel is formed within the bulk solid, the slope cannot be predicted.

In order to state the occuring flow mode, two cameras were used. One camera overhanging on one side, aimed at observing the particles movements at the free surface. The second camera was placed on the side, and fixed on a metallic rolling structure, gifted by man-hand with the same motion as the free surface from top to down; this second camera is aimed at observing the motion of the upper peripheral particles. These two cameras are identical (25 images/sec) and are equipped with a zoom. The experiment is carried out in a blind room and an appropriate lighting is displayed. The video images taken by the cameras are stored in a Pentium 4 PC of 2.6 GHz, equipped with a 3-way station of image acquisition and analysis "Alliance Vision", 2.4 GHz. To analyze the images, two softwares were used: MaStudio for measuring the velocity and Photoshope for performing image by image treatment.

The average velocity of the free surface of a mass flow was analyzed using a "Levelflex" micro impulse system, a compact level transmitter designed to take continuous measurements in loose solids and liquids. This system is connected to a 24VDC / 4-20 mA power supply, and an echograph is used to interpret and record the measurements. Thanks to this system, the free surface level of a mass flow at an arbitrary frequency can be measured and recorded and, as a result, the vertical velocity can be derived. In funnel flow or semi-mass flow, this system can no longer be used since the stick has a transversal velocity. The levelflex setup was cross-validated with manual measurements of velocity. The mass flow rate was measured manually with a stop watch and scales, and compared with mass flow rate obtained using the levelflex equipment. It was also compared with the hereafter method and with the Beverloo formula as well.

IV EXPERIMENTAL STUDY

A Newtonian fluid discharging from a vertical silo is characterized by a horizontal free surface and a flow rate directly related to the fluid height inside the silo, providing the Torricelli formula for non viscous fluids and a viscometer for a viscous fluid. For a granular material, when the aperture is opened, the lower particles start falling and a surface develops in the form of a "V" (an inverted cone for a 3D hopper), moving downwards as the discharge proceeds ; the particles tumble down from this surface's periphery, and its periphery extends outward feeding the strong downward flow underneath the top surface's center. Therefore, a complete characterization of a silo discharge flow should provide not only the flow rate, but also the free surface profile. Degouet [13] classifies the main non intrusive methods used to measure the velocity flied. The paper by Steingart and Evans [14] is the only work concerned with the location and profile of the free surface, and uses a method based on particle image velocimetry (PIV). Their technique requires a transparent silo and therefore cannot be implemented on an industrial silo. In this work, a non intrusive measurement system based on a series of laser planes was developed which determine the evolution of the free surface's position and shape during the silo discharge, and, therefore, the instantaneous flow rate.

4.1 The measurement system

Laser sources (LD-LS0140 Laser levels) are positioned vertically to the silo. A prism-shaped device is placed on each laser source to rotate the laser ray in a plane. Laser planes are uniformly spaced in (y, z) planes parallel to the back wall of the silo. A top-view CCD sensor (camera) records the silo emptying. The CCD sensor must be positioned so that the silo can be completely viewed. Five lasers were used for the measurements. This number of lasers is a tradeoff solution to ensure the process is precise and compatible. If the laser planes are too close, the free surface receives too much light and it is difficult to extract the data. The experimental mechanism is shown in figure 1.

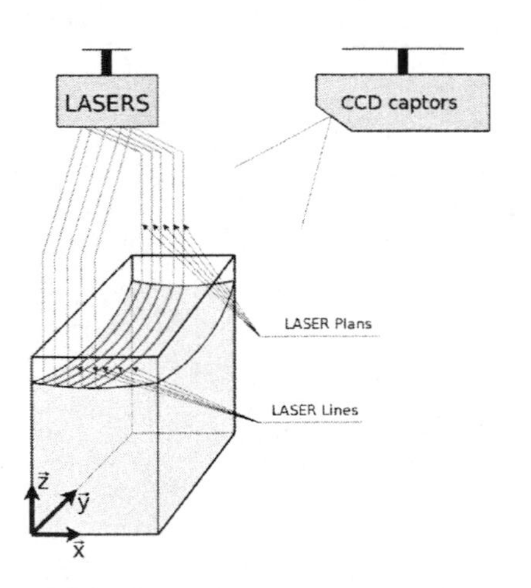

Figure 4: The laser measuring system.

The film shot shows the evolution of the trace of the laser planes on the free surface during the emptying. A program was developed, which processes the film and measures the height of the free surface during the emptying of the silo. The CCD sensor used here allowed to take 25 pictures per second in a 720×576 pixels format. A computer processed the data acquired from the film.

4.2 Calibration

For each laser trace on the free surface, a matrix of values coupling spatial coordinates to each pixel of the picture had to be worked out. To do this, a graduated wood plate was moved in the model. Using the CCD sensor, a series of pictures were taken from different known heights. Then, the graduated marks of each picture allowed matching a value of the coordinates (x, y, z) with the corresponding pixels. By interpolation, height, width and depth values could be matched to each pixel in a picture.

4.3 Picture processing

The emptying of a silo is filmed and gives the positions of the laser traces on the free surface over time. The processing consists of extracting by "thresholding" the pixels corresponding to the traces. The process called thresholding assumes that all the pixels with a red level below a given level are equal to zero and conversely, all the pixels with a red level above the same given value are equal to the same value. The thresholding result is thus a binary picture composed of black and white pixels. The blue and green levels of each pixel are not taken into account. This program automatically carried out this operation on all the individual pictures in the film. Thanks to the collation with the matrices obtained from the calibration, we were able to obtain the coordinates (x, y, z) of each pixel. The process was repeated for each laser trace on the free surface and for each picture of the film. By interpolating between the laser traces, the free surface's position and shape were obtained.

Thanks to this method, we were able to visualize the evolution of the free surface during different emptyings. We were also able to get an insight into the variations of the free surface's shape in real time (within the CCD sensor's acquisition frequency limits) at different locations in the silo and at different moments during the silo's emptying. In addition, due to the size of the particles used, the free surface was not completely smooth. Hence, a filtering by moving averages was performed to filter out the granulous aspect of the free surface. Finally, the instantaneous flow rate could be determlined from an analysis of two successive pictures.

4.4 Results

The flow mode depends on the grain and hopper characteristics (Munch-Andersen and Nielsen [15]). For the "flat plate" model, all the grains tested (soya, colza and rye seeds) undergo a semi-mass flow. For the "single and double hopper" model, all these granular materials studied undergo a mass flow. In addition, the Beverloo formula is valid for all the discharge flows we studied.

As shown above, the shape of the free surface does not change during a mass flow. The microimpulse system was used to determine the height of the ensiled material ; then the free surface's velocity was derived as the ratio between the the difference between two given heights versus the corresponding elapsed time, and finally the mass flow rate after multiplying it by the surface of the silo section and the density of the ensiled material was derived. Figure 5 shows a typical variation of the height versus time during a discharge flow of colza for a model silo equipped with a single hopper, while Figure 6 presents the corresponding variation of the instantaneous velocity. For all flow modes, that is mass, semi-mass or funnel flow, the laser technique could be used.

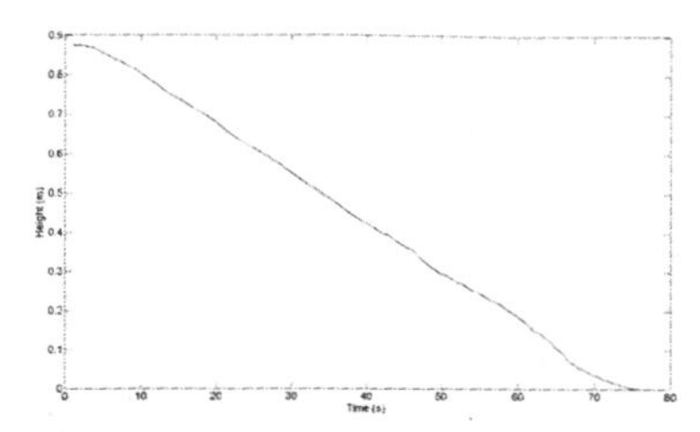

Figure 5: Height vs time variation provided by the microimpulse system during the discharging of ensiled colza seeds in a single hopper.

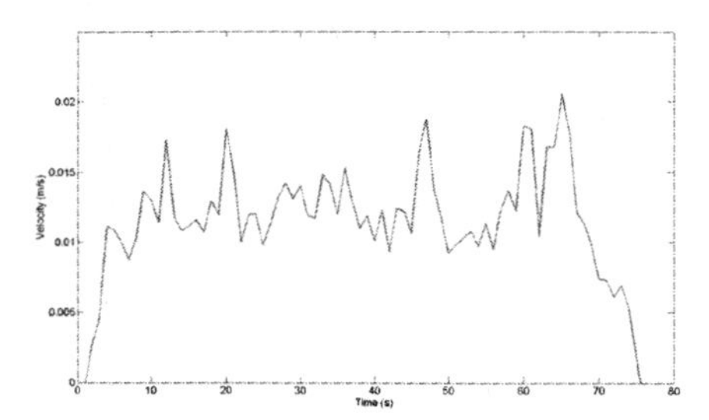

Figure 6: Time variation of instantaneous velocity provided by the microimpulse system during the discharge of ensiled colza seeds using a single hopper.

For all flow modes, that is mass, semi-mass or funnel flow, the laser technique could be used. Figure 7 shows a typical graph of the time variation of the free surface profile and the instantaneous mass flow rate.

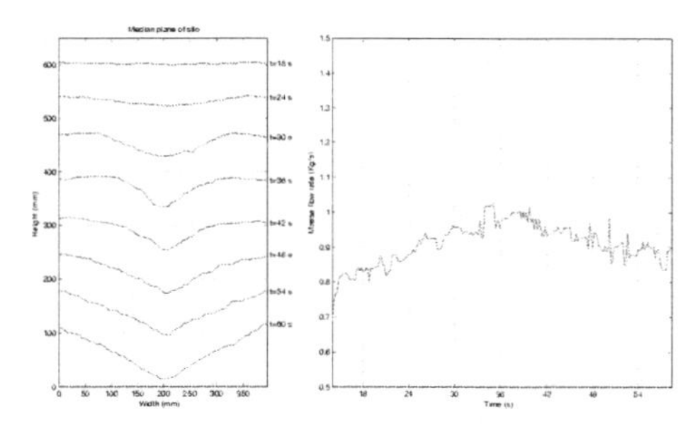

Figure 7 : Time variation of free surface profile and mass flow rate by the laser measuring system during the discharge of ensiled soya seeds using a flat bottom

Tables 3-5 show good agreement between the measurements taken by the laser system developed in this work and those obtained with the microimpulse system, the direct measurement and the Beverloo formula.

	Soya	Colza	Rye
Direct measurement	0.82 ± 0.02	0.98 ± 0.03	0.95 ± 0.02
Micro-impulse	0.89 0.06	1.03 ± 0.06	1.01 ± 0.07
Laser technique	0.91 ± 0.06	1.11 ± 0.08	1.01 ± 0.07
Beverloo et al. [9]	0.84 ± 0.02	0.97 ± 0.02	0.89 ± 0.02

Table 3: Average mass flow rate of silo discharge with the flat bottom.

	Soya	Colza	Rye
Direct measurement	0.69 ± 0.01	0.79 ± 0.02	0.74 ± 0.01
Micro-impulse	0.67 0.06	0.77 ± 0.05	0.71 ± 0.06
Laser technique	0.69 ± 0.05	0.87 ± 0.06	0.072 ± 0.05
Beverloo et al. [9]	0.65 ± 0.05	0.77 ± 0.05	0.69 ± 0.05

Table 4: Average mass flow rate of silo discharge with the single hopper.

	Soya	Colza	Rye
Direct measurement	1.82 ± 0.05	2.04 ± 0.06	1.95 ± 0.05
Micro-impulse	1.85 0.09	2.01 ± 0.09	1.62 ± 0.09
Laser technique	1.9 ± 0.1	2.1 ± 0.1	1.6 ± 0.1
Beverloo et al. [9]	1.9 ± 0.01	2.0 ± 0.01	1.9 ± 0.01

Table 5: Average mass flow rate of silo discharge with double hopper exit.

Furthermore, these tables show that for similar silo exit equipment, a good agreement exists with the flowability characteristics of the diverse studied matters.

Then, the technique was tested with different other types and sizes of openings on the model silo equipped with a flat bottom and filled with soya seeds: circular openings of 6.1 cm, 4.6 cm and 2.7 cm; rectangular openings of 2 cm, 4 cm, 6 cm, 8 cm and

10 cm wide.

Figures 8 and 9 hereafter show the time evolution of the shape and position of the free surface of the ensiled granular material for the circular and the rectangular openings, respectively. The comparison is made at $t = 0s$ and at every fifth of the total discharge duration. The different discharge durations are given in table 6.

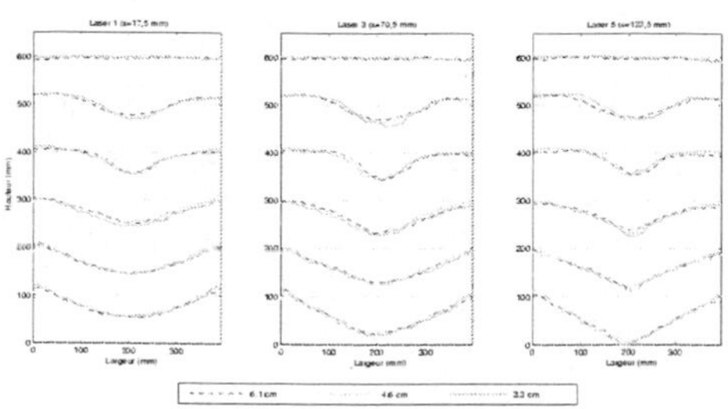

Figure 8. Time evolution of free surface shape and position during a discharge with circular openings

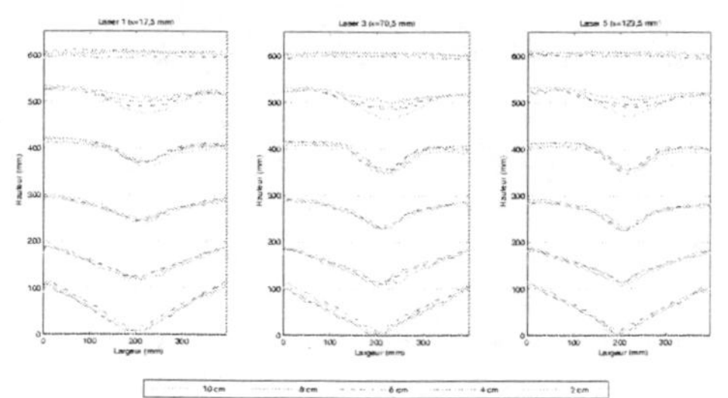

Figure 9. Time evolution of free surface shape and position during a discharge with rectangular openings

Figures 8 and 9 show that the diameter of a circular opening tested, has no influence on the shape of the free surface during a discharge process. While for the rectangular openings tested, the smaller the opening is, the faster the "V" forms on the free surface.

Finally, the flow rate was measured both with circular and rectangular openings and compared with the value obtained manually with a stop watch and scales and with the Beverloo formula as well.
Tables 6 and 7 show a satisfactory agreement between the mass flow rate obtained with the "laser technique" described in this paper and a direct manual measurement using circular and rectangular openings, respectively. Also, these tables give the discharge durations.
Tables 8 and 9 give the mass flow rate calculated using the Beverloo's formula [8] for the different circualr and rectangular openings respectively. It can be noted that hereafter shows that the "laser" measuring technique presented in this paper agrees well with the Beverloo formula for all the circular openings tested. While for the rectangular openings tested, the wider the opening is, the better the "laser" measuring technique presented in this paper agrees with the Beverloo formula [8].

	Discharge duration	Q "direct" (kg/s)	Q "laser" (kg/s)
Opening 6.1cm	47 ± 2	0.82 ± 0.04	0.89 ± 0.08
Opening 4.6cm	118 ± 2	0.33 ± 0.01	0.35 ± 0.04
Opening 3.3cm	342 ± 2	0.11 ± 0.01	0.11 ± 0.02

Table 6: Discharge duration and mass flow rate from direct measurement and from "laser technique" with different circular openings

	Discharge duration	Q "direct" (kg/s)	Q "laser" (kg/s)
Opening 2cm	63 ± 1	0.62 ± 0.01	0.63 ± 0.07
Opening 4cm	14.0 ± 0.4	2.84 ± 0.08	2.9 ± 0.1
Opening 6cm	6.3 ± 0.4	6.4 ± 0.4	6.6 ± 0.3
Opening 8cm	4.1 ± 0.4	10 ± 1	10.7 ± 0.4
Opening 10cm	2.7 ± 0.7	15 ± 2	16.8 ± 0.5

Table 7: Discharge duration and mass flow rate from direct measurement and from "laser technique" with different rectangular openings

	Q (kg/s): Beverloo et al. [8]
Opening: 6.1cm	0.84 ± 0.02
Opening: 4.6cm	0.36 ± 0.01
Opening: 3.3cm	0.12 ± 0.01

Table 8: Mass flow rate calculated using the Beverloo's formula [8] for different circular openings

	Q (kg/s): Beverloo et al. [8]
Opening: 2cm	1.77 ± 0.04
Opening: 4cm	4.54 ± 0.07
Opening: 6cm	8.0 ± 0.1
Opening: 8cm	11.8 ± 0.2
Opening: 10cm	15.8 ± 0.2

Table 9: Mass flow rate calculated using the Beverloo's formula [8] for different rectangular openings

V CONCLUSION

In this paper, an innovative non intrusive technique was developped for the flow rate measurement of a silo discharge of dry food grains.

Using the flow properties of the studied grains (soya, colza and rye seeds), and the characteristics of the grain/wall friction, a slender and completely transparent model silo was designed with diverse possible exit equipments, that is, a flat bottom, a single hopper and a double hopper. The flow mode (mass flow, semi-mass flow and funnel flow) depends on the nature of the studied grain and on the exit equipment used. In a semi-mass flow or in a funnel flow, the free surface of the ensiled material is not horizontal as in a Newtonian fluid ; the surface develops into the form of a "V" and the granular material can be viewed as a complex fluid.

Classical measurement systems were used to investigate the flow rate: microimpulse system for mass flow and direct measurement system (stop watch and scales) and the Beverloo formula were used for all configurations to cross-validate the results obtained with an original system developed in this work. Based on laser technique, this non intrusive method has an advantadge over the PIV method because it does not need transparent walls and consequently it can be implemented on industrial silos. The measured average mass flow rates, for the diverse silo equipments (flat bottom, single hopper and double hopper) agreed with the flowability characteristics of the studied materials (soya, colza and rye seeds).

ACKNOWLEDGEMENT

This work was funded by a PRIR contract, teaming "Port de Commerce de Brest", "Région Bretagne" and "Université de Bretagne Occidentale/IUT de Brest (LIME)".

REFERENCES

[1] A. Reimbert, and M. Reimbert. *Silos*, eds. eyrolles, Paris *(in French)*, 1956.

[2] A. Reimbert, and M. Reimbert. *Silos*, eds eyrolles, Paris *(in French)*, 1979.

[3] Jenike, A.W. (1964). Storage and flow of solids. **Bull. n°123**, *Eng. Exp. Station*, Univ. Utah, Salt lake City.

[4] Brown, C.J. and Nielsen, J. (1998). "Silos. Fundamentals of theory, behaviour and design", Eds. E & FN SPON, London.

[5] Teunou, E. and Fitzpatrick, J.J. (2000). Effect of storage time and consolidation on food powder flowability, *J. Food Eng.*, **43**, 97-101.

[6] Neddermann, R.M. The measurement of the velocity profile in a granular material discharging from a conical hopper, *Chemical Engineering Science*, **43 (7)**, 1988, pp. 1507–1516.

[7] Tüzün, U., Houlsby, G. T., Nedderman, R. M. and Savage, S. B. The flow of granular materials ii, velocity distributions in slow flow, *Chemical Engineering Science*, **37 (12),** 1691–1709.

[8] Beverloo, W. A., Leniger, H. A. and Van de Velde, J. (1961). The flow of granular solids through orifices, *Chemical Engineering Science,* **15** , 260–269.

[9] Bideau, D. and Ammi, M. (2001). Ecoulements gravitaires: sabliers et silos, in « Mécanique des milieux granulaires », pp. 75-114, Eds. Jack Lanier, Hermes Science europe Ltd, Paris, *(in French).*

[10] Rose, H.F. and Tanaka, T. (1959). "Rate of discharge of granular materials from bins and hoppers", The Engineer, London.

[11] Degouet, C., Nsom, B., Lolive, E. and Grohens, A. (2006). Physical and mechanical characterization of soya, colza and rye seeds, *Accepted Appl. Rheol.*

[12] Nielsen, J. and Askegaard, V. (1977). Scale errors in model tests on granular media with special reference to silo models, *Powder Technology*, **16**, 123-130.

[13] C. Degouet. *Caractérisation de matériaux granulaires et de leurs écoulements dans les silos verticaux*, Ph. D. Thesis, Brest, France, 2005 *(in French).*

[14] D. A. Steingart and J. W. Evans. "Measurements of granular flows in two-dimensional hoppers by particle image velocimetry. part i: experimental method and results", *Chemical Engineering Science*, **60**, pp. 1043–1051, 2005.

[15] J. Munch-Andersen and J. Nielsen "Size effects in slender grain silos", *The International Journal of Storing and Handling of Bulk Materials*, **6(5)**, pp. 885-889, 1986.

A Blind Watermarking Algorithm Based on Modular Arithmetic in the Frequency Domain

Cong Jin[1], Zhongmei Zhang[1], Yan Jiang[1], Zhiguo Qu[1], Chuanxiang Ma[2]
1 Department of Computer Science,
Central China Normal University,
Wuhan 430079, P.R.China
2 School of Mathematics and Computer Science,
Hubei University,
Wuhan 430062, P.R. China

***Abstract*-Robustness is the important issue in watermarking, robustness at the same time with blind watermark recovering algorithm remains especially challenging. This paper presents a combined DWT and DCT still image blind watermarking algorithm. The two-level DWT are performed on the original image, the low-frequency sub-band is divided into blocks, the DCT is performed on the every block, the DCT coefficients of every block are sorted using Zig-Zag order, the DCT low-frequency coefficient is selected as embedding watermarking. The watermarking signals are embedded into the selected embedding points using the modular arithmetic. The watermark recovering is the inverse process of the watermark embedding, according to the answer of the modular arithmetic, we can estimate the value of embedded the watermark. The algorithm is compared with a pure DWT-based scheme Experiment results shown that the proposed algorithm is robust to many attacks such as JPEG compression, addition noise, cropping, JPEG compression, median filter, rotation, and resize etc. Proposed algorithm is shown to provide good results in term of image imperceptibility, too.**

I. Introduction

With the development of the computer network and the multimedia technology, protecting the information on the Internet is a very important issue at this time. Digital watermarking is a useful technique for copyright protection. By adding an invisible signal into the original media, it is hoped that the signal will provide an evidence of legal ownership or at least help the owner to detection copyright Violations [1].

Many watermarking algorithms, mainly focusing on the invisibility and robustness, have proposed. How to keep their balance is the research issue. The first group of techniques work in the spatial domain, for example, changing the LSB of some pixels, recording the difference between randomly selected pairs of points, etc [2]. These techniques can suffer from signal compression and hostile attacks. Another group of techniques work in the spatial-frequency domain and add a watermark by manipulating various frequency elements. Frequency domain techniques are much more robust against compression and geometrical transformations than spatial domain techniques.

In many frequency domain techniques, DWT and DCT attract researcher's attention because of their good performances. DWT provides a multi-resolution image representation and has become one of the most important tools in image analysis and coding over the last two decades [3] Wavelet-based watermarking methods exploit the frequency information and spatial information of the transformed data in multiple resolutions to gain robustness. DCT is a technique for converting a signal into elementary frequency components. The DCT is often used in signal and image processing, especially for data compression, because it has a strong "energy compaction" property: most of the signal information tends to be concentrated in a few low-frequency components of the DCT [4]. DWT and DCT are adopted respectively by many watermarking researcher in the last two decades.

An excellent watermarking algorithm is robust against some image processing such as median filtering, JPEG compression, cropping, adding noise etc. Because the good performance of DWT and DCT in image processing, we use the multi-resolution that DWT provided and the strong energy compaction that DCT provided at same time in a watermarking algorithm to gain robustness.

This paper proposed a new watermarking algorithm combining DWT and DCT. The two-level DWT is performed on the original image at first; the DCT is performed on the low-frequency sub-band in the next step, selecting the watermark embedding point in the DCT coefficients. The watermarking signals are embedded by using the modular arithmetic at the selected embedding points. Recovering watermark is the inverse process of the watermark embedding, the whole process need not original image, and it is a blind watermarking algorithm.

The watermarking process includes watermark embedding and watermark recovering. A brief knowledge about the modular arithmetic is introduced in section 2, the process of watermark embedding is described in section 3, watermark recovering is described in section 4, and the experimental results will be shown in section 5.

II. Modular Arithmetic of Real Number

Modular (often also Modulo) arithmetic is an unusually versatile tool discovered by K.F.Gauss. It is a modified system of arithmetic for integers, sometimes referred to as "clock arithmetic", where numbers "wrap around" after they reach a

K. Elleithy (ed.), Advances and Innovations in Systems, Computing Sciences and Software Engineering, 543–547.

certain value (the modulus). For example, whilst 8+6 equals 14 in conventional arithmetic, modulo 12 and the answer is two, as two is the remainder after dividing 14 by the modulus 12.

In this paper, modular arithmetic is defined between real number α and the natural number γ ($\gamma > 1$). The calculation rule is defined as follow: if α is a positive real number, the answer is the residue of α being divided by γ ; if α is a negative real number, firstly, η is the value of γ decreasing absolute value of α , the modular is the residue of η being divided by γ .

III. Watermarking Embedding

In most of the watermark approaches, the watermark is the binary signal and random signal, 8-bit gray image is hardly selected as the watermark signal because the computational complexity is high. But the 8-bit gray image is selected as the Watermarking signal, the embedded watermark signal capacity is high, it is also an important issue that more watermark signals are embedded into the image while achieving invisibility and robustness of image. In this paper, the cover image and the watermarking image are 8-bit gray image. The size of cover image is $M\times M$.

At first, the original image is decomposed into its high pass and low pass sub-bands by using DWT with dauchier-1 filter, daubechier-1 filter is also called the Haar filter, it is popular in image processing. A two-level image with seven sub-bands is gained, which include a low pass sub-band *LL* and six high pass sub-bands.

Many proposed algorithms based DWT embedded watermark image into the high pass sub-bands in order to gain more robust image [6], in this paper, the watermark image is embedded into the low pass sub-band.

Secondly, the *LL* sub-band is considered as a small image, the small image is divided into blocks $B_t(t=1,2,\ldots,n\times n)$, the size of each block is $m\times m$. The DCT of B_t is computed, the DCT coefficients of each block are reordered into a zig-zag scan. Eight points are selected as the watermark embedded points, the locations from the $(L+1)$th to $(L+8)$th in B_t are selected to embed watermark image in every block $(0<L+1<L+8\le m\times m)$.

Thirdly, the each pixel $W(x,y)$ of 8-bit gray watermarking image is converted into a 8-bit binary sequence $F_{x,y}(a)(a=1,2,\ldots,8)$, one bit is embedded into one point, (x,y) is the coordinate of the pixel in the watermarking image.

Fourth, the DCT coefficient of $B_t(i,j)$ selected as the embedding point is modified according to the following rule:

$$s=\text{mod}(B_t(i,j),2\times\beta) \quad (1)$$

$$c=\begin{cases}\beta/2-s & if \quad F_{x,y}(a)=0\\ 3\beta/2-s & if \quad F_{x,y}(a)=1\end{cases} \quad (2)$$

$$B'_t(i,j)=B_t(i,j)+c \quad (3)$$

Which $B'_t(i,j)$ is the DCT coefficient of embedded watermarking signal, t is the block number, (i,j) is the coordinate of embedding point. β is a natural number, and β is 14 and the experiment results are best in the algorithm proposed in this paper.

Finally, B_t is transformed by IDCT, and the watermarked image is obtained.

IV. Watermarking Recovering

The process of the watermark recovering is the inverse process of the watermark embedding. The concrete step is as follow.

Firstly, we begin with DWT and DCT operation on the watermarked image similar to the embedding process.

Secondly, the DCT coefficient of each block are reordered by zig-zag scan, selecting coefficient from the $(L+1)th$ to $(L+8)th$ $(0<L+1<L+8\le m\times m)$, and the operations as following are performed for the selected coefficient $B'_t(i,j)$.

$$p=\text{mod}(B'_t(i,j),2\times\beta) \quad (4)$$

$$\begin{cases}F_{x,y}(a)=0 & if \quad 0\le p\le\beta\\ F_{x,y}(a)=1 & if \quad \beta\le p\le 2\beta\end{cases} \quad (5)$$

Finally, the 8-bit binary sequence is converted into an integer range from 0 to 255; the integer is the coefficient of the coordinateof the extracted watermark image W'.

V. Experiment Results Analysis

Using the watermark embedding and recovering process explained in Section 3 and 4, five images are tested. The five images are all grayscale 8-bit image and size is 512×512 , the watermarking is also a grayscale 8-bit image of size 32×32 . The factors evaluated are watermark embedding imperceptibility and robustness. The algorithm is compared with a pure DWT-based scheme [6].

A. Imperceptibility Evaluation

The imperceptibility quality of watermarked image is measured by the PSNR of the image [7][8], and PSNR is calculated using equation (6). In this paper, the value of the PSNR for each image is as follows in Fig.1-6.

$$PSNR=10\lg\frac{255^2}{\frac{1}{MN}\sum_{i=1}^{M}\sum_{j=1}^{N}(I(i,j)-I'(i,j))} \quad (6)$$

Where $M\times N$ is the image size, I is the original image, I' is the watermarked image.

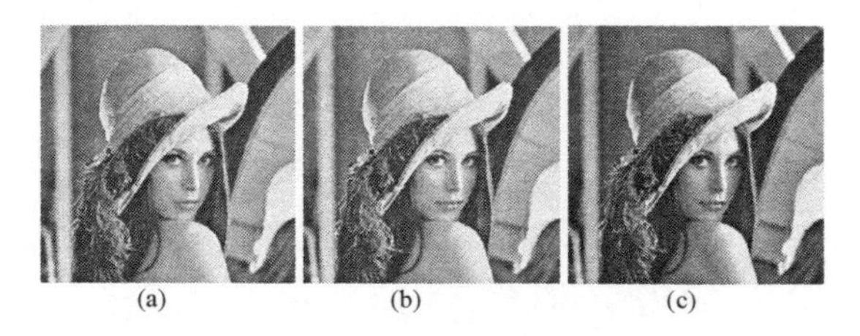

(a) (b) (c)

Fig. 1 (a) is the original image *"lena"*; (b) is the watermarked image by using the algorithm proposed in this paper; the PSNR is 41.6333. (c) is the watermarked image by using the algorithm proposed in [6], the PSNR is 39.8219.

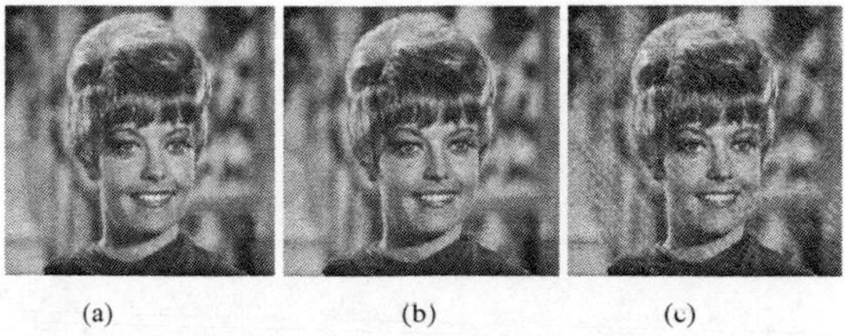

(a) (b) (c)

Fig. 2 (a) is the original image *"zelda"*; (b) is the watermarked image by using the algorithm proposed in this paper; the PSNR is 39.2513. (c) is the watermarked image by using the algorithm proposed in [6], the PSNR is 39.5750.

(a) (b) (c)

Fig. 3 (a) is the original image *"sailboat"*; (b) is the watermarked image by using the algorithm proposed in this paper; the PSNR is 41.6602. (c) is the watermarked image by using the algorithm proposed in [6], the PSNR is 398219.

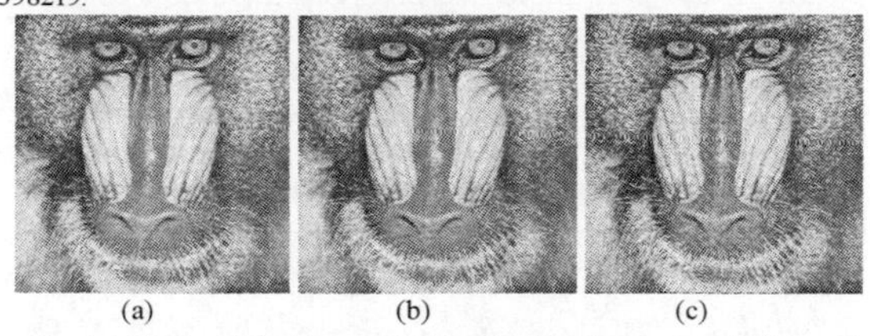

(a) (b) (c)

Fig. 4 (a) is the original image *"baboon"*; (b) is the watermarked image by using the algorithm proposed in this paper; the PSNR is 41.5832. (c) is the watermarked image by using the algorithm proposed in [6], the PSNR is 37.5395.

(a) (b) (c)

Fig. 5 (a) is the original image *"elain"*; (b) is the watermarked image by using the algorithm proposed in this paper; the PSNR is 41.7581. (c) is the watermarked image by using the algorithm proposed in [6], the PSNR is 42.4230.

These figures are shown that the proposed algorithm was used in different kinds of grayscale 8-bit images, although these images are different from each other in background luminance and texture, the imperceptibility quality of watermarked image are all good.

B. Robustness Evaluation

Robustness tests are carried out with conventional attacks such as JPEG compression, addition Salt & Pepper noise, addition Gaussian noise, cropping, resize the watermarked image etc. The robustness is estimated by the value of NMSE, NMSE considers the normalized mean square error between the original watermark and the recovered watermark. NMSE is calculated using equation (7).

$$NMSE = \sum_{i=1}^{M}\sum_{j=1}^{N}(I(i,j) - I'(i,j))^2 / \sum_{i=1}^{M}\sum_{j=1}^{N} I^2(i,j) \tag{7}$$

a) Salt & Pepper Noise

Salt & Pepper noise is data drop-out noise. The noise is caused by errors in the data transmission. The corrupted pixels are either set to the maximum value or have single bits flipped over. In some cases, single pixels are set alternatively to zero or to the maximum value, giving the image a `Salt & Pepper' like appearance. The noise is usually quantified by the percentage of pixels, which are corrupted [9]. The Salt & Pepper noise density is 0.001, 0.002, 0.003, 0.005 respectively. Fig.6 shows some of the recovered watermarking image taking the image "*Lena*" as an example. Table 1 shows the partial value of NMSE with two embedding algorithm when adding Salt & Pepper noise into each watermarked image.

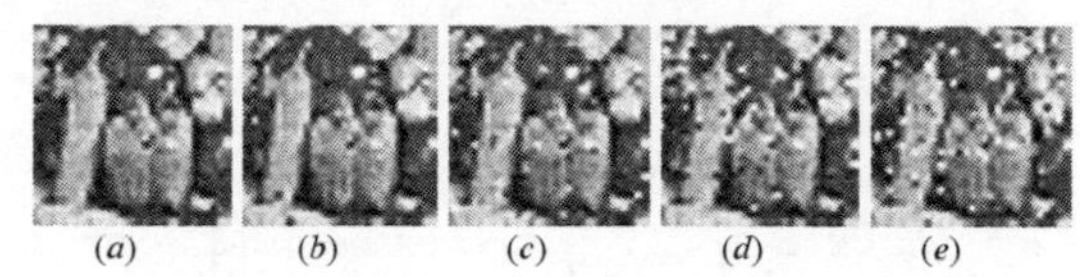

(a) (b) (c) (d) (e)

Fig. 6. (a) is the recovered watermark image; (b) is the recovered watermark image (noise density is 0.001); (c) is the recovered watermark image (noise density is 0.002); (d) is the recovered watermark image (noise density is 0.003); (e) is the recovered watermark image (noise density is 0.005).

TABLE I
THE VALUE OF NMSE WITH TWO WATERMARKING WHEN ADDING SALT & PEPPER NOISE INTO WATERMARKED IMAGE

Image	Salt and pepper noise					
	Algorithm proposed in paper			Algorithm proposed in [6]		
	Noise density			Noise density		
	0.002	0.003	0.005	0.002	0.003	0.005
lena	0.9173	0.9037	0.8445	0.8826	0.8315	0.8011
sailboat	0.9230	0.8172	0.8054	0.8589	0.8043	0.7691
baboon	0.9460	0.8990	0.8368	0.8749	0.8328	0.8362
zelda	0.8571	0.8265	0.7932	0.8749	0.8248	0.7770
elain	0.9474	0.8661	0.8548	0.8976	0.8487	0.8100

b) Gaussian Noise

Gaussian noise is a type of signal processing operation. The amount of noise is controlled by its mean and variance [8]. Gaussian white noise of mean is 0 and variance is 0.0005, 0.0003, 0.0001, 0.0002 respectively. Fig.7 shows some of the recovered watermarking image taking the image "*Baboon*" as an example. Table 2 shows the partial value of NMSE with two embedding algorithm when adding Gaussian noise into each watermarked image.

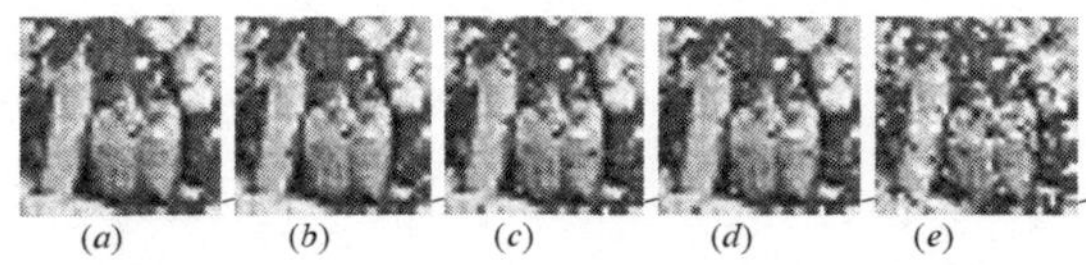

Fig.7. (a) is the recovered watermark image; (b) is the recovered watermark image(variance is 0.0001); (c) is the recovered watermark image (variance is 0.0002) ; (d) is the recovered watermark image (variance is 0.0003); (e) is the recovered watermark image (variance is 0.0005).

TABLE II
THE VALUE OF NMSE WITH TWO WATERMARKING ALGORITHM WHEN ADDING GAUSSIAN NOISE INTO WATERMARKED IMAGE

Image	Gaussian noise insertion					
	Algorithm proposed in paper			Algorithm proposed in [6]		
	variance			variance		
	0.0002	0.0003	0.0005	0.0002	0.0003	0.0005
lena	0.7936	0.7485	0.6668	0.7829	0.7564	0.7000
sailboat	0.7930	0.7264	0.6442	0.6621	0.6451	0.6098
baboon	0.7929	0.7381	0.7339	0.7691	0.7429	0.7042
zelda	0.7946	0.7781	0.6880	0.7702	0.7838	0.6655
elain	0.8693	0.8476	0.7555	0.8237	0.8020	0.7167

c) JPEG Compression

JPEG compression is one of the common compression attacks on digital watermarking. In the experiment, JPEG compression quality factor is 75%, 60%, 50%, 40%, respectively. Fig.8 shows some of the recovered watermarking image taking the image "*Elain*" as an example. Table 3 shows the partial value of NMSE with two embedding algorithm in JPEG compression. Table 3 shows the value of NMSE with two watermarking algorithms after watermarked image JPEG compressed.

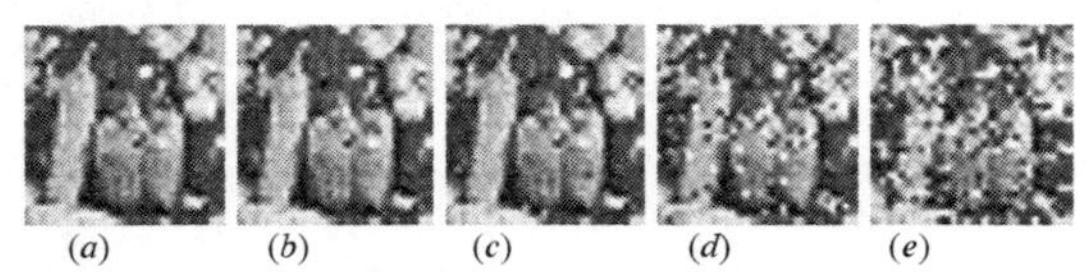

Fig. 8. (a) is the recovered watermark image; (b) is the recovered watermark image (JPEG compression quality factor 75%); (c) is the recovered watermark image (JPEG compression quality factor 60%) ; (d) is the recovered watermark image (JPEG compression quality factor 50%) ; (e)is the recovered watermark image (JPEG compression quality factor 75%).

TABLE III
THE VALUE OF NMSE WITH TWO WATERMARKING ALGORITHM AFTER WATERMARKED IMAGE JPEG COMPRESSED

Image	JPEG compression					
	Algorithm proposed in paper			Algorithm proposed in [6]		
	quality factor			quality factor		
	75%	60%	40%	75%	60%	40%
lena	0.8612	0.8085	0.6916	0.8120	0.7538	0.5016
sailboat	0.8539	0.8142	0.6562	0.6485	0.5631	0.3231
baboon	0.8529	0.7788	0.6562	0.8521	0.7560	0.6621
zelda	0.8640	0.8252	0.7000	0.8089	0.7825	0.5916
elain	0.8577	0.8156	0.6967	0.7804	0.7546	0.5188

d) Cropping

Cropping represents data reduction attack [6]. In experiment 8×8 , 16×16 , 32×32 ,64×64 squares at watermarked image are cropped out at image center, Fig.9 shows some of the recovered watermarking image taking the image "*Sailboat*" as an example. Table 4 shows the partial value of NMSE with two embedding algorithm in cropping.

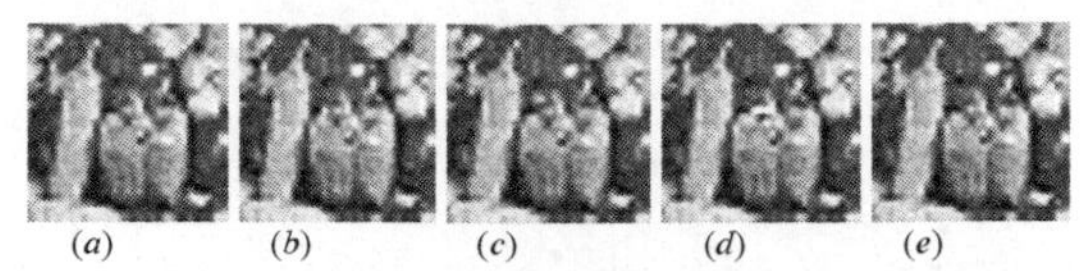

Fig. 9. (a) is the recovered watermark image; (b) is the recovered watermark image(the window is 8×8); (c) is the recovered watermark image (the window is 16×16); (d) is the recovered watermark image (the window is 32×32); (e) is the recovered watermark image (the window is 64×64).

TABLE IV
THE VALUE OF NMSE WITH TWO WATERMARKING ALGORITHM AFTER CROPPING

Image	Cropping					
	Algorithm proposed in paper			Algorithm proposed in [6]		
	window			window		
	16×16	32×32	64×64	16×16	32×32	64×64
lena	0.966	0.9384	0.9209	0.9318	0.9264	0.9126
sailboat	0.9802	0.9727	0.9619	0.9232	0.9113	0.9054
baboon	0.9857	0.9684	0.9509	0.9228	0.9142	0.9139
zelda	0.9557	0.9383	0.9279	0.9134	0.9086	0.9020
elain	0.9645	0.9538	0.9212	0.9167	0.9030	0.9008

e) Median Filter and Global Geometrical Distortion

Median filtering is a type of non-linear filtering that produces a "smoother" image. Global geometrical distortion such as rotation is a big challenge. A small degree of rotation usually retains visual appearance while damaging watermark information. Normally, correlation-based watermark detection is vulnerable to such attack [6]. A 2D median filtering using 3× 3 neighborhood and 60 degrees rotation at image centre are employed in the experiment. Table 5 shows the partial value of NMSE with two embedding algorithm in median filter and global geometrical distortion.

TABLE V
THE VALUE OF NMSE WITH TWO WATERMARKING ALGORITHM AFTER MEDIAN FILTER AND GLOBAL GEOMETRICAL DISTORTION

Image	Median filter and global geometrical distortion			
	Algorithm proposed in paper		Algorithm proposed in [6]	
	Median filter (3×3)	Rotation (60°)	Median filter (3×3)	Rotation (60°)
lena	0.7330	0.6553	0.4091	0.4005
sailboat	0.6359	0.6412	0.2296	0.3254
baboon	0.5941	0.6608	0.3902	0.4141
zelda	0.8132	0.5989	0.1024	0.3038
elain	0.7771	0.6371	0.1786	0.2685

f) Resize

Resize is an attack of changing the image size. Watermarked image is magnified 1.5 times, 2.5 times, and then reduced 2/3, 2/5 in the experiment. Table 6 shows the value of NMSE with two embedding algorithm in resizing the watermarked image.

TABLE VI
THE VALUE OF NMSE WITH TWO WATERMARKING ALGORITHM AFTER RESIZING THE WATERMARKED IMAGE

Image	Resize image			
	Algorithm proposed in paper		Algorithm proposed in [6]	
	reduce2/3 magnify1.5	magnify1.5 reduce2/3	reduce2/3 magnify 1.5	magnify1.5 reduce2/3
lena	0.5973	0.9865	0.4235	0.9326
sailboat	0.4972	0.9865	0.2296	0.9326
baboon	0.5014	0.9865	0.4710	0.9326
zelda	0.6382	0.9865	0.2935	0.9326
elain	0.5384	0.9865	0.3538	0.9326

The proposed algorithm are experimented using many attacks, according to comparing with the pure DWT algorithm, the experimental results shown that the algorithm combining DWT and DCT was better than the pure DWT watermarking algorithm in robustness.

VI. CONCLUSION

An efficient embedding technique is demonstrated. The watermarking algorithm used the multi-resolution of DWT and the energy compaction of DCT, combined DWT and DCT at same time to gain robustness, experiment results shown that the watermarking algorithm is robust for the JPEG compression, noise insertion, cropping, median filter, resizing and rotation etc. Watermark signal is 8-bit grayscale image, embedded watermark capacity is more while achieving imperceptibility and robustness of image. Although the embedding process is simply, the watermark image recovering need not original image, it is a blind watermarking technique. The proposed embedding technique promise good performances in terms of imperceptibility and robustness.

ACKNOWLEDGMENTS

This research was supported by the National Foundation of China under Grant No. 60603069.

REFERENCES

[1]. Tao B. and Dickinson B, "Adaptive Watermarking in the DCT Domain," *Proceedings IEEE International Conference on Acoustics, Speech and Signal Processing, Munich, German*, vol..4 , pp.21-24, 1997.

[2] Mayank Vatsa, Richa Singh, "Robust Biometric Image Watermarking for Fingerprint and Face Template Protection," *IEICE Electronics Express*,. vol.3, pp. 23-28, 2006.

[3] Grgic S., Grgic M., Zovko-Cihlar B, "Performance analysis of image compression using wavelets," *IEEE Transactions on Industrial Electronics*, vol.48, pp.682-695,2002

[4] Wattson A. B,. "Image Compression using the Discrete Cosine Transform," *Mathematics Journal*, vol 4, pp 81-88, 1994.

[5] Nikolaidis N, Pitas I, "Robust Image Watermarking in the Spatial Domain," *Signal Processing*, vol. 66, pp. 385-403, 1998.

[6] Chaw-Seng Woo, Jiang Du, and Binh Pham, "Performance Factors Analysis of a Wavelet based Watermarking Method," *Third Australasian Information Security Workshop (AISW2005), 30 January-3 February, Newcastle, Australia*, vol. 44, pp. 89-97, 2005

[7] Voloshynovskiy S., Pereira S., Iquise V., and Pun T, "Attack Modeling Towards a Second Generation Watermarking," *Benchmark. Signal Processing*, vol. 81, pp. 1177-1214, 2001.

[8] Watson A.B., Yang G.Y., Solomon J.A., Villasenor J, "Visibility of Wavelet Quantization Noise," *IEEE Transactions on Image Processing*,, vol. 6, pp. 1164-1175, 1997.

[9] Boukerrou, Kamel, Kurz, and Ludwik, "Suppression of 'Salt and Pepper' Noise Based on Youden Designs," *Information Sciences*, vol. 110, pp. 217-235, 1998.

Determination of Coordinate System in Short-Axis View of Left Ventricle

Gaurav Sehgal[1], Dr. Gabrielle Horne[2], Dr. Peter Gregson[3]
[1,3]Department of Electrical & Computer Engineering,
C-Building, 1360 Barrington St., Sexton Campus,
Dalhousie University, Halifax, NS - B3J 1Z1, Canada.
[2]Department of Medicine (Cardiology) and Biomedical Engineering,
Dentistry Building, 5981 University Avenue,
Dalhousie University, Halifax, NS - B3H 1W2, Canada.

***Abstract*-With the increasing rate of myocardial infarction (MI) in men and women, it is important to develop a diagnosis tool to determine the effect of MI on the mechanics of the heart and to minimize the effect of heart muscle damage on overall cardiac performance. After a myocardial infarct, the left ventricle of the heart enlarges to compensate for a weak heart muscle. The enlarged and weakened heart gives rise to the clinical syndrome of heart failure. In order to maximize the mechanical performance of the weakened heart, regional ventricular loading and contraction must be understood. To isolate regional wall mechanics, a floating centroid for the left ventricle must to be calculated. This is easy in the normal heart where the left ventricle approximates a single radius of curvature; however in heart failure there are irregular shape changes that complicate this calculation. The conventional method used for centroid calculation employs a center of mass (COM) determination of the whole left ventricle. This method has many shortcomings when applied to an enlarged and irregular left ventricle. This paper proposes a new algorithm for centroid calculation based on iterative majorization to locate the centroid.**

I. Introduction

The heart is one of the most important organs in the human body, with complex functionality and architecture. The heart is primarily considered to be like a shell, with four open cavities or spaces that fill with blood. Two of these are called atria while the other two are called ventricles. The left side of the heart houses one atrium and one ventricle, while the right side houses the other two. A wall called the septum separates the right and left side of the heart. The average heart muscle called the cardiac muscle, contracts and relaxes about 70 to 80 times per minute [1]. As the heart works the hardest of all the organs in the human body, it needs nourishment too. While the circulatory system is busy providing oxygen and nourishment to every cell of the body, the coronary circulatory system is responsible for nourishment of the heart. Coronary circulation refers to the movement of blood through the tissues of the heart [1].

In order to understand the functioning of the heart, an imaging technique called "Cardiac Ultrasonography (Echocardiography)" is used. Echocardiography uses very high frequency (>20,000 Hz) sound waves to visualize the details of heart anatomy [2]. 2D echocardiographic images are most widely used for examining the working of the heart. 2D echocardiography allows a plane of tissue (both depth and width) to be imaged in real time. In this paper, echocardiographic images of the short-axis view of the heart at the level of left ventricle are used for algorithm development and testing as shown in Fig. 1.

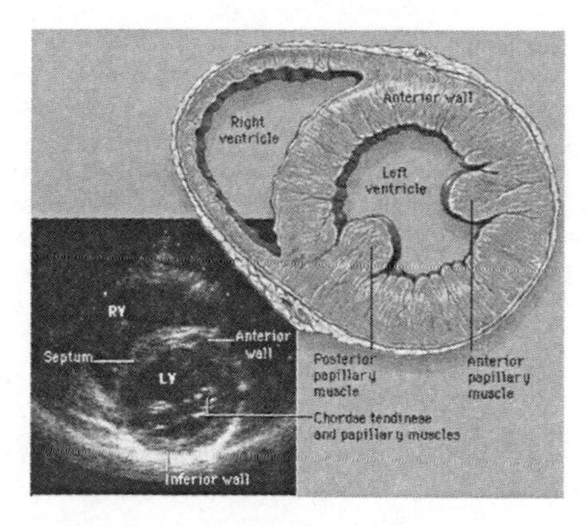

Fig. 1. Echocardiograph image of short-axis view of human heart at left ventricle level [3].

A. Heart Failure

Heart failure is a condition in which the heart has lost the ability to pump enough blood to the body's tissues. With too little blood being delivered, the organs and other tissues do not receive enough oxygen and nutrients to function properly [4]. One of the most common causes of heart failure is atherosclerosis [5], which is the blockage of coronary arteries that supply blood and oxygen to heart tissues. Blockage in arteries occurs due to various factors like smoking, high blood pressure and other mechanical and biochemical forces. After heart failure, the left ventricle enlarges to compensate for the weak heart muscles [5] as shown in Fig. 2.

K. Elleithy (ed.), Advances and Innovations in Systems, Computing Sciences and Software Engineering, 549–554.

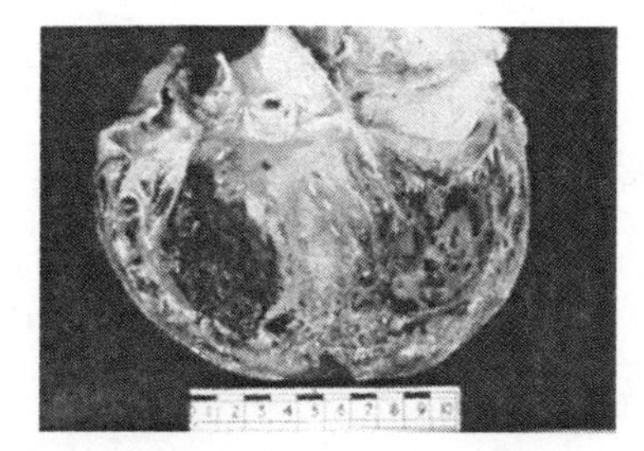

Fig. 2. A heart image after heart failure with the left ventricle split open [4].

B. *The Need for a Coordinate System*

In order to maximize the mechanical performance of a weakened heart, regional ventricular loading and contraction must be understood. Septum movement also needs to be understood with reference to right ventricular pressure. Fig. 3 shows the septum movement in terms of chord shortening in patients with high right-ventricular pressure (JVP) and patients with low right ventricular pressure (N-JVP) [6]. It is evident that there is more septum movement in the case of N-JVP as compared to JVP.

This septum movement plays an important role in patients with heart failure, as it compensates for weak heart muscles [6, 7]. In order to isolate regional wall mechanics and septum movement, a floating center point or centroid for the ventricle must be determined [6]. This centroid can be used to produce 32 radial chords along which the amount of septum movement is determined as shown in Fig. 4. These are the distances that are shown in Fig. 3 for patients with JVP and N-JVP cases. These radial chords help in determining the approach to be taken for increasing or decreasing septum wall movement [6, 7, 8], by varying right ventricular pressure, so that the septum can compensate for weak heart muscles and increase the life span of patients with heart failure.

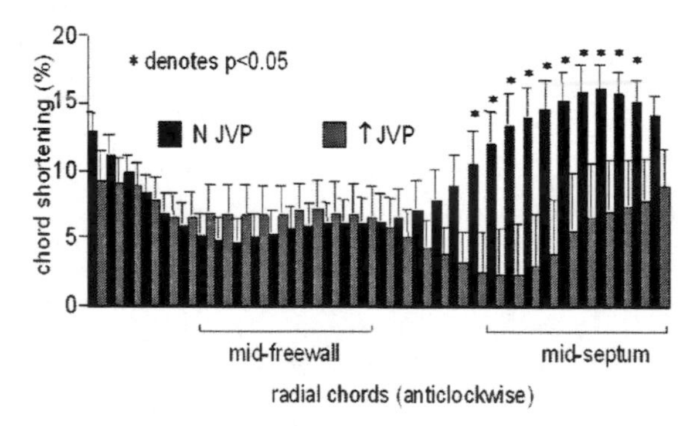

Fig. 3. Septum movement [6].

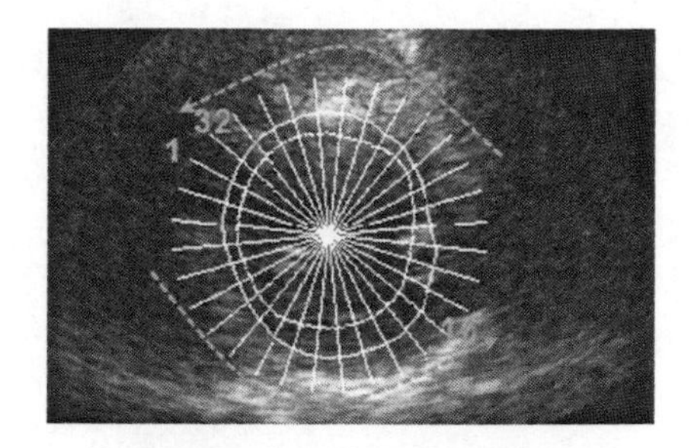

Fig. 4. 32 radial chords drawn in anticlockwise direction on the endocardial region of the left ventricle.

II. Conventional Method

The endocardial region of the left ventricle consists of the septum wall and the freewall region as shown in Fig. 5. The outlining of the endocardial region of the left ventricle is a semi-automated process. The user clicks on the points on the endocardial boundary and the points are connected with one another using spline interpolation. The conventional method used for determining a floating centroid employs a center-of-mass calculation for the entire endocardial region [9, 10]. This method, although easy and requiring less computation, is not always correct. The reason for this is that there is septum movement during each cardiac cycle which makes centroid calculation erroneous.

Fig. 6 shows the center of mass for the left ventricle when the septum is relaxed, whereas Fig. 7 shows the center of mass for the left ventricle with the septum flattened. It can be seen that there is displacement of the center of mass. Although the center of mass calculated is correct according to the definition, this does not serve the purpose since center of mass is used as the centroid for calculating septum movement. As there is movement of the center of mass, the estimated septum movement will be erroneous.

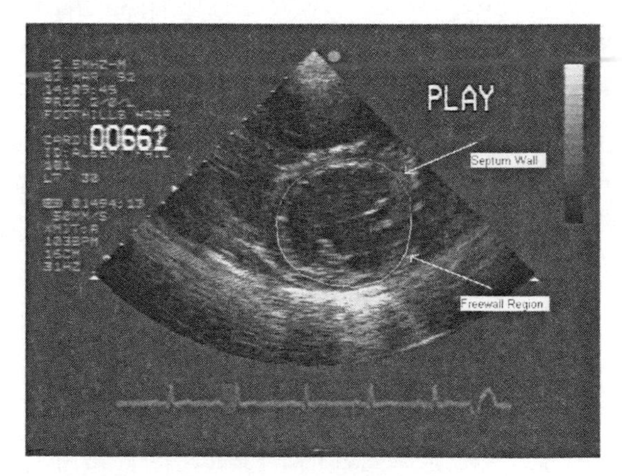

Fig. 5. Septum and Freewall Regions in the left ventricle.

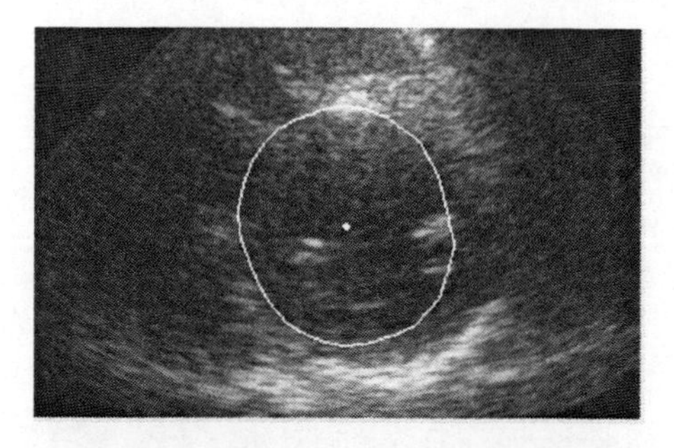

Fig. 6. Center of mass of the left ventricle with the septum relaxed.

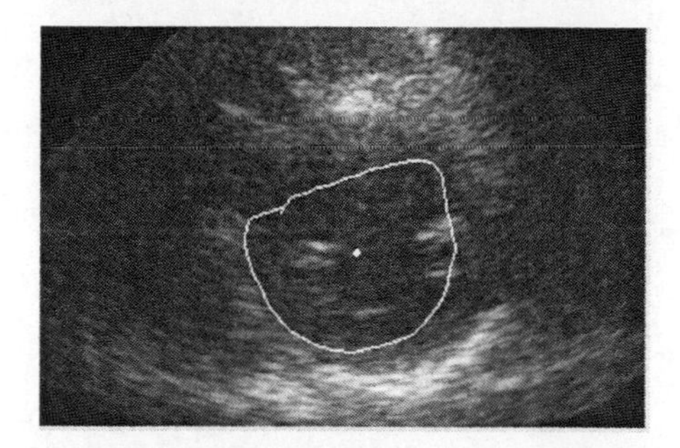

Fig. 7. Center of mass of the left ventricle with the septum flattened.

A. Disadvantages – Conventional Method

The main disadvantage of the conventional method is that it utilizes the whole endocardial region for centroid calculation. Therefore, any variation in the septum wall leads to displacement in the centroid position. This, in effect, gives erroneous septum-wall displacements.

What is needed for centroid calculation is a method that only considers the freewall region in the calculation. As the freewall is almost constant throughout the heart cycle, the centroid position will be fixed, irrespective of septum-wall movement. This centroid then will permit actual septum-wall displacement measurement.

III. Image Preprocessing

The concept for the new algorithm is based on a minimum variance calculation of all the distances from an arbitrary starting point inside the left ventricle to the selected points on the arc located on the freewall region of the left ventricle. These selected points are called circumferential points. Therefore, before we start to shape the algorithm, it is important to select the arc on the freewall region and fix some circumferential points.

A. Arc selection on the freewall of the left ventricle

This process of arc selection is semi-automated. The user clicks on two points on the outline of the endocardial region to define the boundary of the arc. The line joining the two points is called the major axis of the arc. The lower portion of the outline of the endocardial region, delimited by its points of intersection with the major axis, constitutes the arc of interest as shown in Fig. 8. The minor axis is drawn perpendicular to the major axis at the mid-point of the major axis. The major and minor axes are used to determine the eccentricity of the left ventricle. This eccentricity is used in testing the algorithms for different cases.

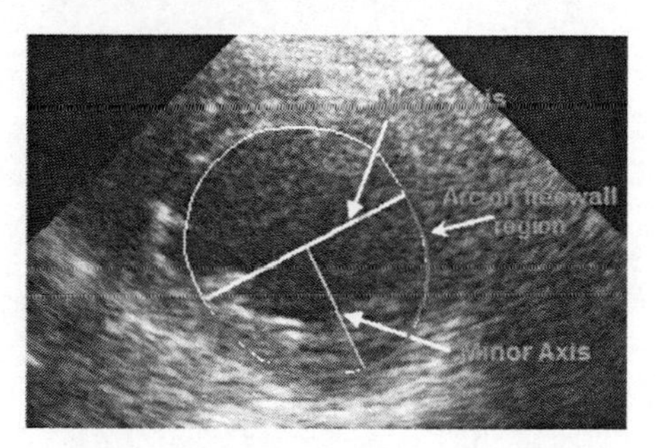

Fig. 8. Selected arc on the freewall region of the left ventricle.

B. Selection of circumferential points

To obtain circumferential points on the arc, an angle is drawn from the point of intersection of the major and minor axes, symmetrical about the minor axis. The angle value (input by the user) gives the spread of circumferential points on the arc. The angle is then divided (using line segments) into a number of divisions equal to the number of circumferential points desired on the arc. Line segments used for angle division have length greater than the minor axis so that they can intersect the arc selected on the endocardial outline. The intersection of each line segment with the arc on the endocardial outline gives the circumferential points as shown in Fig. 9.

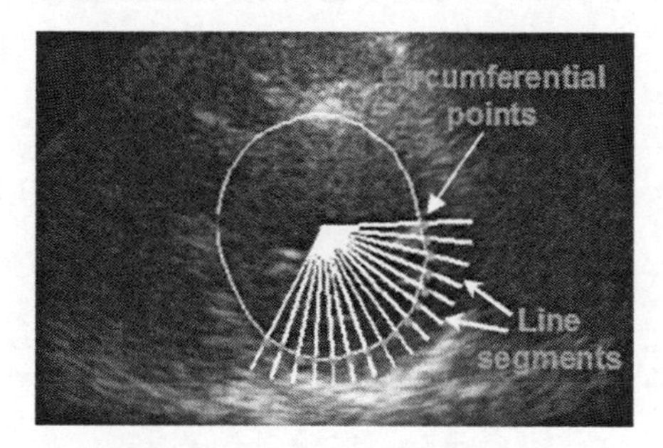

Fig. 9. Circumferential point selection.

IV. New Algorithm

Consider the situation that a circle needs to be fitted through a given set of points where the coordinates of n points are given by the rows of the n x 2 matrix Z and row i is given by the row vector z_i. If all z_i lie exactly on a circle, then there exists a center point x and a radius R such that all the distances d_i between x and z_i equals R. Therefore, the center $L(x)$ is obtained by minimizing (1)

$$L(x,R)=\sum_{i=1}^{n}\left(d_i-R\right)^2 \tag{1}$$

To fit a circle, d_i needs to be defined as a generalized distance

$$d_i=\left(\sum_{s=1}^{m}|u_{is}|^p\right)^{1/p} \tag{2}$$

(p =2 for the Euclidean distance) with p called the Minkowski parameter, m the dimensionality (taken as m = 2), and

$$u_{is}=x_s-z_{is} \tag{3}$$

Different shapes can be fitted for different choices of p. The shape corresponding to p = 2 is the well known circle. For p = 1 it is a diamond, and $p\rightarrow\infty$ it is a square.

Drezner, Steiner, and Wesolowsky (2002) [11] showed that minimizing (1) is equivalent to minimizing

$$\begin{aligned} L(x)&=\sum_{i=1}^{n}\left(d_i-\frac{1}{n}\sum_i d_i\right)^2 \\ &=\sum_{i=1}^{n}d_i^2-\frac{1}{n}\left(\sum_{i=1}^{n}d_i\right)^2 \end{aligned} \tag{4}$$

Equation (4) [11] is the variance equation for finding the centroid of a circle. To generalize (4) for any shape, we propose a new algorithm, called the *"Iterative Majorization Algorithm"* [12].

A. Iterative Majorization

Iterative majorization is a guaranteed descent method that has been mostly developed in the area of psychometrics and statistics [13, 14]. The main idea is to replace, in each iteration, the objective function by an auxiliary function (called the majorizing function) that is easier to minimize such as a quadratic function [15]. Then, an update of the majorizing function is easy to find. By choosing the majorizing function in a special manner, the important property of iterative majorization, that is, the property that the function value will never increase in subsequent iterations, can be proven. In other words, for almost all practical applications, iterative majorization guarantees that the function values decrease.

Let the majorizing function be denoted by $g(x,x_0)$, where x_0 is called the supporting point which is the current known estimate of x. The majorizing function has to have the following properties:

1. The majorizing function $g(x,x_0)$ should be easier to minimize than the original function f(x).
2. The majorizing function is larger than or equal to the original function, that is, $f(x)\leq g(x,x_0)$ for all x.
3. The majorizing function touches the original function at the supporting point x_0, that is, $g(x_0,x_0)=f(x_0)$.

B. Obtaining a Majorization Function for L(x)

The first term in $L(x)$ consist of a sum of d_i^2. The second term of $L(x)$, $-n^{-1}\left(\sum_{i=1}^{n}d_i\right)^2$ can be seen as n^{-1} times the negative square of $\left(\sum_{i=1}^{n}d_i\right)$. Developing the inequality $c(t-t_0)^2\geq 0$ gives

$$\begin{aligned} c(t^2+t_0^2-2tt_0)&\geq 0 \\ -ct^2&\leq -2ctt_0+ct_0^2 \end{aligned} \tag{5}$$

where c is a positive value. Substituting $t=\sum_i d_i$ and $t_0=\sum_i d_{i0}$ gives

$$L(x)\leq\sum_{i=1}^{n}d_i^2-\frac{2}{n}\left(\sum_{i=1}^{n}d_i\right)\left(\sum_{i=1}^{n}d_{i0}\right)+k_1. \tag{6}$$

with d_{i0} the distance based on x_0 (the previous estimation of x) and $k_1=n^{-1}\left(\sum_i d_{i0}\right)^2$. Using the majorizing function for d_i and d_i^2 given by Groenen, Heiser, and Meulman (1999) [16] (6) can be written as

$$\begin{aligned} L(x)&\leq\sum_{i=1}^{n}d_i^2-\frac{2}{n}\left(\sum_{i=1}^{n}d_i\right)\left(\sum_{i=1}^{n}d_{i0}\right)+k_1. \\ &\leq\sum_{is}a_{is}u_{is}^2-2\sum_{is}u_{is}b_{is}+k_2. \end{aligned} \tag{7}$$

where

$u_{is}=x_s-z_{is}$

u_{is0} = distance based on x_0 (previous estimation of x)

ε = convergence criteria (taken as small positive number)

$$b_{is} = \left[\left(n^{-1}\sum_i d_{i0}\right)b_{is}^{(1)} + b_{is}^{(2)}\right]u_{is0}$$

$$k_2 = k_1 + \sum_{is} c_{is}$$

$$a_{is} = \begin{cases} |u_{is0}|^{p-2} / d_{i0}^{p-2} & \text{if } |u_{is0}| > \varepsilon \text{ and } 1 \le p \le 2 \\ |\varepsilon|^{p-2} / d_{i0}^{p-2} & \text{if } |u_{is0}| \le \varepsilon \text{ and } 1 \le p \le 2 \\ 2(p-1) & 2 < p < \infty \end{cases}$$

$$b_{is}^{(1)} = \begin{cases} |u_{is0}|^{p-2} / d_{i0}^{p-1} & \text{if } |u_{is0}| > 0 \text{ and } 1 \le p \le \infty \\ 0 & \text{if } |u_{is0}| = 0 \text{ and } 1 \le p \le \infty \end{cases}$$

$$b_{is}^{(2)} = \begin{cases} 0 & \text{if} 1 \le p \le 2 \\ a_{is} - |u_{is0}|^{p-2} / d_{i0}{}^{p-2} & \text{if } |u_{is0}| > 0 \text{ and } 2 < p < \infty \\ 0 & \text{if } |u_{is0}| = 0 \text{ and } 2 < p < \infty \end{cases}$$

$$c_{is} = \begin{cases} 0 & \text{if } 1 \le p \le 2 \\ \sum_s a_{is}u_{is0}^2 - d_{i0}^2 & \text{if } 2 < p < \infty \\ \sum_s \left(2b_{is}^2 - a_{is}\right)u_{is0}^2 + d_{i0}^2 & \text{if } p \to \infty \end{cases}$$

Inserting $x_s - z_{is}$ for u_{is} in the right hand part of (7) gives

$$\begin{aligned} L(x) &= \sum_{is} a_{is}(x_s - z_{is})^2 - 2\sum_{is}(x_s - z_{is})b_{is} + k_2 \\ &= \sum_s x_s^2 \sum_i a_{is} - 2\sum_s x_s \sum_i (a_{is}z_{is} + b_{is}) + k_3 \end{aligned} \tag{8}$$

where

$$k_3 = k_2 + \sum_{is} a_{is}z_{is}^2 + 2\sum_{is} z_{is}b_{is}$$

Equation (8) shows that the majorizing function on the right hand is indeed quadratic in x_s. The minimum of a quadratic function can easily be obtained by setting the derivative equal to zero and solving for x_s, giving the update

$$x_s^+ = \frac{\sum_i (a_{is}z_{is} + b_{is})}{\sum_i a_{is}} \tag{9}$$

Thus, the iterative majorization algorithm is given by:

1. Choose initial $x_{(0)}$. Set iteration counter k = 0 and set the convergence criterion ε_{conv} to a small positive constant.
2. k = k+1.
3. Calculate the update x^+ using (9), with $x_0 = x^{(k-1)}$.
4. $x^{(k)} = 2x^+ - x^{(k-1)}$.
5. Check the convergence: If $\left\|x^{(k-1)} - x^{(k)}\right\| \ge \varepsilon_{conv}$ go to Step 2.

As a rational starting point for $x^{(0)}$, we use the intersection of the major and minor axes. The result of applying the iterative majorization algorithm to the left ventricle is shown in Fig. 10. In the figure, the white dot is the center of mass, the yellow dot is the starting point $x^{(0)}$ and the red dot is the centroid obtained by using iterative majorization.

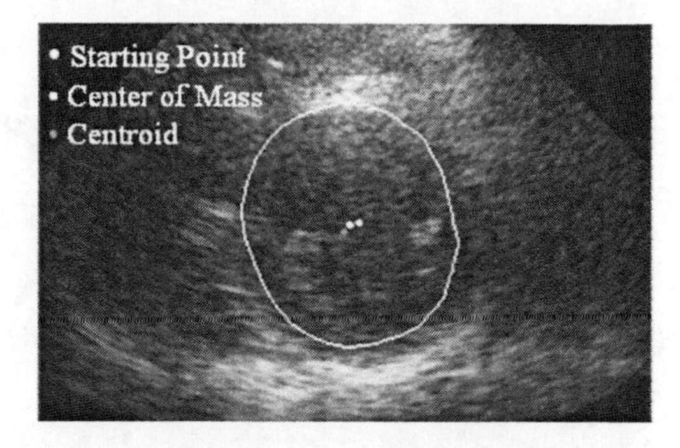

Fig. 10. Iterative majorization result.
Convergence criterion ε_{conv} = 0.1 pixels.
Number of iterations = 10.

V. Experimental Results

The performance of the iterative majorization algorithm is checked by comparing it with the conventional method. Testing was performed on 17 patient data sets using Matlab 6.5. Each data set has 2 cases S1 and S2, where

- S1 → Normal case (no external pressure applied on the heart)
- S2 → Lower body negative pressure

Each case has 6 images, 3 systolic and 3 diastolic. Thus, the total number of images on which algorithm performance is tested is: 17 x 2 x 6 = 204 images. The results of the comparison are shown in Fig. 11.

In the results it can be seen that as the septum flattens there is also a movement of the center of mass (COM) i.e., when the centroid is calculated using the conventional method, whereas with the centroid calculated using the iterative majorization algorithm it remained constant irrespective of septum movement. This centroid measurement will thus provide accurate septum displacement.

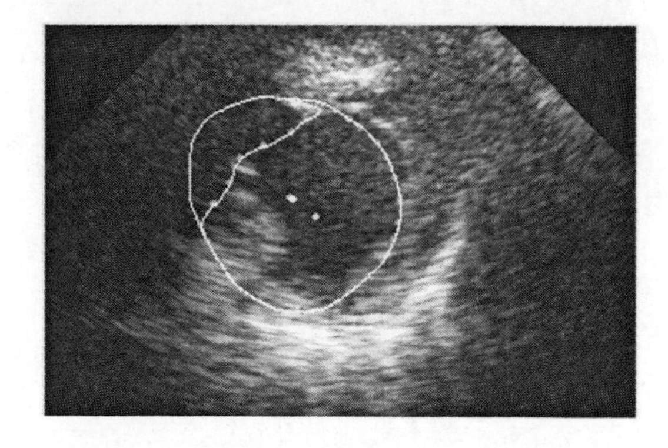

a.

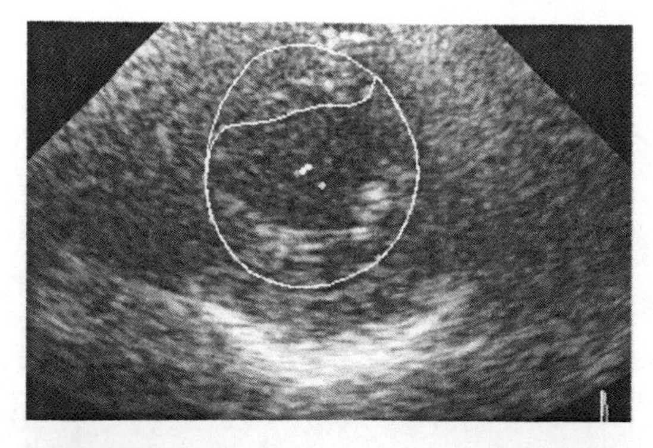

b.

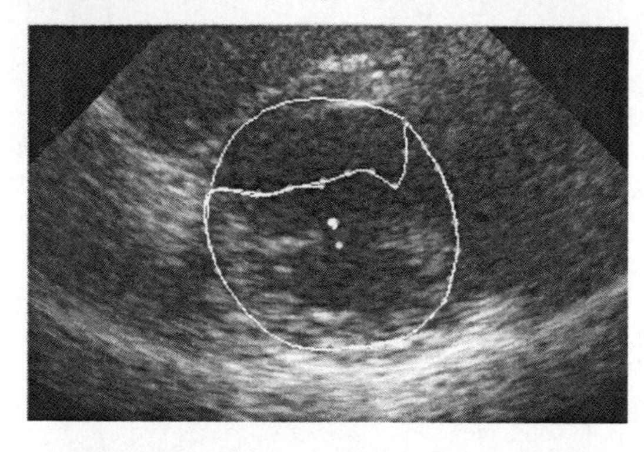

c.

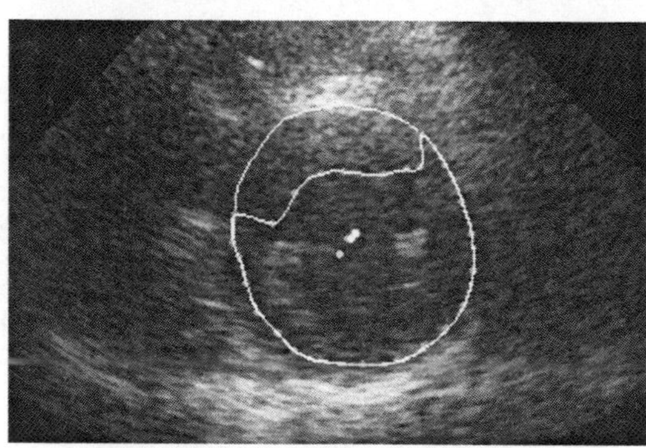

d.

where

- Outer Contour
- Inner Contour with flat septum
- COM of outer contour
- COM of inner contour
- Centroid of outer contour
- Centroid of inner contour

Fig. 11. Results of comparison.

VI. Conclusion and Future work

This paper presents a new algorithm for establishing a coordinate system for images of the weak heart. In weak heart, septum wall motion leads to erroneous estimates of ventricular coordinates and thus erroneous measurements of the septum position when conventional methods are used. The new algorithm was compared to the conventional method used to date. The comparison showed the strength of the iterative majorization algorithm in finding the centroid for any shape variation in the heart.

Further work needs to be done to analyze the long-axis view of the heart.

References

[1] H.R. Schwinger, M. Bohm and A. Koch, "The failing human heart is unable to use the Frank-Starling mechanism," *Am J Physiol*, vol. 74, pp. 959-969, 1994.

[2] B.G. Myers, A.H. Klein and I.E.B. Stofer, "Correlation of echocardiographic and pathologic findings in anteroseptal infarction," *Am Heart J*, vol. 36, pp. 535-575, 1948.

[3] C.J. Dillon, S. Chang and H. Feigenbaum, "Echocardiographic manifestations of left bundle branch block," *J Am Coll Cardiol*, vol. 49, pp. 876-880, 1974.

[4] J.B.N. Ingels, G.T. Daughters, B.E. Stinson and L.E. Alderman, "Evaluation of methods for quantitating left ventricular segmental wall motion in man using myocardial markers as a standard," *Can j Cardiol*, vol. 61, pp. 966-972, 1980.

[5] C Koilpillai, A.M. Quinones and B. Greenberg, "Relation of ventricular size and function to heart failure status and ventricular dysrhythmia in patients with severe left ventricular dysfunction," *Am J Cardiol*, vol. 77, pp. 606-611, 1996.

[6] G.S. Horne, N.R. Anderson, I. Burwash and R.E. Smith, "Difference in ventricular septal motion between subgroups of patients with heart failure," *Can J Cardiol*, vol. 16, pp. 1377-1384, Nov 2000.

[7] I. Kingma, V.J. Tyberg and R.E. Smith, "Effects of diastolic transseptal pressure gradient on ventricular septal position and motion," Am *J Physiol*, vol. 68, pp.1304-1314, 1983.

[8] E.R. Kerber, F.W. Dipple and M.F. Abboud, "Abnormal motion of the interventricular septum in right ventricle volume overload," *Experimental and clinical echocardiographic studies*, vol. 48, pp. 86-96, 1973.

[9] D.J. Pearlman, D.R. Hogan, S.P. Wiske, D.T. Franklin and E.A. Weyman, "Echocardiographic definition of the left ventricular centroid. I. Analysis of methods for centroid calculation from a single tomogram," *J AM Coll Cardiol*, vol. 4, pp. 986-982, Oct 1990.

[10] N.H. Sabbah, T. Kono and S. Goldstein, "Left ventricular shape changes during the course of evolving heart failure," *Am J Physiol*, vol. 263, pp. H266-270, 1992.

[11] Z. Denzner, S. Steiner and O.G. Wesolowsky, "On the circle closest to a set of points," *Comput. Oper. Res.*, vol. 29, pp. 637-650, 2002.

[12] K.V. Deun and F.J.P. Groene, "Majorization algorithms for inspecting circles, ellipses, squares rectangles and rhombi," *Econometric Institute Report EI*, pp. 2003-2035, 2000.

[13] J. De Leeuw, *Information Systems and Data Analysis*, 3rd ed., Berlin: Springer-Verlag, 1994.

[14] J.W. Heiser, *Recent Advances in Descriptive Multivariate Analysis*, Oxford: Oxford University Press, 1995.

[15] K. Lange, R.D. Hunter and I. Yang, "Optimization transfer using surrogate objective functions," *J. Comput. Graph*, vol. 9, pp. 1-20, 2000.

[16] F.J.P. Groene, J.W. Heiser, J.J. Meulman, "Global optimization in least-squares multidimensional scaling by distance smoothing," *J. Classif*, vol. 16, pp. 225-254, 1999.

On-line Modeling for Real-Time, Model-Based, 3D Pose Tracking

Hans de Ruiter, Beno Benhabib
Computer Integrated Manufacturing Laboratory
Department of Mechanical and Industrial Engineering
University of Toronto
5 King's College Road, Toronto, Ontario, M5S 3G8, Canada
deruiter@mie.utoronto.ca

***Abstract*— Model-based object-tracking can provide mobile robotic systems with real-time 6-dof pose information of a dynamic target object. However, model-based trackers typically require the model of the target to be known *a-priori*. This paper presents a novel method capable of building an approximate 3D geometric model of a target object in an on-line mode, fast enough for real-time use by a model-based object tracker. The algorithm constructs a 3D tessellated model and uses projective texture mapping to model the target object's surface features.**

***Index Terms*—Modeling, Pose tracking, Computer Vision.**

I. Introduction

Model-based object-tracking systems can provide mobile robots with real-time sensory feedback. This would allow such robots to respond/interact with objects in their environment with greater ease and accuracy. Ideally, the full 6-dof pose (position and orientation) should be tracked. An autonomous robot, for example, may need to rendezvous with a target from a specific angle and not simply at a specific position.

We have previously developed a real-time, model-based, 6-dof object-tracking algorithm ([1]-[2]). This technique tracks a target-object's pose via a novel combination of 3D model projection and optical flow. Use of depth information generated during the projection process enables estimation of full 6-dof pose, even with only a single camera.

Other effective model-based 6-dof trackers exist ([3]-[9]), though subject to drawbacks. For example, Jurie and Dhome's technique in [3] involves off-line training that is specific to a particular object, thus, precluding the possibility of extending their technique to tracking *a-priori* unknown objects. The other methods ([4]-[9]) track local features such as edges, thus, limiting their systems to use a small sub-set of the available visible data (e.g., corners, line-segments, etc.). Moreover, complex objects and patterns cannot be modeled effectively. Our tracking methodology, on the other hand, uses all available visible surface features of the target object.

All of the aforementioned object-trackers have the same limitation. Namely, the model must be known *a-priori*, thus limiting the range of objects that can be tracked. This paper extends our existing object tracker with a novel, fast model-building algorithm, in order to be able to track a target object whose model is not known *a-priori*. It builds a tessellated 3D model (i.e., a mesh) from images captured using multiple cameras.

Surface features (texture) are mapped onto the 3D model using projective texture-mapping [10]. As the model rotates, parts that were visible disappear from view and new parts become visible. Therefore, the model is rebuilt periodically.

Constructing models from multiple images has been researched extensively and various techniques have been developed. These include space-carving algorithms [11], silhouette-based methods [12], space-curve fitting algorithms [13], and, shape-from-motion algorithms [14].

The key challenge, when building a model for on-line object-tracking, is not building an accurate model, rather, building an approximate visual model in *real-time*. There exists a tradeoff between accuracy, obtained with off-line modelers, and speed. A distributed approach to real-time modeling, requiring a computing cluster, was proposed by Franco et al. [15]. Lee et al. [16] developed a feature-based, real-time modeler. In contrast, our object-tracker requires a visual model, built out of texture-mapped surfaces.

II. Object-Tracker Overview

The proposed system achieves 6-dof pose tracking via a novel 3D model-based approach. The object's pose is estimated in real-time by projecting a visual 3D model of the target object onto the image plane for comparison with the input data. Motion between the object's predicted pose and its actual pose is calculated using optical flow between the *projected* image and the *real* target object image. This motion is, then, used to correct the predicted pose, producing the final pose estimate.

Motion prediction is used to minimize the difference between the estimated and actual poses of the target object. This minimizes the possibility of losing track of the target whilst maximising the accuracy. Since optical flow is a linearization procedure, minimizing the initial error minimizes

K. Elleithy (ed.), Advances and Innovations in Systems, Computing Sciences and Software Engineering, 555–560.

errors introduced due to non-linearities [17].

Projection is performed using 3D graphics hardware via OpenGL [18], relieving the main processor from this task, and hence, significantly increasing the performance of the system. OpenGL also provides a depth-map for the projected image and a mask indicating which pixels belong to the target object. The depth-map provides the missing depth information required to calculate 6-dof motion; the mask is used to restrict the optical-flow calculation to only those pixels that belong to the target object.

In tests performed, this algorithm operated at 45-70 fps and could track objects under various lighting conditions. A full explanation of the tracking algorithm can be found in our previous papers ([1]-[2]).

III. Proposed Modeling Methodology

An overview of the proposed modeling system for object-tracking is shown in Figure 1. The modeling process can be divided into three parts: depth-map extraction, mesh generation, and, texture extraction. Depth-map extraction provides the 3D data required by the mesh generator. A tessellated surface, or mesh, of triangles is built from the depth-map. Texture-extraction extracts the object's surface features, such as coloured markings. The texture and mesh are combined to produce a visual 3D model of the target object.

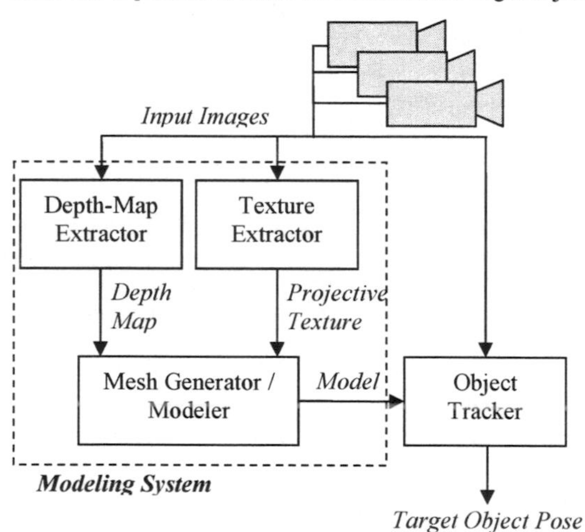

Figure 1: Block diagram of modeling system for object-tracking.

The tessellated model is built by incrementally adding points to the mesh using Delaunay triangulation, as it ensures that the generated mesh is "optimal" [19]. Location for the points in 3D space is calculated by solving the camera-projection equation for $\boldsymbol{p}$:

$$\boldsymbol{p'} = \mathbf{M}_{int}\mathbf{M}_{ext}\boldsymbol{p},\ \boldsymbol{p'} = \begin{bmatrix} p'_x \\ p'_y \\ p'_z \end{bmatrix},\ \text{and}\ \boldsymbol{p} = \begin{bmatrix} p_x \\ p_y \\ p_z \\ 1 \end{bmatrix}, \tag{1}$$

where $\boldsymbol{p'}$ is a (3×1) vector denoting the projected 2D position of $\boldsymbol{p}$ in homogeneous coordinates.[1] The (3×4) matrix $\mathbf{M}_{int}$ is the camera's internal parameters (focal length, scaling factors, etc.). $\mathbf{M}_{ext}$ is a (4×4) matrix that transforms 3D coordinates between the world reference-frame and the camera's reference frame. The variable $\boldsymbol{p'}$ is constructed using the point's 2D on-screen coordinates, $(p'_x/p'_z, p'_y/p'_z)$, and its depth-value for that pixel, (p'_z).

A. Depth-Map Extraction

Depth maps, rather than feature-point based 3D extraction, is used herein so that shapes with no strong feature-points can also be modeled. For example, objects with smooth curved edges do not have well-defined feature-points. The specific Depth-Map-Extraction (DME) algorithm used is not critical to modeling, provided that it operates in real-time. Thus, since the focus of this paper is on the modeling algorithm, the specific DME algorithm used will not be covered in depth.

The specific DME algorithm used bears some similarity to Yang et al.'s algorithm [20] in that it uses graphics hardware and projects the images from the camera onto a candidate depth map. It extracts depth using two or more cameras. The difference lies in that the candidate depth-map in our DME is non-planar and is iteratively improved rather than using Yang et al.'s plane-sweeping algorithm. A smoothness constraint has also been added in order to reduce errors. Our DME algorithm provides a more accurate depth-map whilst still operating in real-time (i.e., at 100 fps when processing five 256×256 images using a Radeon X800 graphics-card).

B. Mesh Generation

1) Selecting Mesh Vertices

Appropriate vertex selection is key to forming an approximate model. Poor selection will make the resulting model a *poor* approximation of the target object. Ideally, a set of principal points would be chosen that minimizes the number of triangles in the resulting mesh.

Finding in real-time appropriate *principal* points in a depth-map poses a few challenges. There is insufficient computation time available to carefully select principal points in the depth-map to provide a good surface approximation with a minimal number of triangles in the mesh. Moreover, depth-maps extracted in real-time from input images generally suffer from spikes/pits and other errors, as can be seen in Figure 2(b). This prevents the use of the depth-map itself for principal-point selection using techniques for approximating range-images, such as has been proposed by Pedrini [21].

The proposed technique selects points based on the input image associated with the depth-map, rather than the depth-

[1] Homogeneous coordinates are used in projective geometry and are defined up to a scale-factor.

map itself. It is based on two principles. Namely:

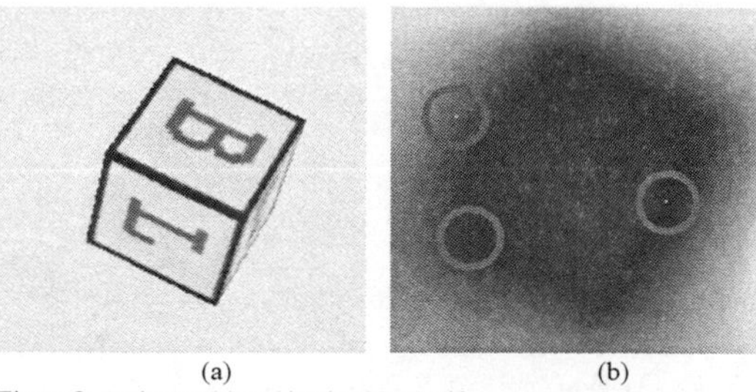

(a) (b)

Figure 2: An image (a) and its depth-map (b)

- Sharp edges, or feature-points, in an image often coincide with principal points of 3D objects, and
- Depth can only be measured in textured image regions (i.e., regions with non-zero intensity-gradients). Smooth image regions have depth-values estimated based on smoothness constraints and adjacent textured regions.

Feature-points alone may not produce meshes that are good approximations of the target object. This is due to smooth curved edges not having definable feature-points. Also, sparsely spaced feature-points can cause large errors at locations between points. In order to solve this problem, additional points are added using the second principle.

In Pedrini's meshing algorithm [21], the mesh is refined iteratively by inserting additional points via a comparison of the error between the depth-map and the mesh. However, for real-time modeling, such computation would be costly. Hence, rather than adding points based on an error measure, points are added in a grid-like fashion. This ensures that enough points are available to produce a reasonable approximation of the target object.

Before discussing the grid-based point insertion scheme, the concept of "texturedness" is defined here. Texturedness is a measure of intensity variations within the image, defined by:

$$T(x,y) = \sum_{i=R,G,B} \left|\nabla I_i(x,y)\right|^2, \quad (2)$$

where $\nabla I_i(x,y)$ denotes the gradient of image I for colour channel i. Thus, regions of uniform colour have zero texturedness. The texturedness value for a pixel increases as the intensity gradients increase.

For simplicity, one could prefer to add points to the mesh on a regular grid. However, true depth information is only present in regions of the image with non-zero texturedness values (smooth areas are estimated using smoothness constraints). In order to solve this problem, the image is divided into a grid of boxes. If a box in the grid contains feature-points, those are added. Otherwise, the pixel closest to the box's centre that has a texturedness value over threshold τ_T (i.e., $T(x,y) > \tau_T$) is selected as the point to add, for that box in the grid. If the box has no pixels with $T(x,y) > \tau_T$, then, no point is added.

The principal-point selection algorithm is, then:

- Find all feature-points using a Harris feature-point detector [22],
- Calculate the texturedness image T (note, this can be combined with the Harris feature detector into one step),
- Perform non-maximal suppression on T using a 3×3 window, and
- For each box in an m×n grid:
  - o If there are any feature-points within this box, add those points to the mesh,
  - o Else, if the box contains pixels for which $T(x,y) > \tau_T$, then:
    - ▪ Find the closest point to the box's centre with $T(x,y) > \tau_T$, and
    - ▪ Add this point to the mesh.
  - o Else, move on to the next box in the grid.

Non-maximal suppression is an operation that sets pixels to zero if they do not have the maximum value within a window (3×3 in this case). Its purpose is to force the selection of points with the largest local $T(x,y)$. Such points are likely to have the most reliable depth values. Figure 3(b) shows an example mesh generated from images taken by five cameras, such as the image in Figure 3(a). This novel technique allows the generation of an approximate target object model at a rate of several frames per second, fast enough for use in real-time object tracking.

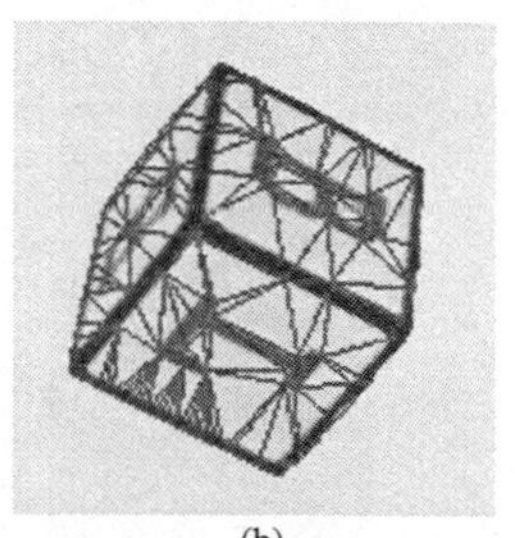

(a) (b)

Figure 3: An input image (a) and the resulting mesh (b).

2) Filtering Noisy Depth-Maps

As abovementioned, depth-maps may suffer from spikes/pits and other errors. Such errors can cause parts of the resulting model to deviate significantly from the target object's true shape. Consequently, the object-tracker may be unable to track such a poorly modeled object. It is, therefore, necessary to try and minimize these errors as much as possible within the constraints of real-time operation. Figure 4(a) shows a depth-map containing spikes. Typically these spikes are only a few pixels in size. A 3×3 median filter is used to eliminate these spikes, as can be seen in Figure 4(b).

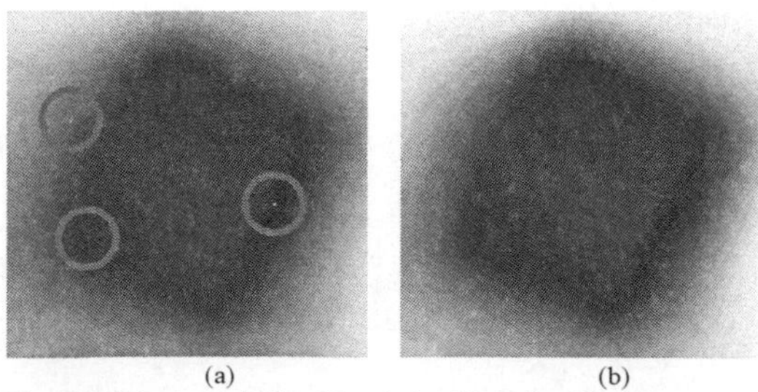

Figure 4: A depth map before (a) and after (b) median filtering.

Despite the median filter removing many of the spikes/pits, some large errors in depth may remain. As a final guard against extraneous points, any points selected that are too far from the target object's origin are discarded. Discarding such points ensures that there are no large outliers capable of significantly affecting the object-tracker's motion calculations.

C. *Texture-Map Extraction*

The 3D model used by the object tracker requires more than just the object's 3D geometry. The object's surface features, such as coloured markings, are just as important. Without both the 3D geometry and surface features tracking cannot proceed. These features are modeled herein using a texture-map. A texture-map is an image that is mapped onto the model's surface, allowing complex surface features to be compactly represented.

Rather than generating a set of standard texture maps for the model, a technique called projective texture mapping [10] is used. This method effectively projects a texture onto a 3D surface in the same manner that a real slide/film projector projects an image onto a screen (Figure 5). This allows the use of one of the input images as a texture without requiring any additional computation at all.

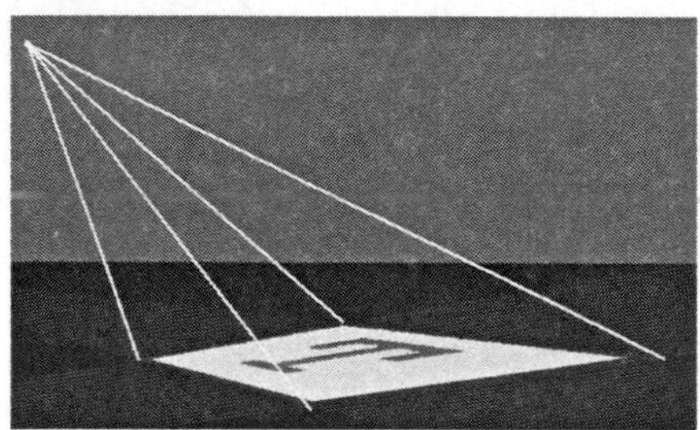
Figure 5: A texture mapped onto a plane using projective texture-mapping.

The texture-extraction procedure is as follows:

- Copy the image from the reference camera into the object's texture map, and
- Record the reference-camera's parameters and pose relative to the target-object's reference frame. These are the texture-projector's parameters and pose.

Using the data obtained above, the object tracker can project the target-object's model onto the image plane including its surface texture. The texture is projected onto the model surface from the viewpoint of the camera that the input image was captured by, i.e., from the texture projector. Note that, the texture-projector's pose is fixed relative to the target object's reference frame. Otherwise, the surface features would move relative to the surface itself.

IV. IMPLEMENTATION

The modeling algorithm described above was implemented in C++ and OpenGL's Shading Language (GLSL). Support for projective texture-mapping was added to the object-tracker's image rendering module using GLSL. The modeler and object-tracker were, then, combined to form one object-tracking system. The modeler builds a model of the target object every n frames. This model is used by the object-tracking algorithm in order to estimate the target-object's pose over time.

Reference-Frame Selection

The output of the object tracker is the target object's reference frame estimated over time. Theoretically, this reference frame can be placed anywhere relative to the object's surface. The optimal selection of reference frame is application dependent. In some applications, the geometric centre may be most appropriate; for others, the centre of mass, or even some other point within the object. When tracking a generic object, whose model is not known *a priori*, it is not possible to find such an optimal reference frame. Instead, a reference frame must be selected based on the 3D mesh built from the input data.

For the purpose of testing the accuracy of tracking using models built from input data, it is best to use the true reference frame of the target object. This allows a direct comparison between the tracker's pose estimate and the actual pose at any time-step. Hence, in this paper, the object-tracking system is provided with the known initial pose of the target object.

V. EXPERIMENTAL RESULTS

Synthetic motion sequences were generated of a 96×96×96 mm cube moving under a predefined trajectory. There were five simulated cameras all facing in the same direction, placed in a plus configuration. The left and right cameras were separated from the central camera by 40 mm whilst the top and bottom cameras were placed 20 mm above and below the central camera, respectively. All cameras had a horizontal field-of-view of 34°. Images measured 512×256 pixels as graphics cards are optimised for image dimensions that are a power-of-two.

Sample frames from a motion sequence are shown in Figure 6. In this sequence, the cube starts its motion at a distance of 1000 mm from the cameras and moves to the right at 60 mm/s (2 mm/frame) and away from the camera at 30 mm/s (1 mm/frame). A blank background has been used because image-segmentation (i.e., separation of the target object from the background) has not yet been addressed. Image segmentation will be addressed in future research.

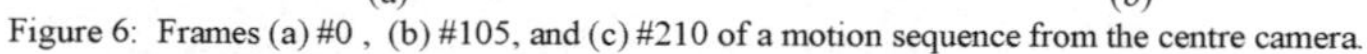

(a) (b) (c)

Figure 6: Frames (a) #0 , (b) #105, and (c) #210 of a motion sequence from the centre camera.

When analyzing the accuracy of vision-based tracking algorithms, it would be beneficial to examine what kind of accuracies can be expected. Accuracy is affected by an object's size, distance from the camera, location on screen, and, the camera's parameters. In order to obtain an approximate idea of what performance to expect, we examine here the accuracies for an object lying on the optical axis at the maximum distance from the camera that occurred in our experiments.

The maximum distance from cameras along the central camera's optical-axis was 1210 mm. Under these conditions, using Equation (1), an on-screen positional error of 1 pixel corresponds to a 1.4 mm error in estimated 3D position parallel to the image plane (i.e., x– or y–axis error). An error of ±2 pixels in on-screen size (caused by a 1 pixel error in the position of each outer edge) produces an error of +34 or −32 mm along the optical-axis, respectively. On-screen errors can also cause errors in the orientation estimate. Let us examine a rotation such that the target-object's outer edges are displaced by 1 pixel. For the x–, y–, and z–axes (in camera coordinates), these errors would be 14°, 14°, and 1.2°, respectively. Cross-coupling between axes due to the non-linear nature of projection can further increase the errors.

Figure 7 shows the tracking errors obtained by the object-tracking system for the sequence shown in Figure 6. The errors correspond approximately to localizing the target to within the nearest pixel on-screen. The modeler updated the target-object's model every 30 frames. In order to measure the tracking errors directly, the system was provided with the target-object's initial pose. The tracker estimated the target-object's pose to within 30 mm and orientation to within 3.5° along each axis.

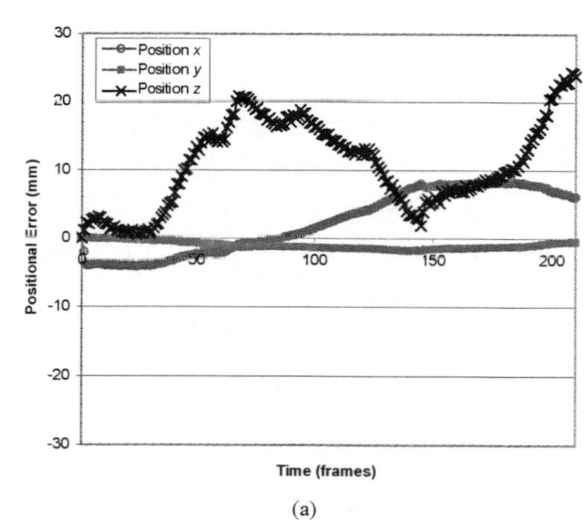

(a)

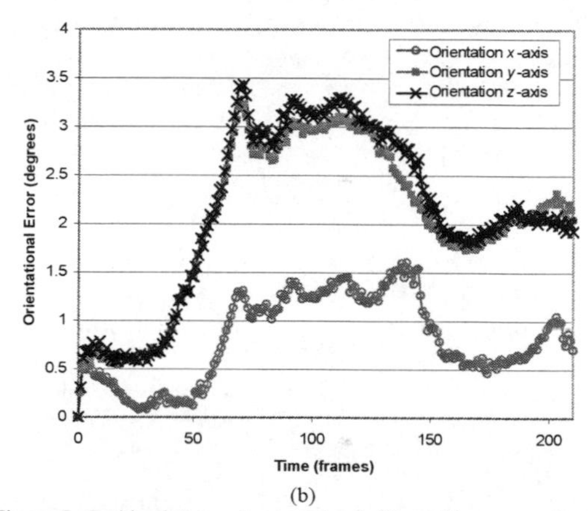

(b)

Figure 7: Positional (a) and orientational (b) tracking errors for the sequence in Figure 6.

The importance of updating the model over time is illustrated in Figure 8, where a case of the object being tracked utilizing only the model built in the first image of the motion-sequence is shown. In this case, as the object rotates, errors present in the model become more apparent. Eventually, the modeled part of the target object rotates out of view and the object tracker loses it.

In regard to computation time, it was noted that the modeler is capable of building a model with 126 triangles at a rate of about 13 fps. The object-tracker, on the other hand, is capable to track an object at rates of 45 to 70 fps. Since, most cameras operate at rates of 30 fps, one can note that the target-object's model can be updated once a second (i.e., every 30 frames) and still successfully track an object. (In our experiments, a computer with a 2.0 GHz Athlon 64 CPU and a Radeon X800 pro graphics-card was used).

VI. Future Work

The current implementation does not address the issue of separating the target-object from the background. Hence the background must be blank. Research will focus on adding image segmentation to the modeler in order to allow this

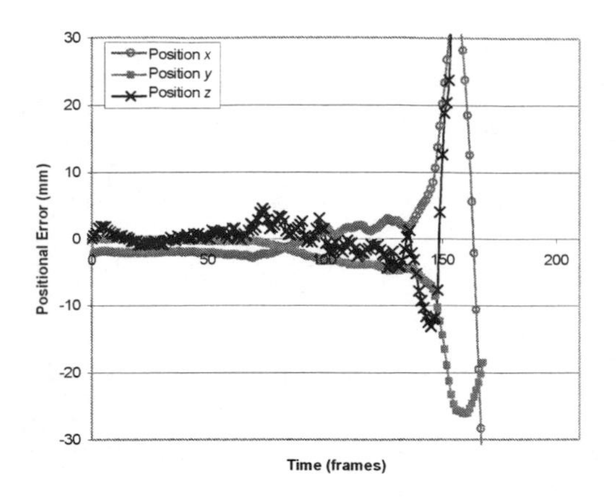

Figure 8: Positional tracking error when tracking without model updating.

object tracking system to work with cluttered backgrounds.

VII. Conclusions

This paper has presented a novel modeler that enables a model-based object-tracker to track the 6-dof pose of an object, whose model is not known *a-priori*, in real-time. The modeler constructs an approximate 3D mesh of the target object by selecting appropriate points on a depth-map built from input images. A novel point-selection algorithm ensures that the 3D mesh is constructed fast enough for use in real-time object tracking. Projective texture-mapping completes the visual model by modeling the object's surface details.

Acknowledgments

This work was supported by the Natural Sciences and Engineering Research Council (NSERC) of Canada.

References

[1] H. de Ruiter and B. Benhabib, "Tracking of rigid bodies for autonomous surveillance," *IEEE Conf. on Mechatronics and Automation*, vol. 2, Niagara Falls, Canada, July 2005, pp. 928–933.

[2] H. de Ruiter and B. Benhabib, "Colour-gradient redundancy for real-time spatial pose tracking in autonomous robot navigation," *Canadian Conf. on Computer and Robotic Vision*, Québec City, Canada, June 2006, pp. 20–28.

[3] F. Jurie and M. Dhome, "Real time robust template matching," *13th British Machine Vision Conf.*, Cardiff, Wales, 2002, pp. 123–132.

[4] E. Marchand, P. Bouthemy, and F. Chaumette, "A 2d-3d model-baed approach to real-time visual tracking," Image and Vision Computing, vol. 19, no. 7, pp. 941–955, November 2001.

[5] T. Drummond and R. Cipolla, "Real-time visual tracking of complex scenes," IEEE Trans. on Pattern Analysis and Machine Intelligence, vol. 24, no. 7, pp. 932–946, July 2002.

[6] A. Comport, E. Marchand, and F. Chaumette, "A real-time tracker for markerless augmented reality," *IEEE and ACM Int. Symposium on Mixed and Augmented Reality*, Tokyo, Japan, October 2003, pp. 36–45.

[7] S. Kim and I. Kweon, "Robust model-based 3d object recognition by combining feature matching with tracking," *Int. Conf. on Robotics and Automation*, vol. 2, Taipei, Taiwan, September 2003, pp. 2123–2128.

[8] V. Kyrki and D. Kragic, "Integration of model-based and model-free cues for visual object tracking in 3d," *Int. Conf. on Robotics and Automation*, Barcelona, Spain, April 2005, pp. 1554–1560.

[9] M. Vincze, M. Schlemmer, P. Gemeiner, and M. Ayromlou, "Vision for robotics: A tool for model-based object tracking," IEEE Robotics and Automation Magazine, vol. 12, no. 4, pp. 53–64, December 2005.

[10] C. Everitt. (2006, October) Projective texture mapping. NVidia. [Online]. Available: http://developer.nvidia.com/object/Projective_Texture_Mapping.html.

[11] K. N. Kutulakos, "Approximate n-view stereo," *7th European Conf. on Computer Vision*, Dublin, Ireland, 2000, pp. 67–83.

[12] A. Mulayim, U. Yilmaz, and V. Atalay, "Silhouette-based 3-d model reconstruction from multiple images," IEEE Trans. on Systems, Man and Cybernetics – Part B: Cybernetics, vol. 33, no. 4, pp. 582–591, August 2003.

[13] F. Kahl and J. August, "Multi-view reconstruction of space curves," *9th IEEE Int. Conf. on Computer Vision*, vol. 2, October 2003, pp. 1017–1024.

[14] P. Simard and F. Ferrie, "Image-based model updating," *13th British Machine Vision Conf.*, Cardiff, UK, 2002, pp. 193–202.

[15] J.-S. Franco, C. Ménier, E. Boyer, and B. Raffin, "A distributed approach for real time 3D modeling," *IEEE Workshop on Real Time 3D Sensors and Their Use*, Washington DC, July 2004.

[16] S. Lee, D. Jang, E. Kim, S. Hong, and J. Han, "A real-time 3d workspace modeling with stereo camera," *Int. Conf. on Robots and Systems*, Edmonton, Canada, August 2005, pp. 2140–2147.

[17] J. Webber and J. Malik, "Robust computation of optical flow in a multi-scale differential framework," *4th Int. Conf. on Computer Vision*, Berlin, Germany, May 1993, pp. 12–20.

[18] (2004, November) Opengl - the industry standard for high performance graphics. OpenGL.org. [Online]. Available: http://www.opengl.org.

[19] M. de Berg, M. van Kreveld, M. Overmars, and O. Schwarzkopf, *Computational Geometry: Algorithms and Applications*, 2nd ed.1em plus 0.5em minus 0.4emBerlin Heidelberg New York: Springer-Verlag, 2000, ch. 9.

[20] R. Yang, G. Welch, and G. Bishop, "Real-time consensus-based scene reconstruction using commodity graphics hardware," *10th Pacific Conf. on Computer Graphics and Applications*, Beijing, China, 2002, pp. 225–234.

[21] H. Pedrini, "Modeling dense range images through fast polygonal approaches," *11th Conf. on Image Analysis & Processing*, Palermo, Italy, September 2001, pp. 448–453.

[22] C. Harris and M. Stephens, "A combined corner and edge detector," *4th Alvey Vision Conf.*, vol. 15, Manchester, UK, September 1988, pp. 141–151.

Grid Enabled Computer Vision System for Measuring Traffic Parameters

Ivica Dimitrovski ✞, Gorgi Kakasevski †, Aneta Buckovska ✞, Suzana Loskovska ✞, Bozidar Proevski ✞
✞ UKIM, Faculty of Electrotechnics and Information Technologies, Skopje, R.Macedonia
† EU, Faculty of Informatics, Skopje, R.Macedonia
✞ ivicad@etf.ukim.edu.mk, † gorgik@etf.ukim.edu.mk

***Abstract*-In this paper, we propose a application for traffic flow analysis including vehicle detection and classification, vehicle speed estimation, and accidents detection. The application is implemented to work in a grid environment with support for data management, job submission and execution and scene recording with remote cameras. The user is allowed to access the Grid infrastructure via a client application and easily manage the system. The management includes selecting stream from different cameras, acquiring results from the analyses and visualization of the obtained results. We implemented a set of algorithms based on image processing for video sequences analyze. Finally, we estimate several aspects of the proposed system in a set of experiments, including an analysis of the usefulness of streaming large video data packets and speed-up possibilities.**

I. Introduction

Image processing techniques have been applied to traffic scenes for a variety of purposes, including: queue detection, incident detection, vehicle classification, vehicle counting and speed detection [1], [2], [3], [4]. Image processing tasks associated to motion detection and estimation algorithms use mathematical techniques dominated by convolution, correlation and filtering. Considering the size of the image (768×576, 416×288, 320×240 pixels), all these tasks have a very high computational cost.

Over the years many vision-based traffic surveillance systems have been developed. The core idea is to have video cameras mounted on poles or other tall structures looking down at the traffic scene. Video is captured, digitalized and transmitted to a central place for processing and analyze. The video material acquired from the cameras can be very large (≈12GB per hour), so the computation cost need for analyze can be extremely high. The best possible solution to overcome this problem is to utilize the grid infrastructure. Grid is a collection of distributed computing resources available over a local or wide area network that appears to an end user or application as one large virtual computing system [6]. The vision is to create virtual dynamic organizations through secure, coordinated resource-sharing among individuals, institutions, and resources. Grid computing is an approach to distributed computing that spans not only locations but also organizations, machine architectures, and software boundaries to provide unlimited power, collaboration, and information access to everyone connected to a grid, in the same way that the Internet integrates resources to form a virtual platform for information. The remainder of the paper is organized as follows. Section 2 gives an overview of the algorithms used for traffic flow measurement, Section 3 introduces our Grid Enabled Computer Vision System (GECVS), and Section 4 describes the implementation of the GECVS. Section 5 presents the experimental results and Section 6 gives a conclusion of the paper.

II. Traffic Flow Measurement Algorithm

The fundamental component of our system is the vehicle detection algorithm which calculates the traffic flow from the sampled images. The input to our tracking algorithm is an image acquired from the traffic camera. In order to use the image pixels to compute real world distances and speeds, we make the following assumptions on the road.

1) The road is flat (all points lie along a single plane)

2) The road is straight (traffic motion is parallel to some axis Fig. 1)

3) Occlusions are minimized (background images separate individual cars)

The assumptions are not rigid, but the closer the data matches the assumptions, the better the results of the algorithm.

The algorithm considers that the pixel brightness is changed by the vehicle presence. For each traffic lane in the scene, we take a group of pixels, comparing their value with a reference without vehicle.

If there are significant differences, the presence of a vehicle is assumed. Because through the day the conditions of brightness change, it is necessary to update the reference background image. Therefore the first step is to find an efficient and accurate way to generate a background image from the given frames. There are several techniques for generating a background image. We used histogram-based method. This technique involves maintaining a color histogram for each pixel over a given set of frames. Assuming a pixel will

Fig. 1. Highway with three lanes

K. Elleithy (ed.), Advances and Innovations in Systems, Computing Sciences and Software Engineering, 561–565.

most frequently have its background RGB value, its color in the final background image corresponds to the most popular value in its histogram. The produced background image can be enhanced with median filter in order to reduce the noise. The main goal of our tracking system is to generate per-lane traffic statistics at real-time speeds, as would be used on a major highway to report traffic flow information. As such, it is reasonable to assume a couple things about the input video sequences. First, we require that the camera is fixed in location, strictly dedicated to analyzing traffic of a road. We also required an overhead view of the traffic in order to prevent the need for tracking multiple layers of traffic. With an overhead view, a system can simply analyze one area of the frame and know that this area belongs to only one lane of traffic. With side views, cars may be occluded by other cars or foreign objects.

In short, we subtract the background reference image from an input frame and create a mask based on some error threshold. We then analyze this mask to track moving objects along the road. We define a region of interest in our input frames, one region for each traffic lane Fig. 1. The region of interest defines the center line of a lane as well as the beginning and end of the observation window.

Once we subtract the background and generate our mask, we are left with a line of 0's and 1's, where 0's indicate background and 1's indicate the presence of a car. Then, from one frame to the next we only need to follow patches of 1's along the line in a consistent manner. For each region of interest (each traffic lane being analyzed), we store the number of cars it has seen, the average speed of the lane, and an array of car structures with an entry for each car currently in the lane. For each lane in each frame, we extract the pixels on the line of interest for the lane and subtract our background information. We then compare the sum of squared differences with a threshold, and generate a mask based on this comparison. The threshold is a parameter of the system and depends on the noise sensitivity of the camera. We perform a median filter on this vector to remove noise in the form of small chunks of black or white that might disrupt the algorithm.

The Fig. 2 shows the tracking process for the middle lane before thresholding, before using the median filter, and after the median filter. White pixels indicate the presence of a car, and black pixels indicate the background.

For vehicle speed detection we used the proposed algorithm in [5]. The image processing algorithms are implemented in C++ and the OpenCV library [7].

Fig. 2. Vehicle detection on middle lane

III. Grid Enabled CV System

Our Grid Enabled CV System is centralized three tier architecture shown in Fig. 3. The first layer is the input layer (cameras connected to computers for acquiring video data), the second layer is the centralized server, the core component of our system, and the third layer is the grid infrastructure.

The cameras are connected directly to the computers with FireWire cable (much simple model is to use USB web cameras with output video in resolution 320x240 or camera with S-Video output). The computers systems access the IEEE 1394 bus directly with Libraw1394 and use Quasar DV codec [8] to decode the video in DV format (defined by IEC61384 and SMPTE314M standards). In our case we use dvgrab, which works with PAL standard (720x576 with 25fps), but also can works with NTSC standard. dvgrab is used to capture DV input video in AVI files from the connected camera on the

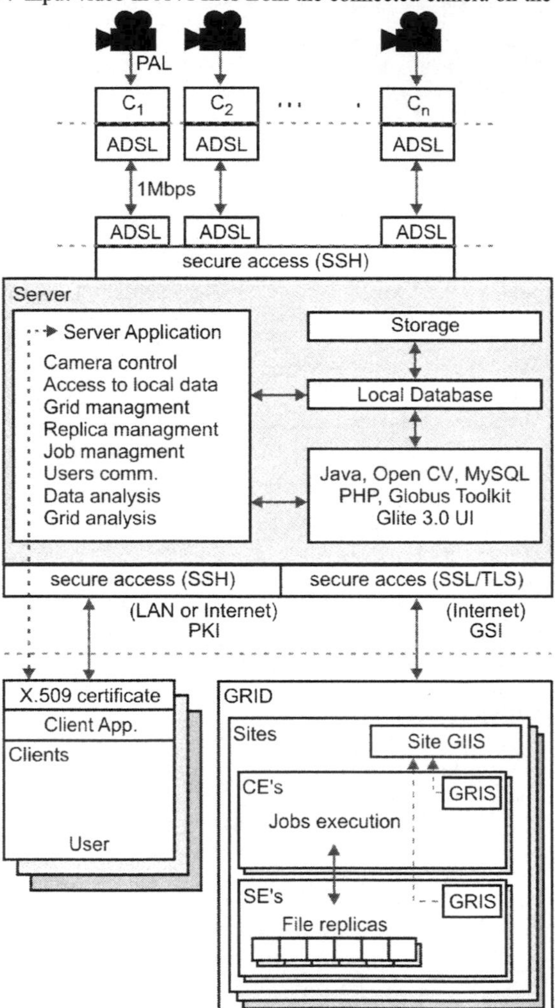

Fig. 3. Grid enabled CV system architecture

FireWire port (ohci1394). The captured video is saved in files. The size of the files is defined by the user of the system. Then the video is converted in MPEG4 format with MEncoder (set option -vf scale=320:240) with 148 Kbps (≈67 MB per hour). The video and additional information (time, size and snapshot) are sent to the server and stored in local database. The original video is kept on the local computer. It will be removed automatically after certain period of time (set by the user), or the user can deleted it immediately.

The connection with the sever is secure connection (SSH). For physical connection we use 1Mbps ADSL with ADSL modem located on the local computer. The daily flow of data is about 1.6 GB per computer.

The users machines communicate with the server through secure connection. On the user machine resides a client application which communicate with the server application. The client application enables the users to use all the functionalities of our GECVS. The client application is implemented in Java and PHP.

In order to use the GECVS the users must have X.509 [11] certificate issued by Certificate Authority (CA) from the Virtual Organization (VO) in which they are members. The server application use standard Grid Security Infrastructure (GSI) implemented in Globus. By default, the underlying communication is based on the mutual authentication of digital certificates and SSL/TLS.

The server contains a storage (HDD's), local database, Java, Open CV libraries, Globus Toolkit, Glite User Interface and the server application. The server application is written in Java with utilization of the popular Java technologies: JMF and JDBC. The local database is implemented in MySQL and the files are stored in the local file system. There are also a few shell scripts written in BASH, TCL and TCL with Expect. We use the library Expect [12] in order to communicate with some processes that need command line interaction with the users. The system is developed under Linux.

IV. IMPLEMENTATION OF GECVS

The described principles were implemented in a complex video system. The main part of this system is the video server and central database subsystem (Fig. 4). The GECVS is intended for various traffic parameters estimation. The server application that resides on the central database subsystem communicates with the clients application, the local database, storage and the grid infrastructure with gLite middleware [9] the successor of LCG middleware. The server application consists of eight main modules:

1) Camera control module from where the user can manage the cameras, change the cameras settings and give commands to the computers.

2) Module for access and management of local data from where the user can access the local database view, change and organized data, get reports and manage files which includes archiving videos on various archive mediums.

3) Grid management module, used for management of different parameters for grid access, testing connections, and view of available resources.

4) In replica management module the user have full control on the files located on the grid Storage Elements (SE). We use high data management tools (lcg_utils) and LCG File Catalog to track the files on grid with lfc file catalog commands and lcg commands for replica management and file catalog interaction. These tools are more practical and easy to use instead of lower level tools (edg-gridftp) where we must work with transport layer and we must know the implementation of SE.

5) From the job management module users can submit jobs to the grid. The jobs are organized in categories based on the algorithm parameters. The user can submit jobs selecting the CE, control job status and get the output. Because the input video file used within the job submission is divided into 24 equal parts (the input video contains information for one day), the type of the jobs is Direct Acyclic Graph (DAG) with two main jobs, one to organize and allocate resources and files on the grid from input XML file given by the server application and one to collect the results and to write the output in XML file. In this DAG job model, there are another 24 slave jobs which are executed in parallel [10]. There is also an option to submit jobs one by one. The server application automatically checks the job status and obtains the results when the job is finished. The results are taken in XML file.

6) Users communication module included for practical reason. This module is used for users communication from where the user can share and organize tasks including sending e-mail messages.

7) In data analysis module the user can view, organize and memorize all obtained results. This module is easily upgradeable. It includes various statistics analysis (daily, weekly, monthly ...). The results are organized according to highway number and camera number for each hour and day separately. The statistics can be made for number of cars,

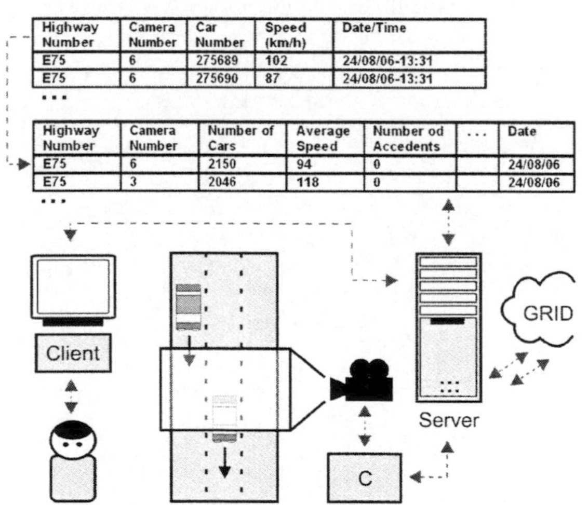

Highway Number	Camera Number	Car Number	Speed (km/h)	Date/Time
E75	6	275689	102	24/08/06-13:31
E75	6	275690	87	24/08/06-13:31

...

Highway Number	Camera Number	Number of Cars	Average Speed	Number od Accedents	...	Date
E75	6	2150	94	0		24/08/06
E75	3	2046	118	0		24/08/06

...

Fig. 4. Grid enabled CV system implementation

number of accidents and speed. The most valuable part of our system is the possibility for traffic flow estimation based on the previously obtained results. This can be very usefully for prediction and elimination of traffic jams and speed violations in different periods of time, different parts of the highway. Very simple output result from our GECVS is shown on Fig. 5.

8) The grid management module is very important for measuring the grid performance. The CEs and SEs resources at the specific site implement an entity called Information Provider, which generates the relevant information about the resource. This information is published by the Grid Resource Information Servers, or GRISes. One or more Grid Index Information Servers (GIISes) can get information from different GRISes and publishes it allowing specialized or global view and search. Local GRIS run on CEs and SEs at each site and report dynamic and static information regarding the status and availability of the services. At each site, a site GIIS, or site BDII, collects the information of all resources given by the GRISes. In order to query directly the IS elements, we use two high level tools: lcg-info and lcg-infosites. With this tools we collect the data from the sites, organize it in local database from where it is presented to the end user. So, user can easy manage jobs activities or set default parameters for the system. All this functionalities are implement with GLUE schema which describe the attributes and the value of the site information CE, SE and Network Monitoring (example WorstRespTime, TotalCPUs, MaxRunningJobs, WaitingJobs, FreeCPUs, CE from GlueCE class) or with using --query attribute of lcg-info tool (example $ lcg-info --list-ce --query 'FreeCPUs=5' --attrs 'FreeCPUs,OS').

While the system is working, it collects various information and parameters of interest which are presented to the user from the grid analysis module. This is very helpful module for users to plan their activities, watch the grid network, CEs and SEs, replicas, successfulness of the jobs. All this data is organized by date, CE and SE (example: user can watch successfulness of CE's and the time for job execution, and the user will submit jobs on CEs with large probability of success).

V. Experimental results

Several measurements were performed in our experiments.

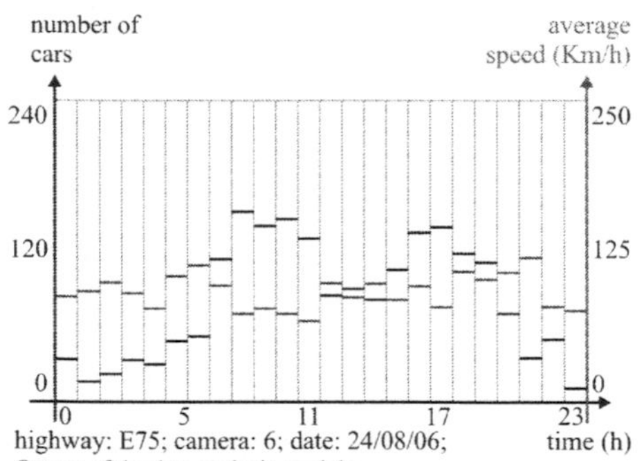

Fig. 5. Output of the data analysis module

First, the initialization time of the grid environment on the computation servers (the "workers") was measured as well as the initialization time due to video file splitting. Second, the duration of streaming the video data to the computation servers was also measured. It should be stated here that is difficult to conduct repeatable experiments in a grid environment since the resources, respectively their availability as well as their computational load can change dynamically (which is an inherent characteristic of a grid). The benefit of streaming large video packets depends on the available network bandwidth which has been measured during our experiments, and the ratio of transmission time and job computation time. The bandwidth (regarding all job submissions in our experiments, more than 500 jobs) for video packet transmission varied between 2,4MBit/s and 8,1 MBit/s. The conducted experiments showed that it is reasonable to stream large video packets to distant servers in order to obtain as many computation servers as possible. The impact of the video packet size was also investigated, respectively how the number of jobs affects the performance. The results demonstrated that the processing time is lower for the larger packet sizes. There are two reasons: First, the splitting process can be conducted faster on the whole video stream if fewer videos packets have to be created. Second, the video streaming produces less overhead since there are fewer streaming connections to establish and to serve due to less computation server request. Thus, it can be concluded that it is necessary to find a compromise between a sufficiently large numbers of job to achieve a high degree of parallelization while limiting this number to avoid the overhead mentioned above.

VI. Conclusion and Future Work

In this paper, we describe the framework of our GECVS and provide the algorithms and implementation results of our current work. We used computer vision based approach to road monitoring and traffic analysis. The algorithms for vehicle tracking, speed measurement and jam detection were implemented in C++ with the OpenCV library. The methods and algorithms are organized in an intelligent video monitoring system with data transferring over grid computing network and archiving in local and central databases. Our system works in grid environment and uses all the functionalities of the grid infrastructure. The system is capable for generating different kind of reports and analysis including various kind of prediction.

Future work will cover complex testing of the system and modification of the used algorithms. Because the system is easily upgradeable we can include more algorithms such as vehicle classifications and road conditions estimation. The existing algorithms can be optimized to give more accurate results in night conditions.

There is also possibility for optimization of the video packet size and thus an optimum number of jobs depending of the number of the available computation elements.

REFERENCES

[1] Image Processing in Road Traffic Analysis, E. Atkociunas, R. Blake, A. Juozapavicius, M. Kazimianec, Nonlinear Analysis: Modelling and Control, 2005, Vol. 10, No. 4, 315–332.
[2] Image Processing Applied to Real Time Measurement of Traffic Flow, Alvaro Soto and Aldo Cipriano, Proceedings of the 28th Southeastern Symposium on System Theory (SSST '96).
[3] An Algorithm to Estimate Mean Traffic Speed Using Uncalibrated Cameras Daniel J. Dailey, Member, IEEE, F. W. Cathey, and Suree Pumrin, IEEE TRANSACTIONS ON INTELLIGENT TRANSPORTATION SYSTEMS, VOL. 1, NO. 2, JUNE 2000.
[4] Visual tracking strategies for intelligent vehicle-highway systems Christopher E. Smith Nikolaos P. Papanikolopoulos Scott A. Brandt Charles Richards, Presented at SPIE's International Symposium on Photonics for Industrial Applications, Intelligent Vehicle Highway Systems, Oct 31 - Nov 4, Boston, MA, 1994.
[5] A Real-time Computer Vision System For Measuring Traffic Parameters, David Beymer, Philip McLauchlan, Benn Coifman, and Jitendra Malik, Proceedings of the 1997 Conference on Computer Vision and Pattern Recognition.
[6] Grid Computing: A Practical Guide to Technology and Applications by Ahmar Abbas ISBN:1584502762 Charles River Media 2004.
[7] http://www.intel.com/technology/computing/opencv/index.htm.
[8] http://libdv.sourceforge.net/, software developed by Charles 'Buck' Krasic, Erik Walthinsen (Omega), 20.10.2003.
[9] http://www.glite.org/, gLite – Lightweight middleware for grid computing, Copyright (c) Members of the EGEE Collaboration 2004.
[10] Patterns: Emerging Patterns for Enterprise Grids, Copyright International Business Machines Corporation 2006.
[11] http://www.ietf.org/rfc/rfc2459.txt, Internet X.509 Public Key Infrastructure Certificate and CRL Profile (Request for Comments: 2459), Network Working Group, Janyary 1999.
[12] Handling Passwords with Security and Reliability in Background Processes, Don Libes, National Institute of Standards and Technology, Proceedings of the Eighth Systems Administration Conference (LISA VIII) San Diego, California, September 19-23, 1994.

Physically Constrained Neural Network Models for Simulation

J. E. Souza de Cursi
INSA de Rouen
Av. de l'Université, BP8
76801 Saint Etienne du Rouvray FRANCE

A. Koscianski
UTFPR
Av. Monteiro Lobato s/n
84016-210 Ponta Grossa BRASIL

***Abstract* — We present a method for combining measurements of a system and mathematical descriptions of its behavior. The approach is the opposite of data assimilation, where data is used in order to correct the results of a model based on differential equations. Here, differential equations are used in order to correct interpolation results. The method may be interpreted as a regularization technique, able to handle the ill-posed character of a neural network regression problem. Significant examples illustrate the numerical behavior and show that the method proposed is effective to calculate.**

***Index Terms*— Simulation models, neural netwoks, inverse problems, regularization, numerical-experimental coupling.**

I. INTRODUCTION

NUMERICAL models for system simulation may be obtained by two different approaches: on the one hand, measurements may be used in order to construct data based models (DBM), such as an interpolating function of the measured data; on the other hand, physical principles may be used in order to derive knowledge based models (KBM), such as a set of partial differential equations describing the behavior of the system. In general, these two approaches are disconnected and are applied separately: comparisons between governing equations issued from physical principles and measurements are principally used in order to identify physical parameters or to provide qualitative analysis.

However, these two approaches are quite complementary and have attractive characteristics: a seductive idea consists in combining their strengths in a mixed one. For instance, data assimilation takes into account experimental data when solving knowledge based models. The opposite approach, which consists in taking into account knowledge based information when constructing data based models has led to the development of Physically Constrained Neural Networks (PCNN), where additional information issued from physics is included in the training equations. This allows the network to better distinguish the error component from the measured data, improving simulation results.

II. SIMULATION MODELS

A. Knowledge based models

KBM generally use physical laws and approximations in order to get a mathematical description of a system under the form of partial or ordinary differential equations. The equations are numerically solved in order to run an experiment and simulate the system.

The main difficulties in the use of KBM are the choice of a correct level of precision, the evaluation of the impact of the simplifications performed and of the lack of knowledge about some aspects of the system. There are several factors that can limit the model reliability [9]:

- a choice of the engineer to disregard effects consider as having minor significance;
- complex interactions between diverse process, that were not considered;
- insufficient knowledge of the internal functioning of the system.

For instance, a column of a bridge may be described as a bar, a beam, a 3D linear solid or a 3D nonlinear solid; it contains different materials and is not perfectly homogeneous. Different KBM are obtained by neglecting some aspects of the mechanical behavior of the column: for instance, if flexion is neglected, a bar model may be used; if the deformation of the cross-section is neglected, a beam model may be used and so on. Each simplification carries a loss of precision. Moreover, a complete model, taking into account non-linearity, non-homogeneity and important geometrical variations requires a deep knowledge about the exact distribution of the materials, their behavior and the mutual influence of these different parts, what leads to expensive computational and experimental costs and non-reproducibility. If other aspects - such as thermal effects, phase transformation, radiation damage, wind and water action - are included, the analysis of the column may become complex and gaps appear in the experimental, numerical and theoretical frameworks. The situation where processes having different time and space scales must be taken into account introduces supplementary complexity [18].

Thus, a complete modeling of a system is generally not realistic and simplifications must be introduced by a judicious

K. Elleithy (ed.), Advances and Innovations in Systems, Computing Sciences and Software Engineering, 567–572.

selection of the parameters and aspects of behavior to be described. Moreover, the computational efficiency is a major aspect of any simulator. Although several numerical schemes exist, there is a strong connection between numerical precision and computational cost (for instance, by the use of a better grid resolution): improving the quality of results generally imply higher computational power [15].

A well known illustration of the relation between computational cost and accuracy of results is the simulation of the heat equation in 1D,

$$u_t = k \cdot u_{xx} \ . \tag{1}$$

Despite the apparent simplicity of the model its numerical implementation is highly sensitive to initial conditions and parameters as time step [3]. For instance, the results presented in Figure 1 required 10000 iterations of a simple finite differences method (forward time, centered space); in this simulation the oscillations were controlled when this number was multiplied by 10.

Complex models as those used in fluid flow and combustion simulation generally have exigencies on CPU power and calculation time that strongly limits software functionalities like interactivity.

B. *Data based models*

The DBM approach tries to replace the lack of knowledge by measured data, in order to obtain a representation of the system. In this framework, interpolation is generally considered as the easiest way of building a simulation model. The mathematical arguments are usually simpler and the associated numerical methods are immediately available through a large number of existing libraries. This approach is often compared to surface reconstruction problems: given a cloud of points we try to recover the original surface from which they have been sampled. A simple approach consists in finding a representation that minimizes the least squares distance to the given data set,

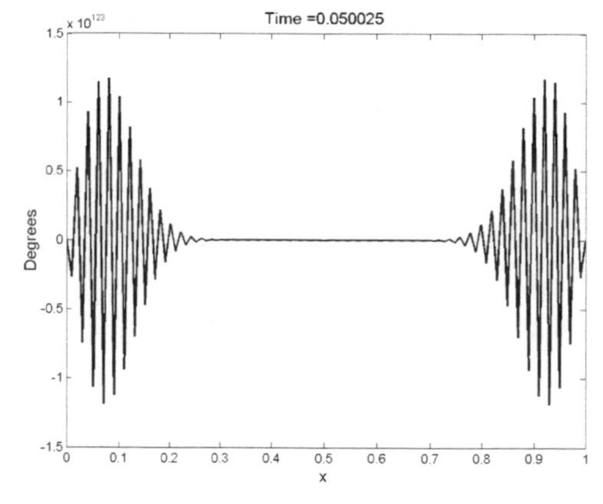

Fig. 1. Simulation of the 1D heat equation showing strong oscillations due to numerical instability.

$$J_D[f] = \sum_{i=0}^{m} \left(f(x_i, \mathbf{w}) - y_i \right)^2 \tag{2}$$

where $f(\cdot)$ is the interpolating function and $\boldsymbol{w}$ is a vector of parameters. It can be shown that solving this equation is equivalent to maximizing a mathematical expectation [8].

The minimum square formulation is widely used and is well adapted to situations where:

- the numerical evaluation of the theoretical model equations requires CPU power which is not available;
- the physics of the phenomenon are incompletely known, precluding the use of a theoretic model;
- the model parameters (e.g. material characteristics) cannot be determined or are affected by large errors;
- experimental data sampled from meaningful situations and with adequate precision is available.

The last of these requirements is rarely fulfilled: most physical phenomena present characteristics which are difficult to observe or are not measurable at all. In many situations, the simple fact of introducing sensors in the system modifies its behavior [9]. For instance, let us consider the measurement of the air pressure over a wing or a sail: the presence of sensors at the wing surface modifies the fluid flow and changes the behavior of the system. Laboratory tests often have difficulties in replicating real operating conditions and the instruments may contaminate the data, resulting biased, noisy measures. These observations show that the solutions of Eq. (2) must be considered with caution: the minimum may not correspond to the best model [1]. In fact, the best representation fitting the available data depends also of other quality criteria, such as precision, smoothness and economy of representation. In particular, the choice of a particular function $f(\cdot)$ must take into account the qualitative behavior of the targeted system. Examples of application are image representation, noise reduction and data compression [4].

Artificial neural networks (ANN) are a particular kind of data based representations, known since some decades; the abilities of ANN to interpolate a given set of data, as well as their limitations and the associated numerical difficulties are documented in the literature. Many of these difficulties are still with us, what makes that some practical aspects of ANN remain empirical, hiding the associated mathematical complexity and the reasons for a possible failure of a model. For instance, the learning procedure often involves the minimization of a non-convex function, what introduces the difficulty of the choice of the initial vector of parameters and of an optimization procedure able to prevent convergence to local minima. Moreover, the regression problem of determination of the ANN parameters is an inverse problem, intrinsically ill-posed. Both these aspects - the minimization of non convex functions and the handling of ill-posed problems - are still active research fields and only partial solutions may be found in the literature. The correct synthesis of a neural network model supposes the resolution of these difficulties.

ANN are a common choice for interpolation. Several types of neural networks have been developed, but probably the most used is the MLP, multi-layer perceptron. A MLP with one hidden layer corresponds to the following equation:

$$f(\mathbf{x},\mathbf{w}) = b_0 + \sum_{i=1}^{N} \rho_i \cdot \sigma(\mathbf{v}_i \mathbf{x} + b_i) \qquad (3)$$

where N is the number of neurons and the values b_i, ρ_i and $\boldsymbol{v}_i$ are assembled in a vector $\boldsymbol{w}$ of network parameters. The function σ (·) is called activation function – for instance, the sigmoidal tangent. Additional layers can be added to the MLP by applying (3) recursively:

$$f_k(\mathbf{x},\mathbf{w}) = b_0 + \sum_{i=1}^{M_k} \rho_i \cdot f_{k-1,i}(\mathbf{x}) \qquad (4)$$

where the subscript k identifies a layer of M_k neurons.

Mathematical results indicate that a MLP with one hidden layer and a sufficient number of neurons can approximate a function with arbitrary accuracy [14]. Defining the minimum number of neurons is however an open problem.

The process of minimizing (2) is called "training" in the literature connected to ANN. The most used procedure is the backpropagation algorithm, which consists of a gradient descent tailored to the MLP in such a way to avoid redundant computations to evaluate the derivatives [16]. The equation is

$$\mathbf{w}_{i+1} = \mathbf{w}_i - \alpha \frac{\partial J}{\partial \mathbf{w}} \ . \qquad (5)$$

The parameter α is usually referred as the learning factor. Equation (3) leads to a non-convex optimization problem and the choice of the initial vector is crucial to a successful minimization. Initialization methods like Nguyen-Widrow show relative success, but it is advisable to perform several minimizations to ensure good results [5]. Perturbed gradients (see for example [19]) and direct methods [20] are some of the techniques that may palliate this difficulty.

A careful choice of the ANN topology is a necessary condition to obtain a fit with the required accuracy. However, matching the data set may fail as the only criteria to choose a representation. To exemplify, Figure 2 shows an interpolation to the heat equation in 1D obtained in Matlab using a MLP of 15 neurons. The circles represent the temperature sampled at the center of a rod. The data were obtained by simulating a theoretical model and adding ± 5% noise. The dotted line is the output of the MLP, achieving an error of the order of 2.489×10^{-27}. Despite the quantitative precision, the solution presents serious flaws; first, it fails completely to extrapolate results in the exterior of the data set, for t < 0.02 and t > 0.5, even if the ANN was correctly prepared to span the entire domain [0.0; 1.0]. Second, the simulation shows an initial increase in temperature, incoherent with the physics of the phenomenon.

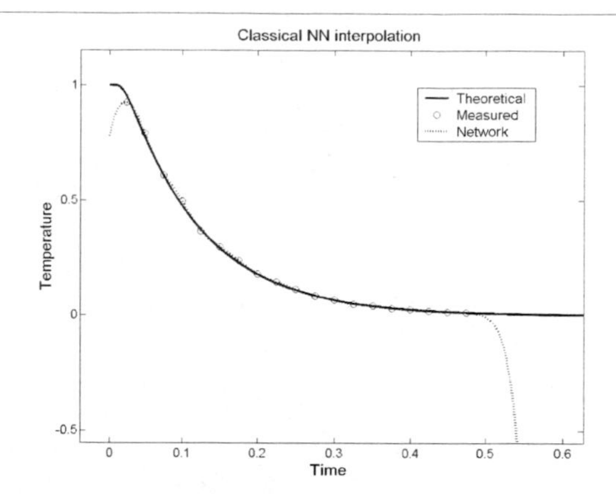

Fig. 2. Neural Network interpolation. The precise agreement with the measures does not validate the NN to simulate the physical phenomenon.

Some alternatives to avoid overfitting would be to reduce the number of neurons or employ techniques like early stopping [5]. Heuristics like these lead to unpredictable results: there is no guarantee that oscillations will not occur and it is not clear how the parameters affect the optimization. The models obtained this way must be simulated and analyzed qualitatively. In problems where the input space is of high dimension this approach is impractical.

C. Comparison

Knowledge based models are able to extrapolate results, allowing the analysis of a system under different conditions. For example, it is possible to run a KBM for different material parameters in order to test alternative designs or to predict the behavior of a given structure under different load conditions. Their main limitation is connected to the complexity of the model: increasing complexity leads to expensive computational cost and loss of accuracy.

Data based models are - in principle - not limited by the complexity of the system physics. The calculations are generally several orders of magnitude faster than the numerical resolution of a set of PDE. However, even if the simulator is able to replicate the data set, it often fails in the extrapolation of the behavior of the system outside of the measured data set. Table 1 highlights these aspects .

TABLE I
COMPARISON OF THE TWO KINDS OF SIMULATION MODELS

Aspect	*KBM*	*DBM*
Regularity	Depends on the equation's properties	Depends on the interpolating function
Extrapolation	Controlled by the equation's properties	Difficulty to control
Computational Cost	Good integration schemes are expensive	Concentrated on training. Cheap Evaluation
Agreement to data	May request parameter identification	Natural result of Interpolations

III. Regularization

Inverse problems arise in many applications in diverse engineering domains, as tomography, geophysics or astronomy. Some examples are: parameter estimation, given the outcome of an experiment; image reconstruction, given a noisy, sparse, set of measures; and model fitting.

A broad representation for the direct problem is given by

$$\mathbf{Af} = \mathbf{d} \tag{6}$$

where $\mathbf{A}$ is an operator related to the physics of the problem, $\mathbf{f}$ is the input domain and $\mathbf{d}$ the image that represents a set of measurements. The inverse problem in the presence of noise is then written

$$\mathbf{f} = \mathbf{A}^{-1}(\mathbf{d}+\varepsilon) \ . \tag{7}$$

The problem (6) is said to be well-posed problem in the sense of Hadamard when three conditions are satisfied: a solution exists; the solution is unique; and the inverse mapping is continuous. The second condition becomes obviously false in a regression problem, given a large class of interpolating functions. In such a case, the sampled data may correspond to different curves - eventually infinitely many ones. Another aspect of ill-posedness may be made clear with the help of basic linear algebra. For instance, suppose $\mathbf{A}$ to be a square matrix of dimension m, positive and semidefinite. The matrix may be decomposed as [21]:

$$\mathbf{A} = \sum_{k=1}^{m} \mu_k \mathbf{v}_k \mathbf{v}_k^T \tag{8}$$

where μ_k and $\mathbf{v}_k$ are respectively the k-th eigenvalue and k-th eigenvector of $\mathbf{A}$. The left side of (6) can now be written as:

$$\mathbf{Af} = \sum_{k=1}^{m} \left(\mu_k \mathbf{v}_k \mathbf{v}_k^T \right) \mathbf{f}_k \tag{9}$$

and:

$$\mathbf{A}^{-1}(\mathbf{d}_k+\varepsilon) = \sum_{k=1}^{m} \left(\frac{1}{\mu_k} \mathbf{v}_k \mathbf{v}_k^T \right) (\mathbf{d}_k+\varepsilon) \ . \tag{10}$$

Equation (10) shows that, in the inverse problem (7), small values μ_k will amplify the noise present in the measures. A similar argument applies for Singular Value Decomposition. In the infinite dimensional case, we may consider as example the linear operator ∂_t present in equation (1). For initial conditions $g(x) = \sin(n\pi x)$, a solution to the heat equation can be written in the form of a series expansion,

$$u(x,t) = \sum_{n=0}^{\infty} \alpha_n \sin(n\pi x) e^{-n^2\pi^2 kt} \ . \tag{11}$$

We observe that the inverse problem associated with (11) exhibits the same difficulty as (10). This can be seen by noting that as $n \to \infty$, the corresponding components in the solution vanish; on the other hand, during parameter estimation, small variations in $u(x, t)$ caused by noise will correspond to arbitrarily large values of n and k and oscillations in $g(x)$.

IV. Model Blending

A. Data Assimilation

Data assimilation techniques are utilized in weather forecasting, a particularly difficulty simulation problem. One of the main problems faced by meteorological simulation is the lack of data. Theoretically, model precision may be increased with the addition of finer levels of detail into the equations; but such models are of less use as we can not provide the required information.

The measures of various atmospheric variables form a sparse grid of data values. Combining this information with models of atmospheric dynamics serves to obtain a finer grid, which can be used in analysis and forecasting. In its simplest form, data assimilation corresponds to equation [13]:

$$\mathbf{x}^a = \mathbf{x}^b + \mathbf{K}\left(\mathbf{y} - \mathbf{H}\mathbf{x}^b\right) \tag{12}$$

where $\mathbf{x}^a$ is known as the analysis vector, $\mathbf{x}^b$ is the background state and contains a guess of the true state of the system. Vector $\mathbf{y}$ contains the measures. Matrix $\mathbf{H}$ is a linear observation operator and $\mathbf{K}$ is the gain matrix.

Equation (12) combines the system physics, represented by $\mathbf{x}^b$, with the measured data $\mathbf{y}$. The result of data assimilation can be interpreted as a calibration of model equations with available data. A weather forecast can then be calculated by running the theoretic model given $\mathbf{x}^a$ as input.

B. Regularization

Data assimilation uses data to correct the system predictions; another possibility is to adopt the symmetric approach and integrate the model equations into the regression problem as a regularization term.

Regularization techniques are employed to handle ill-posed problems, controlling the oscillatory behavior of solutions [7]. A well known form of regularization is Truncated Singular Value Decomposition, which can be understood under the same ideas of eigenvalue decomposition, as exemplified in (9). If the eigenvalues μ_k are sorted in descending order, we may truncate the series preserving only the first $r < m$ terms. As a result, the inverse operator will be less affected by the small noisy components of the measured data.

Another technique is Tikhonov regularization [17]. It proceeds by reducing the solution space to those functions that meet a given smoothness criteria. In the case of data regression, regularization can be done by penalization, adding a term to the least squares formulation. The problem may be written as:

$$J_R[f] = (1-\lambda)J + \lambda R[f] \tag{13}$$

where J is the original minimum squares problem (2), f is the candidate solution, λ is the regularizing parameter and R the regularizing operator. The parameter λ controls the trade-off between data fitting and solution regularization. By choosing $\lambda = 0$ we recover the pure regression problem. On the contrary, when $\lambda \to 1$ the candidate solutions lacking physics behavior are heavily penalized.

One heuristic implementation of this concept for neural networks is to chose $R = \boldsymbol{w}\boldsymbol{w}^{T}$, where $\boldsymbol{w}$ is the parameters vector of the network [12]. The basic assumption here is that, by limiting the magnitude of the weights ρ in equation (2), the output of each neuron tends to small values and this reduces the oscillations. Evidently, this method presents the same difficulties as early stopping. In the literature of inverse problems a common choice for R is:

$$R = \int \frac{\partial^n}{\partial x^n} \quad (14)$$

for some $n > 0$. Higher-order derivatives correspond to more stringent requirements on the solution smoothness. This operator is used for example in the formulation of GRNN, Generalized Regression Neural Networks [6]. Another alternative consists of applying a Fourier transform and filter selected frequency ranges.

C. *Regularization with a physics model*

Classical regularization techniques are able to improve the so called network generalization, but this is done accordingly to arbitrary criteria. As a result, although the method can manage to control oscillation, the functions obtained may still lack a good agreement with the predictions of a theoretic model.

Another possibility is to replace the classical regularizing term with the model equations of the system. This can be done with the aid of a variational formulation, or by penalization with a residual; the resulting optimization problem has the same form as given by (13), but the computed solution in this case is much less prone to random oscillatory effects.

As an example, a regularizing operator corresponding to the heat equation could be:

$$R[f] = \int_0^1 \int_0^1 \left(\frac{\partial}{\partial t} f(t,x,\mathbf{w}) - k \frac{\partial^2}{\partial x^2} f(t,x,\mathbf{w}) \right)^2 dt\, dx . \quad (15)$$

The resulting problem penalizes candidate solutions that deviate from theoretical predictions. A comparison between this new formulation with classical physics-based models and pure data-based models suggest some interesting aspects.

1) Efficiency. The minimization of (13) can be highly time consuming, but this cost is balanced by the gain in simulation speed. Evaluating an ANN is in general much faster than integrating the model equations.
2) Combined sources of information. The term J feeds the network with data that can compensate erroneous theoretic predictions, while the model equations control the ANN tendency to overfit the data and increases its extrapolation capabilities.

This technique can replace classical Neural Networks in those cases where a theoretic model is available and the qualitative aspects of simulation are especially significant.

V. EXPERIMENTS

This section presents two examples of the physical regularization approach. We consider the heat flow problem; and the simulation of the deformation of a structure under impact, employing a model investigated in [11]; the original work used classical neural networks. All the code was written in C++; the derivation of ANN equations as $\partial_{x,w} f(\cdot)$ are shown in [10]. The minimizations were carried out by a combination of Genetic Algorithms and the Nelder-Mead method. Matlab was used to produce the graphical output.

In order to simulate (1) it is necessary to establish parameters values, boundary and initial conditions. In the present case it was chosen $k = 1, f(0, t) = f(1, t) = 0$; and $f(x, 0) = 1$ for $0 < x < 1$. The conditions can be incorporated into (13) by augmenting the equation with new penalization terms, or by a careful choice of the interpolating function. By choosing $f(x, t) = (x^2 - x)\, g(x, t, \boldsymbol{w})$ where $g(x, t, \boldsymbol{w})$ is the neural network, the boundary condition is automatically fulfilled. This restricts the solution space and reduces the search effort. In the present case the initial condition was handled by penalization.

The system was numerically simulated and the values corrupted by 5% noise. A one-hidden layer perceptron with 16 neurons was used. The results are shown in Fig. 3 for two values of λ. By making $\lambda \to 1$ the curve tends to follow more closely the theoretical prediction, as expected.

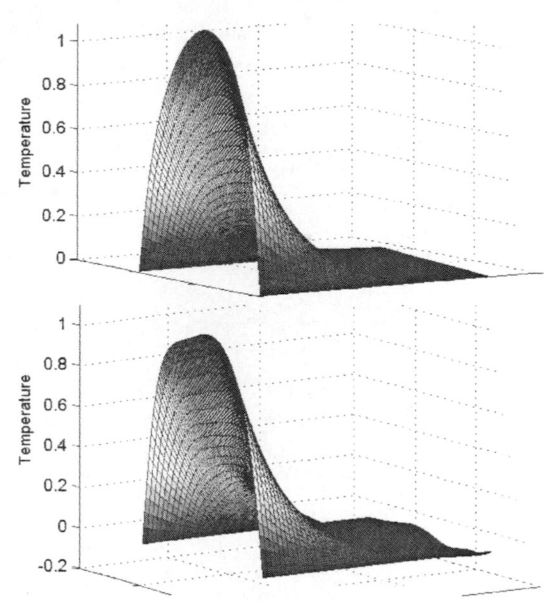

Fig. 3. PCNN simulation of heat flow; on top λ = 0.9; on bottom, λ = 0.1.

The second experiment is the simulation of the deformation of a structure under impact, as described in [11]. The displacement is given by a Maxwell model,

$$m\ddot{u} + c\dot{u} + k(u) \cdot u = 0 \ , \tag{16}$$

where $k\ (t)$ is given by

$$k(x) = k_{max} \cdot e^{-Ax^2} + k_0 \ . \tag{17}$$

The parameters values were k_{max} = 1000 N/m, k_0 = 100 N/m, C = 5.5 kg/s, m = 1 kg, A = 10 and initial velocity v_0 in the range [8 ; 16] m/s. The initial conditions were $u(0) = 0$ and $\dot{u}(0) = v_0$.

The interpolation function was chosen as $f(t, v_0) = t \cdot g(t, v_0, \mathbf{w})$ with $g(\cdot)$ a one-hidden layer perceptron of 25 neurons. The regularizing operator was written as:

$$R[f] = N^{-1} \sum_{k=1}^{N} \left\{ \int_0^T \left(m\ddot{f}(t) + c\dot{f}(t) + k(t) f(t, Vo_k) \right)^2 dt + \left(\frac{\partial}{\partial Vo} f(0, Vo_k) - Vo_k \right)^2 \right\} , \tag{18}$$

where N is the number of sampled trajectories.

Fig. 4 and 5 show simulation results for different values of initial velocity. In both cases the parameter λ was set to 0.9 and the curves correspond to the acceleration of the impact point. The PCNN exhibits a good agreement with the theoretical predictions. The curves correspond to the second derivative $\partial^2 f / \partial t^2$ of the interpolating function, meaning that the displacements and velocities are also available.

A remarkable aspect of the experiment was the robustness of the implementation, when compared to the difficulties reported in [11]: techniques as recurrent networks lead to unpredictable results and were very sensitive to the chosen topology. By contrast the physical regularization makes the whole process simpler. Although tasks as determining the value of λ and the network topology remain empirical, the formulation clearly drives the search to the expected behavior.

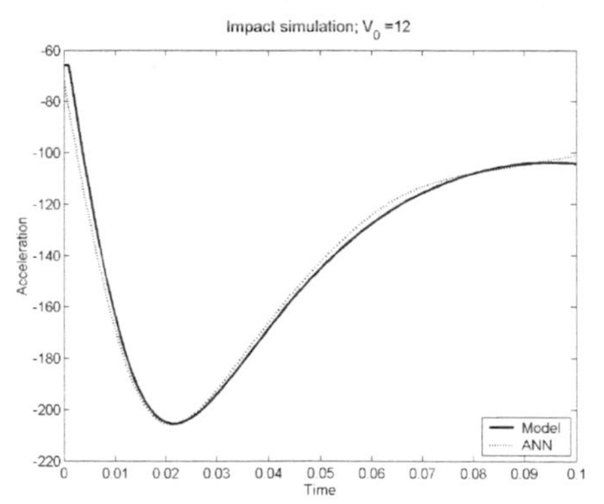

Fig. 4. PCNN simulation of impact, $v_0 = 12$, $\lambda = 0.9$.

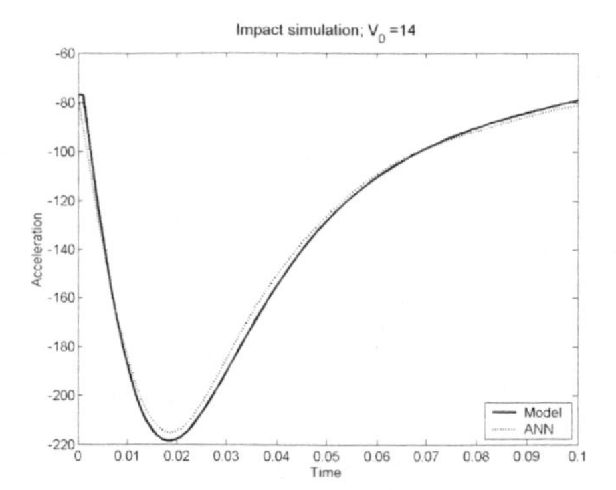

Fig. 4. PCNN simulation of impact, $v_0 = 14$, $\lambda = 0.9$.

REFERENCES

[1] L. A. Aguirre and Billings. S. A. "Dynamical effects of overparametrization in nonlinear models". *Physica D*, 1995.
[2] D. Artiouckhine. "Méthodes d'optimisation et problèmes inverses". *Ecole d'hiver METTI 99*, pp. 55–92, 1999.
[3] J. V. Beck, B. Blackwell and C. R. St. Clair, "Inverse heat conduction: ill-posed problems", Wiley Interscience, New York, 1985.
[4] R. A. DeVore, "Nonlinear approximation". *Acta Numerica*, 1998.
[5] H. Demuth and M. Beale. "Neural network toolbox, users guide", Natick, Mass: The Math Works, 1998.
[6] F. Girosi, M. Jones and T. Poggio, "Regularization theory and neural networks architectures". Neural Computation, 7(2), pp. 219–269, 1995.
[7] P. C. Hansen, "Rank-deficient and discrete ill-posed problems: numerical aspects of linear inversion". *SIAM Monographs on Mathematical Modeling and Computation*, 1997.
[8] S. Haykin, "Neural networks - a comprehensive foundation". Prentice Hall, 2nd edition, 1998.
[9] Y. Jarni and D. Maillet, "Problèmes inverses et estimation de grandeurs en thermique". Ecole d'hiver METTI 99, pp. 1–51, 1999.
[10] A. Koscianski, "Modèles de mécanique pour le temps réel". Doctoral Thesis, INSA de Rouen, Rouen, 2004.
[11] A. Machnick, "Contribution à la simplification de modèles dans un environnement intégré pour le calcul de crash des structures mécaniques". Doctoral Thesis, INSA de Rouen, Rouen, 2005.
[12] D. J. MacKay. "Probable networks and plausible predictions - a review of practical bayesian methods for supervised neural networks". *Network* (IOPP - Institute of Physics Publishing), Bristol, UK, 2000.
[13] N. K. Nichols, "Data assimilation: aims and basic concepts". Proceedings of the NATO Advanced Study Institute on Data Assimilation for the Earth System, Maratea, Italy, June 2002.
[14] A. Pinkus, "Approximation theory of the MLP model in neural networks", *Acta Numerica*, pp. 143–195, 1999.
[15] W. Press, S. A. Teukolsky, W. T. Vetterling, and Brian P. Flannery, "Numerical recipes in C". Cambridge University Press, 1992
[16] D. Rumelhart, G. Hinton and R. Williams, "Learning representations by backpropagating errors". *Nature*, (323), pp. 533–536, 1986.
[17] A. N. Tikhonov and V. Y. Arsenin, "Solution of ill-posed problems". Winston & Sons, Washington, DC, 1977.
[18] A. J. Wallcraft, J. E. Hurlburt, E. J. Metzger, R. C. Rhodes, J. F. Shriver, O. M. Smedstad, "Real-time ocean modeling systems", Computing in Science and Engineering, Vol. 4, No. 2, pp. 50-57, 2002.
[19] M. Solodov. "Convergence analysis of perturbed feasible descent methods". *Journal of Optimization Theory and Applications*, 93(2), pp. 337–353, 1997.
[20] T. G. Kolda, R. M. Lewis, and V. Torczon. "Optimization by direct search: new perspectives on some classical and modern methods". *SIAM Review*, 45(3), pp. 383–482, 2003.
[21] S.M. Tan and Colin Fox, "Inverse problems", The University of Auckland, unpublished.

Index